# THERMODYNAMICS

# THERMODYNAMICS

## THIRD EDITION

## KENNETH WARK

Associate Professor of Mechanical Engineering
Purdue University

**McGRAW-HILL BOOK COMPANY**

New York   St. Louis   San Francisco   Auckland   Bogotá   Düsseldorf
Johannesburg   London   Madrid   Mexico   Montreal
New Delhi   Panama   Paris   São Paulo   Singapore   Sydney   Tokyo   Toronto

**Library of Congress Cataloging in Publication Data**

Wark, Kenneth, date
    Thermodynamics.

    Includes bibliographies and index.
    1.  Thermodynamics.  I.  Title.
QC311.W3  1977        536'.7        76–43276
ISBN 0–07–068280–1

**THERMODYNAMICS**

4567890DODO78321098

This book was set in Times New Roman.
The editors were B. J. Clark and J. W. Maisel;
the cover was designed by Anne Canevari Green;
the production supervisor was Leroy A. Young.
New drawings were done by J & R Services, Inc.
R. R. Donnelley & Sons Company was printer and binder.

*To*
*Lois*
*and*
*Kathy, Karen, and Ron*

# CONTENTS

# PREFACE

This edition is a major revision of the second edition published in 1971. In addition to new examples and problems, nearly half the text contains new or revised material. One major change is the addition of specific examples, problems, and tables in SI (metric) units. This complements similar material in English engineering units ($lb_f$, $lb_m$, ft, s) carried over from the second edition. These are the only two sets of units used in this edition. In general, problems are stated in one set or the other, but units are not mixed. Consequently, the text may be used exclusively in SI or USCS units. In addition, both sets of units may be emphasized by the proper choice of problem assignments. Any example, problem, or table which employs metric data is designated by the symbol M after the number. Metric information usually precedes English engineering data in this edition.

With minor exceptions in this text, the energy in SI units is in kilojoules and the mass is in kilograms. The pressure is in bars rather than a multiple of the pascal unit. The value of the specific volume in tabular presentations is in cubic centimeters per gram rather than cubic meters per kilogram. The latter choice of $v$ frequently leads to values which are very small. However, both units of $v$ are used in problem solutions. The appendix contains separate sets of comparable data in metric and engineering units.

A number of other changes have been made on the basis of classroom use of the second edition over the past five years. In certain areas the subject matter is presented in more distinct modules. This allows for a more flexible choice of topics for course assignments. In addition, the text is adaptable to individualized, or self-paced, instruction. To enhance this latter opportunity, the problems at the end of each chapter are divided or grouped into specific topics and are identified by an appropriate title. This should make it easier for an instructor or student to choose problems for study.

As in the second edition, the introductory concepts of the second law can be approached from either a macroscopic or microscopic viewpoint in this edition.

However, this material no longer appears in a single chapter. The first seven chapters present the basic concepts of first and second laws and properties of matter in a classical or macroscopic context. The statistical viewpoint is given in Chapters 9 and 10. For those wishing to use a statistical introduction to the second law, in place of the macroscopic approach, the first five chapters should be followed by Chapter 9. Chapter 9 is then followed by Chapter 7.

Chapter 2 on the introduction of the closed system energy analysis has been rearranged, and specific work concepts have been broadened in coverage. Various discussions in Chapter 4 on properties of matter have been greatly revised. The control-volume analysis is now developed from a more general (but nonvector) approach. The steady-state, steady-flow analysis is then shown as a special case of engineering importance. In Chapter 7 the development of the concepts of a Carnot heat engine, refrigerator, and heat pump appears immediately after the development of the entropy concept and the directional second law. This strengthens immediately the use of the entropy concept and the TS diagram, without requiring any special equations for computing entropy changes.

There is a greatly expanded section in Chapter 11 on engineering application of air-water vapor mixtures. Chapter 14 now contains a third-law entropy analysis of chemically reacting ideal-gas mixtures. Included also in this chapter is a discussion of the use of the Gibbs function of formation in analyzing reactive systems, including fuel cells. In Chapter 14 the initial analysis of chemical equilibrium in reacting ideal-gas mixtures is now based on the concept of the chemical potential. The analysis of equilibrium for simultaneous chemical reactions is also discussed. Chapter 17 now contains an introductory discussion on absorption refrigeration.

The problems carried over from the second edition have been carefully examined for their appropriateness. Many new problems have been added in USCS units. The problems in SI units, of course, have never before appeared in the textbook. The total number of problems available is extremely large. An instructor should be able to select appropriate problems without repetition over several years.

The author is indebted to Linda N. Shanks for her outstanding job in typing the new portions of the manuscript. Numerous discussions with students and faculty have influenced many of the changes made in this edition. Their input is also gratefully acknowledged.

**Kenneth Wark**

# BASIC CONCEPTS AND DEFINITIONS

## 1-1 THE NATURE OF THERMODYNAMICS

Thermodynamics is a science which comprises the study of energy transformations and of the relationships among the various physical qualities, or properties,
of substances which are affected by these transformations. Predictions of the
physical properties of substances can be carried out either by analyzing the large-
scale (gross) behavior of a substance or by statistically averaging the behavior of
the individual particles which make up the substance. A thermodynamic analysis
which is undertaken without recourse to the nature of the individual particles and
their interactions falls into the domain of *classical thermodynamics*. This is the
macroscopic viewpoint toward matter and its interactions, where the overall,
large-scale effect is the focal point of interest. It requires no hypothesis on the
detailed structure of matter on the atomic scale; consequently, the general laws of
classical thermodynamics are not subject to change as new knowledge on the
nature of matter is uncovered.

Another approach to the study of thermodynamic properties and energy
relationships is based on the statistical behavior of large groups of individual
particles. This method, founded on a microscopic viewpoint, is called *statistical
thermodynamics*. It combines the computational techniques of statistical
mechanics with the findings of quantum theory. The dual purpose of statistical
thermodynamics is to predict and interpret the macroscopic characteristics of
matter in equilibrium situations from their microscopic origins. It postulates that
the values of macroscopic properties (such as pressure, temperature, and density,
among others), which we measure directly or calculate from other measurements,
merely reflect some sort of statistical average of the behavior of a tremendous

number of particles. In present technology, where substances are employed under extreme ranges of temperature and pressure, the predictive methods of statistical thermodynamics are extremely important. In addition, the interpretive role of the microscopic approach is of value. For example, the interpretation in a microscopic context of a thermodynamic property called entropy frequently is helpful in increasing a student's understanding of this concept. This viewpoint of entropy is considered in Chap. 9. This theory also was helpful in the modern development of new, direct energy-conversion methods, such as thermionics and thermoelectrics.

Another independent approach to the analysis of particle behavior is the field of *kinetic theory*. This subject is the study of particle behavior based normally on newtonian mechanics, and requires a detailed knowledge of the interactions between particles. It is extremely useful in developing relations for the transport properties of rate processes, such as viscosity, thermal conductivity, and diffusion coefficients. Since it does not take into account the quantization of energy, it is not successful in predicting thermodynamic properties except in limiting cases. A third microscopic approach is *information theory*. Its application to thermodynamics has been developed since the 1950s. It has made some important contributions to the interpretation of macroscopic state of matter in terms of microscopic behavior.

Emphasis will be placed upon the macroscopic viewpoint in this text, with the statistical viewpoint playing a supporting role. There are several reasons for this. First of all, the solution of a great majority of thermodynamic problems requires an analysis only in terms of macroscopic variables. An understanding of the underlying mechanisms for the process in terms of particle behavior would not be necessary in these cases, nor especially beneficial. Second, classical thermodynamics is an easier, more direct approach to the solution of engineering problems, and often has fewer mathematical complications. It is somewhat more free of abstractions. Finally, it may be pointed out that macroscopic analyses in a number of instances have demonstrated what major area could be studied profitably by the methods of statistical mechanics.

## 1-2 ENERGY AND SOCIETY

One measure of the standard of living in a country is its gross national product (GNP). Data worldwide indicate that the GNP of a country is directly related to the energy consumption per capita in that country. Hence the use of energy by industrialized countries is an important factor in their growth. In the early 1970s the rate of growth of energy usage per capita for the world was around 1.3 percent. In the United States the value was roughly 4 percent. On a 1.3 percent basis a doubling in energy usage would occur in about 50 years, while a 4 percent rate would double the energy usage in about 20 years. These time periods would be valid only if the population and the rate of growth of energy usage were stable. Few countries in the world have a stable or declining population. In addition, the desire of underdeveloped nations to improve their standard of living probably will

**Table 1-1 Estimates of fossil fuel resources in the U.S. in 1970, and the cumulative requirements for these fuels for the 50-year period from 1970 to 2020**

| Fuel | 1970 Resource, kWh | Cumulative requirement, kWh |
|------|--------------------|-----------------------------|
| Coal | $2 \times 10^{15}$ | $0.4 \times 10^{15}$ |
| Oil | $0.3 \times 10^{15}$ | $0.7 \times 10^{15}$ |
| Gas | $0.3 \times 10^{15}$ | $0.6 \times 10^{15}$ |

lead to an increase in the rate of growth of energy usage in the world. Hence a doubling in usage may possibly occur in less than 50 years, worldwide.

Even if one assumes a conservative estimate for the growth of energy usage, the perils that lie ahead for society are fairly apparent. Assume, for example, that the United States desires to be independent of outside sources of fossil fuels. In Table 1-1 are estimates of fossil fuel resources in the United States in terms of coal, oil, and natural gas. Note that coal represents roughly three-fourths of our estimated fossil fuel reserves. Also in the table is an estimate of the cumulative requirements for these fuels for the 50-year period from 1970 to 2020. Assuming that cost does not discourage the purchase of fuel, we find that the life expectancy of oil and natural gas supplies is quite short. If the rate of growth of coal usage remained the same after the year 2020, the life expectancy even for coal would probably not be more than several centuries.

Even though these data are rough estimates, it is obvious that a great deal of activity must occur in the energy field in order to maintain the average standard of living in industrialized countries and to increase that of the underdeveloped nations. One temporary solution is to discover and develop additional sources of oil and gas. Offshore drilling and the further development of underground reserves in Alaska are two options open to the United States. In both cases the expense of obtaining fuel from these sources is considerably higher than that of conventional sources onshore in the continental United States. Another possibility is to continue to develop alternate energy sources. Nuclear fuel for the generation of electrical power has become an important energy source throughout the world, and its use continues to expand. Unfortunately, the $^{235}U$ used in conventional reactors is not in plentiful supply. If nuclear fuel is to remain an important source of energy, it will be necessary to develop breeder reactors which produce additional fuel as well as power during their operation. Commercial operation of such reactors probably will not occur until the end of the present century. Under active investigation is the direct use of solar energy, steam produced from geothermal sources beneath the ground, wind power, and tidal power. The temperature difference between the surface and deeper layers of the ocean is also a potential source for power production.

A third alternative for an industrialized country is to cut the growth rate of energy usage. Such action does not imply a decline in living standards. The use of this alternative does require at least two things, however. One of these is the possible need for reallocation of energy usage. This requires one to establish what products and services are really necessary for the maintenance of the health and welfare of a nation. This action is primarily a political one, and leads to allocation of energy on a priority basis. Second, a serious move must be made to cut the wasteful use of energy in industry, in transportation, and in residential and commercial applications. (In 1972 the allocation of energy to these three basic users in the United States was approximately 43, 24, and 33 percent, respectively.) Such moves are in evidence in a limited number of areas already. The use of greater insulation in the walls and attics of homes and businesses, and the effort to increase the average miles per gallon on gasoline-powered automobiles, are but two examples of energy conservation.

The availability of cheap energy from fossil fuels before the early 1970s encouraged the development of devices which were inefficient in the conversion of energy to more useful forms. With increasing cost and decreasing supply of conventional fossil fuels, it is imperative that engineers look seriously at increasing the efficiency of energy-conversion devices. By efficiency or effectiveness we generally mean the ratio of the desired output to the required input. That is, it is a measure of what is accomplished compared to what it costs.

One important efficiency term for energy conversion devices which transform thermal energy (heat) into work output is the *thermal* efficiency. It is defined as the ratio of the net work output (which is desired) to the heat input (which is costly). Hence,

$$\eta_{\text{thermal}} \equiv \frac{\text{net work output}}{\text{heat input}} \tag{1-1}$$

This definition of thermal efficiency leads to a value which normally lies between 0 and 1. In most instances the ratio in Eq. (1-1) is multiplied by 100, and the thermal efficiency is expressed as a percent. The thermal efficiencies of conventional modern devices are startlingly low. Table 1-2 lists some values for mobile and stationary equipment. One of the larger consumers of fuel in the united States, the spark ignition automotive engine, has essentially the lowest efficiency. In city driving the conversion efficiency probably lies between 10 and 15 percent. The most efficient cyclic device for the production of power is the large steam plant operating on fossil fuels. Even here the efficiency barely exceeds 40 percent. The thermal efficiency of a nuclear-fuel plant is well below this value. There are means of increasing the overall efficiency of a steam power plant, but such changes increase the cost of production of electricity. Eventually the consumer must decide whether the increase in goods and services brought about by expenditure of energy per capita is worth the cost.

The data of Table 1-2 reflect the efficiencies of cyclic devices which convert chemical energy (essentially in the form of fossil fuels) into mechanical energy. Table 1-3 lists the approximate efficiencies of some other types of well-known energy converters. These converters do not operate in a cyclic manner, and have appreciably higher efficiencies. The home furnace is an excellent example of a device where considerable savings in energy consumption might be made.

**Table 1-2 Typical thermal efficiencies, in percent, of some power-producing devices**

| Type | Conditions | Efficiency |
|---|---|---|
| Automobile, spark | Optimum | 25 |
| ignition, gasoline | Steady 60 mi/h | 18 |
| | Steady 30 mi/h | 12 |
| Truck, compression | Full load | 35 |
| ignition, diesel | Half load | 31 |
| Locomotive, diesel | | 30 |
| Gas turbine (100 hp) | | |
| Without regeneration | Optimum | 12 |
| With regeneration | Optimum | 16 |
| Gas turbine(> 7500 kW) | | |
| Without regeneration | Optimum | 25 |
| With regeneration | Optimum | 34 |
| Steam power plant (> 350,000 kW) | Optimum | 41 |

Although only 30 percent of the chemical energy input is lost (as hot gas up the chimney) if the furnace combustion process is optimized, a large number of home furnaces in this country are not checked frequently enough to maintain optimum firing conditions. With dirty fuel nozzles and incorrect air supply, for example, the conversion efficiency can be relatively poor. It has been estimated that the average oil- and gas-heated home may be operating only at the 50 percent level. The significant difference in converter efficiency between a fluorescent and an incandescent lamp should also be noted. The fuel cell listed operates somewhat like a battery, with one major difference. The reactants responsible for the production of electricity from chemical energy are fed continuously to the cell (battery), and the products of the reaction are continuously withdrawn. Unlike a battery, then, the fuel cell theoretically has an unlimited life of operation. The development of the fuel cell for commercial production of electricity is a spin-off of the space age. Fuel cells provide the main source of internal power for space capsules. One

**Table 1-3 Approximate converter efficiencies, in percent, of some well-known devices**

| | |
|---|---|
| (a) Chemical to thermal<br>    Home furnace—70 | (d) Electrical to radiant<br>    Fluorescent lamp—21<br>    Incandescent lamp—7 |
| (b) Chemical to electrical<br>    Storage battery—70<br>    Dry cell battery—90<br>    Fuel cells—70 | (e) Chemical to kinetic<br>    Rockets—45<br>    Jet engines—40 |
| (c) Electrical to mechanical<br>    Electric motor—90 | (f) Potential to mechanical<br>    Hydraulic turbine—95 |

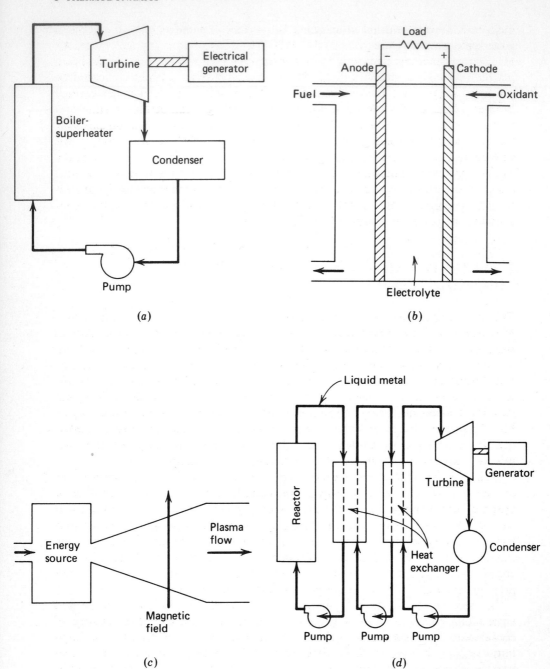

**Figure 1-1** Schematic diagrams of some energy-conversion devices. (*a*) Conventional steam power plant. (*b*) Fuel cell. (*c*) Magneto-hydrodynamic generator (MHD). (*d*) Breeder reactor cycle.

modern converter—not listed in Table 1-3—which may ultimately play a role in large-scale power production is the MHD (magnetohydrodynamic) cycle. An MHD unit generates electricity by passing a very high-temperature gas (called a plasma) through a magnetic field. The high-temperature, ionized gas could be produced through the use of chemical, nuclear, or solar sources. The gas emitted from an MHD unit is still quite hot. Hence the exit gas can be used further as a thermal source for a conventional steam power plant. A similar situation exists for a combined cycle involving a gas turbine–steam power plant combination. In this case the hot gas emitted from the gas turbine is used as the thermal source for the steam power plant. These latter two methods are just two of the possible methods under investigation for increasing the overall conversion efficiency for large electrical power generation units. Figure 1-1 shows schematic diagrams of four conventional or potential energy-conversion systems.

## 1-3 INTRODUCTION TO THE LAWS OF THERMODYNAMICS

This text presents five laws or postulates which govern the study of energy transformations and the relationships among thermodynamic properties. Two of these—the first and second laws— deal with energy, directly or indirectly. Consequently they are of fundamental importance in engineering studies of energy transformations and usage. The remaining three statements—the zeroth law, the third law, and the state postulate—relate to thermodynamic properties. They are important, since the first and second laws lead to equations which involve properties of matter. These latter three statements aid in the evaluation of such properties. In an engineering context, then, the zeroth and third laws and the state postulate are means to an end and not an end in themselves.

The first law of thermodynamics, in its ultimate useful form, is a conservation of energy statement. When energy is transferred from one region to another or changes form within a system, the total quantity of energy is constant. (In this text we shall not consider nuclear transformations of mass to energy.) Most persons have some degree of familiarity with this principle on the basis of their living experience. In addition, restrictive forms of the law appear in many introductory chemistry and physics texts. One of the purposes of this text is to organize these previous experiences into a more general and logical pattern.

The second law of thermodynamics has many ramifications with respect to engineering processes. Among others, it determines the direction of change of equilibrium for a particular system under a given set of constraints. Although important in themselves, these consequences of the second law have no direct bearing on energy transformations. The importance of the second law to society is this: The first law deals with the *quantity* of energy in terms of a conservation rule. The second law deals with the *quality* of energy. It is essentially a nonconservation rule. To speak of the quality of energy may be surprising, since quality implies that some forms of energy are more useful to society than others.

However, such an idea is directly connected with the discussion of energy and societal needs presented in Sec. 1-2. That section stressed the need to optimize the conversion, transmission, and consumption of energy. The concept of optimization of energy usage implies that there are better and worse ways of using energy. It is the potential of various forms of energy to do *useful work* for humanity that determines the quality of those forms. The second law places some restrictions on the transformation of some forms of energy to a more useful type. In addition, we are all familiar with the notion that certain "losses" are associated with energy usage. The presence of friction in a process denotes to most people a loss in performance. Thus certain features of an energy process leads to a "degradation" of energy or, again, a loss in quality. The second law enables the engineer to measure this degradation or change in quality in quantitative terms. That is, the second law provides standards of performance against which actual (real) processes may be compared, in terms of energy usage. If society is to practice conservation in the conversion and consumption of energy, such standards are vitally important. In Sec. 1-2 the values of the thermal efficiency of a number of conventional power-producing devices were tabulated. The values range from roughly 10 to 40 percent. Apparently some energy-conversion devices are much better for converting chemical, solar, or nuclear energy to mechanical work than others. However, an equally important point is whether the value of 40 percent is good, in absolute terms. If a perfect device would yield 45 or 50 percent conversion, then the actual device looks pretty good. On the other hand, if a perfect device is capable of 90 to 95 percent conversion, the 40 percent converter doesn't look too attractive and the 10 percent converter is an extremely poor device. This is one example where the second law provides that standard of performance. Thus, in terms of energy usage, the second law of thermodynamics plays an increasingly important role in the analysis of new processes and in the improvement of old ones.

## 1-4 THERMODYNAMIC SYSTEM, PROPERTY, AND STATE

There are a number of terms used in the study of thermodynamics that are usually introduced to the student in branches of chemistry, physics, and other engineering sciences. Many of these words also are in common usage by the general populace. At this point, and throughout the text, we shall take the opportunity to review concepts and redefine words that should already be a part of the vocabulary of the reader. These concepts and terms are extremely important in thermodynamic studies, hence their meaning must be clear. In some cases it may be necessary to clarify a concept solely because the meaning has been corrupted by sloppy common usage. For example, the word "heat" typically falls into this classification.

A macroscopic thermodynamic *system* is any three-dimensional region of space which is bounded by one or more arbitrary geometric surfaces. The bounding surfaces may be real or imaginary and may be at rest or in motion. The boundary may change its size or shape. The region of physical space which lies outside the arbitrarily selected boundaries of the system is called the *surroundings* or the *environment*. In its usual context the term "surroundings" is restricted to that specific localized region which interacts in some fashion with the system and hence has a detectable influence on the system.

Two examples of macroscopic systems and the representative boundaries are shown in Fig. 1-2. Part (*a*) of the figure illustrates the flow of a fluid through a pipe or duct. The dashed line is a two-dimensional representation of one possible choice of a boundary, fixed in space, which outlines the region of space for which a thermodynamic study is to be made. The inside surface of the pipe may be selected as part of the boundary, and it represents a real obstruction to the flow of matter. However, it is noted that a portion of the boundary is imaginary; i.e., there is no real surface which delineates the position of the boundary at the open ends. These latter boundaries are selected arbitrarily for accounting purposes, and they have no real effect or significance in the actual physical process. Thus we see that it is not necessary for any or all of the boundary to be physically distinguishable when making thermodynamic analyses. Nevertheless, it is extremely important to establish the boundaries of a system clearly before beginning any form of analysis.

In Fig. 1-2*b* a piston-cylinder assembly is shown. Again the selected boundary is given by the dashed line. In this particular case the boundary is physically well established, for the envelope lies just inside the walls of the cylinder and the piston. For this example it should be noted that the shape of the boundary will change when the position of the piston is altered, if the boundary is selected to surround the fluid within the assembly. The change in shape or position of the boundary is always permissible as long as such changes are recognized in subsequent calculations.

A *property* is any characteristic of a system which can in principle be specified by describing an operation or test to which the system is to be subjected. As a result of the particular operation or test, a value can be assigned to the property

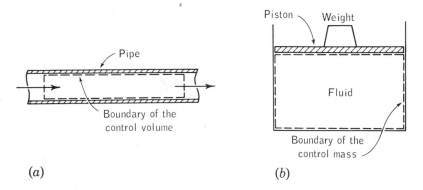

(*a*)    (*b*)

**Figure 1-2** The boundaries of two typical thermodynamic systems. (*a*) Pipe flow; (*b*) piston-cylinder assembly.

under investigation. Properties are defined macroscopically without recourse either to a specific molecular model of matter or to a statistical evaluation of the microscopic behavior of the particles within the system. Examples include pressure, temperature, mass, volume, density, electrical conductivity, acoustic velocity, coefficient of expansion, and many others.

The *state* of a macroscopic system is the condition of the system characterized by the values of its properties. Attention in this text will be directed toward what are known as equilibrium states. The word "equilibrium" is being used in its generally accepted context—the equality of forces, or the state of balance. In future discussion the term "state" will refer to an equilibrium state unless otherwise noted. The concept of equilibrium is an important one, since it is only in an equilibrium state that the thermodynamic properties have any real meaning. By definition:

> A system is in thermodynamic equilibrium if it is not capable of a finite, spontaneous change to another state without a finite change in the state of the surroundings.

There are many types of equilibrium, all of which must be met to fulfill the condition of thermodynamic equilibrium. Included among these are thermal, mechanical, phase, and chemical equilibriums. The criteria for these four types of equilibrium will be examined in some detail in this text. Thermal and mechanical equilibriums are related to the concepts of temperature and pressure, respectively. Phase equilibrium deals with the lack of a tendency for the net transfer of one or more chemical species from one phase to another, when the phases are in contact. A mixture of substances is in chemical equilibrium if there is no tendency for a net chemical reaction to occur.

A *process* is any transformation of a system from one equilibrium state to another. A description of a process typically involves specification of the initial and final equilibrium states, the path (if identifiable), and the interactions which take place across the boundaries of the system during the process. *Path* in thermodynamics refers to the specification of a series of states through which the system passes. Of special significance in thermodynamics is a quasistatic process or path. During such a process the system internally must be infinitesimally close to a state of equilibrium at all times. That is, the path of a *quasistatic* process is a series of equilibrium steps. Although a quasistatic process is an idealization, many actual processes approximate quasistatic conditions closely. It is extremely helpful in engineering analysis if an actual process can be modeled as quasistatic with negligible error. If the path of a process is a nonequilibrium one, it is still essential that the initial and final states be equilibrium states. For nonequilibrium processes, certain intermediate information during the process is missing. Nevertheless, we shall still be able to predict various overall effects, even though a detailed description is not possible.

There are several other processes in thermodynamics of special interest. These are processes during which one property remains constant, and they are designated by the prefix iso- before the property. Table 1-4 is a list of the more

**Table 1-4 A list of special constant-property processes**

| Property held constant | Name of process |
| --- | --- |
| Temperature | Isothermal |
| Pressure | Isobaric (isopiestic) |
| Volume | Isometric (isovolumic) |
| Entropy | Isentropic |
| Enthalpy | Isenthalpic |

*(handwritten annotations: GROUP bracketing Temperature through Entropy; arrow to Entropy; bracket to Enthalpy; "INTERNAL ENERGY" underlined below table)*

important ones found in this text. The last two properties in this list may be unfamiliar to the student. They will be described in detail later in the text. A *cyclic* process is defined as a process with identical end states. The change in the value of any property for a cyclic process is zero.

Any property of a thermodynamic system has a fixed value in a given equilibrium state, regardless of how the system arrives at that state. Therefore the change that occurs in the value of a property when a system is altered from one equilibrium state to another is always the same. This is true regardless of the method used to bring about a change between the two end states. The converse of this statement is equally true. If a measured quantity always has the same value between two given states, that quantity is a measure in the change in a property. This latter assertion will be useful to us in connection with the conservation of energy principle introduced in the next chapter.

The uniqueness of a property value for a given state can be described mathematically in the following manner. The integral of an *exact differential dY* is given by

$$\int_{1}^{2} dY = Y_2 - Y_1 = \Delta Y \tag{1-2}$$

Thus the value of the integral depends solely on the initial and final states. But the change in the value of a property likewise depends only on the end states. Hence the differential change $dY$ in a property $Y$ is an exact differential. Throughout this text the infinitesimal variation of a property will be signified by the differential symbol $d$ preceding the property symbol. For example, the infinitesimal change in the pressure $P$ of a system is given by $dP$. The finite change in a property is denoted by the symbol $\Delta$ (capital delta), for example, $\Delta P$. The change in a property value $\Delta Y$ always represents the final value minus the initial value. This convention must be kept in mind.

Properties are classified as either extensive or intensive. Consider a system arbitrarily divided into a group of subsystems. A property is *extensive* if its value for the whole system is the sum of its values for the various subsystems or parts. If a system is divided into $n$ subsystems, then the extensive property $Y$ for the whole system is given by

$$Y_{\text{system}} = \sum_{i=1}^{n} Y_i$$

where $Y_i$ is the property value for the $i$th subsystem. Examples of extensive properties include volume $V$, energy $E$, and quantity of electric charge $Q_e$. Unlike extensive properties, *intensive* properties have values which are independent of the extent or mass of the system. If a single-phase system in equilibrium is divided arbitrarily into a set of subsystems, the value of a given intensive property will be the same for each macroscopic subsystem. Thus intensive properties have the same value at every small region or point within a system at equilibrium. Examples of intensive properties include temperature, pressure, density, velocity, and concentration. When an extensive property is divided by the total mass of a system, the resulting property is called a *specific* property. For systems in equilibrium a specific property is an intensive property of the system. Examples include specific volume $(v = V/m)$ and specific energy $(e = E/m)$. Capital letters will be used, generally, to denote extensive properties (with mass $m$ a major exception). Lowercase letters will be used for intensive (specific) properties, the most notable exceptions being pressure $P$ and temperature $T$. These latter two properties are always intensive.

## 1-5 DIMENSIONS AND UNITS

*Dimensions* are those names which are used to characterize physical quantities. Common examples of dimensions include length $L$, time $t$, force $F$, mass $M$, electric charge $Q_e$, and temperature $T$. In engineering analysis any equation which relates physical quantities must be dimensionally homogeneous. Dimensional homogeneity requires that the dimensions of terms on one side of an equation equal those on the other side. Such homogeneity also is retained by an equation during any subsequent mathematical operation, and thus is a powerful tool for checking the internal consistency of an equation.

In order to make numerical computations involving physical quantities, there is the additional requirement that units, as well as the dimensions, be homogeneous. *Units* are those arbitrary magnitudes and names assigned to dimensions which are adopted as standard for measurements. For example, the primary dimension of length may be measured in units of feet, miles, centimeters, angstroms, etc. These are all arbitrary lengths which may be related to each other in terms of unit conversion factors, or unitary constants. Unit conversion factors include 12 inches = 1 foot and 60 seconds = 1 minute. A comparable way of writing unit conversion factors is

$$\frac{12 \text{ in}}{1 \text{ ft}} = 1 \qquad \frac{60 \text{ s}}{1 \text{ min}} = 1$$

Terms in equations can always be multiplied by unit conversion factors, since it is always permissible to multiply by unity. A given physical quality may often be measured in several sets of units; it is therefore important to have available tables of common unit conversion factors. These are presented in most engineering and scientific handbooks, such as Marks's Handbook (mechanical engineering),

Perry's Handbook (chemical engineering) the Handbook of Chemistry and Physics, etc. A brief tabulation of unit conversion factors useful for thermodynamic problems met in this text is presented in Table A-1, in the Appendix.

A number of systems of units have been developed over the years. However, in this text we shall consider only two: the SI and the United States Customary system (USCS).

### a The Système Internationale (International System)

The fundamental system of units chosen for scientific work all over the world is the Système Internationale, which is usually abbreviated as SI. The SI employs seven primary dimensions: mass, length, time, temperature, electric current, and luminous intensity, and amount of substance. Table 1-5 lists the SI units which are standards for these dimensions. All these units are defined operationally. The SI unit of length is the meter (m), and is defined as 1,650,763.73 wavelengths of the orange-red line of emission from krypton-86 atoms in vacuum. (Previously the accepted standard was a distance marked on a platinum-iridium bar maintained under prescribed conditions in Sèvres, France.) The unit of time is the second (s), and it is defined as the duration of 9,192,631,770 $\pm$ 20 cycles of a specified transition within the cesium atom. The previous legal standard for time was the mean solar second, which corresponds to 1/86,400 of the mean solar day. The SI unit of mass is the kilogram (kg), and it is represented by the mass of a platinum-iridium cylinder also preserved in Sèvres, France, at the International Bureau of Weights and Measures. The unit of temperature is the kelvin (K), but in this text it will be known as the $°K$. The origin of this unit is discussed more fully in Sec. 1-7. There are also precise operational definitions of the ampere (A), the candela (cd), and the mole (mol).

A mole is the amount of substance containing the same number of particles as there are atoms in 0.012 kg of the pure carbon nuclide $^{12}C$. It has heretofore been frequently used as a unit for mass. For example, the gram-mole used by chemists is that amount of mass which contains as many atoms or molecules of a substance as the number of atoms in 0.012 kg (12 g) of carbon-12. A kilogram-mole of a

### Table 1-5 SI and USCS base units

| Physical quantity | SI Unit and symbol | USCS unit and symbol |
|---|---|---|
| Mass | kilogram (kg) | pound-mass ($lb_m$) |
| Length | meter (m) | foot (ft) |
| Time | second (s) | second (s) |
| Temperature | kelvin (K) | Rankine ($°R$) |
| Electric current | ampere (A) | ampere (A) |
| Luminous intensity | candela (cd) | candle |
| Amount of substance | mole (mol) | |
| Force | | pound-force ($lb_f$) |

substance is 1,000 times as large as a gram-mole. For example, a kilogram-mole of diatomic oxygen, $O_2$, contains 32 kg of oxygen.

All other units in SI are secondary ones, and are derivable in terms of these seven primary or base units. The SI unit of force is the newton (N), and it is derived from Newton's second law, $F = ma$. On the basis of this equation, a force of one newton accelerates one kilogram of mass at one meter per second per second. Since 1 newton = 1 kilogram × 1 meter/(second)$^2$, then

$$\frac{1 \text{ N·s}^2}{\text{kg·m}} = 1 \tag{1-3}$$

Equation (1-3) is a unit conversion factor which relates the derived force unit to the primary mass, length, and time units in the SI. It is useful in converting other secondary units, which are partially defined in terms of force, into a set of units containing only the kilogram, the meter, and the second. For example, the SI unit for pressure is the pascal (Pa), which is defined in force and length units as $1 \text{ N}/m^2$. Employing Eq. (1-3) we find that

$$1 \text{ pa} = \frac{1 \text{ N}}{\text{m}^2} \times \frac{\text{kg·m}}{1 \text{ N·s}^2} = 1 \text{ kg/m·s}^2$$

## Table 1-6  Some SI and USCS derived units

A  SI (Metric)

| Physical quantity | Unit | Symbol | Definition |
|---|---|---|---|
| Force | newton | N | $1 \text{ kg·m/s}^2$ |
| Pressure | pascal | Pa | $1 \text{ kg/m·s}^2$ ($= 1 \text{ N/m}^2$) |
| | bar | bar | $10^5 \text{ kg/m·s}^2$ ($= 10^5 \text{ N/m}^2$) |
| Energy | joule | J | $1 \text{ kg·m}^2/\text{s}^2$ ($= 1 \text{ N-m}$) |
| Power | watt | W | $1 \text{ kg·m}^2/\text{s}^3$ ($= 1 \text{ J/s}$) |
| Electric quantity | coulomb | C | $1 \text{ A·s}$ |
| Electric potential difference | volt | V | $1 \text{ kg·m}^2/\text{A·s}^3$ ($= 1 \text{ A·}\Omega$) |
| Electric resistance | ohm | $\Omega$ | $1 \text{ kg·m}^2/\text{A}^2\text{·s}^3$ ($= 1 \text{ V/A}$) |
| Electric capacitance | farad | F | $1 \text{ A}^2\text{·s}^4/\text{kg·m}^2$ ($= 1 \text{ C/V}$) |

B  USCS (Engineering)

| Physical quantity | Unit | Symbol | Definition |
|---|---|---|---|
| Force | pound-force | $\text{lb}_f$ | $32.174 \text{ lb}_m\text{·ft/s}^2$ |
| Pressure | atmosphere | atm | $68,087 \text{ lb}_m/\text{ft·s}^2$ ($= 14.696 \text{ lb}_f/\text{in}^2$) |
| Energy | foot-pound-force | $\text{ft·lb}_f$ | $32.174 \text{ lb}_m\text{·ft}^2/\text{s}^2$ |
| Power | foot-pound-force/s | $\text{ft·lb}_f/\text{s}$ | $32.174 \text{ lb}_m\text{·ft}^2/\text{s}^3$ ($= 1.82 \times 10^{-3} \text{ hp}$) |

As a result, the secondary unit of pressure, the pascal, can be expressed in terms of its primary equivalent, 1 kg/m·s². Table 1-6 lists some SI secondary or derived units in terms of the primary or base units found in Table 1-5. Special decimal fractions and multiples are designated by certain prefixes in SI units. Among the most important of these are: $10^{-6}$, micro; $10^{-3}$, milli; $10^{-2}$, centi; $10^3$, kilo; and $10^6$, mega. For example, a pressure of 1 bar is equivalent to 100 kilopascals.

A more practical example of the use of the unit conversion factor relating force to mass is the conversion of a set of primary units into a secondary unit. Consider the concept of linear kinetic energy, which is derived from Newton's second law. The familiar result is the term $mV^2/2$. Using primary units and neglecting the factor of 2 in the denominator (since it is unitless),

$$mV^2 = \text{kg} \times \frac{m^2}{s^2} = \text{kg·m}^2/\text{s}^2$$

Kinetic energy expressed in units of kg·m²/s² is perfectly acceptable. However, such usage is unconventional, since the secondary unit for energy in the SI is the joule (J). Conversion to the joule is accomplished through the use of Eq. (1-3). For linear kinetic energy

$$mV^2 = \frac{\text{kg·m}^2}{s^2} \times \frac{1 \text{ N·s}^2}{\text{kg·m}} = \text{N·m} = \text{J}$$

If the mass is expressed in kilograms and the velocity in m/s, the value of $mV^2$ comes out in joules.

In conjunction with Newton's second law it is important to note that *weight* always refers to a force. When it is stated that a body weighs a given amount what we really mean is that this is the force with which the body is attracted toward another body, such as the earth or moon. The acceleration of gravity varies with the distance between two bodies (such as the height of a body above the earth's surface). Thus the weight of a body varies with elevation, while the mass of a body is constant with elevation.

**Example 1-1M** The weight of a piece of metal is 100.0 N at a location where the local acceleration of gravity $g$ is 10.60 m/s². What is the mass of the metal, in kg, and what is the weight of the metal on the surface of the moon, where $g = 1.67$ m/s²?

SOLUTION In this case we write Newton's second law as $F = mg$, since $a = g$. Hence,

$$m = \frac{F}{g} = \frac{100.0(\text{N})}{10.60(\text{m/s}^2)} = 9.434 \frac{\text{N}}{\text{m/s}^2} = 9.434 \text{ kg}$$

The mass of the piece of metal will remain the same regardless of its location. The weight will change, however, with a change in gravitational acceleration. Equating weight to force on the surface of the moon,

$$\text{weight} = F_{moon} = mg = 9.434(\text{kg}) \times 1.67(\text{m/s}^2) = 15.8 \text{ N}$$

Although the mass is the same at the two locations, the weight is significantly different.

## The United States Customary System of Units

An important system of units employed in the United States is the USCS. Unfortunately, it introduces two points of confusion. First of all, pound is used to denote both a unit of force ($lb_f$) and a unit of mass ($lb_m$). If the symbol lb is found in this text without a subscript, it should always be taken as a mass unit. Second, both force and mass are chosen as primary dimensions. Consequently there are four primary dimensions to be used in conjunction with Newton's second law, rather than the three (mass, length, and time) used in the SI. Independent of Newton's second law, the standard gravitational force is defined such that a force of one pound ($lb_f$) will accelerate a mass of one pound ($lb_m$) at a rate of 32.1740 ft/s². Consequently we must write Newton's second law in the form

$$F = \frac{ma}{g_c}$$

when we use the USCS units, where $g_c$ is a proportionality constant. Substituting the definition of a pound force into this modified form of Newton's law, we find that

$$1 \ lb_f = \frac{(1 \ lb_m)(32.174 \ ft/s^2)}{g_c}$$

or that

$$g_c = \frac{32.174 \ lb_m \cdot ft}{lb_f \cdot s^2}$$

Hence, in order to use the USCS units, Newton's second law must be written as

$$F = \frac{ma}{g_c} = \frac{ma}{32.174} \tag{1-4}$$

If, in this equation, the mass is expressed in $lb_m$ and the acceleration in ft/s², then the value of the force will be in $lb_f$. Also keep in mind that weight is a force.

**Example 1–1** The weight of a piece of metal is 220.5 $lb_f$ at a location where the local acceleration of gravity $g$ is 30.50 ft/s². What is the mass of the metal, in $lb_m$, and what is the weight of the metal on the surface of the moon, where $g = 5.48$ ft/s²?

SOLUTION In this case we write Newton's second law as $F = mg/g_c$, since $a = g$. Hence,

$$m = \frac{g_c F}{g} = \frac{32.174(lb_m \cdot ft/lb_f \cdot s^2) \times 220.5(lb_f)}{30.50(ft/s^2)} \doteq 232.6 \ lb_m$$

The mass of the piece of metal will remain the same regardless of its location. Its weight will change, however, with a change in gravitational acceleration. Equating weight to force on the surface of the moon,

$$weight = F_{moon} = \frac{mg}{g_c} = \frac{232.6(lb_m) \times 5.48(ft/s^2)}{32.174(lb_m \cdot ft/lb_f \cdot s^2)} = 39.6 \ lb_f$$

Although the mass is the same at the two locations, the weight is significantly different.

In the USCS we can also use the mole as a unit of mass. In this case it is called the pound-mole (lb-mole). Since the definitions of a mole is the same as noted earlier for the SI units, a pound-mole of diatomic oxygen will contain 32.0 $lb_m$. There are approximately 454 g in a pound-mass. Hence a pound-mole is equivalent to 454 g·mol or 0.454 kg·mol.

The definition of a pound force presented earlier allows us to develop a unit conversion factor relating the pound mass to the pound force. Since

$$1 \ (lb_f) \equiv 32.174 \ \frac{lb_m \cdot ft}{s^2}$$

it follows that

$$\frac{32.174 \ lb_m \cdot ft}{lb_f \cdot s^2} = 1 \qquad (1\text{-}5)$$

Because the term on the left is equal to unity, it can be inserted into any equation without altering the validity of the equation. This unit conversion factor is useful when it is necessary to convert a particular set of units into a more convenient or conventional set. For instance, consider the linear kinetic energy term, $mV^2/2$. If the mass is given in $lb_m$ and the velocity in ft/s (and neglecting the factor 2 which is unitless), then in the USCS (or English engineering) system

$$mV^2 = lb_m \times \frac{ft^2}{s^2} = lb_m \cdot ft^2/s^2$$

Although the above units for $mV^2$ are correct, the more customary units for kinetic energy (KE) are ft·$lb_f$. We achieve the more desirable units by using Eq. (1-5) in the following manner.

$$KE = mV^2 \times \frac{lb_m \cdot ft^2}{s^2} \times \frac{lb_f \cdot s^2}{32.174 \ lb_m \cdot ft} = \frac{mV^2}{32.174} \ (ft \cdot lb_f)$$

Thus, when the primary USCS units given in Table 1-5 are used to evaluate the linear kinetic energy, the resulting value must be divided by the unit conversion factor given by Eq. (1-5) if units of ft·$lb_f$ are desired.

It might be noted that the unit conversion factor of Eq. (1-5) has the same value as $g_c$ in the modified form of Newton's second law, $F = ma/g_c$. Hence a proper format for the linear kinetic energy could be $mV^2/2g_c$ when using USCS units. Older editions of this textbook and some engineering thermodynamics texts by other authors insert the $g_c$ in the denominator whenever the term $mV^2/2$ appears in an equation. This acts as a reminder that the units of $mV^2$ will not be ft·$lb_f$ unless a unit conversion factor is employed. In this edition of the textbook, the term $g_c$ is omitted from all equations. The student should check the units of all terms in an equation, and determine which ones might require some type of unit conversion factor in order to get all quantities in the equation into a consistent set of units. Recall that all terms in an equation must be homogeneous in units.

The kinetic energy quantity is not the only term which frequently requires the use of the unit conversion factor in Eq. (1-5), or $g_c$, in its evaluation. The gravitational potential energy, *mgz*, likewise requires a unit conversion when English engineering units are used and ft·lb$_f$ are the desired units. There are other relationships in the engineering literature, usually based on Newton's second law or on a force balance, which typically require the unit conversion factor given by Eq. (1-5).

## 1-6 DENSITY, SPECIFIC WEIGHT, SPECIFIC VOLUME, AND PRESSURE

Macroscopic thermodynamics deals with systems containing a very large number of particles. It is convenient to consider matter within such systems as a continuum. That is, matter is regarded as distributed continuously throughout space. From a continuum viewpoint it is acceptable to speak of properties at a point. The *density* $\rho$ in the vicinity of a point within a system is defined as the mass per unit volume, $m/V$. The *specific weight* w of a substance is defined as the weight per unit volume. The relation between the specific weight of a body and its density may be obtained from Newton's second law, $F = ma$. If both sides of this equation are divided by the volume of a substance, then $F/V = (m/V)a$, or

$$w = \rho g \qquad (1\text{-}6a)$$

where $g$ is the local acceleration of gravity. When the USCS of units is used, this equation must be modified to

$$w = \frac{\rho g}{g_c} = \frac{\rho g}{32.17} \qquad (1\text{-}6b)$$

The specific weight of a substance is not truly a property of the substance, since the value of w depends upon the local acceleration of gravity. The *specific gravity* of a substance is defined as the ratio of its density to that of water at a specified temperature.

The *specific volume* v is defined as the reciprocal of the density. Thus $v = 1/\rho = V/m$ = volume/mass. The mass units are normally pounds or kilograms. In certain circumstances, however, it is useful to employ the mole as the unit of mass. This may be either a pound-mole or a gram-mole. The number of moles N of a substance is defined as $N = m/M$, where $M$ is the molar mass. The *molar mass* is the mass of a substance which is numerically equal to its molecular weight in any mass unit expressed as mass per mole. For example, the molar mass of helium is 4.003 g/g·mol or 4.003 lb/lb·mol. Thus any property which is partially defined in terms of the mass of the system may be expressed in two ways. As an example, the density might be a mass density or a molar density. (They differ from each other only in that their ratio is a constant.) No special symbol will be used to differentiate between these two ways of expressing properties in terms of mass

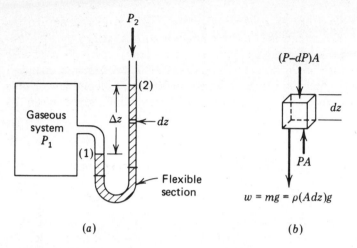

**Figure 1-3** (*a*) Measurement of a pressure differential in terms of a height of a liquid column. (*b*) Force balance on a fluid element within the liquid column.

units. They will be distinguished either by context and usage or by a definite comment as to which is being employed.

The *pressure P* is defined as the normal force per unit area acting on some real or imaginary boundary. Normal forces in static equilibrium will always be considered as compressive; hence the pressure is a positive quantity. In static systems the pressure is uniform in all directions around the vicinity of an elemental volume of fluid. It is well known, however, that the pressure may vary throughout the system for a fluid in the presence of a gravitational field. A familiar example is the variation of the pressure with depth in the common swimming pool.

Pressure differences are frequently measured in terms of the height of a liquid column. Consider the situation shown in Fig. 1-3*a*. A tube containing a liquid is connected to a tank which is filled with a gas at a uniform pressure $P_1$. A pressure $P_2$ exists outside the tank, and this pressure is also applied to the top of the liquid column. The pressure differential $P_1 - P_2$ can be determined from a knowledge of the height $\Delta z$ of the liquid column. The differential height $dz$ shown in Fig. 1-3*a* is shown as a fluid element in Fig. 1-3*b*. Three forces act on the fluid element in the $z$ direction. Two of these are normal compressive forces, and the third is the weight of the element in the gravitational field $g$. On the basis of a force balance for the static situation,

$$\rho(A\,dz)g + PA = (P - dP)A$$

where $A$ is the cross-sectional area normal to the $z$ direction. This equation reduces to $dP = -\rho g\,dz$. If the gravitational acceleration $g$ and the density of the fluid $\rho$ are assumed to be constant, then integration of this expression shows that

$$\Delta P = P_2 - P_1 = -\rho g\,\Delta z = -w\,\Delta z \tag{1-7}$$

The negative sign results from the convention that the height $z$ is measured as positive upward, whereas $P$ decreases in this direction. When English engineering units are employed, the term $w$ must be evaluated by means of Eq. (1-6$b$) if $\Delta P$ is to be expressed as force per unit area.

The actual pressure at a given position in a system is called its *absolute* pressure. The modifying adjective is necessary since, experimentally, most pressure-measuring devices indicate what is known as a gage or vacuum pressure. Absolute pressures must be used in thermodynamic relationships. A positive *gage* pressure is the difference between the absolute pressure and the pressure exerted by the atmosphere at that location. That is

$$P_{gage} = P_{abs} - P_{atm} \tag{1-8}$$

A negative gage pressure (which occurs when the atmospheric pressure is greater than the absolute pressure) is called a *vacuum*. In order to make a vacuum measurement a positive value it is defined as

$$P_{vacuum} = P_{atm} - P_{abs} \tag{1-9}$$

Note that positive and negative gage pressures are pressure differentials. For pressures below atmospheric and slightly above atmospheric this pressure differential is frequently measured by the method shown in Fig. 1-3$a$. The device is known as a manometer, and the liquid it contains may be mercury, alcohol, water, oil, or some other fluid. The pressure differences expressed by Eqs. (1-8) and (1-9) can be evaluated by using Eq. (1-7). The absolute pressure within a system is then determined by combining the evaluation of the pressure difference with an independent measurement of the atmospheric pressure. Figure 1-4 illustrates the relations among the various types of pressures.

Atmospheric pressure is commonly spoken of as the *barometric* pressure. Its value is not fixed, but varies with time and location on the earth. One reference value for pressure frequently used is the *standard atmosphere*. It is defined as the pressure produced by a column of mercury exactly 760 mm in height at $273.15°\,$K

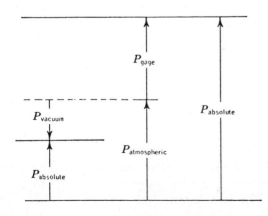

**Figure 1-4** The relationships among the absolute, atmospheric, gage, and vacuum pressures.

or 0°C (32°F) and under standard gravitational acceleration. Some equivalent values in other units are

$$1 \text{ standard atmosphere (atm)} = \begin{cases} 14.696 \text{ lb}_f/\text{in}^2 \text{ (psi)} \\ 29.92 \text{ in of Hg at } 0°C \\ 1.013 \times 10^5 \text{ N/m}^2 = 1.013 \text{ bar} \end{cases}$$

In order to distinguish between absolute and gage pressures in this text, it should be assumed that data are absolute values unless denoted explicitly or implicitly as gage values. In the technical literature the letters $a$ and $g$ are frequently added to abbreviations in order to signify the difference. For example, the absolute and gage pressures in pounds per square inch are denoted by the terms "psia" and "psig," respectively. In some circumstances the symbols "lb/in² gage" or "mbar gage" may be used.

The standard pressure unit in SI is the pascal, which is defined as $1 \text{ N/m}^2$. Although a kilopascal (kPa) is frequently used is a convenient unit of pressure, this text adopts the bar as the SI pressure unit. By definition, a bar equals $10^5 \text{ N/m}^2$. As noted above, it is slightly smaller than one atmosphere.

**Example 1-2M** A barometer reads 735 mm Hg. Determine the atmospheric pressure in mbar by employing Eq. (1-6).

SOLUTION The use of Eq. (1-6) requires data on $\rho$ and $g$. The density of mercury at 0°C is 13.595 g/cm³, and we shall assume this value is relatively independent of temperature. Since the location is not specified, we shall also assume that $g$ is the standard value of 9.807 m/s². With reference to Fig. 1-3, for a barometer $P_1$ is atmospheric pressure and $P_2$ is a complete vacuum, that is, $P_2 = 0$. On the basis of Eq. (1-6),

$$P_{atm} = P_1 = 13.595 \text{ (g/cm}^3) \times 9.807 \text{ (m/s}^2) \times 735 \text{ (mm)} \times 10^{-3} \text{ (m/mm)}$$

$$\times 10^6 \text{ (cm}^3/\text{m}^3) \times 1 \text{ (N·s}^2/\text{kg·m}) \times 10^{-3} \text{ (kg/g)}$$

$$= 0.980 \times 10^5 \text{ N/m}^2 = 0.980 \text{ bar} = 980 \text{ mbar}$$

This value is slightly less than a standard atmosphere, which is 1013 mbar.

**Example 1-2** A barometer reads 28.90 in Hg. Determine the atmospheric pressure in lb$_f$/in² (psi) by employing Eq. (1-6).

SOLUTION The use of Eq. (1-6) requires data on $\rho$ and $g$. The density of mercury at 32°F (which is assumed to be constant) is 13.595 g/cm³. Since the location is not known, we shall assume that $g$ is the standard value of 32.174 ft/s². With reference to Fig. 1-3, for a barometer $P_1$ is atmospheric pressure and $P_2$ is zero, that is, a vacuum. First, on the basis of Table A-1,

$$\rho = 13.595 \text{ (g/cm}^3) \times (1 \text{ lb/ft}^3/0.01602 \text{ g/cm}^3) = 849 \text{ lb/ft}^3$$

Application of Eq. (1-6) then yields

$$P_{atm} = 849 \text{ (lb/ft}^3) \times 32.174 \text{ (ft/s}^2) \times 28.80 \text{ (in)}$$

$$\times \text{ (ft}^3/1728 \text{ in}^3) \times (\text{lb}_f \cdot \text{s}^2/32.174 \text{ lb}_m \cdot \text{ft})$$

$$= 14.20 \text{ lb}_f/\text{in}^2$$

**Example 1-3M** If the barometer reads 735 mm Hg, determine what absolute pressure in bars is equivalent to a vacuum of 280 mm Hg within a system. Neglect effect of temperature on the density of mercury.

SOLUTION A vacuum reading is the difference between the barometric or atmospheric pressure and the absolute pressure, on the basis of Eq. (1-8). In this case the absolute pressure within the system is equal to $735 - 280$, or 455 mm Hg. From Table A-1M it is noted that 760 mm Hg = 1 atm = 1.013 bar at 0°C. Hence

$$P_{abs} = 455 \text{ mm Hg} \times 1.013 \text{ bar}/760 \text{ mm Hg} = 0.606 \text{ bar}$$

**Example 1-3** If the barometer reads 29.1 in Hg, determine what absolute pressure in psia is equivalent to a vacuum of 11 in Hg within a system.

SOLUTION Since a vacuum reading is the difference between the barometric or atmospheric pressure and the absolute pressure, the absolute pressure within the system is equal to $29.1 - 11.0$, or 18.1 in Hg abs. From Table A-1 it is noted that 1 in Hg is equal to 0.491 psi at 32°F. Assuming that the specific gravity of mercury is constant, we, find that

$$P_{abs} = 18.1 \text{ in Hg} \times 0.491 \text{ psia/in Hg} = 8.9 \text{ psia}$$

√ **Example 1-4M** A pressure gage reads 1.60 bars when the barometer has a reading of 755 mm Hg. Determine the absolute pressure of the system in bars.

SOLUTION The absolute pressure is the sum of the gage pressure and the barometric pressure. Using the conversion factors from Table A-1M,

$$P_{bar} = 755 \text{ mm Hg} \times 1.013 \text{ bar}/760 \text{ mm Hg} = 1.01 \text{ bar}$$

$$P_{abs} = P_{gage} + P_{bar} = 1.60 + 1.01 = 2.61 \text{ bar}$$

This is equivalent to 2.58 atm.

**Example 1-4** A pressure gage reads 23.0 psi when the barometer has a reading of 28.9 in Hg. Determine the absolute pressure of the system.

SOLUTION The absolute pressure is the sum of the gage pressure and the barometric pressure. Therefore

$$P_{bar} = 28.9 \text{ in Hg} \times 0.491 \text{ psi/in Hg} = 14.1 \text{ psi}$$

$$P_{abs} = P_{gage} + P_{bar} = 23.0 + 14.1 = 37.1 \text{ psia}$$

## 1-7 TEMPERATURE AND THE ZEROTH LAW

The concept of an absolute temperature is tied in intimately with the second law of thermodynamics. Hence a discussion along these lines must be delayed until later. In this section we shall discuss some important characteristics of temperature. Consider two systems $X$ and $Y$ which initially are individually in equilibrium. If they are brought into contact through a common rigid boundary, there are two possible outcomes with respect to the final states of the systems. One possibility is for the states of $X$ and $Y$ to remain unchanged macroscopically. As a second possibility, both systems will be observed to undergo a change of state until each reaches a new equilibrium state. These changes of state are due to an interaction between $X$ and $Y$. When two systems which are isolated from the local surroundings undergo no further change of state even when in contact through a common rigid boundary, they are said to be in *thermal equilibrium*. It is important to note that both systems change their state as they proceed toward thermal equilibrium.

It is a matter of experience that the value of a single property is sufficient to determine whether systems will be in thermal equilibrium when placed in contact through a common rigid boundary. This property is the *temperature T*. If an interaction occurs, the two systems involved are said to be of unequal temperatures. Such an interaction will continue until the temperatures of the two systems become equal and thermal equilibrium prevails.

Temperature is a property of great importance in thermodynamics, and its value can be obtained easily by indirect measurement with calibrated instruments. The temperature of a system is determined by bringing a second body, a thermometer, into contact with the system and allowing thermal equilibrium to be reached. The value of the temperature is found by measuring some temperature-dependent property of the thermometer. Any such property is called a *thermometric* property. Commonly used properties of materials employed in temperature sensing devices include:

1. Volume of gases, liquids, and solids
2. Pressure of gases at constant volume
3. Electric resistance of solids
4. Electromotive force of two dissimilar solids
5. Intensity of radiation (at high temperatures)
6. Magnetic effects (at extremely low temperatures)

On the basis of experimental observation, it is found that when each of two systems is in thermal equilibrium with a third system, they also will be in thermal equilibrium with each other. This statement is a thermodynamic postulate known as the *zeroth* law. Although it is a statement of common experience, it is not derivable from other laws or definitions. Historically the first and second laws were already numbered before the zeroth law was recognized as an independent postulate. Therefore it is called the zeroth law since in any logical presentation of thermodynamics it precedes the development of the first and second laws.

This law is of importance in the field of thermometry and in the establishment of empirical temperature scales. In practice, the third system in the zeroth law is a thermometer. It is brought into thermal equilibrium with a set of temperature standards and is calibrated. At some later time the thermometer is equilibrated with a system at an unknown temperature, and a value is determined. If thermal equilibrium exists during the calibration process and during the test of the system, the temperature of the system must be the same as that set by the calibration standards, on the basis of the zeroth law.

The absolute temperature scale used by scientists and engineers in the SI is the Kelvin scale. Since 1954 it has been recommended by an international conference that a reference state value of 273.16 be assigned on the Kelvin temperature scale to that state where solid, liquid, and gaseous water all coexist in equilibrium. A state in which three phases of a pure substance coexist in equilibrium is called a triple state of the substance. The triple state of water is 0.01 kelvin ($^\circ$K) higher than the freezing or ice point of water. Thus water freezes at 273.15$^\circ$K. The Celsius

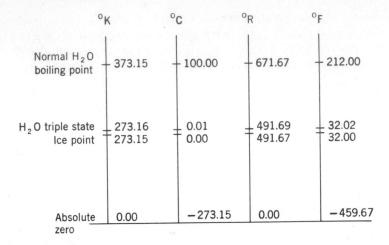

| | °K | °C | °R | °F |
|---|---|---|---|---|
| Normal $H_2O$ boiling point | 373.15 | 100.00 | 671.67 | 212.00 |
| $H_2O$ triple state | 273.16 | 0.01 | 491.69 | 32.02 |
| Ice point | 273.15 | 0.00 | 491.67 | 32.00 |
| Absolute zero | 0.00 | −273.15 | 0.00 | −459.67 |

**Figure 1-5** Comparison of temperature scales.

temperature scale, in °C (formerly called the centigrade scale), is related to the Kelvin scale by the relation

$$T_C = T_K - 273.15 \tag{1-10}$$

Two additional temperature scales are in common usage in this country—the Rankine and the Fahrenheit scale. The temperature in degrees Rankine (°R) is defined arbitrarily at 1.8 times the temperature in kelvins. Consequently,

$$T_R = 1.8 T_K \tag{1-11}$$

In terms of temperature intervals, one sees that

$$1°K = 1.8°R$$

The triple state temperature of water thus is 491.69°R. The Fahrenheit scale $T_F$ (°F) is defined as

$$T_F = T_R - 459.67 \tag{1-12}$$

It is fairly common to round off the last number of Eqs. (1-10) and (1-12) to 273 and 460, respectively, when making engineering calculations. A comparison of these four temperature scales is shown in Fig. 1-5. Familiarity with the interrelationships of these different scales is essential in carrying out thermodynamic calculations. In thermodynamic relationships a symbol $T$ for temperature always implies °K or °R unless specifically stated in terms of the other two scales.

## 1-8 IDEAL-GAS EQUATION OF STATE

An equation of state is a relationship among the properties $y_1$, $y_2$, and $y_3$ such that

$$f(y_1, y_2, y_3) = 0$$

Of particular interest in thermodynamics are equations which relate the variables $P$, $v$, and $T$. On the basis of experimental work carried out originally by Boyle, Charles, and Gay-Lussac, the $PvT$ behavior of many gases at low pressures and moderate temperatures can be approximated quite well by the *ideal (perfect)-gas equation* of state, namely,

$$PV = NR_u T \tag{1-13}$$

or

$$P\bar{v} = R_u T \tag{1-14}$$

where $N$ = number of moles of a gas
$\bar{v}$ = specific volume on a molar basis
$R_u$ = a universal gas constant
(The bar over the $v$ is used only in this section to denote the specific volume on a mole basis. In future sections the units on $v$ should be clear from the context of the equation or discussion.) The values of $R_u$ in several sets of units are

$$R_u = \begin{cases} 0.08315 \text{ bar·m}^3/(\text{kg·mol})(°\text{K}) \\ 8.315 \text{ kJ/(kg·mole})(°\text{K}) \\ 1545 \text{ ft·lb}_f/(\text{lb·mol})(°\text{R}) \\ 0.730 \text{ atm·ft}^3/(\text{lb·mol})(°\text{R}) \\ 1.986 \text{ Btu/(lb·mol})(°\text{R}) \end{cases}$$

These values are also tabulated in Table A-1 in the appendix.

The ideal-gas equation is frequently used with mass units such as pounds and grams instead of pound-moles and gram-moles. In such cases one uses a specific gas constant $R$ in the equation instead of the universal value $R_u$. It is to be noted that

$$R \equiv \frac{R_u}{M}$$

where $M$ is the molar mass. Values of $M$ are given for some elements and common substances in Table A-2. The equivalent forms of the ideal-gas equation become

$$Pv = \frac{R_u T}{M} = RT \qquad PV = mRT \qquad P = \rho RT \qquad PV = \frac{mR_u T}{M} \tag{1-15}$$

where $v$ = specific volume on a mass basis
$\rho$ = density
$m$ = mass of system
Since $R$ depends upon the molar mass of a substance, its value is different for each substance, even when expressed in the same set of units. For example, if the average molar mass of air is taken as 28.97 (or roughly 29), then $R$ for air in the SI is 0.287 J/(g·mol)(°K). The $R$ value for hydrogen in the same set of units is

4.12 J/(g·mol)(°K). It should be carefully noted that the temperature in the ideal-gas equation of state is always expressed in either kelvins or degrees Rankine.

✓ **Example 1-5M** Nitrogen gas, $N_2$, at a pressure of 1.40 bars is maintained at a temperature of 27°C. Determine the specific volume in $m^3/kg$ if it is reasonable to assume that the gas behaves ideally.

SOLUTION From Eq. (1-14) it is seen that $v = R_u T/PM$. The absolute temperature is 273 + 27 = 300°K, the molar mass of nitrogen from Table A-2 is 28.01 kg/kg·mol, and a convenient value of $R_u$ is 0.08315 bar·m³/(kg·mol)(°K). Substitution of these values leads to

$$v = \frac{0.08315 \text{ bar·m}^3}{(\text{kg·mol})(°K)} \frac{300°K}{1.40 \text{ bars}} \frac{\text{kg·mol}}{28.01 \text{ kg}} = 0.636 \text{ m}^3/\text{kg}$$

**Example 1-5** Nitrogen gas at a pressure of 20 psia is maintained at a temperature of 80°F. Determine the specific volume in cu ft/lb$_m$ if it is reasonable to assume that the gas behaves ideally.

SOLUTION From Eq. (1-11) it is seen that $v = R_u T/PM$. Substitution of the above value for $P$ and $T$, and use of 1545 ft·lb$_f$/(lb·mol)(°R) as a proper value for $R_u$ and 28 lb$_m$/(lb·mol) for $M$, leads to

$$v = \frac{1545 \text{ ft·lb}_f}{(\text{lb·mol})(°R)} \frac{(80 + 460)°R}{20 \text{ lb}_f/\text{in}^2} \frac{\text{ft}^2}{144 \text{ in}^2} \frac{\text{lb·mol}}{28 \text{ lb}_m}$$

$$= 10.35 \text{ ft}^3/\text{lb}_m$$

In some calculations involving ideal gases it is not necessary to know the value of the gas constant. We shall find that there are a number of thermodynamic relations which require a knowledge of ratios of a given property rather than a knowledge of the actual values of the property. For example, information on the value of $v_2/v_1$ might be needed. It is easily seen that for a given ideal gas

$$\frac{v_2}{v_1} = \frac{RT_2/P_2}{RT_1/P_1} = \frac{T_2 P_1}{T_1 P_2}$$

To calculate the individual values of $v_2$ and $v_1$ directly from the ideal-gas equation and then take the ratio of the two values involves one in considerably more work than using the above relationship directly. In addition, there is the possibility of mathematical error or errors in the units required when separate calculations are made. Therefore one must keep in mind such shortcut methods for calculations involving ideal gases as that illustrated by the above example.

If a gas were ideal at all pressures, at constant temperature the quantity $Pv$ would be a constant, independent of pressure. A plot of $Pv$ versus $P$ at a given temperature would be a horizontal line. Typical experimental data on such a plot are shown in Fig. 1-6. Note that nitrogen gas approximates ideal behavior over a wide range of pressures, since the line drawn through experimental data is fairly horizontal out to at least 30 atm. Argon gas begins to deviate after about 10 atm. Carbon dioxide, on the other hand, apparently is nonideal at all pressures. Therefore, the quantity $Pv$ for gases generally is a function of both $P$ and $T$. However, as

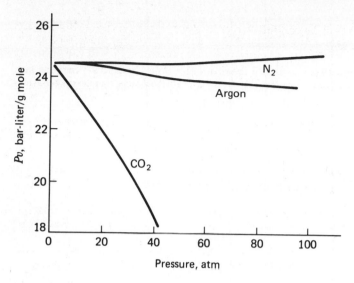

**Figure 1-6** Experimental data showing the variation of $Pv$ with pressure at a given temperature for several gases.

the pressure is lowered, the $Pv$ product approaches the same value, regardless of the nature of the gas. That is, the limiting value of $Pv$ at zero pressure is the same for all gases at the same temperature. Similar plots would result for data at other temperatures, except that the value of the zero pressure intercept differs for every new temperature. Consequently, at zero pressure all gases behave according to the equation $P\bar{v} = R_u T$. As the pressure is raised, some gases deviate from ideal-gas behavior faster than others. Apparently the ideal-gas equation would be considerably in error for $CO_2$ at the temperature used for Fig. 1-6 if the pressure exceeded more than several atmospheres.

As a generalization, the ideal-gas equation of state is only an approximation at best, strictly valid only at zero pressure. Nevertheless, for monatomic and diatomic gases the ideal-gas equation is usually a good approximation up to pressures of 10 to 20 atm at room temperature and above, for errors in accuracy not exceeding several percent. The range of validity of the ideal-gas equation of state for any particular gas depends upon the degree of accuracy desired. In Chap. 4 we shall reexamine the usefulness of this basic equation in the light of further experimental data. A simple method for correcting the equation to give a more accurate $PvT$ relationship for nonideal gases will be shown then.

## 1-9 IDEAL-GAS TEMPERATURE SCALE

Among the thermometric properties employed in the measurement of temperature is the pressure of a gas maintained at constant volume. On the basis of the experimental data shown in Fig. 1-6 we have seen that all gases exhibit the same value of $Pv$ at a given temperature if the pressure is extremely low. Therefore the quantity $Pv$ can be used as a thermometric property to measure the temperature, regardless of the nature of the gas. At extremely low pressures, $Pv$ will vary linearly with $T$ to a high degree of accuracy. If a thermometer contains a low-pressure gas, then when the thermometer is placed in contact with two systems at different temperatures, we should find that in the limit as $P \rightarrow 0$,

$$\frac{T}{T^*} = \frac{Pv}{(Pv)^*}$$

where the asterisk value will be used to represent a reference state. If the gas in the thermometer is maintained at constant volume, the above equation reduces to

$$\frac{T}{T^*} = \frac{P}{P^*} \qquad (1\text{-}16)$$

That is, when one brings a constant-volume-gas thermometer into contact with a system of interest, and then in contact with a system at a reference state, the ratio óf the measured gas pressures enables one to evaluate the temperature of the actual system relative to an assigned value $T^*$ at some reference state. If $T^*$ is chosen to be 273.16, then the Kelvin temperature $T_K$ at any other state as measured by an ideal-gas thermometer is given by

$$T_K = 273.16 \frac{P}{P^*} \qquad (1\text{-}17)$$

Hence the temperature of a given state of a substance may be determined by measuring the value of $P$ when the gas thermometer is in equilibrium with the substance and also measuring $P^*$ when the thermometer is in equilibrium with the triple state of water.

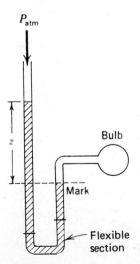

$P_{atm}$

Bulb

Mark

Flexible section

**Figure 1-7** Constant-volume-gas thermometer.

The operation of a constant-volume-gas thermometer is illustrated in Fig. 1-7. The tube on the left is raised or lowered until the mercury level on the right side of the U-tube is at the indicated mark. Then the height $z$ of the mercury column is a measure of the gage pressure of the gas within the thermometer bulb. The height $z$ will vary as the temperature of the gas within the bulb changes.

The empirical temperature scale based on the constant-volume-gas thermometer is commonly referred to as the ideal-gas temperature scale. We assume that the pressure of the gas within the thermometer remains relatively low so that the assumption of ideal-gas behavior is a valid one. It is not an absolute temperature scale, since it depends upon the behavior of a certain class of materials, namely, gases. However, it is a convenient scale due to its linearity, and it can be shown that this scale agrees with an absolute thermodynamic temperature scale developed on the basis of the second law of thermodynamics. Hence the ideal-gas thermometer is a device which measures a truly thermodynamic temperature.

## REFERENCES

Jones, J. B., and G. A. Hawkins: "Engineering Thermodynamics," Wiley, New York, 1960.
Kestin, J.: "A Course in Thermodynamics," Blaisdell, Waltham, Mass., 1966.
Obert, E. F.: "Concepts of Thermodynamics, McGraw-Hill, New York, 1960.
"Temperature in Measurement and Control in Science and Industry," Reinhold, New York, 1962.
Zemansky, M. W.: "Heat and Thermodynamics," 4th ed., McGraw-Hill, New York, 1957.

## PROBLEMS (METRIC)

### Converter efficiencies

**1-1M** An automobile running steadily at 45 km/h for a period of 1.1 h consumes 37.5 l of gasoline. The energy content of the gasoline is 44,000 kJ/kg, and the density is 0.75 g/cm³. The power delivered to the wheels of the car is 64 kW. Determine the thermal efficiency of the engine, in percent, assuming the entire energy content of the fuel is released by combustion.

**1-2M** Consider Prob. 1-1, except that the power is 90 kW, and the fuel consumed is 27 l over a period of 0.53 h. Determine the thermal efficiency.

**1-3M** A small industrial power plant generates 11,600 kg/h of steam, and has an output of 3800 kW. The plant consumes 1450 kg/h of coal, which has an energy content of 29,000 kJ/kg.

(*a*) Determine the overall plant thermal efficiency, in percent.

(*b*) If the energy added to the steam in the steam-generating unit is 2550 kJ/kg, what is the efficiency of the generating unit, in percent?

**1-4M** A steam power plant generates 180,000 kg/h of steam, and has an output of 55,000 kW of power. The plant consumes 19,500 kg/h of coal, which has an energy content of 30,000 kJ/kg.

(*a*) Determine the overall plant thermal efficiency, in percent.

(*b*) If the energy added to the steam in the steam-generating unit of the cycle is 2680 kJ/kg, what is the efficiency of the generating unit, in percent?

**1-5M** A home furnace uses 3.4 l/h of fuel oil, which has a density of 0.88 g/cm³ and an energy content of 46,400 kJ/kg. The furnace is rated at 95,000 kJ/h of heat input to the room air. Determine the chemical to thermal converter efficiency, in percent.

**1-6M** A jet engine consumes 9.00 kg/min of fuel which has a heating value of 41,800 kJ/kg. The velocity of the gas leaving the exit nozzle of the engine is 490 m/s. The ratio of air supplied to fuel consumed on a mass basis is 125 : 1.

(a) Determine the converter efficiency of chemical energy to kinetic energy.

(b) Determine the equivalent power output of the exit gas stream, in kilowatts.

**1-7M** The chemical energy released in a hydrogen-oxygen fuel cell is 15,850 kJ/kg of fuel, and the electrical power produced is 1.2 kW. If the flow rate of fuel to the converter is 0.37 kg/h, determine the converter efficiency, in percent.

### Intensive and extensive properties

**1-8M** Specify whether the following properties are intensive or extensive.

(a) Mass, (b), weight, (c) volume, (d) velocity, (e) density, (f) energy, (g) specific weight, (h) strain, (i) molar density, (j) mass concentration, (k) mole fraction, (l) pressure, (m) temperature, (n) surface area, (o) elevation, (p) potential energy, (q) stress.

**1-9M** Three cubic meters of air at 25°C and 1 bar have a mass of 3.51 kg.

(a) List the values of three intensive and two extensive properties for this system.

(b) If local gravity $g$ is 9.65 m/s$^2$, evaluate another intensive property of the system.

**1-10M** Four cubic meters of water at 25°C and 1 bar have a mass of 3990 kg.

(a) List the values of two extensive and three intensive properties of the system.

(b) If the local gravity $g$ for the system is 9.7 m/s$^2$, evaluate another intensive property.

### Force and mass

**1-11M** A body weighs 48 N in a location where the local acceleration of gravity $g$ is 9.60 m/s$^2$. What force is required to accelerate the body at a rate of 8.0 m/s$^2$?

**1-12M** A 100-g mass is accelerated with a force of (a) 20 N and (b) 33 N. Determine the acceleration of the mass in m/s$^2$.

**1-13M** A body weighs 450 N at a location where the local acceleration of gravity if 7.7 m/s$^2$. What is the mass of the body in kilograms? What force in newtons would be required to impart to this body a horizontal acceleration of 3 m/s$^2$?

**1-14M** A body which has a mass of 5.0 kg is accelerated at a rate of 3.0 m/s$^2$. What total force is required if (a) the body is moving along a horizontal frictionless plane, and (b) the body is moving vertically upward at a location where local gravity is 9.45 m/s$^2$?

**1-15M** The acceleration of gravity as a function of elevation above sea level at 45° latitude is given by $g = 9.807 - 3.32 \times 10^{-6}z$, where $g$ is in m/s$^2$ and $z$ is in meters. Find the height in kilometers above sea level where the weight of a man will have decreased by (a) 2 percent and (b) 4 percent.

**1-16M** (a) A 5-kg mass is weighed on a beam balance under the condition where local gravity is 9.30 m/s$^2$. What is the expected reading, in newtons?

(b) The same mass is weighed with a spring-type balance which has been calibrated for standard gravity. What will be the expected reading, in newtons?

**1-17M** Determine the force on a body of 70-kg mass on the moon where the gravitational acceleration is 1.67 m/s$^2$.

**1-18M** A mass of 2 kg is subjected to a vertical force of 25 N. The local gravity g is 9.60 m/s$^2$ and frictional effects are neglected. Determine the acceleration of the mass if the external vertical force is (a) downward, and (b) upward.

### Density and specific weight

**1-19M** The density of a certain liquid is 0.80 g/cm$^3$. Determine the specific weight in N/m$^3$ where local gravity g is (a) 2.50 m/s$^2$ and (b) 9.50 m/s$^2$.

**1-20M** On the surface of the moon where local gravity $g$ is 1.67 m/s$^2$, 2.2 kg of a gas occupy a volume of 1.2 m$^3$. Determine (a) the specific volume of the gas in m$^3$/kg, (b) the density in g/cm$^3$, and (c) the specific weight in N/m$^3$.

### Pressure

**1-21M** Determine the pressure equivalent to 1 bar in terms of meters of a column of liquid at room temperature, where the liquid is (*a*) water, and (*b*) ethyl alcohol. The specific gravity of ethyl alcohol is 0.789.

**1-22M** If the density of mercury is 13.59 g/cm$^3$, determine the height in centimeters of a liquid column which would be equivalent to 1 bar.

**1-23M** The gage pressure within a system is equivalent to a height of 35 cm of a fluid with a specific gravity of 0.75. If the barometric pressure is 0.980 bar, compute the absolute pressure within the chamber in millibars.

**1-24M** If the barometric pressure is 730 mbar, convert (*a*) an absolute pressure of 2.30 bars to a gage reading in bars, (*b*) a vacuum reading of 500 mbar to an absolute value in bars, (*c*) 0.70 bar absolute to millibars vacuum, and (*d*) a gage reading of 1.30 bars to kilopascals.

**1-25M** A system is evacuated until an attached vacuum gage indicates a reading of 850 mbar. The local barometric pressure is 0.950 bar. Determine the absolute pressure within the tank in mbar.

**1-26M** A manometer reads 640 mbar of vacuum. If the barometric pressure is 970 mbar, determine the absolute pressure in (*a*) millibar, and (*b*) kilopascals.

**1-27M** A tank is divided by a rigid wall into two sections as shown in Fig. P1-27M. Pressure gage *B* reads 1.75 bars, and gage *C* reads 1 10 bars. If the barometric pressure is 0.97 bar, determine the reading on gage *A* in bars.

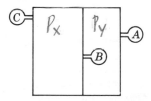

**1-28M** The same as Prob. 1–27M, except that gage *C* reads 5.70 bars, and gage *B* reads 2.20 bars. Determine the reading of gage *A*, and convert this reading to an absolute value.

**1-29M** The same as Prob. 1-27M, except that gage *C* is a vacuum reading of 240 mbar, and gage *B* reads 0.360 bar. Determine the reading on gage *A* in bars.

**1-30M** A mountain climber carries a barometer which reads 950 mbar at the base point of his ascent. During his climb he takes three additional readings which are (*a*) 884 mbar, (*b*) 836 mbar, and (*c*) 787 mbar. Estimate the vertical distance, in meters, he has climbed from the base point if the average air density is assumed to be 1.20 kg/m$^3$. Neglect the effect of altitude on local gravity.

**1-31M** Determine the pressure exerted on a skin diver who has descended to 20 m below the surface of the sea if the barometric pressure is 0.96 bar at sea level and the specific gravity of seawater is 1.03 in this region of the ocean.

**1-32M** If the atmosphere is assumed to be isothermal at 20°C and follows the relationshup $Pv = RT$ (an ideal gas), compute the pressure in bars and the density in kg/m$^3$ at 1000 m above sea level. The pressure and density at sea level are taken to be 1 bar and 1.23 kg/m$^3$, respectively.

**1-33M** Assume that the relationship between the pressure and the specific volume of the air in the atmosphere is given by the equation $Pv^n = $ constant. If the pressure and specific volume at the surface of the earth are 1 bar and 0.844 m$^3$/kg, respectively, determine the height of the atmosphere in meters. The value of the exponent $n$ may be taken as 1.4, and the acceleration of gravity may be assumed to be constant at 9.807 m/s$^2$.

### Temperature

**1-34M** We shall define a new relative temperature scale, called degrees Metric (°M), for which the freezing and boiling points of water are set at 100°M and 1000°M, respectively.

(a) If the scale is linear, derive an equation relating the temperature in degrees Metric ($T_M$) to the corresponding temperature on the Celsius (Centigrade) scale ($T_C$).

(b) What is the value of $T_M$ on this new scale at absolute zero on the Kelvin scale?

**1-35M** A new relative temperature scale, called degrees Thermodynamics (°T), is defined such that the freezing and boiling points of water are given values of 10°T and 210°T, respectively.

(a) If the new scale is linear, derive an equation relating the temperature in degrees Thermodynamic ($T_T$) to the corresponding temperature on the Celsius scale ($T_C$).

(b) Like °C versus °K, the scale in °T can be made into an absolute scale. What is the absolute temperature in °K when it is 0°T on the relative scale?

**1-36M** A new linear absolute temperature scale, called degrees Newton (°N), is defined such that the freezing and boiling points of water are assigned values of 100°N and 200°N, respectively.

(a) Derive an equation relating the temperature in degrees Newton ($T_N$) to the corresponding temperature on the Kelvin scale ($T_K$).

(b) What is the value of $T_N$ on this new scale at absolute zero on the Kelvin scale?

## Ideal Gas

**1-37M** A balloon is filled with methane gas ($CH_4$) at 20°C and 1 bar until the volume is 42.5 m³. Calculate the mass (in kilograms) of methane in the ballon under ideal-gas conditions.

**1-38M** Fifteen hundred kilograms of propane, $C_3H_8$, are to be stored in a gas reservoir at 12°C until needed. The gas is maintained in the reservoir at 3 bars pressure. The molar mass of propane is 44.0. What is the necessary volume of the reservoir, in cubic meters?

**1-39M** A tank contains helium at 6 bars and 40°C. A kilogram of the gas is removed, which causes the pressure and temperature to change to (a) 4 bars and 30°C, and (b) 3.6 bars and 10°C. Determine the volume of the tank in liters.

**1-40M** Two tanks, $A$ and $B$, are connected by suitable pipes through a valve, which initially is closed. Tank $A$ initially contains 0.3 m³ of nitrogen at 6 bars and 60°C, and tank $B$ is evacuated. The valve is then opened, and nitrogen flows into tank $B$ until the pressure in this tank reaches 1.5 bars and the temperature is 20°C. As a result, the pressure in tank $A$ drops to 4 bars and the temperature changes to 50°C. Determine the volume of tank $B$, in cubic meters.

**1-41M** One-half kilogram of helium is placed in a rigid tank 0.5 m³ in volume. If a thermometer in the tank reads 27°C and the barometric pressure is 1 bar, what would be the reading, in bars, of a pressure gage which is inserted into the tank?

**1-42M** A tank contains carbon dioxide at 5 bars and 30°C. A leak occurs in the tank which is not detected until the pressure has dropped to (a) 4 bars and (b) 3.6 bars. If the temperature of the gas at the time of detection of the leak is 20°C, compute the mass of carbon dioxide that has leaked out if the original mass was 25 kg.

**1-43M** Determine the reading on a pressure gage attached to a tank of hydrogen at 22°C when the density of the gas is 0.0798 kg/m³. The barometric pressure is 0.98 bar.

**1-44M** Fifty liters of carbon monoxide are maintained at 1.3 bars at 27°C. Determine the mass of the gas in kilograms.

**1-45M** One kilogram of argon gas (molar mass = 40) is placed in a rigid tank 0.6 m³ in volume. If argon behaves as an ideal gas and the absolute pressure in the tank is 1.4 bars, what should a thermometer read in °C when inserted into the tank?

**1-46M** One kilogram of air (molar mass = 29) is maintained in a 0.6-m³ tank at a gage reading of 500 mbar. If the temperature of the ideal gas is 50°C, what is the atmospheric (barometric) pressure in millibars?

**1-47M** A rigid tank has a volume of 3 m³. It contains a gas having a molar mass of 30 at 8 bars and 47°C. Gas leaks out of the vessel until the pressure is 3 bars at 27°C. What volume is occupied by the gas which escaped if it is at 1 bar and 22°C?

### Constant-volume gas thermometer

**1-48M** A constant-volume-gas thermometer contains helium at a specific volume of 9000 cm$^3$/g. The height of mercury $z$ in the liquid column shown in Fig. 1-7 is (a) −25.0 cm and (b) −20 cm. If the barometric pressure is 970 mbar, what temperature in kelvin is the gas bulb recording?

**1-49M** A constant-volume-gas thermometer is brought into contact with a system of unknown temperature and then into contact with the triple state of water. The mercury column attached to the thermometer has readings of −10.7 and −15.5 cm, respectively. What is the unknown temperature in kelvins. The barometric pressure is 980 mbar, and the thermometer contains hydrogen.

**1-50M** A constant-volume-gas thermometer contains argon at a specific volume of 700 cm$^3$/g. The height of mercury $z$ in the liquid column is Fig. 1-7 is (a) 10 cm, and (b) 20 cm. The density of mercury is 13.6 g/cm$^3$, and the barometric pressure is 1030 mbar. What temperature, in kelvins, is the gas bulb recording?

**1-51M** A constant-volume thermometer is used to measure the temperature of a system. The mass of the gas within the thermometer is varied, and the values of $P$ and $P^*$ are measured. On the basis of the following data:

| $P$, mbar | 1200 | 1000 | 800 | 600 | 400 | 200 |
|-----------|------|------|-----|-----|-----|-----|
| $P/P^*$   | 1.79 | 1.71 | 1.64 | 1.58 | 1.53 | 1.49 |

compute the value of the unknown temperature in kelvins.

# PROBLEMS (USCS)

### Converter efficiencies

**1-1** An automobile running at 30 mph for a period of 1.1 h consumes 10 gal of gasoline. The energy content of the gasoline is 19,000 Btu/lb, and the density of the fuel is 47 lb/ft$^3$. The power delivered to the wheels of the car is 85 hp. Determine the thermal efficiency of the process, in percent, assuming the entire energy content of the fuel is released by combustion.

**1-2** Consider Prob. 1-1, except that the power is 120 hp, and the fuel consumed is 7.2 gal in a period of 0.53 h. Determine the thermal efficiency, in percent.

**1-3** A small industrial power plant generates 25,000 lb/h of steam, and has an output of 3800 kW of power. The plant consumes 3200 lb/h of coal, which has a heating value of 12,500 Btu/lb.
(a) Determine the overall plant thermal efficiency in percent.
(b) If the energy added to the steam in the generating unit is 1100 Btu/lb, what is the efficiency of the steam-generating unit, in percent?

**1-4** A steam power plant generates 400,000 lb/h of steam, and has an output of 55,000 kW of power. The plant consumes 42,800 lb/h of coal, which has a heating value of 13,000 Btu/lb.
(a) Determine the overall thermal efficiency of the plant, in percent.
(b) If the energy added to the steam in the steam-generating unit is 1150 Btu/lb, what is the efficiency of the generating unit, in percent?

**1-5** A home furnace uses 0.90 gal/h of fuel oil, which has a density of 55 lb/ft$^3$ and a heating value of 20,000 Btu/lb. The furnace is rated at 90,000 Btu/h of heat input to the room air which is heated by the furnace. Determine the chemical to thermal conversion efficiency, in percent.

**1-6** A jet engine consumes 2.0 lb/min of fuel which has a heating value of 18,000 Btu/lb. The velocity of the gas leaving the exit nozzle of the engine is 1600 ft/s. The ratio of air supplied to fuel consumed on a mass basis is 125 : 1.
(a) Determine the converter efficiency of chemical energy to kinetic energy.
(b) Determine the equivalent power output of the exit gas stream, in horsepower.

**1-7** The chemical energy released in a hydrogen-oxygen fuel cell is 6830 Btu/lb of fuel, and the electrical power produced is 1.2 kW. If the flow rate of fuel to the converter is 0.81 lb/h, determine the converter efficiency, in percent.

### Intensive and extensive properties

**1-8** Specify whether the following properties are intensive or extensive.
(a) Mass, (b) weight, (c) volume, (d) velocity, (e) density, (f) energy, (g) specific weight, (h) strain, (i) molar density, (j) mass concentration, (k) mole fraction, (l) pressure, (m) temperature, (n) surface area, (o) elevation, (p) potential energy, and (q) stress.

**1-9** Two cubic feet of air at 70°F and 14.6 psia have a mass of 0.149 lb.
(a) List the values of three intensive and two extensive properties.
(b) If the local gravity $g$ is 31.2 ft/s$^2$, evaluate another intensive property of the system.

**1-10** Three cubic feet of water at 60°F and 14.7 psia have a mass of 187 lb.
(a) List the values of two extensive and three intensive properties of the system.
(b) If the local gravity $g$ is 30.8 ft/s$^2$, evaluate another intensive property.

### Force and mass

**1-11** A body weighs 10 lb$_f$ in a location where the local acceleration of gravity is 31.8 ft/s$^2$. What force is required to accelerate the body at a rate of 25 ft/s$^2$?

**1-12** A 100-lb mass is accelerated with a force of 5 lb$_f$. Determine the acceleration of the mass in ft/s$^2$.

**1-13** A body weighs 100 lb$_f$ at a location where the local acceleration of gravity is 25 ft/s$^2$.
(a) What is the mass of the body in pounds-mass?
(b) What force in pounds-force would be required to impart to this body an acceleration of 10 ft/s$^2$?

**1-14** A body which has a mass of 10 lb$_m$ is accelerated at a rate of 10 ft/s$^2$. What total force is necessary if (a) the body is moving along a horizontal frictionless plane, and (b) the body is moving vertically upward at a location where the local effect of gravity is 31.0 ft/s$^2$?

**1-15** The acceleration of gravity as a function of elevation above sea level at 45° latitude is given by $g = 32.17 - 3.32 \times 10^{-6}z$, where $g$ is in ft/s$^2$ and $z$ is in ft. Find the height in miles above sea level where the weight of a man will have decreased by 2 percent.

**1-16** (a) A 10-lb mass is weighed on a beam balance under the conditions where the local acceleration of gravity is 30.5 ft/s$^2$. What will be the expected reading?
(b) The same mass is weighed with a spring-type balance which has been calibrated for standard gravity. What will be the expected reading?

**1-17** Determine the force on a body of 150 lb$_m$ due to the presence of the moon at a point where the gravitational acceleration is one-sixth of the standard gravity on earth.

**1-18** A mass of 5 lb$_m$ is subjected to an external vertical force of 7 lb$_f$. The local gravity $g$ is 31.1 ft/s$^2$, and frictional effects are to be neglected. Determine the acceleration of the mass if the external vertical force is (a) downward, and (b) upward.

**1-19** The density of a certain liquid is 50 lb$_m$/ft$^3$. Determine the specific weight in lb$_f$/ft$^3$ where (a) $g$ is 8.05 ft/s$^2$ and (b) $g$ is 30.0 ft/s$^2$.

**1-20** On the surface of the moon, where the local acceleration of gravity $g$ is 5.47 ft/s$^2$, 5 lb$_m$ of a gas occupy a volume of 40 ft$^3$. Determine (a) the specific volume of the gas in ft$^3$/lb$_m$, (b) the density, and (c) the specific weight in lb$_f$/ft$^3$.

### Pressure

**1-21** Determine the pressure equivalent to 1 atm in terms of feet of a column of liquid at room temperature, where the liquid is (a) water, and (b) ethyl alcohol. The specific gravity of ethyl alcohol is 0.789.

**1-22** If the specific gravity of mercury is 13.59, determine the height, in inches, of a liquid column which would be equivalent to 1 atm.

**1-23** The gage pressure within a system is equivalent to a height of 14 in of a fluid with a specific gravity of 0.75. If the barometric pressure is 29.5 in Hg, compute the absolute pressure within the chamber in psia.

**1-24** If the barometric pressure is 30.55 in Hg, convert (*a*) 35 psia to psig, (*b*) 20 in Hg vac to in Hg abs and to psia, (*c*) 10 psia to in Hg vac, and (*d*) 20 in Hg gage to psia.

**1-25** A system is evacuated until an attached vacuum gage indicates a pressure of 25.5 in Hg. The local barometric pressure is 29.7 in Hg. Determine the absolute pressure within the tank in psia.

**1-26** A manometer reads 20 in of Hg vac. If the barometric pressure is 29.7 in Hg, determine the absolute pressure in (*a*) in Hg and (*b*) psia.

**1-27** A tank is divided by a rigid wall into two sections as shown in Fig. P1.27. Pressure gage *B* reads 24.8 psig and gage *C* reads 15 psig. If the barometric pressure is 29.5 in Hg, determine the reading on gage *A* in psig.

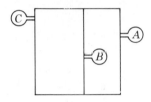

**1-28** Same as Prob. 1-27, except that gage *C* reads 80 psig and gage *B* reads 35 psig. Determine the reading of gage *A* and convert this reading to an absolute value.

**1-29** Same as Prob. 1-27, except that gage *C* has a vacuum reading of 10 psig, and gage *B* reads 18 psig. Determine the reading on gage *A*, is psig.

**1-30** A mountain climber carries a barometer which reads 30.10 in Hg at the base point of his ascent. During his climb he takes three additional readings which are (*a*) 28.90 in Hg, (*b*) 27.74 in Hg., and (*c*) 25.78 in Hg. Estimate the vertical distance he has climbed from the base point, in feet, if the average air density is taken to be 0.074 lb/ft³. Neglect the effect of altitude on local gravity.

**1-31** Determine the pressure exerted on a skin diver who has descended to 65 ft below the surface of the sea if the barometric pressure is 14.5 psia at sea level and the specific gravity of seawater is 1.03 in this region of the ocean.

**1-35** If the atmospheric air is assumed to be isothermal at 60°F and follows the relationship $Pv = RT$ (an ideal gas), compute the pressure in psia and the density in $lb_m/ft^3$ at 3000 ft above sea level. The pressure and density at sea level are taken to be 14.7 psia and 0.077 $lb_m/ft^3$, respectively.

**1-33** Assume that the relationship between the pressure and the specific volume of the air in the atmosphere is given by the equation $Pv^n = $ constant. If the pressure and specific volume at the surface of the earth are 14.7 psia and 13.5 ft³/lb., respectively, determine the height of the atmosphere in feet. The value of $n$ may be taken as 1.4, and the acceleration of gravity may be assumed to be constant at 32.17 ft/sec².

### Temperature

**1-34** A Fahrenheit and a Celsius thermometer are in equilibrium with a fluid. If the reading on the Fahrenheit thermometer is twice that on the Celsius thermometer, what is the absolute Rankine temperature of the fluid?

**1-35** In an experiment, both a Fahrenheit and a Celsius (centigrade) thermometer are in contact with a fluid which is in equilibrium. What is the temperature of the fluid in °K if both thermometers indicate the same numerical value?

**1-36** Carry out the following temperature conversions: (a) 100°F to °C, (b) 45°C to °K, (c) 745°R to °F, (d) 290°K to °R, and (e) 330°K to °F.

### Ideal gas

**1-37** A balloon is filled with methane gas ($CH_4$) at 70°F and 30.0 in Hg abs until the volume is 1500 cu ft. Calculate the mass of methane in the balloon in $lb_m$.

**1-38** It is desired to store methane gas ($CH_4$) underground at a pressure of 200 psia and a temperature of 55°F. Estimate the volume, in cubic feet, required to store 100 $lb_m$ of the gas under these conditions.

**1-39** A tank contains helium at 6 atm and 100°F. A kilogram of the gas is removed, which causes the pressure and the temperature to change to 4 atm and 80°F, respectively. Determine the volume of the tank in cubic feet.

**1-40** Two tanks, A and B, are connected by suitable pipes by a valve, which initially is closed. Tank A initially contains 10 ft³ of nitrogen at 100 psia and 140°F, and tank B is evacuated. The valve is then opened, and nitrogen flows into tank B until the pressure in this tank reaches 20 psia and the temperature is 60°F. As a result, the pressure in tank A drops to 70 psia and the temperature changes to 110°F. Determine the volume of tank B in cubic feet.

**1-41** One pound of helium is placed in a rigid tank 15 ft³ in volume. If a thermometer in the tank reads 170°F and the barometric pressure is 30.5 in Hg, what would be the reading, in psi, of a pressure gage which is inserted into the tank?

1-42 A tank contains carbon dioxide gas at 80 psia and 100°F. A leak occurs in the tank which is not detected until the pressure has dropped to 65 psia. If the temperature of the gas at the time of detection of the leak is 70°F, compute the mass of carbon dioxide that has leaked out if the original mass was 60 $lb_m$.

**1-43** Determine the reading on a pressure gage attached to a tank of hydrogen at 73°F when the density of the gas is 0.00498 $lb_m$/cu ft. The barometric pressure is 29 in Hg.

**1-44** Two cubic feet of carbon monoxide are maintained at 20 psia and 85°F. Determine the mass of the gas in pounds.

**1-45** One pound-mass of argon gas (molar mass = 40) is placed in a rigid tank 10.0 ft³ in volume. If argon behaves as an ideal gas and the absolute pressure in the tank is 20 psia, what should a thermometer read in °C when inserted into the tank?

**1-46** One pound of air (molar mass = 29) is maintained in a 10.0 ft³ tank at a gage reading of 7.0 psig. If the temperature of the ideal gas is 120°F, what is the atmosphere (barometric) pressure in psia?

**1-47** A rigid tank has a volume of 100 ft³. It contains a gas having a molar mass of 30 at 100 psia and 100°F. Gas leaks out of the vessel until the pressure is 40 psia at 80°F. What volume is occupied by the gas which escaped if it is at 29.9 in Hg and 70°F?

### Constant-volume-gas thermometer

**1-48** A constant-volume-gas thermometer contains helium at a specific volume of 145 ft³/lb. The height of mercury z in the liquid column shown in Fig. 1-7 is −10 in, and the density of mercury is 850 lb/ft³. If the barometric pressure is 29.0 in Hg, what temperature in °R is the gas bulb recording?

**1-49** A constant-volume-gas thermometer is brought into contact with a system of unknown temperature and then into contact with the triple state of water. The mercury column attached to the thermometer has readings of −4.20 and −6.10 in, respectively. What is the unknown temperature in °R? The barometric pressure is 29.2 in Hg, and the thermometer contains hydrogen.

**1-50** A constant-volume-gas thermometer contains argon at a specific volume of 11.2 ft³/lb. The height of mercury z in the liquid column, as shown in Fig. 1-7, is (a) 5 in and (b) 10 in. The density of mercury is 850 lb/ft³, and the barometric pressure is 29.90 in Hg. What temperature, in °R, is the gas bulb recording?

**1-51** A constant-volume thermometer is used to measure the temperature of a system. The mass of the gas within the thermometer is varied, and the values of $P$ and $P^*$ are measured. On the basis of the following data, compute the value of the unknown temperature in $°R$.

| $P$, atm | 1.20 | 1.00 | 0.80 | 0.60 | 0.40 | 0.20 |
|---|---|---|---|---|---|---|
| $P/P^*$ | 1.84 | 1.77 | 1.71 | 1.66 | 1.62 | 1.59 |

# TWO

# THE FIRST LAW OF THERMODYNAMICS

As stated earlier, the first law of thermodynamics is concerned with the study of energy transformations. In basic mechanics a few forms of energy, such as gravitational potential and linear kinetic energy, are examined. In electromagnetics other forms of energy, associated with electric and magnetic fields, are introduced. To the chemist the study of the energy associated with atomic and nuclear binding forces is extremely important. Thermodynamics relates these and other forms of energy and describes the change in energy of various types of systems in terms of interactions at the boundaries of the system. Thus thermodynamics is an all-encompassing field which ties together these many diverse areas, which are usually studied independently of one another. The basic laws of thermodynamics are, then, extremely broad in their scope. One of the most important laws of thermodynamics leads to a general conservation of energy principle. The law on which this conservation principle is based is called the first law of thermodynamics.

## 2-1 CLOSED AND OPEN SYSTEMS

The analysis of thermodynamic processes includes the study of the transfer of mass and energy across the boundaries of a system and the effect of these interactions on the state of the system. When the transfer of mass across the bounding surface of a system is prohibited, the system is called a *closed* system. Although the quantity of matter is fixed in a closed system, energy may be allowed to cross its boundaries in the form of work and heat. The matter may also change in chemical composition within the boundaries. The quantity of the matter under observation frequently is called the *control mass*. The control mass within the piston-cylinder assembly in Fig. 1-2b is in a closed system.

It is appropriate to define an *open* system as a system for which mass is permitted to cross the selected boundaries, as well as energy in the form of heat and work. The region of space described in Fig. 1-2a would be analyzed as an open system. In this chapter we shall develop a conservation of energy principle solely for closed systems. Such a conservation statement for open systems will be presented in Chap. 5.

## 2-2 THE CONCEPT OF WORK AND THE ADIABATIC PROCESS

The concept of work is usually introduced in the study of mechanics. Mechanical work is defined as the product of a force $F$ and the displacement $s$ of the force when both are measured in the same direction. The expression for a differential quantity of work $\delta W$ which results from a differential displacement $ds$ is given by

$$\delta W = F \ ds = \mathbf{F} \cdot \mathbf{ds}$$

The last term in this equation is the vector notation for work, which is a scalar quantity. The total work for a finite displacement is obtained from the integration of $F \ ds$. This will require a functional relationship between $F$ and $s$, since $F$ usually will not be a constant.

In thermodynamics it is convenient to consider the concept of work as follows: *Work* is an interaction between a system and its surroundings, and is done by a system if the sole effect external to the boundaries of the system could have been the raising of a weight. This definition leads to a broader interpretation of work than the purely mechanical one. In actual circumstances, effects may appear in the surroundings which do not have associated with them a recognizable force

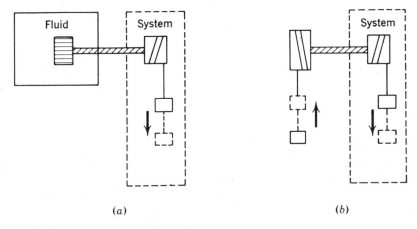

(a)    (b)

**Figure 2-1** Illustration of paddle-wheel work.

moving through a distance. Nevertheless, the real test is whether or not the interaction could have been carried out so that the sole external effect would have been the raising of a weight.

Consider the following example, illustrated in Fig. 2-1a. A closed system contains a pulley-weight system which is attached to a shaft which extends through the boundary of the system. Externally the shaft is attached to a paddle-wheel which rotates in a fluid as the weight is lowered within the system. Is the interaction at the boundary to be called work, in the light of our general definition? Note that the end result of the interaction is a "heating" of the fluid; that is, its temperature increases. One should not infer, however, that the interaction is a heat interaction until the operational test for work has been made. Figure 2-1b shows that the paddle-wheel apparatus could be replaced by another pulley-weight system, and the sole effect external to the system boundary would have been the raising of a weight. Hence the interaction at the boundary in this case must be called work. One should not use the result of an interaction as an intuitive measure of the nature of the interaction itself. Carefully note that a temperature rise of a system does not always imply that a heat interaction has occurred.

The value of a work interaction must necessarily be positive with respect to one system and negative with respect to the other. The sign convention adopted in this edition of the text is that work done *on* a system is positive. Likewise, work done *by* a system is given a negative number. This is a natural convention, in that quantities added are positive and quantities removed are negative.[1] Occasionally the sign on a work quantity will be omitted, and its direction indicated by assigning a subscript "in" or "out" to the symbol for work. The time rate at which work is done on or by a system is defined as *power*. Power is a scalar (nondirectional) quantity.

Any process which involves only work interactions is defined as an *adiabatic* process. In conjunction with this definition, an adiabatic boundary or surface is one which prevents or excludes all interactions except those which can be classified as work effects. Hence, an adiabatic boundary is one which thermally insulates a system, that is, prevents heat interactions.

## 2-3 WORK AS RELATED TO MECHANICAL AND ELECTROSTATIC FORMS OF ENERGY

A combination of the concept of work with Newton's second law of motion leads to the derivation of mechanical forms of energy. Consider the adiabatic change in position of a body of mass $m$ in a gravitational field. The work required to bring about this change of state is obtained by integrating the expression $\delta W = F \, ds$.

---

[1] The sign convention that work done on a system is positive is not a universal convention. There are a number of books in various fields where the author employs the opposite sign convention. The reader must remain aware of the fact that a number of practicing engineers and scientists use the more traditional sign convention, namely, that work done by a system is positive.

The force acting on the body is given by $F = mg$, where $g$ is the local acceleration of gravity. Hence,

$$\delta W_{ad} = mg\ dz$$

where the distance $s$ has been replaced by the coordinate $z$ in the direction of the gravitational field. The subscript "ad" stands for adiabatic. If one assumes that the value of $g$ is constant over the given change in elevation, then integration of the above equation leads to

$$W_{ad} = mg(z_2 - z_1)$$

Thus the work associated with an adiabatic change in position of a system is solely a function of the initial and final positions. In this restricted case, the work required is independent of the path.

Figure 2-2 is an illustration of three different paths along which a body might be moved in a gravitational field such that the body is displaced from position (1) to position (2). Whether the more direct path $A$, or the more devious paths $B$ and $C$, are chosen, the change in the elevation of the body is always $z_2 - z_1$. Hence the quantity of adiabatic work required is the same in all three cases. This also would be true for any other path selected. The quantity $mzg$ is defined as the gravitational potential-energy function, and we note in this special case that

$$\Delta(\text{gravitational potential energy}) = \Delta(\text{PE}) = W_{ad}$$

The above equation relates the adiabatic work done on a system to the change in the gravitational potential-energy function. The actual value of the function is arbitrary, a zero value of the potential-energy function being dependent upon the selection of some elevation where $z$ is taken to be zero.

In a similar manner a system may undergo a change in velocity while other properties of the system, including its elevation, remain constant. The acceleration may be carried out adiabatically, and the work required is again measured by $\delta W = F\ ds$. According to Newton's second law the external force which is used to

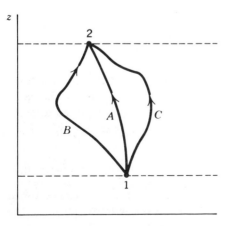

**Figure 2-2** Moving a body along different paths in a gravitational field.

accelerate the system is given by $ma = m(dV/dt)$. For a differential time interval $dt$ the distance $ds$ moved by the system is simply $V\,dt$. Therefore

$$\delta W_{ad} = m\frac{dV}{dt}(V\,dt) = mV\,dV$$

Integration of the above expression yields

$$W_{ad} = \frac{m}{2}(V_2^2 - V_1^2)$$

The quantity $mV^2/2$ is called the linear, or translational, kinetic-energy function, KE. Any change in its value is independent of the path of the process and can be measured by the adiabatic work required to bring about the change in velocity. It is convenient to define the kinetic energy as zero when the velocity is zero relative to some frame of reference.

Recall also from elementary physics that the adiabatic work required to move an electric charge through an electrostatic field from one position to another is independent of the path. Consequently one can define a new function, the electrostatic potential-energy function, such that the adiabatic work done on moving a charge through an electrostatic field is measured by the change in the electrostatic potential-energy function. Symbolically we may write that $\mathbf{F} = Q_e\mathbf{E}$ and $\delta W = \mathbf{F} \cdot d\mathbf{s}$. Therefore

$$W_{ad} = Q_e \int_i^f \mathbf{E} \cdot d\mathbf{s} = \Delta(\text{electrostatic potential energy})$$

where $Q_e$ = charge
  $\mathbf{E}$ = electrostatic field
  $\mathbf{s}$ = distance

The three cases cited above have a number of common features, which are summarized below.

1. The work done on a system in each case has been equated to the change in a specific potential function, and each of the potential functions has been given the name of a specific form of energy.
2. Each form of energy has a value at a given state which is arbitrary, relative to some reference state.
3. Each relationship between work and a change in energy is derived for adiabatic conditions. That is, work is the only type of interaction allowed during the change of state. If these processes were nonadiabatic, properties of the system other than the elevation, velocity, or position would change during the process. In such cases the simple relationships derived above would not generally be valid.

This review of basic concepts from physics reveals that the concept of work frequently is related to specific forms of a quantity called energy. However, the relationships are valid only under adiabatic conditions. We shall next examine the nature of adiabatic changes of state for other types of processes than those discussed above.

## 2-4 THE FIRST LAW OF THERMODYNAMICS

In the preceding section we considered three special cases of adiabatic processes involving work interactions. In each separate case the value of the work interaction was independent of the path between given end states. These cases represented, however, processes with quite different overall changes of state. As a next step, we shall investigate the effect of different types of work interactions on a given system which all lead to the same change of state. To illustrate the point, consider a constant-volume closed system which undergoes a change of state between two given equilibrium states. Two different methods could be employed experimentally to carry out the specified change of state. Process $A$ (see Fig. 2-3$a$) is carried out by allowing a paddle-wheel to rotate in the fluid within the system. If the rotation occurs by dropping a weight, as shown in the figure, then the work done on the system can be measured in terms of the quantity $mg\ \Delta z$, where $m$ is the mass of the weight used and $\Delta z$ is the distance through which the mass moves in the direction of the gravitational field. In process $B$ (see Fig. 2-3$b$) an electrical resistor has been placed within the fluid, and it is connected through the boundaries of the system to an external battery. The amount of electrical work required is the product of the voltage times the charge delivered, or, in an equivalent manner, it can be represented by $\xi It$, where $\xi$ is the cell voltage, $I$ is the current, and $t$ is the time.

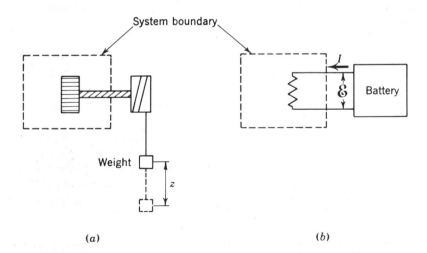

**Figure 2-3** Two different work interactions on the same adiabatic, closed system.

Both processes, *A* and *B*, start at the same initial state and end at the same final state. In addition, one might also consider a process *C* where both paddle-wheel and electrical work are used to bring about the required change of state. It would be found that the work required is always the same for the described adiabatic processes between the same equilibrium states of the closed system. Thus one would find that the value of $mg \, \Delta z$ for process *A* would be identical to the value of $\xi It$ for process *B*, within experimental accuracy, if each process were adiabatic. For process *C* the sum of the combination of paddle-wheel and electrical work would be the same as for the other two processes if we again require an adiabatic process.

On the basis of experimental evidence of this type, which began with the work of Joule in the middle of the nineteenth century, it is possible to make the following broad assertion. This postulate, called the *first law of thermodynamics*, states that

> When a closed system is altered adiabatically, the work is the same for all possible paths which connect two given equilibrium states.

Stated another way, the value of the work carried out on or by a closed adiabatic system is dependent solely upon the end states of the process. This postulate is made regardless of the type of work interaction involved in the process, the type of process, and the nature of the closed system.

The first law of thermodynamics, together with those relationships derivable from it, are so well established that one does not question its validity when properly applied to an extensive range of scientific and engineering problems. Not only is it applicable to simple adiabatic processes such as those which involve paddle-wheel or electrical work in a closed system, but it may also be applied to systems which experience electromagnetic, surface tension, shear, and gravitational field effects. In an expanded formulation it can be applied to closed systems undergoing heat interactions as well as work interactions. Finally, the above generalization can be extended to open systems as well, where mass transfer is permitted across the boundaries of the system.

## 2-5 THE NATURE OF *E*, THE TOTAL ENERGY

It has been postulated that the work done is the same for all adiabatic paths which connect two equilibrium states of a closed system. This statement, the first law of thermodynamics, leads to a general definition of the change in the energy of a closed system between equilibrium states. Recall that the change in the value of a property is fixed by the end states and is independent of the process. If this is true, then specifying the end states will fix the change in values of all properties, regardless of the process between the two specified states. Consequently, it follows logically that any quantity which is fixed by the end states for all processes between those end states must be a measure of the change in the value of a property. Since the adiabatic work is solely a function of the end states for a closed-system process, the quantity of adiabatic work defines or measures the change in a new property function. We shall call this new function the energy $E$ (or total energy) of a closed system.

As a consequence of the first law of thermodynamics we may write

$$W_{ad} \equiv E_2 - E_1 = \Delta E \qquad (2\text{-}1)$$

where 1 and 2 indicate the initial and final states, and $W$ is the work carried out for any adiabatic path between those two states. This equation is an operational definition of the change in the energy of a closed system between two equilibrium states. Equation (2-1) is a fundamental relation which enables one to evaluate the energy of a set of states in terms of adiabatic work interactions. The actual value of $E$ can only be established by arbitrarily assigning a value for a particular substance at a reference state. However, this same procedure was necessary for the mechanical and electrostatic forms of energy discussed earlier. Apparently the derivation and subsequent definition of new forms of energy, such as gravitational potential, kinetic, and electrostatic, are all special cases of Eq. (2-1). Thus the total energy $E$ is the sum of all forms of energy associated with a closed system.

There are numerous forms of energy which constitute the total energy of a macroscopic system. In addition to the gravitational and electrostatic potential energies discussed already with respect to gravitational and electric force fields, a potential energy related to magnetic fields must also be considered. Beside linear kinetic energy related to translational motion of the system as a whole with respect to a reference frame, there is angular or rotational energy of a rigid body. Solids under anisotropic stress have a so-called elastic potential energy, or strain energy, which arises from the work done on a solid material to overcome intermolecular forces while changing its shape. In certain other cases one may wish to acknowledge the energy stored in the surface layer of a liquid, as in the study of interfacial effects between liquid layers.

Even in the absence of the effects of external force fields, motion (linear or rotational), anisotropic shear, and surface tension, a change in the energy of another important type may occur as the result of a work interaction. In the preceding section adiabatic work was added to a closed system in the forms of paddle-wheel work and electrical work dissipated in a resistor (recall Fig. 2-1). In both cases the system underwent the same change of state, but none of the forms of energy outlined above were affected. What was affected in these latter cases was a form of energy called *internal energy*, and it is designated by the symbol $U$ (or $u$ on a unit-mass basis). We have seen that kinetic and potential energies may be associated with macroscopic systems. These same forms of energy may be ascribed to the individual particles of a substance as well. In the gas phase, for instance, particles move at random around the containing vessel as a result of interactions with the walls or with other particles. Based on their velocities relative to the boundaries of the system, there is a translational (linear) kinetic energy associated with each particle. The particles may also have rotational or vibratory motions. Moreover, the particles have a potential energy which arises from the concept of intermolecular forces between the particles of the substance. These forces include the gravitational forces due to mass attraction, electrostatic forces between charges on the particles, nuclear binding forces, etc. The actual nature of these forces on the microscopic level is unimportant to us, since we are interested solely in the gross behavior of a large group of particles. Although internal energy has been described in terms of microscopic characteristics, it is not necessary to acknowledge the particle concept of matter in order to define or evaluate the internal energy. The change in its value can always be established by the quantity of adiabatic work required to change the state of the system. That is, in accordance with Eq. (2-1), when other forms of energy are not affected, $W_{ad} = \Delta U$.

The internal energy described above is related to the internal molecular structure of a substance. If the change in $U$ occurs without a change in chemical composition, we frequently speak of the change in the "sensible" internal energy. The internal energy of a substance may also be altered by changes in the internal atomic structure. The form of internal energy related to changes on the atomic level (but not changes on the subatomic level) might be called "chemical" internal energy. The change in chemical internal energy may be attributed roughly to the change in the energy of the chemical bonds which hold atoms together in molecular form. Familiar examples of internal-energy changes due to chemical reactions include all combustion processes (such as for power generation or propulsion purposes) and the action of a common storage battery when discharging. In this latter case the chemical internal energy, at least in part, is converted directly into electrical energy. There are also well-known internal-energy changes on the subatomic level. Nuclear reactions involving atomic fusion or fission graphically illustrate the tremendous energy which can be released by such changes.

In summary, the total energy $E$ of a system may be broken down into contributions from a number of sources. In general we may write

Total energy = internal energy + kinetic energy + gravitational potential
energy + electrostatic energy + magnetic energy + strain
energy + surface energy + ⋯         (2-2)

The proper inclusion of one or more of the terms on the right side of this expression in a thermodynamic analysis depends primarily upon the type of system under analysis.

## 2-6 A CONSERVATION OF ENERGY PRINCIPLE FOR CLOSED SYSTEMS

The first law of thermodynamics has led to an operational definition of energy. The change in the energy of control mass is equal to the work done on or by the system during an adiabatic process. There are many forms of energy and work which must be considered in the analysis and synthesis of engineering systems. In addition, there is another type of interaction which cannot be classified as a work effect. Consider the stirring of a fluid within an adiabatic, constant-volume system.

It is experimentally known that the same change of state could also be accomplished by bringing the walls in contact with a high-temperature source. The effect which has occurred in the latter case falls into the category of a heat interaction. An adiabatic surface prevents heat transfer. In the case above, the quantity of energy $Q$ transferred in the heat interaction is equal to the change in the energy of the control mass. That is, $Q = \Delta E$. Heat and work are the sole mechanisms by which energy is exchanged with a closed system.

Consider a closed system undergoing a process during which both heat and work effects are present. In this case, the work done will not equal the change in the energy of the system, since the process is nonadiabatic. The difference between the energy change of the system and the work done on the system is defined as a measure of the heat interaction that took place during the process. Mathematically this relationship is expressed by

$$Q \equiv (E_2 - E_1) - W \tag{2-3}$$

or

$$Q + W = \Delta E \tag{2-4}$$

Equation (2-4) is a conservation of energy statement for a closed system, or control mass.[2] Note that heat transfer into a system is taken as positive.

From Eq. (2-4) it is seen that the sum of $Q$ and $W$ is unique between given end states, since $\Delta E$ is fixed in value by the end states. However, $Q$ and $W$ may each take on a wide range of values, as long as their sum equals $\Delta E$. The actual values of $Q$ and $W$ for a set of end states depends upon the path of the process between these states. The differentials of quantities which depend upon the path are known as *inexact* differentials. To emphasize the need for a specification of the path, these quantities will be preceded by the symbol $\delta$ (lowercase delta) for infinitesimal variations in the state of a system. The integration of an inexact differential for a finite change of state does not lead to the use of the symbol $\Delta$. Instead, for finite work and heat interactions we shall write

$$\int_1^2 \delta W = W_{12} \qquad \text{and} \qquad \int_1^2 \delta Q = Q_{12} \tag{2-5}$$

For inexact quantities such as work and heat one does not speak of the change in the quantity, but rather its overall value for the process between states 1 and 2. Note also that since the value of a work interaction (and a heat interaction) is a function of the path, the evaluation of $\delta W$ (or $\delta Q$) around a cycle should not necessarily lead to zero. Finally, it cannot be overemphasized that both work and heat are transient phenomena which occur across the boundary of a system. Thermodynamically, a system never stores heat or work because these phenomena are interactions which cease to exist once the process has ended.

---

[2] As just pointed out in footnote 1 in this chapter, some books use the more traditional sign convention that work done by a system is positive. In such cases the positive sign between Q and W in Eq. (2-4) would be replaced by a negative sign.

Frequently it is convenient to analyze closed systems, or control masses, on a unit-mass basis. If the heat transfer per unit mass $q$ is defined by

$$q \equiv \frac{Q}{m} \tag{2-6a}$$

and the work per unit mass is

$$w \equiv \frac{W}{m} \tag{2-6b}$$

then the conservation of energy principle for a differential change of state is written as

$$\delta q + \delta w = de \tag{2-7}$$

For a finite change of state this becomes

$$q + w = \Delta e \tag{2-8}$$

Equations such as these are important in the thermodynamic analysis of closed systems.

In order to use Eqs. (2-4) and (2-8) in engineering analysis, it is necessary to evaluate, independently, heat-transfer and work quantities and changes in specific forms of energy. A brief description of heat-transfer calculations is presented in Sec. 2-7a. Then Secs. 2-8 and 2-9 are devoted to specific types of work interactions. Lastly, the evaluation of the internal energy of various classes of substances, along with other thermodynamic properties, is covered in Chaps. 3 and 4..

**Example 2-1M** The total energy of a closed system increases by 55.0 kJ during a process as work is done on the system in the amount of 100.0 kJ. How much heat is transferred during the process, and is it added or removed from the system?

SOLUTION The basic starting place in the analysis is the conservation of energy principle represented by Eq. (2-4), $Q + W = \Delta E$. Since work done on a system is positive by convention, the substitution of values for $W$ and $\Delta E$ leads to

$$Q + W = \Delta E$$

$$Q + (+100.0) = +55.0$$

or

$$Q = +55.0 - 100.0 = -45.0 \text{ kJ}$$

Note that the answer is $-45.0$ kJ, and not just $-45.0$. Answers to engineering problems have units, and they *must* be stated specifically. The negative sign on the answer indicates that 45.0 kJ of energy in the form of heat must be removed during the process. Another way of expressing the answer is

$$Q_{out} = 45.0 \text{ kJ}$$

When the subscript "out" or "in" is placed on $Q$ (or $W$), the sign no longer is necessary.

## 2-7 HEAT MEASUREMENT AND ENERGY EQUIVALENTS

### a Measurement of Heat

The concept of heat is defined operationally in the preceding section as a measure of the difference between the change in the energy of a closed system and the work done on the system. Nevertheless, it is a common experience to think of heat transfer as an energy-transfer process which occurs by virtue of a temperature difference between two systems. This is a calorimetric definition of heat transfer. In actual practice the engineer evaluates $Q$ by adopting mechanisms for various types of heat-transfer phenomena. These mechanisms are known as conduction and radiation, in conjunction with convective transfer by fluid motion. They involve temperature as the driving force, and are described mathematically in textbooks and courses in heat transfer. Unfortunately, formal course work in the independent field of heat transfer usually follows an introductory course in thermodynamics. This lack of knowledge of independent heat-transfer calculations places a restriction on thermodynamic problems developed in this text. With regard to $Q$ in such problems, there are three possibilities: (1) it is the unknown in the problem; (2) it is essentially zero and is neglected; and (3) it is a known quantity stated in the problem. With respect to item (2), this occurs when the system is deliberately insulated or the temperature difference between the system and its environment is reasonably small, thus making the process adiabatic. When $Q$ is given, it is implied that independent heat-transfer calculations have been carried out with the resulting reported data.

### b Energy Equivalents in USCS Units

Methods for measuring heat-transfer quantities by calorimetry preceded the development of the first law. As a result, units for heat-transfer quantities have been established independent of units for work interactions. There exists a need, then, to correlate these separate sets of units. This difficulty has been overcome in the SI by using the joule as a measurement of both heat and work. However, in the USCS the basic unit of heat is the British thermal unit (Btu), while work commonly is expressed in ft·lb$_f$. Originally the Btu was defined as the energy required to raise the temperature of one pound of water from 59 to 60°F at one atmosphere pressure. The unit is now defined in terms of mechanical units of energy by the relation

$$1 \text{ Btu} \equiv 778.16 \text{ ft·lb}_f$$

The choice of this value retains the specific heat of water at 60°F at very nearly 1 Btu/(lb$_m$)(°F). A list of energy equivalents appears in Table A-1.

> **Example 2-1** The stored energy of a closed system increases by 55 Btu during a process as work is done on the system in the amount of 77,800 ft·lb$_f$. How much heat is transferred during the process, and is it added or removed from the system?

SOLUTION The basic starting place in the analysis is the conservation of energy principle represented by Eq. (2-4), $Q + W = \Delta E$. Application of this equation to the system of interest requires that the units of energy for $Q$, $W$, and $\Delta E$ must be the same. If we choose to work in Btu, the amount of work done on the system becomes

$$W = 77{,}800 \text{ ft·lb}_f \, \frac{1 \text{ Btu}}{778 \text{ ft·lb}_f} = 100 \text{ Btu}$$

where the conversion factors for the units have been rounded off to three significant figures. We are now in a position to substitute values into Eq. (2-4). Since work done on a system by convention is taken as positive, the substitution of values for $W$ and $\Delta E$ leads to

$$Q + W = \Delta E$$

$$Q + (+100) = +55$$

or

$$Q = +55 \text{ Btu} - 100 \text{ Btu} = -45 \text{ Btu}$$

The negative sign on the answer indicates that 45 Btu of energy in the form of heat must be removed during the process.

## 2-8 QUASISTATIC WORK INTERACTIONS

The equations for work interactions presented in the following section require that the interrelationships between properties be known at all times during the process. This is analogous to the requirement mentioned earlier that $F$ be known as a function of $s$ when evaluating mechanical work by the classical equation $\delta W = \mathbf{F} \cdot d\mathbf{s}$. It is only for equilibrium states that properties are truly defined, and consequently can be expected to be functionally related. Therefore a process must be quasistatic if the following work interactions are to be evaluated purely from a knowledge of the thermodynamic states of the system (independent of information external to the system). A major consequence of quasistatic processes is that the work output is found to be a maximum, and the work input a minimum, for a given process. In the presence of nonequilibrium states the actual measured work output is less than the maximum and the work input is greater than the minimum required. Thus the quasistatic process is a standard to which all other processes may be compared when work interactions are under consideration. A number of equations valid for some particular quasistatic work effects are presented below.

### 1 EXPANSION AND COMPRESSION WORK

As an example of the use of the quasistatic process to calculate the value of work effects, it is interesting to examine the mechanical work done when the boundary of a closed system, or control mass, is moved with respect to the substance as a whole. This common type of mechanical work is usually called expansion, or

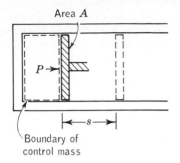

Area $A$

$P \rightarrow$

$\leftarrow s \rightarrow$

Boundary of
control mass

**Figure 2-4** The mechanical work associated with moving the boundary of a piston-cylinder assembly.

boundary work. In some instances it is termed the $P \, dV$ work, for reasons which will become apparent in a moment. An example of this type of work effect is illustrated by a substance enclosed in a piston-cylinder assembly, as shown in Fig. 2-4. The system boundaries are indicated by the dashed line in the diagram. The cross-sectional area of the piston is $A$, and the pressure at the initial equilibrium state is $P$, throughout. The mechanical force exerted uniformly by the substance on the constraining walls of the piston and cylinder is given by $PA$. The mechanical work done on or by the substance is then expressed by

$$\delta W = F \, ds = PA \, ds$$

Since $A \, ds$ is the change in the volume $dV$ due to a deformation of the boundary, the differential quantity of work done when the piston is moved a distance $ds$ is simply

$$\delta W = -P \, dV \tag{2-9}$$

The total boundary work done on or by the substance during a finite change in volume is the sum of the $P \, dV$ terms for each differential change in volume. Mathematically, this is expressed by the relation

$$W_{\text{boundary}} = -\int_{V_i}^{V_f} P \, dV \tag{2-10}$$

where $V_i$ and $V_f$ are the initial and final volumes of the substance. The pressure is expressed in absolute units. It should be noted that the integral of $\delta W$ is simply $W$, and not $\Delta W$. A work interaction is associated with a process and is a function of the path selected.

A plot of a quasistatic process on pressure-volume coordinates is quite useful for describing graphically the boundary work of a process involving a closed system. From integral calculus it is known that the area beneath the curve which represents the quasistatic path of the process is equal to the integral of $P \, dV$ on a pressure-volume diagram. A typical diagram for the evaluation of boundary work is shown in Fig. 2-5. The cross-hatched area on the diagram represents the work done by the gas within the cylinder when the volume is increased by the amount $dV$. The entire area beneath the curve from point 1 to point 2 represents the total

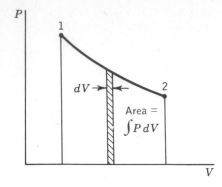

**Figure 2-5** Area representation of boundary work for a quasistatic process.

work done when the gas expands from state 1 to state 2. An indefinitely large number of paths could be drawn between these same two end states, any number of which could be quasistatic. In general, the area beneath each of these paths will be different. This merely emphasizes the fact that work is a path function, and, unlike the change in the value of a property, is not solely dependent upon the end states of the process. (Only in the special case of adiabatic processes does the value of work interactions become independent of the path.)

The integration of the equation for the boundary work requires a knowledge of the functional relationship between $P$ and $V$. This can be found from experimental data or from knowledge of the specific type of process involved and a representative equation of state. An equation of state is an equation which relates $P$ functionally to $V$ and other independent variables. This latter point is best illustrated by examples.

## a $P\,dV$ Work for an Isothermal Ideal-Gas System

A gas which essentially fulfills the condition of an ideal gas, that is, $PV = NR_u T$, is contained within a piston-cylinder assembly. The gas is allowed to expand quasistatically and isothermally from state 1 to state 2. The work done by the gas on the piston face is given by the integral of $-P\,dV$. Noting that $T$ is a constant by definition of an *isothermal* process, we find that

$$W = -\int_{V_1}^{V_2} P\,dV = -\int_{V_1}^{V_2} \frac{NR_u T}{V}\,dV = -NR_u T \ln \frac{V_2}{V_1}$$

By measurement of the temperature of the process and the initial and final volumes, the maximum work done by the system can easily be calculated.

> **Example 2-2M** Two kilograms of nitrogen gas at 27°C and 1.5 bars are compressed isothermally to 3.0 bars. Determine the minimum work of compression, in kilojoules, for the gas.
>
> SOLUTION The minimum work of compression is found when the process is quasistatic. In this case the boundary work is given by the integral of $-P\,dV$. Since the maximum pressure on the diatomic gas is only 3.0 bars, the assumption that nitrogen is an ideal gas under these conditions is a reasonably good one. It is shown above for an isothermal process that the

boundary work for an ideal gas is given by $-NR_u T \ln (V_2/V_1)$. For an ideal gas, also, $P_1 V_1/T_1 = P_2 V_2/T_2$. For an isothermal process of an ideal gas, then, $V_2/V_1 = P_1/P_2$. Consequently, using $R_u$ from Table A-1 and the molar mass from Table A-2,

$$W = -NR_u T \ln (V_2/V_1) = NR_u T \ln (P_2/P_1)$$

$$= 2/28 \text{ kg·mol} \times 8.315 \text{ kJ/(kg·mol)}(°\text{K}) \times 300°\text{K} \times \ln (3.0/1.5)$$

$$= 0.0714(8.315)(300)(0.693) = 123 \text{ kJ}$$

The answer is expected to be positive, since the process is one of compression and work in is positive. The answer could also be expressed as $W_{in} = 123$ kJ. When a work interaction is to be expressed per unit mass, the lowercase letter $w$ is used. In this example, on a unit-mass basis,

$$w = 61.5 \text{ kJ/kg}$$

**Example 2-2** Two pounds of nitrogen gas at 80°F and 20 psia are compressed isothermally to 40 psia. Determine the minimum work of compression in ft·lb$_f$ for the gas.

SOLUTION The minimum work of compression is found when the process is quasistatic, in which case the boundary work is given by the integral of $-P \, dV$. Since the pressure never exceeds 40 psia, the assumption can be made that diatomic nitrogen is an ideal gas during the process. It is shown above for an isothermal process of an ideal gas that the boundary work is given by $-NR_u T \ln V_2/V_1$. Also, for an ideal gas, $P_1 V_1/T_1 = P_2 V_2/T_2$. Under isothermal conditions then, $V_2/V_1$ may be replaced by $P_1/P_2$. Consequently, on the basis of data from Tables A-1 and A-2,

$$W = -NR_u T \ln V_2/V_1 = NR_u T \ln P_2/P_1$$

$$= 2/28 \text{ lb·mol} \times 1545 \text{ ft·lb}_f/(\text{lb·mol})(°\text{R}) \times 540°\text{R} \times \ln 40/20$$

$$- 0.0714(1545)(540)(0.693) - 41,300 \text{ ft·lb}_f$$

The answer is positive, since the process is one of compression. The answer could also be expressed as $W_{in} = 41,300$ ft·lb$_f$. When a work interaction is to be expressed per unit mass, the lowercase letter $w$ is used. In this example, on a unit-mass basis,

$$w = 20,650 \text{ ft·lb}_f/\text{lb}_m$$

## b Boundary Work on Solid and Liquid Phases

The integral of $P \, dV$ may be used to evaluate the quasistatic expansion or compression work for any phase of interest. It is particularly appropriate in this form for the boundary work associated with gaseous systems, since the functional relation between $P$ and $v$ may be established by direct experimental measurements. For solids and liquids, the variation of pressure with volume is more frequently expressed indirectly in terms of a property called the *isothermal coefficient of compressibility* $K_T$. This property is defined as

$$K_T = -\frac{1}{V}\left(\frac{\partial V}{\partial P}\right)_T = -\frac{1}{v}\left(\frac{\partial v}{\partial P}\right)_T \tag{2-11}$$

It is an experimental observation that $(\partial v/\partial P)_T$ is always negative for all phases of matter. The negative sign in the definition of $K_T$ is introduced so that tabulated values of the isothermal compressibility are always positive. It is apparent from the defining equation that $K_T$ will always be given in reciprocal pressure units (for example, $atm^{-1}$). It may be shown that $K_T = 1/P$ for gaseous systems which behave like ideal gases.

To accommodate the use of $K_T$ in the evaluation of boundary work, it is necessary to replace $dv$ in the integral of $P\,dv$ by an equivalent quantity. On the basis that $v$ is solely a function of $T$ and $P$, that is, $v = v(T, P)$, then the total differential is written as

$$dv = \left(\frac{\partial v}{\partial T}\right)_P dT + \left(\frac{\partial v}{\partial P}\right)_T dP \qquad (2\text{-}12)$$

For isothermal processes $(dT = 0)$ Eq. (2-12) reduces to $dv = (\partial v/\partial P)_T\,dP = -vK_T\,dP$. Therefore, the isothermal work of compression per unit mass is given by

$$w_T = -\int_1^2 P\,dv = \int_1^2 vK_T P\,dP$$

The coefficient $K_T$ and the specific volume $v$ frequently are nearly constant for liquids and solids over considerable ranges of temperature and pressure. Under this condition integration of the above expression yields

$$w_T = \frac{vK_T}{2}(P_2^2 - P_1^2) \quad \text{ISOTHERMAL} \qquad (2\text{-}13)$$

This equation provides a means for estimating the work required to compress a solid or liquid under isothermal conditions. Table 2-1 lists $K_T$ values for solid copper and liquid water as a function of temperature. Note that $K_T$ is extremely small with the value in both cases being on the order of $10^{-6}$ $bar^{-1}$.

**Example 2-3M** Water is compressed from 1 to 100 bar while the temperature is maintained at 20°C. Estimate the work required on the closed system in joules.

SOLUTION Equation (2-13) is applicable if we assume that both $K_T$ and $v$ are reasonably independent of pressure. The value of $K_T$ is found in Table 2-1 to be $45.90 \times 10^{-6}$ $bar^{-1}$. The density of water at 20°C is close to 1.0 $g/cm^3$. Hence

$$w_T = \frac{1}{2} \times \frac{1\ cm^3}{g} \times 45.90 \times 10^{-6}\ bar^{-1} \times [(100)^2 - (1)^2]\ bar^2$$

$$= 0.230\ \frac{cm^3\cdot bar}{g} \times \frac{10^5\ N}{bar\cdot m^2} \times \frac{m^3}{10^6\ cm^3} = 0.0230\ J/g$$

This is a small quantity of energy.

**Example 2-3** Water is compressed from 15 to 1500 psia while the temperature is maintained at 100°F. Estimate the work required on the closed system in $ft\cdot lb_f/lb_m$.

SOLUTION Equation (2-13) is applicable if we assume that both $K_T$ and $v$ are reasonably independent of pressure. On the basis of Table 2-1, the value of $K_T$ is roughly $3.06 \times 10^{-6}$ $(psi)^{-1}$. The specific volume of water is essentially 0.01613 $ft^3/lb$. Hence

$$w_T = \frac{1}{2} \times \frac{0.01613\ ft^3}{lb_m} \times \frac{3.06 \times 10^{-6}\ in^2}{lb_f} \times [(1500)^2 - (15)^2]\frac{lb_f^2}{in^4}$$

$$= 0.0555\ lb_f\cdot ft^3/lb_m\cdot in^2 = 8.00\ ft\cdot lb_f/lb_m$$

**Table 2-1 The isothermal compressibility $K_T$ for solid copper and liquid water as a function of temperature, in units of bar$^{-1}$**

|  | (a) Copper | | | | | | |
|---|---|---|---|---|---|---|---|
| Temperature, °K | 100 | 150 | 200 | 250 | 300 | 500 | 800 |
| $K_T \times 10^6$ | 0.721 | 0.734 | 0.749 | 0.763 | 0.778 | 0.839 | 0.922 |
|  | (b) Water | | | | | | |
| Temperature, °C | 0 | 10 | 20 | 30 | 40 | 60 | 80 |
| $K_T \times 10^6$ | 50.89 | 47.81 | 45.90 | 44.77 | 44.24 | 44.50 | 46.15 $\times 10^{-6}$ |

## c Further Considerations of $P\ dV$ Work

Consider now the evaluation of the boundary work for a closed system which undergoes a cyclic, quasistatic process. One possible $Pv$ diagram for the control mass is shown in Fig. 2-6. The path of the cyclic process is from state 1 to states 2, 3, and 4, and finally back to the initial state. For those processes in which the final volume is greater than the initial volume, it is seen that work is done by the system on the surroundings. Hence, for paths 1–2 and 2–3, the work interaction is done by the system. In a similar fashion it will be found that the surroundings do work on the system for path 4–1, since $\Delta V$ is negative. It is apparent that no boundary work is present during path 3–4, because the volume is constant. As a result, the net work done by the system is represented by the area enclosed by the complete cyclic path, and in this case it is a negative value. If the cycle were carried out in the opposite direction, the net work would still be shown by the crosshatched area in the figure, only in this case it would actually be done by the surroundings on the system. Finally, it should be noted that, in general, the area enclosed by any cycle will not be zero. The change in all thermostatic properties will be zero for the cycle, but the net work will be finite.

It has been noted that, in order to evaluate the integral of $P\ dV$, the variation of $P$ in terms of $V$ must be known. If the pressure varied erratically over the volume of the substance, it would be highly unlikely that any simple functional relationship between $P$ and $V$ existed. Thus, in order to determine the boundary

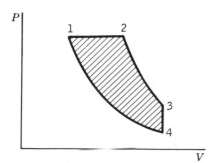

Figure 2-6 The net boundary work during a cyclic process for a closed system.

work for a closed system, the process must be evaluated in terms of quasistatic changes of state. Consider a piston-cylinder assembly filled with a gas; the surfaces of the piston and cylinder are frictionless. If the piston is suddenly accelerated inwardly, the effective pressure at the piston face becomes greater than it would have been if the piston had been moved gradually. From a molecular viewpoint, the gas particles are not able to move away from the approaching piston face fast enough. Consequently, they pile up near the piston and offer a greater resistance to the movement of the piston. As a result, more work is required to compress a gas rapidly than for the corresponding quasistatic process, all other factors being equal. By similar reasoning it would be found that less expansion work is done by the system when the substance is in nonequilibrium, compared with that for a quasistatic expansion. In the case of either nonequilibrium expansion or compression, the value of the work interaction must be measured directly. It could not be calculated from Eq. (2-10) because there is no way of relating $P$ and $V$ thermodynamically for nonequilibrium changes of state. Although the equation for boundary work was derived for volume changes produced by a piston, it should be apparent that this equation applies to any volume change which is carried out quasistatically.

## $\chi$ 2 WORK OF A REVERSIBLE CHEMICAL CELL

A galvanic cell is a device for converting chemical energy into electrical energy by means of a controlled chemical reaction. A reversible cell is one for which the reaction is capable of proceeding in either direction. In order to operate the cell so that it remains close to equilibrium, a slightly lower electric potential must oppose the potential developed by the cell. This could be accomplished by placing a potentiometer in the external circuit which counterbalances the developed potential of the cell. Under these circumstances the ideal potential difference developed is called the electromotive force (emf) and is represented by the symbol $\mathscr{E}$.

The electrical work delivered by the cell for the passage of a differential quantity of charge $dQ_e$ is found in the following way. For a potential difference $\mathscr{E}$, a current $I$ performs work at the rate $\dot{W} = \delta W/dt = \mathscr{E}I$. Also, $I = dQ_e/dt$. Combination of these two expressions yields, for the electrical work,

$$\delta W_{\text{rev cell}} = \mathscr{E}\ dQ_e \tag{2-14}$$

The value of $dQ_e$ is taken to be negative when the cell is discharged. Consequently $\delta W$ will be negative during discharge of the cell, in accordance with our sign convention on work. In accordance with one of Faraday's laws of electrolysis we may write that

$$dQ_e = -j\mathscr{F}\ dN$$

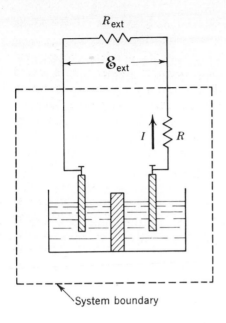

System boundary

**Figure 2-7**  Effect of internal resistance on chemical-cell performance.

where $j$ is the number of moles of electrons transferred per mole of change of the reaction, $dN$ is the differential number of moles that react, and $\mathscr{F}$ is Faraday's constant, which is equal to 96,487 C/mol of electrons. The minus sign results from the convention of making $dN$ positive when a cell is discharged. Thus Eq. (2-14) may also be written in the form

$$\delta W_{\text{rev cell}} = -\mathscr{E} j \mathscr{F} \ dN \tag{2-15}$$

The quantity $j\mathscr{F}$ is the charge (in coulombs) transferred per mole of reacting material. If $\mathscr{E}$ is a constant during charging or discharging, the integration of Eq. (2-15) is straightforward, since $j$ and $\mathscr{F}$ are always constants (although the value of $j$ depends on the specific reaction under consideration).

   If the charge is removed at a finite rate, the potential developed at the terminals of the cell is less than the emf of the cell, which is the maximum potential of the cell. This loss in potential is due to the presence of a finite electric resistance internal to the cell, as shown in Fig. 2-7. The current $I$ causes a voltage drop $\Delta\mathscr{E} = RI$. Therefore, on discharging, $\mathscr{E}_{\text{ext}} = \mathscr{E}_{\text{emf}} - RI$. Upon charging, the applied voltage $\mathscr{E}_{\text{ext}}$ must be greater than the emf by the amount $RI$. Therefore, on charging, $\mathscr{E}_{\text{ext}} = \mathscr{E}_{\text{emf}} + RI$. Hence the maximum work output and the minimum work input for a reversible cell are characterized by quasistatic processes in both directions. For finite rates of charging or discharging, the work associated with a chemical cell would be given by $\delta W_{\text{chem cell}} = \mathscr{E}_{\text{ext}} \ dQ_e$. However, in this case the work must be measured rather than calculated, since $\mathscr{E}_{\text{ext}}$ is not a thermodynamic variable and no unique relation between $\mathscr{E}_{\text{ext}}$ and $Q_e$ is known.

## ✗ 3 WORK IN STRETCHING A LIQUID SURFACE

In heterogeneous systems where gas and liquid (or two liquid phases) may be in contact, the phenomenon of surface tension occurs. For a molecule well inside the liquid phase no resultant force is experienced as it moves around in the fluid because of intermolecular attractions, there being an equal number of molecules immediately surrounding it. However, a molecule near the interface experiences a resultant force directed back into the liquid since the forces of attraction due to gas molecules above it are quite weak compared to those of the liquid phase. Hence additional work must be done on any molecule which is brought from the interior of the liquid to the surface as the surface area is increased. Molecules at the surface may be considered to acquire a surface potential energy over that of the molecules in the interior. This additional energy for the liquid as a whole obviously would be proportional to the surface area between the phases. To increase the surface area of a liquid, then, requires work equal to the increase in the surface potential energy. The quantity of work required is defined in terms of a thermostatic property of the system called the surface tension. The surface tension $\gamma$ of a liquid phase (with respect to the other phase) is defined as the work per unit change in area required to increase the surface area. Consequently, the change in surface energy of a liquid for a differential area change is given by

$$\delta W_{\text{surf tension}} = \gamma \, dA \tag{2-16}$$

The surface tension is given in dimensions of energy per length squared, or force per unit length. Typical values are about $50 \times 10^{-5}$ N/cm, or $3 \times 10^{-3}$ lb$_f$/ft.

## ✗4 WORK DONE ON ELASTIC SOLIDS

In order to change the length of a spring or wire in tension or compression, it is necessary to exert a force $F$ which alters the length $L$. The equation for the differential work required to alter the length from $L$ to $L + dL$ follows directly from the mechanical definition of work, namely,

$$\delta W_{\text{elastic}} = F \, dL \tag{2-17}$$

This expression requires a functional relation between $F$ and $L$ in order to carry out the integration. The force $F$ on a linear elastic spring is commonly related to the displacement $x$ of the spring by the relation $F = k_s x$, where $k_s$ is the spring constant and $x$ is the change in the length of the spring from its unstressed state or condition. That is,

$$F_s = k_s(L_s - L_{s, 0}) \tag{2-18}$$

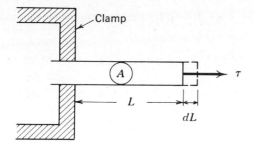

**Figure 2-8** Stretching a wire.

where $L_{s,0}$ is the unstressed length of the spring and $F_s$ is the force on the spring when its length is $L_s$. In this format Eq. (2-17) becomes

$$\delta W_{\text{spring}} = F_s\, dL_s \qquad (2\text{-}19a)$$

Substitution of Eq. (2-18) into Eq. (2-19a) and subsequent integration will yield the work required to compress or stretch an elastic spring. The result is

$$W_{\text{spring}} = k_s(x_2^2 - x_1^2)/2 \qquad (2\text{-}19b)$$

where $x_2 = L_2 - L_0$ and $x_1 = L_1 - L_0$.

It is sometimes more convenient to express the work done on an elastic solid bar or wire in terms of the stress $\sigma$ and the strain $\epsilon$. The stress in the axial direction is defined by $\sigma = F/A$, where $A$ is the cross-sectional area of the solid. The strain for a differential change in length is given by $d\epsilon = dL/L$, as shown in Fig. 2-8. Substitution of these expressions into Eq. (2-17) yields

$$\delta W_{\text{elastic}} = (\sigma A)(L\, d\epsilon) = V_0\, \sigma\, d\epsilon \qquad W = EALE^2 \quad E = \frac{dL}{L} \qquad (2\text{-}20)$$

where $V_0$ is the initial volume of the solid. If the work on the elastic solid is done isothermally, then the relation between stress and strain is given by Young's isothermal modulus, $E_T = \sigma/\epsilon$. When an elastic deformation occurs (the Hooke's law region of a solid), then $E_T$ is a constant.

## ✕ 5 WORK OF POLARIZATION AND MAGNETIZATION

It may be shown from the theory of electromagnetic fields that work is done on a substance contained within an electric or magnetic field when the field is altered. For a dielectric material which lies within an electric field, the work supplied externally to increase the polarization of the dielectric is given by

$$\delta W_{\text{polarization}} = V\mathbf{E} \cdot d\mathbf{P} \qquad (2\text{-}21)$$

where $\mathbf{E}$ is the electric field strength, and $\mathbf{P}$ is the polarization, or electric dipole moment, of the dielectric. In this equation the boldface type emphasizes that $\mathbf{E}$ and $\mathbf{P}$ are both vector quantities, and as such should be multiplied together in accordance with the rules of vector analysis. A similar equation for the work done in increasing the magnetization of a substance due to a change in the magnetic field is expressed by

$$\delta W_{\text{magnetization}} = V \mu_0 \, \mathbf{H} \cdot d\mathbf{M} \tag{2-22}$$

where $\mathbf{H}$ = magnetic field strength
$\quad\quad \mathbf{M}$ = magnetization per unit volume
$\quad\quad \mu_0$ = permeability of free space
$\quad\quad V$ = volume

Additional terms may be added to the above equations for the work required to change either the electric field or magnetic field of free space.

In the preceding sections a number of work interactions were introduced. It is convenient to refer to all types of quasistatic work interactions as a product of a force and a displacement, even though the factors within a given work expression may not bring to mind physical forces and displacements. The intensive property that appears in the equations representing work will be called a *generalized force*, and will be symbolized by $F_k$. In an analogous fashion the extensive property found in these equations will be called a *generalized displacement*, represented by $X_k$. The subscript $k$ simply refers to the $k$th type of a work interaction. The sum of the quasistatic work effects of these types is given by

$$\delta W_{\text{tot}} = \sum_k F_k \, dX_k$$
$$= -P \, dV + F \, dL + \gamma \, dA + V \mu_0 \, \mathbf{H} \cdot d\mathbf{M} + \cdots \tag{2-23}$$

Regardless of whether the generalized work interaction is in or out of the system, the quasistatic value of work is always equal to or less than the actual work effect measured in the environment. That is, $W_{k,\,\text{quasistatic}} \geqslant W_{k,\,\text{measured}}$. In accordance with the general definition of work given earlier, the work $F_k \, dX_k$ (done by the system), when integrated over the path of the quasistatic process, is equivalent solely to the raising of a weight external to the system boundaries.

It is a matter of experience when various work effects should be taken into account. In some instances their contribution may be so small as to be neglected.

**Table 2-2  Generalized quasistatic work interactions**

| System | Generalized force $F_k$ (intensive) | Generalized displacement $X_k$ (extensive) | Equations for work |
|---|---|---|---|
| Linear mechanical | $F$ | $s$ | $F \, ds$ |
| Elastic | $F$ (or $\sigma$) | $L$ (or $\epsilon$) | $F \, dL$ (or $V\sigma \, d\epsilon$) |
| Electrostatic and reversible cell | $\mathscr{E}$ | $Q_e$ | $\mathscr{E} \, dQ_e$ |
| Surface | $\gamma$ | $A$ | $\gamma \, dA$ |
| Capacitor | $E$ | $P$ | $V\mathbf{E} \cdot d\mathbf{P}$ |
| Magnetic | $H$ | $M$ | $V\mu_0 \, \mathbf{H} \cdot d\mathbf{M}$ |
| Boundary | $P$ | $V$ | $-P \, dV$ |

Table 2-2 shows various work interactions in terms of the intensive and extensive factors, as well as the overall differential expressions. This table emphasizes the fact that most generalized forces do not have the dimensions of force, nor do the generalized displacements have the dimensions of length. Nevertheless, the product of the dimensions for any corresponding pair of a generalized force and displacement has the dimensions of work, i.e., energy.

## 2-9 NONQUASISTATIC FORMS OF WORK

Some important forms of work interactions were summarized in Table 2-2. All these interactions can be expressed by an equation of the form $\delta W = F \, dX$. If a process is quasistatic, then $F$ and $X$ are properties of the system and some functional relationship exists between $F$ and $X$ which permits integration of the expression for work. Quasistatic work interactions have the following characteristics:

1. The value of $F_k$ depends only upon the state of the system and is independent of the direction of change of $X$. The change in $X$ can be either positive or negative.
2. Theoretically a system can be returned to its initial state after a quasistatic work interaction merely by reversing the direction of the original work effect.
3. The value of $F_k$ remains finite as $dX$ approaches zero. In each equilibrium state $F_k$ has a fixed and finite value.

Consider, for example, the boundary work done by a piston-cylinder assembly. For a given initial equilibrium state with a pressure $P$, the magnitude of $P$ in the expression $P \, dV$ is the same whether $V$ increases or decreases. After a differential increase in $V$, the system can be returned to its initial state by reversing the direction of the piston movement. Whether for a finite or differential change of state, the value of $P$ remains finite and well defined as long as the process is quasistatic.

There are other important work interactions which do not fall into the quasistatic class. These nonquasistatic forms of work have a number of characteristics which differentiate them from quasistatic interactions. Among these are:

1. The force $F$ depends upon the rate of change of state.
2. The work interaction is unidirectional, and the effect cannot in practice be undone by reversing the original process.
3. The force $F$ approaches zero as the rate of change of state approaches zero. Consequently the work approaches zero for a differential change of state.

A typical example of a nonquasistatic-type work interaction is the case of a fluid undergoing shear deformation due to the rotation of a paddle-wheel or its equivalent. The shear is proportional to the rate of fluid deformation, and as the rate goes to zero, the shear force does likewise. It should be noted that paddle-wheel work may be done on a closed system, but the reverse process is not possible. Thus the interaction is unidirectional. Another example of a unidirectional, nonquasistatic work interaction is the passage of an electric current through a resistor which lies within the system boundaries. The electrical work done on the system is due to some external voltage source, which causes current to pass along conductors connected with the resistor. The usual result of such an interaction is a rise in the temperature of the substance within the system. The reverse process, the spontaneous removal of energy from the system which causes a current to flow through the circuit, never occurs. In cases of both paddle-wheel work and electrical-resistor work, the fluid within the system usually is not in equilibrium during the process. Since the process is nonquasistatic, the work must be measured external to the boundaries of the system. In comparison, a quasistatic form of work is evaluated in terms of changes within the system.

In terms of paddle-wheel work, for example, the rotational mechanical work is evaluated in terms of the external torque $\tau$ transmitted by a rotating shaft. To review the development of the expression relating torque to work, consider the apparatus shown in Fig. 2-9. An external force $F$ acts at a distance $r$ from the center of a shaft. This external force could be represented by a weight attached to the pulley system, for example. The work required to move the force through a differential distance $ds$ is given by $F\,ds$. From the definition of torque we may write that

$$\tau = Fr$$

Also, by definition the angle $d\theta$ in radians is

$$ds = r\,d\theta$$

Hence the relation between rotational mechanical work and torque is simply

$$\delta W = F\,ds = \frac{\tau}{r}(r\,d\theta) = \tau\,d\theta \tag{2-24a}$$

If a constant torque is applied, then

$$W = \tau\theta \tag{2-24b}$$

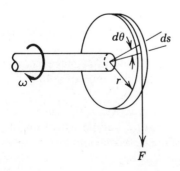

**Figure 2-9** A schematic showing relationship of torque and angular deflection to rotational mechanical work.

where $\theta$ must be expressed in radians, and not degrees. The power associated with rotational work is

$$\dot{P} = \frac{\delta W}{dt} = \tau \frac{d\theta}{dt} = \tau \omega \tag{2-24c}$$

where $\omega$ is the angular velocity in radians per unit time.

The evaluation of the electrical work dissipated within a resistor likewise depends upon external measurements. As noted earlier for the reversible chemical cell, the rate of doing electrical work (or the electrical power) is

$$\text{Power} = \frac{\delta W}{dt} = VI \tag{2-25}$$

When a current of one ampere (A) passes through an electrostatic potential of one volt (V), the power required is defined as one watt (W), which is also 1 J/s. Upon rearrangement, Eq. (2-25) yields the differential work required to pass a current $I$ through a voltage potential $V$ over a differential time period. That is,

$$\delta W_{\text{electrostatic}} = VI \, dt \tag{2-26}$$

Knowledge of current, voltage, and time relations permits the integration of this expression.

## REFERENCES

Hall, N. A., and W. E. Ibele: "Engineering Thermodynamics," Prentice-Hall, Englewood Cliffs, N.J., 1960.

Reif, F.: "Fundamentals of Statistical and Thermal Physics," McGraw-Hill, New York, 1968.

Reynolds, W. C.: "Thermodynamics," McGraw-Hill, New York, 1968.

Zemansky, M. W., and G. C. Van Ness: "Basic Engineering Thermodynamics," 2d ed., McGraw-Hill, New York, 1974.

## PROBLEMS (METRIC)

### Kinetic and potential energy

**2-1M** Determine the work required in kilojoules to accelerate a body of 0.98 kg from a velocity of (a) 100 to 200 m/s and (b) 80 to 180 m/s.

**2-2M** It requires 60 kJ of work to raise a body 90 m in Earth's gravitational field where the local acceleration of gravity is 9.55 m/s².

(a) Find the mass of the body, in kilograms.

(b) If the initial gravitational potential energy of the body was 20 kJ with respect to Earth's surface, determine the final elevation of the body above this surface, in meters.

**2-3M** The acceleration of gravity as a function of elevation above sea level is given by $g = 9.807 - 3.32 \times 10^{-6}z$, where $g$ is in m/s² and $z$ is in meters. A satellite with a mass of 240 kg is boosted 400 km above Earth's surface. What work is required, in kilojoules?

**2-4M** Two hundred kilojoules of work are used to change the kinetic energy of a body by 100 N·m/g.
(a) Determine the mass of the body in kilograms.
(b) If the initial kinetic energy relative to some reference is 40 kJ, determine the final velocity relative to the same reference, in m/s.

### General first law analysis

**2-5M** A closed system undergoes a process during which 9 kJ of heat are removed and 27,000 N m of work are done by the system. The system is then restored to its initial state by adding 6 kJ of heat to it, in addition to work interaction. What is the magnitude in newton-meters and the direction of the work interaction?

**2-6M** During a certain process a closed system does 25 kJ of work while 9 kJ of heat are removed. Then the system is restored to its initial state by means of a process during which 4 kJ are added to it as work. Determine the net heat effect for the cycle in kilojoules, and indicate whether this net effect was on or by the system.

**2-7M** A closed system executes a cyclic process during which there are three separate heat interactions. For the cycle $Q_1$ is 17 kJ, $Q_2$ is $-4$ kJ, and $Q_3$ is $-5$ kJ. During the cycle two work interactions occur, for which $W_1 = 6,000$ N·m. What is the value of the second work interaction $W_2$, and is this quantity of work done on or by the system?

**2-8M** For each of the following cases of a process involving a closed system, fill in the missing data.

| $Q$ | $W$ | $E_1$ | $E_2$ | $\Delta E$ | $Q$ | $W$ | $E_1$ | $E_2$ | $\Delta E$ |
|---|---|---|---|---|---|---|---|---|---|
| (a) 24 | $-13$ | | $-8$ | | (d) 18 | | 26 | | 10 |
| (b) $-8$ | | | 62 | $-18$ | (e) $-8$ | 13 | 28 | | |
| (c) | 17 | $-12$ | | 20 | (f) | $-11$ | | 7 | $-12$ |

**2-9M** A closed system is changed from state 1 to state 2 along path x. It is then returned to the initial state along either of two different paths, y and z. Values of Q and W along the three different paths are listed below for two different systems, A and B. Determine the missing data for Q and W for (a) System A, and (b) System B.

**System A**                                    **System B**

| Path | Process | $Q$ | $W$ | Path | Process | $Q$ | $W$ |
|---|---|---|---|---|---|---|---|
| x | 1 to 2 | 10 | | x | 1 to 2 | | $-7$ |
| y | 2 to 1 | $-7$ | 4 | y | 2 to 1 | $-4$ | 9 |
| z | 2 to 1 | | 8 | z | 2 to 1 | 10 | |

**2-10M** An automobile battery which originally is fully charged gradually discharges while sitting on the shelf at a constant temperature of 30°C and producing no electrical work. However, 1000 kJ of heat are transferred to its environment. If the battery is then recharged slowly to its initial state by a process requiring 440 W·h of work, what is the heat transfer, in kilojoules, to the battery during the charging process?

**2-11M** An experimental energy converter has a heat input of 80,000 kJ/h and a work input of 1.6 kW. The converter produces an electric power output of 18 kW. Calculate the change in the energy of the converter over a time period of 4 min.

### Boundary work

**2-12M** During an expansion process for a closed system the volume of a gas changes from 0.1 to 0.3 m³. The pressure changes during the quasistatic process according to the relation $P = 2.4V + 0.4$, where the units of P and V are bars and cubic meters, respectively.

(a) Find by integration the work done by the gas, in joules.

(b) Check the answer by geometric considerations of the process on a PV diagram.

(c) Find the useful work output if the frictional force between piston and cylinder is 2000 N and the piston area is 0.2 m².

**2-13M** A gas is compressed from a state of 1 bar, 0.30 m³, to a final state of 4 bars. The process equation relating P and V is $P = aV + b$, where $a = -15$ bars/m³. Compute by integration the necessary work, in kilojoules, and sketch the process path on PV coordinates.

**2-14M** A gas is confined in a gastight, frictionless piston-cylinder device surrounded by the atmosphere. Initially, the pressure of the gas is 14 bars and the volume is 0.030 m³. If the gas expands to a final volume of 0.060 m³ and the piston area is 0.1 m², calculate the work done, in newton-meters, against the force F along the shaft connected to the piston. Assume the process is such that $PV^2$ is a constant for the gas at any equilibrium state, and that the atmospheric pressure is 1 bar.

**2-15M** A piston-cylinder device contains 0.96 kg of oxygen ($O_2$) initially at a state of 27°C and 1.5 bars. Heat is supplied to the gas and it expands at constant pressure to a temperature of 627°C. Calculate the quantity of work performed, in kilojoules. Sketch the process on a PV diagram.

**2-16M** One-half kilogram of a gas is contained in a piston-cylinder assembly at initial conditions of 0.02 m³ and 7 bars. The gas is allowed to expand to a final volume of 0.050 m³. Determine the amount of work done in joules for the following processes:

(a) The pressure is constant.

(b) The product PV is a constant.

(c) The product $PV^2$ is a constant.

Compare the different quasistatic processes by use of a PV diagram.

**2-17M** An ideal gas initially in a piston-cylinder assembly at 1.5 bars and 0.03 m³ is first heated at constant pressure until the volume is doubled. It is then allowed to expand isothermally until the volume is again doubled. Determine the total work done by the gas, in newton-meters, and plot the processes on a PV diagram. The initial gas temperature is 500°K.

**2-18M** In many real processes it is found that gases at low pressures fulfill the relationship $PV^n = c$, where the exponent n and c are constants. If n is taken as 1.3, determine the work done, in newton-meters, by air in a piston-cylinder arrangement when it expands quasistatically from 0.01 m³ and 3 bars to a final pressure of 1 bar.

**2-19M** A gas initially at 4 bars is contained in a piston-cylinder assembly. The piston is frictionless and is held against the gas by a spring and the surrounding atmosphere. The gas expands from 0.01 and 0.04 m³. Determine the work done by the gas, in newton-meters, if the force exerted by the spring is (a) directly proportional to the volume of the system, and (b) proportional to the square root of the volume of the system. The pressure of the atmosphere surrounding the assembly is 1 bar.

**2-20M** A quantity of sulfur dioxide ($SO_2$) gas is expanded in a piston-cylinder device. The following experimental data were taken:

| P, bars | 3.45 | 2.75 | 2.07 | 1.38 | 0.69 |
|---|---|---|---|---|---|
| v, m³/kg | 0.125 | 0.150 | 0.187 | 0.268 | 0.474 |

(a) Calculate the work of expansion done per kilogram of $SO_2$.

(b) If the friction between the cylinder and piston was equivalent to 0.15P, what was the work of expansion appearing in the surroundings, in kJ/kg?

**2-21M** A piston-cylinder assembly contains 0.09 m³ of an ideal gas at 1.2 bars and 20°C. The gas is compressed to a final pressure of 3.6 bars and a temperature of 20°C.

(a) Determine the minimum work of compression, in newton-meters, if the process is isothermal.

(b) Consider now that a process occurs between the same two end states in part (a) such that it follows a straight line on PV coordinates. Determine the work of compression for this process, and compare the two processes on a PV diagram.

**2-22M** One kilogram of a gas with a molar mass of 35 is compressed isothermally (constant temperature) at 77°C from a volume of 0.05 m³ to a volume of 0.025 m³. The $PvT$ relationship for the gas is given by $Pv = RT[1 + (c/v^2)]$, where $c = 2.0$ m⁶/(kg·mol)².

(a) Compute the work done on the gas, in newton-meters, if quasistatic.

(b) If $c = 0$, would the work required be greater than, equal to, or less than that calculated in part (a)? Defend your answer in terms of two process lines on a $Pv$ diagram.

**2-23M** One kilogram of a gas with a molar mass of 60 is compressed at a constant temperature of 27°C from a volume of 0.12 m³ to 0.04 m³. The $PvT$ relationship for the gas is given by $Pv = RT[1 + (b/v)]$, where $b$ is 0.012 m³/kg.

(a) Determine the quasistatic work required in newton-meters.

(b) Find the useful work output if the friction force between piston and cylinder is 1000 N and the piston moves 0.5 m.

**2-24M** A gas with a molar mass of 46.0 is compressed from 0.06 to 0.03 m³. The process equation relating $P$ and $V$ is given by $P = 0.001V^{-2} + 0.80$, where $P$ is in bars and $V$ is in cubic meters.

(a) Determine the required work of compression in kilojoules.

(b) Plot the process roughly to scale on a $PV$ diagram, and indicate the area representation of the work input.

**2-25M** A gas is compressed quasistatically in a piston-cylinder device from an initial state of 1.5 bars, 0.4 m³, to a final state of 0.2 m³ in accordance with the following relations: (a) $P(V)^{1/2} = $ constant, (b) $PV = $ constant, and (c) $Pv(\ln v) = $ constant. Determine the work required for the various processes in kilojoules.

**2-26M** A gas is expanded at a constant temperature of 17°C from 1 to 3 m³. The $PvT$ relationship for the gas is given by $[P + (a/v^2)]v = R_u T$, where $a = 15$ bar(m⁶)/(kg·mol)². Compute the work done quasistatically by the gas in newton-meters if the system contains 0.3 kg·mol.

**2-27M** A certain quantity of work is needed to reduce quasistatically the volume of an ideal gas to one-half of its initial value at a constant temperature $T_1$. At what temperature $T_2$ will the same amount of work reduce the volume isothermally to a quarter of the same initial volume if (a) $T_1 = 250°C$, and (b) $T_1 = 350°C$. Give the answer in °K.

**2-28M** Compute the work required to compress copper isothermally at (a) 300°K, and (b) 500°K from 1 bar to 500 bars. Assume the density is 8.90 g/cm³ at both temperatures.

**2-29M** The pressure on 1 kg of water is increased isothermally and quasistatically from 1 bar to 1000 bars. The density of water is 1.0 g/cm³. Estimate the work required in kilojoules if the temperature is (a) 20°C, and (b) 50°C.

**2-30M** The pressure on liquid mercury at 0°C is increased isothermally from 1 bar to 500 bars. Estimate the work required if the isothermal compressibility is $3.87 \times 10^{-6}$ bar⁻¹ and the density is 13.6 g/cm³.

**2-31M** The isothermal compressibility for liquid water at 60°C is given by $K_T = 0.125/v(P + 2740)$, where $K_T$ is in bar⁻¹, $P$ is in bars, and $v$ is in cm³/g. Determine the work required, in joules, to compress 2.5 kg of water isothermally from 1 bar to 600 bars.

**2-32M** Determine the work required, in joules, to compress 20 cm³ of liquid mercury at a constant temperature of 0°C from a pressure of 1 bar to (a) 500 bars, (b) 1000 bars, and (c) 1500 bars. The isothermal compressibility of mercury at the given temperature is given by $K_T = 3.9 \times 10^{-6} - 1.0 \times 10^{-10}P$, where $K_T$ is in bar⁻¹, and $P$ is in bars. The density of mercury may be taken as 13.6 g/cm³.

## Elastic work

**2-33M** The relation between the tension $\tau$, and length $L$, and the temperature $T$ for an elastic substance is given by $\tau = KT(x - x^{-2})$. In this expression $K$ is a constant, $x = L/L_0$, and $L_0$ is the value of $L$ at zero tension and is solely a function of temperature. Determine the work required to stretch such a substance isothermally and quasistatically from $L = L_0$ to (a) $L = 1.2L_0$, (b) $L = 1.4L_0$, and (c) $L = 1.5L_0$.

**2-34M** (*a*) Show that the work required to stretch an initially unstressed wire within the elastic region is given by $W = 0.5ALE\epsilon^2$, where $L$ is the initial length, $A$ is the cross-sectional area, and Young's modulus $E = \sigma/\epsilon$.

(*b*) What is the work required in newton-meters to increase the length of an unstressed steel bar from 10.00 to 10.01 m, if $E = 2.07 \times 10^7$ N/cm$^2$ and $A = 0.3$ cm$^2$?

**2-35M** In Prob. 2-34M use the same values of $E$ and $A$, but stretch the 10-m wire until the force on the bar is (*a*) 8000 N, and (*b*) 50,000 N.

**2-36M** Initially, a vertical steel wire 300 cm long with a cross-sectional area of 0.12 cm$^2$ has the upper end fixed and the lower end hanging free. By placing a load on the free end until the applied force in tension is 5000 N, one finds that the wire has been stretched 0.27 cm above its no-load length. Determine the work required in kilojoules to increase the tension to the 5000-N value.

**2-37M** An elastic spring with a spring constant $k_s$ of 100 N/cm has a length of 20 cm when not under compression. Determine the work required to compress the spring from (*a*) 20 to 19 cm, and (*b*) 19 to 18.5 cm in length.

**2-38M** An elastic linear spring with a spring constant of 144 N/cm is compressed from an initial unconstrained length to a final length of 6 cm. If the work required on the spring is 2.88 J, determine the initial length of the spring.

**2-39M** An elastic linear spring requires 12.6 J of work to compress it from 11 to 10 cm. Find the value of the spring constant $k_s$ in N/cm, if the spring initially is unstressed.

**2-40M** An elastic linear spring with a free length of 15 cm is compressed by a work input of 4.0 J. If the spring constant $k_s$ is 80 N/cm, determine the final length of the spring.

### Electric polarization and magnetization

**2-41M** A parallel-plate condenser in a dc circuit is charged by slowly increasing the voltage across the condenser from 0 to 110 V. Under this condition the voltage $V$ and the charge $Q_e$ are related by the equation $Q_e = kV$, where $k$ is the capacitance of the condenser. Calculate the work in joules necessary to charge the condenser if $k = 2.0 \times 10^{-5}$ C/V.

**2-42M** With respect to the general discussion in Prob. 2-41M, a parallel-plate condenser with a capacitance of $1 \times 10^{-5}$ C/V has a 10-V dc potential suddenly impressed across it and maintained until the condenser is charged.

(*a*) What is the final charge on the condenser?

(*b*) How much work was done on the condenser to charge it?

(*c*) If the same charge is built up by slowly increasing the impressed voltage from zero, how much work, in joules, will be done on the condenser?

(*d*) Why are the answers to (*b*) and (*c*) not the same?

**2-43M** A parallel-plate capacitor is quasistatically charged at room temperature to a potential of 100 V. The capacitor plates are 6 cm square and their separation is 1 mm. The dielectric equation of state for the air between the plates is given by $P = 4.75 \times 10^{-15}E$, where $P$ is in C/m$^2$, and $E$ is in V/m. Determine the work, in joules, required to polarize the air.

**2-44M** Curie's law for paramagnetic substances is given by the relation $M = CH/T$, where $C$ is a constant. For a quasistatic, isothermal change of state, show that the work done per unit volume to change the magnetization is given by

$$w = \mu_0 \frac{T}{2C}(M_i^2 - M_f^2) = \mu_0 \frac{C}{2T}(H_i^2 - H_f^2)$$

### Other work interactions

**2-45M** The drive shaft of an automobile rotates at 3000 revolutions per minute (rpm) and transmits (*a*) 75 kW, and (*b*) 90 kW of power from the engine to the rear wheels. Compute the torque, in newton-meters, developed by the engine.

**2-46M** The torque applied to a shaft is 100 N·m for (a) 10 revolutions, and (b) 25 revolutions. Determine the work input in kilojoules.

**2-47M** A torque of 150 N·m is associated with a shaft rotating at (a) 1500 rpm, and (b) 2500 rpm. Determine the power, in kilowatts, transmitted in each case.

**2-48M** An electric potential of 115 V is impressed on a resistor such that a current of 9 A passes through the resistor over a period of (a) 2 min, and (b) 5 min. Find the amount of electrical work, in kilojoules, done in each case.

**2-49M** A 12-V battery is used to pass a current of (a) 1.5 A, and (b) 4 A through an external resistance for a period of 15 s. Find the amount of electrical work, in kilojoules, done by the battery in each case.

**2-50M** A dc motor operating at 1000 rpm draws a current of 50 A at 24 V. The torque applied to the shaft is 10.5 N·m. What is the rate of heat transfer, in kJ/h?

**2-51M** A 12-V storage battery delivers a current of 10 A for 0.20h. What is the heat transfer, in kilojoules, if the energy of the battery decreases by 104 kJ?

**2-52M** A 12-V battery is charged by supplying a current of 5 A for 40 min. During the charging period a heat loss of 27 kJ occurs from the battery. Find the change in the energy stored in the battery during the specified time period.

### Additional problems

**2-53M** An ideal gas with a molar mass of 50 is contained within a piston-cylinder device initially at 20°C and 0.24 m³/kg. It undergoes an isothermal process to state 2, where the specific volume is 0.12 m³/kg. It is then expanded at constant pressure to state 3, where the specific volume is 0.36 m³/kg. Finally, it is returned to its initial state along a path which is a straight line on $Pv$ coordinates.

(a) Sketch the cyclic process on a $Pv$ diagram.
(b) Determine the values of $P_2$ and $P_3$ in bars.
(c) Calculate the net work of the cycle in newton-meters.

**2-54M** Nitrogen gas initially at 1.2 bars and 30°C is compressed isothermally in a piston-cylinder device to 3.0 bars.

(a) Compute the work required by some outside source and show the area representation on a $Pv$ diagram.

(b) Now consider that atmospheric air at 0.97 bar acts on the back of the piston, in addition to the connecting rod. Compute the net work required by the outside source in this case, for the same change in state of the nitrogen, and the $Pv$ area representation.

(c) Finally, consider in addition to part (b) that friction acts between the piston and cylinder and that the frictional resistance is equivalent to an effective resistive pressure of 0.50 bar which is constant throughout the piston travel. Determine in this case the net work that must be delivered along the connecting rod in N·m/g. Show the area representation of this final net work on a $Pv$ diagram.

**2-55M** A linear spring is compressed from 6 to 4.5 cm with an accompanying heat addition of 3.2 J. The values of $k_s$ and $x_{s,0}$ are 160 N/cm and 6.5 cm, respectively. Determine the internal energy change of the spring in joules.

# PROBLEMS (USCS)

### Kinetic and potential energy

**2-1** Determine the work required in foot-pounds force to accelerate a body of 6.44 $lb_m$ from (a) a velocity of 100 to 300 ft/s, and (b) a velocity of 50 to 150 ft/s.

**2-2** It requires 50,000 ft·lb$_f$ of work to raise a body 280 ft in Earth's gravitational field where the local acceleration of gravity is 31.9 ft/s². 

(a) Find the mass of the body in $lb_m$.

(b) If the initial gravitational potential energy of the body was 15,000 ft·lb$_f$ with respect to Earth's surface, determine the final elevation of the body above this surface, in feet.

**2-3** The acceleration of gravity as a function of elevation above sea level is given by $g = 32.17 - 3.32 \times 10^{-6}z$, where $g$ is in ft/s$^2$ and $z$ is in feet. A satellite with a mass of 520 lb is boosted 240 mi above Earth's surface. What work is required, in ft·lb$_f$?

**2-4** Work in the amount of 160,000 ft·lb$_f$ is used to change the kinetic energy of a body by 40,000 ft·lb$_f$/lb$_m$.

   (a) Determine the mass of the body in lb$_m$.

   (b) If the initial kinetic energy of the body relative to some reference is 30,000 ft·lb$_f$, find the final velocity relative to the same reference, in ft/s.

### General first law analysis

**2-5** A closed system undergoes a process during which 8 Btu of heat are removed and 25 Btu of work are done by the system. The system is then restored to its initial state by adding 7 Btu of heat to it, in addition to a work interaction. What are the magnitude and direction of the work interaction for the second process in Btu?

**2-6** During a certain process a closed system does 30 Btu of work while 10 Btu of heat are removed. Then the system is restored to its initial state by means of a process during which 4 Btu are added to it as work. Determine the net heat effect for the cycle, and indicate whether this net effect was on or by the system.

**2-7** A closed system executes a cyclic process during which there are four separate heat interactions. For the cycle $Q_1 = 20$ Btu, $Q_2 = -3$ Btu, $Q_3 = 5$ Btu, and $Q_4 = -12$ Btu. During the process three work interactions occur, for which $W_2 = -7780$ ft·lb$_f$ and $W_3 = 10,000$ ft·lb$_f$. What is the value of the third work interaction $W_1$, and is this quantity of work done on or by the system during the specific interaction?

**2-8** For each of the following cases of a process involving a closed system, fill in the blank spaces.

| | $Q$ | $W$ | $E_1$ | $E_2$ | $\Delta E$ | | $Q$ | $W$ | $E_1$ | $E_2$ | $\Delta E$ |
|---|---|---|---|---|---|---|---|---|---|---|---|
| (a) | 25 | −10 | | −10 | | (d) | 20 | | | 27 | 10 |
| (b) | −10 | | | 65 | −20 | (e) | −9 | 12 | | 29 | |
| (c) | | 15 | −10 | | 20 | (f) | | −10 | | 6 | −10 |

**2-9** A closed system undergoes a cycle made up of three processes. Fill in the missing data in the tables below for (a) System A, and (b) System B.

### System A

| Process | $Q$ | $W$ | $\Delta E$ |
|---|---|---|---|
| 1–2 | | 0 | 50 |
| 2–3 | 0 | −40 | |
| 3–1 | −30 | | |

### System B

| Process | $Q$ | $W$ | $\Delta E$ |
|---|---|---|---|
| 1–2 | 50 | 0 | |
| 2–3 | 0 | 40 | |
| 3–1 | | −20 | |

### Boundary work

**2-10** During an expansion process for a closed system the volume of a gas changes from 1 to 3 ft$^3$. The pressure changes during the quasistatic process according to the relation $P = 2.4V + 0.4$, where the units of $P$ and $V$ are atmosphere and cubic feet, respectively. Find by integration the work done by the gas in ft·lb$_f$. Check your answer by geometric considerations on a $PV$ diagram.

**2-11** A gas is compressed from a state of 15 psia, 0.5 ft$^3$, to a final state of 60 psia. The process equation relating $P$ and $V$ is $P = aV + b$, where $a = -150$ psia/ft$^3$. Compute by integration the necessary work input in ft·lb$_f$, and sketch the process path to scale on $PV$ coordinates.

**2-12** A gas is confined in a gastight, frictionless piston-cylinder assembly surrounded by the atmosphere. Initially, the pressure of the gas is 200 psia and the volume is 1.0 ft$^3$. If the gas expands to a final volume of 2.0 ft$^3$ and the piston area is 1.5 ft$^2$, calculate the work done in ft·lb$_f$ against the force $F$ along the shaft connected to the piston. Assume the process is such that $PV^2$ is a constant for the gas at any equilibrium state and that the atmospheric pressure is 1 standard atm.

**2-13** A piston-cylinder device contains 0.96 lb of oxygen ($O_2$) initially at a state of 40°F and 30 psia. Heat is supplied to the gas and it expands at constant pressure to a temperature of 1040°F. Calculate the quantity of work performed, in ft·lb$_f$, and sketch the process on a $PV$ diagram.

**2-14** Two pounds-mass of a gas is contained in a piston-cylinder assembly at initial conditions of 2 ft$^3$ and 100 psia. The gas is allowed to expand to a final volume of 5 ft$^3$. Determine the amount of work done in ft·lb$_f$ for the following processes:

    (*a*) The pressure remains constant.
    (*b*) The product $PV$ is a constant.
    (*c*) The product $PV^2$ is a constant.

Compare the different quasistatic processes by the use of a $PV$ diagram.

**2-15** An ideal gas is initially in a piston-cylinder assembly at 20 psia and 1 ft$^3$. The gas is first heated at constant pressure until the volume is doubled. It is then allowed to expand isothermally until the volume is again doubled. Determine the total work done by the gas in ft·lb$_f$/lb·mol of gas, and plot the processes on a $PV$ diagram. The initial gas temperature is 1000°R.

**2-16** In many real processes it is found that gases at low pressures fulfill the relationship $PV^n = c$, where $n$ and $c$ are constants. If $n$ is taken as 1.3, determine the work done by air, in ft·lb$_f$, in a piston-cylinder arrangement when it expands quasistatically from 0.2 ft$^3$ and 60 psia to a final pressure of 20 psia.

**2-17** A gas initially at 60 psia is contained in a piston-cylinder assembly. The piston is frictionless, and is held against the gas by a spring. When the gas expands from 0.2 to 0.8 ft$^3$, determine the work done by the gas in ft·lb$_f$ if the force exerted by the spring is (*a*) directly proportional to the volume of the system, and (*b*) proportional to the square root of the volume of the system. The pressure of the atmosphere surrounding the assembly is 15.0 psia.

**2-18** A quantity of sulfur dioxide gas is expanded in a cylinder equipped with a piston. The following data are taken:

| $P$, psia | $v$, ft$^3$/lb |
|-----------|----------------|
| 50 | 2.0 |
| 40 | 2.4 |
| 30 | 3.0 |
| 20 | 4.3 |
| 10 | 7.6 |

    (*a*) Calculate the work of expansion done per pound of sulfur dioxide.
    (*b*) If the friction between the cylinder and piston was equivalent to 0.2$P$, what was the work of expansion appearing in the surroundings per pound of sulfur dioxide?

**2-19** A piston-cylinder assembly contains 3 ft$^3$ of an ideal gas at 20 psia and 60°F. The gas is compressed to a final pressure of 60 psia and a temperature of 60°F.

    (*a*) Determine the minimum work of compression in ft·lb$_f$ if the process is isothermal.
    (*b*) Consider now that a process occurs between the same two end states as in part (*a*) such that it follows a straight line on $PV$ coordinates. Determine the work of compression for this process, and compare the two processes on a $PV$ diagram.

**2-20** One pound of a gas with a molar mass of 30 is compressed isothermally (constant temperature) at 140°F from a volume of 2 ft$^3$ to a volume of 1 ft$^3$. The $PvT$ relationship for the gas is given by $Pv = RT[1 + (c/v^2)]$, where $c = 450$ ft$^6$/(lb·mol)$^2$.

(a) Compute the work done on the gas in ft·lb$_f$/lb$_m$ if the process is quasistatic.

(b) If $c = 0$, would the work required be greater than, equal to, or less than that calculated in part (a)? Defend your answer in terms of two process lines on a $Pv$ diagram.

**2-21** One-tenth pound of a gas with a molar mass of 60 is compressed at constant temperature of 140°F from a volume of 0.2 ft$^3$ to 0.1 ft$^3$. The $PvT$ relationship for the gas is given by $Pv = RT[1 + (b/v)]$, where $b$ is 0.2 ft$^3$/lb$_m$.

(a) Determine the quasistatic work required in ft·lb$_f$/lb$_m$.

(b) If $b = 0$, would the work input be greater than, equal to, or less than that of part (a)? Defend your answer in terms of a $Pv$ diagram.

(c) What would be the answer to part (b) if the constant $b$ were originally negative, rather than positive?

**2-22** A gas with a molar mass of 46 is compressed from 1.0 to 0.5 ft$^3$. The process equation relating $P$ and $V$ is given by $P = 0.2V^{-2} + 0.8$, where $P$ is expressed in atmospheres and $V$ is in cubic feet.

(a) Determine the required work of compression in foot-pounds force.

(b) Plot the process roughly to scale on a $PV$ diagram, and indicate the area representation of the work input.

**2-23** A gas is compressed in a piston-cylinder device quasistatically from an initial state of 20 psia, 6 ft$^3$, to a final state of 3 ft$^3$, in accordance with the following process relations: (a) $P(V)^{1/2} = $ constant, (b) $PV = $ constant, and (c) $PV(\ln V) = $ constant. Determine the work required for the various processes in ft·lb$_f$.

**2-24** A gas is expanded at a constant temperature of 140°F from 2 to 4 ft$^3$. The $PvT$ relationship for the gas is given by $[P + (a/v^2)]v = R_u T$, where $a = 350$ (atm)(ft$^6$)/(lb·mol)$^2$. Compute the work done quasistatically by the gas, in foot-pounds force, if the system contains 0.2 lb·mol.

**2-25** A certain quantity of work is needed to quasistatically reduce the volume of an ideal gas to one-half of its initial value at a constant temperature $T_1$. At what temperature $T_2$ will the same amount of work reduce the volume isothermally to a quarter of the same initial volume if (a) $T_1 = 400°F$, and (b) $T_1 = 600°F$. Report the answer in °F.

**2-26** Compute the work required to compress copper isothermally at (a) 540°R, and (b) 900°R from 1 to 500 atm. Assume the density of copper is 555 lb/ft$^3$ at both temperatures.

**2-27** The pressure on 1 lb of water is increased quasistatically and isothermally from 1 to 1000 atm. The density of water is 62.3 lb/ft$^3$. Estimate the work required in ft·lb$_f$ if the temperature is (a) 68°F, and (b) 122°F.

**2-28** The pressure on liquid mercury at 32°F is increased isothermally from 1 to 500 atm. Estimate the work required if the isothermal compressibility is $3.92 \times 10^{-6}$ atm$^{-1}$ and the density is 848 lb/ft$^3$.

**2-29** The isothermal compressibility for liquid water at 140°F is given by $K_T = (2.01 \times 10^{-3})/v(P + 2700)$, where $K_T$ is in atm$^{-1}$, $P$ is in atm, and $v$ is in ft$^3$/lb. Determine the work, in ft·lb$_f$, required to compress 4.5 lb of water isothermally from 1 to 600 atm.

**2-30** Determine the work, in ft·lb$_f$, required to compress 1.0 in$^3$ of liquid mercury at a constant temperature of 32°F from a pressure of 1 atm to (a) 500 atm, (b) 1000 atm, and (c) 1500 atm. The isothermal compressibility of mercury at the given temperature is given by $K_T = 3.9 \times 10^{-6} - 1.0 \times 10^{-10}P$, where $K_T$ is in atm$^{-1}$ and $P$ is in atm. The density of mercury may be taken as 850 lb/ft$^3$.

## Elastic work

**2-31** The relation between the tension $\tau$, the length $L$, and the temperature $T$ for an elastic substance is given by $\tau = KT(x - x^{-2})$. In this expression $K$ is a constant, $x = L/L_0$, and $L_0$ is the value of length $L$ at zero tension and is solely a function of the temperature. Determine the work required to stretch such a substance isothermally from $L = L_0$ to (a) $L = 1.1L_0$, (b) $L = 1.2L_0$, and (c) $L = 1.3L_0$.

**2-32** (a) Show that the work required to stretch an initially unstressed wire within the elastic region is given by $W = 0.5ALE\epsilon^2$, where $L$ is the initial length, $A$ is the cross-sectional area, and Young's modulus $E = \sigma/\epsilon$.

(b) What is the work required, in ft·lb$_f$, to increase the length of an unstressed steel wire from 20.00 to 20.01 ft, if $E = 3 \times 10^7$ lb$_f$/in$^2$ and $A = 0.10$ in$^2$?

**2-33** In Prob. 2-32 use the same values of $E$ and $A$, but stretch the 20-ft wire until the force on the wire is (a) 2000 lb$_f$, and (b) 10,000 lb$_f$.

**2-34** Initially, a vertical steel wire 100 in long with a cross-sectional area of 0.050 in$^2$ has the upper end fixed and the lower end hanging free. By placing a load on the free end until the applied force in tension is 1000 lb$_f$, one finds that the wire has been stretched 0.067 in above its no-load length. Determine the work required to increase the tension slowly to 1000 lb$_f$, in ft·lb$_f$.

**2-35** An elastic spring with a spring constant $k_s$ of 100 lb$_f$/in has a length of 10 in when not under compression. Determine the work required to compress the length of the spring from (a) 10 to 9 in, and (b) 9 to 8 in.

**2-36** An elastic spring with a spring constant $k_s$ of 72 lb$_f$/in is compressed from an initial unconstrained length to a final length of 3 in. If the work required on the linear spring is 81 ft·lb$_f$, determine the initial length of the spring.

**2-37** An elastic linear spring requires 252 ft·lb$_f$ of work on it to compress it from 12 to 9 in. Find the value of the spring constant $k_s$, in lb$_f$/in, if the spring initially is unstressed.

**2-38** An elastic linear spring with a free length of 8 in is compressed by a work input of 14 ft·lb$_f$. If the spring constant $k_s$ is 48 lb$_f$/in, determine the final length of the spring.

### Electric polarization and magnetization

**2-39** A parallel-plate condenser in a dc circuit is charged by slowly increasing the voltage across the condenser from 0 to 110 V. Under this condition the voltage $V$ and the charge $Q_e$ are related by the equation $Q_e = kV$, where $k$ is the capacitance of the condenser. Calculate the work in joules necessary to charge the condenser if $k = 2.0 \times 10^{-5}$ F (1 F = 1 C/V).

**2-40** A parallel-plate condenser with a capacitance of $1 \times 10^{-5}$ F has a 10-V dc potential suddenly impressed across it and maintained until the condenser is charged.
(a) What is the final charge on the condenser?
(b) How much work was done on the condenser to charge it?
(c) If the same charge is built up by slowly increasing the impressed voltage from zero, how much work will be done on the condenser?
(d) Why are the answers to (b) and (c) not the same?

**2-41** A parallel-plate capacitor is quasistatically charged at room temperature to a potential of 100 V. The capacitor plates are 3 in square and their separation is 0.040 in. The dielectric equation of state for the air between the plates is given by $P = 4.75 \times 10^{15}E$, where $P$ is in C/m$^2$ and $E$ is in V/m. Determine the work, in joules, required to polarize the air.

**2-42** Curie's law for paramagnetic substances is given by the relation $M = CH/T$, where $C$ is a constant. If a paramagnetic substance undergoes a quasistatic, isothermal change of state, show that the work done by unit volume to change the magnetization is given by

$$w = \mu_0 \frac{T}{2C}(M_i^2 - M_f^2) = \mu_0 \frac{C}{2T}(H_i^2 - H_f^2)$$

### Other work interactions

**2-43** The drive shaft of an automobile rotates at 3000 revolutions per minute (rpm) and transmits (a) 90 hp, and (b) 120 hp from the engine to the rear wheels. Compute the torque developed by the engine in lb$_f$·ft.

**2-44** The torque applied to a shaft is 120 lb$_f$·ft for (a) 10 revolutions, and (b) 25 revolutions. Determine the work input in ft·lb$_f$.

**2-45** A torque of 150 lb$_f$·ft is associated with a shaft rotating at (a) 1500 rpm, and (b) 2500 rpm. Determine the power transmitted in each case in horsepower.

**2-46** An electric potential of 115 V is impressed across a resistor for a period of (a) 2 min, and (b) 5 min. If the current is 1.8 A, find the amount of electrical work, in kilojoules, done in each case.

**2-47** A 12-V battery is used to pass a current of (a) 1.5 A, and (b) 4 A through an external resistance for a period of 15 s. Find the amount of electrical work, in kilojoules, done by the battery in each case.

### Additional problems

**2-48** An automobile battery which originally is fully charged gradually discharges while sitting on the shelf at a constant temperature of 90°F. It produces no electrical work while discharging, but heat transfer of 1000 Btu occurs to its environment. If the battery is then recharged slowly to its initial state by a process requiring 440 Wh of work, what is the heat transfer in Btus to the battery during the charging process?

**2-49** An experimental energy converter has a heat input of 80,000 Btu/h and a work input of 1.2 hp. The converter produces an electric power output of 18 kW. Calculate the change in the energy of the converter in Btu over a time period of 4 min.

**2-50** A dc motor operating at 1000 rpm draws a current of 50 A at 24 V. The torque applied to the shaft is 110 lb$_f$·in. What is the rate of heat transfer in Btu/h?

**2-51** A 12-V storage battery delivers a current of 10 A for 0.20 h. What is the heat transfer if the energy of the battery decreases by 104 Btu?

**2-52** A 12-V battery is charged by supplying a current of 5 A for 40 min. During the charging period a heat loss of 26 Btu occurs from the battery. Find the change in the energy stored in the battery during the specified time period.

# THREE

## STATE POSTULATE AND SOME PROPERTY RELATIONS

An important objective of thermodynamics is to ascertain the functional relationships among certain macroscopic properties of a system at equilibrium. In order to develop these relations, the number of independent variables necessary to fix or establish the state of the system under given conditions must be known. This information is provided by a basic law of thermodynamics known as the state postulate.

## 3-1 THE STATE POSTULATE

Thermodynamics directs its attention to a major class of properties of matter called *thermostatic* or *intrinsic* properties. This class of properties arises from particular characteristics of the mass within the boundaries of the system. They are evaluated by making observations solely on the substance without reference to the surroundings. Examples of thermostatic or intrinsic properties include pressure, temperature, density, and internal energy. From a microscopic viewpoint each thermostatic property characterizes an average of some type of molecular behavior of the particles of the system. For example, the pressure of a gas may be thought of as an average rate of change of momentum of the particles per unit time and per unit area within the system. Since thermostatic properties are all characteristics of microscopic behavior, it is reasonable to expect them to be functionally related. Even though intrinsic properties do not require external datum points for their measurement, many of these properties must be assigned an artificial reference value at one state in order to assign definite numerical values at all other states.

In order to develop functional relationships among the macroscopic intrinsic properties of a substance in equilibrium, we need to know how many intrinsic properties of a substance can be varied independently. For example, how is the internal energy $U$ (used in the conservation of energy equation) related to properties such as pressure, temperature, and others. Consider the general situation where any dependent extensive property $Y_0$ is a function of $n$ other extensive properties. That is, $Y_0 = Y(Y_1, Y_2, \ldots, Y_n)$. Once we have selected values for all the $n$ independent properties, the values of the remaining dependent properties, such as $Y_0$, are fixed. As a simple example, it is known that pure water boils at 212°F (100°C) if the pressure of the substance is adjusted to 1 atm. If the pressure is lowered, such as on a mountain top, the water boils at a lower temperature. Selection of the pressure determines the boiling temperature. Symbolically for this two-phase system, we write that $T = T(P)$.

We seek a general rule that allows us to determine the number of required independent properties for any prescribed physical situation. Experience has shown that the number of independent variables depends upon the nature of the physical situation of interest. On the basis of experimental data for homogeneous (single-phase) substances, the following statement may be made. Known as the *state postulate*, it asserts:

> The number of independent intrinsic properties required to fix the state of a substance is equal to one plus the number of possibly relevant quasistatic work modes.

By "relevant work modes" we mean that they have an appreciable effect on the state of the substance if they are altered during a process. Objects on Earth, for example, are usually under the influence of the natural gravitational, electric, and magnetic force fields of Earth. However, their effects on the outcome of most processes are often negligible. Consequently we ignore their effect when determining the number of independent variables. However, if a strong electric field is applied to a substance containing electric dipoles, we certainly expect its state to be a strong function of the electric field strength. In this case the applied electric field would need to be considered an independent variable. Nonquasistatic work modes are never relevant, in terms of the state postulate.

## 3-2 SIMPLE SYSTEMS

The state postulate asserts that the number of independent thermostatic properties of a substance is related to the number of relevant quasistatic work modes that possibly could alter the state of the substance. When one examines past and present-day engineering and scientific studies, one thing becomes apparent. Seldom does one encounter a system where more than one of the quasistatic work modes listed in Table 2-2 is used to alter the state of the substance. That is, most problem areas involve compressible systems, elastic systems, electrostatic systems, etc., but seldom does one find a combination of these effects influencing the state of the system.

On the basis of this observation, it is convenient to classify various systems as simple systems. By definition, a *simple* system is one for which only one quasistatic work mode might possibly affect the state of the system. On the basis of the state postulate asserted above, the following statement may be made as a state postulate for simple substances:

> The equilibrium state of a simple, homogeneous substance is fixed by specifying the values of any two independent, intrinsic properties.

Once the two independent properties are fixed in value, the values of all remaining intrinsic or thermostatic properties are also fixed. Thus only two intrinsic properties may be independently varied for any simple substance. The state postulate for simple systems is extremely important, since a wide range of scientific and engineering problems involves systems which meet the conditions of a simple system.

The various types of simple systems may be differentiated from each other by speaking of simple compressible systems, simple elastic systems, simple magnetic systems, etc. A *simple compressible* system is defined as one for which the only quasistatic work interaction is boundary $(P\ dV)$ work. For such a substance the effects of capillarity, anisotropic stress, and external force fields are negligible. From a practical viewpoint this means that the system is not influenced by these effects, even though they may be present to some degree. The absence of capillarity, for example, implies that volume effects are much more significant than interfacial surface effects. Due to their importance in engineering studies, property relations for simple compressible substances will be stressed in this and the following chapter.

Any two independent, intrinsic properties are sufficient to fix the state of a simple, homogeneous substance, whether the properties are intensive or extensive in nature. Thus the functional relationship between a set of extensive properties is given by

$$Y_0 = Y(Y_1,\ Y_2) \qquad\qquad (3\text{-}1)$$

and the equivalent relationship for intensive properties is

$$y_0 = y(y_1,\ y_2) \qquad\qquad (3\text{-}2)$$

where the quantities $(Y_1,\ Y_2)$ or $(y_1,\ y_2)$ represent a set of two independent variables. Equation (3-2) follows directly from Eq. (3-1), since any extensive property $Y$ is related to the corresponding intensive value $y$ by the relation $Y = my$. Hence, for a system of fixed mass and chemical composition, the value of the intensive property $y$ is set once the value of $Y$ is fixed. The above equations express some as yet unknown relation between two independent, intrinsic properties and a third dependent property. One of the major tasks of thermodynamics is to develop from theoretical or experimental considerations some explicit expressions for these relationships, especially in terms of intensive properties.

As a familiar example of Eq. (3-2), consider the relationship among the variables $P$, $v$, and $T$. We may write for a simple compressible substance, in accordance with this equation,

$$v = v(P, T)$$

It is well known from elementary chemistry that the specific volume of a gas is a function only of the pressure and the temperature imposed upon the system. The ideal-gas equation of state, $Pv = RT$, is a specific example of a relation commonly used for gases at relatively low pressures. It is also worth noting that for a simple compressible system the only intrinsic form of energy that is altered by a change of state is the internal energy $U$. Therefore the relation $E = E(Y_1, Y_2)$ valid for a simple system reduces to the form $U = U(Y_1, Y_2)$. Although the two independent variables $Y_1$ and $Y_2$ could be chosen from a large set of properties, we shall find it convenient to regard the internal energy primarily as a function of the volume and the temperature of the system. On an intensive basis the functional relationship becomes $u = u(v, T)$. This relationship will be examined shortly.

Similar situations arise for other types of systems where the change in only one generalized displacement is important. For example, in the study of paramagnetic substances the magnetization of the substance is important compared to the volume. By neglecting volume changes, we need only two independent variables for the system. In this case one might seek a relationship among the magnetization $M$, the temperature $T$, and the magnetic field strength $H$. Any two of these can be chosen as the independent properties. In some cases these variables are related by Curie's law, which states that $MT = KH$, where $K$ is Curie's constant. As another example, in the study of elastic substances the elastic work $\sigma \, d\epsilon$ is much more important than boundary work due to volume changes. Again only one quasi-static work interaction is involved and only two independent properties are required. Consequently the variables $(\epsilon, T)$ might be chosen for convenience as the independent variables. In both the above examples, only one external constraint was varied, replacing the volume constraint used with simple compressible systems. If the volume is considered an important parameter for either a magnetic or an elastic substance, then three independent variables would be required. For example, the energy $E$ of an elastic substance could be expressed as $E = E(V, T, \epsilon)$.

The state principle, based on experimental evidence, enables one to predict the number of independent, intrinsic properties required to fix the equilibrium state of a system. On the basis of this principle we may begin to develop various techniques—analytical, graphical, and tabular—for relating and evaluating intrinsic properties. The most important properties to examine at the beginning in terms of their functional relationships are pressure $P$, temperature $T$, specific volume $v$, and internal energy $u$. In addition, we shall find it necessary in the next section to define three other properties of interest, the enthalpy $h$, the specific heat at constant pressure $c_p$, and the specific heat at constant volume $c_v$.

## 3-3 THE CONSERVATION OF ENERGY PRINCIPLE FOR SIMPLE COMPRESSIBLE CLOSED SYSTEMS

The conservation of energy equation for closed systems (control masses) can be expanded into the following form.

$$Q + W = \Delta E$$

$$= \Delta U + \Delta KE + \Delta PE + \Delta(\text{electrostatic energy}) + \cdots \quad (3\text{-}3)$$

where the terms on the right include many possible forms of energy associated with the system. It must also be remembered that the work term $W$ may include several types of work interactions that occur during the specified process. If, in addition to the effects of magnetic and electric fields and surface tension, we elect also to neglect the effects of motion on a substance, the above equation reduces to

$$Q + W = \Delta U \quad (3\text{-}4)$$

where $U$ is a function of two independent intrinsic properties. Other useful forms of Eq. (3-4) are

$$\delta Q + \delta W = dU \quad (3\text{-}5)$$

$$q + w = \Delta u \quad (3\text{-}6)$$

and

$$\delta q + \delta w = du \quad (3\text{-}7)$$

Equations (3-4) to (3-7) apply to simple compressible closed systems which are stationary. The last two of these equations are applied to a substance of unit mass.

In addition to the internal enegy $U$ of a substance, we shall find it convenient to employ in energy balances another intrinsic property called the enthalpy. It is defined by the relation

$$H \equiv U + PV \quad (3\text{-}8)$$

or, per unit mass, by

$$h \equiv u + Pv \quad (3\text{-}9)$$

Since $u$, $P$, and $v$ are properties, $h$ also must be a property. In many cases the enthalpy function has no specific physical interpretation, although it does have the dimension of energy. In a number of thermodynamic equations the quantities $U$ and $Pv$ often appear together. Therefore it is advantageous to write the symbol $h$ in these equations, rather than $u + Pv$. As a result it would also be helpful to tabulate numerical values of the enthalpy along with other property correlations, just as a labor-saving device. In reality, we shall find that the enthalpy function appears in tables and charts more frequently than the internal-energy function. However, the enthalpy $h$ should not be thought of as a specific form of energy, but simply as a defined quantity which is useful in the solution of engineering and scientific problems. The enthalpy function is especially useful in the analysis of open systems.

The enthalpy function plays a minor role in the analysis of closed systems. However, it is a convenient property when analyzing volume changes under the conditions of constant pressure. If a system can be assumed to be a simple compressible system, then the conservation of energy principle reduces to Eq. (3-5), namely,

$$\delta Q + \delta W = dU$$

Since the volume of the control mass is altered, a work interaction is associated with the process. In order to evaluate the work done in moving the boundary, it is necessary to assume that the process is carried out quasistatically. On this basis the work is given by the integral of $-P\,dV$. The energy equation then becomes

$$\delta Q - P\,dV = dU$$

or

$$\delta Q = dU + P\,dV \tag{a}$$

Now, since $H = U + PV$, for a process at constant pressure it is seen that

$$dH = d(U + PV) = dU + d(PV) = dU + P\,dV \tag{b}$$

Substitution of equation (b) into equation (a) leads to an interesting result under the restrictions imposed; that is,

$$\delta Q_P = dH \tag{c}$$

This relation indicates that the quantity of heat transferred to or from a simple compressible system during a quasistatic, constant-pressure process is equal to the change in the enthalpy of the control mass within the boundaries of the system. A knowledge of the enthalpies at the initial and final states would lead to a rapid evaluation of the heat interaction required to bring about the required change of state.

In general, other work interactions may occur during the constant-pressure change of state, in addition to boundary work. For example, if the system were a fluid it might be stirred with a paddle wheel. As another possibility, electrical work might be performed on an electrical resistor within the system. In order to make a broader statement of the special energy equation (c), we may write that

$$\delta Q_P + \delta W' - P\,dV = dU \tag{3-10a}$$

or

$$\delta Q_P + \delta W' = dH \tag{3-10b}$$

where $\delta W'$ in these two equations includes all forms of work other than boundary work. Equations (3-10) are still restricted to quasistatic constant pressure processes.

From the equations developed in this section for simple compressible stationary systems it is apparent that information on the internal energy and the enthalpy over a broad range of conditions will be required in order to analyze and synthesize engineering systems of interest. It will be shown in subsequent sections that the methods of presenting or evaluating $u$ and $h$ depend upon the class or phase of the substance under study.

## 3-4 INTERNAL ENERGY, ENTHALPY, AND RELATED PROPERTIES

The conservation of energy principle developed in Sec. 3-3 requires the knowledge of the internal energy of simple substances. In some special cases we shall find that the enthalpy function $h$, defined as $u + Pv$, is a convenient function to employ in conjunction with this conservation principle. Since the internal energy and the enthalpy of substances are not directly measurable, it is necessary to develop equations for these properties in terms of other measurable properties, such as $P$, $v$, and $T$. For simple compressible systems the internal energy $u$ is a function of two other intensive, intrinsic properties. It is advantageous to select the temperatures and the specific volume as the independent variables. If $u = u(T, v)$, we can write for the total differential of $u$ that

$$du = \left(\frac{\partial u}{\partial T}\right)_v dT + \left(\frac{\partial u}{\partial v}\right)_T dv \qquad (3\text{-}11)$$

The first partial derivative on the right is defined as $c_v$, the specific heat at constant volume. That is,

$$c_v \equiv \left(\frac{\partial u}{\partial T}\right)_v = \frac{1}{m}\left(\frac{\partial U}{\partial T}\right)_v \qquad (3\text{-}12)$$

where $u$ is expressed as the internal energy per unit mass, and $m$ is the total mass of the closed system. It is convenient at times to tabulate and to employ in calculations a molar specific heat $c_v$. This is defined, similarly, by

$$c_v = \left(\frac{\partial u}{\partial T}\right)_v = \frac{1}{N}\left(\frac{\partial U}{\partial T}\right)_v \qquad (3\text{-}13)$$

where $u$ in this case is the internal energy per mole, and $N$ is the number of moles in the system. It should be noted that Eqs. (3-12) and (3-13) are essentially the same, except that different units of mass are employed. The symbol for $c_v$ is the same in either case. The proper units to employ for $c_v$ will be apparent from the context of the problem. The two equations are presented simply because $c_v$ is commonly tabulated in units of both grams (or pounds) and gram-moles (or pound-moles).

Evaluation of the enthalpy function $h$ is usually made in terms of temperature and pressure as the independent variables. By letting $h = h(T, P)$ we may then write that

$$dh = \left(\frac{\partial h}{\partial T}\right)_P dT + \left(\frac{\partial h}{\partial P}\right)_T dP \qquad (3\text{-}14)$$

The first partial derivative on the right is defined as the specific heat at constant pressure $c_P$. Then we have, in a fashion analogous to Eqs. (3-12) and (3-13),

$$c_p \equiv \left(\frac{\partial h}{\partial T}\right)_P = \frac{1}{m}\left(\frac{\partial H}{\partial T}\right)_P \qquad (3\text{-}15)$$

or

$$c_p = \left(\frac{\partial h}{\partial T}\right)_P = \frac{1}{N}\left(\frac{\partial H}{\partial T}\right)_P \qquad (3\text{-}16)$$

Thus we find that derivatives of properties with respect to other properties are themselves state functions of the system. The specific heats $c_v$ and $c_p$ are two of the most important derivative-type properties in thermodynamics.

The dimensions on the values of the specific heats defined above are in terms of energy/(mass)(temperature difference). It is important to note that the temperature quantity in the denominator involves a change in temperature, and not the value of the temperature itself. Consequently, the specific heat may be expressed in terms of either kelvins or degrees Celsius, for example. The symbols °K and °C for these two temperature scales have the same significance in this special case, where only differences in temperature are of interest. The same reasoning obviously applies to degrees Rankine versus degrees Fahrenheit. Common units for specific heats in the SI are kJ/(kg)(°C) [or kJ/(kg)(°K)] and kJ/(kg·mol)(°C). In the USCS the units employed most frequently are Btu/(lb$_m$)(°F) [or Btu/(lb$_m$)(°R)] or Btu/(lb·mol)(°F).

## 3-5 INTERNAL ENERGY, ENTHALPY, AND SPECIFIC-HEAT RELATIONS FOR IDEAL GASES

When gases are at sufficiently low pressures, their $PvT$ behavior is approximated very closely by the relationship $Pv = RT$. In order to make suitable energy balances for processes involving ideal gases, it is necessary to evaluate internal-energy and enthalpy changes for these gases. The internal-energy change for any simple compressible substance is given by Eq. (3-11). When this equation is combined with Eq. (3-12), we find that in general

$$du = c_v \, dT + \left(\frac{\partial u}{\partial v}\right)_T dv \qquad (3\text{-}17)$$

The second coefficient, $(\partial u/\partial v)_T$, is a measure of the change in the internal energy of a substance as the volume is altered at constant temperature. From microscopic considerations it can be argued that the internal energy of an ideal gas should not be a function of the volume of the system. This result is confirmed by macroscopic measurements. As early as the middle of the nineteenth century, Joule carried out a series of experiments which indirectly indicated that the internal energy of gases at low pressures was essentially a function of the temperature only. The fact that $(\partial u/\partial v)_T$ is approximately zero at low pressures for gases is sometimes called Joule's law. Consequently, for substances which approximate ideal-gas behavior, the coefficient of the last term in Eq. (3-17) may be taken as zero. Thus, for ideal gases, we may write

$$du = c_v \, dT \qquad \text{for ideal gases} \qquad (3\text{-}18)$$

for *all* processes, whether constant-volume or not. Hence the internal energy of an ideal gas, unlike that of real gases, is a function of only one independent variable, the temperature. Equation (3-18) is an excellent approximation for many gases up to pressures of several hundred pounds per square inch. This equation is misleading, however, in one respect. The presence of the term $c_v$ in the equation often leads one to misconstrue the proper use of the equation. If a gas behaves essentially as an ideal gas, this expression for $du$ is valid for all processes, regardless of its path. The use of the equation is not restricted to constant-volume processes. The simplicity of the relationship is a consequence of the special nature of an ideal gas.

Integration of Eq. (3-18) for any finite process involving an ideal gas leads to

$$\Delta u = \int c_v \, dT \qquad \text{for ideal gases} \qquad (3\text{-}19)$$

The right-hand side can be evaluated once the empirical data for $c_v$ as a function of temperature are measured.

The extension of these results to the property enthalpy is straightforward. By definition, $h = u + Pv$, and for an ideal gas, $Pv = RT$. Thus we may write

$$dh = du + d(Pv) \qquad \text{and} \qquad d(Pv) = d(RT) = R \, dT$$

The change in the enthalpy for an ideal gas then becomes

$$dh = du + R \, dT \qquad (3\text{-}20)$$

The terms on the right-hand side of Eq. (3-20) are solely a function of temperature for an ideal gas. Consequently, the enthalpy of a hypothetical ideal gas is also only temperature-dependent. Since the specific heat at constant volume is defined for simple substances as $(\partial u/\partial T)_v$, the value of $c_v$ must also be solely a function of temperature for ideal gases. Likewise, the $c_p$ values for ideal gases are a function of temperature only.

To evaluate enthalpy changes of ideal gases, one uses the basic equation represented by Eq. (3-14). When this latter equation is combined with Eq. (3-15) for $c_p$, we find that a general expression for $dh$ for any simple compressible substance is

$$dh = c_p \, dT + \left(\frac{\partial h}{\partial P}\right)_T dP \qquad (3\text{-}21)$$

Since the enthalpy of an ideal gas is solely a function of temperature, the above equation reduces to

$$dh = c_p \, dT \quad \text{for ideal gases} \tag{3-22}$$

or

$$\Delta h = \int c_p \, dT \quad \text{for ideal gases} \tag{3-23}$$

This set of equations is valid for *all* processes of an ideal gas, and is not restricted to constant-pressure processes.

A special relationship between $c_p$ and $c_v$ for ideal gases is obtained by substituting Eqs. (3-18) and (3-22) into Eq. (3-20). This yields

$$c_p \, dT = c_v \, dT + R \, dT$$

or

$$c_p - c_v = R \quad \text{for ideal gases} \tag{3-24}$$

This simple relationship between $c_p$ and $c_v$ for an ideal gas is an important one, since a knowledge of either $c_p$ or $c_v$ allows the other one to be calculated by the above equation. When the specific heats are given as molar values, the value of $R$ in this equation is $R_u$, the universal gas constant.

The integration of Eqs. (3-19) and (3-23) for $\Delta u$ and $\Delta h$ requires a knowledge of the specific-heat variation with temperature at low pressures. The following section discusses the general behavior of the specific heats of gases in terms of their order of magnitude, as well as their functional dependence on both temperature and pressure.

## 3-6 SPECIFIC HEATS OF GASES

The molar specific heats at constant pressure of some common gases are illustrated in Fig. 3-1 as a function of temperature. The data have been extrapolated to zero pressure and are known as zero-pressure or ideal-gas specific heats. The symbols $c_{p,0}$ and $c_{v,0}$ are used frequently to signify values at this state of very low pressure. In Fig. 3-1 it is seen that a monatomic gas such as argon has a value of $c_{p,0}$ which is very close to 5.0 Btu/(lb·mol)(°F) over the entire range of temperature. This value, which is 20.8 kJ/(kg·mol)(°C), is characteristic of all monatomic gases. On the basis of Eq. (3-24) it is found that $c_{v,0}$ for monatomic gases must be approximately 12.5 kJ/(kg·mol)(°C) or 3.0 Btu/(lb·mol)(°F) over a wide range of temperature. As shown in the figure, molecules with two or more atoms do not have a constant value of $c_p$. These more complex molecules exhibit an increase in $c_p$ with increasing temperature at low pressures. The $c_p$ values of the diatomic gases shown in Fig. 3-1 increase as much as 25 percent over a range of 0 to 2000°F (0 to 1100°C). Values of the constant-pressure and constant-volume specific heats are tabulated, in Tables A-4M and A-4, as a function of temperature for a few

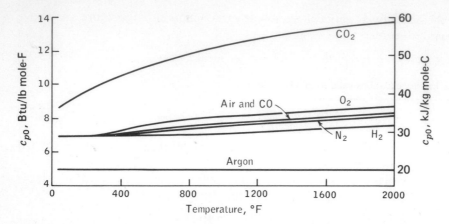

**Figure 3-1** Values of $c_{p,0}$ for seven common gases. (Based on data for NBS Circular 564, 1955.)

common gases at essentially zero pressure. The values of $c_v$ and $c_p$ for monatomic gases quoted above are also repeated in these tables.

Tables A-4M and A-4 also include the variation of the specific-heat ratio $k$ with temperature. This ratio is defined as

$$k \equiv c_p/c_v \qquad (3\text{-}25)$$

The *specific-heat ratio* is an important thermostatic property. Its value for monatomic gases is 1.67 and is essentially constant with temperature. For molecules containing two or more atoms the $k$ value is always less than 1.67 and the values vary considerably with temperature. However, from the definition of $k$ and the fact that $c_p$ is always greater than $c_v$ for an ideal gas $(c_p - c_v = R)$, $k$ is never less than unity. Many common diatomic gases have a specific-heat ratio of, roughly, 1.4 at room temperature and slightly above. The specific-heat ratio is related to either $c_v$ or $c_p$ by the following relations:

$$c_v = \frac{R}{k - 1} \qquad (3\text{-}26)$$

and

$$c_p = \frac{Rk}{k - 1} \qquad (3\text{-}27)$$

These equations are restricted to ideal gases.

In order to integrate Eqs. (3-19) and (3-23) for $\Delta u$ and $\Delta h$, we generally need equations which relate $c_v$ and $c_p$ to temperature. Specific-heat data as a function of temperature are measured directly or evaluated from theory based on a molecular model of matter. Reasonably accurate algebraic equations may then be fitted to the experimental or calculated data. Equations for zero-pressure specific-heat data based on spectroscopic measurements are given in Table 3-1 for a number of

## Table 3-1 Constant-pressure specific-heat equations of various gases at zero pressure

$$c_p/R = a + bT + cT^2 + dT^3 + eT^4$$

$T$ is in kelvins, equations valid from 300 to 1000°K

| Gas | $a$ | $b \times 10^3$ | $c \times 10^6$ | $d \times 10^9$ | $e \times 10^{12}$ |
|---|---|---|---|---|---|
| CO | 3.710 | −1.619 | 3.692 | −2.032 | 0.240 |
| $CO_2$ | 2.401 | 8.735 | −6.607 | 2.002 | |
| $H_2$ | 3.057 | 2.677 | −5.810 | 5.521 | −1.812 |
| $H_2O$ | 4.070 | −1.108 | 4.152 | −2.964 | 0.807 |
| $O_2$ | 3.626 | −1.878 | 7.056 | −6.764 | 2.156 |
| $N_2$ | 3.675 | −1.208 | 2.324 | −6.322 | 0.226 |
| $CH_4$ | 3.826 | −3.979 | 24.558 | −22.733 | 6.963 |
| $C_2H_2$ | 1.410 | 19.057 | 24.501 | 16.391 | −4.135 |
| $C_2H_4$ | 1.426 | 11.383 | 7.989 | −16.254 | 6.749 |

*Source*: Adapted from the data in NASA SP-273, U.S. Government Printing Office, Washington, 1971.

common gases. The units to be employed in the equations must be specified by the author of the equations. In many tables of specific-heat equations the range of temperatures over which an equation is valid will be listed. Also, the maximum percent error of an equation, when used within the prescribed range of temperatures, is frequently noted. The algebraic format of such equations is purely arbitrary.

On the basis of the information on specific-heat data presented above, we are now in a position to evaluate internal-energy and enthalpy changes for ideal gases. In recapitulation, for ideal gases,

$$\Delta u = \int c_v \, dT \tag{3-19}$$

and

$$\Delta h = \int c_p \, dT \tag{3-23}$$

With the use of the types of equations presented in Table 3-1, direct integration of these two equations is possible. Such integrations normally would be carried out for each set of temperature limits of interest. This method of evaluation is acceptable when only a few integrations are anticipated. However, this type of computation is highly undesirable for design purposes if many repetitive calculations are required for specific gases. Consequently the internal-energy and enthalpy data for a few such gases (such as air and other low-molecular-weight gases) have been extensively tabulated over small temperature intervals. These tables are the result of the integration of accurate specific-heat data and they greatly simplify the evaluation of $\Delta u$ and $\Delta h$ when it is not desirable to use computer technology. A set of data for air is shown in Tables A-5M and A-5. In order to obtain the values of $u$ and $h$ shown, the integration process is based on the following modification of Eqs. (3-19) and (3-23), namely,

$$u = u_{ref} + \int_{T_{ref}}^{T} c_v \, dT \quad \text{and} \quad h = h_{ref} + \int_{T_{ref}}^{T} c_p \, dT \tag{3-28}$$

That is, to obtain any $u$ or $h$ value at a given $T$, an arbitrary reference value $u_{ref}$ or $h_{ref}$ is set at a reference temperature, $T_{ref}$. In the ideal-gas tables in the appendix the arbitrary reference value of zero enthalpy is chosen to be at $0°K$. The $h$ and $u$ data in Table A-5M are in kJ/kg (or J/g), while the same properties in Table A-5 are in Btu/lb. Tables A-6M to A-11M list the ideal-gas enthalpies and internal energies of the gases $N_2$, $O_2$, CO, $CO_2$, $H_2O$ and $H_2$, and these $h$ and $u$ data are molar values, that is, kJ/kg mole (or J/g mole). Data for these six gases in USCS units appear in Tables A-6 to A-11, and the molar units in this case are Btu/lb·mol. Other properties that appear in these tables, in addition to $h$ and $u$, will be introduced in a later chapter.

When the temperature interval is relatively small, the specific heats of gases are nearly constant. This can be confirmed by noting data over a several hundred degree interval in either Table A-4M or A-4. Frequently it is convenient in such cases to assume a constant $c_p$ or $c_v$ value, so that Eqs. (3-19) and (3-23) reduce to

$$\Delta u = c_{v, av} \, \Delta T \tag{3-29}$$

and

$$\Delta h = c_{p, av} \, \Delta T \tag{3-30}$$

where $c_{v, av}$ and $c_{p, av}$ in these equations are some sort of average values for the given temperature interval. Usually arithmetic averages are used. This method is especially useful when only tabular specific-heat data are available. The example below illustrates and compares various methods of evaluating the enthalpy change of an ideal gas.

**Example 3-1M** Find the change in enthalpy of 1 kg of air which is heated at low pressure from 300 to $500°K$ by use of (a) empirical specific-heat data, (b) average specific-heat data, and (c) the table for air, Table A-5M.

SOLUTION (a) The constant-pressure specific heat of air may be represented by the equation

$$c_p = 27.43 + 6.180 \times 10^{-3}T - 0.8987 \times 10^{-6}T^2$$

where $c_p$ is measured in kJ/(kg mol)($°K$) and $T$ is in kelvins ($°K$). If the air is assumed to behave as an ideal gas, then $\Delta h = \int c_p \, dT$. Substitution of the above equation for $c_p$ into the expression for $\Delta h$, and subsequent integration, leads to

$$\Delta h = 27.43(T_2 - T_1) + 3.090 \times 10^{-3}(T_2^2 - T_1^2) - 0.2996 \times 10^{-6}(T_2^3 - T_1^3)$$

where $T_2 = 500°K$ and $T_1 = 300°K$. Upon evaluation,

$$\Delta h = 5486 + 494 - 29$$

$$= 5951 \text{ kJ/kg·mol}$$

Since 1 kg·mol of air contains approximately 29 kg,

$$\Delta h = 205 \text{ kJ/kg}$$

Hence the enthalpy change for 1 kg is 205 kJ.

(*b*) On the basis of an average specific heat, we note from Eq. (3-30) that $\Delta h = c_{p,\,av}\,\Delta T$. At 300 and 500°K, from Table A-4M, we note $c_p$ values of 1.005 and 1.029 kJ/(kg)(°K), respectively. Using the arithmetic average, we find that

$$\Delta h = 1.017(200) = 203 \text{ kJ/kg}$$

This answer, based on a linear variation of $c_p$ with temperature over the given temperature range, differs by roughly 1 percent from the integrated value. This close approximation is not unexpected, since the temperature interval is reasonably small.

(*c*) From the tables of properties of dry air, Table A-5M, the *h* values are read directly. Thus,

$$\Delta h = h_2 - h_1 = 503.02 - 300.19 = 203 \text{ kJ/kg}$$

All three answers are substantially the same. This problem illustrates the point that the use of tabular data is much preferred over direct integrations, which are time-consuming. For desk calculations tabular data are extremely useful. In the absence of tabular data and specific-heat equations, use of constant or average specific heat data lead to reasonably accurate answers if the temperature range is small. In fact, in this particular problem, the use of $c_p$ at 300°K (the intial temperature) of 1.005 kJ/(kg)(°C) leads to a $\Delta h$ equal to 201 kJ/kg, which is very close to the other values.

**Example 3-1** Find the change in enthalpy of 1 lb of air which is heated at low pressure from 100 to 500°F by use of (*a*) empirical specific-heat data, (*b*) average specific-heat data, and (*c*) the table for air, Table A-5

SOLUTION (*a*) The constant-pressure specific heat of air may be represented by the equation

$$c_p = 6.557 + 0.8206 \times 10^{-3}T - 0.0663 \times 10^{-6}T^2$$

where $c_p$ is measured in Btu/(lb·mol)(°R) and $T$ is in °R. If the air is assumed to behave as an ideal gas, then $\Delta h = \int c_p \, dT$. Substitution of the above equation for $c_p$ into the expression for $\Delta h$, and subsequent integration, yields

$$\Delta h = 6.557(T_2 - T_1) + 0.4103 \times 10^{-3}(T_2^2 - T_1^2) - 0.0221 \times 10^{-6}(T_2^3 - T_1^3)$$

where $T_2 = 500°F = 960°R$ and $T_1 = 100°F = 560°R$. Upon evaluating,

$$\Delta h = 2623 + 249 - 16$$

$$= 2856 \text{ Btu/lb·mol}$$

Since 1 lb·mol of air contains approximately 29.0 lb,

$$\Delta h = 98.5 \text{ Btu/lb}$$

(*b*) On the basis of an average specific heat we shall employ Eq. (3-30). At 100 and 500°F, from Table A-4, we note $c_p$ values of 0.240 and 0.248 Btu/(lb)(°F), respectively. Using the arithmetic average of $c_p$ leads to

$$\Delta h = c_{p,\,av}\,\Delta T = 0.244(400) = 97.6 \text{ Btu/lb}$$

The use of an arithmetic value of $c_p$ for the temperature range leads to roughly a 1 percent difference, when compared to the integration method. This small difference is not unexpected, since the temperature range is relatively small.

(*c*) The values of *h* at the two temperatures may be read directly from Table A-5 for dry air.

Thus,

$$\Delta h = 231.06 - 133.86 = 97.2 \text{ Btu/lb}$$

Note that all three methods give answers which are in substantial agreement. However, the problem illustrates the point that the use of tabular data is much preferred over repeated direct integrations, which are time-consuming. For desk calculations, tabular data are extremely useful. In the absence of tabular data and specific-heat equations, use of constant or average specific heat data lead to reasonably accurate results if the temperature range is small. In fact, in this particular problem, the use of the initial $c_p$ value at 100°F of 0.240 Btu/(lb)(°F) leads to a $\Delta h$ of 96.0 Btu/lb, which is only in error by 2 percent.

## 3-7 PROBLEM-SOLVING TECHNIQUES

Thermodynamics is a science which aids the engineer in the design of processes and equipment which are useful to mankind. Engineering design involves a great deal of problem solving. Hence, it is important for a student early in his or her academic career to begin to acquire good habits with respect to problem-solving procedures. It is the author's experience that many students may have a reasonable understanding of basic equations and concepts introduced in a course, but do not know how to use them in problem-solving situations. This is especially true when a multiple-step problem arises, where several basic equations and concepts may need to be used. There are certain steps and procedures which are fairly common to most engineering analysis. A general methodology for problem solving involves the following major points.

1. An engineering problem usually begins as a written or verbal statement. The information contained in this statement must be translated into sketches, diagrams, and symbols. Students frequently attempt to put numbers into equations without ever determining the nature of the problem.
2. Sketches of the system, with the appropriate system boundaries indicated, are of great value in approaching a problem in a consistent manner. Input data on the state of the system and on heat and work interactions should be indicated in appropriate places.
3. Process diagrams, such as a $Pv$ diagram, are extremely helpful as an aid in picturing the initial and final states and the path of the process. Such diagrams are useful sometimes in determining what type of tabular data may be needed eventually in the numerical solution.
4. Idealizations or assumptions should be listed that might be necessary in order to solve the problem. For example, is the process necessarily quasistatic, or is the fluid an ideal gas?
5. When one makes use of a relation such as the ideal-gas relation, one has selected an *equation of state*. Equations of state relate properties at a given equilibrium state, but give no information on how these properties vary during a process.
6. The work statement of the problem may indicate a special nature of the path of the process, such as isothermal, constant pressure, adiabatic, etc. These word

statements of how things happen are extremely important and are known as process relations or *process equations*. Make use of this information.

7. Determine what energy interactions are important, and recognize the sign conventions on these terms.

8. Fundamentals should be applied at this point. For example, write a suitable *basic* energy balance. Indicate which terms are zero or negligible.

9. Complete the solution. Watch the units used in the various equations, so that they are consistent. Is the numerical answer reasonable in the light of your common experience? All problems in this text of an engineering nature have been designed to give answers which are consistent with common practice.

Note that you may wish to check a number of things before ever attempting the numerical solution suggested in item 8. Not all of these items may be necessary in a given problem, and no specific order of the items is implied. However, a student's difficulty with a given problem usually arises because one or more of these items have been neglected. This checklist will be employed in the problem examples which follow. The student should look upon the problem examples as a major source for acquiring an understanding of the fundamentals of thermody- namics. In numerical problems it should be assumed that all input data are accurate to three significant figures, even though they frequently will be reported to less than this.

## 3-8 ENERGY ANALYSIS OF CLOSED IDEAL-GAS SYSTEMS

The preceding sections have developed some basic relationships among the properties of an ideal gas. The most useful of these include

$$Pv = RT \qquad (1\text{-}15)$$

$$du = c_v \, dT \qquad (3\text{-}18)$$

$$dh = c_p \, dT \qquad (3\text{-}22)$$

and

$$c_p - c_v = R \qquad (3\text{-}24)$$

In addition, $c_p$ and $c_v$ are solely functions of temperature. For monatomic and diatomic gases these equations can be used up to pressures of several hundred psia and above with reasonable accuracy if the temperature is room temperature and above. If the temperature interval is small (several hundred degrees or less), the internal-energy and enthalpy changes can be estimated accurately by assuming constant specific heats. In this case $\Delta u = c_v \, \Delta T$ and $\Delta h = c_p \, \Delta T$. For some gases the equations for $du$ and $dh$ have been integrated, using accurate specific-heat data, and the results have been presented in tabular form.

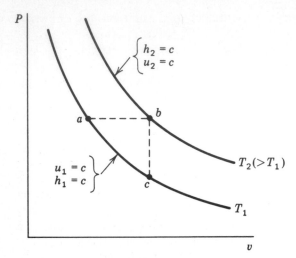

**Figure 3-2** General $Pv$ plot for an ideal gas, showing position of isotherms.

With the availability of data for $u$, $h$, $c_p$, and $c_v$, we are in a position to employ the conservation of energy principle to closed systems containing gases at relatively low pressures. Before proceeding, however, it is important to examine the characteristics of a $Pv$ diagram for an ideal gas. This diagram will be a useful problem-solving device. For an ideal gas we may write that $T = Pv/R$. On a $Pv$ plot this expression is a hyperbola. Figure 3-2 shows a general $Pv$ diagram for an ideal gas. Two isotherms (constant-temperature lines) are shown, where $T_2 > T_1$. The change in temperature along the constant-pressure path $a$–$b$ is also the same along the constant-volume path $c$–$b$. Recall that $u$ and $h$ are solely functions of temperature. Hence the curve along which $a$ and $c$ lie is also a constant-internal-energy and a constant-enthalpy line, $u_1$ and $h_1$. The curve which contains $b$ must also represent $u_2$ and $h_2$, as well as $T_2$. Consequently, the five important properties—$P$, $v$, $T$, $u$, and $h$—are all easily represented on the $Pv$ diagram of an ideal gas.

The following examples illustrate the application of the problem-solving procedure discussed in Sec. 3-7 to ideal-gas systems. The use of the $Pv$ diagram is also demonstrated.

**Example 3-2M** One kilogram of air is compressed slowly in a piston-cylinder assembly from 1 bar, 290°K, to a final pressure of 6 bars. During the process heat is exchanged with the surroundings at a rate sufficient to make the process isothermal. Determine (*a*) the change in internal energy of the air, (*b*)) the work interaction in kJ/kg, and (*c*) the quantity of heat transferred in kJ/kg.

SOLUTION (*a*) The air is considered a simple compressible system which undergoes a constant temperature (isothermal) process. The path of the process is shown on the accompanying $Pv$ diagram, and the system boundary is shown in the sketch of the system. Since the system contains 1 kg of gas, the basic energy equation can be written as

$$q + w = \Delta u$$

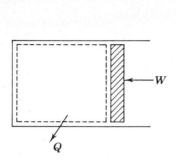

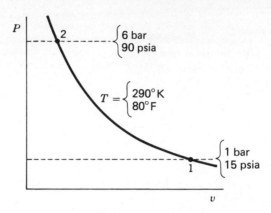

Since air is composed primarily of diatomic gases, it behaves as an ideal gas at 290°K (17°C) even at a pressure of 6 bars. The internal energy of an ideal gas is only a function of temperature. Since the temperature is constant (process equation), the change in internal energy is zero.

(b) On the basis of part (a), the energy equation reduces to $q = -w$. If either of the terms can be evaluated, then the other automatically is found. We have not had any specific equations for computing heat-transfer quantities independently, and therefore we must first evaluate $w$. From the description of the process the only work interaction is boundary work, and it is given by the integral of $-P\,dv$. $P$ and $v$ are functionally related by the ideal-gas equation of state $Pv = RT$ if we assume that the process is quasistatic. Hence,

$$w = -\int P\,dv = -\int RT \frac{dv}{v} = -RT \ln \frac{v_2}{v_1} = RT \ln \frac{P_2}{P_1}$$

The gas constant $R$ is found in Table A-1M. Substitution of the proper data yields

$$w = \frac{8.315}{29} \text{ kJ/(kg)(°K)} \times 290°K \times \ln \frac{6}{1} = 149 \text{ kJ/kg}$$

The positive sign on $w$ indicates that the work is done on the system.

(c) By equating the heat transfer to the negative of the work interaction we note that $q$ equals $-149$ kJ/kg, which is out of the system.

**Example 3-2** One pound of air is compressed slowly in a piston-cylinder assembly from 15 psia, 80°F, to a final pressure of 90 psia. During the process heat is exchanged with the atmosphere at a rate sufficient to make the process isothermal. Determine (a) the change in the internal energy of the gas, (b) the work interaction in ft·lb$_f$/lb$_m$, and (c) the quantity of heat transferred in Btu/lb.

SOLUTION (a) The air is a simple system which undergoes a constant-temperature (isothermal) process. The path of the process is shown on the accompanying $Pv$ diagram, and the system boundary is shown in the sketch of the system. Since the system contains 1 lb of gas, the basic energy equation can be written as

$$q + w = \Delta u$$

Even at a pressure of 90 psia air behaves like an ideal gas at this temperature. The internal energy of an ideal gas is only a function of the temperature. Since the temperature is constant, the change in the internal energy is zero.

(b) On the basis of part (a), the energy equation reduces to $q = -w$. If either of the terms can be calculated, then the other automatically is obtained. We have not had any specific equations or methods introduced for computing heat-transfer quantities independently, and therefore we must tackle the evaluation of $w$. The only work interaction is boundary work, as given by the integral of $-P\,dv$. $P$ and $v$ are functionally related by the ideal-gas equation $Pv = RT$ if we assume that the process is quasistatic. Hence,

$$w = -\int P\,dv = -\int RT\frac{dv}{v} = -RT\ln\frac{v_2}{v_1} = RT\ln\frac{P_2}{P_1}$$

Substitution of proper data yields

$$w = \frac{1545}{29}\,\text{ft·lb}_f/(\text{lb}_m)(°\text{R}) \times 540°\text{R} \times \ln\frac{90}{15} = 51{,}500 \text{ ft·lb}_f/\text{lb}_m$$

(c) By equating the heat transfer to the work interaction and converting to the desired units, we find that

$$q = -w = -\frac{51{,}500}{778} = -66.2 \text{ Btu/lb}$$

The positive value of $w$ indicates that work is done on the system, and the negative value of $q$ indicates that heat is removed from the gas.

**Example 3-3M** An ideal gas has a constant-pressure specific heat of 2.20 kJ/(kg)(°C) and a molar mass of 16.04. Eight kilograms of the gas are heated from 17 to 187°C at constant volume. Determine (a) the work done by the gas, (b) the change in enthalpy of the gas, in kilojoules, and (c) the heat transferred in kilojoules.

SOLUTION (a) The system is constant-volume; therefore no boundary work is done on the simple compressible substance. There is no other work interaction described in the problem; hence $W = 0$. The fact that the boundary work is zero is confirmed by noting that the area under the process curve on the $Pv$ diagram in Fig. 3-3 is zero.

(b) The change in enthalpy of an ideal gas is given by $dH = mc_p\,dT$. There is only a single value of $c_p$ given, so we will assume it is constant. The enthalpy change becomes

$$\Delta H = mc_p\,\Delta T = 8 \text{ kg} \times 2.20 \text{ kJ/(kg)(°C)} \times (187 - 17)°\text{C}$$

$$= 2990 \text{ kJ}$$

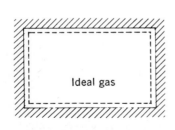
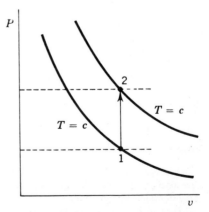

**Figure 3-3** Schematic of system and a $Pv$ diagram of the process for Examples 3-3M and 3-3.

It is important to note that the enthalpy change in this problem is just a number, and has no physical significance.

(c) The quantity of heat transfer is found from the conservation of energy principle. Basically, $Q + W = \Delta U$. Because $W = 0$, $Q = \Delta U$. The internal energy of an ideal gas with a constant specific heat is given by $\Delta U = mc_v \Delta T$. Therefore,

$$Q = mc_v \Delta T$$

The value of $c_v$ is not given in the statement of the problem. However, it is related to $c_p$ for an ideal gas by the relation $c_p - c_v = R$. Use of the universal gas constant from Table A-1M and the molar mass of the gas enables one to compute $c_v$.

$$c_v = c_p - \frac{R_u}{M} = 2.20 - \frac{8.315}{16.04} = 1.68 \text{ kJ/(kg)(°C)}$$

The quantity of heat transferred becomes

$$Q = 8 \text{ kg} \times 1.68 \text{ kJ/(kg)(°C)} \times (187 - 17)°C = 2285 \text{ kJ}$$

The positive value shows that heat is added to the system.

**Example 3-3** An ideal gas has a constant-pressure specific heat of 0.526 Btu/(lb)(°F) and a molar mass of 16.04. Fifteen pounds of the gas are heated from 45 to 345°F at constant volume. Determine (a) the work done by the gas, (b) the change in enthalpy of the gas, and (c) the heat transferred.

SOLUTION (a) The system boundary is rigid; therefore no boundary work is done. There is no other work interaction described in the problem; hence $W = 0$. Note that the area under the process curve on a $Pv$ diagram in Fig. 3-3 is zero.

(b) The change in enthalpy of an ideal gas is given by $dH = mc_p \, dT$. There is only a single value of $c_p$ given, and so we will assume it is constant. The enthalpy change becomes

$$\Delta H = mc_p \Delta T = 15 \text{ lb} \times 0.526 \text{ Btu/(lb)(°F)} \times (345 - 45)°F$$

$$= 2370 \text{ Btu}$$

The enthalpy change in this problem is just a number, and has no physical significance.

(c) The quantity of heat transfer is found from the conservation of energy principle. Basically, $Q + W = \Delta U$. Because $W = 0$, $Q = \Delta U$. The internal energy of an ideal gas with constant specific heats is given by $\Delta U = mc_v \Delta T$. Therefore,

$$Q = mc_v \Delta T$$

The value of $c_v$ is not given in the statement of the problem. However, it is related to $c_p$ for an ideal gas by the relation $c_p - c_v = R$. Use of the universal gas constant and the molar mass of the gas enables one to compute $c_v$.

$$c_v = c_p - \frac{R_u}{M} = 0.526 - \frac{1.986}{16.04} = 0.526 - 0.124 = 0.402 \text{ Btu/(lb)(°F)}$$

The quantity of heat transferred becomes

$$Q = 15(0.402)(345 - 45) = 1810 \text{ Btu}$$

The positive value shows that heat was added to the system.

**Example 3-4M** A vertical piston-cylinder assembly which initially has a volume of 0.1 m³ is filled with 0.1 kg of nitrogen gas. The piston is weighted so that the pressure on the diatomic nitrogen is always maintained at 1.15 bars. Heat transfer is allowed to take place until the volume is 75 percent of its initial value. Determine the quantity and direction of the heat transfer, in kilojoules, and the final temperature of the nitrogen at equilibrium, in °K.

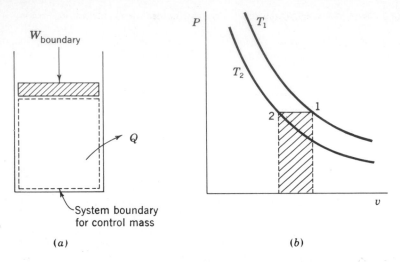

**Figure 3-4** Schematic of system and a $Pv$ diagram of the process for Examples 3-4M and 3-4.

SOLUTION  The system, or control mass, is taken as the mass of nitrogen within the piston-cylinder. A schematic diagram of the system and a process diagram on a $Pv$ plot are shown in Fig. 3-4. Since field effects, etc., are absent, we assume the system to be a simple compressible one. Under this assumption the conservation of energy principle reduces to Eq. (3-5); namely,

$$\delta Q + \delta W = dU$$

Since the volume of the control mass is altered, boundary work is associated with the process. In order to evaluate this form of work, it is necessary to assume that the process is carried out quasistatically. On this basis the work is given by the integral of $-P\,dV$. The energy equation becomes

$$\delta Q - P\,dV = dU$$

or

$$\delta Q = dU + P\,dV = dH$$

The replacement of $dU + P\,dV$ by $dH$ is possible since the process is one of constant pressure.

The enthalpy change can be determined by noting that at 1.15 bars, which is just slightly above atmospheric pressure, nitrogen gas behaves essentially as an ideal gas. For an ideal gas, $dH = mc_p\,dT$. Hence the heat transfer for the proposed process is evaluated by

$$\delta Q = dH = mc_p\,dT$$

or

$$Q = \int mc_p\,dT$$

Integration of the right-hand side of this result depends upon the functional relation between $c_p$ and $T$. If the temperature change is not too great, an average value for $c_p$ can be chosen and the integration carried out quickly. If the temperature change is large, then tabular data would be appropriate.

In order to obtain the initial temperature $T_i$ of the gas, the ideal-gas equation of state is used.

$$T_i = \frac{P_i V_i}{NR_u} = 1.15 \text{ bars} \times 0.1 \text{ m}^3 \times \frac{28}{0.1 \text{ kg·mol}} \times \frac{1}{0.8315} \frac{(\text{kg·mol})(°\text{K})}{\text{bar·m}^3}$$

$$= 387°\text{K}$$

Application of the ideal-gas relation for a constant-pressure process also shows that $T_f = T_i(V_f/V_i)$. Thus the final temperature $T_f$ is

$$T_f = 387 \frac{0.75}{1.0} = 290°\text{K}$$

In the small range of temperatures from $T_i$ to $T_f$, Table A-4M indicates that $c_p$ is nearly constant at 1.041 kJ/(kg)(°C). Consequently, the quantity of heat transferred is

$$Q = mc_{p,\,av}(T_f - T_i) = 0.1 \text{ kg} \times 1.041 \text{ kJ/(kg)(°C)} \times (290 - 387)°\text{C}$$

$$= -10.1 \text{ kJ}$$

The negative sign reveals that heat must be removed from the system in order to reduce the volume to 75 percent of its initial value at constant pressure.

ALTERNATIVE SOLUTIONS AND DISCUSSION  Since the conservation of energy principle under the stated restrictions reduces to $Q = \Delta H = m(h_f - h_i)$, the numerical answer can also be obtained by using the gas table for nitrogen, Table A-6M, at the calculated temperatures. This table indicates values of roughly 11,260 and 8,432 kJ/kg·mol for the initial and final enthalpies, respectively, when based on linear interpolation. Hence the value of the heat transfer is

$$Q = 0.1 \text{ kg} \times \frac{1 \text{ kg·mol}}{28 \text{ kg}} \times (8,432 - 11,260) \frac{\text{kJ}}{\text{kg·mol}}$$

$$= -10.1 \text{ kJ}$$

Good agreement with the result based on an average $c_p$ value is obtained because the variation of $c_p$ with temperature is negligible.

Another alternative solution is to base the calculation on

$$Q = \Delta U + P\,\Delta V$$

without a further combination of the last two terms into $\Delta H$. The change in $\Delta U$ can be found either by $c_{v,\,av}\,\Delta T$ or from $u$ data in the nitrogen table. Since $\Delta V$ is known explicitly, $P\,\Delta V$ can be computed directly. The reader may wish to confirm that $\Delta U$ equals $-7.2$ kJ by either method, and that $P\,\Delta V$ equals $-2.9$ kJ. The work required is shown by the cross-hatched area on the $Pv$ plot, and is into the system. Again the overall answer is $-10.1$ kJ. This latter method is somewhat more descriptive, since it shows explicitly the sources of the energy which contributed to $Q$. The $\Delta H$ computation is faster, but shows less detailed information on the overall process.

**Example 3-4**  A vertical piston-cylinder assembly which initially has a volume of 1.0 ft$^3$ is filled with 0.1 lb$_m$ of nitrogen. The piston is weighted so that the pressure on the nitrogen is always maintained at 20 psia. Heat transfer is allowed to take place until the volume is 90 percent of its initial value. Determine the quantity and direction of the heat transfer and the final temperature of the nitrogen at equilibrium.

SOLUTION  The system, or control mass, is taken as the mass of nitrogen within the confines of the piston cylinder. A schematic diagram of the system and a process diagram on a $Pv$ plot are shown in Fig. 3-4. The system can be treated as a simple compressible system, since the effects of magnetic and electric fields, etc., are absent. Under this assumption the conservation of energy which principle reduces to Eq. (3-5); namely,

$$\delta Q + \delta W = dU$$

Since the volume of the control mass is altered, a work interaction is associated with the process. In order to evaluate the work done in moving the boundary, it is necessary to assume that the process is carried out quasistatically. On this basis the work is given by the integral of $P\, dV$. The energy equation then becomes

$$\delta Q - P\, dV = dU$$

or

$$\delta Q = dU + P\, dV = dH$$

The enthalpy change can be determined by noting that at 20 psia, that is, just slightly above atmospheric pressure, nitrogen gas behaves essentially as an ideal gas. For an ideal gas, $dH = mc_p\, dT$. Hence the heat transfer for the proposed process can be evaluated by

$$\delta Q = dH = mc_p\, dT$$

or

$$Q = \int mc_p\, dT$$

Integration of the right-hand side of this result depends upon the functional relation between $c_p$ and $T$. If the temperature change is not too great an average value for $c_p$ can be chosen and the integration carried out quickly. We have noted already that the specific heats of diatomic gases do not change rapidly with temperature at about room temperature and above. In order to obtain the initial temperature of the gas, the ideal-gas equation of state is used.

$$T_i = \frac{P_i V_i}{N R_u} = 20 \times 144 \ \text{lb}_f/\text{ft}^2 \ \times 1.0 \ \text{ft}^3 \ \times \frac{28}{0.1 \ \text{lb·mol}} \ \frac{1}{} \times \frac{1}{1545} \ (\text{lb·mol})(°R)/\text{ft·lb}_f$$

$$= 522°R = 62°F$$

Application of the ideal-gas relation for a constant-pressure process also shows that, since $P_i V_i/T_i = P_f V_f/T_f$, then $T_f = T_i(V_f/V_i)$. Thus the final temperature is

$$T_f = 522 \ \frac{0.9}{1.0} = 470°R = 10°F$$

In the range of temperatures from $T_i$ to $T_f$ the specific heat at constant pressure is nearly constant at 0.248 Btu/(lb$_m$)(°F). Consequently, the quantity of heat transfer is

$$Q = \int mc_p\, dT = mc_p(T_f - T_i) = 0.1(0.248)(470 - 522) = -1.29 \ \text{Btu}$$

The negative sign reveals that heat must be removed from the control mass in order to reduce the volume to 90 percent of its initial value at constant pressure.

ALTERNATIVE SOLUTION AND DISCUSSION Since the conservation of energy principle for the control mass under the stated restrictions reduces to $Q = \Delta H = m(h_f - h_i)$, the numerical answer can also be obtained by using the gas table for nitrogen, Table A-6, at the calculated temperatures. This table indicates values of 3625 and 3264 Btu/lb·mol for the initial and final enthalpies, respectively, when based on linear interpolation. Hence the value of the heat transfer is

$$Q = 0.1(3264 - 3625)(\tfrac{1}{28}) = -1.29 \ \text{Btu}$$

Good agreement with the result based on an average $c_p$ value is obtained because the variation of $c_p$ with temperature is negligible.

Another alternative solution is to base the calculation on the relation

$$Q = \Delta U + P\,\Delta V$$

without a further combination of the last two terms into $\Delta H$. The change in $\Delta U$ can be found either by $c_{v,\,av}\,T$ or from $u$ data in the nitrogen table. Since $\Delta V$ is known from the input data, $P\,\Delta V$ can be computed directly. The reader may wish to confirm that $\Delta U$ equals $-0.92$ Btu by either method, and that $P\,\Delta V$ equals $-0.37$ Btu. The work required is shown as the cross-hatched area on the $Pv$ diagram, and is into the system. Again, the overall answer is $-1.29$ Btu. This latter method is somewhat slower than using the $\Delta H$ method, but it does show the individual sources of energy which contribute to $Q$.

**Example 3-5M** A rigid insulated tank is divided into two equal volumes by a partition. Inititially, 1 kg of a gas is introduced into one side of the partitioned tank, and the other side remains evacuated. In this equilibrium state the pressure and temperature are 2 bars and 100°C, respectively. The partition is then pulled out, and the gas is allowed to expand into the entire tank. Determine the final pressure and temperature at equilibrium if the gas is ideal.

SOLUTION The boundaries of the closed system will be selected to lie just inside the walls of the entire rigid, insulated tank, including the section which is initially evacuated (see Fig. 3-5). The system may be idealized as a simple compressible one. For such a system the energy equation is

$$Q + W = \Delta U$$

The boundary work associated with a rigid tank is zero, and all other work effects are absent. In addition, $Q$ is zero for the insulated boundary. Because $Q$ and $W$ are zero for this process, we may write for the finite change of state that

$$\Delta U = 0$$

Consequently, the expansion of the gas in this case is one of constant internal energy; that is, $U_2$ equals $U_1$.

For an ideal gas the internal energy is solely a function of the temperature for a given equilibrium state. Since the internal energy is constant, the initial and final temperatures of the

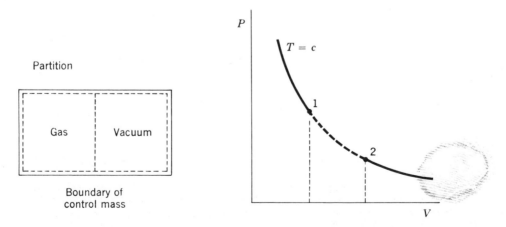

**Figure 3-5** Schematic of system and a $PV$ diagram of the process for Examples 3-5M and 3-5.

ideal gas undergoing this "free expansion" must be the same. Hence $T_2$ is 100°C. For an ideal gas at constant temperature the ideal-gas equation of state leads to the relation $P_2 = P_1(V_1/V_2)$. Since the final volume is twice the initial volume, the final pressure is one-half the original value, or 1 bar.

The process is shown on the $Pv$ diagram. The path between states 1 and 2 is drawn as a dashed line, since the process is nonquasistatic and the temperature is not defined except at the equilibrium end states.

**Example 3-5** A rigid insulated tank is divided into two equal volumes by a partition. Initially, 1 lb of a gas is introduced into one side of the partitioned tank, and the other side remains evacuated. In this equilibrium state the properties of the gas are measured and recorded. The partition is then pulled out, and the gas is allowed to expand into the entire tank. Determine the final pressure and temperature at equilibrium if the gas is ideal and initially at 15 psia and 600°F.

SOLUTION The boundaries of the closed system will be selected to lie just inside the walls of the entire rigid, insulated tank, including the section which is originally evacuated (see Fig. 3-5). The system may be idealized as a simple one. In the absence of all other work effects, the rigid insulated tank assures that the system is an isolated one. For a simple system of fixed composition the energy equation is

$$Q + W = \Delta U$$

However, because $Q$ and $W$ are zero for this process, we may write for the finite change of state that

$$\Delta U = 0$$

Consequently, the expansion of the gas in this case is one of constant internal energy; that is, $U_2$ equals $U_1$.

For an ideal gas it has been shown that the internal energy is solely a function of the temperature for a given equilibrium state. Since the internal energy is constant, the initial and final temperatures of the ideal gas undergoing this "free expansion" must be the same. Hence $T_2$ is 600°F. For an ideal gas at constant temperature the ideal-gas equation leads to the relation $P_2 = P_1(V_1/V_2)$. Since the final volume is twice the initial volume, the final pressure is one-half the original value, or 7.5 psia.

The process is shown on the $Pv$ diagram. The path is drawn as a dashed line, since the process is nonquasistatic and the temperature is not defined except at the equilibrium end states.

# REFERENCES

Joint Army, Navy, and Air Force (JANAF) Thermochemical Tables, NSRDS-NBS-37, June, 1971.

Jones, J. B., and G. A. Hawkins: "Engineering Thermodynamics," Wiley, New York, 1960.

Keenan, J. H., and J. Kaye: "Gas Tables," Wiley, New York, 1945.

Reynolds, W. C.: "Thermodynamics," McGraw-Hill, New York, 1968.

Sears, F. W.: "Thermodynamics, The Kinetic Theory of Gases, and Statistical Mechanics," Addison-Wesley, Reading, Mass., 1955.

Zemansky, M. W., and H. C. Van Ness: "Basic Engineering Thermodynamics," McGraw-Hill, New York, 1966.

# PROBLEMS (METRIC)

### General energy considerations

**3-1M** A piston-cylinder assembly contains a gas which undergoes a series of quasistatic processes which make up a cycle. The processes are as follows: 1–2, adiabatic compression; 2–3, constant pressure; 3–4, adiabatic expansion; 4–1, constant volume. The table below gives data at the beginning and end of each process.

| State | P, bars | V, cm³ | T, °C | U, kJ |
|-------|---------|--------|-------|-------|
| 1 | 0.95 | 5700 | 20 | 1.47 |
| 2 | 23.9 | 570 | 465 | 3.67 |
| 3 | 23.9 | 1710 | 1940 | 11.02 |
| 4 | 4.45 | 5700 | 1095 | 6.79 |

Sketch the cycle on $PV$ coordinates and determine the work and heat transfer (both in kilojoules) for each of the four processes.

**3-2M** An insulated piston-cylinder assembly containing a fluid has a stirring device operated externally. The piston is frictionless, and the force holding it against the fluid is due to standard atmospheric pressure and a coil spring. The spring constant is 7200 N/m. The stirring device is turned 1000 revolutions with an average torque of 0.68 N-m. As a result the piston, which is 0.3 m in diameter, moves outward 0.10 m. Find the change in the internal energy of the fluid in kilojoules.

**3-3M** A piston-cylinder assembly is equipped with a paddle-wheel driven by an external motor and is filled with a gas. The walls of the cylinder, the piston, and the paddle-wheel and its shaft are all made of a material which is a very good insulator. The system contains 50 g of gas. Initially the gas is in state 1 (see table below). The paddle-wheel is then operated, but the piston is allowed to move to keep the pressure constant. Then the paddle-wheel is stopped, and the system is found to be in state 2. There is very little friction between the piston and cylinder walls.

| State | P, bars | v, cu cm/g | u, J/g | h, J/g |
|-------|---------|------------|--------|--------|
| 1 | 35 | 7.11 | 22.75 | 47.64 |
| 2 | 35 | 19.16 | 97.63 | 164.69 |

Determine the energy transfer in joules along the paddle-wheel shaft.

**3-4M** A heavily insulated piston-cylinder device contains a gas which is initially at 6 bars and 177°C and occupies 0.05 m³. The gas undergoes a quasistatic process according to the equation $PV^2 = $ constant. The final pressure is 1.5 bars. Determine (a) the work done in newton-meters, and (b) the change in internal energy in kilojoules.

**3-5M** A closed cylinder with its axis vertical is fitted with a piston in its upper end. The piston is loaded by a weight so that a constant pressure of 3 bars is exerted on the 0.8 kg of gas within the cylinder. The gas decreases in volume from 0.1 to 0.03 m³, and the internal energy decreases by 60 kJ/kg. If the process is quasistatic, determine (a) the work done on or by the gas in kilojoules, (b) the amount of heat added or removed in kilojoules, and (c) the enthalpy change in kJ/kg.

**3-6M** A piston-cylinder device is maintained at a constant pressure of 5 bars and contains 1.4 kg of a gas. During a process the heat transfer out is 50 kJ, while the volume changes from 0.15 to 0.09 m³. Find the change in the internal energy in kJ/kg.

**3-7M** A vertical piston-cylinder assembly contains a gas which is compressed by a frictionless piston weighing 3000 N. During an interval of time a paddle-wheel within the cylinder does 6800 N·m of work on the gas If the heat transfer out of the gas is 10 kJ, and the change in the internal energy is $-1$ kJ, determine the distance the piston moves in meters. The area of the piston is 52 cm$^2$ and the atmospheric pressure acting on the outside of the piston is 1.0 bar.

**3-8M** A piston-cylinder device contains 0.12 kg of an ideal gas initially at 2 bars and 123°C. During a quasistatic, isothermal process heat is removed in the amount of 20 kJ, and electrical work is carried out on the system in the amount of 1.75 W h. Determine the ratio of the final to the initial volume.

**3-9M** On the basis of the equation in Table 3-1 for $c_p$ as a function of temperature at zero pressure for oxygen, calculate the change in the enthalpy when the temperature of oxygen is increased from (a) 300 to 500°K, (b) 300 to 700°K, and (c) 400 to 800°K. Compare these answers to those obtained by use of $h$ data from Table A-7M. What percent error will be introduced by using the arithmetic mean specific heat from Table A-4M to evaluate $\Delta h$ between the given temperature limits?

**3-10M** Same as Prob. 3-9M, except that the gas is carbon dioxide.

**3-11M** Using the empirical equation for $c_p$ of oxygen stated in Table 3-1, calculate the internal energy change when the temperature of oxygen is increased from (a) 300 to 600°K and (b) 300 to 800°K. Compare these answers with those obtained by use of $u$ data from Table A-7M. What percent error will be introduced by using the arithmetic mean specific heat $c_v$ from Table A-4M to evaluate $\Delta u$ between the given temperature limits?

**3-12M** Calculate the change in (a) the internal energy and (b) the enthalpy of helium for the temperature range from 100 to 600°C, in kJ/kg.

**3-13M** Calculate the change in (a) the internal energy and (b) the enthalpy of argon for the temperature range from 300 to 700°K, in kJ/kg.

**3-14M** One-half kilogram of air is compressed isothermally in a closed system from a pressure of 1 bar and a temperature of 20°C to a final pressure of 6 bars.
(a) What is the change in internal energy in kilojoules?
(b) How much work is done in newton-meters?
(c) What is the heat transferred in kJ/kg?

**3-15M** Nitrogen gas at 1 bar and 27°C is contained within a piston-cylinder device. A paddle-wheel within the cylinder is turned until 9500 N·m of energy have been added to the 45 g of gas (neglect mass of paddle-wheel). If the process is adiabatic and at constant pressure, determine the final temperature in °C.

**3-16M** One kilogram of air is contained in a rigid vessel with insulated walls. A paddle-wheel of negligible mass in the vessel is driven by an external motor. During a process the temperature rises from 27 to 127°C. Determine (a) the change in internal energy in kilojoules, (b) the change in the enthalpy, and (c) the work done in newtons-meters.

**3-17M** One-tenth cubic meter of nitrogen at 1 bar and 25°C is contained in a piston-cylinder assembly. A paddle-wheel within the cylinder is turned until 25,000 newton-meters of energy have been added. If the process is adiabatic and at constant pressure, determine the final temperature of the gas in °C.

**3-18M** One-tenth cubic meter of carbon monoxide at 1.3 bars and 27°C is contained in a piston-cylinder device. Electric current flows through a resistor within the cylinder until 0.0157 kWh of electrical energy has entered the gas. If the process is constant pressure and the final volume is 0.2 m$^3$, determine the magnitude and the direction of any heat transfer in kilojoules.

**3-19M** Carbon dioxide is expanded isothermally in a nonflow process from initial conditions of 1.3 bars, 150°C, and 0.1 m$^3$ to a final volume of 0.2 m$^3$. Calculate (a) the work done in N·m/g, and (b) the heat transfer in kJ/kg.

**3-20M** A gram of nitrogen contained in a rigid tank at 27°C is heated until the pressure is doubled. Compute the quantity of heat transferred in kJ/kg.

**3-21M** Two-tenths of a kilogram of air initially at 3 bars and 325°K expands isothermally and quasistatically until the volume is doubled. Evaluate the quantity of heat added or removed in kJ and the change in the internal energy of the air.

**3-22M** In a closed system 0.1 kg of air expands quasistatically from 2 bar, 325°K, until its volume is doubled. The expansion is isothermal. During the process heat is added in the amount of 40 kJ/kg, and electrical work is also done on the system. Determine this quantity of electrical work in watt-hours.

**3-23M** One-tenth cubic meter of air is contained in a piston-cylinder assembly at 1.5 bar and 27°C. It is first heated at constant volume (process 1–2) until the pressure has doubled. Then it is expanded at constant pressure (process 2–3) until the volume has tripled. Find the total heat added in kilojoules, using the air table.

**3-24M** One-tenth of a kilogram of carbon dioxide undergoes a quasistatic, constant-pressure process in a closed cylinder. If the internal-energy change of the system is 17 kJ, determine how much work was done in kilojoules.

**3-25M** A piston-cylinder assembly which initially has a volume of 1 m$^3$ contains 0.16 kg of hydrogen at 2.1 bars. Heat transfer occurs until the final volume is 0.82 m$^3$. If the process occurs at constant pressure, determine (a) the final temperature in °C, and (b) the heat transfer in kilojoules.

**3-26M** Air is contained within a piston-cylinder device at initial conditions of 4 bars and 17°C. A paddle-wheel within the gas is turned until 67.6 N·m/g of work is performed on the gas. During the process the temperature remains constant while the volume is doubled. Determine the magnitude (in kJ/kg) and the direction of any heat transfer.

**3-27M** One-hundredth of a kilogram-mole of helium is contained within an adiabatic piston-cylinder device and is maintained at a constant pressure of 2 bars. A paddle-wheel is operated within the gas until the volume of the gas has increased by 25 percent. Compute (a) the quantity of paddle-wheel work required in newton-meters, and (b) the net work done on the system, if the piston is frictionless and the initial temperature of the gas is 20°C.

**3-28M** A system consists of 0.88 kg of carbon dioxide within a constant-pressure piston-cylinder device which is frictionless. The initial conditions are 7 bars and 17°C. Heat is added until the volume is doubled. On the basis of data in Table A-9M determine (a) the change in internal energy in kilojoules, (b) the change in enthalpy in kilojoules, (c) the work performed by or on the gas in newton-meters, and (d) the quantity of heat supplied in kilojoules.

**3-29M** A rigid tank with a volume of 1 m$^3$ contains oxygen initially at 127°C and 1.5 bars. During a process paddle-wheel work amounts to 2000 N·m, and the temperature drops to 77°C. Determine the magnitude (in kilojoules) and direction of any heat transfer.

**3-30M** Two kilograms of nitrogen are compressed isothermally and frictionlessly in a closed system from 1 bar and 27°C to a final pressure of 1.7 bars. Determine (a) the work required in newton-meters, (b) the heat transfer in kilojoules, and (c) the internal-energy change in kilojoules.

**3-31M** A rigid tank initially contains 0.80 g of air at 295°K and 1.5 bar. An electrical resistor within the tank is energized by passing a current of 0.6 A for a period of 30 s from a 12.0-V source. At the same time a heat loss of 126 J occurs.

(a) Determine the final temperature of the gas in kelvins.
(b) Find the final pressure in bars.

**3-32M** An adiabatic piston-cylinder device contains 0.3 kg of air initially at 300°K and 3 bars. Energy is added to the gas by passing a current of 5 A through a resistor within the system for a time period of 30 s. The gas expands quasistatically at constant pressure until the volume is doubled. Determine the size of the resistor in ohms. Recall that the power dissipated in a resistor is $I^2R$.

**3-33M** Two-tenths kilogram of carbon monoxide in a closed system is heated at constant pressure of 1.8 bars from 27 to 127°C. The measured work output is 5.2 kJ.

(a) Calculate the magnitude in kilojoules and the direction of any heat transfer.
(b) What would be the quantity of heat transferred if the above process were quasistatic and frictionless?

**3-34M** Carbon dioxide is contained in a piston-cylinder device at an initial state of 0.01 m$^3$, 2 bars, and 177°C. It is compressed at constant pressure until the temperature is 77°C. The work required on the gas is measured to be 500 N·m.

(a) Calculate the magnitude in kilojoules and the direction of any heat transfer.
(b) What would be the quantity of heat transfer if the above process were quasistatic and frictionless?

**3-35M** One and one-half kilograms of air in a closed system are heated at a constant pressure of 1.3 bars from 17 to 77°C. The measured work output is 12,000 N·m.

(a) Calculate the heat transfer in kilojoules and determine its direction.

(b) What would be the quantity of heat transferred if the above process were frictionless and quasistatic?

**3-36M** Carbon monoxide gas is contained within a piston-cylinder device at 1 bar and 27°C.

(a) In process A the gas is heated at constant volume until the pressure has doubled. It is then expanded at constant pressure until the volume is three times its initial value.

(b) In process B the same gas in the same initial state is first expanded at constant pressure until the volume has tripled, then the gas is heated at constant volume until it reaches the same final pressure as in process A.

For these two overall processes determine the net heat effect, the net work effect, and the changes in internal energy, in kJ/kg. Compare the results for the two processes.

**3-37M** A piston-cylinder arrangement contains 1 kg of air initially at 2 bars and 77°C. Two processes are involved: a constant-volume process followed by a constant-pressure process. During the first process heat is added in the amount of 50,670 J. During the second process, which follows the first, heat is added at constant pressure until the volume is 2.0 m³. If the processes are quasistatic, evaluate the total work of expansion in kilojoules for the overall system change.

## Specific heat evaluation

**3-38M** One-tenth of a kilogram of an ideal gas is enclosed in a rigid tank at a pressure of 1.2 bars and a temperature of 30°C. A paddle wheel within the tank does 520 N·m of work on the substance, and 810 J of heat are added at the same time. The temperature of the gas, which has a molecular weight of 48, rises 25°C during the process. Compute the average specific heat $c_v$ for the gas, in kJ/(kg)(°C).

**3-39M** Ten grams of an ideal gas having a molar mass of 32 undergoes a quasistatic expansion at constant pressure from 1.3 bars, 20°C, to 80°C. During the process 550 J of heat are added to the closed system. Compute the average value of $c_v$ for the gas, in kJ/(kg)(°C).

**3-40M** One-tenth of a kilogram of an ideal gas (molecular weight = 40) expands quasistatically at constant pressure in a closed system from 1 bar, 40°C to 150°C, and 5200 J of heat are added. Determine the average value of $c_v$, in kJ/(kg)(°C).

**3-41M** One-half kilogram of a gas is contained within a vessel made of rigid walls which do not absorb or transmit heat. 10,000 N·m of work are done on the gas by rotating a paddle-wheel of negligible mass within the system. In addition, 15,000 J are supplied as electrical energy to a resistance heater of negligible mass located in the system. During the process the temperature of the gas increases 50°C. Determine an average $c_v$ value for the gas, in kJ/(kg)(°C).

## Additional problems

**3-42M** Electrical energy is supplied to a mass of 224 g of cadmium at a constant rate of 17 W and at a constant pressure. The cadmium is thermally insulated, and temperature readings are taken at certain time intervals as follows:

| Time, s | Temperature, °C | Time, s | Temperature, °C |
|---------|-----------------|---------|-----------------|
| 0 | 39 | 285 | 137 |
| 15 | 45 | 345 | 155 |
| 45 | 57 | 405 | 172.3 |
| 105 | 80 | 465 | 191 |
| 165 | 100 | 525 | 208 |
| 225 | 119.2 | | |

Draw a graph with time as abscissa and temperature as ordinate. The atomic mass of cadmium is 112. Estimate $c_p$ values at (a) 39°C, (b) 100°C, and (c) 200°C.

**3-43M** A rigid, insulated cylinder is divided into two parts by an uninsulated, frictionless piston which initially is held in a fixed position. One part of the cylinder contains one-half kilogram of a gas at 4 bars and 30°C, while the other part contains one-half kilogram of the same gas at 1.2 bars and 30°C. The piston is then released, and equilibrium is established between the two parts.

(a) What is the final equilibrium temperature in °C?

(b) What is the final equilibrium pressure in bar? Assume that the specific heats, $c_v$ and $c_p$, are constant.

**3-44M** In a tank 0.60 kg of nitrogen are stored at 2 bars and 50°C. Attached to this tank through a suitable valve is a second tank which is 0.50 m³ in volume and completely evacuated. Both tanks are thoroughly insulated. If the valve is opened and equilibrium is allowed to be reached, determine the final pressure in bars.

**3-45M** Two insulated tanks are connected by a valve. One tank contains 0.5 kg of an ideal gas at 80°C, and the other tank contains 1.0 kg of the same gas at 50°C. The valve is opened, and the gases are allowed to mix until equilibrium is reached. All specific heats may be regarded as constant.

(a) Calculate the mixture temperature in °C.

(b) Calculate the total enthalpy change for the two-tank system. Show that this answer is the same regardless of the mass and the temperature of the gas in each tank initially.

**3-46M** In a tank, 1 kg of oxygen is stored at 1 bar and 60°C. Attached to this tank through a suitable valve is a second tank which contains 1 kg of oxygen at 2 bars and 7°C. Both tanks are insulated. If the valve is opened and equilibrium is allowed to be reached, determine the final pressure in bars.


# PROBLEMS (USCS)


### General energy considerations

**3-1** A piston-cylinder assembly contains a gas which undergoes a series of quasistatic processes which make up a cycle. The processes as as follows: 1–2, adiabatic compressions; 2–3, constant pressure; 3–4, adiabatic expansion; 4–1, constant volume. The table below gives data at the beginning and end of the various processes.

| State | P, psia | v, ft³ | T, °F | U, Btu |
|-------|---------|--------|-------|--------|
| 1 | 14 | 0.20 | 70 | 1.39 |
| 2 | 352 | 0.02 | 870 | 3.48 |
| 3 | 352 | 0.06 | 3530 | 10.45 |
| 4 | 65 | 0.20 | 2000 | 6.44 |

Sketch the cycle of $Pv$ coordinates and determine the work and heat transfer (both in Btus) for each of the four processes.

**3-2** An insulated piston-cylinder assembly containing a fluid has a stirring device operated externally. The piston is frictionless, and the force holding it against the fluid is due to standard atmospheric pressure and a coil spring. The spring constant is 500 lb$_f$/ft. The stirring device is turned 10,000 revolutions with an average torque of 0.50 lb$_f$·ft. As a result the piston, which is 2.0 ft in diameter, moves outward 2 ft. Find the change in the internal energy of the fluid in Btus.

**3-3** A piston-cylinder assembly is equipped with a paddle-wheel driven by an external motor and is filled with a gas. The assembly is heavily insulated, and the paddle-wheel is made of a material which is a poor conductor of heat. The system contains 0.1 lb of gas. Initially the gas is in state 1 (see table on p. 104). The paddle-wheel is then operated, but the piston is allowed to move to keep the pressure constant. When the paddle-wheel is stopped, the system is found to be in state 2. There is very little friction between the piston and cylinder walls.

| State | $P$, atm | $v$, in$^3$/lb | $u$, Btu/lb | $h$, Btu/lb |
|-------|----------|----------------|-------------|-------------|
| 1 | 35 | 197 | 9.80 | 20.65 |
| 2 | 35 | 531 | 42.06 | 71.34 |

Determine the energy transfer in the form of paddle-wheel work in ft·lb$_f$.

**3-4** A heavily insulated piston-cylinder device contains a gas which is initially at 100 psia and 350°F, and occupies 1 ft$^3$. The gas undergoes a quasistatic process according to the equation $PV^2 = $ constant. The final pressure is 25 psia. Determine (a) the work done in ft·lb$_f$, and (b) the change in internal energy in Btus.

**3-5** A closed cylinder with its axis vertical is fitted with a piston in its upper end. The piston is loaded by a weight so that a constant pressure of 50 psia is exerted on the 2 lb of gas within the cylinder. The gas decreases in volume from 3 to 1 ft$^3$, and the internal energy decreases by 30 Btu/lb. Assuming the process to be quasistatic, determine (a) the work done on or by the gas in ft-lb, (b) the amount of heat added or removed in Btus, and (c) the enthalpy change in Btu/lb.

**3-6** A piston-cylinder device maintained at a constant pressure of 80 psia contains 3 lb of a gas. During a cooling process the heat transfer out is 50 Btu, while the volume changes from 5 to 3 ft$^3$. Find the change in the internal energy in Btu/lb.

**3-7** A vertical piston-cylinder assembly contains a gas which is compressed by a frictionless piston weighing 684 lb$_f$. During an interval of time a paddle-wheel within the cyclinder does 5000 ft·lb$_f$ of work on the gas. If the heat transfer out of the gas is 10 Btu, and the change in the internal energy is $-1.0$ Btu, determine the distance the piston moves in feet. The area of the piston is 8.0 in$^2$ and the atmospheric pressure acting on the outside of the piston is 14.5 psia.

**3-8** A piston-cylinder device contains 0.27 lb of air initially at 35 psia and 240°F. During a quasistatic, isothermal process heat is removed in the amount of 20 Btu, and electrical work is carried out on the system in the amount of 1.76 Wh. Determine the ratio of the final to the initial volume.

**3-9** An empirical equation for $c_p$ at zero pressure for oxygen may be expressed as $c_p = 11.515 - 172T^{-1/2} + 1530T^{-1}$, where $T$ is in °R, and $c_p$ has units of Btu/(lb·mol)(°R). Calculate the change in the enthalpy when the temperature of oxygen is increased from (a) 100 to 1000°F, (b) 100 to 2000°F, and (c) 300 to 1200°F. Compare these answers to those obtained by use of $h$ data from Table A-7. What percent error will be introduced by using the arithmetic mean specific heat from Table A-4 to evaluate $\Delta h$ between the given temperature limits?

**3-10** Same as Prob. 3-9, except the gas is carbon dioxide. In this case the empirical equation for $c_p$ may be expressed as $c_p = 16.2 - 6.53 \times 10^3 T^{-1} + 1.41 \times 10^6 T^{-2}$, where $T$ is in °R and $c_p$ has units of Btu/(lb·mol)(°R).

**3-11** Using the empirical equation for $c_p$ of oxygen stated in Prob. 3-9, calculate the internal energy change when the temperature of oxygen is increased from (a) 100 to 1000°F and (b) 100 to 2000°F. Compare these answers with those obtained by use of $u$ data from Table A-7. What percent error will be introduced by using the arithmetic mean specific heat $c_v$ from Table A-4 to evaluate $\Delta u$ between the given temperature limits?

**3-12** Calculate the change in (a) the internal energy, and (b) the enthalpy of helium for the temperature range from 100 to 1000°F, in Btu/lb.

**3-13** Calculate the change in (a) the internal energy, and (b) the enthalpy of argon for the temperature range from 80 to 800°F, in Btu/lb.

**3-14** One pound of air is compressed isothermally in a closed system from 15 psia, 60°F, to 90 psia.
(a) What is the change in internal energy in Btus?
(b) How much work is done in ft·lb$_f$?
(c) What is the heat transfer in Btu?

**3-15** One-tenth pound of oxygen is heated in a 0.5 ft$^3$ rigid vessel from 100 to 200°F. Determine (a) the final pressure in psia, and (b) the heat transferred in Btus.

**3-16** One pound of air is contained in a rigid vessel with thermally insulated walls. A paddle-wheel of negligible mass in the vessel is driven by an external motor. During the process the temperature rises from 70 to 170°F. Determine: (a) the change in internal energy, (b) the change in enthalpy, and (c) the work done, all in Btus.

**3-17** A cubic foot of nitrogen at 15 psia and 80°F is contained in a piston-cylinder assembly. A paddle-wheel within the cylinder is turned until 7000 ft·lb$_f$ of energy have been added. If the process is adiabatic and at constant pressure, determine the final temperature of the gas in °F.

**3-18** Three cubic feet of carbon monoxide at 20 psia and 50°F are contained in a piston-cylinder assembly. Electric current flows through a resistor within the cylinder until 0.0096 kwh of electrical energy has entered the gas. If the process is constant pressure and the final volume is 5 ft³, determine the magnitude and the direction of any heat transfer in Btus.

**3-19** Carbon dioxide is expanded isothermally in a nonflow process from initial conditions of 20 psia, 340°F, and 2.0 ft³ to a final volume of 4.0 ft³. Calculate (a) the work done in ft·lb$_f$/lb$_m$, and (b) the heat transfer in Btu/lb.

**3-20** A pound of nitrogen contained in a rigid tank at 100°F is heated until the pressure is doubled. Compute the quantity of heat transferred in Btu/lb.

**3-21** One-tenth pound of air initially at 40 psia and 140°F expands isothermally and quasistatically until the volume is doubled. Evaluate the quantity of heat added or removed in Btu/lb and the change in the internal energy of the air.

**3-22** In a closed system 0.27 lb of air expands quasistatically from 30 psia, 140°F, until its volume is doubled. The expansion is isothermal. During the process heat is added in the amount of 18.5 Btu/lb, and electrical work is also done on the system. Determine this quantity of electrical work in kWh/lb.

**3-23** One cubic foot of air is contained in a piston-cylinder assembly at 20 psia and 40°F. It is first heated at constant volume (process 1–2) until the pressure has doubled. Then it is expanded at constant pressure (process 2–3) until the volume has tripled. Find the total heat added in Btu, using the air table A-5.

**3-24** One-half pound of carbon dioxide undergoes a quasistatic, constant-pressure process in a closed cylinder. If the internal-energy change of the system is 16 Btu, determine how much work was done in Btus.

**3-25** A piston-cylinder assembly which initially has a volume of 1 ft³ contains 0.01 lb of hydrogen at 30 psia. Heat transfer is allowed to take place until the final volume is 0.82 ft³. If the process occurs at constant pressure, determine (a) the final temperature in °F, and (b) the heat transfer in Btu.

**3-26** Air is contained within a piston-cylinder assembly at initial conditions of 4 atm and 60°F. A paddle-wheel within the gas is turned until 22,500 ft·lb$_f$/lb$_m$ of work are performed on the gas. During the process the temperature remains constant while the volume is doubled. Determine the magnitude in Btu/lb and the direction of any necessary heat transfer.

**3-27** One-hundredth of a pound-mole of helium is contained within an adiabatic piston-cylinder assembly and is maintained at a constant pressure of 30 psia. A paddle-wheel is operated within the cylinder until the volume of the gas has increased by 25 percent. Compute (a) the quantity of paddle-wheel work, and (b) the net work done on the system, if the piston is frictionless and the initial temperature of the gas is 60°F.

**3-28** A system consists of 0.88 lb of carbon dioxide within a constant-pressure piston-cylinder device which is frictionless. The initial conditions are 100 psia and 40°F. Heat is added until the volume is doubled. On the basis of data in Table A-9 determine (a) the change in internal energy in Btus, (b) the change in enthalpy in Btus, (c) the work performed by or on the system in ft·lb$_f$, and (d) the quantity of heat supplied in Btus.

**3-29** A rigid tank with a volume of 1 ft³ contains oxygen initially at 200°F and 1.5 atm. During a process paddle wheel work amounts to 1500 ft·lb$_f$, and the temperature drops to 100°F. Determine the magnitude in Btus and direction of any heat transfer.

**3-30** Four pounds of nitrogen are compressed isothermally and frictionlessly in a closed system from

15 psia and 80°F to a final pressure of 25 psia. Determine (a) the work required, (b) the heat transferred, and (c) the change in the internal energy.

**3-31** A rigid wall tank contains 0.23 lb·mol of nitrogen at a pressure of 3 atm and a temperature of 100°F. A resistance coil in the nitrogen receives 0.02 kWh of electrical energy. If during this process 125 Btu of heat are removed, determine (a) the final temperature in °F, and (b) the final pressure in atm.

**3-32** An adiabatic piston-cylinder device contains 0.66 lb of air initially at 80°F and 3 atm. Energy is added to the gas by passing a current of 5 A through a resistor within the system for a time period of 30 s. The gas expands quasistatically at a constant pressure until the volume is doubled. Neglecting energy storage in the resistor, determine the size of the resistor in ohms. Recall that power dissipated in a resistor is $I^2R$.

**3-33** Two-tenths pound of carbon monoxide in a closed system is heated at constant pressure of 25 psia from 80 to 180°F. The measured work output is 890 ft·lb$_f$.
(a) Calculate the magnitude in Btus and the direction of any heat transfer.
(b) What would be the quantity of heat transferred if the above process were frictionless and quasistatic?

**3-34** Carbon dioxide is contained in a piston-cylinder device at an initial state of 0.1 ft$^3$, 25 psia, and 340°F. It is compressed at constant pressure until the temperature is 140°F. The work required on the gas is measured to be 120 ft·lb$_f$.
(a) Calculate the magnitude in Btus and the direction of any heat transfer.
(b) What would be the quantity of heat transfer if the above process were frictionless and quasistatic?

**3-35** Three pounds of air in a closed system are heated at a constant pressure of 20 psia from 65 to 170°F. The measured work output is 12.0 Btu. Calculate the net quantity of heat transferred in Btus during the process, and its direction. What would be the quantity of heat transferred if the above process were frictionless?

**3-36** Carbon monoxide gas is contained within a piston-cylinder device at 15 psia and 60°F.
(a) In process $A$ the gas is heated at constant volume until the pressure has doubled. It is then expanded at constant pressure until the volume is three times its initial value.
(b) In process $B$ the same gas in the same initial state is first expanded at constant pressure until the volume has tripled; then the gas is heated at constant volume until it reaches the same final pressure as in process $A$.
For these two processes determine the net heat effect, the net work effect, and the change in internal energy, all in Btu/lb. Compare the results for the two processes.

**3-37** Air initially at 60 psia and 1 ft$^3$ and with a mass of 0.1 lb expands at constant pressure to a volume at 3 ft$^3$. It then changes state at constant volume to a pressure of 15 psia. If the processes are quasistatic, find the total work done in Btus, the total heat transferred in Btus, and the overall change in internal energy in Btus.

### Specific-heat evaluation

**3-38** One-tenth pound of an ideal gas is enclosed in a rigid tank at a pressure of 18 psia and a temperature of 80°F. A paddle-wheel within the tank does 390 ft·lb$_f$ of work on the substance, and 0.77 Btu of heat is added at the same time. If the temperature of the gas, which has a molecular mass of 48, rises 80°F during the process, compute the average specific heat $c_v$ for the gas in Btu/(lb)(°F).

**3-39** Ten grams of an ideal gas having a molar mass of 32 undergoes a quasistatic expansion at constant pressure from 20 psia, 65°F, to 170°F. During the process 0.51 Btu of heat are added to the closed system. Compute the average value of $c_v$ for the gas in Btu/(lb$_m$)(°F).

**3-40** Two-tenths pound of an ideal gas (molecular weight = 40) expands quasistatically at constant pressure in a closed system from 15 psia, 140°F, to 340°F, and 4.9 Btu of heat is added. Determine the average value of $c_v$ in Btu/(lb)(°F).

**3-41** Ten pounds of a gas are contained within a vessel made of rigid walls which do not absorb or transmit heat. 77,800 ft·lb$_f$ of work are done on the gas by rotating a paddle-wheel of negligible mass within the system. In addition, 150 Btu are supplied as electrical energy to a resistance heater of negligible mass located in the system. During the process the temperature of the gas increases 100°F. Determine an average $c_v$ value for the gas in Btu/(lb)(°F).

**Additional problems**

**3-42** A rigid, insulated cylinder is divided into two parts by an uninsulated, frictionless piston which initially is held in a fixed position. One part of the cylinder contains 1 lb of a gas at 60 psia and 80°F, while the other part contains 1 lb of the same gas at 20 psia and 80°F. The piston is then released, and equilibrium is established between the two parts.

(a) What is the final equilibrium temperature in °F?

(b) What is the final equilibrium pressure in psia? Assume that the specific heats, $c_v$ and $c_p$, are constant.

**3-43** A vertical cylinder made of an insulating material is closed at its lower end by a noninsulated cylinder head and is fitted with a gastight, frictionless piston. The piston is held in its initial position by a spring in such a manner that the spring exerts an initial force at 1000 lb$_f$ on the top of the piston. The piston area is 100 in$^2$, and the weight of the piston may be neglected. The initial volume of the gas within the cylinder is 1 ft$^3$, and the initial temperature is 80°F. Barometric pressure is 29.4 in Hg. During an expansion process the spring is compressed within its elastic limit until the pressure of the gas becomes 50 psia and the volume of the gas has increased to 2 ft$^3$. During this process 15 Btu of work are used to rotate a paddle-wheel in the gas. At the same time 35 Btu of electrical energy are supplied to a resistance wire located in the cylinder. Consider the gas in the cylinder as the system.

(a) Calculate the work done by the system on the piston in ft·lb$_f$.

(b) Calculate the net amount of work done by or on the gas in Btus.

(c) Calculate the temperature rise of the system in °F if it contains air.

(d) Calculate the heat added or removed through the cylinder head in Btus.

**3-44** In a tank 0.130 lb of nitrogen is stored at 30 psia and 140°F. Attached to this tank through a suitable valve is a second tank which is 2.0 ft$^3$ in volume and completely evacuated. Both tanks are thoroughly insulated. If the valve is opened and equilibrium is allowed to be reached, determine the final pressure in psia.

**3-45** Two insulated tanks are connected by a valve. One tank contains 10 lb$_m$ of an ideal gas at 100°F and the other tank contains 20 lb$_m$ of the same gas at 200°F. The valve is opened, and the gases allowed to mix until equilibrium is reached. All specific heats may be regarded as constant.

(a) Calculate the mixture temperature in °F.

(b) Calculate the total enthalpy change for the two-tank system. Show that this answer is the same regardless of the mass and the temperature of the gas in each tank initially.

**3-46** A piston-cylinder machine contains air initially at 20 psia, 240°F, and 0.5 ft$^3$. The piston moves slowly and with negligible friction until the pressure rises to 100 psia. The process is described by the equation $V = 0.6 - 0.005P$, where $V$ is in ft$^3$ and $P$ is in psia.

(a) What is the final temperature?

(b) What mass of air is present?

(c) Determine the work done in ft·lb$_f$.

(d) Determine the heat transfer in Btus.

# FOUR

## PHYSICAL PROPERTIES OF A PURE SUBSTANCE

It is meaningful to examine at this time some basic properties of matter. Of primary importance is the acquisition of a qualitative feeling for the relationships among the basic properties of simple substances. Examination of the available data reveals a consistent pattern in the behavior of substances over a wide range of conditions. It is on this consistency that we wish to focus our attention. In addition, the methods for compiling experimental data will be examined, including graphs, tables, and equations with constants fitted from empirical data. These methods are well accepted; hence it is important for the student to be familiar with their general form and proper use. At the same time, one should gain a quantitative feeling for the order of magnitude of various intrinsic properties. The following discussion will be directed toward the behavior of substances of a single component, or a single chemical species, in a simple, compressible system.

## 4-1 THE $PvT$ SURFACE

In the preceding chapter the state postulate indicated that any intensive, thermostatic property of a simple, homogeneous system is solely a function of two other intrinsic properties. That is, $y_1 = y(y_2, y_3)$, where $y_1$ in general is any thermostatic property. From a mathematical viewpoint, any equation involving two independent variables (such as $y_2$ and $y_3$) can be represented in a rectangular three-dimensional space as a surface. Consequently, the equilibrium states of any simple system can be represented as a surface in space, where the geometric coordinates are intrinsic properties of interest for that particular simple system.

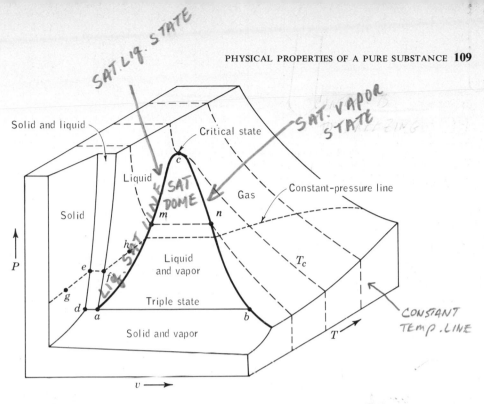

**Figure 4-1** A $PvT$ surface for a substance which contracts on freezing.

In this section we shall focus our attention on simple, compressible substances. Although any number of thermostatic properties might be chosen as
coordinates, it is expedient to examine initially only those diagrams which clearly
exhibit the basic structure of matter in a general fashion. It would also be helpful if
the properties employed are easily measurable and have a fairly simple physical
significance. For these reasons the $PvT$ diagram is extremely important. A typical
example of this diagram for a substance which contracts on freezing is shown in
Fig. 4-1. (Note that water is an exception to this rule.) It should be emphasized
that this figure is not to scale. The solid, liquid, and gas phases appear as surfaces.
In addition to these single-phase conditions, the process of a phase change is well
known. Two phases are coexistent during any phase change, e.g., melting, vaporization, or sublimation. (Sublimation is the transformation of a solid directly into a
gas.) The single-phase regions on the surface are necessarily separated by two-
phase regions. For example, the solid- and liquid-phase regions are seen to be
separated by another surface which represents a two-phase mixture of solid and
liquid. The surfaces of liquid-vapor and solid-vapor mixtures are also shown. (The
words vapor and gas are used interchangeably at this point.)

Any state represented by a point lying on a line separating a single-phase
region from a two-phase region in Fig. 4-1 is known as a *saturation* state. These
are states where a discontinuity occurs in the isotherms (constant-temperature
lines), such as states $m$ and $n$ in the figure. It is significant to note that a change of
phase occurs without a change of pressure or temperature. The curved line which

separates the liquid region from the liquid-vapor region, line *a-m-c*, is referred to as the liquid-saturation line. Any state represented by a point on this line between *a* and *c* is known as a saturated-liquid state. Similarly, the states represented on the curve *c-n-b* are saturated-vapor states.

Since the figure is not to scale and no values are listed on the coordinates, we must rely on actual experimental data to describe more quantitatively the general trends for the majority of substances. The spacing between particles (and hence the volume) undergoes only a very small change during a solid-liquid phase transformation. This volume change is given, for example, by the horizontal distance along the constant-pressure line connecting points *e* and *f* in Fig. 4-1. Since the volume change is small compared with the specific volume of either single phase at that pressure, the solid-liquid region would be quite thin if the diagram were to scale.

In the transition from the liquid to the gas phase, the change in volume generally is very much greater. This would be more readily apparent if the volume axis were more nearly to scale. It has been greatly foreshortened. For most substances the change in specific volume between states *a* and *b* in Fig. 4-1 would be thousands of times greater than the change between states *a* and *d*. It is important to notice, however, that as one moves up the liquid-vapor surface to higher pressures (and temperatures), the volume change for the process of boiling becomes smaller, and eventually becomes nonexistent. Beyond certain pressure-temperature conditions the process of vaporization (or the inverse process of condensation) cannot occur. The state which is the limit beyond which a liquid-vapor transformation is not possible is called the *critical state.* On a *PvT* diagram it appears as a point on the general surface. Associated with it are certain property values which are commonly signified by the subscript *c*. The three properties of present interest will be denoted by $P_c$, $v_c$, and $T_c$ at the critical state.

The critical state may be defined in a negative sense in the following way: If a substance is at a temperature which is higher than its critical temperature, it will not be capable of undergoing condensation to the liquid phase no matter how high a pressure is exerted. Beyond the critical state ($P > P_c$, $T > T_c$) it is impossible to distinguish between the liquid and gas (vapor) phases. For it is seen in Fig. 4-1 that these single-phase surfaces merge into each other. All known substances exhibit this typical behavior. Similarly critical states for other mixture regions, e.g., the solid-liquid surface, have not yet been observed experimentally. The existence of critical behavior demonstrates that the distinction between liquid and gas phases is not clear-cut, if not impossible, in certain situations. It is not really important what name is applied to a substance when it exists in this region beyond the critical state between the liquid and gas. Some prefer merely to refer to it as a fluid. When the pressure is greater than the critical pressure, the state is frequently referred to as a supercritical state. One may wonder if extreme conditions of pressure and temperature are necessarily associated with the critical states of common substances. It turns out that many familiar substances have fairly high critical pressures but critical temperatures which are below normal atmospheric conditions. For example, the critical pressures of hydrogen and oxygen are

**Table 4-1  Critical constants**

| Substance | Temperature | | Pressure | |
|---|---|---|---|---|
| | °K | °R | bars | psia |
| Ammonia ($NH_3$) | 405.5 | 729.8 | 112.8 | 1636 |
| Carbon dioxide ($CO_2$) | 304.2 | 547.5 | 73.9 | 1071 |
| Carbon monoxide (CO) | 133 | 240 | 35.0 | 507 |
| Helium (He) | 5.3 | 9.5 | 2.29 | 33.2 |
| Hydrogen ($H_2$) | 33.3 | 59.9 | 13.0 | 188.1 |
| Nitrogen ($N_2$) | 126.2 | 227.1 | 33.9 | 492 |
| Water ($H_2O$) | 647.3 | 1165.2 | 220.9 | 3204 |

approximately 190 and 730 psia (13.1 and 50.3 bar), respectively, and the corresponding critical temperatures are $-400$ and $-181°F$ ($-240$ and $-118°C$). The critical temperature of carbon dioxide is $88°F$ ($31°C$), whereas that of water is quite high, $705°F$ ($374°C$). Hence no generalization can be made for the range of critical temperatures commonly found. The critical pressures of most common substances, however, are above one atmosphere. Critical data for a few substances are given in Table 4-1, taken from a longer list of substances included in Table A-3. A practical example of a process operating beyond the critical state may be given. A few modern steam power plants operate so that the water during part of its cycle is in those states beyond the critical state which are better characterized simply as fluids. In this particular case this means water not only at temperatures above $650°K$ or $705°F$ (not unusual for any modern high-capacity steam power station), but also at pressures above 220 atm.

Lines *ge* and *fh* in Fig. 4-1 indicate that, at a given pressure, the volume of the solid or liquid phase changes very little as the temperature is increased. This is true when compared with even the small volume change for the solid-liquid phase transition. This is in general agreement with the physical notion that solids and liquids are observed not to undergo any significant volume change during heating or cooling.

An additional unique state of matter may be noted from Fig. 4-1. This is represented by the line parallel to the *Pv* plane marked as the *triple* state. As the term implies, this is a state in which three phases coexist in equilibrium. The triple state in this figure is for equilibrium between a solid, a liquid, and a gas phase. For water this state appears at 0.0061 atm and $0.01°C$ ($32.02°F$). Recall that the triple state of water is used as the reference point for establishing the Kelvin temperature scale. The triple state of water is assigned a temperature value of $273.16°K$. As another example, carbon dioxide exists in these three phases simultaneously at roughly 5 atm and $-57°C$ ($-70°F$). If a piece of solid carbon dioxide (Dry ice) is placed in the open atmosphere, it is noticed that the solid goes directly into the gas or vapor phase. The reason is that at 1 atm carbon dioxide is considerably below the minimum pressure at which its liquid phase can exist; consequently, it sublimes at this pressure. Paradichlorbenzene (used in moth crystals) and iodine

## Table 4-2 Triple-state data

| Substance | $T$, °K | $P$, atm | $T$, °F |
|---|---|---|---|
| Helium 4 ($\lambda$ point) | 2.17 | 0.050 | −456 |
| Hydrogen, $H_2$ | 13.84 | 0.070 | −435 |
| Oxygen, $O_2$ | 54.36 | 0.0015 | −362 |
| Nitrogen, $N_2$ | 63.18 | 0.124 | −346 |
| Ammonia, $NH_3$ | 195.40 | 0.061 | −108 |
| Carbon dioxide, $CO_2$ | 216.55 | 5.10 | −70 |
| Water, $H_2O$ | 273.16 | 0.006 | 32 |

crystals are two other common examples of substance undergoing a direct solid-vapor phase change under atmospheric pressure conditions. Some triple-state data appear in Table 4-2.

Water is an anomalous substance in that it expands upon freezing. Thus the specific volume of the solid phase is greater than that of the liquid phase. The $PvT$ surface modified to take this feature into account is shown in Fig. 4-2.

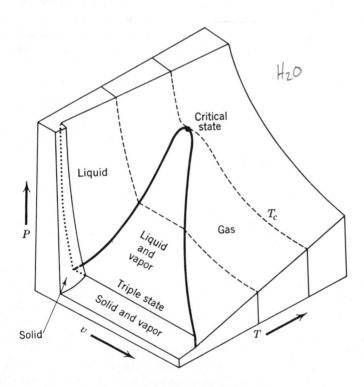

**Figure 4-2** A $PvT$ surface for a substance which expands on freezing.

$$V_{SOLID} > V_{Liq}$$

## 4-2 THE PRESSURE-TEMPERATURE DIAGRAM

Three-dimensional diagrams for the equilibrium states of simple systems are extremely useful in introducing the general relationships between the three phases of matter normally under consideration. The relation of two-phase regions to the single-phase regions is clearly shown, as well as the significance of the critical and triple states of matter. Nevertheless, it will be found to be more convenient in the thermodynamic analysis of simple, compressible systems to work with two-dimensional diagrams. All two-dimensional diagrams may be thought of simply as projections of a three-dimensional surface. For example, the surface presented in Fig. 4-1 may be projected on a $PT$, $Pv$, or $Tv$ plane. Only the first two of these will be explored in any depth at the present time.

A projection of a $PvT$ surface upon the $PT$ plane is commonly termed a phase diagram. It is an empirical fact that both the temperature and the pressure remain constant during a phase change. This is well known for the boiling of water or melting of ice. Consequently, the surfaces in Fig. 4-1 which represent two phases in equilibrium are parallel to the $v$ axis. Hence these surfaces appear as lines when projected onto the $PT$ plane. A typical pressure-temperature plot (based on the general characteristics of Fig. 4-1) is presented in Fig. 4-3. This figure carries the same restriction as before; i.e., it is for a substance which contracts on freezing.

The liquid-vapor surface which appears as a line on the $PT$ diagram is called the liquid-vapor saturation line, since in this case the saturated-liquid and

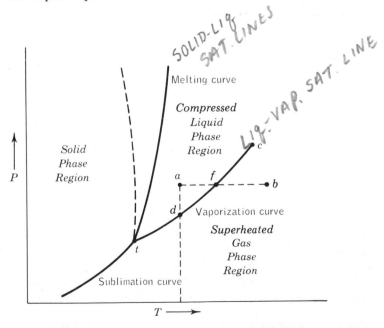

**Figure 4-3** Phase ($PT$) diagram for a substance that contracts on freezing.

saturated-vapor lines project on top of each other. It is also known as the vaporization curve. Similarly, the solid-liquid and solid-vapor surfaces of a $PvT$ diagram are shown as the freezing (or melting) curve and the sublimation curve, respectively, in Fig. 4-3. These are also known as the solid-liquid and the solid-vapor saturation lines. The critical and triple states are designated by the points marked $c$ and $t$. Whereas two-phase systems appear as lines, single-phase systems are represented by areas. The dashed line in the figure represents the fusion (freezing) curve for a substance such as water which expands on freezing. All other specific properties, except the volume, decrease during the freezing of water. Note that for ice an increase in pressure decreases the freezing-point temperature.

The single-phase liquid and gas regions, represented by labeled areas in Fig. 4-3, are sometimes given special names. It is seen from the figure that, for any state in the area marked liquid (such as point $a$), the temperature $T_a$ is below the saturation temperature $T_f$ for the same value of pressure. Such a state is said to be that of a *subcooled* liquid, since it may be achieved by cooling the liquid below its saturation temperature at a given pressure. On the other hand, the pressure $P_a$ of state $a$ is above the saturation pressure $P_d$ for the same temperature. Hence state $a$ is also called the state of a *compressed* liquid, since it may be achieved by compressing the liquid above its saturation pressure at a given temperature. The terms "subcooled" and "compressed" are thus synonymous. In a similar fashion, if the temperature $T_b$ of a substance at state $b$ is above the vapor-liquid saturation temperature $T_f$ for a given pressure, the substance is in a *superheated-vapor* state. The process of superheating generally is defined as one for which the temperature of a vapor is increased at constant pressure.

It is apparent from Fig. 4-3 that for each saturation pressure there is only one saturation temperature. For a liquid-vapor system the saturation pressure is known as the *vapor* pressure. Although there are two independent intensive properties for a single phase of a simple, compressible substance, this number is reduced to one for a single-component, two-phase system. If the pressure is fixed for this latter system, the temperature is fixed and cannot be varied independently. All other intensive properties of each of the two phases in equilibrium are also fixed when one intensive property such as the pressure is fixed. This change in the number of independent variables from two to one has a major influence on the method of presenting tabular property data for single-phase versus two-phase systems. In the case of a three-phase system of a single-component there are no independent variables. Thus the intensive properties at the triple state cannot be varied arbitrarily, but are limited to a single set of values for a given compound. There is only one pressure and one temperature for a substance at its triple state. This is true also for all other intensive properties of the three phases. Hence the triple state of single-component systems is especially well defined in a thermodynamic sense, and is easily reproduced. It is this latter fact that makes the triple state of water an excellent reference state for the establishment of a temperature scale.

Although Fig. 4-3 is the characteristic phase diagram for most substances, there are other substances which exhibit anomalous behavior. It has already been

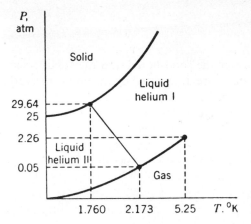

Figure 4-4 Phase diagram for helium 4.

mentioned that the melting curve for water, for example, has a negative slope. Helium is another substance which displays some interesting properties. The $PT$ diagram for $He^4$ is shown in Fig. 4-4 for the low end of the temperature scale. Helium is the only known substance that most probably remains a liquid down to absolute zero temperature. As a result, a pressure greater than 25 atm must be applied in order to form solid $He^4$. Note also that $He^4$ does not have a triple state where solid, liquid, and gas coexist. At 1 atm the substance boils at 4.215°K, and the critical state is at 2.26 atm and 5.25°K. One of the most unusual characteristics of $He^4$ is the existence of two distinct liquid phases, which have quite different thermodynamic properties. Helium II, which exists below 2.173°K, has an extremely low viscosity and is known as a superfluid. The vapor pressure of helium is used as an international standard for the measurement of temperature in the range of, roughly, 0.2 to 5°K.

## 4-3 THE PRESSURE-VOLUME DIAGRAM

Some of the fundamental characteristics of the two-phase surfaces that appear in Fig. 4-1 are masked by the projection of these surfaces onto the $PT$ plane. Hence the projection of the $PvT$ surface onto the $Pv$ plane is of interest. Again, use is made of Fig. 4-1, which is valid for substances which contract on freezing. The resultant projection onto a plane parallel to the $Pv$ axes is shown in Fig. 4-5. Both single- and two-phase regions appear as areas on this new diagram. The saturated-liquid line represents the states of the substance such that any further infinitesimal addition of energy to the substance at constant pressure will change a small fraction of the liquid into vapor. Similarly, removal of energy from the substance at any state which lies on the saturated-vapor line results in partial condensation of the vapor, whereas addition of energy makes the vapor superheated. The two-phase region labeled saturated liquid and vapor, which lies between the saturated-liquid and saturated-vapor lines, is commonly called the *wet region*, or the wet

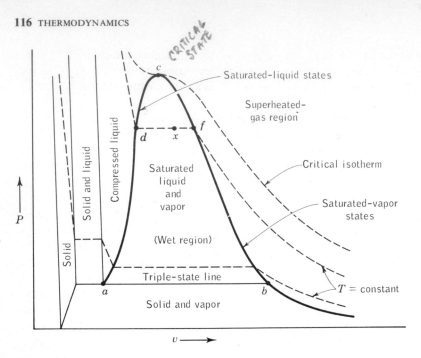

**Figure 4-5** A *Pv* diagram for a substance which contracts on freezing.

dome. The state at the top of the wet region indicated by point *c* is again the critical state.

A state represented by a point in the liquid-vapor region (wet region), such as *x* in Fig. 4-5, is a mixture of saturated liquid and saturated vapor. The specific volumes of these two phases must be points on the saturation line, such as points *d* and *f* in the figure. Hence the specific volume of state *x* simply represents the *average* property value for the two phases in equilibrium. In order to find this average value of *v* (or any other specific property of a liquid-vapor mixture, such as *u* and *h*), we need to know the proportions of vapor and liquid in the saturated vapor-liquid mixture. To achieve this we define the *quality*, usually represented by the symbol *x*, as the fraction of the total mixture which is vapor, based on mass (or weight). That is,

$$\text{Quality} = x = \frac{m_{\text{vapor}}}{m_{\text{total}}} = \frac{m_g}{m_g + m_f} \tag{4-1}$$

In this equation the subscript *g* applies to the saturated vapor state, while subscript *f* denotes the saturated liquid state. It is common to speak of quality also as a percentage, in which case the value of *x* defined above is multiplied by 100. For example, a system composed of a saturated liquid alone may be referred to as having a quality of 0 percent ($x = 0.00$), and a saturated vapor alone has a quality of 100 percent ($x = 1.00$). A saturated vapor alone is frequently called a dry saturated vapor. The word *dry* in this case indicates there is no liquid in the saturated state. The quality is limited to values between zero and unity (or 0 to 100 percent). The use of the term is restricted solely to saturated liquid-vapor mix-tures. Later in this chapter we shall develop equations for computing intensive properties of two-phase mixtures. The quality is an important parameter in such computations.

Extension of the saturated-vapor line below the triple-state line is valid. If energy were now removed, the saturated vapor represented by points along this lower portion of the saturated-vapor curve would condense out into a solid phase rather than a liquid one. The solid, solid-liquid, and compressed-liquid regions may be considered similarly. One final comment is appropriate to the $Pv$ diagram shown in Fig. 4-5. It has been noted earlier that other families of curves represent-ing various properties could be shown on any of the two-dimensional or three-dimensional diagrams. For this reason a family of constant-temperature lines is also shown on the diagram. The general trends shown by these lines can be verified by referring back to the $PvT$ diagram of Fig. 4-1. It is important for the student to have some general concept of the position of constant-temperature lines on a $Pv$ diagram since this diagram will be used in future problem analysis.

It is interesting to compare the $PvT$ behavior of an ideal gas with that of a real gas, in terms of the $PvT$ surface. The general $PvT$ characteristics of a real gas were shown by the gas-phase region of Fig. 4-1. The $PvT$ surface of an ideal gas is shown in Fig. 4-6. Since the $PvT$ relationship for any equilibrium state is given by $Pv = RT$ for an ideal gas, the isotherms (constant-temperature lines) along the surface will appear as hyperbolas. Curves of constant pressure or constant volume are straight lines. Although the ideal-gas equation is strictly valid only in the limit of zero pressure, we have noted that experimental data indicate that many gases nearly fulfill the relation $Pv = RT$ under low-pressure conditions. It is also gen-erally true that most gases behave more ideally as the temperature is raised at constant pressure. Consequently, the conditions of low pressure and high temper-ature usually lead to states of gases which are approximately ideal-gas states. It would be expected, then, that for real gases the equilibrium states on the far

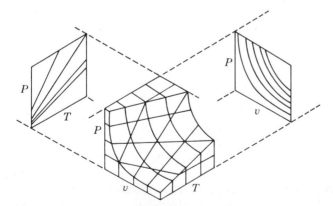

**Figure 4-6** The $PvT$ surface and the $PT$ and $Pv$ projections for ideal-gas behavior.

right-hand side of the *PvT* surface in Fig. 4-1 would approximate ideal-gas behavior.

Figure 4-6 also shows the *PT* and *Pv* projections of the *PvT* surface of an ideal gas. The hyperbolic nature of constant-temperature lines on the *Pv* plane, and the straight-line nature of constant-volume lines on the *PT* plane, are clearly shown.

## 4-4 TABLES OF PROPERTIES OF PURE SUBSTANCES

The preceding sections were presented as a brief introduction to the *PvT* behavior of simple compressible systems containing a pure substance. In addition to the acquisition of the necessary nomenclature required for future discussions, a general qualitative feeling for the *PvT* characteristics of substances has been gained. Some numerical values were also cited to establish the range of interest for some of the various thermodynamic variables. However, in order to carry out quantitative calculations, it is necessary to investigate the more established methods of assembling or storing physical data so that they are readily accessible and in a convenient format for calculation purposes. Such data are primarily either obtained directly from experimental measurements or calculated from equations which employ the preceding experimental measurements. In other instances particular molecular models are chosen, and the experimental work is directed toward the measurement of molecular characteristics. By suitable statistical analysis the average values of macroscopic properties may then be obtained.

The relationships among thermodynamic data are frequently presented in the form of tables. Tabular data represent information taken from direct physical measurements or values calculated from these measurements. They are listed at convenient increments of the independent variables. The following discussion is devoted to an explanation of the tabular formulation of some basic intrinsic properties. For specific examples it will be necessary to refer to only one particular chemical species, since all tables are constructed essentially in the same manner.

For our initial study the following intensive intrinsic properties will be of primary interest: pressure $P$, specific volume $v$, temperature $T$, specific internal energy $u$, specific enthalpy $h$, and specific entropy $s$. (The property entropy has not yet been defined. For the present discussion one is asked to accept the existence of this property without proof. It will not be used to any extent until after it is defined in a later chapter.

### a Superheat tables

In a single-phase region, such as the superheat region, two intensive properties are required to fix or identify the equilibrium state. Variables such as $v$, $u$, $h$, and $s$ are usually tabulated in superheat tables as a function of $P$ and $T$, since the latter

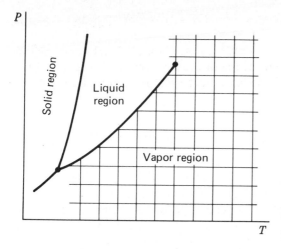

**Figure 4-7** Sketch of *PT* diagram indicating method of tabulating superheat data.

are conveniently measurable properties. The format of a superheat table is readily apparent if we refer again to a *PT* diagram. Figure 4-7 shows the superheated vapor region of a *PT* diagram divided into a grid. The lines of the grid represent integral values of the pressure and the temperature. A superheat table then reports values of $v$, $h$, and $s$ (and frequently $u$) at the grid points of the figure. The values of $u$, $h$, and $s$ are completely arbitrary, each being based on an assigned value at some reference state.

Table 4-3 illustrates one possible format for tabulation of data in the superheat region. A more complete compilation of superheated steam data in metric units is found in Table A-14M, while Table A-14 presents data in USCS units. The format of both tables is identical. The units for specific properties are usually listed at the top of a table, under the heading. With respect to Table 4-3 and Fig. 4-7, note that the data begin with saturated *vapor* data and then proceed at a given pressure to higher integral temperatures. Some superheat tables in the literature will not, however, carry the data down to the saturated vapor state. The temperature of the saturated vapor data is indicated by the value in parentheses after the pressure value. For example, from Table 4-3, at 1.0 bar pressure the saturation temperature is 99.63°C and the specific volume is 1694 cm³/g.

Many scientific and engineering problems involve states of matter which do not fall on the grid of data available for that substance. Interpolation of data then becomes necessary. The intervals for the matrix of data found in unabridged superheat tables are usually chosen so that *linear* interpolation leads to reasonable accuracy. Although the superheat tables in the appendix are condensed versions of the original works, we shall still assume linear interpolation is valid. For professional use one should always employ the complete tables as they appear in the literature, unless high accuracy is not desired. Superheat tables for refrigerant 12 (dichlorodifluoromethane) in metric and engineering units appear as Tables A-18M and A-18. The following examples illustrate the use of superheat tables.

## Table 4-3 Properties of superheated steam ($H_2O$)

A. Metric: $v$, cm³/g; $u$, kJ/kg; $h$, kJ/kg: $s$, kJ/(kg)(°K)

| Temp. °C | $v$ | $u$ | $h$ | $s$ |
|---|---|---|---|---|
| | | | 1.0 bar (99.63°C) | |
| Sat. | 1694.0 | 2506.1 | 2675.5 | 7.3594 |
| 100 | 1696.0 | 2506.7 | 2676.2 | 7.3614 |
| 120 | 1793.0 | 2537.3 | 2716.6 | 7.4668 |
| 160 | 1984.0 | 2597.8 | 2796.2 | 7.6597 |
| 200 | 2172.0 | 2658.1 | 2875.3 | 7.8343 |
| | | | 10.0 bars (179.91°C) | |
| Sat. | 194.4 | 2583.6 | 2778.1 | 6.5865 |
| 200 | 206.0 | 2621.9 | 2827.9 | 6.6940 |
| 240 | 227.5 | 2692.9 | 2920.4 | 6.8817 |
| 280 | 248.0 | 2760.2 | 3008.2 | 7.0465 |
| 320 | 267.8 | 2826.1 | 3093.9 | 7.1962 |

*Source:* Abstracted from Table A-14M.

B. USCS: $v$, ft³/lb; $u$, Btu/lb; $h$, Btu/lb; $s$, BTU/(lb)(°R)

| Temp. °F | $v$ | $u$ | $h$ | $s$ |
|---|---|---|---|---|
| | | | 60 psia (292.7°F) | |
| Sat. | 7.17 | 1098.3 | 1178.0 | 1.6444 |
| 300 | 7.26 | 1101.3 | 1181.9 | 1.6496 |
| 400 | 8.35 | 1140.8 | 1233.5 | 1.7134 |
| 500 | 9.40 | 1178.6 | 1283.0 | 1.7678 |
| 600 | 10.43 | 1216.3 | 1332.1 | 1.8165 |

*Source:* Abstracted from Table A-14.

**Example 4-1M** Determine the internal energy of superheated water vapor at (*a*) 1.0 bar and 110°C, and (*b*) 6 bars and 220°C.

SOLUTION (*a*) The value of $u$ at 1 bar and 110°C can be obtained from Table 4-3 or the more complete data of Table A-14M. With reference to either of these tables, we find that at 1 bar

$$u = 2537.3 \text{ kJ/kg at } 120°C$$

$$u = 2506.7 \text{ kJ/kg at } 100°C$$

Linear interpolation between these two values leads to an internal energy value of 2522.0 kJ/kg at 110°C.

(*b*) In order to find the value of $u$ at 6 bars and 220°C we must turn to Table A-14M in the appendix. In Table A-14M we find data only at 5 and 7 bars and only at 200 and 240°C. To find the desired value of $u$ we must rely on double interpolation. By linearly interpolating, first, with respect to temperature,

$$\text{At 5 bars and } 220°C: \quad u = 2675.2 \text{ kJ/kg}$$

$$\text{At 7 bars and } 220°C: \quad u = 2668.3 \text{ kJ/kg}$$

Consequently, at 6 bars and 220°C the internal energy is the average of these two latter values, or approximately 2671.8 kJ/kg.

**Example 4-1** Determine the enthalpy of superheated water vapor at (a) 60 psia and 450°F, and (b) 110 psia and 550°F.

SOLUTION (a) The value of $h$ at 60 psia and 450°F can be obtained from Table 4-3 or Table A-14. With reference to these tables, we find that at 60 psia

$$h = 1233.5 \text{ Btu/lb at } 400°F$$

$$h = 1283.0 \text{ Btu/lb at } 500°F$$

Linear interpolation between these two values leads to an enthalpy value of 1258.2 Btu/lb at 450°F and 60 psi.

(b) In Table A-14 data are reported only at 100 and 120 psia, and only at 500 and 600°F. To find the desired value at 110 psia and 550°F we must rely on double interpolation. By linearly interpolating first with respect to temperature,

$$\text{At 100 psia and } 550°F: \quad h = 1304.2 \text{ Btu/lb}$$

$$\text{At 120 psia and } 550°F: \quad h = 1302.4 \text{ Btu/lb}$$

Consequently, at 110 psia and 550°F the enthalpy is the average of these latter two values, or approximately 1303.3 Btu/lb.

**Example 4-2M** Determine the pressure of superheated refrigerant 12 at a state of 40°C and an enthalpy of (a) 214.44 kJ/kg, and (b) 214.70 kJ/kg.

SOLUTION This problem illustrates the point that *any* two independent properties fix the state of a simple compressible substance. In this case we seek $P$ as a function of $T$ and $h$.

(a) The superheat data for refrigerant 12 are tabulated in Table A-18M, with the temperature running from top to bottom at any pressure. Starting at 40°C and the lowest pressure (0.6 bar), we note that the enthalpy starts at 216.77 kJ/kg and decreases as the pressure increases for the same temperature. If we continue to read to higher pressures at 40°C, we finally find that $h = 214.44$ kJ/kg at 2.4 bars.

(b) The value of 214.70 kJ/kg for the enthalpy at 40°C lies between 2.0 and 2.4 bars. By linear interpolation we estimate that the pressure corresponding to an enthalpy of 214.70 kJ/kg at 40°C is 2.2 bars.

**Example 4-2** Determine (a) the temperature of superheated refrigerant 12 at a state of 50 psia and an enthalpy of 90.953 Btu/lb, and (b) the pressure at a state of 180°F and an internal energy of 93.800 Btu/lb.

SOLUTION This problem illustrates the point that *any* two independent properties fix the state of a simple compressible substance.

(a) In this case, we seek $T$ as a function of $P$ and $h$. The superheat data for refrigerant 12 are found in Table A-18, with the temperature running from top to bottom for any pressure. At 50 psia the saturated vapor value is 81.249 Btu/lb, and the enthalpy increases with increasing temperature. The desired value of 90.953 Btu/lb for the enthalpy is found at 100°F.

(b) In this case, we seek $P$ as a function of $T$ and $u$. Starting at 180°F and the lowest pressure (10 psia), we note that the internal energy starts at 94.367 Btu/lb and decreases as the pressure increases for the same temperature. If we continue to read $u$ at higher pressures and 180°F, we finally find that $u = 93.800$ Btu/lb at 40 psia.

The preceding examples have illustrated the use of the superheat tables for determining property values of several substances. In general, superheat tables in the literature are arranged like those found in the Appendix, although you will find that the position of the temperature and pressure coordinates are frequently reversed. The metric data of Table 4-3 at 1 bar and 120°C, 160°C, and 200°C could be reported in the reversed format shown in Table 4-4. In this particular case only $v$, $h$, and $s$ data are given. Whether $u$ data are reported in a given table is a matter of choice. Although interpolation is frequently necessary in order to determine an accurate value for a property, the technique is straightforward.

**Table 4-4 An alternate format for reporting properties of superheated steam**

| Pressure, bars | | Temperature, °C | | |
|---|---|---|---|---|
| | | 120 | 160 | 200 |
| 0.7 | | | | |
| | $v$ | 1793. | 1984. | 2172. |
| 1.0 | $h$ | 2716.6 | 2796.2 | 2875.3 |
| | $s$ | 7.4668 | 7.6597 | 7.8343 |
| 1.5 | | | | |

### b. Saturation tables

An intensive property of a two-phase liquid-vapor mixture may be determined by suitable averaging of the property values for each separate phase. The liquid and vapor phases are denoted by the subscript $f$ and $g$, respectively. For any extensive property $Y$ (such as $V$, $U$, or $H$) the specific value $y_x$ (per unit mass of mixture) is found by adding the extensive values for each phase and dividing by the total mass. Hence

$$Y = Y_f + Y_g$$

$$m = m_f + m_g$$

and

$$y_x = \frac{Y}{m} = \frac{Y_f + Y_g}{m_f + m_g}$$

However, the extensive property value for each phase may be expressed in terms of its intensive value and its mass. That is, $Y_f = m_f y_f$ and $Y_g = m_g y_g$, by definition. Substitution of these two definitions into the above expression for $y_x$ yields

$$y_x = \frac{m_f y_f + m_g y_g}{m_f + m_g} = y_f \left( \frac{m_f}{m_f + m_g} \right) + y_g \left( \frac{m_g}{m_f + m_g} \right)$$

However, from the definition of quality $x$ given by Eq. (4-1), it is known that

$$x = \frac{m_g}{m_f + m_g} \qquad \text{and} \qquad 1 - x = \frac{m_f}{m_f + m_g}$$

Consequently, the expression for $y_x$ can be written as

$$y_x = (1 - x)y_f + xy_g \qquad (4\text{-}2)$$

Alternatively, if we designate the difference between the saturated-vapor and saturated-liquid intensive properties by the symbol $y_{fg}$, that is

$$y_{fg} \equiv y_g - y_f$$

then on the basis of Eq. (4-2) we may also write

$$y_x = y_f + x(y_g - y_f) = y_f + xy_{fg} \qquad (4\text{-}3)$$

Equations (4-2) and (4-3) are equivalent. These two equations emphasize the fact that a state in the wet region represents only a hypothetical state which has properties that are average characteristics of both phases. The actual state is the state of the two separate phases which are in contact.

Rearrangement of Eq. (4-3) gives

$$x = \frac{y_x - y_f}{y_{fg}}$$

If this particular equation is applied to the specific volume, we may write $x = (v_x - v_f)/v_{fg}$. This result is useful with respect to the $Pv$ diagram in Fig. 4-5. It is seen that the quality of any state of a liquid-vapor mixture is given by the ratio of horizontal distances within the two-phase region on the diagram. A state of 50 percent quality lies midway between the saturated-liquid and saturated-vapor lines along any constant-pressure line. The above equations are valid for any specific property of a two-phase mixture, such as volume, internal energy, enthalpy, etc.

Intensive properties of a two-phase mixture are readily calculated from equations of the type of (4-2) and (4-3). These equations require a knowledge of the property data for the saturation states of the two phases involved and the quality of the mixtures. Only one intensive property is required to identify the intensive state of the two phases in equilibrium when it is known that a saturated mixture exists. Again it is preferable to tabulate the various intensive properties of each phase of the mixture against either the pressure or the temperature. We have already noted, especially by the use of Fig. 4-3, that fixing the saturation temperature automatically fixes the saturation pressure, and vice versa, for two-phase systems. In many instances two saturation tables may be given, one with temperature as the independent variable and the other with pressure as the independent variable. In these tables the independent variable usually appears only in integral values. Although this may appear as a duplication of data, it is quite convenient when interpolation of either $T$ or $P$ is required.

Table 4-5 is an abridgment of the properties of saturated water as a function of integral values of temperature, in both metric and USCS engineering units. More complete data for saturated water are given in Tables A-12M and A-12 as a function of temperature, and in Tables A-13M and A-13 as a function of pressure. Saturation tables for refrigerant 12 appear as Tables A-16M, A-16, A-17M, and A-17. In some of these tables there are columns of data listed as $h_{fg}$ and $s_{fg}$. As noted earlier, the difference between saturated-vapor and saturated-liquid properties is symbolized by the subscript $fg$. The quantity $h_{fg}$ is called the *enthalpy of vaporization*, or the latent heat of vaporization. It represents the quantity of energy required to vaporize a unit mass of a saturated liquid at a given temperature or pressure. Hence, it is an important thermodynamic property. In these tables $s_{fg}$ is the entropy of vaporization. Both $h_{fg}$ and $s_{fg}$ become zero at the critical state.

The following examples illustrate data retrieval from the various saturation tables for water and refrigerant 12.

## Table 4-5 Properties of saturated liquid and vapor for water

A. Metric: $v$, cm³/g; $u$, kJ/kg; $h$, kJ/kg; $s$, kJ/(kg)(°K)

| Temp. °C $T$ | Press. bars $P$ | Specific volume | | Enthalpy | | Entropy | |
|---|---|---|---|---|---|---|---|
| | | Sat. liquid $v_f$ | Sat. vapor $v_g$ | Sat. liquid $h_f$ | Sat. vapor $h_g$ | Sat. liquid $s_f$ | Sat. vapor $s_g$ |
| 20 | 0.02339 | 1.0018 | 57791 | 83.96 | 2538.1 | 0.2966 | 8.6672 |
| 40 | 0.07384 | 1.0078 | 19523 | 167.57 | 2574.3 | 0.5725 | 8.2570 |
| 60 | 0.1994 | 1.0172 | 7671 | 251.13 | 2609.6 | 0.8312 | 7.9096 |
| 80 | 0.4739 | 1.0291 | 3407 | 334.91 | 2643.7 | 1.0753 | 7.6122 |
| 100 | 1.014 | 1.0435 | 1673 | 419.04 | 2676.1 | 1.3069 | 7.3549 |

*Source:* Abstracted from Table A-12M.

B. USCS: $v$, ft³/lb; $u$, Btu/lb; $h$, Btu/lb; $s$, Btu/(lb)(°R); $T$, °F; $P$, psia

| $T$ | $P$ | $v_f$ | $v_g$ | $h_f$ | $h_g$ | $s_f$ | $s_g$ |
|---|---|---|---|---|---|---|---|
| 60 | 0.2563 | 0.01604 | 1207 | 28.08 | 1087.7 | 0.0555 | 2.0943 |
| 70 | 0.3632 | 0.01605 | 867.7 | 38.09 | 1092.0 | 0.0746 | 2.0642 |
| 80 | 0.5073 | 0.01607 | 632.8 | 48.09 | 1096.4 | 0.0933 | 2.0356 |
| 90 | 0.6988 | 0.01610 | 467.7 | 58.07 | 1100.7 | 0.1117 | 2.0083 |
| 100 | 0.9503 | 0.01613 | 350.0 | 68.05 | 1105.0 | 0.1296 | 1.9822 |

*Source:* Abstracted from Table A-12.

**Example 4-3M** Determine the volume change when 1 g of saturated liquid water is completely vaporized at (a) 100°C, and (b) 300°C.

SOLUTION The volume change during vaporization is given by $v_{fg} = v_g - v_f$.
(a) These values can be found for 100°C from either Table 4-5 or Table A-12M. The result is

$$v_{fg} = 1673 - 1.04 = 1672 \text{ cm}^3/\text{g}$$

(b) At 300°C the data are found only in Table A-12M in the Appendix. At this state,

$$v_{fg} = 21.67 - 1.40 = 20.27 \text{ cm}^3/\text{g}$$

This problem demonstrates the vast change in $v_{fg}$ as the temperature approaches the critical state, which in this case is 374°C. The volume change differs by over a factor of 80. These data indicate that the saturated-vapor line on Fig. 4-5 is considerably out of scale. To agree with the above data, this line should fall much more slowly with decreasing pressure. This trend is typical of most substances.

**Example 4-3** Determine the volume change when 1 lb of saturated liquid water is completely vaporized at (a) 100°F, and (b) 600°F.

SOLUTION The volume change during vaporization is given by $v_{fg} = v_g - v_f$.
(a) These values for 100°F can be found from either Table 4-5 or Table A-12. The result is

$$v_{fg} = 350.0 - 0.016 = 350 \text{ ft}^3/\text{lb}$$

(b) At 600°F the data are found only in Table A-12 in the Appendix. From this table,

$$v_{fg} = 0.2677 - 0.0236 = 0.2441 \text{ ft}^3/\text{lb}$$

These data demonstrate the significant change in $v_{fg}$ as the temperature approaches the critical state, which in this case is 705°F. The volume change differs by over a factor of 1000. The data indicate that the saturated-vapor line on Fig. 4-5 is considerably out of position. To account for the wide variation in $v_{fg}$ with temperature (or pressure), the saturated-vapor line should fall much more slowly with decreasing pressure. This trend is typical of most substances.

**Example 4-4M** Two kilograms of water substance at 200°C are contained in a 0.2-m³ vessel. Determine (a) the pressure, (b) the enthalpy in kJ/kg, and (c) the mass and volume of the vapor within the vessel.

SOLUTION (a) On the basis of the mass and volume data, the overall specific volume is found to be 0.1 m³/kg or 100 cm³/g. From Table A-12 at 200°C it is seen that the water must be a liquid-vapor mixture since $v$ is greater than $v_f$ of 1.1565 cm³/g, but less than $v_g$ of 127.4 cm³/g at this temperature. Thus the pressure must be the corresponding saturation vapor pressure at 200°C. This is 15.54 bars, found in column two of Table A-12M.
(b) The specific enthalpy of the mixture may be found from a form of Eq. (4-3), namely,

$$h_x = h_f + xh_{fg} = h_f + x(h_g - h_f)$$

However, this equation requires a knowledge of the quality. This may be determined from information on $v_x$, which is 100 cm³/g. Note that another form of Eq. (4-3) is

$$v_x = v_f + x(v_g - v_f)$$

Substitution of the values for $v_f$ and $v_g$ at 200°C from Table A-12M leads to

$$100 = 1.1565 + x(127.4 - 1.1565)$$

$$x = \frac{98.84}{127.3} = 0.776$$

The specific enthalpy, then, is found to be

$$h_x = 852.45 + 0.776(1940.7) = 2358 \text{ kJ/kg}$$

(c) The mass of the vapor is $0.776(2) = 1.55$ kg. Hence, the volume occupied by the vapor is

$$V_g = m_g v_g = 1550 \text{ g} \times 127.4 \text{ cm}^3/\text{g} = 197{,}470 \text{ cm}^3 = 0.1975 \text{ m}^3$$

Thus the vapor fills nearly all the volume of the vessel.

**Example 4-4** Five pounds of water substance at 400°F are contained in a 7.0-ft³ vessel. Determine (a) the pressure, (b) the enthalpy in Btu/lb, and (c) the mass and volume of the vapor within the vessel.

SOLUTION From the given data on the total volume and mass we find the specific volume to be 7/5 or 1.4 ft³/lb. From Table A-12 at 400°F it is found that $v$ is greater than $v_f$ of 0.01864 ft³/lb but less than $v_g$ of 1.866 ft³/lb. Hence, the water must be a liquid-vapor mixture. Thus, the pressure must be the corresponding saturation vapor pressure at 400°F. This is 247.1 psia, which is found in column two of Table A-12.

(b) The specific enthalpy of the mixture may be found from a form of Eq. (4-3), namely,

$$h_x = h_f + xh_{fg} = h_f + x(h_g - h_f)$$

The quantities $h_f$ and $h_{fg}$ may be found from the saturation tables, but the quality $x$ is unknown. This latter value is determined from the value of $v_x$, which we found to be 1.4 ft³/lb. We may write Eq. (4-3) as

$$v_x = v_f + x(v_g - v_f)$$

Substitution of the values of $v_f$ and $v_g$ at 400°F from Table A-12 leads to

$$1.4 = 0.01864 + x(1.866 - 0.01864)$$

$$x = \frac{1.38}{1.847} = 0.747$$

With a knowledge of the quality, the specific enthalpy is now found to be

$$h_x = 375.1 + 0.747(826.8) = 992.7 \text{ Btu/lb}$$

(c) The mass of the vapor, based on the quality, is simply $0.747(5) = 3.735$ lb. Hence the volume occupied by the vapor is

$$V_g = m_g v_g = 3.735 \text{ lb} \times 1.866 \text{ ft}^3/\text{lb} = 6.97 \text{ ft}^3$$

Thus the vapor fills nearly all the volume of the vessel.

**Example 4-5M** Three kilograms of saturated liquid water is contained in a constant-pressure system at 5 bars. Energy is added to the fluid until it has a quality of 60 percent. Determine (a) the initial temperature, (b) the final pressure and temperature, and (c) the volume change and the enthalpy change of the water substance.

SOLUTION (a) In a saturation state there is a unique temperature for a given pressure. From the saturation pressure table, Table A-13M, for water we find the saturation temperature corresponding to 5 bars to be 151.9°C.

(b) Since the fluid is not completely vaporized, the pressure and temperature remain equal to their initial values of 5 bars and 151.9°C.

(c) The volume and enthalpy changes are computed from the relations $\Delta V = m(v_2 - v_1)$ and $\Delta H = m(h_2 - h_1)$, where 1 and 2 represent the initial and final states. The initial specific volume and specific enthalpy are read directly from Table A-13M, in terms of $v_f$ and $h_f$. These values are

$$v_1 = v_f = 1.0926 \text{ cm}^3/\text{g} \qquad \text{and} \qquad h_1 = h_f = 640.23 \text{ kJ/kg}$$

The values of $v_2$ and $h_2$ must be calculated on a basis of a liquid-vapor mixture of 60 percent quality. Hence

$$v_2 = v_f + xv_{fg} = 1.09 + 0.60(374.9 - 1.1) = 225.4 \text{ cm}^3/\text{g}$$

$$h_2 = h_f + xh_{fg} = 640.2 + 0.60(2108.5) = 1905 \text{ kJ/kg}$$

Consequently,

$$\Delta V = 3000 \text{ g} \times (225.4 - 1.1) \text{ cm}^3/\text{g} = 672,900 \text{ cm}^3 = 0.673 \text{ m}^3$$

$$\Delta H = 3.0 \text{ kg} \times (1905 - 640) \text{ kJ/kg} = 3800 \text{ kJ}$$

The enthalpy change in this case is equal to the energy added to the system.

**Example 4-5** Three pounds of saturated liquid water are contained in a constant-pressure system at 30 psia. Energy is added to the fluid until it has a quality of 70 percent. Determine (a) the initial temperature, (b) the final pressure and temperature, and (c) the volume change and the enthalpy change of the water substance.

SOLUTION (a) In a saturation state there is only one saturation temperature for a given pressure. From the saturation pressure table, Table A-13, for water the saturation temperature corresponding to 30 psia is 250.34°F.

(b) Since the fluid is not completely vaporized, the pressure and temperature remain equal to the initial values of 30 psia and 250.34°F.

(c) The volume and enthalpy changes are computed from the basic relations, $\Delta V = m(v_2 - v_1)$ and $\Delta H = m(h_2 - h_1)$, where 1 and 2 represent the initial and final states. The initial specific volume and specific enthalpy are read directly from Table A-13, in terms of $v_f$ and $h_f$. These values are

$$v_1 = v_f = 0.017 \text{ ft}^3/\text{lb} \qquad \text{and} \qquad h_1 = h_f = 218.93 \text{ Btu/lb}$$

The values of $v_2$ and $h_2$ must be calculated on a basis of a liquid-vapor mixture of 70 percent quality. Hence

$$v_2 = v_f + xv_{fg} = 0.017 + 0.70(13.75 - 0.017) = 9.63 \text{ ft}^3/\text{lb}$$

$$h_2 = h_f + xh_{fg} = 218.9 + 0.70(945.4) = 880.7 \text{ Btu/lb}$$

Consequently,

$$\Delta V = 3 \text{ lb} \times (9.63 - 0.02) \text{ ft}^3/\text{lb} = 28.8 \text{ ft}^3$$

$$\Delta H = 3 \text{ lb} \times (880.7 - 218.9) \text{ Btu/lb} = 1985 \text{ Btu}$$

The enthalpy change in this case is equal to the energy added to the system during the process.

When searching the literature for property values of a substance, one usually finds both saturation and superheat tables. On the basis of some given property data, it frequently is difficult for those unfamiliar with the tables to know which of these two tables contains the desired information. For example, if the enthalpy and temperature of refrigerant 12 are 212.25 kJ/kg and 40°C, what is the corresponding pressure? When temperature (or pressure) and another property value (such as $v$, $u$, $h$, or $s$) are given, the best method usually is to check in the saturation tables first. By using Table A-16M (saturation-temperature table for refrigerant 12) it is found that $h_f = 74.59$ kJ/kg and $h_g = 203.20$ kJ/kg at 40°C. If the given value had been between 74.59 and 203.2, the substance would be a two-phase mixture, and its quality could be calculated for the saturation pressure of 9.6066 bars. Since the actual value of $h$ is 212.25 kJ/kg, which is greater than $h_g$, the state we seek must be in the superheat region. Now using Table A-18M, we find that the pressure corresponding to the given $h$ and $T$ values is 4.0 bars. (The reader should confirm this value of the pressure.) As a general rule, examination of data in the saturation tables first will shorten the time required to find other property values if the condition of the substance is not known.

### c Compressed- or subcooled-liquid table

There is not a great deal of tabular data for compressed or subcooled liquids in the literature. However, since water is used as the fluid in fossil-fuel power plants, considerable data are available for this substance in the liquid region. The data shown in Table 4-6 are taken from more extensive compilations found in Tables A-15M and A-15. The first row in each set of data is the saturated-liquid data at that temperature. The variation of properties of a compressed liquid with pressure is seen to be slight. By using data for water in the saturated-liquid state of 80°C and 0.4739 bar, for example, as an approximation for the compressed-liquid state of 80°C and 100 bars, errors of 0.45, 0.68, 2.3, and 0.61 percent are made in the values of $v$, $u$, $h$, and $s$, respectively.

When compressed-liquid data based on experimental values are available, they should be used in engineering calculations. However, in the frequent absence of such data the above comparison indicates a general approximation rule. Compressed-liquid data in most cases can be approximated closely by using the

### Table 4-6 Properties of compressed liquid water

A. Metric: Data at 80°C

| P, bars | $v$, cm³/g | $u$, kJ/kg | $h$, kJ/kg | $s$, kJ/(kg)(°K) |
|---|---|---|---|---|
| 0.474 (sat.) | 1.0291 | 334.86 | 334.91 | 1.0753 |
| 50 | 1.0268 | 333.72 | 338.85 | 1.0720 |
| 100 | 1.0245 | 332.59 | 342.83 | 1.0688 |

B. USCS: Data at 150°F

| P, psia | $v$, ft³/lb | $u$, Btu/lb | $h$, Btu/lb | $s$, Bru/(lb)(°R) |
|---|---|---|---|---|
| 3.722 (sat.) | 0.01634 | 117.95 | 117.96 | 0.2150 |
| 500 | 0.01632 | 117.66 | 119.17 | 0.2146 |
| 1000 | 0.01629 | 117.38 | 120.40 | 0.2141 |

*Source:* Abstracted from Tables A-15M and A-15.

property values of the saturated liquid state at the given *temperature*. This simply implies that compressed-liquid data are more temperature-dependent than pressure-dependent. If we designate a compressed-liquid state by the subscript $c$, then in general $y_c \approx y_f$ at the given temperature, where $y$ is any specific property.

**Example 4-6M** Find the change in the specific internal energy of water for a change of state from 40°C, 25 bars, to 80°C, 75 bars by (a) use of the compressed-liquid table, and (b) use of the approximation rule using saturated data.

SOLUTION (a) The data for compressed liquid water appear in Table A-15M. From this table we determine that

$$u_2 - u_1 = 333.15 - 167.25 = 165.90 \text{ kJ/kg}$$

(b) As an approximation method, the values of $u_f$ are used at the specified temperatures, and the pressure data are ignored. Use of Table A-12M leads to

$$u_2 - u_1 = 334.86 - 167.56 = 167.30 \text{ kJ/kg}$$

In the absence of a compressed-liquid table the use of a saturation-temperature table produces an answer which is 0.84 percent in error.

**Example 4-6** Find the change in the specific enthalpy of water for a change of state from 50°F, 500 psia, to 100°F, 1500 psia, by (a) use of the compressed-liquid table, and (b) use of the approximation rule using saturated data.

SOLUTION (a) The data for compressed liquid water appear in Table A-15. From this table we determine that

$$h_2 - h_1 = 71.99 - 19.50 = 52.49 \text{ Btu/lb}$$

(b) As an approximation method, the values of $h_f$ are used at the specified temperatures, and the pressure data are ignored. Use of Table A-12 leads to

$$h_2 - h_1 - 68.05 - 18.06 = 49.99 \text{ Btu/lb}$$

In the absence of a compressed-liquid table the use of a saturation-temperature table produces an answer which is 4.76 percent in error. A much smaller percent error would occur for changes in $v$, $u$, and $s$ for the same initial and final states.

## d Representations of $Pv$ and $PT$

In Sec. 3-7 the use of diagrams as an aid in problem solving was emphasized. As we supply tabular data for the superheat, compressed-liquid, and saturation regions to problems involving the conservation of energy principle, it will be helpful to use $Pv$ and $PT$ diagrams in the overall analysis. The following example lists some processes which are then plotted on these two diagrams. In some cases it is necessary to examine tabular data in order to ascertain the position of the end states. Further problems of this type appear at the end of the chapter.

**Example 4-7M** Plot the following processes on $Pv$ and $PT$ diagrams.
 (a) Superheated vapor is cooled at constant pressure until liquid just begins to form
 (b) A liquid-vapor mixture with a quality of 60 percent is heated at constant volume until its quality is 100 percent.

(c) A liquid-vapor mixture of water with a quality of 50 percent is heated at a constant temperature of 200°C until its volume is 4.67 times the initial volume.

(d) Refrigerant 12 at 8 bars is a saturated liquid. It is heated at constant pressure until its enthalpy has increased by a factor of 2.

SOLUTION The $Pv$ and $PT$ diagrams for processes (a) and (b) are fairly self-evident, as shown on the accompanying figure. For process (c) the final state is not known. It is fixed, however, by the temperature and the specific volume. For the initial state,

$$v_1 = v_f + xv_{fg} = 1.16 + 0.5(127.4 - 1.16) = 64.3 \text{ cm}^3/\text{g}$$

Since the final volume is 4.70 times the initial value,

$$v_2 = 4.67(64.3) = 300.3 \text{ cm}^3/\text{g}$$

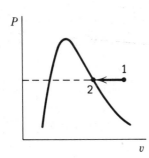
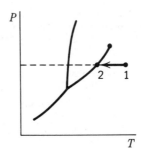

(a)

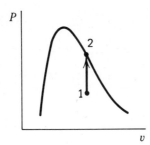
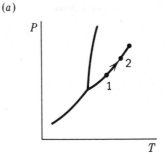

(b)

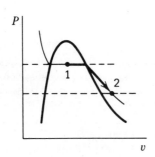
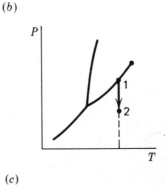

(c)

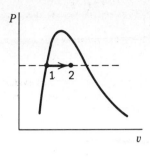
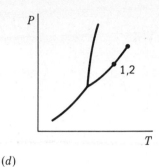

(d)

At 200°C the value of $v_g$ is 127.4 cm³/g. Thus the final state is located far into the superheat region at 200°C. From Table A-14M it is found that $v = 299.9$ cm³/g at 200°C and 7.0 bars. Hence the final pressure is close to 7.0 bars, while the initial saturation pressure is 15.54 bars. The final state of process (d) is determined by the enthalpy at 8 bars. The initial enthalpy, from Table A-17M, is found to be 67.30 kJ/kg. The final enthalpy then is twice this, or $h_2 = 134.6$ kJ/kg. This value is less than $h_g$ at 8 bars; therefore the final state is a wet mixture. To find the final quality,

$$h_2 = 134.6 = h_f + xh_{fg} = 67.3 + x(133.33)$$

$$x = \frac{67.3}{133.33} = 0.505$$

Hence the final state of process (d) lies roughly halfway across the wet region at 8 bars.

**Example 4-7** Plot the following processes on $Pv$ and $PT$ diagrams.
(a) Superheated vapor is cooled at constant pressure until liquid just begins to form.
(b) A liquid-vapor mixture with a quality of 60 percent is heated at constant volume until its quality is 100 percent.
(c) A liquid-vapor mixture of water with a quality of 50 percent is heated at a constant temperature of 400°F until its volume is 5.09 times the initial value.
(d) Refrigerant 12 at 100 psia is a saturated liquid. It is heated at constant pressure until its enthalpy has increased by a factor of 2.

SOLUTION The $Pv$ and $PT$ diagrams for processes (a) and (b) are fairly self-evident, as shown on the accompanying figure. For process (c) the final state is not known. It is fixed, however, by the temperature and the specific volume. For the initial state,

$$v_1 = v_f + xv_{fg} = 0.019 + 0.5(1.143) = 0.591 \text{ ft}^3/\text{lb}$$

Since the final volume is 5.09 times the initial value,

$$v_2 = 5.09(0.591) = 3.008 \text{ ft}^3/\text{lb}$$

At 400°F the value of $v_g$ is 1.162 ft³/lb. Thus the final state is located in the superheat region. From Table A-14 it is found that $v = 3.007$ ft³/lb at 400°F and 160 psia, which is the final state. The final state of process (d) is determined by the final enthalpy at 100 psia. The initial enthalpy, from Table A-17, is 26.542 Btu/lb. Since this value is less than $h_g$ at 100 psia, the final state is a wet mixture. To find the final quality,

$$h_2 = 2(26.542) = h_f + xh_{fg} = 26.542 + x(58.809)$$

$$x_2 = \frac{26.542}{58.809} = 0.451$$

Hence the final state of process (d) lies roughly halfway across the wet region at 100 psia.

## 4-5 TABULAR DATA AND CLOSED-SYSTEM ENERGY ANALYSIS

The preceding section described the format used in the general literature for presenting basic property data in tabular form for simple, compressible substances. This discussion covered the superheat, compressed-liquid, and saturation states. We are now in a position to employ these tables in the solution of problems involving the conservation of energy principle for closed systems. Before proceeding, the reader may wish to review the major points introduced in Sec. 3-7 with respect to problem-solving techniques. Among these techniques is the use of property diagrams, which were also discussed in the preceding section. The following examples illustrate the general approach, and it is important that the reader follow closely the development of each solution.

### a Use of Superheat Data

**Example 4-8M** One kilogram of water substance is maintained in a weighted piston-cylinder assembly at 30 bars and 240°C. The substance is slowly heated at constant pressure until the temperature reaches 320°C. Determine (a) the work required to raise the weighted piston, and (b) the required heat input, in kJ/kg.

SOLUTION The system is the mass of water within the assembly. The initial and final states are determined by first checking the saturation table, A-13M. At 30-bar pressure the saturation temperature is 233.9°C. Since the system temperature is always above this value, the fluid is superheated vapor throughout. However, the initial state lies close to the saturation line. A $Pv$ diagram is shown as Fig. 4-8a, which illustrates the path of the quasistatic heat-addition process.

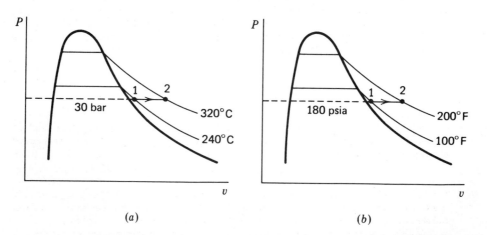

(a)                                        (b)

**Figure 4-8** The $Pv$ diagrams for processes discussed in (a) Example 4-8M, and (b) Example 4-8.

(a) For a slow heat addition, the work during the quasistatic process is given by the integral of $-P\,dv$. Because the pressure is constant, the work effect is given simply by $-P(v_2 - v_1)$. Taking the specific volume values from the superheat table, A-14M, we find that

$$w = -30 \text{ bar} \times (85.0 - 68.2) \text{ cm}^3/\text{g} \times 10^{-1} \text{ N·m/cm}^3\text{·bar}$$

$$= -50.4 \text{ N·m/g} = -50.4 \text{ kJ/kg}$$

The negative sign indicates that work is done by the system on the surroundings.

(b) The heat transfer is given by the relation, $q = \Delta u - w$. Since $w = -P\,\Delta v$, the heat transfer for this quasistatic, constant-pressure process is also given by $q = \Delta h$. Again making use of Table A-14M, the heat transfer is

$$q = h_2 - h_1 = 3043.4 - 2824.3 = 219.1 \text{ kJ/kg}$$

The positive value indicates that heat is transferred into the system. It would be equally correct in this case to evaluate $\Delta u$ from the tables, and subtract from it the value of $w$ found in part a. If this is done, $q = u_2 - u_1 - w = 2788.4 - 2619.7 - (-50.4) = 219.1$ kJ/kg, as before.

If we assume that water vapor is an ideal gas under these conditions, then we can use $c_{p,0}$ data for the given temperature range to calculate $\Delta h$. An average $c_{p,0}$ value for water between 240 and 320°C is approximately 1.984 kJ/(kg)(°C). (The literature source for this value is the *JANAF Thermochemical Tables*.) Hence, for an ideal gas,

$$q = \Delta h = c_{p,0}(T_2 - T_1) = 1.984(80) = 158.7 \text{ kJ/kg}$$

This value constitutes an error of over 27 percent when compared to the result based on tabular (experimental) data. Thus, great care must be taken when analyzing gaseous systems because in many cases the assumption of an ideal gas leads to considerable error.

**Example 4-8** One pound of water substance is maintained in a weighted piston-cylinder assembly at 450 psia and 500°F. The substance is slowly heated at constant pressure until the temperature reaches 600°F. Determine (a) the work required to raise the weighted piston, and (b) the required heat input, in Btu/lb.

SOLUTION The system is the mass of water within the assembly. The initial and final states are determined by first checking the saturation table, A-13. At 450 psia the saturation temperature is 456.4°F. Since the system temperature is always above this value, the fluid is superheated vapor throughout. However, the initial state lies close to the saturation line. A $Pv$ diagram is shown as Fig. 4-8b, which illustrates the path of the quasistatic heat-addition process.

(a) For a slow heat addition, the work during the quasistatic process is given by the integral of $-P\,dv$. Because the pressure is constant, the work interaction is given simply by $-P(v_2 - v_1)$. Taking the specific volume values from the superheat table, A-14, we find that

$$w = -450(144) \text{ lb}_f/\text{ft}^2 \times (1.300 - 1.123) \text{ ft}^3/\text{lb}_m$$

$$= -11,470 \text{ ft·lb}_f/\text{lb}_m$$

The negative sign indicates that work is done by the system on the surroundings.

(b) The heat transfer is given by the relation, $q = \Delta u - w$. Since $w = -P\,\Delta v$, the heat transfer for this quasistatic, constant-pressure process is also given by $q = \Delta h$. Again making use of Table A-14, the heat transfer is

$$q = h_2 - h_1 = 1302.5 - 1238.5 = 64.0 \text{ Btu/lb}$$

The positive value indicates that heat is transferred into the system. It would be equally correct in this case to evaluate $\Delta u$ from the tables, and subtract from it the value of $w$ found in part a. If this is done, $q = u_2 - u_1 - w = 1194.3 - 1145.1 - (-11,470/778) = 64.0$ Btu/lb, as before.

If we assume that water vapor is an ideal gas under these conditions, then we can use $c_{p,0}$ data for the given temperature range to calculate $\Delta h$. An average $c_{p,0}$ value for water between 500 and 600°F is approximately 0.476 Btu/(lb)(°F), based on the *JANAF Thermochemical Tables.* Hence, for an ideal gas,

$$q = \Delta h = c_{p,0} \, \Delta T = 0.476(100) = 47.6 \text{ Btu/lb}$$

This value constitutes an error of over 25 percent when compared to the result based on tabular (experimental) data. Thus, great care must be taken when analyzing gaseous systems because in many cases the assumption of an ideal gas leads to considerable error.

**Example 4-9M** Refrigerant 12 is contained within a rigid tank at an initial state of 2.8 bars and 100°C. Heat is removed until the pressure drops to 2.4 bars. During the process a paddle-wheel within the tank is turned with a constant torque of 6 N·m for 30 revolutions. If the system contains 0.1 kg, compute the required heat transfer in kJ.

SOLUTION Considering the refrigerant within the tank as the system, we write the energy balance in the form

$$q + w = \Delta u = u_f - u_i$$

where $f$ and $i$ stand for the final and initial states. In the absence of boundary work (because of the rigid tank),

$$q = u_f - u_i - w_{\text{paddle}}$$

The saturation temperature at 2.8 bars, from Table A-17M, is $-2.93°C$. Hence the initial state of 100°C is well into the superheat region. From the superheat table, A-18M, we find that

$$u_i = 228.89 \text{ kJ/kg} \qquad \text{and} \qquad v_i = 89.24 \text{ cm}^3/\text{g}$$

The value of $v_i$ is extremely important since, for a rigid system, $v_f = v_i$. Consequently the final state has a pressure of 2.4 bars and a specific volume of 89.24 cm³/g. At 2.4 bars, from the saturated data of Table A-17M, we find that $v_g$ is only 60.76 cm³/g. Thus the final state is also in the superheat region. Looking up the 2.4-bar data in Table A-18M, one sees that a $v$ value of 89.24 cm³/g corresponds very closely to a state of 50°C. Therefore, $u_f = 199.51 \text{ kJ/kg}$. The $Pv$ diagram shows the vertical path above the saturated vapor line (see Fig. 4-9a).

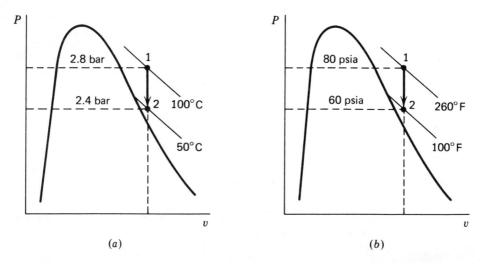

(a)     (b)

**Figure 4-9** The $Pv$ diagrams for processes discussed in (a) Example 4-9M, and (b) Example 4-9.

To complete the analysis we need a value for the paddle-wheel work. Based on the discussion in Chap. 2, rotational mechanical work is given by $\tau\theta$ when $\tau$ is a constant; $\theta$ must be expressed in radians. Hence,

$$W_{paddle} = 6 \text{ N·m} \times 30(2\pi) = 1131 \text{ N·m}$$

Substitution of the proper values into the energy equation yields

$$q = (199.51 - 228.89) \text{ kJ/kg} + \frac{-1131 \text{ N·m}}{100 \text{ g}}$$

$$= -29.38 - 11.31 = -40.69 \text{ kJ/kg}$$

Finally,

$$Q = mq = 0.1 \text{ kg} \times -40.69 \text{ kJ/kg} = -4.07 \text{ kJ}$$

The total heat interaction must account not only for a decrease in the internal energy of the system, but also for the energy added in the form of paddle-wheel work.

**Example 4-9** Refrigerant 12 is contained within a rigid tank at an initial state of 80 psia and 260°F. Heat is removed until the pressure drops to 60 psia. During the process a paddle-wheel within the tank is turned with a constant torque of 3.0 $\text{lb}_f$·ft for 600 revolutions. If the system contains 1.2 lb, compute the required heat transfer in Btu.

SOLUTION Considering the refrigerant within the tank as the system, we write the energy balance in the form

$$q + w - u = u_f - u_i$$

where $f$ and $i$ stand for the final and initial states. In the absence of boundary work (because of the rigid tank),

$$q = u_f - u_i - w_{paddle}$$

The saturation temperature at 80 psia, from Table A-17, is 66.21°F. Hence the initial state of 260°F is well into the superheat region. From the superheat table, A-18, we find that

$$u_i = 104.919 \qquad \text{and} \qquad v_i = 0.7654 \text{ ft}^3/\text{lb}$$

The value of $v_i$ is extremely important since, for a rigid system, $v_f = v_i$. Consequently the final state has a pressure of 60 psia and a specific volume of 0.7654 ft³/lb. At 60 psia, from the saturated data of Table A-17, we find that $v_g$ is only 0.6701 ft³/lb. Thus the final state also is in the superheat region. Looking up the 60-psia data in Table A-18, one sees that a $v$ value of 0.7654 ft³/lb corresponds very closely to a state of 100°F. Therefore, $u_f = 82.024$ Btu/lb. The $Pv$ diagram shows the vertical path above the saturated-vapor line (see Fig. 4-9b).

To complete the analysis we need a value for the paddle-wheel work. Based on the discussion in Chap. 2, rotational mechanical work is given by $\tau\theta$ when $\tau$ is a constant; $\theta$ must be expressed in radians. Hence,

$$W_{paddle} = 3 \text{ lb}_f\text{·ft} \times 600(2\pi) = 11,310 \text{ ft·lb}_f$$

Substitution of the proper values into the energy equation yields

$$q = (82.024 - 104.919) \text{ Btu/lb} - \frac{(11,310/778) \text{ Btu}}{1.2 \text{ lb}}$$

$$= -22.895 - 12.1 = -35.0 \text{ Btu/lb}$$

Finally,

$$Q = mq = 1.2 \text{ lb} \times -35.0 \text{ Btu/lb} = -42.0 \text{ Btu}$$

## b Use of Saturation data

**Example 4-10M** Refrigerant 12 is condensed in a closed system from an initial state of 6 bars and 60°C to a final state of saturated liquid at the same pressure.

(a) Determine the value of the work interaction in N·m/g.
(b) Compute the value of the heat interaction in kJ/kg.

SOLUTION (a) A sketch of the equipment and a $Pv$ process diagram are shown in Fig. 4-10. A piston-cylinder assembly is used to keep the fluid at a constant pressure at 6 bars. At 6 bars the saturation temperature is 22.0°C (Table A-17M). Therefore the initial state of 60°C is a super-

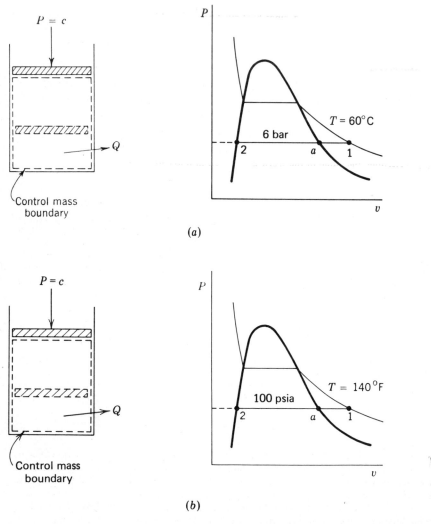

(a)

(b)

**Figure 4-10** Schematic sketches and $Pv$ diagrams for processes represented in Examples 4-10M and 4-10.

heated one, as shown on the $Pv$ diagram. The only work interaction is boundary work, and it is given by the integral of $-P\,dv$ if we assume that the process is quasistatic. In addition, the pressure is held constant. Consequently a knowledge of the initial and final specific volumes is sufficient to evaluate $w$. From the superheat table (A-18M), $v_1 = 34.89$ cm³/g, and from the pressure-saturation table (A-17M), $v_2 = v_f = 0.7566$ cm³/g at 6 bars. Hence,

$$w = -P(v_2 - v_1) = -6 \text{ bars} \times (0.7566 - 34.89) \text{ cm}^3/\text{g} \times 10^{-1} \text{ N·m/cm}^3 \text{·bar} = 20.5 \text{ N·m/g}$$

(b) For a simple, compressible substance the conservation of energy principle reduces to $q + w = \Delta u$. The value of $q$ can be found directly from this equation because the internal-energy change is fixed by the end states and these are known. From Table A-18M, $u_1 = 202.34$ kJ/kg and from Table A-17M, $u_2 = u_f = 56.35$ kJ/kg. Therefore,

$$q = u_2 - u_1 - w = 56.35 - 202.34 - 20.5 = -166.5 \text{ kJ/kg}$$

The answer is negative, as expected, since heat must be removed in order to condense the vapor.

Since the process is at constant pressure with boundary work and assumed quasistatic, the energy equation can be written as $q = \Delta h$. By recognizing this special form of the energy balance one does not need to calculate $w$ and $\Delta u$ separately. For this particular problem, $q = h_2 - h_1 = 56.80 - 223.27 = -166.5$ kJ/kg, as before. It might be pointed out that the heat removed between states $a$ and 2 shown on the $Pv$ diagram is equal to $-h_{fg}$, the negative of the enthalpy of vaporization. Since our energy equation shows that $\Delta h = q$ for a liquid-vapor phase change, it is easy to see why $h_{fg}$ is also called the latent "heat" of vaporization.

**Example 4-10** Refrigerant 12 is condensed in a closed system from an initial state of 100 psia and 140°F to a final state of saturated liquid at the same pressure.
 (a) Determine the value of the work interaction in ft·lb$_f$/lb$_m$.
 (b) Compute the value of the heat interaction in Btu/lb$_m$.

SOLUTION (a) A sketch of the equipment and a process diagram are shown in Fig. 4-10. A piston-cylinder assembly is used to keep the system at a constant pressure of 100 psia. At 100 psia the saturation temperature of refrigerant 12 is 80.76°F (Table A-17). Therefore the initial state of 100°F is a superheated one, as shown on the $Pv$ diagram. The only work interaction is boundary work, and it is given by the integral of $-P\,dV$ if we assume that the process is quasistatic. In addition, the pressure is held constant. Consequently a knowledge of the initial and final specific volumes is sufficient to evaluate $w$. From the superheat table (A-18), $v_1 = 0.4788$ ft³/lb, and from the pressure saturation table (A-17), $v_2 = v_f = 0.0123$ ft³/lb at 100 psia. Hence,

$$w = -P(v_2 - v_1) = -100(144) \text{ lb}_f/\text{ft}^2 \times (0.0123 - 0.4788) \text{ ft}^3/\text{lb}_m$$

$$= 6,710 \text{ ft·lb}_f/\text{lb}_m$$

(b) For a simple, compressible substance the conservation of energy principle reduces to $q + w = \Delta u$. The value of $q$ can be found directly from this equation because the internal-energy change is fixed by the end states and these are known. From Table A-18, $u_1 = 86.647$ Btu/lb and from Table A-17, $u_2 = u_f = 26.31$ Btu/lb. Therefore

$$q = u_2 - u_1 - w = 26.31 - 86.65 - \frac{6,710}{778} = -69.0 \text{ Btu/lb}$$

The answer is negative, as expected, since heat must be removed in order to condense the vapor.

Since the process is constant pressure with boundary work and assumed quasistatic, the energy equation can be written as $q = \Delta h$. (See Sec. 3-3.) By recognizing this special form of the energy balance one does not need to calculate $w$ and $\Delta u$ separately. For this particular problem, $q = h_2 - h_1 = 26.54 - 95.507 = -69.0$ Btu/lb, as before. It might be pointed out that the heat removed between states $a$ and 2 shown on the $Pv$ diagram is equal to $-h_{fg}$, the negative of the enthalpy of vaporization. Since our energy balance shows that $\Delta h = q$ for a liquid-vapor phase change, it is easy to see why $h_{fg}$ is also called the latent "heat" of vaporization.

**Example 4-11M** One-tenth of a kilogram of water at 3 bars and 76.3 percent quality is contained in a rigid tank which is thermally insulated. A paddle-wheel inside the tank is turned by an external motor until the substance is a saturated vapor. Determine the work necessary to complete the process, and the final pressure and temperature of the water.

SOLUTION The boundaries of the control mass are chosen to lie just inside the tank. The fluid is assumed to be a simple, compressible substance, and during the process the water is in fluid shear and is not in equilibrium. However, for this process a knowledge of the end states only is sufficient. For this system the energy equation again is

$$Q + W = \Delta U$$

No heat transfer or boundary work occurs, because of the restrictions on the system (insulated and rigid). Work is done on the system, though, by the action of the paddle-wheel, so that the above equation reduces to

$$W_{paddle} = \Delta U = U_2 - U_1 = m(u_2 - u_1)$$

The work requirements are determined, then, from an evaluation of the initial and final specific internal energies. The initial value of $u$ is found from data in the saturation-pressure table, A-13M, in conjunction with the equations for intensive properties expressed in terms of the quality. That is, for 3 bars and $x = 0.763$, we find that

$$u_1 = u_f(1 - x) + u_g x = 561.2(0.237) + 2543.6(0.763)$$

$$= 2074 \text{ kJ/kg}$$

To find the final state one recognizes that the final specific volume is the same as the initial value, since the tank is rigid and closed. The initial specific volume is

$$v_1 = v_f(1 - x) + v_g x = 1.073(0.237) + 605.8(0.763)$$

$$= 462.5 \text{ cm}^3/\text{g}$$

Thus the final state is a saturated vapor with a specific volume of 462.5 cm³/g. From the accompanying $Pv$ diagram (see Fig. 4-11) it is noted that in order to proceed at constant $v$ from the wet region at 3 bars and 76.3 percent quality to the state of a saturated vapor, the final pressure must be greater than 3 bars. From the saturation-pressure table, A-13M, for steam it is found that a value of $v_g = 462.5$ cm³/g corresponds to a pressure of 4 bars and a temperature of 143.6°C.

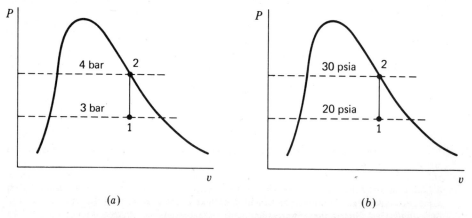

(a)                                                        (b)

**Figure 4-11** The $Pv$ diagrams for processes discussed in (a) Example 4-11M, and (b) Example 4-11.

At this final state the specific internal energy is 2553.6 kJ/kg. The paddle-wheel work that must be supplied, then, is

$$W_{paddle} = m(u_2 - u_1) = 0.1 \text{ kg} \times (2553.6 - 2074) \text{ kJ/kg}$$

$$= 48.0 \text{ kJ}$$

The value of the paddle-wheel work is positive, since the work is done on the system by the rotating shaft.

**Example 4-11** One-tenth of a pound of steam at 20 psia and 68 percent quality is contained in a rigid tank which is thermally insulated. A paddle-wheel inside the tank is turned by an external motor until the substance is a saturated vapor. Determine the work necessary to complete the process and the final pressure and temperature of the steam.

SOLUTION The boundaries of the control mass are chosen to lie just inside the tank. In the initial and final equilibrium states the system can be assumed to be a simple system. During the process the water is in fluid shear and is not in equilibrium. However, we shall find that, for this process, a knowledge of the end states only is sufficient. For a simple, compressible system the energy equation is again

$$Q + W = \Delta U$$

No heat transfer or boundary work occurs, because of the restrictions on the system (insulated and rigid). Work is done on the system, though, by the action of the paddle-wheel, so that the above equation reduces to

$$W_{paddle} = \Delta U = U_2 - U_1 = m(u_2 - u_1)$$

The work requirements are determined, then, from an evaluation of the initial and final sensible internal energies. The initial value of $u$ is found by employing data from the steam table A-13, in conjunction with the equations developed earlier for intensive properties expressed in terms of the quality. That is, for 20 psia and $x = 0.68$, we find that

$$u_x = u_f(1 - x) + u_g x = 196.2(0.32) + (1082.0)(0.68) = 798.2 \text{ Btu/lb}$$

Since the tank is rigid and closed, the final specific volume is the same as the initial value. The initial specific volume is

$$v_1 = v_x = v_f(1 - x) + u_g x = 0.017(0.32) + 20.09(0.68) = 13.67 \text{ cu ft/lb}$$

Thus the final state is a state of saturated vapor where the specific volume is 13.67 ft³/lb. From the accompanying $Pv$ diagram (Fig. 4-11) it is noted that, in order to proceed from the wet region at 20 psia and 68 percent quality to the state of a saturated vapor, the final pressure must be above 20 psia. From the saturation table for steam. Table A-13, it is found that the value of $v_g$ corresponds closely to this at a pressure of 30 psia and a temperature of 250°F. At this state the sensible internal energy is 1088.0 Btu/lb. The paddle-wheel work that must be supplied, then, is

$$W_{paddle} = (1088.0 - 798.2)(0.1)$$

$$= 29.0 \text{ Btu}$$

$$= 22,600 \text{ ft·lb}_f$$

The value of the paddle-wheel work must be positive because of our choice of a sign convention for work interactions.

**Example 4-12M** A piston-cylinder assembly with an initial volume of 0.01 m³ is filled with saturated refrigerant 12 vapor at 16°C. The substance is compressed until a state of 9 bars and 60°C is reached. During the compression process the heat loss amounts to 0.4 kJ. Compute the boundary work required in kJ.

SOLUTION The conservation of energy principle for the simple, compressible substance can be written in the form

$$W = \Delta U - Q = m(u_2 - u_1) - Q$$

The mass of the system is found from the basic relation, $m = V/v$, when evaluated at the initial state. From Table A-16M we find that $v_1 = v_g = 34.42$ cm$^3$/g at 16°C. Therefore

$$m = \frac{V}{v} = \frac{0.01 \text{ m}^3}{34.42 \text{ cm}^3/\text{g}} \times \frac{10^6 \text{ cm}^3}{\text{m}^3} = 291 \text{ g} = 0.291 \text{ kg}$$

The value of $u_1$ is $u_g$ at 16°C, or 176.78 kJ/kg. The saturation temperature at 9 bars, from Table A-17M, is only 37.37°C. Thus at 9 bars and 60°C the substance is superheated. From Table A-18M we find that $u_2 = 199.56$ kJ/kg.

Substitution of the given and acquired data yields

$$W = m(u_2 - u_1) - Q = 0.291 \text{ kg} \times (199.56 - 176.78) \text{ kJ/kg} - 0.4 \text{ kJ}$$

$$= 6.63 - 0.4 = 6.23 \text{ kJ}$$

The heat loss in this case is roughly only 6 percent of the work required. The accompanying $Pv$ sketch of the process (Fig. 4-12) shows the decrease in volume as the vapor is compressed and becomes superheated. Since $P$ as a function of $v$ is not known, the work cannot be evaluated by the integral of $-P \, dV$.

**Example 4-12** A piston-cylinder assembly with an initial volume of 600 in$^3$ is filled with saturated refrigerant 12 vapor at 60°F. The substance is compressed until a state of 120 psia and 140°F is reached. During the compression process the heat loss amounts to 0.5 Btu. Compute the work required in ft·lb$_f$.

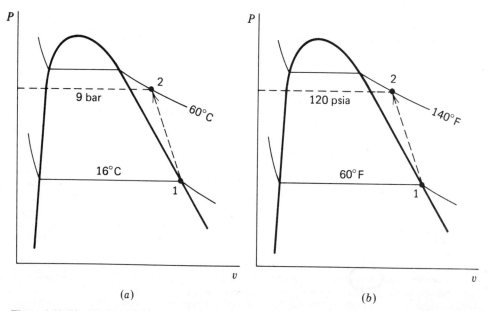

(a)

(b)

**Figure 4-12** The $Pv$ diagrams for processes discussed in (a) Example 4-12M, and (b) Example 4-12.

SOLUTION The conservation of energy principle for the simple, compressible substance can be written as

$$W = \Delta U - Q = m(u_2 - u_1) - Q$$

The mass of the system is found from the basic relation, $m = V/v$, when evaluated at the initial state. From Table A-16 we find that $v_1 = v_g = 0.5584$ ft$^3$/lb at 60°F. Therefore

$$m = \frac{V}{v} = \frac{600 \text{ in}^3}{0.5584 \text{ ft}^3/\text{lb}} \times \frac{1 \text{ ft}^3}{1278 \text{ in}^3} = 0.622 \text{ lb}_m$$

The value of $u_1$ is $u_g$ at 60°F, or 75.92 Btu/lb. The saturation temperature at 120 psia, from Table A-17, is only 93.29°F. Thus at 120 psia and 140°F the substance is superheated vapor. From Table A-18, then, we find that $u_2 = 86.098$ Btu/lb.
  Substitution of the given and acquired data yields

$$W = m(u_2 - u_1) - Q = 0.622 \text{ lb} \times (86.098 - 75.92) \text{ Btu/lb} - 0.5 \text{ Btu}$$

$$= 6.33 - 0.5 = 5.83 \text{ Btu} = 4540 \text{ ft·lb}_f$$

The heat loss in this case is roughly only 9 percent of the work required. The accompanying $Pv$ sketch of the process (Fig. 4-12) shows the decrease in volume as the vapor is compressed and becomes superheated. Since $P$ as a function of $v$ is not known, the work cannot be evaluated by the integral of $-P \, dV$.

## c Use of Compressed-liquid Data

**Example 4-13M** Liquid water is a compressed-liquid state of 75 bars and 40°C is heated quasistatically at constant pressure until it becomes a saturated liquid. Compute the heat input required in kJ/kg.

SOLUTION The path of the process is shown on the accompanying $Pv$ diagram (Fig. 4-13), where the region close to the saturated-liquid line has been greatly enlarged. At constant pressure the quasistatic work is given by $-P \, \Delta v$. Hence the heat input becomes

$$q = \Delta u - w = \Delta u + P \, \Delta v = \Delta h$$

The initial enthalpy at 75 bars and 40°C is found from Table A-15M to be 174.18 kJ/kg. In the final saturated-liquid state at 75 bars the temperature is 290.59°C and the enthalpy is 1292.2 kJ/kg (see Table A-15M). Consequently

$$q = h_2 - h_1 = 1292.2 - 174.2 = 1118.0 \text{ kJ/kg}$$

This large value for $q$ is due to the fact that the temperature changes by over 250°C during the process.

**Example 4-13** Liquid water in a compressed-liquid state of 1000 psia and 100°F is heated quasistatically at constant pressure until it becomes a saturated liquid. Compute the heat input required in Btu per pound.

SOLUTION The path of the process is shown on the accompanying $Pv$ diagram (Fig. 4-13), where the region close to the saturated-liquid line has been greatly enlarged. At constant pressure, the quasistatic work is given by $P \, \Delta v$. Hence the heat input becomes

$$q = \Delta u + w = \Delta u + P \, \Delta v = \Delta h$$

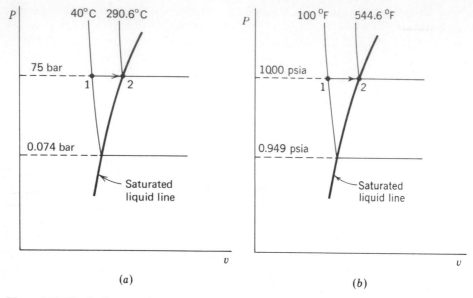

**Figure 4-13** The $Pv$ diagrams for processes discussed in ($a$) Example 4-13M, and ($b$) Example 4-13.

The initial enthalpy at 1000 psia and 100°F is found from Table A-15 to be 70.68 Btu/lb. In the final saturated-liquid state at 1000 psia the temperature is 544.61°F and the enthalpy is 542.4 Btu/lb. Consequently

$$q = 542.4 - 70.7 = 471.7 \text{ Btu/lb}$$

## 4-6 THE COMPRESSIBILITY FACTOR AND CORRESPONDING STATES

Unless the pressure is reasonably low and the temperature relatively high, gases do not exhibit a $PvT$ behavior which can be represented accurately by the ideal-gas equation, $Pv = RT$. One method of retaining an equation format for the $PvT$ relationship of non-ideal gases and still retain reasonable accuracy is through the use of a compressibility-factor correction. The departure from ideal-gas behavior may be characterized by a compressibility factor $Z$, which is defined as

$$Z \equiv \frac{Pv}{RT} \tag{4-4}$$

Since $RT/P$ is the ideal-gas specific volume $(v_{ideal})$ of a gas, the compressibility factor may be considered a measure of the ratio of the actual specific volume to the ideal-gas specific volume. That is, $Z = v_{actual}/v_{ideal}$. For the hypothetical ideal gas the compressibility factor is unity, but for actual gases it can be either less than or greater than unity. Hence, the compressibility factor measures the deviation of an actual gas from ideal-gas behavior.

The specific volume of any gas could be estimated at any desired pressure and temperature if the compressibility factor were known for the gas as a function of $P$ and $T$, since $v = ZRT/P$. This information is found through application of what is known as the *principle of corresponding states*. The principle predicts that the $Z$ factor for all gases is approximately the same when the gases have the same reduced pressure and temperature. The reduced pressure $P_R$ and reduced temperature $T_R$ are defined by

$$P_R \equiv \frac{P}{P_c} \quad \text{and} \quad T_R \equiv \frac{T}{T_c} \qquad (4\text{-}5)$$

Absolute pressures, and temperatures in kelvins or degrees Rankine, must be used in these equations. Thus the critical pressure and temperature of a substance are used to define a reduced state. The validity of such a principle must be based on experimental evidence. When reduced isotherms $T_R$ are plotted on a $ZP_R$ chart, the average deviation of experimental data for a number of gases is somewhat less than 5 percent. Figure 4-14 shows a correlation of actual data for 10 gases for a limited number of reduced isotherms. When the best curves are fitted to all the data, a more complete plot, such as Fig. 4-15, results. This generalized compressibility chart is for $P_R$ values from 0 to 1.0 and $T_R$ values from 0.60 to 5.0. Once the

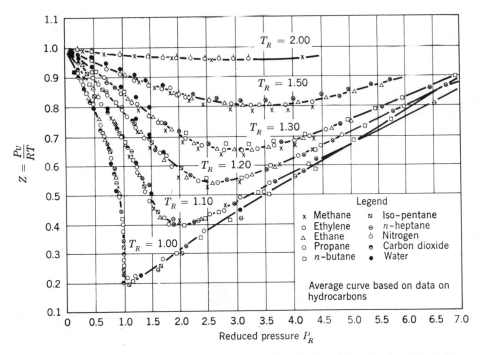

**Figure 4-14** Experimental data correlation for a generalized $Z$ chart. [Gour-Jen Su: Modified Law of Corresponding States, *Ind. Eng. Chem.* (Intern. Edition), **38**:803 (1946).]

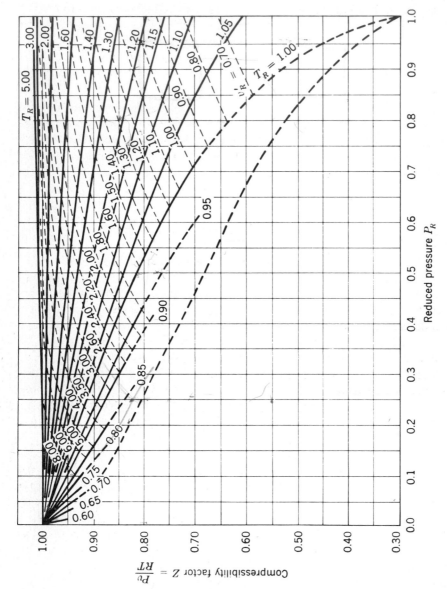

**Figure 4-15** Generalized compressibility chart [L. C. Nelson and E. F. Obert: Generalized Compressibility Charts, *Chem. Eng.*, **61**:203 (1954).]

144

chart has been drawn from data for a limited number of substances, it is assumed to be generally applicable to all gases. As we mentioned above, the generalized chart is only an approximation. However, it is remarkably good for many engineering design problems. The main virtue of the generalized compressibility chart is that it requires only a knowledge of critical pressures and temperatures to predict approximately the specific volume of a real gas. It must be emphasized that the generalized compressibility chart should not be used as a substitute for accurate experimental $PvT$ data. The major role of a generalized compressibility chart is to provide reasonable estimates of $PvT$ behavior in the absence of accurate measurements.

The compressibility factor can also be found when $vT$ data or $vP$ data are given. For correlation purposes it has been found best to use a pseudocritical volume in the definition of the reduced volume, rather than the actual critical volume. If we define a pseudocritical volume $v_c'$ by the quantity $RT_c/P_c$, then the pseudoreduced volume $v_R'$ is

$$v_R' = \frac{vP_c}{RT_c} \tag{4-6}$$

Note that again only a knowledge of $T_c$ and $P_c$ is required. Lines of constant $v_R'$ are also shown on Fig. 4-15, with values running from 0.70 to 8.00.

From the critical data shown in Table A-3 one finds that $P_c$ generally is larger than 30 atm. Standard atmospheric pressure then corresponds to a reduced pressure of about 0.03 or less for most substances. From Fig. 4-9 one sees that the maximum deviation from the ideal-gas law for any gas at 1 atm, regardless of temperature, will be 5 percent or less. When $T_R$ is 1.0 or greater at 1 atm, the deviation will be 1 percent or less. On the other hand, the effect of the critical temperature can be quite pronounced. Consider water, for example, with a critical temperature of 647°K (1165°R). At temperatures from 150 to 550°C (300 to 1000°F) the reduced temperatures range roughly from 0.70 to 1.25. Figure 4-9 indicates that water vapor would deviate greatly from ideal-gas behavior in this temperature range, especially as $P_R$ approaches unity. A $P_R$ value of 1.0 for water corresponds to a pressure of 218 atm.

If one considers nitrogen or argon instead of water, a different situation exists. The critical temperatures of these two gases are 126 and 151°K, respectively. Consequently, for room temperature or above the reduced temperature would always be greater than 2. For such values the compressibility factor is close to unity and is relatively independent of pressure. The assumption of ideal-gas behavior is quite good in these cases ($\pm 2$ percent). Therefore, critical data are extremely useful in estimating whether or not a given gaseous substance will approach ideal-gas behavior at a specified temperature and pressure. In addition, a generalized compressibility chart allows one to predict fairly accurately the deviation from ideality.

**Example 4-14M** Determine the specific volume of water vapor in $cm^3/g$ at 200 bars and 520°C by (a) the ideal-gas equation of state, (b) the principle of corresponding states, and (c) the experimental value in the superheat table.

*200 bars, 520°C      find v (cm³/g)?*

SOLUTION (a) On the basis of the ideal-gas equation, $v = RT/P$. The gas constant $R$, based on Table A-1, is

$$R = \frac{R_u}{M} = \frac{0.08315 \text{ bar·m}^3/(\text{kg·mol})(°\text{K})}{18 \text{ kg/kg·mol}} \times \frac{10^6 \text{ cm}^3}{\text{m}^3} \times \frac{\text{kg}}{10^3 \text{ g}}$$

$$= 4.62 \text{ bar·cm}^3/(\text{g})(°\text{K})$$

Hence,

$$v = \frac{RT}{P} = 4.62 \frac{\text{bar·cm}^3}{\text{g·°K}} \times \frac{793°\text{K}}{200 \text{ bars}} = 18.3 \text{ cm}^3/\text{g}$$

(b) The specific volume based on the principle of corresponding states is given by $v_{actual} = Zv_{ideal}$. The Z factor is found by computing $T_R$ and $P_R$ and then using Fig. 4-15 to find Z. The critical data are found in either Table 4-1 or Table A-3. Based on these data

$$T_R = \frac{793°\text{K}}{647°\text{K}} = 1.23 \qquad P_R = \frac{200 \text{ bars}}{218.3 \text{ atm}} \times \frac{0.987 \text{ atm}}{1 \text{ bar}} = 0.904$$

From Fig. 4-15 the Z value is approximately 0.83. Therefore

$$v = 0.83(18.3) = 15.2 \text{ cm}^3/\text{g}$$

(c) The tabulated value, based on experimental data, is found in Table A-14M to be 15.51 cm³/g. In comparison to the tabulated value, the ideal-gas equation is in error by nearly 20 percent. However, the corresponding states principle leads to an error of only about 2 percent. This chosen state is typical of superheater conditions in large, modern, steam power plants. Based on its Z factor, water vapor in this state is far from behaving as an ideal gas.

**Example 4-14** Determine the specific volume of water vapor in ft³/lb at 1000°F and 3000 psia by (a) the ideal-gas equation of state, (b) the principle of corresponding states, and (c) the experimental value in the superheat table.

SOLUTION On the basis of the ideal-gas equation, $v = RT/P$. The gas constant $R$, based on Table A-1, is

$$R = \frac{R_u}{M} = \frac{10.73 \text{ psi·ft}^3/(\text{lb·mol})(°\text{R})}{18 \text{ lb/lb·mol}} = 0.596 \frac{\text{psi·ft}^3}{\text{lb·°R}}$$

Hence,

$$v = \frac{RT}{P} = 0.596 \frac{\text{psi·ft}^3}{\text{lb·°R}} \times \frac{1460°\text{R}}{3000 \text{ psi}} = 0.290 \text{ ft}^3/\text{lb}$$

(b) The specific volume based on the principle of corresponding states is given by $v_{actual} = Zv_{ideal}$. The Z factor is found by computing $T_R$ and $P_R$ and then using Fig. 4-15 to find Z. The critical data are found in either Table 4-1 or Table A-3. Based on these data,

$$T_R = \frac{1460°\text{R}}{1165°\text{R}} = 1.25 \qquad P_R = \frac{3000 \text{ psi}}{3204 \text{ psi}} = 0.936$$

From Fig. 4-15 the Z value is approximately 0.83. Therefore

$$v = 0.83(0.290) = 0.241 \text{ ft}^3/\text{lb}$$

(c) The tabulated value, based on experimental data, is found in Table A-14 to be 0.2485 ft³/lb. In comparison to the tabulated value, the ideal-gas equation is in error by nearly 17 percent. However, the corresponding states principle leads to an error of only about 3 percent. This chosen state is typical of superheater conditions in large modern steam power plants. Based on its Z factor, water vapor in this state is far from behaving as an ideal gas.

The preceding examples illustrate that $PvT$ data can be estimated to a fair accuracy by employing the principle of corresponding states. It should be emphasized, however, that the $Z$ chart should never be used if experimental data are available.

## 4-7 PROPERTY RELATIONS FOR INCOMPRESSIBLE SUBSTANCES

On the basis of the $PvT$ surface for simple compressible substances (see Fig. 4-1), it is noted, that for many solids and liquids there are wide regions on the $PvT$ surface of equilibrium states where the variation in the specific volume is negligible. Therefore, the assumption that the specific volume (and the density) is constant in the region of interest is often a good approximation to reality, and it leads to no serious error in computations. The equation of state for these two phases often can be represented then by

$$v = \text{constant} \quad \text{or} \quad \rho = \text{constant} \tag{4-7}$$

By definition, a substance of constant density is said to be *incompressible*. Geometrically, on a $PvT$ surface, an incompressible phase would be represented by a plane which is perpendicular to the $v$ axis. The boundary or $P\,dV$ work associated with the change of state of an incompressible substance must always be zero. However, for a simple, compressible substance the only quasistatic work interaction permitted is boundary work. Consequently the only ways to change the internal energy of a simple, incompressible material are by a heat interaction or by nonquasistatic work interactions. For example, stirring of an incompressible liquid by means of a paddle-wheel would be permissible. We noted in Sec. 3-4 that the internal energy of a simple substance can be expressed as $u = u(T, v)$. The total differential becomes

$$du = \left(\frac{\partial u}{\partial T}\right)_v dT + \left(\frac{\partial u}{\partial v}\right)_T dv \tag{3-11}$$

The first partial derivative is $c_v$. In addition, $dv$ is zero for an incompressible substance. Therefore, we may write for a simple, incompressible material that

$$du = c_v\,dT \quad \text{incompressible} \tag{4-8}$$

That is, the internal energy of an incompressible substance is solely a function of temperature. Based on the definition of $c_v$, it is also clear that $c_v$ for an incompressible material is solely a function of the temperature. Hence,

$$u_2 - u_1 = \int_1^2 c_v\,dT \quad \text{incompressible} \tag{4-9}$$

While the internal energy of an incompressible material depends only on the temperature, this is not true of the enthalpy of an incompressible substance. From the definition of the enthalpy function $h = u + Pv$ we see that

$$h_1 - h_1 = u_2 - u_1 + v(P_2 - P_1) \quad \text{incompressible} \tag{4-10}$$

Hence the enthalpy of an incompressible substance is a function of both the temperature and the pressure.

The function $c_p$ is defined as $(\partial h/\partial T)_p$. If we differentiate the relation $h = u + Pv$ with respect to temperature at constant pressure and assume incompressibility, then

$$c_p = \left(\frac{\partial h}{\partial T}\right)_p = \left(\frac{\partial u}{\partial T}\right)_p$$

In addition, because $u$ is solely a function of $T$, we can write that

$$c_v = \left(\frac{\partial u}{\partial T}\right)_v = \left(\frac{\partial u}{\partial T}\right)_p$$

That is, the subscript on $(\partial u/\partial T)$ is immaterial. Since the final quantities in both expressions are equal, we realize that

$$c_p = c_v = c \qquad (4\text{-}11)$$

For an incompressible substance we need not distinguish between $c_p$ and $c_v$, and therefore both can be represented by the symbol $c$. Consequently Eqs. (4-9) and (4-10) can be written as

$$u_2 - u_1 = \int_1^2 c \, dT \qquad (4\text{-}12)$$

and

$$h_2 - h_1 = \int_1^2 c \, dT + v(P_2 - P_1) \qquad (4\text{-}13)$$

for materials that can be approximated as incompressible.

Equation (4-12) supports an approximation introduced in Sec. 4-5 regarding compressed- or subcooled-liquid data. Because subcooled liquids are essentially incompressible, the internal energy of subcooled liquids according to Eq. (4-12) should be only temperature-dependent. For a given temperature, the integral of $c \, dT$ must be zero. Therefore $u_2$ of the subcooled liquid would equal $u_1$ of the saturated liquid at the same temperature. This equality is not true for the enthalpy of a subcooled liquid, since the term $v(P_2 - P_1)$ is present in Eq. (4-13). However, the value of $v$ is usually so small that the contribution of this additional term is quite small compared to $h_1$ for the saturated liquid.

The equations developed in this section are valid if a substance is incompressible. It must be remembered, however, that incompressibility is an idealization, much like that of an ideal gas. Hence these equations yield values which are approximations to real behavior. Appreciable errors may result if the equations are used for regions of the $PvT$ surface where the assumption of constant density is not appropriate.

X **Example 4-15M** Determine the values of $u$ and $h$ for subcooled water at 20°C and 1 bar on the basis of an incompressible fluid.

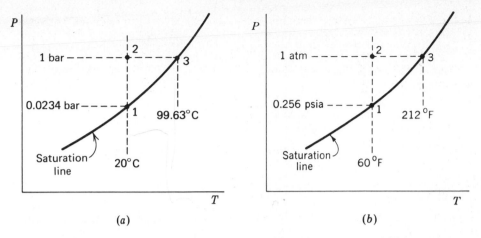

**Figure 4-16** The $PT$ diagrams for processes discussed in (a) Example 4-15M, and (b) Example 4-15.

SOLUTION The accompanying $PT$ diagram (Fig. 4-16) shows state 2 lying in the subcooled- (or compressed-) liquid region at 20°C and 1 bar. We shall use Eq. (4-12) to evaluate $u$ at state 2 on the diagram in terms of state 1 on the saturation line at the same temperature. Since path 1–2 is at constant temperature, the integral in Eq. (4-12) is zero, and $u_2 = u_1 = u_f$ at 20°C. From the saturated data in Table A-12M we find

$$u_2 = u_f \text{ at } 20°C = 83.95 \text{ kJ/kg}$$

The enthalpy at state 2 on the diagram is approximated by employing Eq. (4-13) for the assumed incompressible fluid. Again, the evaluation is along the constant-temperature path. As a result, Eq. (4-13) reduces to

$$h_2 = h_1 + v(P_2 - P_1)$$

The values of $v$, $P_1$, and $h_1$ to use are the saturated-liquid values at 20°C from Table A-12M. Consequently

$$h_2 = 83.96 \text{ kJ/kg} + 1.0018 \text{ cm}^3/\text{g} \times (1 - 0.0234) \text{ bar} \times \frac{\text{J}}{10 \text{ cm}^3 \cdot \text{bar}} \times \frac{1000 \text{ g}}{\text{kg}}$$

$$= 83.96 + 0.10 = 84.06 \text{ kJ/kg}$$

Note that the enthalpy in the subcooled region in this case differs by only roughly 1 percent from the saturated-liquid value at the same temperature.

The enthalpy at state 2 could also be found by selecting path 3–2 at a constant pressure of 1 bar. The enthalpy at state 3 is $h_f$ at 1 bar, which is found in Table A-13M to be 417.46 J/g. Equation (4-13) in this case requires knowledge of $c$. The specific heat $c$ of water between 20 and 99.6°C is relatively constant at 4.184 J/(g)(°C). Hence

$$h_2 - h_3 = c(T_1 - T_3) = 4.184(20 - 99.63) = -332.7 \text{ J/g}$$

$$h_2 = 417.46 - 333.71 = 84.29 \text{ J/g}$$

This method gives about the same answer. However, its accuracy is somewhat in doubt since the use of four significant figures in the calculation is questionable.

**Example 4-15** Determine the values of $u$ and $h$ for subcooled water at 60°F and 1 atm on the basis of an incompressible fluid.

SOLUTION The accompanying $PT$ diagram (Fig. 4-16) shows state 2 lying in the subcooled- (or compressed-) liquid region at 60°F and 1 atm. Equation (4-12) is used to evaluate $u$ at state 2 on the diagram in terms of state 1 on the saturation line at the same temperature. Since path 1–2 is a constant-temperature path, the integral in Eq. (4-12) is zero, and $u_2 = u_1 = u_f$ at 60°F. From the saturated data in Table A-12 we find

$$u_2 = u_f \text{ at } 60°F = 28.08 \text{ Btu/lb}$$

The enthalpy at state 2 on the diagram is approximated by employing Eq. (4-13) for the assumed incompressible fluid. Again, the evaluation is along the constant temperature path. As a result, Eq. (4-13) reduces to

$$h_2 = h_1 + v(P_2 - P_1)$$

The values of $v$, $P_1$, and $h_1$ to use are the saturated-liquid values at 60°F from Table A-12. Consequently

$$h_2 = 28.08 \text{ Btu/lb} + 0.01604 \frac{\text{ft}^3}{\text{lb}_m} \times (14.70 - 0.256)(144) \frac{\text{lb}_f}{\text{ft}^2} \times \frac{\text{Btu}}{778 \text{ ft-lb}_f}$$

$$= 28.08 + 0.04 = 28.12 \text{ Btu/lb}_m$$

Note that the enthalpy in the subcooled region in this case differs only slightly from the saturated-liquid value at the same temperature.

The enthalpy at state 2 could also be found by selecting path 3–2 at a constant pressure of 1 atm. The enthalpy at state 3 is $h_f$ at 1 atm, which is found in Table A-13 to be 180.15 Btu/lb. Equation (4-13) in this case requires knowledge of $c$. The specific heat $c$ of water between 60 and 212°F is relatively constant at 1.0 Btu/(lb)(°F). Hence

$$h_2 - h_3 = c(T_2 - T_3) = 1.0(60 - 212) = -152.0 \text{ Btu/lb}$$

$$h_2 = 180.15 - 152.0 = 28.15 \text{ Btu/lb}$$

This method gives about the same answer. However, its accuracy is somewhat in doubt due to the lack of accuracy for the value of $c$.

Equations (4-12) and (4-13) for the internal-energy and enthalpy changes of an incompressible substance require the use of specific-heat data for solids and liquids. The specific heat of a simple compressible substance theoretically is a function of two independent intensive variables, such as temperature and pressure. Experimental data indicate that the specific heats of solids are primarily a function of the temperature only. This small dependency on pressure is also generally exhibited by the specific heats of liquids. On the other hand, the variation of the specific heats of solids with temperature usually is quite large, whereas liquids are little affected.

The effect of temperature on the specific heat of elemental solids is shown in Fig. 4-17, which is a plot of $c_v$ versus temperature at atmospheric pressure. One can see that there is a significant variation of $c_v$ with temperature. The value approaches zero as the temperature is lowered toward absolute zero. In addition, $c_v$ for all elements approaches a value of roughly 25 kJ/(kg·mol)(°K) as the temperature increases. Common metals like lead, platinum, and aluminum reach this value at room temperature (300°K) or below. Those elements of high melting temperature, low molecular weight, and low molar specific volume follow the behavior of diamond and beryllium in that they exhibit a relatively slow increase

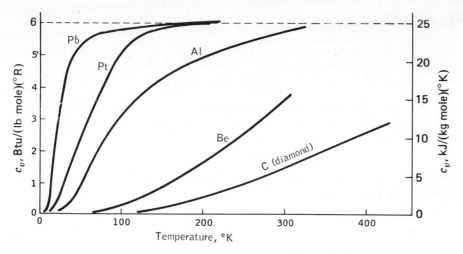

**Figure 4-17** The molar specific heat as a function of temperature for a few common elements.

of $c_v$ with increasing temperature. Experimental data also indicate that $c_v$ rises slowly above 25 kJ/(kg·mol)(°K) [or 6 Btu/(lb·mol)(°F)] at a very high temperatures, but the increase is not particularly significant for many applications. Data for various solids are tabulated in Tables A-19M and A-19.

The variation of $c_p$ from the solid to the liquid phase for water is shown in Fig. 4-18. Again, the rapid rise in the specific heat with temperature is noted for the solid phase (ice) up to the melting point. The value of $c_p$ then doubles as the substance becomes a liquid. The $c_p$ of liquid water remains relatively constant up to the boiling point at 1 atm. The relative temperature-independence of the specific heats of liquids is quite common. The specific heats of some common liquids are found in Tables A-19M and A-19, where the small variation with temperature again may be noted. Although not indicated, the effect of pressure is slight. The value of $c_p$ for liquid water undergoes a change of less than 3 percent at 35°C (100°F) when the pressure is raised from 1 to 400 atm.

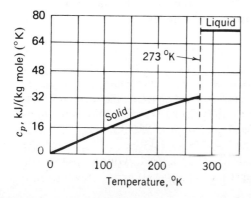

**Figure 4-18** Variation of $c_p$ of water with temperature.

# REFERENCES

Jones, J. B., and G. A. Hawkins: "Engineering Thermodynamics," Wiley, New York, 1960.

Keenan, J. H., et al.: "Thermodynamic Properties of Steam," Wiley, New York, 1969.

Marks, L. S. (Ed.), rev. by T. Baumeister: "Mechanical Engineers' Handbook," 7th ed., McGraw-Hill, New York, 1967.

Perry, R. H., C. H. Chilton, and S. D. Kirkpatrick (Eds.): "Chemical Engineers' Handbook," 5th ed., McGraw-Hill, New York, 1973.

Zemansky, M. W.: "Heat and Thermodynamics," 5th ed., McGraw-Hill, New York, 1968.

# PROBLEMS (METRIC)

### *PT* and *Tv* diagrams

**4-1M** Sketch a *Tv* diagram which includes the compressed-liquid, wet, and superheated-vapor regions of a substance. Include lines of constant pressure which pass through all three regions.

**4-2M** On a phase diagram (*PT*) for water, show approximately the location of constant-volume lines in the three single-phase regions.

### Superheat and saturation data

**4-3M** Fill in the data omitted in the table, for water substance.

| | Pressure bars | Temperature °C | Specific volume, cm³/g | Enthalpy kJ/kg | Internal energy kJ/kg | Quality (if appropriate) |
|---|---|---|---|---|---|---|
| (a) | | 150 | 392.8 | | | |
| (b) | 30 | 400 | | | | |
| (c) | | 100 | | 2200 | | |
| (d) | 60 | | 25.0 | | | |
| (e) | 75 | 80 | | | | |
| (f) | 20 | 400 | | | | |
| (g) | 5 | | | | | 0.90 |
| (h) | | 190 | 156.5 | | | |
| (i) | | 120 | | | 2000 | |
| (j) | | 150 | | 632.2 | | |
| (k) | 6 | | | | 669.9 | |

**4-4M** Determine the required data for water for the following specified conditions: (a) the pressure and specific volume of saturated liquid at 30°C, (b) the temperature and enthalpy of saturated vapor at 7 bars, (c) the specific volume and internal energy at 10 bars and 320°C, (d) the temperature and specific volume at 5 bars and a quality of 80 percent, (e) the specific volume and enthalpy at 80°C and 100 bars, (f) the pressure and enthalpy at 200°C and 70 percent quality, (g) the temperature and internal energy at 15 bars and an enthalpy of 2992.7 kJ/kg, (h) the quality and specific volume at 250°C and an enthalpy of 1943.5 kJ/kg, (i) the internal energy and specific volume at 100°C and an enthalpy of 2676.1 kJ/kg, (j) the pressure and enthalpy at 280°C and an internal energy of 2766.9 kJ/kg, and (k) the temperature and specific volume at 200 bars and an enthalpy of 185.16 kJ/kg.

**4-5M** Complete the following table of properties of refrigerant 12.

| | T. °C | P, bars | x, % | v, cm³/g | h, kJ/kg | u, kJ/kg |
|---|---|---|---|---|---|---|
| (a) | | 12.0 | | 14.41 | | | *Sup heat vap* |
| (b) | 30 | | | | | 188.56 *0.5* *Sup. vap* |
| (c) | | 6.0 | | | 126.7 | *Sup. vap* |
| (d) | 24 | | | | 197.34 | |
| (e) | | 8.0 | | 0.7802 | | *Sat liq* |
| (f) | 60 | 7.0 | | | | *Sup heat vap* |
| (g) | −10 | | 70 | | | *x = 0.70* |
| (h) | 28 | | | | | 62.09 *Sat liq* *vap* |
| (i) | | 5.0 | | | 210.81 | *Sup-htt vap* |
| (j) | 32 | | | 0.7785 | | *Sat. liq* |
| (k) | | 10.0 | | | | 175.2 *x = 0.8957* |

**4-6M** Determine the required data for refrigerant 12 at the following specified conditions: (a) the pressure and specific volume of saturated liquid at 20°C, (b) the temperature and enthalpy of saturated vapor at 8 bars, (c) the specific volume and internal energy at 7.0 bars and 50°C, (d) the temperature and specific volume at 4.0 bars and a quality of 30 percent, (e) the approximate specific volume and enthalpy at 4°C and 10 bars, (f) the pressure and enthalpy at −10°C and 20 percent quality, (g) the temperature and internal energy at 9.0 bars and an enthalpy of 234.03 kJ/kg, (h) the quality and specific volume at 44°C and an enthalpy of 179.4 kJ/kg, (i) the internal energy and specific volume at 28°C and an enthalpy of 198.87 kJ/kg, (j) the pressure and enthalpy at 40°C and an internal energy of 192.23 kJ/kg, and (k) the approximate enthalpy and specific volume at 15 bars and 20°C.

**4-7M** A container holds a mixture of 1 kg of liquid water and 1 kg of steam in equilibrium at 7 bars.
(a) What is the temperature of the mixture?
(b) The pressure is held constant and heat added until the temperature reaches 320°C. Determine the change in volume for the process in cubic meters.
(c) Show the process on a Pv diagram.

**4-8M** Steam is contained in a piston-cylinder device initially at 7 bars and 357.4 cm³/g. It is compressed at constant pressure until it becomes a saturated vapor.
(a) What is the initial temperature in °C?
(b) What is the final temperature?
(c) If the process is quasistatic, how much work input is required in N·m/g?
(d) Determine the initial internal energy in J/g.

**4-9M** Use the ideal-gas equation to calculate the specific volume of steam at the states listed below, and compare your answer with data from the steam tables. (a) 0.06 bar, 80°C; (b) 10 bars, 200°C; (c) 10 bars, 540°C; (d) 120 bars, 400°C; and (e) 120 bars, 740°C.

**4-10M** Steam at 7 bars has a specific volume of 507 cm³/g.
(a) What is its temperature in °C?
(b) Steam at 15 bars has a specific volume of 100 cm³/g. What is its enthalpy in kJ/kg?

**4-11M** Refrigerant 12 at 1.4 bars and 40 percent quality is heated until it becomes a saturated vapor at the same pressure. What is the change in enthalpy of the substance in kJ/kg?

**4-12M** Initially, a 0.3-m³ rigid tank contains saturated water vapor at 1.5 bars. Heat transfer to the surroundings results in a drop in pressure to 1 bar.
For the final equilibrium state determine (a) the temperature in °C, and (b) the ratio of the mass of liquid to the mass of vapor. (c) Show the process on a Pv diagram.

**4-13M** A wet mixture of water substance is maintained in a closed rigid tank at 1 bar.
(a) What must be the quality of the mixture initially if, on heating through a series of equilibrium states, the final state will be the critical state?
(b) What must be the initial ratio of volume of vapor to liquid?

**4-14M** Refrigerant 12 has a specific volume of 40.0 cm$^3$/g. Determine the superheat temperature or the quality, whichever is appropriate, if the pressure is (a) 1.6 bars, (b) 2.8 bars, and (c) 6.0 bars.

**4-15M** A closed tank holds 0.3 m$^3$ of dry, saturated steam at 2.5 bars. Determine the enthalpy of the steam in joules.

**4-16M** Determine the internal energy of 0.1 m$^3$ of refrigerant 12 at $-10°C$ if it is known that the specific volume is 50.0 cm$^3$/g in that state.

**4-17M** Refrigerant 12 is condensed from an initial state of 8 bars and 50°C to a final state of a saturated liquid at the same pressure. Determine (a) the change in the enthalpy in kJ/kg, and (b) the change in volume in cm$^3$/g.

**4-18M** A 1-m$^3$ tank contains a mixture of liquid and vapor at 200°C. The saturated liquid water fills (a) 10 percent, and (b) 5 percent of the tank, and the vapor fills the remaining volume. What is the quality of the mixture?

**4-19M** Dry, saturated steam at 20 bars (state 1) is contained within a piston-cylinder device which initially has a volume of 0.03 m$^3$. The steam is cooled at constant volume until the temperature reaches 200°C (state 2). The system is then expanded isothermally until the volume in state 3 is twice the initial value. Determine (a) the pressure at state 2, (b) the pressure at state 3, (c) the quality or temperature of superheat at state 3, and (d) the change in the internal energy for the two processes $u_2 - u_1$ and $u_3 - u_2$. Plot the two processes on a $Pv$ diagram.

**4-20M** Determine the enthalpy and the specific volume of water for a temperature of 80°C and a pressure of (a) 60 bars, (b) 90 bars, and (c) 125 bars.

**4-21M** Water is maintained at 75 bars and 140°C. What error would be introduced in computing the volume per unit mass if the correction for a compressed liquid were neglected?

**4-22M** A rigid vessel contains saturated water at 270°C. If the correct proportions of liquid and vapor are initially in the vessel, the system will pass through the critical state when heated sufficiently. Determine the initial proportion of liquid and vapor necessary (a) in percent by volume, and (b) in percent by weight.

**4-23M** Water vapor at 20 bars and 300°C is cooled at constant volume until the pressure is 10 bars.
  (a) Determine the quality in the final state.
  (b) Sketch the process on a $Pv$ diagram.

**4-24M** A piston-cylinder device contains 0.2 kg of steam at 7 bars and 220°C. The substance is cooled at constant pressure until three-quarters of it condense.
  (a) What is the change in volume of the system, in liters?
  (b) What is the final temperature in °C?
  (c) Sketch the process on a $Pv$ diagram.

**4-25M** A wet mixture of water undergoes a process such that the enthalpy is maintained constant. The final state is 1 bar and 120°C. If the initial pressure is 20 bars, determine the quality of the initial mixture, in percent.

**4-26M** Same as Prob. 4-25M, except the final state is 0.7 bar and 100°C and the initial pressure is 7 bars.

**4-27M** A cylinder having an initial volume of 2 m$^3$ initially contains steam at 10 bars and 200°C (state 1). The vessel is cooled at constant temperature until the volume is 56.4 percent of the initial volume (state 2). The constant-temperature process is followed by a constant-volume process which ends with a pressure in the cylinder of 30 bars (state 3).
  (a) Determine the pressure in bars and the enthalpy in kJ/kg at state 2.
  (b) Determine the temperature in °C and the enthalpy at state 3.
  (c) Sketch the two processes on a $Pv$ diagram with respect to the vapor dome.

**4-28M** Refrigerant 12 at a pressure of 5 bars has a specific volume of 25.0 cm$^3$/g. It expends at constant temperature until the pressure falls to 2.8 bars.
  (a) Calculate the final specific volume.
  (b) Calculate the final internal energy.
  (c) Show the path of the process on a $Pv$ diagram.

**4-29M** A cylinder fitted with a piston contains steam initially at 30 bar, 540°C, and a volume of 0.1 m³. The steam is compressed without friction until the volume is 0.05 m³. If the pressure remains constant throughout the process, determine the work input in kJ/kg.

**4-30M** A piston-cylinder assembly contains 1.2 kg of steam. Starting with an initial specific volume of 503.4 cm³/g, the gas is slowly compressed until the specific volume reaches 120.0 cm³/g. During the process heat is removed to keep the temperature constant at 280°C. Determine the work input in kilojoules.

**4-31M** Estimate an average $c_p$ of steam at 15 bars and (a) 240°C, and (b) 400°C in kJ/(kg)(°C).

**4-32M** Estimate an average $c_p$ in kJ/(kg)(°C) for steam at 320°C and (a) 0.06 bar, (b) 20 bars, and (c) 60 bars.

## Compressed-liquid data

**4-33M** Saturated liquid water at 80°C is compressed at constant temperature to (a) 50 bars, (b) 100 bars, and (c) 200 bars. Determine the enthalpy change on the basis of experimental compressed-liquid data. In each case, what would be the answer if saturated data were used as an approximation?

**4-34M** Saturated liquid water at 40°C is compressed to 140°C and (a) 50 bars, (b) 100 bars, and (c) 200 bars. Determine the enthalpy change on the basis of compressed-liquid data. In each case, approximate the answer also on the basis of saturated data. What percent error is involved in the second answer?

**4-35M** Same as Prob. 4-34M, except find the change in the specific volume instead.

**4-36M** Same as Prob. 4-34M, except find the change in the internal energy instead.

## Process sketches on Pv diagrams

**4-37M** Steam is compressed isothermally from 1.5 bars, 160°C, to a final specific volume of (a) 250 cm³/g, and (b) 600 cm³/g. Sketch the processes on a Pv diagram, relative to the saturation line.

**4-38M** Refrigerant 12 undergoes a change of state at constant pressure from 3.2 bars, 20°C, to (a) a final temperature of 50°C, and (b) a final specific volume of 40 cm³/g. Sketch the processes on a Pv diagram, relative to the saturation line.

**4-39M** Steam undergoes a change of state at constant pressure from 30 bars and 280°C to (a) a final temperature of 400°C, and (b) a final enthalpy of 2400 kJ/kg. Sketch the processes on a Pv diagram, relative to the saturation line.

**4-40M** Refrigerant 12 undergoes a change of state at constant volume from 7 bars and 60°C to (a) a final pressure of 5 bars, and (b) a final temperature of 100°C. Sketch the processes on a Pv diagram, relative to the saturation line.

**4-41M** Steam undergoes a change of state at constant volume from 15 bars and 162.7 cm³/g to (a) a final temperature of 360°C, and (b) a final pressure of 10 bars. Sketch the processes on a Pv diagram, relative to the saturation line.

**4-42M** Refrigerant 12 undergoes a change of state at constant temperature from 40°C and 9 cm³/g to (a) a final internal energy of 73.82 kJ/kg, and (b) a final pressure of 7 bars. Sketch the processes on a Pv diagram, relative to the saturation line.

## Energy analysis, saturation and superheat data

**4-43M** A cylinder fitted with a gastight, frictionless piston contains 0.5 kg of dry saturated steam at 5 bars. Heat is added until the final pressure is 5 bars and the final temperature is 280°C.
    (a) Determine the work done, in kilojoules.
    (b) Determine the heat supplied, in kilojoules.
    (c) Compute the change in internal energy, in kilojoules.

**4-44M** One and one-half kilograms of dry saturated steam at 3 bars are contained in a piston-cylinder device. Heat is added in the amount of 600 kJ, and a 96-volt dc electrical source supplies current to a resistor within the vapor for a period of 8 min. If the final temperature is 500°C and the pressure remains constant, determine the current required in amperes.

**4-45M** A piston-cylinder device contains refrigerant 12 initially at 2.8 bars and 40°C, and the volume is 0.1 m³. The piston is kept stationary and there is heat transfer to the gas until its pressure rises to 3.2 bars. Then additional heat transfer occurs from the gas during a process in which the volume varies but the pressure is constant. This latter process ends when the temperature reaches 50°C. Assume the processes are quasistatic and find (a) the mass in the system in kilograms, (b) the heat transfer in kilojoules during the constant-volume process, and (c) the heat transfer for the constant-pressure process in kilojoules.

**4-46M** In a constant-volume system steam is cooled from 30 bars and 400°C until it is a saturated vapor. Calculate the heat transfer in kJ/kg.

**4-47M** A rigid vessel with a volume of 0.05 m³ is initially filled with saturated steam at 1 bar. The contents are cooled to 50°C.

    (a) Sketch the process on Pv coordinates with respect to the saturation line.
    (b) What is the final pressure in bars?
    (c) Find the heat transferred from the steam in kilojoules.

**4-48M** A drink contains 25 g of crushed ice. How long will the ice last in an insulated blender agitated by a 0.10-kW motor? (Hint: 335 J are required to melt 1 g of ice.)

**4-49M** Refrigerant 12 at 6 bars undergoes a constant-volume process ending at 4 bars as dry, saturated vapor. Find (a) the initial specific volume and temperature, and (b) the quantity of heat removed in kJ/kg.

**4-50M** One-half kilogram of dry, saturated steam at 1.5 bars is contained in a piston-cylinder device. Heat is added in the amount of 150 kJ, and some work is done on the steam by means of a paddle-wheel until the steam is at 1.5 bars and 400°C. Determine the amount of paddle-wheel work in kilojoules, if the pressure is constant.

**4-51M** Saturated steam at (a) 3 bars, and (b) 5 bars is heated at constant pressure until the tempera-ture reaches 200°C. In addition to 30 kJ/kg added in the form of heat, energy is also added as electrical work by passing current through a resistor which is inside the system. Determine the quantity of electrical work added in kJ/kg.

**4-52M** Refrigerant 12 is condensed from an initial state of 7 bars and 180°C to a final state of saturated liquid at the same pressure. Determine the heat interaction and the work interaction, in kJ/kg.

**4-53M** Water substance is contained in a piston-cylinder device at 10 bars and 200°C. It is compressed isothermally until its quality is 50 percent. The work required to carry this out is 120 kJ/kg.

    (a) What is the final pressure in bars?
    (b) Determine the magnitude in kJ/kg and the direction of any heat transfer.
    (c) Sketch the process path relative to the saturation line on a Pv diagram.

**4-54M** Same as Prob. 4-53M, except that the temperature is 240°C and the work required is 180 kJ/kg.

**4-55M** One kilogram of water substance initially at 10 bars and 200°C is altered isothermally ($T = c$) until its volume is 40 percent of the initial value. During the process the boundary work is 170 kJ/kg, and paddle-wheel work in the amount of 79 N·m/g also takes place.

    (a) Determine the magnitude (in kilojoules) and direction of any heat transfer.
    (b) Sketch the process on a Pv diagram relative to the saturation line.

**4-56M** One-tenth kilogram of refrigerant 12 initially is a wet mixture with a quality of 50 percent at 40°C. It expands isothermally to a pressure of 6 bars. The measured work output due to the expansion is 17 N·m/g.

    (a) Determine the magnitude (in kilojoules) and direction of any heat transfer.
    (b) Sketch the process on a Pv diagram relative to the saturation line.

**4-57M** One-tenth kilogram of water at a pressure of 3 bars occupies a volume of 0.0303 m³ in a piston-cylinder arrangement weighted to maintain a constant pressure. Heat in the amount of 122 kJ is transferred to the water. Find (a) the final temperature in °C, and (b) the work output in kilojoules. Also sketch the path on a Pv diagram relative to the saturation line.

**4-58M** Steam at 1.5 bars and 160°C is contained within a closed rigid tank which has a volume of 0.05 m³. Determine the heat, in kilojoules, that would need to be removed to reduce the pressure to 1.0 bar.

**4-59M** One-tenth kilogram of dry, saturated steam at 25 bars is contained in a closed rigid tank. Work in the amount of 10 kJ is done on the steam by a paddle-wheel, while heat is transferred. The final pressure is 10 bars.

(a) Sketch the path of the process on a $Pv$ diagram.

(b) Calculate the quantity and direction of the heat transfer in kilojoules.

(c) Determine the final temperature in °C.

**4-60M** A closed container maintained at 1 bar pressure is subdivided into two sections by an insulated partition. One section contains 10 g of ice at 0°C, while the other contains dry, saturated steam at 100°C. Determine the amount of steam present if, on removing the partition, it is just sufficient to melt all the ice. (The enthalpy of melting of water is 335 kJ/kg.)

**4-61M** A closed, rigid tank contains 0.5 kg of dry, saturated steam at 4 bars. Heat is added in the amount of 50 kJ, and some work is done by means of a paddle-wheel until the steam is at 7 bars. Calculate the work required in kilojoules.

**4-62M** One kilogram of dry, saturated steam within an adiabatic piston-cylinder assembly is maintained at a constant pressure of 3 bars. A paddle-wheel is operated within the cylinder until the volume of the steam has increased by 29 percent. Compute the paddle-wheel work added in kilojoules.

**4-63M** In a nonflow system, dry, saturated steam at 8 bars is heated at constant volume until its pressure is 10 bars. It is then expanded isothermally to 7 bars by adding 60 kJ/kg of heat. Finally, it is cooled to the saturation temperature by a constant-pressure process. Sketch these processes on a $Pv$ diagram. Calculate the work associated for the total series of processes in kilojoules per kilogram.

**4-64M** One kilogram of dry, saturated steam at 3 bars is contained in a piston-cylinder assembly. Heat is added in the amount of 225 kJ, and some electrical work is done by passing a current of 1.5 A through a resistor in the fluid for 0.5 h. If the final temperature of the steam is (a) 280°C, and (b) 400°C and the process is constant-pressure, determine the necessary voltage, in volts, of the battery which supplied the potential for the current.

**4-65M** Twenty grams of refrigerant 12 in a closed system is initially at 7 bars and 90 percent quality. It undergoes a frictionless process until the temperature reaches 40°C. Compute the amount of heat transferred if the process had been at (a) constant volume, and (b) constant pressure.

**Energy analysis, compressed-liquid data**

**4-66M** One kilogram of water at 100 bars and 40°C is heated at constant pressure to a temperature of (a) 100°C, and (b) 180°C. Determine the heat required using experimental compressed liquid data, in kilojoules. Repeat the calculation using saturation data as an approximation. What percent error is involved with the second method?

**4-67M** Same as Prob. 4-66M, except that the pressure is 200 bars.

**Compressibility factor**

**4-68M** Water vapor exists at 360°C and at (a) 5 bars, and (b) 120 bars. Determine for these pressures the specific volume in $cm^3/g$ based on (1) the ideal-gas equation, and (2) the corresponding states principle. The tabulated values for (a) and (b) are 579.6 and 18.11 $cm^3/g$, respecitively.

**4-69M** Refrigerant 12 is maintained at 60°C and (a) 1 bar, and (b) 14 bars. Determine for both pressures the specific volume in $cm^3/g$ based on (1) the ideal-gas equation, (2) the principle of corresponding states, and (3) experimental data.

**4-70M** Determine the pressure in bars of water vapor at 360°C and 30.89 $cm^3/g$ on the basis of (a) the ideal-gas equation, (b) the principle of corresponding states, and (c) the experimental value.

**4-71M** Refrigerant 12 at a temperature of 100°C has a specific volume of 13.37 $cm^3/g$. Determine the pressure of the gas on the basis of (a) the ideal-gas equation, (b) the corresponding states principle, and (c) the experimental value.

**4-72M** Determine the temperature in °C of carbon dioxide at 60 bars and a specific volume of 10.03 $cm^3/g$, on the basis of (a) the ideal-gas equation, and (b) the compressibility chart. Compare to the experimental value of 100°C.

**4-73M** (a) Estimate how many kilograms of carbon dioxide ($CO_2$) are contained in a cylindrical tank 3 m in diameter and 4 m long if the pressure is 1 bar and the temperature is 30°C.

  (b) Estimate the number of kilograms if the pressure is 60 bars.

**4-74M** Determine the temperature in °C of water vapor at 80 bars and 26.82 cm³/g on the basis of (a) the corresponding states principle, and (b) the experimental value.

**4-75M** Determine the temperature in °C of refrigerant 12 at 14 bars and 14.25 cm³/g on the basis of (a) the corresponding states principle, and (b) the experimental value.

### Incompressible substances

**4-76M** A 26-kg mass of copper is dropped into an insulated tank which contains 0.1 m³ of water at 20°C. What is the initial temperature of the copper if the final equilibrium temperature of the copper and water at 1 atm is (a) 20.8°C, and (b) 21.4°C. Assume the specific heat of copper at 310°K as an average value.                     37°C

**4-77M** An unknown mass of copper at 77°C is dropped into an insulated tank which contains 0.2 m³ of water at 30°C. If the equilibrium final temperature is (a) 31°C, and (b) 32°C, find the mass of copper in kg.

**4-78M** Consider three liquids A, B, and C which have specific heats at constant pressure of 5.0, 10.0, and 12.0 kJ/(kg)(°C), respectively. The three liquids occupy three compartments of an insulated tank, but the walls separating the compartments conduct heat. Initially, one section contains 5 kg of A at 100°C, and the other compartments contain 2 kg of B and 1 kg of C, respectively, each initially at 50°C. Find the temperature of the three fluids at thermal equilibrium, if the overall process is one of constant pressure.

**4-79M** Same as Prob. 4-78M, except the initial temperatures of A, B, and C are 20°C, 60°C, and 100°C, respectively. Find the final equilibrium temperature in °C.

**4-80M** A 2-kg mass of lead at −100°C is brought into contact with an unknown mass of aluminum initially at 0°C. If the final equilibrium temperature of the two metals is (a) −60°C, and (b) −40°C, determine the mass of aluminum in kilograms.

**4-81M** Reconsider Prob. 4-34M. Evaluate the enthalpy change when the final pressure is (a) 50 bars, (b) 100 bars, and (c) 200 bars on the basis that the fluid is an incompressible substance. What percent error is involved in assuming an incompressible substance, in comparison to compressed-liquid data?

**4-82M** Reconsider Prob. 4-36M. Evaluate the internal energy change when the final pressure is (a) 50 bars, (b) 100 bars, and (c) 200 bars on the basis that the fluid is an incompressible substance. What percent error is involved in assuming an incompressible fluid, in comparison to compressed-liquid data?

**4-83M** A paddle-wheel is used to agitate 6 kg of liquid water in a closed, insulated tank. If a 210-W motor is used to drive the paddle-wheel, calculate the temperature rise of the water after a period of (a) 0.5 h, and (b) 1 h, given that the initial temperature is 20°C and the pressure is 1 bar.

## PROBLEMS (USCS)

**4-1** Sketch a $Tv$ diagram which includes the compressed-liquid, wet, and superheated-vapor regions of a substance. Include lines of constant pressure which pass through all three regions.

**4-2** On a $PT$ diagram for water, show approximately the location of constant-volume lines in the three single-phase regions.

### Superheat and saturation data

**4-3** Fill in the data omitted in the table, for water.

| | Pressure psia | Temperature °F | Specific volume ft³/lb | Enthalpy Btu/lb | Quality, if appropriate |
|---|---|---|---|---|---|
| (a) | | 444.7 | 1.162 | | |
| (b) | 400 | 700 | | | |
| (c) | | 240 | | 1000 | |
| (d) | 500 | | 0.700 | | |
| (e) | 1000 | 200 | | | |
| (f) | 800 | | | 1338 | |
| (g) | 160 | | | | x = 0.90 |
| (h) | | 350 | 3.346 | | |
| (i) | | 200 | | 900 | |
| (j) | | 400 | | 375.1 | |

**4-4** Determine the required data for water at the specified conditions: (a) the pressure and specific volume of saturated liquid at 100°F, (b) the temperature and enthalpy for saturated vapor at 100 psia, (c) the specific volume and internal energy at 100 psia and 500°F, (d) the temperature and specific volume at 120 psia and a quality of 80 percent, (e) the specific volume and enthalpy at 100°F and 1000 psia, (f) the pressure and enthalpy at 400°F and 70 percent quality, (g) the temperature and internal energy at 180 psia and an enthalpy of 1297.5 Btu/lb, (h) the quality and specific volume at 500°F and an enthalpy of 845 Btu/lb, (i) the internal energy and specific volume at 212°F and an enthalpy of 1150.5 Btu/lb, (j) the pressure and enthalpy at 500°F and an internal energy of 1175.7 Btu/lb, and (k) the temperature and specific volume at 2000 psia and an enthalpy of 172.6 Btu/lb.

**4-5** Complete the following table of properties of refrigerant 12:

| | T, °F | P, psia | x, % | v, ft³/lb | h, Btu/lb | u, Btu/lb |
|---|---|---|---|---|---|---|
| (a) | | 180 | | 0.2228 | | |
| (b) | 100 | | | | | 82.328 |
| (c) | | 90 | | | 53.6 | |
| (d) | 80 | | | | 84.0 | |
| (e) | | 160 | | 0.0130 | | |
| (f) | 140 | 100 | | | | |
| (g) | 10 | | 70 | | | |
| (h) | 100 | | | | | 33.16 |
| (i) | | 70 | | | 90.09 | |
| (j) | 120 | | | 0.01317 | | |
| (k) | | 140 | | | | 75.08 |

**4-6** Determine the required data for refrigerant 12 at the following specified conditions: (a) the pressure and specific volume of saturated liquid at 70°F, (b) the temperature and enthalpy of saturated vapor at 120 psia, (c) the specific volume and internal energy at 100 psia and 140°F, (d) the temperature and specific volume at 60 psia and 30 percent quality, (e) the approximate specific volume and enthalpy at 10°F and 60 psia, (f) the pressure and enthalpy at 0°F and 25 percent quality, (g) the temperature and internal energy at 120 psia and an enthalpy of 94.736 Btu/lb, (h) the quality and specific volume at 100°F and an enthalpy of 73.59 Btu/lb, (i) the internal energy and specific volume at 110°F and an enthalpy of 87.84 Btu/lb, (j) the pressure and enthalpy at 140°F and an internal energy of 84.899 Btu/lb, and (k) the approximate enthalpy and specific volume at 100 psia and 40°F.

**4-7** A container holds a mixture of 3 lb of liquid and 3 lb of water vapor in equilibrium at 200 psia.

(a) What is the temperature of the mixture?

(b) The pressure is held constant and heat is added until the temperature reaches 500°F. Determine the change in volume for the process, in cubic feet.

**4-8** Steam is contained in a piston-cylinder assembly initially at 100 psia and 5.59 ft³/lb. It is compressed at constant pressure until it becomes a saturated vapor.

(a) What is the initial temperature in °F?

(b) What is the final temperature?

(c) If the process is quasistatic, how much work input is required in ft·lb$_f$/lb$_m$?

(d) Determine the initial internal energy.

**4-9** Use the ideal-gas equation to calculate the specific volume of steam at the states listed below, and compare your answer with the steam tables. (a) 1 psia, 200°F; (b) 200 psia, 400°F; (c) 200 psia, 1000°F; (d) 2000 psia, 700°F; and (e) 2000 psia, 1600°F.

**4-10** (a) Steam at 100 psia has a specific volume of 5.3 ft³/lb. What is its temperature in °F?

(b) Steam at 200 psia has a specific volume of 1.50 ft³/lb. What is its enthalpy?

**4-11** Refrigerant 12 at 12 psia and 40 percent quality is evaporated until it reaches the saturated-vapor state at the same pressure. What is the change in the enthalpy of the substance in Btu/lb?

**4-12** Initially, a 1.0 ft³ rigid tank contains dry, saturated steam at 20 psia. Heat transfer to the surroundings results in a drop in pressure to 15 psia. For the final equilibrium state determine (a) the temperature, and (b) the ratio of the mass of liquid to the mass of vapor.

**4-13** A wet mixture of water substance is maintained in a closed rigid tank at 15 psia. What must be the quality of the mixture initially if, on heating through a series of equilibrium states, the final state will be the critical state? What must be the initial ratio of volume of vapor to liquid?

**4-14** Refrigerant 12 has a specific volume of 2.00 ft³/lb. Determine the superheat temperature or the quality of the substance, whichever is appropriate, if the pressure is (a) 12.0 psia, and (b) 20.0 psia.

**4-15** A closed tank holds 10 ft³ of dry, saturated steam at 40 psia. Determine the enthalpy of the steam in Btu.

**4-16** Determine the internal energy of 1 ft³ of refrigerant 12 at 20°F if it is known that the specific volume is 1.00 ft³/lb in that state.

**4-17** Refrigerant 12 is condensed from an initial state where the pressure is 125 psia and the temperature is 130°F to a final state of the saturated liquid at the same pressure. Determine the change in the enthalpy in Btu/lb and the change in the volume per pound.

**4-18** A 2 ft³ tank contains a mixture of liquid and vapor at 500°F. The saturated water fills (a) 10 percent and (b) 5 percent of the tank, and the vapor fills the remaining volume. What is the quality of the mixture?

**4-19** Dry saturated steam at 120 psia (state 1) is contained within a piston-cylinder assembly which initially has a volume of 1 ft³. The steam is cooled at constant volume until the temperature reaches 300°F (state 2). The system is then expanded isothermally until the volume in state 3 is twice the initial value. Determine (a) the pressure at state 2, (b) the pressure at state 3, (c) the quality or superheat temperature at state 3, and (d) the change in the internal energy for the two processes $u_2 - u_1$ and $u_3 - u_2$. Plot the two processes on a $Pv$ diagram.

**4-20** Determine the enthalpy and the specific volume of water at 100°F and a pressure of (a) 800 psia, (b) 1800 psia, and (c) 2500 psia.

**4-21** Water is maintained at 1000 psia and 300°F. What error would be introduced in computing the volume per unit mass if the correction for a compressed liquid were neglected?

**4-22** A rigid vessel contains saturated water at 500°F. If the correct proportions of liquid and vapor are initially in the vessel, the system will pass through the critical state when heated sufficiently. Determine the initial proportions of liquid and vapor necessary (a) in percent by volume and (b) in percent by weight.

**4-23** Water vapor at 300 psia and 500°F is cooled at constant volume until the pressure is 160 psia. Determine the quality in the final state, and sketch the process on a $Pv$ diagram.

**4-24** A piston-cylinder assembly contains 0.2 lb of steam at 100 psia and 600°F. The substance is cooled at constant pressure until three-quarters of it condenses.

(a) What is the change in volume of the system, in $ft^3$?

(b) What is the final temperature in °F?

(c) Sketch the process on a $Pv$ diagram.

**4-25** A wet mixture of water undergoes a process such that the enthalpy is maintained constant. The final state is 14.7 psia and 250°F. If the initial pressure is 450 psia, determine the quality of the initial mixture, in percent.

**4-26** Same as Prob. 4-25, except the final state is 10 psia and 200°F and the initial pressure is 500 psia.

**4-27** A cylinder having an initial volume of 2 $ft^3$ contains steam at 140 psia and 400°F (state 1). The cylinder is cooled at constant temperature until the volume is 37.5 percent of the initial volume (state 2). The constant-temperature process is followed by a constant-volume process which ends with the pressure in the cylinder at 450 psia (state 3).

(a) Determine the temperature in °F and the enthalpy in Btu/lb at state 2.

(b) Determine the temperature in °F and the enthalpy at state 3.

(c) Sketch the two processes on a $Pv$ diagram with respect to the wet region.

**4-28** Refrigerant 12 at a pressure of 120 psia has a specific volume of 0.25 $ft^3$/lb. It expands at constant temperature until the pressure falls to 50 psia.

(a) Calculate the final specific volume.

(b) Calculate the final internal energy.

(c) Show the path of the process on a $Pv$ diagram.

**4-29** A cylinder fitted with a piston contains steam initially at 500 psia, 1000°F, and a volume of 4 $ft^3$. The steam is compressed without friction until the volume is 2 $ft^3$. If the pressure remains constant throughout the process, determine the work input in Btu/lb.

**4-30** A piston-cylinder assembly contains 2.0 lb of steam. Starting with an initial specific volume of 7.8 $ft^3$/lb, the gas is slowly compressed until the specific volume reaches 2.0 $ft^3$/lb. During the process heat is removed at such a rate as to keep the temperature constant at 600°F. Determine the work input in $ft·lb_f$.

**4-31** Estimate an average $c_p$ for steam at 200 psia and (a) 450°F, and (b) 550°F, in Btu/(lb)(°F).

**4-32** Estimate an average $c_p$ in Btu/(lb)(°F) for steam at 800°F and (a) 1 psia, (b) 300 psia, and (c) 3000 psia.

**Compressed liquid data**

**4-33** Saturated liquid water at 150°F is compressed isothermally to (a) 500 psia, (b) 1500 psia, and (c) 3000 psia. Determine the enthalpy change on the basis of experimental compressed-liquid data. In each case what would the answer be if saturated data were used as an approximation?

**4-34** Saturated liquid water at 200°F is compressed to 300°F and (a) 500 psia, (b) 1500 psia, and (c) 3000 psia. Determine the enthalpy change on the basis of compressed-liquid data. In each case also approximate the answer on the basis of saturated data. What percent error is involved in the second answer?

**4-35** Same as Prob. 4-34, except find the change in specific volume instead.

**4-36** Same as Prob. 4-34, except find the change in internal energy instead.

**Process sketches on $Pv$ diagrams**

**4-37** Steam is compressed isothermally from 300°F, 20 psia, to a final specific volume of (a) 4.0 $ft^3$/lb, and (b) 7 $ft^3$/lb. Sketch the processes on a $Pv$ diagram, relative to the saturation line.

**4-38** Refrigerant 12 undergoes a change of state at constant pressure from 50 psia, 60°F, to (a) a final temperature of 140°F, and (b) a final specific volume of 0.70 $ft^3$/lb. Sketch the processes on a $Pv$ diagram, relative to the saturation line.

**4-39** Steam undergoes a change of state at constant pressure from 400 psia and 500°F to (a) a final temperature of 700°F, and (b) a final enthalpy of 1000 Btu/lb. Sketch the processes on a $Pv$ diagram, relative to the saturation line.

**4-40** Refrigerant 12 undergoes a change of state at constant volume from 100 psia and 140°F to (a) a final pressure of 70 psia, and (b) a final temperature of 200°F. Sketch the processes on a $Pv$ diagram, relative to the saturation line.

**4-41** Steam undergoes a change of state at constant volume from 200 psia and 2.724 ft³/lb to (a) a final temperature of 350°F, and (b) a final pressure of 170 psia. Sketch the processes on a $Pv$ diagram, relative to the saturation line.

**4-42** Refrigerant 12 undergoes a change of state at constant temperature from 100°F and 0.15 ft³/lb to (a) a final internal energy of 79.5 Btu/lb, and (b) a final pressure of 70 psia. Sketch the processes on a $Pv$ diagram, relative to the saturation line.

### Energy analysis, saturation and superheat data

**4-43** A cylinder fitted with a gastight, frictionless piston contains 2 lb of dry saturated steam at 100 psia. Heat is added until the final temperature is 500°F and the final pressure is 100 psia.

    (a) Determine the work done in Btu.

    (b) Determine the heat supplied in Btu.

    (c) Compute the change in internal energy in Btu.

**4-44** Three pounds of dry, saturated steam at 40 psia are contained in a piston cylinder. Heat is added in the amount of 600 Btu, and some work is done on the steam by means of a paddle-wheel until a temperature of 1000°F is reached. If the pressure remains constant at 40 psia, determine the amount of paddle-wheel work done on the steam in ft·lb$_f$.

**4-45** A piston-cylinder machine contains refrigerant 12 initially at 30 psia and 20°F, and the volume is 1 ft³. The piston is kept stationary and there is heat transfer to the gas until its pressure rises to 40 psia. Then additional heat transfer occurs from the gas during a process in which the volume varies but the pressure is constant. This latter process terminates when the temperature reaches 100°F. Assume that the processes are quasistatic and determine (a) the mass present, (b) the heat transfer for the constant-volume process, and (c) the heat transfer for the constant-pressure process.

**4-46** In a constant-volume, fixed-mass system, steam is cooled from 300 psia and 800°F until it is dry and saturated. Calculate the heat transfer in Btu/lb.

**4-47** A rigid steel vessel having a volume of 2 ft³ is initially filled with saturated steam at 14.7 psia. The vessel is sealed and chilled to 100°F.

    (a) Sketch the process on $Pv$ coordinates with respect to the saturated-vapor line.

    (b) What is the final pressure in psia?

    (c) Determine the heat transferred from the steam, in Btu.

**4-48** A drink contains $\frac{1}{20}$ lb of crushed ice. How long will the ice last in an insulated blender agitated by a 0.14-hp motor? (144 Btu are required to melt 1 lb of ice.)

**4-49** Steam of 180 psia undergoes a constant-volume nonflow process ending at 140 psia as dry, saturated vapor. Find (a) the initial specific volume and temperature, (b) the quantity of heat removed in Btu/lb, and (c) the quantity of work done in ft·lb$_f$/lb$_m$.

**4-50** Three pounds of dry, saturated steam at 30 psia are contained in a closed piston cylinder. Heat is added in the amount of 600 Btu, and some work is done on the steam by means of a paddle-wheel until the steam is at 30 psia and 1000°F. Determine the amount of paddle-wheel work done on the steam in ft·lb$_f$, if $P$ is constant.

**4-51** Saturated steam at 20 psia is heated at constant pressure until the temperature reaches 400°F. In addition to 15 Btu/lb added in the form of heat, energy is also added as electrical work by passing current through a resistor which is inside the system. Determine the quantity of electric work added to the system during the process, in Btu/lb$_m$.

**4-52** Refrigerant 12 is condensed from an initial state of 100 psia and 140°F to a final state of saturated liquid at the same pressure. Determine the quantity of the heat interaction, in Btu/lb, and the value of the work interaction, in ft·lb$_f$/lb$_m$.

**4-53** Water substance is contained in a piston-cylinder device at 160 psia and 400°F. It is compressed isothermally until its quality is 50 percent. The work required to carry this out is 70,000 ft·lb$_f$/lb$_m$.

(a) What is the final pressure in psia?

(b) Determine the magnitude, in Btu/lb, and the direction of any heat transfer.

(c) Sketch the process path relative to the saturation line on a Pv diagram.

**4-54** Same as Prob. 4-53, except the temperature is 500°F and the work required is 210,000 ft·lb$_f$/lb$_m$.

**4-55** One pound of water substance initially at 140 psia and 400°F is altered isothermally ($T = c$) until its volume is 40 percent of the initial value. During the process the boundary work is 65,000 ft·lb$_f$/lb$_m$, and paddle-wheel work in the amount of 30,000 ft·lb$_f$/lb$_m$ also takes place.

(a) Determine the magnitude (in Btu) and direction of any heat transfer.

(b) Sketch the process on a Pv diagram relative to the saturation line.

**4-56** One-tenth pound of refrigerant 12 initially is a wet mixture with a quality of 50 percent at 100°F. It expands isothermally to a pressure of 90 psia. The measured work output due to the expansion is 5400 ft·lb$_f$/lb$_m$.

(a) Determine the magnitude (in Btu) and direction of any heat transfer.

(b) Sketch the process on a Pv diagram.

**4-57** Two-tenths pound of water at a pressure of 40 psia occupies a volume of 1.05 ft$^5$ in a piston-cylinder arrangement weighted to maintain a constant pressure. Heat in the amount of 107 Btu is transferred to the water. Find (a) the final temperature in °F, and (b) the work output in Btu.

**4-58** Steam at 20 psia and 300°F is contained within a closed rigid tank which has a volume of 1.50 ft$^3$. Determine the heat that would need to be removed to reduce the pressure to 15 psia.

**4-59** One-tenth pound of dry, saturated steam at 400 psia is contained in a closed rigid tank. Work in the amount of 2.0 Btu is done on the steam by a paddle-wheel, while heat is transferred to or from the steam. The final pressure is 45 psia.

(a) Sketch a Pv diagram and show the path of the process.

(b) Calculate the quantity and direction of the heat transfer in Btu.

(c) Determine the final temperature.

**4-60** A closed container maintained at 1 atm pressure is subdivided into two sections by an insulated partition. One section contains 0.02 lb of ice at 32°F, while the other section contains saturated water vapor at 212°F. Determine the amount of steam present if, on removing the partition, it is just sufficient to melt all the ice. (The enthalpy of melting of water is 144 Btu/lb.)

**4-61** A closed rigid tank contains 2 lb of dry, saturated steam at 60 psia. Heat is added in the amount of 100 Btu, and some work is done on the substance by means of a paddle-wheel until the steam is at 100 psia. Calculate the work done in Btu.

**4-62** One pound of dry, saturated steam within an adiabatic piston-cylinder assembly is maintained at a constant pressure of 20 psia. A paddle-wheel is operated within the cylinder until the volume of the steam has increased by 31.0 percent. Compute the paddle-wheel work added, in ft·lb$_f$.

**4-63** In a nonflow system, dry, saturated steam at 120 psia is heated at constant volume until its pressure is 165 psia. It is then expanded isothermally to 100 psia by adding 61 Btu/lb of heat during the process. Finally, it is cooled to the saturation temperature by a constant-pressure process. Sketch these processes on a Pv diagram, drawn approximately to scale, and calculate the net work done per pound of steam for the total series of processes. Indicate an area on the Pv diagram which represents the net work done.

**4-64** One pound of dry, saturated steam at 60 psia is contained in a piston-cylinder assembly. Heat is added in the amount of 101 Btu, and some electrical work is done on the fluid by passing a steady current of 1.5 A through a resistor in the fluid for a period of 0.5 h. If the final temperature of the steam is 700°F and the process was constant-pressure, determine the necessary voltage of the battery which supplied the potential for the flow of electricity in volts.

**4-65** One pound of refrigerant 12 in a closed system is initially at 100 psia and 90 percent quality. It undergoes a frictionless process until the temperature reaches 120°F. Compute the amount of heat transferred if the process had been at (a) constant volume, and (b) constant pressure.

### Energy analysis, compressed-liquid data

**4-66** One pound of water at 1500 psia and 50°F is heated at constant pressure to a temperature of (a) 150°F, and (b) 300°F. Determine the heat required, in Btu, using experimental compressed-liquid data. Repeat the calculation using saturation data as an approximation. What percent error is involved with the second method?

**4-67** Same as Prob. 4-66, except that the pressure is 2000 psia.

### Compressibility factor

**4-68** Water vapor exists at 700°F and at (a) 80 psia, and (b) 2000 psia. Determine for both pressures the specific volume in ft³/lb based on (1) the ideal-gas equation, (2) the corresponding-states principle, and (3) the experimental value.

**4-69** Refrigerant 12 is maintained at 160°F and (a) 15 psia, and (b) 200 psia. Determine for both pressures the specific volume in ft³/lb based on (1) the ideal-gas equation, (2) the principle of corresponding states, and (3) the experimental value.

**4-70** Determine the pressure in psia of water vapor at 700°F and 0.491 ft³/lb on the basis of (a) the ideal-gas equation, (b) the principle of corresponding states, and (c) the experimental value.

**4-71** Refrigerant 12 at a temperature of 300°F has a specific volume of 0.1368 ft³/lb. Compute the pressure of the gas on the basis of (a) the ideal-gas equation, (b) the corresponding-states principle, and (c) the experimental value.

**4-72** Determine the temperature in °R of carbon dioxide at 900 psia and a specific volume of 0.149 ft³/lb, on the basis of (a) the ideal-gas equation, and (b) the compressibility chart. Compare to the value of 660°R based on experiment.

**4-73** (a) How many pounds of carbon dioxide ($CO_2$) are contained in a cylindrical tank 10 ft in diameter and 7 ft long if the pressure is 14.7 psia and the temperature is 88°F?

(b) How many pounds would be contained in the same tank, if the pressure is 590 psia, for the same temperature?

**4-74** Determine the temperature in °F of water vapor at 12 psia and 0.450 ft³/lb on the basis of (a) the corresponding-states principle, and (b) the experimental value.

**4-75** Determine the temperature in °F of refrigerant 12 at 200 psia and 0.2486 ft³/lb on the basis of (a) the corresponding-states principle, and (b) the experimental value.

### Incompressible substances

**4-76** A 50-lb mass of copper is dropped into an insulated tank which contains 4 ft³ of water at 70°F. What is the initial temperature of the copper if the final equilibrium temperature of the copper and water at 1 atm is (a) 71.20°F, and (b) 72.40°F. Assume the specific heat of copper over the required temperature range is 6.1 Btu/(lb·mol)(°F).

**4-77** An unknown mass of copper at 170°F is dropped into an insulated tank which contains 7 ft³ of water at 90°F. If the final equilibrium temperature is (a) 91.60°F, and (b) 93.5°F, find the mass of copper in pounds.

**4-78** Consider three liquids A, B, and C which have specific heats at constant pressure of 1.0, 2.0, and 2.5 Btu/(lb)(°F), respectively. The three liquids occupy three compartments of a tank, and the walls separating the compartments conduct heat. Initially, one section contains 10 lb of A at 200°F, and the other compartments contain 5 lb of B and 2 lb of C, respectively, each initially at 100°F. Calculate the temperature of the three fluids at thermal equilibrium, assuming the tank is adiabatic and the process is one of constant pressure.

**4-79** Same as Prob. 4-78, except that the initial temperatures of A, B, and C are 80°F, 140°F, and 210°F, respectively. Find the final equilibrium temperature in °F.

**4-80** A 4-lb mass of lead at −150°F is brought into thermal contact with an unknown mass of aluminum initially at 30°F. If the final equilibrium temperature of the two metals is (a) −70°F, and (b) −40°F, determine the mass of aluminum in pounds.

**4-81** Reconsider Prob. 4-34. Evaluate the enthalpy change when the final pressure is (*a*) 500 psia, (*b*) 1500 psia, and (*c*) 3000 psia, assuming the fluid is incompressible. What percent error is involved in assuming an incompressible fluid, in comparison to compressed-liquid data?

**4-82** Reconsider Prob. 4-36. Evaluate the internal energy change when the final pressure is (*a*) 500 psia, (*b*) 1500 psia, and (*c*) 3000 psia, assuming the fluid is incompressible. What percent error is involved in assuming an incompressible fluid, in comparison to compressed-liquid data?

**4-83** A paddle-wheel is used to agitate 12.7 lb of liquid water in a closed, insulated tank. If a $\frac{1}{4}$-hp motor is used to drive the paddle-wheel, calculate the temperature rise of the water after a period of (*a*) 0.5 h, and (*b*) 1 h. The initial temperature is 50°F and the pressure is 1 atm.

# FIVE

## CONTROL-VOLUME ENERGY ANALYSIS

In the two preceding chapters we have developed the basic conservation of energy equations for closed systems, or control masses, and in addition have become familiar with a number of basic relationships among properties. In this chapter we shall develop the basic energy equations for open-system analyses. In terms of the control volume, we must account for energy transported across various parts of the boundary due to mass transfer, as well as account for heat and work interactions. Again, we shall find that knowledge of intrinsic properties of matter is extremely important. The uses of and restrictions on the various equations are of prime interest, but the general methodology for analyzing engineering systems in terms of the energy equations will also be emphasized.

## 5-1 IDEALIZATIONS FOR STEADY-STATE CONTROL-VOLUME ANALYSIS

The first law of thermodynamics resulted from experimental observations on closed systems. This principle generally may be applied to any control mass which can be physically identified throughout a given process. It can be seen, however, that such a conservation law would often be inadequate to the analysis of engineering systems of interest. A large number of engineering problems which require a thermodynamics analysis are open systems where matter flows continuously in and out of a certain defined region of space. In most cases it is physically impossible to trace a given control mass, much less measure the heat and work interactions that might occur to the unit of mass. For example, consider an element of fluid passing through a pump. During its passage the velocity, pressure, and other macroscopic characteristics of the element are continually changing. It is impractical to attempt to measure all the interactions that occur between this fluid element and its neighboring elements. Furthermore, in this case we are not interested in the behavior of an individual element, but rather in the average behavior of all the fluid elements. Thus the analysis of flow processes will require an important modification in one's viewpoint of the system. Fortunately, the results developed for a control mass can be extended properly to any general open system. We shall find it appropriate also to include a conservation of mass principle in terms of an open system.

The analysis of flow processes begins by selecting a region of space called a *control volume*. The boundary of the control volume may be in part a well-defined physical barrier, or all of it may be an imaginary envelope. The establishment of the boundary, or *control surface*, is an important first step in the analysis of any open system. The control volume is fixed in position relative to the observer, but the volume itself may change. However, we shall restrict our initial development in that the control volume will be fixed in size and shape as well as position. An energy balance on the control volume necessitates not only measurements of heat and work interactions, but also an accounting for the energy carried in or out of the control volume by mass transferred across the control surface. Since we are frequently interested in the state of matter at positions on the control surface, one necessary requirement is that the properties of the fluid vary in a continuous manner across the portion of the control surface at which mass is transferred. Discontinuities or chaotic motion may exist within the system, but the flow characteristics at the boundary must be continuous in order that the properties of the fluid may be evaluated. This implies that at each open boundary the conditions of static equilibrium are closely met. This requirement does not prevent the properties at a given open boundary from changing with time, as they must in unsteady-state processes. Nevertheless, it means that the flow is sufficiently well behaved so that its properties are known with adequate accuracy at the boundaries. Since a flow system seems quite divorced from the conditions of static equilibrium, several pertinent comments needed to be made to clarify the situation.

In the preceding chapters a large number of relations have been developed which are valid under the conditions of static equilibrium. In such a state the value of each intrinsic property is the same throughout the system. The only restriction is that the incremental volume of the system to which property values will be assigned be large enough so that the statistical average of a certain molecular characteristic undergoes extremely small fluctuations. For example, consider the density of a small volume of a gaseous system. If the volume taken is small enough, the number of particles, and hence the amount of mass within this volume, will fluctuate wildly over a given time interval. At some instances there may be no particles within it, and an infinitesimal time interval later there may be one or more. Obviously, the concept of density has no real macroscopic meaning in this situation. However, as the small volume is made larger so that on the average it contains a large number of particles, say, $10^6$ or $10^{10}$, then although particles move randomly in and out, the fluctuations from the average value are extremely small. In this case the macroscopic variables will appear to have a constant value. The actual incremental volume selected still could be extremely small macroscopically. Gases at room conditions contain, roughly, $10^{10}$ particles in a cube only $10^{-3}$ mm ($4 \times 10^{-4}$ in) on a side. Thus, after the sample size reaches a certain finite size, any macroscopic property of the static system has the same value for all volumes larger than this minimum value. This minimum size is unimportant to us, since macroscopic measurements usually are based on a portion of the system which contains a tremendous number of particles.

When we consider open systems, a new problem arises with regard to the significance of a property value at the control surface. Since the properties of an element of mass crossing the control surface of the control volume may be changing rapidly as it passes the surface, there is a tendency to question what is meant by a property in this case, and what is its relation to properties defined in thermostatic equilibrium. Are the results of the state postulate still valid? In most cases the answer to this last question is yes, if the physical situation fulfills the conditions for what is frequently called *local equilibrium*. By local equilibrium we mean to imply that the change in any property between two small adjacent fluid elements is very small compared with the average value of the property in this region. That is, $dy/y \ll 1$, where $y$ is some intensive property. If an incremental volume at the boundary is selected, the microscopic characteristics of the particles entering the volume are only slightly different from those leaving it. Hence the unit of volume is so close to static equilibrium that the general results of this latter static condition may be applied to the dynamic situation of open-flow systems. Local equilibrium for open systems is somewhat analogous to quasistatic changes in closed systems. The region of interest is extremely close to equilibrium, so that errors in assuming true equilibrium are negligible. Local equilibrium does not exist, for example, in a differential element within a shock wave. Here the criterion that $dy/y$ is very small is violated. In fact, the element is so far from equilibrium that the normal concept of macroscopic properties is meaningless, and the macroscopic-property variations are essentially discontinuous in this region.

In terms of this local-equilibrium theory, we can define the density $\rho$ and the pressure, for example, in the vicinity of a point in the following way. If $n$ is the number of particles (molecules) in the region, and $m'$ is the mass of a particle, then

$$\rho \equiv \lim_{V \to V'} \frac{m'n}{V} = \lim_{V \to V'} \frac{m}{V}$$

In this expression $V'$ is the smallest volume which can be taken and still let the density have a statistical meaning for which fluctuations around the average value are negligible. On the other hand, the value of $V$ should not be too large, since the density will actually vary within the boundaries of the control volume. In terms of pressure, it is necessary to measure the normal component of the force on an area which is very small but still large enough so that the pressure remains the correct statistical average of a large number of particles. Therefore we shall define the pressure $P$ to be the limiting value of $dF_n/dA$ as $dA$ approaches the smallest area $dA'$ for which the fluid may be considered a continuum.

$$P \equiv \lim_{dA \to dA'} \frac{dF_n}{dA}$$

With this material as background, we now turn to the general derivations of some conservation principles. We shall restrict ourselves to one-dimensional-flow problems. By one-dimensional flow we mean that the properties are constant across any area normal to the flow, and thus vary only on the direction of flow. Such an idealized condition is an approximation to actual behavior. For example, the velocities associated with pipe flow usually vary from zero at the wall to a maximum value at the center of the flow channel at any given cross section. Thus two- or three-dimensional configuration should be considered. Hence, the velocity used in the following equations is an average velocity, constant over the cross section of interest, that represents the overall mass rate of flow.

## 5-2 CONSERVATION OF MASS PRINCIPLE FOR A CONTROL VOLUME

### a General development

An equation which represents a conservation of mass principle for open systems in terms of a control volume may be developed by referring to Fig. 5-1. A control mass (CM) will be followed through a control volume (CV) over a time interval $\Delta t$. On the basis of conservation of mass, the control mass at times $t$ and $t + \Delta t$ must be identical. That is,

$$m_{\text{CM},\, t} = m_{\text{CM},\, t + \Delta t} \tag{a}$$

The control volume under consideration is indicated by the dashed line in the figure, which represents the control surface. In addition to the control volume, it is necessary to consider also the two regions, labeled $A$ and $B$, external to the control

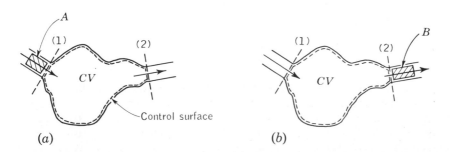

Figure 5-1 The development of conservation principles for a control-volume analysis. (a) Control mass at time $t$; (b) control mass at time $t + \Delta t$.

volume. At some time $t$ we shall consider the control mass to be composed of all the mass within the control volume at that instant plus the mass $\Delta m_A$ within region $A$ (Fig. 5-1$a$). After some time interval $t$ this control mass has shifted position. At time $t + \Delta t$ all the mass within region $A$ has entirely entered the control valume, and some of the control mass initially within the control volume has filled region $B$ (Fig. 5-1$b$). It is not necessary that the masses within regions $A$ and $B$ at times $t$ and $t + \Delta t$, respectively, be of the same magnitude, since the density of matter within the control volume could change with time. Since the control mass initially includes all the mass within the control volume and region $A$,

$$m_{CM,\, t} = m_{CV,\, t} + \Delta m_A \tag{b}$$

Similarly, for the control mass at time $t + \Delta t$, we find that

$$m_{CM,\, t+\Delta t} = m_{CV,\, t+\Delta t} + \Delta m_B \tag{c}$$

Substitution of Eqs. ($b$) and ($c$) into the conservation of mass principle expressed by Eq. ($a$) yields

$$m_{CV,\, t} + \Delta m_A = m_{CV,\, t+\Delta t} + \Delta m_B$$

or, upon rearrangement

$$m_{CV,\, t+\Delta t} - m_{CV,\, t} - \Delta m_A + \Delta m_B = 0 \tag{d}$$

The physical interpretation of this equation is simply that the change in the mass within the control volume over a time interval $\Delta t$ is measured by the difference between the mass within region $A$ which entered and the mass within $B$ which left the control volume during the time interval.

It is now convenient to place the conservation of mass principle on a time basis. This is accomplished by dividing Eq. ($d$) by the time interval $\Delta t$.

$$\frac{m_{CV,\, t+\Delta t} - m_{CV,\, t}}{\Delta t} - \frac{\Delta m_A}{\Delta t} + \frac{\Delta m_B}{\Delta t} = 0 \tag{e}$$

The first term of this equation represents the average rate of change of mass within the control volume during the time interval $\Delta t$. The last two terms are for the average mass flow rates into and out of the control volume in the same time period. In the limit, as $\Delta t \to 0$, these average values become instantaneous rates of change. Consequently, Eq. ($e$) can be written in the differential form

$$\frac{dm_{CV}}{dt} - \frac{dm_A}{dt} + \frac{dm_B}{dt} = 0 \tag{f}$$

The quantity $dm_{CV}/dt$ is the rate of change of the mass within the control volume with respect to time. The second term is the rate of mass leaving region $A$ per unit time, or the rate of mass entering the control volume per unit time at section 1 in the figure. We shall define the quantity $dm/dt$ associated with a given portion of the control surface as the *mass flow rate*, or the mass rate of flow, across the control surface. It will be symbolized by $\dot{m}$, that is,

$$\frac{dm}{dt} \equiv \dot{m} \tag{5-1}$$

Thus the quantity $dm_B/dt$ is the mass flow rate out of the control volume at section 2. The quantity $dm_A/dt$ is equal to $\dot{m}_{\text{in}}$, and $dm_B/dt$ is equal to $\dot{m}_{\text{out}}$. On the basis that $\dot{m}$ is always a positive value, Eq. ($f$) becomes

$$\frac{dm_{\text{CV}}}{dt} - \dot{m}_{\text{in}} + \dot{m}_{\text{out}} = 0$$

or

$$\boxed{\frac{dm_{\text{CV}}}{dt} = \dot{m}_{\text{in}} - \dot{m}_{\text{out}}} \tag{5-2}$$

This equation simply states that the instantaneous rate of change of mass within a control volume is equal to the difference in the mass flow rates in and out of the control volume at that time. It is important to note that Eq. (5-2) is written for the control volume, and not the control mass. Thus we started the development by fixing our attention on a control mass, but have transformed the basic conservation principle for the control mass into one suitable for a control volume. It should be kept in mind that, in general, the values of $\dot{m}$ may vary with time.

Equation (5-2) is a valid expression for a conservation of mass principle on a time basis for any control volume. However, it would be useful only if expressed in terms of measurable thermodynamic quantities. In order to express the quantity $dm_{\text{CV}}/dt$ in terms of properties of the control volume, it is noted that the mass within the control volume at any instant of time is the sum of the masses of each differential-sized element in the region at that time. The mass of any differential element of volume is $\rho \, d\mathscr{V}$. (In the next few paragraphs the volume is represented by a script $\mathscr{V}$ in order to distinguish it from the velocity $V$ of the mass.) The total mass is the sum of the masses of the differential elements. Since the density $\rho$ may be a variable inside the control volume, the total mass is best expressed as the integral of $\rho \, d\mathscr{V}$. Thus

$$m_{\text{CV}} = \int_{\text{CV}} \rho \, d\mathscr{V}$$

where $\mathscr{V}$ is the volume, $\rho$ is the density of any element of volume, and the integration is carried out over the entire region of the control volume. The rate of change of $m_{\text{CV}}$ becomes

$$\frac{dm_{\text{CV}}}{dt} = \frac{d}{dt} \int_{\text{CV}} \rho \, d\mathscr{V} \tag{5-3}$$

The rate of mass transfer into the control volume may be given in terms of properties at the control surface. The volume flow rate across any open boundary of differential area $dA$ is the product of the velocity normal to the area times the area, that is $V_n\, dA$. The subscript $n$ signifies that the velocity component normal to the area is to be used. The mass flow rate through $dA$ is found by multiplying $V_n\, dA$ by the density of the fluid at $dA$. Therefore the rate of mass transfer at a finite boundary of size $A$ is found by integrating over the entire area, or by

$$\dot{m} = \int_A \rho V_n\, dA \qquad (5\text{-}4)$$

One of these terms must be present in Eq. (5-2) for each distinct boundary which permits mass flow. By the substitution of Eqs. (5-3) and (5-4) into Eq. (5-2), the conservation of mass principle for a control volume becomes

$$\frac{d}{dt} \int_{CV} \rho\, dv = \int_A (\rho V_n\, dA)_{in} - \int_A (\rho V_n\, dA)_{out} \qquad (5\text{-}5)$$

This is a general formulation, in nonvector notation, of the conservation of mass principle for a control volume.

## b Steady-state Conservation of Mass Principle

Many engineering problems deal with flow systems which fall under the classification of steady state. A *steady-state* condition for a control volume is one such that the properties at a given position within or at the boundaries of the control volume are invariant with time. The properties of interest include thermostatic ones, such as temperature, pressure, density, and the specific internal energy, as well as those which refer to the overall flow stream, such as velocity. It is the invariance of these properties with respect to time at the control surfaces where mass interactions occur that is of special importance. Keep in mind, however, that the state of the fluid does change as it passes through the control volume. The major purposes of some flow devices is simply to alter the state of the fluid.

Since steady state requires the velocity and the density of the fluid to be constant at each open section of the control surface, the mass flow rate $\dot{m}$ at each open section must also be invariant with time. This is seen by referring to Eq. (5-4) for $\dot{m}$ at a finite boundary of size $A$, namely,

$$\dot{m} = \int \rho V_n\, dA \qquad (5\text{-}4)$$

By fixing the values of $\rho$ and $V_n$ for each element of area $dA$, the total mass flow rate is fixed or steady. If we assume that the density and normal velocity are not only constant at each differential area $dA$, but are also uniform in value across the boundary where mass is being transported, then the mass flow rate is given at any inlet or outlet by

$$\boxed{\dot{m} = \rho V A} \qquad (5\text{-}6)$$

where the velocity $V$ is understood to be the average component of velocity normal to the area.

A further conclusion can be drawn from the restriction of steady state on a process. In general, we have seen that the conservation of mass principle for a control volume is

$$\frac{d}{dt}\int_{cv} \rho \, d\mathcal{V} = \int_A (\rho V_n \, dA)_{in} - \int_A (\rho V_n \, dA)_{out} \qquad (5\text{-}6)$$

Since the properties within the control surface are invariant with time at a given position under steady-state conditions, the rate of change of the mass with respect to time at any position within the control volume must be zero. Hence, in the case of steady state, we see that the left-hand side of the above equation is zero. Therefore the right-hand side reduces, upon rearrangement, to

$$\boxed{\dot m_{in} = \dot m_{out}} \qquad \left(\frac{A\,u_c}{V}\right)_{IN} = \left(\frac{A\,u_c}{V}\right)_{OUT} (5\text{-}7)$$

or in the light of Eq. (5-6),                             CONTINUITY PRINCIPLE

$$(\rho V A)_{in} = (\rho V A)_{out} = \text{constant} \qquad (5\text{-}8a)$$

An equivalent form which frequently is quite useful is

$$* \left(\frac{VA}{v}\right)_{in} = \left(\frac{VA}{v}\right)_{out} = \text{constant} \qquad (5\text{-}8b)$$

where $v$ again is the specific volume of the fluid. The two latter equations are frequently called forms of the *continuity* equation for steady-flow processes. The three equations above are valid only for the case where the control volume has one inlet and one outlet. For the more general case the conservation of mass principle for a control volume under steady-state conditions would be given by

$$\sum \dot m_{in} = \sum \dot m_{out} \qquad (5\text{-}9a)$$

or

$$\sum_{in} (\rho V A) = \sum_{out} (\rho V A) \qquad (5\text{-}9b)$$

An equivalent expression, similar to Eq. (5-8b), would also be valid. These equations simply state that under steady-state conditions there can be no net accumulation of mass within a control surface; hence the sum of the constant rates of flow into the control volume must equal the sum of the constant flow rates out.

Before turning to the development of conservation of energy equations for flow processes, short examples of the use of the continuity equation are presented.

**Example 5-1M** Refrigerant 12 at 8 bars and 50°C flows through a pipe of 1.50 cm internal diameter at a rate of 2 kg/min. Compute the velocity of the fluid in m/s.

SOLUTION At 8 bars the saturation temperature from Table A-17M is found to be 32.74°C. Hence the fluid at 50°C is superheated. From Table A-18M the specific volume at 8 bars and 50°C is given as 24.07 cm³/g. The cross-sectional area of the pipe, normal to the flow velocity, is

$$\boxed{A = \frac{\pi D^2}{4}} = \frac{\pi (1.5)^2}{4} = 1.77 \text{ cm}^2$$

Employing Eq. (5-6), we find that

*[handwritten annotation: density $P = \dfrac{M}{V} = \dfrac{M}{MV} = \dfrac{1}{V}$]*

$$V = \frac{\dot{m}}{\rho A} = \frac{\dot{m}v}{A} = \frac{2 \text{ kg/min} \times 24.07 \text{ cm}^3/\text{g}}{1.77 \text{ cm}^2} \times \frac{10^3 \text{ g}}{\text{kg}}$$

$$= 27,200 \text{ cm/min} = 272 \text{ m/min} = 4.53 \text{ m/s}$$

Velocities in the range of 4 to 40 m/s are quite common for the flow of gases through pipes in commercial processes.

**Example 5-1** Refrigerant 12 at 120 psia and 140°F flows through a pipe of 0.5 in internal diameter at a rate of 4 lb/min. Compute the velocity of the fluid in ft/s.

SOLUTION The saturation temperature of 120 psia from Table A-17 is 93.29°F. Thus the fluid at 140°F is superheated. From Table A-18 the specific volume at 120 psia and 140°F is given as 0.3890 ft³/lb. The cross-sectional area of the pipe, normal to the flow velocity, is

$$A = \frac{\pi D^2}{4} = \frac{\pi (0.5)^2}{4(144)} = 0.001363 \text{ ft}^2$$

Employing Eq. (5-6), we find that

$$V = \frac{\dot{m}}{\rho A} = \frac{\dot{m}v}{A} = \frac{4 \text{ lb/min} \times 0.3890 \text{ ft}^3/\text{lb}}{0.001363 \text{ ft}^2} = 1140 \text{ ft/min}$$

$$= 19.0 \text{ ft/s}$$

Velocities in the range of 10 to 100 ft/s are quite common for the flow of gases through pipes in commercial processes.

**Example 5-2M** Water at 20°C flows through a 1.50-cm rubber hose at a rate of 10 l/min. Compute the velocity of the fluid in m/s and the mass flow rate in kg/min.

SOLUTION The velocity is found by dividing the volume flow rate by the cross-sectional area. The value of $A$ was found in the preceding example to be 1.77 cm². Thus,

*[handwritten annotation: Vol. RATE = VA]*

$$V = \frac{(\text{volume rate})}{\text{area}} = \frac{10 \text{ l/min}}{1.77 \text{ cm}^2} \times \frac{10^3 \text{ cm}^3}{1 \text{ l}}$$

$$= 5650 \text{ cm/min} = 56.5 \text{ m/min} = 0.942 \text{ m/s}$$

The saturation pressure of water at 20°C from Table A-12M is 0.023 bar. Although the pressure is not given, it normally would be above atmospheric conditions. Hence the water is subcooled or compressed. We can approximate the specific volume of the fluid by using the saturated-liquid value at the given temperature. From Table A-12M this is given as 1.002 cm³/g. As a result the mass flow rate becomes

*[handwritten annotation: $\rho = \dfrac{1}{V}$]*

$$m = \rho A V = \frac{(\text{volume rate})}{v} = \frac{10 \text{ l/min}}{1.002 \text{ cm}^3/\text{g}} \times \frac{10^3 \text{ cm}^3}{1 \text{ l}}$$

$$= 9980 \text{ g/min} = 9.98 \text{ kg/min}$$

**Example 5-2** Water at 60°F flows through a ½-in rubber hose at a rate of 3 gal/min. Compute the velocity of the fluid in ft/s and the mass flow rate in lb/min.

SOLUTION There are 7.48 gal in 1 ft³. Thus the volume rate is (3/7.48), or 0.401 ft³/min. Using the saturated-liquid data in Table A-3 as an approximation, we find the specific volume of the fluid at 60°F to be 0.01604 ft$^3$/lb. The cross-sectional area of the hose is 0.001363 ft$^2$, as found in the preceding example. Hence the velocity is

$$V = \frac{(\text{volume rate})}{\text{area}} = \frac{0.401}{0.001363(60)} = 4.9 \text{ ft/s}$$

The mass flow rate becomes

$$\dot{m} = \rho A V = \frac{(\text{volume rate})}{v} = \frac{0.401}{0.01604} = 25.0 \text{ lb/min}$$

Note that although the velocity of the liquid in Example 5-2M (or 5-2) is much less than the gas velocity found in Example 5-1M (or 5-1) for the same area of flow, the mass flow rate of liquid is much greater due to the larger density. Velocities from 0.3 to 3 m/s (or 1 to 10 ft/s) are quite common for liquids in pipe-flow applications.

## 5-3 CONSERVATION OF ENERGY PRINCIPLE FOR A CONTROL VOLUME

### a General Development

The conservation of energy principle may be applied to a control-volume analysis by properly extending the energy equation already developed for a control mass. We shall again consider the control volume shown enclosed within the dashed line in Fig. 5-1a. The control surface at sections 1 and 2 is open to the transfer of mass in or out of the control volume. At some initial time $t$ we focus our attention on the control mass, which is the sum of the mass within the control volume at that instant and the mass $m_A$ within the volume element marked $A$, which lies adjacent to section 1. At time $t + \Delta t$ this control mass has moved to that all the mass originally in region $A$ is now just inside the control volume. In the same time interval, part of the control mass has been pushed out of the control volume into the region marked $B$, which is adjacent to section 2 (Fig. 5-1b). The mass and volume of regions $A$ and $B$ are taken to be small compared with those of the control volume. During the time interval $\Delta t$, heat and work interactions may have occurred to the control mass. These interactions are not shown in the figure. An energy balance on the control mass for the time interval is, simply,

$$Q + W = \Delta E_{CM} = E_{CM, \, t+\Delta t} - E_{CM, \, t} \tag{5-10}$$

Since the masses associated with regions $A$ and $B$ are small compared with the control volume, the heat and work effects on the control mass are essentially the same as those which cross the boundary of the control volume at sections other than 1 and 2. Consequently, whether applied to the control volume or the control mass, the values of $Q$ and $W$ are the same from either viewpoint. At this point, two items must be resolved in order to transform Eq. (5-10) into a relation directly applicable to the control volume itself. First, the right side of this equation is still in terms of the control mass. Second, the work term on the left needs to be split into several terms which represent distinct types of interactions.

The right-hand side of Eq. (5-10) may be expressed in terms of the control volume in the following manner: The energy $E$ of the control mass at time $t$ is the sum of the energies of the mass within the control volume at that instant plus the energy of the mass in region $A$. This may be expressed as

$$E_{CM,\,t} = E_{CV,\,t} + \Delta E_A \tag{a}$$

where the subscripts again indicate the physical significance of the terms and $\Delta E_A$ is the increment of energy associated with the mass which passes from region $A$ into the control volume. In a similar fashion, the energy of the control mass at time $t + \Delta t$ would be

$$E_{CM,\,t+\Delta t} = E_{CV,\,t+\Delta t} + \Delta E_B \tag{b}$$

The quantity $\Delta E_B$ is the increment of energy associated with the mass which passes from the control volume into region $B$. Subtracting Eq. (b) from Eq. (a), we find that

$$E_{CM,\,t+\Delta t} - E_{CM,\,t} = (E_{CV,\,t+\Delta t} - E_{CV,\,t}) - \Delta E_A + \Delta E_B \tag{c}$$

The left side of Eq. (c) is equal to the right side of Eq. (5-10). Substitution of Eq. (c) into Eq. (5-10) yields

$$Q + W = E_{CV,\,t+\Delta t} - E_{CV,\,t} - \Delta E_A + \Delta E_B \tag{d}$$

The conservation of energy principle for a control volume is now placed on a rate basis by dividing Eq. (d) by the time interval $\Delta t$. Hence,

$$\frac{Q}{\Delta t} + \frac{W}{\Delta t} = \frac{E_{CV,\,t+\Delta t} - E_{CV,\,t}}{\Delta t} - \frac{\Delta E_A}{\Delta t} + \frac{\Delta E_B}{\Delta t} \tag{e}$$

The first term on the right represents the average rate of change of energy within the control volume during the time interval $\Delta t$. The last two terms are the average rates of energy transfer into and out of the control volume in the same time period. The limit is now taken as $\Delta t$ approaches zero, so that each term may be expressed as a derivative. As a result

$$\dot{Q} + \dot{W} = \frac{dE_{CV}}{dt} - \frac{dE_A}{dt} + \frac{dE_B}{dt} \tag{5-11}$$

where the following definitions are introduced.

$$\lim_{\Delta t \to 0} \frac{Q}{\Delta t} = \frac{\delta Q}{dt} \equiv \dot{Q} \tag{5-12}$$

and

$$\lim_{\Delta t \to 0} \frac{W}{\Delta t} = \frac{\delta W}{dt} \equiv \dot{W} \tag{5-13}$$

$\dot{Q}$ is the instantaneous rate of heat transfer to the control volume, and $\dot{W}$ is the rate of work done on the system, or the power.

At this point it is desirable to change the last two quantities in Eq. (5-11) to terms identifiable with the mass crossing the control surface. Since the mass in region $A$ would be differentially small for a time interval $dt$, this small mass which lies adjacent to section 1 would essentially have the thermodynamic properties associated with that particular section if the condition of local equilibrium were valid. Consequently, $E_A = e_1 m_1$, or $dE_A/dt = e_1(dm_1/dt)$. A similar expression holds for the transport of energy at section 2. Further, recall that $dm/dt$ is defined as the mass flow rate $\dot{m}$. Then $dE_A/dt = e_{in}\dot{m}_{in}$ and $dE_B/dt = e_{out}\dot{m}_{out}$. Substitution of these two expressions into Eq. (5-11) leads to a conservation of energy principle for a control volume of the following form:

$$\dot{Q} + \dot{W} = \frac{dE_{CV}}{dt} - e_{in}\dot{m}_{in} + e_{out}\dot{m}_{out} \qquad (5\text{-}14)$$

The right-hand side of this final equation is now expressed either in terms of changes in the control volume or in terms of effects at the control surface. The energy $e$ again represents all forms of conceivable energy that might be associated with a unit of mass.

The final task is to recognize the forms of work interactions that can occur in the control mass. The work term is usually separated into two parts. The first of these is called the flow work, or displacement work, and is the work required to push the fluid in or out of the control volume. The second term lumps together all other forms of work, although in conventional problems, where the boundaries of the control volume are rigid, this normally reduces to forms of shaft work. This latter form of work can easily be measured by devices external to the control volume, such as a dynamometer. We therefore turn our attention to the flow, or displacement, work, since it can be expressed in terms of properties of the fluid at the control surface where mass enters the control volume.

Consider the work required to push a differential mass $dm_1$ at section 1 in Fig. 5-2 into the control volume per unit time. In time interval $dt$ region $A$ will pass across section 1, and region $B$ will pass across section 2. The fluid behind region $A$ acts as a piston and pushes the mass within region $A$ into the control volume. At the same time the mass in region $B$ must push the fluid ahead of it out of the way. This fluid also acts with a pistonlike effect to restrain the movement of region $B$ out of the control volume. Thus work is done on the mass of region $A$, while the mass of region $B$ does work on the surroundings.

Since work is the product of a force times a distance, the work per unit time will be the product of the force exerted on the differential mass $dm_1$ times the velocity with which it moves. From mass flow considerations the instantaneous velocity at section 1 is given by $V_1 = v_1 \dot{m}_1/A_1$, which is a form of the continuity-of-flow equation. The force applied is the pressure times the cross-sectional area, or $P_1 A_1$. Multiplying these two terms together, we find that the flow work done on the element of mass $dm_1$ at section 1 per unit time is

$$\dot{W}_{flow,\,1} = P_1 A_1 \left(\frac{v_1 \dot{m}_1}{A_1}\right) = P_1 v_1 \dot{m}_1 = (Pv)_{in}\dot{m}_{in} \qquad (5\text{-}15)$$

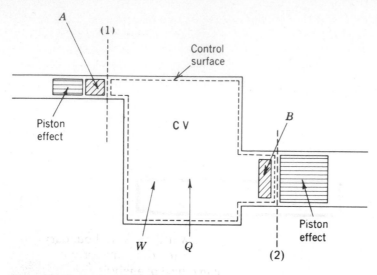

**Figure 5-2** Illustration of displacement or flow work.

For the element of mass $dm_2$ at section 2, work is done by the element on the surroundings as it leaves the control volume. This quantity of work is $-P_2 v_2 \dot{m}_2$, or $-(Pv)_{out} \dot{m}_{out}$. The energy balance on the control volume given by Eq. (5-14) now becomes

$$\dot{Q} + \dot{W}_{net} + (Pv)_{in} \dot{m}_{in} - (Pv)_{out} \dot{m}_{out} = \frac{dE_{CV}}{dt} - e_{in} \dot{m}_{in} + e_{out} \dot{m}_{out} \qquad (5\text{-}16)$$

where the term $\dot{W}_{net}$ represents shaft work and any other forms of work associated with the control volume. It is important to remember that the product $Pv$ is treated as an energy term only in those instances when fluid is crossing the control surface or boundary of an open system. Unlike the quantity $P \, \Delta v$ which results under special circumstances from the integration of $P \, dv$ for closed systems, the quantity $\Delta(Pv)$ found in open system equations is the change in a state function.

By collecting common terms in Eq. (5-16), the energy balance on a rate basis for a control volume with one inlet and one outlet is written as

$$\dot{Q} + \dot{W}_{net} + (e + Pv)_{in} \dot{m}_{in} - (e + Pv)_{out} \dot{m}_{out} = \frac{dE_{CV}}{dt} \qquad (5\text{-}17)$$

All the terms on the left represent instantaneous interactions which add or remove energy from the control volume, and the term on the right is the resultant rate of change of energy of the control volume due to these interactions. Since the instantaneous values on the left side of Eq. (5-17) generally may vary with time, this equation must be integrated over a time period $t$ to give the total energy change of the control volume during this time interval. Thus

$$\int_0^t \dot{Q} \, dt + \int_0^t \dot{W} \, dt + \int_0^t (e + Pv)_{in} \dot{m}_{in} \, dt - \int_0^t (e + Pv)_{out} \dot{m}_{out} \, dt = \Delta E_{CV} \qquad (5\text{-}18)$$

In order to carry out these integrations we will need to know how the various properties and rates vary with time. Another useful formulation of Eq. (5-17) is obtained by multiplying the equation by $dt$. The resulting relationship is valid for a differential change of state for a control volume over a period of time $dt$. Thus

$$\delta Q + \delta W_{shaft} + (e + Pv)_{in}\ dm_{in} - (e + Pv)_{out}\ dm_{out} = dE_{CV} \qquad (5\text{-}19)$$

Since the properties at a boundary may vary with time as mass continues to flow in and out of the control volume, the total energy which crosses a boundary due to mass transfer must is general be found by integration of the $e + Pv$ term over the time interval of interest. Hence Eq. (5-19) may be written as

$$Q_{net} + W_{shaft} + \int (e + Pv)_{in}\ dm_{in} - \int (e + Pv)_{out}\ dm_{out} = \Delta E_{CV} \qquad (5\text{-}20)$$

The integration indicated is in terms of the mass which crosses a boundary over a given time interval. Equations (5-17) to (5-20) represent in one form or another the general energy balance for a control-volume type of analysis. One important characteristic of these quations should be noted. The change of the energy within a control volume over an interval of time is due to two distinct effects. Heat and work interactions occur at boundaries not open to mass flow, and their values are found from measurements made external to the control surface. The remaining energy quantities are evaluated in terms of properties of the control volume at the control surface where mass is transferred.

In general, solution of Eqs. (5-17) to (5-20) requires a knowledge of the functional relationship for the properties of the control volume at each open boundary in terms of the mass crossing the boundaries. In certain special cases the integration and subsequent evaluation of the various terms in these equations are fairly straightforward, for example, the steady-flow process. In other cases, which are generally of the unsteady-flow type, graphical integration or other approximation techniques may become necessary.

## b Steady-state Conservation of Energy Principle

The specification of steady-state operation not only requires that the mass flow rates at each open boundary be constant, but also places a restriction on the heat and work interactions. Because all properties of the fluid within the control volume are invariant with time at a given position, the energy associated with the control volume must also remain constant. The energy of the control volume is affected by the bulk movement of mass across its control surface, as well as by heat and work interactions. The mass-transfer rates are constant in steady-flow processes; consequently, the rate of energy transport associated with the mass crossing every open boundary is also constant. The individual rates for the heat and work interactions are also constant. The sum of the rates of energy transfer into or out of the control volume must, of course, add up to zero. This is clearly seen by applying the general conservation of energy principle given by Eq. (5-17) to a steady-state process. Equation (5-17) is

$$\dot{Q} + \dot{W}_{net} + (e + Pv)_{in} \dot{m}_{in} - (e + Pv)_{out} \dot{m}_{out} = \frac{dE_{CV}}{dt} \tag{5-17}$$

However, the right-hand side of the equation is zero by definition for a steady-state process. Thus we find that, for a steady-state process for a control volume with one inlet and one outlet, the following form of the conservation of energy principle is valid:

$$\dot{Q} + \dot{W}_{shaft} + (e + Pv)_{in} \dot{m} - (e + Pv)_{out} \dot{m} = 0 \tag{5-21}$$

The subscripts have been removed from the mass flow rates since they are equal in magnitude in this case [Eq. (5-7)]. Equation (5-21) symbolizes the statement made above that, for a steady-state operation, the sum of the rates of energy transfer in and out of a control volume must equal zero. If more than one inlet or outlet exists, the conservation of energy principle on a rate basis must be written as

$$\dot{Q} + \dot{W}_{shaft} + \sum_{in} (e + Pv)\dot{m} - \sum_{out} (e + Pv)\dot{m} = 0 \tag{5-22}$$

In this particular equation it must be remembered that the mass flow rates $\dot{m}$ at each open boundary are not necessarily equal. In fact, they normally will not be equal. However, their values must be related by Eq. (5-9a); that is,

$$\sum \dot{m}_{in} = \sum \dot{m}_{out} \tag{5-9a}$$

The analysis of steady-flow systems will frequently require the use of both the conservation of energy and the conservation of mass principles, outlined above.

It is convenient to have the conservation of energy principle for steady-flow processes on the basis of a given time interval, as well as a rate basis. Equation (5-19) was previously developed for a differential time period $dt$ for a control volume undergoing any general process. Under the restriction of steady flow, $dE_{CV}$ is zero; hence this equation for a control volume with only one inlet and one outlet reduces to

$$\delta Q + \delta W + (e + Pv)_{in} \, dm_{in} - (e + Pv)_{out} \, dm_{out} = 0 \tag{5-23}$$

For a control volume with any number of sections open to mass transfer, the following expression would be applicable:

$$\delta Q + \delta W + \sum_{in} (e + Pv) \, dm - \sum_{out} (e + Pv) \, dm = 0 \tag{5-24}$$

Integration of Eq. (5-23) is quite straightforward in the case of a steady-state process. The properties at a given boundary are invariant with time or with respect to the particular element of mass crossing the control surface at a given time increment; hence the quantity $e + Pv$ may be brought directly outside of the integral sign in each term. Integration over a time period during which a unit mass enters and another unit mass leaves the control volume yields

$$q + w_{shaft} + (e + Pv)_{in} - (e + Pv)_{out} = 0 \tag{5-25}$$

Equations (5-21) and (5-25) are extremely important statements of the conservation of energy principle for a control volume under a steady-state condition.

## 5-4 SPECIAL CONSERVATION EQUATIONS

We are now in a position to begin to apply the conservation equations to engineering systems. We recognize that the energy $e$ associated with the mass crossing the boundaries of the control volume is comprised of a number of forms of energy. These include internal energy, kinetic energy, gravitational potential energy, and magnetic and electrical energies. For a large number of problems of practical interest the effects of magnetic and electrical fields are negligible. Thus, it is sufficient to restrict the energy of each unit of mass crossing a boundary to that due to its internal energy, its linear motion, and its position in a gravitational field. In this case:

$$e = u + \frac{V^2}{2} + gz \tag{5-26}$$

Under this restriction the conservation of energy equation per unit mass for a steady-flow process becomes

$$q + w_{\text{shaft}} = \left(u + Pv + \frac{V^2}{2} + gz\right)_{\text{out}} - \left(u + Pv + \frac{V^2}{2} + gz\right)_{\text{in}} \tag{5-27}$$

At this point another important reason for introducing the concept of enthalpy is seen. At every boundary where mass transfer occurs, the sum $u + Pv$ is associated with each unit of mass crossing the control surface. Therefore we may write the conservation of energy principle as

$$\bigstar\, q + w_{\text{shaft}} = \left(h + \frac{V^2}{2} + gz\right)_{\text{out}} - \left(h + \frac{V^2}{2} + gz\right)_{\text{in}} \tag{5-28}$$

Equation (5-28) relates the energy transfer to and from a control volume during a time interval when a unit mass enters and another unit mass leaves the control surface. It should be apparent that the datum points for the velocities and the elevations and the reference state for the enthalpy (or internal energy) must be the same for the inlet and outlet. This will be true for any other form of energy associated with the mass flowing through the control volume.

By rearranging and grouping common terms, the steady-flow energy balance is often found written as

$$q + w_{\text{shaft}} = (h_2 - h_1) + \frac{V_2^2 - V_1^2}{2} + g(z_2 - z_1)$$

$$\bigstar \qquad = \Delta h + \Delta \text{KE} + \Delta \text{PE} \tag{5-29}$$

Subscript 1 designates the inlet, the subscript 2 the outlet. The symbol $\Delta$ in this case signifies outlet conditions minus inlet conditions. This equation, although mathematically correct, is often misinterpreted. It appears as if an energy balance is being made on a unit mass as it passes through the control volume. The fact that an energy balance is being made on a control volume over a certain time interval is made obscure. Keep in mind that during a given interval of time the quantities of mass entering and leaving the control volume are equal in magnitude but they are not the same matter. This equation applies to a control-volume analysis, and not a control-mass analysis.

It is useful also to have Eq. (5-29) on a rate basis. Since

$$\dot{Q} + \dot{W}_{shaft} = (e + Pv)_{out}\,\dot{m} - (e + Pv)_{in}\,\dot{m}$$

by inserting Eq. (5-26), we find that

$$\dot{Q} + \dot{W}_{shaft} = (\Delta h + \Delta KE + \Delta PE)\dot{m} \qquad (5\text{-}30)$$

Although the enthalpy appears in the steady-flow energy balance, the student should not become accustomed to thinking of this quantity always as energy. It is better simply to consider the enthalpy as the convenient sum of two quantities which appear together in the energy balance.

Even though we have restricted ourselves to forms of the conservation of energy equation applicable solely to control volumes with one inlet and outlet, it should be apparent that extension of these basic ideas to systems of three or more flow streams at the control surface is fairly straightforward. If we again restrict the energy of the mass crossing the control surface to those forms given by Eq. (5-26), Eq. (5-30) takes on the more general form,

$$\text{✱}\quad \dot{Q} + \dot{W}_{shaft} = \sum_{out}\left(h + \frac{V^2}{2} + gz\right)\dot{m} - \sum_{in}\left(h + \frac{V^2}{2} + gz\right)\dot{m} \qquad (5\text{-}31)$$

It must be remembered that the $\dot{m}$ terms in the summations are usually different at each open boundary. This is a point that frequently leads to the incorrect analysis of control volumes with several inlets and outlets.

In the analysis of a control mass, or closed system, properties of a fixed quantity of matter were evaluated at the initial and final states of the system. These states were related by some process which occurred over a finite time interval. However, in the case of the control-volume analysis of steady-flow processes, the properties of matter at the boundaries of the control volume are of interest over a time interval irrespective of the fact that the identity of the matter at a boundary continually changes. It should be noted that the viewpoints taken for these two types of analyses are quite different. Since open systems are common in engineering practice, the student must be familiar with the use of the two conservation laws presented in this section—those of mass and energy. In order to familiarize the student with the general method, a number of examples are presented in Sec. 5-5. These problems will attempt to give a feeling of orders of magnitude of various terms, the handling of units, and the importance of various terms in the energy equation when applied to definite types of engineering equipment. Confidence in the use of the control-volume approach to the solution of open-system problems will come only with experience. These solutions require not only the use of one or more of the conservation equations, but also relationships among the properties of matter. Thus we again rely on information from earlier chapters.

## 5-5 ENGINEERING APPLICATIONS INVOLVING STEADY-STATE SYSTEMS

The following examples illustrate the use of the various forms of the conservation of energy equation for a number of control-volume analyses. At this point it is important for the reader to be reminded of a few of the major points that need to be considered in any thermodynamic analysis. These include:

1. Select a suitable control volume for analysis, and sketch the system, indicating the appropriate boundaries.
2. Determine what energy interactions are important, and recognize certain sign conventions on such terms.
3. Write a suitable energy balance for the chosen system.
4. Obtain physical data for the substance under study. Is an equation of state available (such as $Pv = RT$), or must graphical and/or tabular data be employed? What are other property relations for the substance?
5. Determine the path of the process between the initial and final states. Is it isothermal, constant-pressure, quasistatic, adiabatic, etc.?
6. What other idealizations or assumptions are necessary to complete the solution? Is kinetic energy negligible, etc.?
7. Draw suitable diagram(s) for the process, as an aid in picturing the overall problem.
8. Complete the solution for the required item(s) on the basis of the information supplied. Check units in each equation used.

Two of the items in the above list are extremely important, although the neophytes in thermodynamic analysis tend to discount their value. The first item suggests sketching the system under analysis. On the sketch one should indicate all the relevant energy-transfer terms consistent with the boundaries selected for the system. Sketches help the engineer approach a problem in a straightforward and consistent manner. Equally important is the process diagram. By process diagram we mean a drawing of the process on thermodynamic coordinates, such as a $Pv$ plane. The process diagram is especially meaningful in terms of establishing the initial and final states of the process. The value of the process diagram will become more apparent as more properties are introduced, and as problems become more complex. Although the process diagram is listed as item 7, it is frequently one of the first things that is done in a thermodynamic analysis.

Before proceeding with the analysis of some specific steady-state devices, it is appropriate to examine the relative magnitude of kinetic energy and potential energy terms. This will be done for both SI and USCS units.

**ⓐ Kinetic energy and potential energy in SI units** The linear kinetic energy of a unit mass is $V^2/2$. If the velocity is expressed in m/s, then the kinetic energy will be expressed in N·m/kg, or J/kg. Since it is more customary to evaluate energy terms in J/g and kJ/kg, then,

$$KE = \frac{V^2}{2}\frac{m^2}{s^2} \times \frac{N \cdot s^2}{kg \cdot m} \times \frac{kg}{10^3 g} = \frac{V^2}{2000} \text{ kJ/kg (or J/g)}$$

For many engineering applications the values of various energy terms are usually at least 1 kJ/kg in magnitude, and more commonly in the range of 10–100 kJ/kg. For a KE of 1 kJ/kg, the above equation requires a velocity of $(2000)^{1/2}$, or 45 m/s, and a kinetic energy of 10 kJ/kg requires a velocity of 140 m/s. The significant point is that a relatively high velocity is required if the kinetic energy is to be important relative to other terms in an energy balance.

The gravitational potential energy relative to Earth's surface is given by $gz$. If $g$ is in m/s² and the elevation relative to some datum is measured in meters, then the potential energy will be expressed in N·m/kg, or J/kg. Consequently, the value of $gz$ must be divided by 1000, much as in the case of kinetic energy, if the energy is desired in units of kJ/kg (or J/g). The value of $g$ on Earth's surface is roughly 9.8 m/s². For a potential energy of 1 kJ/kg, the elevation required is

$$z = \frac{PE}{g} = 1\frac{kJ}{kg} \times \frac{s^2}{9.8 \text{ m}} \times \frac{kg \cdot m}{N \cdot s^2} \times \frac{1000 \text{ N} \cdot m}{kJ} = 102 \text{ m}$$

For 10 kJ/kg, the height requirement obviously is 1020 m. Since the flow in many industrial devices seldom has an elevation change of this magnitude, the gravitation potential energy change is frequently negligible. Some care must be taken, however, about neglecting the potential energy change when pumping liquids. This will be illustrated by a numerical example later.

**ⓚ Kinetic energy and potential energy in USCS units** The linear kinetic energy of a unit mass is $V^2/2$. If the velocity is expressed in ft/s, then the kinetic energy will be expressed in ft·lb$_f$/lb$_m$ if the term is divided by the unit conversion factor, 32.2 lb$_m$·ft/(lb$_f$)(s²) = 1. In conjunction with tabular data and property equations, we shall find it useful to express energy in Btu/lb. Hence,

$$KE = \frac{V^2}{2}\frac{ft^2}{s^2} \times \frac{lb_f \cdot s^2}{32.2 \text{ lb}_m \cdot ft} \times \frac{Btu}{778 \text{ ft} \cdot lb_f} = \frac{V^2}{50,000} \text{ Btu/lb}_m$$

where the constant 50,000 is approximate. For many engineering applications the values of various energy terms are usually at least 1 to 10 Btu/lb, with some values greater than 100 Btu/lb. For a kinetic energy of 1 Btu/lb, the above equation requires a velocity of $(50,000)^{1/2}$, or 225 ft/s. This is a reasonably high velocity for gas flow in commercial applications. Thus relatively high velocities are required if the kinetic energy of a flow stream is to be important relative to other energy quantities.

The gravitational potential energy relative to Earth's surface is given by $gz$. If $g$ is measured in ft/s² and the elevation relative to some datum is measured in feet, then the potential energy will be expressed in $ft \cdot lb_f / lb_m$ if this term is divided by the unit conversion factor used for the kinetic energy. Again, a factor of 778 for conversion to Btus is also customary. The value of $g$ on the earth's surface is roughly $32.2$ $ft/s^2$. For a potential energy of 1 Btu/lb, the elevation required is

$$z = \frac{PE}{g} = 1\frac{Btu}{lb_m} \times \frac{s^2}{32.2\ ft} \times \frac{32.2\ lb_m \cdot ft}{lb_f \cdot s^2} \times \frac{778\ ft \cdot lb_f}{Btu}$$

$$= 778\ ft$$

Since the flow in many industrial processes seldom has an elevation change of this magnitude, the gravitational potential energy change frequently is negligible. Some care must be taken however, about neglecting this term when pumping liquids through modest elevation changes. This will be illustrated later.

## 1 Nozzles and Diffusers

In many steady-flow processes there is a need to either increase or decrease the velocity of a flow stream. A device which increases the velocity (and hence the kinetic energy) of a fluid at the expense of a pressure drop in the direction of flow is called a *nozzle*. A *diffuser* is a device for increasing the pressure of a flow stream at the expense of a decrease in velocity. These defining conditions apply for both subsonic and supersonic flow. Figure 5-3 is a set of sketches for the general shapes of a nozzle and a diffuser under the conditions of subsonic and supersonic flow. Note that a nozzle is a converging passage for subsonic flow, whereas the passage is diverging for supersonic flow. The opposite conditions hold for a diffuser. We are not in a position to explain the reason for these differences at this point, theoretically; so they should be accepted as a matter of experience. One outgrowth of this which is extremely important in rocket design is that a converging-diverging nozzle must be used if a fluid is to be accelerated from subsonic to supersonic velocities. The actual shapes of these devices in practice are not as shown below, but will probably have curved surfaces in the direction of flow. The true shape is unimportant for the overall energy balances that we wish to make. Since both these devices are merely ducts, it is apparent that no shaft work is involved, and the change in potential energy, if any, would be negligible under

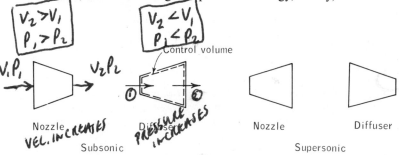

Figure 5-3 General shapes of nozzles and diffusion for subsonic and supersonic flow.

most conditions. For a simple substance the open, steady-flow energy equation per unit mass reduces to

$$q = \frac{V_2^2 - V_1^2}{2} + h_2 - h_1$$

For nozzles, the exit pressure may be imposed or controlled externally. In many cases the heat transfer per unit mass may be quite small compared to the kinetic-energy and enthalpy changes. In some cases the duct may be deliberately insulated. Even without insulation, the fluid velocity may be so large that there is not time enough for the fluid to come to thermal equilibrium with the walls. Also the surface area of the duct is frequently relatively small for effective heat transfer. Thus in many applications the assumptions of an adiabatic process is a good approximation for nozzles and diffusers.

For adiabatic nozzles and diffusers the energy equation reduces to

$$-\Delta h = \Delta KE$$

or

$$-\Delta u - \Delta(Pv) = \Delta KE$$

The latter equation emphasizes the fact that the kinetic-energy change is due to two effects, namely, a change in the internal energy of the fluid and a change in displacement work during the process. Although the use of the enthalpy function is a real convenience, one must not forget the underlying factors which contribute to its change. If we assume, for the purpose of a qualitative illustration, that the fluid is an ideal gas in the nozzle or diffuser, then the last equation can be written as

$$c_v(T_1 - T_2) + R(T_1 - T_2) = \Delta KE$$

Recall that for gases, $c_v > R$, but they are the same order of magnitude. For example, as an approximation for air, $c_v/R = 5/2$. Hence the internal-energy change for air is about twice that of the change in displacement work as the gas is accelerated or decelerated.

In the application of the conservation of energy principle to a nozzle or diffuser, the specification of the inlet conditions is insufficient to determine the outlet state of a fluid. One outlet property or a specification of the type of process is required to fix the final state. The continuity equation may also be useful because of the variable-area nature of these devices.

**Example 5-3M** Water vapor enters a subsonic diffuser at a pressure of 0.7 bar, a temperature of 160°C, and a velocity of 180 m/s. The inlet to the diffuser is 100 cm². During passage through the diffuser the fluid velocity is reduced to 60 m/s, the pressure increases to 1.5 bars, and 19.8 J/g of heat are transferred to the surroundings. Determine (a) the final temperature, (b) the mass flow rate, and (c) the outlet area in cm².

SOLUTION (a) The final temperature is a function of the final pressure, which is 1.5 bars, and one other property. The continuity-of-flow equation is insufficient to help at this point, since both

$A_2$ and $v_2$ are unknown. The energy balance for this steady-state device on a unit-mass basis is, since $w = 0$,

$$q = h_2 - h_1 + \frac{V_2^2 - V_1^2}{2}$$

The potential-energy change is assumed to be zero. It is seen that sufficient information is given to evaluate all the quantities except $h_2$. From the steam table in the superheat region, Table A-14M, it is found that $h_1 = 2798.2$ kJ/kg. Upon substituting the known values, we find that

$$-19.8 = h_2 - 2798.2 + \frac{(60)^2 - (180)^2}{2(1000)}$$

or

$$h_2 = 2798.2 - 19.8 + 14.4 = 2792.8 \text{ kJ/kg}$$

Knowledge of $P_2$ and $h_2$ establishes the values of all other thermostatic properties, such as the temperature $T_2$. The enthalpy of saturated vapor at 1.5 bars is given in Table A-13M as 2693.6 kJ/kg. Since $h_2 = 2792.8$ kJ/kg, the final state is in the superheat region. From the 1.5-bar set of data in Table A-14M, the calculated value of $h_2$ corresponds to a temperature of 160°C. Thus the heat removed has been enough to make the initial and final temperatures the same.

(b) The mass flow rate is found from the continuity equation.

$$\dot{m} = \frac{V_1 A_1}{v_1} = \frac{180 \text{ m/s} \times 100 \text{ cm}^2}{2841 \text{ cm}^3/\text{g}} \times \frac{100 \text{ cm}}{\text{m}} = 634 \text{ g/s} = 0.634 \text{ kg/s}$$

(c) The outlet area is found also by application of the continuity equation. Since $\dot{m}$ is now known, we may find $A_2$ from the relation $\dot{m} = V_2 A_2/v_2 = V_1 A_1/v_1$. From the superheated steam table, A-14M, the value of $v_2$ is given as 1317 cm³/g. Rather than use the value of $\dot{m}$, we shall evaluate the final area by

$$A_2 = \frac{A_1 V_1 v_2}{V_2 v_1} = 100 \text{ cm}^2 \times \frac{180}{60} \times \frac{1317}{2841} = 139 \text{ cm}^2$$

The final area of the diffuser is roughly 40 percent greater than the inlet area. Figure 5-4 illustrates the process on a $Pv$ diagram, which is not to scale.

**Example 5-3** Water vapor enters a subsonic diffuser at a pressure of 10 psia, a temperature of 300°F, and a velocity of 400 ft/s. The inlet to the diffuser is 0.1 ft². During passage through the steady-flow device the fluid velocity is reduced to 175 ft/s, the pressure increases to 20 psia, and 6.6 Btu/lb of heat are transferred to the surroundings. Determine (a) the final temperature in °F, (b) the mass flow rate, and (c) the outlet area in ft².

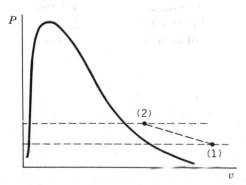

Figure 5-4 Pressure-volume diagram showing pressure increase in a diffuser.

SOLUTION (a) Knowledge of two thermostatic properties is required to find the value of the final temperature. Since $P_2$ is given, one other property value needs to be found. The continuity equation is insufficient to help at this point, since both $A_2$ and $v_2$ are unknown. The energy balance for this device on a unit-mass basis is, since $w = 0$,

$$q = h_2 - h_1 + \frac{V_2^2 - V_1^2}{2}$$

The potential energy change is assumed negligible. It is seen that there is sufficient information given to evaluate all the quantities except $h_2$. From the superheated steam table, A-14, it is found that $h_1 = 1193.7$ Btu/lb. Upon substituting the known values, we find that in units of Btu/lb

$$-6.6 = h_2 - 1193.7 + \frac{(175)^2 - (500)^2}{2(32.2)(778)}$$

or

$$h_2 = 1193.7 - 6.6 + 4.4 = 1191.5 \text{ Btu/lb}$$

Knowledge of $P_2$ and $h_2$ determines $T_2$. The enthalpy of saturated vapor at 20 psia given in Table A-13 is 1156.5 Btu/lb. Since $h_2$ is much larger than this, the final state is a superheated one. From the 20-psia set of data in Table A-14, the calculated value of $h_2$ corresponds to a temperature $T_2$ of 300°F. Thus the initial and final temperatures are the same.

(b) The mass flow rate is found from the continuity equation.

$$\dot{m} = \frac{V_1 A_1}{v_1} = \frac{500 \text{ ft/s} \times 0.1 \text{ ft}^2}{44.99 \text{ ft}^3/\text{lb}} = 1.11 \text{ lb/s}$$

(c) The outlet area is also found by application of the continuity equation. Since $\dot{m}$ is known, we may find $A_2$ from the relation $\dot{m} = V_2 A_2/v_2 = V_1 A_1/v_1$. From the superheated steam table, A-14, the value of $v_2$ is given as 22.36 ft³/lb. Rather than use the value of $\dot{m}$, we shall evaluate the final area by

$$A_2 = \frac{A_1 V_1 v_2}{V_2 v_1} = 0.1 \text{ ft}^2 \times \frac{500}{175} \times \frac{22.36}{44.99} = 0.142 \text{ ft}^2$$

The final area of the diffuser is roughly 40 percent greater than the inlet area. Figure 5-4 illustrates the process on a $Pv$ diagram, which is not to scale.

## 2 Turbines, Pumps, Compressors, and Fans

A turbine, whether the fluid is a gas or a liquid, is a device in which the fluid does work against some type of blade attached to a rotating shaft. As a result, the device produces work which may be used for some purpose in the surroundings. Pumps, compressors, and fans are devices in which work is done on the fluid which results in an increase in pressure of the fluid. Pumps are usually associated with liquids and compressors and fans are employed for gases. The ratio of outlet-to-inlet pressure across a fan will probably be just slightly above 1, whereas for a compressor the ratio will probably be between 3 and 10. For steady flow through any of these devices the energy equation reduces to

$$q + w = h_2 - h_1 + \frac{V_2^2 - V_1^2}{2}$$

The potential-energy change is normally negligible. Two other terms in this expression require comment. As we have noted already, the inclusion of a heat quantity depends upon the mode of operation. If the device is not insulated, the heat gained or lost by the fluid depends upon such factors as whether or not (1) a large temperature difference exists between the fluid and the surroundings, (2) a small flow-velocity exists, and (3) a large surface area is present. In rotating turbomachinery (axial or centrifugal) velocities can be fairly high, and the heat transfer normally is small compared to the shaft work. In reciprocating devices the heat-transfer effects may be reasonably large. Experience and experimental specifications enable the engineer to estimate the relative importance of heat transfer. As a second point, it is found that the change in kinetic energy is usually quite small in these devices, since the velocities at inlet and outlet are frequently less than several hundred feet per second. There are exceptions, of course. In a steam turbine the exhaust velocity can be quite high, due to the large specific volume of the fluid at the low exhaust pressure. On the basis of the continuity equation, velocities may be kept low by selecting large flow areas. However, this may not be a practical choice.

In many cases, the steady-flow energy balance for these work-producing or work-absorbing devices becomes

$$w = h_2 - h_1$$

In this approximate solution it is seen that the enthalpy decreases for a turbine and increases in the direction of flow for compressors and pumps.

**Example 5-4M** Air initially at 1 bar and 290°K is compressed to 5 bars and 450°K. The power input to the air under steady-flow conditions is 5 kW, and a heat loss of 5 kJ/kg occurs during the process. If the changes in potential and kinetic energies are neglected, determine the mass flow rate in kg/min.

SOLUTION The steady-state energy balance given by Eq. (5-29) reduces to

$$q + w_{shaft} = h_2 - h_1$$

All quantities in this equation are known except the shaft work. Since the inlet and exit temperatures are known, the enthalpies can be read from Table A-5M. Therefore,

$$w_{shaft} = h_2 - h_1 - q = 451.8 - 290.2 - (-5) = 166.6 \text{ kJ/kg}$$

Power is the product of the mass flow rate and the shaft work. The mass flow rate is determined, then, by

$$\dot{m} = \frac{\text{Power}}{w_{shaft}} = \frac{5 \text{ kW}}{166.6 \text{ J/g}} \times \frac{1000 \text{ J/s}}{1 \text{ kW}} = 30.0 \text{ g/s} = 1.80 \text{ kg/min}$$

**Example 5-4** Air initially at 15 psia and 60°F is compressed to 75 psia and 400°F. The power input to the air under steady-state conditions is 5 hp, and a heat loss of 4 Btu/lb occurs during the process. If the changes in potential and kinetic energies are neglected, determine the mass flow rate in lb/min.

SOLUTION The steady-state energy balance given by Eq. (5-29) reduces to

$$q + w_{shaft} = h_2 - h_1$$

All quantities in this equation are known except the shaft work. The enthalpies are read from Table A-5 in terms of the initial and final temperatures. Therefore,

$$w_{shaft} = h_2 - h_1 - q = 206.5 - 124.3 - (-4) = 86.2 \text{ Btu/lb}$$

Power is the product of the mass flow rate and the shaft work, that is $\dot{m}w$. The mass flow rate then is

$$\dot{m} = \frac{\text{Power}}{w_{shaft}} = \frac{5 \text{ hp}}{86.2 \text{ Btu/lb}} \times \frac{42.4 \text{ Btu/min}}{1 \text{ hp}} = 2.46 \text{ lb/min}$$

**Example 5-5M** A steam turbine operates with an inlet condition of 30 bars, 400°C, 160 m/s and an outlet state of a saturated vapor at 0.7 bar with a velocity of 100 m/s. The mass flow rate is 1000 kg/min and the power output is 9300 kW. Determine the magnitude and direction of any heat transfer, in kJ/min.

SOLUTION Since the heat transfer is desired on a rate basis, we shall employ Eq. (5-30) as a suitable energy balance under assumed steady-state conditions. That is,

$$\dot{Q} + \dot{W} = \dot{m}\left(h_2 - h_1 + \frac{V_2^2 - V_1^2}{2}\right)$$

Since the saturation temperature at 30 bars is 233.9°C, the initial state of 400°C is superheated. From the superheat table, A-14M, $h_1 = 3230.9$ kJ/kg, and from the saturation pressure table, A-13M, $h_2 = h_g = 2660$ kJ/kg. Consequently,

$$\dot{Q} + (-9300)\text{kW} \times \frac{\text{kJ}}{1 \text{ kW·s}} \times \frac{60 \text{ s}}{\text{min}} = 1000 \text{ kg/min} \times \left[(2660 - 3231) + \frac{(100)^2 - (160)^2}{2(1000)}\right] \text{kJ/kg}$$

$$\dot{Q} = 558,000 + 1000(-571 - 8) = -21,000 \text{ kJ/min}$$

Note that the heat transfer is from the turbine (since the steam is much hotter than the surroundings) and is only about 4 percent of the work interaction. Also, the kinetic-energy change is only 1.4 percent of the enthalpy change. The process is shown on a $Pv$ diagram in Fig. 5-5.

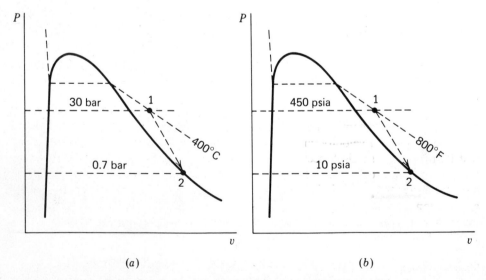

**Figure 5-5** The $Pv$ diagrams for processes represented in Examples 5-5M and 5-5.

**Example 5-5** A steam turbine operates with an inlet condition of 450 psia, 800°F, 450 ft/s and an outlet state of a saturated vapor at 10 psia with a velocity of 250 ft/s. The mass flow rate is 2000 lb/min and the power output is 12,300 hp. Determine the magnitude and direction of any heat transfer in Btu/min.

SOLUTION Since the heat transfer is desired on a rate basis, we shall employ Eq. (5-30) as a suitable energy balance under assumed steady-state conditions. That is,

$$\dot{Q} + \dot{W} = \dot{m}\left[h_2 - h_1 + \frac{V_2^2 - V_1^2}{2}\right]$$

Since the saturation temperature at 450 psia is 456.4°F, the initial state of 800°F is superheated. From Table A-14, $h_1 = 1414.4$ Btu/lb, and from the saturation pressure table, A-13, $h_2 = h_g = 1143.3$ Btu/lb. Consequently,

$$\dot{Q} + (-12,300)\text{hp} \times \frac{42.4 \text{ Btu}}{\text{hp·min}} = 2000 \text{ lb/min} \times \left[(1143.3 - 1414.4) + \frac{(250)^2 - (450)^2}{2(32.2)(778)}\right] \text{Btu/lb}$$

$$\dot{Q} = 521,500 + 2000(-271 - 3) = -26,500 \text{ Btu/min}$$

Note that the heat transfer is from the turbine (since the steam is much hotter than the surroundings) and is only about 5 percent of the work transfer. Also, the kinetic-energy change is only 1 percent of the enthalpy change. The process is shown on a $Pv$ diagram in Fig. 5-5.

# 3 Throttling Devices

Steady-flow systems such as turbines and nozzles produce useful work or increase the kinetic energy of the fluid as it passes through. Accompanying these effects is a drop in pressure. This decrease in pressure is externally controlled. Control over this pressure difference can be accomplished by inserting in the flow system another component called a throttling device. The process of throttling is used for purposes other than control. The main effect, however, is a significant pressure drop without any work interactions or changes in kinetic or potential energy. Flow through a restriction such as a valve or a porous plug fulfills the necessary conditions. A throttling valve is shown in Fig. 5-6. By decreasing the cross-sectional area for flow, a greater flow resistance is introduced. For a given mass flow rate, the greater flow resistance requires a greater pressure drop across the valve.

Although the velocity may be quite high in the region of the restriction, measurements upstream and downstream from the actual valve area will indicate that the change in velocity, and hence the change in kinetic energy, across the restriction is very small. Since the control volume is rigid and no rotating shafts are present, no work interactions are involved, either. The steady-flow energy balance under these restrictions would reduce to $q = h_2 - h_1$. However, in most applications, either the throttling device is insulated or the heat transfer is insignificant. Thus, for a throttling process, the enthalpy change is zero, or

$$h_1 = h_2$$

This statement does not say that the enthalpy is constant for the process, but merely requires that the initial and final enthalpies be the same. The valves in

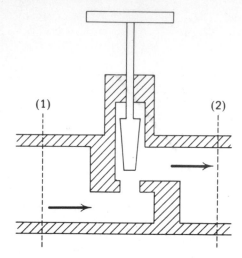

**Figure 5-6** A throttle valve.

water faucets in the home are examples of throttling devices. These devices are also in common usage in most home refrigeration units. A true throttling process which obeys the above equation is frequently called a Joule–Thomson expansion.

**Example 5-6M** Refrigerant 12 is throttled from a saturated liquid at 32°C to a final state where the pressure is 2 bars. Determine the final temperature and the physical state of the fluid at the exit.

SOLUTION From the saturation-temperature table, A-16M, the enthalpy of saturated liquid at 32°F is 66.57 kJ/kg. This also must be the enthalpy of a unit mass after throttling. From the saturation-pressure table, A-17M, at 2 bars the value of $h_f$ is 24.57 kJ/kg and that of $h_g$ is 182.07 kJ/kg. Therefore, the final state must be a mixture of liquid and vapor, and the exit temperature is the corresponding saturation temperature at 2 bars of $-12.53°C$. The quality in the final state is found by

$$h_2 = h_1 = 66.57 = h_f + xh_{fg} = 24.57 + 157.50x$$

or

$$x = \frac{42.0}{157.5} = 0.267$$

The final mixture is roughly 27 percent vapor and 73 percent liquid. The approximate path of the process in the wet region is sketched in Fig. 5-7.

**Example 5-6** Refrigerant 12 is throttled from a saturated liquid at 90°F to a final state where the pressure is 20 psia. Determine the final temperature and the physical state of the fluid at the exit.

SOLUTION From the saturation-temperature table, A-16, the enthalpy of a saturated liquid at 90°F is 28.71 Btu/lb. This also must be the enthalpy of a unit mass after throttling. From the saturation-pressure table, A-17, at 20 psia the value of $h_f$ is 6.77 Btu/lb and that of $h_g$ is 76.40 Btu/lb. Consequently, the final state must be a mixture of liquid and vapor, and the exit

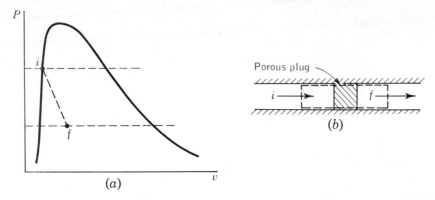

**Figure 5-7** Throttling. (*a*) Process representation; (*b*) control volume.

temperature is the corresponding saturation temperature at 20 psia of $-8.13°F$. The quality in the final state is found by

$$h_2 = h_1 = 28.71 = h_f + xh_{fg} = 6.77 + 69.63x$$

or

$$x = \frac{21.94}{69.63} - 0.315$$

The final mixture is roughly 27 percent vapor and 73 percent liquid. The approximate path of the process in the wet region is sketched in Fig. 5-7.

## 4 Heat Exchangers

One of the most important steady-flow devices of engineering interest is the heat exchanger. These devices serve two useful purposes. Either they are used to remove (or add) energy from some region of space or they are employed to change deliberately the thermodynamic state of a fluid. The radiator of an automobile is an example of a heat exchanger used for heat removal. Modern gas turbines and electrical generators are frequently cooled internally, and their performance is greatly affected by the heat-transfer process. In steam power plants, heat exchangers are used to remove heat from hot combustion gases and subsequently increase the temperature and enthalpy of the steam in the power cycle. This high-enthalpy fluid is then expanded with a resulting large power output. In the chemical industry, heat exchangers are extremely important in maintaining or attaining certain thermodynamic states as chemical processes are carried out. Modern applications of heat exchangers are numerous.

One of the primary applications of heat exchangers is the exchange of energy between two moving fluids. The changes of kinetic and potential energies are usually negligible and no work interactions are present. The pressure drop through a heat exchanger is usually small, hence an assumption of constant pressure is often quite good. A heat exchanger composed of two concentric pipes is

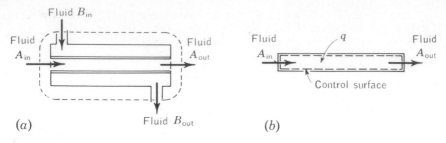

Figure 5-8 Two different control surfaces for a heat exchanger.

shown in Fig. 5-8. One fluid, $A$, flows in the inner pipe, and a second fluid, $B$, flows in the annular space between the pipes. Two different equations arise from energy balances on this equipment, depending upon where the system boundaries are selected. In the first case, the control surface will be placed around the entire piece of equipment, as indicated by the dashed line in Fig. 5-8a. For this system the steady-flow energy balance on a rate basis [Eq. (5-30)] is applicable. We assume that heat transfer external to the device is zero, and $\dot{W}$ is zero since no shaft work exists. In addition, the kinetic- and potential-energy changes of the fluid streams usually are negligible. In terms of the notation shown in Fig. 5-8a, Eq. (5-30) reduces to

$$\dot{m}_A(h_{A1} - h_{A2}) = \dot{m}_B(h_{B2} - h_{B1})$$

Second, it is sometimes desirable to place a set of boundaries around either of the two fluids. In this case not only is there a change in the enthalpy of the fluid, but a heat-transfer term also appears. For example, if the actual heat transfer in the above case is positive from fluid $A$ to fluid $B$, then

$$\dot{Q} = \dot{m}_B(h_{B2} - h_{B1})$$
$$-\dot{Q} = \dot{m}_A(h_{A2} - h_{A1})$$

The heat-transfer rates are, of course, identical. Figure 5-8b shows the system boundaries in this latter case for fluid $A$. Properly used, these equations may be applied to specific types of heat-exchanger equipment called boilers, evaporators, and condensers. As the names suggest, in these types of equipment one of the fluids will change phase. Hence the enthalpy of vaporization will be part of the overall enthalpy change for one of the fluids. Several examples of energy balances on heat-exchanger equipment are given below.

    ✗ **Example 5-7M** A small nuclear reactor is cooled by passing liquid sodium through it. The liquid sodium leaves the reactor at 2 bars and 400°C. It is cooled to 320°C by passing through a heat exchanger before returning to the reactor. In the heat exchanger, heat is transferred from the liquid sodium to water, which enters the exchanger at 100 bars and 40°C and leaves at the same pressure as a saturated vapor. The mass flow rate of sodium is 10,000 kg/h, and its specific heat is constant at 1.25 J/(g)(°C). Determine the mass flow rate of water evaporated in the heat exchanger in kg/h, and the heat-transfer rate between the two fluids in kJ/h.

SOLUTION The mass flow rate can be determined from an energy balance on the entire heat exchanger. It is assumed that the heat loss from the exterior surface of the exchanger is negligible. In terms of the above statement, the energy balance may be written as

$$\dot{m}_{H_2O}(h_{out} - h_{in})_{H_2O} = \dot{m}_{Na}(h_{in} - h_{out})_{Na} = \dot{m}_{Na}c_p(T_{in} - T_{out})_{Na}$$

The inlet water stream is a compressed liquid. Its enthalpy is determined from Table A-15M as 176.38 kJ/kg. The outlet enthalpy of saturated water vapor at 100 bars is found in Table A-13M to be 2724.7 kJ/kg. Substitution of these values into the above equation yields

$$\dot{m}_{H_2O}(2724.7 - 176.4) = 10,000(1.25)(400 - 320)$$

or

$$\dot{m}_{H_2O} = \frac{10,000(1.25)(80)}{2548.3} = 392 \text{ kg/h}$$

The heat-transfer rate from the sodium to the water is the value of either the left- or right-hand side of the energy balance. Using the sodium stream to evaluate the heat-transfer rate, one finds that

$$\dot{Q} = \dot{m}_{Na}c_p(T_{in} - T_{out}) = 10,000(1.25)(80) = 1.0 \times 10^6 \text{ kJ/h}$$

**Example 5-7** A small nuclear reactor is being cooled by passing liquid sodium through it. The liquid sodium leaves the reactor at 25 psia and 725°F. The liquid sodium is cooled to 610°F by passing through a heat exchanger before returning to the reactor. In the heat exchanger, heat is transferred from the liquid sodium to water, which enters the equipment at 1000 psi and 100°F and leaves at the same pressure as a saturated vapor. The mass flow rate of the sodium is 20,000 lb$_m$/h, and its specific heat is constant at 0.3 Btu/(lb)(°F). Determine the mass flow rate of water evaporated in the heat exchanger in lb/h and the heat-transfer rate between the fluids in Btu/h.

SOLUTION The mass flow rate of water can be determined from an energy balance on the entire heat exchanger. It must be assumed that energy leaves only with the fluid streams, so that heat loss from the exterior surface of the heat exchanger is negligible. In actual practice, the external heat loss could be predicted by heat-transfer calculations, which are independent of thermodynamics. In terms of the above problem, the energy balance may be written as

$$\dot{m}_{H_2O}(h_{out} - h_{in})_{H_2O} = \dot{m}_{Na}(h_{in} - h_{out})_{Na} = \dot{m}_{Na}c_p(T_{in} - T_{out})_{Na}$$

The inlet water stream is a compressed liquid. Its enthalpy is determined from Table A-15 as 70.68 Btu/lb. The outlet enthalpy of saturated water vapor at 1000 psia is found in Table A-13 to be 1192.4 Btu/lb. Substitution of these values into the above equation yields

$$\dot{m}_{H_2O}(1192.4 - 70.7) = 20,000(0.3)(725 - 610)$$

or

$$\dot{m}_{H_2O} = \frac{20,000(0.3)(115)}{1121.7} = 615 \text{ lb/h}$$

The heat-transfer rate from the sodium to the water is the value of either the left- or right-hand side of the energy balance. Using the sodium stream to evaluate the heat-transfer rate, one find that

$$\dot{Q} = \dot{m}_{Na}c_p(T_{in} - T_{out}) = 20,000(0.3)(115) = 690,000 \text{ Btu/h}$$

**Example 5-8M** In large steam power plants the use of open feedwater heaters is frequently noted. A sketch of an open feedwater heater, which externally appears as a large rigid tank, is

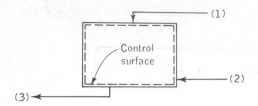

**Figure 5-9** An open feedwater heater.

shown in Fig. 5-9. This feedwater heater may be classified loosely as a heat exchanger, although in this case the fluid streams come into direct contact inside the heater. Compressed liquid water from a feedwater pump in the power-plant cycle enters at section 1. Superheated steam bled from the turbine in the power cycle is shown entering at section 2. After direct mixing within the heater, the resulting fluid, which is a saturated liquid at the heater pressure, leaves at section 3.

For a control volume with three or more flow streams the steady-flow energy equation is given by

$$\cancel{\dot{Q}}^{\,0} - \cancel{\dot{W}_{\text{shaft}}}^{\,0} = \sum_{\text{out}} \left( h + \cancel{\frac{V^2}{2}}^{\,0} + \cancel{gz}^{\,0} \right) \dot{m} - \sum_{\text{in}} \left( h + \cancel{\frac{V^2}{2}}^{\,0} + \cancel{gz}^{\,0} \right) \dot{m}$$

If the heater is essentially adiabatic and the potential- and kinetic-energy changes of the flow streams are negligible, the steady-flow energy balance on the entire heater becomes

$$\dot{m}_1 h_1 + \dot{m}_2 h_2 = \dot{m}_3 h_3$$

A form of the continuity-of-flow equation also shows that

$$\dot{m}_1 + \dot{m}_2 = \dot{m}_3$$

These two equations are sufficient to determine the state and flow rates at the different sections, in conjunction with other data on the power cycle.

Consider now an open feedwater heater operating at 5 bars. The superheated vapor is bled off the turbine at this pressure and 200°C. The compressed liquid coming from the pump is at 40°C. For every kilogram of steam bled from the turbine per unit time, how many kilograms of compressed liquid enter through section 1?

SOLUTION The first task is to evaluate the enthalpies at the three sections. At 5 bars and 200°C the enthalpy from Table A-14M is 2855.4 kJ/kg. At section 1 the compressed liquid enters at 5 bars and 40°C. Since the pressure is not high, we can estimate the enthalpy by using the saturated liquid value at the given temperature. This is found in Table A-12M to be 167.57 kJ/kg. The fluid leaves at section 3 as a saturated liquid at 5 bars. From Table A-13M, $h_3$ is 640.23 kJ/kg. If we let $\dot{m}_2$ be unity, the energy and mass balances lead to

$$2855.4 + \dot{m}_1(167.6) = \dot{m}_3(640.2)$$

or

$$2855.4 + \dot{m}_1(167.6) = (1 + \dot{m}_1)(640.2)$$

The solution to this expression is

$$\dot{m}_1 = \frac{2885.4 - 640.2}{640.2 - 167.6} = 4.75 \text{ kg/unit time}$$

Thus over 80 percent of the mass entering the feedwater heater is compressed liquid, and the remainder comes from steam bled from the turbine.

Example 5-8 In large steam power plants the use of open feedwater heaters is frequently noted. A sketch of an open feedwater heater, which externally appears as a large rigid tank, is shown in Fig. 5-9. This feedwater heater may be classified loosely as a heat exchanger, although in this case the fluid streams come into direct contact inside the heater. Compressed liquid water from a feedwater pump in the power-plant cycle enters at section 1. Superheated steam bled from the turbine in the power cycle is shown entering at section 2. After direct mixing within the heater, the resulting fluid, which is a saturated liquid at the heater pressure, leaves at section 3.

For a control volume with three or more flow streams the steady-flow energy equation is given by

$$\dot{Q} - \dot{W}_{shaft} = \sum_{out} \left(h + \frac{V^2}{2} + gz\right)\dot{m} - \sum_{in} \left(h + \frac{V^2}{2} + gz\right)\dot{m}$$

If the heater is essentially adiabatic and the potential- and kinetic-energy changes of the flow streams are negligible, the steady-flow energy balance on the entire heater becomes

$$\dot{m}_1 h_1 + \dot{m}_2 h_2 = \dot{m}_3 h_3$$

A form of the continuity-of-flow equation also shows that

$$\dot{m}_1 + \dot{m}_2 = \dot{m}_3$$

These two equations are sufficient to determine the state and flow rates at the different sections, in conjunction with other data on the power cycle.

Consider now an open feedwater heater operating at 80 psia. The superheated vapor is bled off the turbine at this pressure and 400°F. The compressed liquid coming from the pump is at 100°F. For every pound of steam bled from the turbine per unit time, how many pounds of compressed liquid enter section 1?

SOLUTION The first task is to evaluate the enthalpies at the three sections. At 80 psia and 400°F the enthalpy from Table A-14 is 1230.6 Btu/lb. At section 1 the compressed liquid enters at 80 psia and 100°F. Since the pressure is not high, we can estimate the enthalpy by using the saturated liquid value at the given temperature. This is found in Table A-12 to be 68.1 Btu/lb. The fluid leaves at section 3 as a saturated liquid at 80 psia. From Table A-13, $h_3$ is 282.2 Btu/lb. If we let $\dot{m}_2$ be unity, the energy and mass balances lead to

$$1230.6 + \dot{m}_1(68.1) = \dot{m}_3(282.2)$$

or

$$1230.6 + \dot{m}_1(68.1) = (1 + \dot{m}_1)(282.2)$$

The solution to this equation is

$$\dot{m}_1 = \frac{1230.6 - 282.2}{282.2 - 68.1} = 4.44 \text{ lb/unit time}$$

Thus a little over 80 percent of the mass entering the feedwater heater per unit time is compressed liquid, and the remainder comes from steam bled from the turbine.

# 5 Pipe Flow

The transport of fluids, either gaseous or liquid, in pipes within a system or between systems is of primary importance to engineers. For liquids a pump usually is inserted somewhere in the piping system, and the pipes may have different diameters at different sections. In addition, the fluid may undergo a considerable elevation change, which is not the case with other pieces of standard equipment. The pipes may be uninsulated, with heat transfer occurring between the fluid and the environment. Therefore, many forms of energy change can be involved. In steady incompressible flow there are frictional losses which are frequently referred to as "head" losses and are represented by $H_f$. By means of experiments it has been found that

$$H_f = f \frac{L}{D} \frac{V^2}{2} \tag{5-32}$$

where $L$ and $D$ are the length and diameter of the pipe, respectively, and $f$ is an experimentally determined "friction factor." Charts which enable one to evaluate $f$ are presented in textbooks on fluid mechanics and in handbooks. For purposes of this book the value of $f$ will be given. For nonhorizontal pipes with an elevation change it is also to be noted that, for incompressible fluids,

$$H_f = \frac{V_1^2 - V_2^2}{2} + \frac{P_1 + P_2}{\rho} + (z_1 - z_2)g \tag{5-33}$$

where $z$ is the elevation with respect to some reference point. This equation dictates that for pipe flow without a pump the internal flow resistance must be overcome by allowing a decrease in kinetic energy, potential energy, or displacement work between the open ends of the control volume.

The conservation of energy principle has the general form

$$q + w_{\text{shaft}} = \Delta u + \Delta(Pv) + \Delta\text{KE} + \Delta\text{PE}$$

By comparing the energy equation to Eq. (5-33) for incompressible fluids, it is seen that in the absence of a work interaction

$$H_f = \Delta u - q \tag{5-34}$$

The above relation provides a physical interpretation of the effects of internal flow resistance on the change in internal energy and heat transfer for the fluid. Thus, forms of energy such as kinetic energy, potential energy, and displacement work are dissipated into a combination of an internal-energy increase and a heat loss.

**Example 5-9M** A pump draws a solution with a specific gravity of 1.50 from a storage tank through a 8-cm pipe. The velocity in the suction (inlet) pipe is 1.2 m/s. The open end of the 5-cm discharge pipe is 15 m above the top of the liquid in the storage tank. Frictional head losses are found to be 25 N·m/kg. What power is required of the pump, in kW?

SOLUTION The top level of the storage tank and the discharge of the 5-cm pipe are designated by states 1 and 2. State $a$ is an intermediate state in the 8-cm pipe. These are shown on the accompanying diagram (see Fig. 5-10). The control volume includes the fluid in the storage tank and within the piping system. The conservation of energy principle for the incompressible fluid can be written in the form

$$w_{\text{shaft}} = H_f + \Delta(Pv) + \Delta\text{KE} + \Delta\text{PE}$$

From the selection of the control volume, $P_1 = P_2$, and for an incompressible fluid $v_1 = v_2$, so

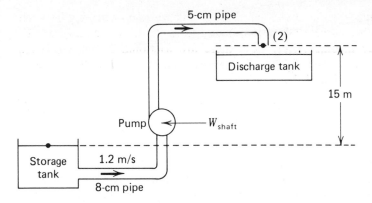

**Figure 5-10** Schematic of flow process for Example 5-9M.

that $\Delta(Pv) = 0$. The initial velocity $V_1$ is zero, and the final velocity is found from the continuity equation.

$$V_2 = V_a \frac{A_a}{A_2} = V_a \left(\frac{D_a}{D_2}\right)^2 = 1.2 \left(\frac{8}{5}\right)^2 = 3.1 \text{ m/s}$$

Substitution of the available data yields

$$w_{shaft} = 25 + \frac{(3.1)^2}{2} + 15(9.8) = 176.8 \text{ N·m/kg}$$

Note that the potential-energy change contributes to over 80 percent of the total value of the work required. The power requirement is obtained from the relation $\dot{W} = \dot{m}w$. The mass rate of flow is

$$\dot{m} = \rho A V = (1.50 \times 1.0)\text{g/cm}^3 \times \frac{\pi(8)^2}{4} \text{ cm}^2 \times 1.2 \text{ m/s} \times \frac{100 \text{ cm}}{\text{m}}$$

$$= 9050 \text{ g/s} = 9.05 \text{ kg/s}$$

In the above calculation the density of water has been taken to be 1.0 g/cm³. Finally,

$$\dot{W} = 177 \text{ N·m/kg} \times 9.05 \text{ kg/s} = 1600 \text{ W} = 1.60 \text{ kW}$$

**Example 5-9** A pump draws a solution with a specific gravity of 1.50 from a storage tank through a 3-in pipe. The velocity in the suction pipe is 4 fps. The open end of the 2-in discharge pipe is 50 ft above the top of the liquid in the storage tank. Frictional losses in the pipe lead to a heat loss from the fluid and an increase in the internal energy of the fluid. Measurements indicate that $\Delta u - q = 10.0$ ft·lb$_f$/lb$_m$. What theoretical horsepower of the pump is required?

SOLUTION The top level of the storage tank and the discharge of the 2-in pipe are designated by states 1 and 2 on the accompanying diagram (see Fig. 5-11). The control volume includes the fluid in the storage tank and within the piping system. The conservation of energy principle can be written in the form

$$w_{shaft} = (\Delta u - q) + \Delta(Pv) + \Delta KE + \Delta PE$$

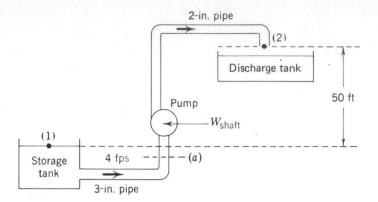

**Figure 5-11** Schematic of flow process for Example 5-9.

From the selection of the control volume, $P_1 = P_2$, and if the fluid is assumed to be incompressible so that $v_1 = v_2$, then $\Delta(Pv) = 0$. The initial velocity is zero, and the final velocity is found from the continuity equation

$$V_2 = V_a \frac{A_a}{A_2} = V_a \left(\frac{D_a}{D_2}\right)^2 = 4\left(\frac{3}{2}\right)^2 = 9 \text{ ft/s}$$

Substitution of the available data yields

$$W_{\text{shaft}} = 10 + \frac{(9)^2}{2(32.2)} + 50 = 61.26 \text{ ft·lb}_f/\text{lb}_m$$

The horsepower requirement is obtained from the equation $\dot{W} = \dot{m}w$. The mass rate of flow is

$$\dot{m} = \rho A V = (62.4 \times 1.50)\frac{\pi(3)^2}{4(144)} 4 = 18.3 \text{ lb}_m/\text{s}$$

Hence

$$\dot{W} = 61.26 \text{ ft·lb}_f/\text{lb}_m \times 18.3 \text{ lb}_m/\text{s} \times \frac{\text{hp·s}}{550 \text{ ft·lb}_f} = 2.04 \text{ hp}$$

**Example 5-10M** Water at 20°C flows through a 0.12-m-diameter pipe for a distance of 600 m. The pressure drop amounts to 25 N/cm² (or 250 kPa) in that distance, while the elevation increases by 8 m. If the friction factor $f$ for the pipe is 0.020, what is the volume flow rate in m³/min?

SOLUTION The volume flow rate depends upon the velocity of the fluid, which is unknown. The velocity appears in both equations for the heat loss, $H_f$; namely, $H_f = f(L/D)(V^2/2)$ and Eq. (5-33),

$$H_f = \frac{V_1^2 - V_2^2}{2} + \frac{P_1 - P_2}{\rho} + (z_1 - z_2)g$$

For incompressible flow through a constant-area duct, $V_1 = V_2$. Thus the first term on the right in the above equation is zero. If we equate the two expressions for $H_f$, we obtain a relationship in which the only unknown is $V$. Hence,

$$f \frac{L}{D} \frac{V^2}{2} = \frac{P_1 - P_2}{\rho} + (z_1 - z_2)g$$

Substitution of values yields

$$0.020 \frac{600 \; V^2 \; \text{m}^2}{0.12 \; 2 \; \text{s}^2} = 25 \frac{\text{N}}{\text{cm}^2} \times \frac{\text{cm}^3}{1 \; \text{g}} \times \frac{1 \; \text{kg·m}}{\text{N·s}^2} \times \frac{\text{m}}{100 \; \text{cm}} \times \frac{1000 \; \text{g}}{\text{kg}} + (-8)\text{m} \times 9.806 \; \text{m/s}^2$$

$$100 \; V^2 = 250 - 78.4$$

$$V = 1.31 \; \text{m/s}$$

The volume flow rate is found by multiplying this velocity by the flow area. Hence

$$\text{Volume rate} = 1.31 \; \text{m/s} \times \frac{\pi(0.12)^2}{4} \; \text{m}^2 = 0.0148 \; \text{m}^3/\text{s} = 0.889 \; \text{m}^3/\text{min}$$

**Example 5-10** Water at 60°F flows through a 6-in-diameter pipe for a distance of 2000 ft. The pressure drop amounts to 39 psi in that distance, while the elevation increases by 20 ft. If the friction factor $f$ for the pipe is 0.020, what is the volume flow rate in gal/min?

SOLUTION The volume flow rate depends on the velocity of the fluid, which is unknown. The velocity appears in both equations for the head loss $H_f$; namely, $H_f = f(L/D)(V^2/2)$ and Eq. (5-33),

$$H_f = \frac{V_1^2 - V_2^2}{2} + \frac{P_1 - P_2}{\rho} + (z_1 + z_2)g$$

For incompressible flow through a constant-area duct, $V_1 = V_2$. Thus the first term on the right in the above equation is zero. If we equate the two expressions for $H_f$, we obtain a relation in which the only unknown is $V$. Hence,

$$f \frac{L}{D} \frac{V^2}{2} = \frac{P_1 - P_2}{\rho} + (z_1 + z_2)g$$

Substitution of values into this equation yields

$$0.020 \frac{2000}{1/2} \frac{V^2}{2(32.2)} = \frac{39(144)}{62.4} + (-20)$$

$$1.242 \; V^2 = 90 - 20 = 70$$

$$V = 7.50 \; \text{ft/s}$$

The volume flow rate equals $V$ times the flow area. Thus

$$\text{Volume rate} = 7.50 \; \text{ft/s} \times \frac{\pi(1/2)^2}{4} \; \text{ft} \times 7.48 \; \text{gal/ft}^3 \times 60 \; \text{s/min}$$

$$= 600 \; \text{gal/min}$$

# 5-6 TRANSIENT FLOW ANALYSIS

There are several prime reasons for studying nonsteady-flow problems. First of all, there are several interesting and useful applications of devices which operate under such conditions. For example, some wind tunnels operate on the discharge of a gas which has been stored under a high pressure in a large, rigid tank. These are known as blowdown wind tunnels. Compressed air stored in a cylinder might also be used to drive a gas turbine coupled with a generator as an auxiliary electric power source in the case of normal power failure. Second, the analysis of nonsteady-flow devices serves to strengthen one's understanding of and one's ability to use the general conservation principles related to mass and energy. The importance of the proper selection of a control volume often becomes apparent in such analyses. As has been pointed out earlier in this text, a control volume may change size and shape, as well as move in space relative to some reference point. In steady-state analyses we usually consider the control volume fixed in space, in size, and in shape. In the following sections we wish to relinquish two of these constraints and allow the size and shape to vary, if need be. However, the control volume will still be considered to be fixed in location. Although the removal of this latter constraint leads to a highly profitable study with important engineering applications, it is more fitting to study such problems in fluid mechanics in conjunction with the conservation of momentum equation.

In the analysis of transient systems the basic energy equations we need are

$$\delta Q + \delta W + (e + Pv)_{\text{in}} \, dm_{\text{in}} - (e + Pv)_{\text{out}} \, dm_{\text{out}} = dE_{\text{CV}} \qquad (5\text{-}19)$$

and

$$Q + W_{\text{net}} + \int (e + Pv)_{\text{in}} \, dm_{\text{in}} - \int (e + Pv)_{\text{out}} \, dm_{\text{out}} = \Delta E_{\text{CV}} \qquad (5\text{-}20)$$

Equation (5-17) on a rate basis is also a useful equation. As we have already noted, the integration and subsequent evaluation of the various terms in these equations is straightforward in the case of a steady-state process. When applied to an unsteady-state problem, the choice of the control volume and subsequent idealizations regarding the process are extremely important. Rather than break down the large number of possible unsteady-state problems into various classes, it is best to approach each problem on its own merits and to analyze each situation beginning with the basic or fundamental equations. This latter consideration, of course, is valid for any type of engineering analysis.

In applying Eqs. (5-19) and (5-20) to transient flow problems, we shall restrict the types of energy associated with the mass within, entering, or leaving a control volume to internal, linear kinetic, and gravitational forms. That is,

$$e = u + \frac{V^2}{2} + zg \qquad (5\text{-}26)$$

Thus we are neglecting any electrical or magnetic field effects. The decision to include the last two terms on the right in any analysis will depend upon the particular circumstances in a given problem. Frequently these latter terms may be omitted, not because their change in value is zero, but because such changes are negligible when compared to other terms in the conservation of energy equation. Within this restriction we can rewrite Eq. (5-19) as

$$\delta Q + \delta W + \left(h + \frac{V^2}{2} + zg\right)_{\text{in}} dm_{\text{in}} - \left(h + \frac{V^2}{2} + zg\right)_{\text{out}} dm_{\text{out}} = dU_{\text{CV}} \qquad (5\text{-}35)$$

In a similar manner Eq. (5-20) becomes

$$Q + W_{net} + \int \left( h + \frac{V^2}{2} + zg \right)_{in} dm_{in}$$

$$- \int \left( h + \frac{V^2}{2} + zg \right)_{out} dm_{out} = \Delta U_{CV} \qquad (5\text{-}36)$$

Although Eqs. (5-35) and (5-36) have been written for a control volume with one inlet and one outlet, it is apparent that a term

$$\left( h + \frac{V^2}{2} + zg \right) dm$$

must be included for every cross section of the control surface through which mass enters or leaves the control volume. Several examples will be presented to illustrate the use of the above equations. One point must be kept in mind. All energy values (and hence enthalpy values as well) must be measured from consistent reference states.

## 5-7 CHARGING AND DISCHARGING RIGID VESSELS

A common and typical example of an unsteady-state flow problem is the charging of a rigid vessel by supplying a fluid from some outside high-pressure source. The rigid vessel may initially be evacuated, or may contain a finite amount of matter within it before the charging process begins. It is not necessary that the fluid used to charge the vessel be identical with the substance already in the vessel. The two fluids will be the same, though, in many practical problems. Generally the mass enters only at one section of the control surface, and no efflux of matter is permitted.

As a more definite illustration of the problem, consider an evacuated, rigid tank connected through a closed valve to a high-pressure line. This situation is shown in Fig. 5-12a. A control volume has been selected, as indicated by the dashed

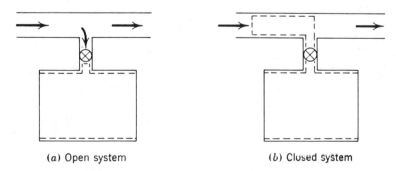

(a) Open system          (b) Closed system

**Figure 5-12** Charging of a vessel.

line in the figure, which includes essentially the volume inside the rigid tank. We shall assume that the volume in the region of the valve and entrance pipe is negligible. Application of Eq. (5-36) to this situation reveals that

$$Q + W_{\text{net}} + \int \left(h + \frac{V^2}{2} + zg\right) dm = \Delta U_{\text{CV}}$$

The net work term is zero in value on the basis of the statement of the process. If the velocity of the flow stream in the high-pressure line is relatively low (e.g., less than 200 ft/s) and the entrance pipe relatively short in the vertical direction, then the kinetic and potential energies of any unit mass on the line may be neglected. Thus the conservation of energy equation for the selected control volume reduces to

$$Q + \int h \, dm = \Delta U_{\text{CV}}$$

At this point one must make a decision about the quantity $Q$. In actual practice the value of the heat transfer could be estimated fairly accurately by independently evaluating the convective and radiative contributions to the overall heat loss or gain. However, if the tank were allowed to fill rapidly, the amount of heat transferred probably would be fairly small even if an appreciable temperature gradient existed between the contents of the tank and the environment. Therefore we shall analyze the process on the basis of negligible heat transfer. Realizing that the overall internal energy change of the control volume can be written as $\Delta(mu)$, we then note that the above energy equation becomes

$$\int h \, dm = m_f u_f - m_i u_i$$

where $i$ and $f$ represent the initial and final states within the control volume. The integration is carried out from zero mass to the final mass which enters from the line.

Frequently the integral of $h \, dm$ is difficult to evaluate because the enthalpy of the mass crossing the control surface varies with respect to time. However, in our particular case we can assume that the quantity of fluid bled from the line is small compared to the quantity flowing through the line. Consequently, the properties of the fluid at the entrance to the valve remain constant. If we designate the line conditions by the subscript $L$, then integration of the above equation leads to

$$h_L(m_L - 0) = m_f u_f - m_i u_i$$

where the terms on the right refer solely to the control volume. For this particular problem $m_i$ is zero (the tank initially was evacuated), and $m_f$ equals $m_L$. Hence

$$h_L m_L = m_L u_f$$

or

$$\text{✳} \quad h_L = u_f$$

This relationship fixes the final state within the control volume. Under the conditions specified, we find that the enthalpy of the fluid in the high-pressure line is numerically equal to the internal energy of the fluid within the control volume at any instant. This latter statement assumes that the fluid equilibrates rapidly within the control volume, so that properties in general are defined and uniform in value at any instant.

The above result is completely general in the sense that the nature of the fluid has never been specified. Though equally applicable to any real gas, the equation above is of special interest when employed for the flow of an ideal gas into the tank. Since $h = c_p T$ and $u = c_v T$ for an ideal gas (if the values of the two specific heats are assumed constant over the required temperature range), then

$$c_p T_L = c_v T_f$$

or

$$T_f = k T_L$$

where $k$ is the specific-heat ratio. This important result shows that $T_f$ is always greater than $T_L$ for flow of an ideal gas into a rigid tank under the following major idealizations.

1. Adiabatic, rigid control volume
2. Equilibrium within the tank at any instant
3. Negligible kinetic energy of the inflowing gas
4. Steady state for the inlet flow

Since $k$ for many gases ranges from 1.25 to 1.40, the absolute temperature rise may be as much as 40 per cent. For real gases, one probably will have to rely on tabular data in order to ascertain the final state of the fluid after it reaches a certain pressure. Such an analysis frequently is one involving iteration.

The problem of charging a tank from a steady-flow line is amenable to attack on the basis of a closed-system analysis as well. It is interesting to pursue this method of analysis, since it provides a striking contrast to the method previously used. The closed system selected for analysis is shown in Fig. 5-12b. Since we have chosen a closed system for analysis, all the mass that eventually enters the tank must initially be included within the system boundaries. Hence the boundaries extend beyond the valve and out into a portion of the line.

The conservation of energy equation in this case is

$$Q + W = \Delta E$$

Again we shall neglect kinetic- and potential-energy terms and assume adiabatic conditions. The energy equation becomes

$$W = \Delta U = m_f u_f - m_i u_i$$

Unlike the control-volume approach, however, neither the work term nor the initial mass term $m_i$ is zero. The work done is the boundary work related to pushing the mass contained within the system, but outside the valve, into the tank. Recall that one idealization employed previously is that steady state exists in the line. Hence the pressure in the line in the vicinity of the valve is constant at some value $P$, and the boundary work is simply $-P\,\Delta V$. Also the initial mass in this closed system, $m_i$, is also that mass contained outside the valve initially, since the tank is evacuated at the start of the process. Consequently the energy equation becomes

$$-P(V_f - V_i) = m_f u_f - m_i u_i$$

or

$$-P(0 - m_L v_L) = m_f u_f - m_L u_L$$

Finally, combining the second term on the right with the term on the left, we find that

$$(P_L v_L + u_L) = m_f u_f$$

$$h_L m_L = m_f u_f$$

and

$$h_L = u_f$$

This result is identical to that obtained by a control-volume analysis. Both were based on comparable idealizations. However, it should be noted that the terms included in the two analyses are quite different. Whether transient flow problems are equally tractable to both control-volume and closed-system analyses depends largely on the particular circumstances. The closed-system approach is usually pertinent only to those situations where the position and state of the mass in transit is well defined.

From a physical viewpoint the discharge of a fluid from a pressurized vessel is similar to the preceding case of the charging of a vessel. However, from a thermodynamic viewpoint there is little similarity in the relations which result from applying the conservation of energy principle to the two processes. We shall again neglect the kinetic and potential energies associated either with the control volume or the mass leaving the vessel through a suitable valve. Thus, $e = u$. The absence of shaft work again is noted. The presence or absence of appreciable heat-transfer effects again depends either upon the nature of the vessel's walls (adiabatic or diathermal), or upon the magnitude of the temperature difference between the system and the environment. In the absence of specific information on heat transfer, we shall assume that the effect is negligible, for one reason or another. If Eq. (5-35) is now applied to the control volume shown in Fig. 5-13$a$, and the above comments are taken into consideration, we find that

$$(u + Pv)\,dm = dU_{\text{cv}}$$

or

$$h\,dm = d(mu)$$

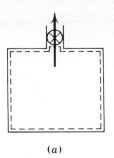

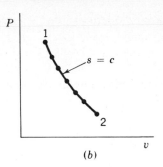

(a)                          (b)

**Figure 5-13** Flow from a pressurized vessel.

This problem is unlike the charging of a vessel from a pressurized line in that the properties of the fluid crossing the control surface are continually changing due to the decrease in the mass within the control volume. For this reason we have elected to write the conservation equation in its differential form. The equation can be rearranged into the following forms.

$$h \, dm = m \, du + u \, dm$$

$$(h - u) \, dm = m \, du$$

and

$$\frac{du}{h - u} = \frac{dm}{m}$$

It should be noted that the specification of property values, such as $h$ and $u$, for the mass within the control volume requires that the fluid be essentially in equilibrium at all times. Hence the process of discharging the gas must be slow enough so that the process is quasistatic.

The equation presented above is completely general. That is, we have not specified the nature of the fluid leaving the pressurized tank. We may introduce two further relationships without losing this generality. First of all, from the definition of the enthalphy function, $h - u = Pv$. Second, since the volume of the control volume remains constant with time, then from the basic relationship $V = mv$ we find that

$$dV = m \, dv + v \, dm = 0$$

Therefore

$$\frac{dm}{m} = -\frac{dv}{v}$$

Substitution of these two basic relations into the conservation of energy equation leads to

$$\frac{du}{Pv} = -\frac{dv}{v}.$$

or

$$du + P\,dv = 0$$

This is a remarkable result, since for a closed system of fixed composition it is realized that

$$T\,ds = du + P\,dv$$

is a valid statement for a simple substance, regardless of the nature of the process. Comparison of these last two equations forces one to conclude that, for the process,

$$ds = 0$$

Therefore, for any fluid which flows from a pressurized vessel under the stated assumptions or idealizations, the specific entropy within the control volume remains constant. The total entropy of the control volume decreases, since the amount of mass decreases. The overall process is highly irreversible, so that the net entropy change of the control volume plus its environment must be positive.

Figure 5-13b illustrates the process path for the discharging of a fluid from a pressurized tank. The path is for the fluid which remains inside the tank. Note that such a process must fulfull the following idealizations:

1. Adiabatic, rigid control volume
2. Equilibrium within the tank at any instant
3. Negligible kinetic- and potential-energy changes

The fact that the process is isentropic is not surprising, since it is adiabatic and quasistatic, and all internal dissipative effects and finite gradients have been ruled out. That is, the process within the tank is adiabatic and internally reversible.

As a specific example of such a process one might consider the flow of an ideal gas from a pressurized tank. The approximate validity of the ideal-gas equation will depend upon the pressure range chosen for the study. Since the process is isentropic, the isentropic relations presented in Sec. 7-9 are valid. For example, the relation $T(v)^{k-1} = $ constant is valid if one assumes constant specific heats. If one wishes to account for variable specific heats, a more accurate theoretical analysis can be made through the use of the $p_r$ and $v_r$ data in the abridged gas tables found in the appendix.

A relationship for the mass remaining in the tank at any instant as a function of the properties of the mass within the control volume can also be obtained for an ideal gas. We have noted that

$$\frac{dm}{m} = -\frac{dv}{v}$$

and from the isentropic relations for an ideal gas

$$\frac{dT}{T} = (1 - k)\frac{dv}{v}$$

Therefore,

$$\frac{1}{k-1}\frac{dT}{T} = \frac{dm}{m}$$

Integration of the above equation yields

$$\frac{m_2}{m_1} = \left(\frac{T_2}{T_1}\right)^{1/(k-1)}$$

If the pressure ratio is known, the comparable equation is

$$\frac{m_2}{m_1} = \left(\frac{P_2}{P_1}\right)^{1/k}$$

A third equation can be derived for the mass as a function of the specific volume, or density.

**Example 5-11M** A tank with a volume of 1.5 $m^3$ is filled with air at a pressure of 7 bars and a temperature of 220°C. Determine the final temperature and the percent of the mass left in the tank if gas is permitted to leave the tank under adiabatic conditions until the pressure drops to 1 bar.

SOLUTION Under the conditions of the problem it is realistic to assume ideal-gas behavior. If the mass within the tank changes its state isentropically, then, since $k$ is roughly equal to 1.4,

$$T_2 = T_1 = \left(\frac{P_2}{P_1}\right)^{(k-1)/k} = 493(1/7)^{0.285} = 283°\text{K} = 10°\text{C}$$

The percent of the mass left in the tank can be found from the relation

$$\frac{m_2}{m_1} = \left(\frac{P_2}{P_1}\right)^{1/k} = (1/7)^{0.715} = 0.249$$

Therefore only 25 percent of the mass is left when a pressure of 1 bar is reached, and the temperature at that instant will be 10°C.

**Example 5-11** A tank with a volume of 50 ft$^3$ is filled with air at a pressure of 100 psia and a temperature of 340°F. Determine the final temperature and the percent of the mass left in the tank for an adiabatic expansion down to the environment pressure of 15 psia.

SOLUTION For a maximum pressure of 100 psia it is realistic to assume ideal-gas behavior at the given temperature. If the mass within the tank changes its state isentropically, then, since $k$ is roughly equal to 1.4,

$$T_2 = T_1 \left( \frac{P_2}{P_1} \right)^{(k-1)/k}$$

$$= 800 \left( \frac{15}{100} \right)^{0.285} = 800(0.5825)$$

$$= 466°R = 6°F$$

The percent of the mass left in the tank can be found from the relation

$$\frac{m_2}{m_1} = \left( \frac{P_2}{P_1} \right)^{1/k} = \left( \frac{15}{100} \right)^{0.715} = 0.257$$

Therefore only 25 percent of the mass is left when pressure equilibrium with the environments is attained, and the temperature at that instant will be 6°F.

There are many other types of transient flow problems of theoretical as well as of practical interest. The solution of each type depends upon the restrictions and idealizations placed on the process. In every case it is best to begin with basic or fundamental equations, and proceed logically from that point. A discussion of a control-volume analysis on which boundary work is present might be helpful at this time.

## 5-8 TRANSIENT-SYSTEM ANALYSIS WITH BOUNDARY WORK

Consider the situation shown in Fig. 5-14. A weighted piston-cylinder assembly initially containing a known quantity of a gas at a known equilibrium state is connected through a valve to a high-pressure line. To simplify the problem, the fluid in the line is the same as that in the piston cylinder, although this is an unnecessary restriction. We are interested in determining the temperature and the quantity of mass within the cylinder after the volume has changed by a known amount.

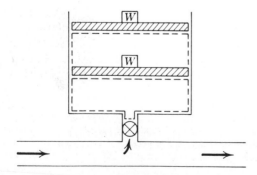

**Figure 5-14** A control volume with a moving boundary.

The control volume under consideration is shown in Fig. 5-14. Since the upper boundary will move, the control volume changes size in this case, and boundary work occurs. It is necessary to place a restriction on the process if the boundary work is to be evaluated. Even though the pressure in the line may be substantially above the pressure within the cylinder (as maintained by the weighted piston), we must assume that the entering gas quickly equilibrates with the gas already present in the control volume so that a uniform pressure $P$ is maintained on the piston by the gas. Then the work done by the gas on the surroundings is simply $-P \, \Delta V$. In the absence of sufficient information we shall assume that the process is adiabatic, which requires that the piston be frictionless. Although the center of mass of the control volume changes in elevation, this potential-energy effect can be neglected. The basic conservation of energy equation for the control volume in this case becomes

$$Q + W_{\text{shaft}} + W_{\text{boundary}} + \int (e + Pv) \, dm = \Delta E_{\text{CV}}$$

This reduces to the relation

$$W_{\text{boundary}} + \int h \, dm = \Delta U_{\text{CV}}$$

In terms of the properties for the initial and final state we find that

$$-P(V_f - V_i) + \int h \, dm = m_f u_f - m_i u_i$$

As a final step, since the flow through the line is steady, we can integrate the quantity $h \, dm$ directly and obtain the equation

$$\text{\Large ✱} \quad -P(V_f - V_i) + h_L m_L = m_f u_f - m_i u_i$$

The unknowns in this equation are $m_L$, $m_f$, and $u_f$. However, the mass quantities are related by

$$\text{\Large ✱} \quad m_L = m_f - m_i$$

Thus we actually have one equation with two unknowns, $m_f$ and $u_f$.

The second required relationship in this case is found from a knowledge of the final volume. Note that $V_f = m_f v_f$. Since $V_f$ is given, the unknowns in this relationship are $m_f$ and $v_f$. Hence another unknown, $v_f$, is introduced. However, $u_f$ and $v_f$ are not independent variables. On the basis of the state principle, for a given pressure within the control volume only $u_f$ or $v_f$ can be varied independently, but not both. Thus a suitable equation of state which uniquely relates $u_f$ and $v_f$ for a given $P$ is the desired third relation which permits the two equations already presented to be satisfied.

↓ **Example 5-12M** A piston-cylinder assembly is attached through a valve to a source of air which is flowing in steady state through a pipe. Initially, inside the cylinder the volume is 0.01 m³, the temperature of the air inside is 40°C, and the pressure is 1 bar. The valve is then opened to the air-supply line which is maintained at 6 bars and 100°C. If the cylinder pressure remains constant, what will be the final equilibrium temperature in °C, and how much mass will have entered through the valve, in grams, when the cylinder volume reaches 0.02 m³.

SOLUTION If the control volume is selected to lie just inside the solid boundaries of the piston-cylinder assembly (see Fig. 5-14) and the gas is assumed to behave ideally, then the conservation of energy equation developed above takes on the form

$$-P(V_f - V_i) + h_L(m_f - m_i) = m_f c_v T_f - m_i c_v T_i$$

$$= m_f c_v \frac{PV_f}{m_f R} - m_i c_v T_i$$

$$= \frac{c_v PV_f}{R} - m_i c_v T_i$$

$$= \frac{PV_f}{k - 1} - m_i c_v T_i$$

Hence by introducing the equation of state for an ideal gas, and the relation $c_v = R/(k-1)$, we find that the resulting equation contains only the single unknown $m_f$. Before substituting numerical values, it is convenient to calculate first the initial mass within the piston-cylinder assembly.

$$m_i = \frac{PV}{RT} = \frac{1 \text{ bar} \times 0.01 \text{ m}^3}{0.08315 \text{ bar·m}^3/(\text{kg·mol})(°K) \times 313°K} \times \frac{29 \text{ kg}}{\text{kg·mol}}$$

$$= 0.0111 \text{ kg} = 11.1 \text{ g}$$

Note also that $h_L = c_p T_L$, and $c_p$, $c_v$, and $k$ values are in Table A-4M. Substitution of known values into the energy equation then yields (expressing each term in newton-meters or joules, and $m_f$ in grams),

$$-1(0.02 - 0.01)10^5 + 1.01(m_f - 11.1)(373) = \frac{1(0.02)(10^5)}{0.4} - 11.1(0.72)(313)$$

$$-1000 + 377m_f - 4180 = 5000 - 2500$$

$$m_f = 7680/377 = 20.4 \text{ g}$$

Thus, the mass entering the assembly during the filling process is

$$m_L = m_f - m_i = 20.4 - 11.1 = 9.3 \text{ g}$$

The final equilibrium temperature is found from the ideal-gas relation:

$$T_f = \frac{PV_f}{m_f R} = \frac{1 \text{ bar} \times 0.02 \text{ m}^3}{20.4 \text{ g} \times 0.08315 \text{ bar·m}^3/(\text{kg·mol})(°K)} \times \frac{29 \text{ kg}}{\text{kg·mol}} \times \frac{1000 \text{ g}}{\text{kg}}$$

$$= 342°K = 69°C$$

Thus, the temperature of the resulting mixture in the assembly increases nearly 30°C above the initial temperature.

**Example 5-12** A piston-cylinder assembly is attached through a valve to a source of air which is flowing in steady state through a pipe. Initially inside the cylinder the volume is 1.0 ft³, the temperature of the air inside is 100°F, and the pressure is 14.8 psia. The valve is then opened to the air-supply line which is at 100 psia and 200°F. If the cylinder pressure remains constant, what will be the final equilibrium temperature in °F, and how much mass will have entered through the valve in pounds when the cylinder volume reaches 2.0 ft³?

SOLUTION If the control volume is selected to lie just inside the solid boundaries of the piston-cylinder assembly (see Fig. 5-14) and the gas is assumed to behave ideally, then the conservation of energy equation developed above takes on the form

$$-P(V_f - V_i) + h_L(m_f - m_i) = m_f c_v T_f - m_i c_v T_i$$

$$= m_f c_v \frac{PV_f}{m_f R} - m_i c_v T_i$$

$$= \frac{c_v PV_f}{R} - m_i c_v T_i$$

$$= \frac{PV_f}{k-1} - m_i c_v T_i$$

Hence by introducing the equation of state for an ideal gas, and the relation $c_v = R/(k-1)$, we find that the resulting equation contains only the single unknown $m_f$. Before substituting numerical values, it is convenient to calculate first the initial mass within the piston-cylinder assembly:

$$m_i = \frac{PV}{RT} = \frac{14.8(144)(1)(29)}{(1545)(560)} = 0.0715 \text{ lb}_m$$

Note that $h = c_p T_L$. Then the substitution of known values into the energy equation yields

$$\frac{-14.8(144)(2-1)}{778} + 0.24(m_f - 0.0715)(660) = \frac{14.8(144)(2.0)}{0.4(778)} - 0.0715(0.171)(560)$$

Then

$$-2.74 + 158.4m_f - 11.32 = 13.7 - 6.84$$

or

$$m_f = 0.132 \text{ lb}_m$$

Thus the mass entering the assembly during the filling process is

$$m_f - m_i = m_L = 0.132 - 0.0715$$

$$= 0.0606 \text{ lb}_m$$

The final equilibrium temperature is found from the ideal-gas relation:

$$T_f = \frac{PV_f}{m_f R} = \frac{14.8(144)(2.0)(29)}{0.132(1545)}$$

$$= 606°R = 146°F$$

If the fluid had been steam (or some other real gas), the solution to this problem would not have been quite as straightforward. In this latter case the property relations are usually not known in an explicit algebraic form. Consequently an iteration process often is required during the numerical evaluation step of the solution.

## PROBLEMS (METRIC)

### Continuity equation

**5-1M** Steam enters a steady-flow device at 40 bars and 320°C with a velocity of 160 m/s. The entrance area is 6 cm$^2$, and the exit area is 21 cm$^2$. If the outlet conditions are 1.5 bars and 160°C determine (a) for a real gas, and (b) for an ideal gas the mass flow rate in kg/s and the exit velocity in m/s. m/s.

**5-2M** Refrigerant 12 enters a steady-flow device at 7 bars, 60°C, with a velocity of 148 m/s. The exit pressure and velocity are 2.8 bars and 120 m/s, respectively. If the mass flow rate is 19.75 kg/s and the exit area is 100 cm$^2$, determine (a) the inlet area in square centimeters, and (b) the exit temperature in °C.

**5-3M** Steam enters a turbine at 60 bars and 500°C with a velocity of 130 m/s, and leaves as a dry, saturated vapor at 0.50 bar. The turbine inlet pipe has a diameter of 0.6 m and the outlet diameter is 3.5 m. Determine (a) the mass flow rate in kg/h, and (b) the exit velocity in m/s.

**5-4M** Steam at a rate of 48,000 kg/h leaves a device at 0.10 bar and 90 percent quality. Calculate the area of the exit in square meters for a mean velocity of 160 m/s.

**5-5M** In a steady-flow device 0.50 kg/min of saturated refrigerant 12 vapor at 5 bars enters with a velocity of 4.0 m/s. The exit area is 0.9 cm$^2$, and the exit temperature and velocity are 60°C and 5 m/s, respectively. Determine (a) the entrance area in square centimeters, and (b) the exit pressure in bars.

**5-6M** Steam enters a turbine at 30 bars pressure, 440°C, at a velocity of 100 m/s and through an area of 0.05 m$^2$. It expands to a pressure of 0.2 bar, a quality of 90 percent, and a velocity of 200 m/s. Determine (a) the mass flow rate through the turbine in steady state, in kg/s, and (b) the exist area, in square meters.

**5-7M** Steam enters a steady-flow device at 1.5 bars and 200°C. At the outlet the pressure is 3 bars, the temperature is 280°C, and the velocity is 120 m/s. If the inlet-to-outlet area ratio for the system is 2.4 : 1, determine the inlet velocity in m/s.

**5-8M** Air initially at 1.7 bars and 80°C flows through a cross-sectional area of 100 cm$^2$ at a rate of 50 kg/min. Downstream at another position the pressure is 3.4 bars, the temperature is 80°C, and the velocity is 15 m/s. Determine (a) the inlet velocity in m/s, and (b) the outlet area in square centimeters.

**5-9M** Air with a density of 1.13 kg/m$^3$ enters a steady-flow device through a duct 0.2 m$^2$ in area with a velocity of 4 m/s. If the outlet density is 2.84 kg/m$^3$, (a) determine the flow rate in kg/s, and (b) determine the area of the discharge duct in square meters if the outlet velocity is 6 m/s.

**5-10M** An ideal gas enters a steady-flow device at 1 bar and 27°C. The inlet to the system has a diameter of 20 cm, and the velocity at that point is 10 m/s. At the outlet the diameter is 7 cm, and the density of the fluid at that section is 4.0 kg/m$^3$. If the molar mass of the gas is 30, determine the velocity in m/s at the outlet and the mass flow rate in kg/s.

**5-11M** An ideal gas with a molar mass of 48 enters a steady-flow device at 3 bars, 157°C, and 180 m/s through an area of 20 cm$^2$. At the exit of the device the pressure is 1.5 bars, the velocity is 120 m/s, and the exit area is 40 cm$^2$. Determine (a) the density of the gas, in kg/m$^3$, (b) the mass flow rate in kg/s, and (c) the exit temperature is °C.

**5-12M** Carbon dioxide enters a steady-flow device at 27°C with a velocity of 25 m/s through an area of 4800 cm$^2$. At the exit of the device the pressure and temperature are 1.4 bars and 47°C, respectively, and the gas moves with a velocity of 9 m/s through an area of 7500 cm$^2$. Determine (a) the mass flow rate in kg/s, and (b) the inlet pressure in bars. Assume ideal-gas behavior.

**5-13M** Water at 20°C flows through a long, plastic tube which does not have a constant diameter. At position 1 in the tube the internal diameter is 0.1 m and the velocity is 3 m/s. At position 2 downstream from position 1 the internal diameter is 0.2 m. The temperature remains constant.
  (a) Determine the mass flow rate in kg/min.
  (b) Determine the velocity at position 2 in m/s.

**5-14M** Water at 15°C flows down a long pipe. At position 1 in the pipe the internal diameter is 20 cm and the velocity is 0.9 m/s. At position 2 downstream from position 1 the velocity is 3.6 m/s.

    (*a*) Determine the mass flow rate in kg/min.

    (*b*) Determine the internal diameter at position 2, in centimeters.

### Diffusers

**5-15M** An adiabatic diffuser is employed to reduce the velocity of a stream of air from 250 m/s to 35 m/s. The inlet pressure is 1 bar, and the inlet temperature is 300°C. Determine the required outlet area in square centimeters if the mass flow rate is 7 kg/s and the final pressure is 1.17 bars.

**5-16M** Air enters a diffuser at 1 bar, 30°C, with a velocity of 150 m/s. The exit temperature is 40°C. If a heat loss of 0.4 kJ/kg occurs, find the exit velocity in m/s.

**5-17M** Air enters a diffuser at 0.7 bar, 57°C, with a velocity of 200 m/s. At the outlet, where the area is 20 percent greater than at the inlet, the pressure is 1 bar. Determine the outlet temperature and the velocity in m/s (*a*) if the process is adiabatic, and (*b*) if the fluid gains 40 kJ/kg in heat transfer as it passes through.

**5-18M** Air enters a diffuser at 0.80 bar and 20°C and leaves at 0.95 bar with a negligible velocity. If the inlet velocity is 300 m/s and the flow is adiabatic at a rate of 25 kg/s, compute (*a*) the exit temperature in °C, and (*b*) the inlet area in square centimeters.

**5-19M** The entrance conditions for an adiabatic diffuser operating on air are 0.8 bar, 27°C, and 250 m/s. The exit pressure is 1 bar, and $A_2/A_1$ is 1.6. Compute the exit velocity in m/s, assuming that the specific heats are constant.

**5-20M** Steam enters an adiabatic diffuser as a saturated vapor at 95°C with a velocity of 350 m/s. At the exit the pressure and temperature are 1 bar and 120°C, respectively. If the exit area is 50 cm², determine (*a*) the exit velocity in m/s, and (*b*) the mass flow rate in kg/s.

**5-21M** Refrigerant-12 enters an adiabatic diffuser as a saturated vapor at 32°C with a velocity of 110 m/s. At the exit the pressure and temperature are 8 bars and 40°C, respectively. If the exit area is 50 cm², determine (*a*) the exit velocity in m/s, and (*b*) the mass flow rate in kg/s.

### Nozzles

**5-22M** Steam enters a nozzle at 30 bars and 320°C and leaves at 15 bars with a velocity of 535 m/s. The mass flow rate is 8000 kg/h. Neglecting the inlet velocity and considering adiabatic flow, compute (*a*) the exit enthalpy in kJ/kg, (*b*) the exit temperature in °C, and (*c*) the nozzle exit area in square centimeters.

**5-23M** Steam at 15 bars and 280°C enters a nozzle with an initial velocity of 125 m/s. The enthalpy of the steam at the exit is 2800 kJ/kg, and the heat loss is 25 kJ/kg. What is the exit velocity in m/s?

**5-24M** Air expands through a nozzle from 5 bars, 157°C, to 1 bar, 47°C. Neglect the inlet velocity, and compute the exit velocity if $q_{in} = 6$ kJ/kg. State any necessary assumptions.

**5-25M** Air is admitted to an adiabatic nozzle at 3 bars, 200°C, and 50 m/s. The exit conditions are 2 bars and 150°C. Determine the ratio of the exit to entrance areas, $A_2/A_1$.

**5-26M** Steam enters an adiabatic nozzle at 5 bars, 280°C, with a velocity of 70 m/s. It expands to a pressure of 1 bar and a temperature of 120°C. Determine the exit velocity in m/s.

**5-27M** Air enters a converging nozzle at 1.8 bars, 67°C, and 40 m/s. At the outlet of the adiabatic passage the pressure is 1 bar and the velocity is six times its initial value. If the inlet area is 100 cm², determine the exit area of the nozzle in square centimeters and the outlet temperature in °C. Use the air table for data.

**5-28M** Steam at 10 bars, 200°C, enters an adiabatic nozzle with a velocity of 20 m/s. The exit conditions are 7 bars and 180°C. Determine the ratio of inlet to exit area, $A_1/A_2$.

### Turbines

**5-29M** Air enters a turbine at 3 bars pressure, a temperature of 47°C, and a velocity of 60 m/s. At the outlet the pressure is 1.2 bars, the temperature is 7°C, and the velocity is 150 m/s. The shaft work delivered by the turbine is 45 kJ/kg. What are the magnitude and direction of the heat transfer in kJ/kg?

**5-30M** Steam enters a turbine at 15 bars and 400°C and exhausts as a saturated vapor at 1 bar. The inlet velocity is 60 m/s, and the exit velocity is 170 m/s. For a mass flow rate of 27,000 kg/h, the power output is 3850 kW. Determine the quantity and direction of the heat transfer in kJ/kg.

**5-31M** Air enters a turbine at 6 bars and 277°C and leaves at 1 bar. The flow rate is 50 kg/min, and the power output is 180 kW. If the heat removed from the air is 30 kJ/kg, compute the exit temperature in °C.

**5-32M** A small gas turbine operating on hydrogen delivers 18 kW to the surroundings. The gas enters the steady-flow device with a velocity of 75 m/s through a cross-sectional area of 0.002 m². The inlet pressure is 2.2 bars and the temperature is 500°K. Although the velocity does not change appreciably, the final pressure drops to 0.8 bar and the final temperature is 380°K. Compute the heat-transfer rate in kJ/min.

**5-33M** Steam flows steadily to a turbine at a rate of 20,000 kg/h, entering at 40 bars, 440°C, and leaving at 0.20 bar with 90 percent quality. A heat loss amounts to 30 kJ/kg. The inlet pipe has a 10-cm diameter, and the exhaust section is rectangular with dimensions of 0.6 m by 0.7 m. Calculate the turbine output power in kilowatts.

**5-34M** Calculate the inlet temperature, in °C, required for a steam turbine which is to develop 10,000 kW of power from a steam flow rate of 42,000 kg/h if the steam enters at 40 bars and leaves at 0.04 bar with a quality of 90 percent. Assume negligible heat transfer and change in kinetic energy.

**5-35M** Air enters a turbine at 6 bars, 740°K, and 120 m/s. The exit conditions are 1 bar, 450°K, and 220 m/s. A heat loss of 15 kJ/kg occurs as the air passes through the turbine. The inlet area is 4.91 cm².
(a) Determine the kinetic energy change in kJ/kg.
(b) Determine the power output in kilowatts.

**5-36M** The inlet conditions for a steam turbine are 40 bars, 600°C, and 50 m/s, while the exit state is 1 bar, 160°C, and 120 m/s. The mass flow rate is 1000 kg/h and a heat loss of 15,000 kJ/h occurs. Calculate (a) the power output in kW, and (b) the ratio of the inlet- to outlet-pipe diameters.

**5-37M** Steam enters a turbine with a negligible velocity at 80 bars and 520°C and leaves at 0.06 bar with 90 percent quality and a velocity of 240 m/s. The turbine delivers 7000 kW, and the cross-sectional area of the exhaust is 0.60 m². Determine the heat transfer from the steam in kJ/kg.

**Compressors**

**5-38M** Calculate the power required by a compressor if air flowing at a rate of 0.8 kg/s enters at 1 bar, 7°C, with a velocity of 70 m/s, and leaves at 2 bars, 77°C, with a velocity of 120 m/s. Heat transferred from the air amounts to 15 kJ/kg.

**5-39M** Air enters a steady-state compressor at 1 bar and 27°C at a rate of 1 kg/min, and leaves at 7 bars and 227°C. The power required to operate the compressor is 3.58 kW. Determine the magnitude (in kJ/h) and direction of any heat transfer.

**5-40M** An air compressor handling 300 m³/min increases the pressure from 1 bar to 2.3 bars, and heat is removed at the rate of 1700 kJ/min. The inlet temperature is 17°C, and at the exit the temperature is 137°C and the area is 200 cm². Find the power input, in kW, necessary to operate the compressor if the inlet velocity is neglected.

**5-41M** A water-cooled compressor changes the state of refrigerant 12 from a saturated vapor at 1 bar to a pressure of 8 bars. The fluid rate is 0.9 kg/min, and the cooling water removes heat at a rate of 140 kJ/min. If the power input to the compressor is 3 kW, determine the final temperature (in °C) of the refrigerant leaving the compressor.

**5-42M** Air at 0.7 bar and 17°C enters a compressor with a velocity of 200 m/s and leaves at a pressure of 6.1 bars, a temperature of 547°C, and a velocity of 250 m/s. The cross-sectional area of the inlet to the compressor is 0.1 m². Determine the power, in kJ/s, required to drive the compressor if the process is adiabatic.

**5-43M** A centrifugal compressor is supplied with dry, saturated steam at 0.2 bar. Five hundred kilograms of steam are compressed per hour to 1 bar and 160°C. During the process 4000 kJ/h are removed from the steam by cooling. What power, in kJ/s, is required to operate the compressor?

**5-44M** A steam compressor is supplied with 50 kg/h of saturated vapor at 0.04 bar and discharges at 1.5 bar and 120°C. The power required is measured to be 2.4 kW. What is the rate of heat transfer from the steam in kJ/min?

**5-45M** Air enters a compressor at 40 m/s through an area of 100 cm² with a pressure and temperature of 1.0 bar and 27°C, respectively. A heat loss of 7.0 kJ/kg occurs during its passage through the compressor, and the outlet conditions are 12.0 bars, 107°C, and 100 m/s.

    (a) Determine the shaft work required (in kJ/kg) in magnitude and direction.
    (b) Determine the mass flow rate through the device, in kg/s.
    (c) Determine the power required, in kilowatts.

**5-46M** A compressor steadily inducts 2000 kg/h of refrigerant 12 at 0.6 bar and 0°C through a pipe with an inside diameter of 7 cm. It discharges the gas at 7 bars, and 140°C through a 2-cm-diameter pipe. During the process, 60,000 kJ/h are lost as heat to the surroundings. Determine the power required to drive the compressor, in kilowatts.

**5-47M** A fan receives air at 970 mbar, 20°C, and 3 m/s and discharges it at 1020 mbar, 22°C, and 18 m/s. If the flow is adiabatic and 50 m³/min enters, determine the required power input in kilowatts.

**Throttling devices**

**5-48M** Steam is throttled from 40 bars to 0.35 bar and 120°C. What is the quality of the steam entering the throttling process? If the inlet and exit velocities are essentially the same, what is the ratio of exit- to inlet-area for the device?

**5-49M** Refrigerant 12 is throttled from the saturated-liquid state at 32°C until the temperature reaches −20°C. What is the final pressure in bars and the final specific volume in cm³/g.

**5-50M** A heavily insulated throttle valve receives steam at 30 bars, 240°C, and exhausts at 7 bars. Velocities are low enough to be neglected.

    (a) Determine the exit temperature in °C.
    (b) If steam were an ideal gas, what would the discharge temperature be?

**5-51M** Liquid refrigerant 12 at 40°C and 10 bars is throttled through a well-insulated capillary tube to (a) 4°C, and (b) −15°C. What is the final pressure in bars and the final quality?

**5-52M** Refrigerant 12 enters a throttling device subcooled 1.31°C below the saturation state at 12 bars. The exit pressure is 2.8 bars. What is the specific volume in the final state, in cm³/g?

**5-53M** Steam flows steadily along a horizontal pipe in which a partially opened valve is located. Upstream of the valve the steam is at 280 bars and 480°C, while downstream of the valve the pressure is 200 bars. Neglecting heat transfer, determine the temperature change of the fluid, in °C.

**5-54M** An incompressible fluid flows adiabatically in a constant-area pipe which contains a throttling valve. Is the internal energy change of the fluid across the valve positive, zero, or negative?

**5-55M** Dry, saturated steam at 8 bars enters a long, insulated pipe. At a point downstream from the entrance where the pressure is 3 bars, determine the quality or the final temperature of the steam, whichever is appropriate.

**Heat exchangers**

**5-56M** Refrigerant 12 with a mass flow rate of 5 kg/min enters a condenser at 14 bars, 80°C, and leaves as a liquid at 13.2 bars, 52°C. Water, which is used as the coolant in the condenser, enters at 12°C and leaves at 24°C. Calculate (a) the heat transfer from the refrigerant 12 in kJ/min, and (b) the mass flow rate of water in kg/min.

**5-57M** Steam enters a heat exchanger at 15 bars and 280°C, where it condenses on the outside of some tubes. The condensed steam leaves the heat exchanger as a saturated liquid at 15 bars with a flow rate of 5000 kg/h. The steam is condensed by passing cool water through the inside of the tubes. The cooling water enters at a temperature of 20°C and experiences a temperature rise of 20°C before leaving the heat exchanger. What flow rate of cooling water is required, in kg/h?

**5-58M** Steam enters a heat exchanger at 40 bars, 500°C, and emerges as a saturated liquid at 15°C. A counterflow of refrigerant 12 enters as a liquid at −40°C, 14.6 bars, and leaves at 36°C, 14.2 bars. If no heat is lost externally from the heat exchanger, determine the flow rate of refrigerant 12 required per kilogram of steam.

**5-59M** Refrigerant 12 enters a condenser at 16 bars, 80°C, and leaves as a liquid at 15.8 bars, 40°C. Water, which is used as the coolant in the condenser, enters under pressure at 10°C and leaves at 30°C. If the rate of refrigerant 12 flow is 5 kg/min, calculate (a) the heat transfer from the refrigerant 12, in kJ/min, and (b) the rate of water required in kg/min.

**5-60M** Five thousand kilograms per hour of a liquid having a mean specific heat of 3.35 kJ/(kg)(°C) are heated from 25 to 80°C by steam which is supplied to the heat exchanger as saturated vapor at 1 bar. The condensed steam leaves the heat exchanger at 95°C and 1 bar. How many kilograms of steam are condensed per hour?

**5-61M** Refrigerant 12 is condensed in a refrigeration cycle from a state of 8 bars and 60°. Air blown across the condenser coils containing the refrigerant fluid experiences a 9°C temperature rise at 1 bar pressure. If the air mass flow rate is 18 kg/min, what is the maximum refrigerant 12 rate permitted in kg/min to assure that the fluid is completely condensed?

## Mixing processes

**5-62M** Water is fed into a pipeline from two different sources. One source delivers steam of 90 percent quality at a rate of 2000 kg/h. The second source delivers steam at a rate of 1750 kg/h at a temperature of 280°C. If the mixing process is adiabatic and at a constant pressure of 10 bars, determine the state of the mixture at equilibrium downstream.

**5-63M** Water is heated in an insulated heat exchanger by mixing it with steam. The water enters at a rate of 100 kg/min at 20°C and 3 bars. The steam enters at 320°C and 3 bars. The mixture leaves the heat exchanger or mixing chamber at 40°C and 3 bars. How much steam is needed, in kg/min?

**5-64M** An open feedwater heater operates at 5 bars pressure. Superheated steam is bled from a turbine and enters the open heater at 200°C. Compressed liquid at 35°C enters the heater at another position. Determine the ratio of the mass flow rate of steam bled from the turbine to the total mass flow rate leaving the heater.

**5-65M** An open feedwater heater operates at 7 bars. Compressed liquid water at 35°C enters at one section. Determine the temperature of the entering steam bled from a turbine if the ratio of the mass flow rate of compressed liquid to that of the superheated steam is 4.5 : 1.

**5-66M** It is desired to have available a flow of liquid water at 3 bars and 70°C. It is proposed to achieve this by mixing in an insulated tank 50 kg/min of water at 3 bars, and 15°C with $m_s$ kg/min of superheated steam at 3 bars and 160°C.

    (a) Determine the required flow rate of superheated steam in kg/min.

    (b) If the exit area is 25 cm², determine the exit velocity in m/s.

## Pipe flow (NOT ON TEST)

**5-67M** Water is contained in a tank 10 m deep, open to the atmosphere at the upper end. At the bottom of the tank is a 1-cm-diameter hole. If the velocity of the stream leaving the hole is uniform across the cross section and no energy transfer occurs to the water flowing out, calculate the velocity of the water flowing out of the hole in m/s, and the flow rate in kg/min. Neglect any temperature change of the fluid.

**5-68M** Liquid water flows over a dam and down into a vertical pipe of 20 cm internal diameter and 160 m long. The temperature of the fluid remains essentially constant, and heat losses are small. Determine the outlet velocity in m/s.

**5-69M** Refrigerant 12 flows steadily through an insulated pipeline 5.0 cm in internal diameter. At a certain cross section in the pipe the pressure is 5 bars and the temperature is 50°C. At another cross section, downstream from the first, the pressure is 4 bars and the specific volume is 50 cm³/g. Find the velocity, in m/s, at the upstream cross section and the mass flow rate in kg/min.

**5-70M** Steam flows through a long, insulated pipe. At one point, where the diameter is 9 cm, the pressure and temperature are 15 bars and 320°C, respectively. At another cross section further downstream, where the diameter is now 7 cm, the pressure is 10 bars and the temperature is 280°C. Determine the mass flow rate of the steam in kg/s and the downstream velocity in m/s.

**5-71M** An ideal gas with constant specific heats $[c_p = 0.86 \text{ J}/(\text{g})(°\text{C})]$ flows through a long, horizontal pipe of constant diameter. The gas enters the pipe at 2.8 bars and 37°C with a velocity of 70 m/s. The gas leaves the pipe at 1.4 bars and 37°C. Determine (a) the exit velocity in m/s, and (b) the heat transfer in kJ/kg.

**5-72M** Steam flows through a long, insulated pipe of constant cross-sectional area. At one point along the pipe the state is 20 bars and 320°C. At another section further downstream the state is 15 bars and 280°C. Determine the velocity at the inlet, in m/s.

**5-73M** A long, horizontal pipe with an internal diameter of 5.25 cm carries refrigerant 12. The fluid enters at 5 bars with a quality of 20 percent and a velocity of 3 m/s. The fluid leaves the pipe at 3.2 bars and 30°C. Determine (a) the mass flow rate, in kg/s, (b) the exit velocity, in m/s, and (c) the heat transfer rate, in kJ/s.

**5-74M** Air flows through a horizontal, insulated pipe of constant diameter. The entrance conditions are 30°C, 3 bars, and 125 m/s. At the exit the velocity is 250 m/s. Determine the exit temperature in °K and the exit pressure in bars.

**5-75M** A pipe 10 cm in diameter contains air at 1.5 bars, 60°C, and 40 m/s at a given cross section. Downstream at another position the pressure is 2.8 bars. If it is desired to maintain the velocity at the same value as before, determine the magnitude, in kJ/kg, and the direction of any required heat transfer.

**5-76M** A steam heating system for a building 150 m high is supplied from boilers 15 m below ground level. Dry, saturated steam is supplied from the boiler at 2.0 bars pressure, and it reaches the 150-m elevation above ground at 1.50 bars. Heat transfer from the supply pipe to the surroundings is 50 kJ/kg. Neglecting kinetic-energy effects, find the state of the steam at the 150-m elevation.

**Pumps** (NOT ON TEST)

**5-77M** Oil with a specific gravity of 0.85 is being pumped from a pressure of 0.7 bar to a pressure of 1.2 bars, and the outlet lies 2 m above the inlet. The fluid flows at a rate of 0.1 m³/s through an inlet cross-sectional area of 0.05 m², and the outlet area is 0.02 m². Determine the power input to the pump in kilowatts.

**5-78M** A pump is employed to deliver 5000 kg/h of water from an elevation 10 m below the pump to an elevation 15 m above the pump. At the lower elevation (state 1) and the upper elevation (state 2) the known data are: $D_1 = 4$ cm, $P_1 = 0.7$ bar, $T_1 = 20°$C, $D_2 = 2$ cm, $P_2 = 5$ bar, and $T_2 = 20°$C.

   (a) Determine the magnitude of the shaft work in kJ/kg if the inlet velocity is 1 m/s and the pipe is insulated.

   (b) Determine the kilowatt rating of the pump.

**5-79M** Water enters a pump at the following conditions: $P_1 = 2$ bars, $T_1 = 15°$C, and $V_1 = 2$ m/s. At the exit, which is 20 m above the inlet, the conditions are: $P_2 = 6$ bars, $T_2 = 15°$C, $V_2 = 8$ m/s, and $D_2 = 2$ cm. The inlet and outlet pipes and the pump are heavily insulated. Determine the required power input to the pump, in kilowatts. Do not neglect any energy term if sufficient information is available to evaluate it.

**5-80M** Water at 20°C is pumped at 1.2 kg/s from the surface of an open tank into a constant-diameter piping system. At the discharge from the pipe the velocity is 10 m/s and the gage pressure is 2.0 bars. The discharge is 15 m above the water surface in the open tank. Neglecting losses in the pipe, determine the power required in kilowatts.

**Transient flow**

**5-81M** A pressurized tank contains 1.5 kg of air at 2 bars and 60°C. Mass is allowed to flow from the tank until the pressure reaches 1 bar. However, during the process heat is added to the air within the tank to keep it at constant temperature. How much heat, in kilojoules, was added during the process? Assume constant specific heats.

**5-82M** An insulated tank with a volume of 0.5 m³ contains air at 1 bar and 25°C. The tank is connected through a valve to a large compressed-air line, which carries air continuously at 7 bars and 120°C. If the valve is opened and air is allowed to flow into the tank until the pressure reaches 5 bars, how much mass, in kilograms, has entered, and what is the final temperature in the tank in °C?

**5-83M** A pressurized tank contains 1 kg of water vapor at 30 bars and 280°C. Mass is allowed to flow from the tank until the pressure reaches 7 bars. However, during the process heat is added to the steam to keep it at constant temperature. How much heat, in kilojoules, was added during the process?

**5-84M** Two adiabatic tanks are interconnected through a valve. Tank $A$ contains 0.2 m³ of air at 40 bars and 90°C. Tank $B$ contains 2 m³ of air at 1 bar and 30°C. The valve is opened until the pressure in $A$ drops to 15 bars. At this instant, $(a)$ what are the temperatures and pressures in both tanks, $(b)$ how much mass has left tank $A$, and $(c)$ what is the total entropy change for the process in kJ/°K? Assume constant specific heats.

**5-85M** A tank with a volume of 3 m³ contains steam at 20 bars and 280°C. The tank is heated until the temperature reaches 440°C. A relief valve is installed on the tank to keep the pressure constant during the process. Evaluate the amount of heat transfer of the steam in kJ and the mass of steam that is bled from the tank.

**5-86M** A rigid insulated tank is initially evacuated. Atmospheric air at 1 bar and 20°C is allowed to leak into the tank until the pressure reaches 1 bar. What is the final temperature of the air within the tank, in °C?

**5-87M** Consider the data of Prob. 5-86M, except that the tank initially contains air at 0.5 bar and 20°C. What is the final temperature, in °C, in this case?

**5-88M** An ideal gas is contained in a rigid tank of volume $V$, initially at $P_1$ and $T_1$. Heat is supplied to the contents until the temperature reaches $T_2$. However, a relief valve allows gas to escape so that the pressure remains constant. Derive an expression for the heat transfer during the process.

**5-89M** Consider the situation in Prob. 5-88M, except that a valve holds the temperature constant while the pressure drops to $P_2$. Derive an expression for the heat transfer in this case?

**5-90M** A tank with a volume of 10 m³ contains air initially at 5 bars, 40°C. Heat is added at a constant rate of 6 kJ/s while an automatic valve allows air to leave the tank at a constant rate of 0.03 kg/s. What is the temperature of the air in the tank $(a)$ 5 min, and $(b)$ 7 min after the initial conditions.

**5-91M** With reference to Prob. 5-90M, how long will it take for the air in the tank to reach $(a)$ 140°C, and $(b)$ 180°C?

**5-92M** With reference to Prob. 5-90M, what is the pressure in the tank after 8 min, starting from the initial conditions?

**5-93M** A container of fixed volume $V$ contains air at pressure $P_1$ and temperature $T_a$. It is surrounded by atmospheric air at a pressure $P_a$ and temperature $T_a$. A valve is opened and atmospheric air is quickly admitted to the container until the pressure reaches the atmospheric value. At that instant the air in the container is at temperature $T_2$. Listing any necessary assumptions, derive an equation for the temperature ratio, $T_R = T_2/T_a$, in terms of the specific-heat ratio $k$ and the pressure ratio, $P_R = P_1/P_a$.

**5-94M** A tank with a volume of 0.5 m³ is half filled with liquid water and the remainder is filled with vapor. The pressure is $(a)$ 20 bars, and $(b)$ 30 bars. Heat is added until one-half of the liquid (by mass) is evaporated while an automatic valve lets saturated water vapor escape at such a rate that the pressure remains constant. Determine the amount of heat transfer in kilojoules.

**5-95M** A pressure vessel with a volume of 0.5 m³ contains saturated water at $(a)$ 250°C, and $(b)$ 300°C. The vessel initially contains 50 percent by volume of liquid. Liquid is slowly withdrawn from the bottom of the tank, and heat transfer takes place so that the contents are kept at constant temperature. How much heat must be added by the time half of the total mass has been removed?

**5-96M** A very large reservoir contains air at 12 bars and an unknown temperature $T_a$. Air flows from this reservoir into a small, insulated tank with a volume of 0.2 m³. The small tank initially contains

0.2 kg of air at 27°C. Then a valve is opened to allow air to flow from the reservoir into the tank until the pressure in the tank becomes 3 bars. At this point the temperature in the tank is observed to be (a) 140°C, and (b) 180°C. What is the temperature of the air in the reservoir, $T_a$? Assume constant specific heats.

**5-97M** A piston-cylinder assembly is attached through a valve to a source of constant-temperature and constant-pressure air. Initially inside the cylinder the volume is 0.1 m³, the temperature of the air is 30°C, and the pressure is 1 bar. The valve is slowly opened to the air-supply line which is at 7 bars and 90°C. The piston moves out as air enters in order to maintain the cylinder pressure at the ambient condition of 1 bar. When the cylinder volume reaches 0.2 m³, (a) what is the temperature inside the cylinder in °C, and (b) how much mass, in kilograms, has entered through the valve?

**5-98M** A tank with a volume of 0.5 m³ contains carbon dioxide at 2 bars and 30°C. Nitrogen at a line pressure and temperature of 8 bars and 150°C, respectively, flows from a pipe into the tank until the pressure reaches 5 bars. If the entire process is adiabatic, determine the final temperature, in °C, of the mixture within the tank.

# PROBLEMS (USCS)

### Continuity of flow

**5-1** Steam enters a steady-flow device at 600 psia and 600°F with a velocity of 550 ft/s. The entrance area is 3.0 in², and the exit area is 26.0 in². If the outlet state is 20 psia and 300°F, determine (a) the mass flow rate in lb/s, and (b) the exit velocity in ft/s.

**5-2** Refrigerant 12 enters a steady-flow device at 100 psia, 160°F, with a velocity of 500 ft/s. The exit pressure and velocity are 40 psia and 400 ft/s, respectively. If the mass flow rate is 40 lb/s and the exit area is 17 in², determine (a) the entrance area in square inches, and (b) the exit temperature in °F.

**5-3** Steam enters a turbine at 1000 psia and 1000°F with a velocity of 440 ft/s and leaves as a saturated vapor at 1 psia. The inlet pipe has a diameter of 2 ft and the outlet diameter as 12 ft.

    (a) Determine the mass flow rate in lb/h.

    (b) Find the exit velocity in ft/s.

**5-4** Steam at a rate of 100,000 lb/h leaves a turbine at 1 psia and 90 percent quality. Calculate the area of the exhaust pipe for a mean velocity of 500 ft/s.

**5-5** In a steady-flow device 1.2 lb/min of saturated refrigerant 12 vapor at 70 psia enters with a velocity of 12 ft/s. The exit area is 0.15 in², and the exit temperature and velocity are 140°F and 16 ft/s, respectively. Determine (a) the entrance area in square inches, and (b) the exit pressure in psia.

**5-6** In a steady-flow device, 100 lb/s of saturated water vapor at 400 psia enters with a velocity of 100 ft/s. The steam exits at a temperature of 900°F and a velocity of 200 ft/s. The exit area is 163 in². Determine (a) the entrance area in square inches, and (b) the exit pressure in psia.

**5-7** Steam enters a steady-flow device at 20 psia and 350°F. At the outlet the pressure is 50 psia, the temperature is 500°F, and the velocity is 400 ft/s. If the inlet-to-outlet area ratio for the system is 2.4 : 1, determine the inlet velocity in ft/s.

**5-8** Air initially at 20 psia and 140°F flows through a cross-sectional area of 0.1 ft² at a rate of 100 lb/min. Downstream from this point the pressure is 40 psia, the temperature is 140°F, and the velocity is 50 ft/s. Determine (a) the inlet velocity in ft/s, and (b) the outlet area in square feet.

**5-9** Air with a density of 0.0702 $lb_m$/ft³ enters a steady-flow system through a duct 2 ft² in area and with a velocity of 10 ft/s. If the outlet density is 0.175 lb/ft³, (a) determine the flow in lb/s, and (b) determine what the area of the discharge duct must be if the outlet velocity is 15 ft/s.

**5-10** An ideal gas enters a steady-flow system at 15 psia and 100°F. The inlet to the system has a diameter of 8 in, and the velocity at that point is 30 ft/s. At the outlet the diameter of the duct is 3 in, and the density of the fluid at that section is 0.25 lb/ft³. If the molar mass of the gas is 30, determine the velocity at the outlet and the mass flow rate in lb/s.

**5-11** An ideal gas with a molar mass of 48 enters a steady-flow device at 3 atm, 310°F, and 600 ft/s through an area of 125 in². At the exit of the device the pressure is 1.5 atm, the velocity is 400 ft/s, and the exit area is 250 in². Determine (a) the density of the gas at the inlet in lb/ft³, (b) the mass flow rate in lb/s, and (c) the exit temperature in °F.

**5-12** Carbon dioxide enters a steady-flow device at 80°F with a velocity of 80 ft/s through an area of 0.050 ft². At the exit of the device the pressure and temperature are 20 psia and 120°F, respectively, and the fluid moves with a velocity of 25 ft/s through an area of 0.080 ft².

    (a) Determine the mass flow rate in lb/s.

    (b) Determine the inlet pressure in psia, assuming the fluid is an ideal gas.

**5-13** Water at 60°F is flowing through a long, plastic tube. The diameter of the tube is not constant. At position 1 in the tube the internal diameter is 1 ft and the fluid velocity is 10 ft/s. At position 2 downstream from position 1 the internal diameter is 2 ft. The temperature of the water remains constant.

    (a) Determine the mass flow rate in lb/h.

    (b) Determine the fluid velocity at position 2 in ft/s.

**5-14** Water at 40°F is flowing through a long pipe. Upstream at position 1 the internal diameter is 16 in and the fluid velocity is 2.5 ft/s. At position 2 downstream the internal diameter is 8 in. Determine (a) the mass flow rate in lb/h, and (b) the fluid velocity at position 2 in ft/s.

### Diffusers

**5-15** An adiabatic diffuser is employed to reduce the velocity of a stream of air from 780 to 120 ft/s. The inlet pressure is 15 psia, and the inlet temperature is 560°F. Determine the required outlet area in in² if the mass flow rate is 15 lb/s and the final pressure is 17.7 psia.

**5-16** Air enters a diffuser at 15 psia, 80°F, with a velocity of 510 ft/s. The exit temperature is 100°F. If a heat loss of 0.2 Btu/lb occurs, find the exit velocity in ft/s.

**5-17** Air enters a diffuser at 10 psia, 140°F, with a velocity of 1000 ft/s. At the outlet, where the area is 28 percent greater than at the inlet, the pressure is 13 psia. Determine the outlet temperature and velocity if (a) the process is adiabatic, and (b) if the fluid gains 2 Btu/lb in the form of heat as it flows through the device.

**5-18** Air enters a diffuser at 10 psia and 60°F and leaves at 15 psia with a negligible velocity. If the inlet velocity is 1095 ft/s and the flow is adiabatic at a rate of 50 lb/s, compute (a) the exit temperature in °F, and (b) the inlet area in ft².

**5-19** The entrance conditions for an adiabatic diffuser operating on air are 10 psia, 100°F, and 800 ft/s. The exit pressure is 14 psia, and $A_2/A_1$ is 1.6. Compute the exit velocity in ft/s assuming the specific heats are constant.

**5-20** Steam enters an adiabatic diffuser as a saturated vapor at 200°F with a velocity of 100 ft/s. At the exit the pressure and temperature are 14.7 psia and 250°F, respectively. If the exit area is 8.0 in², determine (a) the exit velocity in ft/s, and (b) the mass flow rate in lb/s.

**5-21** Refrigerant 12 enters an adiabatic diffuser as a saturated vapor at 80°F with a velocity of 420 ft/s. At the exit the pressure and temperature are 100 psia and 100°F, respectively. If the exit area is 10 in², determine (a) the exit velocity in ft/s, and (b) the mass flow rate in lb/s.

### Nozzles

**5-22** Steam enters a nozzle at 400 psia and 600°F and leaves at 250 psia with a velocity of 1475 ft/s. The flow rate is 18,000 lb/h. Neglecting the inlet velocity and considering adiabatic flow, compute (a) the exit enthalpy, (b) the exit temperature, and (c) the nozzle exit area in ft².

**5-23** Steam at 200 psia and 500°F enters a nozzle with an initial velocity of 400 ft/s. The enthalpy of the steam at the exit is 1120 Btu/lb, and the heat loss from the nozzle is 10 Btu/lb. What is the exit velocity in ft/s?

**5-24** Air expands through a nozzle from 75 psia, 300°F, to 15 psia, 100°F. Neglect the inlet velocity. Compute the exit velocity if $q_{in} = 3.0$ Btu/lb$_m$. State all necessary assumptions.

**5-25** Air expands through a nozzle from 50 psia, 300°F, to 15 psia, 80°F. If the inlet velocity is small and $q_{out} = 2.0$ Btu/lb, compute the exit velocity.

**5-26** Steam enters an adiabatic nozzle at 60 psia, 500°F, with a velocity of 200 ft/s. It expands to a pressure of 15 psia and a temperature of 250°F. Determine the exit velocity in ft/s.

**5-27** Air enters a converging nozzle at 25 psia, 140°F, and 100 ft/s. At the outlet of the adiabatic passage the pressure is 15 psia and the velocity is six times its initial value. If the inlet area is 0.1 ft², determine the exit area of the nozzle in square feet and the outlet temperature of the gas stream in °F. List any necessary assumptions and use the air table A-15 for data.

**5-28** A steady flow of steam enters a nozzle at 1000 psia and 1200°F with a velocity of 300 ft/s. The area of the inlet is 4.53 in² and the exit area is 1.20 in². At the nozzle exit the conditions are 700 psia and 1100°F. Calculate the heat transfer from the steam in Btu/lb.

**Turbines**

**5-29** Air enters a small turbine at a pressure of 45 psia, a temperature of 120°F, and a velocity of 180 ft/s. At the outlet of the turbine the pressure is 18 psia, the temperature is 40°F, and the velocity is 450 ft/s. The shaft work delivered by the turbine is measured to be 15,000 ft·lb$_f$/lb of air passing through the turbine. What are the magnitude and direction of the heat transfer in Btu/lb?

**5-30** Steam enters a turbine at 200 psia and 700°F and exhausts as a saturated vapor at 15 psia. The inlet velocity is 180 ft/s, and the exit velocity is 480 ft/s. For a mass flow rate of 60,000 lb/h, the power output is measured as 4500 hp. Determine the quantity and direction of the heat transfer per pound of steam.

**5-31** Air enters a turbine at 90 psia and 540°F and leaves at 15 psia. The flow rate is 110 lb/min, and the power output is 240 hp. If the heat removed from the air in the turbine amounts to 18 Btu/lb, calculate the exit temperature of the air in °F.

**5-32** A small gas turbine operating on hydrogen delivers 24 hp to the surroundings. The gas enters the steady-flow device with a velocity of 220 ft/s through a cross-sectional area of 0.02 ft². The inlet pressure is 32 psia, and the temperature is 440°F. Although the velocity does not change appreciably, the final pressure drops to 12 psia and the final temperature is 220°F. Compute the heat-transfer rate in Btu/min.

**5-33** Steam enters a turbine at 700 psia, 700°F, and leaves as a dry saturated vapor at 1 psia. The mass flow rate is 300,000 lb/h, and the radiation heat loss from the turbine is 4 Btu/lb. Calculate the power output of the turbine in horsepower.

**5-34** Calculate the inlet (throttle) temperature required for a steam turbine which is to develop 10,000 kW from a steam flow rate of 95,000 lb/h if the steam enters at 600 psia and leaves at 0.5 psia and there is 10 percent moisture in the exhaust steam. Assume negligible heat transfer and change in kinetic energy.

**5-35** A gas turbine power plant takes air from the atmosphere, compresses it, heats it at constant pressure, and then expands it. At the inlet to the power plant the ambient pressure is 14.5 psia and the temperature is 80°F. The inlet velocity of the air stream is 300 ft/s and the inlet area, which lies 25 ft above the floor, is 1.0 ft². The steady-flow process develops 450 hp. At the exit the pressure is 24 psia and the temperature is 340°F. The exit area is 0.8 ft² and lies 10 ft above the floor of the plant. Determine (a) the exit velocity of the flow stream, and (b) the amount and direction of the heat transfer per pound of fluid flowing through the power plant.

**5-36** A turbine is supplied with steam at 400 psia and 700°F through a pipe with an inside diameter of 2.0 in. The turbine exhausts steam at 15 psia and a velocity of 80 ft/s through a pipe with an internal diameter of 5.9 in. The output of the turbine is measured to be 126 hp for a flow rate of 1830 lb$_m$/h. Determine (a) the magnitude and direction of the heat flow in Btu/h if the change in kinetic energy is small, and (b) whether the assumption that the change in kinetic energy is small is justified.

**5-37** Steam enters a turbine with a negligible velocity at 1000 psia and 1000°F and leaves at 1 psia with 10 percent moisture content and a velocity of 700 ft/s. The turbine delivers 9300 hp, and the cross-sectional area of the exhaust is 5.96 ft². Determine the heat loss from the steam passing through the turbine, in Btu/lb.

**Compressors**

**5-38** Calculate the power required by a compressor if air flowing at a rate of 2.0 $lb_m/s$ enters at 15.0 psia, 40°F, with a velocity of 200 ft/s, and leaves at 30.0 psia, 160°F, with a velocity of 400 ft/s. Heat transferred from the air to the cooling water circulating around the compressor casing amounts to 8.0 Btu/lb of air.

**5-39** An air compressor takes in air from the atmosphere at 14.5 psia and 65°F. The pressure ratio is 7.0, and the final specific volume is 3.75 $ft^3/lb_m$. For every pound of air that passes through the compressor, 8.5 Btu are removed as heat. If the changes in the kinetic and potential energies are small, determine the work input in $ft \cdot lb_f/lb$.

**5-40** An air compressor handling 10,000 $ft^3$/min increases the pressure from 15.0 to 35 psia, and heat is removed at the rate of 750 Btu/min. The inlet temperature is 70°F, and at the exit the temperature is 280°F and the area is 0.20 $ft^2$. Find the power input, in horsepower, necessary to operate the compressor under these conditions.

**5-41** A compressor which is cooled by a water jacket in the casing compresses refrigerant 12 from a saturated vapor at 12 psia to a pressure of 125 psia. The refrigerant 12 flows steadily at a rate of 2.0 $lb_m$/min, and the cooling water in the jacket removes heat at the rate of 135 Btu/min. If the power input to the compressor is 4.0 hp, determine the final temperature of the refrigerant 12 leaving the compressor.

**5-42** Air at 10 psia and 60°F enters a compressor of a jet plane with a velocity of 700 ft/s and leaves the compressor at a pressure of 42.8 psia, a temperature of 393°F, and a velocity of 800 ft/s. The cross-sectional area of the inlet to the compressor is 0.9 $ft^2$. Determine the power required to drive the compressor if the process is adiabatic.

**5-43** A centrifugal compressor is supplied with dry, saturated steam at 2 psia. One thousand pounds of steam are compressed per hour to 14.7 psia and 300°F. During the process 3000 Btu are removed per hour from the system by cooling. What horsepower is required to operate the compressor?

**5-44** A steam compressor is supplied with 110 lb/h of saturated vapor at 0.6 psia and discharges at 20 psia and 250°F. The power required is measured to be 3.2 hp. What is the rate of heat transfer from the steam, in Btu/min?

**5-45** Refrigerant 12 enters a compressor as a saturated vapor at 30 psia at a rate of 5 lb/min. The compressor requires an input of 6 hp, and a heat loss of 105 Btu/min occurs. If the exit pressure is 125 psia, determine the exit temperature in °F.

**5-46** A compressor steadily inducts 5000 lb/h of refrigerant 12 at 5 psia and 80°F through a pipe with an inside diameter of 3 in. It discharges the gas at 100 psia and 300°F through a 1-in-diameter pipe. During the process 75,000 Btu/h are lost as heat to the surroundings. Determine the power, in kilowatts, required to drive the compressor.

**5-47** A fan receives air at 14.2 psia, 70°F and 10 ft/s and discharges it at 14.8 psia, 80°F, and 60 ft/s. If the flow is essentially adiabatic and 1800 $ft^3$ of air enter per minute, determine the power input necessary in horsepower.

**Throttling devices**

**5-48** Steam is throttled from 800 psia to 5 psia and 250°F. What is the quality of the steam entering the throttling process? If the change in kinetic energy is small, what is the ratio of exit area to inlet area for the device?

**5-49** Refrigerant 12 is throttled from the saturated-liquid state at 114.5 psia until the temperature reaches $-10$°F. What are the final pressure in psia, the final enthalpy in Btu/lb, and the final specific volume in $ft^3/lb_m$?

**5-50** A heavily insulated throttle valve receives steam at 600 psia, 650°F, and exhausts at 250 psia. Velocities are low enough to be neglected.

    (*a*) Determine the exit temperature in °F.

    (*b*) If steam were an ideal gas, what would the discharge temperature be?

**5-51** Liquid refrigerant 12 at 100°F and 135 psia is throttled through a well-insulated capillary tube to (a) 0°F, and (b) 10°F. What is the final pressure and the final quality?

**5-52** Refrigerant 12 enters a throttling device subcooled 4.35°F below the saturation state at 140 psia. The exit pressure is 30 psia. What is the specific volume in the final state in ft³/lb?

**5-53** Steam flows steadily along a horizontal pipe in which a partially opened valve is located. Upstream of the valve the steam is at 4000 psia and 900°F, while downstream of the valve the pressure is 3000 psia. Neglecting heat transfer, determine the temperature change of the fluid, in °F.

**5-54** An incompressible fluid flows adiabatically in a constant-area pipe which contains a throttling valve. Is the internal-energy change of the fluid across the valve positive, zero, or negative?

**5-55** Dry, saturated steam at 100 psia enters a long, insulated pipe. At a point downstream from the entrance where the pressure is 40 psia, determine the quality or the final temperature of the steam, whichever is appropriate. State any assumptions.

## Heat exchangers

**5-56** Refrigerant 12 with a mass flow rate of 10 lb/min enters a condenser at 200 psia, 180°F, and leaves as a liquid at 190 psia, 120°F. Water, which is used as a coolant in the condenser, enters at 55°F and leaves at 75°F. Calculate (a) the heat transfer from the refrigerant 12 in Btu/min, and (b) the mass flow rate of water in lb/min.

**5-57** Steam enters a heat exchanger at 200 psia and 500°F, where it condenses on the outside of some heat-exchanger tubes. The condensed steam leaves the exchanger as a saturated liquid at 200 psia and at a flow rate of 10,000 lb/h. The steam is condensed by passing cool water through the insides of the tubes. The cooling water enters at a temperature of 50°F and experiences a temperature rise of 45°F before leaving the heat exchanger. What flow rate of cooling water is required in lb/h?

**5-58** Steam enters a heat exchanger at 1000°F, 600 psia, and emerges as a saturated liquid at 60°F. A counterflow of refrigerant 12 enters as liquid at −40°F, 210 psia, and leaves at 100°F, 200 psia. If no heat is lost in the exchanger, calculate the flow rate of refrigerant 12 required per pound of steam.

**5-59** Refrigerant 12 enters a condenser at 200 psia, 120°F, and leaves as a liquid at 200 psia, 70°F. Water, which is used as the coolant in the condenser, enters at 60°F and leaves at 75°F. If the rate of flow of refrigerant 12 is 10 lb/min, calculate (a) the heat transfer from the refrigerant 12 in Btu/min, and (b) the rate of water required in lb/min.

**5-60** Ten thousand pounds per hour of a liquid having a mean specific heat of 0.8 Btu/(lb)(°F) are heated from 80°F to 180°F by steam which is supplied to the heat exchanger as saturated vapor at 14.7 psia. The condensed steam leaves the heat exchanger at 200°F. How many pounds of steam are condensed per hour?

**5-61** Refrigerant 12 is condensed in a refrigeration cycle from a state of 100 psia and 120°F. Air blown across the condenser coil containing the refrigerant fluid experiences a 15°F temperature rise at 1 atm pressure. If the air mass flow rate is 30 lb/min, what is the maximum refrigerant 12 flow rate permitted in lb/min to ensure that the refrigerant fluid is completely condensed?

## Mixing processes

**5-62** Steam is fed into a pipeline from two different sources, both at 100 psia. One source delivers steam of 90 percent quality at a rate of 2100 lb/h. The second source delivers steam at a rate of 4650 lb/h at a temperature of 400°F. If the mixing process is adiabatic and constant-pressure, compute the state of the mixture in the pipeline at equilibrium.

**5-63** Water is heated in an insulated heat exchanger by mixing it with steam. The water enters at a rate of 200 lb/min at 65°F and 50 psia. The steam enters at 600°F and 50 psia. The mixture leaves the exchanger at 100°F and 48 psia. How much steam is needed, in lb/min?

**5-64** An open feedwater heater operates at 60 psia. Superheated steam is bled from a turbine, and enters at one opening with a temperature of 300°F. Compressed liquid at 100°F enters at another section. Determine the ratio of the mass flow rate of steam bled from the turbine to the total mass flow rate leaving the heater.

**5-65** An open feedwater heater operates at 100 psia. Compressed liquid water at 100°F enters at one section. Determine the temperature of the entering steam bled from the turbine if the ratio of the mass flow rate of compressed liquid to that of the superheated steam is 4.5 : 1.

**5-66** It is desired to have available a flow of liquid water at 40 psia and 150°F. It is proposed to achieve this by mixing, in an insulated tank, 100 lb/min of water at 40 psia and 60°F with $m_s$ lb/min of superheated steam at 40 psia and 300°F.

    (a) Determine the required flow rate of steam in lb/min.

    (b) If the exit area is 4.25 in², determine the exit velocity in ft/s.

## Pipe flow

**5-67** Water is contained in a tank 20 ft deep, open to the atmosphere at the upper end. At the bottom of the tank is a $\frac{1}{4}$-in-diameter hole.

    (a) If the velocity of the stream leaving the hole is uniform across the cross section and no energy transfer occurs to the water flowing out, calculate the velocity of the water flowing out of the hole and the flow rate, in lb/min.

    (b) If the effect of friction is to reduce the measured velocity to 95 percent of that calculated in (a), how much does the internal energy of the water change in passing through the hole?

**5-68** Liquid water flows over a dam and down into a vertical pipe of 6 in internal diameter and 500 ft long. The temperature of the fluid remains essentially constant, and heat losses are small. Determine the outlet velocity in ft/s.

**5-69** Refrigerant 12 flows steadily through an insulated pipeline 2.00 in in internal diameter in which there is a pressure drop due to friction. At a certain cross section in the pipe the pressure is 80 psia and the temperature is 120°F. At another cross section, downstream from the first, the pressure is 60 psia and the specific volume is 0.766 ft³/lb. Find the velocity in ft/s at the upstream cross section and the mass flow rate of the fluid in lb/min.

**5-70** Steam flows through a long insulated pipe. At one point along the pipe where the diameter is 4 in, the pressure and temperature are 300 psia and 600°F, respectively. At another cross section farther down the pipe, where the diameter is now 3 in, the pressure is 260 psia and the temperature is 550°F. Determine the mass flow rate of the steam and the velocity at the downstream section of the pipe.

**5-71** An ideal gas with constant specific heats $[c_p = 0.205$ Btu/(lb)(°F)] flows through a long, horizontal pipe of constant diameter. The gas enters the pipe at 40 psia and 100°F and with a velocity of 200 ft/s. The gas leaves the pipe at 20 psia and 100°F. Determine (a) the exit velocity in ft/s, and (b) the heat transfer across the pipe walls, in Btu/lb.

**5-72** Steam flows through a long, insulated pipe of constant cross-sectional area. At one point along the pipe the state is 300 psia and 550°F. At another section further down the pipe the state is 250 psia and 500°F. Determine the velocity $V_1$.

**5-73** A long, horizontal pipe with a constant inside diameter of 2.07 in carries refrigerant 12. The fluid enters at 80 psia with a quality of 20 percent and a velocity of 10 ft/s. The fluid leaves the pipe at 50 psia and 80°F. Determine (a) the mass flow rate in lb/s, (b) the exit velocity in ft/s, and (c) the heat-transfer rate in Btus.

**5-74** Air flows through a horizontal insulated pipe of constant diameter. The entrance conditions are 100°F, 50 psia, and 400 ft/s. At the exit the velocity is 800 ft/s. Determine the exit temperature in °R and the exit pressure in psia.

**5-75** A 4-in-diameter pipe contains air at 20 psia, 140°F, and 120 ft/s at a given cross section. Downstream the pressure reaches 18 psia. If it is desired to maintain the same velocity, determine the magnitude, in Btu/lb, and the direction of any required heat transfer.

**5-76** Water flows through a 12-in-diameter, 1000-ft-long, horizontal pipe at 10 ft/s. The pressure drop due to friction per foot of length of pipe, $\Delta P/L$, is given by the experimental correlation $(P_1 - P_2)/L = 0.02V^2/2v$, where $V$ is the velocity in ft/s, $v$ is the specific volume in ft³/lb, and the constant 0.02 has a unit of $ft^{-1}$.

    (a) If the pipe is insulated, determine the change in internal energy in ft·lb$_f$/lb$_m$.

(b) If the flow is isothermal, determine the magnitude and direction of the heat transfer in $ft \cdot lb_f/lb_m$.

## Pumps

**5-77** Oil with a specific gravity of 0.85 is being pumped from a pressure of 8 in Hg vac to a pressure of 18 psig, and the outlet lies 6 ft above the inlet. The fluid flows at a rate of 3.0 $ft^3/s$ through an inlet cross-sectional area of 0.50 $ft^2$, and the outlet area is 0.30 $ft^2$. Determine the power input to the pump in horsepower.

**5-78** A pump is employed to deliver 11,000 lb/h of water from an elevation 30 ft below the pump to an elevation 50 ft above the pump. At the lower elevation (state 1) and the upper elevation (state 2) the known data are: $D_1 = 2$ in, $A_1 = 0.0233$ $ft^2$, $P_1 = 10$ psia, $T_1 = 60°F$, $D_2 = 1$ in, $A_2 = 0.0060$ $ft^2$, $P_2 = 80$ psia, and $T_2 = 60°F$.

(a) Determine the magnitude and direction of the shaft work in $ft \cdot lb_f/lb_m$ if the inlet velocity is 2.1 ft/s and the pipe is insulated.

(b) Determine the horsepower of the pump.

**5-79** It is necessary to pump liquid water through a positive elevation change of 50 ft. The inlet pipe has a 2-in inside diameter and the inlet velocity is 10 ft/s. At the outlet the pipe is 1 in in diameter. The temperature of the water remains essentially at 60°F, and the gage pressure at inlet and outlet are also the same. Determine the horsepower requirement of the pump.

**5-80** Water at 70°F is pumped at 2.5 lb/s from the surface of an open tank into a constant-area piping system. At the discharge from the pipe the velocity is 30 ft/s and the gage pressure is 25 psig. The discharge is 48 ft above the water surface in the open tank. Neglecting losses in the pipe, determine the power required, in horsepower.

## Transient flow

**5-81** A pressurized tank contains 1 lb of air at 30 psia and 140°F. Mass is allowed to flow from the tank until the pressure reaches 15 psia. However, during the process heat is added to the air within the tank to keep it at constant temperature. How much heat, in Btu, was added during the process? Assume constant specific heats.

**5-82** An insulated tank with a volume of 2 $ft^3$ contains 0.15 lb of air at 15 psia, 80°F. The tank is connected through a valve to a large compressed-air line, which carries air at 100 psia, 240°F. If the valve is opened and air is allowed to flow into the tank until the pressure reaches 80 psia, how much mass has entered and what is the final temperature in the tank?

**5-83** A pressurized tank contains 1 lb of water vapor at 500 psia and 500°F. Mass is allowed to flow from the tank until the pressure reaches 100 psia. However, during the process heat is added to the steam to keep it at constant temperature. How much heat, in Btu, was added during the process?

**5-84** Two adiabatic tanks are interconnected through a valve. Tank A contains 1 $ft^3$ of air at 500 psia and 200°F. Tank B contains 10 $ft^3$ of air at 10 psia and 100°F. The valve is opened until the pressure in A drops to 300 psia. At this instant (a) what are the temperatures and pressures in both tanks, (b) how much mass has left tank A, and (c) what is the total entropy change for the process in Btu/°R? Assume constant specific heats.

**5-85** A tank with a volume of 90 $ft^3$ contains steam at 300 psia and 500°F. The tank is heated until the temperature reaches 800°F. A relief valve is installed on the tank to keep the pressure constant during the process. Evaluate the amount of heat transfer to the steam, in Btu.

**5-86** A rigid-insulated tank is initially evaluated. Atmospheric air at 1 atm and 70°F is allowed to leak into the tank until the pressure reaches 1 atm. What is the final temperature of the air within the tank?

**5-87** Consider the data of Prob. 5-86 except that the tank initially contains air at 0.5 atm and 70°F. What is the final temperature in this case?

**5-88** An ideal gas is contained in a rigid tank of volume $V$, initially at $P_1$ and $T_1$. Heat is supplied to the contents until the temperature reaches $T_2$. However, a relief valve allows gas to escape so that the pressure remains constant. Derive an expression for the heat transfer during the process.

**5-89** Consider the situation in Prob. 5-88, except that a valve holds the temperature constant while the pressure drops to $P_2$. Derive an expression for the heat transfer in this case.

**5-90** A tank of 100 ft$^3$ contains air initially at 80 psia, 100°F. Heat is added at a constant rate of 6 Btu/s while an automatic valve allows air to leave the tank at a constant rate of 0.06 lb/s. What is the temperature of the air in the tank 5 min after the initial conditions?

**5-91** With reference to Prob. 5-90, how long will it take for the air in the tank to reach 350°F?

**5-92** With reference to Prob. 5-90, what is the pressure in the tank after 8 min, starting from the initial conditions?

**5-93** A container of fixed volume $V$ contains air at pressure $P_1$ and temperature $T_a$. It is surrounded by atmospheric air at a pressure $P_a$ and temperature $T_a$. A valve is opened and atmospheric air is quickly admitted to the container until the pressure reaches the atmospheric value. At that instant the air in the container is at temperature $T_2$. Listing any necessary assumptions, derive an equation for the temperature ratio, $T_R = T_2/T_a$, in terms of the specific-heat ratio $k$ and the pressure ratio, $P_R = P_1/P_a$.

**5-94** A tank with a volume of 10 ft$^3$ is half filled with liquid water and the remainder is filled with vapor. The pressure is 500 psia. Heat is added until one-half of the liquid (by mass) is evaporated while an automatic valve lets saturated vapor escape at such a rate that the pressure remains constant. Determine the amount of heat transfer, in Btu.

**5-95** A pressure vessel with a volume of 50 ft$^3$ contains saturated water at 600°F. The vessel initially contains 50 percent by volume liquid. Liquid is slowly withdrawn from the bottom of the tank, and heat transfer takes place so that the contents are kept at constant temperature. How much heat must be added by the time half of the total mass has been removed?

**5-96** A very large reservoir contains air at 200 psia and an unknown temperature $T_a$. Air flows from this reservoir into a small insulated tank with a volume of 2 ft$^3$. The small tank initially contains 0.1 lb of air at 80°F. Then a valve is opened to allow air to flow from the reservoir into the tank until the pressure in the tank becomes 50 psia. At this point the temperature in the tank is observed to be 340°F. What is the temperature of the air in the reservoir, $T_a$? Assume constant specific heats.

**5-97** A piston-cylinder assembly is attached through a valve to a source of constant-temperature and constant-pressure air. Initially inside the cylinder the volume is 1 ft$^3$, the temperature of the air is 100°F, and the pressure is 14.7 psia. The valve is slowly opened to the air-supply line which is at 100 psia and 200°F. The piston moves out as air enters in order to maintain the cylinder pressure at the ambient condition of 14.7 psia. When the cylinder volume reaches 2 ft$^3$, (*a*) what is the temperature inside the cylinder, in °F, and (*b*) how much mass, in lb, has entered through the valve?

# SIX

# A MACROSCOPIC VIEWPOINT OF THE SECOND LAW

Historically, the study of the second law of thermodynamics was developed by persons such as Carnot (a French engineer), Clausius, and Kelvin in the middle of the nineteenth century. This development was made purely on a macroscopic viewpoint, and it is referred to as the classical approach to the second law. A study based on this viewpoint does not require the existence of an atomic theory of matter.

## 6-1 INTRODUCTION

The first law of thermodynamics is a conservation law for energy transformations. Regardless of the types of energy involved in processes—thermal, mechanical, electrical, elastic, magnetic, etc.—the change in the energy of a system is equal to the difference between energy input and energy output. All conservation laws are expressed mathematically this way; that is, as equalities. The first law allows free convertibility from one form of energy to another, as long as the overall quantity is conserved. This law, for example, places no restriction on the conversion of work into heat, or on its counterpart—of heat into work. The unrestricted conversion of work into heat is well known to most persons. Frictional effects are frequently associated with mechanical forms of work which result in a temperature rise of the bodies in contact. Subsequent heat transfer in that area tends then to distribute the energy away from the region of contact. The work interaction, aided by heat transfer, eventually appears as a change in the internal energy of the materials involved in the process. It was noted in Chap. 2 that electrical work is commonly transformed, through the presence of electrical resistance, into a temperature rise of the resistor. If the resistor is uninsulated, heat transfer will then occur to the surroundings. Again, work in some form has ultimately been transformed into internal energy, with heat acting as the intermediate transfer process. Now let us look at the inverse process, the transformation of heat into work.

In nations with a developed or developing technological society, the ability to produce energy in the form of work becomes of prime importance. Work transformations are necessary to transport people and goods, drive machinery, pump liquids, compress gases, and provide energy input to so many other processes that are taken for granted in highly developed societies. Much of the work output in such societies is available in the basic form of electrical energy, which is then converted to rotational mechanical work. Although some of this electrical energy (work) is produced by hydroelectric power plants, by far the greatest part of it is obtained from fossil or nuclear fuels. These fuels allow the engineer to produce a relatively high-temperature gas or liquid stream which acts as a thermal (heat) source for the production of work. Hence the study of the conversion of heat to work is extremely important, especially in the light of developing shortages of fossil and nuclear fuels.

In Chap. 1 it was pointed out that thermal efficiency is the parameter that is used by engineers to measure the effectiveness of the heat-work conversion process. In general, a reasonable measure of performance of any device is the ratio of the desired output to the required (costly) input. For a heat-to-work converter, the thermal efficiency $\eta_{\text{th}}$ is defined as

$$\eta_{\text{th}} \equiv \frac{W_{\text{net}}}{Q_{\text{in}}} \tag{6-1}$$

In the first chapter we noted that the thermal efficiency of common energy-conversion devices typically range from 10 to 40 percent. The important point to note is that the first law places no restriction on this conversion process. A 100 percent conversion is possible in terms of energy conservation alone. One of the contributions of the second law is that it places a restriction on heat-to-work transformations by cyclic devices. The establishment of this restriction is one of the goals of this chapter.

The brief discussion of energy-conversion devices above leads to several other second-law considerations. One of these is the concept that energy has quality, as well as quantity. If work is 100 percent convertible into heat, but the reverse situation is not possible (which we must eventually prove), then work is a more valuable form of energy than heat. Although not as obvious, it can also be shown through second-law arguments that heat has quality in terms of the temperature at which it is discharged from a system. The higher the temperature at which heat transfer occurs, the greater the possible energy transformation into work. Thus thermal energy stored at high temperatures generally is more useful to humanity than that available at lower temperatures. (While there is an immense quantity of energy stored in the oceans, for example, its present availability to us for performing useful tasks is quite low.) This implies, in turn, that thermal energy is *degraded* when it is transferred by means of heat transfer from one temperature to a lower one. Other forms of energy degradation include energy transformations due to frictional effects and electrical resistance, among others. Such effects are highly undesirable if the use of energy for practical purposes is to be maximized. The second law provides some means of measuring this energy degradation.

There are other well-known phenomena which fall under second-law analysis. Consider the placement of any hot object in a cold environment, or vice versa. It is our experience that objects of different temperatures, when brought into thermal contact, tend to reach thermal equilibrium, or equality of temperature. The first law requires that the energy given up by one object is acquired by the other (assuming no losses elsewhere). However, consider now two objects initially at the same temperature. The first law places no restriction on the possibility that one will become cooler while the other becomes warmer, as long as the energy given up by the cooler object is the same as that gained by the warmer system. As a second example, consider a paddle-wheel that stirs a fluid within an insulated container. The paddle-wheel might be rotated by some pulley-weight mechanism. As a result, the potential energy of the weight decreases and the internal energy of the fluid increases, in light of the conservation of energy principle. However, at some later time we do not expect to see the energy of the fluid decrease and the weight return to its initial position, spontaneously. These and many other possible examples illustrate the fact that processes, of their own accord, have a preferred direction of change, irrespective of the first law. Also, after sufficient time, such processes reach an equilibrium state.

In summary, there are a number of phenomena which cannot be explained by conservation principles of any type. Hence we seek another law which, through its generality, will provide guidelines to the understanding and analysis of diverse effects. Among other considerations, the second law is extremely helpful to the engineer in the following ways:

1. It provides the means of measuring the quality of energy.
2. It establishes the criteria for the "ideal" performance of engineering devices.
3. It determines the direction of change for processes.
4. It establishes the final equilibrium state for spontaneous processes.

Since the second law is used to examine the direction of change of processes, it is expressed mathematically as an inequality. This inequality means that the second law is a nonconservation law. A new thermostatic property, the entropy, will be deduced from the general second-law statement. It is this fundamental property which is not conserved in real processes.

## 6-2 EQUILIBRIUM AND THE SECOND LAW

The concept of equilibrium has been extremely important in the preceding discussions of properties of matter and the conservation of energy principle. In the analysis of various processes it was either stated or implied that the initial and final states were equilibrium states. This is necessary since the intrinsic properties of matter are commonly defined only for equilibrium states. Recall also that in the evaluation of various work interactions it is necessary to assume a quasistatic process in order to functionally relate the two parameters, such as $P$ and $V$ in boundary work. That is, a series of equilibrium steps must be assumed in order to integrate such expressions as $P\ dV$. The development of the second law of thermodynamics also relies heavily upon the concept of equilibrium states.

By definition, a system is in equilibrium if a change of state cannot occur while the system is not subject to interactions. If a system is isolated from the environment, and we wait a sufficient time, it is common experience to note no further change in the state of the system. We usually infer then that the system is in equilibrium. No further changes of state will occur unless interactions take place across the boundary. If a system is in equilibrium, a finite change in the state of the system requires a permanent and finite change in the state of the environment.

It is our everyday experience that systems of all types tend to reach a state of equilibrium. If a marble is placed halfway down the side of a bowl and released, it is expected that the marble eventually will come to rest at the bottom without any change in the environment. If two systems of different temperatures are placed in contact through a nonadiabatic wall, but they are insulated from the environment, the two systems will reach a final common temperature. They are said to be in thermal equilibrium. When the two plates of a charged capacitor are shorted out, it is soon found that the plates reach a common electrical potential. When a gas expands from one tank into another evacuated tank, both of which are isolated from the surroundings, it is soon found that there is a common temperature, pressure, density, etc., throughout the two tanks. If two miscible liquids, such as alcohol and water, are placed in contact by removing a partition which originally separated them, we expect the liquids to mix until a uniform state is reached. Many other examples could be provided. There is no way we can prove that systems left to themselves eventually reach an apparent state of equilibrium. It is simply a matter of common experience. In many cases equilibrium is reached very rapidly, while in other cases the process may be rate-limited, so that equilibrium is attained only after a considerable length of time. Nevertheless, we expect things to reach equilibrium in the absence of interactions between the system of interest and its environment.

In the absence of any proof of this generally found behavior, we must postulate that such behavior is to be expected. The following postulate is one form of the *second law* of thermodynamics.

Any system having certain specified constraints and having an upper bound in volume can reach from any initial state a stable equilibrium state with no net effect on the environment.

This statement postulates the existence of stable equilibrium states. By constraints we mean those barriers within or outside of the system which restrict its behavior. These include internal partitions, external conservative force fields, rigid impermeable walls, etc.

Before proceeding, it is important to note that the second law is a directional or limiting law. Processes tend to proceed in one direction and may not be reversed, or processes attain a final equilibrium state which is limited in some sense. For example, a marble placed on the side of a bowl rolls to the bottom. One does not observe it spontaneously rolling back up the side once it has come to rest. If a system of high temperature and a system of low temperature are placed in thermal contact, the former does not get hotter and the latter colder. When a chemical reaction occurs, it is quite possible that an equilibrium state is reached before the reaction is complete. That is, a sizable amount of the reactants may still exist, although no further change of state is observed. The system reaches equilibrium, but the extent of the reaction appears to be limited. When we postulate that any system can reach an equilibrium state, in the absence of interactions, we usually imply that one particular equilibrium state will be reached. Thus the approach to equilibrium is quite directional in its nature.

## 6-3 HEAT ENGINES

We shall now apply the statement of the second law to a special class of devices known as heat engines. A heat engine is a device which operates continuously, or cyclically, and produces work while exchanging heat across its boundaries. The restriction to continuous or cyclic operation implies that the matter within the device is returned to its initial state at regular intervals. For example, in a steam power plant the steam is contained within a closed loop. It is heated, expanded, cooled, and compressed as it circulates around the loop in a continuous manner, returning to its initial state at the same point in the loop. During the circuit it expands and produces power to drive an electrical generator. In addition it exchanges heat with hot combustion gases, as well as with cooling water at the temperature of the environment. Another example of a continuous heat engine is a thermoelectric device such as that shown in Fig. 6-1. Part (a) of the figure illustrates two dissimilar electrical conductors with common junctions maintained at two different temperatures. An electric potential is developed as a result of the temperature difference, and work is produced as a result of the current passing through the motor. If the junction temperatures and the load are maintained constant, the device will produce work continuously while heat is exchanged at the two junctions. Figure 6-1b illustrates a modern version of a thermoelectric device. The dissimlar electrical conductors are semiconductor elements which are placed in series, with alternating junctions in contact with either a high- or low-temperature body. Again, the device operates as a continuous heat engine.

As an example of a cyclic heat engine, consider the piston-cylinder assembly shown in Fig. 6-2a. A piston and weight rest on some supporting pegs at position

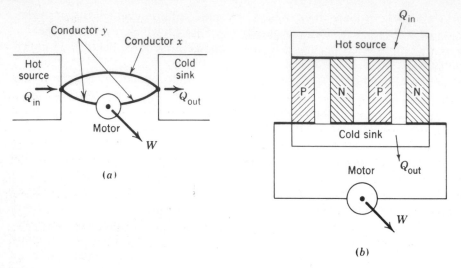

Figure 6-1 A thermoelectric device for power production.

1, and a gas is contained in the volume below the piston. Initially the pressure of the gas is less than the equivalent pressure exerted by the piston-weight-atmosphere combination above the gas. Heat is then added from a high-temperature source until the gas pressure just balances this opposing pressure. The process to this new state 2 from the original state is shown on a $PV$ diagram

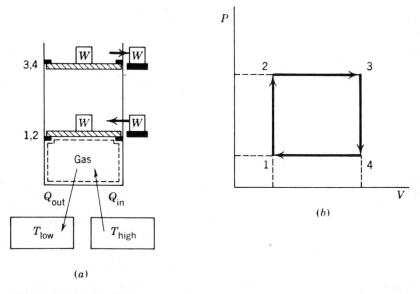

Figure 6-2 A simple, cyclic heat engine.

in Fig. 6-9*b*. If additional heat transfer occurs from the high-temperature source, the gas will expand at constant pressure until the piston hits an upper set of pegs. The heat addition is halted at this point, and the weight is moved horizontally off the piston. The state of the gas is now designated by point 3 on the *PV* diagram. Heat is now rejected from the gas to a low-temperature sink. Until the gas pressure drops so that it just balances the weight of the piston and the atmospheric pressure, the volume does not change. This is state 4. Additional heat removal will lower the piston at constant pressure until the piston reaches its initial position on the lower set of pegs. At this point the cycle has been completed. If another weight were added on at the lower level, the cycle could be repeated. The net work produced by the cycle is measured by the enclosed area on the *PV* diagram. In addition, heat was exchanged between the system and two bodies at different temperatures. Hence the words "continuous" and "cyclic" are descriptive of the nature of processes which alter the state of the matter within a heat engine. In either case the working substance is returned to its initial state at regular intervals.

The three heat-engine devices mentioned above can be represented in a general way by the block diagram of Fig. 6-3. For the heat engine as the closed system, the conservation of energy principle is

$$\sum Q + \sum W = \Delta U$$

For an integral number of cycles, the value of $\Delta U$ is zero. For a continuous system in steady state, the total internal energy of the system is constant and the change in the internal energy is also zero. Thus for either type of heat engine we see that

$$|Q_H| - |Q_L| = W_{net}$$

where $W_{net}$ is always positive for heat engines. Since the heat added and removed from the heat engine have different signs by convention, it is more convenient to use absolute-value signs in the energy equation and to insert a negative sign directly into the equation for the $Q_L$ term. The object of a heat engine is to produce work from the energy added as heat to the system. Therefore a reasonable

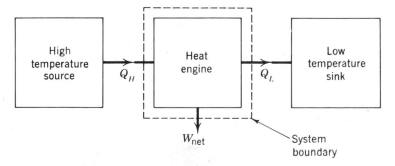

**Figure 6-3** Simple schematic of a heat engine.

measure of its performance is the thermal efficiency $\eta_{th}$ of the engine. That is, as previously defined,

$$\eta_{th} = \frac{W_{net}}{Q_H} \tag{6-1}$$

If we supply 100 units of energy to a heat engine and find that 70 units are rejected, then the net work output is 30 units and the thermal efficiency is 0.30 (or 30 percent). Note that the thermal efficiency will be unity (100 percent) when $Q_L$ is zero. That is, a cyclic heat engine with an efficiency of 100 percent requires no heat rejection to the environment. In the next section we seek to establish whether there is theoretically any limit to the value of the thermal efficiency of *any* heat engine.

## 6-4 PERPETUAL-MOTION MACHINES

It is a matter of experience that the thermal efficiency of all practical heat engines is less than 100 percent. Thus some portion of the heat supplied from a high-temperature source is always rejected to a sink. A basic question with regard to heat engines is whether it is *theoretically* possible to have a device that is 100 percent efficient, even if actual losses due to friction and other factors are neglected. The statement of the second law presented in Sec. 6-2 is sufficient to answer this question. We shall propose the following theorem.

> It is not possible to construct a heat engine which produces no other effect than the exchange of heat from a single source initially in an equilibrium state and the production of work.

Hence we propose to prove that no heat engine can be 100 percent efficient. This corollary to the second law is known as the *Kelvin-Planck* statement of the second law.

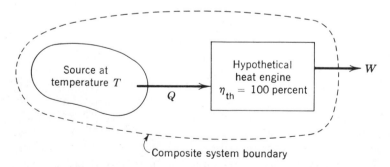

**Figure 6-4** Schematic for the proof of the impossibility of a 100 percent efficient heat engine.

The proof of the theorem is more easily seen with the aid of the diagram in Fig. 6-4. A hypothetical heat engine of 100 percent thermal efficiency undergoes an integral number of cycles. During the process the engine receives a quantity of heat $Q$ from a source at temperature $T$, and produces an equivalent amount of work $W$. The heat engine undergoes a cycle, so that its state is not changed during the process. The source of energy starts at a stable equilibrium state and proceeds to a new state, while the sole effect in the environment is the production of work. The work produced may now be used to further alter the state of the heat source. For example, the heat source might be lifted to a new position or given a finite velocity. As a result of this series of effects, the heat engine is unaltered, the rest of the environment is unchanged, and the heat source has undergone a finite change of state from its original equilibrium state. But, by definition, a system initially in equilibrium cannot undergo a finite change of state without a finite change in its environment. Therefore some part of the proposed process is impossible. The only assumption made in the analysis was that a hypothetical heat engine of 100 percent efficiency could exist. Such an assumption leads to an invalid conclusion. Hence it is not possible to construct a heat engine of 100 percent efficiency if we accept the existence of stable equilibrium states. Therefore all heat engines must reject some portion of the heat supplied to another system (usually called a sink).

Most large fossil- and nuclear-fueled power plants reject heat directly to the environment. Usually the heat is rejected to cooling water which is taken from a river or lake. Since the thermal efficiency of such plants is 40 percent or less, at least 60 percent of the energy released from the fuel must appear in the environment. That portion which is released to the cooling water is what constitutes *thermal pollution* from such plants. For fossil-fueled plants a portion of the 60 percent which is rejected appears as energy in the hot gas stream emitted from the stack of the plant. Nevertheless, approximately one-half of the rejected energy heats the water stream. In a nuclear plant all the rejected energy must appear in the coolant stream, since no combustion gases are associated with such a plant. One method of alleviating thermal pollution in our streams is the use of "dry" cooling towers. In such a system the energy rejected from the power cycle is released to the atmosphere, instead of to a liquid stream. Since the atmosphere is so massive compared to a river, for example, the effects of thermal pollution are greatly diminished.

By definition, a heat engine which exchanges heat with a single system in a stable state and produces work is called a perpetual-motion machine of the second kind, or a PMM2. It is one of the second kind since it violates the second law of thermodynamics. A device of this type would be extremely useful if it could be constructed, since it could refrigerate a region while simultaneously producing work. The concept of a PMM2 is quite useful in proving subsequent theorems in this chapter. A perpetual-motion machine of the first kind (a PMM1) is a device which creates energy and thus violates the first law of thermodynamics. Any process which creates a PMM1 or a PMM2 is impossible.

## 6-5 REVERSIBLE AND IRREVERSIBLE PROCESSES

In the preceding section we found that PMM2-type heat engines are impossible. Therefore all heat engines, theoretical and real, must reject a portion of their heat input to a sink, which usually is part of the environment. However, no limiting maximum value has been set on thermal efficiency. Besides showing that a PMM2 is impossible, thermodynamics also provides the engineer with a numerical upper limit on theoretical thermal efficiency. In order to establish the general relationship, however, it is necessary to describe first what is meant by an "ideal" heat engine which would have maximum theoretical efficiency. An ideal heat engine is one which is reversible. Hence the concept of a reversible process must be described in some detail.

In general, a process commencing from an initial equilibrium state is called *reversible* if at any time during the process both the system and environment can be returned to their initial states. The concept of reversibility by its definition requires restorability. But this requirement of restorability is quite strong, applying to both the system and its surroundings. Normally, interest is centered during a given process solely on the system. For example, a quasistatic process requires equilibrium conditions within and at the boundaries of a system, but it places no restriction on the effects which occur in the surroundings. However, a reversible process requires something of the environment. It is the nature of a reversible process that all heat and work interactions which occurred across the boundaries during the original (forward) process are equal in magnitude but reversed in direction during the reverse process. Thus no net history is left in the surroundings when the system regains its initial state.

In view of our discussions of work interactions in Chap. 2, any quasistatic form of a work interaction (such as boundary work, electric or magnetic field work, etc.) should be carried out in a series of equilibrium steps. Only in this circumstance will the work output by the system equal the work input for the return path. Consider, as an example, a piston-cylinder assembly which contains a gas. The piston will be taken as frictionless, and both the cylinder and the piston are perfect insulators. If the force exerted externally on the gas is decreased by a very small amount, work is performed by the gas on the surroundings. Hypothetically, this amount of work might be stored in the form of the rotational kinetic energy of a flywheel. If the expansion were carried out in a series of small decreases in the external resisting force, the pressure, temperature, and other intrinsic properties of the gas would change uniformly throughout the gas. Hence the work output would be equal to the integral of $P \, dV$, and an equivalent amount of energy would be added to the flywheel. Now the process could be reversed by removing energy from the flywheel through some suitable mechanism. If the gas is now compressed by a series of small increases in the pressure, it can be returned to its initial state. It would be found that the pressure of the gas was exactly the same during the compression as during the expansion for a given position of the piston. Hence the work of compression will equal the work of expansion if the initial and final states of the gas are the same. In addition, all the energy stored in the surroundings (the flywheel) during the expansion will be exactly expended in returning the system to its initial state. Since the system and surroundings are both back in their initial states at the end, the process described above is a reversible one.

The concept of reversibility can also be applied to open systems. As a second example, consider the steady flow of a fluid through an adiabatic frictionless nozzle. The application of the conservation of energy principle shows that the enthalpy of the fluid decreases as the kinetic energy increases. If the nozzle is now followed by an adiabatic, frictionless diffuser, the fluid can be returned to a state identical to that at the entrance of the nozzle. The diffuser increases the enthalpy of the fluid at the expense of a decrease in its kinetic energy. Since the surroundings were not involved, the process described is a reversible one. There are other common examples of reversible processes, and included among these are the ideal pendulum and the pure capacitive-inductive circuit. (See Fig. 6-5.) Both the frictionless pendulum and the ideal capacitive-inductive circuit are forms of an oscillator which undergoes cyclic variations. Another example of a reversible device is an oscillator which interchanges the elastic energy of an ideal spring with the kinetic energy or potential energy of a mass. Other combinations of gravitational potential energy, translational kinetic energy, rotational kinetic energy, elastic energy, magnetic- and electric-field energies, and the frictionless compression and expansion of a gas in a piston cylinder, which would constitute reversible devices, could be considered. Consequently, many of the basic concepts of introductory

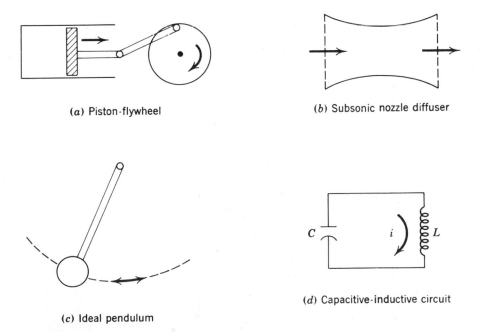

(*a*) Piston-flywheel

(*b*) Subsonic nozzle diffuser

(*c*) Ideal pendulum

(*d*) Capacitive-inductive circuit

**Figure 6-5** Some reversible processes.

physics may be employed to devise reversible processes. Among the processes that are frequently idealized as reversible processes are

1. Restrained expansion or compression
2. Frictionless motion
3. Elastic stretching of a solid
4. Electric circuits of zero resistance
5. Polarization and magnetization effects
6. Restrained discharge of a battery

The reversible process is an idealization or fiction. It is a concept which can be approximated very closely at times by actual devices, but never matched. The reason is that the word "ideal" appeared throughout the foregoing discussion. In the analysis of these ideal processes, we required the absence of friction at the bearing surfaces of the piston cylinder, the flywheel, and the pendulum. The pendulum, in addition, could not be acted upon by air resistance. The capacitive-inductive circuit could not contain a resistive element when the current passed through the circuit. Any springs used had to be elastic ones; i.e., they had to follow Hooke's law. In many actual cases the effects of friction, electrical resistance, and inelasticity can be substantially reduced, but their complete elimination is usually not found. These effects are frequently called *dissipative* effects, since in all cases a portion of the energy originally in the system is converted or dissipated into a less useful form. Only in the absence of dissipative effects can certain forms of energy be converted into other forms without any apparent loss in the capabilities of the system. In such cases the system is truly reversible.

One other criterion, in addition to the absence of dissipative effects, must be met if an isolated system is to be considered reversible. In the discussion of the piston-cylinder and flywheel combination, it was necessary to carry out the expansion and compression of the gas in a series of equilibrium steps. Otherwise the work output and work input would not be the same, and the process would be irreversible. The requirement that a process be quasistatic, in order to be reversible, is a general one. That is, only infinitesimal unbalanced forces are present during a reversible process.

When a process is such that the system and the surroundings cannot be returned to their initial states, the process is said to be irreversible. We have now seen that irreversibilities arise from two sources:

1. Inherent dissipative effects that stem from the nature of the substance itself. These effects include friction of any type, electrical resistance, magnetic hysteresis, and inelasticity.
2. Absence of mechanical, thermal, or chemical equilibrium during a process, i.e., a nonquasistatic process.

The presence of either class of effects is sufficient to make a process nonideal, or irreversible. Since all actual processes include such effects, the reversible process is a limiting process toward which all actual processes may approach asymptotically in performance, but never succeed in realization. Nevertheless, we shall find reversible processes to be appropriate starting places on which to base engineering design calculations. The usefulness of the concept of the reversible process will only become apparent by repeated application of the concept to different situations in the light of the second law of thermodynamics.

From a practical viewpoint, whether or not a given process is reversible is probably best recognized by ascertaining whether irreversibilities occur during the change of state. As a broad generalization, all irreversibilities have this in common: they leave a history, or an imprint, of their presence. Any system which is returned to its initial state after experiencing an irreversible process will leave a history in the surroundings due to the irreversibilities. A partial list of effects which constitute irreversibilities is presented below. Most of these effects fall into the category of common experience. Irreversibilities include:

1. Electrical resistance
2. Inelastic deformation
3. Shock waves
4. Hysteresis effects
5. Viscous flow of a fluid
6. Internal damping of a vibrating system
7. Solid–solid friction
8. Unrestrained expansion of a fluid
9. Fluid flow through valves and porous plugs (throttling)
10. Spontaneous chemical reactions
11. Mixing of dissimilar gases or liquids
12. Osmosis
13. Dissolution of one phase into another phase
14. Mixing of the same two fluids initially at different pressures and temperatures

The foregoing list was made fairly long—not for the purpose of being memorized, although some degree of familiarity is important, but to illustrate two points. First, the common processes that are part of one's experience are all irreversible. Second, these processes cover a diversity of physical and chemical effects.

A gradient within a system, or between two systems, is a source of an irreversibility. Heat transfer is irreversible if its occurrence is due to a finite temperature difference between the system and its surroundings. The transfer of heat is reversible if the temperature difference is made infinitesimally small. In addition, reversibility requires that the quantity of heat transferred on the reversed path be equal in magnitude, but opposite in direction, to that transferred during the original process. This "equal in magnitude, but opposite in sign" concept also applies to any quasistatic work interaction carried out during a reversible process.

We have pointed out that there are either inherent dissipative effects or non-equilibrium effects which make a process irreversible. Absence of these effects in the system and its environment lead to the concept of a *totally reversible* process.

In the thermodynamic analysis of engineering systems it frequently is advantageous to consider processes for which irreversibilities are absent within the system of interest, but not necessarily absent from the environment. Such processes are called *internally reversible*. For simple compressible systems, the only work interaction permitted is quasistatic boundary work if the process is to be internally reversible. Work effects associated with paddle-wheels or electrical resistors within the system are not allowed, for obvious reasons. As a more general statement, only quasistatic work interactions, and not nonquasistatic ones, can occur during internally reversible processes. Finally, an *externally reversible* process is one for which irreversibilities may be present within the boundaries of the system of interest, but the surroundings which interact with the system must undergo only reversible changes.

## 6-6 HEAT AND WORK RESERVOIRS

For the development of further consequences of the second law, it is convenient to define a special heat source or sink known as a heat reservoir. By definition, a heat reservoir is a closed system with the following three characteristics:

1. The only interactions permitted across its boundaries are heat interactions, to the exclusion of work effects.
2. The system equilibrates rapidly, so that it remains essentially in equilibrium during the heat-transfer process.
3. Its temperature remains constant during a finite addition or removal of energy in the form of heat.

Thus energy added or removed from it merely alters its internal energy. The name "heat" reservoir is a misnomer, since it implies that heat is stored in the system. The name really means that this particular system simply acts as a source or sink for the transfer of heat to or from various other systems. Some authors have called this type of reservoir a "thermal energy" reservoir (TER), which might be more appropriate.

A heat reservoir has two important qualities. First, there is no restriction, theoretically, on the phase or chemical composition of the reservoir. Consequently, the only significant property of a heat reservoir is its temperature, which must remain constant. Second, the system changes quasistatically and there are no dissipative effects within the reservoir. Therefore any change in state of a heat reservoir occurs in an *internally reversible* manner.

In practice, a heat reservoir can be achieved in several ways. One method is to have a second system much larger than the system with which it interacts. Any energy in the form of heat added to or removed from the second system will be a small fraction of its total energy. Hence its temperature will tend to remain constant. Large bodies of water such as lakes and oceans, and the atmosphere around the earth, behave essentially as heat reservoirs. Another example of a practical heat reservoir is a two-phase system. Although the ratio of the masses of the two phases will change during heat addition or removal, the temperature will remain fixed as long as both phases coexist.

Another concept of usefulness in thermodynamics is that of a work reservoir, which is called a "mechanical energy" reservoir (MER) by some authors. Similar to a heat reservoir, it is a system which acts as a source or sink for the exchange of work with another system of interest. No heat interactions are associated with it. All motions within a work reservoir are assumed to be frictionless, i.e., nondissipative, and the forces acting across its boundaries are independent of direction or rate of change of the energy-transfer process. Thus a change of state of a work reservoir also occurs in an *internally reversible* manner. Any idealized mechanical, electrical, magnetic, or elastic system could operate as a work reservoir. Examples are pulley-weight systems, rotating flywheels, and elastic springs, among others.

## 6-7 THERMAL EFFICIENCY OF REVERSIBLE AND IRREVERSIBLE ENGINES

Having established the fact that the thermal efficiency of any heat engine is always less than unity, we now turn our attention to a special class of heat engines. With the introduction of the concepts of reversibility and of heat reservoirs, it is possible to classify heat engines either as totally reversible or irreversible. As the name implies, a totally reversible heat engine is one which is free of dissipative or nonequilibrium effects during its operation. These effects must be absent not only within the engine, but also with respect to changes in the environment connected with the operation of the engine. Since a heat engine depends upon heat transfer with at least one source and one sink, it is necessary that all heat interactions be reversible. We consider the transfer of heat to be reversible if the temperature difference between two systems is made infinitesimally small. For a totally reversible heat engine, a temperature difference $dT$ must exist between the working fluid of the engine and the source or sink with which it exchanges energy in the form of heat. If irreversibilities of any kind exist within the engine or result from interactions between the heat engine and its environment, the engine is classified as irreversible.

In this section we shall prove the following theorems regarding the thermal efficiencies of reversible and irreversible heat engines.

1. The efficiency of an irreversible heat engine is always less than the efficiency of a totally reversible heat engine when both are operating between the same heat reservoirs.
2. The efficiencies of two totally reversible heat engines operating between the same heat reservoirs are equal.

These two statements compose what is frequently called Carnot's principle.

The proof of statement 1 is based on the apparatus shown in Fig. 6-6. A totally reversible engine $R$ and an irreversible engine $I$ are operating between the

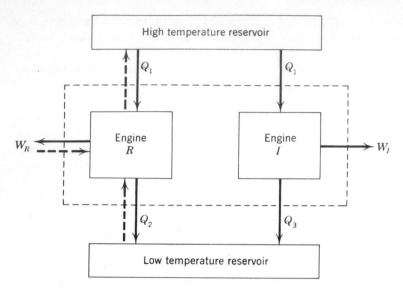

**Figure 6-6** Sketch of apparatus for proof that $N_I < N_R$.

same two heat reservoirs. Both engines receive the same quantity of heat $Q_1$ during an integral number of cycles for each, and the totally reversible engine produces work $W_R$ while the irreversible engine produces work $W_I$. We shall assume that

$$W_I \geqslant W_R$$

in violation of statement 1. Since engine $R$ is totally reversible, its direction of operation may be reversed. In the reverse direction the magnitudes of $Q_1, Q_2$, and $W_R$ remain the same, but their signs are changed (as shown by the dashed arrows in Fig. 6-6). The net result of this new operation is that the high-temperature heat reservoir receives no net energy. In addition, the system comprised of engines $R$ and $I$ (shown by the dashed box in the figure) now exchanges a net amount of heat with a single reservoir. If

$$W_I - W_R > 0$$

then a net amount of work will be produced by the composite system (recall that $W_R$ is now into the system when it operates in the reverse direction). This situation of net work production and net heat transfer with a single reservoir is a PMM2, and is not possible. Therefore $W_I$ cannot be greater than $W_R$.

If, on the other hand,

$$W_I - W_R = 0$$

then from the conservation of energy principle $Q_2$ equals $Q_3$ in magnitude, but the quantities are opposite in direction. Hence the low-temperature reservoir in this situation receives no net heat. But this means the process is reversible and engine $I$

is operating just like engine $R$. Therefore engine $I$ is reversible when $W_I = W_R$. This violates our original assumption, and so $W_I$ cannot equal $W_R$. Since neither the "greater than" sign nor the "equals" sign is permissible, we have proved statement 1. An irreversible engine is always less efficient than a totally reversible engine when both operate between the same two reservoirs.

The proof of the second statement of Carnot's principle follows immediately from the preceding proof. If the efficiencies of two totally reversible engines were not the same when operating between the same reservoirs, then a PMM2 could be set up by operating one in reverse. This would be true no matter which engine were chosen to be the more efficient. Therefore the only way to avoid setting up a PMM2 is to have the same efficiency for both heat engines.

## 6-8 THE THERMODYNAMIC TEMPERATURE SCALE

One of the most noteworthy results of Carnot's principle is the statement that all totally reversible heat engines have the same thermal efficiency when operating between the same two heat reservoirs. In the proof of this statement, it is not necessary to specify the nature of the heat engine itself. That is, the construction and manner of operation of the engine does not affect the thermal efficiency if the engine is totally reversible. In addition, the thermal efficiency does not depend upon the nature of the fluid or matter within the engine. The only other possible influence on the thermal efficiency is the nature of the heat reservoirs. But the only significant property of a heat reservoir is its temperature. Therefore the only parameters that fix the thermal efficiency are the temperatures of the two reservoirs.

The very fact that the efficiency is solely dependent on the two reservoir temperatures provides us with a means of establishing an absolute thermodynamic temperature scale. A brief discussion of temperature scales was made in Chap. 1. Various thermometric properties of substances were suggested for use in measuring the temperature. In each case the nature of the substance influences the value of the temperature determined in a measurement. All ideal gases in a constant-volume device give the same reading in the limit as the pressure is made very small within the thermometer. Nevertheless, the reading still depends upon a special class of substances, the ideal gas. The concept of a reversible engine provides us with the chance to establish a thermodynamic temperature scale which is independent of the properties of the substance used for the measurement.

Now consider a totally reversible heat engine operating between two reservoirs at $T_H$ and $T_L$. According to the Carnot principle we may write that

$$\eta_{\text{th}} = f(T_H, T_L)$$

In addition, the definition of the thermal efficiency leads to the relation

$$\eta_{\text{th}} = \frac{W}{Q_{\text{in}}} = \frac{|Q_H| - |Q_L|}{|Q_H|} = 1 - \frac{|Q_L|}{|Q_H|}$$

If we now arbitrarily select some $f(T_H, T_L)$, then measurements of $Q_H$ and $Q_L$ for *any* totally reversible engine operating between those reservoirs will provide a relation between $T_H$ and $T_L$, since

$$f(T_H, T_L) = 1 - \frac{|Q_L|}{|Q_H|}$$

It has been found convenient and useful to select, arbitrarily, the function of $T_H$ and $T_L$ so that

$$\frac{T_H}{T_L} = \frac{|Q_H|}{|Q_L|} \tag{6-2}$$

This relationship helps define an absolute thermodynamic Kelvin temperature scale. Since it only defines a ratio of temperatures, we assign the thermodynamic temperature $T^*$ of a heat reservoir at the triple state of water to be 273.16°K. If we now operate a totally reversible heat engine between a reservoir at the triple state of water and another reservoir at an unknown temperature $T$, then this latter temperature is related to $T^*$ by

$$T = 273.16 \frac{Q}{Q^*} \tag{6-3}$$

$Q$ is the heat received by the engine from the reservoir at temperature $T$, and $Q^*$ is the heat rejected to the sink at 273.16°K.

Although a totally reversible heat engine is an idealization, available extrapolation techniques enable us to achieve fairly good practical results for a temperature scale based upon this concept. In addition, it can be shown that the temperature scale based on the constant-volume, ideal-gas thermometer is compatible with the absolute thermodynamic temperature scale based on a reversible heat engine. That is, they measure equivalent temperatures. In this latter respect, note the similarity between Eqs. (6-3) and (1-17).

## 6-9 THE CARNOT EFFICIENCY

On the basis of Eq. (6-1) and the first law for a closed system we may write that

$$\eta_{th} = \frac{W_{net}}{Q_{in}} = \frac{|Q_{in}| - |Q_{out}|}{|Q_{in}|} = 1 - \frac{|Q_{out}|}{|Q_{in}|}$$

Combination of Eq. (6-2) with this relation yields

$$\eta_{th, \text{Carnot}} = 1 - \frac{T_L}{T_H} \tag{6-4}$$

The efficiency given by this equation is called the *Carnot* efficiency. According to the Carnot principle, this is the *maximum* efficiency that any heat engine could have when operating between heat reservoirs with temperature $T_H$ and $T_L$. This is an upper limit, since actual engines are always less efficient than reversible engines. To improve the efficiency of a reversible heat engine it is necessary to raise $T_H$, lower $T_L$, or both.

This equation is extremely important with respect to efficiency data given earlier in Table 1-2. Common power-producing devices have efficiencies ranging from 10 to 40 percent. These values are low relative to 100 percent. However, Eq. (6-4) dictates that efficiencies should not be compared to 100 percent, but to some lower theoretical value. For example, consider a steam power plant with a real efficiency of 40 percent. We shall assume that $T_H$ is 800°K (980°F) and $T_L$ is 300°K (80°F). For a heat engine receiving and rejecting heat at these temperatures, the theoretical thermal efficiency according to Eq. (6-4) would be 62.5 percent. Against this value, an actual value of 40 percent does not appear quite so low. Hence, one must recognize that the theoretical values will be considerably less than 100 percent for cyclic heat-to-work energy converters. Also, one major method of increasing the thermal efficiency is by increasing the temperature at which heat is supplied. Unfortunately, this value is restricted by the metallurgical limits of the materials in the heat engine.

## 6-10 THE CLAUSIUS INEQUALITY

As noted earlier, the second law of thermodynamics is concerned with the directional quality of nature. This was expressed initially by the statement that the stable state of a system can be reached from other states, but a system in a stable state will not spontaneously change to other states. This led to a discussion of heat engines. We have shown that irreversible heat engines are always less efficient than totally reversible heat engines when operating between the same heat reservoirs. Thus we find that the second law leads to expressions involving inequalities. Conservation laws, such as the first law, involve equalities. Directional laws involve inequalities.

Another important inequality in thermodynamics which stems from the second law is known as the Clausius inequality. It states that

The cyclic integral of the quantity $\delta Q/T$ for a closed system is always equal to or less than zero.

The proof of this theorem is based on the apparatus shown in Fig. 6-7. A simple, closed system at temperature $T$ has a quantity of heat $\delta Q$ supplied to it while producing a quantity of work $\delta W$. The temperature $T$ is allowed to vary as the state of the system changes. The source of the thermal energy is a heat reservoir at temperature $T_R$, and $T_R \neq T$. In order to eliminate any source of irreversibility outside of the system, heat is transferred from the reservoir to the system through an intermediary, a totally reversible heat engine $R$. This engine receives heat $\delta Q_R$ from the reservoir and supplies heat $\delta Q$ to the system. As a consequence, the engine must deliver a quantity of work $\delta W_E$ to the environment. During the change of state of the system it delivers a quantity of work $\delta W$. The direction of

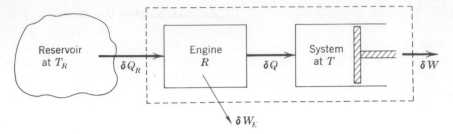

**Figure 6-7** Proof of the Clausius inequality.

transfer of energy for the various parts could be that shown or could be in the opposite direction. It is immaterial to the proof. For the purposes of this discussion, we shall consider the direction of the various heat and work terms to be those shown in the figure. For this theoretical development we shall assume that all work effects are delivered to or from work reservoirs, such as a pulley-weight system. Recall that such systems, as well as the heat reservoir at $T_R$, are reversible in concept. Consequently the only place irreversibilities can occur for the described process is within the system itself. That is, changes to it may be internally reversible or internally irreversible.

The proof of the Clausius inequality is based on two first-law and two second-law statements about the composite system described above. In the equations used below the heat and work terms are absolute values. The signs on the various quantities have already been taken into account. First, an energy balance on the heat engine is

$$\delta W_E = \delta Q_R - \delta Q$$

and for the system

$$\delta W = \delta Q - dU$$

Hence the total work done by the composite of the system and the engine (shown by the dashed line in Fig. 6-7) is

$$\delta W_T = \delta W_E + \delta W = \delta Q_R - dU$$

For the totally reversible heat engine we also know that $T/T_R = \delta Q/\delta Q_R$. This expression may be used to eliminate $\delta Q_R$ from the above equation. Therefore

$$\delta W_T = \frac{T_R}{T} \delta Q - dU$$

Now, consider the situation where the system undergoes a cycle, while the heat engine undergoes an integral number of cycles. The total work done by the composite system for this cyclic process is found by integrating the above equation. Using a cyclic integral notation, we find that

$$\oint \delta W_T = T_R \oint \frac{\delta Q}{T} - \oint dU$$

The $T_R$ has been brought outside the integral sign since it is a constant. The composite of the system and the heat engine is equivalent to a device which operates in a cycle while exchanging heat with a single reservoir. Therefore the composite device cannot produce work, or a PMM2 would result. However, there is no restriction against the situation where the work is equal to or less than zero. Thus we can state the second-law requirement on cyclic heat engines that exchange heat with a single reservoir in the following form:

$$\oint \delta W_T \leqslant 0$$

In addition, the cyclic integral of $dU$ is zero. As a result, the preceding equation becomes

$$T_R \oint \frac{\delta Q}{T} \leqslant 0$$

$T_R$ is always positive. Consequently,

$$\oint \frac{\delta Q}{T} \leqslant 0 \tag{6-5}$$

This is the Clausis inequality. It applies equally well to a truly cyclic device or a continuous, steady-state device.


## 6-11 ENTROPY

In the preceding proof of the Clausius inequality, all changes external to the system were made reversible. However, changes within the system could be reversible or irreversible, since the type of process associated with the system was not specified. The possibility of either internally reversible or irreversible processes accounts for the presence of the "equal to" and "less than" signs found in Eq. (6-5). First of all, let us consider that the changes within the system are internally reversible during a specified process. Since the external effects are already specified as reversible, the process is totally reversible. In this case the overall process can be reversed in direction. The magnitudes of heat and work terms will remain the same, with only a change in sign. In particular, for the original direction we may write that

$$\oint \frac{\delta Q}{T} \leqslant 0$$

If the cyclic path is now reversed, we must write that

$$\oint \frac{\delta Q}{T} \geqslant 0$$

since the $\delta Q/T$ terms are equal in magnitude but opposite in sign for the reversed direction. However, the greater than sign in this latter inequality violates the

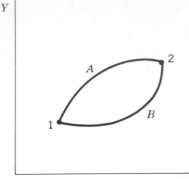

$X$     **Figure 6-8** Cyclic path of a reversible process.

Clausius inequality, which must be valid for any cyclic process. Hence the inequality sign cannot apply to a closed system which undergoes an internally reversible, cyclic process. But use of the equality sign is permitted for each direction of the cycle, without any second-law violation. Consequently, for all internally reversible processes involving closed systems

$$\oint \frac{\delta Q}{T} = 0 \qquad \text{internally reversible} \qquad\qquad (6\text{-}6)$$

The presence of internal irreversibilities leads to the inequality

$$\oint \frac{\delta Q}{T} < 0 \qquad \text{internally irreversible} \qquad\qquad (6\text{-}7)$$

In this latter case the presence of dissipative or nonequilibrium effects leads to the inequality. Once these relations are derived they become general in nature. That is, they are valid independent of whether the surroundings are reversible or not. The reason for this is that $\delta Q$ and $T$ refer solely to the system, and what happens in the surroundings is irrelevant to the evaluation of $\delta Q/T$.

We shall now turn our attention to the internally reversible case, as governed by Eq. (6-6). Consider a closed system undergoing a cyclic process, as shown in Fig. 6-8. The coordinates $X$ and $Y$ represent any two pertinent intrinsic properties, such as $P$ and $V$. For the internally reversible cycle consisting of paths $A$ and $B$ we may write that

$$\oint \frac{\delta Q}{T} = 0 = \int_1^2 \left(\frac{\delta Q}{T}\right)_A + \int_2^1 \left(\frac{\delta Q}{T}\right)_B$$

This equation can be rearranged into the form

$$\int_1^2 \left(\frac{\delta Q}{T}\right)_A = \int_1^2 \left(\frac{\delta Q}{T}\right)_B$$

Therefore if the process is internally reversible, the integral of $\delta Q/T$ is the same along paths $A$ and $B$. Since $A$ and $B$ were chosen arbitrarily, the internal of $\delta Q/T$ should lead to the same value for any reversible path between states 1 and 2.

In Chap. 2 we accepted the fact (based on experiment) that the adiabatic work between two given states of a closed system is always the same, regardless of the path chosen. This leads to the definition of the change in energy of a system, that is,

$$W_{ad} \equiv \Delta E$$

Now we have a similar situation. Between two given end states, the integral of $\delta Q/T$ always has the same value for any internally reversible path between those end states. Therefore this integral defines the change in a new property that is analogous to the change in energy. We shall call this integral a measure of the change in the entropy of the closed system. The function itself will be represented by the symbol $S$. Hence,

$$\Delta S = S_2 - S_1 \equiv \int \left(\frac{\delta Q}{T}\right)_{int\ rev} \tag{6-8}$$

or

$$dS = \left(\frac{\delta Q}{T}\right)_{int\ rev} \tag{6-9}$$

If a process is carried out in an internally reversible manner, then only a knowledge of the transient heat effects and of the temperature of the system at each stage of the process is required in order to evaluate $\Delta S$. The restriction of internal reversibility is inherent in the derivation and cannot be removed from this particular equation.

Note that Eq. (6-8) can also be written, on an intensive basis, as

$$s = \int \left(\frac{\delta q}{T}\right)_{int\ rev} + s_0$$

This equation shows that the integration of $\delta Q/T$ gives the entropy only within an arbitrary constant. As long as we work with pure substances, or nonreacting mixtures, a knowledge of the constant $s_0$ is unimportant because only a change in the entropy is usually required. Therefore tables and graphs may be constructed for which the base, or reference, value of $s$ is completely arbitrary. For example, in the saturation-temperature tables for water, A-12M and A-12, the specific entropy is chosen to be zero for saturated liquid water at the triple state (0.01°C or 32.02°F). Other reference states are chosen for other substances.

The dimensions of the entropy function are energy/(absolute temperature). Units in SI for the specific entropy are kJ/(kg)(°K) or kJ/(kg·mol)(°K), while USCS units commonly are Btu/(lb)(°R) or Btu/(lb·mol)(°R). Other sets of units may also be found in the literature.

## 6-12 INCREASE IN ENTROPY PRINCIPLE

It has been demonstrated that the entropy function is related uniquely to the value of $\delta Q/T$ for internally reversible processes. An additional relationship between $\Delta S$ and the integral of $\delta Q/T$ can be obtained by considering an internally irreversible process between two given end states of a closed system. Note the cyclic process shown in Fig. 6-8. We shall now retain the statement that the path $B$ is internally reversible. However, path $A$ will be defined as an internally irreversible process. For the entire cycle we write that

$$\oint \frac{\delta Q}{T} = \int_1^2 \left(\frac{\delta Q}{T}\right)_A + \int_2^1 \left(\frac{\delta Q}{T}\right)_B = \int_1^2 \left(\frac{\delta Q}{T}\right)_A - \int_1^2 \left(\frac{\delta Q}{T}\right)_B$$

The limits on the integral along path $B$ can be reversed, since in this case the integral represents the change in a property. In addition, for an internally irreversible process, the Clausius inequality requires that

$$\oint \frac{\delta Q}{T} < 0 \qquad\qquad (6\text{-}7)$$

Also, because path $B$ is internally reversible,

$$\int_1^2 \left(\frac{\delta Q}{T}\right)_B = S_2 - S_1$$

Combining these three expressions, we find that

$$\int_1^2 \left(\frac{\delta Q}{T}\right)_A - (S_2 - S_1) < 0$$

Path $A$ is *any* internally irreversible path from 1 to 2. Hence, for internally irreversible processes between equilibrium states

$$S_2 - S_1 > \int \left(\frac{\delta Q}{T}\right)_{\text{irrev}} \qquad\qquad (6\text{-}8)$$

In general, for any type of process involving a closed system,

$$S_2 - S_1 \geqslant \int \frac{\delta Q}{T} \qquad\qquad (6\text{-}9a)$$

or

$$dS \geqslant \frac{\delta Q}{T} \qquad\qquad (6\text{-}9b)$$

For this latter pair of equations, the equality sign applies for internally reversible processes, and the inequality for internally irreversible processes.

From Eq. (6-9b) two extremely important thermodynamic relations can be deduced. First, in the absence of heat transfer any closed system must fulfill the condition that

$$dS_{\text{adia}} \geqslant 0 \qquad\qquad (6\text{-}10)$$

For a finite change of state,

$$\Delta S_{adia} \geqslant 0 \tag{6-11}$$

The entropy function always increases in the presence of internal irreversibilities for an adiabatic, closed system. In the limiting case of an internally reversible adiabatic process the entropy will remain constant.

Although Eqs. (6-10) and (6-11) are extremely useful in the analysis of engineering processes, there are a number of processes which are not adiabatic. In such situations these equations are not directly applicable. However, an equivalent inequality in terms of the entropy function can be developed for cases where heat transfer occurs. Rather than studying just the closed system, however, it is necessary to include in the second-law analysis every part of the surroundings which is affected by changes in the system. In effect, we must look at the total changes, and not just at the system change. In practice, these changes outside of the system occur in regions reasonably close to the system. If we enclose in an arbitrary set of boundaries all those parts which are being affected, then no other changes are observed outside of this boundary. Such an enlarged system, which excludes further interactions across its boundaries, is frequently called an *isolated* system. No work or heat interactions or mass transfers occur across the boundaries of an isolated system.

Since no heat or mass transfer takes place across its boundaries, every isolated system fulfills the conditions of being closed and adiabatic. As a result, Eq. (6-10) applies also to an isolated system. Therefore,

$$dS_{\text{isolated system}} \geqslant 0 \tag{6-12}$$

For a finite change of state,

$$\Delta S_{\text{isolated system}} \geqslant 0 \tag{6-13}$$

To emphasize the fact that for nonadiabatic systems we must take into account all parts of physical space that are being affected by a change in the state of a particular system, Eq. (6-13) may be written as

$$\Delta S_{\text{total}} = \sum \Delta S_{\text{subsystems}} \geqslant 0 \tag{6-14}$$

That is, the algebraic sum of the entropy changes of *all* the subsystems participating in a process is zero or positive. If all the parts or subsystems of an overall system undergo reversible changes, and all interactions between these parts are reversible, then the total entropy of the overall (isolated) system will remain the same. If irreversibilities occur within any of the parts or are associated with any interaction between two or more of the subsystems, the entropy of the overall system must increase.

Equations (6-10) to (6-14) are the basic governing rules stemming from the second law of thermodynamics. Although other properties and further theorems may be developed which are more useful in certain types of engineering analysis, we begin with the entropy function and the " increase in entropy principles " stated above. Before proceeding, there are several important comments which must be made with respect to these principles.

1. The increase in entropy principles are *directional* statements. They limit the direction in which processes can proceed. A decrease in entropy is not possible for a closed system which is adiabatic, or for a composite of systems which interact among themselves.
2. The entropy function is a *nonconserved* property, and the increase in entropy principles are nonconservation laws. Only in the case of the reversible processes is entropy conserved. Reversible processes do not create entropy.
3. The second law states that all systems, left to themselves, are capable of reaching a state of equilibrium. The law now requires, in addition, that the entropy continually increase as that state of equilibrium is approached. Mathematically, the state of equilibrium must be attained when the entropy function reaches its *maximum* possible value, consistent with the constraints on the system. (Since a maximization process is involved, the techniques of the differential calculus are useful in evaluating the final equilibrium state.)
4. The increase in entropy principles are intimately connected with the concept of *irreversibility*. It is the presence of irreversibilities that leads to the increase in entropy. We shall find that the greater the magnitude of the irreversibilities, the greater the overall entropy change. Hence the entropy function can be used quantitatively by the engineer to measure dissipative effects or losses associated with any type of process. It may also be used to establish performance criteria.

The last item above is extremely important with respect to the analysis of engineering systems. To conserve or make better use of energy supplies, it is necessary to have some measure of current or predicted performance. The entropy function, or other properties based on this function, is extremely valuable in providing guidelines in this direction.

## 6-13 THE ENTROPY CHANGE OF HEAT AND WORK RESERVOIRS

At this point it is appropriate to consider a special application of the integral of $\delta Q/T$. A heat reservoir has been defined as a closed system which undergoes only internally reversible changes at constant temperature, while exchanging heat with another system. As a consequence of being an internally reversible system, the integral of $\delta Q/T$ applies directly to a heat reservoir. The value of the integral is the entropy change of any heat reservoir. Since the temperature is constant, integration leads to

$$\Delta S_{\text{heat reservoir}} = \int \left(\frac{\delta Q}{T}\right)_{\text{int rev}} = \frac{Q}{T} \tag{6-15}$$

The proper sign of $Q$ must be taken into account when numerically evaluating $Q/T$. When heat is added to a heat reservoir, its entropy must increase, while the removal of heat always decreases the entropy of the reservoir.

A work reservoir is a device which has the capability of delivering or absorbing energy in the form of work in a reversible manner. No heat effect is associated with its operation. The entropy change of a work reservoir is also easily determined. Such devices are described as internally reversible and adiabatic. Under these conditions the entropy change of reversible work sources or sinks must be zero. That is,

$$\Delta S_{\text{work reservoir}} = \int \left(\frac{\delta Q}{T}\right)_{\text{int rev}} = 0 \tag{6-16}$$

A qualitative explanation for this is as follows. When an idealized pulley-weight or flywheel device is used, for example, only its external state is altered, such as its overall gravitational potential energy or rotational kinetic energy. The thermostatic or intrinsic state of each, represented by variables such as temperature, pressure, or internal energy, is not changed. Since entropy is a thermostatic property, its value for these idealized work storage devices also is not changed.

## REFERENCES

Fast, J. D.: "Entropy," McGraw-Hill, New York, 1963.

Hatsopoulos, G. N., and J. H. Keenan: "General Principles of Thermodynamics," Wiley, New York, 1968.

Howerton, M. T.: "Engineering Thermodynamics," Van Nostrand, Princeton, N.J., 1962.

Reif, F.: "Fundamentals of Statistical and Thermal Physics," McGraw-Hill, New York, 1965.

Tribus, M.: "Thermostatics and Thermodynamics," Van Nostrand, Princeton, N.J., 1961.

## PROBLEMS (METRIC)  *(DO ALL)*

### Carnot heat engine

**6-1M** A totally reversible heat engine operates between temperatures of (a) 427 and 17°C, and (b) 537 and 27°C. Compute the ratio of the heat absorbed from the source to the work output.

**6-2M** A Carnot engine receives 10 kJ of heat from a reservoir at the normal boiling point of water (100°C) and rejects heat to another reservoir maintained at the triple state of water (0.01°C). Compute the work done in kilojoules and the thermal efficiency of the engine.

**6-3M** At what temperature is heat supplied to a Carnot engine that rejects 1000 kJ/min of heat at 7°C and produces (a) 40 kW of power, and (b) 50 kW of power?

**6-4M** A Carnot engine operates between 37 and 717°C. It is proposed to increase the temperature of the high-temperature source to 1027°C. A counterproposal is to lower the sink temperature to achieve the same new thermal efficiency as would have been attained by raising the source temperature to 1027°C. What would be the new required sink temperature in °C?

**6-5M** The efficiency of a Carnot engine discharging heat to a cooling pond at 27°C is 30 percent. If the cooling pond receives $10^6$ J/min, what is the power output of the engine in kilowatts? What is the temperature of the high-temperature source?

**6-6M** In a Carnot engine it is found that an efficiency of 50 percent occurs, with 500 kJ/cycle taken from the source reservoir. Calculate the sink temperature in °C and the heat rejected to the sink per cycle if the source temperature is 307°C.

**6-7M** A patent application claims that a heat engine which receives heat at 160°C and rejects it to a sink at 5°C is capable of delivering 0.12 kW·h of energy for every 1000 kJ received by the engine. Is this a valid claim?

**6-8M** A Carnot engine with an efficiency of 40 percent receives 4000 kJ/h from a high-temperature source and rejects heat to a sink at 25°C. What is the power output in kilowatts and the temperature of the source in °C?

### Clausius inequality

**6-9M** In a given power cycle the working fluid receives 3150 kJ/kg of heat at an average temperature of 440°C and rejects 1950 kJ/kg of heat to cooling water at 20°C. If there are no other heat interactions to or from the fluid, does the cycle satisfy the Clausius inequality?

**6-10M** In a given power cycle the working fluid receives 3340 kJ/kg of heat at an average temperature of 480°C and rejects 2020 kJ/kg of heat to the environment at 15°C. In the absence of other heat interactions, does the cycle satisfy the Clausius inequality?

### Entropy changes of heat reservoirs

**6-11M** With reference to Prob. 6-2M, compute the entropy changes of the heat source and heat sink in kJ/°K.

**6-12M** With reference to Prob. 6-3M, compute the entropy changes of the heat source and heat sink in kJ/(°K)(min) for part (a) and for part (b).

**6-13M** With reference to Prob. 6-6M, compute the entropy changes of the heat source and heat sink in kJ/°K/cycle.

**6-14M** A reversible heat engine receives 1000 kJ of heat from a reservoir at 377°C and rejects heat to another reservoir at 27°C. Compute the entropy changes of the two reservoirs in kJ/°K.

## PROBLEMS (USCS)

### Carnot heat engine

**6-1** A totally reversible heat engine operates between temperatures of 660 and 100°F. Compute the ratio of the heat absorbed from the source to the work output.

**6-2** A Carnot engine develops 2 hp and rejects 7500 Btu/h to a sink at 60°F. Determine the source temperature in °F.

**6-3** At what temperature is heat supplied to a Carnot engine that rejects 1000 Btu/min of heat at 40°F and produces 50 hp?

**6-4** A Carnot engine operates between 120 and 1280°F. It is proposed to increase the high-temperature source to 1860°F. A counterproposal is to lower the sink temperature to achieve the same thermal efficiency as would have been attained by raising the source temperature to 1860°F. What would be the new necessary sink temperature in °F?

**6-5** The efficiency of a Carnot engine discharging heat to a sink at 80°F is 30 percent. If the sink receives 700 Btu/min, what is the power output of the engine? What is the temperature of the high-temperature source?

**6-6** In a Carnot engine it is found that an efficiency of 50 percent occurs, with 500 Btu/cycle taken from the source. Calculate the sink temperature and the heat rejected to the sink per cycle if the source temperature is 560°F.

**6-7** A patent claims that an engine which receives heat at 320°F and rejects it to a sink at 40°F is capable of delivering 0.10 kWh of energy for every 1000 Btu received by the engine. Is this a valid claim?

**6-8** A Carnot engine with an efficiency of 40 percent receives 4000 Btu/h from a high-temperature source and rejects heat to a sink at 80°F. What is the power output in horsepower and the temperature of the source in °F?

### Clausius inequality

**6-9** In a power cycle the working fluid receives 1350 Btu/lb of heat at an average temperature of 820°F and rejects 840 Btu/lb of heat to cooling water at 70°F. If there are no other heat interactions to or from the fluid, does the cycle satisfy the Clausius inequality?

**6-10** In a power cycle the working fluid receives 1440 Btu/lb of heat at an average temperature of 900°F and rejects 870 Btu/lb of heat to the environment at 60°F. In the absence of other heat interactions, does the cycle satisfy the Clausius inequality?

### Entropy change of heat reservoirs

**6-11** With reference to Prob. 6-2, compute the entropy changes of the heat source and heat sink in Btu/°R.

**6-12** With reference to Prob. 6-3, compute the entropy changes of the heat source and sink in Btu/(°R)(min).

**6-13** With reference to Prob. 6-6, compute the entropy changes of the heat source and sink in Btu/°R/cycle.

**6-14** A reversible heat engine receives 1000 Btu of heat from a reservoir at 540°F and rejects heat to another reservoir at 60°F. Compute the entropy change of the two reservoirs in Btu/°R.

## SOME CONSEQUENCES OF THE SECOND LAW

A new intrinsic property—entropy—was introduced in the preceding chapter. It is intimately connected with the second law of thermodynamics, which is concerned with the directional or limiting quality of phenomena and the degradation of performance due to the presence of irreversibilities. These items of directionality and performance will be examined closely in this chapter. In addition, it is necessary to examine in some detail the methods of evaluating entropy changes of common substances.

### 7-1 THE TEMPERATURE-ENTROPY DIAGRAM

The use of property diagrams as an aid in problem solving was emphasized earlier in this text when the conservation of energy principle was applied to closed and open systems. In conjunction with the second law of thermodynamics, it is extremely helpful to be able to plot processes on diagrams for which one of the coordinates is entropy, or for which entropy lines appear on the diagram. One of the most fundamental coordinate systems for second-law analysis is the $TS$ diagram. The equation for the entropy change of a fixed mass,

$$dS = \left(\frac{\delta Q}{T}\right)_{\text{int rev}}$$

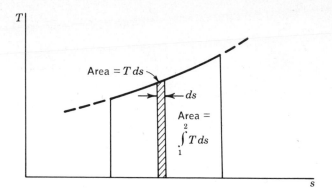

**Figure 7-1** $TS$ diagram for an internally reversible process.

can be rearranged to show that

$$\delta Q_{\text{int rev}} = T\, dS \tag{7-1}$$

or
$$Q_{\text{int rev}} = \int T\, dS \tag{7-2}$$

For a process which is carried out in an internally reversible manner, the heat transferred during the process is represented by an area on a $TS$ diagram. The temperature must be on an absolute scale. Figure 7-1 illustrates this point for some arbitrary internally reversible process. The value of $Q$ may be determined by actual measurement of the area or by integration if the relationship between $T$ and $S$ is known.

The temperature-entropy diagram is obviously quite useful for processes which are at either constant $T$ or $S$, since these paths may be shown as straight lines. In many cases a $TS$ diagram is useful, if only for the qualitative information it provides. In the analysis of simple systems, sketches of processes on $TS$ coordinate plots are often extremely informative and highly recommended.

The general characteristics of a $TS$ diagram are shown in Fig. 7-2. Only the gas and liquid regions of the pure substances appear. A more detailed $Ts$ diagram for $CO_2$ which includes all three phase regions is presented in Fig. A-26. The student should become familiar with the placement of constant-volume and constant-pressure lines on a $TS$ diagram. It should be observed that in any single-phase region, constant-volume lines have a steeper slope than constant-pressure lines through any given state. A general mathematical proof of this is given as a problem at the end of the chapter. One should also recognize that $TS$ diagrams are really projections of a three-dimensional surface of equilibrium states upon the $TS$ plane. The third dimension might be conveniently chosen to be either pressure or volume, although other intrinsic properties could be selected.

The $Pv$ diagram is extremely useful in the analysis of processes involving boundary ($P\, dv$) work. It is helpful at this point to indicate (without proof until later) the position of constant entropy lines on a $Pv$ plot. Figure 7-3 illustrates this

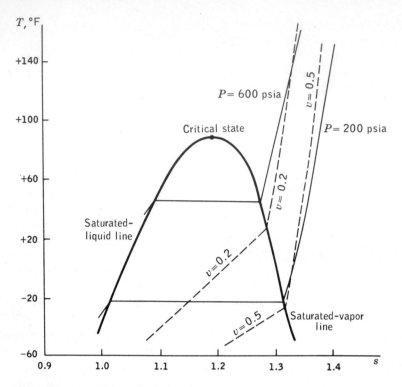

**Figure 7-2** A $TS$ diagram for $CO_2$.

situation for the gas phase, and the position of isothermal (constant temperature) lines are also shown for comparison. Processes during which the entropy remains constant are called *isentropic*. Note that isentropic lines have a steeper slope than isothermal lines in the gas region of a $Pv$ diagram.

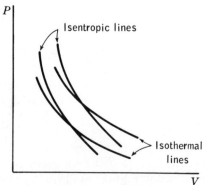

$V$     **Figure 7-3** Process lines on a pressure-volume diagram.

## 7-2 THE CARNOT HEAT ENGINE AND THE REVERSED CARNOT CYCLE

It has been demonstrated that the maximum thermal efficiency of any heat engine operating between two heat reservoirs is the theoretical Carnot efficiency. It is given by

$$\eta_{Carnot} = (T_H - T_L)/T_H = 1 - (T_L/T_H)$$

These are several theoretical cycles, composed of a series of different processes, which have efficiencies equal to the Carnot efficiency. One of the best known of these, which operates in a totally reversible manner between fixed-temperature reservoirs, is called the Carnot cycle. Such an engine can operate as either a closed or open system. Basically, the Carnot cycle requires that the working medium of the engine undergo a series of four processes. These processes include two isothermal reversible ones and two adiabatic reversible ones. Figure 7-4 shows a $PV$ diagram for a Carnot cycle. In this case the working medium might be considered, for example, a gas contained within a piston-cylinder device.

A heat reservoir at a temperature $T_H$ supplies heat to the engine which is at a temperature $T_H - dT$. Since there is a differential temperature difference between the heat reservoir and the engine, the process is externally reversible. In order to maintain an infinitesimal temperature difference during the heat-addition process, the working medium of the engine is allowed to expand isothermally. This isothermal expansion is shown as process 1-2. The engine produces a net work output during the expansion. From state 2, the working medium is allowed to expand reversibly and adiabatically to a new state 3. During process 2-3 additional work is produced. State 3 corresponds to a temperature of $T_L + dT$. The system is now compressed isothermally to state 4. During the compression process, heat is necessarily removed to maintain the system at $T_L + dT$. The heat

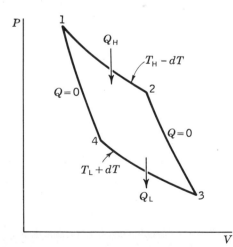

**Figure 7-4** Pressure-volume diagram for a Carnot engine.

is rejected to a heat reservoir at a temperature $T_L$. Again this process is externally reversible since a temperature difference of $dT$ is maintained across the boundaries of the engine. State 4 is so selected that, by a final reversible, adiabatic compression, the working medium is returned to the initial state. This is process 4-1 in Fig. 7-4. During processes 3-4 and 4-1 work is performed on the system. Overall, heat is exchanged between the working medium of the engine and two fixed-temperature reservoirs, and net work is produced during each complete cycle. The area enclosed on the $PV$ diagram is a measure of the net work output for the Carnot cycle, when boundary work is the major type of work under consideration.

The term "working medium" has been used in this discussion to emphasize the fact that many different types of systems can undergo a Carnot cycle. Although commonly associated with gases, a Carnot cycle can be executed by liquids, rubber bands (and other elastic media), electric cells, paramagnetic substances, liquid-vapor mixtures, etc. In most of these latter cases some other forms of reversible work will be important, rather than $P\,dV$ work. Consequently, some coordinate system other than $PV$ will be useful in illustrating the overall cycle. The area enclosed by the Carnot cycle on these other plots of an intensive generalized force against an extensive generalized displacement will again represent the net reversible work output for the cycle.

The heat supplied to and rejected from a Carnot engine is easily shown on a $TS$ diagram. Figure 7-5 illustrates this point, where the Carnot cycle appears as a rectangular area. The four states numbered in Fig. 7-5 correspond to the same states in Fig. 7-4. For an isothermal reversible process $Q$ equals $T\,\Delta S$. Therefore, the area beneath the horizontal line connecting states 1 and 2 represents the heat supplied $Q_H$. In a similar fashion the heat rejected $Q_L$ is given by the area beneath the line 3-4. From an energy balance on the heat engine the difference between $Q_H$ and $Q_L$ is the net work produced by the engine during a cycle. Hence the area enclosed by two isothermal, reversible lines and two isentropic lines on a $TS$ plot also is a measure of the net work of a Carnot cycle. When a consistent set of units is employed, the enclosed areas for the cycle on $PV$ and $TS$ diagrams should be the same for the same end states of the cycle. Figure 7-5 demonstrates quite well the effect of the values of $T_H$ and $T_L$ on the Carnot efficiency. By either increasing

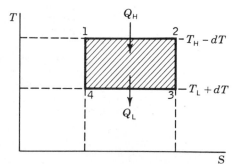

Figure 7-5 A $TS$ diagram for a Carnot engine.

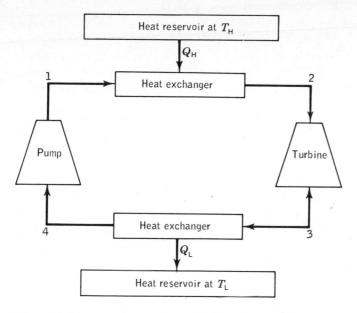

**Figure 7-6** Schematic diagram of a steady-flow Carnot engine.

$T_H$ or decreasing $T_L$ the crosshatched area becomes a larger fraction of the total area beneath the line 1-2. Hence the ratio $W_{net}/Q_{in}$ increases under these circumstances.

The use of the Carnot cycle is not restricted to closed systems. If the working medium is a gas or liquid, it may execute a Carnot cycle in a suitable series of steady-flow devices. A common scheme for a steady-flow Carnot cycle is shown in Fig. 7-6. The isothermal heat-addition and heat-rejection processes (1-2 and 3-4) are accomplished by heat exchangers. The adiabatic expansion and compression processes (2-3 and 4-1) require a turbine and a pump, respectively. A steady-flow Carnot cycle with a liquid-vapor mixture as the working substance is considered below.

**Example 7-1M** Steam at 30 bars and 50 percent quality is heated in a steady-flow heat exchanger until it becomes saturated vapor. It is then expanded isentropically through a turbine to 0.080 bar. This is followed by cooling in a second heat exchanger and then isentropic compression to the initial state. Determine the Carnot efficiency for the cycle and net work output.

SOLUTION The efficiency of the cycle is found most directly by determining the temperature of the fluid in the two isothermal heat exchangers. Since the heating and cooling processes involve wet mixtures, the temperatures required are simply the saturation temperatures corresponding to the pressures in the heat exchangers. A sketch of the cyclic process is shown on a $Ts$ diagram. From steam table A-13M, these temperatures are found to be 233.9 and 41.51°C, or 507.1 and 314.7°K. Therefore

$$\eta_{Carnot} = \frac{507.1 - 314.7}{507.1} = 0.379 \text{ (or } 37.9\%)$$

The net work output is found from the thermal efficiency by $w_{net} = \eta_{th} q_{in}$. The heat input during the constant-pressure evaporation process is simply $h_2 - h_1$ if the changes in kinetic and potential energy are neglected. From the data in Table A-13M,

$$h_1 = 0.5(1008.4) + 0.5(2804.2) = 1906.3 \text{ kJ/kg}$$

$$h_2 = h_g \text{ at 30 bars} = 2804.2 \text{ kJ/kg}$$

$h_1 = h_f + x h_{fg}$

$h_1 = 1008.4 + .5(1795.7)$

$h_1 = 1906.3$

Hence

$$w_{net} = 0.379(2804.2 - 1906.3) = 340.3 \text{ kJ/kg}$$

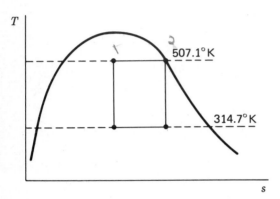

(a) Example 7-1M

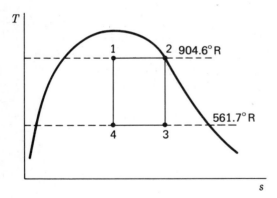

(b) Example 7-1

**Example 7-1** Steam at 400 psia and 50 percent quality is heated in a steady-flow heat exchanger until it becomes saturated vapor. It is then expanded isentropically through a turbine to 1 psia. This is followed by cooling in a second heat exchanger and then isentropic compression to the initial condition. Determine the Carnot efficiency for the cycle and the net work output.

SOLUTION The efficiency of the cycle is found most directly by determining the temperature of the fluid in the two isothermal heat exchangers. Since the heating and cooling processes involve wet mixtures, the temperatures required are simply the saturation temperatures corresponding to the pressures in the heat exchangers. From steam table A-4, these are found to be 444.6 and 101.7°F, or 904.6 and 561.7°R, respectively. Therefore

$$\eta_{th} = \frac{904.6 - 561.7}{904.6} = 0.379 (\text{or } 37.9\%)$$

The net work output by definition of the thermal efficiency is given by

$$w_{net} = \eta_{th} q_{in}$$

The heat input during the constant-pressure evaporation process is simply $h_2 - h_1$ if the changes in kinetic and potential energy are neglected. From the tabular data in steam table A-4,

$$h_1 = 0.5(424.2) + 0.5(1205.5) = 814.9 \text{ Btu/lb}$$

$$h_2 = h_g \text{ at } 400 \text{ psia} = 1205.5 \text{ Btu/lb}$$

Hence

$$w_{net} = 0.379(1205.5 - 814.9) = 148.0 \text{ Btu/lb}$$

A sketch of this cyclic process is shown on a $Ts$ diagram.

The Carnot cycle is of particular importance since its thermal efficiency is the maximum value for any heat engine operating between two given heat reservoirs. This efficiency is a theoretical one, because a reversible engine cannot be attained in practice. Nevertheless, the efficiency of a Carnot cycle represents a standard to which all actual heat-engine cycles may be compared.

A Carnot heat-engine cycle has been described as a totally reversible cycle. The engine itself is reversible and heat transfer to or from it occurs through differential temperature differences. Because of its total reversibility, it may truly be operated in the reverse direction. The end result is a device which acts as a refrigerator or heat pump. Heat is absorbed in this case from a low-temperature heat source, and heat is rejected to a high-temperature sink during the cycle. It is essential that work be supplied *to* the device from an external source. This is because of a corollary of the second law known as the *Clausius* statement of the second law, which states:

It is impossible to operate a cyclic device in such a manner that the sole effect external to the device is the transfer of heat from one heat reservoir to another at a higher temperature.

The Clausius corollary is also an obvious consequence of the increase in entropy principle. Let us assume that a certain cyclic device can bring about the transfer of heat from a cold to a hot body without any further external effects. The entropy change of the device is zero, since it is cyclic. The only other possible entropy change is that of the reservoirs. If H represents the heat reservoir of higher temperature and L the lower one, then for the transfer of a differential quantity of heat $\delta Q$,

$$dS_{res} = \frac{\delta Q}{T_H} - \frac{\delta Q}{T_L} = \delta Q \frac{T_L - T_H}{T_H T_L}$$

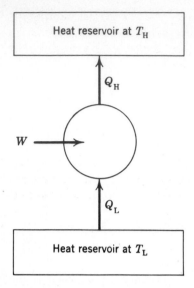

Heat reservoir at $T_H$

$Q_H$

$W$

$Q_L$

Heat reservoir at $T_L$

**Figure 7-7** Schematic of a Carnot refrigerator.

But $T_H$ is greater than $T_L$, so that $dS$ of the universe is negative. This is impossible, in view of our second-law statement; consequently, the original premise is incorrect and the Clausius statement is validated.

Figure 7-7 is a schematic of a refrigerator or heat pump operating between fixed-temperature levels. The difference between such a device as a heat pump or a refrigerator is merely one of definition. The purpose of a refrigerator is to maintain a low-temperature source of finite size at a predetermined temperature by removing heat from it. A heat pump maintains a high-temperature sink at a given level by supplying it with heat taken from a low-temperature source. For example, a home or business office can be supplied with heat taken from the ground or air on a cold winter day. Due to the work input to the reversed engine, the first law dictates that $Q_H$ is larger than $Q_L$. Even more important is the fact that $Q_H$ can be considerably larger than $W_{in}$. Hence, in using a heat pump to heat a building, the costly (electrical) work input may be only a fraction of the thermal energy delivered.

The concept of thermal efficiency is not useful when applied to a reversed cycle. When dealing with refrigerators and heat pumps the analogous term usually applied is the coefficient of performance, which is abbreviated COP, In refrigeration processes the object is to remove the maximum quantity of heat per unit of net work input. Therefore

$$\text{COP}_{\text{refrig}} \equiv \frac{Q_L}{W_{\text{net}}} = \frac{Q_{\text{in}}}{Q_{\text{out}} - Q_{\text{in}}} \tag{7-3}$$

The purpose of a heat pump is to supply the maximum quantity of heat to a high-temperature sink per unit of required net work input. Hence

$$\text{COP}_{\text{heat pump}} \equiv \frac{-Q_H}{W_{\text{net}}} = \frac{Q_{\text{out}}}{Q_{\text{out}} - Q_{\text{in}}} \tag{7-4}$$

The coefficient of performance may be expressed in terms of temperatures as well as of heat quantities. For heat pumps and refrigerators heat is rejected to the high-temperature sink and heat is removed from a low-temperature source. Hence the entropy change for a differential cycle of the device is

$$dS_{tot} = dS_H + dS_L + dS_{device} = \frac{\delta Q_H}{T_H} - \frac{\delta Q_L}{T_L} + 0$$

where the signs on the heat quantities have already been assumed. Consequently, only the absolute values of $\delta Q_H$ and $\delta Q_L$ should be used in these two equations. The maximum performance of refrigerators and heat pumps, as with heat engines, is attained when $dS_{tot}$ is zero. In this circumstance we see that the relationship between temperatures and heat quantities is exactly the same as for heat engines, that is, $\delta Q_H/T_H = \delta Q_L/T_L$, or

$$\frac{|Q_H|}{|Q_L|} = \frac{T_H}{T_L} \tag{6-2}$$

Consequently, for a reversed Carnot cycle, we find that substitution of Eq. (6-2) into Eq. (7-3) yields

$$COP_{refrig} = \frac{T_L}{T_H - T_L} \qquad \text{reversed Carnot} \tag{7-5}$$

In a similar manner Eq. (7-4) for a heat pump becomes

$$COP_{heat\ pump} = \frac{T_H}{T_H - T_L} \qquad \text{reversed Carnot} \tag{7-6}$$

It should be recognized that, although the limiting value of the thermal efficiency is unity, the coefficient of performance theoretically can be much larger

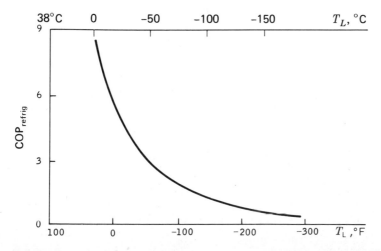

**Figure 7-8** Variation of the coefficient of performance with $T_L$ for a fixed value of $T_H$ of 25°C or 77°F.

than 1. Figure 7-8 illustrates the variation of $\text{COP}_{\text{refrig}}$ as $T_L$ is changed for a fixed value of the rejection (sink) temperature $T_H$ of 25°C or 77°F. It is noted that the coefficient of performance increases rapidly as the temperature difference between source and sink is decreased. The colder the low-temperature source of a refrigerator must be maintained, the larger the work input must be per unit of heat $Q_L$ removed.

**Example 7-2M** A Carnot refrigerator is maintaining foodstuffs in a refrigerated region at 2°C by rejecting heat to the atmosphere at 27°C. It is desired to maintain some frozen foods at $-17$°C with the same sink temperature of 27°C. What percent increase in work input will be required for the frozen-food unit over the refrigerated unit for the same quantity of heat $Q_L$ removed?

SOLUTION For each Carnot refrigerator two equations are valid, namely, $Q_H/Q_L = T_H/T_L$ and $Q_H - Q_L = W$, where the $Q$ and $W$ terms are absolute values. The combination of these equations gives

$$W_{in} = Q_L\left(\frac{T_H}{T_L} - 1\right)$$

Denoting the work input for the freezer operating between $-17$ and 27°C by a prime, we find that

$$W_{in} = Q_L\left(\frac{300}{275} - 1\right) = 0.0909Q_L$$

and

$$W'_{in} = Q_L\left(\frac{300}{256} - 1\right) = 0.1719Q_L$$

The percent increase of $W'$ over $W$ is given by

$$\frac{W' - W}{W} = \frac{0.1719Q_L - 0.0909Q_L}{0.0909Q_L} = 0.891 \text{ (or } 89.1\%)$$

Hence, by changing the temperature difference between source and sink from 25 to 44°C, the required theoretical work input is nearly doubled.

**Example 7-2** A Carnot refrigerator is maintaining foodstuffs in a refrigerator area at 40°F by rejecting heat to the atmosphere at 80°F. It is desired to maintain some frozen foods at 0°F with the same sink temperature of 80°F. What percent increase in work input will be required for the frozen-food unit over the refrigerated unit for the same quantity of heat removed?

SOLUTION For each Carnot refrigerator two sets of equations are valid, that is, $Q_H/Q_L = T_H/T_L$ and $Q_H - Q_L = W$. The combination of these equations gives

$$W_{in} = Q_L\left(\frac{T_H}{T_L} - 1\right)$$

Denoting the work input for the freezer operating between 0 and 80°F by a prime, we see that

$$W_{in} = Q_L(\tfrac{540}{500} - 1) = 0.080Q_L$$

and

$$W'_{in} = Q_L(\tfrac{540}{460} - 1) = 0.174Q_L$$

The percent increase of $W'$ over $W$ is given by

$$\frac{W' - W}{W} = \frac{0.174Q_L - 0.080Q_L}{0.080Q_L} = 1.175 (\text{or } 117.5\%)$$

Hence, by changing the temperature difference between source and sink from 40 to 80°F, the required work input is more than doubled.

One of the important contributions of the second law of thermodynamics to engineering analysis is the establishment of performance standards. In this section the thermal efficiency and the coefficient of performance have been reviewed for cyclic devices operating between heat reservoirs. These theoretical equations give values for the best performance under the stated restrictions. These values for $\eta_{th}$ and COP are the maximum possible, and may be used as comparative standards for the actual performance of heat engines, heat pumps, and refrigerators.

## 7-3 EFFECTS OF REVERSIBLE AND IRREVERSIBLE HEAT INTERACTIONS

In an engineering context irreversibilities degrade performance, and the degree or magnitude of the irreversibility is measured by the total entropy change of the interacting systems. An example of this is the effect of irreversible heat transfer associated with a real heat engine on its output. Before developing a quantitative analysis of this effect, we shall review the concept of reversible heat transfer in terms of the second law.

Essentially there are two methods of transferring heat reversibly between two systems. One of these is to maintain only an infinitesimal temperature difference between the two systems. To illustrate this, consider the exchange of a quantity of heat $Q$ between two heat reservoirs at temperatures $T$ and $T - \Delta T$. The total entropy change is simply

$$\Delta S_{total} = \frac{-Q}{T} + \frac{Q}{T - \Delta T} = \frac{Q}{T}\left(\frac{\Delta T}{T - \Delta T}\right)$$

Since $\Delta T$ always has to be equal to or less than $T$, the last term is always equal to or greater than zero, in accordance with the increase in entropy principle. As the two temperatures approach each other, $\Delta T$ approaches zero and the total entropy change also approaches zero. In the limit as $\Delta S_{total}$ becomes zero, the process by definition becomes reversible. However, we must have a temperature difference in order for heat transfer to occur. We can consider heat transfer to be reversible if the temperature difference between the two systems is made infinitesimally small.

The foregoing discussion indicates that as the temperature difference between two reservoirs (or any two systems in general) increases, the total entropy change increases for the heat exchange process. This point is also illustrated through the use of a $TS$ diagram for such a process. As an example, consider two systems H and L with absolute temperatures $T_H$ and $T_L$ of 1000 and 500 (in units of either °K

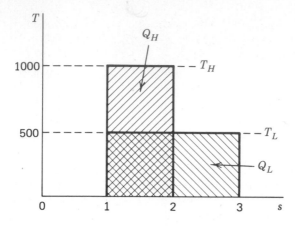

**Figure 7-9** Area representation of $Q$ for an irreversible heat-transfer process.

or °R). One thousand units of heat are transferred from H to L. The entropy change of H, an internally reversible heat reservoir, is simply $Q/T_H$, which in this case is

$$\Delta S_H = \frac{Q}{T_H} = \frac{-1000}{1000} = -1 \text{ entropy units}$$

Since the reservoir is internally reversible, the values of $Q$, $T_H$, and $S_H$ can be shown on a $TS$ diagram. Figure 7-9 shows the area representation of $Q$ for $T_H$ of 1000 and a $\Delta S$ of 1 entropy unit.

Now, the heat $Q$ received by L is the same as that given up by H. On the $TS$ diagram of Fig. 7-9 this heat quantity received by L must have the same area are previously shown. But it must also lie below the temperature line of 500. The cross-hatched area below $T_L$ again represents $Q$. Since in this case $Q = T_L \Delta S_L$, and $T_L$ is one-half that of $T_H$, $\Delta S_L$ must be twice as large as $\Delta S_H$, or 2 units. It is easily seen that as $T_L$ approaches $T_H$, these two cross-hatched areas must approach the same position on the diagram, and the total entropy change must become zero in the limit as $T_L$ reaches $T_H$. As $T_L$ becomes smaller, $\Delta S_L$ must become increasingly larger to maintain the same rectangular area on the $TS$ plot. As a consequence $\Delta S$ of the composite system of H and L steadily increases. This is a basic example of the increase in entropy principle.

When the temperature difference between two systems is finite, it is still possible to transfer heat between them in a reversible manner. Based on the discussion in the preceding section, this may be accomplished theoretically by inserting a reversible heat engine between the two systems. If a Carnot engine is used, then $\Delta S_H = Q_H/T_H$ and $\Delta S_L = Q_L/T_L$, where $Q_H \neq Q_L$. However, for a Carnot engine $Q_H/T_H = Q_L/T_L$. Since the heat-transfer quantities are of opposite sign, the total entropy change is zero, which indicates the reversibility of the process.

If we now compare the two preceding topics—irreversible heat transfer across a finite temperature difference and reversible heat transfer by insertion of a reversible heat engine—then we have one of many possible examples of irreversibilities leading to degradation of energy. When a heat engine is inserted between two systems which have a finite temperature difference between them, work is produced as a result of the heat exchange. For the same $\Delta T$ between the systems, direct irreversible heat transfer leads to no work production at all. Consequently, any time direct heat transfer occurs between two systems, the capability of producing some work from the interaction is lost. Since work is a more valuable form of energy in transition than heat, this loss of work capability is equivalent to a degrading or loss of *quality* of the energy involved, where quality is a measure of its potential usefulness.

The preceding discussion indicates that heat transfer through large finite temperature differences should be avoided, if possible, since a large "work potential" is lost. The maximum work output or the work potential obtainable from a quantity of heat $Q$ is given by

$$W_{max} = Q_{in}\eta_{Carnot} = Q_{in}\left(1 - \frac{T_L}{T_H}\right) \tag{7-7}$$

where the source and sink for heat transfer are at constant temperature. For most heat engines the environment acts as the sink; therefore $T_L$ is not a variable under control of the engineer. As $T_H$ increases, there exists the possibility that a larger fraction of $Q$ might be converted into work. That is, a given quantity of heat has a higher potential for conversion into work if it is supplied at a higher temperature to a reversible engine. This implies that heat energy itself has quality as well as quantity. The higher the temperature at which thermal energy is available to the user, the higher its quality at that moment. There is a tremendous amount of energy in the ocean, for example, but its low temperature gives it a low work potential, or quality.

## 7-4 THE $T\,dS$ EQUATIONS

The basic relation for the evaluation of the entropy change of any closed system is the integral of $\delta Q/T$ along an internally reversible path. Under isothermal conditions this integration is straightforward. For heat reservoirs, for example, the entropy change is simply $Q/T$. When $T$ is a variable, it is necessary to find some functional relationship between $\delta Q$ and $T$ before integrating. The formal procedure for doing this is outlined in this section.

When a closed system is restricted to internally reversible changes of state, the only work interactions which are permitted are of the quasistatic type discussed in Chap. 2. The sum of all the generalized quasistatic work interactions may be represented by the term $\sum_k F_{k,\,eq}\,dX_k$. Therefore the conservation of energy equation for a control mass undergoing internally reversible changes is

$$\delta Q = dU + \sum_k F_{k,\,eq}\,dX_k$$

If this relation is substituted into the quantity $\delta Q/T$, we find that

$$dS = \frac{1}{T}\left(dU + \sum_k F_{k,\,eq}\, dX_k\right)$$

or $\quad\quad T\, dS = dU + P\, dV - \sigma\, d\epsilon - \gamma\, dA - E\, dP - \mu_0 VH\, dM + \cdots \quad\quad (7\text{-}8)$

For a simple, compressible system Eq. (7-8) takes on the abbreviated form

$$T\, dS = dU + P\, dV \quad\quad\quad\quad (7\text{-}9)$$

This equation is called the first $T\, dS$ or Gibbsian equation for simple systems. Since $dH = dU + P\, dV + V\, dP$, the two preceding equations can be written as

$$T\, dS = dH - V\, dP + \sum_{k-1} F_{k,\,eq}\, dX_k \quad\quad\quad (7\text{-}10)$$

and $\quad\quad\quad\quad T\, dS = dH - V\, dP \quad\quad\quad\quad\quad (7\text{-}11)$

In Eq. (7-10) the summation is now over $k - 1$ work effects; that is, the $P\, dV$ term is excluded.

On a unit-mass basis the $T\, dS$ equations for a simple, compressible substance become

$$T\, ds = du + P\, dv \quad\quad\quad\quad (7\text{-}12)$$

and $\quad\quad\quad\quad T\, ds = dh - v\, dP \quad\quad\quad\quad\quad (7\text{-}13)$

It is these equations which will be used to evaluate the specific entropy change of several classes of substances.

The set of Eqs. (7-8) to (7-13) was derived on the basis of the process being carried out in an internally reversible manner. However, the integration of these equations will lead to the correct entropy change between two equilibrium states whether the actual process is reversible or irreversible. The change in entropy between two states is independent of the path because entropy is a point function. Integration of the above equations requires a knowledge of the functional relationship among properties. The restrictions, or regions of applicability, for the $T\, dS$ equations are important. They are summarized as follows:

1. They are in general applied to homogeneous systems. They may be applied to heterogeneous systems if the composition of each phase remains fixed and the pressure and temperature are the same for every phase. For example, a two-phase, single-component system would fulfill the requirement over a wide range of temperatures and pressures. However, a two-phase system composed of air could not be used since the composition of each phase differs for each equilibrium temperature.
2. They are valid only for chemically invariant systems. No chemical reactions are permitted within the system boundaries. The equations must be modified to permit their usage for chemically variant systems.
3. They are employed only between equilibrium states.
4. They are equally valid for simple or nonsimple systems as long as the equations are modified to include the proper quasistatic work interactions.

The $T\ dS$ equations are extremely useful, since they allow one to calculate the entropy changes of substances once the functional relationships among the properties involved are known. Shortly we shall examine several special classes of substances for which the integration of the $T\ dS$ equations is quite simple. At the same time we should recognize that the entropy values we find in tables have resulted from the complex integration of equations directly derivable from the $T\ dS$ equations. It is equally important to recognize that these equations are also valid for a mass passing through an open steady-flow system, since we have already seen that the property entropy is solely a function of the intrinsic state of matter. Its value should not be influenced by extrinsic changes of state, such as changes in velocity or position of the mass. The $T\ dS$ equations are extremely valuable in the further development of thermodynamic relationships.

## 7-5 ENTROPY CHANGES OF IDEAL GASES

The $T\ dS$ equations are important because they allow one to calculate the entropy changes of substances once the functional relationships among the properties are known. A simple and useful example is the application of the $T\ dS$ equations to systems containing ideal gases. As indicated in the preceding section, the basic place to start is with the $T\ dS$ equations on a unit-mass basis, namely,

$$T\ ds = du + P\ dv \tag{7-12}$$

and
$$T\ ds = dh - v\ dP \tag{7-13}$$

For an ideal gas, $du = c_v\ dT$, $dh = c_p\ dT$, and $Pv = RT$. Therefore

$$ds = \frac{du + P\ dv}{T} = c_v\ \frac{dT}{T} + R\ \frac{dv}{v} \tag{7-14}$$

and
$$ds = \frac{dh - v\ dP}{T} = c_p\ \frac{dT}{T} - R\ \frac{dP}{P} \tag{7-15}$$

Integration of the last term in these equations requires no further information since $R$ is a constant. We have seen, however, that both $c_v$ and $c_p$ are functions of temperature alone for ideal gases.

If the temperature range is small, the arithmetically averaged specific heat may be used with negligible error. Table A-4 shows that the variation of $c_p$ and $c_v$ over several hundred degrees is small in many cases. Integration of Eqs. (7-14) and (7-15) for constant specific heats shows that, for ideal gases,

$$\Delta s = c_v\ \ln \frac{T_2}{T_1} + R\ \ln \frac{v_2}{v_1} \tag{7-16}$$

and
$$\Delta s = c_p\ \ln \frac{T_2}{T_1} - R\ \ln \frac{P_2}{P_1} \tag{7-17}$$

These equations provide an easy method for evaluating entropy changes of ideal gases over fairly small temperature ranges.

**Example 7-3M** Air is compressed from 1 bar, 27°C, to 3.5 bars, 127°C, in a closed system. Determine the entropy change in kJ/(kg)(°K) or J/(g)(°K).

SOLUTION For the small temperature range under consideration, from 300 to 400°K, the values of $c_v$ and $c_p$ may be considered constant. From Table A-4M it is found that their average values in this range are 0.722 and 1.009 kJ/(kg)(°K), respectively. The value of $R$ needed can be found from the universal values found in Table A-1, or by recalling that for ideal gases $R = c_p - c_v$. From this latter relation we find that $R = 0.287$ kJ/(kg)(°K). Since both temperature and pressure data are given, Eq. (7-17) is an appropriate equation to use to evaluate $\Delta s$. Hence

$$\Delta s = 1.009 \ln (400/300) - 0.287 \ln (3.5/1) = -0.0693 \text{ kJ/(kg)(°K)}$$

The negative answer is not in conflict with the second law, since the process is not stated to be adiabatic. Figure 7-10 shows the process on a $Ts$ diagram. As described in Sec. 7-1, constant-pressure lines run from lower left to upper right on a $Ts$ plot, and the lines tend to diverge slightly as the temperature increases.

**Example 7-3** Air is compressed from 15 psia, 80°F, to 50 psia, 240°F, in a closed system. Determine the entropy change per pound.

SOLUTION The accompanying sketch (see Fig. 7-10) shows the process on a $Ts$ diagram. For the small temperature range under consideration, the values of $c_v$ and $c_p$ may be considered constant, and they have average values of 0.173 and 0.241 Btu/(lb)(°F), respectively. Since $R = c_p - c_v$ for ideal gases, $R$ is 0.068 Btu/(lb)(°F). Employing Eq. (7-17), one finds that

$$\Delta s = 0.241 \ln \tfrac{700}{540} - 0.068 \ln \tfrac{50}{15} = -0.0199 \text{ Btu/(lb)(°R)}$$

The negative answer is not in conflict with the increase in entropy principle for a closed system, since the process is not stated to be adiabatic.

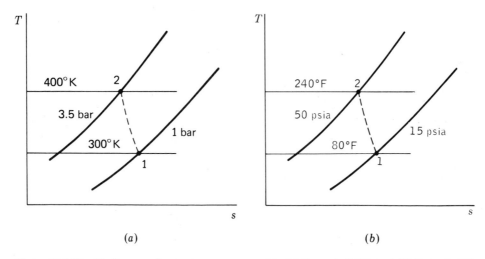

(a)                                        (b)

**Figure 7-10** The $Ts$ diagrams for processes represented in (a) Example 7-3M, and (b) Example 7-3.

**Example 7-4M** A rigid, insulated tank contains 1.2 kg of nitrogen gas at 350°K and 1 bar. A paddle-wheel inside the tank is driven by a pulley-weight mechanism. During an experiment 25,000 N·m of work is expended on the gas through the pulley-weight mechanism. Determine the entropy change for the nitrogen in kJ/°K.

SOLUTION The paddle-wheel work constitutes an internal irreversibility. Nevertheless, the $T\,ds$ equations will still lead to a correct evaluation of $\Delta S$. In order to use these equations, however, we must determine the final state of the gas. The initial state already is completely specified. Since energy information is given, the application of an energy balance on the system probably will provide additional information on the final state. The tank is insulated and rigid, so boundary work and heat transfer are zero. However, paddle-wheel work is present. Thus,

$$Q + W_{\text{paddle}} = \Delta U = mc_v\,\Delta T$$

$$0 + 25,000 \text{ N·m} = 1200\text{g} \times 0.744 \text{ J/(g)(°C)} \times (T_2 - 350)°\text{K}$$

$$T_2 = 350 + 28.0 = 378°\text{K}$$

A $c_v$ value at the initial temperature is used, since the final temperature is not yet known. Since $\Delta T$ is only 28°C, the use of the initial $c_v$ value is quite appropriate in this case.

Since the tank is rigid, the initial and final specific volumes are equal. Therefore Eq. (7-16) is more appropriate to use than Eq. (7-17), because the last term in Eq. (7-16) is zero for the process. As a consequence,

$$\Delta S = m\,\Delta s = mc_v\,\ln\,(T_2/T_1) = 1200(0.744)\,\ln\,(378/350)$$

$$= 68.7 \text{ J/°K} = 0.0687 \text{ kJ/°K}$$

The positive change in entropy is brought about by the irreversible work interaction. Since the process is adiabatic and irreversible, the entropy change must be positive in accordance with the increase in entropy principle for adiabatic, closed systems.

A $TS$ plot of the process is shown in Fig. 7-11a. From the ideal-gas equation, $P_2/P_1 = T_2/T_1$. Based on the above data, the pressure increased during the process. We noted in Sec. 7-1 that constant-volume lines have a steeper positive slope than constant-pressure lines on a $TS$ plot. Hence the diagram requires that the entropy change by positive.

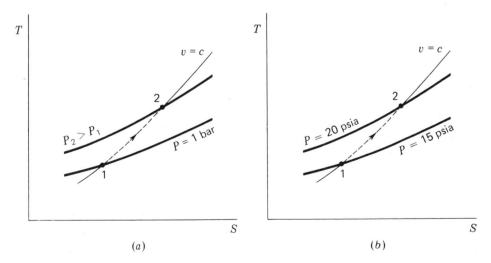

**Figure 7-11** The $TS$ diagrams for processes represented in (a) Example 7-4M, and (b) Example 7-4.

**Example 7-4** A rigid, insulated tank contains 2 lb of nitrogen gas at 140°F and 15 psia. A paddle-wheel inside the tank is driven by a pulley-weight mechanism. During an experiment 10,000 ft·lb$_f$ of work is expended on the gas through the pulley-weight mechanism. Determine the entropy change for the nitrogen in Btu/°R.

SOLUTION The paddle-wheel work constitutes an internal irreversibility. Nevertheless, the $T\,ds$ equations will still lead to a correct evaluation of $\Delta S$. In order to use these equations, however, the initial and final states must be determined. In this case the initial state already is completely specified, but only the volume of the final state is known. Since energy information is given in the problem statement, the application of an energy balance probably will provide additional information on the final state. The tank is insulated and rigid, so boundary work and heat transfer are zero. However, paddle-wheel work is present. Thus,

$$Q + W_{\text{paddle}} = \Delta U = mc_v\,\Delta T$$

If Btu are the units of each term, then

$$0 + \frac{10,000}{778} = 2(0.178)(T_2 - 140)$$

$$T_2 = 140 + 36.1 = 176.1°F = 636.1°R$$

A $c_v$ value at the initial temperature, rather than an average value, is used, since the final temperature is not known until the energy balance is made. Since the temperature rise is only 36°F, use of the initial $c_v$ value is appropriate.

Since the tank is rigid, the initial and final specific volumes are equal. Hence the entropy change can be found most easily by using Eq. (7-16) because the last term is zero. Therefore

$$\Delta S = m\,\Delta s = mc_v\,\ln\,(T_2/T_1) = 2(0.178)\,\ln\,(636.1/600)$$

$$= 0.0208\ \text{Btu/°R}$$

The positive change in entropy is brought about by the irreversible work interaction. The entropy change must be positive for this process in accordance with the increase in entropy principle for irreversible, adiabatic closed-system processes.

A $TS$ plot of the process is shown in Fig. 7-11b. From the ideal-gas equation, $P_2/P_1 = T_2/T_1$. Therefore the pressure increased during the process. We noted in Sec. 7-1 that constant-volume lines have a steeper slope than constant-pressure lines on a $TS$ plot. Hence the diagram requires that the entropy change be positive.

Examples 7-4M and 7-4 illustrate an interesting point with regard to the increase in entropy principle and nonquasistatic work interactions such as paddle-wheel work. Assume that the reverse process occurs, so that energy is removed from the gas and paddle-wheel work is delivered to the surroundings. For this adiabatic process, the entropy change of the closed system would be equal in magnitude, but opposite in sign, to the original stirring process. But a negative entropy change is not permitted for adiabatic closed systems. Thus processes like paddle-wheel and electrical-resistor work are truly nonreversible.

**Example 7-5M** A weight piston-cylinder assembly maintained at 1.5 bars contains air initially at 500°K. Heat transfer occurs until the air reaches a temperature of 400°K. Determine the specific entropy change of the air in kJ/(kg)(°K).

SOLUTION The closed system is chosen to be the air within the assembly. The constant-pressure process is shown as path $A$ from state 1 to state 2 on the accompanying $PV$ plot (see

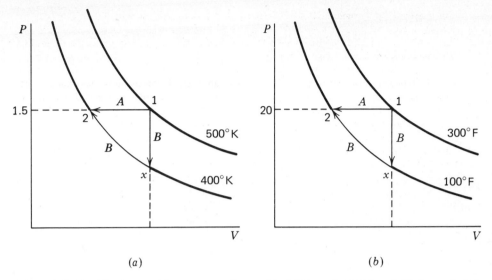

**Figure 7-12** The PV diagrams for processes discussed in (a) Example 7-5M, and (b) Example 7-5.

Fig. 7-12). If we choose Eq. (7-17) to evaluate the entropy change, then the pressure term in the equation drops out, since $P = $ constant. From Table A-4M we find the average $c_p$ value to be 1.021 kJ/(kg)(°K). Therefore

$$\Delta s = c_{p,\text{av}} \ln\ (T_2/T_1) = 1.021\ \ln\ (400/500) = -0.228\ \text{kJ/(kg)(°K)}$$

The entropy change is negative since heat is removed during the process in order to lower the temperature.

ALTERNATE SOLUTION Although it appears convenient, it is not necessary to evaluate $\Delta s$ along path $A$ by means of Eq. (7-17). Equation (7-16) involves a temperature term and a specific-volume term. This equation, in reality, represents a path such as $B$ shown on the $PV$ diagram. The term $c_v \ln\ (T_2/T_1)$ evaluates the entropy change at constant volume for a change in temperature. In this case this is process 1-x. The term $R \ln\ (v_2/v_1)$ is the change in entropy at constant temperature for a change in volume. This is process x-2 of path $B$ on the diagram. Physically, starting from state 1, the piston is clamped in position to fix the volume. Heat is then removed until the temperature drops from 400 to 300°K. Then the piston is released and the volume is allowed to decrease isothermally. During this second quasistatic step the pressure increases from $P_x$ to $P_2$.

For the process 1-x the volume is constant, and the entropy change for this step is

$$s_x - s_1 = c_{v,\text{av}} \ln\ (T_x/T_1) = c_{v,\text{av}} \ln\ (T_2/T_1)$$

$$= 0.734\ \ln\ (400/500) = -0.164\ \text{kJ/(kg)(°K)}$$

For the process x-2 the temperature is constant, and the entropy change for this second step is

$$s_2 - s_x = R \ln\ (v_2/v_x) = R \ln\ (v_2/v_1)$$

Use of the ideal-gas equation reveals that $v_2/v_1 = P_1 T_2/P_2 T_1 = T_2/T_1$. Consequently,

$$s_2 - s_x = R \ln\ (T_2/T_1) = \frac{8.315}{29}\ \text{J/(g)(°K)} \times \ln\ (400/500)$$

$$= -0.0640\ \text{kJ/(kg)(°K)}$$

For the overall process 1-x-2,

$$s_2 - s_1 = -0.164 + (-0.064) = -0.228 \text{ kJ/(kg)(°K)}$$

This answer is the same as that calculated for the direct path 1-2. This is not unexpected. Entropy is a property, and the change in a property depends solely upon the end states. This example does point out, however, that from the evaluation standpoint some paths lead to simpler calculations than others.

**Example 7-5** A weight piston-cylinder assembly maintained at 20 psia contains air initially at 300°F. Heat transfer occurs until the air reaches a temperature of 100°F. Determine the specific entropy change of the air in Btu/(lb)(°R).

SOLUTION The closed system is chosen to be the air within the assembly. The constant-pressure process is shown as path $A$ from state 1 to state 2 on the accompanying $PV$ plot (see Fig. 7-12). If we choose Eq. (7-17) to evaluate the entropy change, then the pressure term in the equation drops out, since $P = $ constant. From Table A-4 we find the average $c_p$ value to be 0.241 Btu/(lb)(°R). Therefore

$$\Delta s = c_{p,\,av} \ln (T_2/T_1) = 0.241 \ln (560/660) = -0.0737 \text{ Btu/(lb)(°R)}$$

The entropy change is negative since heat is removed during the process in order to lower the temperature.

ALTERNATE SOLUTION Although it appears convenient, it is not necessary to evaluate $\Delta s$ along path $A$ by means of Eq. (7-17). Equation (7-16) involves a temperature term and a specific-volume term. This equation, in reality, represents a path such as $B$ shown on the $PV$ diagram. The term $c_v \ln (T_2/T_1)$ evaluates the entropy change at constant volume for a change in temperature. In this case this is process 1-x. The term $R \ln (v_2/v_1)$ is the change in entropy at constant temperature for a change in volume. This is process x-2 of path $B$ on the diagram. Physically, starting from state 1, the piston is clamped in position to fix the volume. Heat is then removed until the temperature drops from 760 to 560°R. Then the piston is released and the volume allowed to decrease isothermally. During this second quasistatic step the pressure increases from $P_x$ to $P_2$.

For the process 1-x the volume is constant, and the entropy change for this step is

$$s_x - s_1 = c_{v,\,av} \ln (T_x/T_1) = c_{v,\,av} \ln (T_2/T_1)$$

$$= 0.173 \ln (560/760) = -0.0528 \text{ Btu/(lb)(°R)}$$

For the process x-2 the temperature is constant, and the entropy change for this second step is

$$s_2 - s_x = R \ln (v_2/v_x) = R \ln (v_2/v_1)$$

Use of the ideal-gas equation reveals that $v_2/v_1 = P_1 T_2/P_2 T_1 = T_2/T_1$. Consequently,

$$s_2 - s_x = R \ln (T_2/T_1) = \frac{1.986}{29} \text{ Btu/(lb)(°R)} \times \ln (560/760)$$

$$= -0.0209 \text{ Btu/(lb)(°R)}$$

For the overall process 1-x-2,

$$s_2 - s_1 = -0.0528 + (-0.0209) = -0.0737 \text{ Btu/(lb)(°R)}$$

This answer is the same as that calculated for the direct path 1-2. This is not unexpected. Entropy is a property, and the change in a property depends solely upon the end states. This example does point out, however, that from the evaluation standpoint some paths lead to simpler calculations than others.

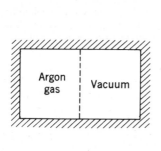

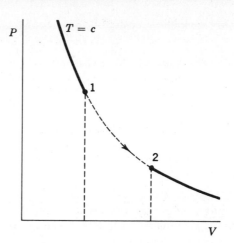

**Figure 7-13** Schematic and $PV$ diagram for process discussed in Examples 7-6M and 7-6.

**Example 7-6M** Argon gas is contained in an insulated tank at 25°C and 2 bars. A partition is punctured, and the gas flows into another insulated tank of the same volume. The second portion of the tank is initially evacuated. Determine the entropy change.

SOLUTION Argon may be considered to be an ideal gas. The two portions of the tank, together, are selected as the boundaries of the closed system. Figure 7-13 shows a schematic and a $PV$ diagram. Since the tanks are rigid, the work interactions are zero. The process is adiabatic and highly irreversible. On the basis of the increase in entropy principle we expect a positive change in entropy.

To evaluate $\Delta s$ we need more information on the final state. Since both $Q$ and $W$ are zero, $\Delta U$ is also zero. The internal energy of an ideal gas is solely a function of temperature; consequently, the temperature remains at 25°C during the free expansion. From the ideal-gas equation, $P_1 V_1/T_1 = P_2 V_2/T_2$. Hence the final pressure is one-half the initial value, or 1 bar. Either Eq. (7-16) or Eq. (7-17) may be used to evaluate $\Delta s$ since in both cases the temperature term is zero. Choosing the former, we find that

$$\Delta s = R \ln (v_2/v_1) = \frac{8.315}{40} \frac{\text{kJ}}{(\text{kg})(°\text{K})} \times \ln 2 = 0.225 \text{ kJ/(kg)(°K)}$$

This problem illustrates why the restriction of internal reversibility is so important on the integral of $\delta Q/T$. Since $Q$ is zero, the integral of $\delta Q/T$ is zero. However, $\Delta S$ is not zero, as shown by our calculation. Thus we see that the integral of $\delta Q/T$ has no relation to $\Delta S$ if the process to which it is applied is internally irreversible.

**Example 7-6** Argon gas is contained in an insulated tank at 100°F and 30 psia. A partition is punctured, and the gas flows into another insulated tank of the same volume. The second portion of the tank is initially evacuated. Determine the entropy change.

SOLUTION Argon may be considered to be an ideal gas. The two portions of the tank, together, are selected as the boundaries of the closed system. Figure 7-13 shows a schematic and a $PV$ diagram. Since the tanks are rigid, the work interactions are zero. The process is adiabatic and highly irreversible. On the basis of the increase in entropy principle we expect a positive change in entropy.

To evaluate $\Delta s$ we need more information on the final state. Since both $Q$ and $W$ are zero, $\Delta U$ is also zero. The internal energy of an ideal gas is solely a function of temperature; consequently, the temperature remains at 100°F during the free expansion. From the ideal-gas equation, $P_1 V_1/T_1 = P_2 V_2/T_2$. Hence the final pressure is one-half the initial value, or 15 psia. Either Eq. (7-16) or (7-17) may be used to evaluate $\Delta s$ since in both cases the temperature term is zero. Choosing the former, we find that

$$\Delta s = R \ln (v_2/v_1) = \frac{1.986}{40} \text{ Btu/(lb)(°R)} \times \ln 2 = 0.0344 \text{ Btu/(lb)(°R)}$$

This problem illustrates why the restriction of internal reversibility is so important on the integral of $\delta Q/T$. Since $Q$ is zero, the integral of $\delta Q/T$ is zero. However, $\Delta S$ is not zero, as shown by our calculation. Thus we see that the integral of $\delta Q/T$ has no relation to $\Delta S$ if the process to which it is applied is internally irreversible.

More accurate evaluation of Eqs. (7-14) and (7-15) is attained by the use of analytical expressions for the specific heats as a function of temperature. Table 3-1 is an example of such data. By programming equations for specific heat data on a computer, rapid evaluations of $\Delta s$ may be made. Unless many such calculations are needed, however, it may not be desirable to use a high-speed computer. An alternative is to tabulate the integration of Eqs. (7-14) and (7-15) in some manner similar to the tabulations of $h$ and $u$ in the ideal-gas tables in the Appendix. In devising tables for the accurate evaluation of $\Delta s$, Eq. (7-15) is usually used. Integrating the last term of this equation in the usual manner, we note that

$$\Delta s = \int_1^2 c_p \frac{dT}{T} - R \ln \frac{P_2}{P_1}$$

Now we shall define a function $s^0$ such that its value relative to a reference state can be found by

$$s_1^0 - s_{\text{ref}}^0 = \int_{\text{ref}}^1 c_p \frac{dT}{T}$$

The determination of $s_1^0 - s_{\text{ref}}^0$ only requires a knowledge of $c_p$ as a function of $T$. In a similar manner for another temperature $T_2$,

$$s_2^0 - s_{\text{ref}}^0 = \int_{\text{ref}}^2 c_p \frac{dT}{T}$$

If we now subtract the first equation for $s^0$ from the second one,

$$s_2^0 - s_1^0 = \int_1^2 c_p \frac{dT}{T}$$

This is the first term on the right of the specific entropy equation. Consequently, in general for an ideal gas,

$$s_2 - s_1 = s_2^0 - s_1^0 - R \ln \frac{P_2}{P_1} \qquad (7\text{-}18)$$

The units of $s^0$ and $R$ in this equation must be consistent.

Referral to Tables A-5M to A-11M and Tables A-5 to A-11 will show that the function $s^0$ has been tabulated solely as a function of temperature for these ideal gases. For air the value of $s^0$ in Table A-5M has units of kJ/(kg)(°K), while in Table A-5 the units are Btu/(lb)(°R). For the remaining gases the metric data have units of kJ/(kg·mol)(°K), while the USCS data have units of Btu/(lb·mol)(°R). The superscript on $s$ indicates a standard-state value: the gas is an ideal gas at 1 atm pressure. The following two examples illustrate data acquisition from these tables.

**Example 7-7M** Air is compressed from 1 bar, 27°C, to 3.5 bars, 127°C. Determine the entropy change employing the air tables, A-5M.

SOLUTION On the basis of Eq. (7-18), $\Delta s = s_2^0 - s_1^0 - R \ln (P_2/P_1)$. From Table A-5M at 300 and 400°K, $s_1^0 = 1.70203$ and $s_2^0 = 1.99194$ kJ/(kg)(°K). Therefore

$$\Delta s = 1.99194 - 1.70203 - \frac{8.315}{29} \ln (3.5/1) = -0.0693 \text{ kJ/(kg)(°K)}$$

This is in good agreement with Example 7-3M since the temperature range is fairly small.

**Example 7-7** Air is compressed from 15 psia, 80°F, to 50 psia, 240°F. Determine the entropy change by using the air table, A-5.

SOLUTION $\Delta s = s_2^0 - s_1^0 - R \ln (P_2/P_1)$. From Table A-5, at 540 and 700°R, $s_1^0 = 0.6008$ and $s_2^0 = 0.6632$ Btu/(lb)(°R). Therefore

$$\Delta s = 0.6632 - 0.6008 - \frac{1.986}{29} \ln (50/15) = 0.0200 \text{ Btu/(lb)(°R)}$$

This is in good agreement with Example 7-3 since the temperature range is fairly small.

**Example 7-8M** Nitrogen gas maintained at 2 bars is heated from 300°K to (a) 500°K, and (b) 800°K. Evaluate the entropy change for both sets of temperature increments by means of (1) an average $c_p$ value and (2) the nitrogen table, A-6M.

SOLUTION (a) Since the pressure is constant, Eq. (7-17) reduces to $\Delta s = c_{p, \text{av}} \ln (T_2/T_1)$ and Eq. (7-18) becomes $\Delta s = s_2^0 - s_1^0$. On the basis of data from Tables A-4M and A-6M we find that

$$\Delta s = c_{p, \text{av}} \ln (T_2/T_1) = 1.047(28) \ln (500/300) = 14.98 \text{ kJ/(kg·mol)(°K)}$$

$$\Delta s = s_2^0 - s_1^0 = 206.630 - 191.682 = 14.95 \text{ kJ/(kg·mol)(°K)}$$

(b) For the second increment of temperature we find that

$$\Delta s = 1.08(28) \ln (800/300) = 29.66 \text{ kJ/(kg·mol)(°K)}$$

$$\Delta s = 220.907 - 191.682 = 29.23 \text{ kJ/(kg·mol)(°K)}$$

For the 300 to 500°K interval, use of an average $c_p$ value leads to a 0.2 percent error over the integrated value. For the 300 to 800°K interval, this error increases to 1.5 percent.

**Example 7-8** Nitrogen gas maintained at 30 psia is heated from 80°F to (a) 400°F, and (b) 1000°F. Evaluate the entropy change for both sets of temperature intervals by means of (1) an average $c_p$ value and (2) the nitrogen table, A-6.

SOLUTION (a) Since the pressure is constant, Eqs. (7-17) and (7-18), respectively, reduce to,

$$\Delta s = c_{p, \text{av}} \ln T_2/T_1$$

$$\Delta s = s_2^0 - s_1^0$$

On the basis of data from Tables A-4 and A-6 we find that

$$\Delta s = 0.250(28) \ln (860/540) = 3.258 \text{ Btu/(lb·mol)(}^\circ\text{R)}$$

$$\Delta s = 49.031 - 45.781 = 3.250 \text{ Btu/(lb·mol)(}^\circ\text{R)}$$

(b) For the second temperature interval,

$$\Delta s = 0.259(28) \ln (1460/540) = 7.213 \text{ Btu/(lb·mol)(}^\circ\text{R)}$$

$$\Delta s = 52.867 - 45.781 = 7.086 \text{ Btu/(lb·mol)(}^\circ\text{R)}$$

For the 80 to 400°F interval, use of an average $c_p$ value leads to a 0.25 percent error over the integrated value. For the 540 to 1000°F interval, this error increases to 1.8 percent.

When tables of $s^0$ values are available, they provide a fast and accurate means of evaluating the effect of temperature on the ideal-gas entropy. In addition, the use of average-specific-heat data lead to an increasing error as the temperature interval increases. It must be kept in mind, however, that regardless of the method used to evaluate the $T \, ds$ equations for an ideal gas, the result is equally applicable to a closed system and to a unit mass passing through a steady-state control volume.

## 7-6 ENTROPY CHANGES INVOLVING REAL GASES AND SATURATION STATES

In Chap. 4 tables of data representing superheated-vapor, saturation, and compressed-liquid states were introduced. The emphasis then was on the properties $P$, $V$, $T$, $u$, and $h$, since we were concerned at that point with first-law analyses. In order to apply the second law to real gases and saturated fluids we must reconsider these tables in the light of the entropy function. There are no simple algebraic relationships (as there are for ideal gases and incompressible substances) between the entropy $s$ and the pressure and temperature for real gases, saturation states, and compressed liquids. The values of $s$ (relative to an arbitrary reference state) in these cases are determined from fairly complex numerical techniques based on experimental $P$, $v$, $T$, and specific-heat data. These values are correct only to within the accuracy permitted by the experimental measurements and numerical computations. Reevaluation of tabular data is frequently necessary as improved experimental data become available.

The value of the entropy at a given state is tabulated in exactly the same manner as the properties $v$, $u$, and $h$. In the superheat region, where there are two independent properties, the entropy is tabulated along with the other properties as a function of temperature and pressure. For the saturation states, the values of $s_f$ and $s_g$ are given as a function of either temperature or pressure. For wet mixtures the quality of the mixture must be known in order to ascertain its specific entropy. Finally, if compressed-liquid data are available, the entropy function again is tabulated against the temperature and pressure of the fluid. In the absence of compressed-liquid data, the value of $s_c$ in that region can be estimated by using $s_f$ at the given temperature.

Several examples below illustrate the basic use of tables in terms of the entropy function. The practical use of tabulated entropy data will be shown in the remaining sections of this chapter, and in numerous later chapters.

**Example 7-9M** Steam at 40 bars and 280°C is changed at constant volume to a state of 9 bars. Determine the entropy change for the process in kJ/(kg)(°K).

SOLUTION The saturation temperature at 40 bars from Table A-13M is 250.4°C. Therefore the initial state is superheated vapor. From Table A-14M the entropy and the specific volume are read directly, and are found to be

$$s_1 = 6.2568 \text{ kJ/(kg)(°K)}$$

$$v_1 = 55.46 \text{ cm}^3/\text{g}$$

The final state is determined or fixed by the values of $P_2$ and $v_2$, which is the same as $v_1$. From Table A-13M, it is found that at 9 bars $v_2$ lies between $v_f$ and $v_g$. Consequently, the final state is a wet mixture. The quality of the state is found by

$$v_2 = v_x = v_f + xv_{fg}$$

$$55.46 = 1.12 + x(215.0 - 1.1)$$

$$x = 54.34/213.9 = 0.254 \text{ (or } 25.4\%)$$

The knowledge of the quality enables us to evaluate the entropy.

$$s_2 = s_f + sx_{fg} = 2.0946 + 0.254(6.6226 - 2.0946) = 3.245 \text{ kJ/(kg)(°K)}$$

Hence the entropy change for the process is

$$\Delta s = s_2 - s_1 = 3.245 - 6.257 = -3.012 \text{ kJ/(kg)(°K)}$$

The process diagram is shown on the $Ts$ plot of Fig. 7-14. The negative sign is not significant, other than to indicate that the gas is cooled during the process.

**Example 7-9** Steam at 500 psia and 600°F is changed at constant volume to a state of 100 psia. Determine the entropy change for the process, in Btu/(lb)(°R).

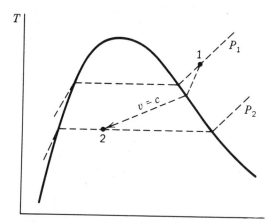

**Figure 7-14** Process diagram for Examples 7-9M and 7-9.

SOLUTION The saturation temperature at 500 psia is roughly 467°F; therefore, the initial state is a superheated vapor. From Table A-14 the entropy is read directly, and is found to be

$$s_1 = 1.5585 \text{ Btu/(lb)(°R)}$$

The final state is determined by the pressure and the volume. Initially the specific volume is 1.158 ft$^3$/lb. This is also the final specific volume at a pressure of 100 psia. From Table A-13, it is found that at this pressure this value of $v$ lies between $v_f$ and $v_g$. Consequently the final state is a wet mixture. The quality at this state is found by

$$v_x = v_f + xv_{fg}$$

$$1.158 = 0.01774 + x(4.432 - 0.01774)$$

$$x = 1.14/4.414 = 0.258$$

The knowledge of the quality enables us to evaluate the entropy.

$$s_2 = s_f + xs_{fg} = 0.4740 + 0.258(1.1286) = 0.7652 \text{ Btu/(lb)(°R)}$$

Hence the entropy change is

$$\Delta s = s_2 - s_1 = 0.7652 - 1.5585 = -0.7933 \text{ Btu/(lb)(°R)}$$

The negative sign on the answer is not significant, other than to indicate that the gas is cooled during the process. The process diagram is shown on the accompanying $Ts$ plot (Fig. 7-14).

**Example 7-10M** One-tenth kilogram of water substance initially at 3 bars, and 200°C is contained within a closed system. During a process a heat removal of 7.70 kJ and a work input of 17,500 N·m occur, resulting in a final pressure of 15 bars. Determine the entropy change of the water substance in kJ/°K.

SOLUTION If the steam tables are to be used to evaluate data, a state must be explicitly defined. The initial pressure and temperature are sufficient in this case to fix the initial state. However, only the pressure is known for the final state. Therefore, we need one more property of a simple, compressible substance in order to find the entropy of the final state. Since the values of the energy interactions are known for the process, it is reasonable to suspect that an energy balance on the process will yield more information on the final state.
For a closed system,

$$Q + W = \Delta U = m(u_2 - u_1)$$

The only unknown in this expression is the final specific internal energy $u_2$. The saturation temperature at 3 bars is 133.55°C. Hence the initial temperature of 200°C indicates that the initial state is a superheated vapor. From Table A-14M

$$u_1 = 2650.7 \text{ kJ/kg} \quad \text{and} \quad s_1 = 7.3115 \text{ kJ/(kg)(°K)}$$

Use of the energy balance then yields

$$u_2 = u_1 + \frac{Q + W}{m} = 2650.7 + \frac{-7.70 + (17,500/1000)}{0.1} = 2748.7 \text{ kJ/kg}$$

The internal energy of a saturated vapor $u_g$ at 15 bars is approximately 2595 kJ/kg. Thus the final state is also a superheated vapor. At 15 bars and a specific internal energy of 2748.7 kJ/kg, we find from Table A-14M that

$$T_2 = 280°C \quad \text{and} \quad s_2 = 6.8381 \text{ kJ/(kg)(°K)}$$

Consequently the total entropy change of the system is

$$\Delta S = m\,\Delta s = 0.1 \text{ kg} \times (6.8381 - 7.3115) \text{ kJ/(kg)(°K)}$$

$$= -0.04734 \text{ kJ/°K}$$

The negative sign on the entropy change does not violate the increase in entropy principle, since the closed system is not adiabatic.

**Example 7-10** One-tenth pound of water substance initially at 40 psia and 300°F is contained within a closed system. During a given process a heat removal of 3.40 Btu and a work input of 7.53 Btu occur, resulting in a final pressure of 200 psia. Determine the entropy change of the water substance in Btu/°R.

SOLUTION If the steam tables are to be used to evaluate data, a state must be explicitly defined. The initial pressure and temperature are sufficient to fix the initial state. However, only the pressure is known for the final state. Therefore we need one more property in order to find the entropy of the final state. Since the values of the interactions are known during the process, it is reasonable to suspect that an energy balance on the process will yield more information on the final state.

For a closed system,

$$Q + W = \Delta U = m(u_2 - u_1)$$

The only unknown in this expression is the final specific internal energy $u_2$. The saturation temperature at 40 psia is roughly 267°F. Hence an initial temperature of 300°F at this pressure indicates that the initial state is a superheated vapor. From the superheat table, Table A-14, we find that

$$u_1 = 1105.1 \text{ Btu/lb} \qquad \text{and} \qquad s_1 = 1.6993 \text{ Btu/(lb)(°R)}$$

Use of the energy balance then yields

$$u_2 = u_1 + \frac{Q + W}{m} - 1105.1 + \frac{-3.40 + 7.53}{0.1} = 1146.4 \text{ Btu/lb}$$

The internal energy of a saturated vapor $u_g$ at 200 psia is approximately 1114 Btu/lb. Thus the final state is also a superheated vapor. At 200 psia and an internal energy of 1146.4 Btu/lb, we find that

$$T_2 = 450°F \qquad \text{and} \qquad s_2 = 1.5938 \text{ Btu/(lb)(°R)}$$

Consequently the total entropy change of the system is

$$\Delta S = m\,\Delta s = 0.1 \text{ lb} \times (1.5938 - 1.6993) \text{ Btu/(lb)(°R)}$$

$$= -0.01055 \text{ Btu/°R}$$

The negative entropy change found here does not violate the second law, per se, since the system chosen is neither adiabatic or isolated.

**Example 7-11M** Refrigerant 12 initially at 2 bars and 30°C is expanded isothermally and internally reversibly to 1 bar in a steady-flow process. In the absence of work effects, determine the change in kinetic energy in kJ/kg.

SOLUTION Assuming the potential-energy change to be negligible, the steady-flow energy equation reduces to $q = \Delta h + \Delta KE$. Since the process is isothermal and reversible, $q$ is given by $T\,\Delta s$. Therefore the change in kinetic energy becomes

$$\Delta KE = T\,\Delta s - \Delta h = T(s_2 - s_1) - (h_2 - h_1)$$

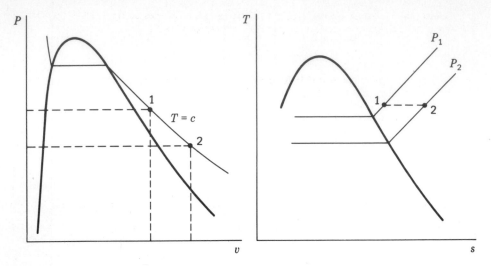

**Figure 7-15** The $Pv$ and $Ts$ process diagrams for Examples 7-11M and 7-11.

A sketch of the process is shown in Fig. 7-15 on $Pv$ and $Ts$ diagrams. The saturation temperature is $-12.53°C$ at 2 bars and $-30.10°C$ at 1 bar. Since the process occurs at $30°C$, the fluid is in the superheat region throughout the process. From Table A-18M:

$$P_1 = 2 \text{ bars} \qquad\qquad P_2 = 1 \text{ bar}$$
$$T_1 = 30°C \qquad\qquad T_2 = 30°C$$
$$s_1 = 0.7978 \text{ kJ/(kg)(°K)} \qquad s_2 = 0.8488 \text{ kJ/(kg)(°K)}$$
$$h_1 = 208.60 \text{ kJ/kg} \qquad\quad h_2 = 210.02 \text{ kJ/kg}$$

Substitution of these values into the preceding equation yields

$$\Delta KE = 303(0.8488 - 0.7978) - (210.02 - 208.60)$$

$$= 15.45 - 1.42$$

$$= 14.03 \text{ kJ/kg (or J/g)}$$

Note that in order to maintain the system isothermal, a total of 15.45 J/g of heat must be added to the system.

**Example 7-11** Refrigerant 12 initially at 20 psia and 80°F is expanded isothermally and internally reversibly to 15 psia in a steady-flow process. In the absence of work effects, determine the change in kinetic energy in Btu per pound.

SOLUTION Assuming the potential-energy change is negligible, the steady-flow energy balance reduces to $q = \Delta h + \Delta KE$. However, for isothermal, reversible flow, $q$ is given by $T \Delta s$. Therefore the change in kinetic energy becomes

$$\Delta KE = T \Delta s - \Delta h = T(s_2 - s_1) - (h_2 - h_1)$$

A sketch of the process is shown in Fig. 7-15 on $Pv$ and $Ts$ diagrams. The saturation temperature at 20 psia is $-8.13°F$ and $-20.75°F$ at 15 psia. Since the process occurs at 80°F, the fluid is a superheated vapor throughout the process. From Table A-18:

$$P_1 = 20 \text{ psia} \qquad\qquad P_2 = 15 \text{ psia}$$
$$T_1 = 80°F \qquad\qquad T_2 = 80°F$$
$$s_1 = 0.1955 \text{ Btu/(lb)(°R)} \qquad s_2 = 0.2005 \text{ Btu/(lb)(°R)}$$
$$h_1 = 89.168 \text{ Btu/lb} \qquad h_2 = 89.383 \text{ Btu/lb}$$

Substitution of these values into the preceding equation yields

$$\Delta KE = 540(0.2005 - 0.1955) - (89.383 - 89.168)$$

$$= 2.70 - 0.22$$

$$= 2.48 \text{ Btu/lb}$$

In order to maintain the system isothermal, a total of 2.7 Btu/lb of heat must be added to the fluid.

# 7-7 ENTROPY CHANGE OF AN INCOMPRESSIBLE SUBSTANCE

A substance of constant density (or specific volume) was defined in Sec. 4-7 as an incompressible substance. Significantly, the internal energy of such a substance is given by

$$du = c_v \, dT \qquad\qquad (4\text{-}8)$$

and the specific heats at constant volume and constant pressure are equal; that is,

$$c_p = c_v = c \qquad\qquad (4\text{-}11)$$

For any simple, compressible, substance we recognize that $T \, ds = du + P \, dv$. However, the last term is zero for an incompressible substance, by definition. Therefore, employing the relations above, we see that the entropy change of an incompressible substance is given by

$$ds = \frac{du}{T} = \frac{c_v \, dT}{T} = \frac{c \, dT}{T} \qquad \text{incompressible} \qquad (7\text{-}19)$$

In many cases it is reasonable to assume that the specific heat is constant over a given temperature range, in which case we find that

$$s_2 - s_1 = c \ln \frac{T_2}{T_1} \qquad\qquad (7\text{-}20)$$

In those cases where the specific heat $c$ varies significantly with temperature, it would be necessary to integrate Eq. (7-19) by first inserting a functional relationship for $c$ as a function of $T$. This would be especially true for solids at very low temperatures (close to absolute zero).

**Example 7-12M** Two separated blocks of copper, $A$ and $B$, have masses of 1 kg and 3 kg and initial temperatures of 100 and 300°C, respectively. They are brought into contact and allowed to come to thermal equilibrium at atmospheric pressure while insulated from the surroundings. Determine the entropy change of each block and the total entropy change.

SOLUTION An energy balance on a given block indicates that at constant pressure $Q = \Delta H = mc_p \, \Delta T$. Within an accuracy of several percent, the specific heat may be assumed constant. All the heat removed from the hot block goes to the cold block. By equating the two heat quantities,

$$[mc_p(T_2 - T_1)]_A = -[mc_p(T_2 - T_1)]_B$$

Upon cancellation of the $c_p$ quantities, we find that the only unknown in the energy balance is $T_2$.

$$1(T_2 - 100) = -3(T_2 - 300)$$

$$T_2 = 250°C$$

The entropy change is found by application of Eq. (7-20). Since $\Delta S = m \, \Delta s$,

$$\Delta S = mc \ln (T_2/T_1)$$

From Table A-19M, the average $c_p$ value in the overall temperature range is 25.9 kJ/(kg·mol)(°K) or 0.407 kJ/(kg)(°K). Hence,

$$\Delta S_A = 1(0.407) \ln (523/373) = 0.138 \text{ kJ/°K}$$

$$\Delta S_B = 3(0.407) \ln (523/573) = -0.111 \text{ kJ/°K}$$

The surroundings are not affected by the process, so the total entropy change is just the sum of $A$ and $B$.

$$\Delta S_{\text{total}} = 0.138 + (-0.111) = 0.027 \text{ kJ/°K}$$

The increase in entropy is due to the irreversible heat transfer between the two solids. A $TS$ diagram for the process is sketched in Fig. 7-16. The areas under the constant-pressure line must be equal, since $Q = \int T \, dS$ for internally reversible processes. In addition, both areas have a common state, $T_2$. Because of the position of the constant-pressure line on a $TS$ diagram, the absolute value of the entropy change of $B$ must be less than that of $A$ if the areas are to be equal.

**Example 7-12** Two separated blocks of copper, $A$ and $B$, have masses of 0.1 lb and 0.3 lb and initial temperatures of 100 and 300°F, respectively. They are brought into contact and allowed to come to thermal equilibrium at atmospheric pressure while insulated from the surroundings. Determine the entropy change of each block and the total entropy change.

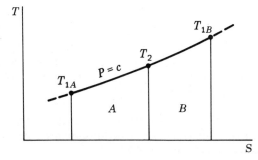

**Figure 7-16** Process diagram for Examples 7-12M and 7-12.

SOLUTION An energy balance on a given block indicates that at constant pressure $\delta Q = dH = mc_p \, dT$. Within an accuracy of several percent, the specific heat may be assumed to be constant at 0.0973 Btu/(lb)(°F). Since the heat removed from the hot block all goes to the cold block,

$$[mc_p(T_2 - T_1)]_A = -[mc_p(T_2 - T_1)]_B$$

Upon cancellation of the $c_p$ quantities, we find that the only unknown in the energy balance is $T_2$.

$$0.1(T_2 - 100) = -0.3(T_2 - 300)$$

$$T_2 = 250°F$$

The entropy change is found from Eq. (7-20). Since $\Delta S = m \, \Delta s$,

$$\Delta S = mc \ln (T_2/T_1)$$

$$\Delta S_A = 0.1(0.0973) \ln (710/560) = 0.00231 \text{ Btu/°R}$$

$$\Delta S_B = 0.3(0.0973) \ln (710/760) = -0.00199 \text{ Btu/°R}$$

The surroundings are not affected by the process, so the total entropy change is just the sum of $A$ and $B$.

$$\Delta S_{\text{total}} = 0.00231 + (-0.00199) = 0.00032 \text{ Btu/°R}$$

The increase in entropy is due to the irreversible heat transfer between the two solids. A $TS$ diagram for the process is sketched in Fig. 7-16. The areas under the constant-pressure line must be equal, since $Q = \int T \, dS$ for internally reversible processes. In addition, both areas have a common state, $T_2$. Because of the position of the constant-pressure line on the $TS$ diagram, the absolute value of the entropy change for $B$ must be less than that of $A$ if the areas are to be equal.

**Example 7-13M** Saturated liquid water at 40°C is heated and compressed to 80°C and 75 bars. Estimate the entropy change by (a) assuming an incompressible fluid, (b) using saturation data for the liquid, and (c) using the compressed-liquid table.

SOLUTION (a) For an incompressible fluid, $\Delta s = c \ln (T_2/T_1)$. From Table A-19M, the specific heat in this range of temperatures is roughly 4.187 kJ/(kg)(°C). Therefore

$$\Delta s = 4.187 \ln (353/313) = 0.504 \text{ kJ/(kg)(°K)}$$

(b) In the absence of compressed-liquid data for state 2, saturated-liquid data may be used at the given temperature. Employing the data in Table A-12M, we find that

$$\Delta s = s_{f, \, 80} - s_{f, \, 40} = 1.0753 - 0.5725 = 0.5028 \text{ kJ/(kg)(°K)}$$

(c) The entropy of state 2 in part (b) is found more accurately by use of the compressed-liquid table, A-15M. Hence

$$\Delta s = 1.0704 - 0.5725 = 0.4979 \text{ kJ/(kg)(°K)}$$

This last answer is the most accurate, since it is based directly on experimental data. Use of saturated data leads to a 1 percent error, while the assumption of an incompressible fluid leads to only a 1.2 percent error. Thus, the use of the incompressible-fluid relation for $\Delta s$ is a reasonable approximation when tabular data are unavailable for the liquid state.

**Example 7-13** Saturated liquid water at 32°F is heated and compressed to 200°F and 1000 psia. Determine the entropy change by (a) assuming an incompressible fluid, (b) using saturated data for the liquid, and (c) using the compressed-liquid table.

SOLUTION (a) For an incompressible fluid, $\Delta s = c \ln (T_2/T_1)$. From Table A-19, the specific heat of water in this temperature range is roughly 1.0 Btu/(lb)(°F). Therefore

$$\Delta s = 1.0 \ln (660/492) = 0.294 \text{ Btu/(lb)(°R)}$$

(b) In the absence of compressed-liquid data for state 2, saturated-liquid data may be used at the given temperature. Employing the data in Table A-12, we find that

$$\Delta s = s_{f,\,200} - s_{f,\,32} = 0.2940 - (-0.0003) = 0.2940 \text{ Btu/(lb)(°R)}$$

(c) The entropy of state 2 in part (b) is found more accurately by use of the compressed-liquid table, A-15. Hence

$$\Delta s = 0.29281 - (-0.0003) = 0.2928 \text{ Btu/(lb)(°R)}$$

This last answer is the most accurate since it is based directly on experimental data. Use of saturated data leads to a 0.4 percent error, while the assumption of an incompressible fluid leads to the same percent of error in this case. Thus the use of an incompressible fluid relation for $\Delta s$ is a reasonable approximation when tabular data are unavailable for the liquid state.

# 7-8 SOME RELATIONSHIPS FOR WORK INTERACTIONS

In the discussion of quasistatic work interactions for closed systems in Chap. 2, it was pointed out that such interactions lead to the maximum work output and the minimum work input. With the second law as background, we are now in a position to prove the above statement. Although the following analysis is generally valid for any type of quasistatic work effect, we shall restrict this discussion to the study of simple, compressible substances. The extension to other types of systems is readily apparent. For a simple, compressible substance in a closed system we may write two basic equations:

$$T \, dS = dU + P \, dV \tag{7-9}$$

and

$$\delta Q_{\text{act}} + \delta W_{\text{act}} = dU \tag{3-5}$$

Subscripts have been written on the $\delta Q$ and $\delta W$ terms to indicate that these are actual values, and apply to either internally reversible or irreversible changes. Consequently, $\delta W_{\text{act}}$ is measured external to the system. That is, the work is the measured input or output. If we use Eq. (3-5) to eliminate $dU$ from Eq. (7-9), then

$$T \, dS = \delta Q_{\text{act}} + P \, dV + \delta W_{\text{act}}$$

or

$$dS = \frac{\delta Q_{\text{act}}}{T} + \frac{1}{T}(P \, dV + \delta W_{\text{act}}) \tag{7-21}$$

In this equation, $P \, dV$ is the work carried out during an internally reversible process. In addition, we have developed from the second law that $dS \geqslant \delta Q_{\text{act}}/T$. Therefore

$$\frac{1}{T}(\delta W_{\text{act}} - \delta W_{\text{rev}}) = dS - \frac{\delta Q_{\text{act}}}{T} \geqslant 0$$

Since the absolute temperature $T$ is positive by concept, we find that

$$\delta W_{\text{act}} \geqslant \delta W_{\text{rev}} \tag{7-22}$$

This inequality is valid regardless of the direction of change of the closed system. That is, it applies whether the work interaction is in or out. Noting that the inequality sign on any of the preceding equations denotes the presence of irreversibilities, we have now shown that irreversibilities within a closed system reduce the work output or increase the work input. To maximize the work output, we must attempt to minimize irreversibilities within a system.

The $T\,ds$ equations are also useful in deriving a basic relationship for shaft work associated with a control volume. The conservation of energy principle for a differential mass of a simple, compressible substance passing through a steady-state device is

$$\delta w_{\text{shaft}} = dh + d\text{KE} + d\text{PE} - \delta q$$

If the process is internally reversible, then, for the same unit mass

$$\delta q = T\,ds = dh - v\,dP \tag{7-13}$$

Substitution of this latter relationship into the energy equation yields

$$\delta w_{\text{shaft}} = dh + d\text{KE} + d\text{PE} - (dh - v\,dP)$$
$$= v\,dP + d\text{KE} + d\text{PE} \tag{7-23}$$

For flow over a finite distance through a steady-state device the frictionless mechanical work associated with a unit mass is

$$w_{sf,\text{ rev}} = \int v\,dP + \Delta\text{KE} + \Delta\text{PE} \tag{7-24}$$

If the changes in the kinetic and potential energies are negligible, Eq. (7-24) reduces to

$$w_{sf,\text{ rev}} = \int v\,dP \tag{7-25}$$

Equations (7-24) and (7-25) are useful in a quantitative sense only if the functional relationship between $v$ and $P$ is known. A specific example of this is shown in the next section. The reader should be careful to distinguish between the nonflow and steady-flow work equations for simple compressible substances. Confusion often arises from the similarity between $P\,dv$ and $v\,dP$. By sketching a process on a $Pv$ diagram, the student can easily distinguish the difference between these two work expressions in terms of areas on the diagram.

Since Eqs. (7-23) to (7-25) are developed under internally reversible conditions, they should lead to the maximum work output or minimum work input to steady-flow devices such as turbines, compressors, and pumps. By an analysis similar to that given for closed systems at the beginning of this section, it can be shown that a relation of the type

$$\delta W_{sf,\,act} \geqslant \delta W_{sf,\,rev} \tag{7-26}$$

applies to these steady-flow work devices.

## 7-9 ISENTROPIC PROCESSES

An internally reversible process has a special significance to engineers. As pointed out in the last section, such a process leads to the maximum work output or minimum input for devices such as turbines and compressors. Although it is not as obvious, we shall soon show that internally reversible processes also lead to the best performance of nozzles, diffusers, and other non-work-producing devices. Consequently, the internally reversible process can be used as a standard to which all real processes may be compared, whether applied to closed or open systems. In addition, it was emphasized in Chap. 5 during the discussion of control-volume analysis that many devices are essentially adiabatic, for various reasons. Consequently, the adiabatic, internally reversible process is frequently chosen as a standard to which actual processes may be compared.

The basic relation for the entropy change of any mass of material undergoing an internally reversible process is $dS = \delta Q/T$. It is immediately seen that adiabatic, internally reversible processes must be ones of constant entropy. When the entropy does not change, a process is called *isentropic*. Isentropic processes are used by the engineer in the theoretical analysis of open and closed systems, since such processes are the limit of extrapolation of real adiabatic processes. This situation is illustrated on a $Ts$ diagram in Fig. 7-17 for a process which involves an increase in pressure, for example.

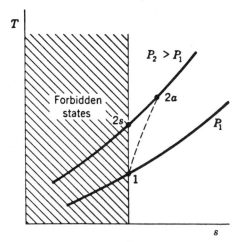

**Figure 7-17** Forbidden states for irreversible-adiabatic processes.

We have already noted that constant-pressure lines have positive slopes on a $Ts$ diagram. If an internally reversible process occurs adiabatically, the final state $2s$ lies directly above the initial state 1. However, if irreversibilities are present, the final state $2a$ must lie to the right of state 1 on the $P_2$ line; that is, $s_2 > s_1$. This increase in entropy is dictated by the general statement, $dS_{adia} \geqslant 0$. (In this discussion and others later in this chapter, the symbol $s$ stands for an isentropic final state, while the symbol $a$ represents an actual final state which occurs due to irreversibilities.) The position of state $2a$ on line $P_2$ depends upon the extent of the irreversibilities. State $2a$ cannot lie to the left of the vertical line through state 1, since all states to the left of this line are forbidden states for an adiabatic process. Consequently, it is seen that the isentropic process is the limiting process as irreversibilities are reduced under adiabatic conditions. As a limiting condition, the isentropic process is a standard of performance against which real processes may be compared.

In previous sections of this chapter, it was shown that special equations can be developed for the entropy change of several specific classes of substances. These special equations may now be used to derive isentropic relations for these same substances.

## 1 Isentropic Ideal-Gas Relations

In Sec. 7-5 the $T\,ds$ equations were applied to ideal gases. In the case where the specific heats are assumed constant, or an average value over the given temperature interval is used, the following equations are valid.

$$\Delta s = c_v \ln \frac{T_2}{T_1} + R \ln \frac{v_2}{v_1} \tag{7-16}$$

and

$$\Delta s = c_p \ln \frac{T_2}{T_1} - R \ln \frac{P_2}{P_1} \tag{7-17}$$

For an isentropic process, $\Delta s = 0$. If Eqs. (7-16) and (7-17) are set equal to zero, the following relations result.

$$\frac{T_2}{T_1} = \left(\frac{v_1}{v_2}\right)^{k-1} \qquad \text{Isentropic process} \tag{7-27a}$$

and

$$\frac{T_2}{T_1} = \left(\frac{P_2}{P_1}\right)^{(k-1)/k} \qquad \text{Isentropic process} \tag{7-28a}$$

In these expressions $k$ is the specific heat ratio, $c_p/c_v$. It appears in the above equations by use of the identities, $c_v/R = 1/(k-1)$ and $c_p/R = k/(k-1)$, for ideal gases. If the ideal-gas equation is substituted into either Eq. (7-27) or Eq. (7-28) so that $T$ is eliminated as a variable, then

$$\frac{P_2}{P_1} = \left(\frac{v_1}{v_2}\right)^{k} \qquad \text{Isentropic process} \tag{7-29a}$$

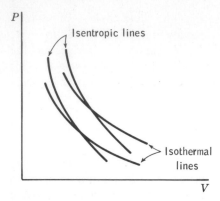

$P$

Isentropic lines

Isothermal lines

$V$

**Figure 7-18** Process lines on a pressure-volume diagram.

Another way of expressing these relationships among $P$, $v$, and $T$ is as follows:

$$T(v)^{k-1} = \text{constant} \qquad (7\text{-}27b)$$

$$T^k P^{1-k} = \text{constant} \qquad (7\text{-}28b)$$

$$P(v)^k = \text{constant} \qquad (7\text{-}29b)$$

Equations (7-27) to (7-29) represent process equations for the isentropic change of state for ideal gases. As a word of warning, note that the three constants in the above set of equations are not equal in value or dimensions. Typical values of $k$ are given in the tables of specific heats, Tables A-4M and A-4.

In Fig. 7-18 isothermal and isentropic lines are plotted on a $PV$ diagram for comparative purposes. Isentropic lines have a larger negative slope than isothermal lines through the same state point on a $PV$ diagram for ideal gases. It is important to note that the above set of relations is valid for any fixed mass of an ideal gas which undergoes a process at constant entropy. The mass may be in a closed system or flowing through a steady-state control volume.

**Example 7-14M** Air enters an adiabatic compressor at 17°C and is compressed through a pressure ratio of 8.6 : 1. If the compression is assumed to be internally reversible, determine the enthalpy change based on inlet conditions.

SOLUTION The enthalpy change for constant specific heats is given by $\Delta h = c_p(T_2 - T_1)$ for an ideal gas. The final temperature is found from the isentropic relation

$$\frac{T_2}{T_1} = \left(\frac{P_2}{P_1}\right)^{(k-1)/k}$$

At the inlet condition of 17°C, the $k$ value from Table A-4M is found to be 1.40; hence

$$T_2 = 290(8.6)^{0.286} = 290(1.85) = 537°\text{K}$$

Therefore, using the inlet $c_p$ value of 1.005 kJ/(kg)(°C),

$$\Delta h = 1.005(537 - 290) = 248 \text{ kJ/kg}$$

**Example 7-14** Air enters an adiabatic compressor at 40°F and is compressed through a pressure ratio of 8.6 : 1. For an internally reversible compression process, determine the enthalpy change based on inlet conditions.

SOLUTION The enthalpy change for a constant $c_p$ value is given by $\Delta h = c_p(T_2 - T_1)$ for an ideal gas. The final temperature is found from the isentropic relation

$$\frac{T_2}{T_1} = \left(\frac{P_2}{P_1}\right)^{(k-1)/k}$$

At the inlet condition of 40°F, the $k$ value from Table A-4 is found to be 1.40; hence

$$T_2 = 500(8.6)^{0.286} = 500(1.85) = 925°R$$

Therefore, using the inlet $c_p$ value of 0.240 Btu/(lb)(°F),

$$\Delta h = 0.240(925 - 500) = 102 \text{ Btu/lb.}$$

In some isentropic calculations it is necessary to account for the variation of $c_p$ with temperature, if reasonable accuracy is desired. There are two methods by which this may be accomplished. The most fundamental way is based on the direct use of Eq. (7-18). For an isentropic process this equation reduces to

$$0 = s_2^0 - s_1^0 - R \ln \frac{P_2}{P_1}$$

Recall that the quantity $s_2^0 - s_1^0$ is a measure of the integral of $c_p \, dT/T$ between states 1 and 2. Therefore the above equation does account for the variation of specific heat with temperature. Also, we have already seen that $s^0$ values are tabulated solely as a function of temperature. Such data are available for hundreds of ideal gases, and have been tabulated over a wide range of temperatures in the *JANAF* Tables. (These tables, produced by the Dow Chemical Company, are distributed by the U.S. Government Printing Office.) In this text, the $s^0$ values for air are given on a unit-mass basis (gram or pound), while for all other gases the values are on a molar basis.

If the initial state is known, then the final state of an isentropic process of an ideal gas is found by using one of the two following equations:

$$s_2^0 = s_1^0 + R \ln \frac{P_2}{P_1} \qquad (7\text{-}30)$$

and

$$P_2 = P_1 \exp \frac{s_2^0 - s_1^0}{R} \qquad (7\text{-}31)$$

Equation (7-30) permits the calculation of $T_{2s}$ if $P_2$ is known, while Eq. (7-31) evaluates $P_{2s}$ if $T_2$ is known. The subscript $s$ again symbolizes an isentropic state. The use of these expressions is illustrated below.

**Example 7-15M** Reconsider the data of Example 7-14M. Use variable-specific-heat data to determine the enthalpy change of air when it is compressed through an 8.6 : 1 pressure ratio from an initial temperature of 17°C.

SOLUTION The final temperature for the compression process is found by using Eq. (7-30). In Table A-5M, the value of $s_1^0$ at 290°K is found to be 1.66802 kJ/(kg)(°K). Substitution of the known data yields

$$s_2^0 = 1.66802 + (8.315/29) \ln 8.6 = 2.285 \text{ kJ/(kg)(°K)}$$

In Table A-5M, this value of $s_2^0$ is found to lie between 530 and 540°K. By linear interpolation $T_{2s}$ is 533°K. By further interpolation of $h$ data, $h_{2s}$ is 536.8 kJ/kg, and $h_1$ at 290°K is 290.16 kJ/kg. Hence

$$\Delta h = 536.8 - 290.2 = 246.6 \text{ kJ/kg}$$

This is about 0.6 percent smaller than that obtained from constant-specific-heat data in Example 7-14M. Although the answers in this case differ very little, the above method is quick and always more accurate, since the variation in specific heats is automatically included.

**Example 7-15** Reconsider the data of Example 7-14. Use variable-specific-heat data to determine the enthalpy change of air when it is compressed through an 8.6 : 1 pressure ratio from an initial temperature of 40°F.

SOLUTION The final temperature for the compression process is found by using Eq. (7-30). In Table A-5, the value of $s_1^0$ at 500°R is noted to be 0.58233 Btu/(lb)(°R), while $h_1$ is 119.48 Btu/lb. Substitution of the known data into Eq. (7-30) yields

$$s_2^0 = 0.58233 + (1.986/29) \ln 8.6 = 0.7297 \text{ Btu/(lb)(°R)}$$

In Table A-5, this value of $s_2^0$ falls very close to 920°R. At this temperature, $h_2$ is 221.18 Btu/lb. Hence

$$\Delta h = 221.2 - 119.5 = 101.7 \text{ Btu/lb}$$

This answer is fairly close to that given by assuming constant specific heats. Nevertheless, the above method is quick and always provides more accuracy, since the variation in specific heats is automatically included.

The use of $s^0$ data, in conjunction with Eqs. (7-30) and (7-31), provides a general method of evaluating isentropic changes for ideal gases. In some special cases it is useful to extend the method one step further. For example, the preceding method is an iterative (trial and error) process when the volume ratio is known rather than the pressure ratio. This makes it more difficult to evaluate expansion and compression processes in piston-cylinder devices such as automotive engines. Second, for some gases such as air, numerous repetitive calculations may be required for a particular design. In such cases it is helpful, but not essential, to generate a new set of isentropic data.

This set of data is based on Eq. (7-31), namely,

$$P_2 = P_1 \exp \frac{s_2^0 - s_1^0}{R} \tag{7-31}$$

Solving for the pressure ratio under isentropic conditions,

$$\left(\frac{P_2}{P_1}\right)_s = \exp \frac{s_2^0 - s_1^0}{R} = \frac{\exp\left(s_2^0/R\right)}{\exp\left(s_1^0/R\right)}$$

where the subscript $s$ on $(P_2/P_1)$ emphasizes the isentropic restriction. The quantity $\exp\left(s^0/R\right)$ is solely a function of the temperature, and is defined as the *relative pressure* $p_r$. Therefore, for an isentropic process of an ideal gas,

$$\left(\frac{P_2}{P_1}\right)_s = \frac{p_{r2}}{p_{r1}} \tag{7-32}$$

The values of the relative pressure $p_r$ may be tabulated for a given gas along with $u$, $h$, and $s^0$ as a function of the temperature. Equation (7-32) serves the same purpose that Eq. (7-28a) does for ideal gases of constant specific heat.

In some cases it is necessary to work with volume ratios instead of pressure ratios for isentropic processes. A relative volume $v_r$ can be defined in a manner similar to that for the relative pressure. By the use of the ideal-gas relation and Eq. (7-32),

$$\frac{v_2}{v_1} = \frac{P_1 T_2}{P_2 T_1} = \frac{P_{r1} T_2}{P_{t2} T_1} = \frac{T_2 \, p_{r1}}{p_{r2} \, T_1}$$

The function $T/p_r$ is solely a function of temperature at a given state, and is defined as the *relative volume* $v_r$. Hence

$$\left(\frac{v_2}{v_1}\right)_s = \frac{v_{r2}}{v_{r1}} \tag{7-33}$$

Equation (7-33) for variable specific heats is equivalent to Eq. (7-27a) for constant specific heats.

It is convenient, then, to tabulate the functions $p_r$ and $v_r$ versus temperature for ideal gases. This has been done over a wide range of temperatures for a limited number of gases of engineering importance. In this text these two functions appear only in the air tables, A-5M and A-5.

**Example 7-16M** Rework Example 7-15M by using relative-pressure data instead of $s^0$ data.

SOLUTION In Table A-5M, at 290°K we find $p_{r1}$ to be 1.2311, and $h_1$ again is 290.16 kJ/kg. The final state is found by

$$P_{r2} = p_{r1}(P_2/P_1) = 1.2311(8.6) = 10.59$$

In Table A-5M, at 530 and 540°K the values of $p_r$ are 10.37 and 11.10, respectively. By linear interpolation $T_{2s}$ is 533°K, which agrees with the previous calculation, using $s^0$ data, in Example 7-15M. As a result the enthalpy change again is 246.6 kJ/kg.

**Example 7-16** Rework Example 7-15 by using relative-pressure data instead of $s^0$ data.

SOLUTION In Table A-5, at 500°R we find that $p_{r1}$ is 1.0590. The final state is found by

$$P_{r2} = p_{r1}(P_2/P_1) = 1.0590(8.6) = 9.106$$

In Table A-5 for air, at 920°R $p_r = 9.102$. Therefore $T_{2s}$ is close to 920°R, in agreement with Example 7.15. As a result the enthalpy change again is 101.7 Btu/lb.

A further example of the use of isentropic relations for ideal gases is in conjunction with equations previously developed for reversible work effects for closed and open systems. For closed-system boundary work

$$w = -\int P \, dv \tag{2-10}$$

and for steady-flow mechanical work

$$w = \int v \, dP + \Delta KE + \Delta PE \tag{7-24}$$

In both of these cases a functional relationship between $P$ and $v$ is required. For an isentropic process where the variation of the specific heats with temperature is small,

$$P(v)^k = \text{constant} \qquad (7\text{-}29b)$$

This latter relation provides a means of evaluating Eqs. (2-10) and (7-24) under adiabatic conditions for ideal gases. By direct substitution of Eq. (7-29b) into these expressions followed by integration, it can be shown that

$$w_{\text{nonflow}} = \frac{P_2 v_2 - P_1 v_1}{k - 1} \qquad (7\text{-}34)$$

$$w_{\text{steady-flow}} = \frac{k(P_2 v_2 - P_1 v_1)}{k - 1} + \Delta\text{KE} + \Delta\text{PE} \qquad (7\text{-}35)$$

These equations give the maximum work output or minimum work input under adiabatic operation. It can also be shown that these equations are in agreement with the respective energy balances for closed and open systems.

## 2 Isentropic Relations for Incompressible Substances

For incompressible substances the entropy change is given by Eq. (7-19), namely, $ds = c\, dT/T$. If the specific heat is relatively constant, the integration of this equation leads to Eq. (7-20), that is

$$\Delta s = c \ln \frac{T_2}{T_1}$$

Consequently, an isentropic process involving an incompressible substance is also isothermal, so that $T_2 = T_1$. This is the only simplification in property relations which results from an isentropic restriction. As an example, the flow of liquids such as water through insulated pipes is close to isothermal. Since pipe flow is not truly isentropic, due to frictional effects, a slight but nearly undetectable temperature rise does occur.

> **Example 7-17M** Reconsider the data of Example 5-9M. However, assume the flow to be adiabatic and internally reversible, so that the frictional head loss is zero. Determine the required power to the pump, in kilowatts.
>
> SOLUTION The figure in Example 5-9M shows a schematic of the flow system and important input data. The conservation of energy principle for the control volume, which includes the fluid in the storage tank and within the piping system, is
>
> $$q + w_{\text{shaft}} = \Delta u + \Delta Pv + \Delta\text{KE} + \Delta\text{PE}$$
>
> The value of $q$ is zero, and since the process is isentropic the temperature is constant if the fluid is assumed incompressible. But the internal energy change of an incompressible fluid is solely a function of temperature. Hence, $\Delta u = 0$. The energy equation then reduces to
>
> $$w_{\text{shaft}} = \Delta Pv + \Delta\text{KE} + \Delta\text{PE}$$

In Example 5-9M, however, it is shown that $\Delta Pv = 0$, as well. Using the values of the kinetic- and potential-energy changes found in Example 5-9M,

$$w_{shaft} = \frac{(3.1)^2}{2} + 15(9.8) = 151.8 \text{ N·m/kg}$$

Since the mass flow rate was found to be 9.05 kg/s, the power required is

$$\dot{W} = 151.8 \text{ N·m/kg} \times 9.05 \text{ kg/s} = 1374 \text{ W} = 1.37 \text{ kW}$$

This is 86 percent of that required when friction is taken into account.

**Example 7-17** Reconsider the data of Example 5-9. However, assume the flow to be adiabatic, internally reversible, and incompressible. Determine the required power to the pump, in horsepower.

SOLUTION The figure in Example 5-9 shows a schematic of the flow system and important input data. The conservation of energy principle for the control volume, which includes the fluid in storage tank and within the piping system, is

$$q + w_{shaft} = \Delta u + \Delta Pv + \Delta KE + \Delta PE$$

The value of $q$ is stated to be zero, and for isentropic flow of an incompressible fluid, the temperature is constant. This requires that $\Delta u$ also be zero. Hence the energy equation reduces to

$$w_{shaft} = \Delta Pv + \Delta KE + \Delta PE$$

In Example 5-9, however, it was determined that $\Delta Pv$ is also zero, by choice of the control volume. Consequently, using the values the $\Delta KE$ and $\Delta PE$ found in Example 5-9,

$$w_{shaft} = \frac{(9)^2}{2(32.2)} + 50 = 51.3 \text{ ft-lb}_f/\text{lb}_m$$

Since the mass flow rate was found to be 18.3 $\text{lb}_m/\text{s}$, the power required is

$$\dot{W} = 51.3 \text{ ft·lb}_f/\text{lb}_m \times 18.3 \text{ lb}_m/\text{s} \times \frac{\text{hp·s}}{550 \text{ ft·lb}_f} = 1.71 \text{ hp}$$

This is 84 percent of that required when friction is taken into account.

# 3 Isentropic Process Evaluation Using Superheat and Saturation Data

There are no special relations for the evaluation of isentropic changes for fluids in the superheat or saturation region, other than $s_1 = s_2$. However, this process information, in conjunction with data on the initial and final states, is usually sufficient. An appropriate use of an energy balance or the continuity-of-flow equation may also be necessary to complete the solution. Several examples of this type of evaluation are provided below.

**Example 7-18M** Refrigerant 12 passes adiabatically through a nozzle in steady flow until its pressure reaches 2 bars. At the inlet to the nozzle the pressure and temperature are 9 bars and 100°C, respectively. Under reversible-flow conditions in the nozzle, determine the exit velocity in m/s if the inlet velocity is small.

SOLUTION If the flow is reversible and adiabatic, it is also isentropic. Thus the inlet and outlet entropy values are equal. The final velocity may be found from an energy balance on the nozzle; that is, since $q$ and $w$ are zero,

$$h_2 - h_1 = \frac{V_1^2 - V_2^2}{2}$$

$V_1$ is small and will be neglected. $V_2$ can be found from the equation if $h_2$ can be found. From Table A-18M, the value of $h_1$ is 248.54 kJ/kg and that of $s_1$ is 0.8190 kJ/(kg)(°K). The final state is determined from the values of $P_2$ and $s_2$. At 2 bars the value of $s_g$ is 0.7035 kJ/(kg)(°K), so the final state is superheated vapor also. From Table A-18M, it is found that at 2 bars $s = 0.8184$ kJ/(kg)(°K) at 40°C. This is close to the value of $s_1$ which equals $s_2$. At this state of 40°C and 2 bars, $h_2 = 214.97$ J/g. Substitution of these $h$ values into the energy balance leads to

$$214.97 - 248.54 = -\frac{V_2^2}{2(1000)}$$

where the factor 1000 converts kilograms to grams. Thus

$$V_2 = (67,140)^{1/2} = 259 \text{ m/s}$$

**Example 7-18** Refrigerant 12 passes adiabatically through a nozzle in steady flow until its pressure reaches 30 psia. At the inlet to the nozzle the pressure is 120 psia and the temperature is 220°F. Under reversible-flow conditions in the nozzle, determine the exit velocity in ft/s.

SOLUTION If the flow is reversible and adiabatic, it is also isentropic. Thus the inlet and exit entropy values are equal. The final velocity can be found from the energy balance on the nozzle; that is,

$$h_2 - h_1 = \frac{V_1^2 - V_2^2}{2}$$

Since no information is given on the inlet velocity, we shall assume that it is negligible. The final and initial enthalpies can be determined by knowing any two intrinsic properties at those states. From Table A-18, the refrigerant 12 superheat table, it is found that $h_1 = 108.509$ Btu/lb and $s_1 = 0.1993$ Btu/(lb)(°R). The final state is determined from the values of $P_2$ and $s_2$. At 30 psia and $s_2 = s_1$, it is found that the final temperature is very close to 120°F. As an approximation, the final enthalpy is then 94.843 Btu/lb. Substitution of these values in the energy balance leads to

$$94.843 - 108.509 = -\frac{V_2^2}{2(32.2)(778)}$$

or

$$V_2 = 827 \text{ ft/s}$$

**Example 7-19M** Steam enters a turbine at 30 bars, 500°C, and 70 m/s. It expands through a pressure ratio of 10:1 and leaves at a velocity of 140 m/s. If the process is adiabatic and internally reversible, calculate the work output in kJ/kg.

SOLUTION A steady-flow energy balance on the adiabatic turbine shows that

$$w = h_2 - h_1 + \frac{V_2^2 - V_1^2}{2}$$

The potential-energy change, if any, is neglected. Since the velocities are given, the problem is to evaluate $h_2 - h_1$. The value of $h_1$ is found from knowledge of $P_1$ and $T_1$. From steam table A-14M, for the initial state:

$$P_1 = 30 \text{ bars} \qquad h_1 = 3456.5 \text{ kJ/kg}$$
$$T_1 = 500°C \qquad s_1 = 7.2338 \text{ kJ/(kg)(°K)}$$

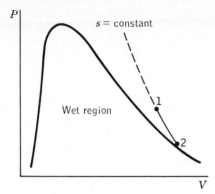

$P$

$s = \text{constant}$

Wet region

1

2

$V$

**Figure 7-19** A $PV$ process diagram for Examples 7-19M and 7-19.

The process is adiabatic and reversible and hence, isentropic. The specific entropy of a unit mass flowing through the turbine is constant. For the final state, then, the pressure is 3 bars and the entropy is 7.2338 kJ/(kg)(°K). The saturated-vapor entropy at 3 bars is given as 6.9919 kJ/(kg) × (°K). Therefore the fluid is slightly superheated on leaving the turbine. Entering Table A-14M again, we find that at 3 bars we must interpolate between 160 and 200°C. Such interpolation shows a turbine outlet temperature of roughly 183°C with an enthalpy of 2830 kJ/kg. Consequently the work done for the isentropic process is

$$w = 2830 - 3456 + \frac{(140)^2 - (70)^2}{2(1000)}$$

$$= -626 + 7.4 = -619 \text{ kJ/kg}$$

The process is illustrated on the $Pv$ diagram in Fig. 7-19. This is the maximum work output for an adiabatic turbine operating under the specified conditions.

**Example 7-19** Steam enters a turbine at 600 psia, 900°F, and 250 ft/s. It expands through a pressure ratio of 10 : 1 and leaves at a velocity of 500 ft/s. If the process is adiabatic and internally reversible, calculate the work output per pound of steam.

SOLUTION A steady-flow energy balance on the turbine shows that

$$w = h_2 - h_1 + \frac{V_2^2 - V_1^2}{2}$$

Heat transfer is zero, and the potential-energy change, if any, is negligible. Since the velocities are given, the problem is to evaluate $h_2 - h_1$. Two properties are required to find a third property in the superheat table. From steam table A-4, for the initial state:

$$P_1 = 600 \text{ psia} \qquad h_1 = 1462.9 \text{ Btu/lb}$$
$$T_1 = 900°F \qquad s_1 = 1.6766 \text{ Btu/(lb)(°R)}$$

The process is adiabatic and reversible and hence, isentropic. The specific entropy of each unit mass flowing through the turbine is constant. For the final state, pressure is 60 psia and the entropy is 1.6766 Btu/(lb)(°R). The saturated-vapor entropy at 60 psia (293°F) is given as 1.6443 Btu/(lb)(°R). Therefore the vapor is slightly superheated on leaving the turbine. Entering Table A-14 again, one finds that, at 60 psia and 350°F, $s$ is listed as 1.6830 Btu/(lb)(°R). At 60 psia and

300°F $s$ is 1.6496 Btu/(lb)(°R). Interpolation shows a turbine outlet temperature of 340°F with an enthalpy of 1203.0 Btu/lb. Consequently the work done per pound for the isentropic process is

$$w = 1203.0 - 1462.9 + \frac{(500)^2 - (250)^2}{2(32.2)(778)}$$

or
$$w = -256.2 \text{ Btu/lb}$$

The process is illustrated on the accompanying $PV$ diagram (Fig. 7-19).

The discussion in this section has centered on isentropic behavior of substances in various closed and open devices. Although no process in practice is isentropic, it is an extremely important concept with respect to devices that are essentially adiabatic. Since internally reversible processes by concept are free of nonequilibrium and dissipative effects, the isentropic performance of a device may be used as a standard to which real behavior is compared. Isentropic conditions lead to a *maximization* or *minimization* of important system variables, such as work effects or final state properties such as velocity, temperature, or pressure. In a later section of this chapter we shall examine the effect of irreversibilities on adiabatic devices. In this context the isentropic behavior of devices is also important.

## 7-10 THE ENTHALPY-ENTROPY DIAGRAM

In the analysis of steady-flow processes an enthalpy-entropy diagram, as well as a $Ts$ plot, is quite useful. The enthalpy is the important thermostatic property in the steady-flow energy balance, and entropy is the principal property of concern with respect to the second law. Thus the coordinates of an $hs$ diagram represent the two major properties of interest in the first- and second-law analysis of open systems. The vertical distance between two states on this diagram is a measure of $\Delta h$. The enthalpy change, in turn, is related through the adiabatic steady-flow energy balance to the work and/or kinetic-energy changes for turbines, compressors, nozzles, etc. The horizontal distance between two states $\Delta s$ is a measure of the degree of irreversibility for an adiabatic process. As a consequence we shall find the $hs$ diagram helpful in visualizing process changes for control-volume analyses.

In addition to its use in process visualization, the $hs$ diagram is also a means of presenting data. Such data can be read with reasonable accuracy if the diagram is drawn to a suitable size or scale. An $hs$ diagram for steam is included in the Appendix as Fig. A-25. A schematic of an $hs$ diagram, commonly called a *Mollier* diagram, is shown in Fig. 7-20. On an $hs$ plot, constant-pressure lines and constant-temperature lines are straight in the wet (liquid-vapor) region. Lines of constant quality within the wet region lie roughly parallel to the saturated-vapor line. (On some Mollier charts lines of constant quality are marked as percent-moisture lines. Percent moisture means the percent of liquid in the liquid-vapor mixture. The saturated-vapor line, for example, is a 0 percent–moisture line.) Another feature of this diagram is that constant-temperature lines become hori-

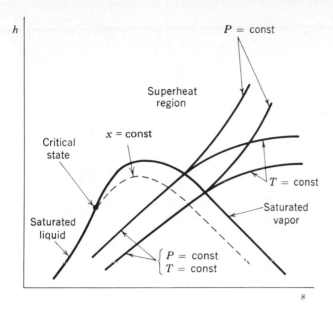

**Figure 7-20** Schematic of a Mollier diagram.

zontal in the superheat region, which lies on the far right side of the plot. As the pressure is lowered on a gas at constant temperature, the gas behaves more like an ideal gas. For ideal gases the temperature and enthalpy are directly proportional. Hence, at low enough pressures, the temperature lines should lie parallel to the enthalpy lines. As an example, below roughly 40°C or 100°F steam behaves as an ideal gas at all pressures up to the saturation pressure for a given temperature. Finally, note that constant-volume lines are not normally shown on this type of diagram.

It has been pointed out that in certain parts of the superheated-vapor region on an $hs$ diagram the temperature and enthalpy lines are parallel. This means that an $hs$ and a $Ts$ diagram for an ideal gas may be superimposed upon each other. This technique of letting the ordinate of a plot represent both $T$ and $h$ is quite useful for visualization of steady-flow processes. This is especially true when one is analyzing in a qualitative sense the effect of irreversibilities on performance.

## 7-11 EFFICIENCIES OF SOME STEADY-FLOW DEVICES

Dissipative effects inherently accompany the flow of real fluids through steady-flow devices. These irreversibilities downgrade the performance of the device and lead to an increase in the entropy of the fluid. The magnitude of the entropy change is in proportion to the degree of the irreversibility. From an engineering viewpoint it is desirable to have parameters which measure the performance of individual types of equipment in terms of energy degradation. For heat engines, which are cyclic devices, the thermal efficiency is a practical measure of performance. However, values of such efficiencies are meaningless unless they are compared to the theoretical idealized performance. In this case we found that the Carnot thermal efficiency was one standard against which real cyclic heat engines might be compared.

In order to develop parameters for the comparative performance of specific types of steady-flow devices, we must recognize that the actual flow through many of these devices is approximately adiabatic. Therefore the idealized performance of such equipment occurs when the flow is adiabatic and internally reversible, that is, isentropic. An appropriate measure of their actual performance is a comparison of the actual desired output to the theoretical isentropic output. This ratio is defined as the *adiabatic* or *isentropic* efficiency of the steady-flow device.

The purpose of a turbine is to produce useful work output. Therefore the isentropic or adiabatic turbine efficiency $\eta_T$ is defined as

$$\eta_T \equiv \frac{w_a}{w_s} \qquad \text{Turbine} \qquad (7\text{-}36)$$

where the subscripts $a$ and $s$ again represent the actual and the isentropic flows, respectively. Under adiabatic conditions and with a negligible potential energy change, the conservation of energy principle for a unit mass passing through a control volume is

$$w_{\text{shaft}} = \Delta h + \Delta \text{KE} \qquad (7\text{-}37)$$

When the kinetic-energy change across a turbine is small, the turbine efficiency may be approximated by

$$\eta_T = \frac{(h_1 - h_2)_a}{(h_1 - h_2)_s} \qquad \text{Turbine} \qquad (7\text{-}38)$$

where 1 and 2 represent the inlet and outlet states. The value of $\eta_T$ for actual machines ranges from 80 to 90 percent.

A nozzle is a flow channel constructed to accelerate the fluid to a higher velocity. Therefore, the adiabatic or isentropic nozzle efficiency $\eta_N$ is defined as

$$\eta_N \equiv \frac{V_a^2/2}{V_s^2/2} \qquad \text{Nozzle} \qquad (7\text{-}39)$$

where $V_a^2/2$ is the actual kinetic energy at the nozzle exit, and $V_s^2/2$ is the kinetic energy at the nozzle exit for isentropic flow to the same exit pressure. Nozzle efficiencies generally range upward from 90 percent. For converging nozzles used in subsonic flow, efficiencies of 0.95 or higher are quite common. If the inlet nozzle velocity is small compared to $V_a$ or $V_s$ at the exit, then Eq. (7-37) permits the nozzle efficiency to be expressed also by

$$\eta_N = \frac{(h_1 - h_2)_a}{(h_1 - h_2)_s} \qquad \text{Nozzle} \qquad (7\text{-}40)$$

This expression is analogous to Eq. (7-38) for a turbine.

Although turbines typically are insulated to prevent a heat loss from the working fluid, in some cases a deliberate attempt is made to cool the fluid passing through a compressor. For conditions where the heat loss is essentially zero, the adiabatic or isentropic compressor efficiency $\eta_C$ is defined as the ratio of the isentropic work required to the actual work required for the same pressure rise. That is,

$$\eta_C \equiv \frac{w_s}{w_a} \qquad \text{Compressor} \qquad (7\text{-}41)$$

When the kinetic-energy change across the compressor is negligible, Eq. (7-37) permits us to write Eq. (7-41) as

$$\eta_C = \frac{(h_2 - h_1)_s}{(h_2 - h_1)_a} \qquad \text{Compressor} \qquad (7\text{-}42)$$

where 1 and 2 represent the inlet and outlet states. The value of $\eta_C$ ranges roughly from 75 to 85 percent for actual gas compressors.

## 7-12 IRREVERSIBLE ADIABATIC PROCESSES IN STEADY-STATE DEVICES

One of the major engineering uses of the second law is the prediction of the performance of steady-state flow devices when internal irreversibilities are present. It has already been demonstrated in Sec. 7-8 that the performance of adiabatic shaft-work devices is degraded by the presence of irreversibilities. Isentropic processes lead to the maximum work output or minimum work input. It is essential for us to show that isentropic processes are also standards of performance for adiabatic nonwork devices. In addition, the effects of irreversibilities on changes in various fluid properties need to be established. The following discussion is divided into two classes. First we shall analyze incompressible-flow problems, and this will be followed by a discussion of compressible flow in steady-state devices. In the latter case, attention will be specifically directed toward the use of ideal gases, since the property relationships are relatively simple and the results are generally applicable to real gases.

## 1 Incompressible Fluids

The basic conservation of energy principle for control volumes operating in steady state is

$$q + w_{\text{shaft}} = \Delta h + \Delta KE + \Delta PE \qquad (5\text{-}29)$$

In the applications to be studied here, the device is assumed to be adiabatic and the effect of gravitational potential-energy changes is assumed negligible. The enthalpy and entropy changes of an incompressible fluid are given by

$$\Delta h = c\,\Delta T + v\,\Delta P \tag{4-13}$$

and
$$\Delta s = c\ln\frac{T_2}{T_1} \tag{7-20}$$

The energy equation for the control volume becomes

$$w_{\text{shaft}} = c\,\Delta T + v\,\Delta P + \Delta\text{KE} \tag{7-43}$$

Equations (7-20) and (7-43) will now be applied to various flow devices. It is also true that $\Delta S_{\text{adia}} \geq 0$.

**a Turbines** In the analysis of incompressible flow through turbines, it is acceptable as a first approximation to neglect kinetic-energy changes. Hence the energy equation in this case reduces to

$$w_{\text{shaft}} = c\,\Delta T + v\,\Delta P \tag{7-44}$$

In the limiting case of isentropic flow, Eq. (7-20) dictates that $T_2 = T_1$, since $\Delta s = 0$. If the process is isothermal, then the internal-energy term $c\,\Delta T$ is zero. As a result of adiabatic, internally reversible flow, the shaft work is equal to the change in the displacement work. That is,

$$w_{\text{shaft}} = v(P_2 - P_1)$$

This same result could have been obtained from the expression for steady-flow mechanical work developed in Sec. 7-8. Recall that for frictionless processes,

$$w_{\text{shaft}} = \int v\,dP + \Delta\text{KE} + \Delta\text{PE} \tag{7-24}$$

If we again neglect kinetic- and potential-energy changes, then the shaft work is the integral of $v\,dP$. But $v$ is a constant, and so the integral becomes equal to $v(P_2 - P_1)$, which agrees with the previous derivation.

In the case of adiabatic, internally irreversible flow the entropy must increase. Equation (7-20) now requires that $T_2 > T_1$. The effect of this temperature rise on the work output is easily seen if we rearrange Eq. (7-44) to the form

$$v(P_1 - P_2) = -w_{\text{shaft}} + c(T_2 - T_1)$$

The pressure drop $(P_1 - P_2)$ is positive and is fixed by the inlet and outlet conditions. The shaft-work output is positive, and the internal-energy change is positive due to the temperature rise required by the second law. With irreversibilities present, the change in the displacement work goes to increasing the temperature of the fluid as well as producing new work output. Greater irreversibilities within the control volume lead to a larger temperature rise, with a concomitant decrease in the net work output. This is an example of mechanical dissipation of energy into a less useful form, an internal-energy increase.

**b Pumps** The basic equations for the analysis of incompressible flow through a pump are exactly the same as for a turbine, namely,

$$\Delta s = c \ln \frac{T_2}{T_1} \tag{7-20}$$

and

$$w_{\text{shaft}} = c \, \Delta T + v \, \Delta P \tag{7-44}$$

For isentropic flow the temperature remains constant and the shaft work again is related solely to the displacement work. The given shaft-work input goes to increasing the pressure of the fluid, with no other effect.

If the process is adiabatic and internally irreversible, the increase in entropy leads to a rise in temperature of the fluid, as given by Eq. (7-20). The immediate effect on the pressure rise of the fluid is seen from Eq. (7-44). For a given work input, the added energy is split between an increase in internal energy and an increase in displacement work. For the same work input, the pressure rise is greater for the isentropic case than for the irreversible process.

**Example 7-20M** Water enters a pump at 1 bar and 30°C. Shaft work is done on the fluid in the amount of 4500 N·m/kg.
(a) Determine the pressure rise if the process is isentropic.
(b) Determine the pressure rise if the fluid temperature rises by 0.1°C during the process.

SOLUTION (a) For an isentropic process, $w_{\text{shaft}} = v \, \Delta P$ for an incompressible substance. Solving for the pressure rise, we find that

$$\Delta P = \frac{w_{\text{shaft}}}{v} = \frac{4500 \text{ N·m}}{\text{kg}} \times \frac{1 \text{ g}}{\text{cm}^3} \times \frac{10 \text{ cm}^3 \cdot \text{bar}}{\text{N·m}} \times \frac{\text{kg}}{1000 \text{ g}} = 45 \text{ bars}$$

(b) For the irreversible process with a temperature rise of 0.1°C, Eq. (7-44) must be employed. Substitution of the proper values yields, when each term is expressed in N·m/g,

$$w_{\text{shaft}} = c \, \Delta T + v \, \Delta P$$

$$\frac{4500}{1000} = 4.18(0.1) + \frac{1.0(P)}{10}$$

$$\Delta P = 10(4.50 - 0.42) = 40.8 \text{ bars}$$

When water is compressed from atmospheric conditions to around 40 or 50 bars, every 0.1°C rise in temperature due to irreversibilities results in roughly a 10 percent reduction in the pressure rise under adiabatic flow.

**Example 7-20** Water enters a pump at 15 psia and 100°F. Shaft work is done on the fluid in the amount of 1550 ft·lb$_f$/lb$_m$.
(a) Determine the pressure rise if the process is isentropic.
(b) Determine the pressure rise if the fluid temperature rises by only 0.2°F.

SOLUTION (a) For an isentropic process, $w_{\text{shaft}} = v \, \Delta P$. Solving for the pressure rise, we find that

$$\Delta P = \frac{w_{\text{shaft}}}{v} = \frac{1550 \text{ ft·lb}_f}{\text{lb}_m} \frac{\text{lb}_m}{0.01613 \text{ ft}^3} \frac{\text{ft}^2}{144 \text{ in}^2} = 667 \text{ psi}$$

(b) For the irreversible process with a temperature rise of $0.2°F$, Eq. (7-44) must be employed. Substitution of the proper values yields

$$w_{shaft} = c\,\Delta T + v\,\Delta P$$

$$1550 \text{ ft·lb}_f/\text{lb}_m = 1.0 \text{ Btu}/(\text{lb}_m)(°F) \times 0.2°F \times 778 \text{ ft·lb}_f/\text{Btu} + 0.01613 \text{ ft}^3/\text{lb}_m \times \Delta P$$

$$\Delta P = \frac{1550 - 155}{0.01613} = (667 - 66.7) \text{ psi} = 600 \text{ psi}$$

The results of this calculation are quite revealing. When water is compressed from atmospheric conditions to around 600 to 700 psia, every $0.2°F$ rise in temperature due to irreversibilities results in a 10 percent reduction in the pressure rise under adiabatic conditions.

**c Nozzles** For flow through adiabatic nozzles, the basic equations for an incompressible fluid are

$$-\Delta h = \Delta KE$$

or

$$(u_1 - u_2) + v(P_1 - P_2) = \Delta KE \tag{7-45}$$

and

$$s_2 - s_1 = c \ln \frac{T_2}{T_1} \tag{7-20}$$

Under the conditions of isentropic flow the temperature change is zero. Hence the internal-energy change is also zero, and the energy equation for the control volume reduces to

$$v(P_1 - P_2) = \Delta KE$$

This result simply states that the desired kinetic-energy change through the nozzle is due solely to the change in displacement work if the flow is incompressible and isentropic.

In the case of internal irreversibilities, which lead to a rise in temperature of the incompressible fluid, the change in kinetic energy is expressed by a rearranged form of Eq. (7-45), i.e.,

$$v(P_1 - P_2) = \Delta KE + (u_2 - u_1)$$

$$= \Delta KE + c(T_2 - T_1)$$

We see that the increase in internal energy decreases the kinetic-energy change for a given inlet and outlet pressure. Dissipation of some of the displacement work increases the internal energy of the fluid. As a result, the kinetic-energy change for irreversible flow is less than that for isentropic flow.

**d Diffusers** The purpose of a diffuser is to decelerate the flow and increase the pressure of the fluid. The basic equations are the same as for a nozzle, and are given by Eqs. (7-20) and (7-45). If we assume that the final velocity is negligible compared to the initial velocity, then Eq. (7-45) can be written as

$$c(T_1 - T_2) + \frac{V_1^2}{2g_c} = v(P_2 - P_1)$$

In isentropic flow the first term on the left is zero. Therefore $P_2$ is determined solely by the values of $P_1$ and $V_1$. When internal irreversibilities are present, $T_2 > T_1$. Hence the first term on the left is negative, and the pressure rise will be less than that for isentropic flow. Part of the available kinetic energy at the inlet appears as a temperature rise instead of a pressure rise.

**Example 7-21M** Determine the maximum pressure rise in a diffuser if liquid water enters at 1 bar, 30°C, and 7 m/s, and leaves at a negligible velocity.

SOLUTION The maximum pressure rise is obtained if the process is isentropic. For isentropic flow of an incompressible fluid, $T_1 = T_2$, and the energy equation for the control volume is

$$\frac{V_1^2}{2} = v(P_2 - P_1)$$

Therefore, when both terms are expressed in N·m/g,

$$\frac{(7)^2}{2(1000)} = \frac{1}{10} \Delta P$$

$$\Delta P = 0.245 \text{ bar}$$

**Example 7-21** Determine the maximum pressure rise in a diffuser if water enters at 15 psia, 100°F, and 20 ft/s, and leaves at a negligible velocity.

SOLUTION The maximum pressure rise is obtained if the process is isentropic. Under this condition $T_1 = T_2$ for an incompressible fluid, and the energy equation for the control volume becomes

$$\frac{V_1^2}{2} = v(P_2 - P_1)$$

Therefore, when each term is expressed in ft lb$_f$/lb$_m$,

$$\frac{(20)^2}{2(32.2)} = 0.01613(144) \Delta P$$

or

$$\Delta P = 2.67 \text{ psi}$$

**e Throttling devices** A throttling device has been described earlier as a flow restriction which leads to a drop in pressure of the fluid. No interactions occur to the fluid, and kinetic- and potential-energy changes are negligible. Therefore $h_1 = h_2$ if the inlet and outlet are selected to be relatively far upstream and downstream from the flow restriction. No proof of the drop in pressure across these devices has yet been shown. The proof stems from the second law. Splitting the enthalpy function back into its two contributions of internal energy and displacement work, we see that the statement $h_1 = h_2$ can be written as

$$u_2 - u_1 = v(P_1 - P_2)$$

or

$$c(T_2 - T_1) = v(P_1 - P_2)$$

For incompressible fluids,

$$\Delta s = c \ln \frac{T_2}{T_1}$$

If the flow is assumed to be isentropic, then $T_2 = T_1$ on the basis of the $\Delta s$ expression. The energy relation then requires that $P_1 = P_2$. Hence no properties change in the direction of flow, and flow itself cannot occur. The flow must be irreversible. In this case $T_2 > T_1$ since the entropy increases. The energy relation now requires that $P_2 < P_1$, which is the result we expected but had not yet proved. The flow accelerates and the pressure increases as the fluid reaches the restriction. At the outlet of the restriction the fluid decelerates to the original velocity, so that the original enthalpy is recovered. However, the original pressure is not recovered. This is offset by a rise in temperature of the fluid.

## 2 Compressible Fluids

The basic conservation of energy statement for a control volume operating in steady state is

$$q + w_{shaft} = \Delta h + \Delta KE + \Delta PE$$

We shall assume that heat interactions and potential-energy changes are negligible. In addition we shall assume that the fluid behaves as an ideal gas. This provides us with some fairly simple property relationships, while at the same time leading to results that are typical of nonideal gases as well. The energy equation will have the form

$$w_{shaft} = \Delta h + \Delta KE \tag{7-37}$$

Ideal-gas relationships of interest to us include

$$Pv = RT$$

$$\Delta u = c_v \, \Delta T$$

$$\Delta h = c_p \, \Delta T$$

$$\Delta s = c_p \ln \frac{T_2}{T_1} - R \ln \frac{P_2}{P_1} \tag{7-17}$$

Note that we shall also assume constant specific heats.

**a Turbines** For adiabatic turbines involving ideal gases the conservation of energy equation, Eq. (7-37), reduces to

$$w_{shaft} = c_p(T_2 - T_1)$$

if we neglect the kinetic-energy changes for a first approximation. The equation for the entropy change can be written as

$$\Delta s + R \ln \frac{P_2}{P_1} = c_p \ln \frac{T_2}{T_1}$$

The fixed conditions for a turbine include $P_1$, $T_1$, and $P_2$. When $\Delta s$ is zero (isentropic), $T_2$ is fixed, which in turn fixes the shaft-work output. The isentropic

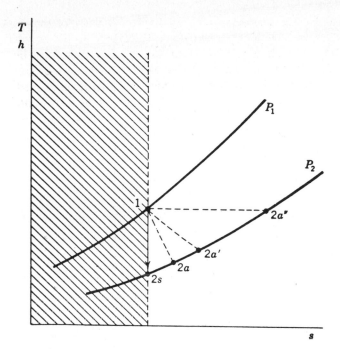

**Figure 7-21** Effect of irreversibilities on adiabatic turbine performance.

process from state 1 to state 2s is shown in Fig. 7-21. Note that the ordinate on the diagram is either $T$ or $h$. This is possible since $h$ is proportional to $T$ for an ideal gas. Thus we are plotting a $Ts$ and an $hs$ diagram simultaneously. Recall that $w_{shaft} = \Delta h$. Therefore the vertical distance between any two states on the diagram represents the shaft work on an $hs$ diagram.

For the internally irreversible process, $\Delta s$ is positive. From the preceding equation for $\Delta s$ it is seen that as $\Delta s$ increases, $T_2$ must increase for a given $T_1$, $P_1$, and $P_2$. Therefore the final state $2a$ for an irreversible process lies up the $P_2$ line at a higher $T$ and higher $s$. This also means that state $2a$ has a greater enthalpy than state $2s$. Consequently the shaft-work output is less. State $2a'$ represents the final state for a more irreversible process than that represented by $2a$. In the limit, state $2a''$ would be the final state, and for the given pressure drop $\Delta h$ would be zero. The shaft work also would be zero. In this extreme case the turbine acts like a throttling valve. This would be a highly undesirable method of operation.

**Example 7-22M** For the problem in Example 7-19M, consider the case where the internal irreversibilities reduce the work output to 85 percent of the reversible case. For the same pressure ratio determine (a) the final temperature, and (b) the final specific entropy.

SOLUTION The actual work output for the irreversible expansion is simply the reversible work times the turbine efficiency. From Example 7-19M, $w_{rev} = -619$ kJ/kg

$$w_{irr} = \eta_T w_{rev} = 0.85(-619) = -526 \text{ kJ/kg}$$

Although it is not quite correct, since the outlet specific volume will change for the irreversible case, we shall assume that the final velocity remains the same. Hence $\Delta KE$ is still 7.4 kJ/kg. The new exit enthalpy is found from the first law; i.e., $w = \Delta h + \Delta KE$. Thus

$$-526 = (h_{2a} - 3456.5) + 7.4$$

Therefore,

$$h_{2a} = 3456.5 - 526 - 7.4 = 2923 \text{ kJ/kg}$$

From Table A-14M, at 3 bars this enthalpy value is found to lie between 200 and 240°C. By linear interpolation the final temperature is roughly 228°C and the specific entropy is 7.429 kJ/(kg)(°K). For the adiabatic and irreversible expansion the final temperature is 45°C higher and the specific entropy is 0.195 kJ/(kg)(°K) higher than for the reversible process in Example 7-19M. This increase in entropy for the adiabatic irreversible process is in accord with the second law.

**Example 7-22** For the problem in Example 7-19 consider the case where internal irreversibilities reduce the work to 85 percent of the reversible case. For the same pressure ratio determine (a) the final temperature and (b) the final specific entropy.

SOLUTION The actual work for the irreversible expansion is simply the reversible work times the turbine efficiency. So

$$w_{irr} = w_{rev}\eta_T$$

$$= -256.2(0.85) = -217.8 \text{ Btu/lb}$$

If the final velocity is assumed the same, the new enthalpy drop across the turbine is found from the first law; i.e.,

$$-217.8 = (h_2 - 1462.9) + 3.7$$

Therefore

$$(h_2)_{irr} = 1462.9 - 217.8 - 3.7 = 1241.4 \text{ Btu/lb}$$

By interpolation in superheat table A-14 for $P_2 = 60$ psia and $h_2 = 1241.4$ Btu/lb, the final temperature is, roughly, 415°F and the specific entropy is 1.723 Btu/(lb)(°R). For the adiabatic irreversible expansion the final temperature is 75°F higher and the specific entropy is 0.046 Btu/(lb)(°R) higher than for the reversible work process. The second law requires that the specific entropy increase for an adiabatic irreversible process on a fixed mass, as was demonstrated by this specific example.

**b Compressors** The basic equations for the study of compressor operation are the same as those used for turbine analysis. However, there are two different modes of operation we need to examine. Consider first the case where the shaft-work input to the compressor is fixed. We wish to determine the compressor outlet pressure in this situation. The energy equation, neglecting kinetic-energy changes, is

$$w_{shaft} = \Delta h = c_p(T_2 - T_1)$$

We note for a given compressor inlet temperature and given shaft-work input that the outlet temperature is fixed. This is true whether irreversibilities are present or not. The outlet pressure is determined from second law considerations. Rearranging Eq. (7-17), we find that

$$R \ln \frac{P_2}{P_1} = c_p \ln \frac{T_2}{T_1} - \Delta s$$

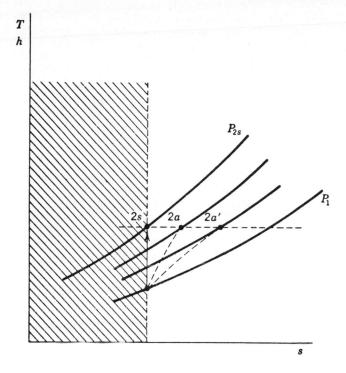

**Figure 7-22** Effect of irreversibilities on adiabatic compressor-outlet pressure for fixed work input.

Under isentropic conditions $P_2$ is uniquely determined. This state, designated by $2s$, is shown in Fig. 7-22. The isentropic process, $1$-$2s$, again is shown on a composite $Ts$ and $hs$ diagram. The vertical distance on the $hs$ plot represents the shaft-work input.

Now let us consider irreversibilities. $T_2$ remains the same from energy considerations. Since $\Delta s$ is positive, we see from the above equation that for a given $T_1$, $P_1$, and $T_2$ the value of $P_2$ is less than for the isentropic case. On Fig. 7-22 several lines of pressure lower than $P_{2s}$ have been plotted. State $2a$ represents one possible final state when irreversibilities are present. As the extent of irreversibilities increases, $\Delta s$ increases and $P_2$ decreases. State $2a'$ is the result of greater irreversibilities.

A second important type of compressor operation is the case where a definite final pressure $P_2$ is required. In refrigeration cycles this final pressure determines the temperature at which the refrigerant condenses. Thus close tolerances must be maintained on the compressor exit pressure. In this case $P_1$, $P_2$, and $T_1$ are given data. If we again rearrange Eq. (7-17), then

$$c_p \ln \frac{T_2}{T_1} = R \ln \frac{P_2}{P_1} + \Delta s$$

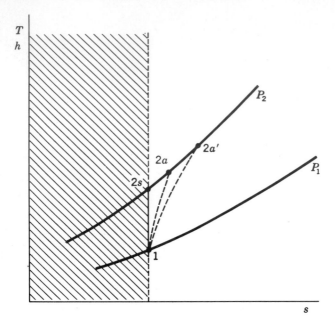

**Figure 7-23** Effect of irreversibilities.

In the isentropic situation, $T_2$ is fixed by the given conditions. The isentropic process is shown in Fig. 7-23 where again $T$ and $h$ appear on the ordinate. State $2s$ again is the isentropic final state.

For the irreversible case, the preceding equation shows that $T_2$ increases as the entropy increases. States $2a$ and $2a'$ show this effect. The major consequence, however, is the large change in shaft work required. Based on $w_{shaft} = \Delta h$, we see that there are sizable increases in shaft work as the degree of irreversibility increases. More shaft work is required to attain the same required pressure. The isentropic process is an obvious standard of performance.

**Example 7-23M** Nitrogen is compressed adiabatically from 1.20 bar, 290°K, to a final pressure of 5.18 bars.
(a) If the final temperature is 480°K, how many degrees of the actual temperature rise is due to internal irreversibilities within the compressor?
(b) What is the adiabatic efficiency of the compressor?

SOLUTION (a) In order to find the $\Delta T$ due to irreversibilities, we must first find the temperature rise in the absence of irreversibilities. Isentropic compression will give the minimum temperature rise, as shown on the $Ts$ diagram in Fig. 7-24.
$T_{2s}$ can be found by employing the $s^0$ data in Table A-6M. At $T_1 = 290°K$, $s_1^0 = 190.695$ kJ/(kg·mol)(°K). Therefore,

$$s_2^0 = s_1^0 + R \ln \frac{P_2}{P_1} = 190.695 + 8.315 \ln (5.18/1.20)$$

$$= 202.856 \text{ kJ/(kg·mol)(°K)}$$

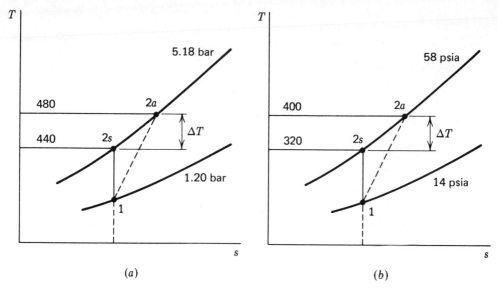

**Figure 7-24** Process diagrams for (a) Example 7-23M, and (b) Example 7-23.

In Table A-6M this value of $s^0$ corresponds closely to a temperature of 440°K. Hence the temperature rise due to internal irreversibilities is $480 - 440 = 40°C$.

An alternate solution would be to make use of the isentropic relation $T_2/T_1 = (P_2/P_1)^{(k-1)/k}$. This approach requires a guess as to a proper value of $k$ to use, since $T_2$ is unknown. The method used above avoids this difficulty.

(b) Neglecting kinetic-energy changes, shaft work equals $\Delta h$. The enthalpies for the problem are $h_1 = 8432$ kJ/kg·mol, $h_{2s} = 12,811$ kJ/kg·mol, and $h_{2a} = 13,988$ kJ/kg·mol. Thus

$$\eta_C = \frac{h_{2s} - h_1}{h_{2a} - h_1} = \frac{12,811 - 8,432}{13,988 - 8,432} = 0.788 \text{ (or 78.8 percent)}$$

Due to irreversibilities, the compressor-work input is 27 percent greater than for the ideal isentropic process.

**Example 7-23** Nitrogen is compressed adiabatically from 14 psia, 60°F, to a final pressure of 58 psia.

(a) If the final outlet temperature is 400°F, how many degrees of the temperature rise is due to internal irreversibilities within the compressor?

(b) What is the compressor adiabatic efficiency?

SOLUTION (a) In order to find the $\Delta T$ due to irreversibilities, we must first find the temperature rise in the absence of irreversibilities. Isentropic compression will give the minimum temperature rise, as shown on the $Ts$ diagram in Fig. 7-24. $T_{2s}$ can be found by employing the $s^0$ data in Table A-6. At $T_1 = 520°R$, $s_1^0 = 45.519$ Btu/(lb·mol)(°R). Therefore,

$$s_2^0 = s_1^0 + R \ln \frac{P_2}{P_1} = 45.519 + 1.986 \ln (58/14)$$

$$= 48.34 \text{ Btu/(lb·mol)(°R)}$$

In Table A-6 this value of $s^0$ corresponds to a temperature of $780°R$. Hence the temperature rise due to internal irreversibilities is $400 - 320 = 80°F$.

An alternate solution would be to make use of the isentropic relation $T_2/T_1 = (P_2/P_1)^{(k-1)/k}$. This approach requires a guess as to the proper value of $k$ to use, since $T_2$ is as yet unknown. The method above avoids this difficulty.

(b) Neglecting kinetic-energy changes, shaft work equals $\Delta h$. The enthalpies for the problem are $h_1 = 3611$ Btu/lb·mol, $h_{2s} = 5424$ Btu/lb·mol, and $h_{2a} = 5986$ Btu/lb·mol. Thus

$$\eta_C = \frac{h_{2s} - h_1}{h_{2a} - h_1} = \frac{5424 - 3611}{5986 - 3611} = 0.763 \text{ (or 76.3 percent)}$$

Due to internal irreversibilities, the compressor-work input is 31 percent greater than for the ideal isentropic process.

**c Nozzles** Since the shaft work is zero, and we are neglecting heat transfer and potential-energy changes, the conservation of energy statement for a nozzle is

$$\Delta KE = -\Delta h = c_p(T_1 - T_2)$$

The entropy change for an ideal gas is still governed by

$$\Delta s = c_p \ln \frac{T_2}{T_1} - R \ln \frac{P_2}{P_1}$$

The independent variables are $T_1$, $P_1$, and $P_2$. For isentropic flow the above equation shows that $T_2$ is fixed by the three variables. Since $T_{2s}$ is fixed, the energy equation dictates that $\Delta KE$ is fixed. Hence the acceleration of a gas through a nozzle is determined solely by the three variables if the flow is isentropic.

The effect of internal irreversibilities is most clearly seen by rearranging the equation for $\Delta s$ into the form

$$c_p \ln \frac{T_2}{T_1} = R \ln \frac{P_2}{P_1} + \Delta s$$

Keep in mind that $T_1$, $P_1$, and $P_2$ are fixed for a given mode of operation. As $\Delta s$ increases with increasing irreversibilities, $T_{2a}$ likewise increases. This is shown in Fig. 7-25. That is, $T_{2a} > T_{2s}$. Since $\Delta KE = c_p(T_1 - T_2)$, we see that isentropic expansion gives a larger $\Delta T$, and thus a larger $\Delta KE$. An isentropic expansion through a nozzle is more effective in accelerating the flow than an irreversible expansion.

**d Diffusers** A diffuser decelerates the flow and increases the pressure. For an ideal gas flowing through the control volume, the energy equation simplifies to

$$c_p \Delta T = -\Delta KE$$

If the final velocity is small compared to the initial value, then $V_1$ fixes the kinetic-energy change, which in turn fixes $\Delta T$. Hence the temperature change is independent of whether irreversibilities are present or not. The property that is affected by

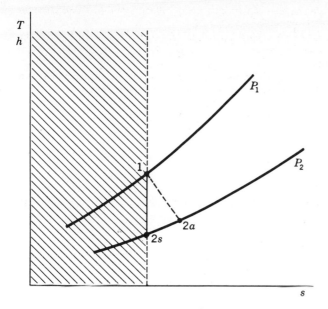

**Figure 7-25** Effect of irreversibilities on nozzle performance.

irreversibilities is the pressure. For an ideal gas with a fixed temperature ratio $T_2/T_1$ it can be shown that

$$\Delta s + R \ln \frac{P_{2a}}{P_1} = R \ln \frac{P_{2s}}{P_1}$$

$P_{2s}$ is fixed by the values of $P_1$, $T_1$, and $T_2$. Therefore as $\Delta s$ increases, the above equation shows that $P_{2a}$ decreases for a given $\Delta KE$. This effect is illustrated in Fig. 7-26. Higher values of $s_2$, due to irreversibilities, require lower values of $P_{2a}$ for a fixed value of $T_2$.

**e Throttling devices** In a throttling process $h_1 = h_2$. If we continue to restrict ourselves to ideal-gas flow, then $T_1 = T_2$, since $h$ is only a function of $T$. As a result,

$$\Delta s = c_p \ln \frac{T_2}{T_1} - R \ln \frac{P_2}{P_1} = -R \ln \frac{P_2}{P_1}$$

In the discussion of incompressible flow through throttling devices it was shown that $\Delta s$ must be greater than zero. In order for this to be so, the above equation requires that $P_2 < P_1$. This is the same result as for incompressible flow. Throttling always appears as a horizontal line on an $hs$ diagram. The process line also must always be from higher to lower pressure.

A practical example of the use of the throttling process is in conjunction with the determination of the quality of a real fluid in a flow system. If a wet vapor is flowing through a duct, the measurement of its pressure and temperature is

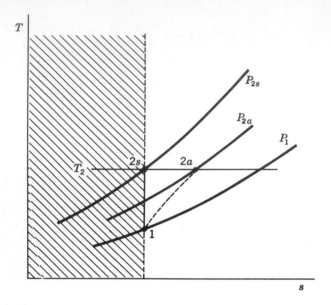

**Figure 7-26** Effect of irreversibilities on diffuser-outlet pressure.

insufficient to determine its state. This is true since $P$ and $T$ are not independent properties for a wet saturation state. However, if the pressure of the wet vapor is lowered sufficiently by throttling, the substance will become a superheated vapor. This may be accomplished in this case by bleeding off a portion of the wet vapor and allowing it to flow first through a narrow constriction in the flow passage. Then it passes into a larger chamber, where its pressure and temperature are measured. A throttling calorimeter is sketched in Fig. 7-27$a$. If the calorimeter is insulated (adiabatic) and the kinetic-energy change is negligible, a steady-flow energy balance on the calorimeter indicates that

$$h_{\text{steady-flow}} = h_{\text{calorimeter}}$$

Note that the process is not a constant-enthalpy one, but merely one in which the initial and final enthalpies are equal. Thus the path of the throttling process shown in Fig. 7-27$b$ appears as a dashed line on the $hs$ plot, since the actual path is not known.

    The problem can be solved numerically in two ways: use of tabular data and use of a Mollier diagram. When tabular data are used, the final enthalpy is determined from the values of pressure and temperature at the exit of the calorimeter. This enthalpy value and knowledge of the initial pressure (or temperature) in the pipe is sufficient to determine the quality of the fluid in the initial state. If a reasonably accurate Mollier chart of the substance is available, then one proceeds as follows: The initial quality is found by starting on the $hs$ diagram at the measured calorimeter pressure and temperature and moving to the left (at constant $h$) until one crosses the initial saturation-pressure (or -temperature) line for

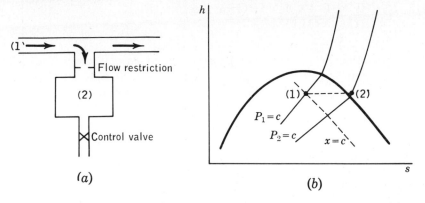

**Figure 7-27** (*a*) Schematic of a throttling calorimeter. (*b*) An *hs* diagram for a throttling process.

the flow system. The quality is then read directly from the Mollier diagram at this point.

**Example 7-24M** Wet steam at 15 bars is throttled into a calorimeter. The exit conditions of the calorimeter are measured as 1 bar and 100°C. What is the quality of the steam?

SOLUTION From the superheat table A-14M the exit enthalpy is found to be 2676.2 kJ/kg. This must also be the enthalpy at 15 bars as a wet vapor. The quality may be found by employing the relation $h_x = h_f + x h_{fg}$. From the saturation table A-13M, at 15 bars $h_f$ is 844.9 kJ/kg and $h_{fg}$ is 1947.3 kJ/kg. Hence

$$x = \frac{2676.2 - 844.9}{1947.3} = 0.940 \text{ (or 94.0 percent)}$$

**Example 7-24** Wet steam which flows through a 6-in pipe is bled off at 200 psia into a throttling calorimeter. The exit conditions of the calorimeter are 16 psia and 260°F. What is the quality of the steam at 200 psia?

SOLUTION A The quality will first be determined from the Mollier diagram for steam, Fig. A-25. By moving horizontally from the intersection of the 16-psia and 260°F lines until the 200-psia line is crossed, one reads the quality directly as approximately 3 percent liquid, or 97 percent vapor.

SOLUTION B The quality may also be determined by computations. From the superheat table A-14, at 16 psia and 260°F the enthalpy is found to be 1173.4 Btu/lb. This must also be the enthalpy at 200 psia as a wet vapor. The quality may be found by employing the relation $h_x = h_f + x h_{fg}$. From the saturation table A-13, at 200 psia, $h_f$ is 355.6 Btu/lb and $h_{fg}$ is 843.7 Btu/lb. Consequently,

$$x = \frac{1173.4 - 355.6}{843.7} = 0.970$$

A quality of 97 percent agrees with the value obtained in Solution A.

## 7-13 THE INCREASE IN ENTROPY PRINCIPLE FOR A CONTROL VOLUME

The second law of thermodynamics may be expressed as an increase in entropy principle for adiabatic closed and isolated systems. In each case attention is focused on a particular mass or set of masses within a given boundary. We now wish to extend this principle to a region of space—the control volume. The approach is similar to that used for the derivation of the conservation of energy principle in Chap. 5. Figure 7-28 shows a control mass relative to a control volume at times $t$ and $t + \Delta t$. At time $t$,

$$S_{CM,\,t} = S_{CV,\,t} + S_A \tag{a}$$

and at time $t + \Delta t$,

$$S_{CM,\,t+\Delta t} = S_{CV,\,t+\Delta t} + S_B \tag{b}$$

Subtraction of Eq. ($a$) from Eq. ($b$) yields

$$S_{CM,\,t+\Delta t} - S_{CM,\,t} = S_{CV,\,t+\Delta t} - S_{CV,\,t} + S_B - S_A \tag{c}$$

or

$$\Delta S_{CM} = \Delta S_{CV} + s_2 m_2 - s_1 m_1 \tag{d}$$

where $S_B = s_2 m_2$ and $S_A = s_1 m_1$ and the $\Delta$ in front of $S_{CM}$ and $S_{CV}$ represents the change over a time period $\Delta t$.

The entropy change of any control mass is generally given by the relation $dS \geqslant \delta Q/T$. Since in this case the temperature varies across the control mass, we are required to evaluate the quantity $Q_i/T_i$ at each boundary where heat enters or leaves. The subscript $i$ denotes a location at the control surface of temperature $T_i$. Hence

$$\Delta S_{CM} \geqslant \Sigma \frac{Q_i}{T_i} \tag{e}$$

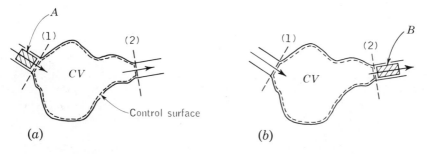

$(a)$ $(b)$

**Figure 7-28** Development of an entropy relation for a control volume. ($a$) Control mass at time $t$; ($b$) control mass at time $t + \Delta t$.

Substituting Eq. (*d*) into Eq. (*e*) we find that

$$\Delta S_{CV} + s_2 m_2 - s_1 m_1 - \Sigma \frac{Q_i}{T_i} \geqslant 0 \qquad (7\text{-}46)$$

For an infinitesimal time interval $dt$ the above equation for multiple outlets and inlets becomes

$$dS_{CV} + \sum_{\text{out}} (s \, dm) - \sum_{\text{in}} (s \, dm) - \Sigma \frac{\delta Q_i}{T_i} \geqslant 0 \qquad (7\text{-}47)$$

Finally, on a rate basis for the control volume,

$$\frac{dS_{CV}}{dt} + \sum_{\text{out}} (s\dot{m}) - \sum_{\text{in}} (s\dot{m}) - \Sigma \frac{\dot{Q}_i}{T_i} \geqslant 0 \qquad (7\text{-}48)$$

In these equations the equality is associated with internally reversible processes and the inequality with internally irreversible processes. For an irreversible process the sum of the terms in the above three equations is positive, and this sum is a measure of the entropy "production" due to the irreversibilities.

Under steady-state conditions, the quantity $dS_{CV}$ (or $\Delta S_{CV}$) must be zero in Eqs. (7-46) to (7-48). In addition, if a process is adiabatic, the sum of the $Q_i/T_i$ terms is zero. When both are valid restrictions, Eq. (7-48) becomes

$$\sum_{\text{out}} (s\dot{m}) - \sum_{\text{in}} (s\dot{m}) \geqslant 0 \qquad (7\text{-}49)$$

This expression relates the flow of entropy across the boundaries of an adiabatic, steady-state control volume. When internal irreversibilities are present within the control volume, the sum of the terms on the left side is positive and measures the rate of entropy production, $dS_{\text{prod}}/dt$.

> **Example 7-25M** Consider the open feedwater heater discussed in Example 5-8M. The device is steady-state, with superheated steam (flow stream 1) entering at 1 kg/unit time at 5 bars and 200°C. Subcooled water (flow stream 2) enters at 4.75 kg/unit time at 5 bars and 40°C. Saturated liquid water (flow stream 3) at 5.75 kg/unit time leaves at 5 bars. Determine the rate of entropy production for the irreversible mixing process in kJ/(°K)(unit time).
>
> SOLUTION Equation (7-49) will be applied to determine the rate of entropy production. The equation may be written as
>
> $$\frac{dS_{\text{prod}}}{dt} = s_3 \dot{m}_3 - s_1 \dot{m}_1 - s_2 \dot{m}_2$$
>
> The specific entropies at states 1, 2, and 3 are found from Tables A-14M, A-12M, and A-13M to be 7.0592, 0.5725, and 1.8607 J/(g)(°K), respectively. Substitution of these values yields
>
> $$\frac{dS_{\text{prod}}}{dt} = 1.8607(5.75) - 7.0592(1) - 0.5725(4.75)$$
>
> $$= 10.70 - 7.06 - 2.72 = 0.92 \text{ kJ/(°K)(unit time)}$$
>
> The entropy production is positive, in accordance with the second law, and is due to the irreversible mixing of flow streams at different temperatures.

**Example 7-25** Consider the open feedwater heater discussed in Example 5-8. The device is steady-state, with superheated steam (flow stream 1) entering at 1 lb/unit time at 80 psia and 400°F. Subcooled water (flow stream 2) enters at 4.44 lb/unit time at 80 psia and 100°F. Saturated liquid water (flow stream 3) at 5.44 lb/unit time leaves at 80 psia. Determine the entropy production of the irreversible mixing process in Btu/(°R)(unit time).

SOLUTION Equation (7-49) will be applied to determine the rate of entropy production.

$$\frac{dS_{prod}}{dt} = s_3 \dot{m}_3 - s_1 \dot{m}_1 - s_2 \dot{m}_2$$

The entropies at states 1, 2, and 3 are found from the steam tables to be 1.6790, 0.1295, and 0.4534 Btu/(lb)(°R), respectively. Substitution of these values yields

$$\frac{dS_{prod}}{dt} = 0.4534(5.44) - 1.6790(1) - 0.1295(4.44)$$

$$= 2.465 - 1.679 - 0.575 = 0.211 \text{ Btu}/(°R)(\text{unit time})$$

The entropy production is positive, and is due to the irreversible mixing of flow streams at different temperatures.

# REFERENCES

Fast, J. D.: "Entropy," McGraw-Hill, New York, 1963.
*Joint Army, Navy and Air Force (JANAF) Thermochemical Tables*, NSRDS-NBS-37, June, 1971.
Keenan, J. H., and J. Kaye: "Gas Tables," New York, 1945.

# PROBLEMS (METRIC)

**Carnot heat engines**

**7-1M** A Carnot heat engine operates between temperatures of 927 and 27°C. For every kilowatt of power output, calculate the heat supplied and the heat rejected in kJ/h, and the thermal efficiency.

**7-2M** At what temperature in °C is heat supplied to a Carnot engine that rejects 1000 kJ/min of heat at (a) 7°C, and (b) 37°C and produces 40 kW of power.

**7-3M** A Carnot engine operates between 1000 and 300°K. The change in entropy of the supply or source reservoir is 0.5 kJ/°K. Find (a) the heat added in kilojoules, and (b) the network output in kilojoules.

**7-4M** A reversible heat engine exchanges heat with three reservoirs and produces work in the amount of 400 kJ. Reservoir A has a temperature of 500°K and supplies 3000 kJ to the engine. If reservoirs B and C have temperatures of 400 and 300°K, respectively, how much heat, in kilojoules, does each exchange with the engine and what is the direction of the heat exchanges?

**7-5M** Same as Prob. 7-4M, except that the work output is 600 kJ and the temperatures of reservoirs A, B, and C are 600, 400, and 300°K, respectively.

**7-6M** Two Carnot engines A and B are operated in series. The first one (A) receives heat at 627°C and rejects it to a reservoir at temperature T. The second engine (B) receives the heat rejected by the first engine, and in turn rejects heat to a reservoir at 27°C. Calculate the temperature T in °C for the situation where (a) the work outputs of the two engines are equal, and (b) the efficiencies of the two engines are equal.

**7-7M** Same as Prob. 7-6M, except that the temperatures of the heat reservoirs are 827 and 127°C.

**7-8M** A Carnot engine which operates on air as the working medium has a pressure of 15 bars and a specific volume of 0.075 m$^3$/kg at the beginning of the adiabatic expansion. Twenty kJ/kg of heat are added during each cycle, and the sink temperature is 7°C. Compute the thermal efficiency of the engine and the work output in kJ/kg.

**7-9M** A Carnot engine which produces 10 kJ of work for one cycle has a thermal efficiency of 50 percent. The working fluid is 0.5 kg of air, and the pressure and volume at the beginning of the isothermal expansion are 7 bars and 0.11 m$^3$, respectively. Find (a) the heat transfer and work for each of the four processes, (b) the temperature at the end of each process, and (c) the volume at the end of the isothermal expansion process.

**7-10M** A Carnot engine operates on air as the working fluid between temperatures of 700 and 300°K. At the beginning of the isothermal heat addition the pressure is 6.5 bars. During the heat-addition process the volume is doubled. Find the net work of the cycle per kilogram of air.

**7-11M** The pressure, volume, and temperature in a Carnot engine using air as the working medium are, at the beginning of the isothermal expansion, 5 bars, 550 cm$^3$, and 260°C, respectively. During the isothermal expansion 0.3 kJ of heat is added, and the maximum volume is 3300 cm$^3$. Determine (a) the volume after the isothermal expansion, (b) the sink temperature in °C, (c) the heat rejected per cycle, and (d) the thermal efficiency.

**7-12M** A Carnot cycle operates with steam as the working medium. At the end of the adiabatic compression the pressure is 15 bars and the quality is 20 percent. Heat is added during the isothermal expansion until the steam becomes a saturated vapor. The steam then expands adiabatically until the pressure is 1 bar. Determine for this cycle (a) the efficiency, and (b) the quantity of work per cycle, in kJ/kg.

**7-13M** A Carnot engine operates with steam as the working fluid in a closed system. At the end of the adiabatic compression the pressure is 10 bars and the quality is 10 percent. At the end of isothermal expansion the steam is a saturated vapor. During the isothermal compression the pressure is 0.3 bar. Determine (a) the efficiency of the cycle, and (b) the amount of work done per cycle.

**7-14M** A Carnot engine contains 0.1 kg of water. At the beginning of the heat-addition process the fluid is a saturated liquid, and at the end of this process it is a saturated vapor. Heat addition is at 120 bars and heat rejection is at 0.3 bar. Determine (a) the thermal efficiency, and (b) the heat added and net work done per cycle.

**7-15M** A Carnot cycle, using steam as the working fluid, operates between 270 and 40°C. At the beginning of heat addition the quality is 50 percent and at the end of this process the pressure is 40 bars.

(a) Determine the efficiency of the engine.

(b) If the engine contains 0.01 kg of steam and operates at 3000 cycles/min, what is the power output in kilowatts?

**7-16M** In a Carnot cycle, heat is added in process 1-2 and rejected in process 3-4. For the cycle, $T_1 = 600°K$, $T_4 = 300°K$, and $S_2 - S_1 = 0.1$ kJ/°K.

(a) Determine the thermal efficiency.

(b) Now suppose process 2-3 is not reversible. It is still adiabatic but, as a result of irreversibilities, the entropy increases 0.02 kJ/°K during the process. Determine the thermal efficiency in this latter case.

**7-17M** A steady-flow type Carnot engine operates between the temperature limits of 400 and 30°C. During the isothermal-expansion process 350 kJ of heat are added per kilogram of fluid. However, irreversibilities in the turbine cause an increase in entropy through the adiabatic turbine amounting to 0.04 kJ/(kg)(°K). Calculate the heat rejected and the ratio of the actual work output to the maximum possible work output if the expansion had been reversible. All other parts of the cycle are assumed to be reversible. Sketch the irreversible cycle on a $Ts$ diagram.

### Carnot refrigerators and heat pumps

**7-18M** A Carnot heat engine which operates between temperature levels of 927 and 33°C rejects 30 kJ to the low-temperature sink. The work output of this engine is used to drive a heat pump. The heat pump removes 270 kJ of heat from a low-temperature reservoir and rejects it to the surroundings at 33°C. Determine the temperature, in °C, of the low-temperature reservoir for the heat pump.

**7-19M** A Carnot heat pump is used for heating a building. The outside air at $-8°C$ is the cold reservoir, the building at 27°C is the hot reservoir, and 200,000 kJ/h are required for heating. Find (a) the heat taken from the outside in kJ/h, and (b) the power input required, in kilowatts.

**7-20M** A reversible refrigerating machine absorbs 200 kJ/min from a cold space and requires 1.5 kW to drive it. If the machine is reversed and receives 800 kJ/min from the hot source, how much power, in kilowatts, does it produce?

**7-21M** A Carnot heat pump operating between the temperatures of $-7$ and 29°C has a power input of 3.5 kW. Determine the coefficient of performance.

**7-22M** If the thermal efficiency of a Carnot engine is 1/6, find the coefficient of performance of (a) a Carnot refrigerator, and (b) a Carnot heat pump operating between the same temperature limits.

**7-23M** A computer room is to be kept at 20°C. Assume that the computer transfers 1500 kJ/h to the room air. If the environmental temperature is 35°C, what is the minimum power input required for the cooling process?

**7-24M** A Carnot heat pump is to be used to maintain a home with a heat loss of 80,000 kJ/h at 22°C. If the makeup heat is to be supplied by a heat pump from the outside air at $-5°C$, find the power input required, in kilowatts. If electricity costs 3.5 cents per kilowatthour, what would be the operating cost for 1 day?

**7-25M** A reversed Carnot engine is to be used to produce ice at 0°C. The heat-rejection temperature is 30°C, and the enthalpy of freezing is 335 kJ/kg. How many kilograms of ice can be formed per hour per kilowatt of power input?

**7-26M** A ton of refrigeration is defined as a heat-absorption rate of 211 kJ/min from a cold source. It is desired to operate a reversed Carnot cycle between temperature limits of $-20$ and $+35°C$ so that 8 ton of refrigeration are produced. Calculate (a) the number of kilowatts required to operate the cycle, and (b) the coefficient of performance.

**7-27M** A heat pump operates on a reversed Carnot cycle, removes heat from a low-temperature source at $-15°C$, and rejects heat to a sink at 26°C. If electricity costs 3.9 cents per kilowatthour, determine the cost of operation for supplying a home with 50,000 kJ/h.

**7-28M** A Carnot refrigerator removes heat from a sink at $-8°C$ and rejects heat to the atmosphere at 15°C. The refrigerator is coupled to the output of a Carnot engine which receives heat at 577°C and also rejects to the atmosphere. Determine the ratio of the heat supplied to the engine to the heat removed by the refrigerator.

**7-29M** A Carnot heat engine is used to drive a Carnot refrigerator. The heat engine receives $Q_1$ at $T_1$ and rejects $Q_2$ at $T_2$. The refrigerator removes a quantity of heat $Q_3$ from a source at $T_3$ and rejects a quantity of heat $Q_4$ at $T_4$. Develop an expression for the ratio $Q_3/Q_1$ in terms of the various temperature of the heat reservoirs.

**7-30M** A reversed Carnot engine employs air as the working fluid. At the beginning of isothermal expansion the air is at 1 bar and 5°C; it fills a volume of 0.02 m³. During the isothermal expansion the volume is doubled. At the end of the adiabatic compression the temperature is 383°C. Calculate, in kilojoules, the heat added and the heat rejected during the cycle.

**7-31M** A reversed Carnot engine is used as a refrigerator. Heat is rejected at 35°C from refrigerant 12, which acts as the working substance. At the beginning of the isothermal expansion, the pressure is 1.2 bars and the quality is 30 percent. Sufficient heat is added from the low-temperature source to ensure that the refrigerant at the end of the isothermal heat-addition process is saturated vapor. Determine (a) the coefficient of performance for the cycle, and (b) the power input in kilowatts if the mass flow rate is 1.1 kg/min.

**7-32M** Initially, two identical bodies of constant specific heat are maintained at the same temperature. These two bodies of finite size are then used as reservoirs for a refrigerator. Heat is removed from one body and rejected to the other body. As a result, the temperature of one body continues to decrease, and that of the other continually increases. The process is at constant pressure, and it is assumed that neither body undergoes a phase change. Show that the minimum work required to decrease the temperature of the colder body to some value $T_f$ which is less than $T_i$, the initial temperature of both bodies, is

$$W_{\min} = mc_p \left( \frac{T_i^2}{T_f} + T_f - 2T_i \right)$$

**Other reversible cycles**

**7-33M** A reversible heat engine employing an ideal gas with constant specific heats undergoes the following cyclic process: constant-volume heating from state 1 to state 2, adiabatic expansion from state 2 to state 3, and then constant-pressure compression from state 3 to state 1. Prove that the thermal efficiency of the cycle is

$$\eta_{\text{th}} = 1 - k \, \frac{V_3/V_1 - 1}{P_2/P_1 - 1}$$

**7-34M** A piston-cylinder device contains air initially at 3.5 bars and 307°C. The gas undergoes the following cyclic process: 1-2, constant-volume heating until the pressure has doubled; 2-3, adiabatic expansion to a temperature of 702°C; 3-1, constant-pressure compression to the initial state. Use the air table to find (a) the heat input in kJ/kg, (b) the heat output in kJ/kg, and (c) the thermal efficiency.

**7-35M** A piston-cylinder device initially contains air at a low temperature and 1 bar. In order to produce a refrigeration effect, the following cycle is proposed: 1-2, adiabatic, reversible compression to a state of 4.75 bars and 390°K; 2-3, constant-pressure cooling to the initial temperature; 3-1, isothermal expansion with heat addition to the original pressure of 1 bar. The heat removed in process 2-3 is 140.8 kJ/kg. Use the air table.

    (a) Sketch $Pv$ and $Ts$ diagrams of the cycle.
    (b) Determine the original temperature in °K.
    (c) Determine the heat added during the isothermal expansion process in kJ/kg.
    (d) Compute the coefficient of performance.

**7-36M** One kilogram of air is contained in a piston-cylinder device at 1 bar and 7°C. The gas undergoes the following cycle: 1-2, constant-volume heating until the pressure is 4 bars; 2-3, adiabatic, reversible expansion to the initial pressure and a temperature of 790°K; 3-1, constant-pressure cooling to the initial state.

    (a) Sketch $Pv$ and $Ts$ diagrams for the cycle.
    (b) Determine the thermal efficiency of the cycle.

**7-37M** Determine the ratio of the efficiencies for the two cycles shown in Fig. P7-37(a and b) in terms of the temperatures $T_1$ and $T_2$. What is the limiting value of the ratio as $T_2$ approaches infinity?

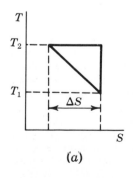

(a)

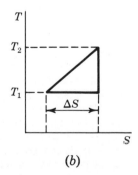

(b)

### Irreversible heat transfer and quality of energy

**7-38M** One thousand kilojoules of heat are transferred from a heat reservoir at 850°K to a second reservoir at (a) 550°K, and (b) 330°K.
  (1) Calculate the entropy change of each reservoir in kJ/°K.
  (2) Is the sum of the entropy changes of the reservoirs in agreement with the second law?

**7-39M** One hundred kilojoules of heat are transferred from an isothermal system at (a) 550°K, and (b) 450°K to the environment at 7°C.
  (1) Calculate the entropy change of the system and the environment in kJ/°K.
  (2) Is the sum of the entropy changes of the reservoirs in agreement with the second law?

**7-40M** Consider a heat engine which receives 1000 kJ of heat from a reservoir at (a) 1100°K, and (b) 1200°K. The engine, which is internally reversible, produces 690 kJ of work and rejects heat reversibly to a sink at 27°C. Determine (1) the temperature in °C at which the engine isothermally received the heat from the source reservoir, and (2) the total entropy change of the process, in kJ/°K.

**7-41M** A heat engine receives 100 kJ of heat reversibly from a source reservoir at 800°K. The engine is internally reversible and produces 60 kJ of work. However, the heat-rejection process to a sink temperature of (a) 35°C, and (b) 20°C, occurs across a finite temperature difference. Determine (1) the temperature in °C at which the fluid within the engine rejects heat, and (2) the total entropy change for the process, in kJ/°K.

**7-42M** Which quantity of heat theoretically has the higher quality, 1000 kJ at 800°K or 3000 kJ at 380°K? The environmental temperature is 7°C.

**7-43M** Which quantity of heat theoretically has the higher quality, 2000 kJ at 1100°K or 5000 kJ at 530°K? The environmental temperature is 22°C.

**7-44M** Which quantity of energy theoretically has the higher quality, 1000 kJ of shaft work or 3000 kJ of heat at 500°K? The environmental temperature is 20°C.

**7-45M** Which quantity of energy theoretically has the higher quality, 530 kJ of shaft work or 1000 kJ of heat at 480°K. The environmental temperature is 27°C.

### Entropy change of an ideal gas

**7-46M** Two kilograms of nitrogen gas at 1.3 bars and 27°C are heated to 227°C in a constant-volume process. Compute the entropy change in kJ/°K.

**7-47M** Nitrogen, an ideal gas, is heated in a rigid, closed container from 60°C, 1 bar, to 72°C. What is the entropy change in kJ/(kg·mol)(°K).

**7-48M** A piston-cylinder assembly contains air at 350°K and 1 bar. The pressure is increased to 1.3 bars, but during the process heat transfer occurs so that the temperature remains constant. Determine the entropy change in kJ(kg)(°K).

**7-49M** A vertical piston-cylinder assembly contains nitrogen gas at a constant pressure of 1 bar. It is cooled from 150 to 40°C. Compute (a) the heat removed, in kJ/kg, (b) the work done, in kJ/kg, and (c) the change in entropy, in kJ/(kg)(°K).

**7-50M** A rigid tank contains 50 g of carbon monoxide gas at 0.95 bar and 30°C. Heat is added until the pressure reaches 1.3 bars. Compute (a) the heat transferred, in kilojoules, and (b) the change in entropy, in kJ/°K.

**7-51M** A gas at 27°C and 0.95 bar is compressed isothermally from 10 to 2 m³. Is the entropy increased or decreased, and by how much, if the gas is (a) air, or (b) nitrogen.

**7-52M** Nitrogen gas, initially at 6 bars and 207°C, is cooled in a rigid tank until the temperature reaches 37°C. Calculate the change in entropy, in kJ/(kg)(°K).

**7-53M** Calculate the change in entropy of 1 kg of He which is heated at constant pressure from 0.8 bar and 20°C to a final temperature of 60°C in a closed system.

**7-54M** A piston-cylinder assembly contains air at 350°K and 1 bar. The pressure is increased quasi-statically to 1.6 bars, but during the process the temperature remains constant. Determine the entropy change in kJ/(kg)(°K).

**7-55M** One kilogram of air at 27°C is heated at constant pressure to a temperature of 527°C. Determine the change in entropy in kJ/°K.

**7-56M** An ideal monatomic gas undergoes an irreversible process from 1 bar, 17°C, to 1.8 bars, 57°C. Determine the entropy change of the system in kJ/(kg·mol)(°K).

### Energy balances and entropy changes of ideal gases

**7-57M** One-fourth kilogram of air in a closed system undergoes an irreversible process from 1.3 bars, 27°C, to a final pressure of 2.7 bars. Measurements during the process indicate that 3.5 kJ of heat were removed from the air and the work done on the air was 10.7 kJ. Compute the entropy change for the irreversible process in kJ/°K.

**7-58M** Air is contained in a rigid adiabatic tank in an initial state of 0.95 bar and 27°C. A paddle-wheel within the system, which is driven by an external motor, stirs the air until the pressure rises to 1.4 bars. Calculate (a) the work done on the gas in kJ/kg, and (b) the entropy change in kJ/(kg)(°K).

**7-59M** A monatomic gas initially at 15°C and 1 bar is contained in a rigid, adiabatic tank. Paddle-wheel work in the amount of 75 kJ/kg is done on the system. Determine the entropy change in kJ/(kg)(°K) if the gas is (a) helium, and (b) argon.

**7-60M** Five kilograms of diatomic hydrogen at 6 bars and 17°C are contained in a constant-pressure device. Heat in the amount of 7980 kJ is added to the contents. Compute the entropy change of the hydrogen, in kJ/°K.

**7-61M** A rigid tank with a volume of 0.03 m³ contains oxygen at an initial state of 87°C and 1.5 bar. During a process paddle-wheel work is carried out by applying a torque of 13 N·m for 25 revolutions, and a heat loss of 3.71 kJ occurs. Determine the entropy change of the gas, in kJ/°K.

**7-62M** A piston-cylinder device contains 10 kg of air at 2 bars and 327°C. During a constant-pressure process 3770 kJ of heat are removed from the air. Using air table data, determine the entropy change of the air, in kJ/°K.

**7-63M** One-half kilogram of air is compressed isothermally and quasistatically from 1 bar, 270°K, to a final state which requires 63.9 kJ of boundary-work input to the piston-cylinder device. Find (a) the final pressure in bars, and (b) the change in entropy in kJ/°K.

**7-64M** A piston-cylinder assembly contains 0.01 kg of hydrogen at 2 bars and 320°K. During a certain experiment, 7.4 kJ of heat are removed from the gas at constant pressure. Determine the entropy change of the hydrogen, in kJ/°K.

**7-65M** One-half kilogram of air in a closed system undergoes an irreversible process at constant volume from an initial state of 0.1 m³ and 27°C. Measurements during the process indicate that 9.5 kJ of heat were removed and paddle-wheel work was carried out by applying a torque of 15.7 N·m to the shaft for 400 revolutions.

    (a) Find the final temperature of the air, in °C.
    (b) Compute the entropy change of the air, in kJ/°K.

**7-66M** One kilogram of air in a closed system undergoes an irreversible process from 1.4 bars and 27°C, to the same final pressure. Measurements during the process indicate that 16.2 kJ of heat were removed and the net work done on the air was 40.1 kJ.

    (a) Determine the final temperature, in °C.
    (b) Compute the entropy change of the air, in kJ/°K.

**7-67M** An ideal gas is throttled from a high pressure to a lower pressure through a well-insulated flow restriction. Prove that such a process is internally irreversible.

**7-68M** Oxygen is throttled from 2 bars and 600°K to (a) 1.4 bars, and (b) 1.2 bars. Find the entropy change, in kJ/(kg)(°K).

**7-69M** The entropy increase of 0.5 kg of nitrogen is (a) 0.1406 kJ/°K, and (b) 0.1987 kJ/°K, when it is throttled adiabatically from 8 bars. Determine the final pressure in bars.

**7-70M** Air flows isothermally and adiabatically at 47°C through a variable-area duct such that the entropy increase due solely to internal irreversibilities is (a) 0.0640 kJ/(kg)(°K), and (b) 0.1464 kJ/(kg)(°K).

(1) If the initial pressure is 5 bars, determine the final pressure in bars.

(2) Determine the ratio of inlet to outlet area, if the device is horizontal.

### Entropy tabulations for superheat and saturation states

**7-71M** Find the specific entropy, in kJ/(kg)(°K), for water substance in parts (a) through (j) listed in Prob. 4-3M.

**7-72M** Find the specific entropy, in kJ/(kg)(°K), for water substance in parts (a) through (k) listed in Prob. 4-4M.

**7-73M** Find the specific entropy, in kJ/(kg)(°K), for refrigerant 12 in parts (a) through (k) in Prob. 4-5M.

**7-74M** Find the specific entropy, in kJ/(kg)(°K), for refrigerant 12 in parts (a) through (k) in Prob. 4-6M.

### Entropy changes in the superheat and saturation regions

**7-75M** One kilogram of steam at 30 bars and 320°C undergoes a reversible, isothermal expansion in a cylinder having a gastight piston until the pressure reaches 10 bars. Calculate the heat transfer in J/g and sketch the process on a $Ts$ diagram.

**7-76M** One kilogram of steam at 15 bars and 320°C is contained in a closed system at constant pressure. During an adiabatic process 100,000 N·m of work are performed on the steam by stirring. The steam is then cooled at constant pressure by the transfer of 12,700 J of heat to the surroundings at 17°C.

    (a) Sketch the process on a $Ts$ diagram.

    (b) What is the change in the entropy of the steam, in kJ/°K?

    (c) What is the entropy change of the surroundings, in kJ/°K?

**7-77M** Steam is originally at 60 bars, 500°C.

    (a) It is expanded isothermally and reversibly to 15 bars in a closed system. Calculate the volume change and the work done, in kJ/kg.

    (b) If the same expansion to 15 bars is carried out in a steady-flow device, calculate the work done, in kJ/kg.

**7-78M** Saturated liquid water at 150°C is heated isothermally and reversibly in a closed system. Calculate the work done, in kJ/kg, when 400 kJ/kg of heat are added.

**7-79M** Steam enters a turbine at 40 bars and 500°C. The entropy increase as it passes through the turbine is 0.85 kJ/(kg)(°K), and the steam exhausts at 0.060 bar. A mass flow rate of 50,000 kg/h produces a power output of 13,000 kW. If the inlet and exit velocities are 70 and 150 m/s, respectively, determine the quantity and direction of the heat transfer, in kJ/kg.

**7-80M** Saturated liquid refrigerant 12 at 7 bars flows through a long, insulated capillary tube until the pressure reaches 1.8 bars. Find the entropy change, in kJ/(kg)(°K).

**7-81M** Refrigerant 12 at 2.8 bars and 60°C is compressed in a closed system to 14 bars. If the process is isothermal, find the heat transfer and work done, in kJ/kg.

**7-82M** Refrigerant 12 flows through a well-insulated capillary tube. The entering state is 8 bars, 40°C, and the exit temperature is 24°C. What is the change in entropy, in kJ/(kg)(°K)?

**7-83M** A rigid vessel contains 0.2 kg of steam at 7 bars and 280°C. Heat is transferred out of the system until the temperature drops to 200°C.

    (a) Determine the entropy change, in kJ/°K.

    (b) Sketch the process on the $Ts$ plane.

**7-84M** A piston-cylinder device initially contains 0.1 kg of steam at 30 bars and 500°C. During a constant-pressure process a heat loss of 22.6 kJ occurs.

    (a) Determine the entropy change, in kJ/°K.

    (b) Sketch the process on a $Ts$ plot, relative to the saturation line.

**7-85M** A piston-cylinder assembly contains 4 kg of steam at 30 bars and 360°C. During a certain experiment, 1340 kJ of heat are removed from the steam at constant pressure.

(a) Determine the entropy change of the steam, in kJ/°K.

(b) Illustrate the process on a $Ts$ diagram.

**7-86M** A piston-cylinder device initially contains refrigerant 12 at 3.2 bars and 80°C. It is compressed quasistatically and at constant pressure, with a boundary-work input of 14.73 kJ/kg. At the same time, a heat loss of 127.4 kJ/kg occurs.

(a) Determine the entropy change, in kJ/(kg)(°K).

(b) Sketch the process on a $Ts$ plot.

**Entropy changes for incompressible substances**

**7-87M** Fifty kilograms of water at 1 bar and 20°C are mixed with 20 kg of water initially at 1 bar and 90°C. If the mixing process is adiabatic and at constant pressure, determine the total change in entropy for the 70 kg of water, in kJ/°K.

**7-88M** An 8-kg iron casting at 550°C is quenched in a tank containing 0.03 $m^3$ of water initially at 25°C. Assuming no heat loss to the surroundings and that the specific heat of the iron averages 0.45 kJ/(kg)(°C), determine the change in entropy of (a) the water, (b) the casting, and (c) the total process.

**7-89M** A 1-kg block of copper at 150°C and a 1-kg block of aluminum at 50°C initially are isolated from each other and from the local surroundings. They are then placed in thermal contact with each other, but are still isolated from the surroundings. Calculate $\Delta S$ for each block if they attain thermal equilibrium. The specific heat of copper averages 0.385 kJ/(kg)(°C), and for aluminum the value is 0.90 kJ/(kg)(°C). Report the answers in kJ/°K.

**7-90M** A small solid object of constant heat capacity $C$ equal to 0.70 kJ/°K is initially at 580°K. The object is thrown into a large lake which is at a temperature of 290°K. Calculate the total entropy change for this process, in kJ/°K. (NOTE: heat capacity $C$ equals mass times specific heat, $mc$.)

**7-91M** A kilogram of ice at 0°C is mixed with 7 kg of water at 30°C. Compute the change in entropy of the composite system if the process is adiabatic and the pressure is maintained at 1 bar. The enthalpy of melting of ice is 335 kJ/kg. Report the answer in kJ/°K.

**Increase in entropy principle**

**7-92M** An open feedwater heater mixes liquid and vapor water from two different sources. Super-heated steam enters at one section at 5 bars and 240°C. Compressed liquid at the same pressure and 35°C enters at another inlet. The mixture of these two streams leaves at the same pressure as a saturated liquid. If the heater is adiabatic, calculate the increase in entropy per kilogram of mixture leaving the heater.

**7-93M** One kilogram of saturated water vapor is condensed at 240°C. The heat removed is used to vaporize a quantity of saturated-liquid refrigerant 12 maintained at 30°C.

(a) Calculate the entropy change, in kJ/°K, for the water vapor which condenses.

(b) Calculate the entropy change, in kJ/°K, for the refrigerant 12 which vaporizes.

(c) Is the sum of parts (a) and (b) compatible with the second law?

**7-94M** Ten grams of nitrogen initially are maintained at 1 bar and 27°C. The gas is compressed irreversibly and isothermally to 3 bars. All the heat lost by the system, which amounts to 1250 J, appears in the surroundings at a temperature of 27°C. Compute the entropy change of the system, the surroundings, and the universe, in J/°K.

**7-95M** A resistor of 20 Ω, initially at 27°C, is maintained at a constant temperature while a current of 5 A is allowed to flow for 2 s.

(a) Determine the entropy change of the resistor and of the universe, in J/°K.

(b) Now, insulate the resistor and carry out the same experiment. For this latter case, determine the entropy change of the resistor and the universe. The specific heat of the resistor is 1.05 J/(g)(°C), and the mass of the resistor is 8.0 g.

**7-96M** An electrical resistor of 30 Ω is maintained at a constant temperature of 17°C. A current of 5 A is maintained for a time interval of 3 s. Determine, if the temperature of the surroundings is 17°C, (a) the entropy change of the resistor, and (b) the entropy change of the universe. Record both answers in J/°K.

**7-97M** Air is initially at 2.3 bars and 62°C, and expands to 1.4 bars and 22°C. Can this change of state be carried adiabatically? Why?

**7-98M** Refrigerant 12 flows at a low velocity through a well-insulated capillary tube. The initial state is a saturated vapor at 8 bars, and the final state is 1 bar. Determine the entropy change of the fluid, in kJ/(kg)(°K). Does this value agree qualitatively with the second law? What is the ratio of the final velocity to the initial velocity?

**7-99M** A turbine takes in air at 6 bars and 597°C, and expands adiabatically to 1 bar and 297°C; KE and PE are negligible.
(a) Using the air table, compute the work output, in kJ/kg.
(b) Determine whether the process is reversible or irreversible.
(c) Again using the air table, determine the final temperature if the process is reversible.

**7-100M** One kilogram of air in a closed system expands from 2.6 bars and 267°C to 1.3 bars and 137°C, producing 50 kJ of work. The environmental temperature is 27°C.
(a) Determine the entropy change of the system plus environment.
(b) Draw the path of the process on a $Ts$ diagram.

**7-101M** Ten kilograms of liquid water at 20°C are mixed adiabatically at 1 bar pressure with 6 kg of water at 100°C.
(a) Compute the entropy change for each quantity of water, and the total change, in kJ/°K.
(b) Sketch the separate processes that each quantity of water undergoes on the same $TS$ diagram.

**7-102M** Carbon monoxide at 1 bar and 37°C is compressed in a closed system to 3 bars and 147°C. The required work input during the process is 88 kJ/kg, and the surrounding temperature is 25°C. Determine the total entropy change for the CO and the environment, in kJ/°K per kilogram of CO.

**7-103M** Five kilograms of nitrogen in a closed system expand from 3 bars and 157°C to 1 bar and 17°C. During the process 230 kJ of work are produced, and heat is transferred to the environment at 7°C.
(a) Determine the magnitude and direction of the heat transfer, in kJ.
(b) Prove whether the process is reversible, irreversible, or impossible.

**7-104M** One kilogram of air initially at 6.5 bars and 330°K is contained in a weighted piston-cylinder arrangement which maintains a constant pressure. During a process, 23.4 kJ of heat are transferred from the air to the surroundings, which are at 25°C, while 5.3 kJ of work are performed on the system by the weighted piston.
(a) Calculate the entropy change of the air in the cylinder, in kJ/°K.
(b) Determine the entropy change of the surroundings, in kJ/°K.
(c) Does the overall process satisfy the second law of thermodynamics? Why?

**7-105M** Refrigerant 12 enters a constant-diameter pipe at 5 bars and 50°C and leaves at 4 bars and 30°C. The surrounding temperature is 25°C.
(a) Determine the entropy change of the fluid, in kJ/(kg)(°K).
(b) Find the entropy change of the surroundings, in kJ/°K per kilogram of fluid.
(c) Does the overall process satisfy the second law? Why?

## Isentropic processes of ideal gases

**7-106M** Air is being compressed adiabatically and without friction from 1 bar and 20°C to 3 bars in a closed system.
(a) Calculate the work input and the enthalpy change, in kJ/kg.
(b) For the same conditions, repeat the calculation for air which is cooled externally so that the process is isothermal and frictionless.

**7-107M** Nitrogen at 3 bars and 400°K is allowed to expand adiabatically and without friction in a closed system to a final pressure of 1.7 bars. Compute the work done, in kJ/kg.

**7-108M** One-half kilogram of air is compressed adiabatically and without friction from an initial pressure of 1.3 bars and a volume of 0.3 m³ to a final volume of 0.06 m³. Calculate the change in internal energy of the air, in kilojoules.

**7-109M** One-tenth kilogram of air initially at 1 bar and 27°C is compressed adiabatically and without friction to a final pressure of 1.9 bars. Making use of the air table, compute the change in the internal energy, in kJ.

**7-110M** Air enters a nozzle at 11 bars and 57°C. For a frictionless, adiabatic process, what is the velocity of the air at the exit of the nozzle where the pressure is 4 bars? Neglect the inlet velocity.

**7-111M** Air expands adiabatically and without friction through a nozzle from 8 bars and 227°C to the exit, where the pressure is 3 bars and the area is 8 cm². If the inlet velocity is negligible, compute the exit velocity and the mass flow rate.

**7-112M** Argon, initially at 6.4 bars and 280°C, flows at a rate of 5 kg/s through an insulated, frictionless nozzle. The initial kinetic energy is negligible, and the outlet pressure is 1.4 bars. Determine (a) the final temperature in °C, (b) the final velocity in m/s, and (c) the exit area in square centimeters.

**7-113M** Air enters a diffuser at 0.7 bar and 7°C with a velocity of 300 m/s. The outlet velocity is 70 m/s, and the process is adiabatic and frictionless.
    (a) What is the final temperature in °C?
    (b) What is the final pressure in bars?

**7-114M** Air enters a diffuser at 0.6 bar, −3°C, and 260 m/s. The air stream leaves the diffuser at a velocity of 130 m/s. For adiabatic, frictionless flow, (a) what are the temperature and pressure at the outlet, and (b) what is the ratio of the outlet area to the inlet area?

**7-115M** A heavily insulated turbine receives helium at 8.4 bars and 550°C and exhausts at 1.4 bars. Neglect kinetic- and potential-energy changes. Assuming isentropic flow, determine the exhaust temperature, in °C, and the work output, in kJ/kg.

**7-116M** The inlet conditions for an air turbine are 6 bars and 327°C. The air expands adiabatically and without friction to 1 bar. For a flow rate of 40 kg/min, compute the power output, in kilowatts.

**7-117M** Air is compressed adiabatically and without friction from 0.9 bar and 17°C to 2.7 bars. It is discharged at a velocity of 180 m/s through a duct area of 0.05 m². If the inlet velocity is small, determine the power input required, in kilowatts.

**7-118M** Nitrogen gas flows through a compressor at the rate of 5 kg/s and undergoes an isentropic change from 1 bar and 17°C, to 2.7 bars. The measured kinetic-energy change is 5 kJ/kg. Compute the power input to the gas, in kilowatts.

**7-119M** Argon at a rate of 0.04 kg/s enters a well-insulated, frictionless compressor at 1.6 bars and 25°C. The exit pressure is 6.2 bars, and the kinetic- and potential-energy changes are negligible. Determine (a) the exit temperature, in °K, and (b) the power input, in kilowatts.

**Isentropic processes in superheat and saturation regions**

**7-120M** Refrigerant 12 enters a nozzle and expands from 6 bars and 120°C to a pressure of 1 bar. The nozzle is adiabatic, and the inlet velocity may be neglected. Calculate the exit velocity in m/s if the flow is reversible.

**7-121M** Steam at 3 bars and 200°C expands isentropically through a nozzle to 0.35 bar.
    (a) If the inlet velocity is negligible, what is the discharge velocity, in m/s?
    (b) For a flow rate of 40 kg/min, what is the exit area of the nozzle in square centimeters?

**7-122M** Steam, at 5 bars and 320°C, expands through a nozzle to 3 bars of pressure at a rate of 50 kg/min. If the process is isentropic and the initial velocity is 50 m/s, calculate (a) the velocity at the exit, in m/s, and (b) the exit area in square centimeters.

**7-123M** Refrigerant 12, at 0.6 bar and 20°C, enters a diffuser with a velocity of 200 m/s. The process is isentropic and the exit temperature is 50°C. Determine (a) the exit pressure, in bars, and (b) the exit velocity, in m/s.

**7-124M** One gram of dry, saturated steam at 7 bars expands in a reversible, adiabatic process to a pressure of 1.5 bars. Calculate the work done, in kilojoules, if the system is closed.

**7-125M** Steam undergoes an isentropic expansion from 40 bars and 440°C to 0.1 bar. If the change in kinetic energy is negligible and the mass flow rate is 20,000 kg/h, determine (a) the kilowatt output, and (b) the approximate quality of the steam at the exhaust.

**7-126M** Refrigerant 12 enters a steady-flow turbine at 6 bars and 120°C. It passes through isentropically and leaves at 1 bar. Neglecting potential- and kinetic-energy changes, determine the work output, in kJ/kg.

**7-127M** A steam turbine is limited to a maximum inlet temperature of 540°C. The exhaust pressure is 0.10 bar, and the moisture in the turbine is not to exceed 12.6 percent. What is the maximum allowable inlet turbine pressure if the turbine is adiabatic and reversible?

**7-128M** Refrigerant 12 is compressed from a saturated vapor at $-5$°C to a final pressure of 8 bars in a steady-flow process.
  (a) What is the lowest final temperature possible if the process is adiabatic?
  (b) What is the minimum shaft work required in kJ/kg?
  (c) If the actual final temperature is 12°C higher than the minimum temperature due to irreversibilities, what is the percent increase in shaft work required?

**7-129M** Dry, saturated steam at 1.5 bars is compressed in a reversible, adiabatic process to 7 bars.
  (a) Determine the final temperature, in °K.
  (b) Calculate the work required, in kJ/kg, if the process is steady-flow.
  (c) If the mass flow rate is 10,000 kg/h, what power input is required, in kilowatts?

## Isentropic processes with incompressible substances

**7-130M** Oil with a specific gravity of 0.85 is being pumped from a pressure of 0.7 bar to a pressure of 1.2 bars, and the outlet lies 2 m above the inlet. The fluid flows at a rate of 0.1 $m^3$/s through an inlet cross-sectional area of 0.05 $m^2$, and the outlet area is 0.02 $m^2$. Determine the power input to the pump, in kilowatts.

**7-131M** A pump is used to deliver 5000 kg/h of water from an elevation 10 m below the pump to 15 m above the pump. The inlet conditions are: $D_1 = 4$ cm, $P_1 = 0.7$ bar, and $T_1 = 20$°C, while exit conditions are: $D_2 = 2$ cm and $P_2 = 5$ bar.
  (a) Determine the shaft work, in kJ/kg, if the inlet velocity is 1 m/s and the pipe is insulated.
  (b) Determine the power required for the pump, in kilowatts.

**7-132M** Water at 20°C is pumped at 1.2 kg/s from the surface of an open tank into a constant-diameter piping system. At the discharge from the pipe, the velocity is 10 m/s and the gage pressure is 2.0 bars. The discharge is 15 m above the water surface in the open tank. For reversible, adiabatic flow, determine the power, in kilowatts, required for the pump.

## Steady-flow mechanical work

**7-133M** Carbon monoxide flows frictionlessly and isothermally through a 75-$cm^2$ constant-area duct. Upstream the pressure is 1.6 bars and the temperature is 100°C. After flowing against an impeller which produces shaft work, the gas reaches a downstream position where the pressure is 1 bar. If the mass flow rate is 1.8 kg/s, determine (a) the shaft work, and (b) the heat transfer; express both answers in kJ/kg.

**7-134M** Air is compressed isothermally from 0.96 bar and 7°C to 4.8 bars. Flow through the compressor is steady at 0.95 kg/s. Kinetic and potential energies are negligible. Calculate (a) the power input, in kilowatts, and (b) the rate of heat removal, in kJ/s, if the process is frictionless.

**7-135M** A fluid flows frictionlessly through an open system which requires work. The entrance conditions are $P_1 = 3.4$ bars, $v_1 = 0.6$ $m^3$/kg, and $A_1 = 0.1$ $m^2$. At the exit $P_2 = 13.6$ bars and $A_2 = 0.05$ $m^2$. The mass flow rate is 7 kg/s, and for the process $Pv$ is constant. Find the work required, in kJ/kg.

**7-136M** Helium with a density of 1.1 kg/$m^3$ enters a frictionless, steady-flow system at a pressure of 1.4 bars through a 48-cm-diameter duct with a velocity of 20 m/s. It leaves with a density of 4.8 kg/$m^3$ through a 12-cm-diameter duct. The outlet is located 10 m below the inlet. Calculate the work done on the helium if the process is carried out adiabatically.

**7-137M** Air enters an adiabatic turbine at 6 bars and 227°C. The changes in kinetic and potential energy are negligible, and the final pressure is 1.2 bars. Determine the maximum possible work output, in kJ/kg.

### Adiabatic efficiencies

**7-138M** The nozzle of a turbojet engine received gas from the turbine exhaust, at 1.8 bars and 707°C, with a velocity of 70 m/s. The nozzle expands the gas adiabatically to a pressure of 0.7 bar. Determine the discharge velocity if the nozzle efficiency is 90 percent. Assume the gas is air, and use the air table data. Plot the process on a $Ts$ diagram.

**7-139M** Air, at 1.6 bars and 57°C, discharges through a converging nozzle. The final pressure is 1 bar, and the initial velocity is negligible. If the nozzle efficiency is 92 percent, determine the exhaust velocity, in m/s.

**7-140M** Steam passes through a turbine from an initial state of 100 bars, 520°C, to a final pressure of 15 bars. If the flow is essentially adiabatic and the measured outlet temperature is 280°C, determine the isentropic efficiency of the turbine.

**7-141M** Air expands in a adiabatic turbine from 3 bars and 117°C to 1 bar.
(a) Determine the maximum work output, in kJ/kg.
(b) If the actual outlet temperature is 30°C, determine the turbine isentropic efficiency.

**7-142M** Steam at 20 bars, 440°C, and 80 m/s enters an adiabatic turbine and expands to 0.7 bar at a rate of 10,000 kg/h.
(a) Determine the turbine inlet area, in square centimeters.
(b) If the isentropic efficiency of the turbine is 80 percent, determine the steam outlet tempera-ture, in °C.
(c) Find the power output, in kilowatts.

**7-143M** Air is compressed adiabatically in steady flow from 1 bar and 17°C, to 6 bars. If the compres-sor efficiency is 82 percent, determine (a) the outlet temperature, in °C, (b) the temperature rise, in °C, due to irreversibilities, and (c) the actual work input, in kJ/kg.

**7-144M** Air is compressed in a steady-state device from 1 bar and 27°C to 5 bars. If the outlet temperature is (a) 250°C, and (b) 240°C, determine the adiabatic efficiency of the compressor.

**7-145M** Refrigerant 12 is compressed from a saturated vapor at −10°C to a final pressure of 8 bars. If the process is adiabatic and the compressor efficiency is 78 percent, determine the outlet temperature, in °C.

## PROBLEMS (USCS)

### Carnot heat engines

**7-1** A Carnot engine operates between temperatures of 1440 and 60°F. For every horsepower of output, calculate (a) the heat supplied and the heat rejected in Btu/h, and (b) the thermal efficiency.

**7-2** At what temperature in °F is heat supplied to a Carnot engine that rejects 1000 Btu/min of heat at 40°F and produces 50 hp?

**7-3** A Carnot heat engine operates between 1000 and 500°R. The change in entropy of the supply or source reservoir is 0.5 Btu/°R. Find (a) the cycle efficiency, (b) the heat added, in Btu, and (c) the net work output, in Btu.

**7-4** A reversible heat engine exchanges heat with three reservoirs and produces work in the amount of 400 Btu. Reservoir $A$ has a temperature of 800°R and supplies 2400 Btu to the engine. If reservoirs $B$ and $C$ have temperatures of 600 and 400°R, respectively, how much heat, in Btus, does each exchange with the engine, and what is the direction of the heat exchanges?

**7-5** Same as Prob. 7-4, except that the work output is 600 Btu, and the temperatures of reservoirs $A$, $B$, and $C$ are 1200, 800, and 600°R, respectively.

**7-6** Two Carnot engines $A$ and $B$ are operated in series. The first one ($A$) receives heat at 1100°F and rejects heat to a reservoir at temperature $T$. The second engine ($B$) receives the heat rejected by the first engine and in turn rejects it to a heat reservoir at 80°F. Calculate the temperature $T$ in °F for the situation where (a) the work outputs of the two engines are equal, and (b) the efficiencies of the two engines are equal.

**7-7** Same as Prob. 7-6, except that the temperatures of the heat reservoirs are 1500 and 100°F.

**7-8** A Carnot engine which operates on air as the working medium has a pressure of 200 psia and a specific volume of 1.20 ft$^3$/lb at the beginning of the adiabatic expansion. Nine Btu/lb of heat are added during each cycle, and the sink temperature is 40°F. Compute the efficiency of the engine and the work output in ft·lb$_f$/lb$_m$.

**7-9** A Carnot engine which produces 10 Btu of work for one cycle has a thermal efficiency of 50 percent. The working fluid is 1.0 lb of air, and the pressure and volume at the beginning of the isothermal expansion process are 100.0 psia and 4.0 ft$^3$, respectively. Find (a) the heat transfer and work for each of the four individual processes; (b) the temperature at the end state of each process; (c) the volume at the end of the isothermal expansion process.

**7-10** The thermodynamic fluid in a Carnot engine is 1 lb of air. The engine operates between an energy source at 1200°R and an energy sink at 500°R. At the beginning of the isothermal heat addition the pressure is 100 psia. During the heat-addition process the volume is doubled. Find the net work of the cycle per pound of air.

**7-11** The pressure, volume, and temperature in a Carnot engine (using air as the medium) are, at the beginning of the isothermal expansion, 75 psia, 0.2 ft$^3$, and 500°F, respectively. During the isothermal expansion, 3.0 Btu are added. The maximum volume is 1.2 ft$^3$. Determine (a) the volume after the isothermal expansion, (b) the sink temperature in °F, (c) the heat rejected per cycle, and (d) the thermal efficiency of the cycle.

**7-12** A Carnot cycle operates with steam as the working medium. At the end of the adiabatic compression the pressure is 250 psia and the quality is 0.20. Heat is added during the isothermal expansion until the steam becomes dry and saturated. The steam then expands adiabatically until the pressure is 15 psia. Determine for this cycle (a) the efficiency, and (b) the quantity of work per cycle, in Btu/lb.

**7-13** A Carnot cycle operates with steam as the working fluid in a closed system. At the end of adiabatic compression the pressure is 200 psia and the quality is 20 percent. Heat is added during the isothermal expansion until the steam is a saturated vapor. The fluid then expands adiabatically to 10 psia.
  (a) Determine the thermal efficiency of the cycle.
  (b) Determine the heat input to the cycle, in Btu/lb.

**7-14** A Carnot engine contains 1 lb of water. At the beginning of the heat-addition process the fluid is a saturated liquid and at the end of the process it is a saturated vapor. Heat addition is at 640°F and heat rejection at 40°F.
  (a) Sketch $Pv$ and $Ts$ diagrams of the process.
  (b) Determine the thermal efficiency.
  (c) Determine the heat added and the net work done per cycle, in Btu/lb.

**7-15** A Carnot engine, using steam as the working fluid, operates between 500 and 100°F. At the beginning of the heat-addition process the quality is 50 percent, and at the end of this process the pressure is 400 psia.
  (a) What is the thermal efficiency of the engine?
  (b) The engine contains 0.01 lb of steam and operates at 3000 cycles per minute. What is the horsepower output?

**7-16** In a Carnot engine cycle heat is added in process 1-2 and rejected in process 3-4. For the cycle $T_1 = 1080°R$, $T_4 = 540°R$, and $S_2 - S_1 = 0.1$ Btu/°R.
  (a) Determine the thermal efficiency and the work done in one cycle.
  (b) Now suppose process 2-3 is not reversible. It is still adiabatic, but, as a result of friction, the entropy increases 0.02 Btu/°R during the process. Determine the thermal efficiency in this latter case.

**7-17** A steady-flow-type Carnot heat engine operates between the temperature limits of 740 and 100°F. During the isothermal expansion process 180 Btu of heat are added per pound of fluid. However, irreversibilities in the turbine cause an increase in entropy through the adiabatic turbine amounting to 0.02 Btu/(lb)(°R). Calculate the heat rejected and the ratio of the actual work output to the maximum possible work output if the expansion had been reversible. All other parts of the cycle are assumed to be reversible. Sketch the irreversible cycle on a $Ts$ diagram, indicating the areas which represent heat and work.

## Carnot refrigerators and heat pumps

**7-18** A Carnot heat engine which operates between temperature levels of 1190 and 90°F rejects 30 Btu to the low-temperature sink. The work output of this engine is used to drive a heat pump. The heat pump removes 270 Btu of heat from a low-temperature reservoir and rejects heat to the surroundings at 90°F. Determine the temperature of the low-temperature reservoir for the heat pump, in °F.

**7-19** A Carnot heat pump is used for heating a building. The outside air at 22°F is the cold body, the building at 72°F is the hot body, and 200,000 Btu/h is required for heating. Find (a) heat taken from the outside per hour, and (b) power required.

**7-20** A reversible refrigerating machine absorbs 200 Btu/min from a cold space and requires 2 hp to drive it. If the machine is reversed and receives 800 Btu/min from the hot body, how much power does it produce?

**7-21** A Carnot heat pump operating between the temperatures of 20 and 80°F has a power input of 4 hp. Determine the coefficient of performance.

**7-22** If the thermal efficiency of a Carnot engine is $\frac{1}{6}$, calculate the coefficient of performance of (a) a Carnot refrigerator and (b) a Carnot heat pump.

**7-23** A computer room is to be kept at 70°F. Assume that the air in the room is dry and pure. The computer transfers 1500 Btu/h to the room air. If the surrounding temperature is 105°F, what is the minimum power required to accomplish this refrigeration?

**7-24** A Carnot heat pump is to be used to maintain a home with a heat loss of 80,000 Btu/h at 74°F. If the heat is removed by the heat pump from the outside air at 20°F, find the required horsepower input. If electricity costs 4.5 cents per kilowatthour, what would the operating cost be for 1 day?

**7-25** A reversed Carnot engine is to be used to produce ice at 32°F. The heat-rejection temperature is 81°F, and the enthalpy of fusion of water is 144 Btu/lb. Compute the pounds of ice that can be formed per hour per horsepower of power input.

**7-26** A ton of refrigeration is defined as the heat-absorption rate of 200 Btu/min from the cold source. It is desired to operate a reversed Carnot cycle between the temperature limits of $-10$ and $+85°F$ so that 8 tons of refrigeration are produced. Calculate the power required to operate the refrigeration plant and the coefficient of performance.

**7-27** A heat pump operates on a reversed Carnot cycle, removes heat from a low-temperature source at $-20°F$, and rejects heat to a sink at 80°F. If electricity costs 4.5 cents per kilowatthour, determine the cost of operation for supplying a home with 50,000 Btu/h.

**7-28** A Carnot refrigerator removes heat from a sink at $-20°F$ and rejects heat to the atmosphere at 50°F. The refrigerator is powered by the output of a Carnot engine which receives heat from a source at 1070°F and also rejects to the atmosphere. Determine the ratio of the heat supplied to the engine to the heat removed by the refrigerator.

**7-29** A Carnot heat engine is employed to drive a Carnot refrigerator. The heat engine receives $Q_1$ at $T_1$ and rejects $Q_2$ at $T_2$. The refrigerator removes a quantity of heat $Q_3$ from a source at $T_3$ and rejects a large quantity of heat $Q_4$ at $T_4$. Develop an expression for the ratio $Q_3/Q_1$ in terms of the various temperatures of the heat reservoirs.

**7-30** A reversed Carnot engine employs air as the working fluid. At the beginning of the isothermal expansion the air is at 15 psia, 40°F, and it fills a volume of 2.0 ft³. During the isothermal expansion the volume is doubled. At the end of the adiabatic compression the temperature is 240°F. Calculate the heat added and the heat rejected during the cycle, in Btus.

**7-31** A reversed Carnot engine is being used as a refrigerator. Heat is rejected at 100°F from refrigerant 12, which acts as the working substance. At the beginning of the isothermal expansion, the pressure is 12 psia and the quality is 30 percent. Sufficient heat is added from the low-temperature source to ensure that the refrigerant at the end of the isothermal heat-addition process is dry and saturated. Determine (a) the coefficient of performance for the cycle, and (b) the power input if the mass flow rate is 2.0 lb/min.

**7-32** Initially, two identical bodies of constant specific heat are maintained at the same temperature. These two bodies of finite size are then used as reservoirs for a refrigerator. Heat is removed from one body and rejected to the other body. As a result, the temperature of one body continues to decrease, and that of the other continually increases. The operation is at constant pressure, and it is assumed that neither body undergoes a phase change. Show that the minimum work required to decrease the temperature of the colder body to some value $T_f$ which is less than $T_i$, the initial temperature of both bodies, is

$$W_{min} = mc_p\left(\frac{T_i^2}{T_f} + T_f - 2T_i\right)$$

### Other reversible cycles

**7-33** A reversible heat engine employing an ideal gas with constant specific heats undergoes the following cyclic process: constant-volume heating from state 1 to state 2, adiabatic expansion from state 2 to state 3, and then constant-pressure compression from state 3 to state 1. Prove that the thermal efficiency of the cycle is

$$\eta_{th} = 1 - k\frac{V_3/V_1 - 1}{P_2/P_1 - 1}$$

**7-34** A piston-cylinder assembly contains air initially at 52 psia and 580°F. The gas undergoes the following cyclic process: 1-2, isentropic expansion to 14 psia and 260°F; 2-3, isothermal compression to the original pressure; 3-1, constant-pressure heating to the initial state. Use the air tables to compute (*a*) the heat input in Btu/lb, (*b*) the heat output, and (*c*) the thermal efficiency. Sketch $Pv$ and $Ts$ diagrams for the cycle.

**7-35** A piston-cylinder assembly initially contains air at a low temperature and 1 atm. In order to produce a refrigeration effect, the following cycle is proposed: 1-2, isentropic compression to a state of 3.1 atm and 120°F; 2-3, constant-pressure cooling of 38.4 Btu/lb to the original temperature; 3-1, isothermal expansion with heat addition to the original pressure of 1 atm. Use the air tables.

    (*a*) Sketch $Pv$ and $Ts$ diagrams of the cycle.
    (*b*) Determine the original temperature $T_1$.
    (*c*) Determine the heat added during the isothermal expansion process.
    (*d*) Compute the coefficient of performance of the cycle.

**7-36** One pound of nitrogen is contained in a piston-cylinder arrangement at 15 psia and 40°F. The gas undergoes the following cyclic process: 1-2, constant-volume heating until $P_2$ is 60 psia; 2-3, isentropic expansion to the initial pressure and a temperature of 940°F; 3-1, constant-pressure cooling to the initial state.

    (*a*) Sketch $Pv$ and $Ts$ diagrams for the cycle.
    (*b*) Determine the thermal efficiency of the cycle.

**7-37** Determine the ratio of the efficiencies for the two cycles shown in Fig. P7-37*a* and *b* in terms of the temperatures $T_1$ and $T_2$. What is the limiting value of the ratio as $T_2$ approaches infinity? Explain the significance.

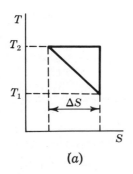

(*a*)

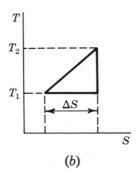

(*b*)

### Irreversible heat transfer and quality of energy

**7-38** One thousand Btu of heat are transferred from a heat reservoir at 1040°F to a second reservoir at (a) 540°F, and (b) 140°F.

(1) Calculate the entropy change of each reservoir in Btu/°R.

(2) Is the sum of the entropy changes of the reservoirs compatible with the second law?

**7-39** One hundred Btu of heat are transferred from an isothermal system at (a) 540°F, and (b) 340°F to the environment at 40°F.

(1) Calculate the entropy change of the system and the environment in Btu/°R.

(2) Is the sum of the entropy changes in agreement with the second law?

**7-40** Consider a heat engine which receives 1000 Btu of heat from a reservoir at (a) 1500°F, and (b) 1700°F. The engine, which is internally reversible, produces 654 Btu of work and rejects heat reversibly to a sink at 80°F. Determine (1) the temperature in °F at which the engine isothermally receives the heat from the source reservoir, and (2) the total entropy change of the process, in Btu/°R.

**7-41** A heat engine receives 100 Btu of heat reversibly from a source reservoir at 1000°F. The engine is internally reversible and produces 58 Btu of work. However, the heat-rejection process to a sink reservoir at (a) 100°F, and (b) 60°F, occurs across a finite temperature difference. Determine (1) the temperature in °F at which the fluid within the engine rejects heat and (2) the total entropy change for the process, in Btu/°R.

**7-42** Which quantity of heat theoretically has the higher quality, 1000 Btu at 1000°F or 3000 Btu at 200°F? The environmental temperature is 40°F.

**7-43** Which quantity of heat theoretically has the higher quality, 2000 Btu at 1500°F or 5000 Btu at 500°F? The environmental temperature is 70°F.

**7-44** Which quantity of energy theoretically has the higher quality, 1000 Btu of shaft work or 3000 Btu of heat at 500°F? The environmental temperature is 60°F.

**7-45** Which quantity of energy theoretically has the higher quality, 40,000 ft$\cdot$lb$_f$ of shaft work or 100 Btu of heat at 400°F? The environmental temperature is 80°F.

### Entropy change of an ideal gas

**7-46** Two pounds of nitrogen gas at 20 psia and 100°F are heated to 400°F in a constant-volume process. Compute the entropy change in Btu/°R.

**7-47** Air is heated in a rigid closed container from 140°F, 15 psia, to 176°F. What is the entropy change of the air in Btu/(lb·mol)(°R)?

**7-48** Find the entropy change of methane ($CH_4$) when the temperature is increased from 80 to 240°F, while the pressure increases by a factor of 2. Use data from Table 3-1.

**7-49** A vertical piston-cylinder assembly contains nitrogen gas at a constant pressure of 1 atm. It is cooled from 300 to 100°F. Compute (a) the work done, in Btu/lb, and (b) the change in entropy, in Btu/(lb)(°R).

**7-50** A rigid tank contains 0.1 lb of carbon monoxide at 14.5 psia and 100°F. Heat is added until the pressure reaches 18.6 psia. Compute (a) the heat transferred, in Btu, and (b) the change in entropy, in Btu/°R.

**7-51** Fifty cubic feet of a gas at 80°F and 14 psia are compressed isothermally to 10 ft³. Is the entropy increased or diminished, and by how much, if the gas is (a) air, or (b) nitrogen?

**7-52** Nitrogen gas initially at 100 psia and 400°F is cooled quasistatically in a closed, rigid tank until the temperature reaches 100°F. Calculate the changes in the internal energy and the entropy in Btu/lb and Btu/(lb)(°R).

**7-53** Calculate the change in entropy of 1 lb of helium which is heated quasistatically at constant pressure from 10 psia, 60°F, to a final temperature of 140°F in a closed system. The molar value of $c_p$ may be taken as 5.0 Btu/(lb·mol)(°F).

**7-54** A piston-cylinder assembly contains air at 630°R and 15 psia. The pressure is increased quasistatically to 20 psia, but during the process heat transfer occurs so that the temperature remains constant. Determine the entropy change in Btu/(lb)(°R).

**7-55** One pound of air at 80°F is heated quasistatically at constant pressure in a closed system to a temperature of 980°F. Determine the change in entropy in Btu/(lb)(°R).

**7-56** An ideal monatomic gas undergoes an irreversible process from 1 atm, 60°F, to 1.8 atm, 130°F. Determine the entropy change of the system in Btu/(lb·mol)(°R).

**7-57** One-half pound of air in a closed system undergoes an irreversible process from 20 psia, 80°F, to a final pressure of 40 psia. Measurements during the process indicate that 3.5 Btu of heat were removed from the air and the work done on the air was 8.2 Btu. Compute the entropy change of the air for the irreversible process, in Btu/°R.

**7-58** Air is contained in a rigid, adiabatic tank in an initial state of 14 psia and 80°F. A paddle-wheel within the system, which is driven by an external motor, stirs the air until the pressure rises to 21 psia. Calculate the work done on the gas in ft·lb$_f$/lb$_m$ and the entropy change in Btu/(lb)(°R).

**7-59** A monatomic gas is stirred by a paddle-wheel in a rigid tank until its temperature changes from 40 to 540°F. If the tank is insulated and the pressure is initially 15 psia, determine (a) the work added, in Btu/lb·mol, and (b) the entropy change, in Btu/(lb·mol)(°R). Assume $c_P = 5.0$ Btu/(lb·mol)(°F) and $c_v = 3.0$ Btu/(lb·mol)(°F).

**7-60** Five pounds of gaseous hydrogen at 100 psia and 40°F are contained in a constant-pressure device; 5145 Btu of heat are added reversibly, and then the hydrogen is stirred by a propeller until the final temperature is 440°F. Compute the total change in entropy of the hydrogen, in Btu/°R.

**7-61** A rigid tank with a volume of 1 ft³ contains oxygen at an initial state of 200°F and 1.5 atm. During a process, paddle-wheel work on the system amounts to 1500 ft·lb$_f$, and a heat loss of 3.52 Btu occurs. Determine the entropy change of the gas, in Btu/°R.

**7-62** A piston-cylinder assembly contains 10 lb of air at 2 atm and 620°F. During a constant-pressure process, 1270 Btu of heat are removed from the air. Determine the entropy change for the air, in Btu/°R, using air table data.

**7-63** One pound of air is compressed isothermally and quasistatically from 15 psia, 500°R, to a final state which requires 55.1 Btu of work input to a piston-cylinder assembly. Find (a) the final pressure, in psia, and (b) the change in entropy, in Btu/°R.

**7-64** A piston-cylinder assembly contains 0.01 lb of hydrogen at 30 psia and 105°F. During a certain experiment, 3.5 Btu of heat are removed from the gas at constant pressure. Determine the entropy change of the hydrogen, in Btu/°R.

**7-65** One pound of air in a closed system undergoes an irreversible process from 10 ft³ and 80°F at constant volume. Measurements during the process indicate that 9.0 Btu of heat were removed and paddle-wheel work was carried out by applying a torque of 23.9 lb$_f$·ft to the shaft for 100 revolutions.
  (a) Calculate the final temperature of the air, in °F.
  (b) Compute the entropy change of the air, in Btu/°R.

**7-66** One pound of air in a closed system undergoes an irreversible process from 20 psia and 80°F to the same final pressure. Measurements during the process indicate that 7.0 Btu of heat were removed and the net work done on the air was 17.3 Btu.
  (a) Determine the final temperature, in °F.
  (b) Compute the entropy change of the air, in Btu/°R.

**7-67** An ideal gas is throttled from a high pressure to a lower pressure through a well-insulated flow restriction. Prove that such a process is internally irreversible.

**7-68** Oxygen is throttled from 30 psia and 1000°R to (a) 20 psia, and (b) 15 psia. Find the change in entropy, in Btu/(lb)(°R).

**7-69** The entropy increase of 1 lb of nitrogen is (a) 0.0362 Btu/(lb)(°R), and (b) 0.0253 Btu/(lb)(°R), when it is throttled adiabatically from 100 psia. Determine the final pressure, in psia.

**7-70** Air flows isothermally and adiabatically at 100°F through a variable-area duct such that the entropy increase due solely to internal irreversibilities is (a) 0.0153 Btu/(lb)(°R), and (b) 0.0350 Btu/(lb)(°R).
  (1) If the initial pressure is 50 psia, determine the final pressure.
  (2) Determine the ratio of inlet to outlet areas, if the device is horizontal.

### Entropy tabulations for superheat and saturation states

**7-71** Find the specific entropy, in Btu/(lb)(°R), for parts (a) through (j) listed in Prob. 4-3 for water substance.

**7-72** Find the specific entropy in Btu/(lb)(°R) for parts (a) through (k) listed in Prob. 4-4 for water substance.

**7-73** Find the specific entropy in Btu/(lb)(°R) for parts (a) through (k) in Prob. 4-5 for refrigerant 12.

**7-74** Find the specific entropy in Btu/(lb)(°R) for parts (a) through (k) listed in Prob. 4-6 for refrigerant 12.

### Energy analysis and entropy changes from tabular data

**7-75** One pound of steam at 200 psia and 400°F undergoes a reversible isothermal expansion in a cylinder having a gastight piston until the pressure reaches 20 psia. Calculate the heat transfer and show the process on a $Ts$ diagram.

**7-76** One pound of refrigerant 12 vapor at 100°F and 125 psia is contained in a closed system at constant pressure. During an adiabatic process 29,820 ft·lb$_f$ of work are performed on the fluid by stirring. The fluid is then cooled at constant pressure by the transfer of 20.8 Btu of heat to the surroundings at 60°F.

    (a) Sketch the complete process on a $Ts$ diagram.

    (b) What is the change in the entropy of the refrigerant, in Btu/°R?

**7-77** Steam is originally at 1200°F, 1000 psia.

    (a) It is expanded isothermally and reversibly to 200 psia in a closed system. Calculate the volume change and work done.

    (b) If the same expansion to 200 psia is carried out in a steady-flow system, calculate the work done in Btu/lb.

**7-78** Refrigerant 12 initially at 50 psia and 100°F undergoes a reversible isothermal expansion to 30 psia in a steady-flow process. No work is done during the process. Compute the change in the kinetic energy of the fluid in Btu/lb.

**7-79** Steam enters a turbine at 600 psia and 800°F. The entropy increase through the turbine is 0.206 Btu/(lb)(°R), and the steam exhausts at 1.0 psia. A mass flow rate of 108,000 lb/h produces a power output of 15,000 hp. If the inlet and outlet velocities are 200 and 400 ft/s, respectively, determine the quantity and direction of the heat transfer in Btu/lb.

**7-80** Saturated liquid refrigerant 12 at 100 psia flows through a long, insulated capillary tube until the pressure reaches 20 psia. Find the entropy change in Btu/(lb)(°R).

**7-81** Refrigerant 12 at 40 psia and 140°F is compressed in a closed system to 200 psia. If the process is isothermal, calculate the heat transfer and work done in Btu/lb.

**7-82** Refrigerant 12 flows through a well-insulated capillary tube. The entering state is 125 psia, 100°F, the exit temperature is 80°F, and the process is adiabatic. What is the change in entropy per pound?

**7-83** A rigid vessel contains 0.5 lb of steam at a pressure of 100 psia and a temperature of 500°F. Heat is transferred out of the system until the temperature drops to 400°F. Determine the entropy change in Btu/°R.

**7-84** A piston-cylinder device contains 0.2 lb of steam initially at 450 psia and 900°F. During a constant-pressure process a heat loss of 33.2 Btu occurs. Determine the entropy change of the steam in Btu/°R.

**7-85** A piston-cylinder device contains 10 lb of steam at 400 psia and 700°F. During a certain experiment 1570 Btu of heat is removed from the steam as the pressure remains constant. Determine the entropy change for the steam, in Btu/°R.

**7-86** A piston-cylinder device contains refrigerant 12 at 50 psia and 120°F. It is compressed quasistatically and at constant pressure, with a boundary-work input of 4450 ft·lb$_f$. At the same time there is a heat loss of 34.08 Btu. If the system contains 1 lb, determine the entropy change in Btu/(lb)(°R).

**Entropy changes of incompressible substances**

**7-87** One hundred pounds of water at 1 atm and 60°F are mixed with 40 lb of water initially at 1 atm and 200°F. If the mixing process is adiabatic and at constant pressure, determine the total charge in entropy for the 140 lb of water, in Btu/°R.

**7-88** A 5-lb iron casting at 950°F is quenched in a tank containing 1 ft$^3$ of water initially at 75°F. Assuming no heat loss to the surroundings and that the specific heat of the iron casting averages 0.11 Btu/(lb)(°F), determine the change in entropy of (a) the water, (b) the casting, and (c) the universe.

**7-89** A 1-lb block of copper at 250°F and a 1-lb block of aluminum at 120°F initially are isolated from each other and from the local surroundings. They are then placed in thermal contact with each other, but are still isolated from the surroundings. Calculate $\Delta S$ for each block if they attain thermal equilibrium. Specific-heat data for pure metals are given in Table A-19

**7-90** A small, solid object of constant heat capacity $C$ equal to 12 Btu/°R is initially at 580°F. The object is then thrown into a large lake which is at a temperature of 60°F. Calculate the total entropy change for this process in Btu/°R. (*Note:* heat capacity $C$ equals mass times specific heat, $mc$).

**7-91** A pound of ice at 32°F is mixed with 6.5 lb of water at 70°F. Compute the change in entropy of the composite system if the process is adiabatic and the pressure is maintained at 14.7 psia. (Enthalpy of melting of water is 144 Btu/lb.)

**Increase in entropy principle**

**7-92** An open feedwater heater mixes liquid water and water vapor from two different sources. In one such device superheated steam enters at one opening at 60 psia and 300°F. Compressed liquid at the same pressure and 100°F enters at another inlet. The mixture of these two streams leaves at the same pressure as a saturated liquid. Calculate the increase in entropy per pound of mixture leaving the heater if the heater is adiabatic.

**7-93** One pound of saturated water vapor is condensed at 340°F. The heat removed is used to vaporize a quantity of saturated liquid refrigerant 12 maintained at 120°F.

    (a) Calculate the entropy change for the water vapor in Btu/°R.

    (b) Calculate $\Delta S$ for the refrigerant 12 in Btu/°R.

    (c) Is the sum of parts (a) and (b) compatible with the second law?

**7-94** Ten grams of nitrogen initially are maintained at a pressure of 1 atm and a temperature of 60°F. It is compressed irreversibly and isothermally to 3 atm. All the heat lost by the system, which amounts to 1.2 Btu appears in the surroundings at a temperature of 60°F. Compute the entropy change of the system, the surroundings, and the universe.

**7-95** A resistor of 20 Ω, initially at 80°F, is maintained at a constant temperature while a current of 5 A is allowed to flow for 2 sec.

    (a) Determine the entropy change of the resistor and of the universe in Btu/°R.

    (b) Then the resistor is insulated and the same experiment is carried out. For the latter case determine the entropy change of the resistor and the universe. The specific heat of the resistor is known to be 1.05 Btu/(lb)(°F), and the mass of the resistor is 0.020 lb.

**7-96** An electrical resistor of 30 Ω is maintained at a constant temperature of 70°F. A current of 5 A is maintained for a time interval of 2 s. Determine, if the temperature of the surroundings is 70°F, (a) the entropy change of the resistor in Btu/°R, and (b) the entropy change of the universe in Btu/°R.

**7-97** Air is initially at 40 psia, 140°F, and expands to 20 psia, 80°F. Can this be a single adiabatic process?

**7-98** Refrigerant 12 flows at a low velocity through a well-insulated capillary tube. The initial state is a saturated vapor at 120 psia, and the final state is 15 psia. Determine the entropy change in Btu/(lb) × (°R). Does this value agree qualitatively with the second law? Sketch the path on a $Ts$ diagram. What is the ratio of the final velocity to the initial velocity?

**7-99** A gas turbine takes in air at 90 psia, and 1100°F and expands it adiabatically to 15 psia and 540°F. KE and PE are negligible.

    (a) Compute the work output using the air tables.

    (b) Determine whether the process is reversible or not.

**7-100** One pound of air in a closed system expands from 40 psia and 500°F to 20 psia and 280°F, producing 22.8 Btu of work. The environmental temperature is 80°F. Determine the entropy change of the system plus environment. Draw the path of the process on a $Ts$ diagram.

**7-101** Twenty pounds of water at 180°F are mixed adiabatically at 1 atm pressure with 12 lb of water at 60°F.
    (a) Compute the entropy change for each quantity of water and the total change in Btu/°R.
    (b) Sketch the separate processes that each quantity of water undergoes on the same $TS$ diagram.

**7-102** Carbon monoxide at 1 atm and 100°F is compressed in a closed system to 3 atm and 300°F. The required work input during the process is 38.0 Btu/lb, and the surrounding temperature is 80°F. Determine the total entropy change of the CO system and the environment, in Btu/°R per pound of CO.

**7-103** Ten pounds of nitrogen in a closed system expand from 45 psia and 320°F to 15 psia and 60°F. During the process 200 Btu of work are produced, and heat is transferred to the environment at 40°F.
    (a) Determine the magnitude, in Btus, and the direction of the heat transfer.
    (b) Show whether the process is reversible, irreversible, or impossible.

**7-104** One pound of air initially at 100 psia and 600°R is contained in a weighted piston-cylinder device which maintains a constant pressure. Heat in the amount of 25.63 Btu is transferred from the air to the atmosphere which is at 80°F, while 5.06 Btu of work are performed on the system by the piston.
    (a) Calculate the final air temperature in °R.
    (b) Determine the entropy change of the air, in Btu/°R.
    (c) Determine the entropy change of the atmosphere, in Btu/°R.
    (d) On the basis of the overall entropy change, does the process satisfy the second law?

**7-105** Refrigerant 12 flows down a horizontal, constant-diameter pipe from an initial state of 70 psia and 120°F to a final state of 60 psia and 80°F. The changes in kinetic and potential energy are negligible. If the environmental temperature is 80°F, find the total entropy change of the process in Btu/°R per pound of refrigerant.

### Isentropic processes of ideal gases

**7-106** Air is being compressed adiabatically and without friction from 14 psia and 60°F to 40 psia in a closed system. Calculate the work input and the enthalpy change in Btu/lb.

**7-107** Nitrogen at 45 psia and 240°F is allowed to expand adiabatically and without friction in a closed system to a final pressure of 25 psia. Compute the work done by each pound of nitrogen in ft·lb$_f$.

**7-108** A mass of 5.5 lb of air is compressed adiabatically and without friction from an initial pressure of 20 psia and volume of 100 ft$^3$ to a final volume of 20 ft$^3$. Calculate the change in internal energy of the air.

**7-109** One pound of air initially at 15 psia and 40°F is compressed adiabatically and without friction to a final pressure of 28.5 psia. Compute the change in the internal energy of the gas in Btu/lb, making use of the air table.

**7-110** Air enters a nozzle at 160 psia and 120°F. For a frictionless, adiabatic process, what is the velocity of the air at the exit of the nozzle where the pressure is 50 psia? Neglect the entrance velocity.

**7-111** Air expands adiabatically and without friction through a nozzle from 120 psia and 440°F, to the exit, where the pressure is 20 psia and the area is 1.0 in$^2$. The inlet velocity is negligible. Compute the exit velocity and the mass flow rate.

**7-112** Argon, initially at 100 psia and 540°F, flows at a rate of 10 lb/s through an insulated, frictionless nozzle. The initial kinetic energy is negligible, and the outlet pressure is 20 psia. Determine (a) the final temperature in °F, (b) the final velocity, and (c) the exit area in in$^2$.

**7-113** Air enters a diffuser at 10 psia and 40°F with a velocity of 900 ft/s. The outlet velocity is 200 ft/s, and the process is adiabatic and frictionless.
    (a) What is the final temperature in °F?
    (b) What is the final pressure in psia?

**7-114** The diffuser of a jet engine draws in air at a pressure of 8.0 psia, a temperature of 20°F, and a velocity of 800 ft/s. The air stream leaves the diffuser at a velocity of 400 ft/s. If the flow is adiabatic and quasistatic, what are the temperature and the pressure at the outlet? What is the ratio of outlet area to inlet area?

**7-115** A heavily insulated turbine receives helium at 100 psia and 1000°F and exhausts at 20 psia. The process is frictionless, and potential- and kinetic-energy changes are small. Determine the exhaust temperature in °F and the work done in Btu/lb.

**7-116** The inlet conditions for a gas turbine operating on air are 90 psia and 620°F. The air expands adiabatically and without friction to a final pressure of 15 psia. The flow rate is 100 lb/min. Compute the power output in horsepower if the change in kinetic energy is small.

**7-117** Air is being compressed adiabatically and without friction from 14 psia and 60°F to 40 psia, where it is discharged at a velocity of 550 ft/s through a duct area of 0.40 ft². If the inlet velocity is small, determine the power required in horsepower.

**7-118** Nitrogen gas flows through a compressor at the rate of 10 lb/s and undergoes an isentropic change from 15 psia and 60°F, to 40 psia. The measured kinetic-energy change is 2.5 Btu/lb. Compute the power input to the gas.

**7-119** Argon at a rate of 0.1 lb/s enters a well-insulated, frictionless compressor at 20 psia and 80°F. The exit pressure is 100 psia, and the kinetic- and potential-energy changes are negligible. Determine the exit temperature in °R and the power input in horsepower.

**Isentropic processes in the superheat and saturation regions**

**7-120** Refrigerant 12 enters a nozzle and expands from 90 psia and 240°F to 15 psia. The nozzle is adiabatic, and the inlet kinetic energy may be neglected. Calculate the exit velocity in ft/s if the flow is reversible.

**7-121** Steam at 60 psia and 350°F expands isentropically in a nozzle to 5 psia.
    (a) If the inlet velocity is negligible, determine the discharge velocity in ft/s.
    (b) For a flow rate of 100 lb/min, what is the required exit area in in²?

**7-122** Steam at 60 psia and 600°F expands through a nozzle to 40 psia at a rate of 100 lb/min. If the process is isentropic and the initial velocity is 100 ft/s, calculate (a) the velocity at the exit of the nozzle in ft/s and (b) the exit area in ft².

**7-123** Steam at 40 psia and 300°F enters a diffuser with a velocity of 1500 ft/s. If the process is isentropic and the exit velocity is 450 ft/s, determine the exit pressure and temperature in psia and °F.

**7-124** One pound of dry saturated steam at 100 psia expands in a reversible adiabatic process to a pressure of 20 psia. Calculate the work done in Btu/lb if the system is closed.

**7-125** Steam undergoes an isentropic expansion from 600 psia and 800°F to 1 psia. If the change in kinetic energy is negligible and the mass flow rate is 40,000 lb/h, determine (a) the power output in horsepower, and (b) the approximate quality of the steam exhausting from the turbine.

**7-126** Refrigerant 12 enters a steady-flow turbine at 90 psia and 240°F. The flow is isentropic and the exit pressure is 30 psia. Neglecting kinetic- and potential-energy changes, determine the work output in Btu/lb.

**7-127** A steam turbine is limited to a maximum inlet temperature of 1000°F. The exhaust pressure is 1 psia, and the moisture in the turbine exhaust stream is not to exceed 10 percent. What is the maximum allowable inlet turbine pressure if the turbine is adiabatic and reversible?

**7-128** Refrigerant 12 is compressed from a saturated vapor at −20°F to a final pressure of 125 psia in a steady-flow process.
    (a) What is the lowest final temperature possible?
    (b) What is the minimum shaft work required in Btu/lb?
    (c) If the actual final temperature is 20°F higher than the minimum temperature due to the presence of internal irreversibilities, what is the percent increase in the shaft work required? Show the processes on a Ts diagram.

**7-129** Saturated water vapor is compressed in a reversible, adiabatic process from 20 psia and 300°F, to 120 psia.

(a) Find the final temperature in °F.

(b) If the system is steady-flow, calculate the work required in Btu/lb.

(c) If the mass flow rate is 20,000 lb/h, compute the required power input in horsepower.

### Isentropic processes with incompressible substances

**7-130** Oil with a specific gravity of 0.85 is being pumped from a pressure of 8 in Hg vacuum to a pressure of 18 psig. The outlet of the piping system lies 18 ft above the inlet. The fluid flows adiabatically at a rate of 3 ft$^3$/s through an inlet cross-sectional area of 0.50 ft$^2$, and the outlet area is 0.30 ft$^2$. If the inlet temperature is 60°F, determine the minimum horsepower required for the pump.

**7-131** A pump is used to deliver 11,000 lb/h of water from an elevation 30 ft below the pump to an elevation 50 ft above the pump. At the lower elevation (state 1) and the upper elevation (state 2) the known data are: $A_1 = 0.0233$ ft$^2$, $P_1 = 10$ psia, $T_1 = 60$°F, $A_2 = 0.0060$ ft$^2$, and $P_2 = 80$ psia. If the inlet velocity is 2.1 ft/s and the pipe is insulated, determine the minimum pump work required, in horsepower.

**7-132** Water at 70°F is pumped adiabatically at 2.5 lb/s from the surface of an open tank into a constant-area piping system. At the discharge from the pipe, which is 48 ft above the water surface of the open tank, the velocity is 30 ft/s and the gage pressure is 25 psig. Determine the minimum horsepower required to drive the pump.

### Steady-flow mechanical work

**7-133** Carbon monoxide flows frictionlessly and isothermally through a constant-area duct of 12.0 in$^2$. Upstream at position (1) the pressure is 20 psia and the temperature is 200°F. After flowing against an impeller which produces shaft work, the gas reaches a downstream position where the pressure is 15 psia. If the mass flow rate is 3.95 lb/s, determine (a) the shaft work in ft·lb$_f$/lb$_m$, and (b) the heat transfer in the same units.

**7-134** Air is compressed isothermally from 14 psia, and 40°F, to 70 psia. Flow through the compressor is steady at 2.0 lb/s. Kinetic and potential energies are negligible. Assuming that air is an ideal gas, calculate the power input to the compressor and the heat removed from the air, both in Btu/s, if the process is frictionless.

**7-135** Fluid flows frictionlessly in an open system. Conditions at the entrance of the system are $P_1 = 50$ psia, $v_1 = 10$ ft$^3$/lb, and $A_1 = 1.0$ ft$^2$. At the exit of the system $P_2 = 200$ psia and $A_2 = 0.5$ ft$^2$. The mass rate of flow is 10 lb/s, and, for the process, $Pv = $ constant. Determine the mechanical work for the system.

**7-136** Helium with a density of 0.07 lb/ft$^3$ enters a frictionless steady-flow system at a pressure of 20 psia through a 24-in-diameter duct with a velocity of 20 ft/s. It leaves with a density of 0.3 lb/ft$^3$ through a 6-in-diameter duct. The outlet of the system is located 20 ft below the inlet. Calculate the work done on the helium if the process is carried out adiabatically.

**7-137** Air enters an adiabatic turbine at 100 psia and 440°F. The changes in kinetic and potential energy are negligible, and the final pressure is 20 psia. Determine the maximum possible work output in Btu/lb.

### Adiabatic efficiencies

**7-138** The nozzle of a turbojet engine receives gas from the turbine exhaust at 27 psia and 1300°F, with a velocity of 200 ft/s. The nozzle expands the gas adiabatically to a pressure of 10.5 psia. Determine the discharge velocity from the nozzle if the nozzle efficiency is 90 percent. Assume the gas is air, and use variable specific heat data. Plot the process on a $Ts$ diagram.

**7-139** Air at 24.2 psia and 600°R discharges through a converging nozzle. The final pressure is 14.7 psia, and the initial velocity is negligible. If the nozzle efficiency is 89 percent, determine the exhaust velocity in ft/s.

**7-140** Steam expands through a turbine from 800 psia and 800°F to 14.7 psia and 250°F. Determine the adiabatic efficiency in percent.

**7-141** Air expands through a turbine from 45 psia and 240°F, to 15 psia and 90°F. Determine the adiabatic efficiency in percent.

**7-142** Steam expands through a turbine from 2000 psia and 1000°F to 10 psia and 93 percent quality. Determine the turbine adiabatic efficiency.

**7-143** Air is compressed adiabatically from 14.5 psia and 40°F to 98 psia. If the compressor efficiency is 82 percent, find the outlet temperature in °F and the work required in Btu/lb for a steady-flow process.

**7-144** Air is compressed from 15 psia and 60°F to 90 psia and 480°F in an adiabatic, steady-flow process. Determine the adiabatic efficiency of the compressor.

# EIGHT

## AVAILABLE ENERGY AND AVAILABILITY

The first law of thermodynamics provides a means of interrelating various forms of energy in terms of their quantities. However, various forms of the same quantity of energy may differ in their usefulness to society. This implies that energy has quality in addition to quantity. The second law provides several means for assessing the quality of energy. Among the concepts used for this purpose are the parameters known as available energy and availability. This chapter presents the basic development of these concepts for closed and open systems.

## 8-1 AVAILABLE AND UNAVAILABLE ENERGY

In the operation of any reversible heat engine a portion of the heat supplied from a constant-temperature source must be rejected to a sink at a lower temperature. We have noted that the amount of heat rejected for a given heat input is reduced by lowering the sink temperature $T_L$. For all practical purposes there is a minimum value of $T_L$ which can be considered in most cases. This normally corresponds to the temperature of the atmospheric air, or of a large body of water, or even of the earth at that particular time. Although this ambient temperature is a variable, its range of variation is quite small. Since any practical heat sink is quite large, its temperature may be considered to be constant during any finite addition of heat to it. The lowest practical sink temperature for a given application will be designated as $T_0$.

A reversible engine such as the Carnot engine operates between two heat reservoirs of fixed temperature. This engine has the maximum thermal efficiency of all engines operating between two temperature extremes. However, many applications of interest involve a heat source of variable temperature. This is true since most energy sources are of limited size. Hence any energy in the form of heat removed from them causes their temperature to drop. It is important, then, to define terminology and derive relationships which would prove useful in determining the effectiveness of heat engines operating with heat supplied from a source of variable temperature. The effects of irreversible heat transfer on the ultimate production of work will also be examined.

The simple concepts of available and unavailable energy are useful to clarify the situation. These are defined in the following way: *Available energy* is that portion of the heat added to a system which, instead, could have been converted into work if it is added to a reversible heat engine which rejects to a reservoir at $T_0$. *Unavailable energy* is simply that portion of a quantity of heat which is not converted into work if it is added to a reversible engine which rejects to a reservoir at $T_0$. These quantities of energy will be represented by $Q_A$ and $Q_U$, respectively. If a quantity of heat $Q$ is added to the heat engine, then $Q = Q_A + Q_U$. In the case of a Carnot engine, it is apparent that, in terms of the nomenclature previously used, $Q = Q_H$, $Q_A = W$, and $Q_U = Q_L$.

A more general problem is illustrated in Fig. 8-1. Heat is added reversibly to a system the temperature of which rises from an initial state $i$ to a final state $f$ during the process. The question arises as to how much of this heat quantity is available energy. In other words, if this same quantity of heat were added directly to a reversible heat engine instead of to the system, what would be the maximum amount of work that the engine could produce? Since the temperature level varies for each increment of heat added to the engine, it is necessary, first, to consider any differential increment of heat $\delta Q_x$ added to the heat engine at a temperature $T_x$. Because the temperature of this small quantity of heat is essentially constant, the

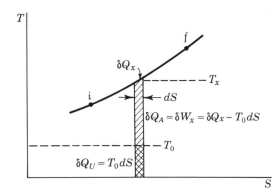

**Figure 8-1** Available and unavailable energy for a differential addition of heat to a system.

heat engine operates as a Carnot engine between the temperature levels of $T_x$ and $T_0$. The maximum work delivered by the engine will be

$$\delta W_{max} = (T_x - T_0)\frac{\delta Q_x}{T_x} = \delta Q_x - \frac{T_0\,\delta Q_x}{T_x} \tag{8-1}$$

where $\delta Q_x$ is the increment of heat added to the engine. The unavailable energy is given by

$$\delta Q_U = \delta Q_x - \delta W_{max} = \frac{T_0\,\delta Q_x}{T_x} \tag{8-2}$$

For the finite process from state $i$ to state $f$ one must integrate the above equations. This results in

$$W_{max} = Q - T_0\int_i^f \frac{\delta Q_x}{T_x}$$

and

$$Q_U = T_0\int_i^f \frac{\delta Q_x}{T_x}$$

However, the integral of $\delta Q_x/T_x$ is simply the entropy change of the system when heat is added reversibly to it. Consequently,

$$W_{max} = Q - T_0(S_f - S_i) = Q - T_0\,\Delta S \tag{8-3}$$

and

$$Q_U = T_0(S_f - S_i) = T_0\,\Delta S \tag{8-4}$$

Since the heat was added reversibly, we may also write

$$\delta Q_A = \delta W_x = T\,dS - T_0\,dS = (T - T_0)\,dS \tag{8-5}$$

Equation (8-5) may not be integrated directly to find $Q_A$ unless the relationship between $T$ and $S$ is known along the complete path. Figure 8-2 shows the avail-

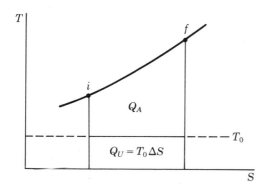

**Figure 8-2** Available and unavailable energy quantities.

able and unavailable portions of the total heat supplied to the engine. The unavailable energy is always determined by $T_0 \, \Delta S$ and is represented by a rectangular area on a $TS$ plot. The quantity $Q$ is usually found from the conservation of energy principle.

**Example 8-1M** Air in a closed system at 1 bar and 27°C is heated in a reversible manner at constant pressure until its temperature reaches 227°C. How much of the heat added is available energy, in kJ/kg of air, if the lowest sink temperature is 7°C?

SOLUTION The problem reduces to this question: If the air is now cooled reversibly to its initial state by removing heat, and this heat is supplied to a series of reversible heat engines, what is the maximum work that can be produced? Figure 8-2 again illustrates the general problem, where the line passing through states $i$ and $f$ is now a constant-pressure line. The heat removed from the air is found from the first law for the constant-pressure process. If the air is an ideal gas, and using the air table,

$$q_{\text{out}} = \Delta u - w = \Delta h = h_2 - h_1$$

$$= 503.0 - 300.2 = 202.8 \text{ kJ/kg}$$

The absolute value of the change in entropy of the air is

$$\Delta s = s_2^0 - s_1^0 - R \ln P_2/P_1 = 2.21952 - 1.70203 = 0.5175 \text{ kJ/(kg)(°K)}$$

Hence

$$q_U = T_0 \, \Delta s = 280(0.5175) = 145 \text{ kJ/kg}$$

The available-energy part of the total heat supplied, then, is

$$q_A = q - q_U = 202.8 - 145 = 57.8 \text{ kJ/kg}$$

When the 202.8 kJ of heat was added reversibly to each kilogram of air, a certain portion of this heat was still available for producing work at a later time. However, the available energy in this case represents only 29 percent of the total heat added.

**Example 8-1** Air at 15 psia and 80°F is heated in a reversible manner at constant temperature until its temperature reaches 400°F. How much of the heat added is available energy per pound of air heated if the lowest sink temperature is 40°F?

SOLUTION The problem really reduces to this: If the air is now cooled reversibly to its initial state by removing heat, and this heat is supplied to a series of reversible engines, what is the maximum work that can be gained by such a device? A series of engines is required since the temperature level of each successive increment of heat is less. Figure 8-2 again illustrates the general problem, where the line passing through the states $i$ and $f$ is now a constant-pressure line. The heat removed from the air is found from the first law for the constant-pressure process. If the air behaves ideally,

$$q_{\text{out}} = \Delta u - w = \Delta h = c_p(T_f - T_i)$$

$$= 0.240(400 - 80) = 76.8 \text{ Btu/lb}$$

The change in entropy of the air during cooling is

$$\Delta s = \int \frac{\delta q}{T} = \int c_p \frac{dT}{T} = c_p \ln \frac{T_f}{T_i}$$

$$= 0.240 \ln \frac{860}{540} = 0.1118 \text{ Btu/(lb)(°R)}$$

Hence

$$q_U = T_0 \, \Delta s = 500(0.1118) = 55.9 \text{ Btu/lb}$$

The available-energy part of the total heat supplied, then, is

$$q_A = q - q_U = 76.8 - 55.9 = 20.9 \text{ Btu/lb}$$

When the 76.8 Btu of heat was added in a reversible manner to each pound of air, a certain portion of this heat was still available for producing work at a later time. However, the available energy in this case represents only 27 percent of the total heat added per pound.

In the preceding example the heat transfer to the gaseous system occurred reversibly. Since each increment of heat remained at the same level of temperature, the available energy was the same whether the total energy was stored in the gas or in the original undefined energy source. The significance of the concept of available energy is more clearly seen if one considers the same quantity of heat now transferred to the gaseous system across a finite temperature difference. The following examples illustrate the point.

**Example 8-2M** Air in a closed system at 1 bar and 27°C is heated at constant pressure until its temperature reaches 227°C. The heat is supplied from a constant-temperature reservoir. Determine how much of the heat removed from the reservoir is available and unavailable energy for reservoir temperatures of (a) 550°K, and (b) 800°K. The lowest surrounding temperature is 280°K. Compare the results with those of the preceding example (8-1M).

SOLUTION (a) The total heat removed from the heat reservoir at 550°K is again 202.8 kJ/kg of air. However, in this new case the heat might have been supplied to a single Carnot heat engine since the supply temperature is now constant. Therefore

$$q_U = q \, \frac{T_0}{T_H} = 202.8 \, \frac{280}{550} = 103 \text{ kJ/kg}$$

and

$$q_A = w_{max} = q - q_U = 202.8 - 103 = 100 \text{ kJ/kg}$$

From the previous example (8-1M) we have seen that, if the same quantity of heat is run through a series of heat engines by cooling air from 500 to 300°K, the available and unavailable energies are 57.8 and 145 kJ/kg, respectively. Thus it is seen that heat is more useful to us when it is supplied from a higher temperature level. When heat is transferred irreversibly from the reservoir at 550°K to the air, a theoretical loss of 42.2 kJ/kg of available energy results. This is a 42 percent loss.

The relative change in the available and unavailable energies due to irreversible heat transfer is conveniently shown on a $TS$ diagram. Since both the heat reservoir and the air may be made to undergo internally reversible processes, the heat quantities are areas on the diagram. In Fig. 8-3 the available and unavailable energies for Example 8-1 and part a of this example are illustrated. The total area under each curve must equal 202.8 kJ.

(b) A further increase in the supply-reservoir temperature increases the loss of available energy due to the large irreversibilities of the heat-transfer process. It can be shown that when the high-temperature reservoir is at 800°K, then $q_A$ and $q_U$ become 131.8 and 71.0 kJ/kg, respectively. The result of the irreversible heat transfer of 202.8 kJ/kg of air from 800°K to the heated air would be the loss of 74.0 kJ/kg of available energy, or 56 percent of the maximum.

**Example 8-2** Air at 15 psia and 80°F is heated at constant pressure until its temperature reaches 400°F. The heat is supplied from a constant-temperature reservoir. Determine how much of the heat removed from the reservoir is available and unavailable energy for the reservoir

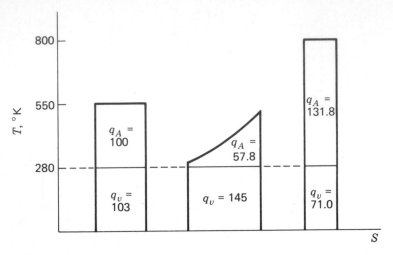

**Figure 8-3** The effect of heat transfer across a finite temperature difference on the available energies in Example 8-2M.

temperatures of (*a*) 540°F and (*b*) 1040°F. The lowest surrounding temperature is 40°F. Compare the results with those of the preceding example (8-1).

SOLUTION (*a*) The total heat removed from the heat reservoir at 540°F is again 76.8 Btu/lb of air. However, in this new case the heat might have been supplied to a single Carnot engine since the supply temperature is now constant. Therefore

$$q_U = q \frac{T_0}{T_H} = 76.8 \frac{500}{1000} = 38.4 \text{ Btu/lb}$$

and
$$q_A = w_{net} = q - q_U = 38.4 \text{ Btu/lb}$$

We have seen (Example 8-1) that, if the same quantity of heat is run through a series of heat engines by cooling air from 400 to 80°F, the available and unavailable energies are 20.9 and 55.9 Btu/lb, respectively. Thus it is seen that heat is more useful to us when it is supplied from a higher temperature level. When heat is transferred irreversibly from the reservoir at 1000°R to the air (in a range from 540 to 960°R), a theoretical loss of 17.5 Btu-lb of available energy, or a 45.6 percent loss, results.

The relative change in the available and unavailable energies due to irreversible heat transfer is conveniently shown on a *TS* diagram. Since both the heat reservoir and the air may be made to undergo internally reversible processes, the heat quantities are areas on the diagram. In Fig. 8-4 the available and unavailable energies for Example 8-1 and part *a* of this example are illustrated. The total area under each curve must equal 76.8 Btu.

(*b*) A further increase in the supply-reservoir temperature increases the loss of available energy due to the larger irreversibilities of the heat-transfer process. It can be shown that when the high-temperature reservoir is at 1500°R, then $q_A$ and $q_U$ become 51.2 and 25.6 Btu/lb, respectively. The result of the irreversible heat transfer of 76.8 Btu/lb of air from 1500°R to the heated air would be the loss of 30.3 Btu/lb of available energy, or 59 percent of the maximum.

Example 8-2M (and Example 8-2) has illustrated the highly undesirable characteristic of irreversible heat transfer in terms of available energy. It also points out a deficiency of the first law, which recognizes heat and work as forms of

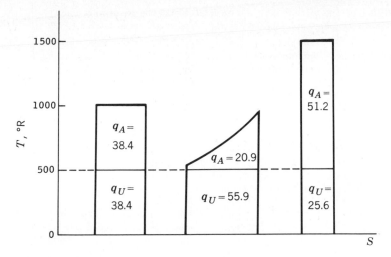

**Figure 8-4** The effect of heat transfer across a finite temperature difference on the available and unavailable energies in Example 8-2.

energy which are related to each other only in terms of the change in the energy of a closed system during a process. However, the second law recognizes not only that heat is not completely convertible into work by a cyclic process, but also that heat has a sense of quality about it. The quality of heat is with reference to the temperature at which it is available for use in a heat engine. The higher the temperature of a heat reservoir, the higher the quality of a unit of energy stored therein. Energy at high temperatures has a higher quality because a greater portion of it is convertible into available energy than the same quantity of energy stored in heat reservoirs at lower temperatures. The following examples further illustrate the usefulness of the concept of available and unavailable energy.

**Example 8-3M** The conversion of work into less useful forms of energy due to irreversible processes is often unavoidable. A prime example is the expenditure of paddle-wheel work. Consider a closed, insulated piston-cylinder device maintained at a pressure of 2 bars. Inside the cylinder, initially, is 1 kg of saturated liquid water. A paddle-wheel inside stirs the contents until 80 percent of the water is vaporized. How much of the work added can be converted back into work if the surrounding temperature is 20°C?

SOLUTION Work can be produced by removing the insulation and cooling the water until it all becomes liquid again. Since the condensation process is a constant-temperature one, the maximum work obtainable is found by supplying the heat removed from the water to a Carnot heat engine. The temperature limits for the engine would be the saturation temperature of 120.2°C (at 2 bars) and 20°C. The heat added to the engine is simply $xh_{fg} = 0.80(2201.9) = 1762$ kJ/kg. Since the heat rejected, or the unavailable energy, is $T_0 \, \Delta s$,

$$q_U = T_0 \, \Delta s = T_0 x s_{fg} = 293(0.80)(5.597) = 1312 \text{ kJ/kg}$$

For a paddle-wheel-work addition of 1762 kJ/kg, the available energy regained is

$$q_A = q - q_U = 1762 - 1312 = 450 \text{ kJ/kg}$$

This represents only 26 percent of the original expended work. The use of work to carry out a process which could have been accomplished by the addition of heat is never desirable if it is avoidable.

If, instead, the water had been vaporized by a heat addition from a reservoir at 450°K, a considerably smaller loss of work would have occurred. In this case

$$q_U = q\frac{T_0}{T_H} = 1762\frac{293}{450} = 1147 \text{ kJ/kg}$$

and

$$q_A = q - q_U = 1762 - 1147 = 615 \text{ kJ/kg}$$

Therefore the loss of possible work due to heat addition would have been only 165 kJ/kg (615 − 450) compared to 1312 kJ/kg (1762 − 450) when paddle-wheel work was employed. Hence a work effect should never be used where a heat interaction could accomplish the same result, if irreversibilities are to be kept to a minimum.

**Example 8-3** The conversion of work into less useful forms of energy due to irreversible processes is often unavoidable. A prime example of this is the expenditure of paddle-wheel work. Consider a closed, insulated piston-cylinder assembly maintained at a pressure of 25 psia. Inside the cylinder, initially, is 1 lb of saturated liquid water. A paddle-wheel inside stirs the contents until the water is 80 percent vaporized. How much of the work added can be converted back into work if the surrounding temperature is 80°F?

SOLUTION Work can be produced by removing the insulation and cooling the system until all the water becomes saturated liquid again. Since the condensation process is a constant-temperature one, the maximum work obtainable is found by supplying the heat removed from the water to a Carnot engine. The temperature limits for this engine would be the saturation temperature of 240°F (at 25 psia) and 80°F. The heat added to the engine is simply $xh_{fg} = 0.80(952.1) = 761.7$ Btu/lb. Since the heat rejected, or the unavailable energy, is $T_0\,\Delta s$.

$$q_U = T_0\,\Delta s = T_0\,xs_{fg} = 540(0.80)(1.3606) = 587.8 \text{ Btu/lb}$$

For a paddle-wheel-work addition of 761.7 Btu/lb, the available energy regained is

$$q_A = 761.7 - 587.8 = 173.9 \text{ Btu/lb}$$

This represents only 23 percent of the original expended work. The use of work to carry out a process which could have been accomplished by the addition of heat is never desirable, if it is avoidable. If the water, instead, had been vaporized by a heat addition from a reservoir at 1000°R, a considerably smaller loss of work would have occurred. In this case

$$q_U = q\frac{T_0}{T_H} = 761.7\frac{540}{1000} = 411.3 \text{ Btu/lb}$$

and

$$q_A = q - q_U = 761.7 - 411.3 = 350.4 \text{ Btu/lb}$$

Therefore the loss of possible work due to heat addition would have been only 176 Btu/lb (350.4 − 173.9) compared with 588 Btu/lb (761.7 − 173.9) when paddle-wheel work was employed. It is important to note that in this case that the temperature of the supply-heat reservoir should be as low as possible, and yet allow heat transfer to the water. Since the water was maintained at 700°R throughout the process, the supply reservoir could well have been as low as 750°R, instead of the 1000°R used in the problem. This would further reduce the loss of possible work due to irreversible heat transfer from 176 Btu/lb to, roughly, 39 Btu/lb. Hence a work effect should never be used where a heat interaction could accomplish the same result, if irreversibilities are to be kept to a minimum.

**Example 8-4M** One kilogram of saturated liquid water at 3 bars is vaporized by cooling air at 1 bar pressure from 540 to 320°C. Determine the change in available energy due to the irreversible heat transfer, and draw a $TS$ plot for the process. The lowest surrounding temperature is 5°C.

SOLUTION First the available energy obtained by cooling the air and supplying this heat directly to a reversible engine will be calculated. The heat removed is the enthalpy of condensation at 3 bars or 133.6°C. Table A-15M lists this as 2163.8 kJ/kg. The quantity of air cooled is found by applying an energy balance on the heat-transfer process.

$$Q(\text{air cooled}) = Q(\text{water evaporated})$$

$$(mc_p\,\Delta T)_{\text{air}} = (mh_{fg})_{\text{water}}$$

For an average $c_p$ value of 1.07 J/(g)(°C), $m_{\text{air}}$ is found to be 9.19 kg. The total entropy change of the air during the constant-pressure cooling process is

$$\Delta S = mc_p\,\ln\,\frac{T_2}{T_1}$$

$$= 9.19(1.07)\,\ln\,\frac{593}{813} = -3.10 \text{ kJ/}°\text{K}$$

The heat rejected, or the unavailable energy, may now be calculated by using the absolute value of $\Delta S$ determined above. Thus

$$Q_U = T_0\,\Delta S = 278(3.10) = 862 \text{ kJ}$$

The available energy is

$$Q_A = Q - Q_U = 2164 - 862 = 1302 \text{ kJ}$$

After the irreversible heat-transfer process to the water occurs, some available energy still exists. The evaporated water may be cooled to its initial state, and the heat removed may be used in a heat engine. Since the heat removed would all be at 133.6°C (the saturation temperature), the available energy would be that obtainable from a single Carnot engine.

$$Q_A = Q\left(1 - \frac{T_0}{T_H}\right) = 2164(1 - 278/407) = 686 \text{ kJ}$$

The available energy has been reduced 47 percent $[(1302 - 686)/1302]$ by the irreversible heat transfer from the hot air to the boiling water. Figure 8-5 is a $TS$ diagram for the heat-transfer process described above. The areas for the cooled air and the heated water are superimposed, to emphasize the large increase in unavailable energy. The lined area in the lower left-hand corner is the unavailable energy when the heat from the air is used directly in a heat engine. The cross-hatched area is the additional unavailable energy created when the energy from the air is used to heat the water. The reduction in the available energy is quite apparent.

**Example 8-4** One pound of saturated liquid water at 50 psia is vaporized by cooling air at 1 atm pressure from 1000 to 600°F. Determine the change in available energy due to the irreversible heat transfer, and draw a $TS$ plot for the process. The lowest surrounding temperature is 40°F.

SOLUTION First the available energy obtained by cooling the air and supplying this heat directly to a reversible engine will be calculated. The heat removed is the latent heat of vaporization at 50 psia and 281°F. Steam table A-15 lists this as 924.2 Btu/lb. The quantity of air cooled is found by applying an energy balance on the evaporation process.

$$Q(\text{air cooled}) = Q(\text{water evaporated})$$

$$(mc_p\,\Delta T)_{\text{air}} = (mh_{fg})_{\text{water}}$$

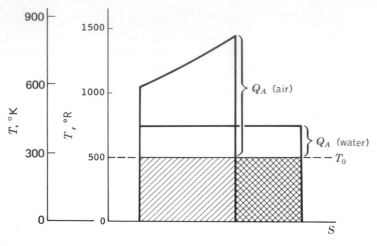

**Figure 8-5** The effect of irreversible heat transfer from hot gases to evaporating water on $Q_A$ and $Q_U$.

For an average $c_p$ value of 0.257 Btu/(lb)(°F), $m_{air}$ is found to be 8.99 $lb_m$. The total entropy change of the air during the constant-pressure cooling process is

$$\Delta S = \int \frac{\delta Q}{T} = \int mc_p \frac{dT}{T} = mc_p \ln \frac{T_2}{T_1}$$

$$= 8.990(0.257) \ln \frac{1060}{1460} = -0.737 \text{ Btu/°R}$$

The heat rejected, or the unavailable energy, may now be calculated by using the absolute value of $\Delta S$ determined above. Thus

$$Q_U = T_0 \ \Delta S = 500(0.737) = 368.5 \text{ Btu}$$

The available energy is

$$Q_A = Q - Q_U = 924.2 - 368.5 = 555.5 \text{ Btu}$$

After the irreversible heat-transfer process to the water occurs, some available energy still exists. The evaporated water may be cooled to its initial state, and the heat removed may be used in a heat engine. Since the heat removed would all be at 281°F (the saturation temperature), the available energy would be that obtainable from a single Carnot engine.

$$Q_A = Q \left(1 - \frac{T_0}{T_H}\right) = 924.2(1 - 500/741) = 301 \text{ Btu}$$

The available energy has been reduced 46 percent [(555.5 − 301)/555.5] by the irreversible heat transfer from the hot air to the boiling water.

Figure 8-5 is a $TS$ diagram for the heat-transfer process described above. The areas for the cooled air and the heated water are superimposed, to emphasize the large increase in unavailable energy. The lined area in the lower left-hand corner is the unavailable energy when the heat from the air is used directly in a heat engine. The crosshatched area is the additional unavailable energy created when the energy from the air is used to heat the water. The reduction in the available energy is quite apparent.

The losses associated with irreversible heat transfer processes, as described above, are typical of the situation found in steam power plants. Energy from hot combustion gases is radiated or convected as heat to the boiler tubes. The percent of loss in available energy is great. The fraction of the heat liberated by the burning of coal which is thrown away in the flue gases in a steam power plant is relatively small (15 to 20 percent). However, its temperature level is quite high. Therefore the loss in available energy in the flue gases is significant. A much larger fraction of the heat liberated by combustion eventually appears in the condenser cooling water at the turbine outlet. The temperature level of this energy is so low, however, that most of it is unavailable energy. When the production of work through the use of heat engines is of prime interest, the quality (as measured by the temperature level) of the heat available may be more important that its quantity.

The introduction of the concepts of available and unavailable energy has been used to emphasize more strongly the differences between heat and work. Once heat is transferred across a finite temperature difference, or work of any type is expended irreversibly, its usefulness at a later time is diminished. When heat is transferred from a high to a low temperature, some available energy is always lost. For a given quantity of heat, the available energy associated with it decreases as the supply temperature decreases. Unavailable energy can never be partially converted into further work unless a colder sink or reservoir is established. Note that available and unavailable energy are not properties of a system. Their values depend upon the sink temperature $T_0$. The concepts introduced in this section can be extended further by introducing the concepts of availability and irreversibility.

## 8-2  CLOSED-SYSTEM AVAILABILITY

The concept of available energy introduced in the preceding section is quite useful in engineering analysis. It is one method of measuring the potential of a process to produce work, or of measuring the loss in capability of producing work. However, there are a number of situations for which we need a broader criterion for establishing the maximum work output that any system might ultimately produce. A major criterion is the maximum useful work output that a system, proceeding between equilibrium states, could produce while exchanging heat solely with the local environment.

Consider a closed system $X$ with equilibrium properties $P$, $V$, $T$, $S$, and $U$ (and any other properties of interest) which is surrounded by a local environment or atmosphere of constant temperature $T_0$ and constant pressure $P_0$. Since the system and its environment are not in equilibrium (thermal or mechanical), these two regions may interact and produce work. This situation is illustrated in Fig. 8-6. The work delivered to other systems external to the given system $X$ and its environment is symbolized by $\delta W_X$. We now wish to determine the maximum useful work output that can be delivered by the composite system–environment when the system passes between two given equilibrium states. During this change of state, heat may be exchanged only with the local atmosphere. (This restriction

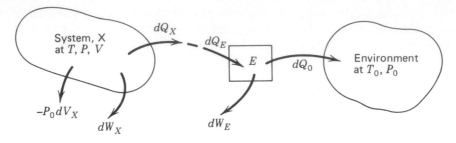

**Figure 8-6** Composite system-environment producing work.

with regard to heat transfer will be removed later.) In addition, the system itself may be deformable. Thus boundary work may be associated with the change of state. However, we shall assume that the local environment is so large that its outer boundary is fixed.

The quantity $\delta W_X$ is the total work delivered by the system during its change of state. However, *useful work* is defined as the work done by the system which does not include that portion done on the atmosphere. Since the atmosphere pressure $P_0$ is constant, then, for the system $X$,

$$\delta W_{X,\,useful} = \delta W_X - (-P_0\, dV_X)$$

This quantity of useful work is not the maximum useful work that can be produced during the process. To maximize the work output, the closed-system process must be totally reversible. This requires that any heat transfer between the system and the environment must be reversible. To accomplish this, it is necessary that a reversible heat engine be inserted between the system and the environment, as shown in Fig. 8-6. This engine will produce an additional quantity of work, $\delta W_E$. For the overall process, the maximum useful work, then, is

$$\delta W_{useful,\,max} = \delta W_X - (-P_0\, dV_x) + \delta W_E \qquad (a)$$

We now wish to transform Eq. $(a)$ so that only thermodynamic properties will appear on the right side of the equation. This requires an evaluation of the quantities $\delta W_X$ and $\delta W_E$.

An energy balance on the cyclic heat engine is of the form

$$\delta Q_E + \delta Q_0 + \delta W_E = dU_E = 0 \qquad (b)$$

In addition, the second law applied to the engine yields

$$\frac{\delta Q_E}{T} = \frac{-\delta Q_0}{T_0} \qquad (c)$$

A combination of Eqs. $(b)$ and $(c)$ leads to

$$\delta W_E = -\delta Q_E - \delta Q_0 = -\delta Q_E + \delta Q_E \frac{T_0}{T} \qquad (d)$$

Furthermore, the first law for the closed system leads to an expression for $\delta W_X$. If we assume that the energy of the system may be represented solely by the internal energy $U$ then

$$dU_X = \delta W_X + \delta Q_X \tag{e}$$

Substitution of Eqs. (d) and (e) into Eq. (a) yields

$$\delta W_{u, \, max} = -\delta Q_E + \delta Q_E \frac{T_0}{T} + dU_X - \delta Q_X + P_0 \, dV_X \tag{f}$$

However, $\delta Q_E$ and $\delta Q_X$ are the same quantity, except that they have opposite signs; $\delta Q_E$ is into the heat engine, but $\delta Q_X$ is taken as out of the system. Hence, $\delta Q_E = -\delta Q_X$. If this relation is placed in Eq. (f), then

$$\delta W_{u, \, max} = \delta Q_X - \delta Q_X \frac{T_0}{T} + dU_X - \delta Q_X + P_0 \, dV_X$$

$$= dU_X + P_0 \, dV_X - T_0 \frac{\delta Q_X}{T} \tag{g}$$

Since the system change is internally reversible, the quantity $\delta Q_X / T$ in the above equation is equal to $dS_X$. Therefore,

$$\delta W_{u, \, max} = dU_X + P_0 \, dV_X - T_0 \, dS_X \tag{h}$$

Integration of this expression between states 1 and 2 of the closed system leads to the desired result.

$$W_{useful, \, max} = \Delta U + P_0 \, \Delta V - T_0 \, \Delta S$$

$$= (U_2 - U_1) + P_0(V_2 - V_1) - T_0(S_2 - S_1) \tag{8-6}$$

On a unit-mass basis this may be written as

$$w_{useful, \, max} = \Delta u + P_0 \, \Delta v - T_0 \, \Delta s$$

$$= (u_2 - u_1) + P_0(v_2 - v_1) - T_0(s_2 - s_1) \tag{8-7}$$

These two equations represent the maximum useful work output, since the process was assumed to be totally reversible. If the process is possibly irreversible, we would write on the basis of the second law that $\delta Q_X / T \leqslant dS_X$. When this is substituted into Eq. (g) and the equation is then integrated, the result is

$$w_{useful} \leqslant \Delta u + P_0 \, \Delta v - T_0 \, \Delta s \tag{8-8}$$

A similar expression may be written for $W_{useful}$. Equation (8-8) is a general relationship for the useful work associated with the change of state of a simple, compressible, closed system which exchanges heat solely with the environment. When the process is totally reversible, the equality sign applies, and $w$ in this case is the maximum useful work output.

The preceding development applies to a closed system which undergoes a change of state between specified end states. In some cases it is of interest to determine the maximum useful work when a system proceeds from a given state to a state of thermal and mechanical equilibrium with the environment. The final temperature and pressure of the system in this case would be $T_0$ and $P_0$, respectively. When the system and the atmosphere are in equilibrium with each other, the system is said to be in the *dead state*. The maximum useful work output that may be obtained from a system-atmosphere combination as the system proceeds from a given state to the dead state while exchanging heat solely with the atmosphere is defined as the *availability* of the system in the given initial state. The availability of a closed system is given by the symbol $\Phi$ for the total mass and by $\phi$ for a unit mass. In terms of Eqs. (8-6) and (8-7) we find that

$$\Phi = (U_0 - U) + P_0(V_0 - V) - T_0(S_0 - S)$$

$$= (U_0 + P_0 V_0 - T_0 S_0) - (U - P_0 V - T_0 S) \qquad (8\text{-}9)$$

and

$$\phi = (u_0 - u) + P_0(v_0 - v) - T_0(s_0 - s)$$

$$= (u_0 + P_0 v_0 - T_0 s_0) - (u + P_0 v - T_0 s) \qquad (8\text{-}10)$$

where the symbols without subscripts refer to the initial state of the closed system. Although we speak of the availability of a system in a given state, note that the availability is a property which is a function of both the state of the system and the local atmosphere. Its value depends upon $T_0$ and $P_0$ as well as the properties of the system.

By applying Eq. (8-10) to states 1 and 2 of a closed system we find that

$$\Delta\phi = \phi_2 - \phi_1 = -[(u_2 - u_1) + P_0(v_2 - v_1) - T_0(s_2 - s_1)]$$

When this result is inserted into Eq. (8-8) we obtain the result,

$$w_{\text{useful}} \leqslant -\Delta\phi \qquad (8\text{-}11a)$$

or, for the total mass within a closed system,

$$W_{\text{useful}} \leqslant -\Delta\Phi \qquad (8\text{-}11b)$$

Again, the equality sign applies when the overall process is carried out in a totally reversible manner. In this latter case, $W$ is the optimum useful work. When irreversibilities are present, the useful work will always be less than the negative value of $\Delta\Phi$.

**Example 8-5M** Steam is contained in a piston-cylinder device. The initial conditions before expansion are 10 bars, 280°, and a volume of 3000 cm³. After expansion the pressure is 1.5 bars and the volume is 14,400 cm³. During the expansion process the heat transfer to the environment is 0.50 kJ. Assume the atmosphere to be at 1 bar and 22°C.

(*a*) Determine the actual useful work done by the system, in kilojoules.

(*b*) Determine the change in the availability of the steam during the process, in kilojoules.

SOLUTION (*a*) The actual useful work is the difference between the total work and the boundary of $P\,dV$ work done against the atmosphere. The total work can be found from the conservation of energy equation, and the boundary work is $-P_0(V_2 - V_1)$. Hence we may write that

$$W_u = W_{\text{tot}} - W_{\text{atm}}$$

$$= m(u_2 - u_1) - Q - (-P_0)(V_2 - V_1)$$

From tabular data for steam in the superheat region at 10 bars and 280°C, we find that $u_1 = 2760$ kJ/kg, $v_1 = 248$ cm³/g, and $s_1 = 7.0465$ kJ/(kg)(°K). In addition, the mass of the system is $m = V_1/v_1 = 3000/248 = 12.1$ g $= 0.0121$ kg. To find the final state, we note that $v_2 = v_1(V_2/V_1) = 248(14,400/3000) = 1190$ cm³/g. From the superheat table at 1.5 bars, the above value of $v_2$ corresponds closely to the value of $v$ found at a temperature of 120°C. Hence $u_2 = 2533$ kJ/kg and $s_2 = 7.2693$ kJ/(kg)(°K).

Hence

$$W_u = 0.0121(2533 - 2760) - (-0.5) + \frac{1(0.014 - 0.003)(10^5)}{10^3}$$

$$= -2.75 + 0.5 + 1.10 = -1.15 \text{ kJ}$$

(b) The change in availability, or the maximum useful work associated with the process, is found on the basis of Eq. (8-6):

$$W_{u, \text{max}} = (U_2 + P_0 V_2 - T_0 S_2) - (U_1 + P_0 V_1 - T_0 S_1)$$

$$= m(u_2 - u_1) + P_0(V_2 - V_1) - T_0(S_2 - S_1)$$

$$= -2.75 + 1.10 - 295(0.0121)(7.2693 - 7.0465)$$

$$= -2.75 + 1.10 - 0.80 = -2.45 \text{ kJ}$$

For the stated process the actual useful work is roughly 47 percent of the maximum useful work.

**Example 8-5** Steam is contained in a piston-cylinder device. The initial conditions before expansion are 160 psia, 500°F, and a volume of 0.1 ft³. After expansion the pressure is 20 psia and the volume is 0.65 ft³. During the expansion process the heat transfer to the environment is 0.50 Btu. Assume the atmosphere to be at 14.7 psia and 80°F.

(a) Determine the actual useful work done by the system, in Btus.

(b) Determine the change in the availability of the steam during the process, in Btus.

SOLUTION (a) The actual useful work is the difference between the total work and the boundary or $P \, dV$ work done against the atmosphere. The total work can be found from the conservation of energy equation, and the boundary work is $-P_0(V_2 - V_1)$. Hence we may write

$$W_u = W_{\text{tot}} - W_{\text{atm}}$$

$$= m(u_2 - u_1) - Q - (-P_0)(V_2 - V_1)$$

From tabular data for steam in the superheat region at 160 psia and 500°F, we find that $u_1 = 1171.2$ Btu/lb, $v_1 = 3.44$ ft³/lb, and $s_1 = 1.6518$ Btu/(lb)(°R). In addition, the mass of the system is $m = V_1/v_1 = 0.1/3.44 = 0.0291$ lb. To find the final state we note that $v_2 = v_1(V_2/V_1) = 3.44(0.65/0.1) = 22.36$ ft³/lb. From the superheat table at 20 psia, the above value of $v_2$ corresponds to a value of $v$ at 300°F. Hence $u_2 = 1108.7$ Btu/lb and $s_2 = 1.7805$ Btu/(lb)(°R). Hence

$$W_u = 0.0291(1108.7 - 1171.2) - (-0.5) + \frac{14.7(0.65 - 0.1)(144)}{778}$$

$$= -1.82 + 0.50 + 1.50 = -0.82 \text{ Btu}$$

(b) The change in availability, or the maximum useful work associated with the process, is found on the basis of Eq. (8-6):

$$W_{u, \text{max}} = (U_2 + P_0 V_2 - T_0 S_2) - (U_1 + P_0 V_1 - T_0 S_1)$$

$$= m(u_2 - u_1) + P_0(V_2 - V_1) - mT_0(s_2 - s_1)$$

$$= -1.82 + 1.50 - .0291(540)(1.7805 - 1.6518)$$

$$= -1.82 + 1.50 - 2.02 = -2.34 \text{ Btu}$$

For the stated process the actual useful work is roughly 35 percent of the maximum useful work.

The preceding analysis has dealt with a closed system which exchanges heat solely with the environment or atmosphere. In more general circumstances the system may also exchange heat with some other reservoir at a temperature $T_R$. Equation (8-11) can be modified to fit this new situation. A thermal-energy (heat) reservoir is a closed system. Hence, as it exchanges heat $\delta Q_R$ with the system of interest, the reservoir undergoes an availability change $d\Phi_R$. For the overall process, then

$$\delta W_u \leqslant d\Phi + d\Phi_R \qquad (8\text{-}12)$$

It is convenient to transform the last term of Eq. (8-12) into a more familiar format. On the basis of Eq. (8-9),

$$d\Phi_R = dU_R + P_0 \, dV_R - T_0 \, dS_R$$

In addition, recall that a heat reservoir undergoes only internally reversible changes of state and that no work is performed. As a result, $dV_R = 0, \delta Q_R = dU_R$, and $dS_R = \delta Q_R/T_R$. When these expressions are substituted into the above equation,

$$d\Phi_R = dQ_R - T_0 \frac{\delta Q_R}{T_R} = \delta Q_R\left(1 - \frac{T_0}{T_R}\right) \qquad (8\text{-}13)$$

where $Q_R$ is measured with respect to the reservoir. Therefore, for the situation where heat may also be exchanged with a reservoir at $T_R$, we find that, generally

$$\delta W_u \leqslant (dU + P_0 \, dV - T_0 \, dS) + \delta Q_R\left(1 - \frac{T_0}{T_R}\right)$$

$$= -d\Phi + \delta Q_R\left(1 - \frac{T_0}{T_R}\right) \qquad (8\text{-}14)$$

or

$$W_u \leqslant (\Delta U + P_0 \, \Delta V - T_0 \, \Delta S) + Q_R\left(1 - \frac{T_0}{T_R}\right)$$

$$= -\Delta\Phi + Q_R\left(1 - \frac{T_0}{T_R}\right) \qquad (8\text{-}15)$$

If heat is transferred from the reservoir to the system, the quantity $Q_R$ is negative when substituted into Eqs. (8-14) and (8-15). A set of equations similar to the above expressions can be written for $w_u$ on a unit-mass basis.

**Example 8-6M** A tank of air at 2 bars and 300°K has a volume of 3 m³. Heat is transferred to the air from a reservoir at 1000°K until the temperature of the air in the tank is 600°K. The surrounding atmosphere is at 17°C and 1 bar. Calculate (a) the initial and final availability of the air, in kilojoules, and (b) the maximum useful work associated with the process, in kilojoules.

SOLUTION (a) Determination of the availability of a state requires information on $u$, $v$, and $s$. For the initial state

$$m = \frac{PV}{RT} = \frac{2(3)(29)}{0.08315(300)} = 6.98 \text{ kg}$$

$$v_0 = \frac{RT}{P} = \frac{0.08315(290)}{1(29)} = 0.832 \text{ m}^3/\text{kg} \qquad v_1 = \frac{V}{m} = \frac{3}{6.98} = 0.430 \text{ m}^3/\text{kg}$$

$$s_0 - s_1 = c_p \ln \frac{T_0}{T_1} - R \ln \frac{P_0}{P_1} = 1.01 \ln \frac{290}{300} - \frac{8.315}{29} \ln \frac{1}{2}$$

$$= -0.0342 + 0.1987 = 0.165 \text{ kJ}/(\text{kg})(^\circ\text{K})$$

Hence

$$\Phi_1 = m[(u_0 - u_1) + P_0(v_0 - v_1) - T_0(s_0 - s_1)]$$

$$= 6.98[0.718(290 - 300) + 1(0.832 - 0.430)(100) - 290(0.165)]$$

$$= 6.98(-7.18 + 40.2 - 47.85) = -103.5 \text{ kJ}$$

For the final state, $v_2 = v_1$, since the tank is assumed rigid. Also, $P_2 = P_1(T_2/T_1) = 2(600/300) = 4$ bars. In addition,

$$s_0 - s_2 = 1.02 \ln \frac{290}{600} - \frac{8.315}{29} \ln \frac{1}{4} = -0.742 + 0.397 = -0.345 \text{ kJ}/(\text{kg})(^\circ\text{K})$$

Therefore the availability of the final state is

$$\Phi_2 = 6.98[0.740(290 - 600) + 1(0.832 - 0.430)(100) - 290(-0.345)]$$

$$= 6.98(-229.4 + 40.2 + 100.1) = -621.9 \text{ kJ}$$

(b) The maximum useful work is given by Eq. (8-15), namely,

$$W_{\text{useful, max}} = -\Delta\Phi + Q_R\left(1 - \frac{T_0}{T_R}\right)$$

The value of $Q_R$ is found from an energy balance on the air in the tank. Since the volume of the tank is constant,

$$Q = \Delta U - W = m(u_2 - u_1) = 6.98(0.740)(600 - 300) = 1550 \text{ kJ}$$

The value of $Q_R$ with respect to the reservoir is the negative of $Q$, or $-1550$ kJ. Consequently,

$$W_{\text{useful, max}} = -(-621.9 + 103.5) + (-1550)\left(1 - \frac{290}{1000}\right)$$

$$= 518 - 1100 = -582 \text{ kJ}$$

Thus a maximum useful output of 582 kJ is associated with the stated process.

**Example 8-6** A tank of air at 25 psia and 100°F has a volume of 20 ft$^3$. Heat is transferred to the air from a reservoir at 1500°R until the temperature of the air in the tank is 660°F. The surrounding atmosphere is at 60°F and 14.7 psia. Calculate (a) the initial and final availability of the air, in Btus, and (b) the maximum useful work associated with the process, in Btus.

SOLUTION (a) Determination of the availability of a state requires information on $u$, $v$, and $s$. For the initial state

$$m = \frac{PV}{RT} = \frac{25(20)(29)}{10.73(560)} = 2.41 \text{ lb}$$

$$v_0 = \frac{RT}{P} = \frac{10.73(520)}{29(14.7)} = 13.1 \text{ ft}^3/\text{lb} \qquad v_1 = \frac{V}{m} = \frac{20}{2.41} = 8.30 \text{ ft}^3/\text{lb}$$

$$s_0 - s_1 = c_p \ln \frac{T_0}{T_1} - R \ln \frac{P_0}{P_1} = 0.240 \ln \frac{520}{560} - \frac{1.986}{29} \ln \frac{14.7}{25}$$

$$= -0.0178 + 0.0364 = 0.0186 \text{ Btu}/(\text{lb})(^\circ\text{R})$$

Hence the availability for the initial state is

$$\Phi_1 = m[(u_0 - u_1) + P_0(v_0 - v_1) - T_0(s_0 - s_1)]$$

$$= 2.41[0.171(520 - 560) + \frac{14.7(144)}{778}(13.1 - 8.30) - 520(0.0186)]$$

$$= 2.41(-6.84 + 13.06 - 9.67) = -8.31 \text{ Btu}$$

For the final state, $v_2 = v_1$, since the tank is assumed rigid. Also, $P_2 = P_1(T_2/T_1) = 25(1120/560) = 50$ psia. In addition,

$$s_0 - s_2 = 0.246 \ln \frac{520}{1120} - \frac{1.986}{29} \ln \frac{14.7}{50} = -0.1887 + 0.0838$$

$$= -0.105 \text{ Btu/(lb)(°R)}$$

Therefore the availability of the final state is

$$\Phi_2 = 2.41[0.177(520 - 1120) + \frac{14.7(144)}{778}(13.1 - 8.30) - 520(-0.105)]$$

$$= 2.41(-106.2 + 13.1 + 54.6) = -92.8 \text{ Btu}$$

(b) The maximum useful work is given by Eq. (8-15), namely,

$$W_{\text{useful, max}} = -\Delta\Phi + Q_R\left(1 - \frac{T_0}{T_R}\right)$$

The value of $Q_R$ is found from an energy balance on the air in the tank. Since the volume of the tank is constant,

$$Q = U - W = m(u_2 - u_1) = 2.41(0.177)(660 - 100) = 239 \text{ Btu}$$

The value of $Q_R$ with respect to the reservoir is the negative of $Q$, or $-239$ Btu. Consequently,

$$W_{\text{useful, max}} = -(92.8 + 8.3) + (-239)\left(1 - \frac{520}{1500}\right)$$

$$= 84.5 - 156 = -71.5 \text{ Btu}$$

Thus a maximum output of 126 Btu is associated with the stated process.


## 8-3 AVAILABILITY IN STEADY-FLOW PROCESSES

The maximum work delivered by an open system is the sum of that delivered by the system and that produced by a reversible heat engine as heat is exchanged between the system and the environment. Figure 8-7 illustrates the arrangement. The maximum work is given by

$$\delta W_{\text{max, sf}} = \delta W_{\text{rev, sf}} + \delta W_{\text{eng}}$$

$$= \delta W_{\text{rev, sf}} + \delta Q - \frac{T_0 \, \delta Q}{T}$$

$$= \delta W_{\text{rev, sf}} + \delta Q - T_0 \, dS \qquad (8\text{-}16)$$

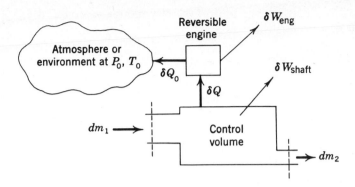

**Figure 8-7** Composite control-volume–environment producing maximum work.

The first term on the right represents shaft work delivered by the flow device. The conservation of energy statement for a steady-state control-volume analysis is

$$\delta Q + \delta W_{\text{sf}} = (h + \text{KE} + \text{PE})_2 \, dm - (h + \text{KE} + \text{PE})_1 \, dm$$

if any magnetic, electrical, or surface-tension effects are neglected. Substitution of this relationship into the preceding one yields

$$\delta W_{\text{max, sf}} = (h + \text{KE})_2 \, dm - (h + \text{KE})_1 \, dm - T_0 \, dS$$

$$= (h + \text{KE} - T_0 s)_2 \, dm - (h + \text{KE} - T_0 s)_1 \, dm \qquad (8\text{-}17)$$

where the potential-energy term has been omitted. If heat is exchanged also with a reservoir at temperature $T_R$, we must modify the above equation in a manner analogous to that used in obtaining Eq. (8-14). Thus

$$\delta W_{\text{max, sf}} = (h_2 + \text{KE}_2 - T_0 s_2) \, dm - (h_1 + \text{KE}_1 - T_0 s_1) \, dm$$

$$+ \delta Q_R \left(1 - \frac{T_0}{T_R}\right) \qquad (8\text{-}18)$$

On the basis of a unit mass entering and leaving the control volume,

$$w_{\text{max, sf}} = (h_2 + \text{KE}_2 - T_0 s_2) - (h_1 + \text{KE}_1 - T_0 s_1)$$

$$+ q_R \left(1 - \frac{T_0}{T_R}\right)$$

$$= \Delta h + \Delta \text{KE} - T_0 \, \Delta s + q_R \left(1 - \frac{T_0}{T_R}\right) \qquad (8\text{-}19)$$

Similar to the closed-system analysis presented above, the actual work output is always less than the sum of the terms on the right of Eq. (8-19) if irreversibilities are present. Hence a general statement for the work associated with a steady-flow process with heat exchanged between the control volume and the reservoir at $T_R$ or the surroundings at $T_0$ is

$$w_{sf} \leqslant \Delta h + \Delta \text{KE} - T_0 \, \Delta s + q_R \left( 1 - \frac{T_0}{T_R} \right) \qquad (8\text{-}20)$$

The equality holds for totally reversible processes, in which case the work interaction has its optimum value. The sign on $q_R$ is taken with respect to the reservoir.

The *stream availability* of a fluid in steady flow is defined as the maximum work output which can be obtained as the fluid is changed reversibly to the dead state while exchanging heat solely with the atmosphere. The stream availability will be given the symbol $\psi$ for a unit mass and $\Psi$ for the total mass, and $\Psi = m\psi$. On the basis of Eq. (8-19),

$$\psi \equiv (h_0 - T_0 s_0) - (h_1 + \text{KE}_1 - T_0 s_1) \qquad (8\text{-}21)$$

A potential-energy term may also be added to both terms on the right, if desired. Note that the dead state implies not only thermal and mechanical equilibrium of the fluid with the atmosphere at $T_0$ and $P_0$, but also that the kinetic energy at the dead state is zero. The definition of $\psi$ applies to a simple, compressible fluid where electrical, magnetic, and surface-tension effects are negligible. It is important to distinguish the stream availability $\psi$ from the closed-system availability $\phi$ defined in the preceding section. Equation (8-20) can now be rewritten in terms of the stream availability. In general,

$$w_{sf} \leqslant -\Delta\psi + q_R \left( 1 - \frac{T_0}{T_R} \right) \qquad (8\text{-}22)$$

where $q_R$ again is measured with respect to the reservoir. The stream availability of a fluid in steady flow can be increased only by transferring heat to it from a source at a temperature other than $T_0$ or by doing work on the fluid.

**Example 8-7M** An adiabatic turbine operates with air initially at 5 bars, 400°K, and 150 m/s. The exit conditions are 1 bar, 300°K, and 70 m/s. Determine the actual work output and compare this with the maximum work obtainable. Assume that $c_p$ is constant at 1.01 kJ/(kg)(°K) and that the atmosphere is at 1 bar and 17°C.

SOLUTION The actual work output is found from the conservation of energy equation for a control volume around the turbine:

$$w_{act} = h_2 - h_1 + \frac{V_2^2 - V_1^2}{2}$$

$$= 1.01(300 - 400) + \frac{(70)^2 - (150)^2}{2(1000)}$$

$$= -101 - 8.8 = -109.8 \text{ kJ/kg}$$

The maximum work possible is calculated from Eq. (8-19):

$$w_{sf, max} = (h_2 - h_1) - T_0(s_2 - s_1) + \frac{V_2^2 - V_1^2}{2}$$

$$= w_{act} - T_0(s_2 - s_1)$$

$$= -109.8 - 290\left(1.01 \ln \frac{300}{400} - \frac{8.315}{29} \ln \frac{1}{5}\right)$$

$$= -109.8 - 49.6 = -159.4 \text{ kJ/kg}$$

The presence of irreversibilities within the turbine have led to a loss of 49.6 kJ/kg of work output, which is a sizable fraction of the maximum possible work output.

**Example 8-7** An adiabatic turbine operates with air initially at 75 psia, 240°F, and 450 ft/s. The exit conditions are 15 psia, 80°F, and 200 ft/s. Determine the actual work output and compare this with the maximum useful work obtainable. Assume that $c_p$ is constant at 0.240 Btu/(lb)(°F) and that the atmosphere is at 1 atm pressure and 70°F.

SOLUTION The actual work output is found from the conservation of energy equation for a suitable control volume around the turbine:

$$w_{act} = h_2 - h_1 + \frac{V_2^2 - V_1^2}{2}$$

$$= 0.24(80 - 240) + \frac{(200)^2 - (450)^2}{2(32.2)(778)}$$

$$= -38.4 - 3.3 = -41.7 \text{ Btu/lb}$$

The maximum work possible is calculated from Eq. (8-19):

$$-w_{max, sf} = (h_1 - h_2) + T_0(s_2 - s_1) + \frac{V_1^2 - V_2^2}{2g_c}$$

$$= -w_{act} + T_0(s_2 - s_1)$$

$$= 41.7 + 530\left(0.24 \ln \frac{540}{700} - \frac{53.3}{778} \ln \frac{15}{75}\right)$$

$$= 41.7 + 25.3 = 67.0 \text{ Btu/lb}$$

The presence of irreversibilities within the turbine have led to a loss of 25.3 Btu/lb of work output, which is a sizable fraction of the maximum work possible from this steady-flow device.

**Example 8-8M** Nitrogen gas initially at 3.6 bars and 27°C is throttled through a well-insulated valve to a pressure of 1.1 bars. The atmospheric temperature is 15°C. Determine the maximum work output under steady-flow conditions.

SOLUTION Equation (8-19) is the basis for the calculation. Throttling is a process for which $\Delta h = 0$. Since $\Delta h$ for an ideal gas equals $c_p \Delta T$, the temperature of the nitrogen remains constant at 27°C. Also, the kinetic-energy change is neglected. Hence the expression for the maximum work output becomes

$$w_{sf, max} = \Delta h + \Delta KE - T_0 \, \Delta s$$

$$= -T_0 \left( c_p \ln \frac{T_2}{T_1} - R \ln \frac{P_2}{P_1} \right) = RT_0 \ln \frac{P_2}{P_1}$$

$$= \frac{8.315(288)}{29} \ln \frac{1.1}{3.6} = -97.9 \text{ kJ/kg}$$

For the actual process, of course, the work output is zero.

**Example 8-8** Nitrogen gas initially at 50 psia and 100°F is throttled through a well-insulated valve to a pressure of 15 psia. The atmospheric temperature is 60°F. Determine the maximum work output under steady-flow conditions.

SOLUTION Equation (8-19) is the basis for the calculation. Throttling is a process for which $\Delta h = 0$. Since $\Delta h$ for an ideal gas equals $c_p \, \Delta T$, the temperature of the nitrogen remains constant at 100°F. Also, kinetic-energy changes are neglected. Hence the expression for the maximum work output becomes

$$w_{sf, max} = \Delta h + \Delta KE - T_0 \, \Delta s$$

$$= -T_0 \left( c_p \ln \frac{T_2}{T_1} - R \ln \frac{P_2}{P_1} \right) = RT_0 \ln \frac{P_2}{P_1}$$

$$= \frac{1.986(520)}{29} \ln \frac{15}{50} = -42.9 \text{ Btu/lb}$$

For the actual process, of course, the work output is zero.

## 8-4 IRREVERSIBILITY FOR CLOSED AND OPEN SYSTEMS

In the analysis of either closed or open systems, the conservation of energy principle, in conjunction with auxiliary equations or data, is usually sufficient to ascertain the actual work delivered to or by a particular device. For engineering purposes it is important to have some standard to which actual performance can be compared. The concept of availability has been introduced for this purpose in the preceding sections. Maximum work output or minimum work input is associated with processes that are totally reversible. The presence of irreversibilities leads to a decrease in the amount of work which could be produced and delivered to devices external to the system and its local environment. This loss of work capability will be defined as the irreversibility $I$ of any process, and will be taken to be a positive quantity. Hence the irreversibility is the difference between the actual work delivered and the maximum work obtainable for a given change of state for a system. Thus

$$I \equiv W_{act} - W_{max} \tag{8-23}$$

In determining $W_{max}$ it is necessary that the system undergo a process between the same end states and exchange the same quantities of heat with various reservoirs.

In addition, the system will probably also have to exchange heat with the environment. Equation (8-23) may also be written in terms of the irreversibility per unit mass $i$, where $i = I/m$.

In the analysis of the irreversibility associated with a closed-system process, it is convenient to express the work quantities in terms of useful work rather than total work. Note that the boundary work done on the environment is the same in any reversible process as in an actual process. If this latter quantity is subtracted from both $W_{act}$ and $W_{max}$, then

$$I = W_{u, act} - W_{u, max} \tag{8-24}$$

In conjunction with Eq. (8-15), the above equation may also be expressed as

$$I = W_{u, act} + \Delta\Phi - Q_R\left(1 - \frac{T_0}{T_R}\right) \tag{8-25}$$

The quantity $W_{u, act}$ normally would be found from an energy balance on the actual process.

The quantity $I$ can be expressed in terms of the entropy function. For a closed system we may substitute the equality form of Eq. (8-14) into a differential form of Eq. (8-24). Thus

$$\delta I = \delta W_{u, act} - \delta W_{u, max}$$

$$= \delta W_{tot, act} + P_0 \, dV - (dU + P_0 \, dV - T_0 \, dS) - \delta Q_R\left(1 - \frac{T_0}{T_R}\right)$$

$$= \delta W_{tot, act} - dU + T_0 \, dS - \delta Q_R + \delta Q_R \frac{T_0}{T_R}$$

From the conservation of energy principle the quantity $\delta W_{tot, act}$ may be replaced by $dU - \delta Q$, where $\delta Q$ is the sum of all heat interactions. In this particular case, $-\delta Q$ is equal to $\delta Q_R + \delta Q_0$. Substitution of these quantities yields

$$\delta I = \delta Q_0 + T_0 \, dS + \delta Q_R \frac{T_0}{T_R}$$

In addition, $\delta Q_0 = T_0 \, dS_0$ and $\delta Q_R = T_R \, dS_R$, because any heat reservoir by concept is considered to be internally reversible. Thus the equation for the irreversibility $\delta I$ of a process for a closed system becomes

$$\delta I = T_0 \, dS_0 + T_0 \, dS + T_0 \, dS_R$$

$$= T_0 \, (dS)_{total} \tag{8-26a}$$

or 

$$I = T_0 \, (\Delta S)_{total} \tag{8-26b}$$

The direct proportionality between the quantity $I$ and the total entropy change associated with a process is not unexpected. The total increase in entropy of the parts of an isolated system is due solely to the presence of irreversibilities within the overall system.

An equation equivalent to Eq. (8-26) can also be developed for open systems in steady flow. On the basis of Eq. (8-18) we find that

$$\delta W_{sf,\,max} = (h + KE - T_0 s)_2 \; dm - (h + KE - T_0 s)_1 \; dm + \delta Q_R\left(1 - \frac{T_0}{T_R}\right)$$

In addition, the steady-flow energy equation for a control volume is

$$\delta W_{sf,\,act} = -\delta Q + (h + KE)_2 \; dm - (h + KE)_1 \; dm$$

The potential-energy terms have been omitted from both of the preceding equations. As a result, the differential form of Eq. (8-23) becomes

$$\delta I_{sf} = -\delta Q + T_0(s_2 - s_1) \; dm - \delta Q_R + \delta Q_R \frac{T_0}{T_R}$$

However, $-\delta Q = \delta Q_0 + \delta Q_R$, $\delta Q_0 = T_0 \, dS_0$, and $\delta Q_R = T_R \, dS_R$. With these substitutions the above equation becomes

$$\delta I_{sf} = T_0 \; dS_0 + T_0(s_2 - s_1) \; dm + T_0 \; dS_R$$

$$= T_0 \; dS_{isol} \tag{8-27}$$

Per unit mass of fluid flowing through the control volume,

$$i_{sf} = T_0[\Delta s_0 + (s_2 - s_1) + \Delta s_R] \tag{8-28}$$

where $\Delta s_0$ and $\Delta s_R$ are the entropy changes of the local surroundings and a heat reservoir at $T_R$, respectively. Thus the steady-flow analysis leads to an equation analogous to Eq. (8-26) for closed systems.

On the basis of the increase in entropy principle, which states that $dS_{isol} \geqslant 0$, we can conclude that

$$\delta I \geqslant 0 \tag{8-29}$$

or, for a finite change of state,

$$I \geqslant 0 \tag{8-30}$$

The equality sign in either of these two equations applies only in the case of a totally reversible process. For an irreversible process the irreversibility is a measure of the decrease in energy which can be converted into work.

**Example 8-9M** Calculate the irreversibility of the process considered previously in Example 8-6M, in kilojoules.

SOLUTION The irreversibility of the closed system is

$$I = W_{u,\,act} - W_{u,\,max}$$

In this particular problem the tank is rigid and the actual useful work is zero. The maximum useful work was previously found to be $-582$ kJ. Therefore the irreversibility of the process is simply 582 kJ. This value can also be calculated from the expression

$$I = T_0(\Delta S)_{tot} = T_0(\Delta S_{sys} + \Delta S_{res})$$

The entropy change of the system is given by

$$\Delta S_{sys} = mc_v \ln \frac{T_0}{T_1} = 6.98(0.740) \ln \frac{600}{300} = 3.58 \text{ kJ/}^\circ\text{K}$$

The entropy change of the reservoir is simply $Q_R/T_R = -1550/1000 = -1.55$ kJ/°K. Therefore the total entropy change is $3.58 + (-1.55) = 2.03$ kJ/°K, and

$$I = 290(2.03) = 589 \text{ kJ}$$

This answer agrees with the preceding evaluation within the accuracy of the calculation.

**Example 8-9** Calculate the irreversibility, in Btus, of the process considered previously in Example 8-6.

SOLUTION The irreversibility of the closed system is

$$I = W_{u, \text{act}} - W_{u, \text{max}}$$

In this particular problem the tank is rigid and the actual useful work is zero. The maximum useful work was previously found to be $-71.5$ Btu. Therefore the irreversibility of the process is simply 71.5 Btu. This value can also be calculated from the expression

$$I = T_0(\Delta S)_{tot} = T_0(\Delta S_{sys} + \Delta S_{res})$$

The entropy change of the system is given by

$$\Delta S_{sys} = mc_v \ln \frac{T_2}{T_1} = 2.41(0.177) \ln \frac{1120}{560} = 0.296 \text{ Btu/}^\circ\text{R}$$

The entropy change of the reservoir is simply $Q_R/T_R = -239/1500 = -0.159$ Btu/°R. Therefore the total entropy change is $0.296 - 0.159 = 0.137$ Btu/°R, and

$$I = 520(0.137) = 71.2 \text{ Btu}$$

This answer agrees with the preceeding evaluation within the accuracy of the calculation.

**Example 8-10M and 8-10*** Consider the expansion of nitrogen gas through a throttling valve under the conditions expressed in either Example 8-8M or 8-8. Determine the irreversibility of the process.

SOLUTION Per unit mass of nitrogen passing through the valve, $i = T_0 \Delta s_{tot}$. Under steady-state conditions the only entropy change is that of the fluid as it passes through the valve. As a result, the irreversibility is equal in magnitude, but opposite in sign, to the steady-flow maximum work calculated in Examples 8-8M and 8-8. That is, $i = -w_{sf, \text{max}}$. This result is not surprising, since the capability of work output associated with a finite pressure drop is completely lost when a fluid is throttled.

# PROBLEMS (METRIC)

**Available energy**

**8-1M** A constant-volume tank contains saturated water vapor at 90°C. Heat is added to the water until the pressure in the tank reaches (a) 1.5 bars, and (b) 1 bar. For a sink temperature of 10°C, compute how much of the heat added is available energy.

---

* This example can be read with no concern for units.

**8-2M** A kilogram of air at 1.8 bars is colled at constant volume from 450 to 300°K. All the heat which leaves the system appears in the surroundings at (a) 12°C, and (b) 27°C. Determine how much of the heat removed is available energy, in kilojoules. Draw a $Ts$ diagram for the process, and label the areas which represent available and unavailable energy.

**8-3M** Ten kilograms of air at 1 bar and 350°K are in a closed, rigid tank. Heat is transferred to the air from a thermal-energy reservoir that has a constant temperature of 500°K. The ambient temperature is 300°K.

(a) How much heat, in kilojoules, can be transferred to the air in the tank from the energy source?

(b) How much of the heat transferred to the air is available energy?

**8-4M** Consider air in a closed system. The mass of air is 0.1 kg; the initial conditions are 1 bar and 22°C. An irreversible process takes place during which 1.7 kJ of heat are removed and 2.6 kJ of work are done on the gas.

(a) Find the change in entropy for this process if the final pressure is 2 bars.

(b) If the sink temperature is 17°C, determine the available portion of the heat transfer.

**8-5M** Defend or refute the following argument: From a theoretical viewpoint it is claimed that 1200 kJ of heat available from a thermal energy reservoir at 827°C are more useful than 1600 kJ of heat available at 482°C.

**8-6M** A kilogram of air at 1 bar is cooled at constant pressure from 440 to 300°K, with all the heat given up by the system appearing in the environment, which has a temperature of 290°K. Determine how much of the heat removed is available energy. Compute the entropy change for the overall process, in kJ/kg of air.

**8-7M** In a steam-power plant, saturated water vapor leaving the turbine at 0.08 bar is condensed to saturated liquid. If 7°C is the temperature of the sink reservoir, what fraction of the heat removed is available energy?

**8-8M** Steam at 3 bars and (a) 240°C, and (b) 280°C is allowed to cool at constant pressure in a closed system until it reaches thermal equilibrium with the surroundings at 20°C. Compute the loss in available energy in kJ/kg.

**8-9M** Referring to Example 8-3M, calculate the loss of possible work due to heat addition when the temperature of the heat reservoir is at (a) 400°K, (b) 1200°K, (c) 2000°K, and (d) infinity. Compare with the loss of work when paddle-wheel work is used to evaporate the water.

**8-10M** Heat for a heat engine is available at a constant temperature of 1427°C, but actually is transferred to the working medium which is at a temperature of 500°K. The lowest available environmental temperature is 300°K. What fraction of the heat supplied becomes unavailable because the heat transferred is received by the working fluid at 500°K rather than 1427°C?

**8-11M** A rigid tank contains 1 kg of carbon monoxide at 0.9 bar and 27°C. A paddle-wheel stirs the contents of the tank adiabatically until the temperature has been raised to 367°C. If the gas is now returned to its initial state by rejecting heat to the surroundings at 17°C, determine (a) the paddle-wheel work done on the gas initially, in kilojoules, and (b) the portion of the heat removed in the second process that is available energy.

**8-12M** One kilogram of carbon dioxide is heated at constant pressure from 300 to 500°K. Calculate (a) the heat supplied, in kilojoules, (b) the change in entropy, and (c) the maximum work that could be obtained from the heat quantity found in part (a) if the minimum sink temperature is 7°C.

**8-13M** A fluid is evaporated in a boiler by transferring heat from hot gases which are at 1400°K. If the fluid boils at 1000°K, what fraction of the available energy is lost due to the heat transfer? On the same basis, if the fluid boils at 700°K, what fraction is lost? Sketch the calculated quantities on a $Ts$ diagram. $T_0 = 25$°C.

**8-14M** A hot oil vapor at 327°C condenses in a heat exchanger. The heat removed from the oil is used to evaporate water at 300°C on the other side of the heat-exchanger tubes. The steam formed in the heat exchanger is then used in a reversible power cycle which has a rejection temperature of 17°C.

(a) Compute the fraction of the available energy initially in the condensing oil vapor which is lost because of the irreversible heat transfer within the heat exchanger.

(b) What fraction of the heat transfer from the oil eventually appears as useful work output?

**8-15M** A hydrocarbon oil is to be cooled in a heat exchanger from 400 to 320°K by exchanging heat with water which enters the exchanger at 20°C at a rate of 2000 kg/h. The oil flows at a rate of 750 kg/h, and has an average specific heat of 2.30 kJ/(kg)(°K). Compute (a) the total change in entropy due to the heat-transfer process, in kJ/°K per kg of oil, and (b) the loss in available energy due to the process if the surrounding temperature is 17°C.

**8-16M** In a steam-power plant, most of the energy released by the combustion of fuel which does not appear as work output appears either as moderate-temperature thermal energy in the flue gases or as low-temperature thermal energy in the steam leaving the turbine. The energy in the flue gas is rejected to the atmosphere, and the energy removed from the steam in the condenser is rejected to cooling water. Consider the following data: For every 100 kJ of energy released by combustion of fuel, 32 kJ appear as work output, 46 kJ as heat rejected at 40°C to cooling water at 20°C, and 22 kJ as heat rejected from the flue gases at an average temperature of 200°C to the ambient air at 20°C. Determine the available energy (a) of the heat removed from the flue gases, and (b) of the heat rejected from the condensing steam. Compare the two answers with each other and with the actual work output.

**8-17M** In a steam-power-plant boiler, the temperature of the hot gas is 1450°K. The mixture of water and steam inside the boiler tubes is at (a) 427°C, and (b) 527°C. The lowest sink temperature for heat rejection is 23°C. What fraction of the heat transferred cannot be used to produce work, theoretically, because of the irreversible heat transfer from the hot gas to the relatively cool water?

**8-18M** Nitrogen gas flows at a rate of 2 kg/s through a steady-flow device, reversibly and at constant pressure. The inlet conditions are 1 bar and 300°K, and the final temperature is 400°K. The temperature of the surroundings is 10°C. Compute per unit mass (a) the heat added during the process, and (b) the available energy associated with the heat added to the gas.

**8-19M** Water enters a heat exchanger at 30 bars and 500°C and leaves at 30 bars and 280°C. What fraction of the heat removed from the steam is available energy? $T_0 = 17°C$.

## Availability for closed systems

**8-20M** A tank with a volume of 0.3 m$^3$ contains air at 6 bars and 300°K. The surrounding atmosphere is at 0.95 bar and 300°K. Determine the availability of the air, in kilojoules.

**8-21M** What is the maximum useful work that can be done by the air in Prob. 8-20M if it is supplied with 150 kJ of heat from a reservoir at 177°C?

**8-22M** Let the air in Prob. 8-20M undergo a free expansion until its volume is (a) doubled, and (b) tripled. What is the change in the availability for the process, in kilojoules?

**8-23M** What is the maximum useful work associated with 50 kg of water at 0°C if the surrounding atmosphere is at 0.95 bar and 20°C?

**8-24M** How much useful work can be done by 1 kg of steam in a closed system at 80 bars and 400°C? The environment is at 1 bar and 25°C.

**8-25M** One kilogram of air initially at 1 bar and 27°C is contained in a well-insulated tank. An impeller inside the tank is turned by an external mechanism until the pressure is 1.2 bars. Determine (a) the actual work required, in kilojoules, (b) the maximum useful work associated with the change of state, in kj, and (c) the irreversibility of the process. Let $T_0 = 27°C$ and $P_0 = 1$ bar.

**8-26M** A piston-cylinder device contains 0.4 kg of air at 1 bar and 27°C. Determine the minimum work input, in kilojoules, required to compress the air to 4 bars and 127°C. $T_0 = 20°C$ and $P_0 = 1$ bar.

**8-27M** A storage battery is capable of delivering 1 kWh of energy. What volume of air stored in a tank at 27°C and (a) 20 bars, and (b) 40 bars, is required to theoretically have the same work capability? The state of the environment is 27°C and 1 bar.

**8-28M** Determine the availability of a unit mass of an ideal gas in a closed system at temperature $T$ (different from $T_0$ of the surroundings), but at a pressure $P$ which is the same as $P_0$ of the surroundings. By using your knowledge of property relations of ideal gases, express the answer in terms of $T_0$, $P_0$, $T$, and any required constants of the gas.

**8-29M** Determine the availability of a unit mass of an ideal gas in a closed system at temperature $T_0$

which is the same as that of the surroundings, but at a pressure $P$ which is different from $P_0$ of the surroundings. By using your knowledge of property relations of ideal gases, express your answer in terms of $T_0$, $P_0$, $P$, and any required constants of the gas.

**8-30M** One-fourth kilogram of an ideal gas initially at 1.4 bars and 25°C is in an insulated tank. An impeller within the tank is turned by an external motor until the pressure is 1.8 bars. Determine the irreversibility of the process if the gas is (a) nitrogen, (b) hydrogen, and (c) carbon dioxide. The atmosphere is at 0.96 bar and 22°C.

**8-31M** A tank of air at 12 bars and 227°C has a volume of 0.8 m$^3$. The air is cooled by heat transfer until the temperature is 27°C. The surroundings are at 1 bar and 27°C.
  (a) Calculate the initial and final availability, in kilojoules.
  (b) Calculate the irreversibility of the process, in kilojoules.

**8-32M** Steam is contained in a piston-cylinder device. Before expansion the state is 10 bars, 280°C, and 0.01 m$^3$. After expansion the pressure and volume are 1.5 bars and 0.06 m$^3$. The heat transfer during the expansion is $-0.8$ kJ. Calculate (a) the availability of the initial and final states, in kilojoules, and (b) the irreversibility of the process, in kilojoules. $T_0 = 20$°C and $P_0 = 1$ bar.

**8-33M** A piston-cylinder device expands air from 6 bars, 77°C, and 0.06 m$^3$ to 3.5 bars and 0.15 m$^3$. During the process 35 kJ of heat are added to the air from a source at 500°K. The atmosphere is at 1 bar and 300°K. Using the air table, determine (a) the initial and final availability of the air, in kilojoules, and (b) the irreversibility of the process, in kilojoules.

### Availability in open systems

**8-34M** Air enters a turbine at 3 bars and 480°K and exhausts at 1 bar. The surroundings are at 1 bar and 20°C. Compute the maximum shaft work possible, in kJ/kg.

**8-35M** Steam enters a turbine at 30 bars and 400°C and expands to 1 bar, 120°C, in a steady-flow, adiabatic process. The ambient pressure and temperature are 1 bar and 27°C. Disregard changes in kinetic and potential energy. Determine (a) the actual work delivered, in kJ/kg, (b) the maximum possible work for the given conditions, and (c) the work if the expansion was isentropic from the initial conditions to the given final pressure.

**8-36M** Air enters a compressor in steady flow at 1.4 bars, 17°C, and 70 m/s. It leaves the compressor at 3.5 bars, 127°C, and 110 m/s. Determine the minimum work required and the irreversibility of the process. $T_0 = 7$°C; $P_0 = 1$ bar.

**8-37M** Steam enters a turbine at 80 bars and 560°C at a rate of 50,000 kg/h. Partway through the turbine 25 percent of the flow is bled off at 20 bars and 440°C. The rest of the steam leaves the turbine at 0.10 bar as a saturated vapor. Determine (a) the availability of the three streams of interest, (b) the maximum power output possible, in kilowatts, and (c) the actual power output in kilowatts if the flow is adiabatic. The environment is at 1 bar and 20°C.

**8-38M** Air is compressed adiabatically in steady flow from 1 bar and 17°C to 4 bars. Neglect changes in kinetic and potential energy. The required work input is 180 kJ/kg. Determine the irreversibility of the process, in kJ/kg. Let $T_0 = 17$°C.

**8-39M** Saturated liquid refrigerant 12 enters an expansion valve at 6 bars and leaves at 2 bars. Determine the irreversibility of the process if (a) it is adiabatic, and (b) the fluid receives 4 kJ/kg of heat from the atmosphere. Atmospheric conditions are 1 bar and 27°C.

**8-40M** Steam which is a saturated vapor at 30 bars is throttled to 7 bars. The change in kinetic energy is negligible, and the atmospheric temperature is 12°C. Calculate the irreversibility of the process in kJ/kg.

**8-41M** Air enters a counterflow heat exchanger at 2.6 bars and 27°C and is heated to 127°C. The air flow rate is 1.2 kg/s. Heat is transferred to the air stream from a hot, flue-gas stream which enters the exchanger at 1.2 bars and 230°C and leaves at 95°C. The values of $c_v$ and $c_p$ for the flue-gas stream are 0.63 and 0.84 kJ/(kg)(°K), respectively. Both streams pass through the heat exchanger with negligible change in pressure and kinetic energy. Assume no heat loss to the environment, which is at 0.98 bar and 22°C. Determine the irreversibility of the process in kilojoules per kilogram of air.

# PROBLEMS (USCS)

### Available energy

**8-1** A constant-volume tank contains dry, saturated water vapor at 200°F. Heat is added to the water until the pressure in the tank reaches 20 psia. For a sink temperature of 50°F, compute how much of the heat added is available energy.

**8-2** A pound of air at 25 psia is cooled at constant volume from 340 to 40°F. All the heat which leaves the system appears in the surroundings at 20°F. Determine how much of the heat removed is available energy, in Btus. Draw a $Ts$ diagram for the process, and label the areas which represent available and unavailable energy.

**8-3** Ten pounds of air at 15 psia and 600°R are in a closed, rigid tank. Heat is transferred to the air from an energy reservoir that has a constant temperature of 900°R. The ambient temperature is 500°R.

    (a) How much heat can be transferred to this air from the energy source?
    (b) How much of the heat transferred to the air is available energy?
    (c) How much of the heat transferred from the source is unavailable energy?

**8-4** Consider air in a closed system. The mass of air is 0.1 lb; the initial conditions are 15 psia and 40°F. An irreversible process takes place during which 1.7 Btu of heat are removed and 2.6 Btu of work are done on the air.

    (a) Find the change in entropy for this process if the final pressure is 30 psia.
    (b) If the sink temperature is 50°F, determine the available portion of the heat transfer.

**8-5** Defend or refute the following argument: From a theoretical viewpoint it is claimed that 1200 Btu of heat available from a reservoir at 1500°F are more useful than 1600 Btu of heat available at 900°F. The ambient temperature is 60°F.

**8-6** A pound of air at 15 psia is cooled at constant pressure from 340 to 80°F, with all the heat given up by the system appearing in the environment, which has a temperature of 60°F. Determine how much of the heat removed is available energy. Compute the entropy change for the universe, in Btu/lb.

**8-7** In a steam-power plant, saturated dry vapor leaving the turbine at 1 psia is condensed to saturated liquid. If 40°F is the lowest-temperature reservoir to reject heat to, what fraction of the heat removed is available energy?

**8-8** Steam at 1 atm and 500°F is allowed to cool at constant pressure in a closed system until it reaches thermal equilibrium with the surroundings at 70°F. Compute the loss in available energy, in Btu/lb.

**8-9** Referring to Example 8-3, calculate the loss of possible work due to heat addition when the temperature of the heat reservoir is at (a) 700°R, (b) 2000°R, (c) 5000°R, (d) infinity. Compare with the loss of work when paddle-wheel work is employed to evaporate the water.

**8-10** Heat for a heat engine is available at a constant temperature of 2140°F, but is actually transferred to the working medium at a constant temperature of 440°F. The lowest available environmental temperature is 80°F. How many Btu per 100 Btu supplied become unavailable because the heat transfer occurs at 440 instead of 2140°F?

**8-11** A rigid tank contains 1 lb of carbon monoxide at 14 psia and 80°F. A paddle-wheel stirs the contents of the tank adiabatically until the temperature has been raised to 240°F. If the gas is now returned to its initial state by rejecting heat to the surroundings at 50°F, determine (a) the paddle-wheel work done on the gas initially, and (b) the portion of the heat removed in the second process that is available energy.

**8-12** One pound of carbon dioxide is heated at constant pressure from 80 to 440°F. Calculate (a) the heat supplied, (b) the change in entropy, (c) the unavailable energy, and (d) the maximum work that could be obtained from the heat quantity found in part (a) if the minimum sink temperature is 40°F.

**8-13** Fluid is evaporated in a boiler by transferring heat from hot gases which are at 2500°R. If the fluid boils at 2000°R, how much of the heat transferred is available energy per 100 Btu of heat transferred? On the same basis, if the fluid boils at 1500°R, how much of it is available energy? Sketch these calculated quantities as areas on a $Ts$ diagram.

**8-14** A hot oil vapor at 640°F condenses in a heat exchanger. The heat removed from the oil is used to evaporate water at 540°F on the other side of the heat-exchanger tubes. The steam formed in the heat exchanger is then used in a reversible power cycle, which has a rejection temperature of 50°F. Compute the fraction of the available energy initially in the condensing oil vapor which is lost owing to irreversible heat transfer within the heat exchanger.

**8-15** A hydrocarbon oil is to be cooled in a heat exchanger from 260 to 120°F by exchanging heat with water which enters the exchanger at 50°F at a rate of 4000 lb/h. The oil flows at a rate of 1500 lb/h, and has an average specific heat of 0.55 Btu/(lb)(°F). Compute (a) the total change in entropy due to the heat-transfer process in Btu/°R per pound of oil, and (b) the loss in available energy due to the process, based on the fact that the oil might have been cooled directly by using the heat removed to run a Carnot engine which rejects heat to the surroundings at 60°F.

**8-16** In a steam-power plant most of the energy released by the combustion of fuel which does not appear as work output appears either as high-temperature thermal energy in the hot flue gases or as low-temperature thermal energy in the steam leaving the turbine. The energy in the hot flue gases is rejected to the ambient atmosphere, and the energy removed from the steam in the condenser of the power plant is rejected to cooling water. Consider the following data: For every 100 Btu of energy released by the combusion of fuel, 24 Btu appears as work output, 54 Btu as heat rejected at 100°F to cooling water at 70°F, and 22 Btu as heat rejected from hot flue gases at an average temperature of 750°F to the ambient air at 60°F. Determine the available energy (a) of the heat removed from the hot flue gases, and (b) of the heat rejected from the condensing steam at 100°F. Compare the two answers.

**8-17** In a steam-power-plant furnace, the temperature of the fire is 2200°F. The mixture of water and steam inside the boiler tubes is at 400°F. The lowest temperature at which heat can be dissipated is 100°F. How many Btu (out of each 1000 transferred) cannot be used to produce work because the heat is transferred to the water at 400 instead of 2200°F?

**8-18** Nitrogen gas flows at a rate of 2.0 lb/s through a steady-flow device reversibly and at constant pressure. The inlet conditions to the control volume are 15 psia and 80°F, and the final temperature is 280°F. The temperature of the surroundings is 50°F. On the basis of 1 lb of nitrogen, compute (a) the heat added during the process, (b) the work done by or on the nitrogen, and (c) the available energy associated with the heat added to the gas.

**8-19** Water enters a heat exchanger at 500 psia and 900°F and leaves at 500 psia and 400°F. What fraction of the heat removed from the steam is available energy?

## Availability for closed systems

**8-20** A tank with a volume of 10 ft$^3$ contains air at 100 psia and 70°F. The surrounding atmosphere is at 14.5 psia and 70°F. Determine the availability of the air, in Btus.

**8-21** What is the maximum useful work, in Btus, that can be done by the air in the preceding problem if it is supplied with 150 Btu of heat from a reservoir at 300°F?

**8-22** Suppose that the air in Prob. 8-60 is allowed to undergo a free expansion until its volume becomes 20 ft$^3$. What is the change in the availability for the process, in Btus?

**8-23** What is the maximum useful work associated with a 50-lb cake of ice at 32°F if the surrounding atmosphere is at 14 psia and 60°F?

**8-24** How much useful work can be done by 1 lb of steam in a closed system at 1000 psia and 800°F? The environment is at 14.7 psia and 70°F.

**8-25** One pound of air initially at 15 psia and 90°F is contained in a well-insulated tank. An impeller inside the tank is turned by an external mechanism until the pressure is 18 psia. Determine (a) the actual work for the process in Btu/lb, (b) the maximum useful work associated with the change of state in Btu/lb, and (c) the irreversibility of the process.

**8-26** A piston-cylinder assembly contains 1 lb of air at 15 psia and 70°F. Determine the minimum shaft work, in Btu, required to compress the air to 60 psia and 250°F.

**8-27** A storage battery is capable of delivering 1000 Wh of energy. What volume of air stored in a tank at 500 psia and 80°F is required to have the same capability? $T_0 = 80°F$; $P_0 = 14.7$ psia.

**8-28** A tank of air at 200 psia and 360°F has a volume of 30 ft³. The air is cooled by heat transfer until the temperature is 80°F. The surroundings are at 14.7 psia and 80°F.

(a) Calculate the initial and final availability, in Btus.

(b) Calculate the irreversibility of the process, in Btus.

**8-29** Steam is contained in a piston-cylinder device. Before expansion the state is 160 psia, 500°F, and 0.1 ft³. After expansion the state is 20 psia and 0.6 ft³. The heat transfer during the expansion is −0.8 Btu. Calculate (a) the availability of the initial and final states, in Btus, and (b) the irreversibility of the process, in Btus.

**8-30** A piston-cylinder device expands air from 100 psia, 140°F, and 2 ft³ to 60 psia and 5 ft³. During the process 75 Btu of heat is added to the air from a source at 1000°R. The atmosphere is at 14.7 psia and 500°R. Using the air table, determine (a) the initial and final availability of the air, in Btus, and (b) the irreversibility of the process, in Btus.

## Availability in open systems

**8-31** Air enters a turbine at 45 psia and 400°F and exhausts at 15 psia. The surroundings are at 14.7 psia and 70°F. Compute the maximum shaft work possible, in Btu/lb.

**8-32** Steam enters a turbine at 400 psia and 700°F and expands to 14.7 psia and 250°F in a steady-flow, adiabatic process. The ambient pressure and temperature are 14.7 psia and 80°F. Disregard increments of kinetic and potential energy. Determine (a) the actual work delivered, in Btu/lb, (b) the maximum possible work for the given conditions, and (c) the work if the expansion was isentropic from the initial conditions to the same final pressure.

**8-33** Air enters a compressor in steady flow at 20 psia, 50°F, and 200 ft/s. It leaves the compressor at 50 psia, 260°F, and 350 ft/s. Determine the maximum useful work and the irreversibility, assuming the process is adiabatic. $T_0 = 40°F$; $P_0 = 1$ atm.

**8-34** Steam enters a turbine at 1000 psia and 1100°F at a rate of 100,000 lb/h. Partway through the turbine, 25 percent of the flow is bled off at 300 psia and 800°F. The rest of the steam leaves the turbine at 1 psia as a saturated vapor. Determine (a) the availability of the three streams of interest in Btu/lb, (b) the maximum work, in horsepower, and (c) the actual work output in horsepower if the flow is adiabatic. The environment is at 14.7 psia and 70°F.

**8-35** Air is compressed adiabatically in steady flow from 15 psia and 60°F to 60 psia. Neglect changes in kinetic and potential energy. The required work input is 80 Btu/lb. Determine the irreversibility of the process, in Btu/lb.

**8-36** Saturated liquid refrigerant 12 enters an expansion valve at 100 psia and leaves at 30 psia. Determine the irreversibility of the process if (a) it is adiabatic, and (b) the fluid receives 4 Btu/lb of heat from the atmosphere. The atmospheric conditions are 14.7 psia and 70°F.

**8-37** Steam which is a saturated vapor at 400 psia is throttled to 100 psia. The change in kinetic energy is negligible and the atmospheric temperature is 50°F. Calculate the irreversibility of the process, in Btu/lb.

**8-38** Air enters a counterflow heat exchanger at 40 psia and 80°F and is heated to 260°F. The air flow rate is 1.2 lb/s. Heat is transferred to the air stream from a hot, flue-gas stream which enters the exchanger at 14.9 psia, 450°F, and leaves at 200°F. The values of $c_v$ and $c_p$ for the flue-gas stream are 0.150 and 0.200 Btu/(lb)(°F), respectively. Both streams pass through the heat exchanger with negligible change in pressure and kinetic energy. Assume no heat loss to the environment, which is at 70°F and 14.6 psia. Determine the irreversibility of the process in Btus per pound of air.

# NINE

## A STATISTICAL VIEWPOINT OF ENTROPY AND THE SECOND LAW

In the last half of the nineteenth century, through the work of Maxwell, Clausius, Boltzmann, and others, the concept of the entropy function was developed in terms of microscopic parameters of a macroscopic system. The modern development of the subject, begun by Planck, Einstein, and others at the beginning of the twentieth century, requires the use of quantum mechanics and various techniques of statistical analysis. Since a statistical analysis of the behavior of a large number of particles is involved, this microscopic approach to thermodynamics is called statistical thermodynamics. It is based on its own set of postulates, which are independent of those of classical or macroscopic thermodynamics.

## 9-1 QUANTIZATION OF ENERGY

Before 1900, two major concepts regarding the nature of matter and energy had been established. One of these was the particle nature of matter. A quantity of mass was conceived of as being composed of atoms and molecules, and each particle in theory was a separate entity and identifiable. Also, the wave nature of electromagnetic radiation was well entrenched and in common usage. In the early 1900s, the concepts of energy and matter underwent a profound change. The work of Planck, Einstein, and others proposed that energy was particle-like. Light energy was considered to be absorbed or emitted in discrete quantities called quanta, or photons. In 1901, Planck derived an equation for the spectral distribution of blackbody radiation associated with electromagnetic radiation. He assumed that the energies corresponding to the various standing electromagnetic waves could take on only discrete values. These values were integral multiples of $hv$, where $v$ is the frequency of the wave and $h$ is constant known as Planck's constant. In 1905, Einstein employed the concept of light quanta introduced by Planck to explain the anomalous photoelectric effect. It had been discovered earlier that electrons were ejected from solids when radiation fell upon them. It was also known that the maximum energy of the ejected electrons was solely a function of the frequency of the incident radiation. Einstein assumed that radiation consisted of quanta of size $hv$, which could be absorbed by the electrons. A portion of this energy was expended as the electron escaped from the solid. Thus the photoelectric effect was explained by acknowledging the discrete nature of radiant energy.

Then, in the 1920s, de Broglie suggested that matter, especially electrons, might exhibit wave characteristics. The de Broglie equation relating the momentum $p$ of a particle to its characteristic wavelength $\lambda$ is $p = h/\lambda$. In 1927, the experimental work of Davisson and Germer showed that electrons are diffracted when they are passed through a crystal lattice. Hence these small particles exhibit a characteristic (diffraction) usually associated with wave phenomena. These and other experimental results have demonstrated the duality of nature. Both matter and energy, under various circumstances, exhibit the properties either of waves or of discrete quantities.

An example of the discreteness of energy values is provided by the emission of light from substances such as sodium vapor when they are heated in a gas flame. The measured wavelength of light emitted from a sodium source is 0.5890 $\mu$m. Electrons orbiting the nucleus have been thermally excited. When an excited electron "falls" or "decays" from the excited state to the normal or "ground" state of energy, the energy difference between these states of the electron is emitted as light. The radiant energy is always emitted at the same wavelength, and it is assumed to be emitted in the form of a single photon with an energy $hv$. This energy difference is given by the Einstein-Planck equation,

$$\epsilon_1 - \epsilon_0 = hv$$

where $\epsilon_1$ = energy of excited state
$\epsilon_0$ = ground state of energy
$h = 6.625 \times 10^{-34}$ J·s

For sodium emission at 0.5890 $\mu$m, this energy-level difference is 2.10 eV.

One of the important tasks of wave, or quantum, mechanics is to relate mathematically the wave nature of matter to the discrete energy values that are associated with individual particles or with large groups of particles. Schrödinger in the mid-1920s suggested an equation which represents the wave nature of matter in terms of a set of standing waves. The Schrödinger wave equation is taken as a basic postulate of quantum theory. Solution of this equation for the various energy modes leads to a set of quantum, or energy, levels for each mode.

Recall that an energy mode is simply one of the physically distinguishable means by which a particle can have energy. The discrete or allowed energy states of particles are determined by (1) the nature of the particles, and (2) the circumstances of the system. By circumstances of the system we mean such constraints as the volume of the system or the strength of the applied magnetic field. For the present we shall simply accept the notion that energy is discrete, without any further examination of its quantitative aspects.

## 9-2 THE ENTROPY FUNCTION

A group of particles that constitutes an isolated thermodynamic system has an overall energy $U$. This energy is shared by the individual particles. However, each particle is not required to have the same energy $\epsilon$. The results of quantum mechanics indicate that a whole series of energy values must be considered. Moreover, only certain discrete values of energy are allowed or need to be regarded. These discrete values are set by certain constraints on the system. For example, it is found that the translational (linear) kinetic energies of gas particles are inversely proportional to the volume to the two-thirds power. It is common practice to use the term *energy level* when referring to one of the allowed energy values. The series of allowed energy levels begins at some minimum value called the *ground level* (or zero level) of energy. The series then progresses through distinct steps to larger and larger energy values. Such a series of energy levels exists for each energy mode. For example, one would need to consider for a diatomic gas at low pressures at least those series of energy levels for the translational, rotational, and vibrational modes.

The term "energy level" allows one to present the concept of discrete energy values very simply. A typical diagram for energy levels is shown in Fig. 9-1. The symbols $\epsilon_0$, $\epsilon_1$, $\epsilon_2$, ... represent the different energies per particle that a particle may have when occupying the ground level, the first level, etc. By occupying a level it is meant that a particle is considered to have the energy value associated with that particular level. In Fig. 9–1 the vertical distance of any line above the base line is proportional to the energy difference between that level and the ground level. The convenience of a diagram of this type as an aid in picturing

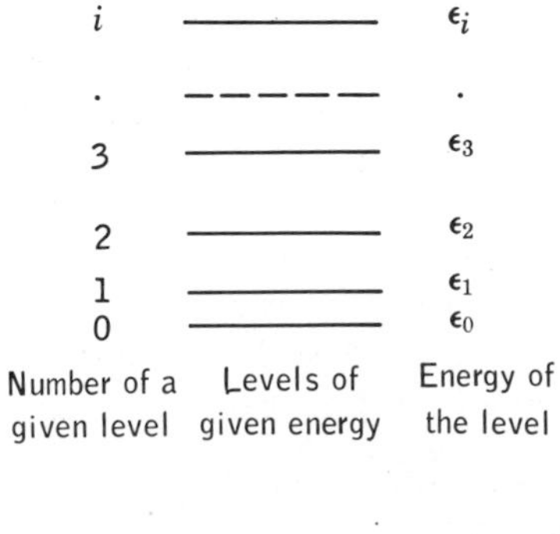

| $i$ | ———— | $\epsilon_i$ |
| . | — — — — | . |
| 3 | ———— | $\epsilon_3$ |
| 2 | ———— | $\epsilon_2$ |
| 1 | ———— | $\epsilon_1$ |
| 0 | ———— | $\epsilon_0$ |
| Number of a given level | Levels of given energy | Energy of the level |

**Figure 9-1** A set of discrete energy levels.

the complex and chaotic nature of a thermodynamic system on a microscopic scale will soon become apparent. The actual values of the discrete energy levels for the possible energy modes are those given by solutions of the Schrödinger wave equation. The spacing between levels in Fig. 9-1 is shown to increase as the energy $\epsilon_i$ increases. This is in agreement with the analytical results obtained for an energy mode such as rotation. An energy mode such as vibration has equally spaced energy levels.

The microscopic picture of matter is one of an endless array of interactions or encounters between the various particles of the system. In the gas phase the particles move through the vessel undergoing encounters with other particles at the rate of, roughly, $10^{33}$ s$^{-1}$/cm$^3$ under normal atmospheric pressure and temperature conditions. As a result of these encounters energy is transferred between the interacting bodies. The mechanism of the energy transfer is not presently of concern. What is of vital interest is that, at each new instant of time, the microscopic representation of any isolated system has changed. The exchange of energy by molecular encounters is restrained only by the conditions that the resultant energy values associated with the various energy modes of the particles afterward satisfy the values allowed by quantum restrictions and that the total energy be conserved. Thus any system, internally in equilibrium, has a microscopic structure which leads to definite macroscopic properties such as pressure, temperature, specific volume, specific internal energy, etc. However, a host of different descriptions on a particle basis will exist over a given time interval. A major task to be resolved is to ascertain what connection exists, if any, between the multitude of microscopic descriptions which are possible and the values we assign to properties based on macroscopic observations.

Consider an isolated system which contains $n$ particles and has a total energy $U$. If the particles are considered independent of one another, then at any given time the energy of the $i$th molecule is given by $\epsilon_i$. The total internal-energy requirements of the system must be fulfilled by the relationship

$$U = \sum_i n_i \epsilon_i \tag{9-1}$$

where $n_i$ represents the number of molecules whose energy is $\epsilon_i$. The values of $\epsilon_i$ are, of course, restricted to those values allowed by quantum theory. The summation is taken over all the levels which may be expected to contain one or more particles. (The largest value of $\epsilon_i$ in the summation obviously never can be greater than the value of $U$; that is, one particle has associated with it all the energy of the system.) It has been noted already that the energy $\epsilon_i$ possessed by a particle is relative to some ground-level energy $\epsilon_0$. For simplicity in the following examples, the energy (quantum) levels will be regarded as equally spaced and the value of the ground-level energy will be taken as zero. First an isolated system will be chosen which contains four particles and a total energy of 6 units. Only one energy mode is considered, and the quantum levels for this energy mode are spaced 1 unit apart, beginning at zero. The following question naturally arises: In what manner may particles be distributed among the various energy levels and still satisfy the condition that the total energy be 6 units? Obviously, only the first seven allowed

**Table 9-1 The nine possible macrostates of four particles having a total energy of 6 units, for equally spaced energy levels**

| Energy level | Distribution | | | | | | | | |
|---|---|---|---|---|---|---|---|---|---|
| | $A$ | $B$ | $C$ | $D$ | $E$ | $F$ | $G$ | $H$ | $I$ |
| $\epsilon_0$ | 3 | 2 | 2 | 1 | 1 | 2 | | | 1 |
| $\epsilon_1$ | | 1 | | 2 | 1 | | 3 | 2 | |
| $\epsilon_2$ | | | 1 | | 1 | | | 2 | 3 |
| $\epsilon_3$ | | | | | 1 | 2 | 1 | | |
| $\epsilon_4$ | | | | 1 | 1 | | | | |
| $\epsilon_5$ | | 1 | | | | | | | |
| $\epsilon_6$ | 1 | | | | | | | | |
| $W$ | 4 | 12 | 12 | 12 | 24 | 6 | 4 | 6 | 4 |

energy levels need be taken into account, since higher-valued ones could not possibly be occupied. A summary of the possible distributions of particles among energy levels for the chosen example is given in Table 9-1. The nine possibilities appear in the vertical columns labeled $A$ to $I$. The values in the horizontal rows give the number of particles of energy $\epsilon_i$ for each distribution. The student may wish to verify that the vertical columns represent the only possible distributions.

We shall define a *macrostate* as any possible microscopic state of an assembly of particles described in terms of the number of particles in each energy level at a given instant of time. The description of a macrostate does not require enumeration of which particles have a given energy, but only an accounting of how many particles have a given energy. In Table 9-1 the nine distributions labeled $A$ to $I$ will now be identified as the nine possible macrostates for a system composed of four particles and having a total energy of 6 units. Each of these macrostates satisfies the macroscopic description of the state of the system.

Further analysis of Table 9-1 leads to the concept of a microstate if we assume that the particles are distinguishable. Consider any one of the nine macrostates, e.g., column $A$. Here three particles have a ground level of energy $\epsilon_0$ and the remaining one is in the sixth quantum level of energy $\epsilon_6$. Because of the random exchange of energy, the particle having 6 units of energy could be any one of the four particles in the system. Over a sufficient period of time macrostate $A$ will occur a number of times. However, each of the four distinguishable particles will share in occupying the sixth level of energy for this particular macrostate. Thus the macrostate $A$ really consists of four microstates all of which are equally likely to occur over a sufficiently long period of time. A *microstate*, then, is a description of which particles have certain energies. Of course, we cannot specify which microstate exists at a given instant of time, or even which macrostate. Microstates do exist, however, and they are continually changing with time. The other distributions, $B$ to $I$, in Table 9-1 likewise may be formed in a number of ways, depending upon the placement of certain particles in various energy levels. The last horizontal row in the table, marked $W$, lists the number of microstates for each macrostate. We shall define the number of microstates $W$ for a given macrostate as the *thermodynamic probability* of that macrostate. Unlike the usual mathematical probability which varies from zero to unity, the thermodynamic probability is never less than unity and is usually an extremely large number of thermodynamic systems. The value of $W$ in the case of Table 9-1 is obtained from a permutation formula

$$W = \frac{n!}{n_1!n_2! \cdots n_k!}$$

This equation gives the number of arrangements (microstates) possible for $n$ items where the identity of each item as residing in a particular group is recognized. However, the order of items within a group is immaterial. The symbols $n_1, n_2, \ldots$ represent the number of items in groups 1, 2, etc. As an example, four particles are placed in three groups containing 2, 1, and 1 particles, respectively. For this situation the above equation gives for the possible arrangements that

$$W = \frac{4!}{2!1!1!} = 12$$

Hence distributions $B$, $C$, and $D$ all have 12 possible internal arrangements, or microstates. The other values of $W$ in Table 9-1 are easily verified. It is seen from Table 9-1 that the total number of microstates is 84. Since all the microstates will be assumed to occur with equal likelihood, over a long period of time each microstate would be formed $\frac{1}{84}$ of the time.

In actual thermodynamic systems, the accounting of macrostates and microstates is even more complicated than the simple examples discussed above. First of all, quantum theory indicates that frequently one must consider energy states that are distinct, but of equal or nearly equal energy. When $g_i$ quantum states of the same energy $\epsilon_i$ exist, the system of particles is said to be *degenerate*. We then speak of $g_i$ degenerate states within an energy level. The values of $g_i$ may be different for the various energy levels. The degeneracy of a system of particles does not alter the number of macrostates, but greatly increases the number of microstates. For example, in the case of distinguishable particles the number of microstates per energy level increases by the factor $g_i^{n_i}$. Therefore the total possible arrangements for given values of $n_i$ and $g_i$ for distinguishable particles becomes

$$W = n! \prod \frac{g_i^{n_i}}{n_i!}$$

A second important factor is the assumption of distinguishability of the particles. Distinguishability was assumed to provide a model that permitted an easy explanation of macrostates and microstates. Modern quantum statistics is based on indistinguishable particles. Therefore the equation above for $W$ is incorrect for this latter model, and other equations for $W$ must be developed. Nevertheless, regardless of the model, the concept of macrostates and microstates is a pertinent one for the development of a new thermodynamic property.

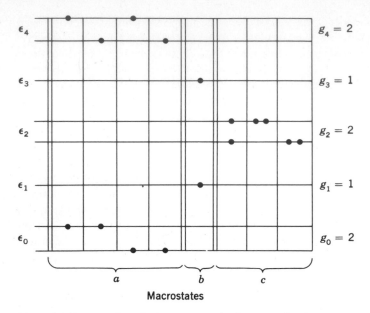

Figure 9-2 Enumeration of macrostates and microstates for degenerate system of indistinguishable particles.

Consider the following degenerate system of two indistinguishable particles which shares a total of 4 units of energy. The energy spacing again is 1 unit, the ground-state value is zero, and the degeneracies of the levels, starting from the ground level, are 2, 1, 2, 1, and 2. Figure 9-2 summarizes the macrostates and microstates for these data. The macrostates $a$, $b$, and $c$ contain 4, 1, and 3 microstates, respectively. Note that interchange of the two particles does not constitute a new microstate, since the particles are indistinguishable. Each macrostate has a specified number of particles per energy level, but there is no further required specification as to which degenerate state within a level has particles associated with it.

For a system of $n$ particles and total energy $U$, there is a finite fixed number of macrostates for each set of allowed energy levels. Since the number of macrostates is fixed in each case, the number of microstates is also set. The total number of microstates for a given thermodynamic state is represented by $W_{tot}$. This simply is the sum of the $W$ values for each possible macrostate. In our first example $W_{tot}$ would be 84, and in the second example the number would be 8. The number of microstates $W_{tot}$ thus is a unique and identifiable number (at least in theory) for every equilibrium state. If $W_{tot}$ has a certain fixed value for each equilibrium state, it partially fulfills the concept of a property in at least one respect. There is one and only one value for each equilibrium state. If a system is carried through a cyclic process, the value of $W_{tot}$ will necessarily be the same for the initial and final states of the cycle.

Although the value of $W_{tot}$ is fixed for a given equilibrium state, it is not a useful measure of a thermodynamic property. If any new property is to be conveniently tabulated and correlated with other properties, its value should vary in direct proportion to its mass; i.e., it should be extensive in nature. The thermodynamic probability $W$ is not an extensive variable. The theory of combinations and permutations states that the total arrangements of two independent groups taken together are the product of the arrangements of each individual group. In terms of thermodynamic probability, if system $A$ can be arranged in 3 ways and system $B$ in 2 ways, there are 6 possible arrangements that might occur when both systems are considered simultaneously. For two thermodynamic systems considered simultaneously, $W_{AB} = W_A W_B$. Hence the property $W$ is multiplicative in nature.

Although $W$ itself is multiplicative, there exists a mathematical function of $W$ such that one of its basic characteristics is its additivity. In fact, the only function of $W$ which has this feature is a logarithmic function. We have already seen that for a composite system the value of $W$ is

$$W_{AB} = W_A W_B$$

By taking the logarithm of each side, we find that

$$\log W_{AB} = \log W_A + \log W_B$$

Therefore we shall define a new thermodynamic property in terms of the thermodynamic probability of the system. This property is called the entropy of the system, and is designated by the symbol $S$. In general, $S$ is related to $W_{tot}$ by the definition,

$$S = k \ln W_{tot} \tag{9-2}$$

where $k$ is a proportionality constant. It is necessary to choose the logarithm to the base $e$ in order to make the numerical values of the entropy function agree with values calculated from macroscopic measurements. This equation is known as the Boltzmann-Planck equation. Its general validity has been well established. The constant $k$, which is called the *Boltzmann constant*, is important for two reasons: It determines both the magnitude and the units for the entropy function, since the $\ln W_{tot}$ term is a pure number. Quantitatively, it is given by

$$k = \frac{R_u}{N_A}$$

where $R_u$, again, is the universal gas constant, and $N_A$ is Avogadro's number. In the metric system of units the value of $k$ is $1.380 \times 10^{-23}$ J/($^\circ$K)(molecule). This value is also listed in Table A-1, for reference.

The development of this new thermodynamic function, which we have called the entropy of a system, may be made by two approaches on the microscopic level. The method of statistical mechanics based on the Boltzmann-Planck equation is well accepted in the modern literature. Recently, the work of Tribus, based on earlier papers by Shannon and Jaynes, has shown that an equivalent approach to the entropy function in thermodynamics is possible by extending the methods of information theory. This latter theory has been used primarily in the field of communications. In this text the statistical development of the entropy function will be based on the older of these methods, statistical mechanics. The reader may wish to refer to the work of Tribus for the parallel development of statistical concepts.

## 9-3 THE SECOND LAW OF THERMODYNAMICS

On the basis of microscopic concepts a new macroscopic property of matter, the entropy, has been introduced. It has been defined in terms of the number of microscopic arrangements, the thermodynamic probability $W$, of an isolated system of given energy $U$ and volume $V$ which contains $n$ particles. There is a certain value of $W_{tot}$ for each equilibrium state of a macroscopic system, and therefore a unique value of $S$ associated with each equilibrium state. As any system undergoes a change of state, it generally is expected that $W_{tot}$ will change and therefore that the system will have a new value for its entropy. There is also the possibility, however, that $W_{tot}$ may not change in some cases. Hence the chance that the entropy will not be altered by some changes of state should not be ruled out. This would simply mean that the total number of arrangements did not change from one equilibrium state to another.

The reason for introducing this new macroscopic property is that it is intimately connected with a basic characteristic of nature. It is a matter of experience that there is a directional quality about physical phenomena. For example, if two masses at different temperatures are brought into thermal contact (and they are insulated from the environment), it is observed that they always approach a common final temperature. The conservation of energy principle requires that the energy given up by one mass must equal that energy taken on by the other mass, but it does not require that the final temperatures be the same. In the study of chemical reactions it is found that many reactions do not proceed to completion under certain experimental conditions, such as a specified temperature and pressure of the system. For example, consider the reaction of CO and $H_2O$ to form $CO_2$ and $H_2$ at 1500°K and 1 atm. The theoretical equation shows that $CO + H_2O \rightarrow CO_2 + H_2$. That is, for every mole of CO and $H_2O$ introduced into a reaction vessel, theoretically 1 mole each of $CO_2$ and $H_2$ should be formed. However, the yield of these products is considerably less than this at the stated conditions. The conservation of energy principle places no restriction on the yield. As a third example, consider the mixing of two different gases by breaking a partition which initially separates them. After a sufficient period of time the gases are found to be uniformly mixed, within experimental accuracy. The spontaneous separation of the two gases, such that at a later time one gas is detected wholly on one side of the vessel and the other gas solely on the other side, is not observed. The example may appear trivial, for our experience tells us this separation process will not occur. However, it is important to be able to explain our experiences on the basis of basic laws. There certainly is no first-law restriction on the unmixing process.

As a final example, take the case of cyclic devices which are used to convert heat released from various sources into work output. Internal-combustion engines and nuclear steam-power stations, for example, fall into this general classification. Although great strides have been made in the design of such devices by engineers, the complete conversion of energy from fuels to work output has not been even closely approached for practical engines. (The conversion seldom approaches 40 percent.) Is this the fault of the designer or is there some theoretical reason for the observed limitations? The first law, on the other hand, permits the complete conversion of thermal energy into work by devices such as these. The above limited list of examples could be extended greatly, each example being illustrative of a process which satisfies the conservation of energy principle but whose final state is restricted. This restriction is either in a directional sense or a quantitative (extent of process) sense. Although more examples could be taken from widely diversified fields, it is found that a single postulate of thermodynamics is sufficient to explain all the observed results.

Before a postulate known as the second law of thermodynamics is presented, it may be helpful if we examine in a qualitative manner several examples of physical phenomena which illustrate the general principle. It must be kept in mind, however, that these examples are not proof of the validity of the statement to be made later. Laws are generalizations based on limited experimental evidence. Their validity is generally based on the ability of the laws, and corollaries of the laws, to successfully predict future behavior. The three examples below, then, are presented simply as common illustrations of behavior which lead to the general postulate. This postulate, the second law of thermodynamics, will be expressed in terms of the behavior of an isolated system. In each case we shall study an isolated system which changes state due to the removal of constraint. If the constraint is removed, various macroscopic (and microscopic) parameters change in value until a new equilibrium state is reached. We shall find that the final equilibrium state of the isolated system is intimately connected to the behavior of $W_{tot}$ (and hence $S$) during the change of state.

First consider the free expansion of a gas from one-half of an insulated, rigid vessel into the other half of the vessel, which is initially evacuated. If the inside of the entire vessel walls is chosen as the boundary of the system, then the system is an isolated one, since both $Q$ and $W$ are zero and no mass crosses the boundaries. All changes of state occur within the vessel, and the environment is unaffected. On the basis of the conservation of energy principle, the energy of the gas does not change during the expansion process, (when the internal partition is removed). However the values of the permitted energy levels for the particles do change. As noted earlier, for a simple substance the energy levels for the translational kinetic energy of gases are inversely proportional to the volume to the two-thirds power. When the gas fills the entire vessel, the energy spacings $\Delta\epsilon$ are somewhat smaller than before. Thus we have the situation of a system of $n$ particles with fixed energy $U$ where the particles now redistribute themselves over allowed energy levels which are more closely spaced. It should be apparent that in this situation the value of $W_{tot}$ must increase. That is, there are many more ways to distribute a given amount of energy among a given number of particles if the energy spacings decrease. For example, recall the data from Table 9-1, where four distinguishable particles having a total energy of 6 units were distributed over energy levels 1 unit apart. Nine macrostates resulted. If the energy levels become $\frac{1}{2}$ unit apart (due to a volume change), 30 macrostates would be possible. The total number of arrangements would increase from 84 to 373. The important result is that the change in the total number of arrangements, and hence the entropy change, during a free expansion is always *positive*.

As a second example, consider the sliding of a block down an inclined plane. To eliminate the consideration of a change in the kinetic energy of the block as it slides down the plane, we shall assume that the solid–solid friction between block and plane is sufficient so that the block moves very slowly. After a finite time interval the block comes to rest at the bottom of the plane. To further simplify the analysis, we may assume that the heating effect at the solid–solid interface affects only the block and plane. That is, no energy in the form of heat is transferred to the environment. Our isolated system then is simply the block and plane. The conservation of energy principle for this isolated system reduces merely to $\Delta E = 0$, or $\Delta U + \Delta PE = 0$. The decrease in the potential energy of the block appears as an increase in the internal energy of both the block and the plane. Assuming that the particles of a solid act independently of one another and that the allowed energies of the particles are not altered by the process, the number of microscopic arrangements will increase for a solid as its energy increases. Recall again the data for Table 9-1. A system of four distinguishable particles with a total energy of 6 units has 9 macrostates and 84 microstates (for equally spaced energy levels). If the total energy is decreased to 4 units and all other data are the same, then the number of macrostates is reduced to 5, and the number of microstates is reduced from 84 to 35. A significant increase in microstates occurs, then, with an increase in energy. Therefore the value of $W_{tot}$ for both the block and the plane increases during the frictional process. The end result is that the presence of solid–solid friction always increases the total entropy of those parts of an isolated system affected by the process.

As a final example, consider an isolated system consisting of two subsystems, $X$ and $Y$. Initially they are isolated from each other by means of a rigid, thermally insulating partition. The partition acts as a constraint on the energy of each subsystem. If the total energy of the composite isolated system is $U_C = U_X + U_Y$, then the energy of $X$ must remain constant, and likewise for subsystem $Y$. Associated with each subsystem are thermodynamic probabilities $W_X$ and $W_Y$, and for the composite isolated system, $W_C = W_X W_Y$. The two subsystems are now brought into thermal contact through the rigid partition. Since the volume of each is fixed, no boundary work is possible. However, an energy exchange in the form of heat is permitted. Now the total energy $U_C$ is shared by the two subsystems. On a microscopic basis there is no real limitation to the manner in which the energy is shared between $X$ and $Y$. As the energy of each subsystem varies, the value for $W$ for each subsystem changes.

As an illustrative example, let $\epsilon_0 = 0$ in both subsystems, with $\Delta\epsilon = 1$. Since

**Table 9-2 Microstates for two systems in thermal contact**

| $U_X$ | $U_Y$ | $W_X$ | $W_Y$ | $W_C$ | $P_i$ |
|-------|-------|-------|-------|-------|-------|
| 0 | 8 | 1 | 45 | 45 | 0.015 |
| 1 | 7 | 4 | 36 | 144 | 0.048 |
| 2 | 6 | 10 | 28 | 280 | 0.093 |
| 3 | 5 | 20 | 21 | 420 | 0.140 |
| 4 | 4 | 35 | 15 | 525 | 0.175 |
| 5 | 3 | 56 | 10 | 560 | 0.186 |
| 6 | 2 | 84 | 6 | 504 | 0.168 |
| 7 | 1 | 120 | 3 | 360 | 0.120 |
| 8 | 0 | 165 | 1 | 165 | 0.055 |
|   |   |   |   | 3003 | 1.000 |

the volumes of $X$ and $Y$ are constant, the energy-level values do not change for gases contained within the subsystems. Subsystem $X$ will contain four distinguishable particles, subsystem $Y$ will contain three distinguishable particles, and the total energy shared by the two subsystems will be 8 units. Table 9-2 summarizes the various possible microscopic descriptions, where each level is nondegenerate.

No matter what the original split in the total energy had been, the final value of 3003 for $W_{tot}$ would have been much larger than the initial value once the constraint between the two subsystems was removed. Although the illustration involves distinguishable particles and small numbers, the same qualitative result would occur for indistinguishable particles and large degenerate systems. Apparently the removal of a thermal constraint, so that an isolated system changes to a new equilibrium state, always leads to an increase in $W_{tot}$ and hence to an increase in the entropy of the isolated system.

(The preceding example simply illustrates the point that the removal of a thermal constraint brings about an increase in the thermodynamic probability of an isolated system. However, even though the total energy may now be divided between the two subsystems in nine different ways, we would not expect to find the energy of either subsystem to vary significantly at equilibrium if the subsystems are relatively large. That is, at macroscopic equilibrium all the thermodynamic properties, including each subsystem's energy, should be constant on the basis of experimental measurements. Therefore many of the possible splits in energy are never seen, even though they are theoretically possible. We shall use the data from Table 9-2 in a following section to illustrate another important point regarding the macroscopic equilibrium state.)

The preceding discussion can be summarized in the following way. If a constraint of some type is removed from an isolated system initially in equilibrium, the number of internal arrangements $W_{tot}$ will either increase or remain the same. That is $W_{tot, final} \geqslant W_{tot, initial}$. Although all our examples were ones during which $W_t$ increased, the possibility that the thermodynamic probability of an isolated system will not change must be included. Experience simply dictates that the number of arrangements never decreases. Therefore we shall postulate the following statement as the *second law of thermodynamics.*

> The entropy of an isolated system always either increases or remains the same when the system changes from one equilibrium state to another.

This statement frequently is referred to as the principle of increase of entropy. The second law may be expressed mathematically for any differential change of state by

$$dS_{\text{isol sys}} \geq 0 \qquad (9\text{-}3a)$$

For a finite process it must be true that

$$\Delta S_{\text{isol sys}} \geq 0 \qquad (9\text{-}3b)$$

We must infer from these two equations not only that the overall change in entropy must be positive (or zero), but also that every differential step of the overall process must have an entropy change that is positive (or zero).

The general statement that the entropy of an isolated system always either increases or remains the same leads to the concept of reversible and irreversible processes. Consider first an isolated system which undergoes a change of state during which its entropy increases. This is permissible in the light of the second law. It is then proposed to reverse the direction of the original forward process in order to return the isolated system to its initial state. Since the entropy is a property of a system, the reverse process requires that the entropy of the isolated system decrease. But the second law states that the entropy cannot decrease if the system is isolated. Consequently the reverse process is impossible, and the original forward process is called *irreversible.* Any process is irreversible if the sum of the entropy changes for all subsystems involved in the process is positive.

There are some processes for which the entropy change of an isolated system will be zero. As a result, the entropy change would be zero for the reversed process. This latter reversed process does not violate the second law. Thus there are processes involving isolated systems which may proceed in either direction, since $\Delta S$ remains zero at all times. Such processes are called *reversible.* The entropy function then is a test for reversibility. Reversibility is an extremely important concept to the engineer, since the degree of irreversibility in a process determines, in part, its performance. The role of reversible and irreversible processes in engineering design will be amply illustrated throughout this text.

One of the most important features of the second law is its formulation in terms of reversible and irreversible processes. Another important feature of the second law is the recognition that it is a *nonconservation* law. Many of the basic laws studied in physics are conservation laws, and include the familiar ones involving mass, energy, momentum, and charge. A conservation principle is extremely useful in any field, since it provides a method of accounting for various changes which occur in a system. The second law, on the other hand, is the antithesis of a conservation principle. It states that in all those processes which might be termed irreversible, the entropy of the isolated system always increases. The decrease in entropy for any type of change of an isolated system is not possible. Hence entropy is not a conserved property. It is this characteristic that makes the second law a *directional* or *limiting* law. That is, it establishes a criterion for the direction toward or limit to which all real processes can proceed. If the entropy of an isolated system cannot decrease during any process, a large number of equilibrium states are excluded from the possibility of being the final equilibrium state. This may help to explain why certain phenomena are observed in nature, whereas other effects never occur. The second law of thermodynamics, expressed here in terms of an increase in entropy principle, has a broad application in providing guidelines to the understanding of natural phenomena and the design of engineering systems.

One final characteristic of the second law should be noted. Even though the entropy of an isolated system increases when a process is irreversible, it cannot increase indefinitely. The system must finally reach a new equilibrium state that is consistent with the constraints on the system. Therefore it may also be stated that the entropy of an isolated system always tends toward a maximum value, and when it reaches that maximum value, the system is characterized by a new set of equilibrium properties. The entropy function thus can be used as a *criterion for equilibrium* in an isolated system. The entropy increases as a process progresses, but in the limit as equilibrium is approached $dS$ likewise approaches zero. Mathematically, $dS = 0$ is the criterion for equilibrium for an isolated system.

In summary, we have defined a new extensive property, the entropy, by the Boltzmann-Planck equation, $S = k \ln W_{tot}$. The constant $k$ is Boltzmann's constant, and $k = R_u / N_A$, where $N_A$ is Avogadro's number. The entropy is a measure of the number of microscopic arrangements in which a system could be found, consistent with the constraints on the system. The second law of thermodynamics then postulates that the entropy of an isolated system either increases or remains the same during a change of state; that is, $dS_{isol} \geq 0$. The inequality sign applies to that class of processes called irreversible, while the equality sign is valid for reversible processes. Thus the second law is a directional law, since it prohibits certain processes from ever occurring. An isolated system reaches a new equilibrium state, after the removal of a constraint, when the entropy reaches a maximum value. In a broad sense, then, the second law of thermodynamics is concerned with three subjects: (1) the direction of change when spontaneous processes occur, (2) the criteria for equilibrium in thermodynamic systems, and (3) the effect of irreversibilities on performance.

At this point a number of questions remain unanswered. What do we really imply in physical terms when we say a process is reversible or irreversible? If the entropy function is so important, how does one evaluate it by purely macroscopic techniques? How do we use the entropy function and the second law in the solution of scientific and engineering problems? Can the second law be applied in a modified form to closed systems, regardless of the environment? It is to these related questions that we must now turn.

## 9-4 REVERSIBLE AND IRREVERSIBLE PROCESSES

The reader at this point is directed to reread Sec. 6-5, and then follow this with Sec. 9-5.

## 9-5 MICROSCOPIC IMPLICATIONS OF WORK AND HEAT

Heat and work effects have been investigated extensively in a macroscopic sense in earlier chapters. In such considerations it was not necessary to acknowledge the existence of atomic particles or energy levels. However, it is important for us to examine these interactions on a microscopic basis, especially with regard to internally reversible processes. We shall find that the effects of heat and work interactions on changes at the microscopic level are quite different. In addition, the results of such an examination will be quite helpful in developing an expression for the entropy change of a closed system in terms of macroscopic variables.

In an earlier section it was noted that the discrete values of energy allowed for a given energy mode are set by the value of one of the generalized coordinates (displacements) of the system or by the nature of the substance. As an example, consider the translational kinetic energies of gas particles maintained in a box of volume $V$. The values of the $\epsilon_i$'s are found to be proportional to $1/V^{2/3}$. That is, the volume of the system determines the spacing of the energy levels for this particular form of energy. Other generalized displacements will determine the magnitude of the energy-level spacing for other energy modes. Consequently the energy-level spacing for a certain energy mode will change only if the corresponding generalized displacement is altered. Let us consider a simple system containing a monatomic gas. The volume is slowly reduced by a compression process. As the volume diminishes, the energy levels become farther apart. For each energy level, $\epsilon_i \rightarrow \epsilon_i + d\epsilon_i$, where $d\epsilon_i$ is positive for a compression process. Each level contains $n_i$ particles. Therefore an increment of energy $n_i \, d\epsilon_i$ must be supplied at each level of energy. If the only energy added to all levels is due to boundary work, i.e., if the process is internally reversible, then

$$\delta W_{\text{int rev}} = \sum_i n_i \, d\epsilon_i \qquad (9-4)$$

Internal reversibility is required, since any dissipative effects present (such as friction) would negate the statement that the generalized work interaction exactly equaled the change in the energy of the closed system. The above equation is not valid for nonquasistatic types of work interactions. For example, it has been pointed out that paddle-wheel effects, the flow of electric current through a resistor, and work done against frictional forces do not necessarily involve a change in a generalized displacement. Consequently these effects do not contribute to the change in the $\epsilon_i$ values of the particles. For simple systems the discrete energy values can only be changed by volume changes. It is also important to note that the $n_i$ values do not change if the only interaction is the internally reversible

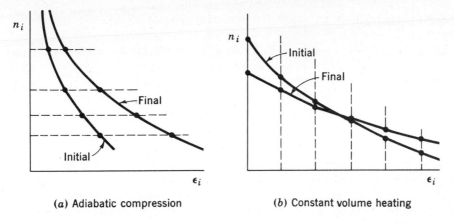

(a) Adiabatic compression  (b) Constant volume heating

**Figure 9-3** Shifting of $\epsilon_i$ or $n_i$ by work or heat effects during an internally reversible process.

change of a generalized displacement. The addition of energy to a closed system due to internally reversible work effects is not to be associated with a change in the particle distribution. The energy of each level changes with a constant population of particles for a given macrostate. Figure 9-3a illustrates a hypothetical change in volume of a system so that the energy spacing is doubled. Note that the $n_i$ values are constant.

On a microscopic basis we have seen that $U = \sum_i n_i \epsilon_i$. Therefore, for a differential change of state,

$$dU = \sum_i n_i \, d\epsilon_i + \sum_i \epsilon_i \, dn_i$$

According to this equation any change in the energy $U$ must be associated with two effects: (1) the population of each energy level is held fixed while the energy values themselves change, and (2) the values of the energy levels are held fixed while the distribution of particles among the levels changes. In addition, the conservation of energy principle for closed systems states that

$$dU = \delta Q + \delta W \tag{3-5}$$

For an internally reversible change of state, $W = -\sum_i n_i \, d\epsilon_i$. Therefore it follows from the above equations that the macroscopic heat effects during an internally reversible process bring about a microscopic change in the state of the system given by

$$\delta Q_{\text{int rev}} = \sum_i \epsilon_i \, dn_i \tag{9-5}$$

Thus the population of each level changes by $dn_i$ when a differential heating effect occurs. Since $\sum_i dn_i = 0$ for a closed system, some of the levels become more highly populated, whereas other levels decrease in their population. Figure 9-3b shows the hypothetical case of heat addition in the absence of work interactions.

Equation (9-5) requires an internally reversible process because we placed this same restriction on Eq. (9-4).

The macroscopic conservation of energy principle for a closed system has been described in terms of microscopic changes within the system. However, such descriptions of $\delta Q$ and $\delta W$ are valid only if the system is free of irreversibilities. Note that the absence of irreversibilities refers only to the system. Hence the restriction on Eqs. (9-4) and (9-5) is internal reversibility, not total reversibility.

## 9-6 ENTROPY AND THE MOST PROBABLE MACROSTATE

The entropy function, as defined by the Boltzmann-Planck equation, is given by the relations $S = k \ln W_{\text{tot}}$. In practice, however, it is much more convenient to evaluate $S$ by using a modified form of the Boltzmann-Planck equation. The validity of this modification can be justified by examining one further characteristic of systems on the microscopic level. Recall the examples presented in Sec. 9-2 which illustrated the concepts of macrostates and microstates. The first system considered had four distinguishable particles and a total energy of 6 units, which resulted in nine macrostates (see Table 9-1). With each macrostate is associated a certain number of microstates. All microstates are tacitly assumed to be equally likely in occurrence in a random molecular system. Macrostate $E$ has 24 possible microstates out of the 84 total. This is 28 percent of the total number of arrangements. The macrostate which has the largest number of microstates is defined as the *most probable macrostate*. Hence macrostate $E$ is the most probable macrostate for the given system. Associated with every thermodynamic system is a macrostate which can be called the most probable one.

The most probable macrostate defines the energy distribution which would be observed more frequently than any other distribution. For macroscopic thermodynamic systems it turns out that this distribution is overwhelmingly more probable than any other. Therefore it is the characteristics of the most probable macrostate that determine the equilibrium thermodynamic properties. The most probable distribution is, in a sense, the equilibrium distribution. Although there are innumerable macrostates possible for a given thermodynamic state, experience dictates that all of them can be ignored, except the most probable one, in computing thermodynamic properties.

The importance of the most probable macrostate in computing thermodynamic properties can be illustrated by the following example. Recall from Sec. 9-3 the situation where two subsystems $X$ and $Y$ were brought into thermal contact and allowed to share a total energy of 8 units. Table 9-2 summarized the total number of arrangements (3003 microstates) for each possible division of the energy between the two subsystems. It is seen that the most probable split of energy occurs when subsystem $X$ has 5 units of energy. The probability of occurrence $p_i$ is 18.6 percent, since this description has 18.6 percent of the total number of microstates. The probability of occurrence of a situation where subsystem $X$

has no energy is seen to be quite unlikely (1.5 percent), and this certainly fits our intuition. However, note that the probability of subsystem $X$ having 4 or 6 units of energy is almost as probable as the case where $X$ has 5 units, which is the most probable state. Thus fluctuations from the most probable state are quite probable, and this does not fit our expectation. In the final equilibrium state we do not anticipate any property of either subsystem, including the energy, to vary to any measurable extent. The reason this example contradicts our experience is that both subsystems are too small to be considered macroscopic systems.

A more meaningful analysis of the above situation is made when one studies two subsystems each of which has an extremely large number of particles and the total shared energy is large. When $n$ is very large, it should be realized that $W$ for a given subsystem is an extremely rapidly increasing function of its energy. In this case, since the total energy is shared, as subsystem $X$ acquires an increasing amount of energy, the energy of $Y$ decreases. Therefore, $W_X$ increases extremely rapidly while $W_Y$ decreases extremely rapidly as $X$ acquires an increasing share of the total energy. Let us represent the thermodynamic probability of the composite system by $W_C$. Since $W_C = W_X W_Y$, the product of $W_X$ and $W_Y$ exhibits an extremely sharp maximum at some value of $U_X$. Therefore a plot of the probability of various microscopic descriptions versus $U_X$ must also show the same sharp maximum. This state of maximum $W_C$ is, of course, the most probable split in energy, and we might symbolize the energy of subsystem $X$ in this state by $U_{X, mp}$. The general characteristics of a plot of $p_i$ (the probability of a certain split in the shared energy) versus $U_X$ is shown in Fig. 9-4. The region $\Delta U$ in the figure where $p_i$ has a significant value is such that $\Delta U \ll U_{X, mp}$. We see that when two large systems are in thermal equilibrium, they do not exhibit a widely varying energy for either one, in a macroscopic sense. The probability that $U_X$ will be appreciably different from $U_{X, mp}$ is negligible. On the microscopic level we see that wide variations are possible, but highly improbable. Hence our instruments never

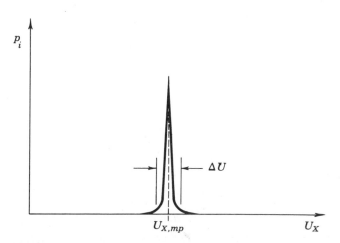

**Figure 9-4** Probability of occurrence of subsystem $X$ when its energy is $U_X$.

detect any fluctuations. The most probable arrangement and those only slightly different from it in their arrangement predominate over all other possibilities.

The preceding discussion was based on two systems which shared a common quantity of energy. The same general conclusions can be drawn for a single isolated system of energy $U$. Although the energy is fixed, there is a huge number of macrostates that might exist. However, for any large system of particles, only a small fraction of the possible macrostates account for a large fraction of the total number of microstates. In addition, the characteristics of the energy distributions of this small number of macrostates are not very different from the most probable distribution. That is, the probability of occurrence of a macrostate falls off extremely fast for very small changes in the $n_i$ values associated with the most probable macrostate. For example, consider a degenerate system of distinguishable particles. Small changes $\delta n_i$ are made at various energy levels, starting with the system in the most probable macrostate. The value of $W_{mp}$ is altered to a new value $(W_{mp} + \delta W)$. A mathematical analysis of this situation leads to the following approximation:

$$\frac{W_{mp} + \delta W}{W_{mp}} \approx \exp\left[-\frac{1}{2}\sum_i \frac{(\delta n_i)^2}{n_i}\right]$$

Note that whether the $\delta n_i$ values are positive or negative, the new macrostate always has fewer arrangements than the most probable one. Hence the equation is consistent with the definition of the most probable macrostate. In most systems $n_i$ will usually be large, such as $10^{15}$ to $10^{20}$. Therefore, even when $\delta n_i$ is a fraction of a percent, the bracketed term is extremely large. This is due to the fact that the $\delta n_i$ terms are all squared. Consequently small changes in $\delta n_i$ lead to a drastic reduction in the number of microstates when starting from the most probable macrostate. Obviously only those macrostates which are nearly like the most probable one in their distribution of particles need to be considered. For this reason it is reasonable to base macroscopic properties solely on the particle distribution of the most probable macrostate.

If the most probable macrostate is a sufficient model for the prediction of thermodynamic properties, how is the concept of the entropy function affected, since it is based on $W_{tot}$, and not $W_{mp}$. Actually, since $W_{mp}$ has a fixed value for any equilibrium state, it could serve as the basis for the definition of the entropy function, as well as $W_{tot}$. However it is not necessary to modify our definition of $S$ as given by the Boltzmann-Planck equation. As a result of the mathematics used when dealing with systems of a large number of particles, it turns out that for all practical purposes $\ln W_{mp} = \ln W_{tot}$. For evaluating the entropy, then, we use the most probable macrostate to find $W$, rather than account for all macrostates. The reasonableness of this approximation is illustrated by the following example. It is necessary to limit the example to a study of relatively small systems, but the extrapolation to larger systems is quite obvious.

Table 9-3 lists the arrangement of distinguishable particles among three equally spaced energy levels for systems $x$, $y$, and $z$. The specific macrostate shown in each case has been chosen to be the most probable one. The only difference among the systems is that the values of $n$ and $U$ have been varied. (Note that none

**Table 9-3**

| $n$ | $W_{mp}$ | $W_{tot}$ | $W_{mp}/W_{tot}$ | $\log_{10} W_{mp}$ | $\log_{10} W_{tot}$ | Log ratio |
|------|----------|-----------|------------------|---------------------|---------------------|-----------|
| 62   | $1.4 \times 10^{14}$ | $4.25 \times 10^{16}$ | $3.3 \times 10^{-3}$ | 14.154 | 16.623 | 0.851 |
| 310  | $5.95 \times 10^{75}$ | $1.13 \times 10^{80}$ | $5.25 \times 10^{-5}$ | 75.783 | 80.053 | 0.948 |
| 1550 | $4.4 \times 10^{386}$ | $5.7 \times 10^{395}$ | $7.7 \times 10^{-10}$ | 386.642 | 395.757 | 0.978 |

of the systems is large enough to be considered macroscopic in size.) In each case the average energy per particle is exactly the same. For these three systems we wish to compare the values of $W_{mp}$ to $W_{tot}$ and $\ln W_{mp}$ to $\ln W_{tot}$. The results are summarized as shown. Note that the values of $W_{mp}$ and $W_{tot}$ are extremely large, even for these relatively small systems. As the number of particles increases, the ratio $W_{mp}/W_{tot}$ becomes very small. More importantly, as the number of particles increases, it appears that $\log W_{mp}$ approaches $\log W_{tot}$. (*Note*: logarithms to the base 10 have been used in the table for convenience. The ratio of logarithms would be the same regardless of the base used.) If the example were extrapolated to a system of $10^{23}$ particles, the ratio of logarithms would be unity for all practical purposes. Although we have used an example involving a system of distinguishable particles with a limited size in all cases, the conclusion reached is applicable to real systems as well. That is, $\ln W_{mp} = \ln W_{tot}$ to any desired accuracy if the system is large. Hence the evaluation of the entropy function by use of the Boltzmann-Planck equation can be made in terms of a modified equation,

$$S = k \ln W_{mp} \tag{9-6}$$

In evaluating the entropy function, we need not bother with the details of all the possible macrostates that could possibly exist for a particular equilibrium state. Attention is focused simply on the most probable macrostate.

In summary, for macroscopic thermodynamic systems, the most probable macrostate and those nearly identical to it predominate. Therefore the macroscopic properties of a system can be based entirely on the microscopic characteristics of the most probable macrostate. Calculation of properties based on this approach will agree with experimental measurements. (Exceptions to this rule must be permitted, since in a few cases fluctuations from the mean are significant.) It is of particular interest that the entropy function can be correctly evaluated in terms of the Boltzmann-Planck equation by employing the logarithm of the thermodynamic probability associated with the most probable macrostate.

## 9-7 ENTROPY CHANGES IN TERMS OF MACROSCOPIC VARIABLES

The entropy function $S$ has been introduced and defined in terms of microscopic concepts. However, our day-to-day experiences, such as our observations and measurements, are macroscopic. Hence it is highly desirable to relate the entropy function to macroscopic phenomena. Recall that in the development of the conservation of energy principle we found it necessary to consider only changes in the internal energy and enthalpy functions. Absolute values of $u$ and $h$ were never required. We shall also find it necessary to consider only changes in the value of the entropy function. For an infinitesimal process the change in the entropy of a closed system is found by differentiating Eq. (9-6). Thus, in general,

$$dS = k \, d(\ln W_{mp}) \qquad (9\text{-}7)$$

That is, the entropy change is determined by the change in the number of arrangements of the most probable distribution as the equilibrium state is altered. Our goal is to relate this change in $\ln W_{mp}$ to macroscopically measurable quantities.

We have already noted in preceding examples of isolated systems of given degeneracy that the thermodynamic probability $W$ for any macrostate depends solely upon the $n_i$ values associated with each energy level. Since the logarithm of $W$ would also depend solely on the $n_i$ values for that macrostate, we may denote functionally that

$$\ln W = f(n_1, n_2, n_3, \ldots, n_i, \ldots)$$

If a differential change of state occurred, the change in $\ln W$ for any macrostate would be given by the total differential,

$$d \ln W = \frac{\partial \ln W}{\partial n_1} dn_1 + \frac{\partial \ln W}{\partial n_2} dn_2 + \cdots = \sum_i \frac{\partial \ln W}{\partial n_i} dn_i$$

This equation is valid for any macrostate, including the most probable macrostate. Thus we may write specifically that

$$d \ln W_{mp} = \sum_i \frac{\partial \ln W_{mp}}{\partial n_i} dn_i \qquad (9\text{-}8)$$

The change in the entropy function for an infinitesimal change of state between two equilibrium states then becomes, if we substitute Eq. (9-8) into Eq. (9-7),

$$dS = k \sum_i \frac{\partial \ln W_{mp}}{\partial n_i} dn_i \qquad (9\text{-}9)$$

The evaluation of the quantity $\partial \ln W_{mp}/\partial n_i$ is fairly straightforward, although there is a considerable amount of unavoidable mathematics. Since the most probable macrostate is, by definition, the one with the largest number of arrangements, we may approach the problem as one requiring the maximization of a function through the techniques of the differential calculus. One additional complication must be considered, however. In addition to the general equation relating the dependent variable, $\ln W$, to the independent variables, the $n_i$'s, there are two more equations relating the $n_i$'s which must be satisfied during the maximization process. These are

$$\sum_i n_i = n \qquad (9\text{-}10)$$

and

$$\sum_i \epsilon_i n_i = U \qquad (9\text{-}1)$$

The problem we have, then, is one maximization in the presence of constraining equations.

The details of the maximization process are given in Appendix A-2. The major result is an equation relating $\partial \ln W_{mp}/\partial n_i$ to the value of the energy level where particles have been added or removed. For any energy level where $dn_i$ particles are shifted when the system changes from one equilibrium state to another,

$$\frac{\partial \ln W_{mp}}{\partial n_i} = A + B\epsilon_i \tag{9-11}$$

The two quantities $A$ and $B$ represent two constants, as yet undetermined. This relationship is the one we sought in order to evaluate $dS$ through the use of Eq. (9-9). If we combine Eqs. (9-9) and (9-11), then

$$dS = k \sum_i \frac{\partial \ln W_{mp}}{\partial n_i} \, dn_i = k \sum_i (A + B\epsilon_i) \, dn_i$$

$$= kA \sum_i dn_i + kB \sum_i \epsilon_i \, dn_i \tag{9-12}$$

The terms $A$ and $B$ have been brought outside the summation signs, since they are constants. The first summation on the right side of Eq. (9-12) is zero, for the following reason. In a system of fixed number of particles,

$$\sum_i n_i = n \tag{9-10}$$

Differentiation of this equation yields

$$\sum_i dn_i = 0$$

Consequently Eq. (9-12) reduces to the form

$$dS = kB \sum_i \epsilon_i \, dn_i \tag{9-13}$$

Equation (9-13) represents the change in entropy due to a differential change of state between two equilibrium states, expressed in terms of the shifting of particles among the energy levels. The equation is completely general in this form. The change of state is not restricted to any particular type of process and may involve both heat and work interactions.

Equation (9-13) contains two quantities that must be replaced by their counterparts in macroscopic thermodynamics, namely, $B$ and $\sum_i \epsilon_i \, dn_i$. The first of these is the constant which appears in the derivation of Eq. (9-11). This constant is evaluated in statistical thermodynamics by employing the constraining equations in a particular fashion. This type of evaluation is mathematically lengthy, and need not concern us here. Instead, a more qualitative and physical interpretation will be given.

Consider once more two simple, closed systems $X$ and $Y$ which are brought into thermal contact through a rigid wall. The volume of each system is constant, and the total energy of the composite system $X + Y$ is constant. The total entropy of the composite system is the sum of the entropies of the two individual systems. If the two systems are not initially in mutual equilibrium, a heat interaction will occur until a new equilibrium state is reached. On the basis of the second law of thermodynamics, the entropy of the composite system will increase until a maximum value is attained. As the maximum is reached, $dS = 0$. During the process the change in the entropy for either system can be expressed in the following way. For any simple system, there are two independent properties. In this case it is convenient to let these two properties be the energy and the volume. Hence, $S = S(U, V)$. For a differential change of state

$$dS = \left(\frac{\partial S}{\partial U}\right)_V dU + \left(\frac{\partial S}{\partial V}\right)_U dV$$

In our particular example the volume is being held fixed on each system, and therefore the second term on the right is zero for each system. Consequently the total entropy change for the composite system $C$ may be written as

$$dS_C = dS_X + dS_Y = \left(\frac{\partial S}{\partial U}\right)_V\bigg|_X dU_X + \left(\frac{\partial S}{\partial U}\right)_V\bigg|_Y dU_Y \tag{9-14}$$

As the composite system approaches equilibrium, $dS_C = 0$. The variation of the composite entropy with energy $U_X$ is shown in Fig. 9-5. In addition, the total energy of the composite system is a constant given by $U_C = U_X + U_Y$. Thus $dU_X = -dU_Y$. These two facts in combination with Eq. (9-14) indicate that at the state of equilibrium

$$\left(\frac{\partial S}{\partial U}\right)_V\bigg|_X = \left(\frac{\partial S}{\partial U}\right)_V\bigg|_Y \tag{9-15}$$

The derivatives expressed in this equation are properties of the systems $X$ and $Y$. However, the only property of each system that necessarily must be the same at

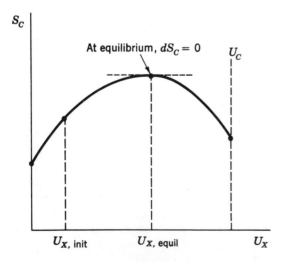

Figure 9-5 Variation of composite entropy with energy shared by systems $X$ and $Y$.

the state of thermal equilibrium is the temperature. Therefore it is to be expected that the partial derivative $(\partial S/\partial U)_V$ is related in some way to the temperature of the system. The dimensions of the partial derivative give a clue to this relationship. From Eq. (9-6) it is seen that the dimensions of $S$ are those of the Boltzmann constant $k$, which are energy/(absolute temperature). When divided by the dimension of $U$, which is energy, we see that the dimension of the partial derivative is the reciprocal of the absolute temperature $T^{-1}$. Hence it is appropriate to define a thermodynamic temperature which is the reciprocal of the derivative found in Eq. (9-15). Consequently,

$$T \equiv \frac{1}{(\partial S/\partial U)_V} \quad \text{or} \quad \left(\frac{\partial S}{\partial U}\right)_V = \frac{1}{T} \tag{9-16}$$

It can be shown that this thermodynamic temperature is the same as that given by the ideal-gas temperature scale.

As a final step, the partial derivative in Eq. (9-16) can be related to the constant $B$. Recall that in general the change in $U$ is given by $dU = \sum_i \epsilon_i \, dn_i + \sum_i n_i \, d\epsilon_i$. In our example involving systems $X$ and $Y$ the volume of each is constant. Hence the term $\sum_i n_i \, d\epsilon_i$ is zero in the expression for $dU$, since a change in $\epsilon$ depends upon a change in $V$. As a result we find that $dU_V = \sum_i \epsilon_i \, dn_i$. If we substitute this result into Eq. (9-13) for $dS$, then for this specific example

$$dS = kB \, dU \qquad \text{at constant volume}$$

or
$$\left(\frac{\partial S}{\partial U}\right)_V = kB \tag{9-17}$$

If Eqs. (9-16) and (9-17) are now equated, we find that

$$kB = \frac{1}{T} \tag{9-18}$$

Thus the constant $B$ which appeared originally as one of the constants in Eq. (9-11) is inversely proportional to the thermodynamic temperature or ideal-gas temperature, measured either in degrees Rankine or degrees Kelvin. Although derived for a special process, Eq. (9-18) is of general validity. Direct substitution of the above relation into Eq. (9-13) leads to the expression

$$dS = \frac{1}{T} \sum_i \epsilon_i \, dn_i \tag{9-19}$$

All that remains now is the replacement of the summation term by an equivalent macroscopic quantity.

In Sec. 9-5 it was demonstrated that a heat interaction affects the number of particles associated with each energy level if the process is internally reversible. That is,

$$\delta Q_{\text{int rev}} = \sum_i \epsilon_i \, dn_i \tag{9-5}$$

If this relation is substituted into Eq. (9-19), we obtain the result that

$$dS = \left(\frac{\delta Q}{T}\right)_{\text{int rev}} \tag{9-20}$$

If a process is carried out in an internally reversible manner, then only a knowledge of the transient heat effects and of the temperature of the system at each stage of the process is required in order to evaluate $\Delta S$. The restriction of internal reversibility is inherent in the derivation and cannot be removed from this particular equation. Although internally reversible processes are limiting ones, Eq. (9-20) could be used to evaluate entropy changes from experimental measurements of heat transfer and temperature if irreversibilities within the system were negligible.

Integration of Eq. (9-20) leads to the relationship, on an intensive basis, that

$$s = \int \left(\frac{\delta q}{T}\right)_{\text{int rev}} + s_0 \tag{9-21}$$

This equation shows that integration of $\delta q/T$ gives the entropy only within an arbitrary constant. As long as we work with pure substances, or nonreacting mixtures, a knowledge of the constant $s_0$ is unimportant because only a change in the entropy is usually required. Therefore tables and graphs may be constructed for which the base, or reference, value of $s$ is completely arbitrary. For example, in the saturation-temperature tables for water, A-12M and A-12, the specific entropy is chosen to be zero for saturated liquid water at the triple state ($0.01°C$ or $32.02°F$). Other reference states are chosen for other substances.

The dimensions of the entropy function are energy/(absolute temperature). The SI units for the specific entropy are kJ/(kg)($°K$) or kJ/(kg·mol)($°K$), while USCS units commonly are Btu/(lb)($°R$) or Btu/(lb·mol)($°R$). Other sets of units may also be found in the literature.

## 9-8 HEAT AND WORK RESERVOIRS

The reader at this point is directed to read Secs. 6-6 and 6-13, and then follow these sections with Sec. 9-9.

## 9-9 APPLICATION OF THE SECOND LAW TO CLOSED SYSTEMS

Our fundamental statement of the second law of thermodynamics has been made in terms of the behavior of isolated systems. In many cases, however, it is either desirable or convenient to focus our attention on the behavior of a closed system. The restatement of the second law in terms of a closed system may be made in the following manner.

Consider a closed system which, in general, may be affected by work and heat interactions. Any process it undergoes may be either reversible or irreversible. In

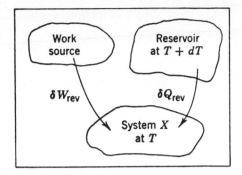

**Figure 9-6** A closed system undergoing reversible interactions with its surroundings.

order to focus attention on the effect of irreversibilities within the closed system, all changes external to the system which affect the state of the system will be made reversible. For example, any energy in the form of work supplied to the system must be from a reversible source, such as a flywheel, pulley-weight assembly, etc. In addition, any heat interaction must occur through an infinitesimal temperature difference and will be supplied from a heat reservoir. Figure 9-6 shows a closed system $X$ at a temperature $T$ undergoing reversible work and heat interactions. There is no restriction on the type of change within the system $X$.

The composite of the system and other systems with which it interacts is an isolated system to which we may apply the second law. The entropy change $dS_C$ of the composite system is

$$dS_C = dS_X + dS_{\text{heat reservoir}} + dS_{\text{work source}} \geqslant 0$$

Since the temperature difference between the heat reservoir and the closed system is an infinitesimal $dT$, the entropy change of the reservoir can be expressed in terms of system conditions. That is,

$$dS_{\text{reservoir}} = -\frac{\delta Q}{T}$$

In addition, the entropy change of a reversible work source is zero, as noted in the preceding section. Therefore the inequality stated above reduces to

$$dS_X + \left(-\frac{\delta Q}{T}\right) \geqslant 0$$

or, upon rearrangement,

$$dS_{\text{closed}} \geqslant \left(\frac{\delta Q}{T}\right)_{\text{act}} \tag{9-22}$$

In Eq. (9-22) the quantity $\delta Q$ represents the actual heat transfer to or from the system. The equality sign applies when the closed-system change is internally reversible, as we have noted previously. The inequality sign is valid if irreversibilities are present within the system.

Equation (9-22) leads to an important relation for any type of closed system. In the absence of heat transfer, the entropy change for any closed system must fulfill the condition that

$$dS_{\text{closed, adia}} \geq 0 \qquad\qquad (9\text{-}23)$$

Thus, for adiabatic changes the entropy must increase in the presence of internal irreversibilities. For internally reversible adiabatic processes, $dS = 0$. Note that Eq. (9-23) is another mathematical statement of a directional law.

## 9-10 HEAT ENGINE CONCEPTS

From an engineering viewpoint the interrelationships of heat and work as required by the first and second laws are extremely important in the analysis of work-producing devices. Thermal energy (heat) is available from a number of sources including fossil fuels, nuclear fission, and solar radiation. The conversion of heat into a more useful form of energy, work, is accomplished by a class of devices known as heat engines. A heat engine is a device which operates continuously or cyclically and produces work while exchanging heat across its boundaries. The restriction to continuous or cyclic operation implies that the substance within the device is returned to its initial state at regular intervals. For example, in a steam-power plant the steam is contained within a closed loop. It is heated, expanded, cooled, and compressed as it circulates around the loop in a continuous manner, returning to its initial state at the same point in the loop. A cyclic heat engine might be one of the nature of a piston-cylinder assembly.

The second law places a severe restriction on the operation of a heat engine. A corollary to the second law states that

> It is impossible to operate in any manner a cyclic device that will produce work while exchanging heat with a single heat reservoir.

This corollary is known as the *Kelvin-Planck* statement of the second law. The proof of this statement follows directly from the increase in entropy principle for an isolated system. Three factors need to be checked in terms of entropy changes. Since the engine is cyclic in nature, it will undergo no net entropy change. The production of work in itself involves no creation of entropy if the energy equivalent to the work effect is stored in the surroundings in a reversible manner, as in a work reservoir. The third effect is the heat transfer from a heat reservoir to the cyclic heat engine. The entropy change of the reservoir is simply $Q/T$, and must be negative, since $Q$ is negative with respect to the reservoir. Consequently the total entropy change of the cyclic device and its surroundings is negative. This is a violation of the second law as expressed by Eq. (9-3). Therefore the Kelvin-Planck statement is a direct consequence of the second-law statement given previously.

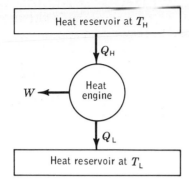

**Figure 9-7** Schematic of a heat engine operating between two heat reservoirs at $T_H$ and $T_L$.

Machines which would operate in violation of the Kelvin-Planck statement are known as perpetual-motion machines of the second kind (PMM2). The existence of such machines would be a tremendous help to the world. Work could be derived from the enormous low-temperature heat sources such as the oceans in practically unlimited quantities. Unfortunately, such an operation is impossible. Note that neither the first nor second law restricts the inverse operation, that is, the complete conversion of work into heat.

On the basis of the Kelvin-Planck statement, all cyclic heat engines must reject a portion of the heat received to another heat reservoir. A schematic of a heat engine operating between reservoirs of temperatures $T_H$ and $T_L$ is shown in Fig. 9-7. Basically, the heat engine operates under these conditions:

1. Heat $Q_H$ is supplied to the heat engine from a heat reservoir at a temperature $T_H$ (high temperature).
2. The engine produces a net quantity of work $W$ during the cyclic process.
3. Heat $Q_L$ is rejected from the engine during the cycle to a heat reservoir at a temperature $T_L$ (low temperature), where $T_L < T_H$.

For the heat engine as the closed system of interest, the conservation of energy principle is

$$\sum Q - \sum W = \Delta U$$

For a cyclic device, $\Delta U$ is zero. Thus for the engine shown in Fig. 9-7 we see that

$$|Q_H| - |Q_L| = W_{net}$$

where $W_{net}$ is always positive for heat engines. Since the heat added and the heat removed from the heat engine have different signs by convention, it is convenient to use absolute-value signs in the energy equation and insert a negative sign directly into the equation for the $Q_L$ term. The object of a heat engine is to produce work from the energy added as heat. A measure of its performance is the ratio of work output to heat input. This ratio is defined as the thermal efficiency $\eta_{th}$ of the engine. That is,

$$\eta_{th} \equiv \frac{W_{net}}{Q_{in}} = \frac{W_{net}}{Q_H} \qquad (9\text{-}24)$$

If we supply 100 units of energy to a heat engine and find that 70 units are rejected, then the net work output is 30 units and the thermal efficiency is 0.30 (or 30 percent).

We now wish to establish the maximum value of the thermal efficiency that any heat engine could achieve. If we apply the first law to a heat engine for a differential cycle, then

$$\delta Q_H - \delta Q_L - \delta W_{net} = 0 \qquad (a)$$

Since $\delta Q_L$ is out of the system, it is preceded by a negative sign in Eq. $(a)$. When numerical values are substituted into this equation, only absolute values should be used, since the sign convention on the heat quantities has already been included. In addition, the total entropy change for the cyclic process is

$$dS_{tot} = dS_H + dS_L + dS_{engine} \qquad (b)$$

The entropy change of the heat engine is zero because it is a cyclic device. The values of $dS_H$ and $dS_L$ for the heat reservoirs are $-\delta Q_H/T_H$ and $\delta Q_L/T_L$, respectively. The first term is negative since heat leaves the high-temperature reservoir. Consequently, Eq. $(b)$ may be written as

$$dS_{tot} = -\frac{\delta Q_H}{T_H} + \frac{\delta Q_L}{T_L} \qquad (c)$$

Substitution of the term $\delta Q_L$ from Eq. $(a)$ into Eq. $(c)$ yields

$$dS_{tot} = -\frac{\delta Q_H}{T_H} + \frac{\delta Q_H}{T_L} - \frac{\delta W}{T_L} \qquad (d)$$

Upon rearrangement, the thermal efficiency is found to be given by

$$\eta_{th} \equiv \frac{\delta W}{\delta Q_H} = \frac{T_H - T_L}{T_H} - \frac{T_L\, dS_{tot}}{\delta Q_H} \qquad (9\text{-}25)$$

The first term on the right-hand side always lies between 0 and 1, and it is solely a function of the temperatures of the heat reservoirs. The second term on the right is always positive or zero, since the second law requires that $dS_{tot} \geq 0$. Hence the effect of the last term always is to reduce the thermal efficiency or in the limit to leave it unchanged for a reversible process. Equation (9-25) indicates that the maximum thermal efficiency is attained when $dS_{tot}$ is zero. Under this condition Eq. $(c)$ requires that $-\delta Q_H/T_H + \delta Q_L/T_L = 0$. For a finite totally reversible cycle this becomes

$$\frac{|Q_H|}{T_H} = \frac{|Q_L|}{T_L} \qquad (9\text{-}26)$$

The maximum efficiency of a heat engine operating between two heat reservoirs is achieved when the heat quantities and reservoir temperatures satisfy the above relation. Under these circumstances the thermal efficiency is called the Carnot efficiency, and it is given simply by

$$\eta_{Carnot} = \frac{T_H - T_L}{T_H} \tag{9-27}$$

Equation 9-27 holds for any totally reversible heat engine operating between two reservoirs of fixed temperature. The temperature values must be expressed, of course, in either degrees Kelvin or degrees Rankine.

The efficiency of any heat engine operating between two constant-temperature reservoirs is increased by either employing supply reservoirs of very high temperature or rejecting to heat reservoirs of extremely low temperature. The latter condition is difficult to accomplish since temperatures below atmospheric conditions must be artifically acquired. The use of heat from high-temperature sources is often restricted by the availability of materials that can withstand such extreme conditions. Much work is being carried on at the present time in the field of materials science to develop high-temperature-resistant solids, so that the theoretical thermal efficiencies of operating devices can be substantially increased. The maximum efficiency of unity is reached when $T_H$ is infinite or $T_L$ is absolute zero. It is important to recognize that, for reasonable values of $T_H$ and $T_L$ (for example, 1000 and 300°K, respectively, for continuously operating engines), the thermal efficiency never approaches unity. When one considers that practical engines may have a heat supply the temperature of which varies considerably below the maximum value and that mechanical inefficiencies must also be included, it is not too surprising that modern heat engines have overall efficiencies of only 25 to 40 percent. It is essential to understand, however, that even under optimum theoretical conditions, a heat engine is relatively inefficient in converting heat into work by means of a cyclic process. This severe limitation has led to a renewed interest in work-producing devices which are not limited to the Carnot efficiency, such as fuel cells. This latter device has other inherent advantages, including its adaptability to unique environments and silent operation.

It has been demonstrated that the maximum efficiency of any heat engine operating between two fixed heat reservoirs occurs when the entropy change of the universe during the cyclic process is zero. From this we inferred earlier that the entire cycle must be carried out in a reversible manner. Several important consequences of this may be shown. These are usually known as the Carnot corollaries, or the *Carnot principle*. Normally, at least two parts, or statements, are included:

1. No heat engine is more efficient than a reversible engine which operates between the same heat reservoirs.
2. All reversible heat engines which operate between the same heat reservoirs will have the same thermal efficiency.

The first point has already been proved by use of Eq. (9-25). A more efficient engine would require that $dS$ of the universe be negative, in violation of the second law. The point can also be shown in the following way. Equation ($c$) for a finite cycle becomes

$$\Delta S_{tot} = -\frac{Q_H}{T_H} + \frac{Q_L}{T_L} \geqslant 0$$

This can be rewritten in the form

$$\frac{Q_L}{Q_H} \geqslant \frac{T_L}{T_H}$$

where the inequality sign applies to an irreversible cycle. It is also generally true for an irreversible cycle and a reversible cycle that

$$\eta_{irrev} = 1 - \frac{Q_L}{Q_H}$$

and

$$\eta_{Carnot} = 1 - \frac{T_L}{T_H}$$

Combination of these two efficiency expressions leads to

$$\eta_{irrev} + \frac{Q_L}{Q_H} = \eta_{Carnot} + \frac{T_L}{T_H}$$

But it was noted above that for an irreversible cycle $Q_L/Q_H > T_L/T_H$. Therefore, as a general statement for heat engines

$$\eta_{irrev} < \eta_{Carnot} \tag{9-28}$$

The inequality in the second law requires that an irreversible engine be less efficient than a reversible engine when operating between the same heat reservoirs. The second part of Carnot's principle is self-apparent from Eq. (9-27).

One final point, which often is included in the Carnot principle, needs to be emphasized. The proof that a reversible engine has the maximum efficiency was accomplished without reference to the nature of the working medium or the operation of the engine itself. Whether the reversible engine operates on hydrogen gas, air, water, or mercury is of no consequence. Likewise, the sequence of processes which make up the cycle is not important.

## REFERENCES

Fast, J. D.: "Entropy," McGraw-Hill, New York, 1963.

Hatsopoulos, G. N., and J. H. Keenan: "General Principles of Thermodynamics," Wiley, New York, 1968.

Howerton, M. T.: "Engineering Thermodynamics," Van Nostrand, Princeton, N.J., 1962.

Lee, J. F., F. W. Sears, and D. L. Turcotte: "Statistical Thermodynamics," Addison-Wesley, Reading, Mass., 1963.

Reif, F.: "Fundamentals of Statistical and Thermal Physics," McGraw-Hill, New York, 1965.

Sonntag, R. E., and G. J. Van Wylen: "Fundamentals of Statistical Thermodynamics," Wiley, New York, 1966.

Tribus, M.: "Thermostatics and Thermodynamics," Van Nostrand, Princeton, N.J., 1961.

# PROBLEMS

**9-1** What is the frequency of a photon of radiation which has an energy of 1 eV?

**9-2** The frequency of a photon of radiation is $0.1 \times 10^{15}$ s$^{-1}$. What is the energy associated with this photon in (a) joules, and (b) electronvolts?

### Particle distributions

**9-3** Determine the number of ways that seven distinguishable particles can be arranged into three groups so that the groups contain (a) 3, 2, and 2 particles, and (b) 4, 2, and 1 particle(s).

**9-4** Determine the number of ways that eight distinguishable particles can be arranged into three groups so that the groups contain (a) 3, 3, and 2 particles, and (b) 4, 2, and 2 particles.

**9-5** Four distinguishable balls are dropped at random into two boxes. After repeated tests, what fraction of the time will we expect to find that two balls are in each box if (a) the boxes are indistinguishable, and (b) the boxes are distinguishable?

**9-6** Five distinguishable balls are dropped at random into three boxes. After repeated tests, what fraction of the time will we expect to find (a) that the distribution of balls is 3 : 1 : 1 if the boxes are indistinguishable, and (b) that 3 balls are in the first box, 1 ball is in the second box, and 1 ball is in the third box.

**9-7** Consider the following two distributions ($A$ and $B$) of particles among five energy levels:

| Energy of level | $A$ | $B$ |
|---|---|---|
| 0 | 12 | 13 |
| 1 | 6 | 6 |
| 2 | 3 | 1 |
| 3 | 2 | 2 |
| 4 | 1 / | 2 |

Determine (a) the total number of particles in each distribution, (b) the total energy of each distribution, and (c) the ratio of the number of microstates of the most probable one to the other distribution.

**9-8** An isolated system contains three particles with a total energy of 3 units. The energy levels for the energy mode under consideration are equally spaced one unit apart, and the ground level of energy is taken as zero.

(a) How many macrostates are possible for this system? How many microstates, if the particles are distinguishable? (Use a diagram similar to those used in the text to help develop your answers.)

(b) Is there a most probable macrostate in part (a)?

**9-9** Consider the following three distributions, $A$, $B$, and $C$:

| Energy of level | $A$ | $B$ | $C$ |
|---|---|---|---|
| 0 | 15 | 16 | 15 |
| 1 | 8 | 8 | 7 |
| 2 | 4 | 3 | 3 |
| 3 | 2 | 1 | 3 |
| 4 | 1 | 2 | 2 |

If the number of arrangements per distribution is given by the permutation formula $W = n!/\pi n_i!$, determine which distribution is most probable.

**9-10** A system of three indistinguishable particles has a total energy of 3 units. The ground level is taken to be zero, and the spacing between levels is 1 unit. The degeneracy for the bottom four levels, starting from the ground level, is 1, 2, 2, and 2. Determine the number of macrostates and microstates.

**9-11** Consider a system of three indistinguishable particles. The energies of each are restricted to values of 0, 1, 2, 3, and 4. Determine the number of macrostates and microstates if each of the energy levels has a degeneracy of unity. The total energy is 6.

**9-12** Consider a system of three distinguishable particles with a total energy of 9 units. The degeneracy of each level is unity, and the particles are restricted to energy values of 0, 1, 2, 3, and 4. Determine the number of macrostates and microstates.

**9-13** An isolated system contains three particles with a total energy of 5 units. The energy levels are equally spaced 1 unit apart, and the ground level is taken as zero. How many macrostates and microstates are possible if the particles are distinguishable?

**9-14** Consider a system of four distinguishable particles with a total energy of 6 units. The ground level is taken to be zero, the spacing between levels is 1 energy unit, and the levels are nondegenerate. Determine the number of macrostates and microstates.

**9-15** Consider a system of ten distinguishable particles for which the energy levels are quantized according to the relation $\epsilon_i = i(i + 1)$, where $i = 0, 1, 2, \ldots$ The total energy is 18 units.

(a) How many macrostates are possible?
(b) How many microstates are possible?
(c) How many microstates are there for the most probable macrostate?
(d) How many particles are in the various energy levels, beginning at the ground level, for the most probable distribution?

**9-16** Same as Prob. 9-15, except that there are 12 particles and the total energy is 24 units.

**9-17** Consider a system of six distinguishable particles with a total energy of 12 units. The energy levels are quantized according to the relation $\epsilon_i = i + \frac{1}{2}$, where $i = 0, 1, 2, \ldots$

(a) How many macrostates and microstates are there?
(b) Is there a most probable macrostate?

**9-18** Same as Prob. 9-17, except that there are eight particles and the total energy is 14 units.

### Effect of heat addition on particle distribution

**9-19** Consider a system of three particles with a total energy of 4 units, initially. The ground level is taken to be zero, and the spacing between levels is 1 unit. A heat interaction occurs, and the total energy increases to 5 units. With the heat interaction, the spacing of energy levels remains the same.

(a) What is the number of macrostates initially and finally?
(b) If the particles are assumed to be distinguishable, what is the number of microstates initially and finally?

**9-20** Same as Prob. 9-19, except that the total energy initially is 5 units and finally is 6 units.

### Lagrangian technique

**9-21** Find the values of $x$, $y$, and $z$ which will make the function $R = x^2 + y^2 + z^2$ a minimum, with the restriction that $x + y + z = 9$ and $x + 2y + 3z = 20$.

**9-22** Find the dimensions for a box with an open top, the volume of which is 500 in$^2$, such that the total surface area is a minimum.

**9-23** Find the least and greatest distances of a point on the ellipse

$$\frac{x^2}{9} + \frac{y^2}{4} = 1$$

from the straight line $x + 2y = 8$.

**9-24** Find the values of $x$, $y$, and $z$ which make the function $F = x^2 + 4y^2 + 4z^2$ a minimum, subject to the restrictions that $x + y + z = 6$ and $x + 2y + 4z = 32$.

**9-25** Determine the values of $x$, $y$, and $z$ that make the function $u = xy^2z^3$ a minimum, subject to the relation $x + y + z = 36$.

**9-26** Find the dimensions of an open box (no top) that has a total surface area of 108 in$^2$, such that the total volume is a maximum.

**9-27** A system of 1000 distinguishable particles has three allowed energy levels which have values of 1, 2, and 3 energy units. By the method of Lagrangian multipliers determine the number of particles found in each of the three energy levels for the most probable macrostate if the total energy of the system is (a) 1600 units, (b) 2000 units, and (c) 2400 units.

**9-28** A system of 10,000 distinguishable particles has three allowed energy levels. The energy associated with the levels is 0, 1, and 2 energy units. By the method of Lagrangian multipliers determine the number of particles found in each of the three levels for the most probable macrostate if the total energy of the system is (a) 7000 units, (b) 11,000 units, and (c) 14,000 units.

### Carnot heat engine*

**9-29** A totally reversible heat engine operates between temperatures of 427 and 17°C. Compute the ratio of the heat absorbed by the engine from the source to the work output.

**9-30** A Carnot engine develops 1.5 kW of power output and rejects 7500 kJ/h to a sink at 20°C. Determine the source temperature, in °C.

**9-31** At what temperature is heat supplied to a Carnot engine that rejects 1000 kJ/min of heat at 7°C and produces (a) 40 kW of power, and (b) 50 kW of power?

**9-32** A Carnot engine operates between 37 and 717°C. It is proposed to increase the temperature of the high-temperature source to 1027°C. A counterproposal is to lower the sink temperature to achieve the same new thermal efficiency as would have been achieved by raising the source temperature to 1027°C. What would be the new required sink temperature, in °C?

**9-33** The efficiency of a Carnot engine discharging heat at 27°C is 30 percent. If the sink receives one million J/min, what is the power output of the engine, in kilowatts? What is the temperature of the high-temperature source?

**9-34** In a Carnot engine it is found that an efficiency of 50 percent occurs, with 500 kJ/cycle taken from the source reservoir. Calculate the sink temperature, in °C, and the heat rejected to the sink in kJ/cycle if the source temperature is 307°C.

**9-35** A patent application claims that a heat engine which receives heat at 160°C and rejects to a sink at 5°C is capable of delivering 0.10 kWh of energy for every 1000 kJ received as heat by the engine. Is this a valid claim?

**9-36** A Carnot engine with an efficiency of 40 percent receives 4000 kJ/h from a high-temperature source and rejects heat to a sink at 25°C. What is the power output, in kilowatts, and the temperature of the source, in °C?

**9-37** With reference to Prob. 9-30, compute the entropy changes of the heat source and heat sink in kJ/(°K)(h).

**9-38** With reference to Prob. 9-31, compute the entropy changes of the heat source and heat sink in kJ/(°K)(min) for part (a) and for part (b).

**9-39** With reference to Prob. 9-34, compute the entropy changes of the heat source and heat sink in kJ/°K/cycle.

**9-40** A reversible heat engine receives 1000 kJ of heat from a reservoir at 377°C and rejects heat to another reservoir at 27°C. Compute the entropy changes of the two reservoirs in kJ/°K.

* Although the following problems are all expressed in metric units, the units are not essential in comprehending the principles exemplified.

# TEN

# APPLICATIONS OF
# QUANTUM-STATISTICAL MECHANICS

The modern theories on the structure of matter and the mathematical laws which govern the evaluation of properties of matter are sufficiently well developed to permit the statistical computation of macroscopic properties to a high degree of accuracy. In this chapter we shall analyze a number of widely different systems, which require the application of Maxwell-Boltzmann, Bose-Einstein, or Fermi-Dirac statistics. The subject matter includes a study of the properties of an ideal gas, the photon-gas or blackbody radiation, the specific heat of a solid, and the electron gas in a metal. These are some of the many uses of quantum-statistical mechanics which lead to a broader understanding of the nature of matter. The material developed here is based on the fundamental ideas presented in Chap. 9.

It has been postulated that the entropy function for an equilibrium state can be approximated to a high degree of accuracy when evaluated on the basis of the most probable macrostate. Since the most probable macrostate by definition is the one with the largest number of microstates, the thermodynamic probability $W$ will have a maximum value when it is described in terms of the $n_i$ values for the most probable macrostate. The evaluation of the analytical expressions for $n_i$ thus requires a maximization of the thermodynamic probability associated with a particular physical system. When the appropriate equations for $n_i$ are inserted into the equations for $\ln W$, they lead to the determination of the entropy function for that system.

## 10-1 SIMPLE KINETIC THEORY OF A GAS

The kinetic theory of gases permits the evaluation of macroscopic properties of matter based on the behavior of individual particles, or molecules, within the system. Several basic assumptions, or idealizations, are involved. Primarily, such a study is based on a model of a particle which behaves as a smooth, rigid, perfectly elastic sphere. The result of this idealization is that the properties of interest are unaffected by the internal structure of the particle. For example, the only form of internal energy which need be considered is the translational, or linear, kinetic energy of the particles. In addition, the particles are assumed to be distributed uniformly within the containing vessel, and all directions of molecular motion are equally probable.

Based on Newton's second law, the pressure is a measure of the rate of change of momentum per unit time and per unit area. From a kinetic viewpoint, the pressure can be evaluated from the rate of change of the momentum of the particles as they collide with a unit surface of the vessel. If the velocity of a particle is represented by $c$, then, for any particle,

$$c^2 = u^2 + v^2 + w^2$$

where $u$, $v$, and $w$ are the velocities in the $x$, $y$ and $z$ directions, respectively. Now consider collisions with a wall perpendicular to the $x$ axis. Since all collisions are assumed to be perfectly elastic, the change in momentum in the $x$ direction of a particle of mass $m'$ due to a collision is $2m'u$. The pressure exerted on the wall is found by multiplying this change in momentum per collision by the number of collisions per unit time and area.

The evaluation of the number of impacts per unit time and area is aided by referring to Fig. 10-1. Consider for the moment only those particles with a given velocity $c$ which collide with an area $dA$ in the time interval $dt$. During the time interval, only those particles which initially were within an oblique cylinder of base $dA$ and slant height $c\,dt$ would reach the base area. The perpendicular distance from the top of the cylinder to the surface is $u\,dt$, where $u$ now may be considered the speed in the positive $x$ direction. Hence the volume of this cylinder

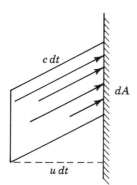

**Figure 10-1** Evaluation of molecular collisions per unit time and area.

is $u \, dt \, dA$. If the density of particles of speed $u$ is given by $n_u$, the number of particles within the cylinder of speed $u$ which will reach the surface $dA$ in the time $dt$ is $n_u u \, dt \, dA/2$. The factor of 2 is necessary since only one-half of the particles with a given speed along the $x$ axis are moving toward the surface. The number of particles of velocity $c$ and speed $u$ in the direction that will reach the surface per unit time and area is simply $n_u u/2$. Multiplication of this value by the momentum change per particle yields the pressure exerted in the $x$ direction by the particles of speed $u$. This is $2m'u(n_u u/2) = n_u m'u^2$. Since all positive values of $u$ are possible, the total pressure is found by summing up the quantity $n_u m'u^2$ for all values of $u$. Thus

$$P_x = \sum n_u m'u^2 = m' \sum n_u u^2$$

The value of $n_u$, the number of particles per unit volume which have a speed $u$ in the $x$ direction, is a function of $u$, and hence must remain inside the summation sign. At this point the functional relationship is unknown; nor is it required.

A mean-squared velocity $\overline{u^2}$ can now be defined by the following equation:

$$n\overline{u^2} \equiv \sum n_u u^2$$

Consequently, the pressure exerted in the $x$ direction (and similarly in the $y$ and $z$ directions, since the derivation is independent of the choice of coordinates) is

$$P_x = m'n\overline{u^2} \qquad P_y = m'n\overline{v^2} \qquad P_z = m'n\overline{w^2} \qquad (10\text{-}1)$$

It is noted from the above equations that the mean-squared velocity is the same in all three directions, since $P_x = P_y = P_z$ for a substance in static equilibrium. Thus

$$\overline{c^2} = \overline{u^2} + \overline{v^2} + \overline{w^2} = 3\overline{u^2}$$

Substitution of this result into Eqs. (10-1) yields the expression for the pressure exerted by the particles. Hence

$$P = \frac{m'n\overline{c^2}}{3} \qquad (10\text{-}2)$$

where $n$ is the total number of particles per unit volume.

Since the quantity $m'n$ is the mass of the particles per unit volume of the substance, then $m'n = m/V$. Hence Eq. (10-2) becomes, upon rearrangement,

$$PV = \frac{2}{3} \frac{m\overline{c^2}}{2}$$

But the quantity $m\overline{c^2}/2$ is the average translational, or linear, kinetic energy of the system of particles $U_{tr}$. Thus

$$PV = \tfrac{2}{3}U_{tr} \qquad (10\text{-}3)$$

On the basis of kinetic theory, the pressure is proportional to the average translational kinetic energy per unit volume. Note that all gases at the same pressure and volume have the same total translational kinetic energy.

A kinetic-theory interpretation of temperature may also be obtained by comparing Eq. (10-3) with the empirical ideal-gas relation

$$PV = \frac{m}{M} R_u T$$

where $M$ again is the molecular weight, or molar mass, of the gas. Such a comparison has merit, since we have neglected intermolecular forces between particles in the above derivation. When the $PV$ quantities are compared, we find that

$$U_{tr} = \frac{3}{2} \frac{m}{M} R_u T = \tfrac{3}{2} mRT \qquad (10\text{-}4)$$

On a unit-mass basis this becomes

$$u_{tr} = \tfrac{3}{2} RT \qquad (10\text{-}5)$$

where $u_{tr}$ is the translational energy per unit mass. Since $R_u \equiv kN_A$, where $k$ is Boltzmann's constant and $N_A$ is Avogadro's number, we may also write

$$\tfrac{1}{2} m' \overline{c_2} = \tfrac{3}{2} kT = \epsilon_{tr} \qquad (10\text{-}6)$$

where $\epsilon_{tr}$ is the average translational energy per particle. Either of the latter two equations indicates that the temperature of an ideal gas is proportional to the mean kinetic energy of the particles. The above development is especially meaningful for a monatomic ideal gas, for which the translational energy is the only significant form over a wide range of temperatures.

Although the mean kinetic energy of all particles is the same at the same temperature, independent of their masses and the pressure, this is not true regarding the mean velocity of the particles. If one defines the root-mean-square velocity $c_{rms}$ by the quantity $(\overline{c^2})^{1/2}$, it is found that

$$c_{rms} = \left(\frac{3kT}{m'}\right)^{1/2} = \left(\frac{3R_u T}{M}\right)^{1/2} \qquad (10\text{-}7)$$

Note that the rms velocity is inversely proportional to the molar mass of gas. For a given temperature, as the molar mass increases, the rms velocity of the gas decreases.

**Example 10-1** Determine the rms velocity of (a) helium and (b) nitrogen particles at a temperature of 27°C, based on kinetic theory.

SOLUTION (a) Employing Eq. (10-7) for helium, which has a molar mass of 4, we find that

$$c_{rms,\,He} = \frac{[3(1.38 \times 10^{-16})(300)(6.025 \times 10^{23})]^{1/2}}{4^{1/2}}$$

$$= 1.37 \times 10^5 \text{ cm/sec} = 4550 \text{ ft/s}$$

(b) Since the diatomic nitrogen molecule is very close to seven times as heavy as a helium atom, its rms velocity at the same temperature is easily found to be

$$c_{rms,\,N_2} = \frac{4550}{7^{1/2}} = 1720 \text{ ft/s} = 524 \text{ m/s}$$

Thus light molecules move faster than heavier ones. As a note of comparison, the speed of sound through air at this temperature is, roughly, 1100 ft/s or 350 m/s. Finally, it should be remembered that, although the rms velocity of a system of particles is always positive, the average velocity is always zero.

The kinetic theory of gases is also useful in providing some insight on the specific heats of monatomic and diatomic gases. The approach presented here is not particularly useful for gases more complex than these. If one begins by examining monatomic species, the only form of molecular motion is a translational one. Since the molecular potential energy of an ideal gas may be assumed to be zero, the total internal energy is composed of linear kinetic energy. It was shown above that such energy on the basis of a simple kinetic theory is given by

$$u_{tr} = \tfrac{3}{2}RT \tag{10-5}$$

When the internal energy is expressed per mole, the universal gas constant $R_u$ should be employed in the above equation. If we use the definition of $c_v$ provided by Eq. (3-12), the molar specific heat at constant volume for particles with only translational kinetic energy becomes

$$c_v = \left(\frac{\partial u}{\partial T}\right)_v = \tfrac{3}{2}R_u \tag{10-8}$$

Furthermore, the molar specific heat at constant pressure can now be found by employing Eq. (3-24):

$$c_p = c_v + R_u = \tfrac{3}{2}R_u + R_u = \tfrac{5}{2}R_u \tag{10-9}$$

On a molar basis $R_u = 8.315$ J/(g·mol)(°K) or 1.986 Btu/(lb·mol)(°R). As an approximation, then, we find that $c_v$ is roughly 12.5 J/(g·mol)(°K) or 3 Btu/(lb·mol)(°R), and $c_p$ is 20.8 J/(g·mol)(°K) or 5 Btu/(lb·mol)(°R) for a monatomic gas such as helium or argon. These values are in substantial agreement with the experimental values presented in Sec. 3-6. Thus kinetic theory gives an accurate evaluation of the specific heats of monatomic gases. One advantage of kinetic theory is apparent when compared with classical thermodynamics. This latter approach is unable to predict actual numerical values of properties; it merely provides relationships among the various properties. Numerical values come from experimental results. However, kinetic theory gives quantitative answers directly. It must be kept in mind, though, that these quantitative values from kinetic theory are the result of assuming a molecular model before starting any calculations.

In order to extend this discussion to diatomic molecules, it is important, first, to consider what is meant by a degree of freedom with relation to any energy mode. *Energy modes* are those physically distinguishable means by which a particle can have energy, e.g., in translation, in rotation, in vibration, etc. A *degree of freedom* is defined as any independent quantity which must be specified in order to determine the energy of a molecule. For example, if one specifies the translational velocity of a particle in terms of its components in three rectilinear coordinate directions, the translational kinetic energy of a particle is fixed by specifying the translational kinetic energies in these three directions. Hence one speaks of 3 translational degrees of freedom. The average translational kinetic energies in the three directions are all equal. Since the total kinetic energy on the average is $\frac{3}{2}R_u T$ as given by Eq. (10-5), the average kinetic energy in each coordinate direction must be $\frac{1}{2}R_u T$. This result is a special case of the equipartition of energy theorem developed in classical and statistical mechanics. The equipartition of energy theorem demonstrates that if the energy associated with any degree of freedom is a quadratic function of a coordinate or a component of velocity or momentum, the mean value of the corresponding energy is $\frac{1}{2}R_u T$. The internal energy per mole for particles with $f$ quadratic terms is thus given by

$$u_{tot} = \frac{f}{2} R_u T$$

The equations for $c_v$, $c_p$, and $k$ become

$$c_v = \left(\frac{\partial u}{\partial T}\right)_v = \frac{f}{2} R_u$$

$$c_p = c_v + R_u = \frac{f}{2} R_u + R_u = \frac{f+2}{2} R_u$$

$$k = \frac{f+2}{f}$$

Note that $c_v$, $c_p$, and $k$ arc independent of temperature according to kinetic theory.

For diatomic molecules, rotational energy must also be considered. If the atoms of the molecule are point masses and are assumed to be rigidly connected together, then 2 rotational degrees of freedom exist for rotation about the two axes that are perpendicular to the axis connecting the atoms. The energy of rotation for the axis of symmetry is zero, since the moment of inertia for point masses in this direction would be zero. In addition to rotation, the atoms may also vibrate along the axis which connects them. There are 2 vibrational quadratic terms for this single vibratory motion because vibrational energy consists of both kinetic and potential forms of energy. These are specified by the velocity and position of the atoms. Hence each mode of harmonic vibrational motion contributes $R_u T$ units of energy to the diatomic molecule, and not $\frac{1}{2}R_u T$. In summary, the total, or maximum, degrees of freedom for a diatomic molecule, according to simple kinetic theory, comprise 3 in translation, 2 in rotation, and 2 in vibration. Thus we expect a maximum of 7 degrees of freedom for a diatomic molecule.

The main drawback to the kinetic theory of specific heats is that these quantities are found to be independent of temperature, as noted above. We have already seen from experimental data, however, that these properties are quite temperature-dependent in the gas phase. In spite of this disagreement, the values supplied by kinetic theory are useful in a special sense. At elevated temperatures (considerably above room temperatures) the specific heats of all diatomic gases approach the maximum, or classical, values. In this case $c_v$ would be $\frac{7}{2}R_u$ and $c_p$ would be $\frac{9}{2}R_u$. For a number of diatomic gases it is found that, in the vicinity of room temperature, vibrational motion is essentially zero. Thus there are only 5 degrees of freedom and $c_v$ and $c_p$ should be $\frac{5}{2}R_u$ and $\frac{7}{2}R_u$, respectively, while $k$ would be $\frac{7}{5}$, or 1.4. Finally, at very low temperatures, rotational motion stops, and the energy is entirely translational. In this case, diatomic molecules behave as monatomic ones from an energy standpoint. This is substantiated experimentally at low temperatures. Thus, in certain temperature ranges, the results from kinetic theory are valid if it is permissible to neglect various degrees of freedom. Kinetic theory, then, provides some guidelines in the evaluation of the internal energy and the specific heats of monatomic and diatomic ideal gases. In order to assess the variation of these properties more correctly, it is necessary to turn to statistical methods based on quantum theory.

## 10-2 BOSE-EINSTEIN, FERMI-DIRAC, AND MAXWELL-BOLTZMANN STATISTICS

It is a matter of experience that three different equations for $W_{mp}$ in terms of $n_i$ are necessary to satisfy different physical situations. Hence some care must be taken in selecting the proper equation for $W_{mp}$ when the entropy function is to be evaluated. The three "statistics" differ mainly in the manner in which the arrangements, or microstates, are to be counted. Maxwell-Boltzmann (MB) statistics were developed in the late 1800s, before the advent of quantum mechanics. Therefore this method of evaluation is not restricted to the rules of quantization of energy. In addition, MB statistics assume that particles are distinguishable (localized particles). In this latter respect these statistics are useful in evaluating the entropy of an idealized solid, for which the particles have relatively fixed positions in space and hence can be considered as distinguishable. In the case of gases, however, MB statistics lead to an erroneous result for the evaluation of the entropy, since gas molecules certainly are not distinguishable from one another.

Since 1920 quantum-statistical mechanics has led to two other sets of results, based on what are known as Bose-Einstein (BE) and Fermi-Dirac (FD) statistics. These two forms of statistics are valid for a system of identical, independent, indistinguishable (nonlocalized) particles. The necessity of two different types of statistics arises from a consideration of the symmetry of the wave function $\psi$ introduced in quantum mechanics. A fundamental study of this function is well beyond the requirements of this text. The important fact is that nuclei, atoms, and molecules which contain an odd number of elementary particles (electrons, protons, and neutrons) follow the quantum statistics of Fermi-Dirac. These statistics would be applicable to gases such as deuterium (D), helium-3 ($He^3$), ammonia ($NH_3$), and nitric oxide (NO). If the number of elementary particles is even (an even mass number), BE statistics apply. In this class are such gases as atomic hydrogen (H), molecular hydrogen ($H_2$), nitrogen ($N_2$), and helium-4 ($He^4$). It is an experimental fact, for example, that at low temperatures the properties of the two helium isotopes referred to above are quite different.

Based on the preceding discussion, it is important to ascertain the mathematical relations for the thermodynamic probability $W_{tot}$ for systems which follow Bose-Einstein, Fermi-Dirac, and Maxwell-Boltzmann statistics. These equations can be derived from the theory of combinations. However, this factor must be considered: Based on quantum mechanics, it is frequently found that several quantum states have the same, or nearly the same, energy $\epsilon_i$. These states are then grouped together as belonging to the same energy level, and the number of these states of equal energy is known as the *degeneracy* of the level, $g_i$. Although the energies of the states are equal, they still are distinct states, and shifting of particles from one state to another on the same level constitutes a new arrangement. The values of the degeneracy (also known as the multiplicity, or statistical weight) for the levels of each energy mode are determined from quantum mechanics. In addition, the problem is complicated by the fact that the particles in some cases are considered distinguishable, whereas in others they are considered indistinguishable. A final restriction is that in FD statistics only one particle is allowed in a given energy state, whereas in Be and MB statistics there is no limit on the number of particles per energy state. The restriction in FD statistics is based on Pauli's exclusion principle. A combination of these various restrictions leads to the different methods of accounting.

## 1 Bose-Einstein Statistics

For this combinatorial method the particles are considered indistinguishable (as in a gas phase) and there is no limit on the number of particles which can occupy a given energy state. The basic question which arises is, what is the number of ways that $n_i$ particles assigned to the $\epsilon_i$ energy level can be distributed over the $g_i$ states associated with that level? To illustrate the method of computation, we shall assign the letters $a, b, c, \ldots$ to the $g_i$ energy states that have the same energy $\epsilon_i$. Each of these states may contain any number of particles, beginning with zero, although the total number of particles for the energy level is restricted to the value of $n_i$. The particles themselves will carry tags numbered consecutively from 1 to $n_i$. In actuality, of course, the particles cannot be so numbered, since they are assumed to be indistinguishable. The numbering scheme is merely for purposes of deriving the desired equation, and it has no physical significance. Now we shall arrange all the letters and all the numbers in a horizontal row. When numbers follow a letter, our convention will be that particles so numbered are associated with the particular state designated by the letter. For example, consider an energy level which has four energy states $a, b, c,$ and $d$ and six particles $1, 2, 3, 4, 5,$ and $6$; that is, $g_i = 4$ and $n_i = 6$. One possible arrangement of these six particles in the four states of equal energy might be

$$a12b345cd6$$

This implies that state $a$ contains particles 1 and 2, state $b$ contains particles 3, 4, and 5, state $c$ is empty, and state $d$ contains particle 6.

There are several important factors to consider now. The first of these is that all possible sequences similar to the one above must begin with a letter; otherwise the beginning of the sequence would have no meaning in the context for which the system of numbering and lettering was established. There are apparently $g_i$ ways in which a sequence could begin, since any letter for a state could be used. The pattern of letters and numbers which follow the initial letter is immaterial since our accounting system would not be violated. The question which now needs answering is this: For a given initial letter, how many different ways could the remaining letters and numbers be arranged? The answer is given by the permutations of $y$ objects in a sequence, which is simply $y!$ (factorial $y$). For example, three objects can be arranged in six $(1 \times 2 \times 3)$ ways. For a given starting letter there are $n_i + g_i - 1$ remaining numbers and letters; hence there are $(n_i + g_i - 1)!$ possible arrangements for a given starting letter. For all possible starting letters the number of arrangements would simply be

$$g_i(n_i + g_i - 1)!$$

However, although the above expression satisfies the requirement that any number of particles occupy a given energy state, it does not account for the indistinguishability of either the particles or the states. For example, in the simple illustration given above, it is not significant in BE statistics whether the arrangement is written as $a12b345cd6$ or $a21b345cd6$. That is, interchange of the order in which the particles 1 and 2 are placed into state $a$ should not be counted as two distinct sequences, since particles 1 and 2 are in reality indistinguishable. The same kind of reasoning applies to the arrangement of the states. The two sequences $a12b345cd6$ and $b345ca12d6$ are identical for accounting purposes, since rearranging the order in which the states are filled does not lead to a new microstate. The particles can be arranged in $n_i!$ different ways, which are really the same microstate, and the order of filling states with a predetermined number of particles can be done in $g_i!$ ways. Thus the number of microstates in BE statistics for the $i$th energy level is found by dividing the total number of arrangements obtained earlier by the factorial quantities just discussed. Hence

$$W_{i,\text{BE}} = \frac{g_i(n_i + g_i - 1)!}{(g_i)!\,(n_i)!} = \frac{(n_i + g_i - 1)!}{(g_i - 1)!\,(n_i)!}$$

This equation gives the number of arrangements for the $i$th level of energy for any given number of particles $n_i$. The value of $W_i$ for each level is independent of that for other levels once the values of $n_i$ for all the levels are assigned. We have already noted that if an event $X$ can occur in a total of $x$ ways and if a different event $Y$ can occur in $y$ ways, the simultaneous event can occur in $xy$ ways. Consequently, the total number of different arrangements for a certain set of $n_i$ values is found as the product of the different arrangements for each level. That is,

$$W_{\text{BE}} = \prod W_{i,\text{BE}} = \prod \frac{(n_i + g_i - 1)!}{(g_i - 1)!\,(n_i)!} \qquad (10\text{-}10)$$

The symbol $\prod$ indicates multiplication analogous to the symbol $\sum$ used to indicate addition. Different values of $n_i$ assigned to each of the levels lead to different values of $W_{BE}$. When the maximum value of $W_{BE}$ is used in conjunction with the Boltzmann-Planck equation, we should obtain the entropy of particles which fulfill the restrictions customarily assumed for BE statistics. This value should agree with that computed by macroscopic techniques for systems which contain a large number of particles.

## 2 Fermi-Dirac Statistics

This method of enumerating the microstates or the thermodynamic probability of a system again requires that the particles be considered indistinguishable. In addition, and in contradiction to BE statistics, the Pauli exclusion principle is applied to a system of particles which follow FD statistics. Thus only one particle per energy state is permitted, or zero. Under such restrictive conditions we should expect that the equation for the thermodynamic probability $W$ for FD statistics would differ from BE statistics. Again consider the $i$th level, where there are $g_i$ energy states and $n_i$ particles available which may be placed in the $g_i$ states. Now, however, $n_i$ must be equal to or less than $g_i$, and the number of particles in a given state must be unity, or zero. As an example we shall consider a system of four energy states and two particles for a given energy level. The two unoccupied states will be denoted by the symbols $x$ and $y$, and the two states occupied by a particle, by the symbols $a$ and $b$. Permutation of the occupied and unoccupied states leads to the following 24 possible arrangements since the permutation of $g_i$ distinguishable items is simply $g_i!$, and $4! = 24$.

$$
\begin{array}{cccc}
abxy & baxy & abyx & bayx \\
axby & bxay & aybx & byax \\
axyb & bxya & ayxb & byxa \\
xaby & xbay & yabx & ybax \\
xayb & xbya & yaxb & ybxa \\
xyab & xyba & yxab & yxba
\end{array}
$$

However, the interchange of either two occupied states or two unoccupied states does not lead to a new arrangement, since in reality the occupied states are indistinguishable from each other, as well as the unoccupied states. In general, there are $n_i$ occupied states and $g_i - n_i$ unoccupied states. Thus the total number of arrangements based on distinguishable states, $g_i!$, must be divided by the permutation of $n_i$ states and $g_i - n_i$ states. In the example shown above, the total number of permutations, 24, should be divided by $(2!)(2!)$ in order to account for indistinguishability. Hence the number of arrangements which really should be counted is only six. All four vertical columns listed above satisfy the requirements for FD statistics. However, only one may be chosen, and not all four. Finally, we find, then, that the number of arrangements of a system of particles which follow FD statistics is given by

$$W_{i,\,FD} = \frac{g_i!}{n_i!(g_i - n_i)!}$$

for each energy level. Since the number of arrangements of each level is independent of the other levels, again the thermodynamic probability is given as the product of the number of arrangements for each level. Hence

$$W_{FD} = \prod \frac{g_i!}{n_i!\,(g_i - n_i)!} \tag{10-11}$$

A different value for $W_{FD}$ exists for every possible set of $n_i$ values for the allowed energy levels. Once the $n_i$ values are found which maximize $W_{FD}$, the result (10-11) is useful in determining the entropy for those thermodynamic systems composed of independent, indistinguishable particles which obey the Pauli exclusion principle.

### 3 Maxwell-Boltzmann Statistics

Before the advent of quantum mechanics, a scheme had been developed for statistically evaluating the properties of matter. The main defect in this approach stems from the assumption that the particles one deals with are distinguishable in nature. As a consequence, the determination of the entropy of an ideal gas, for example, is in error unless a correction is applied to the final answer. The fact that it was derived, originally, before the acceptance of quantum ideas is not important, for the general method may be easily adapted to particles which have energy that is quantized.

In MB statistics there is no limit on the number of particles which can occupy a given energy state. However, for a given energy level, if two particles in two different energy states for that level are exchanged, this constitutes a new microstate, since the particles are assumed to be distinguishable. Such an exchange would not lead to a new arrangement in BE statistics. Consider a system of $n$ particles and $i$ energy levels. The total number of arrangements for placing $n_1$ particles in the first level, $n_2$ particles in the second level, etc., is given by

$$W = \frac{n!}{n_1!\,n_2! \cdots n_i!} = \frac{n!}{\prod (n_i)!}$$

where $n_i$ represents the number of particles in the $i$th level. However, the number of arrangements is actually larger than this value, because we have not accounted for different arrangements within a given energy level. It can be shown that $n_i$ particles may be distributed over $g_i$ states in $g_i^{n_i}$ different ways. For example, if the states are numbered 1 and 2 and the distinguishable particles are lettered $a$, $b$, and $c$, the eight possible arrangements for a given level are:

| 1 | $a$ | $b$ | $c$ | $ab$ | $ac$ | $bc$ | $abc$ | |
|---|-----|-----|-----|------|------|------|-------|-----|
| 2 | $bc$ | $ac$ | $ab$ | $c$ | $b$ | $a$ | | $abc$ |

Since the number of arrangements within a given energy level is independent of the other levels, the total number of arrangements of the system due solely to rearrangement of particles within the levels is the product of the $g_i^{n_i}$ terms for all the levels. The total number of distributions, or the thermodynamic probability, for MB statistics for a given set of $n_i$ values, then, is the product of the two relations developed above. Thus

$$W_{MB} = \frac{n!}{\prod (n_i)!} \prod (g_i)^{n_i} = n! \prod \frac{(g_i)^{n_i}}{(n_i)!} \tag{10-12}$$

The relationship (10-12) would be valid for identical, independent, distinguishable particles. The number of particles per state is not limited, but the exchange of particles between states of a given level constitutes new arrangements.

## 4 Corrected Boltzmann Statistics

The evaluation of the entropy of an ideal gas is based on the indistinguishability of particles. For such particles two equations are available: the Bose-Einstein and Fermi-Dirac. It is interesting to examine these two expressions in the limiting case when $g_i \gg n_i$. The FD statistics are given by

$$W_{FD} = \prod \frac{g_i!}{n_i!(g_i - n_i)!} \tag{10-11}$$

This equation can also be written as

$$W_{FD} = \prod \frac{g_i(g_i - 1)(g_i - 2) \cdots (g_i - n_i + 1)}{n_i!} \tag{10-13}$$

The basic equation for BE statistics is

$$W_{BE} = \prod \frac{(g_i + n_i - 1)!}{n_i!(g_i - 1)!} \tag{10-10}$$

This expression can be written in an equivalent manner as

$$W_{BE} = \prod \frac{g_i(g_i + 1)(g_i + 2) \cdots (g_i + n_i - 1)}{n_i!} \tag{10-14}$$

In the special case when $g_i \gg n_i$, it is noted that both Eqs. (10-13) and (10-14) reduce to the form

$$W_{FD} = W_{BE} = \prod \frac{(g_i)^{n_i}}{n_i!} \tag{10-15}$$

For most gases at relatively low pressure or high temperature, the number of quantum states $g_i$ available at any energy level $\epsilon_i$ is much larger than the number of particles $n_i$ found in that level. That is,

$$g_i \gg n_i \tag{10-16}$$

is a valid statement of ideal-gas behavior. When the result of Eq. (10-15) is compared to Eq. (10-12) for $W_{MB}$, we find that

$$W_{BE} = W_{FD} = \frac{W_{MB}}{n!} = W_{MB, corr}$$

where $W_{MB, corr}$ is known as the corrected Boltzmann statistics. By definition,

$$W_{MB, corr} \equiv \prod \frac{(g_i)^{n_i}}{n_i!} \tag{10-17}$$

In these cases where Eq. (10-16) is a valid approximation, we find that both Bose-Einstein and Fermi-Dirac statistics reduce to a common expression for the thermodynamic probability, as given by Eq. (10-17). This model of corrected Boltzmann statistics will be used in the subsequent sections which describe macroscopic properties of an ideal gas based on molecular behavior.

## 10-3   THE BE, MB, AND FD DISTRIBUTION LAWS

In the preceding section brief derivations for the thermodynamic probability of systems of particles which follow the BE, FD, MB, and corrected Boltzmann statistics were presented. Since we have seen that the entropy of a macroscopic system can be approximated to a high degree of accuracy through the knowledge of the thermodynamic probability of the most probable macrostate, the determination of the $n_i$ values which lead to a maximum value of $W$ is the next logical step. There will be three different equations for $n_i$, depending upon the statistical model employed. (Both the MB and the corrected Boltzmann statistics lead to the same equation for the most probable distribution.) Only the equations for $n_i$ for the MB and BE models will be derived completely. In deriving expressions for $n_i$, it is important to keep in mind that the values of $\epsilon_i$ and $g_i$ for each energy level are constant and presumably known from quantum-mechanical calculations.

## 1 Stirling's Approximation Formula

Before proceeding to the development of the distribution laws, it is necessary to introduce an approximation formula for factorial quantities, such as $(x!)$. An expression for $x!$ when $x$ is a large integer is given by the infinite-series expansion

$$x! = (2\pi x)^{1/2}\left(\frac{x}{e}\right)^x\left(1 + \frac{1}{12x} + \frac{1}{288x^2} + \cdots\right)$$

When $x$ is very large, all the terms inside the parentheses except the first term may be neglected. Then

$$x! = (2\pi x)^{1/2}\left(\frac{x}{e}\right)^x$$

or
$$\ln x! = \tfrac{1}{2}\ln(2\pi x) + x\ln x - x$$

However, $\ln x$ is much smaller than $x$ when $x$ is large; so the first term on the right of the last equation is also negligible. The final result is known as *Stirling's approximation* formula for $\ln x!$; namely,

$$\ln x! = x\ln x - x \tag{10-18}$$

When $x$ is very large, say, $10^{20}$, the error introduced is negligible.


## 2 Maxwell-Boltzmann Statistics

The distribution law for a system of independent, distinguishable particles where there is no restriction placed on the number of particles in a given quantum state is found by use of MB statistics. The relationship between $n_i$, $g_i$, and $\varepsilon_i$ in the most probable macrostate may be found by maximizing the function $\ln W_{MB}$. In this case we wish to maximize

$$\ln W_{MB} = \ln\left[n!\prod\frac{(g_i)^{n_i}}{n_i!}\right]$$

Since the logarithm of a product of terms is equal to the sum of the logarithms of the terms,

$$\ln W_{MB} = \ln n! + \sum_i (n_i \ln g_i - \ln n_i!) \tag{10-19}$$

Maximization of $\ln W_{MB}$ requires that it be differentiated with respect to $n_i$. To avoid differentiation of factorial quantities, we may replace $\ln n!$ and $\ln n_i!$ by use of Stirling's formula. As a result, Eq. (10-19) becomes

$$\ln W_{MB} = n\ln n - n + \sum_i (n_i \ln g_i - n_i \ln n_i + n_i)$$

which reduces to

$$\ln W_{MB} = n\ln n + \sum_i (n_i \ln g_i - n_i \ln n_i) \tag{10-20}$$

This equation is now differentiated with respect to $n_i$, with $n$ and $g_i$ remaining constant. The derivative of the first term on the right is zero. Note, however, that it is this term which is present in MB statistics but is removed in corrected MB statistics. Consequently, the derivative of either $\ln W_{MB}$ or $\ln W_{MB, corr}$ will lead to the same expression. This is

$$d(\ln W_{MB}) = \sum_i (\ln g_i - \ln n_i - 1) \, dn_i \qquad (10\text{-}21)$$

However, since the total number of particles must be conserved,

$$\sum_i n_i = n \qquad (9\text{-}10)$$

and for a differential change in state,

$$\sum_i dn_i = 0 \qquad (10\text{-}22)$$

Therefore Eq. (10-21) reduces to

$$d(\ln W_{MB}) = \sum_i \ln \left(\frac{g_i}{n_i}\right) dn_i = 0 \qquad (10\text{-}23)$$

Equation (10-23) is set equal to zero to maximize $\ln W_{MB}$.

The variations of $dn_i$ are not independent of each other. As indicated by Eq. (10-22), the conservation of the number of particles places a constraint on the variation of the $dn_i$ terms. The total energy $U$ also must be conserved during a differential variation in the molecular state of the system. Since

$$\sum_i n_i \epsilon_i = U \qquad (9\text{-}1)$$

then

$$\sum_i \epsilon_i \, dn_i = 0 \qquad (10\text{-}24)$$

The problem resolves itself into one requiring the maximization of a function for which two constraining equations also exist. One method suitable for the solution of such a problem is known as the method of lagrangian undetermined multipliers. This method is outlined and several simple examples are presented in Sec. A-1 in the Appendix. The reader may wish to refer to the Appendix before continuing with this section.

To carry out the lagrangian method, we shall multiply Eqs. (10-22) and (10-24) by the arbitrary Lagrangian multipliers $\ln A$ and $B$, respectively. Upon subtraction of these resulting equations from Eq. (10-23), we find that

$$\sum_i \left(\ln \frac{g_i}{n_i} - \ln A - B\epsilon_i\right) dn_i = 0 \qquad (10\text{-}25)$$

At this point only $(i - 2)$ of the $dn_i$ terms are independent. But by selecting $\ln A$ and $B$ to have certain values, the remaining $dn_i$ terms become independent. Hence their coefficients may be set equal to zero. All of these coefficients have the same format; namely,

$$\ln \frac{g_i}{n_i} - \ln A - B\epsilon_i = 0 \qquad (10\text{-}26)$$

for all values of $i$. Solving for $n_i$, we find that

$$n_i = \frac{g_i}{Ae^{B\epsilon_i}} \qquad \text{(Maxwell-Boltzmann)} \qquad (10\text{-}27)$$

This equation is the distribution law for uncorrected or corrected Maxwell-Boltzmann statistics (or simply Boltzmann statistics). It is the most probable distribution of particles among energy states for independent, distinguishable particles. Note the exponential nature of the distribution. The value of $n_i/g_i$ falls off rapidly as the value of $\epsilon_i$ increases for given values of $A$ and $B$. The constants $A$ and $B$ will be examined in detail in later sections.

## 3 The Bose-Einstein Distribution Law

The distribution law for a system of independent, indistinguishable particles where no restriction is placed on the number of particles in a given quantum state is found by the use of BE statistics. The relationship between $n_i$, $g_i$, and $\epsilon_i$ is found by maximizing the function $\ln W_{BE}$. From differential calculus it is known that the maximum value of a well-behaved function can be found by differentiating the function and setting the derivative equal to zero. In this particular case, the function we wish to maximize is given by the logarithm of Eq. (10-10), which was of the form

$$W_{BE} = \prod \frac{(n_i + g_i - 1)!}{(g_i - 1)! \, n_i!} \qquad (10\text{-}10)$$

The logarithm of Eq. (10-10) can be written in a more convenient form by recalling that the logarithm of a product of terms is equal to the sum of the logarithms of the terms. Consequently,

$$\ln W_{BE} = \ln \left[ \prod \frac{(n_i + g_i - 1)!}{(g_i - 1)! \, n_i!} \right] = \sum_i \ln \frac{(n_i + g_i - 1)!}{(g_i - 1)! \, n_i!} \qquad (10\text{-}28)$$

If differentiation is carried out on Eq. (10-28), it will be necessary to differentiate factorial quantities; hence this form of the equation for $\ln W$ is difficult to handle mathematically. However, a valid approximation for factorials is given by Stirling's formula, Eq. (10-18).

Use of Stirling's approximation allows us to transform Eq. (10-28) into an expression where the factorials will be absent. Therefore, after suitable cancellation, an approximate expression for $\ln W_{BE}$ is

$$\ln W_{BE} = \sum_i [(n_i + g_i - 1) \ln (n_i + g_i - 1) - n_i \ln (n_i) - (g_i - 1) \ln (g_i - 1)]$$

This form will now be differentiated with respect to the variable $n_i$ and set equal to zero. The final result, which the student may wish to confirm, is

$$d(\ln W_{BE}) = \sum_i \ln \left( \frac{n_i + g_i - 1}{n_i} \right) dn_i = 0 \qquad (10\text{-}29)$$

Equation (10-29) gives the change in the function $\ln W_{BE}$ as the values for the number of particles $n_i$ associated with each level are altered. If the variations of each of the $n_i$ values are independent of each other for all energy levels, the solution to Eq. (10-29) is found by setting the coefficients in front of each of the $dn_i$ terms equal to zero. However, the values of $dn_i$ are not independent of each other. In this particular case there are two constraining equations which must also be fulfilled. The system under consideration contains $n$ particles and has a total energy given by $U$. The total number of particles in the system must equal the total of those occupying each energy level. This is expressed by

$$\sum_i n_i = n \qquad (9\text{-}10)$$

In addition, the energy of the system must be conserved as particles take on different energies in encounters with other particles. This conservation of energy may be expressed by

$$\sum_i n_i \epsilon_i = U \qquad (9\text{-}1)$$

The problem resolves itself into one requiring the maximization of a function for which two constraining equations also exist, namely, Eqs. (9-10) and (9-1). One method suitable for the solution of such a problem is known as the method of lagrangian undetermined multipliers. This method is outlined and several simple examples are presented in Sec. A-1 in the Appendix. Application of this method to the present problem is shown below. The reader may wish to refer to the Appendix before continuing with this section.

Differentiation of the two constraining equations with respect to $n_i$ leads to two additional equations containing the variable $dn_i$. When these two equations are put together with Eq. (10-29), the following set of equations results:

$$\sum_i \ln \frac{n_i + g_i - 1}{n_i} dn_i = 0 \qquad (10\text{-}29)$$

$$\sum_i dn_i = 0 \qquad (10\text{-}22)$$

$$\sum_i \epsilon_i \, dn_i = 0 \qquad (10\text{-}24)$$

First we shall multiply. Eqs. (10-22) and (10-24) by the arbitrary Lagrangian multipliers $\ln A$ and $B$. Upon subtraction of these resulting equations from Eq. (10-29), we find that

$$\sum_i \left( \ln \frac{n_i + g_i - 1}{n_i} - \ln A - B\epsilon_i \right) dn_i = 0 \qquad (10\text{-}30)$$

By selecting $\ln A$ and $B$ to be such values that the remaining $dn_i$ terms become independent, the coefficients of the terms in Eq. (10-30) may be set equal to zero. Hence

$$\ln \frac{n_i + g_i - 1}{n_i} - \ln A - B\epsilon_i = 0$$

The values for $n_i$ to be assigned to each of the $i$ energy levels which lead to the maximum value of $W$ may now be found. Solving for $n_i$, we find that

$$n_i = \frac{g_i - 1}{Ae^{B\epsilon_i} - 1} \tag{10-31}$$

In a number of cases it is found from quantum mechanics that the values of $g_i$ for every energy level are much greater than unity. Under this restriction the preceding equation reduces to the following form, known as the BE distribution law:

$$n_{i,\,BE} = \frac{g_i}{Ae^{B\epsilon_i} - 1} \tag{10-32}$$

It should be recalled again that such an equation is valid for independent, indistinguishable particles where no restriction is placed on the number of particles which occupy a given energy state. Since $A$ and $B$ are constants, it is seen that the number of particles associated with a given energy level under the above restrictions for a given equilibrium state is solely a function of the value of the energy level itself and the degeneracy of the level. The constants $A$ and $B$ will be examined in Secs. 10-4 and 10-5.

## 4 A Comparison of the Three Distribution Laws

The distribution law for Fermi-Dirac statistics can be determined in a similar manner by employing Eq. (10-11) for the thermodynamic probability of such a system. The two constraining equations used in conjunction with MB and BE statistics are still valid. The resulting equations for the three statistics are summarized below:

Bose-Einstein: $\qquad\qquad n_i = \dfrac{g_i}{Ae^{B\epsilon_i} - 1} \qquad\qquad (10\text{-}32)$

Fermi-Dirac: $\qquad\qquad n_i = \dfrac{g_i}{Ae^{B\epsilon_i} + 1} \qquad\qquad (10\text{-}33)$

Maxwell-Boltzmann: $\qquad\quad n_i = \dfrac{g_i}{Ae^{B\epsilon_i}} \qquad\qquad\quad (10\text{-}27)$

Fermi-Dirac statistics apply to independent, indistinguishable particles, and the maximum number of particles in an energy state is unity. Maxwell-Boltzmann statistics were derived for independent, distinguishable particles, with no limitation on the number of particles allowed in a given quantum state. Although we have particularly stressed systems of independent particles, in all fairness it must be pointed out that the methods of statistical thermodynamics are applicable to systems of dependent particles as well.

We shall examine the significance of the constants $A$ and $B$ shortly. Before doing this, however, it is pertinent to examine the above equations when written in a slightly different form. Rearrangement of these expressions leads to the following set of equations:

Bose-Einstein:
$$\frac{g_i}{n_i} + 1 = Ae^{B\epsilon_i}$$

Fermi-Dirac:
$$\frac{g_i}{n_i} - 1 = Ae^{B\epsilon_i}$$

Maxwell-Boltzmann:
$$\frac{g_i}{n_i} = Ae^{B\epsilon_i}$$

Consequently, these three equations are approximately identical when the value of $g_i/n_i$ is much greater than unity. Under such circumstances both of the equations for the newer quantum statistics reduce to the classical, or Boltzmann, distribution. It is a matter of experience that, for reasonable ranges of temperature and pressure, the value of $g_i/n_i$ for nonlocalized gas particles is much greater than unity. That is, the number of quantum states available to the particles of the system is very much greater than the number of particles. For example, this approximation is valid for the translational energies of helium gas at 1 atm pressure down to about $10°K$. The approximation improves as the temperature is increased. Hence the MB equation for the equilibrium distribution of particles among the allowed translational energy levels is a sufficiently valid relation for determining the values of $n_i$ for ideal gases. This is true in spite of the fact that the equation for $W_{MB}$ originally was derived for distinguishable, or localized, particles. Thus, when dealing with gases over a wide range of conditions, the three distribution laws lead to practically identical results for the values of $n_i$.

The evaluation of equations for $n_i$ enables us to determine the entropy of a thermodynamic system, except that these equations for $n_i$ still contain the unknown lagrangian multipliers $A$ and $B$. It is necessary, first, to ascertain their significance in terms of pertinent thermodynamic variables, before the equations for $\ln W_{mp}$ can serve in determining the entropy of a system.

## 10-4 EVALUATION OF THE LAGRANGIAN MULTIPLIER $A$

In deriving the distribution functions for the most probable macrostate in Sec. 10-3, we employed the differential forms of the expressions for the thermodynamic probability $W$ and the two constraining equations. However, the two constraining equations in the form of Eqs. (9-10) and (9-1) still have not been used. These two equations provide the additional information necessary to determine the values of $A$ and $B$ in terms of other thermodynamic parameters. The first of these two equations, namely,

$$\sum_i n_i = n \tag{9-10}$$

may now be employed to evaluate the parameter $A$ (or $e^A$). The distribution function for a gas at reasonable pressures and temperatures, according to the preceding section, may be taken as

$$\frac{n_i}{g_i} = \frac{e^{-B\epsilon_i}}{A} \tag{10-34}$$

Substitution of Eq. (10-34) for $n_i$ into Eq. (9-10) leads to

$$\frac{\sum g_i e^{-B\epsilon_i}}{A} = n$$

or

$$A = \frac{1}{n} \sum_i g_i e^{-B\epsilon_i} \tag{10-35}$$

This equation is not explicitly an expression for $A$ in terms of thermodynamic parameters, since the equation still contains the lagrangian multiplier $B$. Once $B$ is evaluated, a more useful form of Eq. (10-35) can be presented. Again it is assumed that $\epsilon_i$ and $g_i$ are known from quantum mechanics. It should be noted that the multiplier $A$ is a dimensionless number. However, the parameter $B$ must have the dimensions of reciprocal energy.

The substitution of Eq. (10-35) for $A$ into the expression for $n_i$ yields the following expressions for the $i$th level:

$$n_i = \frac{g_i e^{-B\epsilon_i}}{A} = \frac{n g_i e^{-B\epsilon_i}}{\sum g_i e^{-B\epsilon_i}} \tag{10-36}$$

This equation expresses the equilibrium distribution of the most probable macrostate for independent particles in a gas phase where $g_i \gg n_i$. In practice, such an equation will be valid for gases which behave as ideal gases, for example. The remaining step is to determine the thermodynamic significance of the lagrangian multiplier $B$.

## 10-5 THE LAGRANGIAN MULTIPLIER $B$

The evaluation of the Lagrangian multiplier $B$ requires the use of the second constraining equation, namely,

$$\sum n_i \epsilon_i = U = n\bar{\epsilon} \tag{9-1}$$

where $\bar{\epsilon}$ is the average energy per particle for a particular energy mode. If we restrict ourselves to systems of gaseous molecules which approximate closely the

MB distribution law, then, upon combining Eqs. (10-8) and (10-17), we find for the average energy per particle that

$$\bar{\epsilon} = \frac{\sum n_i \epsilon_i}{n} = \frac{\sum \epsilon_i g_i e^{-B\epsilon_i}}{\sum g_i e^{-B\epsilon_i}} \tag{10-37}$$

For the translational energy mode the linear kinetic energy of a particle can be expressed in terms of the velocity components $u$, $v$, and $w$ in the three rectangular coordinate directions $x$, $y$, and $z$. This is given by

$$\epsilon_i = \tfrac{1}{2}mc^2 = \tfrac{1}{2}mu^2 + \tfrac{1}{2}mv^2 + \tfrac{1}{2}mw^2$$

The degeneracy $g_i$ of each translational level may be taken as unity. Hence substitution of the equation for $\epsilon_i$ for the translational mode into Eq. (10-37) leads to

$$\bar{\epsilon} = \frac{\sum\sum\sum (\tfrac{1}{2}mu^2 + \tfrac{1}{2}mv^2 + \tfrac{1}{2}mw^2)\exp\left[-B(\tfrac{1}{2}mu^2 + \tfrac{1}{2}mv^2 + \tfrac{1}{2}mw^2)\right]}{\sum\sum\sum \exp\left[-B(\tfrac{1}{2}mu^2 + \tfrac{1}{2}mv^2 + \tfrac{1}{2}mw^2)\right]} \tag{10-38}$$

In this equation the summations are taken over all values of the velocities in the three coordinate directions. It is assumed in classical statistical mechanics that the distribution of allowed translational velocities is continuous; that is, all velocities are allowed. Quantum mechanics dictates that, although the allowed energy values are discrete, the translational energy levels are extremely close together in spacing. From either approach, then, the process of summation in Eq. (10-38) may be replaced by one of integration. In addition, the integration will be carried out from $-\infty$ to $+\infty$, although no particle can have a velocity greater than the speed of light. Mathematically, it may be shown that the contribution of extremely high energy values to the value of the integral is negligible; consequently, they may be included in the integration without affecting the result. Therefore the average translational energy is now closely approximated by

$$\bar{\epsilon} = \frac{\displaystyle\int\limits_{-\infty}^{+\infty}\!\!\!\iint (\tfrac{1}{2}mu^2 + \tfrac{1}{2}mv^2 + \tfrac{1}{2}mw^2)\exp\left[-B(\tfrac{1}{2}mu^2 + \tfrac{1}{2}mv^2 + \tfrac{1}{2}mw^2)\right] du\,dv\,dw}{\displaystyle\int\limits_{-\infty}^{+\infty}\!\!\!\iint \exp\left[-B(\tfrac{1}{2}mu^2 + \tfrac{1}{2}mv^2 + \tfrac{1}{2}mw^2)\right] du\,dv\,dw}$$

$$\tag{10-39}$$

Since the variables $u$, $v$, and $w$ are independent of each other, the triple integral may be separated into the product of three single integrals. The numerator of Eq. (10-39), denoted by $I$, may be expressed by three terms of the form

$$I_u = \tfrac{1}{2}m \int_{-\infty}^{+\infty} u^2 e^{-Bmu^2/2}\,du \int_{-\infty}^{+\infty} e^{-Bmv^2/2}\,dv \int_{-\infty}^{+\infty} e^{-Bmw^2/2}\,dw$$

where $I = I_u + I_v + I_w$. The values of $I_u$, $I_v$, and $I_w$ will be the same. The solution for $I_u$ involves recognition of two standard integrals, namely,

$$\int_{-\infty}^{+\infty} x^2 e^{-ax^2}\,dx = \frac{1}{2}\left(\frac{\pi}{a^3}\right)^{1/2}$$

and

$$\int_{-\infty}^{+\infty} e^{-ax^2}\,dx = \left(\frac{\pi}{a}\right)^{1/2}$$

In our case, $a = Bm/2$. The evaluation of $I_u$ becomes

$$I_u = \frac{m}{2}\frac{1}{2}\left(\frac{8\pi}{B^3m^3}\right)^{1/2}\left(\frac{2\pi}{Bm}\right)^{1/2}\left(\frac{2\pi}{Bm}\right)^{1/2} = \frac{\pi}{2B}\left(\frac{8\pi}{B^3m^3}\right)^{1/2}$$

The value of the denominator of Eq. (10-39) is found from the second of the standard integrals listed above. The result is $(2\pi/Bm)^{3/2}$. Therefore the average translational energy is simply

$$\bar{\epsilon} = 3\frac{\pi}{2B}\left(\frac{8\pi}{B^3m^3}\right)^{1/2}\left(\frac{Bm}{2\pi}\right)^{3/2} = \frac{3}{2B}$$

The MB distribution law used in this development is valid for gases at low densities which approximate the ideal-gas state. However, in Sec. 10-1, it was shown by kinetic theory that the average translational energy of a gas molecule is $\frac{3}{2}kT$ [Eq. (10-6)]. Therefore

$$B = \frac{1}{kT} \tag{10-40}$$

where $k$ again is Boltzmann's constant, and $T$ is the ideal-gas temperature. Although we have chosen to derive relationship (10-40) between the lagrangian multiplier $B$ and the absolute temperature $T$ on the basis of an ideal gas, the result is general. In Chap. 1 temperature was presented as an empirical concept in terms of heat-transfer and thermal-equilibrium concepts. According to kinetic theory, it is also a measure of the average translational energy of molecules in an equilibrium state. Now it is important to recognize that the macroscopic property temperature has a statistical counterpart. It is the parameter which establishes the shape of the equilibrium distribution function for the $n_i$ values for a given set of allowed energy levels and degeneracies. As in the case of all statistical properties, the concept of temperature is of no consequence when the number of particles becomes extremely small.

Before returning to the main theme in this chapter, it is interesting to observe the effect of the temperature upon the most probable distribution of particles among the allowed energy levels. This is most easily seen by graphical techniques. To simplify the problem for the moment, we shall assume that the degeneracies are all unity, so that only one energy state exists for each energy level. Also, the ground level of energy, $\epsilon_0$, is taken to be zero in value, and the spacing of the energy levels is uniformly one energy unit apart. Under these restrictions, the MB distribution law for $n_i$ reduces to

$$n_i = \frac{ne^{-\epsilon_1/kT}}{1 + e^{-\epsilon_1/kT} + e^{-\epsilon_2/kT} + \cdots} \tag{10-41}$$

where $B$ has been replaced by its macroscopic counterpart, $1/kT$. For numerical computation involving this hypothetical system, the values of $B$ (or $1/kT$) are chosen to be $1.0, 0.5$, and $0.2$. The discrete distribution given by Eq. (10-41) for this particular example is shown in Fig. 10-2. It should be kept in mind that as $B$ decreases, the temperature increases. The effect of $B$ (or $T$) on the $n_i$ values is quite pronounced. At high positive values of $B$ (or low temperatures), the distribution

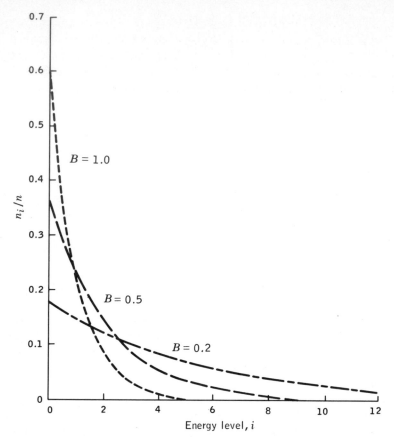

**Figure 10-2** The ratio $n_i/n$ associated with the discrete energy levels based on Eq. (10-41), where $\Delta \epsilon = 1.0$, $\epsilon_0 = 0$, and $B$ has values of 1.0, 0.5, and 0.2.

curve falls rapidly, indicating that most of the particles are in the lower levels. At high temperatures (or as $B$ approaches zero), the particles are nearly evenly spread among the allowed energy levels. Thus we see that particles preferentially exist in the lower energy states, but as the temperature is increased, more particles begin to occupy the higher energy states. This result is somewhat intuitive, if one loosely associates an increase in temperature with a corresponding increase in energy of the system. With the increase in energy, more particles have to take on larger energy values, in order not to violate Eq. (9-1) and still maintain an exponential distribution for the equilibrium system.

## 10-6 THE MOLECULAR PARTITION FUNCTION AND ITS RELATION TO THE MACROSCOPIC PROPERTIES OF AN IDEAL GAS

Earlier in this text the concept of entropy was introduced on the basis of the quantization of energy. The entropy of a system may be found from the BP equation, $S = k \ln W_{\text{tot}}$. In this equation $W_{\text{tot}}$ is the total number of arrangements, or microstates, associated with the equilibrium state. A valid approximation to the BP equation for systems containing a large number of particles is the relation $S = k \ln W_{mp}$, where $W_{mp}$ is the number of arrangements, or the thermodynamic probability, of the most probable macrostate. Three methods of evaluating the thermodynamic probability of any macrostate were introduced in the preceding sections. The basis for the BE, FD, and MB statistics was the distinguishability feature of particles and the Pauli exclusion principle. As a result, three distribution laws are available for evaluating $n_i$ for the most probable macrostate. These equations for $n_i$ are primarily a function of $\epsilon_i$, $g_i$, and $T$. In summary, they are

Bose-Einstein: $$n_i = \frac{g_i}{A e^{B\epsilon_i} - 1} \qquad (10\text{-}32)$$

Fermi-Dirac: $$n_i = \frac{g_i}{A e^{B\epsilon_i} + 1} \qquad (10\text{-}33)$$

Maxwell-Boltzmann: $$n_i - \frac{g_i}{A e^{B\epsilon_i}} \qquad (10\text{-}27)$$

The constants $A$ and $B$ are lagrangian multipliers, and $B = 1/kT$, where $k$ is the Boltzmann constant. When the quantity $g_i/n_i$ is much greater than unity, both of the equations for the newer quantum statistics reduce essentially to the classical, or MB, distribution. It is found, generally, that the Boltzmann distribution function will be applicable to all ordinary gases, even at low temperatures. Exceptions to this general rule are the light gases at low temperatures, especially $H_2$, He, and Ne. The use of the unmodified BE equations would be applicable to these gases at low temperatures if the molecule were made up of an even number electrons, protons, and neutrons. For an odd number of elementary particles the Fermi-Dirac statistics would be applied.

On the basis of the foregoing discussion, the MB distribution law given by Eq. (10-27) will be applicable to the study of the translational, rotational, and vibrational energy modes of an ideal gas. The evaluation of the lagrangian multiplier $A$ was carried out in Sec. 10-4 for the MB distribution. By employing the constraining equation $\sum_i n_i = n$, it was found that

$$A = \frac{1}{n} \sum_i g_i e^{-B\epsilon_i} \qquad (10\text{-}35)$$

In any equilibrium state the values of $\epsilon_i$, $g_i$, and $T$ (or $B$) are known for each energy mode, at least in theory. Consequently, the summation quantity which appears in Eq. (10-35) has a known value for a given energy mode. This summation is called the *molecular partition* function, and we shall give it the symbol $z$. Therefore

$$z \equiv \sum_i g_i e^{-B\epsilon_i} \qquad (10\text{-}42)$$

It should be noted that the partition function is a dimensionless number. In terms of the lagrangian multiplier $A$, it is clearly seen from Eq. (10-35) that $A = z/n$. The partition function is important in that it is a fundamental function required for calculating the statistical properties of a system of independent particles. It will be demonstrated in the following sections that macroscopic properties may be expressed in terms of the molecular partition function and other macroscopic variables, such as the temperature, the volume, and the universal gas constant $R_u$. The MB distribution law expressed in terms of $z$ becomes

$$n_i = \frac{ng_i e^{-B\epsilon_i}}{\sum\limits_i g_i e^{-B\epsilon_i}} = \frac{ng_i e^{-B\epsilon_i}}{z} \tag{10-43}$$

This equation is extremely important for expressing macroscopic properties of ideal gases in terms of the partition function.

The internal energy associated with a given energy mode is given by

$$U = \sum\limits_i n_i \epsilon_i \tag{9-1}$$

If Eq. (10-43) for $n_i$ in terms of the partition function is substituted into the above expression, it becomes

$$U = \frac{\sum\limits_i ng_i \epsilon_i e^{-B\epsilon_i}}{z} \tag{10-44}$$

A further refinement of this equation is possible by comparing it with the derivative of $\ln z$ with respect to $B$. The values of $\epsilon_i$ are held fixed during the differentiation. Based in Eq. (10-42), the logarithm of $z$ is

$$\ln z = \ln \sum\limits_i g_i e^{-B\epsilon_i}$$

If this expression is differentiated with respect to $B$ at constant $\epsilon_i$, we find that

$$\left(\frac{\partial \ln z}{\partial B}\right)_V = -\frac{\sum\limits_i g_i \epsilon_i e^{-B\epsilon_i}}{z} \tag{10-45}$$

Comparing Eq. (10-45) with Eq. (10-44), we obtain the following equation for $U$:

$$U = -n\left(\frac{\partial \ln z}{\partial B}\right)_V = nkT^2\left(\frac{\partial \ln z}{\partial T}\right)_V = R_u T^2\left(\frac{\partial \ln z}{\partial T}\right)_V \tag{10-46}$$

The transformation of the second quantity to the third is based on the relation $B = 1/kT$, and hence $dB = -dT/kT^2$. If the number of particles is $N_A$, Avogadro's number, then $N_A k$ may be replaced by $R_u$. Hence the last quantity above is valid for a mole of a substance. Equation (10-46) shows that the contribution of any energy mode to the internal energy may be found from a knowledge of the quantity $(\partial \ln z/\partial B)_V$ or its equivalent.

The evaluation of the entropy of an ideal gas is based on the indistinguishability of particles. In the special case when $g_i \gg n_i$, we have found that Bose-

Einstein and Fermi-Dirac statistics reduce to corrected Maxwell-Boltzmann statistics, where

$$W_{BE} = W_{FD} = W_{MB,\,corr} = \prod \frac{(g_i)^{n_i}}{n_i!} \tag{10-15}$$

With this approximation for $W$, the entropy of an ideal gas becomes

$$S = k \ln W = k(\sum n_i \ln g_i - \sum \ln n_i!)$$

Introducing Stirling's approximation for $\ln n_i!$, one obtains

$$S = k\left(\sum n_i \ln \frac{g_i}{n_i} + n\right) \tag{10-47}$$

At this point the relation for $g_i/n_i$ may be substituted as given by Eq. (10-43). Since $g_i/n_i = z \exp(B\epsilon_i)/n$, then

$$\ln \frac{g_i}{n_i} = \ln \frac{z}{n} + B\epsilon_i$$

The substitution of this relation into Eq. (10-47) yields

$$S = kn\left(\ln \frac{z}{n} + \frac{BU}{n} + 1\right)$$

When $U$ is replaced by Eq. (10-46), then

$$S_{approx} = kn\left[\ln z - \ln n + 1 + T\left(\frac{\partial \ln z}{\partial T}\right)_V\right] \tag{10-48}$$

An approximate equation for the molar entropy of an ideal gas is

$$S_{approx} = R_u\left[\ln z - \ln N_A + 1 + T\left(\frac{\partial \ln z}{\partial T}\right)_V\right] \tag{10-49}$$

or
$$S_{approx} = R_u(\ln z - \ln N_A + 1) + \frac{U}{T} \tag{10-50}$$

The equations for the entropy of indistinguishable particles are restricted to those cases for which the inequality $g_i \gg n_i$ is valid and Stirling's approximation holds. It is conventional to lump the $+1$ and the $-\ln N_A$ terms in the preceding two equations with the translational entropy contribution. Hence these two terms do not appear with the entropy contributions for any other energy mode.

Differentiation of Eq. (10-46) for the internal energy leads directly to the constant-volume molar specific heat $c_v$.

$$c_v = \left(\frac{\partial U}{\partial T}\right)_V = \frac{\partial}{\partial T}\left(R_u T^2 \frac{\partial \ln z}{\partial T}\right)_V \tag{10-51}$$

Expressing the results of the above differentiation in terms of both $T$ and $B$, one finds that

$$c_v = \frac{R_u}{T^2}\left[\frac{\partial^2 \ln z}{\partial(1/T)^2}\right]_V = R_u B^2\left(\frac{\partial^2 \ln z}{\partial B^2}\right)_V \qquad (10\text{-}52)$$

Since Eq. (10-52) is valid for an ideal gas, the constant-pressure specific heat $c_p$ may be found by adding $R_u$ to this equation. The enthalpy function is determined, of course, from the basic definition, $H = U + PV$.

The equations developed above permit the evaluation of pertinent thermodynamic properties in terms of the partition function $z$ of a given energy mode, the temperature and volume of the equilibrium state, and the constants $R_u$ and $N_A$. The molecular partition function of each energy mode depends in turn upon the allowed energy levels for that mode. Hence we need, next, to investigate the methods for determining the values of the discrete energy levels. These are obtained from suitable solutions of the Schrödinger wave equation, which is a basic postulate of wave mechanics.

One of the important tasks of wave, or quantum, mechanics is to relate mathematically the wave nature of matter to the discrete energy values that are associated with individual particles. A set of discrete, or allowed energy values may be assigned to all modes of energy storage. As in Planck's work, where only discrete frequencies were allowed for the standing waves of electromagnetic radiation, Schrödinger, in the mid-1920s, suggested that matter could be represented by a set of standing waves. This description of the wave nature of matter leads to the allowed quantum levels of energy associated with various energy modes. The complete formulation of the quantum-mechanical laws of motion for a particle is beyond the requirements of this text.

In thermodynamics one is primarily concerned with particles which are restrained to certain regions of space. In this case standing waves are set up which are not time-dependent. Schrödinger presented a solution for the time-independent wave function $\psi$. This wave function in quantum mechanics is the counterpart of the amplitude in classical mechanics. The wave equation itself can be solved fairly easily if the system is restricted to one of independent particles. Such a system is one of separable energy states. This separation occurs in two ways: First, the potential energy of interaction is constant and is taken to be zero. Each particle then has its own definite energy. This condition is met by an ideal gas, where intermolecular forces are not considered. Second, we require separability of the energy modes of the individual particles, e.g., among the translational, rotational, and vibrational modes of energy storage. This idea of separation allows the Schrödinger wave equation to be broken down into a group of simpler differential equations which are easier to solve.

# 10-7 THE TRANSLATIONAL ENERGY LEVELS OF A PARTICLE

Based on the concept of the separability of Schrödinger's wave equation, the translational energy of a particle in a box may now be determined. The wave equation for the translational mode takes on the form

$$\frac{\partial^2 \psi_x}{\partial x^2} + \frac{\partial^2 \psi_y}{\partial y^2} + \frac{\partial^2 \psi_z}{\partial z^2} = -\frac{8\pi^2 m}{h^2} \epsilon_{tr} \psi_{tr} \tag{10-53}$$

where $\epsilon_{tr}$ is the total translational energy of the particle. The potential energy of a particle has been taken to be zero. Moreover, this equation can be further reduced since

$$\psi_{tr} = \psi_x \psi_y \psi_z$$

where $\psi_x$, $\psi_y$, and $\psi_z$ are functions only of the coordinates $x$, $y$, and $z$, respectively. Thus the Schrödinger wave equation applied to the one-dimensional component of translation is simply an ordinary differential equation given by

$$\frac{d^2 \psi_x}{dx^2} = \frac{-8\pi^2 m}{h^2} \epsilon_x \psi_x \tag{10-54}$$

Similar equations apply in the $y$ and $z$ directions.

The solution to this equation is of the form

$$\epsilon_x = \frac{h^2}{8m} \left(\frac{k_x}{a}\right)^2 \tag{10-55}$$

where $k_x$ is any positive integer starting with unity. That is,

$$k_x = 1, 2, 3, \ldots$$

The term $k_x$ is known as the translational quantum number in the $x$ direction. Similar equations are valid in the $y$ and $z$ directions. The total translational energy becomes

$$\epsilon_{tr} = \frac{h^2}{8m} \left[ \left(\frac{k_x}{a}\right)^2 + \left(\frac{k_y}{b}\right)^2 + \left(\frac{k_z}{c}\right)^2 \right] \tag{10-56}$$

The range of values for the $k$'s is not unbounded, but is limited by the total energy of the system. The quantities $a$, $b$, and $c$ are the dimensions in the $x$, $y$, and $z$ directions.

Since the values of the translational quantum numbers begin at unity, the smallest translational energy of a particle in a cubic box is $\epsilon_0 = 3h^2/8mL^2$. In classical statistical mechanics (predating the conception of quantum theory), the ground level was taken as zero. Although not zero, the value of $\epsilon_0$ is very small. The order of magnitude is, roughly, $10^{-18}$ J/g·mole. The average translational energy of a mole of a gas is 3740 J at 300°K, according to kinetic theory. Hence, not only is the ground-level energy extremely small, but also the energy differences between levels are so small that the energy distribution closely approximates that of a continuum. From a mathematical viewpoint, the allowed translational energies appear to be continuous rather than discrete.

## 10-8 THE ROTATIONAL ENERGY OF A RIGID LINEAR ROTATOR

All diatomic gases and a number of triatomic gases (for example, $CO$, $CO_2$, $CS_2$, $N_2O$) are linear molecules. The atoms of the molecule lie in a straight line. As a first approximation for these gases it may be assumed that the interatomic distances are fixed. The assumption of a rigid rotator is valid generally at not too high temperatures. Under these conditions the solution of the wave equation shows that the energy of the $j$th level is

$$\epsilon_{rot} = \frac{j(j+1)h^2}{8\pi^2 I} \tag{10-57}$$

where $I$ is the moment of inertia of the molecule. Some values of $I$ for common diatomic gases are found in Table 10-1. The energy $\epsilon_{rot}$ represents the total energy of both rotational degrees of freedom. The rotational quantum number $j$ can have integral values beginning at zero; i.e.,

$$j = 0, 1, 2, 3, \dots$$

**Table 10-1 Moments of inertia and vibrational wave numbers of some diatomic molecules**

| Molecule | $I$, g·cm$^2$ | $w$, cm$^{-1}$ |
|---|---|---|
| $H_2$ | $0.462 \times 10^{-40}$ | 4405 |
| HF | 1.34 | 4138 |
| HCl | 2.65 | 2990 |
| HBr | 3.32 | 2650 |
| HI | 4.31 | 2309 |
| $N_2$ | 14.1 | 2360 |
| CO | 14.6 | 2170 |
| NO | 16.5 | 1904 |
| $O_2$ | 19.4 | 1580 |
| $B_2$ | 22.8 | 1061 |
| $F_2$ | 31.4 | 923 |
| $Cl_2$ | 116.7 | 561 |
| $Br_2$ | 347 | 323 |
| $I_2$ | 752 | 215 |

*Source:* JANAF Thermochemical Tables, NSRDS-NBS-37, 1971.

Since the $j$ value of zero gives a ground level of energy equal to zero, all other levels are relative to a zero base value.

The order of magnitude for the spacing of the rotational quantum levels is much larger than that for the translational mode. The general range is from 5 to 500 J/g·mol. Excitation of the rotational levels then would occur at much higher temperatures than for translation, since larger quantities of energy will be required.

In addition to the energy-level spacings, we need information on the number of energy states per level. Quantum mechanics demonstrates that there are $2j + 1$ molecular quantum states which correspond approximately to the same magnitude of the rotational energy. Therefore, in rotation,

$$g_j = 2j + 1 \tag{10-58}$$

The degeneracy of a rotational level thus increases linearly with the rotational quantum number $j$. The larger the energy, the larger is the degeneracy.

## 10-9 THE VIBRATIONAL ENERGY OF A LINEAR HARMONIC OSCILLATOR

The Schrödinger wave equation for a one-dimensional harmonic oscillator is of the form

$$\frac{d^2\psi}{dx^2} + \frac{8\pi^2 m}{h^2}(\epsilon - \tfrac{1}{2}kx^2)\psi = 0 \tag{10-59}$$

The general solution of this differential equation is given by an infinite series. The allowed vibrational energy levels are quantized, and given by

$$\epsilon_{\text{vib}} = (v + \tfrac{1}{2})hv \tag{10-60}$$

where $v = 0, 1, 2, 3, \ldots$. The symbol $v$ is the vibrational quantum number, and $v$ is the fundamental frequency of oscillation in $\sec^{-1}$. The frequency $v$ lies in the infrared region of the electromagnetic spectrum for diatomic gases. Recall that the equation for the rotational energy indicates that the ground-level energy is zero. However, in the ground level, the vibrational energy is $\tfrac{1}{2}hv$. The quantity $hv$ is referred to as a quantum of energy.

In some instances it is convenient to suppress the ground-level energy $\epsilon_{\text{vib, 0}}$ and then correct for this later. Equation (10-60) may be modified to

$$\epsilon_{\text{vib}} - \epsilon_{\text{vib, 0}} = vhv \tag{10-61}$$

or
$$\epsilon_{\text{vib}} - \epsilon_{\text{vib, 0}} = vhcw \tag{10-62}$$

In Eq. (10-62) $c$ is the speed of light and $w$ is the vibrational wave number in reciprocal centimeters. The vibrational wave numbers of some common diatomic gases are found in Table 10-1. These are usually obtained from spectroscopic data. If the total energy is ever required in a calculation, the ground-level energy equal to $\tfrac{1}{2}N_A hv$ per mole must be added to the results.

The energy spacing of the vibrational levels is extremely large, not unusually from 4000 to 40,000 J/g·mol. Because of these large values it will be shown later that even at several thousand degrees of temperature, most particles can have only the vibrational energies associated with the first few levels. According to Eq. (10-60), the vibrational levels of a harmonic oscillator are equally spaced. Quantum mechanics also shows that the degeneracy of a one-dimensional harmonic oscillator is unity. That is, $g_{vib} = 1$.

## 10-10 THE MOLECULAR PARTITION FUNCTION FOR THE TRANSLATIONAL ENERGY MODE

In an earlier section, the translational energy of an independent particle was determined in terms of the integer quantum numbers. The energy $\epsilon_{tr}$ is given by

$$\epsilon_{tr} = \frac{h^2}{8m} \left( \frac{k_x^2}{a^2} + \frac{k_y^2}{b^2} + \frac{k_z^2}{c^2} \right) \tag{10-56}$$

In order to evaluate the translational partition function, the above equation is substituted into the expression for $z$.

$$z_{tr} = \sum e^{-B\epsilon_{tr}} = \sum_{k_x} \sum_{k_y} \sum_{k_z} \exp \left[ \frac{-Bh^2}{8m} \left( \frac{k_x^2}{a^2} + \frac{k_y^2}{b^2} + \frac{k_z^2}{c^2} \right) \right] \tag{10-63}$$

The value of $z$ depends upon the summation over all available levels of energy. Under normal conditions, if the dimensions of the volume of the box are large, the intervals in the above summations are very close together. This is especially true since $h^2$ is such a very small number. Therefore the summation process may be replaced by integration. Accepting this approximation for the moment, the summation on the right-hand side of Eq. (10-63) may be evaluated as the product of three integrals of similar form. Considering the $x$ direction only,

$$z_x = \sum \exp \left( -\frac{Bh^2 k_x^2}{8ma^2} \right) = \int_0^\infty \exp \left( -\frac{Bh^2 k_x^2}{8ma^2} \right) dk_x$$

$$= \int_0^\infty \exp \left( -C k_x^2 \right) dk_x$$

where $C = Bh^2/8ma^2$. From standard integral tables it is found that

$$\int_0^\infty \exp \left( -C k_x^2 \right) dk_x = \frac{1}{2} \left( \frac{\pi}{C} \right)^{1/2}$$

Consequently,

$$z_x = \frac{1}{2} \left( \frac{8\pi ma^2}{Bh^2} \right)^{1/2} = \left( \frac{2\pi ma^2}{Bh^2} \right)^{1/2}$$

or
$$z_x = \frac{a}{h} \left( \frac{2\pi m}{B} \right)^{1/2} \tag{10-64}$$

Since $z_{tr} = z_x z_y z_z$ according to Eq. (10-63), the molecular partition function for the total translational energies of independent particles becomes

$$z_{tr} = \frac{abc}{h^3} \left(\frac{2\pi m}{B}\right)^{1/2} \tag{10-65}$$

Recognizing again that $B = 1/kT$ and $abc$ is the volume of the system, the translational partition function can also be written as

$$z_{tr} = \frac{V}{h^3} (2\pi m k T)^{3/2} \tag{10-66}$$

For a given chemical species, the translational partition function is primarily a function of the volume and temperature of the system.

## 10-11 THE MOLECULAR PARTITION FUNCTION OF A RIGID ROTATOR

In Sec. 10-8 the energy for level $j$ in rotation for a rigid diatomic molecule was given as

$$\epsilon_{rot} = \frac{j(j + 1)h^2}{8\pi^2 I} \tag{10-57}$$

where $I$ is the moment of inertia. Another form of this equation is

$$\epsilon_{rot} = j(j + 1)k\theta_R \tag{10-67}$$

where the constant $\theta_R$ is defined as

$$\theta_R \equiv \frac{h^2}{8\pi^2 I K} \tag{10-68}$$

The quantity $\theta_R$ has the dimension of temperature, and is known as the characteristic rotational temperature. Since the degeneracy of each rotational energy level is $2j + 1$, the rotational partition function of a rigid rotator becomes

$$z_{rot} = \sum_{j=0} g_j e^{-\epsilon_j/kT} = \sum_{j=0} (2j + 1)e^{[-j(j + 1)\theta_R]/T} \tag{10-69}$$

Although the energy levels are more widely spaced in rotation than in translation, a continuum of energy often is a reasonable assumption. The summation of $z_{rot}$ under this approach becomes an integration. The evaluation of the rotational partition function by integration may be justified when $\theta_R \ll T$. Hence

$$z_{rot} = \int_0^\infty (2j + 1)e^{[-j(j + 1)\theta_R]/T} \, dj \tag{10-70a}$$

Note that $(2j + 1) \, dj$ is equivalent to $d(j^2 + j)$. The above expression can thus be written as

$$z_{rot} = \int_0^\infty e^{[-(j^2 + j)\theta_R]/T} \, d(j^2 + j) = \frac{T}{\theta_R} \int_0^\infty e^{-x} \, dx \tag{10-70b}$$

where $x = [(j^2 + j)\theta_R]/T$. The value of the integral in Eq. (10-70b) is unity. Consequently the rotational partition function, when $\theta_R \ll T$, is simply

$$z_{rot} = \frac{T}{\theta_R} = \frac{8\pi^2 I k T}{h^2} \tag{10-71}$$

This equation is valid for a heteronuclear diatomic molecule, assuming a continuum for the energy levels and behavior as a rigid rotator.

A heteronuclear molecule is one composed of two or more dissimilar atoms. A diatomic molecule of this type would be carbon monoxide. On the other hand, a homonuclear molecule would be one composed of like atoms, such as diatomic nitrogen or hydrogen. In polyatomic molecules additional configurations are possible by rotation of the molecule around its axes. That is, equivalent orientations are possible because of rotation alone in space. For homonuclear molecules these orientations are indistinguishable. A symmetry number $\sigma$ is necessary in the denominator of the rotational partition function; otherwise we would account for too many configurations of the molecule which are really indistinguishable. For homonuclear diatomic molecules the symmetry number is 2, and for heteronuclear ones the value of $\sigma$ is unity. Therefore Eq. (10-71) should be revised into the general form

$$z_{rot} = \frac{8\pi^2 I k T}{\sigma h^2} = \frac{T}{\sigma \theta_R} \tag{10-72}$$

The approximation of integration, instead of direct summation, which led to Eq. (10-72) is generally valid when $8\pi^2 I k T/h^2$ is greater than about 5. Consequently, the approximate equation for the rotational partition function is valid for a majority of diatomic molecules well below $100°K$.

When the energy levels are not closely spaced it may be necessary to carry out the actual summation of terms in Eq. (10-69), rather than use an integration technique. This is accomplished through the Euler-Maclaurin summation formula. The result is

$$z_{rot} = \frac{T}{\sigma \theta_R} \left[ 1 + \frac{1}{3}\left(\frac{\theta_R}{T}\right) + \frac{1}{15}\left(\frac{\theta_R}{T}\right)^2 + \frac{4}{315}\left(\frac{\theta_R}{T}\right)^3 + \cdots \right] \tag{10-73}$$

Note that when $T \gg \theta_R$, this equation reduces to Eq. (10-72).

## 10-12 VIBRATIONAL PARTITION FUNCTION OF A LINEAR HARMONIC OSCILLATOR

The vibrational partition function may now be obtained by the usual summation process. Integration is not possible in this case since the spacing between energy levels is so large. The multiplicity, or degeneracy factor, of the vibrational levels is unity. Hence

$$z_{vib} = \sum g_v \exp\left(-\frac{\epsilon_v}{kT}\right) = \sum_{v=0} \exp\left(-\frac{vh\nu}{kT}\right) \tag{10-74}$$

The vibrational energy $\epsilon_{vib}$ used above is that given by Eq. (10-61), where the ground-level energy has been suppressed. Hence, in the partition function above, a constant multiplicative factor, $\exp{(-hv/2kT)}$, has likewise been suppressed.

Replacing $hv/kT$ by $x$ in Eq. (10-74), the series expansion in $x$ can be reduced to a simpler form:

$$z_{vib} = \sum_{v=0} \exp{(-vx)} = 1 + e^{-x} + e^{-2x} + e^{-3x} + \cdots$$

$$= (1 - e^{-x})^{-1} \qquad (10\text{-}75)$$

The equation for the partition function of a harmonic oscillator is then

$$z_{vib} = \left(1 - \exp{\frac{-hv}{kT}}\right)^{-1} \qquad (10\text{-}76)$$

When $x > 5$, $z_{vib}$ is essentially unity.


## 10-13 PROPERTIES OF AN IDEAL MONATOMIC GAS

The internal energy, entropy, and heat capacities of an ideal monatomic gas may now be determined by the use of the partition function. Since monatomic gases may be considered as point masses, the only mode of energy storage of importance at temperatures up to several thousands of degrees is the translational one.

The molar translational energy $U$ is given by Eq. (10-46) in the form of

$$U = R_u T^2 \left(\frac{\partial \ln z}{\partial T}\right)_V$$

On the basis of Eq. (10-66), the partial derivative of $\ln z$ is found to be $3/2T$. Hence the molar translational energy is

$$U = R_u T^2 \frac{3}{2T} = \frac{3}{2} R_u T \qquad (10\text{-}77)$$

The molar entropy of ideal monatomic gases is found by use of the expression for indistinguishable particles:

$$S = R_u \ln \frac{z}{N_A} + R_u T \left(\frac{\partial \ln z}{\partial T}\right)_V + R_u \qquad (10\text{-}49)$$

Substitution of the translational partition function and its derivative gives

$$S = R_u \ln \left[\frac{V}{N_A} \left(\frac{2\pi mkT}{h^2}\right)^{3/2}\right] + R_u T \frac{3}{2T} + R_u$$

$$= R_u \ln \left[\frac{V}{N_A} \left(\frac{2\pi mkT}{h^2}\right)^{3/2}\right] + \frac{5}{2} R_u \qquad (10\text{-}78)$$

Expression (10-78) is known as the Sackur-Tetrode equation. It was derived independently by these two men by somewhat different methods around 1912. The molar volume in the equation may be replaced by $RT/P$ when convenient. By substituting for $V$ and by expressing the mass of the atom in terms of the molar mass $M$ of the substance, the molar entropy of a monatomic gas may be reduced to a function of the system temperature (in degrees Kelvin) and pressure (in atmospheres), the molar mass, and numerical constants.

$$S = R_u(\tfrac{5}{2} \ln T - \ln P + \tfrac{3}{2} \ln M - 1.164) \qquad (10\text{-}79)$$

The molar heat capacity at constant volume, $c_v$, is found most easily by noting that $c_v = (\partial U/\partial T)_V$. Differentiation of Eq. (10-77) shows that, for an ideal monatomic gas,

$$c_v = \tfrac{3}{2}R_u \qquad (10\text{-}80)$$

Hence the evaluations of the molar internal energy and molar heat capacity at constant volume obtained by statistical methods are in excellent agreement with both the kinetic theory and experimental results.

**Example 10-2** Determine the translational contribution to the entropy of monatomic chlorine as an ideal gas at 1000°K and 1 bar by employing Eqs. (10-78) and (10-79).

SOLUTION We shall first evaluate various portions of Eq. (10-78). The quantity $V/N_A$ is found as follows:

$$V/N_A = NR_uT/PN_A = 1(0.08315)(1000)/(1)(6.023 \times 10^{26})$$

$$= 1.38 \times 10^{-25} \text{ m}^3$$

The molar mass of monatomic chlorine is approximately 35.5 kg/kg·mol. Hence

$$\frac{2\pi mkT}{h^2} = \frac{2\pi(35.5)(1.38 \times 10^{-23})(1000)}{(6.625 \times 10^{-34})^2} = 1.164 \times 10^{22} \text{ m}^{-2}$$

On the basis of Eq. (10-78), then,

$$S_{tr} = 8.315[\ln (1.38 \times 10^{-25})(1.164 \times 10^{22})^{3/2} + 2.5]$$

$$= 8.315(18.97 + 2.5) = 178.5 \text{ kJ/(kg·mol)(°K)}$$

On the basis of Eq. (10-79) we find that

$$S_{tr} = 8.315[(\tfrac{5}{2}) \ln 1000 - \ln 1.013 + (\tfrac{3}{2}) \ln 35.5 - 1.164]$$

$$= 8.315(17.27 - 0.01 + 5.35 - 1.16)$$

$$= 178.4 \text{ kJ/(kg·mol)(°K)}$$

The answers are slightly different due to round-off errors.

## 10-14 CONTRIBUTIONS BY THE ROTATIONAL MODE TO THE PROPERTIES OF DIATOMIC GASES

Diatomic molecules will also have molar translational energies equal to $\tfrac{3}{2}RT$. In addition, energy may be stored in the energy modes internal to the molecule rotation and vibration. The evaluation of the rotational contribution to the internal energy and of other properties depends upon the rotational partition function. In Sec. 10-11 it was noted that, for most diatomic molecules above 100°K, this partition function could be approximated by

$$z_{rot} = \frac{8\pi^2 I k T}{\sigma h^2} \tag{10-72}$$

The $(\partial \ln z/\partial T)_V$ is $1/T$. The rotational internal energy is then

$$U_{rot} = R_u T^2 \left(\frac{\partial \ln z}{\partial T}\right)_V = R_u T^2 \frac{1}{T} = R_u T \tag{10-81}$$

This value is in agreement with the classical kinetic-theory value given earlier.

At low temperatures it is necessary to compute the rotational contribution to the internal energy by direct summation. The energy levels are sufficiently far apart so that a continuum of energy is no longer acceptable. By the substitution of

$$\epsilon_i = \epsilon_{rot} = \frac{j(j+1)h^2}{8\pi^2 I} \tag{10-57}$$

and

$$g_i = 2j + 1$$

into the expression for $U$, the rotational internal energies may be determined at various temperatures. To keep the results perfectly general, it is convenient to plot $U/RT$ versus $8\pi^2 I k T/h^2$ since both are dimensionless. The quantity $h^2/8\pi^2 Ik$ is the characteristic rotational temperature $\theta_R$ for the substance. A plot of $U/RT$ against $T/\theta_R$ for a rigid rotator is shown in Fig. 10-3. It is seen that, as $T/\theta_R$ goes beyond 5.0, the value of $U/RT$ approaches the classical value of 1. Thus, at room temperatures, many diatomic gases are fully excited in the rotational energy mode. Some typical values of $\theta_R$ are listed in Table 10-2.

The specific heat at constant volume for a diatomic rigid rotator must be handled similarly to the internal energy. At sufficiently high values of the temperature and the moment of inertia the value of the $c_v$ is found by differentiating Eq. (10-81).

$$c_v = \left(\frac{\partial U}{\partial T}\right)_V = R_u$$

This is the classical value given by kinetic theory. At low values of $8\pi^2 I k T/h^2$, the specific heat must be found by direct summation also. The results of this summation are shown also in Fig. 10-3. The curve has a maximum value of about 1.1 when $T/\theta_R$ is 0.8. It then levels off quickly to the classical value of 1.0 for values of $T/\theta_R$ greater than 2. Hence all diatomic gases except hydrogen have a fully excited rotational heat capacity for temperatures greater than 50°K.

As noted in Sec. 10-6, the entropy contributions of the internal modes of energy storage are found by neglecting two terms in Eq. (10-49), since they have already been accounted for in the translational mode. This is equivalent to con-

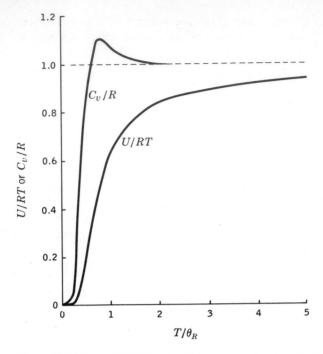

**Figure 10-3** Values of $U/RT$ and $c_v/R$ for the rotational mode of a diatomic rigid rotator.

**Table 10-2 Characteristic rotational temperatures $\theta_R$ for some diatomic gases**
$(\theta_R = h^2/8\pi^2 Ik)$

| Gas | $\theta_R$, °K | $\theta_R$, °R |
|-----|------|------|
| $H_2$ | 87.2 | 157 |
| HF | 30.1 | 54.1 |
| HCl | 15.2 | 27.4 |
| HBr | 12.1 | 21.8 |
| HI | 9.35 | 16.9 |
| $N_2$ | 2.86 | 5.15 |
| CO | 2.76 | 4.96 |
| NO | 2.44 | 4.39 |
| $O_2$ | 2.08 | 3.74 |
| $B_2$ | 1.77 | 3.19 |
| $F_2$ | 1.28 | 2.30 |
| $Cl_2$ | 0.345 | 0.621 |
| $Br_2$ | 0.116 | 0.209 |
| $I_2$ | 0.054 | 0.097 |

*Source:* Computed from data in Table 10-1.

sidering the internal modes as consisting of distinguishable elements. For such distinguishable elements

$$S_{dist} = R_u\left[\ln z + T\left(\frac{\partial \ln z}{\partial T}\right)_V\right] \tag{10-82}$$

Since $\ln z_{rot} = \ln (8\pi^2 IkT/\sigma h^2)$, we have

$$\left(\frac{\partial \ln z_{rot}}{\partial T}\right)_V = \frac{1}{T}$$

The molar entropy of a rigid rotator, then, is

$$S_{rot} = R_u \ln \frac{8\pi^2 IkT}{\sigma h^2} + R_u = R_u(\ln z_{rot} + 1) \tag{10-83}$$

The independent variables are $I$, $T$, and $\sigma$. After substituting for the natural constants, this expression reduces to

$$S_{rot} = R_u(\ln I + \ln T - \ln \sigma + 89.40) \tag{10-84}$$

where $I$ is measured in g·cm$^2$ and $T$ in °K.

**Example 10-3** Determine the rotational contribution to the entropy of hydrogen chloride, HCl, as an ideal gas at 100°K.

SOLUTION First it is necessary to determine whether the integration or summation form of the partition function is applicable. From Table 10-2 the value of $\theta_R$ is 15.2°K. Hence $\theta_R/T$ is 0.152 and $T/\theta_R$ is 6.58. On the basis of Eq. (10-73), and noting that $\sigma$ is unity,

$$z_{rot} = \frac{T}{\sigma\theta_R}\left[1 + \frac{1}{3}\left(\frac{\theta_R}{T}\right) + \frac{1}{15}\left(\frac{\theta_R}{T}\right)^2 + \cdots\right]$$

$$= 6.58\left[1 + \frac{1}{3}(0.152) + \frac{1}{15}(0.152)^2 + \cdots\right]$$

$$= 6.58(1 + 0.051 + 0.002 + \cdots) = 6.93$$

This calculation indicates that an error of roughly 5 percent would occur if we used the integrated form given by Eq. (10-73). In order to evaluate the entropy by means of Eq. (10-82) we must find $(\delta \ln z/\delta T)$. Differentiation of Eq. (10-73) yields

$$\left(\frac{\partial \ln z}{\partial T}\right)_{rot} = \frac{\frac{1}{\sigma T}\left[\frac{T}{\theta_R} - \frac{1}{15}\left(\frac{\theta_R}{T}\right)\right]}{z_{rot}}$$

$$= \frac{0.01[6.58 - 0.0666(0.152)]}{6.93} = 0.00949$$

Therefore

$$S_{rot} = R_u \ln z + T\left(\frac{\delta \ln z}{\delta T}\right)_V$$

$$= 8.315[\ln 6.93 + 100(0.00949)]$$

$$= 8.315(1.936 + 0.949) = 23.99 \text{ kJ/(kg·mol)(°K)}$$

If, instead, we assume the validity of Eq. (10-84),

$$S_{rot} = R_u(\ln I + \ln T - \ln \sigma + 89.40)$$

$$= 8.315(\ln 2.65 \times 10^{-40} + \ln 100 + 89.40)$$

$$= 8.315(-91.13 + 4.605 + 89.40) = 23.91 \text{ kJ/(kg·mol)(°K)}$$

Although $z$ varies by 5 percent for the two methods, the error in $S$ in this case is small if the integrated format is chosen. This occurs because $\ln z$ is involved in evaluating the entropy, and not $z$ itself.

## 10-15 CONTRIBUTIONS BY THE VIBRATIONAL MODE TO THE PROPERTIES OF DIATOMIC GASES

The partition function of a group of harmonic oscillators is

$$z_{vib} = \left(1 - \exp\frac{-hv}{kT}\right)^{-1} \tag{10-76}$$

This simple form is valid at all temperatures for any gas. Therefore the development of equations for the vibrational contribution to various properties is much more straightforward than for the rotational mode.

The vibrational energy $U_{vib}$ of a mole of gas can be found by again applying the relation

$$U_{vib} = R_u T^2 \left(\frac{\partial \ln z_{vib}}{\partial T}\right)_V \tag{10-46}$$

Since the derivative of the logarithm of $z_{vib}$ is

$$\frac{\partial \ln z_{vib}}{\partial T} = \frac{-1}{1 - \exp(+hv/kT)} \frac{hv}{k} \frac{1}{T^2}$$

the energy stored in a system of harmonic oscillators is given by

$$U_{vib} - U_{vib, 0} = \frac{R_u T}{\exp(+hv/kT) - 1} \frac{hv}{kT} = \frac{R_u Tx}{e^x - 1} \tag{10-85}$$

where $x = hv/kT$. Equation (10-85) excludes the ground-level energy, which has been shown to equal $\frac{1}{2}N_A hv$.

The contribution of the vibrational energy mode to the constant-volume specific heat may now be determined by differentiation of Eq. (10-85) with respect to temperature at constant volume. The result is

$$c_{v, vib} = \frac{R_u \exp(-hv/kT)}{[\exp(-hv/kT) - 1]^2} \left(\frac{hv}{kT}\right)^2 = \frac{R_u e^x x^2}{(e^x - 1)^2} \tag{10-86}$$

The mathematical functions $x(e^x - 1)$ and $e^x x^2/(e^x - 1)^2$ are known as Einstein functions, and have been tabulated in standard sources as a function of $x$.

At relatively high temperatures (dependent upon the value of $v$, the frequency) the factor $hv/kT$, or $x$, becomes quite small. The limit of $x/(e^x - 1)$ as $x \to 0$ is unity. Therefore the limiting vibrational-energy storage per mode is $R_u T$. The corresponding contribution to $c_v$ is obviously $R_u$. This is reached for $c_v$ when $1/x$ reaches approximately 2.0. Hence, at sufficiently high temperatures, the contribution of the vibrational mode to the specific heat reaches the classical value based on the equipartition of energy theorem.

The dimensionless parameters $U/RT$ and $c_v/R$ for a harmonic oscillator system are shown in Fig. 10-4. The abscissa is the dimensionless ratio $T/\theta_v$, where $\theta_v$ is the characteristic vibrational temperature defined by $h\nu/k$. In terms of the vibrational wave number $w$, in $cm^{-1}$, the abscissa is also equivalent to $T/1.438w$. These dimensionless plots are quite important since only one curve is required to express the complete range of possible values of such variables as $I$, $\nu$, and $w$, which are characteristic of the particular substances. Representative values of $\theta_v$ are given in Table 10-3.

The contribution of the vibrational mode to the molar entropy is found by the proper application of the entropy equation for distinguishable elements,

$$S_{\text{dist}} = R\left[\ln z + T\left(\frac{\partial \ln z}{\partial T}\right)_v\right]$$
(10-82)

By use of Eq. (10-76) for the vibrational partition function of a harmonic oscillator it may be shown that

$$S_{\text{vib}} = R_u\left[\frac{x}{e^x - 1} - \ln\left(1 - e^{-x}\right)\right]$$
(10-87)

where again $x = h\nu/kT$.

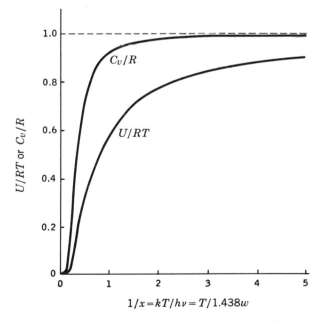

**Figure 10-4** The dimensionless internal energy and specific heat of a system of harmonic oscillators.

## Table 10-3
### Characteristic vibrational temperatures $\theta_v$ for some diatomic gases
$(\theta_v = hcw/k)$

| Gas | $\theta_v$, °K | $\theta_v$, °R |
|-----|------|--------|
| $H_2$ | 6340 | 11,420 |
| HF | 5950 | 10,720 |
| HCl | 4300 | 7,740 |
| HBr | 3810 | 6,860 |
| HI | 3320 | 5,970 |
| $N_2$ | 3390 | 6,100 |
| CO | 3120 | 5,610 |
| NO | 2740 | 4,940 |
| $O_2$ | 2270 | 4,080 |
| $B_2$ | 1530 | 2,750 |
| $F_2$ | 1330 | 2,400 |
| $Cl_2$ | 807 | 1.450 |
| $Br_2$ | 465 | 837 |
| $I_2$ | 310 | 558 |

*Source:* Computed from data in Table 10-1.

**Example 10-4** Determine the constant-pressure specific heat of carbon monoxide gas at 300, 600, 1200, and 1800°K and at a pressure of 1 atm, assuming ideal-gas behavior.

SOLUTION Since ideal-gas behavior is assumed, the $c_p$ values may be obtained by first determining $c_v$ for the three energy modes and then adding $R_u$ to the sum of these contributions. The value of $c_v$ for the 3 translational degrees of freedom totals $1.5R_u$ at all temperatures. The moment of inertia of CO is given by Table 10-1 as $14.6 \times 10^{-40}$ g·cm². At 300°K the value of $8\pi^2 IkT/h^2$ would be, roughly, 110. Since this value is an order of magnitude greater than 5, the use of integration to evaluate the partition function is allowable. Thus the rotational contribution to $c_v$ is $R_u$ at all the desired temperatures.

The vibrational contribution to $c_v$ is evaluated by use of Eq. (10-86). Table 10-1 lists the vibrational wave number for CO as 2170 cm$^{-1}$. Recall that $x = hv/kT = 1.438w/T$. The values of $x$ corresponding to the four given temperatures are then 10.41, 5.20, 2.60, and 1.73, respectively. The results for all three energy modes are summarized below.

| $T$, °K | $c_{v, vib}$ | $c_{v, tot}$ | $c_p$(calc.) | $c_p$(tabl.) |
|---------|----------|----------|----------|----------|
| 300 | $0.003R_u$ | $2.503R_u$ | 29.13 | 29.14 |
| 600 | $0.151R_u$ | $2.651R_u$ | 30.36 | 30.44 |
| 1200 | $0.586R_u$ | $3.086R_u$ | 33.98 | 34.17 |
| 1800 | $0.784R_u$ | $3.284R_u$ | 35.62 | 35.91 |

The $c_p$ values are listed above in J/(g·mol)(°K). The calculated values compare favorably with the tabulated ones, which are based on the JANAF Thermochemical Tables.

**Example 10-5** Determine the entropy of carbon monoxide gas at 300 and 1800°K if the gas is assumed ideal at 1 atm.

SOLUTION The translational entropy at 300°K is found by using Eq. (10-79). Substitution for $T$, $P$, and $M$ leads to

$$S_{tr} = (\tfrac{5}{2} \ln 300 - \ln 1 + \tfrac{3}{2} \ln 28.011 - 1.164)R_u$$

$$= 18.129 R_u$$

The rotational entropy requires knowledge of $I$ and $\sigma$. The symmetry number of heteronuclear diatomic molecules is unity. The moment of inertia again is $14.6 \times 10^{-40}$ g·cm², from Table 10-1. Substitution of these data into Eq. (10-84) yields

$$S_{rot} = (\ln 14.6 \times 10^{-40} + \ln 300 - \ln 1 + 89.40)R_u$$

$$= 5.68 R_u$$

The calculation of the vibrational entropy contribution involves the use of Eq. (10-87). The value of $x$ at 300°K is given by

$$x = \frac{1.438w}{T} = \frac{1.438(2170)}{300} = 10.41$$

Hence

$$S_{vib} = (0.0003 + 0.00003)R_u$$

$$= 0.00033 R_u$$

The vibrational contribution at 300°K is negligible. A summary of these results and of similar computations at 1800°K is presented below:

| $T$, °K | $S_{tr}$ | $S_{rot}$ | $S_{vib}$ | $S$(calc.) | $S$(tabl.) |
|---------|----------|-----------|-----------|------------|------------|
| 300 | $18.13R_u$ | $5.68R_u$ | 0 | 198.0 | 197.7 |
| 1800 | $22.59R_u$ | $7.46R_u$ | $0.57R_u$ | 254.6 | 254.8 |

The last two columns are in units of J/(g·mol)(°K). The final column is based on values tabulated in the JANAF Tables. The comparison in this case is good.

## 10-16 MAXWELL ENERGY-DISTRIBUTION FUNCTION FOR TRANSLATIONAL ENERGIES

The MB distribution function may now be employed, in conjunction with the specific relationship for the discrete energy levels of the translational mode, to determine for this mode the distribution of the energy among the various particles. Of particular interest will be a plot of $p_i$ versus $i$ (or some function of the energy $\epsilon_i$ of each level), where $p_i$ is defined as $n_i/n$. That is, $p_i$ is the probability, or the fraction of the particles, associated with a given energy level. Since the allowed energy values are discrete, a typical presentation of the data would appear as a bar graph. The height of each bar represents the value of $p_i$ at that particular energy level. In the case of translational energies, the separation of the energy levels has been shown to be so small that the assumption of a continuum of energies is a valid one. On this basis the distribution curve for the translational mode can be presented as a continuous function. In place of the plot of $p_i$ versus $i$ for the discrete case, in the continuous case a plot of $dp/d\epsilon$ versus $\epsilon$ might be appropriate. The differential area below the curve and between the values of $\epsilon + d\epsilon$ then would represent the probability of particles having energy in that range. The total area below the curve between $\epsilon = 0$ and any value $\epsilon$ represents the fraction of the total particles whose energies lie between 0 and $\epsilon$. Hence $dp$ stands for the fraction of the total particles which have energy in some $d\epsilon$ range.

The Boltzmann distribution function relates the probability $p_i$ to a given energy state by

$$p_i = \frac{g_i e^{-B\epsilon_i}}{z} \tag{10-43}$$

Since $B$, $\epsilon_i$, and $z$ are constants for a given state of a system, differentiation of this equation gives simply

$$dp_i = \frac{e^{-B\epsilon_i}}{z} dg_i$$

The subscripts $i$ may now be dropped since we wish to apply this equation to a continuous distribution. Thus

$$dp = \frac{e^{-B\epsilon}}{z} dg \tag{10-88}$$

If an expression for $dg$ in terms of $d\epsilon$ can be found for the translational mode and substituted into Eq. (10-88), we shall have the desired result. The number of energy states of approximately the same translational energy is given by those states whose values of $k_x^2 + k_y^2 + k_z^2$ are the same for a system of cubic dimensions. This is easily seen since the translational energy is given by

$$\epsilon_{tr} = \frac{h^2}{8mL^2} (k_x^2 + k_y^2 + k_z^2) \tag{10-56}$$

where $a$, $b$, and $c$ all are of length $L$. Geometrically, the quantum states are represented by points in a three-dimensional plot of the quantum numbers, $k_x$, $k_y$, and $k_z$. Points reside only where the quantum numbers are positive and integral values. Therefore all points in an octant of a sphere which lie at a radius of $r$ from the center have the same energy; $r$ is defined by

$$r = (k_x^2 + k_y^2 + k_z^2)^{1/2} \tag{a}$$

Those states whose energies are between $\epsilon$ and $\epsilon + d\epsilon$ must lie in the differential volume of a spherical shell between $r$ and $r + dr$. The volume of this shell is $4\pi r^2 dr$. Only one octant of the shell volume is meaningful, and this represents differentially the number of states $dg$ which we seek. Therefore

$$dg = \tfrac{1}{2}\pi r^2 dr \tag{b}$$

By combining Eq. (10-56) and Eq. (a) and solving for $r$, one finds that

$$r = \frac{L}{h}(8m)^{1/2}(\epsilon)^{1/2}$$

where $\epsilon$ now is the total translational energy of a state. Differentiation of the above equation gives

$$dr = \frac{L}{h}(8m)^{1/2}(\tfrac{1}{2})\epsilon^{-1/2}\, d\epsilon$$

Substitution of the equations for $r$ and $dr$ into Eq. (b) for $dg$ results in

$$dg = 2\pi\left(\frac{L}{h}\right)^3 (2m)^{3/2}(\epsilon)^{1/2}\, d\epsilon \tag{10-89}$$

Finally, Eq. (10-88) for $dp$ may be expressed in terms of $\epsilon$ by use of the above result.

$$dp = \frac{2\pi V e^{B\epsilon}}{zh^3}(2m)^{3/2}\epsilon^{1/2}\, d\epsilon$$

where $V$ has replaced $L^3$. However, $z$ for the translational mode has been shown to be $(2\pi m k T)^{3/2} V/h^3$. Therefore

$$dp = \frac{2\epsilon^{1/2}e^{-\epsilon/kT}\, d\epsilon}{\pi^{1/2}(kT)^{3/2}} \tag{10-90}$$

This equation gives the fraction of particles with translational energies lying between $\epsilon$ and $\epsilon + d\epsilon$. It was first derived by Maxwell by means of the kinetic theory of gases. By letting $\epsilon/kT$ equal the variable $c$, the preceding equation can be arranged into a more convenient form,

$$dp = \frac{2c^{1/2}e^{-c}\, dc}{\pi^{1/2}} \tag{10-91}$$

Instead of plotting $dp/d\epsilon$ versus $\epsilon$ as originally suggested, $dp/dc$ versus $c$ will be chosen. This approach is used in Fig. 10-5 because both $p$ and $c$ are dimensionless numbers. Thus a selection of $T$ is not required. Any increment of area, such as the crosshatched area on the figure, is a measure of the fraction of the particles whose energies lie in that range of $c$. The fraction of particles whose values of $c$ lie between 1.0 and 1.1 is, roughly, 4.05 percent. It is apparent that the great majority of particles have energies less than $5kT$. In fact, nearly 90 percent have energies less than $3kT$ and nearly 75 percent less than $2kT$. The most probable translational energy is $kT/2$. The mean translational energy has already been shown from kinetic theory to be $1.5kT$, in Sec. 10-1.

It is also of interest to determine the distribution of molecular speeds for an ideal gas at equilibrium. Since $\epsilon = mv^2/2$, then $d\epsilon = mv\, dv$. Substitution of these latter two expressions into Eq. (10-90) yields

$$dp_v = \left(\frac{2}{\pi}\right)^{1/2}\left(\frac{m}{kT}\right)^{3/2} v^2 \exp\left(-\frac{mv^2}{2kT}\right) dv \tag{10-92}$$

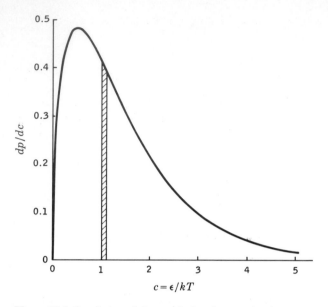

**Figure 10-5** Population of the translational energy levels.

The quantity $dp_v$ represents the fraction of the particles which have speeds in the range from $v$ to $v + dv$. This distribution function is the MB distribution of speeds. A dimensionless plot of this function can be obtained by plotting $(dp_v/dv)/(8m/\pi kT)^{1/2}$ versus $mv^2/2kT$. This plot is shown in Fig. 10-6. The area under the curve represents the total fraction of the particles in the system, and must equal unity. From the figure it is noted that the most probable speed is that

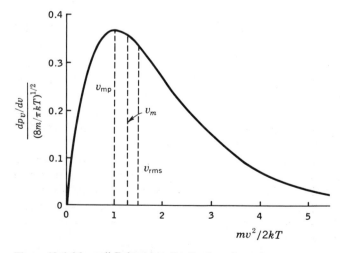

**Figure 10-6** Maxwell-Boltzmann distribution of speeds.

for which the abscissa has a value of unity. Hence the most probable speed for a MB distribution is

$$v_{mp} = \left(\frac{2kT}{m}\right)^{1/2} \tag{10-93}$$

This result can be checked by differentiating the quantity $dp_v/dv$ in Eq. (10-92) with respect to $v$ and setting the result equal to zero. Although an extreme range of speeds is possible, the curve in Fig. 10-6 indicates that a large majority of the particles have speeds fairly close to the most probable speed.

The mean, or average, speed of the particles is defined by

$$v_m \equiv \int_0^\infty v \, dp_v$$

Substitution of Eq. (10-92) for $dp_v$ into this relation gives

$$v_m = \frac{4K^3}{\pi^{1/2}} \int_0^\infty v^3 \exp\left(-K^2 v^2\right) dv$$

where $K = (m/2kT)^{1/2}$. The above integral has a value of $1/2K^4$. Thus the mean, or average, speed of the particles for a MB distribution is

$$v_m = \left(\frac{8kT}{\pi m}\right)^{1/2} \tag{10-94}$$

The rms speed $v_{rms}$ is defined as

$$(v_{rms})^2 = \int_0^\infty v^2 \, dp_v$$

When the Maxwell-Boltzmann distribution is substituted and the integration is carried out, one finds that

$$v_{rms} = \left(\frac{3kT}{m}\right)^{1/2} \tag{10-95}$$

This is in agreement with Eq. (10-7) developed in Sec. 10-1. In this earlier work the rms speed was developed from kinetic theory. A comparison of the three preceding equations will show that the relative magnitudes of the most probable, mean, and rms speeds are in the ratio of $1 : 1.129 : 1.225$. The relative position of these three speeds is shown in Fig. 10-6. It must be pointed out that the MB distribution has been confirmed experimentally in a number of ways.

Finally, the speed distribution function given by Eq. (10-92) can be converted into a velocity distribution function. The quantity $dp_v$ represents the fraction of particles with speeds between $v$ and $v + dv$. Let us symbolize this by $dn_v/n$, where $dn_v$ is the number of particles with speeds between $v$ and $v + dv$. We wish to relate this quantity, $dn_v$, to the number of particles with velocities between $V$ and $V + dV$, as represented by $dn_v$.

Recall that the translational energy $\epsilon_{tr}$ can be represented by a point in a three-dimensional plot of quantum numbers $k_x$, $k_y$, and $k_z$ (see Eq. 10–56). Likewise, the velocity $V$ of any particle can be represented graphically by the vector sum of the three component velocities, $V_x$, $V_y$, and $V_z$. Hence all particles with a velocity of $V$ lie within a cube in velocity space the dimensions of which are $V_x$ to $V_x + dV_x$, $V_y$ to $V_y + dV_y$, and $V_z$ to $V_z + dV_z$. In other words, if the velocity vector is drawn from the origin of $V_x$, $V_y$, $V_z$ space, its point will lie within the cube described above. Now let that velocity vector trace out a spherical shell of inner radius $V$ and outer radius $V + dV$. All particles whose speed $v$ to $v + dv$ is the same as the magnitude of velocity vector $V$ (to $V + dV$) will be represented by points within that spherical shell. That is, the number of particles $dn_v$ in that speed range is related to the number of particles $dn_V$ with that velocity range by

$$dn_v = 4\pi v^2 \, dn_V$$

Therefore we can replace $dp_v$ in Eq. (10-92) by $4\pi v^2 \, dn_V / n$. As a result,

$$\frac{dn_V}{n} = dp_V = \left(\frac{m}{2\pi kT}\right)^{3/2} \exp\left(-\frac{mV^2}{2kT}\right) dV \qquad (10\text{-}96)$$

This is the MB velocity distribution. Note that the velocity vectors are distributed uniformly in the spherical shell between $V$ and $V + dV$ and that the number of points in each shell decreases exponentially as the velocity increases.

In order to plot the velocity distribution function, it is necessary to reduce it to the one-dimensional case. This is easily done by noting that $V^2 = V_x^2 + V_y^2 + V_z^2$ and $dV = dV_x \, dV_y \, dV_z$. Hence the right-hand side of Eq. (10-96) can be written as the product of three terms which are identical except for subscripts. Since the notations $x$, $y$, and $z$ are completely interchangeable when

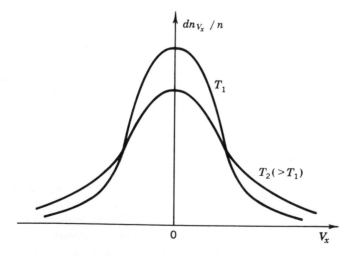

**Figure 10-7** Maxwell-Boltzmann distribution of $x$ component of velocity.

used as subscripts, we find that the velocity distribution function in any coordinate direction ($x$, for example) is

$$\frac{dn_{V_x}}{n} = \left(\frac{m}{2\pi kT}\right)^{1/2} \exp\left(-\frac{mV_x^2}{2kT}\right) dV_x \tag{10-97}$$

A plot of $dn_V/n$ versus $V_x$ is shown in Fig. 10-7. The area under the curve, quantitatively, should equal unity. The function is symmetric around the vertical axis, as expected. Note that increasing the temperature ($T_2 > T_1$) tends to flatten the curve, and significant portions of the distribution extend to larger (positive and negative) values of $V_x$. This is as anticipated. Larger temperatures imply a greater total energy. Hence a greater fraction of particles take on higher energies, or, equivalently, larger velocities.

## 10-17 RADIATION AND THE PHOTON GAS

One of the initial objectives during the development of the quantum theory was the prediction of the energy distribution of radiation as a function of wavelength or frequency. Experimental work had already established the nature of the distribution, but no theoretical analysis was capable of correctly predicting the distribution over the entire range of wavelengths until the work of Planck in the early 1900s. Before turning to the development of the Planck distribution law, it is useful to apply classical thermodynamics to the study of blackbody radiation.

In the study of kinetic theory in Sec. 10-1, it was found that the pressure exerted by particles upon a wall was equal to two-thirds of the energy density of the particles; that is, $PV = \frac{2}{3}U_{tr}$, or

$$P = \frac{2}{3}u'$$

where $u'$ is the energy per unit volume. In addition, the pressure is proportional to the momentum of the particles, and the momentum of a particle equals twice the energy divided by the velocity. In an analogous manner, the pressure exerted by a photon gas is proportional to its momentum; however, the momentum of a photon simply equals its energy divided by its velocity. Consequently, the pressure exerted by a photon gas should be only one-half that of a particle gas, and, in terms of its energy density, would be given by

$$P = \frac{u'}{3} \tag{10-98}$$

Thus the quantity $u'/3$ represents the mechanical pressure exerted on a reflecting surface by a photon gas in equilibrium with an enclosed cavity. It should be kept in mind for electromagnetic radiation in an enclosure that the energy density $u'$ depends upon the temperature of the walls of the enclosure, but is independent of the volume. If the volume of an enclosure is enlarged at constant temperature, more radiation is emitted, so that the energy density remains constant. This is diametric to the situation for an ideal gas. If the volume of an ideal gas is altered at constant temperature, the energy density is like-wise altered.

Consider now a cavity in which all changes are reversible, and the only work interaction is due to a change in the volume. The basic $T \, dS$ equation

$$T \, dS = dU + P \, dV \tag{7-9}$$

is applicable. Also, $U = Vu'$ and $P = u'/3$. By the use of the two latter expressions, one obtains the equation

$$T \, dS = V \, du' + \frac{4}{3} u' \, dV = V \frac{du'}{dT} dT + \frac{4}{3} u' \, dV$$

Division of the above equation by $T$ and application of the test for exactness (Sec. 13-1) to the right-hand side yields

$$\frac{1}{T} \frac{du'}{dT} = \frac{4}{3} \left( \frac{1}{T} \frac{du'}{dT} - \frac{u'}{T^2} \right)$$

Rearrangement leads to

$$\frac{du'}{u'} = 4 \frac{dT}{T}$$

and this integrates to

$$u' = aT^4 \tag{10-99}$$

This result is referred to as Stefan's law, where $a$ is a constant. The major qualitative consequence of this law is that the energy density of radiation rises very rapidly with an increase in temperature.

On the basis of Eq. (10-99), it is easily seen that the following two equations are also valid for electromagnetic radiation:

$$P = \frac{aT^4}{3} \tag{10-100}$$

$$S = \tfrac{4}{3}aVT^3 \tag{10-101}$$

Equation (10-100) indicates that the radiation pressure might become quite significant at high temperatures, e.g., in the vicinity of the sun. The constant $a$ will be evaluated shortly by means of quantum statistics.

In order to determine the energy distribution of a photon gas, it is necessary to find the number of photons which have a frequency between $v$ and $v + dv$. A photon can be represented in classical electromagnetic theory as a standing wave. The question arises as to how many standing waves there are with a frequency between $v$ and $v + dv$ in an enclosure of volume $V$. In a one-dimensional analysis any frequency must be some integral multiple of the fundamental frequency $c/2l$, where $c$ is the speed of light and $l$ is the length of the vessel. The factor $c/2l$ applies equally well in all three coordinate directions. Hence, in general, any frequency $v_i$ can be represented by the relation

$$v_i^2 = \left(\frac{ck_x}{2l_x}\right)^2 + \left(\frac{ck_y}{2l_y}\right)^2 + \left(\frac{ck_z}{2l_z}\right)^2$$

where $k_x$, $k_y$, and $k_z$ are integers starting with unity. As a result of the above relation, the allowed frequencies can be considered as points in a lattice where the spacing of the lattice points is $c/2l_x$ in the $x$ direction, etc. In such a "frequency space" there are on the average $8l_x l_y l_z/c^3$, or $8V/c^3$ points in a unit volume. Consider now the number of points which lie in a spherical shell of radius $v$ to $v + dv$ in the positive octant. The volume of this shell is $\frac{1}{8}(4\pi v^2 \, dv)$. The total number of standing waves $dg_i$ in this octant of frequency space equals the number of points per unit volume times the volume. That is,

$$dg_i = \frac{8V}{c^3}\left(\frac{1}{8}\right)(4\pi v^2 \, dv) = \frac{4\pi V v^2 \, dv}{c^3} \qquad (10\text{-}102)$$

Now if we assume that photons behave as bosons, i.e., they fulfill the Bose–Einstein distribution law, then

$$n_i = \frac{g_i}{e^{hv/kT} - 1} \qquad (10\text{-}32)$$

For a differential range of frequencies, $n_i$ and $g_i$ can be replaced by $dn_i$ and $dg_i$, respectively. Upon substituting the equation for $dg_i$, Eq. (10-102), into the modified BE distribution function, we find that the mean number of photons with frequencies between $v$ and $v + dv$ is given by

$$dn_i = \frac{8\pi V v^2 \, dv}{c^3}\frac{1}{e^{hv/kT} - 1} \qquad (10\text{-}103)$$

The factor 8 appears in Eq. (10-103), rather than 4, since the polarization of light waves leads to twice the number of standing waves.

The energy of a photon is $hv$, and the total radiant energy in a given frequency interval is $hv \, dn_i$. Consequently, the energy density in the same frequency interval is

$$\frac{hv \, dn_i}{V} = u_v' \, dv$$

Thus

$$u_v' = \frac{8\pi v^3 h}{c^3}\frac{1}{e^{hv/kT} - 1} \qquad (10\text{-}104)$$

It is customary to express the radiant density $u_v'$ in terms of the wavelength $\lambda$. Note that $v = c/\lambda$, and $dv = -c \, d\lambda/\lambda^2$. Transformation of the quantity $u_v' \, dv$ to the energy density per unit wavelength interval $u'_\lambda$ leads to the following result (neglecting the negative sign):

$$u'_\lambda = \frac{8\pi hc\lambda^{-5}}{e^{hc/\lambda kT} - 1} \tag{10-105}$$

The famous equation (10-105) is Planck's energy-density distribution law for blackbody radiation. This relation has been experimentally confirmed, and the agreement validates the assumption that photons may be considered as bosons when making quantum-mechanical calculations. In the original experimental work the constant $h$ was an empirical constant. The measurement of the distribution of radiant energy as a function of wavelength provides a means of evaluating Planck's constant. Figure 10-8 is a plot of the energy density per unit wavelength $u'_\lambda$ as a function of wavelength. The energy density maximizes at some wavelength, and then falls off rapidly as the wavelength increases.

The mean, or average, energy density is found by integrating the energy density per unit of wavelength over all possible wavelengths. Thus

$$u' = \int_0^\infty u'_\lambda \, d\lambda = 8\pi hc \int_0^\infty \frac{1}{e^{hc/\lambda kT} - 1} \frac{d\lambda}{\lambda^5}$$

By letting $x = hc/\lambda kT = K/\lambda T$, it may be shown that the above expression reduces to the form

$$u' = 8\pi hc \frac{T^4}{K^4} \int_0^\infty \frac{x^3 \, dx}{e^x - 1} \tag{10-106}$$

The integral itself has a value of $\pi^4/15$. Hence the average energy density becomes

$$u' = \frac{8\pi^5 (kT)^4}{15c^3 h^3} = aT^4 \tag{10-107}$$

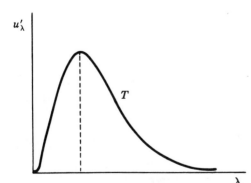

Figure 10-8 Plot of Planck's energy-density distribution law for blackbody radiation.

This equation has the same form as Eq. (10-99). However, we are now in a position to evaluate the constant. On the basis of Eq. (10–107), we find that

$$a = \frac{8\pi^5 k^4}{15c^3 h^3} = 7.5 \times 10^{-20} \text{ N·cm/(cm}^3)(^\circ K)^4 \tag{10-108}$$

Equations (10-105) and (10-107), obtained on the basis of BE statistics, are extremely useful in the study of blackbody radiation.

## 10-18 THE EINSTEIN AND DEBYE MODELS OF A MONATOMIC SOLID

By the early 1900s experimental evidence had accumulated which indicated that the specific heats of monatomic solids fell off rapidly as the temperature approached absolute zero and that the specific heat became zero at $0^\circ K$. A number of empirical relations can be proposed which enable one to fit the experimental data with suitable equations. One of the earliest attempts to predict theoretically the variation of the specific heat of monatomic solids with temperature was made by Einstein. As an idealization, we consider the crystal as a perfect cubic lattice, and the oscillations of each atom in the lattice may be resolved into the three orthogonal coordinates. The amplitudes of the three vibrations are assumed to be simple harmonic, and all have the same frequency. Hence a crystalline solid consisting of $n$ atoms is represented by $3n$ linear harmonic oscillators which are independent and are distinguishable by means of their position in the lattice structure. The distinguishability implies the applicability of the MB distribution law.

The specific-heat contribution by a system of linear harmonic oscillators of frequency $v$ has already been evaluated in Sec. 10-15 in terms of Eq. (10-86). Since we are dealing with $3n$ oscillators in the present case, the molar specific heat of a monatomic solid under the above idealizations becomes

$$c_v = \frac{3R_u e^x x^2}{(e^x - 1)^2} \tag{10-109}$$

where $x = hv/kT$. As $T$ increases, this function approaches the limiting value of $3R_u$, which is in good agreement with experimental evidence. In addition, the value of $c_v$ approaches zero as $T$ approaches zero. However, the variation of $c_v$ falls off too rapidly as the temperature is lowered when a comparison is made between the theoretical equation and experimental data. Nevertheless, the Einstein formula was one of the earliest attempts to use quantum theory to predict the behavior of matter.

To formulate a more correct picture of a real solid, one must recognize that the oscillators are neither harmonic nor alike. Instead, the theory of crystal vibration predicts a broad spectrum of frequencies which vary from the lowest value $v_1$

to the maximum value $v_m$. The distribution of frequencies varies with the temperature. Debye proposed an approximate expression for this distribution by treating a solid as an elastic medium. The lowest $3n$ stable modes of vibration of an elastic, continuous solid were taken as the required independent modes. On this basis Debye was able to derive the following expression for the internal energy of vibration of a solid medium:

$$U = 9RT\left(\frac{T}{\theta_D}\right)^3 \int_0^{x_m} \frac{x^3 \, dx}{e^x - 1} \qquad (10\text{-}110)$$

where $x = hv/kT$, $\theta_D = hv_m/k$, and $v_m$ again is the maximum allowed frequency. The specific heat at constant volume is found from the definition $c_v = (\partial U/\partial T)_v$. Carrying out the required differentiation, we obtain

$$c_v = 9R\left(\frac{T}{\theta_D}\right)^3 \int_0^{x_m} \frac{x^4 e^x \, dx}{(e^x - 1)^2} = f\left(\frac{T}{\theta_D}\right)$$

Through integration by parts, one is able to transform the above integral into two terms. The result is

$$c_v = \frac{36R}{x_m^3} \int_0^{x_m} \frac{x^3 \, dx}{e^x - 1} - \frac{9Rx_m}{e^{x_m} - 1} \qquad (10\text{-}111)$$

The integral in the first term, which also appears in the expression for $U$, is frequently termed Debye's integral. The value of the integral cannot be determined except by numerical calculation. $\theta_D$ is known as the Debye temperature of a solid.

Table 10-4 lists values of $c_v/3R_u$ as a function of $\theta_D/T$, which result from Eq. (10-111). As $T/\theta_D$ becomes large, the value of $c_v$ approaches the classical value

**Table 10-4 Debye specific heat function $c_v/3R$**

| $\theta_D/T$ | $c_v/3R$ | $\theta_D/T$ | $c_v/3R$ |
|---|---|---|---|
| 0 | 1.0000 | 5.5 | 0.3133 |
| 0.5 | 0.9876 | 6.0 | 0.2656 |
| 1.0 | 0.9517 | 6.5 | 0.2251 |
| 1.5 | 0.8960 | 7.0 | 0.1909 |
| 2.0 | 0.8254 | 7.5 | 0.1622 |
| 2.5 | 0.7459 | 8.0 | 0.1382 |
| 3.0 | 0.6628 | 9.0 | 0.1015 |
| 3.5 | 0.5807 | 10.0 | 0.0758 |
| 4.0 | 0.5031 | 11.0 | 0.0577 |
| 4.5 | 0.4320 | 13.0 | 0.0354 |
| 5.0 | 0.3686 | 15.0 | 0.0231 |

*Note:* When $\theta_D/T > 16$, $c_v/3R = 77.93(T/\theta_D)^3$.

**Table 10-5 Debye temperatures $\theta_D$ for a few elements, in degrees Kelvin**

| | | | |
|---|---|---|---|
| Aluminum | 390 | Lead | 90 |
| Beryllium | 900 | Magnesium | 290 |
| Carbon (diamond) | 1860 | Nickel | 375 |
| Copper | 315 | Platinum | 225 |
| Iron | 450 | Silver | 212 |

of $3R_u$. At extremely low temperatures $[(T/\theta_D) < 0.1]$ the second term in Eq. (10-111) approaches zero. The integral in the first term becomes the same as that which appeared in the discussion of blackbody radiation in Sec. 10-17. Since the value of the integral with limits from zero to infinity is $\pi^4/15$, the equation for $c_v$ at small temperatures is approximated by

$$c_v = \frac{36\pi^4 R}{15x_m^3} = 233.8R \left(\frac{T}{\theta_D}\right)^3 \tag{10-112}$$

This expression is called Debye's $T$-cubed law, since the specific heat of a monatomic solid at low temperatures is approximated fairly accurately as a cubic function of the absolute temperature. As noted above, this equation is generally valid when $T < 0.1\theta_D$. Typical values of $\theta_D$ for some elements are given in Table 10-5. On the basis of these data it is seen that temperatures less than $30°$K are frequently required to validate the use of the Debye $T$-cubed law.

From experimental measurements it has been found that many materials satisfactorily follow the Debye equation, including elements and simple compounds. The $T$-cubed law has proved invaluable for the extrapolation of $c_v$ data of crystals from temperatures in the liquid-hydrogen region ($20°$K) down to absolute zero. However, a large number of discrepancies do exist, as might be expected. The type of specific crystalline lattice structure would affect the frequency spectrum for example. Also, the presence of an electron-gas specific heat (Sec. 10-19) must be taken into account at temperatures close to absolute zero. Nevertheless, the Debye model for evaluating the internal energy and the specific heat of a crystalline solid is an excellent example of the use of quantum theory to predict the properties of a substance.

## 10-19 THE ELECTRON GAS

The free, or conducting, electrons in a metal behave essentially as a gas. Consequently, it is common to speak of them collectively as an electron gas. If these particles are free to move about, their behavior is similar to that of the translational motion of an ideal monatomic gas. Such a direct analogy to an ideal gas is misleading, however, since these classical particles would contribute the amount $3R/2$ to the specific heat of a metal, if the vibrational and electronic energies were separable. This would double the Dulong-Petit value of $3R/2$ for metals, which is contrary to experience. If the conducting electrons were subject to quantum theory, however, we should have a quite different situation. In place of the classical MB statistics we now assume that free electrons fulfill FD statistics. That is, any energy state can have associated with it no more than one electron. Nevertheless, the allowed energies and degeneracies of electrons will still be those associated with translational motion.

$$\epsilon_i = (k_x^2 + k_y^2 + k_z^2)\frac{h^2}{8mL^2} = \frac{r^2 h^2}{8mL^2} \tag{10-56}$$

In addition, the degeneracy of the translational energy mode is found from

$$dg = 4\pi\left(\frac{L}{h}\right)^3 (2m)^{3/2}(\epsilon)^{1/2}\,d\epsilon \tag{10-89}$$

In the equation for $dg$ it should be noted that an extra factor of 2 has been included to account for the two possible spins of an electron for a given energy state. The FD statistics are represented by the relation

$$n_i = \frac{g_i}{Ae^{B\epsilon_i} + 1} \tag{10-33}$$

where $B = 1/kT$. Replacing $n_i$ and $g_i$ by $dn$ and $dg$, we obtain, for the number of particles $dn_\epsilon$ within the energy range from $\epsilon$ to $\epsilon = d\epsilon$,

$$dn_\epsilon = \frac{\pi}{2}\frac{(8m)^{3/2}\,V}{h^3}\frac{\epsilon^{1/2}\,d\epsilon}{Ae^{B\epsilon_i} + 1} \tag{10-113}$$

The parameter $A$ is determined by noting that the integral of expression (10-113) leads to the total number of particles $n$. Hence

$$n = \frac{\pi}{2}\frac{(8m)^{3/2}V}{h^3}\int_0^\infty \frac{\epsilon^{1/2}\,d\epsilon}{Ae^{B\epsilon_i} + 1} \tag{10-114}$$

Although this equation relates the constant $A$ to the total number of particles, it is difficult to carry out the indicated integration.

A convenient method of approximating the constant $A$ is as follows: Consider the situation where all the energy states are filled from the ground state up to a maximum value $\epsilon^*$. This maximum energy is called the Fermi energy, corresponding to the Fermi level. (The energy states must be filled with two electrons at a time, because of the spin factor.) To evaluate $\epsilon^*$ it is necessary simply to integrate Eq. (10-89) for $dg$ from $\epsilon = 0$ to $\epsilon = \epsilon^*$. The integral of $dg$ between these limits is, of course, the total number of particles $n$, since all the energy states are filled by definition of the Fermi level. Consequently, from Eq. (10-89), we find that

$$n = \frac{\pi}{2}\frac{(8m)^{3/2}V}{h^3}\int_0^{\epsilon^*} \epsilon^{1/2}\,d\epsilon \tag{10-115}$$

and therefore

$$\epsilon^* = \frac{h^2}{8m}\left(\frac{3n}{\pi V}\right)^{2/3} \tag{10-116}$$

Note that the Fermi energy depends upon the density of electrons in the metal, and is independent of the temperature. As a typical value, consider that copper at room temperature has one free electron per atom. The Fermi energy of such a material would be

$$(\epsilon^*)_{Cu} = \frac{(6.624 \times 10^{-27})^2}{8(9.1 \times 10^{-28})} \left[ \frac{3(6.025 \times 10^{23})(8.9)}{3.14(63)} \right]^{2/3}$$

$$= 1.13 \times 10^{-18} \text{ J} = 7.07 \text{ eV}$$

The quantities $n/V$ found in Eqs. (10-114) and (10-116) may now be equated. This leads to the relation

$$1 = \frac{3}{2(\epsilon^*)^{3/2}} \int_0^\infty \frac{\epsilon^{1/2} \, d\epsilon}{A e^{B\epsilon} + 1}$$

The constant $A$ is now related to the Fermi energy $\epsilon^*$. One further reduction in this equation may be made by substituting the defined quantity $x = \epsilon/kT$, and $d\epsilon = kT \, dx$. Finally,

$$1 = \frac{3}{2} \left( \frac{kT}{\epsilon^*} \right)^{3/2} \int_0^\infty \frac{x^{1/2} \, dx}{A e^x + 1} \tag{10-117}$$

To integrate expression (10-117) it is necessary to make some sort of approximation. As a first approximation we shall assume that $A e^x \ll 1$. When this is done, the denominator of Eq. (10-117) reduces to unity and the integration becomes quite simple. The value of the integrand approaches zero as $x$ becomes large. Hence we shall take the upper limit of the integral to be $\ln(1/A)$ and assume that the value of the integrand above this limit is negligible. As a result, the integral in Eq. (10-117) reduces to $\frac{2}{3}[\ln(1/A)]^{3/2}$, and the entire equation reduces to

$$A = e^{-\epsilon^*/kT} \tag{10-118}$$

For this first approximation to be good, it is necessary that $\epsilon^* \gg \epsilon$. That is, most of the electrons must have energies which are considerably less than that of the Fermi level.

When the first approximation for $A$ is substituted into the energy-distribution equation, Eq. (10-113) becomes

$$\frac{dn_\epsilon}{d\epsilon} = \frac{\pi(8m)^{3/2} V \epsilon^{1/2}}{2h^3(e^{B(\epsilon - \epsilon^*)} + 1)} \tag{10-119}$$

Now, as the temperature approaches absolute zero, the exponential term in the denominator approaches $-\infty$ for all cases where $\epsilon < \epsilon^*$. Under this circumstance the quantity $dn_\epsilon/d\epsilon$ varies as $\epsilon^{1/2}$. However, when $\epsilon > \epsilon^*$, the exponential term is $+\infty$ for a zero absolute temperature, and $dn_\epsilon/d\epsilon$ is zero. The behavior of the energy distribution at absolute zero is shown in Fig. 10-9. The shape of the curve indicates a significance of the Fermi level of energy. It is the maximum energy associated with any electron when the material is held at absolute zero temperature. It is important to note that the energies of electrons are neither zero nor constant at $0°K$, but vary from zero to a maximum value at the Fermi level. Since

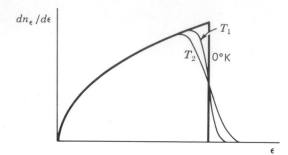

**Figure 10-9** Energy distribution in an electron gas.

at this temperature all the electrons have an energy less than or equal to the Fermi energy, the approximation for the constant $A$ given by Eq. (10-118) is quite good. It is of interest also to calculate the average energy of the electron gas at absolute zero. This is found by employing the relationship

$$\epsilon_{avg} = \frac{\int \epsilon \, dn_\epsilon}{\int dn_\epsilon} = \frac{\int_0^{\epsilon*} \epsilon^{3/2} \, d\epsilon}{\int_0^{\epsilon*} \epsilon^{1/2} \, d\epsilon} = \frac{3}{5} \epsilon* \tag{10-120}$$

This average energy of an electron in an electron gas at $0°K$ is extremely high. A monatomic, ideal gas would have to be heated to thousands of degrees to have the same average energy.

As the temperature rises, these simplified results are no longer valid. A few electrons with energies close to $\epsilon*$ at $T = 0°K$ are excited to energies slightly above the Fermi level as the temperature increases. Hence the distribution curve experiences a slight rounding off at these higher temperatures, as indicated by the lighter lines $T_1$ and $T_2$ in Fig. 10-9. Nevertheless, the number of electrons which move up to higher energies is so small that the electron-gas specific heat is hardly affected, as we shall soon see.

Sommerfeld found by means of a better approximation for the constant $A$ at higher temperatures that the average energy of the electron gas could be expressed by

$$\epsilon_{avg} = \frac{3}{5} \epsilon* \left[ 1 + \frac{5\pi^2}{12} \left( \frac{kT}{\epsilon*} \right)^2 + \cdots \right]$$

The total internal energy is $n\epsilon_{avg}$; consequently, the specific heat at constant volume contributed by the electron gas would be

$$c_{v, \, el} = \left( \frac{\partial U}{\partial T} \right)_v = \frac{n\pi^2 k^2 T}{2\epsilon*}$$

This approximation to the electron-gas specific heat assumes that there is one free electron per atom. For 1 mole of the substance, the specific heat becomes $R_u \pi^2 kT/2\epsilon*$. It is frequently convenient to define a Fermi characteristic temperature $\theta*$ by the quantity $\epsilon*/k$. As a result the molar specific heat of an electron gas based on the Sommerfeld approximation is

$$c_{v,\,el} = \frac{\pi^2}{2}\left(\frac{T}{\theta*}\right)R_u \tag{10-121}$$

On the basis that the Fermi energy of copper is, roughly, 7 eV, the Fermi characteristic temperature for copper is about 80,000°K. Thus Eq. (10-121), when applied to copper, becomes

$$c_{v,\,el} = (6 \times 10^{-5}T)R_u$$

At room temperature the electron-gas contribution to the specific heat of copper would be, roughly, $0.02R_u$. This compares with a specific heat of the vibrating atoms of $\frac{3}{2}R_u$; consequently, the contribution by the electron gas is negligible in this temperature range. At extremely low temperatures, of the order of 10°K or less, the effect indicated by Eq. (10-121) must be added to the Debye $T$-cubed term found in Sec. 10-18, since both are of the same order of magnitude.

## REFERENCES

Crawford, F. H.: "Heat, Thermodynamics, and Statistical Physics," Harcourt, New York, 1963.
Davidson, N.: "Statistical Mechanics," McGraw-Hill, New York, 1962.
Lee, J. F., F. W. Sears, and D. L. Turcotte: "Statistical Thermodynamics," Addison-Wesley, Reading, Mass., 1963.

## PROBLEMS

**10-1** Compute the arithmetic mean speed and the rms speed for the following groups of particles:

|  |  | $c_i$, cm/sec | | | | |
|---|---|---|---|---|---|---|
|  |  | 0 | 1 | 2 | 3 | 4 |
| (a) | $n_i$ | 0 | 2 | 4 | 5 | 3 |
| (b) | $n_i$ | 1 | 3 | 5 | 7 | 3 |

**10-2** Determine the mean translational kinetic energy of a gas molecule, in joules, (a) at 300°K, and (b) at 3000°K.

**10-3** Compute the rms speed, in m/s, of argon atoms (a) at 100°K and (b) at 900°K.

**10-4** Compute the rms speed, in m/s, at 300°K for the gases (a) hydrogen and (b) mercury.

**10-5** Determine the temperature, in °K, at which the mean linear kinetic energy of a particle will be equal to that of a single charged particle of the same mass that has been accelerated from rest through a potential difference of 1 V.

**10-6** Derive the FD distribution law by starting with the BP equation and employ Eq. (10-11) as a valid expression for the thermodynamic probability for indistinguishable particles restricted to the Pauli exclusion principle. Maximize ln $W$, and reduce it to a usable form by employing the Stirling approximation.

**10-7** How many macrostates would there be for Prob. 9-8 if FD statistics were applicable? Why?

**10-8** Sketch the Maxwell-Boltzmann, Bose-Einstein, and Fermi-Dirac macrostates for a collection of three particles distributed among three energy levels, where the highest level has two states associated with it.

**10-9** Two particles are to be distributed among four energy states of a given energy level. Sketch the possible distributions for FD, BE, and MB statistics, and check the results against the equations for $W$.

**10-10** Two particles are to be distributed among three energy states of a given energy level. Sketch the possible energy distributions for FD, BE, and MB statistics, and then check against the equations for $W$ presented for these three cases.

**10-11** Derive a statistical equation for the enthalpy $H$ in terms of the variables $z$, $T$, and $V$.

**10-12** Determine the magnitude of the spacing of the lower energy levels of the translational mode for helium gas under standard conditions, in J/g·mol, if $L$ is 1 m.

**10-13** Determine the value of $k_x$ (the translational quantum number) for the average energy of translation of an argon atom at 300°K in a box 1 m long.

**10-14** Determine the approximate magnitude of the spacing of the first three energy levels for rotation, in J/g·mol, for (a) oxygen, (b) hydrogen, and (c) chlorine.

**10-15** Determine the approximate magnitude of the spacing of the first two energy levels for vibration, in J/g·mol, for (a) oxygen, (b) hydrogen, and (c) chlorine.

**10-16** Determine the fraction of the molecules in the lowest translational level at 300°K and 1 atm for a molecule whose mass is 1 g/g·mol, if $L$ is 1 m.

**10-17** Determine the maximum value of $n_i/g_i$ for $He^4$ gas at 1 atm and 5°K. Show that MB statistics would not be applicable in this temperature range.

**10-18** Determine the translational partition function for helium at (a) 300°K, and (b) 1000°K and contained in a box with a volume of 1 m³.

**10-19** Determine the rotational partition function for $Cl_2$ and HI gases at (a) 300°K, and (b) 1000°K at 1 atm.

**10-20** Calculate the rotational partition function of HBr gas at 100°K. Is Eq. (10-72) valid under this circumstance?

**10-21** Calculate the vibrational partition function for HI gas at (a) 900°K, and (b) 1200°K and 1 atm, assuming behavior as a harmonic oscillator.

**10-22** Calculate the rotational partition function of gaseous diatomic chlorine at (a) 300°K, (b) 600°K, and (c) 900°K.

**10-23** Plot the distribution of molecules among the various rotational levels for carbon monoxide, CO, at (a) 100°K, and (b) 200°K.

**10-24** For gaseous HI, find the temperature at which exactly 50 percent of the molecules have a rotational quantum number greater than 3.

**10-25** Prove analytically that the maximum of the curve in Fig. 10-5 occurs when $\epsilon = kT/2$.

**10-26** Calculate the relative populations in the rotational mode for the $j = 4$ and $j = 5$ levels at (a) 300°K, (b) 1000°K, and (c) 3000°K, for HBr.

**10-27** Determine the values of $n_i/n$ for the first seven vibrational levels of diatomic chlorine at (a) 300°K, (b) 600°K, and (c) 1000°K.

**10-28** Determine the probabilities of the first seven vibrational energy levels of diatomic iodine gas at (a) 300°K, and (b) 500°K.

**10-29** Determine the values of $n_i/n$ among the various vibrational levels of diatomic bromine gas at (a) 300°K, (b) 700°K, and (c) 1500°K.

**10-30** The ground level of energy of the chlorine atom has a degeneracy of 4. The next highest level of energy has a degeneracy of 2 and lies 881 cm⁻¹ above the ground level. Calculate the ratio of the number of atoms in the highest level to those in the ground level (a) at 100°K, (b) at 1000°K, and (c) at 5000°K.

**10-31** For diatomic oxygen, determine the fraction of the particles that have a vibrational quantum number less than 4, at (a) 2500°K, and (b) 3000°K.

**10-32** Calculate the most probable values of $n_i/n$ for the first four energy levels of vibration of diatomic chlorine at (a) 1000°K, (b) 1500°K, and (c) 2000°K.

**10-33** Plot the most probable values of $n_i/n$ for the rotational energy levels of CO at 300°K, using only even values of the quantum number $j$, up to $j = 20$.

**10-34** Calculate and plot the rotational internal energy versus temperature for HCl from 0 to 100°K, using the appropriate method for this temperature range. (Use 20°K intervals.)

**10-35** Compute the various contributions to $c_p$ and $s$ at (a) 300°K and (b) 2000°K for diatomic chlorine at 1 atm. Also compute the total value for each property in $J/(g \cdot mol)(°K)$.

**10-36** Calculate $c_p$ for $Br_2$ at temperatures of 200°K and 700°K.

**10-37** Compute the translational, rotational, and vibrational contributions to $c_p$ for CO at (a) 100°K, (b) 300°K, and (c) 1000°K.

**10-38** Calculate the molar entropy of gaseous argon at 1 atm and (a) 300°K and (b) 1000°K.

**10-39** Calculate the translational molar entropy of nitrogen at 800°K and 1 atm.

**10-40** Determine the rotational and vibrational contributions to the entropy of nitrogen gas at 1 atm and 800°K. Compare with results given in Probl. 10-39.

**10-41** Calculate the entropy of HI gas at 1 atm and 1100°K in $J/(g \cdot mol)(°K)$.

**10-42** Determine the vibrational contribution to $c_v$ in $J/(g \cdot mol)(°K)$ for (a) $H_2$ and (b) $Cl_2$ gas at 300, 1000, and 3000°K.

**10-43** At what temperature would the vibrational specific heat of $H_2$ approach 90 percent of its classical value?

**10-44** Estimate the values of the constant-volume specific heat and the entropy at 500°K and 1 atm pressure for carbon monosulfide (CS). The moment of inertia is $3.41 \times 10^{-39}$ g·cm$^2$, and the vibrational wave number is 1285.1 cm$^{-1}$.

**10-45** Estimate the constant-pressure specific heat and the entropy of carbon disulfide ($CS_2$) at 1000°K and 1 atm pressure by assuming behavior as a rigid rotator-harmonic oscillator. The moment of inertia of the linear molecule is $25.68 \times 10^{-39}$ g·cm$^2$, and the fundamental vibrational wave numbers are 396.8, 396.8, 658, and 1532.5 cm$^{-1}$.

**10-46** Derive Eq. (10-93) by differentiation techniques.

**10-47** Derive Eq. (10-95) for the rms speed.

**10-48** What percent of the classical value of $c_v$ for a solid is obtained then $T/\theta_D$ is unity, using the Debye function?

**10-49** The experimental value for $c_v$ of diamond at 293°K is 6.03 $J/(g \cdot mol)(°C)$. Calculate $c_v$ at this temperature based on the Debye function.

**10-50** Determine the temperature for carbon (diamond) at which its $c_v$ value is (a) 60 percent, (b) 90 percent, and (c) 98 percent of its classical value.

**10-51** The specific heat of aluminum is 0.439 $J/(g \cdot atom)(°C)$ at 23°K. Calculate the value at 273°K, and compare with the experimental value of 23.47 $J/(g \cdot atom)(°C)$.

**10-52** The specific heat of aluminum is 18.93 $J/(g \cdot atom)(°C)$ at 173°K. What is the characteristic Debye temperature based on this experimental value? Compare it with that calculated from data in Prob. 10-51.

**10-53** Calculate the Fermi energy for (a) silver, and (b) aluminum, in electronvolts.

**10-54** To what temperature must argon gas be heated so that the particles in translation have the same average energy as an electron gas in copper at absolute zero temperature?

**10-55** Calculate the vibrational specific heat and the electron-gas specific heat of (a) aluminum, and (b) silver at 5°K in $J/(g \cdot atom)(°C)$.

**10-56** Assuming one conduction electron per atom and a metal density of 10.5 g/cm$^3$, evaluate the lattice and electronic contributions to $c_p$ of silver at (a) 10°K, (b) 30°K, and (c) 300°K.

# ELEVEN

## NONREACTIVE IDEAL-GAS MIXTURES

The basic thermodynamic laws introduced so far are of general validity. In the application of these laws to closed and open systems, however, we have dealt primarily with systems containing a single chemical species. Analytical expressions, tables, and graphs have been presented which relate such properties as $P$, $v$, $T$, $u$, $h$, $s$, $c_v$, and $c_p$ for systems of a single component. In view of the fact that many engineering applications involve multicomponent systems, it is vital to have some understanding of methods of evaluating the properties of such systems.

## 11-1 SOME BASIC DEFINITIONS FOR GAS MIXTURES

A complete description of a multicomponent system requires a specification not only of two properties such as the pressure and the temperature, but also of the composition. It is apparent that properties such as $u$, $h$, $v$, and $s$ of a mixture are different for every different composition. Consequently, the tabulation of properties of mixtures is not too fruitful, because of the tremendous quantity of data required. One method of overcoming this difficulty is to devise rules for averaging the properties of the individual components, so that the resulting value is representative of the overall composition. This method has the advantage that the properties of mixtures are based on the properties of the individual chemical species, which in general will be known. Hence much of the data we have already investigated for single components will now be applicable to evaluating properties of mixtures. We shall restrict the initial investigation to the properties of ideal-gas mixtures.

In the study of thermodynamic systems there are two general bases for measuring the mass within the system. One of these involves the use of units like kilograms or pounds. The other basis for mass is the mole unit, i.e., kilogram-moles or pound-moles. Thus, in dealing with gas mixtures, we may analyze systems from either viewpoint. When the analysis of a gas mixture is made on the basis of mass or weight, it is called a *gravimetric* analysis. For a nonreacting gas mixture it is apparent that the total mass of the mixture $m_m$ is the sum of the masses of each of the $k$ components. That is,

$$m_m = m_1 + m_2 + m_3 + \cdots + m_k = \sum_{i=1}^{k} m_i \tag{11-1}$$

The mass fraction $mf_i$ of the $i$th component is defined as

$$mf_i \equiv \frac{m_i}{m_m} \tag{11-2}$$

The sum of the mole fractions of all the components in a mixture also is unity. If an analysis of a gas mixture is based on the number of moles of each component present, the analysis is termed a *molar* analysis. The total number of moles $N_m$ for a mixture is given by

$$N_m = N_1 + N_2 + N_3 + \cdots + N_k = \sum_{i=1}^{k} N_i \tag{11-3}$$

and the mole fraction of any component $y_i$ is defined as

$$y_i \equiv \frac{N_i}{N_m} \tag{11-4}$$

The sum of the mole fractions of all the components in a mixture also is unity. From the definition of the molar mass (or molecular weight) $M_i$, the mass of a component is related to the number of moles of that component by

$$m_i = N_i M_i \tag{11-5}$$

If Eq. (11-5) is substituted into Eq. (11-1) for each of the components, then

$$m_m = N_1 M_1 + N_2 M_2 + N_3 M_3 + \cdots + N_k M_k = N_m M_m \tag{11-6}$$

where $M_m$ is an average, or apparent, molar mass, or molecular weight, for the mixture. The solution of Eq. (11-6) for $M_m$ yields, in terms of $y_i$,

$$M_m = \sum_{i=1}^{k} y_i M_i \tag{11-7}$$

The average molecular weight of a gas mixture, then, is the sum over all the components of the mole fraction times the molecular weight.

As an example of this latter relationship, the average molar mass of atmospheric air can be approximated in the following way. If we neglect the presence of other gas components, air is composed of roughly 78.1 percent $N_2$, 21.0 percent $O_2$, and 0.9 percent argon. Substitution of these values into Eq. (11-7) yields

$$M_{air} = 0.781(28.0) + 0.210(32.0) + 0.009(40) = 28.95$$

When other gases are included in the calculation for the average composition of air, the commonly quoted value of 28.97 results.

## 11-2 $PvT$ Relationships for Mixtures of Ideal Gases

The $PvT$ relationship for a mixture of gases is usually based on two rules. The first of these is known as *Dalton's law* of additive pressures. This rule states that the total pressure exerted by a mixture of gases is the sum of the component pressures $p_i$, each measured alone at the temperature and volume of the mixture. It is inferred from this rule that each gas acts independently of the others present in the mixture. Hence Dalton's rule can be written in the form

$$P = p_1 + p_2 + p_3 + \cdots + p_k = \sum_i p_i \tag{11-8}$$

where $p_i$ is the component pressure of the $i$th component. A physical representation of the additive-pressure rule is shown in Fig. 11-1 for the case of two gases $A$ and $B$.

It would be expected that ideal gases fulfill Dalton's rule exactly, since the concept of an ideal gas implies that intermolecular forces are negligible, and thus the gases do act independently of one another. The pressure exerted by an ideal gas in a gas mixture, in view of Dalton's rule, can be expressed as

$$p_i = \frac{N_i R_u T}{V} \tag{11-9}$$

where $T$ and $V$ are the absolute temperature and volume of the mixture, respectively. If the component pressures of each chemical species are now substituted into Eq. (11-8), we find that

$$P = \frac{N_1 R_u T}{V} + \frac{N_2 R_u T}{V} + \cdots + \frac{N_i R_u T}{V}$$

$$= (N_1 + N_2 + \cdots + N_i)\frac{R_u T}{V} = \frac{N_m R_u T}{V} \tag{11-10}$$

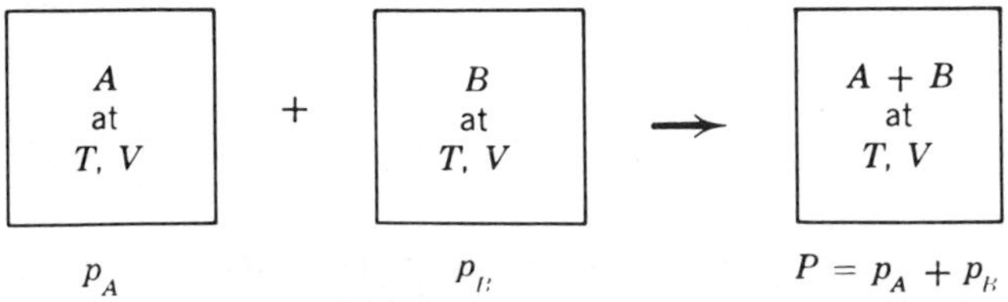

| $A$ at $T, V$ | $+$ | $B$ at $T, V$ | $\longrightarrow$ | $A + B$ at $T, V$ |
|---|---|---|---|---|
| $p_A$ | | $p_{I\!:}$ | | $P = p_A + p_{I\!:}$ |

**Figure 11-1** Schematic representation of Dalton's law of additive pressure.

Thus the gas mixture also follows the ideal-gas equation, as would be expected. Equation (11-10) has been confirmed experimentally for gases at relatively low pressures, which verifies Dalton's additive-pressure rule for mixtures of ideal gases. Note that the total pressure depends upon the total number of moles of gas present, but not on their identity. Dalton's rule may also be applied to mixtures of real gases as an approximation technique. However, the results will not necessarily agree with experiment, due to the influence of intermolecular forces between real-gas particles.

Recall from Eq. (11-6) that $N_m$ equals $m_m/M_m$. Making this substitution into the above equation, we obtain the following relation:

$$P = \frac{m_m}{M_m} \frac{R_u T}{V} = \frac{m_m T}{V} \frac{R_u}{M_m} = \frac{m_m R_m T}{V} \tag{11-11}$$

where $R_m$ is an apparent gas constant for the gas mixture on a mass basis and is defined as

$$R_m = \frac{R_u}{M_m} \tag{11-12}$$

The value of $M_m$ is obtained from Eq. (11-7). Equation (11-11) is the ideal-gas relation for a gas mixture on a mass basis, rather than the molar basis given by Eq. (11-10).

A relationship between the component pressure $p_i$ of an ideal-gas mixture and its mole fraction $x_i$ is found by dividing Eq. (11-9) by Eq. (11-10). This leads to

$$\frac{p_i}{P} = \frac{N_i R_u T/V}{N_m R_u T/V} = \frac{N_i}{N_m} = y_i$$

or

$$p_i = y_i P \tag{11-13}$$

In the thermodynamic literature the product $y_i P$ is frequently defined as the *partial* pressure $p_i'$ of a gas. Therefore, for a mixture of ideal gases, the component pressure $p_i$ and the partial pressure $p_i'$ are equal in value. For nonideal-gas mixtures in general this is not true. The partial pressure always has a value by definition, but this value is not necessarily the actual pressure $p_i$ exerted by the component of interest if the gases are not ideal.

Another interpretation or description of gas mixtures is that based on the *Amagat-Leduc law* of additive volumes. This law states that the total volume of a mixture of gases is the sum of the volumes that each gas would occupy if measured individually at the pressure and the temperature of the mixture. This may be expressed by the relation

$$V = V_1 + V_2 + V_3 + \cdots + V_k = \sum_{i=1}^{k} V_i \tag{11-14}$$

where $V_i$ is the volume of the $i$th component measured at the pressure and the temperature of the mixtures. This relation may be applied to both ideal-gas and real-gas mixtures, although in the latter case it is only approximately valid. When Amagat's law is applied to ideal gases in a mixture, the ideal-gas equation for each component becomes $PV_i = N_i R T$. Substitution of this equation into the Amagat-Leduc law results in an expected form, namely,

$$V = \frac{N_1 R_u T}{P} + \frac{N_2 R_u T}{P} + \cdots + \frac{N_k R_u T}{P}$$

$$= (N_1 + N_2 + \cdots + N_k) \frac{R_u T}{P}$$

$$= \frac{N_m R_u T}{P} \tag{11-15}$$

Now if the equation for $V_i$ is divided by Eq. (11-15), we find that

$$\frac{V_i}{V} = \frac{N_i R_u T/P}{N_m R_u T/P} = \frac{N_i}{N_m} = y_i \tag{11-16}$$

A physical interpretation of the component volume $V_i$ can be obtained for ideal-gas mixtures in the following way: Consider a mixture of two ideal gases in a system of volume $V$ with a total pressure $P$ and a temperature $T$. Hypothetically, the gases could be separated so that one species occupied a certain part of the volume, and the other species filled the remaining volume. The temperature and pressure of each separate gas would still be identical. The actual volume filled by a component volumes is the total volume of the closed system. A schematic representation of the additive-volume rule is shown in Fig. 11-2 for the case of two gases.

**Example 11-1M** A rigid tank contains 0.5 kg of oxygen at 3 bars and 40°C. Nitrogen is to be added to the tank until the pressure reaches 4 bars at the same temperature. Determine the mass of nitrogen required, in kilograms.

SOLUTION On the basis of Eq. (11-9) we may write, since $V$ and $T$ are the same for the two gases,

$$P_{O_2}/P_{N_2} = N_{O_2}/N_{N_2}$$

or $$N_{N_2} = N_{O_2}(p_{N_2}/p_{O_2})$$

All the quantities on the right are known. The value of the moles of oxygen is

$$N_{O_2} = m/M = (0.5 \text{ kg})/(32 \text{ kg/kg·mol}) = 0.0156 \text{ kg·mol}$$

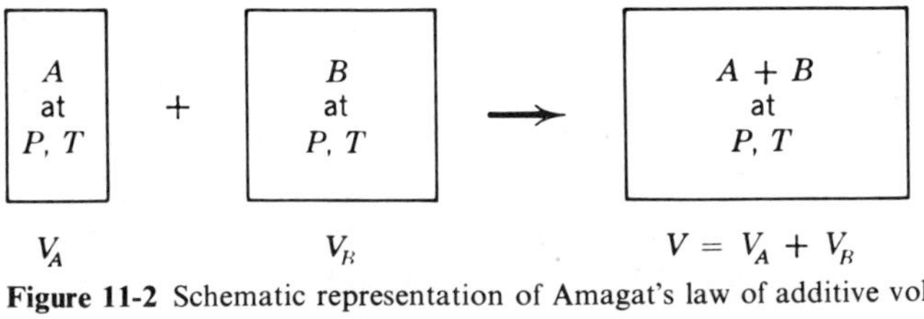

| A at $P$, $T$ | + | B at $P$, $T$ | → | A + B at $P$, $T$ |

$V_A$        $V_B$        $V = V_A + V_B$

**Figure 11-2** Schematic representation of Amagat's law of additive volumes.

The pressure of the oxygen does not change when nitrogen is added, therefore its partial pressure is 3 bars. Using Eq. (11-8),

$$p_{N_2} = P - p_{O_2} = 4 - 3 = 1 \text{ bar}$$

Hence,

$$N_{N_2} = 0.0156(1/4) = 0.0390 \text{ kg·mol}$$

As a result the mass of nitrogen added is

$$m_{N_2} = NM = 0.00390(28) = 0.109 \text{ kg}$$

**Example 11-1** A rigid tank contains 0.8 lb of oxygen at 30 psia and 75°F. Nitrogen is added to the tank until the pressure reaches 40 psia at the same temperature. Determine the mass of nitrogen added, in pounds.

SOLUTION On the basis of Eq. (11-9) we may write, since $V$ and $T$ are the same for the two gases, for the final state that

$$p_{O_2}/p_{N_2} = N_{O_2}/N_{N_2}$$

or

$$N_{N_2} = N_{O_2}(p_{N_2}/p_{O_2})$$

All the quantities on the right are known. The value of the moles of oxygen is

$$N_{O_2} = m/M = (0.8 \text{ lb})/(32 \text{ lb/lb·mol}) = 0.0250 \text{ lb·mol}$$

The pressure of the oxygen does not change when nitrogen is added, therefore its partial pressure is 30 psia. Using Eq. (11-8),

$$p_{N_2} = P - p_{O_2} = 40 \quad 30 = 10 \text{ psia}$$

Hence,

$$N_{N_2} = 0.0250(10/40) = 0.00625 \text{ lb·mol}$$

As a result the mass of nitrogen added is

$$m_{N_2} = NM = 0.00625(28) = 0.175 \text{ lb}$$

On the basis of Eqs. (11-13) and (11-16) it is evident that

$$\frac{p_i}{P} = y_i = \frac{N_i}{N_m} = \frac{V_i}{V} \tag{11-17}$$

Hence, for ideal-gas mixtures, the mole fraction, the volume fraction, and the ratio of component pressure to total pressure are all equal. The direct relationship between mole fractions and volume fractions enables one to convert volumetric analyses to mass analyses. The next two examples are illustrations of such conversion calculations. Note that a tabular type of calculation is quite convenient, where the calculations proceed from left to right.

**Example 11-2M** A gaseous fuel is composed of 20 percent $CH_4$, 40 percent $C_2H_6$, and 40 percent $C_3H_8$, where all percentages are by volume. Determine the gravimetric analysis of the fuel, the apparent molecular weight of the mixture, and the apparent gas constant for the mixture.

SOLUTION It is convenient to select 100 moles as a basis for the subsequent calculation. Since the volume fraction is also the mole fraction, if it is assumed that the gases are ideal, then the percentages given here are also the number of moles of each component per 100 mol of mixture. These values are listed in the second column of the accompanying table. The third column lists the molecular weight, or molar mass, of each. Recall that the mass of a component is given by $N_i M_i$ [Eq. (11-5)]. Consequently, the mass of each component per 100 mol of mixture is found by multiplying the value in column 2 by the value in column 3 for each species. The result is shown in column 4. The number at the bottom is the summation for the masses. The gravimetric analysis, then, is found simply by dividing the mass of each component by the total mass, as given by column 5. It should be noted that the gravimetric analysis shown by column 5 is significantly different from the volumetric, or molar, analysis of the fuel mixture. The sum of column 4 is the mass of mixture per 100 mol of mixture. Therefore the mass per mole must be $\frac{3280}{100} = 32.80$. But this is the definition of the apparent molar mass of a mixture. The gas constant on a mass basis may now be found.

(a) Metric: $R_m = \dfrac{R_u}{M_m} = \dfrac{0.08315}{32.80} = 2.54 \times 10^{-3} \text{ bar·m}^3/(\text{kg·mol})(^\circ\text{K})$

(b) USCS: $R_m = \dfrac{R_u}{M_m} = \dfrac{0.730}{32.80} = 0.0223 \text{ atm·ft}^3/(\text{lb·mol})(^\circ\text{R})$

Other values of $R_m$ in metric and USCS units may be evaluated by using other values of $R_u$ found in Table A-1.

| (1) Component | (2) Moles per 100 moles of mixture | (3) Molar mass | (4) Mass per 100 moles of mixture | (5) Mass analysis, percent |
|---|---|---|---|---|
| $CH_4$ | 20 | 16 | 320 | 9.76 |
| $C_2H_6$ | 40 | 30 | 1200 | 36.59 |
| $C_3H_8$ | 40 | 44 | 1760 | 53.65 |
| Total | | | 3280 | 100.00 |

**Example 11-2** See Example 11-2M for the appropriate calculations.

**Example 11-3M** An ideal-gas mixture consists of 10 percent hydrogen, 48 percent oxygen, and 42 percent carbon monoxide by mass. Determine the volumetric analysis in percent and the apparent molar mass.

SOLUTION In the table below the first three columns list the chemical species, the mass fraction, and the molar mass. Column 4 is the moles of each per unit mass of mixture, found by

| (1) Component | (2) Mass fraction, percent | (3) Molar mass | (4) Moles per unit mass of mixture | (5) Mole fraction, percent | (6) Volume analysis, percent |
|---|---|---|---|---|---|
| $H_2$ | 0.10 | 2 | 0.050 | 0.6250 | 62.50 |
| $O_2$ | 0.48 | 32 | 0.015 | 0.1875 | 18.75 |
| CO | 0.42 | 28 | 0.015 | 0.1875 | 18.75 |
| Total | | | 0.080 | 1.0000 | 100.00 |

dividing column 2 by column 3. The mole fraction of each constituent in column 5 is then determined by dividing each value in column 4 by the total in column 4. Since the volume fraction is equal to the mole fraction, the volumetric analysis in column 6 is based directly on the values in column 5.

Note that hydrogen, which is present in the smallest amount in the gravimetric analysis, has the largest percentage on a volumetric basis. The apparent molar mass is obtained from the last value in column 4.

$$M_m = \frac{1}{\text{moles/unit mass of mixture}} = \frac{1}{0.080} = 12.5$$

**Example 11-3** See Example 11-3M as an appropriate problem.

# 11-3 PROPERTIES OF MIXTURES OF IDEAL GASES

In a system of ideal gases the temperature $T$ applies to all the gases within the system when occupying a volume $V$ at a total pressure $P$. The component pressures are given, of course, by the quantity $y_i P$. Other thermodynamic properties of the individual gases and of the mixture may be obtained by the application of the *Gibbs-Dalton law*, which is simply a generalization of Dalton's rule of additive pressures. It states that in a mixture of ideal gases each component of the mixture acts as if it were alone in the system at the volume $V$ and the temperature $T$ of the mixture. Consequently all extensive properties of the multicomponent mixture could be found by summing the contributions made by each gas component.

On the basis of the Gibbs-Dalton law the total internal energy of the mixture $U_m$ is given by

$$U_m = U_1 + U_2 + U_3 + \cdots + U_k = \sum_{i=1}^{k} U_i \qquad (11\text{-}18)$$

The total internal energy of each component may be expressed by $m_i u_i$ on a mass basis or by $N_i u_i$ on a mole basis, where the units on $u_i$ are different for the two cases. Hence we may write two relationships for the internal energy in the following manner:

$$m_m u_m = m_1 u_1 + m_2 u_2 + m_3 u_3 + \cdots + m_k u_k \qquad (11\text{-}19)$$

and
$$N_m \bar{u}_m = N_1 \bar{u}_1 + N_2 \bar{u}_2 + N_3 \bar{u}_3 + \cdots + N_k \bar{u}_k \qquad (11\text{-}20)$$

where $u_m$ in Eq. (11-19) is the specific internal energy of the mixture on a mass basis, and $\bar{u}_m$ in Eq. (11-20) is the molar specific internal energy. The internal energy $u_i$ of any component is equal to that of the pure component at the temperature and volume of the mixture; that is, $u_i = u(T, V)$. However, recall that the internal energy of an ideal gas is solely a function of the temperature. Therefore $u_i = u(T)$ is a valid expression for the internal energy of every component of an ideal-gas mixture.

If Eq. (11-20) is divided through by $N_m$, the molar specific internal energy can also be found by

$$\bar{u}_m = y_1 \bar{u}_1 + y_2 \bar{u}_2 + y_3 \bar{u}_3 + \cdots + y_k \bar{u}_k = \sum_{i=1}^{k} y_i \bar{u}_i \tag{11-21}$$

where $\bar{u}_i$ is the molar specific internal energy of the $i$th component when pure at the temperature $T$ of the mixture. It must be kept in mind that the internal energy of an ideal gas is established by selecting an arbitrary value for $u_i$ at some reference temperature $T_0$. Consequently, the internal energy $U_m$ of a gas mixture depends upon the arbitrary reference states chosen for each of the gases in the mixture. Frequently, the value of the internal energy of an ideal gas is computed on the basis that $u_i$ is zero at a reference state of $T_0$ equal to $0°K$ (or $0°R$).

An equation for the constant-volume specific heat of a mixture, $c_{v, m}$, on either a mass or a molar basis, is developed by differentiating Eqs. (11-19) and (11-20) with respect to temperature. First, considering Eq. (11-19), we find that

$$m_m \left(\frac{\partial u_m}{\partial T}\right)_v = m_1 \left(\frac{\partial u_1}{\partial T}\right)_v + m_2 \left(\frac{\partial u_2}{\partial T}\right)_v + \cdots + m_k \left(\frac{\partial u_k}{\partial T}\right)_v$$

After dividing through by $m_m$ and replacing $(\partial u_i / \partial T)_v$ by $c_{v, i}$, this equation becomes

$$c_{v, m} = \frac{m_1 c_{v, 1} + m_2 c_{v, 2} + \cdots + m_k c_{v, k}}{m_m} = \sum_{i=1}^{k} mf_i c_{v, i} \tag{11-22}$$

A similar development for the molar specific heat of a mixture leads to

$$\bar{c}_{v, m} = y_1 \bar{c}_{v, 1} + y_2 \bar{c}_{v, 2} + \cdots + y_k \bar{c}_{v, k} \tag{11-23}$$

where each $\bar{c}_{v, i}$ term must be expressed in molar units.

The enthalpy of a mixture of ideal gases will simply be the sum of the enthalpies of the individual chemical species. Therefore

$$\begin{aligned} H_m &= H_1 + H_2 + H_3 + \cdots + H_k \\ &= m_1 h_1 + m_2 h_2 + m_3 h_3 + \cdots + m_k h_k \\ &= N_1 \bar{h}_1 + N_2 \bar{h}_2 + N_3 \bar{h}_3 + \cdots + N_k \bar{h}_k \end{aligned} \tag{11-24}$$

On the second line of this equation the specific enthalpies must be on a mass basis, and on the third line the specific enthalpies employed must be the molar values. The enthalpies of ideal gases, similar to the internal energies, are based on some arbitrary reference state. Recall that the enthalpy of an ideal gas is only a function of the temperature of the system. In a manner analogous to that shown above, the differentiation of Eq. (11-24) with respect to temperature leads to equations for the constant-pressure specific heat of an ideal-gas mixture. As a result,

$$c_{p, m} = \frac{m_1 c_{p, 1} + m_2 c_{p, 2} + \cdots + m_k c_{p, k}}{m_m} \tag{11-25a}$$

and

$$\bar{c}_{p, m} = y_1 \bar{c}_{p, 1} + y_2 \bar{c}_{p, 2} + \cdots + y_k \bar{c}_{p, k} \tag{11-25b}$$

The selection of arbitrary reference states for values of $u$ and $h$ has no influence on the values of $c_{v,m}$ and $c_{p,m}$, since the specific heats are derivatives, in a mathematical sense. The calculation of some properties of a mixture of ideal gases is shown below.

**Example 11-4M** Consider the mixture of three ideal gases used in Example 11-3M at a temperature of 300°K and a total pressure of 2 bars. Determine (a) the partial pressure of each component, (b) the constant-volume specific heat of the mixture, and (c) the specific enthalpy of the mixture.

SOLUTION A summary of the necessary data for the calculations is shown in the accompanying table. The gravimetric and volumetric analyses are taken from Example 11-3M, and the values of $c_v$ and $h$ for each gas are taken from Tables A-4M, A-7M, A-8M, and A-11M.

| Component | Gravimetric analysis, % | Volumetric analysis, % | $c_v$ at 300°K, kJ/(kg)(°K) | $h$ at 300°K, kJ/kg·mol |
|-----------|-------------------------|------------------------|------------------------------|---------------------------|
| $H_2$ | 10 | 62.50 | 10.183 | 8523 |
| $O_2$ | 48 | 18.75 | 0.658 | 8736 |
| CO | 42 | 18.75 | 0.744 | 8723 |

(a) The component pressure of each gas is given by $y_i P$. Thus,

$$p(H_2) = 0.6250(2) = 1.250 \text{ bars}$$

$$p(O_2) = 0.1875(2) = 0.375 \text{ bar}$$

$$p(CO) = 0.1875(2) = 0.375 \text{ bar}$$

$$\overline{\phantom{P = 2.000 \text{ bars}}}$$

$$P = 2.000 \text{ bars}$$

(b) The constant-volume specific heat on a mass basis is determined from Eq. (11-22). Upon substitution of the values listed above,

$$c_{v,m} = 0.10(10.183) + 0.48(0.658) + 0.42(0.744)$$

$$= 1.65 \text{ kJ/(kg)(°K)}$$

(c) The evaluation of the specific enthalpy on a molar basis is found by first modifying Eq. (11-24) into the form

$$\overline{h}_m = y_1 \overline{h}_1 + y_2 \overline{h}_2 + y_3 \overline{h}_3 + \cdots$$

Substitution of the values from the table above leads to

$$\overline{h}_m = 0.6250(8523) + 0.1875(8736) + 0.1875(8723)$$

$$= 8600 \text{ kJ/kg·mol}$$

It should be noted that this enthalpy value is arbitrary, and depends upon the reference states for the enthalpies of the pure gases.

**Example 11-4** Consider the mixture of three ideal gases used in Example 11-3 at a temperature of 100°F and a total pressure of 14.5 psia. Determine (a) the partial pressures of each component, (b) the constant-volume specific heat of the mixture, and (c) the specific enthalpy of the mixture.

SOLUTION  A summary of the necessary data for the calculation is shown in the table below. The gravimetric and volumetric analyses are taken from Example 11-3, and the values for $c_v$ and $h$ for each gas are taken from Tables A-7, A-8, and A-11.

| Component | Gravimetric analysis, % | Volumetric analysis, % | $c_v$ at 100°F, Btu/(lb)(°F) | $h$ at 100°F, Btu/mol |
|---|---|---|---|---|
| $H_2$ | 10 | 62.50 | 2.441 | 3798.8 |
| $O_2$ | 48 | 18.75 | 0.158 | 3886.6 |
| CO | 42 | 18.75 | 0.178 | 3889.5 |

(a) The component pressure of each gas equals the mole fraction times the total pressure. Since the mole percent and the volumetric analysis are equivalent, the component pressures for each gas are

$$p(H_2) = 0.6250(14.5) = 9.06 \text{ psia}$$

$$p(O_2) = 0.1875(14.5) = 2.72 \text{ psia}$$

$$p(CO) = 0.1875(14.5) = 2.72 \text{ psia}$$

$$\overline{\phantom{xxxxxxxx}}$$

$$P = 14.50 \text{ psia}$$

(b) The constant-volume specific heat on a mass basis is determined from Eq. (11-22) and the values for $c_{v,i}$ listed above. Upon substitution,

$$c_{v,m} = 0.10(2.441) + 0.48(0.158) + 0.42(0.178)$$

$$= 0.395 \text{ Btu/(lb)(°F)}$$

(c) The evaluation of the specific enthalpy on a molar basis is found by modifying Eq. (11-24). Since $H_m = N_m h_m$, Eq. (11-24) can be rearranged in the form

$$h_m = y_1 h_1 + y_2 h_2 + y_3 h_3 + \cdots$$

Substitution of the values from the table above leads to

$$h_m = 0.6250(3798.8) + 0.1875(3886.6) + 0.1875(3889.5)$$

$$= 3831.3 \text{ Btu/mol}$$

It should be remembered that this enthalpy value is arbitrary, and depends upon the reference states for the enthalpies of the pure gases.

The entropy of a mixture of ideal gases can also be determined on the basis of the Gibbs-Dalton law. Since each gas behaves as if it alone occupied the volume $V$ of the system at the mixture temperature $T$, we may write

$$S_m = S_1(T, V) + S_2(T, V) + \cdots + S_k(T, V) \qquad (11\text{-}26)$$

If the entropy of an ideal gas is to be expressed as a function of temperature and pressure, then recall that

$$ds = c_p \frac{dT}{T} - R \frac{dp}{p} \qquad (7\text{-}15)$$

The symbol $p$ has been used for the pressure in this case, since, at the temperature and the volume of the mixture, the pressure of any gas is measured by its partial, or component, pressure, and not the total pressure $P$ of the system. The change in entropy of an ideal gas which is part of a mixture, in terms of its partial pressure $p_i$, is given by

$$\Delta s_i = c_{p_i} \ln \frac{T_2}{T_1} - R_i \ln \frac{p_{i,2}}{p_{i,1}} \qquad (11\text{-}27)$$

Recall that in the gas tables the integral of $c_p \, dT/T$ is given by $s_2^0 - s_1^0$. Consequently, the entropy change of the $i$th component in a mixture of ideal gases may also be found from

$$\Delta s_i = s_{i,2}^0 - s_{i,1}^0 - R_i \ln \frac{p_{i,2}}{p_{i,1}} \qquad (11\text{-}28)$$

if the gas tables are available for the component. The total change of entropy for a mixture of gases is determined from the sum of the entropy changes of the individual constituents. That is,

$$\Delta S_m = N_1 \, \Delta s_1 + N_2 \, \Delta s_2 + \cdots + N_k \, \Delta s_k \qquad (11\text{-}29)$$

The use of the entropy function in calculations for processes involving ideal-gas mixtures will now be illustrated by several examples.

**Example 11-5M** A mixture of gases consisting of 0.2 kg of nitrogen and 0.3 kg of argon is compressed from 2 bars and 300°K to 6 bars adiabatically and reversibly in a closed system. Determine (a) the final temperature, (b) the work required, in kilojoules, and (c) the entropy change of the nitrogen and the argon in kJ/°K.

SOLUTION (a) We shall assume ideal-gas behavior. The process is isentropic; hence the final temperature can be estimated from the relationship $T_2/T_1 = (P_2/P_1)^{(k-1)/k}$. The specific-heat ratio $k$ must be the average for the mixture, that is, $k_m = c_{p,m}/c_{v,m}$. Equations (11-22) and (11-25a) are used to evaluate $c_{p,m}$ and $c_{v,m}$. Since $T_2$ is unknown, we must use the specific-heat values for nitrogen in Table A-4M for the initial temperature. The $c_v$ and $c_p$ values for argon are the same for all monatomic gases on a molar basis, as noted in Table A-4M. On a mass basis, the $c_v$ and $c_p$ values for argon are 0.313 and 0.520 J/(g)(°K), and are essentially independent of temperature. Therefore

$$c_{v,m} = \frac{0.2(0.743) + 0.3(0.313)}{0.2 + 0.3} = 0.485 \text{ J/(g)(°K)}$$

$$c_{p,m} = \frac{0.2(1.039) + 0.3(0.520)}{0.2 + 0.3} = 0.728 \text{ J/(g)(°K)}$$

Therefore the specific-heat ratio $k_m$ for the mixture is $0.728/0.485 = 1.50$. The final temperature $T_2$, then, is

$$T_2 = T_1 \left(\frac{P_2}{P_1}\right)^{(k-1)/k} = 300 \left(\frac{6}{2}\right)^{(1.50-1)/1.50} = 433°K$$

Due to the relatively small temperature change, data based on the initial temperature probably are of sufficient accuracy.

(b) The work required for the adiabatic-compression process of a closed system is simply

$$W = U_2 - U_1 = mc_{v,m}(T_2 - T_1)$$

Although an average $c_{v,m}$ value should be used, we shall use one based on the initial temperature, which was calculated above.

$$W = 0.5 \text{ kg} \times 0.485 \text{ kJ/(kg)(°K)} \times (433 - 300)°K = 32.3 \text{ kJ}$$

(c) The entropy changes of the individual gases in the mixture are found by applying Eq. (11-27). Note, however, that for this process the mole fraction of either gas does not change. Consequently, the ratio of component pressures in this equation may be replaced by the ratio of the total pressure of the system, $P_2/P_1$. If the specific heats at the average temperature of the process are now used, then

$$(\Delta s)_{N_2} = 1.043 \ln \frac{433}{300} - \frac{8.315}{28} \ln \frac{6}{2} = 0.0565 \text{ kJ/(kg)(°K)}$$

$$(\Delta s)_{Ar} = 0.520 \ln \frac{433}{300} - \frac{8.315}{40} \ln \frac{6}{2} = -0.0376 \text{ kJ/(kg)(°K)}$$

The total entropy change for each gas, then, is

$$(\Delta S)_{N_2} = 0.2(0.0565) = 0.0113 \text{ kJ/°K}$$

$$(\Delta S)_{Ar} = 0.3(-0.0376) = -0.0113 \text{ kJ/°K}$$

Since the overall process is isentropic, the sum of $\Delta S$ for the two gases should be zero. Note, however, that each gas undergoes a definite entropy change of its own. For a mixture of two gases, this change is always equal in magnitude but opposite in sign.

**Example 11-5** A mixture of gases consisting of 2 lb of nitrogen and 3 lb of argon is compressed from 14.5 psia and 60°F to 58 psia adiabatically and reversibly in a closed system. For this process determine (a) the final temperature, (b) the work required in Btu/lb of mixture, and (c) the entropy change of the argon and the entropy change of the nitrogen in Btu/°R.

SOLUTION (a) Since the pressures are relatively low, the gases are assumed to behave as ideal gases. The process is isentropic; hence the final temperature can be computed from the isentropic ideal-gas relationship $T_2/T_1 = (P_2/P_1)^{(k-1)/k}$. The specific-heat ratio $k$ to be used must be the average for the mixture. Equations (11-22) and (11-25a) can first be used to evaluate the constant-volume and constant-pressure specific heats. Then $k$ is determined from its definition, namely, $k = c_p/c_v$. A more accurate evaluation of $T_2$ is made if the $k$ value is the average value for the process. However, since $T_2$ is yet unknown, it is impossible to know what average should be taken. In this case we shall use the specific-heat values for nitrogen in Table A-4 for the initial temperature of the mixture. Since the $c_p$ and $c_v$ values do not vary rapidly with temperature, this is a fair first approximation. The $c_v$ and $c_p$ values for argon are 0.0745 and 0.124 Btu/(lb)(°F) and are roughly independent of temperature. Substituting values into Eqs. (11-22) and (11-25a), we find that

$$c_{v,m} = \frac{2(0.178) + 3(0.0745)}{5} = 0.116 \text{ Btu/(lb)(°F)}$$

and

$$c_{p,m} = \frac{2(0.248) + 3(0.124)}{5} = 0.174 \text{ Btu/(lb)(°F)}$$

Therefore the specific-heat ratio $k_m$ for the mixture is $0.174/0.116 = 1.50$. The final temperature $T_2$, then, is

$$T_2 = T_1\left(\frac{P_2}{P_1}\right)^{(k-1)/k} = 520\left(\frac{58}{14.5}\right)^{(1.50-1)/1.50} = 825°R = 365°F$$

At 365°F the value of $k_m$ is roughly the same. For our present purposes we shall assume that a $k$ value based on the initial temperature is of sufficient accuracy.

(b) The work required for the compression process can be found directly from the conservation of energy principle for closed systems. Because the process is adiabatic, the work is given simply by

$$w = u_2 - u_1 = c_{v,m}(T_2 - T_1)$$

Again, it would be appropriate to use, in general, an average $c_{v,m}$ value for the temperature range of the process. As a first approximation, though, we shall use the $c_{v,m}$ value for the initial state, which was calculated above. The required work then becomes

$$w = 0.116(825 - 520) = 35.4 \text{ Btu/lb of mixture}$$

(c) The entropy changes of the individual gases in the mixture are found from Eq. (11-27); namely, $\Delta s_i = c_{p,i} \ln (T_2/T_1) - R_i \ln (p_{i,2}/p_{i,1})$. The mole fraction of either gas does not change during the process; consequently, the ratio of partial pressures in this equation is the same as the ratio of the total pressure of the system, $P_2/P_1$. If the specific heats at the average temperature of the process are used, then

$$(\Delta s)_{N_2} = 0.249 \ln \frac{825}{520} - \frac{1.986}{28} \ln \frac{58}{14.5} = +0.0165 \text{ Btu/(lb)(°R)}$$

$$(\Delta s)_{Ar} = 0.124 \ln \frac{825}{520} - \frac{1.986}{40} \ln \frac{58}{14.5} = -0.0115 \text{ Btu/(lb)(°R)}$$

The total entropy change for each gas, then, is

$$(\Delta S)_{N_2} = 2(+0.0165) = +0.0330 \text{ Btu/°R}$$

$$(\Delta S)_{Ar} = 3(-0.0115) = -0.0345 \text{ Btu/°R}$$

Since the overall process is isentropic, the sum of $\Delta S$ for nitrogen and argon should be zero. The slight discrepancy is due to round-off error and the fact that $k_m$ was based on the initial temperature of the mixture. The important thing to note here is that, although the total entropy change is zero, both gases undergo a definite entropy change of their own. For a mixture of two gases, this change is always equal in magnitude but opposite in sign.

## 11-4 MIXING PROCESSES INVOLVING IDEAL GASES

When an ideal gas mixture undergoes a process change without a change in composition, the component pressure of each species remains the same. Consequently, the value of $p_{i,2}/p_{i,1}$ is the same as $P_2/P_1$ for the total mixture. However, when two or more pure gases are mixed, or two gas mixtures come into contact, the component or partial pressure may change significantly. This change must be accounted for, especially when evaluating the entropy change of the individual gases.

Consider a rigid tank divided into $k$ compartments by partitions. Different ideal gases fill each of the $k$ compartments, and the total pressure and temperature of each pure gas initially are the same. The tank is insulated, and then the partitions are pulled from the tank. Eventually, each gas spreads into the total volume of the tank, and a new equilibrium state is attained. In the absence of heat and work interactions, the energy equation for the closed system reduces to $\Delta U = 0$.

In terms of the individual components this takes the form

$$U_{\text{initial}} = U_{\text{final}}$$

or

$$\sum N_{i,1} u_{i,1} = \sum N_{i,2} u_{i,2} \tag{11-30}$$

where $u_{i,1}$ and $u_{i,2}$ are the initial and final molar internal energies of the $i$th component, and they are solely a function of the temperature for ideal gases. The conservation of energy principle requires that

$$U_2 = \sum N_{i,2} u_{i,2} = \text{constant}$$

The values of $N_i$ are constant, since chemical reactions are not under consideration. In addition, it must be noted that $u_i$ for any component increases with increasing temperature. If the temperature increases upon mixing, each of the $u_{i,2}$ values would be larger. Such an increase in the $u_{i,2}$ values would increase the overall energy $U_2$. However, the conservation of energy principle requires that $U_2$ be constant. Therefore, the final temperature must be equal to the initial value for the case under consideration. If the initial temperatures are not the same, then the first law may be used to determine the final temperature. For a two-component system of species $A$ and $B$ with constant specific heats, the first law leads to

$$T_2 = \frac{N_A c_{v,A} T_{A,1} + N_B c_{v,B} T_{B,1}}{N_A c_{v,A} + N_V c_{v,B}} \tag{11-31}$$

Equation (11-27) could also be used to find $T_2$ by employing $u$ data from the appropriate gas tables.

The effect of mixing of two gases initially at the same pressure and temperature on the final pressure may be found by using the ideal-gas equation. For any component $i$ we may write

$$\frac{p_{i,2} V_{i,2}}{p_{i,1} V_{i,1}} = \frac{N_i R_u T}{N_i R_u T}$$

or

$$\frac{p_{i,2}}{p_{i,1}} = \frac{V_{i,1}}{V_{i,2}} = \frac{V_{i,1}}{V} = y_i$$

where $V$ is the total volume of the mixture. Hence $p_{i,2} = y_i p_{i,1} = y_i P_{\text{initial}}$. If we sum over all components,

$$P_{\text{final}} = \sum p_{i,2} = \sum y_i P_{\text{initial}} = P_{\text{initial}}$$

Therefore the pressure also does not change upon mixing in this special case. When the components initially are at different pressures and temperatures, the final pressure is determined directly from the ideal-gas equation, after an energy balance has been made to find the final temperature.

The mixing of two or more gases is highly irreversible, and under adiabatic conditions a positive entropy change would be expected. The entropy change of each component is given by Eq. (11-28) multiplied by the moles of that species.

$$(\Delta S)_i = N_i \left( s_{i,2}^0 - s_{i,1}^0 - R_u \ln \frac{p_{i,2}}{p_{i,1}} \right)$$

If we again consider the special case of mixing of gases initially at the same temperature and pressure, then $T_2 = T_1$. Consequently the first two terms on the right cancel. Also, $p_{i,1} = P$ and $p_{i,2} = y_i P$, since the initial and final total pressures are equal. Therefore for the $i$th species,

$$(\Delta S)_i = -N_i R_u \ln \frac{y_i P}{P} = -N_i R_u \ln y_i \tag{11-32}$$

The sum of the $(\Delta S)_i$ terms for the mixture is the total entropy change for the adiabatic mixing process.

$$\Delta S_m = -R_u \sum N_i \ln y_i \tag{11-33}$$

Since the $y_i$ values are always less than unity, the entropy change given by Eq. (11-33) is always positive. In addition, the total entropy change is independent of the composition of the gases, and depends solely on the number of moles of each gas involved in the mixing process. Finally, the entropy change for the mixing of gases developed above is valid only if all the gases are distinguishable from one another. The entropy change is zero when the same gas is mixed at constant pressure and temperature.

**Example 11-6M** A rigid, insulated tank is divided into two compartments by a partition. Initially 0.02 kg-mol of nitrogen fills one compartment at 2 bars and 100°C. The other compartment contains 0.03 kg mol of carbon dioxide at 1 bar and 20°C. The partition is removed and the gases allowed to mix. Determine the temperature and pressure of the mixture and the entropy change for the mixing process.

SOLUTION For the low pressures involved the gases are assumed to behave as ideal gases. Since both $Q$ and $W$ are zero, then the first law for a closed system dictates that $\Delta U = 0$, or $U_{init}$ equals $U_{final}$. On the basis of Eq. (11-30)

$$[Nc_v(T_2 - T_1)]_{N_2} + [Nc_v(T_2 - T_1)]_{CO_2} = 0$$

The final temperature $T_2$ is the same for each gas, and probably lies between 20 and 100°C. Based on Table A-4M we shall let $c_v$ for nitrogen and carbon dioxide be constant and have values of 0.744 and 0.680 kJ/(kg)(°K), respectively. These are 20.8 and 29.9 kJ/(kg·mol)(°K) on a molar basis. Substitution of numerical values into the above equation yields

$$0.02(20.8)(T_2 - 100) + 0.03(29.9)(T_2 - 20) = 0$$

Therefore

$$T_2 = 45.3°C$$

The final pressure is determined from the ideal-gas relation, $PV = NR_u T$. The total volume is the sum of the volumes of the original compartments, which may also be determined from the same equation. Hence

$$V_{N_2} = \frac{NR_u T}{P} = \frac{0.02(0.08315)(373)}{2} = 0.310 \text{ m}^3$$

$$V_{CO_2} = \frac{0.03(0.08315)(293)}{1} = 0.731 \text{ m}^3$$

$$V_{tot} = 0.310 + 0.731 = 1.041 \text{ m}^3$$

As a result,

$$P_2 = \frac{N_m R_u T_2}{V_m} = \frac{0.05(0.08315)(318)}{1.041} = 1.27 \text{ bars}$$

The entropy change is the sum of the entropy changes for the individual components. If we use Eq. (11-27), we will need $c_p$ data. Recall that for ideal gases $c_p = c_v + R$. Hence for $N_2$ and $CO_2$ the $c_p$ values are 29.1 and 38.2 kJ/(kg·mol)(°K), respectively. Keep in mind that each gas exerts only its component pressure at the total volume and temperature of the system. For the two gases we find that

$$(\Delta S)_{N_2} = 0.02 \left[ 29.1 \ln \frac{318}{373} - 8.315 \ln \frac{0.4(1.27)}{2} \right]$$

$$= 0.02(-4.64 + 11.40) = 0.135 \text{ kJ/°K}$$

$$(\Delta S)_{CO_2} = 0.03 \left[ 38.2 \ln \frac{318}{293} - 8.315 \ln \frac{0.6(1.27)}{1} \right]$$

$$= 0.03(3.13 + 2.26) = 0.162 \text{ kJ/°K}$$

Therefore, for the entire system,

$$\Delta S_{tot} = 0.135 + 0.162 = 0.297 \text{ kJ/°K}$$

The entropy change for each gas could also be calculated from Eq. (11-28), which employs the $s^0$ values from the gas tables A-6M and A-9M. For example, the entropy change for nitrogen would be

$$(\Delta S)_{N_2} = 0.2 \left[ 193.377 - 198.027 - 8.315 \ln \frac{0.4(1.27)}{2} \right]$$

$$= 0.2(-4.65 + 11.40) = 0.135 \text{ kJ/°K}$$

This agrees with the previously calculated value, because of the small temperature range involved.

**Example 11-6** A rigid, insulated tank is divided into two compartments by a partition. Initially 0.2 mol of nitrogen is introduced into one compartment at a pressure of 30 psia and a temperature of 140°F. At the same time 0.3 mol of carbon dioxide is introduced into the other compartment at 15 psia and 60°F. The partition is then removed and the gases are allowed to mix. Determine the temperature and pressure of the mixture at equilibrium and the entropy change for the mixing process. The constant-volume molar specific heats for nitrogen and carbon dioxide will be assumed to be constant for the process, and are 5.0 Btu/(mol)(°R) and 6.9 Btu/(mol)(°R), respectively.

SOLUTION For the low pressures involved, the gases behave essentially as ideal gases. We shall assume that the mixture obeys the Gibbs-Dalton law. The internal energy of ideal gases is solely a function of temperature. Thus the final equilibrium temperature depends upon the internal energy of the mixture. There are no work or heat interactions for the process if the system is chosen to include the entire tank. Therefore the internal energy of the system does not change. In terms of the mixing process, this means that $U_{init}$ equals $U_{final}$, or $\Delta U = 0$. Consequently, the sum of the changes in internal energy of the two gases must equal zero. That is,

$$[Nc_v(T_2 - T_1)]_{N_2} + [Nc_v(T_2 - T_1)]_{CO_2} = 0$$

The final temperature $T_2$ is the same for each gas at equilibrium. Substitution of the proper numerical values into the equation yields

$$0.2(5.0)(T_2 - 140) + 0.3(6.9)(T_2 - 60) = 0$$

Therefore

$$T_2 = 86°F$$

The final pressure may now be determined from the ideal-gas relation, $PV = NR_uT$. The total volume of the mixture is the sum of the volumes of the original compartments, which may also be determined from the same equation. Hence

$$V_{N_2} = \frac{NR_uT}{P} = \frac{0.2(1545)(600)}{30(144)} = 42.9 \text{ ft}^3$$

$$V_{CO_2} = \frac{0.3(1545)(520)}{15(144)} = 111.6 \text{ ft}^3$$

$$V_{tot} = 42.9 + 111.6 = 154.5 \text{ ft}^3$$

$$P_2 = \frac{N_m R_u T_2}{V_2} = \frac{(0.2 + 0.3)(1545)(460 + 86)}{154.5(144)} = 19.0 \text{ psia}$$

On the basis of the Gibbs-Dalton law, the entropy change is the sum of the entropy changes for the individual components. We may apply Eq. (11-27) to each constituent, keeping in mind that a gas exerts only its component pressure at the volume and temperature of the system. The $c_p$ data required for Eq. (11-27) may be calculated from the ideal-gas relation, $c_p = c_v + R$. Using the $c_v$ data given in the statement of the problem, we find that $c_p$ values for nitrogen and carbon dioxide are approximately 7.0 and 8.9 Btu/(lb·mol)(°R), respectively. On this basis

$$(\Delta S)_{N_2} = N\left(c_p \ln \frac{T_2}{T_1} - R_u \ln \frac{P_2}{P_1}\right)$$

$$= 0.2\left[7.0 \ln \frac{546}{600} - 1.986 \ln \frac{0.4(19.0)}{30}\right]$$

$$= 0.2(-0.659 + 2.74) = 0.416 \text{ Btu/°R}$$

$$(\Delta S)_{CO_2} = 0.3\left[8.9 \ln \frac{546}{520} - 1.986 \ln \frac{0.6(19.0)}{15}\right]$$

$$= 0.3(0.434 + 0.554) = 0.988 \text{ Btu/°R}$$

Therefore, for the entire system,

$$\Delta S_{tot} = 0.416 + 0.988 = 1.404 \text{ Btu/°R}$$

The entropy change for each gas could also be calculated from Eq. (11-28), which employs the $s^0$ values from gas tables A-6 and A-9. For example, the entropy change for nitrogen on this basis would be

$$(\Delta S)_{N_2} = 0.2\left[45.858 - 46.514 - 1.986 \ln \frac{0.4(19.0)}{30}\right]$$

$$= 0.2(-0.656 + 2.74) = 0.416 \text{ Btu/°R}$$

This agrees with the previously calculated value, because of the small temperature range involved.

Mixing of ideal gases also occurs in steady-flow, open systems. The calculations are essentially the same as for the closed system analysis shown above, except a different energy equation must be used. In addition, there is no general way to evaluate the final pressure unless some additional data on flow conditions entering and leaving the control volume are given.

## 11-5 PROPERTIES OF A MIXTURE OF AN IDEAL GAS AND A VAPOR

Although the relationships developed in the preceding sections for the properties of ideal-gas mixtures are of general usefulness, there is one additional complication that must be recognized when dealing with ideal-gas mixtures. There is always the possibility that one or more of the ideal gases may exist in a state that is close to a saturation state for the given component. Recall that, according to the Gibbs-Dalton law, each component exists as if it were alone at the system temperature and volume. We have seen that each gas exerts a pressure which is equal to its component pressure. But the component pressure can never be greater than the saturation pressure for that component at the mixture temperature. Any attempt to increase the component pressure beyond the saturation pressure results in partial condensing of the vapor. For example, consider increasing the total pressure of an ideal-gas mixture at constant temperature. Since the mole fractions of every component gas are fixed (at least temporarily), the component pressure of each increases in direct proportion to the increase in the total pressure. However, if the component pressure of any constituent eventually exceeds its saturation pressure for that temperature, the gas will begin to condense out as the pressure is increased further. This process is illustrated on a $Pv$ diagram in Fig. 11-3a. The gas which condenses under this circumstance is usually spoken of as a vapor. Consequently, the ideal-gas mixtures are referred to as gas-vapor mixtures.

A similar situation also occurs when a gas-vapor mixture is cooled at constant pressure. In this case the partial pressure of the vapor remains constant (up to the point of condensation), but the temperature eventually is lowered sufficiently to equal the saturation temperature for the given partial pressure. As the temperature is lowered still further, the saturation pressure corresponding to the temperature becomes less than the actual partial pressure, and hence some of the vapor must condense. The effect of lowering the temperature of a mixture containing

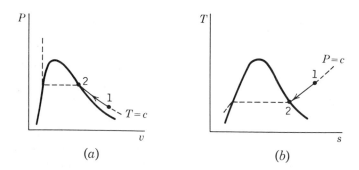

(a)                                    (b)

**Figure 11-3** The effect of raising the total pressure or lowering the temperature of a gas mixture containing a vapor.

condensable vapor is shown on a $Ts$ diagram in Fig. 11-3$b$, where state 2 is the state of incipient condensation. A well-known example of a gas mixture containing a condensable vapor is the air in the atmosphere. The condensation of water from the air as the temperature is lowered, forming dew, is a common experience. Although there are a number of systems of engineering interest which involve the use of gas-vapor mixtures, we shall direct our attention in this chapter to air–water-vapor mixtures. Such a mixture is frequently referred to as atmospheric air. The study of the basic properties of such a mixture is quite important, since it is realized that the composition of atmospheric air plays a decisive role in the proper functioning of the human body.

We have already assumed that the gases which make up the atmospheric air are ideal gases and that the Gibbs-Dalton law is valid. Several other properties of the components of the mixture should be summarized. In dealing with the conditioning of air, we normally are interested in a temperature range from, roughly, $-10$ to $40°C$ or 20 to $110°F$. Under this circumstance the constant-pressure specific heat of dry air is roughly a constant, and will be taken to be $1.00\ \mathrm{J/(g)(°C)}$ or $0.240\ \mathrm{Btu/(lb)(°F)}$. It must also be recognized that, although the water vapor is in either the dry, saturation state or superheated-vapor state, its enthalpy is solely a function of temperature. This is necessarily true, since we have assumed the water vapor to behave as an ideal gas. The enthalpy of the water vapor in a dry-air–water-vapor mixture for all practical purposes equals the enthalpy of water vapor in the vapor-saturation state at the temperature of the mixture. This point is confirmed by referring to either a $Ts$ or an $hs$ diagram for water based on actual experimental data. It will be seen that constant $h$ and $T$ lines coincide in the low-pressure range of the superheat region. This approximation is not too good for temperatures much above $40°C$ or $100°F$.

The temperature of the mixture as measured by a conventional thermometer is called the *dry-bulb* temperature, such as $T_1$ in Fig. 11-3. In Fig. 11-3$b$ a process was illustrated for which the temperature of a gas-vapor mixture was lowered and the total pressure of the mixture remained constant. Until the saturation state of the vapor is reached, the partial pressures of the constituents also remain constant. The temperature at which the mixture becomes saturated, or condensation begins, when a mixture of dry air and water vapor is cooled at constant pressure from an unsaturated state, is called the *dew-point* temperature. Hence the temperature at state 2 on the $Ts$ diagram is the dew point of any mixture for which the partial pressure of the water vapor is represented by the constant-pressure line. The dew-point temperature is the saturation temperature of water which corresponds to the partial pressure of the water vapor actually in the atmospheric air. A dry-air–water-vapor mixture which is saturated with water vapor is frequently called saturated air.

For air–water-vapor mixtures which are not saturated, we need to be able to denote the quantity of water vapor present at a given state of the mixture. This is done conventionally in two ways: The *relative humidity* $\phi$ is defined as the ratio of the partial pressure of the vapor in a mixture to the saturation pressure of the vapor at the same temperature and pressure of the mixture. If $p_v$ represents the actual vapor pressure and $p_g$ represents the saturation pressure at the same temperature, then

$$\phi \equiv \frac{p_v}{p_g} \tag{11-34}$$

The pressures used to define the relative humidity are shown in the $Ts$ diagram of Fig. 11-4, which is an extension of the data shown in Fig. 11-3b. State 1 is the initial state of the water vapor in the mixture, and its vapor pressure at this state is $p_1$. If this same vapor were present in saturated air at the same temperature, its pressure would necessarily have to be that given at state 3, which is the saturation pressure $p_g$ for that temperature. In terms of the figure, $\phi = p_1/p_3$. The pressure $p_g$ is, of course, greater than $p_v$ since the values of constant-pressure lines on a $Ts$ diagram in the superheat region increase as one proceeds from lower right to upper left on the diagram. Consequently, the relative humidity $\phi$ is always less than or equal to unity, as one would expect. It might be noted that since the relative humidity is defined solely in terms of the vapor in the mixture, it is independent of the state of the air in the mixture. Based on the assumption that both the dry air and the water vapor in the mixture behave as ideal gases, the equation for the relative humidity can be expressed in terms of specific volumes (or densities) as well as partial pressures. That is,

$$\phi = \frac{p_v}{p_g} = \frac{RT_v/v_v}{RT_g/v_g} = \frac{v_g}{v_v} = \frac{\rho_v}{\rho_g} = \frac{m_v}{m_{sat}} \tag{11-35}$$

since the temperatures and the gas constants in the relation are equal. It is seen that saturated air is atmospheric air with a relative humidity of 1.0 (or 100 percent).

The *humidity ratio* (or specific humidity) $\omega$ describes the quantity of water vapor in a mixture in terms of the amount of dry air present. It is formally defined as the ratio of the mass of water vapor present, $m_v$, to the mass of dry air, $m_a$. These masses are on a kilogram or a pound basis. The humidity ratio is not a measure of the mass fraction of the water vapor in the mixture, as should be carefully noted. Such a measure is not useful in dealing with gas mixtures which

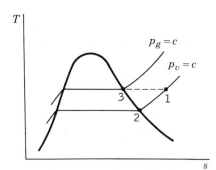

**Figure 11-4** $Ts$ diagram for water vapor in atmospheric air.

change composition during a process, as will be shown shortly. In equation form, the humidity ratio is

$$\omega \equiv \frac{m_v}{m_a} \tag{11-36}$$

Recall that the specific volume is defined as $v = V/m$. Since both the water vapor and the dry air occupy the same total volume, the mass ratio can be replaced by the ratio of specific volumes. Hence

$$\omega = \frac{V/v_v}{V/v_a} = \frac{v_a}{v_v} = \frac{\rho_v}{\rho_a} \tag{11-37}$$

Going one step further, the humidity ratio can also be expressed in terms of the partial pressures of the two constituent gases. The specific volume of an ideal gas in a gas mixture is given by $v_i = RT/p_i = R_u T/p_i M_i$, where $p_i$ is the partial pressure and $M_i$ is the molar mass of the component. By substituting this equation into Eq. (11-37), we find that

$$\omega - \frac{R_u T/p_a M_a}{R_u T/p_v M_v} = \frac{M_v p_v}{M_a p_a}$$

The ratio of molar masses for water to air is 0.622, and $p_a = P - p_v$; thus

$$\omega = 0.622 \frac{p_v}{p_a} = 0.622 \frac{p_v}{P - p_v} \tag{11-38}$$

Finally, we are able to relate the humidity ratio and the relative humidity of a mixture by combining Eqs. (11-34) and (11-38). This yields

$$\phi = \frac{p_v}{p_g} = \frac{p_a \omega}{0.622 p_g} \tag{11-39}$$

The evaluation of the above quantities and their use in a simple problem are illustrated in the example that follows.

**Example 11-7M** An air–water-vapor mixture at 25°C and 1 bar has a relative humidity of 50 percent. Compute (a) the humidity ratio, (b) the dew point, and (c) the amount of water added or removed per kilogram of dry air if the mixture undergoes a process during which its state is changed to 20°C and 40 percent relative humidity.

SOLUTION (a) The humidity ratio is evaluated from Eq. (11-38). From steam table A-12M it is found that the saturation pressure $p_g$ at 25°C is 0.0317 bar. Therefore the actual vapor pressure $p_v$ is $\phi p_g = 0.50(0.0317) = 0.0159$ bar. Then

$$\omega_1 = \frac{0.622(0.0159)}{1.00 - 0.0159} = 0.01005 \text{ kg water/kg dry air}$$

(b) The dew point by definition is the temperature at which the actual vapor pressure becomes the saturation pressure. From Table A-12M the vapor pressure at 10°C is 0.01228 bar and at 15°C it is 0.01705 bar. By linear interpolation, the dew point is roughly 13.8°C.

(c) The amount of water added or removed from the mixture per kilogram of dry air is the difference between the final and initial humidity ratios. At 20°C the saturation vapor pressure is

0.02339 bar; so the actual vapor pressure is 0.40(0.02339) = 0.00936 bar. For the same mixture pressure, the final humidity ratio becomes

$$\omega_2 = \frac{0.622(0.00936)}{1.00 - 0.00936} = 0.00588 \text{ kg water/kg dry air}$$

Therefore,

$$\omega_2 - \omega_1 = 0.00588 - 0.01005 = -0.00417 \text{ kg water/kg dry air}$$

The negative sign indicates that water was removed during the process.

**Example 11-7** An air–water-vapor mixture at 75°F and 14.7 psia has a relative humidity of 50 percent. Compute (a) the humidity ratio, (b) the dew point, and (c) the amount of water added or removed per pound of dry air if the mixture undergoes a process during which its state is changed to 60°F and 40 percent relative humidity.

SOLUTION (a) The humidity ratio is evaluated from Eq. (11-38). From steam table A-12 it is found that the saturation pressure $p_g$ at 75°F is 0.4300 psia. Therefore the actual vapor pressure $p_v$ is $\phi p_g = 0.50(0.4300) = 0.2150$ psia. Then

$$\omega_1 = \frac{0.622(0.2150)}{14.7 - 0.2150} = 0.00925 \text{ lb water/lb dry air}$$

Since the humidity ratio is usually a very small number, it is frequently tabulated or referred to on the basis of grains of water per pound of dry air. There are 7000 gr in a pound by definition; hence, for the initial state,

$$\omega_1 = 0.00925(7000) = 64.75 \text{ gr water/lb dry air}$$

(b) The dew point by definition is the temperature at which the actual vapor pressure becomes the saturation pressure. From Table A-3 the vapor pressure at 55°F is 0.2141 psia. Consequently, the dew-point temperature of the initial mixture is close to 55°F, since the actual vapor pressure is 0.2150 psia.

(c) The amount of water added or removed from the mixture per pound of dry air is simply the difference between the final and initial humidity ratios At 60°F the saturation pressure is 0.256 psia; so the actual vapor pressure would be 0.40(0.256) = 0.1024 psia. For the same mixture pressure, the final humidity ratio becomes

$$\omega_2 = \frac{0.622(0.1024)}{14.7 - 0.1024} = 0.00437 \text{ lb water/lb dry air}$$

Since the final humidity ratio is smaller than the initial value, the quantity of water removed is 0.00925 − 0.00437 = 0.00488 lb/lb of dry air.

**Example 11-8M** From the initial data in Example 11-7M, determine the quantity of water condensed if the mixture is simply cooled at constant pressure to a final temperature of 10°C.

SOLUTION Since the dew point of the mixture is, roughly, 13.8°C, water must condense when the mixture is cooled at 10°C. In the final state the mixture will be saturated. At 10°C the saturation vapor pressure is 0.01228 bar. Therefore the final specific humidity is

$$\omega_2 = \frac{0.622(0.01228)}{1.00 - 0.01228} = 0.00773 \text{ kg water/kg dry air}$$

The water removed is 0.01005 − 0.00773 = 0.00272 kg water/kg dry air.

**Example 11-8** From the initial data in Example 11-7, determine the quantity of water condensed if the mixture is simply cooled at constant pressure to a final temperature of 50°F.

SOLUTION Since the dew point of the mixture is, roughly, 55°F, water must condense when the mixture is cooled to 50°F. In the final state the mixture will be saturated. At 50°F the saturation vapor pressure is 0.178 psia. Therefore the specific humidity is

$$\omega_2 = \frac{0.622(0.178)}{14.7 - 0.178} = 0.00763 \text{ lb water/lb dry air}$$

The water removed is $0.00925 - 0.00763 = 0.00162$ lb/lb of dry air.

## 11-6 THE ADIABATIC-SATURATION AND WET-BULB TEMPERATURES

In the preceding section the definitions of relative humidity and humidity ratio were introduced as useful quantities in the analysis of problems dealing with the conditioning of water-vapor–dry-air mixtures. Such quantities are not of value, however, unless they can be measured experimentally. In the preceding examples we have presupposed a knowledge of their numerical values. We must therefore investigate methods for evaluating either the relative humidity or the humidity ratio from easily measured thermodynamic parameters.

One method of evaluating the humidity ratio of a mixture is based on the technique of adiabatic saturation. The experimental apparatus is sketched in Fig. 11-5a, and the $Ts$ diagram for the process is shown in Fig. 11-5b.

Referring to Fig. 11-5a, we see that an unsaturated mixture of gases enters a steady-flow channel with a dry-bulb temperature $T_1$ and a relative humidity which is less than 100 percent. If the channel is long enough, the mixture will pick up additional moisture as it passes over the liquid water in the bottom of the channel, and it will leave the device as a saturated mixture at some temperature $T_2$. The device is normally insulated, so that the final temperature achieved when completely saturated is known as the *adiabatic-saturation* temperature. The adiabatic-saturation temperature is always less than the dry-bulb temperature $T_1$ since the evaporation of water into the mixture requires energy, which comes from both the air mixture passing through and the liquid water in the channel. Consequently, the gas mixture is cooled as it becomes saturated. A steady flow of liquid water is

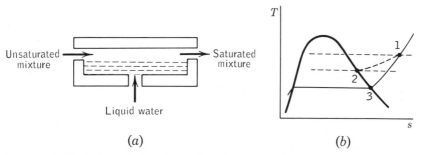

(a)                                        (b)

**Figure 11-5** Physical description of the adiabatic-saturation process and its representation on a $Ts$ diagram.

added at temperature $T_2$ to make up for the evaporation of water into the gas stream. The process itself for the mixture is shown by the dashed line from state 1 to state 2 in Fig. 11-5b. The temperature at state 3 is the dew-point temperature for the mixture, which is lower than both the initial dry-bulb temperature and the adiabatic-saturation temperature.

In most cases the kinetic- and potential-energy changes of the air stream are negligible; hence the conservation of energy principle for the steady-flow process under these conditions reduces simply to $H_{in} = H_{out}$. The enthalpy brought into the control volume which lies inside the channel consists of two parts, that due to the entering gas stream and that due to the liquid makeup water. The exit enthalpy is associated solely with the mixture stream leaving the control volume. On the basis of the Gibbs-Dalton law, the enthalpy of a dry-air–water-vapor mixture is the sum of the enthalpies of the individual components. Thus, per unit mass of dry air,

$$h_{mixture} = h_{dry\ air} + h_{water\ vapor}$$

$$= c_{pa} T_a + \omega h_v \tag{11-40}$$

where $h_v$ is the enthalpy of the water vapor. In atmospheric air problems, $h_v$ may be evaluated by $h_g$ for a saturated vapor at the given air temperature.

On the basis of a unit mass of dry air entering and leaving the control volume in a given time interval for the adiabatic-saturation process, the energy balance becomes

$$h_{a1} + \omega_1 h_{v1} + (\omega_2 - \omega_1)h_{f2} = h_{a2} + \omega_2 h_{v2}$$

where $h_f$ is the enthalpy of saturated liquid water. On the basis of Eq. (11-40) this equation becomes

$$c_{pa} T_{a1} + \omega_1 h_{v1} + (\omega_2 - \omega_1)h_{f2} = c_{pa} T_{a2} + \omega_2 h_{v2}$$

Noting that $h_{v2} - h_{f2} = h_{fg2}$ and rearranging the equation, we obtain the following expression for $\omega_1$.

$$\omega_1 = \frac{c_{pa}(T_2 - T_1) + \omega_2 h_{fg2}}{h_{v1} - h_{f2}} \tag{11-41}$$

Every quantity of the right-hand side of the equation is known, once the temperatures $T_1$ and $T_2$ are measured. The humidity ratio at $T_2$ is known, since this state is a saturation state. Thus the humidity ratio (and hence the relative humidity) of an unsaturated gas-vapor mixture could be ascertained by measuring the inlet and outlet temperatures and the total pressure of the mixture which undergoes a process of adiabatic saturation.

Although the technique of adiabatic saturation leads to the desired results, it is difficult in practice to attain a saturated state by this method without employing an extremely long flow channel, which is impractical. In lieu of this, a temperature equivalent to the adiabatic-saturation temperature for water-vapor–air mixtures, known as the *wet-bulb* temperature, is used. This temperature is easily measured by the following technique. The bulb of an ordinary thermometer is covered with a wick which has been moistened with water. The unsaturated atmospheric air of undetermined humidity ratio is then passed over the wetted wick until dynamic equilibrium is attained and the temperature of the wick (and hence the thermometer) reaches a stable value. It is found that, for air–water-vapor mixtures at normal temperatures and pressures, the wet-bulb temperature, the determination of which relies on heat- and mass-transfer rates, is close in value to the adiabatic-saturation temperature. Thus the temperature $T_2$ used in Eq. (11-41) to obtain the

**Table 11-1 Properties of saturated water as a function of temperature**

| A. Metric data: $h$, kJ/kg | | | | B. USCS data: $h$, Btu/lb | | | |
|---|---|---|---|---|---|---|---|
| Temp. °C $T$ | Press. bars $P$ | Sat. liquid $h_f$ | Sat. vapor $h_g$ | Temp. °F $T$ | Press. lb/in$^2$ $P$ | Sat. liquid $h_f$ | Sat. vapor $h_g$ |
| 6 | 0.00935 | 25.2 | 2512.4 | 40 | 0.1217 | 8.0 | 1078.9 |
| 8 | 0.01072 | 33.6 | 2516.1 | 42 | 0.1315 | 10.0 | 1079.8 |
| 10 | 0.01228 | 42.0 | 2519.8 | 44 | 0.1420 | 12.0 | 1080.7 |
| 11 | 0.01312 | 46.2 | 2521.6 | 46 | 0.1532 | 14.0 | 1081.5 |
| 12 | 0.01402 | 50.4 | 2523.4 | 48 | 0.1652 | 16.1 | 1082.4 |
| 13 | 0.01497 | 54.6 | 2525.3 | 50 | 0.1780 | 18.1 | 1083.3 |
| 14 | 0.01598 | 58.8 | 2527.1 | 52 | 0.1917 | 20.1 | 1084.2 |
| 15 | 0.01705 | 63.0 | 2528.9 | 54 | 0.2064 | 22.1 | 1085.1 |
| 16 | 0.01818 | 67.2 | 2530.8 | 56 | 0.2219 | 24.1 | 1085.9 |
| 17 | 0.01938 | 71.4 | 2532.6 | 58 | 0.2386 | 26.1 | 1086.8 |
| 18 | 0.02064 | 75.6 | 2534.4 | 60 | 0.2563 | 28.1 | 1087.7 |
| 19 | 0.02198 | 79.8 | 2536.2 | 62 | 0.2751 | 30.1 | 1088.6 |
| 20 | 0.02339 | 84.0 | 2538.1 | 64 | 0.2952 | 32.1 | 1089.4 |
| 21 | 0.02487 | 88.1 | 2539.9 | 66 | 0.3165 | 34.1 | 1090.3 |
| 22 | 0.02645 | 92.3 | 2541.7 | 68 | 0.3391 | 36.1 | 1091.2 |
| 23 | 0.02810 | 96.5 | 2543.5 | 70 | 0.3632 | 38.1 | 1092.0 |
| 24 | 0.02985 | 100.7 | 2545.4 | 72 | 0.3887 | 40.1 | 1092.9 |
| 25 | 0.03169 | 104.9 | 2547.2 | 74 | 0.4158 | 42.1 | 1093.8 |
| 26 | 0.03363 | 109.1 | 2549.0 | 76 | 0.4446 | 44.1 | 1094.7 |
| 27 | 0.03567 | 113.3 | 2550.8 | 78 | 0.4750 | 46.1 | 1095.5 |
| 28 | 0.03782 | 117.4 | 2552.6 | 80 | 0.5073 | 48.1 | 1096.4 |
| 29 | 0.04008 | 121.6 | 2554.5 | 82 | 0.5414 | 50.1 | 1097.3 |
| 30 | 0.04246 | 125.8 | 2556.3 | 84 | 0.5776 | 52.1 | 1098.1 |
| 31 | 0.04496 | 130.0 | 2558.1 | 86 | 0.6158 | 54.1 | 1099.0 |
| 32 | 0.04759 | 134.2 | 2559.9 | 88 | 0.6562 | 56.1 | 1099.9 |
| 33 | 0.05034 | 138.3 | 2561.7 | 90 | 0.6988 | 58.1 | 1100.7 |
| 34 | 0.05324 | 142.5 | 2563.5 | 92 | 0.7439 | 60.1 | 1101.6 |
| 35 | 0.05628 | 146.7 | 2565.3 | 94 | 0.7914 | 62.1 | 1102.4 |
| 36 | 0.05947 | 150.9 | 2567.1 | 96 | 0.8416 | 64.1 | 1103.3 |
| 38 | 0.06632 | 159.2 | 2570.7 | 98 | 0.8945 | 66.1 | 1104.2 |

*Source:* Data abridged from "Steam Tables, Metric" and "Steam Tables, English," by J. H. Keenan, F. G. Keyes, P. G. Hill, and J. G. Moore, Wiley, New York, 1969.

initial humidity ratio is normally the wet-bulb temperature, and this leads to an answer of sufficient accuracy. This method does not usually lead to correct answers if the pressure of air–water-vapor mixtures is quite different from the usual atmospheric pressures, or for any other gas-vapor mixtures.

To expedite computations in this section and in Sec. 11-8, Table 11-1 contains saturated-water data as a function of temperature over smaller temperature intervals than appear in Tables A-12M and A-12. Examples in the remaining portion of this chapter will use data primarily from Table 11-1, instead of those in the Appendix.

**Example 11-9M** A sample of atmospheric air at 1 bar has a dry-bulb temperature of 24°C and a wet-bulb temperature of 16°C. Determine (a) the humidity ratio, (b) the relative humidity, and (c) the enthalpy of the mixture per kilogram of dry air.

SOLUTION (a) The humidity ratio, or specific humidity, is calculated from Eq. (11-41). The value of $\omega_2$ in this equation is determined from Eq. (11-38), where in this special case $p_v$ at state 2 is the saturation pressure at 16°C. Therefore,

$$\omega_2^{\cdot} = \frac{0.622(0.01818)}{1.0 - 0.01818} = 0.0115 \text{ kg water/kg dry air}$$

The enthalpies of the water in the liquid and vapor phases are found in Table 11-1. Hence, using Eq. (11-41),

$$\omega_1 = \frac{1.0(16 - 24) + 0.0115(2463.6)}{2545.4 - 67.2} = 0.00820 \text{ kg water/kg dry air}$$

(b) The relative humidity is now computed from Eqs. (11-38) and (11-39). First,

$$\omega_1 = 0.00820 = \frac{0.622 p_v}{1.0 - p_v}$$

The solution of this equation yields a value of $p_v = 0.0130$ bar. Thus the relative humidity, by Eq. (11-39), is

$$\phi = p_v/p_g = 0.0130/0.02985 = 0.436 \text{ (or 43.6\%)}$$

(c) The enthalpy of the mixture is the sum of the enthalpies of the individual components. Per unit mass of dry air this is

$$h_m = h_a + \omega h_v$$

where $h_v$ again is the enthalpy of saturated vapor, and $h_a$ of dry air is given by $c_p T$. We obtain, then,

$$h_1 = 1.0(24) + 0.00820(2545.4) = 44.87 \text{ kJ/kg dry air}$$

It is important to note the arbitrariness of this enthalpy value. The enthalpy of air is chosen to be zero at 0°C, and the enthalpy of the water vapor, on the basis of the steam tables, is zero as a saturated liquid at the triple state of 0.01°C. If differences in enthalpies are to be calculated, the enthalpy at each state must be based on the same reference values.

**Example 11-9** It is found from measurements that a certain sample of atmospheric air at 14.7 psia has a dry-bulb temperature of 74°F and a wet-bulb temperature of 59°F. Determine (a) the humidity ratio, (b) the relative humidity, and (c) the enthalpy of the mixture per pound of dry air.

SOLUTION (a) The humidity ratio, or specific humidity, is found from Eq. (11-41). The value of $\omega_2$ in this equation is determined from Eq. (11-38), where in this special case $p_v$ at state 2 is the saturation pressure at 59°F. Consequently,

$$\omega_2 = 0.622(0.247)/(14.7 - 0.247) = 0.0106 \text{ lb water/lb dry air}$$

The enthalpies of the water in the liquid and vapor phases are found from the steam tables in the appropriate places. Therefore

$$\omega_1 = \frac{0.24(59 - 74) + 0.0106(1060.5)}{1094.1 - 27.1}$$

$$= 0.00716 \text{ lb water/lb dry air} = 50.1 \text{ gr water/lb dry air}$$

(b) The relative humidity is now computed from Eqs. (11-38) and (11-39). First,

$$\omega_1 = 0.00716 = \frac{0.622p_v}{14.7 - p_v}$$

The solution of this equation yields a value of $p_v$ equal to 0.167 psia. Then the relative humidity $\phi = p_v/p_g = 0.167/0.415 = 0.402$, or roughly 40 percent.

(c) The enthalpy of the mixture is the sum of the enthalpies of the individual components on the basis of the Gibbs-Dalton law. Per pound of dry air this is given by $h_a + \omega h_v$, where $h_v$ again is the enthalpy of saturated water vapor at the given temperature, and the enthalpy of dry air is given by $c_p T$. One obtains, then,

$$h_1 = 0.24(74) + 0.00716(1094.1) = 25.60 \text{ Btu/lb dry air}$$

It is important to note the complete arbitrariness of this enthalpy value. The enthalpy of the air was chosen to be zero at 0°F, and the enthalpy of the water vapor, on the basis of the steam tables, is zero as a saturated liquid at 32°F. If differences in enthalpies are to be calculated for engineering purposes, the enthalpy at each state must be based on the same reference values.

It is convenient to have values of the specific volume of the gas-vapor mixture on the basis of a unit mass of dry air. Consider a sample of atmospheric air with a volume such that it contains unit mass of dry air, plus water vapor. The volume $v$ of the mixture per unit mass of dry air is the same as the volume of the dry air per unit mass of dry air. This latter quantity is calculated simply by $v_a = R_a T/p_a$, where $p_a$ is the component pressure of the dry air. Hence the value $v$ we seek is also given by $R_a T/p_a$. With the equations and definitions introduced in the preceding material, we are now in a position to analyze engineering processes involving gas–water-vapor mixtures. No further concepts or definitions are necessary other than the first law of thermodynamics. To make such analyses easier, though, it is convenient to plot the important parameters of such mixtures on a diagram known as a psychrometric chart. A brief discussion of this chart is presented in the following section.

## 11-7 THE PSYCHROMETRIC CHART

To facilitate computations of process changes for dry-air–water-vapor mixtures, it is convenient to plot some of the important parameters on a diagram known as a psychrometric chart. The evaluation of the humidity ratio (specific humidity) requires a specification of the total pressure; consequently, this chart is usually based on a total pressure of one standard atmosphere. Normally, the humidity ratio appears on the ordinate and the dry-bulb temperature on the abscissa. Both of these appear as linear scales. Equation (11-38) shows that the humidity ratio is directly a function of the vapor pressure of the water, at a given total pressure. Therefore the vapor pressure is also frequently plotted on the ordinate, in psia, but this scale will not be linear. We have also seen from Eq. (11-41) that the humidity ratio is determined from a knowledge of the wet-bulb and dry-bulb temperatures for the mixture. Hence the main parameter on the psychrometric chart which relates the humidity ratio and the dry-bulb temperature is the wet-bulb temperature. Lines of wet-bulb temperatures run from the upper left to the lower right of the chart. An outline of a psychrometric chart is shown in Fig. 11-6. The wet-bulb temperatures begin at a 100 percent saturation line, and at this saturation line the wet-bulb and dry-bulb temperatures have the same value. Inasmuch as the relative humidity is directly related to the humidity ratio, lines of relative humidity are also conveniently plotted on this diagram. The saturation line is, of course, a line of 100 percent relative humidity. Other relative-humidity lines follow the same general shape to the right of the saturation line.

   In addition to these parameters, it is also useful to have lines of constant enthalpy and constant specific volume on the psychrometric chart. It will be noted that lines of constant enthalpy (Btu per pound of dry air) lie approximately parallel to lines of constant wet-bulb temperatures, although in reality they are not exactly parallel. During a process of adiabatic saturation the wet-bulb temperature remains the same and the total enthalpy of the stream remains very nearly constant, except for the slight enthalpy change due to the addition of water to the air stream. As a first approximation, the wet-bulb temperatures and the enthalpies are

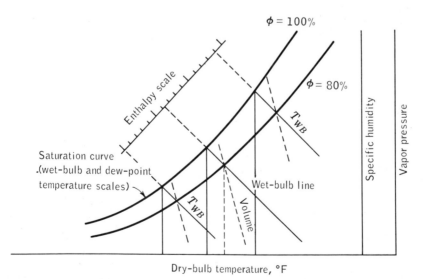

**Figure 11-6** Outline of a psychrometric chart.

frequently plotted as parallel lines, if great accuracy is not desired. Other charts take this slight deviation into account by providing separate scales for the two properties. Lines of constant enthalpy appear as straight lines on the chart. In SI units the enthalpy values tabulated on the chart are based on a zero value for the dry air and for the water vapor at 0°C. On the chart using USCS units the enthalpy values are based on a zero value for the dry air at 0°F and a zero value for the water vapor at 32°F. Finally, lines of constant specific volume also appear. In SI units the values are reported in cubic meters per kilogram of dry air, while in engineering units the values are in terms of cubic feet of mixture per pound of dry air. Correction tables are frequently attached to the psychrometric chart, which allow the user to correct for different total pressures. These corrections are normally small, and will be neglected here. The psychrometric chart provides a convenient and fast method for ascertaining the properties of dry-air–water-vapor mixtures. A psychrometric chart in SI and in USCS units is given in Figs. A-20M and A-20 in the Appendix.

**Example 11-10M** The dry-bulb and wet-bulb temperatures of atmospheric air at a total pressure of 1 bar are 23°C and 16°C, respectively. From the psychrometric chart, Fig. A-20M, determine (a) the humidity ratio, (b) the relative humidity, (c) the vapor pressure, in bars, (d) the dew point, (e) the enthalpy, and (f) the specific volume in m$^3$/kg.

SOLUTION The properties are found from the chart by first finding the point at the intersection of the vertical dry-bulb-temperature line and the sloping wet-bulb-temperature line.

(a) The humidity ratio is read from the ordinate at the right to be about 0.0087 kg water/kg dry air.

(b) It is found that the 50 percent–relative-humidity line runs through the point selected.

(c) The vapor pressure is also read from the ordinate scale at the right, and is, roughly, 0.0140 bar.

(d) The dew point is the temperature at which condensation would just begin if the mixture were cooled at constant pressure. Since the vapor pressure and the humidity ratio remain constant until condensation begins, the dew-point temperature is found by moving horizontally to the left from the initial state until the saturation line is reached. The temperature at this point is close to 12°C, which is the dew point.

(e) The enthalpy is found by following the wet-bulb line (which is also an enthalpy line) from the initial state up to the enthalpy scale. The value read is approximately 45.3 kJ/kg of dry air.

(f) At the initial state a constant-specific-volume line of 0.86 m$^3$/kg passes just to the left of the point. The actual value is close to 0.861 m$^3$/kg of dry air.

**Example 11-10** The dry-bulb and wet-bulb temperatures of atmospheric air at a total pressure of 14.7 psia are 74°F and 59°F, respectively. From the psychrometric chart, Fig. A-20, determine (a) the humidity ratio, (b) the relative humidity, (c) the vapor pressure, in psia, (d) the dew point, (e) the enthalpy, and (f) the specific volume per pound of dry air.

SOLUTION The properties are found from the chart by first finding the point at the intersection of the vertical dry-bulb-temperature line and the sloping wet-bulb-temperature line.

(a) The humidity ratio is read from the ordinate to be slightly more than 50 gr/lb of dry air.

(b) It is found that the 40 percent–relative-humidity line runs through the point selected.

(c) The vapor pressure is also read from the ordinate, and is, roughly, 0.167 psia.

(d) The dew point is the temperature at which condensation would just begin if the mixture were cooled at constant pressure. Since the vapor pressure and the humidity ratio remain constant until condensation begins, the dew point is found by moving horizontally to the left from the initial state until the saturation line is reached. The dry-bulb temperature at the intersection of the constant-humidity ratio line and the saturation line is found to be 48°F, which is the dew-point temperature.

(*e*) The enthalpy of the mixture is found by following the wet-bulb line (which is also an enthalpy line) from the initial state up to the enthalpy scale in the upper left of the diagram. The enthalpy is read to be approximately 25.9 Btu/lb dry air.

(*f*) At the initial state a constant-specific-volume line equal to 13.6 ft$^3$/lb dry air passes through the point on the diagram.

# 11-8 AIR-CONDITIONING PROCESSES

A person generally feels more comfortable when the air within a building is maintained in a fairly limited range of temperatures and relative humidities. However, due to mass- and heat-transfer between the inside of the building and the local environment, and due to internal effects such as cooking, baking, and clothes washing in the home, the temperature and the relative humidity frequently reach undesirable levels. To achieve values of $T$ and $\phi$ within the desired ranges (the comfort zone), it is usually necessary to alter the state of the air. As a result, equipment must be designed to raise or lower the temperature and the relative humidity, individually or simultaneously. In addition to altering the state of a specific air stream by heating, cooling, humidifying, or dehumidifying, a change in state can also be attained by mixing directly the internal or building air with another air stream from, for example, outside the building. Thus there are a number of basic processes to be considered with respect to the conditioning of atmospheric air.

The basic relations available for the evaluation of such processes are three in number. An energy balance on the flow stream(s) may be necessary, as well as mass balances on the water vapor and dry air. In addition, of course, property data for dry air and water substance must be known. In this section all metric examples will be solved by analytical expressions and steam-table data. Examples in USCS units will be solved numerically by means of the psychrometric chart found in the Appendix. However, in either case the psychrometric chart will be used to describe qualitatively various process designs. The chart is extremely helpful as an aid in visualizing changes of state brought about by process equipment.

## 1 Dehumidification with Heating

A fairly common condition within industrial and residential buildings, especially in the summer, is the tendency toward high temperatures and high relative humidities. The discomfort of the body in this situation is well known. A major method of lowering both $T$ and $\phi$ simultaneously is illustrated in Fig. 11-7a. The air to be treated is passed through a flow channel which contains cooling coils. The fluid inside the coils might be, for example, relatively cold water or a refrigerant which has been cooled in a vapor-compression refrigeration cycle. The initial state of the

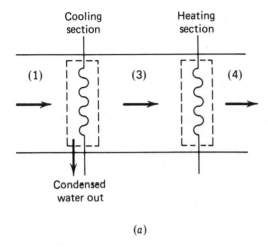

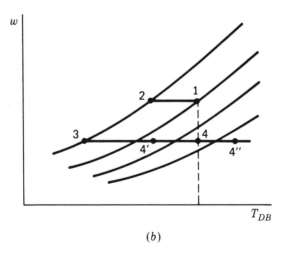

**Figure 11-7** Dehumidification process with heating. (*a*) Equipment; (*b*) psychrometric chart process diagram.

air stream is shown as state 1 on the sketch of a psychrometric chart in Fig. 11-7b. As the air passes through the cooling coil its temperature decreases and its relative humidity increases, at constant specific humidity. If the air remains in contact with the cooling coil sufficiently, the air stream will reach its dew point, indicated by state 2 in Fig. 11-7b. Further cooling requires the air to remain saturated, and its state follows the 100 percent–relative-humidity line to the left toward state 3. During this latter process water condenses out from the air, and its specific humidity is lowered. Hence, by sufficient contact with the coil, both the temperature and water content of the air are lowered. In many cases this conditioned-air stream flows directly back and mixes with the air in the building.

However, in some cases the conditioned air may be at too low a temperature. This is overcome by then passing the air stream leaving the cooling coil section through a heating section. By proper choice of the temperature of the fluid within the heating coil, the temperature of the air stream leaving the overall equipment may be adjusted to the desired value. Three possibilities are shown as states 4, 4', and 4" in Fig. 11-7b. By proper adjustment of the amount of cooling (which controls the position of state 3) and the amount of heating, a suitable state 4 may be attained. The following examples illustrate the overall process.

**Example 11-11M** Outside atmospheric air at 32°C and 70 percent relative humidity is to be conditioned so it enters a home at 22°C and 45 percent relative humidity. First the air passes over cooling coils. The air is cooled below its dew point, and water condenses from the air stream until the desired humidity ratio is reached. The air then passes over a heating coil until the temperature reaches 22°C. Determine (a) the amount of water removed, in kg/kg dry air, (b) the heat removed by the cooling system, in kJ/kg dry air, and (c) the quantity of heat added in the final section, in kJ/kg dry air.

SOLUTION The heat-transfer quantities are given by $q = h_{out} - h_{in}$ and the water removed is measured by $\Delta\omega$ for process 2-3 on Fig. 11-7b. Therefore the $h$ and $\omega$ values are the important properties to evaluate for the overall process. The basic expressions are

$$h_m = c_p T_{air} + \omega h_g$$

$$\omega = 0.622 p_v / (P - p_v)$$

From Table 11-1 the vapor pressures at 32 and 22°C are 0.04759 and 0.02645 bar, respectively. We assume $P = 1$ bar, and recall that $c_p$ for air is 1.00 kJ/(kg)(°C). Therefore

$$\omega_1 = \omega_2 = \frac{0.622(0.70)(0.04759)}{1.00 - 0.70(0.04759)} = 0.0214 \text{ kg water/kg dry air}$$

$$\omega_3 = \omega_4 = \frac{0.622(0.45)(0.02645)}{1.00 - 0.45(0.02645)} = 0.00749 \text{ kg water/kg dry air}$$

At states 1 and 4 the enthalpies are

$$h_1 = 1.00(32) + 0.0214(2559.9) = 86.78 \text{ kJ/kg dry air}$$

$$h_4 = 1.00(22) + 0.00749(2541.7) = 41.04 \text{ kJ/kg dry air}$$

To find $h_2$ and $h_3$, we need information on $T_2$ and $T_3$; $T_2$ is the dew-point temperature and is found from the fact that $p_{v,\,2} = p_g$ at $T_2 = 0.7(0.04759) = 0.0333$ bar. From Table 11-1 a pressure of 0.0333 bar falls between saturation temperatures of 25 and 26°C. By linear interpolation, $T_2 = 25.8$°C. Thus

$$h_2 = 1.00(25.8) + 0.0214(2548.7) = 80.34 \text{ kJ/kg dry air}$$

Finally, $p_{v,\,3} = p_g$ at $T_3 = 0.45(0.02645) = 0.01190$ bar. This pressure falls between 8 and 10°C in Table 11-1. By linear interpolation, $T_3 = 9.5$°C. Hence

$$h_3 = 1.00(9.5) + 0.00749(2518.9) = 28.37 \text{ kJ/kg dry air}$$

(a) The quantity of water removed is given by the difference in the humidity ratios between states 2 and 3. Consequently,

$$\Delta\omega = 0.00749 - 0.0214 = -0.01391 \text{ kg water/kg dry air}$$

The negative sign indicates water was removed from the flow.
(b) The heat removed in the cooling-coil section is found from an energy balance on the section. Neglecting kinetic-energy changes, the steady-flow equation per kilogram of dry air is

$$q = (1)h_3 + (\omega_3 - \omega_1)h_{f,\,3} - (1)h_1$$

$$= 1(28.37) + (-0.01391)(39.9) - 1(86.78)$$

$$= 28.37 - 0.56 - 86.78 = -58.97 \text{ kJ/kg dry air}$$

Note that the energy removed by the condensed-liquid stream is extremely small, and might be neglected as a first approximation.
(c) The heat added in the final section is simply the enthalpy change of the air stream, if it is assumed that the process is adiabatic. Hence

$$q_{\text{in}} = h_4 - h_3 = 41.04 - 28.37 = 12.67 \text{ kJ/kg dry air}$$

The data for state 2 are not necessary for the solution, but were listed merely to indicate the dew-point properties of the mixture. The values calculated for the enthalpies and the humidity ratios at the various states should now be checked by means of the psychrometric chart in the Appendix.

**Example 11-11** Outside atmospheric air at 88°F and 60 percent relative humidity is to be conditioned so that it enters a home at 72°F and 40 percent relative humidity. First the air passes over cooling coils. The air is cooled below its dew point, and water condenses from the air stream until the desired humidity ratio is reached. The air then passes over a heating coil until the temperature reaches 72°F. Determine (a) the amount of water removed per pound of dry air, (b) the heat removed by the cooling system, in Btu/lb dry air, and (c) the quantity of heat added in the heating section, in Btu/lb dry air.

SOLUTION The heat-transfer quantities are given by $q = h_{\text{out}} - h_{\text{in}}$ and the water removed is measured by $\Delta\omega$ for process 2-3 in Fig. 11-7b. Therefore the $h$ and $\omega$ values are the important properties for evaluating the overall process. The basic expressions are

$$h_m = c_p T_{\text{air}} + \omega h_g$$

$$\omega = 0.622 p_v/(P - p_v)$$

From Table 11-1 the vapor pressures at 88 and 72°F are 0.6562 and 0.3887 psia, respectively. We shall assume that $P = 14.7$ psia, and recall that $c_p$ for air is 0.240 Btu/(lb)(°F). Therefore

$$\omega_1 = \omega_2 = \frac{0.622(0.60)(0.6562)}{14.7 - 0.60(0.6562)} = 0.0171 \text{ lb water/lb dry air}$$

$$= 120 \text{ gr water/lb dry air}$$

$$\omega_3 = \omega_4 = \frac{0.622(0.40)(0.3887)}{14.7 - 0.40(0.3887)} = 0.00665 \text{ lb water/lb dry air}$$

$$= 46.6 \text{ gr water/lb dry air}$$

At states 1 and 4 the enthalpies are

$$h_1 = 0.24(88) + 0.0171(1099.9) = 39.9 \text{ Btu/lb dry air}$$

$$h_4 = 0.24(72) + 0.00665(1092.9) = 24.6 \text{ Btu/lb dry air}$$

To find $h_2$ and $h_3$ we need information on $T_2$ and $T_3$; $T_2$ is the dew-point temperature for state 1 and is found from the fact that $p_{v,2} = p_g$ at $T_2 = 0.6(0.6562) = 0.3937$ psia. From Table 11-1 a pressure of 0.3937 psia falls between saturation temperatures of 72 and 74°F. By linear interpolation, $T_2 = 72.4$°F. Thus

$$h_2 = 0.24(72.4) + 0.0171(1093.1) = 36.1 \text{ Btu/lb dry air}$$

Finally, $p_{v,3} = p_g$ at $T_3 = 0.4(0.3887) = 0.1555$ psia. This pressure falls between 46 and 48°F in Table 11-1. By linear interpolation, $T_3 = 46.4$°F. Hence

$$h_3 = 0.24(46.4) + 0.00665(1081.7) = 18.3 \text{ Btu/lb dry air}$$

(a) The quantity of water removed is given by the difference in the humidity ratios between states 2 and 3. Consequently

$$\Delta\omega = 46.6 - 120.0 = -73.4 \text{ gr water/lb dry air}$$

$$= -0.01045 \text{ lb water/lb dry air}$$

The negative sign indicates water was removed from the flow stream.
(b) The heat removed in the cooling-coil section is found from an energy balance on the section. Neglecting kinetic-energy changes, the steady-flow equation per pound of dry air is

$$q = (1)h_3 + (\omega_3 - \omega_1)h_{f,3} - (1)h_1$$

$$= 1(18.3) + (-0.01045)(14.0) - 1(39.9)$$

$$= 18.3 - 0.1 - 39.9 = -21.7 \text{ Btu/lb dry air}$$

Note that the energy removed by the condensed-liquid stream is extremely small, and might be neglected as a first approximation.
(c) The heat added in the final section is simply the enthalpy change of the air stream between states 3 and 4, if it is assumed that the process is adiabatic. Hence

$$q_{in} = h_4 - h_3 = 24.6 - 18.3 = 6.3 \text{ Btu/lb dry air}$$

The data for state 2 are not necessary, but were listed merely to indicate the dew-point properties of the initial air. The values calculated for the enthalpies and the humidity ratios at the various states should now be checked by means of the psychrometric chart in the Appendix.

## 2 Evaporative cooling

In desert climates the air in the atmosphere is frequently hot and dry (very low relative humidity). Rather than pass the air through a refrigerated cooling section, which is costly, it is possible to take advantage of the low humidity to achieve cooling. This is accomplished by passing the air stream through a water-spray section, as shown in Fig. 11-8a. (The equivalent effect may be carried out by passing the air through a filter bed of some type, through which water is allowed to trickle. This provides reasonably good air–liquid-stream contact.) Due to the low relative humidity, part of the liquid-water stream evaporates. The energy for the evaporation process comes from the air stream, so that it is cooled. The overall effect is a cooling and a humidification of the air stream. Since the air is so dry to begin with, the additional moisture added to the air does not particularly make the living conditions uncomfortable. This process is essentially equivalent to the adiabatic-saturation process discussed in Sec. 11-6. Consequently the path of the process follows a constant-wet-bulb or constant-enthalpy line on a psychrometric chart, as shown in Fig. 11-8b. There is a minimum temperature which can be achieved by such a process, which is the saturation state 2′ on the figure. The examples below illustrate the analysis of such a process.

**Example 11-12M** Desert air at 36°C and 10 percent relative humidity passes through an evaporative cooler.
(a) If the final temperature is 20°C, how much water is added per kilogram of dry air and what is the final relative humidity?
(b) What is the minimum temperature that could be achieved by this process? The total pressure is 1 bar.

SOLUTION (a) The amount of water added is $\Delta\omega$. For the initial state, using Eq. (11-38) and Table 11-1,

$$\omega_1 = \frac{0.622(0.10)(0.05947)}{1.00 - 0.10(0.05947)} = 0.00372 \text{ kg water/kg dry air}$$

Since the process is one of constant enthalpy,

$$c_p T_1 + \omega_1 h_{g,\,1} = c_p T_2 + \omega_2 h_{g,\,2}$$

$$1.00(36) + 0.00372(2567.1) = 1.00(20) + \omega_2(2538.1)$$

$$\omega_2 = (36.0 + 9.55 - 20.0)/2538.1 = 0.0101 \text{ kg water/kg dry air}$$

Therefore the amount of water added is

$$\omega_2 - \omega_1 = 0.0101 - 0.00372 = 0.00638 \text{ kg water/kg dry air}$$

Then, from Eq. (11-38), the vapor pressure is

$$0.0101 = 0.622 p_v/(1 - p_v) \qquad p_v = 0.01598 \text{ bar}$$

At 20°C the saturation vapor pressure is 0.02339 bar. Hence

$$\phi_2 = p_v/p_g = 0.01598/0.02339 = 0.683 \text{ (or 68.3 percent)}$$

The values calculated above may be checked on a psychrometric chart.

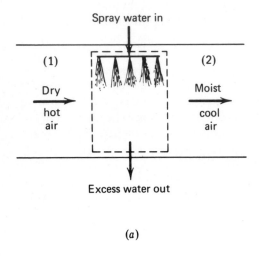

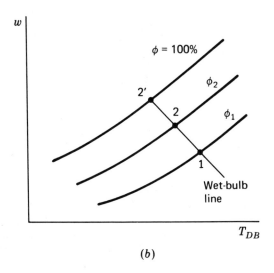

**Figure 11-8** Evaporation cooling. (a) The process equipment; (b) the process on a psychrometric chart.

(b) The minimum temperature is the adiabatic-saturation value, as represented by $T_2$ in Eq. (11-41). Unfortunately, this equation cannot be solved directly for $T_2$. A trial-and-error solution is necessary. Based on known data, Eq. (11-41) becomes

$$0.00372 = \frac{1.00(T_2 - 36) + \omega_2 h_{fg2}}{2567.1 - h_{f2}}$$

In addition, the equation for $\omega_2$ at the final saturation state is

$$\omega_2 = \frac{0.622 p_{g2}}{1.0 - p_{g2}}$$

To solve these two equations, we assume first that $T_2 = 16°C$. Then

$$\omega_2 = \frac{0.622(0.01818)}{1.0 - 0.01818} = 0.0115 \text{ kg water/kg dry air}$$

Finally, the right-hand side of Eq. (11-41) becomes

$$\frac{1.0(16 - 36) + 0.0115(2463.6)}{2567.1 - 67.2} = 0.00333$$

When 17°C is used as an estimate, the right-hand side of Eq. (11-41) is equal to 0.00451. Therefore the minimum temperature lies between 16 and 17°C, and this state is denoted by state 2' on Fig. 11-8b. This minimum temperature can also be found on a psychrometric chart by following a constant-enthalpy line from the initial state to a state of 100 percent relative humidity.

**Example 11-12** Desert air at 98°F and 10 percent relative humidity passes through an evaporative cooler.
(a) If the final temperature is 70°F, how much water is added per pound of dry air and what is the final relative humidity?
(b) What is the minimum temperature that could be achieved by this process?

SOLUTION (a) The amount of water added is $\Delta\omega$. For the initial state, using Eq. (11-38) and Table 11-1,

$$\omega_1 = \frac{0.622(0.10)(0.8945)}{14.7 - 0.10(0.8945)} = 0.00381 \text{ lb water/lb dry air}$$

$$= 26.7 \text{ gr water/lb dry air}$$

The energy associated with the liquid spray can be neglected. As a result the process is essentially one of constant enthalpy, and the energy balance becomes

$$c_p T_1 + \omega_1 h_{g, 1} = c_p T_2 + \omega_2 h_{g, 2}$$

$$0.24(98) + 0.00381(1104.2) = 0.24(70) + \omega_2(1092.0)$$

$$\omega_2 = (23.5 + 4.2 - 16.8)/1092.0 = 0.00998 \text{ lb water/lb dry air}$$

$$= 69.9 \text{ gr water/lb dry air}$$

Therefore the amount of water added is

$$\omega_2 - \omega_1 = 0.00998 - 0.00381 = 0.00617 \text{ lb water/lb dry air}$$

$$= 43.2 \text{ gr water/lb dry air}$$

Then, from Eq. (11-38), the final vapor pressure is

$$0.00998 = 0.622 p_v/(14.7 - p_v) \qquad p_v = 0.232 \text{ psia}$$

At 70°F the saturation vapor pressure is 0.3632 psia. Hence

$$\phi_2 = p_v/p_g = 0.232/0.363 = 0.639 \text{ (or 63.9 percent)}$$

The values calculated above can be checked on a psychrometric chart.

(b) The minimum temperature is the adiabatic-saturation value, which is $T_2$ in Eq. (11-41). Unfortunately, this equation cannot be solved directly for $T_2$. A trial-and-error solution is necessary. Based on known data, Eq. (11-41) becomes

$$0.00381 = \frac{0.24(T_2 - 98) + \omega_2 h_{fg2}}{1104.2 - h_{f2}}$$

In addition, the equation for $\omega_2$ at the final saturation state is

$$\omega_2 = \frac{0.622 p_{g2}}{14.7 - p_{g2}}$$

To solve these two equations, we shall first let $T_2 = 62°F$. Hence

$$\omega_2 = \frac{0.622(0.2751)}{14.7 - 0.2751} = 0.0119 \text{ lb water/lb dry air}$$

Then the right-hand side of Eq. (11-41) above becomes

$$\frac{0.24(62 - 98) + 0.0119(1058.5)}{1104.2 - 30.1} = 0.00364$$

This result is very close to the desired value of 0.00381. Hence 62°F is the minimum possible temperature and this state is denoted by state 2′ on Fig. 11-8b. This minimum temperature value can also be found on a psychrometric chart by following a constant-enthalpy line from the initial state to a state of 100 percent relative humidity.

## 3 Heating with Humidification

In winter or at high altitudes the air in the atmosphere frequently is dry (low relative humidity) and cold. Thus the engineering problem is one of increasing both the water content and the temperature of any inlet air into a building. One method of achieving humidification with heating is shown in Fig. 11-9a. The air stream passes first over a heating coil, and then through a spray section. Process 1-2 is well defined for the heating section, but process 2-3, shown in Fig. 11-9b, has a number of possible end states. State 3 is a function of the temperature of the water which enters the air stream. Usually the water will be nearly the same temperature as the air stream. Thus process 2-3 will be evaporative cooling, as described in the preceding section. As another possibility, steam may be introduced in place of liquid water. This results in humidification with additional heating, as shown by state 3′ in Fig. 11-9b. The processes could be reversed, of course, with water introduction followed by heating. Heating with humidification is illustrated by the following examples.

**Example 11-13M** An air stream at 8°C and 30 percent relative humidity is first heated to 32°C and then passed through an evaporative cooler until the temperature reaches 26°C. Determine (a) the heat added, in kJ/kg dry air, and (b) the final relative humidity. Total pressure is 1 bar.

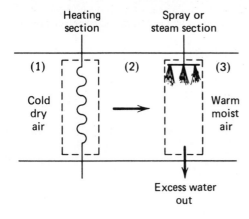

(a)

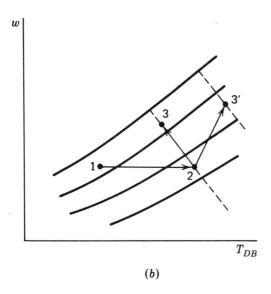

(b)

**Figure 11-9** Heating with humidification. (a) Schematic of the process equipment; (b) process diagram on a psychrometric chart.

SOLUTION (a) The initial enthalpy and humidity ratio are evaluated as follows from Eqs. (11-38) and (11-41):

$$\omega_1 = \frac{0.622(0.30)(0.01072)}{1.00 - 0.30(0.01072)} = 0.00201 \text{ kg water/kg dry air}$$

$$h_1 = 1.00(8) + 0.00201(2516.1) = 13.05 \text{ kJ/kg dry air}$$

Since the humidity ratio is constant during heating,

$$h_2 = 1.00(32) + 0.00201(2559.9) = 37.15 \text{ kJ/kg dry air}$$

Therefore the heat added is

$$q = h_2 - h_1 = 37.15 - 13.05 = 24.1 \text{ kJ/kg dry air}$$

(b) The humidity ratio at state 3 is determined by equating the initial and final enthalpies for the spray process. Thus

$$c_p T_2 + \omega_2 h_{g,\,2} = c_p T_3 + \omega_3 h_{g,\,3}$$

$$1.00(32) + 0.00201(2559.9) = 1.00(26) + \omega_3(2549.0)$$

$$\omega_3 = (32.0 + 5.15 - 26.0)/2549.0 = 0.00437 \text{ kg water/kg dry air}$$

This value of $\omega_3$ is used to find the vapor pressure at state 3 through the use of Eq. (11-38).

$$0.00437 = 0.622 p_v/(1 - p_v) \qquad p_v = 0.00700 \text{ bar}$$

At 26°C the vapor pressure from Table 11-1 is 0.03363 bar. Hence the relative humidity at state 3 is

$$\phi_3 = p_v/p_g = 0.00700/0.03363 = 0.208 \text{ (or 20.8 percent)}$$

**Example 11-13** An air stream at 50°F and 40 percent relative humidity is first heated to 96°F and then passed through an evaporative cooler until the temperature reaches 70°F. Determine (a) the heat added, in Btu/lb dry air, and (b) the final relative humidity. The total pressure is 14.7 psia.

SOLUTION (a) The initial enthalpy and humidity ratio are evaluated as follows from Eqs. (11-38) and (11-41):

$$\omega_1 = \frac{0.622(0.40)(0.178)}{14.7 - 0.40(0.178)} = 0.00303 \text{ lb water/lb dry air}$$

$$h_1 = 0.24(50) + 0.00303(1083.3) = 15.28 \text{ Btu/lb dry air}$$

Since the humidity ratio is constant during heating,

$$\omega_2 = 0.24(96) + 0.00303(1103.3) = 26.2 \text{ Btu/lb dry air}$$

Therefore the heat added is

$$q = h_2 - h_1 = 26.2 - 15.3 = 10.9 \text{ Btu/lb dry air}$$

(b) The humidity ratio at state 3 is determined by equating the initial and final enthalpies for the spray process. Thus

$$c_p T_2 + \omega_2 h_{g,\,2} = c_p T_3 + \omega_3 h_{g,\,3}$$

$$0.24(96) + 0.00303(1103.3) = 0.24(70) + \omega_3(1092.0)$$

$$\omega_3 = (26.2 - 16.8)/1092 = 0.00855 \text{ lb water/lb dry air}$$

This value of $\omega_3$ is next used to find the vapor pressure at state 3 through the use of Eq. (11-38).

$$0.00855 = \frac{0.622\phi p_g}{14.7 - \phi p_g} = \frac{0.622(0.363)\phi}{14.7 - 0.363\phi}$$

$$\phi = 0.55 \text{ (or 55 percent)}$$

The values calculated above can be checked on the psychrometric chart.

## 4 Adiabatic Mixing of Two Streams

Another important application in air conditioning is the mixing of two dry air–water-vapor streams, as shown in Fig. 11-10a. Three basic relations can be written for the overall control volume on a rate basis:

1. mass balance on dry air

$$\dot{m}_{a1} + \dot{m}_{a2} = \dot{m}_{a3} \tag{11-42a}$$

2. mass balance on water vapor

$$\dot{m}_{a1}\omega_1 + \dot{m}_{a2}\omega_2 = \dot{m}_{a3}\omega_3 \tag{11-42b}$$

3. energy balance for adiabatic mixing

$$\dot{m}_{a1}(h_{a1} + \omega_1 h_{g1}) + \dot{m}_{a2}(h_{a2} + \omega_2 h_{g2}) = \dot{m}_{a3}(h_{a3} + \omega_3 h_{g3}) \tag{11-42c}$$

When properties of two of the flow streams are known, these three equations are sufficient to evaluate the properties of the third stream.

When the properties of the two inlet streams are known, the three equations above may be used to evaluate $\omega_3$ and $h_3$ of the exit stream. If Eqs. (11-42b) and (11-42c) are divided by $\dot{m}_{a3}$, then

$$\frac{\dot{m}_{a1}}{\dot{m}_{a3}}\omega_1 + \frac{\dot{m}_{a2}}{\dot{m}_{a3}}\omega_2 = \omega_3 \tag{11-43}$$

and

$$\frac{\dot{m}_{a1}}{\dot{m}_{a3}}h_1 + \frac{\dot{m}_{a2}}{\dot{m}_{a3}}h_2 = h_3 \tag{11-44}$$

The ratios $\dot{m}_{a1}/\dot{m}_{a3}$ and $\dot{m}_{a2}/\dot{m}_{a3}$ represent the fractions of the total flow which enter the mixing process at states 1 and 2. Thus knowledge of the inlet states and the inlet mass flow rates of dry air is sufficient to determine the exit humidity ratio and enthalpy. These latter two properties fix all the other properties of the exit stream, such as its dry-bulb and wet-bulb temperatures and its relative humidity.

The mixing process also has an interesting interpretation on a psychrometric chart. For this purpose it is useful to combine Eqs. (11-42a) through (11-42c) to form two additional expressions. When Eqs. (11-42a) and (11-42b) are combined so that $\dot{m}_{a3}$ is eliminated, then

$$\frac{\dot{m}_{a1}}{\dot{m}_{a2}} = \frac{\omega_2 - \omega_3}{\omega_3 - \omega_1} \tag{11-45}$$

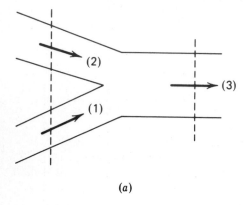

(a)

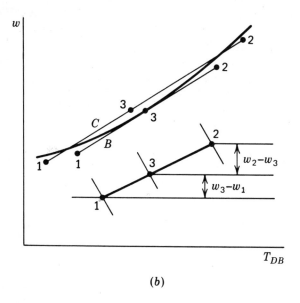

(b)

**Figure 11-10** Adiabatic mixing of two air streams. (a) Schematic of mixing process; (b) process on a psychrometric chart.

If Eq. (11-42a) is substituted into Eq. (11-42c), the result is

$$\frac{\dot{m}_{a1}}{\dot{m}_{a2}} = \frac{h_2 - h_3}{h_3 - h_1} \tag{11-46}$$

Both Eqs. (11-45) and (11-46) have a geometric interpretation with respect to the psychrometric chart. For example, note process line A on Fig. 11-10b. State 3 will have a humidity ratio which lies between that of states 1 and 2. Equation (11-45) dictates that the vertical distances between states 2 and 3 and between states 3 and 1 on the psychrometric chart are in proportion to the ratio of the mass flow rates of dry air for streams 1 and 2. A similar analysis may be made for Eq. (11-46) in terms of the constant-enthalpy lines on the chart.

Figure 11-10b illustrates three possible situations with respect to state 3. For process line A, states 1 and 2 are situated so that state 3 must lie below the 100 percent relative humidity line. In this case $\phi_3 < 1$, regardless of the mass ratio, $\dot{m}_{a1}/\dot{m}_{a2}$. Process line B indicates that state 3 may be saturated ($\phi = 1$). For given positions of states 1 and 2, this requires a definite value for $\dot{m}_{a1}/\dot{m}_{a2}$. Finally, if states 1 and 2 lie close to or on the 100 percent relative humidity line, then state 3 may lie to the left of the saturation line. In this case, water will condense during the mixing process, and frequently will remain suspended as foglike droplets in the exit flow stream. Generally, this would be an undesirable condition if the flow stream goes directly into a home or business area. The following examples illustrate the adiabatic mixing process of two flow streams.

**Example 11-14M** An air stream (1) enters an adiabatic mixing chamber at a rate of 150 m³/min at 10°C and $\phi = 0.80$. It is mixed with another stream (2) at 32°C and $\phi = 0.60$ at a rate of 100 m³/min. Determine the final temperature and relative humidity of the exit stream, if the total pressure is 1 bar.

SOLUTION Equations (11-45) and (11-46) require a knowledge of the mass flow rates of dry air. The volume rates given are for the total flow, including water vapor. However, $\dot{m}_a$ can be found by dividing the volume rate by the specific volume of the air. It was noted in Sec. 11-6 that $v = R_a T/p_a$. Thus, for state 1

$$p_{v1} = \phi_1 p_g = 0.8(0.01228) = 0.0098 \text{ bar}$$

$$p_{a1} = P - p_{v1} = 1.0 - 0.0098 = 0.9902 \text{ bar}$$

$$v_1 = \frac{0.08315(283)}{29(0.9902)} = 0.819 \text{ m}^3/\text{kg dry air}$$

$$\dot{m}_{a1} = 150/0.819 = 183 \text{ kg dry air/min}$$

$$\omega_1 = \frac{0.622(0.0098)}{0.9902} = 0.00616 \text{ kg water/kg dry air}$$

$$h_1 = 1.0(10) + 0.00616(2519.8) = 25.5 \text{ kJ/kg dry air}$$

Similarly for state 2,

$$p_{v2} = \phi_2 p_g = 0.6(0.04759) = 0.0286 \text{ bar}$$

$$p_{a2} = P - p_{v2} = 1.0 - 0.0286 = 0.9714 \text{ bar}$$

$$v_2 = \frac{0.08315(305)}{29(0.9714)} = 0.900 \text{ m}^3/\text{kg dry air}$$

$$\dot{m}_{a2} = 100/0.900 = 111 \text{ kg dry air/min}$$

$$\omega_2 = \frac{0.622(0.0286)}{0.9714} = 0.0183 \text{ kg water/kg dry air}$$

$$h_2 = 1.0(32) + 0.0183(2559.9) = 78.8 \text{ kJ/kg dry air}$$

From these data we can evaluate $\omega_3$ and $h_3$ from Eqs. (11-45) and (11-46).

$$\frac{183}{111} = \frac{\omega_3 - 0.0183}{0.00616 - \omega_3}$$

$$\omega_3 = 0.0108 \text{ kg water/kg dry air}$$

$$\frac{183}{111} = \frac{h_3 - 78.8}{25.5 - h_3}$$

$$h_3 = 45.64 \text{ kJ/kg dry air}$$

The temperature $T_3$ is found from the value of $h_3$, since

$$h_3 = 45.64 = 1.0(T_3) + 0.0108(h_{v3})$$

The solution is by trial and error, since Table 11-1 must be used to find $h_{v3}$ at guessed values of $T_3$. At 18°C, $h_3$ is calculated by the above equation to be 45.37 kJ/kg dry air, and at 20°C the value is 47.41 kJ/kg dry air. Since we seek a value of 45.64, the actual exit temperature is close to 18°C. From the $\omega_3$ value the vapor pressure may be found.

$$\omega_3 = 0.0108 = 0.622 p_v/(1 - p_v) \qquad p_v = 0.01707 \text{ bar}$$

Since the saturation pressure at 18°C is 0.02064 bar, then

$$\phi = p_v/p_g = 0.01707/0.02064 = 0.827 \text{ (or 82.7 percent)}$$

**Example 11-14** An air stream (1) enters an adiabatic mixing chamber at a rate of 1500 ft$^3$/min at 50°F and $\phi = 0.80$. It is mixed with another stream (2) at 90°F and $\phi = 0.60$ entering at a rate of 1000 ft$^3$/min. Determine the final temperature and relative humidity of the exit stream, if the total pressure is 14.7 psia.

SOLUTION The values of $T_3$ and $\phi_3$ can be read directly from the psychrometric chart if we can find the position of line $A$ in Fig. 11-10b, and either $h_3$ or $\omega_3$. The mass flow rates of dry air at states 1 and 2 are found in the following manner: The specific volumes of air at states 1 and 2 read from the psychrometric chart are 12.96 and 14.24 ft$^3$/lb dry air, respectively. Therefore the mass flow rates are

$$\dot{m}_{a1} = 1500/12.96 = 115.7 \text{ lb dry air/min}$$

$$\dot{m}_{a2} = 1000/14.24 = 70.2 \text{ lb dry air/min}$$

The psychrometric chart also indicates that

$$\omega_1 = 43 \text{ gr water/lb dry air} \qquad \omega_2 = 128 \text{ gr water/lb dry air}$$

$$h_1 = 18.5 \text{ Btu/lb dry air} \qquad h_2 = 41.8 \text{ Btu/lb dry air}$$

These values, in conjunction with Eqs. (11-42), (11-43), and (11-44), can be used to determine $h_3$ and $\omega_3$. From Eq. (11-42),

$$\dot{m}_{a3} = 115.7 + 70.2 = 185.9 \text{ lb dry air/min}$$

Next, from Eq. (11-43),

$$\omega_3 = \frac{115.7(43) + 70.2(128)}{185.9} = 75.1 \text{ lb water/lb dry air}$$

Finally, from Eq. (11-44),

$$h_3 = \frac{115.7(18.5) + 70.2(41.8)}{185.9} = 27.3 \text{ Btu/lb dry air}$$

These two values fix state 3. However, this state must also lie on the line which connects states 1 and 2 on the psychrometric chart, as a check. From the chart we find that

$$T_3 = 65°F \qquad \phi_3 = 82 \text{ percent}$$

and these values do lie on the connecting line.

## 5 Wet Cooling Tower

At fossil- and nuclear-fueled power plants a considerable portion of the energy released by the fuel must be rejected to the environment. Cooling water from natural sources such as rivers and lakes is commonly used to carry off some or all of the rejected energy. Due to environmental concern, there is a limitation on the temperature of the cooling-water effluent from the plant which is discharged back to the natural source. (The resulting temperature rise is known as thermal pollu-tion.) Since the air of the atmosphere is so large in size, it can also be used as a sink for the energy rejected to the environment. The engineering problem is the transfer of this energy from the cooling water to the atmospheric air.

One method of transferring energy from any water stream to the atmosphere is through the use of a wet cooling tower. A schematic of the device is shown in Fig. 11-11. The warm cooling water is sprayed from the top of the tower and falls downward under gravitational force. Atmospheric air is inducted at the bottom of the tower by a fan and flows upward and in counterflow to the falling water droplets. The water and air streams are thus brought into intimate contact, and a small fraction of the water stream evaporates into the air stream. The evaporation process requires energy, and this energy transfer results in a cooling of the remain-ing water stream. The cooled water stream is then returned to the power plant to pick up additional waste energy. Since a portion of the cooling water is eva-porated into the air stream, an equivalent amount of water must be added as makeup somewhere in the flow cycle.

To make an energy analysis of a wet cooling tower, a control volume is drawn around the entire tower, as shown by the dashed line in Fig. 11-11. The process is assumed to be adiabatic, the fan work is neglected, and changes in kinetic and potential energy are negligible. Consequently, the basic energy balance given by Eq. (5-31) reduces to

$$\sum (\dot{m}h)_{\text{out}} - \sum (\dot{m}h)_{\text{in}} = 0$$

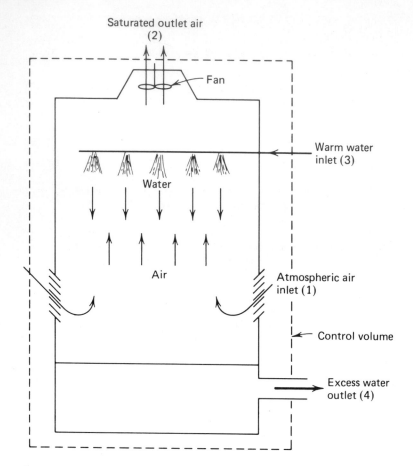

**Figure 11-11** Schematic of a wet cooling tower.

In terms of Fig. 11-11 this may be written as

$$\dot{m}_{a1}(h_{a1} + \omega_1 h_{v1}) + \dot{m}_{w3} h_{w3} = \dot{m}_{a2}(h_{a2} + \omega_2 h_{v2}) + \dot{m}_{w4} h_{w4} \qquad (11\text{-}47)$$

For atmospheric air, the enthalpy of the water vapor $h_v$ may be approximated by $h_g$ at the given temperature, and for the water stream enthalpy $h_w$ may be evaluated by $h_f$ at the given temperature. Also recall that $h_a = c_{pa} T$, and note that $\dot{m}_{a1} = \dot{m}_{a2} = \dot{m}_a$. In this new symbolism, then, Eq. (11-47) becomes

$$\dot{m}_a[c_{pa}(T_1 - T_2) + \omega_1 h_{g1} - \omega_2 h_{g2}] = \dot{m}_{w4} h_{f4} - \dot{m}_{w3} h_{f3} \qquad (11\text{-}48)$$

As a final step, it is convenient to replace $\dot{m}_{w4}$ in terms of $\dot{m}_{w3}$. This can be done by making a mass balance on the water passing through the control volume. This mass balance is

$$\dot{m}_{w4} = \dot{m}_{w3} - \dot{m}_a(\omega_2 - \omega_1) \qquad (11\text{-}49)$$

Substitution of Eq. (11-49) into Eq. (11-48) yields

$$\dot{m}_a[c_{pa}(T_1 - T_2) + \omega_1 h_{g1} - \omega_2 h_{g2} + (\omega_2 - \omega_1)h_{f4}] = \dot{m}_{w3}(h_{f4} - h_{f3}) \qquad (11\text{-}50)$$

In many uses of cooling towers the value of $\dot{m}_{w3}$ is known, as well as the temperature drop needed for the water stream. The above equation is then used to calculate the mass flow rate of air required on a dry basis. To obtain the total mass flow of air, the dry rate is multiplied by $(1 + \omega_1)$.

The mass flow rate of dry air may also be calculated directly from Eq. (11-47) if data from a psychrometric chart are to be used. If Eqs. (11-47) and (11-49) are combined, the result is

$$\dot{m}_a h_{m1} + \dot{m}_{w3} h_{f3} = \dot{m}_a h_{m2} + h_{f4}[\dot{m}_{w3} - \dot{m}_a(\omega_2 - \omega_1)]$$

or $\qquad \dot{m}_a[h_{m1} - h_{m2} + h_{f4}(\omega_2 - \omega_1)] = \dot{m}_{w3}(h_{f4} - h_{f3})$ $\qquad$ (11-51)

where $h_m$ is the total enthalpy of dry air plus water vapor recorded on the psychrometric chart.

**Example 11-15M** Water enters a cooling tower at 40°C and leaves at 25°C. The tower receives atmospheric air at 20°C and 40 percent relative humidity. The air leaves the tower at 35°C and 95 percent relative humidity. Find the mass flow rate of dry air, kg/min, passing through the tower if the water rate at the inlet is 12,000 kg/min.

SOLUTION The main data needed, in addition to steam-table information, are the humidity ratios for the inlet and exit air streams. These are obtained as follows:

$$\omega_1 = \frac{0.622(0.40)(0.02339)}{1.0 - 0.40(0.02339)} = 0.00603 \text{ kg water/kg dry air}$$

$$\omega_2 = \frac{0.622(0.95)(0.05628)}{1.0 - 0.95(0.05628)} = 0.0351 \text{ kg water/kg dry air}$$

Other data include:

$$h_{g1} = 2538.1 \text{ kJ/kg} \qquad h_{f3} = 167.6 \text{ kJ/kg}$$

$$h_{g2} = 2565.3 \text{ kJ/kg} \qquad h_{f4} = 104.9 \text{ kJ/kg}$$

With these data, and recalling that $c_{pa} = 1.00 \text{ kJ/(kg)(°C)}$, we find that Eq. (11-50) yields

$$\dot{m}_a[1.0(20 - 35) + 0.00603(2538.1) - 0.0351(2565.3) + (0.0351 - 0.00603)(104.9)]$$

$$= 12,000(104.9 - 167.6)$$

$$\dot{m}_a = \frac{12,000(-62.7)}{-15 + 15.3 - 90.0 + 3.0} = 8,680 \text{ kg dry air/min}$$

To find the mass flow rate of atmospheric air required, the above answer should be multiplied by $(1 + \omega_1)$.

**Example 11-15** Water enters a cooling tower at 100°F and leaves at 80°F. The tower receives atmospheric air at 70°F and 40 percent relative humidity, and the air leaves at 95°F and 95 percent relative humidity. Find the mass flow rate of dry air, in lb/min, required if the water rate at the inlet is 25,000 lb/min.

SOLUTION The main data needed, in addition to steam-table information, are the humidity ratios for the inlet and outlet air stream. From the psychrometric chart, A-20, the value of $\omega_1$ is found to be 44 gr/lb or 0.00629 lb water/lb dry air. The exit condition falls off the chart, and must be calculated.

$$\omega_2 = \frac{0.622(0.95)(0.816)}{14.7 - 0.95(0.816)} = 0.0346 \text{ lb water/lb dry air}$$

Other data needed include:

$$h_{g1} = 1092.0 \text{ Btu/lb} \qquad h_{f3} = 68.1 \text{ Btu/lb}$$

$$h_{g2} = 1102.9 \text{ Btu/lb} \qquad h_{f4} = 48.1 \text{ Btu/lb}$$

With these data, and recalling that $c_{pa} = 0.24$ Btu/(lb)(°F), we find that Eq. (11-50) yields

$$\dot{m}_a[0.24(70 - 95) + 0.00629(1092.0) - 0.0346(1102.9) + (0.0346 - 0.00629)(48.1)]$$

$$= 25,000(48.1 - 68.1)$$

$$\dot{m}_a = \frac{25,000(-20.0)}{-6.0 + 6.9 - 38.2 + 1.4} = 14,000 \text{ lb dry air/min}$$

To find the mass flow rate of atmospheric air needed, the above answer should be multiplied by $(1 + \omega_1)$.

# PROBLEMS (METRIC)

### Mass, volume, pressure, temperature relations

**11-1M** A gas mixture has the following volumetric analysis: $N_2$, 60 percent; $CO_2$, 33 percent; $O_2$, 7 percent.
  (a) Determine the gravimetric analysis.
  (b) What is the mass in kilograms of 30 m³ of this gas at 1 bar and 27°C?

**11-2M** Calculate the specific volume, in cm³/g, of an ideal-gas mixture composed of 16 percent $H_2$, 28 percent CO, and 56 percent $N_2$, by weight, at 1 bar and 77°C.

**11-3M** The volumetric analysis of a mixture of gases in $N_2$, 70 percent; $CO_2$, 20 percent; $O_2$, 10 percent.
  (a) Determine the gravimetric analysis.
  (b) Compute the component pressure of $CO_2$ if the mixture is at 1 bar and 60°C.
  (c) How many kilograms of the mixture would be contained in 15 m³ at the conditions specified in part (b)?

**11-4M** A gas mixture has the following gravimetric analysis: $O_2$, 32 percent; CO, 56 percent; He, 12 percent. Determine (a) the volumetric analysis, and (b) the apparent gas constant, in kJ/(kg)(°K).

**11-5M** A gas mixture contains 0.28 kg of CO, 0.16 kg of $O_2$, and 0.66 kg of $CO_2$ at 1.4 bars and 37°C. Compute (a) the partial pressure of each component, in millibars, and (b) the volume occupied by each component measured, in cubic meters, at its own partial pressure and mixture temperature.

**11-6M** A 0.1-m³ tank contains 0.7 kg of $N_2$ and 1.1 kg of $CO_2$ at 27°C. Compute (a) the component pressure of each species, in bars, (b) the component volume of each species, in cubic meters, (c) the total pressure of the mixture in bars, and (d) the gas constant for the mixture, in kJ/(kg)(°K).

**11-7M** A gas mixture is made up of 20 percent $O_2$ and 80 percent Ar by weight. The total pressure is 3 bars, and the temperature is 27°C. Determine (a) the partial pressure of argon, in bars, (b) the apparent molar mass in kg/kg·mol, and (c) the gas constant, in kJ/(kg)(°K).

**11-8M** A mixture of ideal gases has the following volumetric analysis: $CO_2$, 50 percent; $N_2$, 30 percent; $CH_4$, 20 percent.
  (a) Calculate the gravimetric analysis of the mixture and the apparent molar mass.
  (b) A tank of 22.4 l capacity contains 0.08 kg of this gas at 7°C. What is the pressure in the tank, in bars?

**11-9M** A mixture at 300°K and 1.8 bars has a volumetric analysis as follows: $O_2$, 60 percent; $CO_2$, 40 percent. Compute (a) the mass analysis, (b) the partial pressure of $O_2$ in bars, and (c) apparent molar mass of the mixture in kg/kg·mol.

**11-10M** In what molar ratio should helium and krypton be mixed so that the gas mixture may have the same average molar mass as pure argon?

**11-11M** In what molar ratio should helium and carbon dioxide be mixed so that the gas mixture may have the same average molar mass as pure diatomic oxygen?

**11-12M** A mixture of ideal gases at 1.1 bars and 77°C has the following volumetric analysis: 87.5 percent He and 12.5 percent $N_2$. Calculate (a) the partial pressure of He, (b) the percent $N_2$ by weight, (c) the apparent gas constant in bar·m³/(kg)(°K), and (d) the specific heat at constant volume in kJ/(kg)(°K).

**11-13M** A mixture of ideal gases at 1.2 bars and 127°C has the following volumetric analysis: 25 percent CO and 75 percent Ar. Calculate (a) the partial (component) pressure of CO, (b) the percent Ar by weight, (c) the apparent gas constant in kJ/(kg)(°K), and (d) the specific heat at constant pressure in kJ/(kg)(°K) for the mixture.

**11-14M** A gas sample has the following volumetric analysis: $H_2$, 4 percent; CO, 12 percent; $CO_2$, 35 percent, and $N_2$, 49 percent. Determine (a) the gravimetric analysis, (b) the component pressure of CO and $CO_2$, and (c) the constant-pressure specific heat of the mixture in kJ/(kg)(°K) if the temperature is 77°C and the total pressure is 1 bar.

**11-15M** A rigid tank contains 2.0 kg of $H_2O$, 4.0 kg of $O_2$, and 7.0 kg of CO at 77°C. If the total pressure is 200 mbar, calculate the partial pressure of CO, in millibars, and the enthalpy of the mixture, in kJ/kg, based on tabular data in the Appendix.

**11-16M** For the mixture described in Prob. 11-2M determine at 350°K (a) the enthalpy of the mixture in kJ/(g·mol), and (b) $c_p$ of the mixture in kJ/(kg·mol)(°C).

**11-17M** The mixture described in Prob. 11-3M is cooled as it passes through a heat exchanger in steady state from 550 to 325°K. Compute the heat transferred, in kJ/kg·mol of mixture.

**11-18M** For the mixture described in Prob. 11-5M, evaluate (a) $c_v$, and (b) $c_p$ for the mixture, in kJ/(kg)(°K) at 77°C.

**11-19M** For the mixture described in Prob. 11-5M, determine (a) the internal energy, and (b) the enthalpy of the mixture, in kJ/kg·mol, based on tabular data at 127°C.

**11-20M** For the mixture described in Prob. 11-6M, determine (a) the enthalpy, and (b) the internal energy of the mixture in kJ/kg·mol (relative to 298°K), based on tabular data.

**11-21M** Based on the data of Prob. 11-7M, determine the constant-pressure specific heat, in kJ/(kg·mol)(°C), of the mixture.

**11-22M** For the mixture described in Prob. 11-1M, compute the value of (a) $c_p$, and (b) $c_v$ for the mixture in J/(g·mol)(°C) at 27°C.

**11-23M** Referring to Prob. 11-9M, determine (a) the enthalpy change of the mixture, in kJ/kg·mol, between 300 and 400°K, and (b) $c_p$ of the mixture at 350°K, in kJ/(kg·mol)(°K).

**Energy analysis of ideal-gas mixtures**

**11-24M** An equimolar mixture of two gases flows steadily through a heat exchanger. The initial and final temperatures of the mixture are 300 and 1000°K, respectively. If the total flow rate is 10 g·mol/s, what is the rate of heat transfer to the gas in kJ/s for a mixture of (a) $N_2$ and $CO_2$, (b) CO and $CO_2$, and (c) $N_2$ and Ar.

**11-25M** A mixture of three gases with component pressures in the ratio 1 : 2 : 5 is expanded isentropically from 600°K and 10 bars to 1 bar in a piston-cylinder assembly. For constant chemical composition how much work is done, in kJ/kg·mol of mixture, for mixtures of (a) $H_2O$, CO, and $N_2$, and (b) CO, Ar, and $N_2$.

**11-26M** A gas mixture consisting of $N_2$, $CO_2$, and $H_2O$ in a molar ratio of 4 : 1 : 1 enters a turbine at 1000°K. The gases expand isentropically through an 8 : 1 pressure ratio. Compute the work output, in kJ/kg·mol, of the mixture.

**11-27M** A piston-cylinder assembly contains 0.06 kg · mol of $N_2$ and 0.04 kg · mol of $O_2$. The mixture is compressed isothermally at 500°K from 1 bar to 3 bars. What is the required work input, in kilojoules?

**11-28M** One gram·mole of $CO_2$ initially at 2 bars and 27°C is mixed adiabatically with two gram·moles of $O_2$ initially at 5 bars and 152°C. During the constant-volume mixing process electrical energy equivalent to 0.67 kJ/g·mol of mixture is added. Determine the final temperature of the mixture, in °C.

**11-29M** One kilogram·mole of $CO_2$, initially at 2 bars and 202°C, is mixed adiabatically in a closed system at constant pressure with two kilogram·moles of $N_2$, initially at 2 bars and 27°C. During the process, 1000 kJ are added in the form of electrical work. Determine the final temperature of the mixture, in °C.

**11-30M** Two cubic meters of $N_2$ at 1.4 bars and 127°C are allowed to mix with five cubic meters of $H_2$ at 3 bars and 27°C by removing a partition between the two tanks of gas. If the final temperature is 77°C, determine the magnitude and direction of the heat transfer, in kilojoules.

**11-31M** Two kilograms of nitrogen, initially in a rigid tank at 7 bars and 177°C, are mixed with one kilogram of oxygen, initially in another tank at 3 bars and 27°C, by opening a valve connecting the two tanks. Determine (a) the heat transfer, in kilojoules, and (b) the final pressure, in bars, if the final temperature is 102°C.

### Entropy changes of ideal-gas mixtures

**11-32M** Hydrogen and nitrogen in separate streams enter an adiabatic mixing chamber in a 3 : 1 molar ratio. At the inlet the hydrogen is at 2 bars and 77°C, and the nitrogen is at 2 bars and 277°C. The mixture leaves at 1.9 bars.
 (a) Determine the final temperature of the mixture, in °K.
 (b) Compute the entropy change of the hydrogen gas, in kJ/(kg·mol)(°K).

**11-33M** An insulated 0.3-m³ tank is divided into two sections by a partition. One section is 0.2 m³ in volume and initially contains hydrogen gas at 2 bars and 127°C. The remaining section initially holds nitrogen gas at 4 bars and 27°C. The adiabatic partition is then removed, and the gases are allowed to mix. Determine (a) the temperature of the equilibrium mixture, in °K, (b) the pressure of the mixture, and (c) the entropy change for each component and the total value, in kJ/°K.

**11-34M** A system consists of two tanks interconnected with a pipe and valve. One tank with a volume of 0.83 m³ contains 2 kg of Ar at 1.5 bars and 27°C. The other tank has a volume of 0.33 m³ and holds 1.6 kg of $O_2$ at 5 bars and 127°C. The valve is opened and the gases mix. The atmospheric temperature is 17°C; the final temperature of the mixture is 77°C. Determine (a) the heat transfer, in kilojoules, (b) the final pressure, in bars, (c) the change in the entropy of the Ar and the $O_2$, in kJ/°K, and (d) the entropy change of the atmosphere, in kJ/°K.

**11-35M** An equimolar mixture of $N_2$ and $CO_2$ is compressed from 1 bar and 27°C, to 4 bars and 127°C. Compute the entropy change of the mixture, in kJ/(kg)(°K).

**11-36M** An equimolar mixture of $N_2$ and $CO_2$ at 1.5 bars and 127°C is cooled at constant pressure to 27°C. Compute the change in enthalpy, internal energy, and entropy of the mixture per kilogram · mole of mixture, using the gas tables.

**11-37M** A quantity, 1 kg·mol, of a mixture containing 0.5 kg·mol of $N_2$ and 0.5 kg·mol of $H_2$ is reversibly and isothermally compressed in a cylinder from 3 to 6 bars.
 (a) Calculate the specific entropy change for each species, in kJ/(°K).
 (b) Compute the mixture entropy change, in kJ/°K.

**11-38M** A tank with a volume of 0.3 m³ contains $O_2$ at 6 bars and 352°C. A second tank, with a volume of 0.5 m³, contains He at 4 bars and 7°C. The tank walls are heavily insulated. The two gases are allowed to mix through a connecting pipe. Determine (a) the final temperature, in °C, (b) the final pressure, in bars, and (c) the total entropy change, in kJ/°K.

### Properties of air-water vapor mixtures

**11-39M** A tank contains 15 g of dry air and 50 g of saturated water vapor at 70°C. Determine (a) the volume of the tank, in liters, and (b) the total pressure, in millibars.

**11-40M** A tank contains 7 g of dry air and 70 g of dry, saturated steam at a temperature of 80°C. Determine (a) the volume of the tank, in liters, and (b) the total pressure, in millibars.

**11-41M** If the partial pressure of water vapor in atmospheric air is 30 mbar at 30°C, determine (a) the relative humidity, and (b) the approximate dew-point temperature, using the steam tables.

**11-42M** The partial pressure of water vapor in atmospheric air is 25 mbar at 25°C. Determine (a) the relative humidity, (b) the dew point, in °C, and (c) the specific humidity, in g/kg, for 980 mbar total pressure. Use the steam tables.

**11-43M** An atmospheric air mixture at 960 mbar contains 2 percent water by volumetric analysis. If the dry-bulb temperature is 25°C, determine (a) the dew-point temperature, (b) the humidity ratio, in g/kg, and (c) the enthalpy, in kJ/kg, (based on $h = 0$ at 0°C for both dry air and water). Use steam-table data.

**11-44M** Atmospheric air at 30°C and 1 bar has a relative humidity of 60 percent. What is (a) the partial pressure of dry air in the mixture, in millibars, (b) the dew-point temperature, in °C, and (c) the specific humidity, in g/kg. Use steam-table data.

**11-45M** The dry-bulb temperature of atmospheric air is 25°C, and the relative humidity is 30 percent. Find the values of (a) the partial pressure of the water vapor, (b) the humidity ratio if the total pressure is 1 bar, and (c) the dew point. Use steam-table data.

**11-46M** Atmospheric air with a relative humidity of 50 percent is held at 35°C and 970 mbar. Calculate (a) the specific humidity, in g/kg, and (b) the enthalpy, in kJ/kg, where $h = 0$ at 0°C. Use steam-table data.

**11-47M** Air at a barometric pressure of 960 mbar and a dry-bulb temperature of 20°C has a relative humidity of 40 percent. Determine, through the use of the steam tables, (a) the specific humidity, in g/kg, (b) the dew point, in °C, and (c) the enthalpy, in J/g (based on $h = 0$ at 0°C).

**11-48M** Arrange the temperature values 19, 35, and 8°C in the following order: dry-bulb, dew-point, wet-bulb.

**11-49M** Atmospheric air at 1 bar has a dry-bulb temperature of 25°C and a wet-bulb temperature of 20°C. Determine, through the use of steam-table data and Eq. (11-41), (a) the humidity ratio, in g/kg, (b) the relative humidity, and (c) the enthalpy, in J/g (based on $h = 0$ at 0°C).

**11-50M** Atmospheric air at 970 mbar has a dry-bulb temperature of 30°C and a wet-bulb temperature of 20°C. Determine, through the use of steam-table data and Eq. (11-41), (a) the humidity ratio, in g/kg, (b) the relative humidity, and (c) enthalpy, in J/g (based on $h = 0$ at 0°C).

**11-51M** Same as Prob. 11-49M, except that the dry-bulb and wet-bulb temperatures are 27 and 21°C, respectively.

**11-52M** Condensation on cold-water pipes often occurs in warm, humid rooms. The outside of a pipe may reach a minimum of 10°C, and the room temperature is kept at 25°C.

(a) What is the maximum limit of relative humidity in the room to avoid condensation at normal atmospheric pressure?

(b) How high must the minimum temperature be raised if condensation is avoided and the room relative humidity is 54 percent at 25°C.

**11-53M** On a cold winter's day the inside surface of a wall in a home is found to be 15°C and the air within the room is at 22.5°C.

(a) What is the maximum relative humidity that the air can have without the occurrence of condensation of water on the wall?

(b) If added insulation in the wall raises the maximum permissible relative humidity to 75 percent, what is the new, permissible inside-wall temperature, in °C?

**Property evaluation from a psychrometric chart**

**11-54M** The dry-bulb and wet-bulb temperatures for atmospheric air are 30 and 23°C, respectively. Determine (a) the partial pressure of the water vapor, (b) the humidity ratio, (c) the specific volume per mass of dry air, (d) the relative humidity, and (e) the enthalpy, in kJ/kg, of dry air. Employ the psychrometric chart, and assume the total pressure is 1 bar.

**11-55M** The humidity ratio of atmospheric air is 0.010 kg/kg of dry air and the relative humidity is 60 percent. Determine, from the psychrometric chart, (a) the dry-bulb temperature, (b) the wet-bulb temperature, (c) the partial pressure of water vapor, and (d) the enthalpy per kilogram of dry air.

**11-56M** Atmospheric air has a dry-bulb temperature of 32°C and a wet-bulb temperature of 26°C. From the psychrometric chart, determine (a) the humidity ratio or specific humidity, (b) the dew point, (c) the enthalpy, (d) the specific volume, and (e) the relative humidity.

**11-57M** Atmospheric air at a pressure of 1 bar and 28°C has a specific humidity of 0.0090 kg/kg. Determine, from the psychrometric chart, (a) the relative humidity, (b) the vapor pressure, (c) the enthalpy, (d) the specific volume, and (e) the dew-point temperature.

**11-58M** Air which originally exists at 36°C, 1 bar, and a relative humidity of 40 percent is cooled at constant pressure to 24°C. Determine (a) the relative humidity, (b) the humidity ratio, and (c) the dew-point temperature at the final state, employing the psychrometric chart.

**11-59M** Atmospheric air at 1 bar has dry-bulb and wet-bulb temperatures of 25 and 18°C, respectively. From the psychrometric chart, determine (a) the humidity ratio, (b) the dew point, (c) the specific volume, (d) the relative humidity, and (e) the enthalpy.

**Heating, cooling, and dehumidification**

**11-60M** Atmospheric air with dry-bulb and wet-bulb temperatures of 30 and 20°C, respectively, is cooled to 18°C at a constant pressure of 1 bar. If the mass flow rate through the cooling unit is 500 kg/h, determine the heat removed, in kJ/h.

**11-61M** Atmospheric air at 0.98 bar and 26°C, with a relative humidity of 70 percent, is cooled to 10°C.

    (a) How many grams of water vapor are condensed per kilogram of dry air?
    (b) How much heat is removed, in kJ/kg of dry air?

**11-62M** Atmospheric air with dry-bulb and wet-bulb temperatures of 28 and 20°C, respectively, is cooled to 17°C at a constant pressure of 0.99 bar. The volume flow rate entering the cooling unit is 100 m³/h.

    (a) Determine the mass flow rate in kg/h.
    (b) Determine the heat removed, in kJ/h.

**11-63M** Atmospheric air at 28°C and 70 percent relative humidity flows over a set of cooling coils at a rate of 500 m³/min. The liquid which condenses is removed from the system at 10°C. Subsequent heating of the air results in a final temperature of 24°C and a relative humidity of 40 percent. Determine (a) the heat removed in the cooling section, in kJ/min, (b) the heat added in the heating section, in kJ/min, and (c) the amount of vapor condensed, in kg/min.

**11-64M** A 3-m³ storage tank initially contains atmospheric air which has been compressed to 5 bars and 150°C with a relative humidity of 10 percent. The air then is cooled back to the ambient temperature of 17°C. Determine (a) the dew point of the mixture, (b) the temperature at which condensation begins, (c) the amount of water condensed, and (d) the heat transferred from the tank, in kJ.

**11-65M** Atmospheric air at 30°C and 60 percent relative humidity flows over a set of cooling coils at an inlet flow rate of 1500 m³/min. The liquid which condenses is removed from the system at 17°C. Subsequent heating of the air results in a final state of 25°C and 60 percent relative humidity. Determine (a) the heat removed in the cooling section, in kJ/min, (b) the heat added in the heating section, in kJ/min, and (c) the amount of vapor condensed, in kg/min.

**11-66M** Ambient air from outside a store is to be conditioned from a temperature of 29°C and 80 percent relative humidity to a final state of 21°C and 40 percent relative humidity. If the pressure remains constant at 1 bar, calculate (a) the amount of water removed, (b) the heat removed in the cooling process, and (c) the heat added in the final step of the process.

**Evaporative cooling**

**11-67M** Atmospheric air at 34°C and 20 percent relative humidity passes through an evaporative cooler until the final temperature is (a) 24°C, and (b) 20°C. Determine how much water is added to the air and what the final relative humidity is in each case.

**11-68M** Atmospheric air at 28°C and 10 percent relative humidity passes through an evaporative cooler until the final relative humidity is (a) 40 percent, and (b) 60 percent. For each case find the amount of water added to the air and the dew point of the final state.

**11-69M** Atmospheric air at 36°C and 10 percent relative humidity passes through an evaporative cooler.

    (a) What is the minimum temperature that could be achieved by the process?

    (b) If the final humidity ratio is (1) 0.007 kg/kg, and (2) 0.009 kg/kg, how much water is added in the cooler and what is the final temperature?

### Heating and humidification

**11-70M** An air stream at 10°C and 40 percent relative humidity is first heated to 33°C and then passed through an evaporative cooler until the temperature reaches 26°C. Determine (a) the heat added, (b) the water added, and (c) the final relative humidity.

**11-71M** Atmospheric air at 12°C and 30 percent relative humidity is first heated to 35°C and then passed through an evaporative cooler until the temperature reaches (a) 26°C, and (b) 20°C. Determine the heat added in the first section and the water added in the second section.

**11-72M** An air stream at 34°C with a wet-bulb temperature of 15°C passes through an evaporative cooler until the temperature is (a) 21°C, and (b) 17°C. It is then heated to 30°C. Determine the amount of water added in the first section and the heat added in the second section of the equipment.

### Adiabatic mixing

**11-73M** Two streams of air at 1 bar are mixed together in an adiabatic, steady-flow process. One air stream, at 36°C and 40 percent relative humidity, enters at a rate of 5 kg/min, while the second stream, at 5°C and 100 percent relative humidity, enters at (a) 10 kg/min, and (b) 15 kg/min. Determine (1) the exit humidity ratio, (2) the exit stream enthalpy, and (c) the exit temperature, in °C.

**11-74M** One air stream, at 15°C and 30 percent relative humidity, is mixed adiabatically with another air stream, at 30°C and 70 percent relative humidity, in a steady-flow process. If the mass flow rate of the hot air stream is (a) twice, and (b) three times that of the colder stream, determine the final temperature and relative humidity of the final mixture.

**11-75M** In an air-conditioning process 50 m³/min of outside air at 29°C and 80 percent relative humidity are mixed adiabatically with 100 m³/min of inside air initially at 20°C and 30 percent relative humidity. Compute for the resultant mixture (a) the dry-bulb temperature, (b) the wet-bulb temperature, (c) the humidity ratio, and (d) the relative humidity.

### Wet cooling tower

**11-76M** Water enters a cooling tower at 36 m³/min and is cooled from 30 to 20°C. The entering atmospheric air at 1 bar has dry-bulb and wet-bulb temperatures of 21 and 15°C, respectively. The air is saturated when it leaves the tower. Determine (a) the volume rate of inlet air required, in m³/min, and (b) the rate of water evaporated, in kg/h. The exit air temperature is 28°C.

**11-77M** Water enters a cooling tower at 40°C and leaves at 22°C. The tower receives 10,000 m³/min of atmospheric air at 1 bar, 20°C, and 40 percent relative humidity. The air exit condition is 32°C and 90 percent relative humidity. Determine (a) the mass flow rate of dry air passing through the tower, in kg/min, (b) the mass flow rate of entering water, in kg/min, and (c) the amount of water evaporated, in kg/min.

**11-78M** It is desired to cool 1300 kg/min of water from 40 to 26°C. The cooling tower receives 800 m³/min of air at 1 bar with dry-bulb and wet-bulb temperatures of 29 and 21°C, respectively. If the evaporation rate from the water stream is 1600 kg/h, determine the temperature of the exit air stream.

**11-79M** Same as Prob. 11-78M, except that the outlet water rate is 1250 kg/min and the exit air relative humidity is 95 percent. The inlet water rate and the evaporation rate are now unknown.

# PROBLEMS (USCS)

**Mass, volume, pressure, temperature relations**

**11-1** A gas mixture has the following volumetric analysis: $N_2$, 60 percent; $CO_2$, 33 percent; $O_2$, 7 percent.

    (*a*) Determine the analysis on a mass basis.

    (*b*) What is the mass of 1000 ft$^3$ of this gas when the pressure is 15 psia and the temperature is 80°F?

**11-2** Calculate the specific volume of an ideal-gas mixture comprised of 16 percent $H_2$, 28 percent CO, and 56 percent $N_2$, by weight, at 15 psia and 140°F, in ft$^3$/lb.

**11-3** The volumetric analysis of a mixture of gases is as follows: $N_2$, 70 percent; $CO_2$, 20 percent; $O_2$, 10 percent.

    (*a*) Determine the gravimetric analysis.

    (*b*) Compute the component pressure of $CO_2$ if the pressure and temperature of the mixture are 15 psia and 140°F, respectively.

    (*c*) How many pounds of the mixture would be contained in 450 ft$^3$ at the conditions given in part (*b*)?

**11-4** A gas mixture has the following gravimetric analysis: $O_2$, 32 percent; CO, 56 percent; He, 12 percent. Determine (*a*) the volumetric analysis and (*b*) the apparent gas constant.

**11-5** A gas mixture contains 0.28 lb of CO, 0.16 lb of $O_2$, and 0.66 lb of $CO_2$ at 20 psia and 90°F. Compute (*a*) the partial pressure of each component, and (*b*) the volume occupied by each component measured at its own partial pressure and mixture temperature.

**11-6** A 3-ft$^3$ tank contains 0.7 lb of nitrogen and 1.1 lb of carbon dioxide at 90°F. Compute (*a*) the partial pressure of each component, (*b*) the partial volume of each component, (*c*) the pressure of the mixture, and (*d*) the gas constant for the mixture in ft·lb/(lb)(°R).

**11-7** A gas mixture is made up of 20 percent He, 30 percent Ar, and 50 percent $N_2$ by weight. The total pressure is 50 psia, and the temperature is 50°F. Determine (*a*) the partial pressure of the He, and (*b*) the apparent molecular weight and the gas constant per pound of the mixture.

**11-8** A mixture of ideal gases gave the following volumetric analysis: $CO_2$, 50 percent; $N_2$, 30 percent; $CH_4$, 20 percent.

    (*a*) Calculate the gravimetric analysis of the mixture and the apparent molecular weight.

    (*b*) A tank of 359-ft$^3$ capacity contains 32 lb of this gas mixture at 32°F. What is the pressure in the tank?

**11-9** A mixture having a volumetric analysis as follows is cooled under constant-volume conditions from 240°F and 35 psia to a final temperature of 50°F; $O_2$, 60 percent; $N_2$, 40 percent. Compute (*a*) volumetric analysis at 50°F, (*b*) partial pressures at 50°F, (*c*) apparent molecular weight, and (*d*) mass percent of $O_2$ and $N_2$.

**11-10** In what molar ratio should He and Kr be mixed so that the gas mixture may have the same average molar mass as pure Ar?

**11-11** In what molar ratio should He and $CO_2$ be mixed so that the gas mixture may have the same average molar mass as pure diatomic oxygen?

**Internal energy, enthalpy, and specific heats**

**11-12** A mixture of ideal gases at 16 psia and 80°F has the following volumetric analysis: 87.5 percent He and 12.5 percent $N_2$. Calculate (*a*) the partial pressure of He, (*b*) the percent $N_2$ by weight, (*c*) the apparent gas constant, and (*d*) the specific heat at constant volume, in Btu/(lb)(°F).

**11-13** A mixture of ideal gases at 16 psia and 100°F has the following volumetric analysis: 25 percent CO and 75 percent He. Calculate (*a*) the partial pressure of CO, (*b*) the percent He by weight, (*c*) the apparent gas constant, and (*d*) the specific heat at constant pressure, in Btu/(lb)(°F) for the mixture.

**11-14** A gas sample was found to have the following volumetric analysis: $H_2$, 4 percent; CO, 12 percent; $CO_2$, 35 percent; and $N_2$, 49 percent. Determine (a) the gravimetric analysis, (b) the partial pressures, and (c) the constant-pressure specific heat of the mixture if the temperature is 100°F and the total pressure is 14.5 psia.

**11-15** A rigid tank at 80°F contains 2.0 lb of $H_2O$, 4.0 lb of $O_2$, and 7.0 lb of CO. If the total pressure is 3.0 psia, calculate the partial pressure of CO in psia and the enthalpy of the mixture, based on data from gas tables in the Appendix.

**11-16** For the mixture described in Prob. 11-2 determine, at 140°F (a) the enthalpy of the mixture in Btu/lb·mol, and (b) $c_p$ of the mixture, in Btu/(lb·mol)(°F).

**11-17** The mixture described in Prob. 11-3 is cooled as it passes through a heat exchanger in steady state from 540 to 140°F. Compute the heat transferred, in Btu/mol of mixture.

**11-18** For the mixture described in Prob. 11-5, evaluate (a) $c_v$, and (b) $c_p$ for the mixture, in Btu/(lb)(°F).

**11-19** For the mixture described in Prob. 11-5, determine (a) the internal energy, and (b) the enthalpy, in Btu/lb·mol, based on tabular data at 100°F.

**11-20** For the mixture described in Prob. 11-6, determine (a) the enthalpy, and (b) the internal energy, in Btu/lb·mol, for the mixture at 90°F.

**11-21** Based on the data of Prob. 11-7, determine the constant-pressure specific heat of the mixture, in Btu/(lb·mol)(°F).

**11-22** For the mixture described in Prob. 11-1, compute the value of (a) $c_p$, and (b) $c_v$ for the mixture, in Btu/(lb·mol)(°F), at 80°F.

**11-23** Referring to Prob. 11-9, determine (a) the enthalpy change, in Btu/lb·mol, and (b) $c_p$ of the mixture at 140°F, in Btu/(lb·mol)(°F).

### Energy analysis

**11-24** An equimolar mixture of $H_2$ and $CO_2$ flows steadily through a nuclear reactor at constant pressure. No chemical reactions occur. The initial and final temperatures of the mixture are 500 and 2500°R, respectively. If the molar flow rate is 10 lb·mol/s, what is the rate of heat transfer from the reactor to the gas, in Btu/s?

**11-25** A mixture of $H_2O$, CO, and $N_2$, with component pressures in the ratio 1 : 2 : 5, respectively, is expanded isentropically from 2000°R and 10 atm to 1 atm in a piston-cylinder assembly. Assuming constant chemical composition, how much work is done, in Btu/lb·mol?

**11-26** A gas mixture consisting of $N_2$, $CO_2$, and $H_2O$ in a molar ratio of 4 : 1 : 1 enters a turbine at 2500°R. The gases expand isentropically through an 8 : 1 pressure ratio. Compute the work output, in Btu/lb·mol of mixture.

**11-27** A piston-cylinder contains 0.6 lb·mol of $N_2$ and 0.4 lb·mol of $O_2$. The mixture is compressed isothermally at 1000°R from 1 to 3 atm. What is the required work input in foot-pound force?

**11-28** One pound-mole of $CO_2$, initially at 30 psia and 100°F, is mixed adiabatically with two pound-moles of $O_2$, initially at 80 psia and 300°F. During the constant-volume mixing process electrical energy equivalent to 200 Btu/lb·mol of mixture is added. Determine the final temperature of the mixture, in °F.

**11-29** One pound-mole of $CO_2$, initially at 30 psia and 200°F, is mixed adiabatically in a closed system at constant pressure with two pound-moles of $N_2$, initially at 30 psia and 80°F. During the process 400 Btu are added in the form of electrical work. Determine the final temperature of the mixture, in °F.

**11-30** Two cubic feet of nitrogen at 20 psia and 180°F are allowed to mix with five cubic feet of hydrogen at 50 psia and 40°F by removing a partition between the two tanks of gas. If the final temperature is 80°F, determine the magnitude and direction of the heat transfer, in Btus.

**11-31** Two pounds of nitrogen at 100 psia and 140°F in a rigid tank are mixed with one pound of oxygen at 50 psia, 60°F, in another rigid tank by opening a valve connecting the two tanks. The components and mixture are ideal gases, and their specific heats may be assumed to be constant. Determine (a) the heat transfer, in Btus, and (b) the final pressure, in psia, if the final temperature is 95°F.

### Entropy changes of ideal-gas mixtures

**11-32** Hydrogen and nitrogen in separate streams enter an adiabatic mixing chamber in a 3 : 1 molar ratio. At the inlet the hydrogen is at 25 psia, 140°F, and the nitrogen is at 25 psia, 540°F. The mixture leaves the chamber at 24 psia.

(a) Determine the final temperature of the mixture, in °R.

(b) Compute the entropy change of the hydrogen gas, in Btu/(lb·mol)(°R).

**11-33** An insulated 3-ft$^3$ tank is divided into two sections by a partition. One section is 2 ft$^3$ in volume and initially contains hydrogen gas at 30 psia and 110°F. The remaining section initially holds nitrogen gas at 50 psia and 50°F. The adiabatic partition is then removed, and the gases are allowed to mix. Determine (a) the temperature of the equilibrium mixture, (b) the pressure of the mixture, and (c) the entropy change for the process.

**11-34** A system consists of two tanks interconnected with a pipe and valve. One tank with a volume of 37.0 ft$^3$ contains 5.0 lb of argon at 20 psia and 90°F. The other tank has a volume of 8.0 ft$^3$ and holds 4.0 lb of oxygen at 100 psia and 140°F. The valve is opened, and the gases are allowed to mix. The atmospheric temperature is 40°F, and during the mixing process 31 Btu of heat leave the uninsulated system. Determine (a) the final temperature, (b) the final mixture pressure, (c) the apparent molecular weight of the mixture, (d) the change in entropy of the argon, in Btu/°R, and (e) the entropy change of the atmosphere, in Btu/°R.

**11-35** An equimolar mixture of nitrogen and carbon dioxide is compressed from 15 psia and 60°F to 60 psia and 280°F. Compute the entropy change of the mixture, in Btu/(lb)(°R).

**11-36** An equimolar mixture of nitrogen and carbon dioxide at 20 psia and 120°F is cooled at constant pressure to 60°F. Compute the change in enthalpy, internal energy, and entropy of the mixture, per mole of mixture, using the gas tables.

**11-37** One mole of a mixture containing half-mole of $N_2$ and half-mole of $H_2$ is reversibly and isothermally compressed in a cylinder from 50 to 100 psia. Assume the gases to be ideal.

(a) Calculate the specific entropy change for each component.

(b) Calculate the total entropy change for the mixture.

**11-38** A tank with a volume of 10 ft$^3$ contains $O_2$ at 100 psia and 640°F. A second tank, with a volume of 15 ft$^3$, contains He at 80 psia and 40°F. The tank walls are heavily insulated. The two tanks are interconnected, and the gases are allowed to mix. Determine (a) the final temperature, in °F, (b) the final pressure, in psia, and (c) the total entropy change, in Btu/°R.

### Properties of air–water-vapor mixtures

**11-39** A tank contains 0.018 lb of air and 0.12 lb of saturated water vapor at 150°F. Determine (a) the volume of the tank, in cubic feet, and (b) the total pressure, in psia.

**11-40** A tank contains 0.015 lb of air and 0.15 lb of dry, saturated steam at a temperature of 180°F. Determine (a) the volume of the tank, and (b) the total pressure in the tank.

**11-41** If the partial pressure of water vapor in atmospheric air is 0.400 psia at 90°F, calculate (a) the relative humidity and (b) the approximate dew point for the water-vapor–air mixture, using the steam tables.

**11-42** The partial pressure of water vapor in atmospheric air is 0.35 psia at 80°F. Determine (a) the relative humidity, (b) the dew point, in °F, and (c) the specific humidity, in lb/lb when the total pressure is 14.6 psia. Use steam-table data.

**11-43** A dry-air–water-vapor mixture at 14.5 psia contains 2 percent water by volume. If the dry-bulb temperature is 75°F, determine (a) the dew-point temperature, (b) the humidity ratio, and (c) the enthalpy, in Btu/lb of dry air.

**11-44** Atmospheric air at 90°F and 14.7 psia has a relative humidity of 60 percent.

(a) What is the partial pressure of the dry air in the mixture?

(b) What is the dew-point temperature?

**11-45** The dry-bulb temperature of atmospheric air is 80°F, and the relative humidity is 30 percent.

(a) Compute the value of (1) the humidity ratio and (2) the partial pressure of the water vapor if the total pressure is 14.7 psia.

(b) What is the dew point?

**11-46** A mixture of dry air and water vapor with a relative humidity of 50 percent is held at a temperature of 95°F and 14.30 psia. Calculate the specific humidity in gr/lb.

**11-47** Air at a barometric pressure of 14.5 psia and a dry-bulb temperature of 70°F has a relative humidity of 40 percent. Determine, through the use of the steam tables, (a) the specific humidity and (b) the dew point.

**11-48** Arrange the temperature values 66, 97, 46 in the following order: dry-bulb, dew-point, wet-bulb.

**11-49** Atmospheric air at 14.7 psia has a dry-bulb temperature of 78°F and a wet-bulb temperature of 68°F. Determine, through the use of steam-table data and Eq. (11-41), (a) the humidity ratio, in lb/lb, (b) the relative humidity, and (c) the enthalpy, in Btu/lb.

**11-50** Atmospheric air at 14.6 psia has a dry-bulb temperature of 86°F and a wet-bulb temperature of 68°F. Determine, through the use of steam-table data and Eq. (11-41), (a) the humidity ratio, in lb/lb, (b) the relative humidity, and (c) the enthalpy, in Btu/lb.

**11-51** Same as Prob. 11-49, except that the dry-bulb and wet-bulb temperatures are 82°F and 72°F, respectively.

**11-52** Condensation on cold-water pipes often occurs in warm humid rooms. If the water temperature in the pipes may reach a minimum of 45°F and the room temperature is kept at 75°F, what is the maximum limit of relative humidity in the room to avoid condensation at normal atmospheric pressure?

**11-53** On a cold winter day the inside surface of a wall in a home is found to be 58°F. If the air within the room is at 72°F, what is the maximum relative humidity that the air can have without the occurrence of condensation of water on the wall?

### Property evaluation from a psychrometric chart

**11-54** The dry-bulb and wet-bulb temperatures for atmospheric air are 85 and 74°F, respectively. Determine (a) the partial pressure of the water vapor, (b) the humidity ratio, (c) the specific volume per pound of dry air, (d) the relative humidity, and (e) the enthalpy, in Btu/lb of dry air. Employ the psychrometric chart, and assume the total pressure is 14.696 psia.

**11-55** The humidity ratio of atmospheric air is 70 gr/lb of dry air, and the relative humidity is 60 percent. Determine, from the psychrometric chart, (a) the dry-bulb temperature, (b) the wet-bulb temperature, (c) the partial pressure of the water vapor, and (d) the enthalpy per pound of dry air.

**11-56** Atmospheric air has a dry-bulb temperature of 90°F and a wet-bulb temperature of 84.5°F. From the psychrometric chart determine (a) the specific humidity, (b) the dew point, (c) the enthalpy, (d) the specific volume, and (e) the relative humidity.

**11-57** Atmospheric air at a pressure of 14.7 psia and 90°F has a specific humidity of 0.0095 lb water vapor/lb dry air. Determine from the psychrometric chart (a) the relative humidity, (b) the vapor pressure, (c) the total enthalpy, (d) the specific volume, and (e) the dew-point temperature.

**11-58** Air which exists originally at 100°F, 14.7 psia, and a relative humidity of 40 percent is cooled at constant pressure to 70°F. Determine (a) the relative humidity, (b) the humidity ratio, and (c) the dew-point temperature at the second condition, in °F; and (d) sketch the process and the end points on a psychrometric-chart type of plot.

**11-59** Atmospheric air at 14.7 psia has dry-bulb and wet-bulb temperatures of 78 and 64°F, respectively. From the psychrometric chart, determine (a) the humidity ratio, (b) the dew point, (c) the specific volume, (d) the relative humidity, and (e) the enthalpy.

### Heating, cooling, and dehumidification

**11-60** Atmospheric air with dry-bulb and wet-bulb temperatures of 90 and 70°F, respectively, is cooled to 65°F at a constant pressure of 14.7 psia. If the mass flow rate of air through the cooling unit is 500 lb/h, determine the heat removed, in Btu/h.

**11-61** Atmospheric air at 14.2 psia, 80°F, and a relative humidity of 70 percent is cooled to 50°F. If the process is at constant pressure, determine (a) the pounds of water vapor condensed, and (b) the quantity of heat removed, in Btu/lb of dry air.

**11-62** Atmospheric air with dry-bulb and wet-bulb temperatures of 82 and 68°F, respectively, is cooled to 61°F at a constant pressure of 14.7 psia. The volume flow rate entering the cooling unit is 100 ft³/min.

(a) Determine the mass flow rate in lb/min.

(b) Determine the heat removed, in Btu/min.

**11-63** Atmospheric air at 86°F and 60 percent relative humidity flows over a set of cooling coils at an inlet flow rate of 2000 ft³/h. The liquid which condenses is removed from the system at 60°F. Subsequent heating of the air results in a final state of 75°F and 50 percent relative humidity. Determine (a) the heat removed in the cooling section, in Btu/h, (b) the heat added in the heating section, in Btu/h, and (c) the amount of vapor condensed, in lb/h.

**11-64** A 100-ft³ storage tank initially contains atmospheric air which has been compressed to 80 psia and 300°F at a relative humidity of 10 percent. The air then cools back to the ambient temperature of 60°F. Determine (a) the dew point of the mixture, (b) the temperature at which condensation begins, (c) the amount of water condensed per pound of dry air, and (d) the heat transferred to or from the tank during the process, in Btu/lb dry air.

**11-65** Atmospheric air at 84°F and 70 percent relative humidity and a rate of 15,000 ft³/min flows over a set of cooling coils. The liquid which condenses is removed from the system at 50°F. Subsequent heating of the air results in a final temperature of 75°F and a relative humidity of 40 percent. Determine the heat removed in the cooling section and the heat added in the heating section, in Btu/h.

**11-66** Ambient air from outside a store is to be conditioned from a temperature of 85°F and 80 percent relative humidity to a final state of 70°F and 40 percent relative humidity. On the assumption that the process occurs at a constant pressure of 14.7 psia, calculate (a) the amount of water removed per pound of dry air, (b) the heat removed in the cooling process, and (c) the heat added in the final step of the process.

## Evaporative cooling

**11-67** Atmospheric air at 95°F and 20 percent relative humidity passes through an evaporative cooler until the final temperature is (a) 75°F, and (b) 68°F. Determine how much water is added to the air and what the final relative humidity is in each case.

**11-68** Atmospheric air at 85°F and 10 percent relative humidity passes through an evaporative cooler until the final relative humidity is (a) 40 percent, and (b) 60 percent. For each case find the amount of water added to the air and the dew-point temperature of the final state.

**11-69** Atmospheric air at 90°F and 10 percent relative humidity passes through an evaporative cooler.

(a) What is the minimum temperature that could be achieved by the process?

(b) If the final humidity ratio is (1) 0.007 lb/lb and (2) 0.009 lb/lb, how much water is added in the cooler and what is the final temperature?

## Heating and humidification

**11-70** An air stream at 50°F and 40 percent relative humidity is first heated to 90°F and then passed through an evaporative cooler until the temperature reaches 80°F. Determine (a) the heat added, (b) the water added, and (c) the final relative humidity.

**11-71** Atmospheric air at 55°F and 30 percent relative humidity is first heated to 95°F and then passed through an evaporative cooler until the temperature reaches (a) 80°F, and (b) 68°F. Determine the heat added in the first section and the water added in the cooler section.

**11-72** An air stream at 95°F with a wet-bulb temperature of 60°F passes through an evaporative cooler until the temperature is (a) 70°F, and (b) 62°F. It is then heated to 85°F. Determine the amount of water added in the first section and the heat added in the second process.

## Adiabatic mixing

**11-73** Two streams of air at 14.7 psia are mixed together in an adiabatic, steady-flow process. One air stream at 100°F and 40 percent relative humidity enters at a rate of 5 lb/min, while the second stream at 40°F and 100 percent relative humidity enters at 10 lb/min. Determine (a) the exit humidity ratio, (b) the exit stream enthalpy, and (c) the final temperature of the exit stream, in °F.

**11-74** One air stream at 60°F and 30 percent relative humidity is mixed adiabatically with another air stream at 90°F and 70 percent relative humidity in a steady-flow process. If the mass flow rate of the hot air stream is twice that of the colder stream, determine the final temperature and relative humidity of the mixture.

**11-75** In an air-conditioning process 500 ft³/min outside air at 85°F and 80 percent relative humidity are mixed adiabatically with 1000 ft³/min of inside air initially at 70°F and 30 percent relative humidity. Compute for the resultant mixture (a) the dry-bulb temperature, (b) the wet-bulb temperature, (c) the humidity ratio, and (d) the relative humidity.

## Wet cooling tower

**11-76** Water enters a cooling tower at 9000 gal/min and is cooled from 86 to 68°F. The entering atmospheric air at 14.7 psia has dry-bulb and wet-bulb temperatures of 70 and 60°F, respectively. The air is saturated when it leaves the tower. Determine (a) the volume rate of inlet air required, in ft³/min, and (b) the rate of water evaporated, in lb/h. The exit air temperature is 82°F.

**11-77** Water enters a cooling tower at 105°F and leaves at 72°F. The tower receives 300,000 ft³/min of atmospheric air at 14.7 psia, 68°F, and 40 percent relative humidity. The air exit condition is 90°F and 90 percent relative humidity. Determine (a) the mass flow rate of dry air passing through the tower, in lb/min, (b) the mass flow rate of entering water, in lb/min, and (c) the amount of water evaporated, in lb/min.

**11-78** It is desired to cool 3000 lb/min of water from 103 to 80°F. The cooling tower receives 24,000 ft³/min of air at 14.7 psia with dry-bulb and wet-bulb temperatures of 85 and 70°F, respectively. If the evaporation rate from the water stream is 3600 lb/h, determine the temperature of the exit air stream.

**11-79** Same as Prob. 11-78, except that the outlet water rate required is 2800 lb/min and the exit air has a relative humidity of 95 percent. The inlet water rate and the evaporation rate are now unknown.

# TWELVE

## *PvT* BEHAVIOR OF REAL GASES AND REAL-GAS MIXTURES

In Chap. 1 some qualitative characteristics of the *PvT* behavior of real gases were presented in order to arrive at the concept of an ideal gas and an ideal-gas thermometer. The data presented were primarily in the form of $Pv = f(P, T)$. Quantitative data for real gases were presented in Chap. 4 in the form of tables like those which appear in the Appendix. However, in many cases it would be convenient to have analytical expressions for the relationships among these three properties. There is no set method for doing this; consequently, we shall look at a few of the many correlations available for the equations of state of real gases. The accuracy of these *PvT* relations varies with the type of gas and the range of the properties under consideration. The best methods have not necessarily been chosen for analysis in this chapter, since the purpose here is simply to provide some insight into typical correlations. When the occasion arises, a search of the literature will reveal numerous equations of state which might profitably be used for particular investigations.

## 12-1 THE VIRIAL EQUATION OF STATE

For simple, compressible systems we have postulated that two independent properties are required in order to fix the equilibrium state. If $Pv$ is chosen as the dependent thermodynamic property, it may be considered to be a function of two independent properties such as pressure and temperature, or volume and temperature. The data of Fig. 12-1 illustrate a functional relationship of the type

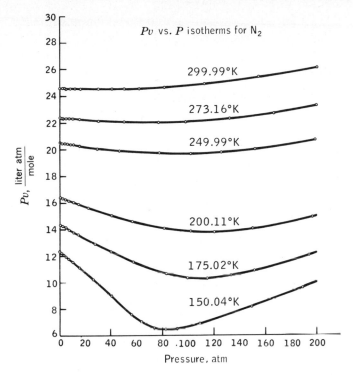

**Figure 12-1** Variation of $Pv$ of nitrogen with pressure at constant temperature. (From M. W. Zemansky, "Heat and Thermodynamics," 4th ed., McGraw-Hill, New York, 1957.)

$Pv = f(P, T)$. This functional relationship may be expressed to any desired accuracy by an infinite-series expansion of the type

$$Pv = a + bP + cP^2 + dP^3 + \cdots \tag{12-1}$$

If the independent variables were selected as $v$ and $T$, the following form of an infinite series would be applicable:

$$Pv = a\left(1 + \frac{b'}{v} + \frac{c'}{v^2} + \frac{d'}{v^3} + \cdots\right) \tag{12-2}$$

Equations of this type, which relate $P$, $v$, and $T$, are known as *virial equations* of state. The coefficients $a$, $b$, $c$, etc., are called the first, second, third, etc., virial coefficients for the respective equations. These coefficients are solely a function of the temperature of the system. They are of particular interest since they can be given a physical significance on a molecular scale. As a result, they can be determined, at least in theory, from statistical mechanics, as well as from direct macroscopic measurements illustrated in Fig. 12-1. The virial coefficients are also, of course, a function of the substance of interest. Since all $Pv T$ equations of state should reduce to $Pv = R_u T$ as the pressure approaches zero, it is seen that the constant $a$, common to both equations, is equal to $R_u T$.

As an example, a virial equation for nitrogen gas at $0°C$ and for pressures up to 200 atm has been developed as follows:

$$Pv = 22{,}414.6 - 10.281P + 0.065189P^2 + 5.1955 \times 10^{-7}P^4$$

$$- 1.3156 \times 10^{-11}P^6 + 1.009 \times 10^{-16}P^8 \quad (12\text{-}3)$$

The volume in this equation is expressed in $cm^3/g \cdot mol$, and the pressure is in atmospheres.

When the pressure is relatively low, it is seen from Eq. (12-3) that the latter terms for $Pv$ contributed little compared with the first several terms. Thus at low pressures a virial equation of the type

$$Pv = RT + bP \quad (12\text{-}4)$$

is often of sufficient accuracy, where $b$ is the second virial coefficient. This coefficient is negative at low temperatures, but increases with increasing temperature and eventually is positive. The temperature for which the second virial coefficient is zero is known as the Boyle temperature. An empirical equation for the second virial coefficient $b$ of nitrogen has the form

$$b = 39.5 - \frac{1.00 \times 10^4}{T} - \frac{1.084 \times 10^6}{T^2} \quad (12\text{-}5)$$

where $b$ is in $cm^3/g \cdot mol$, and $T$ is in degrees Kelvin. We shall find in the following chapter that equations similar to (12-4) and (12-5) are useful not only in correlating $PvT$ data, but also in evaluating other properties of matter.

## 12-2 THE VAN DER WAALS EQUATION OF STATE

In 1873 van der Waals proposed an equation of state which was an attempt to correct the ideal-gas equation so that it would be applicable to real gases. On the basis of simple kinetic theory particles are assumed to be point masses, and there are no intermolecular forces between particles. However, as the pressure increases on a gaseous system, the volume occupied by the particles may become a significant part of the total volume. In addition, the intermolecular attractive forces become important under this condition. To account for the volume occupied by the particles, van der Waals proposed that the specific volume in the ideal-gas equation of state be replaced by the term $v - b$. At the same time the ideal pressure was to be replaced by the term $P + a/v^2$. The constant $b$ is the covolume of the particles, and the constant $a$ is a measure of the attractive forces. Thus the van der Waals equation is

$$\left(P + \frac{a}{v^2}\right)(v - b) = RT \quad (12\text{-}6)$$

Both $a$ and $b$ have units which must be consistent with those employed for $P$, $v$, and $T$. Note that, as the pressure approaches zero and the specific volume approaches infinity, the correction terms are negligible and the equation reduces to $Pv = RT$. Empirically, it is observed that, as the pressure increases, the $a/v^2$ term usually becomes important sooner than the $b$ correction factor.

The van der Waals equation is moderately successful, but it has a fundamental weakness that the constants $a$ and $b$ in actuality vary with temperature. Hence their values should be determined empirically for particular regions of pressure and temperature. One specific method of evaluating the two constants is based on the experimental fact that the critical isotherm on a $Pv$ diagram has a zero slope at the critical state for all gases. In addition, this isotherm goes through a point of inflection at the critical state. This phenomenon is illustrated in Fig. 12-2, which shows some typical isotherms based on the van der Waals equation of state. The curve marked $T_c$ is the critical isotherm, and it is fairly representative of actual experimental data. On the basis that the critical isotherm goes through a point of inflection at the critical state, we may write the following equations which are valid at the critical state:

$$\left(\frac{\partial P}{\partial v}\right)_{T_c} = 0 \quad \text{and} \quad \left(\frac{\partial^2 P}{\partial v^2}\right)_{T_c} = 0$$

These equations permit the evaluation of the constants in any two-constant equation of state. For example, the van der Waals constants $a$ and $b$ may now be determined by this method, It will be found that, on the basis of Eq. (12-6),

$$\left(\frac{\partial P}{\partial v}\right)_{T_c} = \frac{-RT_c}{(v_c - b)^2} + \frac{2a}{v_c^3} = 0$$

and

$$\left(\frac{\partial^2 P}{\partial v^2}\right)_{T_c} = \frac{2RT_c}{(v_c - b)^3} - \frac{6a}{v_c^4} = 0$$

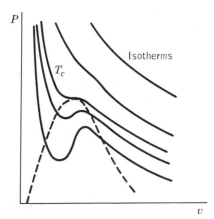

Figure 12-2 Isotherms for a van der Waals gas.

The subscript $c$ indicates that the property must be evaluated at the critical state. By combining these two expressions with the van der Waals equation of state it is found that

$$a = \frac{27}{64} \frac{R^2 T_c^2}{P_c} \qquad b = \frac{R T_c}{8 P_c} \qquad \frac{P_c v_c}{R T_c} = \frac{3}{8}$$

The above equations for $a$ and $b$ allow the determination of these constants from $P_c$ and $T_c$, which are obtained experimentally. However, since it is observed that the values of $a$ and $b$ actually need to be varied in order to relate $P$, $v$, and $T$ accurately, the use of critical data to evaluate the two constants must lead to an equation of state of limited accuracy. Another indication of this is the fact that the quantity $P_c v_c / R T_c$ is 0.375 according to the van der Waals equation. Experimentally, this is found to lie between 0.20 and 0.30 for the majority of gases. Thus the van der Waals equation is an approximation which is sometimes in serious error. Nevertheless, it is of historical interest as one of the first attempts to correct the ideal-gas equation so that real-gas behavior might be predicted. Tables A-3M and A-3 list some typical values for the van der Waals constants based on critical data.

## 12-3 OTHER EQUATIONS OF STATE

Two other two-constant equations of state are those of Berthelot and Dieterici. These have the form

Berthelot:
$$P = \frac{RT}{v - b} - \frac{a}{T v^2} \tag{12-7}$$

Dieterici:
$$P = \frac{RT}{v - b} e^{-a/RTv} \tag{12-8}$$

The Berthelot equation is quite similar to the van der Waals equation, except that the term $a/v^2$ also contains the temperature in the denominator. As a result, the constants $a$ and $b$ in this equation are not the same as those for the van der Waals equation of state, when based on critical data. The Dieterici equation was developed primarily to give better agreement with the quantity $P_c v_c / R T_c$ determined experimentally. It was noted earlier that this quantity based on the van der Waals equation was considerably in error. Like the van der Waals equation, these equations are also of limited accuracy.

Although many other equations of state could be presented, of varying complexity, the Redlich-Kwong equation is of considerable interest. This equation of state contains only two constants, and is empirical in nature. Nevertheless, it appears to have considerable accuracy over a wide range of $PvT$ conditions. On the basis of theoretical and practical considerations, Redlich and Kwong proposed in 1949 the following relationship:

$$P = \frac{RT}{v - b} - \frac{a}{T^{1/2} v (v + b)} \tag{12-9}$$

The constants $a$ and $b$ again can be evaluated from critical data. It is found that $a = 0.4275R^2T_c^{2.5}/P_c$, and $b = 0.0867RT_c/P_c$. One of the primary considerations in the development of this equation was that, at high pressures, the volume of all gases tends to approach the limiting value of $0.26v_c$, and this value is essentially independent of the temperature. Consequently, $b$ also equals $0.26v_c$. As a result, it is found that the equation gives quite good results at high pressures. The equation also appears to be fairly accurate for temperatures above the critical value. When $T$ is less than $T_c$, the equation is found to deviate from experimental data as the temperature is lowered. However, in several instances it has been shown that the Redlich-Kwong equation is as accurate as the eight-constant Benedict-Webb-Rubin equation of state. Consequently, the Redlich-Kwong equation of state has the advantages of simplicity combined with a fair degree of accuracy, especially with respect to other two-constant equations of state.

Since the Redlich-Kwong equation is a cubic equation in $v$, a rearrangement of the equation leads to a more convenient method for finding $v$ when $P$ and $T$ are known. By multiplying the equation by $v/RT$, we find that the Redlich-Kwong equation can be written as

$$\frac{Pv}{RT} = Z_{RK} = \frac{v}{v-b} - \frac{a}{(v+b)RT^{1.5}}$$

$$= \frac{1}{1-k} - \left(\frac{a}{bRT^{1.5}}\right)\left(\frac{k}{1+k}\right) \qquad (12\text{-}10a)$$

where $k$ is defined as $b/v$ and $Z_{RK}$ is a compressibility factor for a Redlich-Kwong gas. (See Sec. 4-6 for an introductory discussion of the compressibility factor $Z$.) Moreover, it has been noted above that $a$ and $b$ are functions of the critical temperature and pressure. As a result, $a/bRT^{1.5} = 4.934(T_c/T)^{1.5}$, and

$$Z_{RK} = \frac{1}{1-k} - \frac{4.934}{(T/T_c)^{1.5}}\left(\frac{k}{1+k}\right) \qquad (12\text{-}10b)$$

Also,

$$k = \frac{b}{v} = \frac{bP}{ZRT} = \frac{0.0867P_R}{ZT_R} \qquad (12\text{-}10c)$$

where $P_R = P/P_c$ and $T_R = T/T_c$. When $P$ and $T$ are given and $P_c$ and $T_c$ are known, Eqs. (12-10b) and (12-10c) contain two unknowns, namely, $Z$ and $k$. The two equations may be solved by iteration by hand or computer techniques. For example, as a first guess let $Z = 1$. Then use this value in Eq. (12-10c) to find a value of $k$. Use of this first estimate of $k$ in Eq. (12-10b) yields a new approximation for $Z$. This in turn is used again in Eq. (12-10c) to find a new value of $k$. This procedure is continued until the iteration produces changes in $Z$ and $k$ which are smaller than some prescribed small values.

**Example 12-1M** Estimate the pressure which would be exerted by 3.7 kg of CO in a 0.030-m³ container at 215°K, employing (a) the ideal-gas equation, (b) the van der Waals equation, and (c) the Redlich-Kwong equation of state.

SOLUTION The specific volume of the gas is $0.030/3.7 = 0.00811$ m$^3$/kg $= 0.227$ m$^3$/kg mole, and $R_u$ is 0.08315 bar·m$^3$/(kg·mol)(°K).

(a) The pressure is computed directly from the ideal-gas equation as

$$P = \frac{RT}{v} = \frac{0.08315(215)}{28(0.00811)} = 78.7 \text{ bars}$$

(b) The constants for the van der Waals equation are found in Table A-3M. For CO we find that $a = 1.463$ bar·m$^6$/(kg·mol)$^2$ and $b = 0.0394$ m$^3$/kg·mol. Substitution of these values in the equation of state leads to

$$\left(P + \frac{a}{v^2}\right)(v - b) = RT$$

$$\left[P + \frac{1.463}{(0.227)^2}\right][0.00811(28) - 0.0394] = 0.08315(215)$$

The solution to this equation gives the pressure to be 66.9 bars.

(c) The specific volume is 0.227 m$^3$/kg·mol and the temperature is 215°K. The constants $a$ and $b$ are found in Table A-21M to be 17.26 bar(m$^6$)(°K)$^{1/2}$/(kg·mol)$^2$ and 0.02743 m$^3$/kg·mol, respectively. Making the proper substitutions, we find that

$$P = \frac{RT}{v - b} - \frac{a}{T^{1/2}v(v + b)}$$

$$= \frac{0.08315(215)}{0.227 - 0.0274} - \frac{17.26}{(215)^{1/2}(0.227)(0.227 + 0.0274)}$$

$$= 69.2 \text{ bars}$$

**Example 12-1** Estimate the pressure which would be exerted by 8.2 lb of CO in a 1-ft$^3$ container at $-78°$F, employing (a) the ideal-gas equation, (b) the van der Waals equation of state, and (c) the Redlich-Kwong equation.

SOLUTION The temperature of $-78°$F is equal to 382°R. The specific volume of the gas is $1/8.2 = 0.122$ ft$^3$/lb.

(a) The pressure is computed directly from the ideal-gas equation as

$$P = \frac{RT}{v} = \frac{1545(382)}{28(0.122)(144)} = 1200 \text{ psia} = 81.7 \text{ atm}$$

(b) The constants for the van der Waals equation are found in Table A-3. For CO we find that $a = 372$ atm·ft$^6$/(lb·mol)$^2$ and $b = 0.63$ ft$^3$/lb·mol. In this system of units the value of $R_u$ will be 0.730 atm·ft$^3$/(lb·mol)(°R). In addition, the specific volume to be used is $0.122(28) = 3.42$ ft$^3$/lb·mol. Substitution of these values in the proper equation leads to

$$\left(P + \frac{a}{v^2}\right)(v - b) = RT$$

$$\left[P + \frac{372}{(3.42)^2}\right](3.42 - 0.63) = 0.73(382)$$

The solution of this equation gives the pressure to be 68.2 atm.

(c) The specific volume is 3.42 ft$^3$/lb·mol and the temperature is 382°R. The constants $a$ and $b$ are found in Table A-21 to be 5870 atm(ft$^6$)(°R)$^{1/2}$/(lb·mol)$^2$ and 0.4395 ft$^3$/lb·mol, respectively.

Making the proper substitutions, we find that

$$P = \frac{RT}{v - b} - \frac{a}{T^{1/2}v(v + b)}$$

$$= \frac{0.73(382)}{3.42 - 0.440} - \frac{5870}{(382)^{1/2}(3.42)(3.42 + 0.440)}$$

$$= 70.8 \text{ atm}$$

One of the inherent limitations in two-constant equations of state is the lack of accuracy over a wide range of conditions. To achieve higher accuracy requires equations of state with a considerably larger number of terms, or constants. An example of a more complex equation which has been successful in predicting the $PvT$ behavior especially of hydrocarbons is the Benedict-Webb-Rubin equation. This equation is

$$P = \frac{RT}{v} + \left(B_0 RT - A_0 - \frac{C_0}{T^2}\right)\frac{1}{v^2} + \frac{bRT - a}{v^3} + \frac{a\alpha}{v^6}$$

$$+ \frac{c}{v^3 T^2}\left(1 + \frac{\gamma}{v^2}\right) \exp\left(-\gamma/v^2\right) \tag{12-11}$$

This equation has eight adjustable constants for a given substance. Values of these constants for five common substances are found in Table A-21M and A-21. The units on these constants must be consistent with those used for $P$, $v$, $R$, and $T$. Another example of a complex equation of state is that suggested by Martin and Hou. It fits polar and nonpolar substances with densities up to 50 percent greater than the critical density. The original form of this equation in 1955 had nine adjustable constants. Since that time (1959), the equation has been modified to include eleven constants.

**Example 12-2M** Estimate the pressure of CO for the conditions expressed in Example 12-1M, on the basis of the Benedict-Webb-Rubin equation of state.

SOLUTION The values of the constants for the B-W-R equation for CO are listed in Table A-21M. The specific volume is 0.227 l/g·mol, and $R_u = 0.08315$ bar(l)/(g·mol)(°K). Making the proper substitutions, we find that

$$P = \frac{0.08315(215)}{0.227} + \left[0.05454(0.08315)(215) - 1.3587 - \frac{8.673 \times 10^3}{(215)^2}\right]\frac{1}{(0.227)^2}$$

$$+ \frac{0.002632(0.08315)(215) - 0.0371}{(0.227)^3} + \frac{0.0371(1.35 \times 10^{-4})}{(0.227)^6}$$

$$+ \frac{1.054 \times 10^3}{(0.227)^3(215)^2}\left[1 + \frac{0.0060}{(0.227)^2}\right] \exp\left[-0.0060/(0.227)^2\right]$$

$$= 78.75 - 11.09 + 0.86 + 0.04 + 1.94 = 70.50 \text{ bars}$$

*NBS Technical Note* 202 reports a pressure of 70.91 bars for the given specific volume and temperature. Thus the B-W-R equation is in error less than 0.6 percent. In comparison, the results of Example 12-1M indicate that the van der Waals equation is 5 percent too low, while the Redlich-Kwong equation is 1.8 percent too low.

**Example 12-2** Estimate the pressure of CO for the conditions expressed in Example 12-1, on the basis of the Benedict-Webb-Rubin equation of state.

SOLUTION The values of the constants for the B-W-R equation for CO are listed in Table A-21. The specific volume is 3.42 ft³/lb·mol and $T$ is 382°R. Making the proper substitutions, we find that

$$P = \frac{0.730(382)}{3.42} + \left[0.8740(0.73)(382) - 344.5 - \frac{7.124 \times 10^6}{(382)^2}\right]\frac{1}{(3.42)^2} + \frac{0.676(0.73)(382) - 150.73}{(3.42)^3}$$

$$+ \frac{150.73(0.5556)}{(3.42)^6} + \frac{1.387 \times 10^7}{(3.42)^3(382)^2}\left[1 + \frac{1.541}{(3.42)^2}\right]\exp\left[-1.541/(3.42)^2\right]$$

$$= 81.54 - 12.79 + 0.94 + 0.05 + 2.36 = 72.1 \text{ atm}$$

Because of the complexity of the B-W-R equation, the above answer is probably a good approximation to the true value. In comparison, the results of Example 12-1 indicate that the van der Waals equation is 5.4 percent too low, while the Redlich Kwong equation is 1.8 percent too low.

## 12-4 GENERALIZED COMPRESSIBILITY CHARTS

If we are to represent the $PvT$ behavior of real gases accurately, we have seen that we require fairly complicated empirical expressions. Each of these equations of state requires its own set of experimentally determined constants. A general method of correlating $PvT$ data, which has a reasonable degree of accuracy combined with an inherent simplicity, is based on the van der Waals principle of corresponding states. This principle was first introduced in Sec. 4-6. The reader should review that section at this point.

In summary, a compressibility factor $Z$ is inserted into the ideal-gas equation as a correction factor for real-gas behavior. That is,

$$Pv = ZRT \tag{12-12a}$$

The compressibility factor in general is a function of two intrinsic properties for each simple substance.

The principle of corresponding states indicates that the thermodynamic properties of all fluids should have approximately the same value when in the same reduced state. In general the reduced state is specified by two reduced properties, where any reduced property is the actual value of the property divided by its value at the critical state. For example,

$$P_R = \frac{P}{P_c} \qquad T_R = \frac{T}{T_c} \tag{12-12b}$$

Experimental data confirm the expectation that $Z$ has approximately the same value for all gases at the same values of $P_R$ and $T_R$. Originally a generalized compressibility chart was presented as Fig. 4-15 for a $P_R$ range from 0 to 1. Two additional generalized charts are presented in Figs. A-25M and A-26M for different ranges of the parameter $P_R$.

It would be very useful also to have reduced-volume lines on a generalized compressibility chart. If such lines were available, $vT$ data or $vP$ data could be employed to predict pressure or temperature data, respectively. For correlation purposes a pseudocritical specific volume has been defined by the relation $v'_c = RT_c/P_c$. This leads to a pseudoreduced volume $v'_R$ defined as

$$v'_R = \frac{v}{v'_c} = \frac{vP_c}{RT_c}$$

Pseudoreduced-volume lines are found on all three generalized compressibility charts found in the text.

A few general characteristics of the compressibility charts should be noted:

1. For reduced temperatures greater than 2.5, the value of $Z$ is greater than unity for all pressures. Under these circumstances the actual volume will always be greater than the ideal-gas volume for the same pressure and temperature.
2. For reduced temperatures below 2.5, the reduced isotherms go through a minimum at fairly low reduced pressures. In this region the volume is less than the ideal-gas volume, and the deviation from ideal-gas behavior is substantial.
3. The $Z$ factor is not obtainable at the critical state from the chart, since the critical reduced isotherm $(T_R = 1)$ is vertical in the region of $P_R = 1$. This is true since $Z_c$ takes on a wide range of values for various substances.
4. In the reduced pressure range of 8 to 10, all gases deviate from the ideal-gas relation about the same amount, regardless of temperature.
5. When $P_R$ is greater than 10, the deviation from ideal-gas behavior can approach several hundred percent.
6. In the limit, of course, as $P_R$ approaches zero, the value of $Z$ approaches unity for all values of the reduced temperature.

The generalized compressibility chart introduced above is extremely useful in estimating the $PvT$ behavior of gases, since only one chart is required for a large number of substances. It has the additional advantage of requiring a knowledge of only two physical properties of a compound, the critical values of the pressure and the temperature. These, usually, are easily measured quantities. The generalized chart is one of many illustrations in engineering and science where it is very helpful to correlate data on graphs by employing dimensionless parameters. For extreme accuracy in predicting $PvT$ data, one must rely on specific equations of state for individual compounds, which are found frequently in the literature. Further improvements even in the principle of corresponding states method may be found in other texts.

**Example 12-3M** Estimate the pressure exerted by CO with a specific volume of 0.227 m³/kg·mol at 215°K, and compare the result with the answers obtained in Examples 12-1M and 12-2M.

SOLUTION The critical pressure and temperature of CO are given in Table A-3M and are 35.0 bars and 1330°K, respectively. The reduced properties are then

$$v_R' = \frac{vP_c}{RT_c} = \frac{0.227(35.0)}{0.08315(133)} = 0.718$$

$$T_R = \frac{T}{T_c} = \frac{215}{133} = 1.62$$

From Fig. A-25M the value of $P_R$ is 2.1, and since $P_c$ equals 35.0 bars, the actual pressure is 2.1(35.0) = 73.5 bars. This compares favorably with the value of 70.5 bars obtained by the Benedict-Webb-Rubin equation in Example 12-2M. In this example we could also calculate the pressure by reading off the value of the compressibility factor of 0.90 and then use the relation $P = ZRT/v$. Since the ideal-gas evaluation gives 78.7 bars, the use of the compressibility factor leads to a pressure of 0.90(78.7), or 70.8 bars. This is in slight disagreement with the value of 73.5 bars found by reading $P_R$ directly from the chart. Both approaches should lead to the same answer, if the chart could be read with greater accuracy.

**Example 12-3** Estimate the pressure exerted by CO with a specific volume of 0.122 ft³/lb at −78°F, and compare the result with the answers obtained in Examples 12-1 and 12-2.

SOLUTION The critical pressure and temperature of CO are given in Table A-3 as 34.5 atm and 239°R, respectively. The reduced properties are then

$$v_R' = \frac{vP_c}{RT_c} = \frac{0.122(34.5)(28)}{0.73(239)} = 0.675$$

$$T_R = \frac{T}{T_c} = \frac{382}{239} = 1.60$$

From Fig. A-25M the value of $P_R$ is 2.1, and since $P_c$ equals 34.5 atm, the actual pressure is 2.1(34.5) = 72.5 atm. This compares favorably with the value of 72.1 atm obtained by the Benedict-Webb-Rubin equation in Example 12-2. The van der Waals value of 68.2 atm is somewhat low, and the ideal-gas value of 81.7 is much too large. In the above example one could also calculate the pressure by reading off the value of the compressibility factor of 0.88 and then use the relation $P = ZRT/v$. Since the ideal-gas evaluation gives 81.7 atm, the use of the compressibility factor leads to a pressure of 0.88(81.7), or 72.0 atm. This agrees, of course, with the value of 72.5 atm found by reading $P_R$ directly from the chart, within the accuracy of the chart.

**Example 12-4M** Estimate the temperature of ethane gas $(C_2H_6)$ when it has a pressure of 68.4 bars and a specific volume of 0.0104 m³/kg.

SOLUTION The critical pressure and temperature of ethane are given in Table A-3M as 48.8 bars and 305°K, respectively. Since the molar mass of ethane is 30, then the reduced properties are

$$v_R' = \frac{vP_c}{RT_c} = \frac{0.0104(30)(48.8)}{0.08315(305)} = 0.600$$

$$P_R = \frac{P}{P_c} = \frac{68.4}{48.8} = 1.40$$

The intersection of a $v_R'$ line of 0.600 and a $P_R$ line of 1.40 leads to the desired values of $T_R$ and $Z$. From Fig. A-25M we find that $Z = 0.70$ and $T_R = 1.20$. Consequently, $T = T_c T_R = 305(1.20) = 366°K$. A similar value may be found by using the $Z$ value in the expression $T = Pv/ZR$.

**Example 12-4** Estimate the temperature of ethane gas ($C_2H_6$) when it has a pressure of 67.5 atm and a specific volume of 0.167 ft³/lb.

SOLUTION The critical pressure and temperature of ethane are given in Table A-3 as 48.2 atm and 550°R, respectively. Since the molar mass of ethane is 30, then the reduced properties are

$$v_R' = \frac{vP_c}{RT_c} = \frac{0.167(30)(48.2)}{0.730(550)} = 0.603$$

$$P_R = \frac{P}{P_c} = \frac{67.5}{48.2} = 1.40$$

The intersection of a $v_R'$ line of 0.603 and a $P_R$ line of 1.40 leads to the desired values of $T_R$ and $Z$. From Fig. A-25M we find that $T_R = 1.20$ and $Z = 0.70$. Consequently, $T = T_c T_R = 550(1.20) = 660°R = 200°F$. A similar answer may be found by using the $Z$ value in the expression $T = Pv/ZR$.

## 12-5 REAL-GAS MIXTURES

In Chap. 11 it was pointed out that the evaluation of the properties of mixtures is a difficult one, since any one property depends not only upon two independent properties such as pressure and temperature, but also upon a specification of the composition of the mixture such as the mole fractions of each component. Hence the tabulation of the properties of mixtures is not too fruitful because of the tremendous quantity of data required. As in the approach for ideal gases presented earlier, one solution is to employ the properties of the individual pure constituents in some sort of mixture rule. The two most common rules are those of additive pressures and additive volumes introduced in the preceding chapter. For example, the total pressure of a mixture of gases might be computed by Dalton's law of additive pressures as the sum of the pressures exerted by the individual constituents. That is,

$$P_m = P_1 + P_2 + P_3 + \cdots]_{T,V} \tag{12-13}$$

where the pressures of the components 1, 2, 3, etc., are computed from more exact methods for each individual gas. Each component pressure is evaluated at the volume and the temperature of the mixture as noted in Eq. (12-13). The real error in this approach is that each gas is assumed to act as if it alone occupied the entire volume and its behavior were not affected by the presence of other chemical species. As a consequence, it might be expected that the law of additive pressures will be more valid at relatively low pressures or densities, and considerable error might accrue at higher pressures.

If each gas in a gas mixture were assumed to follow the van der Waals equation, the equation for component $i$ would be

$$P_i]_{T,V} = \frac{R_u T}{v_i - b_i} - \frac{a_i}{v_i^2}$$

where $v_i$ is the molar volume of the component. However, the molar volume of any species is related by definition to the molar volume $v_m$ of the mixture by $v_m = y_i v_i$. The substitution of this relation into the above equations yields

$$p_i]_{T,V} = \frac{y_i R_u T}{v_m - y_i b_i} - \frac{y_i^2 a_i}{v_m^2}$$

If this general form of the equation is substituted into the additive-pressure law, we find upon rearrangement that

$$P_m = R_u T\left(\frac{y_1}{v_m - y_1 b_1} + \frac{y_2}{v_m - y_2 b_2} + \cdots\right) - \frac{1}{v_m^2}(a_1 y_1^2 + a_2 y_2^2 + \cdots) \tag{12-14}$$

For any mixture of given temperature, volume, and composition, the right-hand side of this equation can be computed directly, assuming that the van der Waals constant for each constituent has already been determined.

Another approach to the use of the additive-pressure law would be to assume that the compressibility factors for the individual components could be used. If an equation of the type $p_i V = Z_i N_i R_u T$ is substituted in the additive-pressure law for each component, the result is

$$P_m = \frac{R_u T}{V}(N_1 Z_1 + N_2 Z_2 + \cdots)$$

However, if an average compressibility factor $Z_m$ for the mixture is defined in terms of the equation $P_m V = Z_m N_m R_u T$, then we see that

$$Z_m N_m = Z_1 N_1 + Z_2 N_2 + \cdots$$

or $\qquad\qquad Z_{m,V,T} = y_1 Z_1 + y_2 Z_2 + \cdots \tag{12-15}$

where $y$ again is the molefraction for that species. The subscripts $V$ and $T$ on $Z_m$ emphasize that the compressibility factors for each component are to be evaluated at the volume and temperature of the mixture. In general it can be stated that the rule of additive pressures in the form of Eq. (12-15) tends to give values of $Z_m$ which are greater than the experimental value at low pressures or densities and to give values which are too low at high pressures. To overcome this difficulty, a method, frequently called the Bartlett rule of additive pressures, may be employed. By this method the same form of Eq. (12-15) is used, but the individual compressibility factors are evaluated, using the temperature and the molar volume of the mixture rather than the molar volume of the component. This approach tends to give better results in the low-pressure region. Other equations of state can be used, of course, in conjunction with the law of additive pressures.

Amagat's rule of additive volumes may also be applied to gas mixtures as an approximation technique. In this case

$$V_m = V_1 + V_2 + V_3 + \cdots]_{P,T} \tag{12-16}$$

where the individual volumes are computed from appropriate equations of state based on the pressure and the temperature of the mixture. If the compressibility-factor methods is employed, then $P_m V_i = Z_i N_i R_u T$. The use of this expression in Eq. (12-16) leads to the following relationship for the overall mixture compressibility factor $Z_m$:

$$Z_{m, P, T} = y_1 Z_1 + y_2 Z_2 + \cdots \tag{12-17}$$

The subscripts $P$ and $T$ emphasize that the individual $Z$ factors are to be evaluated at the pressure and the temperature of the mixture. Although Eqs. (12-15) and (12-17) are identical in form for the additive-pressure and additive-volume rules, the method of evaluation is quite different. As a result, the two methods give different values for a given problem, when applied to real gases. Since Amagat's rule takes into account the total pressure of the system, it effectively is accounting for the influence of intermolecular forces between different chemical species, as well as between like molecules. It is a matter of experience that the rule of additive volumes for real-gas mixtures leads to $Z_m$ values which are too low in the low-pressure region. In general, it appears that Amagat's rule is probably superior to the additive-pressure rule except at relatively low pressures.

For preliminary engineering estimates another mixture rule known as Kay's rule is satisfactory to within, roughly, 10 percent. In this method, pseudocritical temperature and pressure for mixtures are defined in the following manner:

$$T'_c = y_1 T_{c1} + y_2 T_{c2} + y_3 T_{c3} + \cdots \tag{12-18a}$$

and

$$P'_c = y_1 P_{c1} + y_2 P_{c2} + y_3 P_{c3} + \cdots \tag{12-18b}$$

This technique is quite useful since it requires merely a knowledge of the critical temperatures and pressures of the constituent gases. Other types of correlations for real-gas mixtures are available in the literature.

> **Example 12-5** A system contains a mixture of hydrogen and nitrogen which is 75 mole percent hydrogen and 25 mole percent nitrogen at 77°F (25°C). When the specific volume is 1.355 ft³/lb·mol (0.0845 m³/kg·mol), the pressure experimentally is found to be 400 atm. Estimate what the pressure would be on the basis of (a) ideal-gas behavior, (b) additive-pressure rule and van der Waals gases, (c) additive-volume rule and van der Waals gases, (d) additive-pressure rule and compressibility factors, (e) additive-volume rule and compressibility factors, (f) the Bartlett rule of additive pressures, and (g) pseudocritical temperature and pressure approach.
>
> SOLUTION The critical data and the van der Waals constants for hydrogen and nitrogen are as follows:
>
> $$T_{c, H_2} = 59.8 + 14.4 = 74.2°R \qquad a_{H_2} = 62.80 \text{ atm·(mol)}^2/\text{ft}^6$$
>
> $$P_{c, H_2} = 12.8 + 8.0 = 20.8 \text{ atm} \qquad b_{H_2} = 0.426 \text{ ft}^3/\text{lb·mol}$$
>
> $$T_{c, N_2} = 227°R \qquad a_{N_2} = 346 \text{ atm·(mol)}^2/\text{ft}^6$$
>
> $$P_{c, N_2} = 33.5 \text{ atm} \qquad b_{N_2} = 0.618 \text{ ft}^3/\text{lb·mol}$$
>
> For use with the generalized compressibility charts the critical temperature and pressure of hydrogen have been corrected by the factors of 14.4 and 8.0, as indicated in the main text, for greater accuracy. The value of the universal gas constant $R_u$ to use with the above set of units is 0.73 atm·ft³/(lb·mol)(°R).

(a) The pressure based on the ideal-gas equation is

$$P = \frac{NRT}{V} = \frac{0.73(537)}{1.355} = 290 \text{ atm}$$

This value is in error by over 25 percent.

(b) For the additive-pressure rule with the van der Waals equation, we may employ Eq. (12-14). Thus

$$P = 0.73(537) \left[ \frac{0.75}{1.355 - 0.75(0.426)} + \frac{0.25}{1.355 - 0.25(0.618)} \right]$$

$$- \frac{1}{(1.355)^2} [62.80(0.75)^2 + 346(0.25)^2]$$

$$= 392(0.9335) - 0.545(56.9) = 335 \text{ atm}$$

The result is better than the ideal-gas relation, but still quite low compared with the experimental value.

(c) According to the additive-volume rule, the total volume is the sum of the volumes of the components, measured at the pressure and the temperature of the mixture. The pressure is unknown; so we write the following three equations:

$$1.355 = 0.75v_{H_2} + 0.25v_{N_2}$$

$$P_{H_2} = \frac{RT}{v - b} - \frac{a}{v^2} = \frac{0.73(537)}{v - 0.426} - \frac{62.80}{v^2} = \frac{392}{v - 0.426} - \frac{62.80}{v^2}$$

$$P_{N_2} = \frac{392}{v - 0618} - \frac{346}{v^2}$$

For a satisfactory solution the pressures of the two gases must be equal. Employing an iteration process, we assume that $v$ of the hydrogen is 1.38 ft$^3$/lb·mol. Then, from the first equation above,

$$v_{N_2} = \frac{1.355 - 0.75(1.38)}{0.25} = 1.28$$

Substituting these two values into the last two equations, we find that

$$P_{H_2} = \frac{392}{1.38 - 0.426} - \frac{62.80}{(1.38)^2} = 378 \text{ atm}$$

$$P_{N_2} = \frac{392}{1.28 - 0.618} - \frac{346}{(1.28)^2} = 381 \text{ atm}$$

These two answers are sufficiently close together not to warrant another trial. This solution is, roughly, 5 percent lower than the measured value.

(d) The additive-pressure rule and the compressibility chart can be combined into a very rapid method of approximation if pseudoreduced-volume lines are available. Recall that $v'_R = vP_c/RT_c$, where $v$ is the specific volume an individual gas would have if it filled the entire system. On this basis the pseudoreduced volumes and the reduced temperatures for the two gases become

$$v'_{R, H_2} = \frac{1.355(20.8)(2120)}{0.75(74.2)(1545)} = 0.71 \qquad T_{R, H_2} = \frac{537}{74.2} = 7.25$$

$$v'_{R, N_2} = \frac{1.355(33.5)(2120)}{0.25(227)(1545)} = 1.10 \qquad T_{R, N_2} = \frac{537}{227} = 2.37$$

From Figs. A-25M and A-26M the $Z$ factors for hydrogen and nitrogen are found to be 1.06 and 0.99, respectively. Hence

$$Z_m = 0.75(1.06) + 0.25(0.99) = 1.05$$

and
$$P = Z_m P_{ideal} = 1.05(290) = 305 \text{ atm}$$

(e) According to the additive-volume rule, each gas exists at the pressure and the temperature of the mixture. Since the pressure is unknown, it must be assumed under an iteration process. If we first assume that the system pressure is 350 atm, then

$$P_{R, H_2} = \frac{350}{20.8} = 16.8 \qquad T_{R, H_2} = 7.25$$

$$P_{R, N_2} = \frac{350}{33.5} = 10.45 \qquad T_{R, N_2} = 2.37$$

The individual compressibility factors based on these reduced properties are 1.20 and 1.21 for hydrogen and nitrogen, respectively. Therefore

$$Z_m = 0.75(1.20) + 0.25(1.21) = 1.20$$

and
$$P = Z_m P_{ideal} = 1.20(290) = 352 \text{ atm}$$

This value agrees substantially with the assumed value; so no further trial is necessary. In this case the additive-volume rule was a better approximation than the additive-pressure rule, although both give too low a value.

(f) The Bartlett rule of additive pressures is based on using the molar specific volume of the mixture for the specific volume of any component. If the value of 1.355 is used in the pseudo-reduced volume definition, we find that, from the calculation made in part (d),

$$v'_{R, H_2} = 0.75(0.71) = 0.53 \qquad T_{R, H_2} = 7.25$$

$$v'_{R, N_2} = 0.25(1.10) = 0.28 \qquad T_{R, N_2} = 2.37$$

The compressibility factors for hydrogen and nitrogen under these conditions are approximately 1.25 and 1.17, respectively. Thus

$$Z_m = 0.75(1.25) + 0.25(1.17) = 1.23$$

and
$$P = Z_m P_{ideal} = 1.23(290) = 360 \text{ atm}$$

Note that the answer is much improved over that obtained by the unmodified rule of additive pressures.

(g) If, finally, the definitions of the pseudocritical temperature and pressure defined above are employed, then

$$T'_c = y_1 T_{c1} + y_2 T_{c2} = 0.75(74.2) + 0.25(227) = 112.4°R$$

$$P'_c = y_1 P_{c1} + y_2 P_{c2} = 0.75(20.8) + 0.25(33.5) = 24.0 \text{ atm}$$

Consequently,

$$v'_R = \frac{vP'_c}{RT'_c} = \frac{1.355(24.0)}{0.73(112.4)} = 0.397$$

$$T'_R = \frac{T}{T'_c} = \frac{537}{112.4} = 4.79$$

From Fig. A-26M, the $Z$ factor is 1.31, and the corresponding pressure of the system is $P = 1.31(290) = 379 \text{ atm}$. All the methods employed above gave answers which were too low, although some were much better than others.

# REFERENCES

Benedict, O., G. B. Webb, and L. C. Rubin: *J. Chem. Phys.*, **8**(4): 334–345 (1940).
Lewis, G. N., and M. Randall: "Thermodynamics," 2d ed., McGraw-Hill, New York, 1961.
Martin, J. J., and Y. C. Hou: *A.I. Ch. E. J.*, **1**(2): 142–151 (1955).
Martin, J. J., R. M. Kapoor, and N. deNevers: *A.I. Ch. E. J.*, **5**(2): 159–164 (1959).
Obert, E. F.: "Concepts of Thermodynamics," McGraw-Hill, New York, 1960.
Otto, J., A. Michels, and H. Wouters: *Physik. Z.*, **35**(3): 97–100 (1934).
Redlich, O., and J. N. S. Kwong: *Chem. Rev.*, **44**(1): 233–244 (1949).

# PROBLEMS (METRIC)

**12-1M** Determine the expressions for the constants $a$ and $b$ in the Berthelot equation of state in terms of the critical-state values of $P$, $v$, and $T$.

**12-2M** Determine the equations for $a$ and $b$ in the Dieterici equation of state in terms of the critical-state values of $P$, $v$, and $T$.

**12-3M** Determine equations for $a$ and $b$ in the Redlich-Kwong equation of state in terms of the critical values of $P$, $v$, and $T$.

**12-4M** Steam at 100 bars and 360°C is a real gas. Determine the specific volume in $cm^3/g$ by the following methods: (*a*) ideal gas, (*b*) van der Waals, (*c*) Redlich-Kwong, (*d*) compressibility factor, and (*e*) tabulated value.

**12-5M** Carbon dioxide at 313°K and 73 bars has an observed specific volume of 200 $cm^3/g \cdot mol$. Calculate the value based on (*a*) ideal-gas equation, (*b*) Redlich-Kwong equation, (*c*) van der Waals equation, and (*d*) compressibility factor.

**12-6M** Compute the specific volume of steam at 140 bars and 400°C by means of (*a*) ideal-gas equation, (*b*) van der Waals equation, (*c*) Redlich-Kwong equation, (*d*) compressibility factor, and (*e*) tabulated value.

**12-7M** The specific volume of refrigerant 12 at 60°C is 19.41 $cm^3/g$. Determine the pressure of the fluid by means of (*a*) ideal-gas equation, (*b*) van der Waals equation, (*c*) Redlich-Kwong equation, (*d*) compressibility factor, and (*e*) tabulated value.

**12-8M** Same as Prob. 12-6M except that we wish to find the pressure if the given state is 360°C and 30.9 $cm^3/g$.

**12-9M** Same as Prob. 12-7M, except that the state is 80°C and 12.0 $cm^3/g$.

**12-10M** Same as Prob. 12-6M, except that we wish to find the temperature for the given state of 100 bars and 23.3 $cm^3/g$.

**12-11M** Same as Prob. 12-7M, except that we wish to find the temperature for the given state of 9 bars and 22.0 $cm^3/g$.

**12-12M** The specific volume of saturated $CO_2$ vapor at 25°C is 4.19 $cm^3/g$. Estimate the pressure, in bars, on the basis of (*a*) ideal-gas equation, (*b*) Redlich-Kwong equation, (*c*) Benedict-Webb-Rubin equation, (*d*) van der Waals equation, and (*e*) compressibility factor. The tabulated value is 64.0 bars.

**12-13M** Determine the pressure of $N_2$ at $-129$°C if the specific volume is 2.11 $cm^3/g$, employing (*a*) van der Waals equation, (*b*) Redlich-Kwong equation, (*c*) Benedict-Webb-Rubin equation, and (*d*) principle of corresponding states.

**12-14M** Methane gas is maintained at 23 bars and $-82$°C. Compute the specific volume in $m^3/kg$ on the basis of (*a*) Redlich-Kwong equation, (*b*) van der Waals equation, and (*c*) principle of corresponding states.

**12-15M** Calculate the pressure of methane ($CH_4$) at 12°C and 0.0187 $m^3/kg$ on the basis of (*a*) principle of corresponding states, (*b*) Redlich-Kwong equation, and (*c*) Benedict-Webb-Rubin equation.

**12-16M** Calculate the pressure required to compress 100 l of nitrogen at 745 mm Hg and 23°C to 1.0 l at −110°C, using $Z$ data.

**12-17M** On the basis of the Redlich-Kwong equation calculate the specific volume in $cm^3/g$ mole at (a) 350°K and 101 bars, and (b) 400°K and 101 bars, for carbon dioxide.

**12-18M** The constants for the virial equation of state in the form $Pv = A[1 + (b'/v) + (c'/v^2) + \cdots]$ have been determined experimentally for nitrogen at −100°C. The values are: $A = 14.39$, $b' = -0.05185$ $m^3/kg{\cdot}mol$, $c = 0.002125$ $m^6/(kg{\cdot}mol)^2$. Also, $P$ is in bars, $T$ in °K, and $v$ in $m^3/kg{\cdot}mol$.

    (a) Determine the compressibility factor at 68 bars and −100°C from the above equation.

    (b) Compare the above result to that obtained from a generalized compressibility chart.

**12-19M** Calculate the specific volume of $N_2$ in $cm^3/g$ at 627°K and 51 bars, using the compressibility factor method. What would be the percent error in assuming nitrogen to be an ideal gas under these conditions?

**12-20M** Diatomic oxygen is at 100 bars and −73°C. Calculate how many kilograms can be stored in a tank with a volume of 3 $m^3$ if we use (a) the ideal-gas equation, and (b) the corresponding-states concept.

**12-21M** Estimate the temperature of 1 kg of n-octane ($C_8H_{18}$) at 27 bars if the volume is 0.0125 $m^3$.

**12-22M** On the basis of the $Z$ factor estimate the kilograms of propane gas in a 0.3-$m^3$ cylinder if the pressure is 200 bars and the temperature is (a) 207°C, and (b) 283°C.

**12-23M** Nitrogen gas at a pressure of 100 bars and −70°C is contained in a tank of 0.25 $m^3$. Heat is added until the temperature is 37°C. Determine approximately, through the use of the $Z$ factor, (a) the specific volume of the gas, in $m^3/kg{\cdot}mol$, and (b) the final pressure, in bars.

**12-24M** Ethane gas is being transported through a pipeline at a pressure of 95 bars and 55°C. Using the generalized chart, determine the increase in percent, in the velocity of the ethane that is required to deliver the same mass flow rate if the line pressure remains the same but the temperature increases to 116°C.

**12-25M** Diatomic oxygen at a pressure of 101 bars and −27°C is contained in a tank of 0.20 $m^3$. Heat is transferred until the temperature is 5°C. Estimate (a) the specific volume of the gas in $m^3/kg{\cdot}mol$, and (b) the final pressure in bars, using the compressibility chart.

**12-26M** Calculate the specific volume in $m^3/kg{\cdot}mol$ of nitrogen at 102 bars and −45°C by (a) the ideal-gas equation, (b) the compressibility chart, and (c) the virial equation. The virial coefficients for nitrogen at this temperature are $b = -2.34 \times 10^{-2}$, $c = 3.61 \times 10^{-5}$, and $d = 5.18 \times 10^{-7}$, where $P$ is in bars, $T$ is in °K, and $v$ is in $m^3/kg{\cdot}mol$.

**12-27M** Determine the specific volume of nitrogen at 200 bars and 0°C by means of Eq. (12-3). Compare with the answer found by using Eqs. (12-4) and (12-5).

**12-28M** Same as Prob. 12-27M, except the pressure is 170 bars.

**12-29M** On the basis of the generalized chart, determine the kilograms of ethylene ($C_2H_4$) in a 0.5-$m^3$ tank if the temperature is 38°C and the pressure is 82 bars. If the temperature is increased to 123°C, find the final pressure, in bars.

**12-30M** Determine the specific volume of nitrogen at 150°K and 64 bars by means of Eqs. (12-4) and (12-5), and compare with the result based on the generalized compressibility chart.

**12-31M** *NBS Technical Note* 202 (1963) lists the specific volume of carbon monoxide at 200°K and 60 atm to be 0.2356 $m^3/kg{\cdot}mol$. Estimate the value of $P$ at the given $v$ and $T$ by means of (a) the ideal-gas equation, (b) the van der Waals equation, (c) the Redlich-Kwong equation, (d) the Benedict-Webb-Rubin equation, and (e) the generalized compressibility chart.

**12-32M** A mixture consists of 1 kg of $CO_2$ and 1 kg of water vapor at 20 bars and 200°C. Calculate (a) the partial pressure of each component, (b) the component pressure based on the additive-pressure rule and $CO_2$ is an ideal gas, (c) the component pressure based on the additive-pressure rule if $CO_2$ is a real gas, and (d) the volume of the mixture.

**12-33M** On the basis of compressibility factors, determine the isothermal work necessary to compress carbon dioxide from 70 to 250 bars at a temperature of 92°C in a steady-flow process.

**12-34M** Same as Prob. 12-33M, except that the $CO_2$ is at 153°C.

### Real-gas mixtures

**12-35M** Assume air is a mixture of two gases, and the composition is 78 mole percent nitrogen and 22 mole percent oxygen. What is the pressure of the mixture if 14 m³ of it at 20°C and 1 bar is compressed to 0.028 m³ and 37°C? Assume the following: (a) ideal-gas law; (b) van der Waals gases, using the additive-pressure law; (c) van der Waals gases, using the additive-volume law; and (d) reduced-coordinates and additive-volume law.

**12-36M** One kilogram-mole each of two gases $A$ and $B$ exists in a mixture at 260°K and a total pressure of 100 bars. Find the total volume in cubic meters, using (a) the additive-pressure rule, and (b) the additive-volume rule. The critical pressures and temperatures of $A$ and $B$ are: $P_{cA} = 3$ bars, $P_{cB} = 4$ bars, $T_{cA} = T_{cB} = 200$°K.

**12-37M** Same as Prob. 12-36M, except that the mixture temperature is 240°K and the total pressure is 90 bars.

**12-38M** One wishes to prepare a mixture of 60 mole percent acetylene $(C_2H_2)$ and 40 mole percent $CO_2$ at 47°C and 100 bars, in a 1-m³ tank. The tank initially contains acetylene at 47°C and pressure $P_1$. Carbon dioxide is then bled into the tank from a line containing $CO_2$ at 47°C and 100 bars until the tank pressure reaches 100 bars. What is the value of $P_1$ such that when the tank pressure reaches 100 bars the composition within the tank will be 60 mole percent acetylene. Assume the validity of Kay's rule for the mixture.

**12-39M** Calculate the pressure exerted by a mixture of 0.5 kg·mol of methane $(CH_4)$ and 0.5 kg·mol of propane $(C_3H_8)$ for a temperature of 90°C and a volume of 0.48 m³. Use (a) van der Waals equation and the additive-pressure law, (b) the compressibility chart and additive volumes, and (c) the compressibility chart and Kay's rule. Compare with the ideal-gas value. The observed value is 50.6 bars.

## PROBLEMS (USCS)

**12-1** Show that, for an adiabatic reversible expansion of a van der Waals gas, $T(v - b)^{k-1} = $ constant.

**12-2** Determine the equations for $a$ and $b$ in the Dieterici equation of state in terms of the critical-state values of $P$, $v$, and $T$. Then compute the value of $Z_c$ for such a gas, and compare with experimental data for real gases.

**12-3** Steam at 1600 psia and 740°F is a real gas. Determine the specific volume by the following methods: (a) compressibility factor, (b) ideal gas, (c) van der Waals, and (d) tabulated values.

**12-4** Carbon dioxide at 313°K and 72 atm has an observed specific volume of 3.21 ft³/lb·mol. Calculate the value based on (a) principle of corresponding states, (b) ideal gas, and (c) van der Waals.

**12-5** Compute the specific volume of steam at 2000 psia and 800°F by means of (a) the ideal-gas equation and (b) the compressibility chart. (c) Compare with tabulated data.

**12-6** The specific volume of ammonia at 120°F is 1.443 ft³/lb. Determine the pressure of the substance by means of (a) the ideal-gas equation, (b) the compressibility factor, (c) the van der Waals equation, (d) the Redlich-Kwong equation, and (e) the tabulated value.

**12-7** Determine the pressure of nitrogen at $-200$°F if the specific volume is 0.03375 ft³/lb, employing (a) the van der Waals equation, (b) the Redlich-Kwong equation, and (c) the principle of corresponding states.

**12-8** Compute the pressure of methane $(CH_4)$ at a temperature of 200°F and a specific volume of 0.235 ft³/lb by using (a) the compressibility factor, (b) the van der Waals equation, and (c) the Redlich-Kwong equation.

**12-9** Methane gas is maintained at 335 psia and $-115$°F. Compute the specific volume of the gas in ft³/lb on the basis of the principle of corresponding states.

**12-10** Calculate the volume of 1 lb of methane $(CH_4)$ at 40°F and 1000 psia, using compressibility data.

**12-11** Calculate the pressure required to compress 10 ft$^3$ of nitrogen at 14.4 psia and 73°F to 0.1 ft$^3$ at −167°F, employing $Z$ data.

**12-12** (*a*) Evaluate the constants $a$ and $b$ in the Redlich-Kwong equation for carbon dioxide in units compatible with $P$ in atm, $T$ in °R, and $v$ in ft$^3$/lb·mol.

(*b*) Calculate $v$ for 100 atm and 630°R.

(*c*) Calculate $v$ for 100 atm and 1260°R.

(*d*) Compare the results with data for $v$ from the *National Bureau of Standards Circular* 564.

**12-13** The constants for the virial equation of state in the form $Pv = A[1 + (b'/v) + (c'/v^2) + \cdots]$ have been determined experimentally for nitrogen at −100°C. The values are: $A = 0.634$, $b' = -2.3146 \times 10^{-3}$, and $c' = 4.235 \times 10^{-6}$.

(*a*) Determine the compressibility factor at 67 atm and −100°C from the above equation.

(*b*) Compare the above result to that obtained from a generalized compressibility chart.

**12-14** (*a*) Calculate the specific volume in ft$^3$/lb for $N_2$ at 1130°R and 735 psia, using the compressibility factor.

(*b*) What would be the percentage error in assuming nitrogen to be an ideal gas under these conditions?

**12-15** Oxygen ($O_2$) is at 100 atm and −100°F. Calculate how many pounds can be stored in a tank with a volume of 100 ft$^3$ if we (*a*) use the ideal-gas equation and (*b*) use the principle of corresponding states.

**12-16** Estimate the temperature of 1.0 lb of *n*-octane ($C_8H_{18}$) at 27 atm if the volume is 0.20 ft$^3$.

**12-17** Determine the pounds of propane gas in a 10-ft$^3$ cylinder if the pressure is 2900 psia and the temperature is (*a*) 405°F and (*b*) 540°F.

**12-18** Nitrogen gas at a pressure of 1500 psia and −97°F is contained in a tank of 10 ft$^3$. Heat is transferred until the temperature is 100°F. Determine approximately (*a*) the specific volume of the gas in ft$^3$/lb·mol, and (*b*) the final pressure in psia, using the compressibility chart.

**12-19** Ethane gas is being transported through a pipeline at a pressure of 1400 psia and 130°F. Using the generalized compressibility chart, determine, in percent, the increase in velocity of the ethane required to deliver the same mass flow rate if the line pressure remains the same but the temperature increases to 240°F.

**12-20** Calculate the specific volume in ft$^3$/lb·mol of nitrogen at 1500 psia and −50°F by (*a*) the ideal-gas law, (*b*) the compressibility-factor method, and (*c*) the virial equation. (The virial coefficients for nitrogen at this temperature are $b = -3.75 \times 10^{-1}$, $c = 5.86 \times 10^{-4}$, and $d = 8.53 \times 10^{-6}$, where $P$ is in atm and $T$ is in °R.)

**12-21** Determine the specific volume of nitrogen at 200 atm and 0°C by means of Eq. (12-3). Compare with the answer found by using Eqs. (12-4) and (12-5).

**12-22** On the basis of the generalized compressibility chart, determine the pounds of ethylene ($C_2H_4$) in a 10 ft tank if the temperature is 73°F and the pressure is 1225 psia. If the temperature is increased to 252°F, find the final pressure in psia.

**12-23** Determine the specific volume of nitrogen at 150°K and 60 atm by means of Eqs. (12-4) and (12-5), and compare with the result based on the generalized charts, Figs. A-25M and A-26M.

**12-24** Air is a mixture of two real gases. What is the pressure of the mixture if 500 ft$^3$ of it at 70°F are compressed to 1.0 ft$^3$? The original pressure was 1 atm, and the final temperature is 100°F. Assume the following: (*a*) ideal-gas; (*b*) van der Waals gases, using the additive-pressure law; (*c*) van der Waals gases, using additive-volume law; and (*d*) reduced coordinates and additive-volume law.

**12-25** A mixture consists of 1.0 lb of $CO_2$ (an ideal gas) and 1.0 lb of water vapor (a real gas) at 300 psia and 400°F. Calculate (*a*) the partial pressure of each component, (*b*) the component pressure based on the additive-pressure rule, (*c*) the component pressure based on the additive-pressure rule if $CO_2$ is a real gas, and (*d*) the volume of the mixture.

**12-26** On the basis of compressibility factors, determine the isothermal work necessary to compress carbon dioxide from 1000 to 4000 psia at a temperature of 197°F in a steady-flow process.

**12-27** One mole of each of two gases $A$ and $B$ exists in a mixture at 250°R and a total pressure of

100 atm. Find the total volume, in cubic feet, using (a) the additive-pressure rule; and (b) the additive-volume rule. The critical pressures and temperatures of A and B are: $P_{cA} = 3$ atm, $P_{cB} = 4$ atm, $T_{cA} = T_{cB} = 100°R$.

**12-28** One wishes to prepare a mixture of 60 mole percent acetylene ($C_2H_2$) and 40 percent $CO_2$ at 120°F, 1500 psia, in a 1.0 ft³ tank. The tank initially contains acetylene at 120°F and pressure $P_1$. Carbon dioxide is then bled into the tank from a line containing $CO_2$ at 120°F and 1500 psia until the tank pressure reaches 1500 psia. What is the value of $P_1$ such that when the tank pressure reaches 1500 psia the composition within the tank will be 60 percent acetylene? Assume the validity of Kay's rule for the mixture.

**12-29** On the basis of Eqs. (12-4) and (12-5) determine the specific volume of nitrogen in ft³/lb at 1000 psia and 300°R, and compare the result to the experimental value of 0.0828 ft³/lb. What percent error is made if one assumes ideal-gas behavior at this state?

# THIRTEEN

## GENERALIZED THERMODYNAMIC RELATIONSHIPS

In the solution of engineering and scientific problems it is essential to be able to determine the values of thermodynamic properties. Since classical thermodynamics is an experimental science, a great deal of empirical data has been obtained over the past decades. The student must recognize, however, that only a relatively few properties can be evaluated by direct experimentation. Of these, the correlation among pressure, specific volume, and temperature and the relationship between the specific heats and temperature at low pressures are most easily measured. The evaluation of such properties as the internal energy, the enthalpy, and the entropy is made on the basis of calculations involving the directly measurable data mentioned above. Consequently, one of the major tasks of thermodynamics is to provide basic equations which enable one to evaluate properties such as $u$, $h$, and $s$, and others, from measurable-property data.

A second point to recognize is that, in many cases, insufficient data are available to calculate properties, even if the basic mathematical relations have been developed. Hence approximation techniques are needed, since in numerous situations either insufficient time or money may require that alternative methods of evaluating properties be available. A case in point is the use of the generalized compressibility factor $Z$, introduced in the preceding chapter, which provides a means of correlating $PvT$ data in the absence of sufficient direct experimental information. This particular method will be extended to other properties, in this chapter. Before investigating some general thermodynamic relations for the qualitative and quantitative insight they provide, a brief review of some rules of partial differential calculus is in order.

## 13-1 FUNDAMENTALS OF PARTIAL DERIVATIVES

Many of the expressions to be developed in this chapter will involve a dependent variable expressed as a function of two independent variables, since we are primarily interested in simple systems. With this in mind, we shall first consider three thermodynamic variables represented by $x$, $y$, and $z$. Their functional relationship may be expressed either in the form $f(x, y, z) = 0$ or in the form $x = x(y, z)$. In the latter case we could also write that $y = y(x, z)$ or that $z = z(x, y)$. The differential of the dependent variable $x$ is given by the equation

$$dx = \left(\frac{\partial x}{\partial y}\right)_z dy + \left(\frac{\partial x}{\partial z}\right)_y dz \qquad (a)$$

Similar expressions for $dy$ and $dz$ may also be written, depending upon which variable is selected as the dependent variable.

 The reason for not considering a partial derivative as a fraction is easily seen by recalling that the equilibrium states of a simple system may be represented by a three-dimensional surface. Such a surface is shown in Fig. 13-1 for a single-phase region, where the thermodynamic properties are symbolized by $x$, $y$, and $z$. As an example, this surface could represent the superheat region of a $PvT$ diagram similar to that shown in Fig. 4-1. Sections of the surface have been cut away around the equilibrium state $D$, so that the curvature of the surface is more clearly seen. Consider a plane of constant $z$ which intersects the surface. The curve of intersection is labeled with the points $C$, $D$, and $E$ on the diagram. The quantity $(\partial x/\partial y)_z$ is the slope of the surface at any state along this curve of intersection. In particular, at state $D$ a tangent has been drawn to the curve which is the line $AB$.

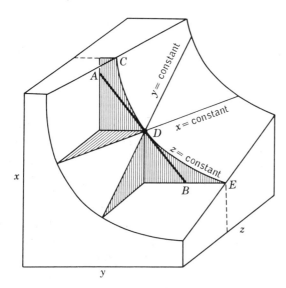

**Figure 13-1** A representation of a partial derivative for a thermodynamic surface.

The value of the slope of this line is also the value of $(\partial x/\partial y)_z$ at state $D$. This partial derivative exists, of course, when any other plane of constant $z$ intersects the equilibrium surface, as long as the surface is continuous in the $z$ direction at the state of interest. It is apparent that similar interpretations of the quantities $(\partial x/\partial z)_y$ and $(\partial y/\partial z)_x$ can be made when planes of constant $y$ and $x$, respectively, intersect the surface of equilibrium states. The parameters $x$, $y$, and $z$ represent any combination of three intrinsic properties of a thermodynamic system.

A mathematical test is available from partial differential calculus to determine whether or not the total derivative of a function is an exact differential. This test is a necessary and sufficient condition for the total derivative to be exact. Since differential equations relating thermodynamic properties are exact, the test for exactness provides a means of developing additional relations among thermostatic properties. This method is especially useful when a property is presented as a function of two independent variables, as in the case of a simple system. Recall Eq. (a): If $x = x(y, z)$, then

$$dx = \left(\frac{\partial x}{\partial y}\right)_z dy + \left(\frac{\partial x}{\partial z}\right)_y dz$$

By denoting the coefficient of $dy$ by $M$ and the coefficient of $dz$ by $N$, the above equation becomes

$$dx = M\,dy + N\,dz \tag{b}$$

Partial differentiation of $M$ and $N$ with respect to $z$ and $y$, respectively, leads to

$$\frac{\partial M}{\partial z} = \frac{\partial^2 x}{\partial y\,\partial z} \quad \text{and} \quad \frac{\partial N}{\partial y} = \frac{\partial^2 x}{\partial z\,\partial y}$$

If these partial derivatives exist, it is known from calculus that the order of differentiation is immaterial, so that

$$\frac{\partial M}{\partial z} = \frac{\partial N}{\partial y} \tag{13-1a}$$

When Eq. (13-1a) is satisfied for any function $x$, then $dx$ is an exact differential. Equation (13-1a) is known as the test for exactness. Its usefulness in terms of thermostatic properties is demonstrated in the next section.

Another expression from partial differential calculus which is useful in relating thermodynamic variables is the *cyclic* relation. This relation, which is not derived in the text, is

$$\left(\frac{\partial w}{\partial x}\right)_y \left(\frac{\partial x}{\partial y}\right)_w \left(\frac{\partial y}{\partial w}\right)_x = -1 \tag{13-1b}$$

Equation (13-1b) is important in interrelating thermodynamic variables. It also illustrates the fallacy of handling partial derivatives as fractions. When the variables held constant on the partial derivatives are different, the quantities within the partials cannot be canceled.

## 13-2 SOME FUNDAMENTAL RELATIONS FOR SIMPLE, COMPRESSIBLE SYSTEMS

In Chap. 7 several important equations were developed for the entropy change of simple, compressible systems of fixed chemical composition by combining the first and second laws of thermodynamics. The first law in this particular case is $\delta q + \delta w = du$. For an internally reversible process, the heat and work interactions may be represented by $T\,ds$ and $-P\,dv$. Substitution of these latter two expressions into the first law yields

$$T\,ds - P\,dv = du \tag{13-2}$$

Since $h = u + Pv$, then $du = dh - P\,dv - v\,dP$. Replacement of $du$ in Eq. (13-2) by this relationship leads to the second $T\,ds$ equation, namely,

$$T\,ds + v\,dP = dh \tag{13-3}$$

Two additional equations of interest may be formed by defining two other properties of matter. The Helmholtz function $a$ is defined by the equation

$$a \equiv u - Ts \tag{13-4}$$

Hence,

$$da = du - T\,ds - s\,dT \tag{13-5}$$

If Eq. (13-2) is substituted into Eq. (13-5), then

$$da = -P\,dv - s\,dT \tag{13-6}$$

The Gibbs function $g$ is defined by the equation

$$g \equiv h - Ts \tag{13-7}$$

Therefore,

$$dg = dh - T\,ds - s\,dT \tag{13-8}$$

Substitution of $dh$ from Eq. (13-3) into Eq. (13-8) yields

$$dg = v\,dP - s\,dT \tag{13-9}$$

In order to summarize these four important relationships among properties of simple systems, they are collected together as a set and presented below.

$$du = T\,ds - P\,dv \tag{13-2}$$

$$dh = T\,ds + v\,dP \tag{13-3}$$

$$da = -P\,dv - s\,dT \tag{13-6}$$

$$dg = v\,dP - s\,dT \tag{13-9}$$

Note that the variables on the right-hand side of these equations include only $P$, $v$, $s$, and $T$. These equations are sometimes referred to as the Gibbsian equations.

They relate the change in various properties of a simple system during a differential change between equilibrium states.

One of the most important sets of thermodynamic relations which arise from the Gibbsian, or $T\,ds$, equations above is obtained by applying the test for exactness to the expressions for $du$, $dh$, $da$, and $dg$. Since these quantities are exact differentials, the test for exactness must lead to an equality. Application of Eq. (13-1a) to the four equations above yields the following set of relations among partial derivatives:

$$\left(\frac{\partial T}{\partial v}\right)_s = -\left(\frac{\partial P}{\partial s}\right)_v \tag{13-10}$$

$$\left(\frac{\partial T}{\partial P}\right)_s = \left(\frac{\partial v}{\partial s}\right)_P \tag{13-11}$$

$$\left(\frac{\partial P}{\partial T}\right)_v = \left(\frac{\partial s}{\partial v}\right)_T \tag{13-12}$$

$$\left(\frac{\partial v}{\partial T}\right)_P = -\left(\frac{\partial s}{\partial P}\right)_T \tag{13-13}$$

This set of equations is referred to as the Maxwell relations. Their importance is not apparent at this point, but a simple example may help illustrate their usefulness. Consider a system at a given equilibrium state. For a differential change of state it is desired to know the rate of change of entropy of the system as the volume is altered isothermally. This could apply, for example, to a gas contained within a piston-cylinder assembly which is expanded isothermally. It is not possible to measure $(\partial s/\partial v)_T$ directly for the process, since entropy variations cannot be evaluated directly. However, Eq. (13-12) states that it is necessary only to measure the rate of change of pressure with temperature at constant volume, since $(\partial P/\partial T)_v = (\partial s/\partial v)_T$. From an experimental viewpoint, it is relatively easy to measure pressure and temperature variations.

## 13-3 GENERALIZED RELATIONS FOR CHANGES IN ENTROPY, INTERNAL ENERGY, AND ENTHALPY FOR SIMPLE, COMPRESSIBLE SYSTEMS

One of the most important functions of thermodynamics is to provide fundamental equations for the evaluation of properties or the change in properties under the most general considerations. These equations, for example, should be independent of the type or the phase of the substance under consideration. Once these " generalized " equations have been developed, it will then be necessary to provide experimental information on specific substances if further numerical evaluation is to be profitable. It would be desirable to have the experimental information in an analytical form, but in many cases it may be necessary to rely on tabular or graphical forms of equations of state. In any event, our present goal is the derivation of these generalized equations. Again we shall restrict ourselves to simple, compressible thermodynamic systems. In this way the number of independent variables will be restricted to two. Of primary interest in this section will be equations for the changes in entropy, internal energy, and enthalpy. Relationships for the specific heats will appear in the following section.

It is most convenient to begin with a derivation of an equation for the change in entropy of a simple, homogeneous system. Once this is accomplished, equations for $du$ and $dh$ may then be obtained directly by employing the $T\,ds$ equations already developed. It is desirable to derive generalized relations in terms of easily measured independent variables. For this reason we shall select the independent properties to be any pair of the group of $(P, v, T)$. Thus three equations for $ds$ can be determined. For practical reasons, we are usually concerned only with the two pairs of variables $(T, v)$ and $(T, P)$. If the entropy is chosen to be a function of $T$ and $v$, we may write that the total derivative of $s$ is given by

$$ds = \left(\frac{\partial s}{\partial T}\right)_v dT + \left(\frac{\partial s}{\partial v}\right)_T dv \tag{13-14}$$

It is now desirable to attempt to express $ds$ solely in terms of measurable quantities. This requires replacing the partial derivatives in Eq. (13-14) with other terms which include only the variables $P$, $v$, $T$ and the specific heats $c_v$ and $c_p$.

The first partial derivative on the right is related to $c_v$ and $T$. If $u = u(T, v)$, then

$$du = \left(\frac{\partial u}{\partial T}\right)_v dT + \left(\frac{\partial u}{\partial v}\right)_T dv$$

$$= c_v\, dT + \left(\frac{\partial u}{\partial v}\right)_T dv \tag{13-15}$$

When $du$ in the above expression is replaced by Eq. (13-2), then, upon rearrangement

$$ds = \frac{c_v\, dT}{T} + \frac{1}{T}\left[\left(\frac{\partial u}{\partial v}\right)_T + P\right] dv \tag{13-16}$$

Equations (13-14) and (13-16) are both valid expressions for $ds$. Hence, the coefficients in front of the $dT$ terms on the right must be equal, and

$$\left(\frac{\partial s}{\partial T}\right)_v = \frac{c_v}{T} \tag{13-17}$$

The second partial derivative is contained in a Maxwell relation:

$$\left(\frac{\partial s}{\partial v}\right)_T = \left(\frac{\partial P}{\partial T}\right)_v \tag{13-12}$$

By making the proper substitutions into Eq. (13-14), we find that

$$ds = \frac{c_v\, dT}{T} + \left(\frac{\partial P}{\partial T}\right)_v dv \tag{13-18}$$

This is the result we sought, since the right-hand side of the equation is now expressed solely in terms of measurable quantities. An equivalent equation for $ds$ in terms of the variables $T$ and $P$ may be obtained by starting with the relation

$$ds = \left(\frac{\partial s}{\partial T}\right)_P dT + \left(\frac{\partial s}{\partial P}\right)_T dP \qquad (13\text{-}19)$$

Again, the second partial derivative may be replaced by a Maxwell relation. In addition, it can be shown in a manner similar to the development of Eq. (13-17) that

$$\left(\frac{\partial s}{\partial T}\right)_P = \frac{c_p}{T} \qquad (13\text{-}20)$$

As a result, the substitution of Eqs. (13-13) and (13-20) into Eq. (13-19) yields the desired result,

$$ds = \frac{c_p\, dT}{T} - \left(\frac{\partial v}{\partial T}\right)_P dP \qquad (13\text{-}21)$$

Integration of Eqs. (13-18) and (13-21) between the same two equilibrium states should lead to the same value for $\Delta s$, since the change in the value of a property is independent of the method employed to calculate it. These equations are generalized relations, since they are not restricted to any particular substance or to any particular phase of a substance. They are restricted, however, to simple systems.

A generalized equation for the sensible internal-energy change is now possible by recalling the first $T\, ds$ equation developed previously.

$$du = T\, ds - P\, dv \qquad (13\text{-}2)$$

The quantity $ds$ is eliminated from this equation by substituting Eq. (13-18) for it. After separation of variables it turns out that

$$du = c_v\, dT + \left[T\left(\frac{\partial P}{\partial T}\right)_v - P\right] dv \qquad (13\text{-}22)$$

The generalized equation for the change in enthalpy is found by employing the second $T\, ds$ equation, namely,

$$dh = T\, ds + v\, dP \qquad (13\text{-}3)$$

Substitution of Eq. (13-21) for $ds$ and subsequent rearrangement leads to

$$dh = c_p\, dT + \left[v - T\left(\frac{\partial v}{\partial T}\right)_P\right] dP \qquad (13\text{-}23)$$

Integration of Eqs. (13-18), (13-21), (13-22), and (13-23) for actual systems requires experimental knowledge of the $PvT$ behavior of the substance in the region of interest, plus experimental information on the relationship between the specific heats and temperature.

The change in enthalpy, for example, of a simple, compressible substance is found by integration of Eq. (13-23). As a result

$$h_2 - h_1 = \int_1^2 c_p \, dT + \int_1^2 \left[ v - T\left(\frac{\partial v}{\partial T}\right)_P \right] dP \tag{13-24}$$

In order to integrate the first term, information is required on the variation of $c_p$ with temperature at a fixed pressure. (Frequently this pressure is chosen to be essentially zero, so that ideal-gas specific-heat data, $c_{p,0}$, are used.) The integration of the second term requires knowledge of the $PvT$ behavior of the substance for the range of pressure desired at a given temperature. Due to the format of the coefficient of $dP$ in Eq. (13-24), it is helpful if the $PvT$ equation of state is explicit in $v$. On the other hand, an equation explicit in $P$ would be advantageous when Eq. (13-22) for $du$ is integrated. Figure 13-2 shows one possible path of integration between two real-gas states, 1 and 2, on $PT$ and $Ts$ diagrams. For this particular path the first term of Eq. (13-24) is integrated at zero pressure between states $x$ and $y$. The second term in Eq. (13-24) must be integrated twice in this case. One integration is at constant temperature $T_1$ between states 1 and $x$, and the other integration is at $T_2$ between states $y$ and 2.

**Examples 13-1M and 13-1** In Chap. 3 it was pointed out that the internal energy of gases at low pressures could be approximated quite successfully by a relation of the type $du = c_v \, dT$. This assumption that the internal energy was solely a function of temperature was based on Joule's law, which stated that $(\partial u/\partial v)_T$ at low pressures was experimentally determined to be zero. Demonstrate that Joule's law holds exactly for ideal gases.

SOLUTION In order to evaluate Joule's law in terms of an ideal gas, $(\partial u/\partial v)_T$ must be expressed in terms of $P$, $v$, and $T$. This is easily done by writing the total derivative of $u$ as a function of $T$ and $v$ and then comparing this equation with Eq. (13-22). These two equations are

$$du = \left(\frac{\partial u}{\partial T}\right)_v dT + \left(\frac{\partial u}{\partial v}\right)_T dv$$

and

$$du = c_v \, dT + \left[ T\left(\frac{\partial P}{\partial T}\right)_v - P \right] dv \tag{13-22}$$

Since the latter expression is of general validity, it is apparent on comparison of these two equations that

$$\left(\frac{\partial u}{\partial v}\right)_T = T\left(\frac{\partial P}{\partial T}\right)_v - P$$

This relation is valid for any simple, compressible system. We shall now apply the ideal-gas relationship to it, namely, $Pv = RT$. It is easily shown that, for an ideal gas, $(\partial P/\partial T)_v = R/v$. Substituting this into the equation for $(\partial u/\partial v)_T$, we find that, for an ideal gas,

$$\left(\frac{\partial u}{\partial v}\right)_T = T\left(\frac{R}{v}\right) - P = P - P = 0$$

Thus an ideal gas is always a substance such that the internal energy is truly only a function of temperature, as given by $du = c_v \, dT$. Since many gases at low pressures approximately follow the ideal-gas law, their internal energies also are only a function of temperature, to a high degree of accuracy. If a similar treatment is carried out for the enthalpy function, it may be shown that

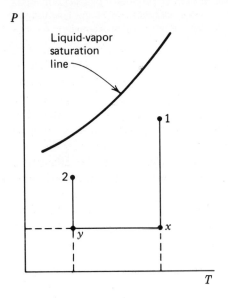

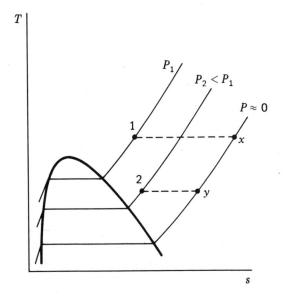

**Figure 13-2** Possible path of integration between two real-gas states.

$(\partial h/\partial P)_T$ is always zero for ideal gases. This proves that, for hypothetical ideal gases, the enthalpy change is always given by $dh = c_p\,dT$.

**Examples 13-2M and 13-2** Find the change in enthalpy and entropy for a real gas along an isothermal path between pressures $P_1$ and $P_2$. Assume for the range of pressures involved that the $PvT$ behavior of the gas is represented with reasonable accuracy by the relation $v = RT/P + b - a/RT$, where $a$ and $b$ are constants.

SOLUTION The change in enthalpy is given by the second term of Eq. (13-24), namely,

$$(h_2 - h_1)_T = \int_1^2 \left[ v - T\left(\frac{\partial v}{\partial T}\right)_P \right] dP$$

The equation of state is explicit in $v$. Hence $(\partial v/\partial T)_P = R/P + a/RT^2$, and

$$v - T\left(\frac{\partial v}{\partial T}\right)_P = \frac{RT}{P} + b - \frac{a}{RT} - T\left(\frac{R}{P} + \frac{a}{RT^2}\right) = b - \frac{2a}{RT}$$

Therefore,

$$(h_2 - h_1)_T = \left(b - \frac{2a}{RT}\right)(P_2 - P_1)$$

To evaluate the entropy change we choose Eq. (13-21), rather than Eq. (13-18), because it requires an equation of state explicit in $v$. Hence

$$(s_2 - s_1)_T = -\int_1^2 \left(\frac{\partial v}{\partial T}\right)_P dP = -\int_1^2 \left(\frac{R}{P} + \frac{a}{RT^2}\right) dP$$

$$= -R \ln \frac{P_2}{P_1} - \frac{a(P_2 - P_1)}{RT^2}$$

# 13-4 GENERALIZED RELATIONS FOR $c_p$ AND $c_v$

Two expressions for the specific heats have already been developed in this chapter:

$$c_v = T\left(\frac{\partial s}{\partial T}\right)_v \tag{13-17}$$

$$c_p = T\left(\frac{\partial s}{\partial T}\right)_P \tag{13-20}$$

The latter two expressions are generalized relations for $c_v$ and $c_p$, and they may be used in any single-phase region where $PvT$ data are available. An alternative method of evaluating specific-heat data is based on the experimental fact that it is relatively easier to measure specific-heat data at low pressures than at elevated pressures. For example, we have already seen from Chap. 3 that much is known about the specific heats of common gases as a function of temperature at low pressures. Such data were called "zero-pressure" specific heats. Consequently, we are primarily concerned with determining in what manner specific-heat values vary with increasing pressure (or decreasing specific volume) at constant temperature. Such an evaluation must again be based solely on the use of measured $PvT$

data in the desired range of equilibrium states. In a mathematical sense what we seek are expressions for the two terms $(\partial c_v/\partial v)_T$ and $(\partial c_p/\partial P)_T$. Generalized equations for these two expressions are obtained by starting with Eqs. (13-18) and (13-21); that is,

$$ds = \frac{c_v\,dT}{T} + \left(\frac{\partial P}{\partial T}\right)_v dv \tag{13-18}$$

and

$$ds = \frac{c_p\,dT}{T} - \left(\frac{\partial v}{\partial T}\right)_P dP \tag{13-21}$$

Since these equations are expressions for exact differentials, the test for exactness may be employed. When applied to Eq. (13-18), the following relation results:

$$\left(\frac{\partial c_v}{\partial v}\right)_T = T\left(\frac{\partial^2 P}{\partial T^2}\right)_v \tag{13-25}$$

This is the desired relation. If we start with Eq. (13-21), it can be shown by an analogous procedure that

$$\left(\frac{\partial c_p}{\partial P}\right)_T = -T\left(\frac{\partial^2 v}{\partial T^2}\right)_P \tag{13-26}$$

In order to obtain the value of $c_p$, for example, at an elevated pressure, Eq. (13-26) is integrated from zero pressure to the desired value. Hence

$$c_p - c_{p,0} = -T\int_0^P \left(\frac{\partial^2 v}{\partial T^2}\right)_P dP \tag{13-27}$$

where $c_{p,0}$ again is the zero-pressure, or ideal-gas, specific heat. The integration of the right-hand side requires a knowledge of the $PvT$ behavior of the substance in either tabular or analytical form.

Another thermodynamic relation of interest is the difference between the constant-pressure and constant-volume specific heats, that is, $c_p - c_v$. One reason for this interest is that $c_p$ values usually are easier to measure than $c_v$ values. Hence $c_v$ values could in theory be evaluated solely from $c_p$ and $PvT$ data. Since the change in any property value is not dependent upon the method of evaluation, we may equate the two equations for $ds$ previously presented. Equating Eqs. (13-18) and (13-21), we find

$$\frac{c_v\,dT}{T} + \left(\frac{\partial P}{\partial T}\right)_v dv = \frac{c_p\,dT}{T} - \left(\frac{\partial v}{\partial T}\right)_P$$

or

$$\frac{c_p - c_v}{T}\,dT = \left(\frac{\partial P}{\partial T}\right)_v dv + \left(\frac{\partial v}{\partial T}\right)_P dP$$

Differentiation with respect to pressure at constant volume yields

$$\frac{c_p - c_v}{T}\left(\frac{\partial T}{\partial P}\right)_v = \left(\frac{\partial v}{\partial T}\right)_P$$

or upon rearrangement,

$$c_p - c_v = T\left(\frac{\partial v}{\partial T}\right)_P \left(\frac{\partial P}{\partial T}\right)_v \tag{13-28}$$

An equivalent form for $c_p - c_v$ can be obtained by replacing $(\partial P/\partial T)_v$ in terms of the cyclic rule $(\partial P/\partial T)_v = -(\partial v/\partial T)_P(\partial P/\partial v)_T$. Use of this expression in Eq. (13-28) leads to the relation

$$c_p - c_v = -T\left(\frac{\partial v}{\partial T}\right)_P^2 \left(\frac{\partial P}{\partial v}\right)_T \tag{13-29}$$

A number of important qualitative results stem from Eq. (13-29).

First of all, on the basis of experimental data it is known that $(\partial P/\partial v)_T$ is always negative for all substances in all phases. Since the first partial in Eq. (13-29) is a squared term, then $c_p - c_v$ must always be positive, or zero. This quantity becomes zero on two occasions. The first of these is, apparently, when $T$ is absolute zero on the thermodynamic scale, if the remaining terms remain finite at this state. Consequently, the specific heats at constant pressure and constant volume are identical at the absolute zero of temperature. The specific heats will also be equal if the value of $(\partial v/\partial T)_P$ is ever zero. This occurs, for example, in the case of liquid water at, roughly, 4°C, where the fluid is at its state of maximum density. It should also be noted that even at temperatures above zero, the difference in specific heats will generally be small for liquids and solids. This is true since the value of $(\partial v/\partial T)_P$ is very small for most equilibrium states, which can be verified by referring to a $PvT$ surface for a substance, such as was shown in Fig. 4-1. Hence one frequently speaks of the specific heat of a liquid or a solid without specifying the type, since the $c_v$ and $c_p$ values are not significantly different in many cases. The data quoted are usually $c_p$ values. The fact that $c_p \geq c_v$ also leads to the generalization that constant-volume lines always have steeper slopes than constant-pressure lines at the same point on a $Ts$ diagram.

**Examples 13-3M and 13-3** Determine the isothermal change in $c_p$ with pressure for the same gas studied in Example 13-2.

SOLUTION The isothermal change in $c_p$ with pressure is given by Eq. (13-26), namely, $(\partial c_p/\partial P)_T = -T(\partial^2 v/\partial T^2)_P$. In Example 13-2 the equation of state was $v = RT/P + b - a/RT$, and the first derivative was found to be

$$\left(\frac{\partial v}{\partial T}\right)_P = \frac{R}{P} + \frac{a}{RT^2}$$

The second partial derivative, then, is

$$\left(\frac{\partial^2 v}{\partial T^2}\right)_P = -\frac{2a}{RT^3}$$

Consequently,

$$(c_{p2} - c_{p1})_T = -\int_1^2 T\left(-\frac{2a}{RT^3}\right) dP = \frac{2a(P_2 - P_1)}{RT^2}$$

The property variation of solids and liquids is often expressed in terms of the volumetric expansion coefficient $\beta$ and the isothermal coefficient of compressibility $K_T$. These two quantities are defined as

$$\beta = \frac{1}{v}\left(\frac{\partial v}{\partial T}\right)_P \tag{13-30}$$

and

$$K_T = -\frac{1}{v}\left(\frac{\partial v}{\partial P}\right)_T \tag{13-31}$$

Substitution of these two equations into Eq. (13-29) leads to

$$c_p - c_v = \frac{vT\beta^2}{K_T} \tag{13-32}$$

The use of $\beta$ and $K_T$ is quite helpful in many calculations since their values may often be assumed to be constant during a given process. The slow variation of these two properties with temperature is illustrated by the data of Table 13-1, which shows data for solid copper.

**Table 13-1** $\beta$ and $K_T$ for copper as a function of temperature

| $T$, °K | $\beta \times 10^6$, (°K)$^{-1}$ | $K_T \times 10^7$, cm$^2$/newton |
|---------|----------|----------|
| 100 | 31.5 | 0.721 |
| 150 | 41.0 | 0.733 |
| 200 | 45.6 | 0.748 |
| 250 | 48.0 | 0.762 |
| 300 | 49.2 | 0.776 |
| 500 | 54.2 | 0.837 |
| 800 | 60.7 | 0.922 |

**Examples 13-4M and 13-4** At 500°K the values of $v$, $\beta$, and $K_T$ for solid copper are 7.115 cm$^3$/g·mol, $54.2 \times 10^{-6}$ °K$^{-1}$, and $0.837 \times 10^{-7}$ cm$^2$/N, respectively.

(a) Determine the value of $(c_p - c_v)$ in J/(g·mol)(°C).

(b) If the value of $c_p$ is 26.15 J/(g·mol)(°C) at this temperature, what percent error would be made in $c_v$ if we assume that $c_p = c_v$?

SOLUTION (a) The difference between $c_p$ and $c_v$ is obtained directly by substituting the appropriate values into Eq. (13-32). Thus,

$$c_p - c_v = \frac{vT\beta^2}{K_T} = \frac{7.115(500)(54.2 \times 10^{-6})^2}{0.837 \times 10^{-7}} \frac{N(\text{cm})}{\text{g·mol}(°K)} \times \frac{m}{10^2 \text{ cm}}$$

$$= 1.249 \text{ J/(g·mol)(°K)}$$

(b) If $c_p = 26.15$ J/(g·mol)(°K), then the actual value of $c_v$ is 24.90 J/(g·mol)(°K). Therefore the percent error in assuming that $c_v$ equals $c_p$ is 1.249/24.90 = 0.050, or 5 percent. As a result, one must be careful not to assume that $c_p$ and $c_v$ are equal for solid materials if the temperature is sufficiently high.

## 13-5 VAPOR PRESSURE AND THE CLAPEYRON EQUATION

The vapor pressures of all liquids vary with saturation temperature in essentially the same manner. The dependency of the saturation pressure on the temperature will now be developed from theoretical considerations. The generalized relationship that results is also valid for solid-gas and solid-liquid phase changes. We shall begin by calculating the entropy change of a simple substance during a phase change. This entropy change in terms of the variables $v$ and $T$ has already been presented in the form of Eq. (13-14), namely,

$$ds = \left(\frac{\partial s}{\partial v}\right)_T dv + \left(\frac{\partial s}{\partial T}\right)_v dT \tag{13-14}$$

However, for any process involving a change in phase, we realize that the temperature is constant during the phase change. Therefore the above equation reduces to

$$ds = \left(\frac{\partial s}{\partial v}\right)_T dv$$

The quantity $(\partial s/\partial v)_T$ can be replaced by the Maxwell relation given by Eq. (13-12), which shows that $(\partial s/\partial v)_T = (\partial P/\partial T)_v$. Hence

$$ds = \left(\frac{\partial P}{\partial T}\right)_v dv$$

The term $(\partial P/\partial T)_v$ is the slope of the saturation curve at a given saturation state, and this quantity is independent of the volume during a change of phase. Consequently, the partial derivative may be written as a total derivative, $dP/dT$, and it may be moved outside of the integral sign during the integration of the above equation. Integration leads to

$$s_2 - s_1 = \frac{dP}{dT}(v_2 - v_1)$$

or

$$\frac{dP}{dT} = \frac{s_2 - s_1}{v_2 - v_1} \tag{13-33}$$

where the subscripts 1 and 2 represent the saturation phases for the process. For example, they may represent the saturated-vapor and saturated-liquid phases during a vaporization process.

The entropy change during a phase change may be evaluated from the first and second laws. From the second law $ds = \delta q/T$, and for a constant-pressure process (such as a phase change) the first law for a closed system is $\delta q = dh$. Thus $ds = dh/T$ and $s_2 - s_1 = (h_2 - h_1)/T$. Equation (13-33) then becomes

$$\frac{dP}{dT} = \frac{h_2 - h_1}{T(v_2 - v_1)} = \frac{\Delta h}{T \, \Delta v} \tag{13-34}$$

Equation (13-34) is called the Clapeyron equation. It is generally valid for any phase change which occurs at constant pressure and temperature. For a liquid-vapor phase change this equation might be written as

$$\frac{dP}{dT} = \frac{h_{fg}}{Tv_{fg}}$$

where the subscripts follow nomenclature introduced in Chap. 4. In general, $\Delta h$ and $\Delta v$ are the enthalpy and volume changes between any two saturation states at the same pressure and temperature. Note that the Clapeyron equation permits the evaluation of enthalpy changes for phase changes from a knowledge of only $PvT$ data.

For liquid-vapor and solid-vapor phase changes, Eq. (13-34) can be further modified by introducing several approximations. For purposes of discussion we shall consider only the first of these, but the results are equally applicable to solid-vapor phase changes. For liquid-vapor phase changes at relatively low pressures, the value of $v_g$ is many times the size of $v_f$. Thus a good approximation is to replace $v_{fg}$ by $v_g$ in the above equation. Also, at these low pressures, the $PvT$ relation for the vapor closely follows that for an ideal gas; that is, $v_g = RT/P$. By making these two successive approximations in Eq. (13-34), we find that

$$\frac{dP}{dT} = \frac{Ph_{fg}}{RT^2}$$

or

$$\frac{dP}{P} = \frac{h_{fg}}{RT^2} \, dT \tag{13-35}$$

Equation (13-35) is frequently called the Clapeyron-Clausius equation. Integration of this equation depends upon the variation of $h_{fg}$ with temperature. If a small variation of pressure (or temperature) is chosen so that the change in $h_{fg}$ over the interval of integration is small, then integration yields

$$\ln P = -\frac{h_{fg}}{R}\left(\frac{1}{T}\right) + C \tag{13-36}$$

where $C$ is a constant of integration. This equation indicates that the vapor pressure of a liquid is very closely an exponential function of the saturation temperature. The general form of the equation is also valid for saturation data below the triple state in the sublimation region.

It should be kept in mind that Eq. (13-36) is only an approximation. A more accurate analytical expression for the variation of the saturation pressure with respect to temperature requires that additional terms be added. For example, a better approximation might be given by an equation of the form

$$\ln P_{\text{sat}} = A + \frac{B}{T} + C \ln T + DT + ET^2 + \cdots \tag{13-37}$$

The constants $A, B, C$, etc., are adjusted to obtain the best fit with the experimental data. Nevertheless, the simple exponential form given by Eq. (13-36) is fairly accurate in many cases.

An equation like Eq. (13-37) is important for the following reason. If the equation represents a precise fit to experimental data, then the derivative of the equation will give an accurate value of $dP/dT$. Substitution of this value of $dP/dT$ into the Clapeyron equation, Eq. (13-34), will lead to an accurate evaluation of $\Delta h$ for a liquid-gas or a solid-gas phase transformation.

The relationship given by Eq. (13-34) is quite useful to demonstrate one other point: The slope of a saturation line on a $PT$ diagram apparently depends upon the signs of $\Delta h$ and $\Delta v$. In most cases, when heat is added to a closed system to bring about a phase change, the volume also increases. Hence $dP/dT$ is usually positive. However, in the case of the melting of water and a few other substances, the volume decreases. The slope of the melting curve on a phase diagram for these few substances must then be negative. This was pointed out in Chap. 4 during the discussion of phase diagrams, but now the Clapeyron equation theoretically substantiates what is empirically observed. The freezing temperature of any substance which expands on freezing is lowered when the pressure is increased.

**Example 13-5M** Estimate the enthalpy of vaporization of water at 200°C using $PvT$ data from Table A-12M.

SOLUTION On the basis of Eq. (13-34) it is seen that

$$\Delta h = T \, \Delta v \, \frac{dP}{dT}$$

The value of $dP/dT$ may be approximated to a high degree by $\Delta P/\Delta T$ in the region of interest. The saturation data at a 10° temperature interval on either side of 200°C are as follows: At 190°C the saturation pressure is 12.54 bars, and at 210°C the pressure is 19.06 bars. Hence $\Delta P/\Delta T$ equals $(19.06 - 12.54)/(210 - 190)$, or 0.326 bars/°C. The change in volume $v_{fg}$ at 200°C is given as 127.3 cm³/g. Substitution of these values into the above equation yields

$$\Delta h = \frac{473(127.3)(0.326)}{10} = 1963 \text{ J/g}$$

where the factor of 10 converts units of cm³·bar to N·m. Table A-12M lists the value of $h_{fg}$ as 1941 J/g. An error of roughly 1 percent results from this approximation technique.

**Example 13-5** Estimate the enthalpy of vaporization of water at 400°F using $PvT$ data from Table A-12.

SOLUTION On the basis of Eq. (13-34) it is seen that

$$\Delta h = T \, \Delta v \, \frac{dP}{dT}$$

The value of $dP/dT$ may be approximated to a high degree by $\Delta P/\Delta T$ in the region of interest. The saturation data at a 10° temperature interval on either side of 400°F are as follows: At 390°F the saturation pressure is 220.20 psia, and at 410°F the pressure is 276.50 psia. Hence $\Delta P/\Delta T$

equals $(276.50 - 220.20/(410 - 390)$, or 2.819 psi/°F. The change in volume $v_{fg}$ at 400°F is given as 1.8474 ft$^3$/lb. Substitution of these values into the above equation yields

$$\Delta h = \frac{860(1.8474)(2.819)(144)}{778} = 827 \text{ Btu/lb}$$

Table A-12 lists the value as 826.8 Btu/lb, so good agreement is obtained by the calculation above.

**Example 13-6M** The saturation pressure and the enthalpy of vaporization of refrigerant 12 at 20°C are found to be 5.673 bars and 140.9 J/g, respectively. Without any additional experimental data, estimate the saturation pressure at 0°C.

SOLUTION Definite integration of Eq. (13-35) leads to

$$\ln \frac{P_2}{P_1} = \frac{h_{fg}(T_2 - T_1)}{RT_1 T_2}$$

Although it would be better to use the average value of $h_{fg}$ between the temperatures of interest, we must use the value at $T_1$, for lack of information. Substituting in the proper values, we find that

$$\ln \frac{P_2}{P_1} = \frac{140.9(121)(0 - 20)}{8.315(293)(273)} = -0.513$$

Therefore $P_2/P_1 = 0.599$, and $P_2 = 0.599(5.673) = 3.398$ bars. The tabulated value at 0°C is 3.086 bars. Although the estimate is in error by 10 percent, it is a fair approximation in the absence of further experimental data.

**Example 13-6** The saturation pressure and the enthalpy of vaporization for referigerant 12 at 70°F are measured and found to be 84.89 psia and 60.31 Btu/lb, respectively. Without any additional experimental data, estimate the saturation pressure at 10°F.

SOLUTION Definite integration of Eq. (13-35) leads to

$$\ln \frac{P_2}{P_1} = \frac{h_{fg}(T_2 - T_1)}{RT_1 T_2}$$

Although it would be better to use the average value of $h_{fg}$ between the temperatures of interest, we must use the value at $T_1$, for lack of more information. Substituting in the proper values, we find that

$$\ln \frac{P_2}{P_1} = \frac{60.31(121)(470 - 530)}{1.986(470)(530)} = -0.885$$

Therefore $P_1/P_1 = 0.413$, and $P_2 = 0.413(84.89) = 35.0$ psia. The tabulated value is, roughly, 29.3 psia at 10°F. Although the estimate is in error by some 20 percent, it is seen that a fair approximation is possible without further experimental data.

## 13-6 THE JOULE-THOMSON COEFFICIENT

Consider the flow of a fluid down a duct which contains a restriction to the flow. This restriction might be some type of porous plug, such as steel wool or cotton. The effect of this porous plug is a significant pressure drop across the restriction. A schematic of the equipment is shown in Fig. 13-3a. The work effects within the control surfaces 1 and 2 are zero, and the flow passage is heavily insulated;

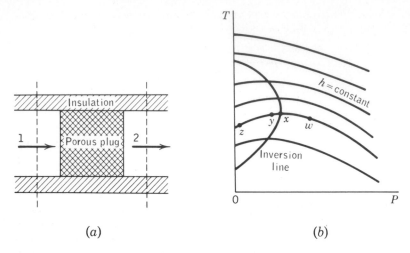

**Figure 13-3** Equipment for Joule-Thomson experiment and a plot of data resulting from typical measurements.

consequently, the heat effect is negligible. The flow of the fluid is adjusted to steady-state conditions. In addition, the changes in the kinetic and potential energies of the flow stream across the porous plug may also be made negligible. The conservation of energy principle for a control volume around the restriction is

$$q + w = \Delta h + \Delta KE + \Delta PE$$

In light of the above statements, this equation reduces merely to

$$h_1 = h_2$$

We have noted previously that a process for which the inlet and outlet enthalpies are the same is called throttling. The effect of throttling has a number of important scientific and engineering applications. Based on some original experiments by Joule and Thomson, the outcome of allowing a fluid to flow through a restriction from a higher to a lower pressure such as throttling is frequently called the Joule-Thomson effect. The importance of the Joule-Thomson effect is severalfold. First of all, other thermostatic properties may be related or evaluated from measurements of the Joule-Thomson effect. These include, for example, specific volumes, specific heats, and enthalpies. Second, we shall find that under certain conditions the result of throttling is a reduction in the temperature of the fluid. Thus low temperatures may be achieved with a device which has no moving parts. In fact, under proper conditions, it is possible that one or more of the components of a gas flow stream might pass into the liquid phase during the throttling process. Such liquefaction might provide a means of separating components of a gas mixture.

It is an experimental fact that throttling of a fluid leads to a final temperature which may be higher or lower than the initial value, depending upon the values of

$P_1$, $T_1$, and $P_2$. A mathematical measure of this effect is given by Joule-Thomson coefficient $\mu_{JT}$, which is defined as

$$\mu_{JT} = \left(\frac{\partial T}{\partial P}\right)_h \tag{13-38}$$

The Joule-Thomson coefficient may be readily determined at various states by plotting experimental data in terms of a family of constant-enthalpy lines on a $TP$ diagram. To obtain this plot the values of $P_1$ and $T_1$ upstream from the restriction are held fixed, and the pressure $P_2$ downstream is varied experimentally. For each setting of $P_2$ the downstream temperature $T_2$ is measured. Under throttling conditions the state for each measurement made downstream has the same enthalpy as the initial state upstream. After making a sufficient number of measurements downstream for a given state upstream, a line of constant enthalpy can be drawn on the $TP$ diagram. Then either the initial pressure or temperature is altered, and the measurement procedure is repeated for this new value of the enthalpy. In this manner a whole family of constant-enthalpy lines on a $TP$ plot may be obtained. A typical result is shown in Fig. 13-3b. The slope of a constant-enthalpy line at any state is a measure of the Joule-Thomson coefficient at that state, i.e., a measure of $(\partial T/\partial P)_h$.

Figure 13-3b shows that a number of the constant-enthalpy lines have a state of maximum temperature. The line shown in the figure which passes through these states of maximum temperature is called the inversion line, and the value of the temperature for that state is the inversion temperature. A pressure line will cut the inversion curve at two different states; hence one speaks of the upper and lower inversion temperatures for a given pressure. This line has an important physical significance. To the right of the inversion line on a $TP$ plot the Joule-Thomson coefficient is negative. That is, in this particular region the temperature will increase as the pressure decreases through the throttling device. A heating effect occurs. On the other hand, to the left of the inversion curve the Joule-Thomson coefficient is positive, which means that cooling will occur for expansions in this region. Hence, on the throttling of a fluid, the final temperature after a porous plug may be greater than, equal to, or less than the initial temperature, depending upon the final pressure for any given set of initial conditions. For example, in Fig. 13-3b a typical initial state might be point $w$. Expansion to the inversion curve (point $x$) results in heating of the fluid. If further expansion to point $y$ is permitted, some cooling will occur, but this is not sufficient to lower the temperature back to that of the initial state. However, if expansion to point $z$ is possible, enough cooling will occur to bring the final temperature to a lower value than that for the initial state.

It should also be noted that, for some initial states, a cooling process is impossible. The upper part of the inversion curve passes through zero pressure at some finite temperature for all substances. Consequently, many enthalpy lines at high temperatures never pass through the inversion line, as seen in Fig. 13-3b. For these enthalpy lines the Joule-Thomson coefficient is always negative throughout the range of pressures. Examples of this are hydrogen and helium, which have negative coefficients at ordinary temperatures and low pressures. Hence, for these two gases, the temperature must be artificially lowered considerably before throttling can be employed for an additional cooling effect. For most substances, however, at ordinary temperatures the Joule-Thomson coefficient is negative at high pressures, and it becomes positive at low pressures. For a given pressure drop it is seen that the maximum cooling effect is attained only if the initial state lies on the inversion line. If the initial state lies to the right of the inversion curve, part of the expansion results in heating, which counters the desired effect.

Another useful diagram is the Joule-Thomson coefficient plotted against temperature for various pressures. This is obtained by measuring the slopes of the constant-enthalpy lines at various states from a diagram such as Fig. 13-3b. Data for the Joule-Thomson coefficient of nitrogen determined from experimental work are shown in Fig. 13-4. Note that the coefficient is zero at approximately $-135$ and $+220°C$ for a pressure of 200 atm. These temperatures correspond to the lower and upper inversion temperatures at this pressure. The following example makes use of this diagram to estimate the cooling effect that can be achieved by the throttling of nitrogen from a high pressure to atmospheric conditions.

**Examples 13-5M and 13-5** Estimate the final temperature which could be obtained by throttling nitrogen gas from $-50°C$ and 100 atm to 1 atm pressure.

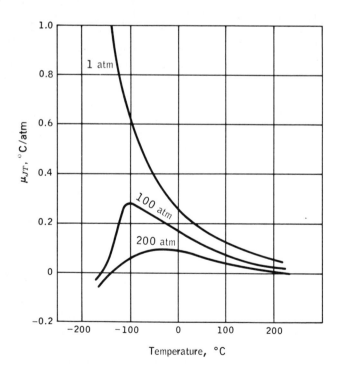

**Figure 13-4** The Joule-Thomson coefficient for nitrogen. [Based on the data of Roebuck and Osterling, *Phys. Rev.*, **48**(5), 450–457. **48**: 450 (1935).]

SOLUTION In order to estimate the final temperature, one needs to know the average Joule-Thomson coefficient in the given pressure range. Two properties are required to evaluate a third property at a given equilibrium state, but the final temperature is unknown. To evaluate the coefficient at the final state we shall first assume that there is no temperature change on throttling. From Fig. 13-4 it is estimated that the average coefficient between 100 atm and 1 atm at $-50°C$ is $0.31°$ C/atm. On this basis a drop of 99 atm in pressure would give a temperature drop of $99(0.31)$, or $31°C$. Thus a better approximation to the final temperature at 1 atm is $-81°C$. Now a new average coefficient may be estimated. At 100 atm, $-50°C$, and 1 atm, $-81°C$, the coefficients are, roughly, 0.23 and $0.51°C/atm$, respectively. The average then is $0.37°C/atm$, and the corresponding temperature drop is $37°C$. Another trial, based on the new final-temperature estimate of $-87°C$, leads to a final temperature of $-89°C$. Further refinement is unnecessary on the basis of the available data. Hence it is estimated that the throttling of nitrogen gas from 100 atm and $-50°C$ to 1 atm will result, approximately, in a temperature drop of $39°C$, or $70°F$.

It is useful to examine the Joule-Thomson coefficient in terms of a generalized equation, that is, its relation to the variables $P$, $v$, and $T$ and the specific heats. This is easily obtained by recalling the generalized equation for the enthalpy, namely,

$$dh = c_p \, dT + \left[ v - T\left(\frac{\partial v}{\partial T}\right)_P \right] dP \qquad (13\text{-}23)$$

By differentiating this equation with respect to the pressure at constant enthalpy, one obtains the following result:

$$\left(\frac{\partial T}{\partial P}\right)_h = \frac{1}{c_p} \left[ T\left(\frac{\partial v}{\partial T}\right)_P - v \right] = \mu_{JT} \qquad (13\text{-}39)$$

Thus the Joule-Thomson coefficient may be calculated from a knowledge of the $PvT$ relationship of the fluid and the specific heat at constant pressure for that state. In practice, one could use the Joule-Thomson coefficient, which is easily measured, to evaluate specific-heat data at elevated pressures. If this generalized relationship is applied to an ideal gas, an interesting result occurs. We find that, since $(\partial v/\partial T)_P$ for an ideal gas is simply $R/P$,

$$\mu_{JT,\ \text{ideal gas}} = \frac{1}{c_p} \left( \frac{RT}{P} - v \right) = 0$$

Hence an ideal gas undergoes no change in temperature upon throttling. This is not surprising since it has already been pointed out in Chap. 4 that the enthalpy of an ideal gas is solely a function of its temperature. If the initial and final enthalpies of a fluid are equal for a throttling process by definition, an ideal gas under this condition would also have the same initial and final temperatures. No actual gas is an ideal gas; however, many gases at low pressures approximate this condition. Consequently, the temperature change upon throttling of real gases at low pressures is often quite small.

## 13-7 GENERALIZED THERMODYNAMIC CHARTS

The principle of corresponding states discussed in Sec. 12-4 is extremely useful in predicting property values other than $P$, $v$, and $T$. These three values were previously correlated through the compressibility factor $Z$ and the reduced properties $P_R$, $v'_R$, and $T_R$. The compressibility factor and reduced coordinates may be used to evaluate such properties as the enthalpy, the entropy, and the specific heat at constant pressure for gases at elevated pressures. The usefulness of such a method will be that only the critical pressure and temperature are required for any given substance. The correlations for these properties again will be presented in graphical form. The method of evaluation involves the generalized equations previously developed in this chapter.

Recall from Sec. 13-3 that the enthalpy of simple homogeneous substances may be evaluated from the generalized equation

$$dh = c_p \, dT + \left[ v - T\left(\frac{\partial v}{\partial T}\right)_P \right] dP \qquad (13\text{-}23)$$

The change in the enthalpy of a gas with temperature is fairly easily computed, since it requires a knowledge only of the variation of $c_p$ with temperature at the desired pressure. Hence the first term on the right of the above equation is not too difficult to evaluate in many cases. However, the variation of $h$ with pressure is not so straightforward because it requires a knowledge of the $PvT$ behavior of each substance of interest. Since detailed data for many compounds will be lacking, a more general method must be employed.

At constant temperature it is noted that the enthalpy change is given by

$$dh = \left[ v - T\left(\frac{\partial v}{\partial T}\right)_P \right] dP$$

If the compressibility relation $Pv = ZRT$ is used, one finds that

$$dh_T = \left[ \frac{ZRT}{P} - \frac{ZRT}{P} - \frac{RT^2}{P}\left(\frac{\partial Z}{\partial T}\right)_P \right] dP = -\frac{RT^2}{P}\left(\frac{\partial Z}{\partial T}\right)_P dP$$

Before integrating this expression it must be transformed into reduced coordinates so that the result will be of general validity. By definition, $T = T_c T_R$ and $P = P_c P_R$. Hence

$$dT = T_c \, dT_R \qquad \text{and} \qquad dP = P_c \, dP_R$$

Substitution of these expressions into the equation for $dh_T$ yields

$$dh_T = -\frac{RT_c^2 T_R^2}{P_c P_R}\left(\frac{\partial Z}{T_c \, \partial T_R}\right)_{P_R} P_c \, dP_R = -RT_c T_R^2\left(\frac{\partial Z}{\partial T_R}\right)_{P_R} d\ln P_R$$

Upon integrating at constant temperature we obtain the expression

$$\frac{\Delta h_T}{R_u T_c} = -\int_i^f T_R^2\left(\frac{\partial Z}{\partial T_R}\right)_{P_R} d\ln P_R \qquad (13\text{-}40)$$

where the subscripts $i$ and $f$ signify the initial and final limits of integration for the reduced pressure. For convenience, the enthalpy should be evaluated from the ideal-gas to a real-gas state at the same temperature. The lower limit on the right, then, is zero pressure, for which state $P_R$ is likewise zero. The enthalpy of an ideal gas will be indicated by an asterisk, that is, $h^*$. The upper limit is the actual real-gas enthalpy $h$ at some elevated pressure $P$. Hence

$$\frac{h^* - h}{R_u T_c} = \int_0^P T_R^2 \left(\frac{\partial Z}{\partial T_R}\right)_{P_R} d \ln P_R \qquad (13\text{-}41)$$

The value of the integral is obtained by graphical integration, employing data from the generalized compressibility chart. The result of integration leads to values of $(h^* - h)/R_u T_c$ as a function of $P_R$ and $T_R$. A plot of these data is called a generalized enthalpy chart, and a typical chart is shown in Fig. A-27M. An example of the use of this chart is given in the following.

**Examples 13-7M and 13-7** Methane gas ($CH_4$) is cooled in a constant-pressure process from $T_1$ to $T_2$. Calculate the heat transfer per unit mass of methane ($a$) using the generalized enthalpy chart, Fig. A-26, and ($b$) assuming ideal-gas behavior.

SOLUTION A sketch of the process from state 1 to state 2 on a $Ts$ diagram is shown in Fig. 13-5. The actual path follows the constant-pressure line. The first law for a closed system at constant pressure reduces to $q = \Delta h$. However, the enthalpy change $h_2 - h_1$ cannot be obtained directly from the generalized enthalpy chart, since the chart always gives the enthalpy difference between an actual state and an ideal-gas state at the same temperature. Hence the sketch also shows a line of approximately zero pressure, which is the ideal-gas state. The generalized enthalpy chart, Fig. A-27M, gives the enthalpy change between such states at 1 and 1*, or 2 and 2*. Since the enthalpy is a state function, we may calculate the change of this property by the path 1-1*-2*-2 as well as by the direct path 1-2. Consequently,

$$q = \Delta h = h_2 - h_1 = (h_1^* - h_1) + (h_2^* - h_1^*) - (h_2^* - h_2)$$

The first and last terms on the right are obtained from Fig. A-27M, the generalized enthalpy chart.

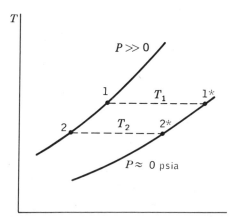

**Figure 13-5** Process diagram for Examples 13-7M and 13-7.

*Metric problem:* (a) The constant-pressure process occurs at 65 bars, and the initial and final temperature are 70 and $-6°C$, respectively. The value of $c_p$ at zero pressure is given by $c_{p,0} = 18.9 + 0.0555T$, where $T$ is in °K and the specific heat is in J/(g·mol)(°K). From the statement of the problem,

$$P_{R1} = 65/46.4 = 1.40 \qquad P_{R2} = 1.40$$

$$T_{R1} = 343/191 = 1.80 \qquad T_{R2} = 267/191 = 1.40$$

At these values we find from the enthalpy chart that $(h^* - h)/R_u T_c$, at states 1 and 2, has values of 0.53 and 0.92, respectively. Hence

$$h_1^* - h_1 = 0.53(8.315)(191) = 842 \text{ J/g·mol}$$

$$h_2^* - h_2 = 0.92(8.315)(191) = 1461 \text{ J/g·mol}$$

The value of $h_2^* - h_1^*$ is found from the integration of the zero-pressure specific-heat data; that is

$$h_2^* - h_1^* = \int_{343}^{267} (18.9 + 0.0555T) \, dT = -1436 - 1286 = -2722 \text{ J/g·mol}$$

Substitution of these values into the first law yields

$$q = 842 + (-2722) - 1461 = -3341 \text{ J/g·mol} = -209 \text{ J/g}$$

(b) If the effect of pressure is neglected, the change in enthalpy for this process is the same as for an ideal gas; that is

$$q = h_2^* - h_1^*$$

This value has already been calculated and found to be $-2722$ J/g·mol, which is equivalent to $-170$ J/g. Thus a 19 percent error will result if the effect of pressure on the enthalpy is neglected.

*USCS problem:* (a) The constant-pressure process occurs at 55 atm, and the initial and final temperatures are 160 and 20°F, respectively. The value of $c_p$ at zero pressure is given by $c_{p,0} = 4.52 + 0.00737T$, where $T$ is in °R and the specific heat is in Btu/(lb·mol)(°R). From the statement of the problem,

$$P_{R1} = 55/45.8 = 1.20 \qquad P_{R2} = 1.20$$

$$T_{R1} = 620/344 = 1.80 \qquad T_{R2} = 480/344 = 1.40$$

At these values we find from the enthalpy chart that $(h^* - h)/R_u T_c$, at states 1 and 2, has values of 0.53 and 0.92, respectively. Hence

$$h_1^* - h_1 = 0.53(1.986)(344) = 362 \text{ Btu/lb·mol}$$

$$h_2^* - h_2 = 0.92(1.986)(344) = 629 \text{ Btu/lb·mol}$$

The value of $h_2^* - h_1^*$ is found from the integration of the zero-pressure specific-heat data; that is

$$h_2^* - h_1^* = \int_{620}^{480} (4.52 + 0.00737T) \, dT = -633 - 567 = -1200 \text{ Btu/lb·mol}$$

Substitution of these values into the first law yields

$$q = 362 + (-1200) - 629 = -1467 \text{ Btu/lb·mol} = -91.7 \text{ Btu/lb}$$

(b) If the effect of pressure is neglected, the change in enthalpy for this process is the same as for an ideal gas; that is

$$q = h_2^* - h_1^*$$

This value has already been calculated and found to be $-1200$ Btu/lb·mol, which is equivalent to $-75.0$ Btu/lb Thus an 18 percent error will result if the effect of pressure on the enthalpy is neglected.

For a number of different processes, including those with heat transfer, it is important to have a generalized entropy chart as well as the enthalpy chart. The entropy chart is based on the generalized equation for the entropy change of a simple substance, namely,

$$ds = \frac{c_p}{T}\,dT - \left(\frac{\partial v}{\partial T}\right)_P dP \tag{13-21}$$

As in the situation for the enthalpy function, we note that the first term on the right requires only specific-heat data for a substance at the required pressure. The term in Eq. (13-21) that is difficult to evaluate in many cases is the second one, simply because sufficient $PvT$ data may not be available for the substance of interest. Therefore a generalized approach is necessary in many cases. Following the procedure for the enthalpy function, we shall integrate Eq. (13-21) from essentially zero pressure to the desired pressure at constant temperature. This is denoted mathematically by

$$(s_P - s_0^*)_T = -\int_0^P \left(\frac{\partial v}{\partial T}\right)_P dP \tag{13-42}$$

Normally, the next procedure would be to insert the definitions of the compressibility factor and the reduced pressure and temperature into this expression. However, Eq. (13-42) is not directly useful since the entropy at the ideal-gas state of zero pressure is infinite in value. This dilemma is circumvented in the following way: We shall apply Eq. (13-21) to an isothermal change between zero pressure and a given finite pressure $P$, but we shall assume that the gas behaves as an ideal gas at all times. Hence

$$(s_P^* - s_0^*)_T = -\int_0^P \left(\frac{\partial v}{\partial T}\right)_P dP = -R \int_0^P \frac{dP}{P} \tag{13-43}$$

The state represented by $s_P^*$ is a fictional state since an ideal gas, for which the equation of state $Pv = RT$ is valid, exists only at zero pressure. However, we may still assign values to the state even if it is nonexistent. If Eq. (13-42) is now subtracted from Eq. (13-43), we obtain

$$(s_P^* - s_P)_T = -\int_0^P \left[\frac{R}{P} - \left(\frac{\partial v}{\partial T}\right)_P\right] dP \tag{13-44}$$

From the definition of the compressibility factor it is found that

$$\left(\frac{\partial v}{\partial T}\right)_P = \frac{RZ}{P} + \frac{RT}{P}\left(\frac{\partial Z}{\partial T}\right)_P$$

The use of this equation permits Eq. (13-44) to be written as

$$(s_P^* - s_P)_T = -R \int_0^P \left[ \frac{1-Z}{P} - \frac{T}{P}\left(\frac{\partial Z}{\partial T}\right)_P \right] dP$$

This latter result can now be expressed in terms of reduced properties,

$$(s_P^* - s_P)_T = -R \int_0^{P_R} \frac{1-Z}{P_R} dP_R + RT_R \int_0^{P_R} \left(\frac{\partial Z}{\partial T_R}\right)_{P_R} \frac{dP_R}{P_R}$$

By comparing the last term of this equation with Eq. (13-41) one finds that this term can be written as a function of $h^* - h$. The final result is

$$\frac{(s_P^* - s_P)_T}{R_u} = \frac{h^* - h}{R_u T_R T_c} - \int_0^{P_R}(1 - Z)\frac{dP_R}{P_R} \tag{13-45}$$

The value of the first term on the right is available from a generalized enthalpy chart. The last term on the right must be evaluated by graphical integration of compressibility data. Equation (13-45) permits evaluation of the departure of the entropy value from the ideal-gas value at the same pressure and temperature. A graphical presentation of $(s_P^* - s_P)_T/R_u$ versus the reduced pressure and temperature is given in Fig. A-28M in the form of a generalized entropy chart. Among other applications, a generalized entropy chart is useful for isentropic processes of real gases.

**Examples 13-8M and 13-8** Carbon dioxide is compressed reversibly and adiabatically from known values of $P_1$ and $T_1$ to $P_2$ in a steady-flow process. Determine the final temperature with the aid of the generalized entropy chart, if the gas at $P_1$ is ideal.

SOLUTION The process is reversible and adiabatic, hence isentropic. Thus we seek the value of $T_2$ for which $s_2 = s_1$. The calculation will require two steps if we make use of the generalized entropy chart. First we shall calculate the change in entropy from the initial to the final state as if the gas were ideal. Then we shall evaluate the entropy change from the ideal-gas to a real-gas state at the final pressure and temperature. The sum of these two changes must be zero if the process is isentropic. Thus

$$s_1 - s_2 = s_1^* - s_2 = (s_1^* - s_2^*) + (s_2^* - s_2) = 0$$

State 1 equals state 1* since the initial state is an ideal-gas state. The quantity $s_2^* - s_2$ is found from Fig. A-28M in terms of $P_{R2}$ and $T_{R2}$. The term $s_1^* - s_2^*$ may be found from the ideal-gas relation

$$s_2^* - s_1^* = \int_1^2 c_p \, d(\ln T) - R \ln \frac{P_2}{P_1} = s_2^0 - s_1^0 - R \ln \frac{P_2}{P_1}$$

where the $s^\circ$ values are found in the ideal-gas table. Therefore the total entropy change becomes

$$s_1 - s_2 = s_1^0 - s_2^0 + R \ln \frac{P_2}{P_1} + (s_2^* - s_2)_{\text{chart}} = 0$$

The values of $s_2^0$ and the chart correction are unknown, however, since the final temperature is not known. Hence the solution becomes one of iteration. A trial temperature is chosen repeatedly until the above equation is satisfied.

*Metric analysis:* The initial pressure and temperature are 1 bar and 220°K, and the final pressure is 40 bars. The following data are available:

$$P_{R2} = 40/73.9 = 0.541 \qquad s_1^0 = 202.966$$

$$R \ln \frac{P_2}{P_1} = 8.315 \ln 40 = 30.67 \text{ kJ/(kg·mol)(°K)}$$

As a first trial we shall assume that $T_2$ is 500°K. On this basis,

$$T_{R2} = 500/304 = 1.64 \qquad (s_2^* - s_2)/R_u = 0.125 \qquad s_2^0 = 234.814$$

Substitution of all the data into the equation for $s_1 - s_2$ yields

$$s_1 - s_2 = 202.97 - 234.81 + 30.67 + 0.125(8.315) = -0.13 \text{ kJ/(kg·mol)(°K)}$$

The assumption of $T_2$ equal to 500°K gives a total entropy change close to zero. If a further assumption of $T_2 = 490°K$ is made, we find that

$$s_1 - s_2 = 202.97 - 233.92 + 30.67 + 0.130(8.315) = 0.80 \text{ kJ/(kg·mol)(°K)}$$

The first approximation leads to a negative $\Delta s$ value, while the second evaluation yields a positive $\Delta s$. Hence the best approximation to the final temperature lies between 490 and 500°K, and is probably around 498°K.

USCS *analysis:* The initial pressure and temperature are 1 atm and 400°R, and the final pressure is 40 atm. The following data are available:

$$P_{R2} = \frac{40}{72.9} - 0.548 \qquad s_1^0 = 48.555 \text{ Btu/(lb·mol)(°R)}$$

$$R \ln \frac{P_2}{P_1} = 1.986 \ln 40 = 7.33 \text{ Btu/(lb·mol)(°R)}$$

As a first trial we shall assume that $T_2$ is 900°R. On this basis,

$$T_{R2} = 900/548 = 1.64 \qquad (s_2^* - s_2)/R_u = 0.125 \qquad s_2^0 = 56.070$$

Substitution of all the data into the equation for $s_1 - s_2$ yields

$$s_1 - s_2 = 48.555 - 56.070 + 7.33 + 0.25 = +0.065 \text{ Btu/(lb·mol)(°R)}$$

The assumption of this temperature gives a total change in entropy close to zero. If a further assumption of $T_2 = 920°R$ is made, we find that

$$s_1 - s_2 = 48.555 - 56.305 + 7.33 + 0.24 = -0.180 \text{ Btu/(lb·mol)(°R)}$$

Consequently, the best approximation to the final temperature is probably around 905°R. As a comparison we can calculate the final temperature based solely on ideal-gas behavior from the gas table A-9. At 400°R the relative pressure $p_{r1}$ is 0.04153. Hence $p_{r2} = 40(0.04153) = 1.661$, which is the relative pressure for a final temperature of about 884°R. The use of the generalized entropy chart predicts a final temperature about 20° higher than this. The entropy correction for real-gas behavior is especially significant at low values of the reduced temperature.

Although only two generalized charts have been presented here, it should be apparent that any number of these might be devised, once a generalized equation is available for a property in terms of the variables $P$ and $T$. For example, a generalized chart which permits the estimation of $c_p$ values at high pressures is available. Chemists and chemical engineers find a generalized chart for the property called fugacity to be quite useful. In the absence of an abundance of $PvT$ data for a substance, generalized charts are powerful tools for predicting the properties of a fluid—gas, or liquid.

## REFERENCES

Hall, N. A., and W. E. Ibele: "Engineering Thermodynamics," Prentice-Hall, Englewood Cliffs, N.J., 1960.

Zemansky, M. W.: "Heat and Thermodynamics," 4th ed., McGraw-Hill, New York, 1957.

## PROBLEMS (METRIC)

### Generalized relations

**13-1M** Prove that the constant-pressure lines in the wet region of an $hs$ diagram are straight and not parallel and that the slope of a constant-pressure line in the superheat region increases with temperature.

**13-2M** From any of the four Maxwell relations, derive the other three, making use of Eq. (13-1b).

**13-3M** Derive expressions for (a) $(\partial u/\partial P)_T$, and (b) $(\partial u/\partial v)_T$ that involve only $P$, $T$, and $v$.

**13-4M** Derive the relation $c_p = T(\partial s/\partial T)_P$.

**13-5M** From Eqs. (13-17) and (13-20), show that the slope of a constant-volume line is greater than that for a constant-pressure line through the same state point in the gas region of a $Ts$ diagram.

**13-6M** Derive the expression $c_p = T(\partial P/\partial T)_s(\partial v/\partial T)_P$.

**13-7M** Derive the expression $c_v = -T(\partial v/\partial T)_s(\partial P/\partial T)_v$.

**13-8M** Approximate the value of $c_p$ for steam at (a) 60 bars and 400°C, and (b) 120 bars and 480°C, by employing the equation derived in Prob. 13-6M. Compare with the value obtained by approximating the derivative, $c_p = (\partial h/\partial T)_P$.

**13-9M** Approximate the value of $c_p$ for refrigerant 12 at (a) 6 bars and 50°C, and (b) 12 bars and 80°C, by employing the equation derived in Prob. 13-6M. Compare with the value obtained by approximating the derivative, $c_p = (\partial h/\partial T)_P$.

**13-10M** Show that the slope of a constant-pressure line in the vapor region of a $Ts$ diagram normally increases with increasing temperature.

**13-11M** Derive the relation $(\partial^2 g/\partial T^2)_P = -c_p/T$.

**13-12M** Derive the relation $(\partial^2 a/\partial T^2)_v = -c_v/T$.

**13-13M** What can be concluded qualitatively about the change in the enthalpy of a fluid during an isentropic compression?

**13-14M** Prove that, for any homogeneous system, $c_p/c_v = (\partial P/\partial v)_s(\partial P/\partial v)_T$.

**13-15M** Check the validity of Eq. (13-13) by finding approximate values of the derivatives on both sides of the equation for steam at (a) 15 bars and 360°C, and (b) 80 bars and 400°C.

**13-16M** Check the validity of Eq. (13-13) by finding approximate values of the derivatives on both sides of the equation for refrigerant 12 at (a) 7 bars and 50°C, and (b) 12 bars and 80°C.

**13-17M** On the basis of the van der Waals equation of state and the generalized relations:

(a) Show that $(h_2 - h_1)_T = (P_2 v_2 - P_1 v_1) + a(1/v_1 - 1/v_2)$.

(b) Show that $(s_2 - s_1)_T = R \ln [(v_2 - b)/(v_1 - b)]$.

(c) Evaluate $(h_2 - h_1)_T$ in kJ/kg·mol for oxygen at 27°C when it is compressed from 1 bar to 100 bars. Compare with the ideal-gas solution to the problem.

**13-18M** Develop an expression for the isothermal change in internal energy for a substance which follows the Redlich-Kwong equation of state. The result should be given in terms of $T$, $v$, $a$, $b$, and a constant.

**13-19M** On the basis of the equation derived in Prob. 13-18M, evaluate with the data in Table A-21M the internal-energy change of steam as it is compressed isothermally at 360°C from 23.31 cm³/g to 11.05 cm³/g. Check the answer against tabular data for steam.

**13-20M** Same as Prob. 13-19M, except that we wish to compress refrigerant 12 isothermally at 80°C from 27.48 cm³/g to 11.98 cm³/g.

**13-21M** Derive expressions for the isothermal change (a) of enthalpy, and (b) of entropy for a substance which follows the Berthelot equation of state.

**13-22M** A pressure-enthalpy diagram is frequently used in the refrigeration industry. Determine an equation for the slope of an isentropic line on this diagram in terms of $PvT$ data only.

**13-23M** Use Eqs. (13-27), (12-4), and (12-5) to predict the $c_p$ value of nitrogen, in kJ/(kg·mol)(°C), at (a) 200°K and 40 bars, and (b) 190°K and 70 bars.

**13-24M** A gas has a compressibility factor $Z$ given by $Z = 1 + aP/T^2$. Derive an expression for the change in enthalpy between two states at the same temperature but different pressures.

**13-25M** Derive an expression for the change in enthalpy in terms of temperature and pressure changes from a state $T_1$ and $P_1$ to a state of higher values of $T_2$ and $P_2$ for a gas whose equation of state is $Pv/RT = 1 + AP/T$ and whose specific heat at a pressure $P_0$ is given by $c_{p,0} = 1 + BT$; $A$ and $B$ are constants, and $P_0$ is less than $P_1$ and $P_2$.

## Use of $\beta$ and $K_T$

**13-26M** Show that the change in volume of a substance can be related to $\beta$ and $K_T$ such that $dv/v = \beta\, dT - K_T\, dP$.

**13-27M** On the basis of the result shown in Prob. 13-26M, estimate the percent change in the volume of copper when it changes state from 200°K and 1 bar to 300°K and 1000 bars.

**13-28M** A 0.1-kg mass of copper is heated from 250 to 500°K and is compressed from 1 bar to 500 bars. Estimate the change in volume in cubic centimeters if the value of $v$ at 500°K is 7.115 cm³/g·mol.

**13-29M** Evaluate (a) the coefficient $\beta$, and (b) the coefficient $K_T$ symbolically for a van der Waals gas.

**13-30M** At 500°K the value of $v$ is 7.115 cm³/g·mol.
    (a) Determine the value of $c_p - c_v$ for solid copper, in kJ/(kg·mol)(°K).
    (b) If $c_p$ is 26.15 kJ/(kg·mol)(°K), what percent error would be made in $c_v$ if we assumed that $c_p = c_v$ at this temperature?

**13-31M** At 300°K the value of $v$ for solid copper is 7.062 cm³/g·mol. Determine the value of $c_p - c_v$, in kJ/(kg·mol)(°K).

**13-32M** At 0°C the values of $v$, $\beta$, and $K_T$ for liquid mercury are 14.67 cm³/g·mol, $174 \times 10^{-6}$ °K$^{-1}$, and $3.79 \times 10^{-7}$ cm²/N, respectively.
    (a) Determine the value of $c_p - c_v$, in kJ/(kg mol)(°K).
    (b) If $c_p$ is 28.0 kJ/(kg·mol)(°K), what percent error would be made in $c_v$ if we assumed that $c_p = c_v$?

## Phase changes

**13-33M** The vapor pressure of carbon tetrachloride at several temperatures is as follows:

| T, °C | 25 | 35 | 45 | 55 |
|---|---|---|---|---|
| P, mbar | 151.7 | 232.5 | 345.1 | 498.0 |

Plot ln $P$ versus $1/T$, and from the slope evaluate the mean enthalpy of vaporization.

**13-34M** Estimate the enthalpy of vaporization of water at (a) 100°C, (b) 150°C, and (c) 200°C by employing (1) the Clapeyron equation, and (2) the Clapeyron-Clausius equation. Compare with data from Table A-12M.

**13-35M** Estimate from tabular data the enthalpy of vaporization of refrigerant 12 at (a) 8°C, (b) 24°C, and (c) 44°C by using (1) the Clapeyron equation, and (2) the Clapeyron-Clausius equation. Compare result with data from Table A-16M.

**13-36M** The specific volumes of liquid water and ice at 0°C are 1.0002 and 1.0911 cm$^3$/g, respectively. Estimate the melting-point temperature of ice at (a) 250 bars, (b) 500 bars, and (c) 750 bars, if the enthalpy of fusion is 333.4 kJ/kg.

**13-37M** The vapor pressure of water can be represented with reasonable accuracy between 5 and 50 bars by the equation, $\ln P = (-4692/T) + 0.0124 \ln T + 12.58$, with $P$ in bars and $T$ in °K. Compute the enthalpy of vaporization at (a) 160°C, (b) 200°C, and (c) 250°C, using values of the specific volume from tables.

**13-38M** The triple state of carbon dioxide is $-56.6$°C and 5.178 bars. Predict the slopes of the three saturation lines in the vicinity of the triple state on a $PT$ diagram on the basis of the following triple-state data.

| Phase | Solid | Liquid | Vapor |
|---|---|---|---|
| $h$, J/g | 181.3 | 380.2 | 728.5 |
| $v$, cm$^3$/g | 0.661 | 0.849 | 72.22 |

**13-39M** Helium 4 boils at 4.22°K at 1 bar, and the enthalpy of vaporization is 83.3 J/g·mol. By producing a vacuum over the liquid phase, the fluid boils at a lower temperature. Estimate what pressure, in millibars, is necessary to produce a temperature of (a) 2°K, (b) 1°K, and (c) 0.5°K.

**13-40M** At the triple state of water the pressure and temperature are 6.12 mbar and 0.010°C. The enthalpy of melting is 333.4 kJ/kg, and the specific volumes of the liquid and solid phases are 1.0002 and 1.0911 cm$^3$/g, respectively. A person skates on ice at 30°F on blades with a contact area of 0.32 cm$^2$. What must the weight and the mass, in N and kg, respectively, of a person be to just melt the ice beneath the blades.

**13-41M** Consider the thermodynamic data given in Prob. 13-40M. A person with a mass of 80 kg is ice-skating on blades which have a total area of 0.25 cm$^2$ in contact with the ice. The temperature of the ice is (a) $-2$°C, and (b) $-3$°C. Will the ice melt under the blades?

**Joule-Thomson coefficient**

**13-42M** Determine the Joule-Thomson coefficient for water at (a) 30 bars and 320°C, (b) 60 bars and 320°C, and (c) 100 bars and 400°C, in °C/bar.

**13-43M** A gas obeys the relation $P(v - b) = RT$, where $b$ is a positive constant. Determine the Joule-Thomson coefficient of the gas. Could this gas be cooled effectively by throttling?

**13-44M** Calculate the Joule-Thomson coefficient for a Dieterici gas, and prove that the inversion temperature for such as gas is $2a(v - b)/Rbv$.

**13-45M** By employing the generalized equation for the Joule-Thomson coefficient and the tabular data for refrigerant 12, estimate the coefficient in °C/bar at (a) 10 bars, and 80°C, (b) 12 bars and 80°C, and (c) 14 bars and 80°C.

**13-46M** The Joule-Thomson coefficient for nitrogen at 40 bars and $-73$°C is approximately 0.4°C/bars on the basis of Fig. 13-4. Determine the value of $c_p$ in J/(g·mol)(°C) if Eqs. (12-4) and (12-5) represent the $PvT$ behavior of nitrogen at this state.

**13-47M** An acceptable equation of state for helium gas is given by $Pv = RT - aP/T + bP$, where $a = 386.7$°K·cm$^3$/g·mol and $b = 15.29$ cm$^3$/g·mol.

(*a*) Compute the Joule-Thomson coefficient at 150 and 15°K.

(*b*) Find the inversion temperature in °K.

(*c*) Estimate the temperature reached in an ideal throttling process from 25 bars and 15°K to 1 bar.

**13-48M** By the use of Eq. (12-5), employ a graphical technique to determine roughly the inversion temperature for nitrogen, in °K.

**13-49M** On the basis of Eq. (12-5) determine the Joule-Thomson coefficient for nitrogen at 1 bar and (*a*) 300°K, (*b*) 500°K, and (*c*) 700°K.

### Generalized charts

**13-50M** Calculate the value, in kJ/kg, of ($h^* - h$) at 40 bars and 320°C for steam by use of (*a*) the generalized chart, and (*b*) steam-table data.

**13-51M** Same as Prob. 13-50M, except for a state of 80 bars and 360°C.

**13-52M** Calculate the value, in kJ/kg, of ($h^* - h$) at 16 bars and 80°C for refrigerant 12 by use of (*a*) the generalized chart, and (*b*) tabular data.

**13-53M** Refrigerant 12 is throttled from 16 bars and 100°C to 1.4 bars. Determine the final temperature based on a generalized chart. Compare the result with that obtained from tabulated data. The zero-pressure specific heat in this temperature range may be taken as 0.65 J/(g)(°C).

**13-54M** Same as Prob. 13-53M, except that the fluid is throttled from 16 bars and 80°C, to 4 bars.

**13-55M** By means of a generalized chart, estimate the change in enthalpy in kJ/kg·mol accompanying the isothermal expansion of ethane from 20 bars, 30°C, to 2 bars.

**13-56M** Methane gas is cooled in a constant-pressure process at 55 bars from 100 to 30°C. The equation for $c_{p,0}$ is given in Table 3-1. Calculate the heat transferred, in kJ/kg.

**13-57M** Ethane is compressed from 30°C and 15 bars to 60°C and 98 bars in a steady-flow process. The molar specific heat in kJ/(kg·mol)(°C) is given by $c_{p,0} = 16.8 + 0.123T$, where $T$ is in °K. Using generalized charts, determine the enthalpy change, in kJ/kg·mol, if (*a*) the gas is an ideal gas, and (*b*) the gas is a real gas.

**13-58M** Nitrogen gas is compressed isothermally from the ideal-gas state to 50 bars and 250°K. Determine the value of $h^* - h$ by employing (*a*) Eq. (12-5) and the generalized equation, and (*b*) the generalized chart.

**13-59M** Methane gas ($CH_4$) is compressed in a steady-flow process from 13.9 bars and 71°C to 186 bars and 300°C. Using generalized charts, determine (*a*) the change in enthalpy, in kJ/kg·mol, (*b*) the change in entropy in kJ/(kg·mol)(°K).

**13-60M** Nitrogen at 100 bars and 200°K is contained in a tank of 0.3 m³. Heat is transferred to the nitrogen until the temperature is 350°K. Determine the heat transferred and the final pressure (*a*) using generalized charts, and (*b*) assuming ideal-gas behavior.

**13-61M** Calculate (*a*) the work of compression, and (*b*) the heat transfer, in kJ/kg, when ethane is compressed reversibly and isothermally from 5 to 98 bars at 30°C in a steady-flow process.

**13-62M** Carbon dioxide is compressed in a reversible, adiabatic, steady-flow process from 5 bars, 30°C, to 40 bars. Determine the work of compression, in kJ/kg.

**13-63M** Ethylene at 67°C and 1 bar is compressed isothermally in a reversible, nonflow process to 255 bars. On the basis of generalized charts, determine (*a*) the change in entropy, in kJ/(kg·mol)(°K), (*b*) the change in internal energy, in kJ/kg·mol, (*c*) the heat transferred, and (*d*) the work required, in kJ/kg·mol.

**13-64M** Ethylene is being expanded in a gas turbine from 180°C and 307 bars to 67°C and 61 bars. The intake capacity is 1 m³/min. A mean $c_p$ value of 50.0 kJ/(kg·mol)(°K) may be assumed at zero pressure. Estimate the shaft-output power, in kilowatts.

**13-65M** Steam at 280 bars and 520°C is expanded adiabatically and reversibly in a steady-flow process to 100 bars. Determine the final temperature, in °C, and the change in enthalpy using (*a*) generalized charts, and (*b*) tabular data.

**13-66M** Propane ($C_3H_8$) enters a pipe at 34.2 bars, 100°C, and a velocity of 40 m/s. The gas flows adiabatically through the pipe until the pressure reaches 10.7 bars. If the zero-pressure value of $c_p$ is relatively constant at 1.76 kJ/(kg)(°C), estimate the exit temperature, in °C, and the exit velocity, in m/s.

**13-67M** Oxygen initially at 40 bars and $-27$°C is compressed adiabatically in a steady-flow process to a final pressure of 200 bars. Determine the minimum work of compression if (a) the generalized charts are used, and (b) the gas is assumed to be ideal.

**13-68M** Carbon dioxide is compressed isentropically in steady flow from 4 bars and 0°C to 60 bars. Determine the work of compression in kJ/kg·mol.

# PROBLEMS (USCS)

### Generalized relations

**13-1** Prove that the constant-pressure lines in the wet region of an $hs$ diagram are straight and not parallel and that the slope of a constant-pressure line in the superheat region increases with temperature.

**13-2** From any of the four Maxwell relations, derive the other three, making use of Eq. (13-1b).

**13-3** Derive expressions for (a) $(\partial u/\partial P)_T$ and (b) $(\partial u/\partial v)_T$ that involve only the pressure, temperature, and specific volume.

**13-4** Derive the relation $c_p = T(\partial s/\partial T)_P$.

**13-5** From Eqs. (13-17) and (13-20), show that the slope of a constant-volume line is greater than that for a constant-pressure line through the same state point on a $Ts$ diagram.

**13-6** Derive the expression $c_p = T(\partial P/\partial T)_s(\partial v/\partial T)_P$.

**13-7** Derive the expression $c_v = -T(\partial v/\partial T)_s(\partial P/\partial T)_v$.

**13-8** Approximate the value of $c_p$ for steam at 800 psia and 650°F by employing the equation derived in Prob. 13-6. Compare with the value obtained directly from the definition of this property.

**13-9** Approximate the value of $c_v$ for steam at 600 psia and 600°F by employing the equation derived in Prob. 13-7.

**13-10** Show that the slope of a constant-pressure line in the vapor region of a $Ts$ diagram normally increases with increasing temperature.

**13-11** Derive the relation $(\partial^2 g/\partial T^2)_P = -c_p/T$.

**13-12** Derive the relation $(\partial^2 a/\partial T^2)_v = -c_v/T$.

**13-13** What can be concluded qualitatively about the change in the enthalpy of a fluid during an isentropic compression?

**13-14** Prove that, for any homogeneous system, $c_p/c_v = (\partial P/\partial v)_s(\partial P/\partial v)_T$.

**13-15** Find the approximate values of the derivatives on both sides of Eq. (13-13) for steam at 200 psia and 500°F in the same set of units.

**13-16** Develop the general relations $(\partial h/\partial v)_T = T(\partial P/\partial T)_v + v(\partial P/\partial v)_T$, and evaluate $(\partial h/\partial v)_T$ for an ideal gas.

**13-17** On the basis of the van der Waals equation of state and the generalized relations:
(a) Show that $(h_2 - h_1)_T = (P_2 v_2 - P_1 v_1) + a(1/v_1 - 1/v_2)$.
(b) Show that $(s_2 - s_1)_T = R \ln [(v_2 - b)/(v_1 - b)]$.
(c) Evaluate $(h_2 - h_1)_T$ in Btu/lb·mol for oxygen at 80°F which is compressed from 1 to 100 atm. Compare with the ideal-gas solution.

**13-18** Derive expressions for the isothermal change of enthalpy and entropy for a substance which follows the Berthelot equation of state.

**13-19** A pressure-enthalpy diagram is frequently employed in the refrigeration industry. Determine an equation for the slope of an isentropic line on this diagram in terms of $PvT$ data only.

**13-20** A gas has a compressibility factor $Z$ given by $Z = 1 + aP/T^2$. Derive an expression for the change in enthalpy between two states at the same temperature but different pressures.

**13-21** Derive an expression for the change in enthalpy in terms of temperature and pressure changes from a state $T_1$ and $P_1$ to a state of higher values $T_2$ and $P_2$ for a gas whose equation of state is $Pv/RT = 1 + AP/T$ and whose specific heat at a pressure $P_0$ is given by $c_{p,0} = 1 + BT$; $A$ and $B$ are constants, and $P_0$ is less than $P_1$ and $P_2$.

**13-22** A gas the equation of state of which is $P(v - b) = RT$ undergoes a change of state from 20 psia and 500°R to 800 psia and 700°R. Compute the enthalpy change if $c_{p,0} = 0.24 + 1.3 \times 10^{-4}T$ and $b = 0.0039$, with $T$ in °R and $P$ in psia.

**13-23** At 900°R the values of $v$, $\beta$, and $K_T$ for solid copper are 51.24 ft³/lb·mol, $30.1 \times 10^{-6}$ °R⁻¹, and $4.003 \times 10^{-12}$ ft²/lb$_f$, respectively.

   (a) Determine the value of $c_p - c_v$ in Btu/(lb mol)(°F).

   (b) If $c_p$ is 6.25 Btu/(lb·mol)(°F), what percent error would be made in $c_v$ if we assumed that $c_p = c_v$?

**13-24** At 32°F and 14,225 psia the values of $v$, $\beta$, and $K_T$ for liquid mercury are 0.235 ft³/lb·mol, $96.7 \times 10^{-6}$ °R⁻¹, and $1686 \times 10^{-12}$ ft²/lb$_f$, respectively.

   (a) Determine the value of $c_p - c_v$, in Btu/(lb·mol)(°F).

   (b) If $c_p$ is 6.69 Btu/(lb·mol)(°F), what percent error would be made in $c_v$ if we assumed that $c_p = c_v$?

## Phase changes

**13-25** The vapor pressure of carbon tetrachloride at several temperatures is as follows:

| $T$, °F | 77 | 95 | 113 | 131 |
|---|---|---|---|---|
| $P$, mm Hg | 113.8 | 174.4 | 258.9 | 373.6 |

Plot log $P$ versus $1/T$, and from the slope evaluate the mean enthalpy of vaporization in the given temperature range.

**13-26** Find the enthalpy of vaporization of water at 400°F by employing (a) the Clapeyron equation, and (b) the Clapeyron-Clausius equation. Compare with tabulated data.

**13-27** From tabular data determine the enthalpy of vaporization of refrigerant 12 at 70°F using (a) the Clapeyron equation, and (b) the Clapeyron-Clausius equation.

**13-28** The effect of pressure on the melting point of sodium has been measured by Bridgman [*Phys. Rev.*, 3:127 (1914)]. Some of the data include:

| $P$, atm | $T$, °F | $\Delta v \times 10^5$, ft³/lb |
|---|---|---|
| 5810 | 288.5 | 30.06 |
| 7740 | 310.6 | 27.46 |
| 9680 | 332.0 | 24.97 |

Determine the approximate enthalpy of fusion at 310.6°F, and compare with the tabulated value of 51.6 Btu/lb.

**13-29** The specific volumes of water and ice at 32°F are 0.01602 and 0.01747 ft³/lb, respectively. Estimate the melting-point temperature of ice at 500 atm if the enthalpy of fusion at 32°F is 143.32 Btu/lb.

**13-30** The vapor pressure of water can be represented by the equation, $\ln P = (-8445/T) + 0.0124 \ln T + 15.252$, with $P$ in psia and $T$ in °R. Compute the enthalpy of vaporization at 400°F using values of the specific volumes from tables.

**13-31** The enthalpy of fusion of water is practically constant at 143.8 Btu/lb. Calculate the freezing-point temperature at a pressure of (a) 10,000 psia and (b) 20,000 psia.

## Joule-Thomson coefficient

**13-32** Using data from Table A-14 and noting that $c_p = 0.48$ Btu/(lb)(°F), estimate the Joule-Thomson coefficient for steam at 100 psia and 700°F.

**13-33** Estimate the Joule-Thomson effect on the temperature of nitrogen which is throttled from 200 atm and 25°C to 1 atm.

**13-34** Determine the Joule-Thomson coefficient for water at 500 psia and 550°F, in °F/psi.

**13-35** Determine the change in the Gibbs function for the vaporization of refrigerant 12 at 10°F.

**13-36** A gas obeys the relation $P(v - b) = RT$, where $b$ is a positive constant. Determine the Joule-Thomson coefficient of the gas. Could this gas be cooled effectively by throttling?

**13-37** By employing the generalized equation for the Joule-Thomson coefficient and the tabular data for refrigerant 12 in Table A-18, estimate the coefficient in °F/atm at (a) 300 psia and 200°F, and (b) 500 psia and 240°F.

**13-38** The Joule-Thomson coefficient for nitrogen at 40 atm and $-100°F$ ($-73°C$) is approximately 0.4°C/atm on the basis of Fig. 13-4. Determine the value of $c_p$ in cal/(g·mol)(°C) if Eqs. (12-4) and (12-5) represents the $PvT$ behavior of nitrogen at this state.

**13-39** Estimate the enthalpy deviation of carbon monoxide at 1500 psia and $-100°F$ in Btu/lb·mol.

**13-40** An acceptable equation of state for helium gas is given by $Pv = RT - aP/T + bP$, where $a = 11.17°R$ ft³/lb·mol and $b = 0.245$ ft³/lb·mol.

    (a) Compute the Joule-Thomson coefficient for helium at 270° and 27°R.

    (b) Find the inversion temperature in °R.

    (c) Estimate the temperature reached in an ideal throttling process from 24 atm and 27°R to 1 atm.

**13-41** By the use of Eq. (12-5), employ a graphical technique to determine approximately the inversion temperature for nitrogen in °K.

**13-42** On the basis of Eq. (12-5), determine the Joule-Thomson coefficient of nitrogen at 1 atm and (a) 300°K, (b) 500°K, and (c) 700°K.

## Generalized charts

**13-43** Carbon dioxide is compressed in a reversible, adiabatic, steady-flow process from 5 atm and 460°R, to 40 atm. Determine the work of compression in Btu/lb.

**13-44** Refrigerant 12 is throttled from 600 psia and 280°F to 100 psia. Determine the final temperature based on generalized charts. The zero-pressure specific heat in this temperature range may be taken as 0.155 Btu/(lb)(°F). Compare with the result obtained from tabulated values of properties.

**13-45** By means of a generalized chart, estimate the change in enthalpy, in Btus, accompanying the isothermal expansion of 1 mol of ethane from 300 atm and 90°F, to 30 atm.

**13-46** Nitrogen at a pressure of 1500 psia and $-100°F$ is contained in a tank of 10 ft³. Heat is transferred to the nitrogen until the temperature is 200°F. Determine the heat transfer and the final pressure of the nitrogen (a) using compressibility data, and (b) assuming ideal-gas behavior.

**13-47** Methane gas is cooled in a constant-pressure process at 800 psia from 200 to 100°F. The molar specific heat at zero pressure is given by $c_{p,0} = 4.52 + 0.00737T$, where $T$ is in °R. Calculate the heat transferred per pound.

**13-48** Ethane is compressed from 90°F and 15 atm to 145°F and 96 atm in a steady-flow process. The molar specific heat at zero pressure is given by $c_{p,0} = 4.01 + 0.01636T$, where $T$ is in °R. Determine the enthalpy change of the gas, in Btu/lb·mol (a) if the gas is an ideal gas, and (b) if the gas is a real gas, using generalized charts.

**13-49** Calculate (a) the work of compression and (b) the heat transfer per pound when ethane is compressed reversibly and isothermally from 70 to 1420 psia at 90°F in a steady-flow process.

**13-50** Nitrogen gas is compressed isothermally from the ideal-gas state to 50 atm and 250°K. Determine the value of $h^* - h$ by (a) employing Eq. (12-5) and the generalized equation, and (h) employing the generalized chart.

**13-51** Calculate the value of $(h^* - h)_T$ in Btu/lb at 200 psia and 140°F for refrigerant 12 by use of (a) the generalized chart and (b) the data of Table A-18.

**13-52** Calculate the value of $(h^* - h)_T$ at 500 psia and 600°F for steam, in Btu/lb, by use of (a) the generalized chart and (b) the data of Table A-14 or the Mollier diagram.

**13-53** Nitrogen gas is compressed isothermally at $-100°F$ from 15 psia to 300 psia. Determine the enthalpy change, in Btu/lb, by employing the generalized equation in conjunction with Eqs. (12-4) and (12-5).

**13-54** The equation of state of a gas is $v = RT/P - aT/(T - b)$, where $a$ and $b$ are positive constants. $T$ is greater than $b$.

(a) Derive an expression for $h_2 - h_1$ between two states at the same temperature but at different pressures.

(b) If this gas flows through a throttling valve, does the temperature decrease, increase, or remain unchanged?

(c) Is $c_p$ a function of $P$ for this gas?

**13-55** Listed below are some data for the values of $b$ and $T(db/dT)$ for air as a function of temperature, on the basis of the virial equation, $Pv = RT + bP$. Compute the entropy change in Btu/(lb·mol)(°R) for an isothermal compression at (a) 360°R and (b) 450°R from 1 to 50 atm.

| $T$, °R | $b$, ft³/lb·mol | $T(db/dT)$, ft³/lb·mol |
|---|---|---|
| 360 | −0.613 | 1.57 |
| 450 | −0.310 | 1.17 |

**13-56** Ethylene at 127°F and 14.7 psia is compressed isothermally in a reversible, nonflow process to 4080 psia. Using the principle of corresponding states, determine (a) the change in the specific entropy, (b) the change in the specific internal energy, (c) the heat added or removed, and (d) the work done per pound.

**13-57** Propane is being pumped a distance of 250 m. At a booster pump a gas is compressed from 73°F and 185 psia to 340°F and 1235 psia. The mean-zero-pressure specific heat $c_p$ is 0.44 Btu/(lb)(°F). It is desired to deliver 100 ft³/min at the compressor discharge. Determine the horsepower required for the compression process.

**13-58** Methane gas initially at 90°F and 400 psia is compressed isothermally and reversibly in a steady-flow device to a pressure of 2700 psia with a flow rate of 800,000 lb/h. Determine the following four quantities for the condition of (a) a gas which follows the principle of corresponding states, and (b) a gas which behaves as an ideal gas at all pressures: (1) the entropy change in Btu/(lb·mol)(°R), (2) the heat transferred in Btu/lb·mol, (3) the work required in Btu/lb·mol, and (4) the required area of the compressor outlet in square inches if the outlet velocity is 100 ft/s.

**13-59** Steam at 4500 psia and 1000°F is expanded adiabatically and reversibly in a steady-flow process to 1600 psia. Determine the final temperature and the change in enthalpy employing (a) the principle of corresponding states, and (b) the experimental data for steam.

**13-60** Ethylene ($C_2H_2$) at 100°F and 1 atm is compressed isothermally in a reversible, nonflow process to 278 atm. Using the principle of corresponding states, determine (a) the change in specific internal energy, (b) the change in entropy per mole, and (c) the work done in Btu/lb·mol.

**13-61** Methane gas ($CH_4$) is compressed in a steady-flow process from 168 psia and 160°F to 2860 psia and 570°F. Using generalized charts, determine (a) the change in enthalpy in Btu/lb·mole, (b) the change in internal energy in Btu/lb·mol, and (c) the change in entropy in Btu/(lb·mol)(°R)

# FOURTEEN

## COMBUSTION AND THERMOCHEMISTRY

In the preceding chapters attention was focused on the thermodynamic analysis of nonreacting systems. There are many practical examples, however, of engineering systems in which chemical reactions play a major role. It is our purpose in this chapter to apply the basic concepts of the first law of thermodynamics to reacting systems.

## 14-1 STOICHIOMETRY OF REACTIONS

Although chemical reactions could involve many different types of reactants, our interest will be directed mainly toward combustion reactions. Combustion normally involves the reaction of fuels containing primarily carbon and hydrogen with oxygen or air to form carbon dioxide ($CO_2$), carbon monoxide (CO), and water ($H_2O$) as the primary products. Other possible products will be discussed later, when the effects of dissociation are introduced.

One basic consideration in the analysis of combustion processes is the *theoretical* or *stoichiometric* reaction for a given fuel. A theoretical reaction requires the complete combustion of carbon, hydrogen, and any other combustible elements in the fuel. For example, all the carbon present is assumed to be burned to carbon dioxide and all the hydrogen is converted into water. In addition, no oxygen is present in the products of combustion. Hence the complete combustion of methane ($CH_4$) with oxygen is written as

$$CH_4 + 2O_2 \rightarrow CO_2 + 2H_2O$$

The theoretical oxygen requirement for a given fuel is the minimum oxygen required for the complete combustion. For the combustion of methane, two moles of $O_2$ are required per mole of fuel. The balanced chemical equation for the complete combustion of a fuel, such as methane, is called the *stoichiometric* equation, and in this case, theoretically, no oxygen will appear in the products of combustion. In general, one might consider the chemical reaction

$$v_A A + v_B B + \cdots \rightarrow v_L L + v_M M + \cdots \tag{14-1}$$

where the uppercase letters $A$, $B$, etc., represent the chemical species of the reactants and $L$, $M$, etc., represent the chemical species of the products. The $v_i$ terms are known as the stoichiometric coefficients of the various species. In the combustion of methane, $v_A = 1$, $v_B = 2$, $v_L = 1$, and $v_M = 2$. In general, any number of reactants or products might need to be considered.

The combustion of fuels in industrial practice is normally accomplished by employing air as the oxidizer. The major components of air are considered to be approximately 21 percent oxygen by volume, 78 percent nitrogen, and 1 percent argon. Small amounts of carbon dioxide and other gases are present, of course. It is convenient to assume that air is composed of 21 percent oxygen and 79 percent nitrogen by volume. Hence there is 21 moles of oxygen to every 79 moles of nitrogen in our assumed composition of atmospheric air. Therefore we may write

$$1 \text{ mol } O_2 + 3.76 \text{ mol } N_2 = 4.76 \text{ mol air}$$

or

$$1 \text{ lb } O_2 + 3.31 \text{ lb } N_2 = 4.31 \text{ lb air}$$

or

$$1 \text{ kg } O_2 + 3.31 \text{ kg } N_2 = 4.31 \text{ kg air}$$

The average molecular weight of air is 28.97, which we shall round off to 29.0 in most calculations.

When air is used as the oxidizer, we speak of the *theoretical air* requirements of a fuel. In the case of the oxidation of methane, the theoretical or stoichiometric reaction with air is written in the form

$$CH_4 + 2O_2 + 2(3.76)N_2 \rightarrow CO_2 + 2H_2O + 7.52N_2$$

The theoretical or chemically correct combustion of propane $(C_3H_8)$ with air is given by the chemical equation

$$C_3H_8 + 5O_2 + 5(3.76)N_2 \rightarrow 3CO_2 + 4H_2O + 18.80N_2$$

In each case no oxygen appears in the products of combustion, and the nitrogen has been assumed to undergo no chemical change.

For the complete combustion of carbon and hydrogen to $CO_2$ and $H_2O$ we may use the term theoretical, stoichiometric, or chemically correct oxygen, or air, requirements. When this quantity is not used in a process, we speak of the *percent theoretical* oxygen, or air, actually used. The stoichiometric quantity is the 100 percent theoretical requirement. When a deficiency is used, the percent theoretical is somewhere between 0 and 100 percent, and an excess of oxygen (or air) means that some value greater than the 100 percent theoretical value was employed.

Thus 200 percent theoretical air means that twice as much air is supplied as is necessary for complete combustion. In such a case oxygen will necessarily appear in the product gases. Other terms in frequent usage are the *percent excess* and the *percent deficiency* of oxygen (or air). As examples, 150 percent theoretical air is equivalent to 50 percent excess air, and 80 percent theoretical air is a 20 percent deficiency of air. When 150 percent theoretical air or 50 percent excess air is supplied, the combustion reaction for propane becomes

$$C_3H_8 + 7.5O_2 + 28.20N_2 \rightarrow 3CO_2 + 4H_2O + 2.5O_2 + 28.20N_2$$

This is a theoretical reaction, since in practice other products such as carbon monoxide may appear in small quantities. In writing chemical equations like the one above, no information about the products of the actual process is necessary.

In addition to the preceding nomenclature, the relationship between the fuel and air supplied to a combustion process is frequently given in terms of the *air-fuel* or *fuel-air* ratio. The air-fuel ratio (AF) is defined as the mass of air supplied per unit mass of fuel supplied. The fuel-air ratio (FA) is the reciprocal of the above definition. For the 100 percent theoretical combustion of propane, for example, the chemical equation shows that 23.80 mol of air (5 mol of $O_2$ + 18.80 mol of $N_2$) are required per mole of fuel. Consequently, the air-fuel ratio for theoretical combustion of this fuel is

$$AF = \frac{23.80 \text{ kg·mol air}}{\text{kg·mol fuel}} \times \frac{29 \text{ kg air}}{\text{kg·mol air}} \times \frac{\text{kg·mol fuel}}{44 \text{ kg fuel}} = 15.7 \frac{\text{kg air}}{\text{kg fuel}}$$

Since the air-fuel ratio is expressed as mass of air per mass of fuel, its value is the same whether expressed as kilograms per kilogram as above, or as pounds per pound. The fuel-air ratio for the same combustion process would be 0.0637, in units of kg fuel/kg air or lb fuel/lb air. Many hydrocarbon fuels from oil or natural gas require an air-fuel ratio in the neighborhood of 15 to 16 for stoichiometric combustion.

Finally, the relationship between the amounts of fuel and air supplied to a combustion process is also given by the *equivalence* ratio, $\phi$. By definition,

$$\phi = \frac{FA_{actual}}{FA_{stoich}}$$

where FA in the numerator represents the fuel-air ratio used under actual combustion conditions and FA in the denominator is the stoichiometric or chemically correct value. The value of $\phi$ is less than 1 when an excess of oxidant (such as air or oxygen) is used. This is also called a lean mixture. A rich mixture is one where $\phi$ is greater than unity, and the fuel is in excess of the stoichiometric requirement. The term equivalence ratio is used frequently with respect to spark ignition and compression ignition engine operation, and to gas turbine analysis.

**Example 14-1M** Determine the volume of air required to burn gaseous propane to completion, per kilogram and per kilogram-mole of fuel. The theoretical quantity of air is supplied and the temperature and pressure of the air are 25°C and 0.98 bar, respectively.

SOLUTION Under the stated conditions the air is assumed to behave as an ideal gas. The volume occupied by each pound of air is found from the ideal-gas equation.

$$v = \frac{RT}{P} = \frac{1545 \text{ ft·lb}_f}{(\text{lb·mol})(°R)} \times 540°R \times \frac{\text{ft}^2}{14.5(144) \text{ lb}_f} \times \frac{\text{lb·mol}}{29 \text{ lb}} = 13.78 \text{ ft}^3/\text{lb}$$

In the discussion preceding this example the air-fuel ratio was found to be 15.7 for the stoichiometric amount of air. Therefore the volume of air required per kilogram of fuel is

$$V = 0.872 \text{ m}^3/\text{kg air} \times 15.7 \text{ kg air/kg fuel} = 13.7 \text{ m}^3/\text{kg fuel}$$

On the basis of 1 kg · mol of fuel, the volume of air is

$$V = 13.7 \text{ m}^3/\text{kg fuel} \times 44 \text{ kg fuel/kg·mol fuel} = 603 \text{ m}^3/\text{kg·mol}$$

**Example 14-1** Determine the volume of air required to burn gaseous propane to completion, per pound and per pound-mole of fuel. The theoretical quantity of air is supplied and the temperature and pressure of the air are 80°F and 14.5 psia, respectively.

SOLUTION Under the stated conditions the air is assumed to behave as an ideal gas. The volume occupied by each pound of air is found from the ideal-gas equation.

$$v = \frac{RT}{P} = \frac{1545 \text{ ft·lb}_f}{(\text{lb·mol})(°R)} \times 540°R \times \frac{\text{ft}^2}{14.5(144) \text{ lb}_f} \times \frac{\text{lb·mol}}{29 \text{ lb}} = 13.78 \text{ ft}^3/\text{lb}$$

In the discussion preceding this example the air-fuel ratio was found to be 15.7 for the stoichiometric amount of air. Therefore the volume of air required per pound of fuel is

$$V = 13.78 \text{ ft}^3/\text{lb air} \times 15.7 \text{ lb air/lb fuel} = 216 \text{ ft}^3/\text{lb fuel}$$

On the basis of 1 lb·mol of fuel, the volume of air is

$$V = 216 \text{ ft}^3/\text{lb fuel} \times 44 \text{ lb fuel/lb·mol fuel} = 9500 \text{ ft}^3/\text{lb·mol}$$

**Example 14-2M** A gaseous fuel contains the following components on a volumetric or mole basis: hydrogen, 2 percent; methane, 10 percent: and nitrogen, 88 percent. Calculate the air-fuel ratio required, kg air/kg fuel, when the fuel is burned with 20 percent excess air.

SOLUTION The first step is to write the chemical reactions for the combustion of the fuel per mole of fuel. In terms of the theoretical or chemically correct oxygen requirements,

$$0.02H_2 + 0.01O_2 \rightarrow 0.02H_2O$$

$$0.10CH_4 + 0.20O_2 \rightarrow 0.10CO_2 + 0.20H_2O$$

As a result we find that 0.21 mole of oxygen are required per mole of fuel for complete combustion. The theoretical air-fuel ratio is

$$AF = \frac{0.21(32)}{0.02(2) + 0.10(16) + 0.88(28)} = \frac{6.72}{26.28} = 0.256$$

For 20 percent excess air the required value is

$$AF = 0.256(1.20) = 0.307 \text{ kg air/kg fuel}$$

This value is extremely low because most of the entering fuel stream is nitrogen, which is essentially an inert gas.

**Example 14-2** Read Example 14-2M. The air-fuel ratio found in the solution may be expressed as lb air/lb fuel as well as kg air/kg fuel.

In the incomplete combustion of carbon in a fuel, the carbon reacts according to the reaction $C + \frac{1}{2}O_2 \rightarrow CO$. Since oxygen has a greater affinity for combining with hydrogen than it does with carbon, all the hydrogen in a fuel normally is converted to water. If there is insufficient oxygen to assure complete combustion, it is always the carbon which is not completely reacted. In actual practice, there is usually CO in the products, even though an excess of oxygen was supplied. This may be attributed either to incomplete mixing during the process or to insufficient time for complete combustion. In addition, the nitrogen in the air and in the fuel partially reacts with oxygen to form oxides of nitrogen. While the quantity formed is small (usually less than 2000 parts per million), oxides of nitrogen are recognized as air pollutants.

The following example involves a combustion process carried out with a deficiency of air. The example illustrates the use of a basic tool in combustion analysis—the principle of the conservation of mass of each of the elements present in the overall reaction. By making appropriate mass balances on each element, we can systematically determine additional information on the initial reactant state or the state of the products. This technique will be used in subsequent examples as well.

**Example 14-3M** Propane gas is allowed to react with 80 percent theoretical air. Determine the theoretical equation for the reaction.

SOLUTION With a deficiency of air it is assumed that all the hydrogen is converted into water, but the carbon is converted into both CO and $CO_2$. The equation for 100 percent theoretical air has been shown to be

$$C_3H_8 + 5O_2 + 18.80N_2 \rightarrow 3CO_2 + 4H_2O + 18.80N_2$$

For 80 percent theoretical air we may write, in general, that

$$C_3H_8 + 4O_2 + 15.04N_2 \rightarrow aCO + bCO_2 + 4H_2O + 15.04N_2$$

The problem is to predict theoretically the quantities of CO and $CO_2$ in the products, as given by the unknowns $a$ and $b$. One of the basic premises for reacting systems of this sort is that all atomic species are conserved. Therefore we can write mass balances for each of the atomic species present in a reaction. Balances on hydrogen and nitrogen are not informative in this particular case. However, a carbon and an oxygen balance lead to the following equations:

C balance:  $\qquad\qquad\qquad\qquad\qquad$ $3 = a + b$

O balance:  $\qquad\qquad\qquad\qquad\qquad$ $8 = a + 2b + 4$

The solution to these two equations is that $a$ equals 2 and $b$ equals 1. Hence the correct theoretical equation for the combustion of propane with 80 percent theoretical air is

$$C_3H_8 + 4O_2 + 15.04N_2 \rightarrow 2CO + 1CO_2 + 4H_2O + 15.04N_2$$

The values of $a$ and $b$ would vary, depending upon the percent theoretical air used. This equation would not be applicable to the actual combustion process, since there may be $O_2$ in the products, as well as other species in small amounts. However, the chemical equation can be used as a first approximation of what should be expected in the way of products for this particular reaction under the given conditions.

**Example 14-3** Read Example 14-3M above.

Up to this point we have not designated the phases of the reactants and the products for a reaction, since we have been interested only in reviewing the chemistry of combustion processes and in introducing new terminology pertinent to the subject. Nevertheless, acknowledgment of the fact that different phases might be present during the reaction is important for several reasons. One of these is the problem of the dew point of the product gases. In the combustion of hydrocarbon fuels, one of the main products is water. In a mixture of ideal product gases, this water vapor has a certain partial pressure. If the partial pressure ever becomes greater than the saturation pressure of water at a given temperature, some of the water will condense out as the temperature is lowered further. The presence of liquid-water droplets in the combustion gases may lead to corrosion problems, for one thing. Consequently, it is useful to be able to predict the dew point of a given product gas. This requires a knowledge of the partial pressure of the water vapor in a gas, which in turn is a function of the mole fraction of the water vapor. The example below illustrates a typical calculation for the dew point of a combustion gas, when dry air is used.

**Example 14-4M** Gaseous propane is burned with 150 percent theoretical air at a pressure of 970 mbar. If the air is dry, determine the mole analysis of the product gas and the dew point of the gas mixture, in °C.

SOLUTION The chemical equation for the reaction of propane ($C_3H_8$) with 150 percent theoretical air has been given previously as

$$C_3H_8 + 7.5O_2 + 28.20N_2 \rightarrow 3CO_2 + 4H_2O + 2.5O_2 + 28.20N_2$$

The total moles of products is 37.7; therefore the molar analysis of the products is simply

$$\text{Mole fraction } CO_2 = \frac{3}{37.7} = 0.0796 (7.96 \text{ percent})$$

$$\text{Mole fraction } H_2O = \frac{4}{37.7} = 0.1061 \ (10.61 \text{ percent})$$

$$\text{Mole fraction } O_2 = \frac{2.5}{37.7} = 0.0663 \ (6.63 \text{ percent})$$

$$\text{Mole fraction } N_2 = \frac{28.20}{37.7} = 0.7480 \ (74.80 \text{ percent})$$

The partial pressure of the water vapor in the product gas is $0.1061(970) = 103$ mbar $= 0.103$ bar. From Table A-12M the saturation temperature corresponding to this pressure is, roughly, 46°C. When the gas mixture is cooled to 46°C at constant pressure, the dew point is reached. For temperatures below this value water will condense out of the product gas.

**Example 14-4** Gaseous propane is burned with 150 percent theoretical air at a pressure of 14.5 psia. If the air is dry, determine the mole analysis of the product gas and the dew point of the gas mixture, in °F.

SOLUTION The evaluation of the mole analysis is found in the solution to Example 14-4M, above. The mole fraction of water vapor is found to be 0.1061; therefore the partial pressure of the water vapor is $0.1061(14.5) = 1.54$ psia. From Table A-12 the saturation temperature corresponding to this pressure is, roughly, 117°F. When the gas mixture is cooled to 117°F at constant pressure, the dew point of the gas mixture is reached. For temperatures below this value water will condense out of the gaseous system.

## 14-2 ACTUAL COMBUSTION PROCESSES

In the preceding section the discussion and examples were based on the premise that complete information was available on the reactants entering into a combustion process. In addition, it was necessary to assume that, in the presence of excess air, all carbon within a fuel would be converted completely into carbon dioxide. However, it is a common experience based on measurements of product gases that carbon monoxide is frequently present in significant quantities, even if an excess of air has been used. As a further point, there are a number of applications for which actual measurement of the air-fuel ratio is difficult to ascertain. The flow of fuel, whether solid, liquid, or gas, into the reaction chamber is normally fairly well known; however, the air flow might be hard to measure accurately. To overcome these problems, an analysis of the gaseous products may be made with a good degree of accuracy. From this analysis a significant amount of information on the overall combustion process is learned.

There are numerous experimental methods that can be used to determine the concentration of various components in the actual gaseous products of combustion. The need for more accurate techniques increased with the passage of strict air-pollution emission standards by the state and federal governments, beginning in the late 1960s. Today there is a large number of manufacturers of suitable equipment. The analysis of combustion gases is usually reported on either a "dry" or "wet" basis. On a dry basis the percent water vapor in the gas stream is not reported. The well-established Orsat analyzer is a typical piece of equipment which reports the overall analysis on a dry basis. Although the mole fraction of water vapor in the original gas sample is not reported by such a measurement, this does not limit the usefulness of the technique.

Typical calculations based on representative "dry" analyses are given below. The general methodology is based on the use of mass balances on the atomic species present in the reacting mixture. This approach was introduced earlier, and illustrated in Example 14-3. We shall now make extensive use of the fact that atomic species are conserved in chemical reactions when nuclear effects are absent.

**Example 14-5M** Propane is reacted with air in such a ratio that an analysis of the products of combustion gives $CO_2$, 11.5 percent; $O_2$, 2.7 percent; and CO, 0.7 percent. What is the percent theoretical air used during the test?

SOLUTION The proper basis for beginning the analysis is 100 mol of dry product gases. From the analysis above the moles of $N_2$ in the dry products must be 85.1 mol. With this in mind, a chemical equation for the reaction might be

$$xC_3H_8 + aO_2 + 3.76aN_2 \rightarrow 11.5CO_2 + 0.7CO + 2.7O_2 + 85.1N_2 + bH_2O$$

A balance on each element enables us to evaluate each of the unknown coefficients. For nitrogen,

N$_2$ balance: $$3.76a = 85.1$$

$$a = 22.65$$

A carbon balance leads directly to the quantity of fuel used.

C balance:
$$3x = 11.5 + 0.7 = 12.2$$

$$x = 4.07$$

The only unknown at this point is the value of $b$, the moles of water in the products. However, two mass balances have not been used, namely, those for oxygen and hydrogen. The additional mass balance provides in this case, where the composition of the fuel is known, an independent check on the preceding calculations. From the oxygen balance,

O balance:
$$2(22.65) = 2(11.5) + 0.7 + 2(2.7) + b$$

$$b = 16.2$$

Then, employing a hydrogen balance, we check the computation.

$$8x = 2b$$

H balance:
$$8(4.07) = 2(16.2)$$

$$32.56 = 32.4$$

This check is fairly good, considering the limited significant figures used for the calculations. The actual reaction, then, is represented by the chemical equation

$$4.07C_3H_8 + 22.65O_2 + 85.1N_2 \longrightarrow 11.5CO_2 + 0.7CO + 2.7O_2 + 85.1N_2 + 16.2H_2O$$

By dividing through by 4.07, the equation could be placed on the basis of 1 mol of fuel. The stoichiometric equation for propane has been shown to be

$$C_3H_8 + 5O_2 + 18.80N_2 \longrightarrow 3CO_2 + 4H_2O + 18.80N_2$$

The moles of oxygen used theoretically per mole of fuel is 5. For the actual combustion, this ratio is $22.65/4.07 = 5.57$. Therefore the percent theoretical air used in the actual combustion process was

$$\text{Percent theoretical air} = \frac{5.57(100)}{5.0} = 111 \text{ percent}$$

or the percent excess air was 11 percent.

**Example 14-5** See Example 14-5M above.

**Examples 14-6M and 14-6** An unknown hydrocarbon fuel, $C_xH_y$, was allowed to react with air. An Orsat analysis was made of a representative sample of the product gases, with the following result: $CO_2$, 12.1 percent; $O_2$, 3.8 percent; and CO, 0.9 percent. Determine (a) the chemical equation for the actual reaction, (b) the composition of the fuel, (c) the air-fuel ratio used during the text, and (d) the excess or deficiency of air used.

SOLUTION  The percentages of the three gases in the Orsat analysis added up to 16.8 percent. If it is assumed that the remaining gas in the sample is nitrogen, the volumetric percent of nitrogen must be 83.2 percent. With this information we may now write a general chemical equation for the reaction in the form

$$C_xH_y + aO_2 + 3.76aN_2 \longrightarrow 12.1CO_2 + 3.8O_2 + 0.9CO + 83.2N_2 + bH_2O$$

The equation actually reverses the technique used in the preceding section. Basically, we ask ourselves what initial composition of fuel and what air-fuel ratio would be required to produce certain known percentages of product gases. The calculation is based on 100 mol of dry product gases as the starting point. The unknown values of $x$ and $y$ determine the fuel composition, and the value of $a$ will establish the air-fuel ratio. In addition, the value of $b$, the moles of water vapor formed, will need to be found if the dew point is desired. A nitrogen balance determines the value of $a$.

$N_2$ balance:

$$3.76a = 83.2$$

$$a = 22.1$$

An oxygen balance enables one to determine the moles of water formed during the reaction.

O balance:

$$2(22.1) = 2(12.1) + 2(3.8) + 0.9 + b$$

$$b = 11.5$$

Now carbon and hydrogen balances may be used to find the values of $x$ and $y$.

C balance:

$$x = 12.1 + 0.9 = 13.0$$

H balance:

$$y = 2b = 2(11.5) = 23.0$$

With a knowledge of the values of the initially unknown quantities, the chemical equation may now be written as

$$C_{13}H_{23} + 22.1O_2 + 83.2N_2 \longrightarrow 12.1CO_2 + 3.8O_2 + 0.9CO + 83.2N_2 + 11.5H_2O$$

The fact that the values of $x$ and $y$ came out to be whole numbers has no significance, since 3-figure accuracy is involved. Usually, the values of $x$ and $y$ for similar problems of this type will not be whole numbers. Also, the formula $C_{13}H_{23}$ should not be thought of as belonging to a single chemical species, but as the average formula for a fuel which probably contains a large number of different compounds.

From the balanced chemical equation above, the air-fuel ratio can be computed. Since 105.3 mol of air were used per mole of fuel, then

$$AF = \frac{105.3 \text{ mol air}}{1 \text{ mol fuel}} \times \frac{29 \text{ lb air}}{1 \text{ mol air}} \times \frac{\text{mol fuel}}{179 \text{ lb fuel}} = 17.1 \text{ lb air/lb fuel (or kg air/kg fuel)}$$

Finally, the excess or deficiency of air can be found by first computing the theoretical AF value for this fuel. The chemical equation for combustion of $C_{13}H_{23}$ with the theoretical-air requirements is

$$C_{13}H_{23} + 18.75O_2 + 3.76(18.75)N_2 \longrightarrow 13CO_2 + 11.5H_2O + 18.75(3.76)N_2$$

Since 18.75 mol of $O_2$ are required and 22.1 mol of $O_2$ were actually used, the percent excess air used during the test is

$$\text{Percent excess} = \frac{22.1 - 18.75}{18.75}(100) = 18 \text{ percent}$$

The preceding discussion has dealt with the chemistry of reactive systems in terms of theoretical and actual considerations. Terminology used in the engineering studies of reactive systems has been reviewed. The presentation has been based primarily on the principle of conservation of atomic species during chemical reactions. We are now in a position to consider reactive systems in the light of the conservation of energy principle.

## 14-3 STEADY-FLOW ENERGY ANALYSIS OF REACTING MIXTURES

The application of the conservation of energy principle to reacting systems is merely an extension of the ideas presented earlier in the text for systems of fixed composition. For a simple, compressible closed system the basic relationship is

$$Q + W = \Delta U \tag{3-4}$$

Engineering applications of reacting systems are generally directed toward steady-state, steady-flow processes. In the absence of magnetic and electric fields, the conservation of energy principle reduces to the familiar equation

$$Q + W_{\text{shaft}} = \Delta H + \Delta KE + \Delta PE \tag{14-2}$$

For a chemically reactive system the $\Delta H$ term in this latter expression may be written as

$$\Delta H = H_{\text{prod}} - H_{\text{reac}}$$
$$= \sum_i (N_i h_i)_{\text{prod}} - \sum_i (N_i h_i)_{\text{reac}} \tag{14-3}$$

where $h_i$ is the molar enthalpy of any product or reactant at the temperature and pressure of the reaction, and $N_i$ is the moles of any product or reactant. An analogous equation is valid for a constant-volume process for a closed system, where all the enthalpy symbols in Eq. (14-3) are replaced by the internal energy.

The evaluation of the $h_i$ quantities in Eq. (14-3) introduces a difficulty unique to reactive systems. In Chap. 3 and 4 methods for evaluating the enthalpy of an ideal or real gas were introduced. A common way of presenting such data is in tabular format. The Appendix contains a number of tables of thermodynamic data for ideal and real gases. Recall, however, that the values of $h$ in these tables are dependent upon the choice of a reference state. The ideal-gas data, for example, are based arbitrarily on a zero value of the enthalpy at absolute zero of temperature. The steam data, on the other hand, are based on a zero reference value for the saturated liquid at the triple state. In the literature we may find tables of these same substances with reference states different from those discussed above, since it is merely an arbitrary choice of the author of a given table. Since reference states are completely arbitrary, different values of $\Delta H$ in Eq. (14-3) will result when tables based on different reference states are employed. We must seek another technique for evaluating the enthalpy function which avoids this difficulty.

The tables of $h$ data discussed above still are useful if we correctly account for another major contribution to the enthalpy of any substance when considering chemical reactions. Energy is associated with forces which bind atoms together into a specific compound. As a result, each reactant and product of a chemical reaction has a definite "bond" energy at a given temperature and pressure. When reactants form products, certain bonds are disrupted and others are formed. The sum of the bond energies of the reactants may be quite different from that of the products of a reaction. This difference in bond energy must be strictly accounted for in an energy balance on the reacting system. In general, the energy associated with the products may be more or less than the energy of the reactants, depending upon (1) the chemical nature of the reactants and products, and (2) the state of the reactants and products.

A consistent accounting of this change in "chemical" energy is accomplished by introducing the concept of the enthalpy of formation $\Delta h_f$ of a substance. The *enthalpy of formation* is defined as the energy released or absorbed when a chemical compound is formed from its elements. In addition, the temperature and pressure of the compound formed by the reaction must be the same as that of the initial reactants (elements). On the basis of Eq. (14-3) this formation process is denoted symbolically by

$$\Delta h_f = h_{\text{compound}} - \sum_i (v_i h_i)_{\text{elements}} \tag{14-4}$$

where $v_i$ again is the stoichiometric coefficient for a given element. If we rearrange this equation, then

$$h_{\text{compound}} = \Delta h_f + \sum_i (v_i h_i)_{\text{elements}} \tag{14-5}$$

Hence the concept of the enthalpy of formation enables us to evaluate the enthalpy of any specific compound in terms of the quantities on the right-hand side of Eq. (14-5).

The actual numerical evaluation of Eq. (14-5) requires the following three considerations:

1. The enthalpy of formation $\Delta h_f$ is determined by laboratory measurement or by statistical (microscopic) evaluation. The methods used are not important to the following discussion. However, it should be noted that values of the enthalpy of formation are available for hundreds of substances. The enthalpy of formation for a number of common substances is given in Tables A-22M and A-22 in the Appendix. These data are reported at 1 atm and 25°C (77°F). The sign convention for $\Delta h_f$ is the same as that for $Q$ and $W$. When energy is released as a compound is formed from its stable elements, the value of $\Delta h_f$ is negative.

2. The term "elements" must be strictly defined. The reactants in this case are taken arbitrarily as the *stable* form of the elements at the specified state. For example, the stable forms of elements such as hydrogen, nitrogen, and oxygen at 1 atm and room temperature are $H_2(g)$, $N_2(g)$, and $O_2(g)$. On the other hand, the stable form of carbon under these same conditions is solid graphite, $C(s)$, and not diamond.

3. The enthalpy of all stable elements at 1 atm and 25°C (77°F) is assigned a value of zero. It is this choice which establishes a consistent and convenient basis for measuring energy transformations in chemically reactive systems.

The state of 25°C (298.15°K) or 77°F (536.7°R) and 1 atm is the *standard reference* state for thermochemical calculations. Properties at this state are frequently symbolized by the superscript °. Hence the values of the enthalpy of formation in Tables A-22M and A-22 are titled as $\Delta h_f^\circ$. If we now apply the three considerations discussed above to Eq. (14-5), then for a formation reaction at 25°C and 1 atm the equation becomes

$$h_{\text{compound}}(\text{at } 25°\text{C and 1 atm}) = \Delta h_f^\circ \qquad (14\text{-}6)$$

Hence the enthalpy of any compound at 298°K and 1 atm is equal to its enthalpy of formation at that state, while the enthalpy of stable elements at this same state is zero.

The final step is to evaluate the enthalpy at a specified temperature and pressure which is different from the standard reference state. To accomplish this, we must add to the value given by Eq. (14-6) the change in enthalpy between the reference state of 25°C and 1 atm and the specified state. This contribution is simply that found from conventional ideal-gas tables and steam tables, for example. If we restrict ourselves to reacting ideal-gas mixtures, then the enthalpy of each gas is independent of pressure. In this case the total enthalpy $h_{i,T}$ of an ideal gas at temperature $T$ is given by

$$h_{i,T} = \Delta h_f^\circ + (h_T - h_{298})_i \qquad (14\text{-}7)$$

where $h_T$ is the enthalpy at the specified temperature $T$ and $h_{298}$ is the enthalpy at the reference temperature of 298°K or 537°R. If tabular data are not available to evaluate the last term in Eq. (14-7), the enthalpy change $h_T - h_{298}$ must be computed from the integral of $c_p\, dT$. Thus the enthalpy of any compound is composed of two parts, that associated with its formation from elements at a given temperature and pressure and that associated with a change of state at constant composition. It is convenient to call the second term on the right of Eq. (14-7) the *sensible* enthalpy change of a substance.

If we now combine Eqs. (14-2), (14-3), and (14-7), the steady-state, steady-flow energy balance for processes which include chemical reactions becomes

$$Q + W_{\text{shaft}} = \sum_{\text{prod}} N_i(\Delta h_f^\circ + h_T - h_{298} + \text{KE})_i - \sum_{\text{reac}} N_i(\Delta h_f^\circ + h_T - h_{298} + \text{KE})_i$$
$$(14\text{-}8)$$

The potential energy change of the flow stream has been neglected in the above expression. In many combustion processes there is no shaft work for the selected control volume and the change in kinetic energy frequently is negligible. Under these additional constraints the energy balance reduces to

$$Q = \sum_{\text{prod}} N_i(\Delta h_f^\circ + h_T - h_{298})_i - \sum_{\text{reac}} N_i(\Delta h_f^\circ + h_T - h_{298})_i \qquad (14\text{-}9)$$

The value of $Q$, of course, may be positive or negative. Chemical reactions which liberate energy in the form of heat are called exothermic, and those which absorb energy are known as endothermic.

**Example 14-7M** In Section 14-1 it was shown that the combustion of methane with the stoichiometric amount of oxygen is given by the chemical equation

$$CH_4(g) + 2O_2(g) \longrightarrow CO_2(g) + 2H_2O(g)$$

where the symbol $(g)$ after the chemical formula indicates the gas phase. Determine the heat released or absorbed if this reaction occurs at 1 atm and 25°C.

SOLUTION Equation (14-9) is applicable to this problem, in conjunction with data from Table A-22M. In this particular case the sensible enthalpy change $h_T - h_{298}$ is zero for reactants and products, since both the initial and final temperatures are the standard reference temperature. Substitution of appropriate data yields

$$Q = \sum_{\text{prod}} N_i \, \Delta h_{f,i}^\circ - \sum_{\text{reac}} N_i \, \Delta h_{f,i}^\circ$$

$$= 1(-393,520) + 2(-241,820) - 1(-74,870) - 2(0)$$

$$= -802,290 \text{ kJ/kg·mol } CH_4$$

It should be noted that the above answer is independent of the amount of oxidant supplied to the reaction. If air is supplied, even in an excess amount, the oxygen and nitrogen enter and leave at the standard reference temperature. Hence the value of $h_T - h_{298}$ is zero for these elements, regardless of the quantity of each.

**Example 14-7** In Sec. 14-1 it was shown that the combustion of methane with the stoichiometric amount of oxygen is given by the chemical equation

$$CH_4(g) + 2O_2(g) \longrightarrow CO_2(g) + 2H_2O(g)$$

where the symbol $(g)$ after the chemical formula indicates the gas phase. Determine the heat released or absorbed if this reaction occurs at 1 atm and 77°F.

SOLUTION Equation (14-9) is applicable to this problem, in conjunction with data from Table A-22. In this particular case the sensible enthalpy change $h_T - h_{537}$ is zero for reactants and products, since both the initial and final temperatures are the standard reference temperature. Substitution of appropriate data yields

$$Q = \sum_{\text{prod}} N_i \, \Delta h_{f,i}^\circ - \sum_{\text{reac}} N_i \, \Delta h_{f,i}^\circ$$

$$= 1(-169,300) + 2(-104,040) - 1(-32,210) - 2(0)$$

$$= -345,170 \text{ Btu/lb·mol } CH_4$$

It should be noted that the above answer is independent of the amount of oxidant supplied to the reaction. If air is supplied, even in an excess amount, the oxygen and nitrogen enter and leave at the standard reference temperature. Hence the value of $h_T - h_{537}$ is zero for these elements, regardless of the quantity of each.

**Example 14-8M** Methane gas initially at 400°K is burned with 50 percent excess air which enters the combustion chamber at 500°K. The reaction, which occurs at 1 atm, goes to completion and the temperature of the product gases is 1800°K. Determine the heat transfer to or from the combustion chamber, in kJ/kg·mol of fuel.

SOLUTION The chemical equation for the complete combustion of methane with 50 percent excess air is

$$CH_4(g) + 3O_2(g) + 3(3.76)N_2(g) \longrightarrow CO_2(g) + 2H_2O(g) + 11.28N_2(g) + O_2(g)$$

The state of the water in the products is indicated as a gas, since the final temperature is well above the dew point. Also, since the partial pressure of the water vapor is only 132 mbar, the vapor may be assumed to be an ideal gas along with the other product gases. The enthalpy data for all the gases except methane are obtained from Tables A-6M through A-10M, and the enthalpy of formation data are listed in Table A-22M. No tabular data are available in this text that permit the calculation of the enthalpy change for methane from 298 to 400°K. This quantity must be determined by integrating $c_p$ data. One equation for $c_p$ of methane is

$$c_p(CH_4) = (1.702 + 9.081 \times 10^{-3}T - 2.164 \times 10^{-6}T^2)R_u$$

where $T$ is expressed in °K. The appropriate value of $R_u$ is 8.315 kJ/(kg·mol)(°K). Therefore the enthalpy change is given by

$$\Delta h(CH_4) = 8.315 \int_{298}^{400} (1.702 + 9.081 \times 10^{-3}T - 2.164 \times 10^{-6}T^2)\, dT$$

$$= 8.315\Big\{1.702(400 - 298) + \frac{9.081 \times 10^{-3}}{2}[(400)^2 - (298)^2]$$

$$- \frac{2.164 \times 10^{-6}}{3}[(400)^3 - (298)^3]\Big\}$$

$$= 8.315(174 + 323 - 27) = 3910 \text{ kJ/kg·mol } CH_4$$

Substitution of the above value and data from tables in the Appendix into Eq. (14-9) yields

$$Q = 1(-393,520 + 88,806 - 9364) + 2(-241,820 + 72,513 - 9904) + 11.28(0 + 57,651 - 8669)$$

$$+ 1(0 + 60,371 - 8682) - 1(-74,870 + 3910) - 3(0 + 14,770 - 8682)$$

$$- 11.28(0 + 14,581 - 8669) = -82,290 \text{ kJ/kg·mol } CH_4$$

In Example 14-7M for the stoichiometric combustion of methane gas at 25°C it was determined that the heat released is 802,290 kJ/kg·mol. When excess air is present and the products are heated to 1800°K, as in this example, roughly 90 percent of the energy released at 25°C is used to heat the products to 1800°K.

**Example 14-8** Methane gas initially at 700°R is burned with 50 percent excess air which enters the combustion chamber at 900°R. The reaction, which occurs at 1 atm, goes to completion and the temperature of the product gases in 3300°R. Determine the heat transfer to or from the combustion chamber in Btu/lb·mol of fuel.

SOLUTION The chemical equation for the complete combustion of methane with 50 percent excess air is

$$CH_4(g) + 3O_2(g) + 3(3.76)N_2(g) \longrightarrow CO_2(g) + 2H_2O(g) + 11.28N_2(g) + O_2(g)$$

The state of the water in the products is indicated as a gas, since the final temperature is well above the dew point. Also, since the partial pressure of the water vapor is only 1.92 psia, the water vapor may be assumed to be an ideal gas along with the other product gases. The enthalpy data for all the gases except methane are obtained from Tables A-6 through A-10, and the enthalpy of formation data are listed in Table A-22. No tabular data are available in this text that permit the calculation of the enthalpy change for methane from 537 to 700°R. This quantity must be determined by integrating $c_p\, dT$. The equation for $c_p$ is

$$c_p(CH_4) = 3.381 + 18.04 \times 10^{-3}T - 4.30 \times 10^{-6}T^2$$

where $T$ is expressed in °R and $c_p$ is in Btu/(lb·mol)(°R). The enthalpy change for methane becomes

$$\Delta h(CH_4) = \int_{537}^{700} (3.381 + 18.04 \times 10^{-3}T - 4.30 \times 10^{-6}T^2) \, dT$$

$$= 3.381(700 - 537) + \frac{18.04 \times 10^{-3}}{2}[(700)^2 - (537)^2]$$

$$- \frac{4.30 \times 10^{-6}}{3}[(700)^3 - (537)^3]$$

$$= 551 + 1819 - 270 = 2100 \text{ Btu/lb·mol } CH_4$$

Substitution of the above value and data from tables in the Appendix into Eq. (14-9) yields

$$Q = 1(-169{,}300 + 39{,}087 - 4028) + 2(-104{,}040 + 31{,}918 - 4258)$$

$$+ 11.28(0 + 25{,}306 - 3730) + 1(0 + 26{,}412 - 3725) - 1(-32{,}210 + 2100)$$

$$- 3(0 + 6338 - 3725) - 11.28(0 + 6268 - 3730) = -27{,}290 \text{ Btu/lb·mol } CH_4$$

In Example 14-7 for the stoichiometric combustion of methane gas at 537°R the heat released was found to be 345,170 Btu/lb·mol of methane. When excess air is present and the products are heated to 3300°R, as in this example, more than 90 percent of the energy released at 537°R is used to heat the products to 3300°R.

Before proceeding with thermochemical calculations, it is useful to introduce some additional concepts and notation commonly found in the literature.

## 14-4 ENTHALPY OF REACTION AND HEATING VALUES

The enthalpy of formation introduced in the preceding section applies to a restricted type of chemical reaction—the formation of a compound from its elements. More generally, the enthalpy change associated with any chemical reaction for which the products are returned to the original temperature of the reactants is called the *enthalpy of reaction*, $\Delta h_R$. Since in a closed system held at constant pressure $Q = \Delta H$, the enthalpy of reaction is frequently called the heat of reaction at constant pressure. A graphical representation of the enthalpy of reaction as a function of temperature for an exothermic reaction is shown in Fig. 14-1. Although the value of $H$ for the sum of the reactant or product species is completely arbitrary (as given by states $a$ and $b$ in the figure), the vertical distance between the lines must equal the enthalpy of reaction at that particular temperature and the specified pressure. The vertical distance between the lines is not constant, since the enthalpy of reaction is a function of temperature for any given pressure. Since the line $ab$ in Fig. 14-1 is drawn at 25°C, the enthalpy of reaction shown is the standard-state value $\Delta h_R^\circ$, if the pressure is 1 atm. If the reactants in the figure are stable elements and the product is a single compound formed from these elements, the vertical distance $ab$ in this case at 25°C and 1 atm would represent $\Delta h_f^\circ$.

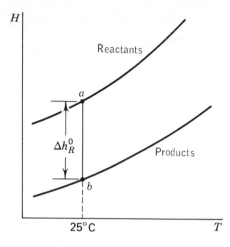

Figure 14-1 The enthalpy of reaction as a function of temperature for an exothermic reaction.

The enthalpy of reaction for any reaction is uniquely related to enthalpy of formation data for the reactants and products. On the basis of Eq. (14-3),

$$\Delta H = \sum_i (N_i h_i)_{prod} - \sum_i (N_i h_i)_{reac} \qquad (14\text{-}3)$$

However, for any reaction which occurs at 25°C and 1 atm, Eq. (14-7) dictates that for both reactants and products

$$h_{i,298} = \Delta h_{f,298} + (h_{298} - h_{298})_i = \Delta h_{f,298}$$

Consequently, for a reaction occurring at the standard reference state of 25°C and 1 atm we find that

$$\Delta h_R^\circ = \sum_i (N_i \, \Delta h_f^\circ)_{prod} - \sum_i (N_i \, \Delta h_f^\circ)_{reac} \qquad (14\text{-}10)$$

Thus the enthalpy of reaction in the standard reference state is solely a function of the enthalpies of formation of the reactants and products in that state. In Examples 14-7M and 14-7 we determined the heat released by the stoichiometric and complete combustion of methane with oxygen at 25°C (or 77°F) and 1 atm. It is now seen that the solution follows the format of Eq. (14-10), since the problem reduced to the use of enthalpy of formation data for the reactants and products. Hence the heat released in that case is also the enthalpy of reaction for the gas-phase combustion process.

**Example 14-9M** The reaction of carbon monoxide with steam in the gas phase is known as the water-gas reaction. The overall reaction is

$$CO(g) + H_2O(g) \longrightarrow CO_2(g) + H_2(g)$$

Determine the enthalpy of reaction at 25°C and 1 atm.

SOLUTION The enthalpy of reaction can be found from Eq. (14-10). In terms of specific components,

$$\Delta h_R^\circ = 1(\Delta h_f^\circ)_{CO_2} + 1(\Delta h_f^\circ)_{H_2} - 1(\Delta h_f^\circ)_{CO} - 1(\Delta h_f^\circ)_{H_2O}$$

The enthalpy of formation data are found in Table A-22M. Hence,

$$\Delta h_R^\circ = 1(-393,520) + 1(0) - 1(-110,530) - 1(-241,830)$$

$$= -41,160 \text{ kJ/kg·mol CO}$$

The quantity of energy released per mole of reactant is relatively small in this case. Compare this answer to the result of Example 14-7M.

**Example 14-9** Read Example 14-9M; however, replace the calculation with data at 77°F in USCS units.

SOLUTION Using data from Table A-22, the enthalpy of reaction for the water-gas reaction becomes

$$\Delta h_R^\circ = 1(-169,300) + 1(0) - 1(-47,550) - 1(-104,040)$$

$$= -17,710 \text{ Btu/lb·mol CO}$$

The quantity of energy released per mole of reactant in this case is relatively small. Compare this answer with the result of Example 14-7.

A common use of the enthalpy of reaction in thermochemical calculations is found by collecting similar terms in Eq. (14-9). This equation may be written in the form

$$Q = \sum_{\text{prod}} N_i \, \Delta h_f^\circ - \sum_{\text{reac}} N_i \, \Delta h_f^\circ + \sum_{\text{prod}} N_i(h_T - h_{298})_i - \sum_{\text{reac}} N_i(h_T - h_{298})_i$$

However, according to Eq. (14-10), the sum of the first two terms on the right-hand side is equal to $\Delta h_R^\circ$. Therefore, Eq. (14-10) can also be expressed as

$$Q = \Delta h_R^\circ + \sum_{\text{prod}} N_i(h_T - h_{298})_i - \sum_{\text{reac}} N_i(h_T - h_{298})_i \qquad (14\text{-}11)$$

This equation is only useful if the enthalpy of reaction for the desired reaction is available in a table. If it is not, then Eq. (14-9) must be used, since enthalpy of formation data are readily available in the literature. The following example illustrates how the enthalpy of reaction leads to a useful graphical interpretation of thermochemical calculations.

**Example 14-10M** Carbon monoxide and the 300 percent theoretical air requirement enter a steady-flow combustor at 400°K and a low pressure. The energy released by the reaction heats the product gases to 1400°K. If the combustion process is complete, determine the heat gained or lost through the walls of the combustor.

SOLUTION The balanced chemical reaction for complete combustion of the specified reactants is

$$CO(g) + 1.5O_2(g) + 5.64N_2(g) \longrightarrow CO_2(g) + O_2(g) + 5.64N_2(g)$$

If we neglect the kinetic-energy change, the quantity of heat transferred is given by $\Delta H$. A graphical interpretation of this enthalpy change is shown in Fig. 14-2. The variation of the composite sensible enthalpy of the reactants as a function of temperature is shown by the upper line. The initial state of 400°K is designated by state 1. Likewise, the variation of the composite enthalpies of the product gases with temperature is shown as the lower line. The final state of

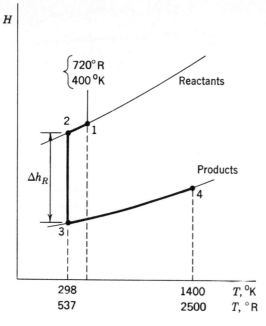

**Figure 14-2** Process diagram for Examples 14-10M and 14-10.

$1400°K$ is marked as state 4. The quantity $Q$ equals $H_4 - H_1$. Since the enthalpy is a property, we may state that

$$Q = \Delta H = (H_2 - H_1) + (H_3 - H_2) + (H_4 - H_3)$$

The quantity $H_2 - H_1$ is the sensible enthalpy change of the reactants from 400 to 298°K, $H_3 - H_2$ is the enthalpy of reaction at 298°K, and $H_4 - H_3$ is the sensible enthalpy change of the products from 298 to 1400°K.
Consequently we can write the energy equation as

$$Q = \sum_{reac} N_i(h_{298} - h_{400}) + \Delta h_{R,\,298} + \sum_{prod} N_i(h_{1400} - h_{298})$$

This expression has the same format as Eq. (14-11). The first quantity is evaluated from Tables A-6M, A-7M, and A-8M.

$$\sum_{reac} = 1(8669 - 11,644) + 1.5(8682 - 11,711) + 5.64(8669 - 11,640)$$

$$= -24,275 \text{ kJ}$$

The enthalpy of reaction is found by substituting data from Table A-22M into Eq. (14-10).

$$\Delta h_{R,\,298} = 1(-393,520) + 1(0) + 5.64(0) - 1(-110,530) - 1.5(0) - 5.64(0)$$

$$= -282,990 \text{ kJ}$$

Finally, the third quantity is evaluated from Tables A-6M, A-7M, and A-9M.

$$\sum_{prod} = 1(65,271 - 9,364) + 1(45,648 - 8,682) + 5.64(43,605 - 8,669)$$

$$= 289,910 \text{ kJ}$$

On the basis of 1 mole of CO, the energy equation becomes $q = -24,275 + (-282,990) + 289,910 - -17,360$ kJ/kg·mol CO Thus heat must be removed so that the product gases may reach 1400°K. This problem could also have been solved directly by using Eq. (14-9). However, the use of Eq. (14-11) lends itself to a graphical interpretation which shows specifically the various contributions to the overall enthalpy change.

**Example 14-10** Carbon monoxide and the 300 percent theoretical air requirement enter a steady-flow combustion chamber at 260°F and a low pressure. The energy released by the reaction heats the product gases to 2040°F. If the combustion process is complete, determine the heat gained or lost through the walls of the combustor, in Btu/lb·mol of CO.

SOLUTION The balanced chemical equation for the complete combustion of the specified reactants is

$$CO_2(g) + 1.5O_2(g) + 5.64N_2(g) \longrightarrow CO_2(g) + O_2(g) + 5.64N_2(g)$$

If we neglect any kinetic-energy change, the quantity of heat transferred is given by $\Delta H$. A graphical interpretation of the overall enthalpy change is shown in Fig. 14-2. The variation of the composite sensible enthalpy of the reactants as a function of temperature is shown by the upper line. The initial state of 720°R (260°F) is designated by state 1. Likewise, the variation of the composite enthalpies of the product gases with temperature is shown as the lower line. The final state of 2500°R is marked as state 4. The quantity $Q$ equals $H_4 - H_1$. Since enthalpy is a property, we may state that

$$Q = \Delta H = (H_2 - H_1) + (H_3 - H_2) + (H_4 - H_3)$$

The quantity $H_2 - H_1$ is the sensible enthalpy change of the reactants from 720 to 537°R, $H_3 - H_2$ is the enthalpy of reaction at 537°R, and $H_4 - H_3$ is the sensible enthalpy change of the products from 537 to 2500°R. Consequently the energy equation takes on the format of Eq. (14-11), namely,

$$Q = \sum_{\text{reac}} N_i(h_{537} - h_{720}) + \Delta h_{R,\,537} + \sum_{\text{prod}} N_i(h_{2520} - h_{537})$$

The first term is evaluated from Tables A-6, A-7, and A-8.

$$\sum_{\text{reac}} = 1(3725 - 5006) + 1.5(3725 - 5023) + 5.64(3730 - 5005)$$

$$= -10,420 \text{ Btu}$$

The enthalpy of reaction is found by substituting data from Table A-22 into Eq. (14-10).

$$\Delta h_{R,\,537} = 1(-169,300) + 1(0) + 5.64(0) - 1(-47,540) - 1.5(0) - 5.64(0)$$

$$= -121,760 \text{ Btu}$$

Finally, the third quantity is evaluated from data in Tables A-6, A-7, and A-9.

$$\sum_{\text{prod}} = 1(27,801 - 4028) + 1(19,443 - 3725) + 5.64(18,590 - 3730)$$

$$= 123,300 \text{ Btu}$$

For 1 mole of CO, the energy equation becomes

$$q = -10,420 + (-121,760) + 123,300 = -8880 \text{ Btu/lb·mol CO}$$

Thus heat is removed so that the product gases may reach 2500°R (2040°F). This problem also could be solved directly by using Eq. (14-9). However, the use of Eq. (14-11) lends itself to a graphical interpretation which shows specifically the various contributions to the overall enthalpy change.

Many engineering applications of chemical reactions involve the combustion of hydrocarbon fuels with oxygen or air. The enthalpy of reaction for a combustion process is frequently called the *enthalpy of combustion,* $\Delta h_c$. Typical values of the enthalpy of combustion are found in Tables A-23M and A-23. Such data are quite useful when using the steady-flow energy balance in the format of Eq. (14-11). The enthalpy of combustion of a fuel is also referred to as the heating value of the fuel. The heating value is defined as energy released from the combustion process at a given temperature and is always positive in value. Therefore the value of $\Delta h_c$ and the heating value for a given fuel are equal in magnitude but opposite in sign.

An important item to consider in the analysis of combustion reactions is the phase of each reactant and product. This is especially true of the fuel used and of the water formed by the reaction. Phase changes require a considerable amount of energy, and this must be accounted for in an energy balance. When water in the product gases is in the vapor state, the energy released is called the *lower heating value.* Normally water is in the vapor state if the temperature of the product gas is reasonably high. In this case the partial pressure of the water will be less than the saturation pressure for that temperature. Thus the dew-point temperature of the product gases, discussed in Sec. 14-1, becomes an important parameter when applying the first law to reacting systems. As the temperature of a mixture of product gases is lowered below the dew point, most of the water will condense. The process of condensation results in an additional release of energy. This enthalpy change due to condensation is accounted for by specifying the *higher heating value* of the fuel, which is based on liquid water in the products of combustion. The example below illustrates that there can be a significant difference between the lower and higher heating values of a given fuel. Thus the notation of whether $H_2O(l)$ or $H_2O(g)$ appears in a chemical equation is quite important from an energy standpoint.

The phase of the fuel used in the combustion process must also be strictly observed in the overall energy released by a reaction is to be correctly evaluated. When a liquid hydrocarbon fuel is used, the energy released is less than for a gaseous fuel, since energy must be supplied to vaporize the fuel. However, the effect of the phase of the fuel on the enthalpy of combustion is usually much less than the phase in which water appears (i.e., as a liquid or vapor) in the product gases.

**Example 14-11M** In Example 14-7M it was noted that the enthalpy of reaction (or combustion) of gaseous methane at 25°C is $-802{,}290$ kJ/kg·mol when water appears in the products as a gas. From a knowledge of the lower heating value, determine the higher heating value at the same temperature.

SOLUTION From Table A-12M the enthalpy of vaporization $h_{fg}$ of water is 2442.2 kJ/kg or 44,010 kJ/kg·mol of water formed at 25°C. The combustion reaction for methane shows 2 mol of water formed per mole of methane. Hence the enthalpy of reaction (or combustion) corresponding to the higher heating value of 25°C and 1 atm is

$$\Delta h_{c,\,higher} = \Delta h_{c,\,lower} - 2h_{fg}$$

$$= -802{,}290 - 2(44{,}010)$$

$$= -890{,}310 \text{ kJ/kg·mol CH}_4$$

In this case the higher heating value is approximately 11 percent higher than the lower heating value. This indicates the significant difference that may exist between higher and lower heating values.

**Example 14-11** In Example 14-7 it was noted that the enthalpy of reaction (or combustion) of gaseous methane at 77°F is $-345,170$ Btu/lb·mol when water appears in the products as a gas. From a knowledge of the lower heating value, determine the higher heating value at the same temperature.

SOLUTION From Table A-12 the enthalpy of vaporization $h_{fg}$ of water is 1050 Btu/lb or 18,920 Btu/lb·mol of water formed at 25°C. The combustion reaction for methane shows 2 mol of water formed per mole of methane. Hence the enthalpy of reaction (or combustion) corresponding to the higher heating value at 25°C and 1 atm is

$$\Delta h_{c,\text{higher}} = \Delta h_{c,\text{lower}} - 2h_{fg}$$

$$= -345,170 - 2(18,920)$$

$$= -383,010 \text{ Btu/lb·mol } CH_4$$

In this case the higher heating value is approximately 11 percent higher than the lower heating value. This indicates the significant difference that may exist between higher and lower heating values.

**Example 14-12M** Liquid butane ($C_4H_{10}$) at 25°C is sprayed into a combustion chamber. In addition, 400 percent theoretical air is supplied at an inlet temperature of 600°K. The gaseous products of combustion leave the chamber at 1100°K. If complete combustion is assumed, because of the large excess of air, find how much heat, per kilogram-mole of fuel, must be removed from the gases in the combustion chamber.

SOLUTION For the fuel under consideration the stoichiometric chemical equation is

$$C_4H_{10}(l) + 6.5O_2(g) + 24.2N_2(g) \longrightarrow 4CO_2(g) + 5H_2O(g) + 24.4N_2(g)$$

For 400 percent theoretical air the equation becomes

$$C_4H_{10}(l) + 26.0O_2(g) + 97.6N_2(g) \longrightarrow 4CO_2(g) + 5H_2O(g) + 19.5O_2(g) + 97.6N_2(g)$$

The heat transfer is found through use of Eq. (14-11), namely,

$$Q = \sum_{\text{prod}} N_i(h_T - h_{298})_i + \Delta h_R{}^\circ - \sum_{\text{reac}} N_i(h_T - h_{298})_i$$

The terms in this equation are illustrated in Fig. 14-3 on an enthalpy-temperature diagram. The vertical distance between states 1 and 2 represents the sensible enthalpy change of the reactants, as given by the last term of the energy balance. In this particular example, the only reactant at an elevated temperature is the 400 percent theoretical air. The enthalpy of reaction (or combustion) at 298°K is the distance between states 2 and 3. Finally, the sensible enthalpy change for the products is given by the vertical distance between states 3 and 4.

The enthalpy of reaction at 25°C is obtained from Table A-23M, under the listing of the enthalpy of combustion. However, the tabulated value of $-2,877,100$ kJ/kg·mol is for $n$-butane in the gaseous state, while this problem involves liquid $n$-butane. To correct the above value, we need the enthalpy of vaporization also shown in the table. Since the fuel must be vaporized, the energy released will be less than the above value. The enthalpy of vaporization is given as 21,060 kJ/kg·mol. Therefore the $\Delta h_c$ value we should use is $-2,856,040$ kJ/kg·mol. The difference in this case is slight, but care should always be taken to recognize the phases of the chemical species. For example, the above value for $\Delta h_c$ (and other values in Table A-23M) are for water as a liquid product. Since water will be gaseous at 1100°K, we must also lower the enthalpy of

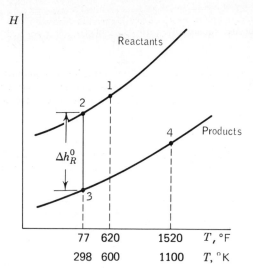

**Figure 14-3** A first-law analysis of a reacting system.

combustion by $5h_{fg} = 5(44,010)$, or 220,050 kJ/kg·mol of fuel. The correct value of $\Delta h_c$ to use, then, is $-2,635,990$ kJ/kg·mol. Note that the correction for the phase change of water is substantial.

The other enthalpy values required for Eq. (14-11) are obtained from Tables A-6M through A-10M in the Appendix. Use of these data and the enthalpy of combustion calculated above yields

$$q = -2,635,990 + 4(48,258 - 9364) + 5(40,071 - 9904) + 19.5(34,899 - 8682)$$

$$+ 97.6(33,426 - 8669) - 26.0(17,929 - 8682) \quad 97.6(17,563 - 8669)$$

$$= -510,540 \text{ kJ/kg·mol } C_4H_{10}$$

ALTERNATIVE SOLUTION Instead of using the enthalpy of combustion data, corrected for phase changes, it is also correct to use the enthalpy of formation data of Table A-22 in conjunc-conjunction with Eq. (14-9). By using the enthalpy of formation of gaseous water from this table, we do not have to correct for $h_{fg}$ of water. The value of $h_{fg}$ for liquid butane still must be used, however. The answer obtained by this method should be in substantial agreement with the preceding solution, except for round-off errors.

**Example 14-12** Liquid butane $(C_4H_{10})$ at $77°F$ is sprayed into a combustion chamber which is operating at a constant pressure of 5 atm. In addition, 400 percent theoretical air is supplied by an air compressor which has compressed the air to 5 atm and $620°F$. The gaseous products of combustion leave the combustion chamber and enter a turbine which has a maximum permissible temperature of $1520°F$. If complete combustion is assumed, because of the large excess of air, how much heat per mole of fuel must be removed from the gases in the combustion chamber to assure no damage to the turbine?

SOLUTION For the particular fuel under consideration the stoichiometric equation is

$$C_4H_{10}(l) + 6.5O_2(g) + 24.4N_2(g) \longrightarrow 4CO_2(g) + 5H_2O(g) + 24.4N_2(g)$$

For 400 percent theoretical air (or 300 percent excess air) the equation becomes

$$C_4H_{10}(l) + 26.0O_2(g) + 97.6N_2(g) \longrightarrow 4CO_2(g) + 5H_2O(g) + 19.5O_2(g) + 97.6N_2(g)$$

The inlet air temperature is 1080°R and the temperature of the product gases is 1980°R. The required heat transfer is ascertained through the use of Eq. (14-11), namely,

$$Q = \sum_{prod} N_i \, \Delta h_f - \sum_{reac} N_i \, \Delta h_f + \sum_{prod} N_i \, \Delta h - \sum_{reac} N_i \, \Delta h$$

$$= \Delta h_{R,\,537} + \sum_{prod} N_i (h_T - h_{537})_i - \sum_{reac} N_i (h_T - h_{537})_i \qquad (14\text{-}11)$$

This equation states that the heat which must be removed equals the energy released by the reaction at 77°F, plus the energy made available by cooling any reactant down to 77°F, minus the energy required to heat the products from 77°F to their final outlet temperature. The terms on the right-hand side of the equation are illustrated on the enthalpy-temperature diagram of Fig. 14-3. The value of $\Delta h_R$ at 77°F is the value of $\Delta H$ between states 2 and 3 in the figure. The value of $\sum N_i(\Delta h)$ for the products is given by the vertical distance between states 3 and 4. Similarly, the sensible enthalpy change of the reactants is represented by the distance between states 1 and 2. In our particular example, the only reactant at an elevated temperature is the 400 percent theoretical air supplied to the combustion chamber.

The enthalpy of reaction at 77°F (537°R) is obtained from Table A-23, under the listing of the enthalpy of combustion. However, the tabulated value of $-1{,}237{,}800$ Btu/lb·mol is for *n*-butane in the gaseous state, while this problem involves liquid *n*-butane. To correct the above value, we need the enthalpy of vaporization data also shown in the table. When a liquid fuel is used, the energy released is less than that for a gaseous fuel at a given temperature, since the fuel must be vaporized. The enthalpy of vaporization in this particular case is 9060 Btu/lb·mol. Therefore the $\Delta h_R$ value we should use is $-1{,}228{,}740$ Btu/lb·mol. The difference in this case is slight, but care should always be taken to recognize the phases of the chemical species involved in any reaction. For example, the values in Table A-23 are for water as a liquid product. Since water will be gaseous at 1980°R, we must also lower the enthalpy of combustion by 5(18,920), or 94,600 Btu/lb·mol of fuel. The correct $\Delta h_R$ value to use, then, is $-1{,}134{,}140$ Btu/lb·mol. Note that the correction for the phase change of water is substantial.

The other enthalpy values required to evaluate Eq. (14-11) are obtained from Tables A-6 through A-10 in the Appendix. Use of these data and the enthalpy of combustion calculated above yields

$$q = -1{,}134{,}140 + 4(20{,}753 - 4028) + 5(17{,}235 - 4258)$$

$$+ 19.5(14{,}995 - 3725) + 97.5(14{,}375 - 3730)$$

$$- 26.0(7697 - 3725) - 97.6(7551 - 3730)$$

$$= -219{,}840 \text{ Btu/lb·mol } C_4H_{10}$$

ALTERNATIVE SOLUTION Instead of using the enthalpy of combustion data, corrected for phase changes, it is also correct to use the enthalpy of formation data of Table A-22 in conjunction with Eq. (14-9). By using the enthalpy of formation of gaseous water from this table, we do not have to correct for $h_{fg}$ of water. The value of $h_{fg}$ of butane, however, still must be used. The answer obtained by this method should be in substantial agreement with the preceding solution, except for round-off errors.

The preceding example illustrates the equivalency of Eqs. (14-9) and (14-11). Thermochemistry problems may be solved by employing enthalpy of formation data or by enthalpy of reaction (combustion) data. Frequently it is a matter of personal choice. Nevertheless, in the absence of enthalpy of reaction data for a specific reaction, it is necessary to rely on enthalpy of formation data. These latter data are readily available.

## 14-5 ADIABATIC-COMBUSTION TEMPERATURE

In the absence of work effects and any appreciable kinetic-energy change of the flow stream, the energy released by a chemical reaction in a steady-flow reactor appears in two forms: heat loss to the surroundings and a temperature rise of the product gases. The smaller the heat loss, the larger the temperature rise becomes. In the limit of adiabatic operation of the reactor, the maximum temperature rise will occur. In a number of engineering applications of reacting systems, such as rocket-propulsion and gas-turbine cycles, it is desirable to be able to predict the maximum attainable temperature by the product gases. This maximum temperature is referred to as the *adiabatic-flame* or *adiabatic-combustion* temperature of the reacting mixture. On the basis of Eq. (14-9) the energy balance for a reacting mixture in steady flow and under adiabatic conditions becomes

$$\sum_{\text{prod}} N_i(\Delta h_f^\circ + h_T - h_{298})_i = \sum_{\text{reac}} N_i(\Delta h_f^\circ + h_T - h_{298})_i \qquad (14\text{-}12)$$

Since the initial temperature and composition of the reactants normally is known, the right-hand side of Eq. (14-12) can be evaluated directly. To obtain the maximum temperature rise of the products, a reaction must go to completion. Hence the $N_i$ values of the products are also known from the chemistry of the reaction. In addition, the values of $\Delta h_f^\circ$ and $h_{298}$ for each of the products is obtainable from tables of thermochemical data. Therefore, the only unknowns in Eq. (14-12) are the $h_T$ values for each of the product gases at the unknown adiabatic-combustion temperature. Since the $h_T$ values are tabulated against temperature in the Appendix, the solution to Eq. (14-12) is one of iteration. That is, a temperature must be guessed, and then the enthalpy values of the product gases at that temperature are found from their respective tables. If the guessed temperature is correct, then the numerical values of the left-hand and right-hand sides of Eq. (14-12) are equal. If not, then another estimate of the final temperature must be tried.

An estimate of the maximum combustion temperature made on this basis normally is conservative. That is, the calculated value frequently will be several hundred degrees higher than a measured value. In actual practice the calculated combustion temperature is not attained because of several effects. First of all, combustion is seldom complete. In addition, heat losses may be minimized but not eliminated. Last, some of the products of combustion will dissociate into other chemical species as a result of the high temperatures present. These dissociation reactions normally are endothermic, and consume some of the energy released by the overall reaction. The next two examples illustrate adiabatic-combustion-temperature calculations.

> **Example 14-13M** Liquid butane ($C_4H_{10}$) at 25°C and 400 percent theoretical air at 600°K react at constant pressure in a steady-flow process. Determine the adiabatic-combustion temperature in °K.

> SOLUTION The equation for the reaction under consideration is the same as in Example 14-12M.

> $$C_4H_{10}(l) + 26O_2(g) + 97.6N_2(g) \longrightarrow 4CO_2(g) + 5H_2O(g) + 19.5O_2(g) + 97.6N_2(g)$$

Equation (14-12) is applicable to this problem, namely,

$$\sum_{\text{reac}} N_i (\Delta h_f^\circ + h_T - h_{298})_i = \sum_{\text{prod}} N_i (\Delta h_f^\circ + h_T - h_{298})_i$$

As noted in the discussion which precedes this example, all the values of the terms in this equation are known, except the $h_T$ values for the products. These values are unknown because the final temperature of the products is not known. Substitution of the known values from tables in the Appendix yields

$$1(-126{,}150 - 21{,}060) + 26(0 + 17{,}929 - 8682) + 97.6(0 + 17{,}563 - 8669)$$

$$= 4(-393{,}520 + h_{T,\text{CO}_2} - 9364) + 5(-241{,}820 + h_{T,\text{H}_2\text{O}} - 9904)$$

$$+ 19.5(0 + h_{T,\text{O}_2} - 8682) + 97.6(0 + h_{T,\text{N}_2} - 8669)$$

In the above equation the value 21,060 in the first term is the enthalpy of vaporization of $n$-butane. Solving for the unknown enthalpy values at the adiabatic-combustion temperature, we find that

$$4h_{T,\text{CO}_2} + 5h_{T,\text{H}_2\text{O}} + 19.5h_{T,\text{O}_2} + 97.6h_{T,\text{N}_2} = 4{,}846{,}800 \text{ kJ}$$

We are now involved in an iteration, or trial-and-error, type of solution. The value of the adiabatic-flame temperature is guessed, which enables us to look up the corresponding $h_T$ values for each of the products. These values permit the evaluation of the left-hand side of the above equation. Several estimates may be required before the approximate answer is obtained.
A reasonably fast convergence on the correct temperature is achieved in the following manner. The chemical equation for the overall reaction shows that most of the combustion product is nitrogen (roughly 77 percent on a mole basis). Hence, as a first approximation, we shall assume the product gas is entirely nitrogen. In this case the energy balance reduces to

$$126.1 h_{T,\text{N}_2} = 4{,}846{,}800 \text{ kJ}$$

or

$$h_{T,\text{N}_2} = 38{,}400 \text{ kJ/kg·mol}$$

From Table A-6M it is found that this enthalpy value corresponds roughly to a temperature of 1240°K. This value will be found to be too high. The actual product gases contain $CO_2$, $H_2O$, and $O_2$ as well as $N_2$. The triatomic gases, $CO_2$ and $H_2O$, have a larger average specific heat $c_p$ than that of nitrogen in the given temperature range. As a result, they absorb more energy per mole for a given temperature change than does nitrogen. The effect of the presence of these triatomic gases, then, is to lower the final temperature from that predicted solely on the basis of nitrogen. The table below shows a summary of the iteration process in the vicinity of the final answer.

|  | 1200°K | 1220°K |
|---|---|---|
| $4h_T(\text{CO}_2)$ | 215,400 | 219,900 |
| $5h_T(\text{H}_2\text{O})$ | 221,900 | 226,300 |
| $19.5h_T(\text{O}_2)$ | 749,700 | 763,700 |
| $97.6h_T(\text{N}_2)$ | 3,589,400 | 3,655,300 |
| $\sum h_T(\text{prod})$ | 4,776,400 | 4,865,200 |

The sum of the enthalpy values for the products of combustion should be 4,846,800 kJ. The data in the table above indicate that the adiabatic-combustion temperature is close to 1220°K.

**Example 14-13** Liquid butane ($C_4H_{10}$) at 77°F and 400 percent theoretical air at 1080°R react at constant pressure in a steady-flow process. Determine the adiabatic-combustion temperature in °R.

SOLUTION The equation for the reaction under consideration is the same as in Example 14-12.

$$C_4H_{10}(l) + 26O_2(g) + 97.6N_2(g) \longrightarrow 4CO_2(g) + 5H_2O(g) + 19.5O_2(g) + 97.6N_2(g)$$

Equation (14-12) is applicable to this problem, namely,

$$\sum_{\text{reac}} N_i(\Delta h_f^\circ + h_T - h_{537})_i = \sum_{\text{prod}} N_i(\Delta h_f^\circ + h_T - h_{537})_i$$

As noted in the discussion which precedes this example, all the values of the terms in this equation are known, except the $h_T$ values for the products. These values are unknown because the final temperature of the products is not known. Substitution of the known values from tables in the Appendix yields

$$1(-54,270 - 9060) + 26(0 + 7697 - 3725) + 97.6(0 + 7551 - 3730)$$

$$= 4(-167,300 + h_{T, CO_2} - 4028) + 5(-104,040 + h_{T, H_2O} - 4258)$$

$$+ 19.5(0 + h_{T, O_2} - 3725) + 97.6(0 + h_{T, N_2} - 3730)$$

In the above equation the value 9060 in the first term is the enthalpy of vaporization of $n$-butane. Solving for the unknown enthalpy values at the adiabatic-combustion temperature, we find that

$$4h_{T, CO_2} + 5h_{T, H_2O} + 19.5h_{T, O_2} + 97.6h_{T, N_2} = 2,084,400 \text{ Btu}$$

We are now involved in an iteration, or trial-and-error, type of solution. The value of the adiabatic-flame temperature is guessed, which enables us to look up the corresponding $h_t$ values for each of the products. These values permit the evaluation of the left-hand side of the above equation. Several estimates may be required before the approximate answer is obtained. A reasonably fast convergence on the correct temperature is achieved in the following manner. The chemical equation for the overall reaction shows that most of the combustion product is nitrogen (roughly 77 percent on a mole basis). Hence, as a first approximation, we shall assume the product gas is entirely nitrogen. In this case the energy balance reduces to

$$126.1h_{T, N_2} = 2,084,400 \text{ Btu}$$

or

$$h_{T, N_2} = 16,530 \text{ Btu/lb·mol}$$

From Table A-6 it is found that this enthalpy value corresponds roughly to a temperature of 2260°R. This value will be found to be too high. The actual product gases contain $CO_2$, $H_2O$, and $O_2$ as well as $N_2$. The triatomic gases, $CO_2$ and $H_2O$, have a larger average specific heat $c_p$ than that of nitrogen in the given temperature range. As a result, they absorb more energy per mole for a given temperature change than does nitrogen. The effect of the presence of these triatomic gases, then, is to lower the final temperature from that predicted solely on the basis of nitrogen. The table below shows a summary of the iteration process in the vicinity of the final answer.

|  | 2180°R | 2220°R |
|---|---|---|
| $4h_T(CO_2)$ | 93,700 | 95,900 |
| $5h_T(H_2O)$ | 96,500 | 98,600 |
| $19.5h_T(O_2)$ | 325,500 | 332,200 |
| $97.6h_T(N_2)$ | 1,559,500 | 1,591,100 |
| $\sum h_T(\text{prod})$ | 2,075,200 | 2,117,800 |

The sum of the enthalpy values for the products of combustion should be 2,084,400 Btu. The data in the table above indicate that the adiabatic-combustion temperature is close to 2180°R.

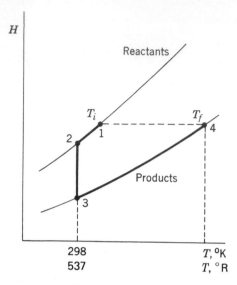

**Figure 14-4** Process diagram for adiabatic combustion.

For the preceding example, Eq. (14-11) could be used as the energy balance, rather than Eq. (14-12). This would necessitate the use of $\Delta h_R{}^\circ$ (or $\Delta h_c$) information, instead of $\Delta h_f$ data. The process of adiabatic combustion can be interpreted in terms of Eq. (14-11) on an $HT$ diagram, as shown in Fig. 14-4. State 1 on the figure represents the initial reactants at some temperature $T_i$ which is not 298°K (or 537°R). The products will be at some final temperature $T_f$. On the basis of an energy balance, the total enthalpy of the products must equal that of the reactants. Consequently, state 4 must lie horizontally to the right of state 1, for the direct process 1-4. For the indirect path 1-2-3-4, the energy available by cooling the reactants from $T_i$ to 298°K (537°R), plus the energy $\Delta h_R{}^\circ$ released by the reaction at 298°K, must equal the enthalpy change of the products from 298°K to $T_f$. Although the use of Eq. (14-12) in terms of $\Delta h_f{}^\circ$ data might be considered a more basic approach, Eq. (14-11) lends itself again to a graphical interpretation of the various energy terms.

**Example 14-14M** Determine the adiabatic-flame temperature for the reaction of liquid butane ($C_4H_{10}$) with the theoretical air requirements at 1 atm if all the reactants enter the combustion chamber at 25°C in steady flow.

SOLUTION The chemical equation for the reaction is

$$C_4H_{10}(l) + 6.5O_2(g) + 24.4N_2(g) \longrightarrow 4CO_2(g) + 5H_2O(g) + 24.4N_2(g)$$

The energy balance for the steady-flow process again reduces to

$$\sum_{reac} N_i(\Delta h_f^\circ + h_T - h_{298})_i = \sum_{prod} N_i(\Delta h_f^\circ + h_T - h_{298})_i$$

The only reactant terms in this case are $\Delta h_f{}^\circ$ and the enthalpy of vaporization for butane. Substitution of tabular data yields

$$1(-126,150 - 21,060) = 4(-393,520 + h_{T,CO_2} - 9364) + 5(-241,820 + h_{T,H_2O} - 9904)$$
$$+ 24.4(h_{T,N_2} - 8669)$$

Therefore,

$$4h_{T, CO_2} + 5h_{T, H_2O} + 24.4h_{T, N_2} = 2,934,500 \text{ kJ}$$

To shorten the iteration process, we again assume that all the products are nitrogen. This leads to the approximation,

$$h_{T, N_2} = 2,934,500/33.4 = 87,860 \text{ kJ/kg·mol}$$

From the nitrogen table, A-6M, the temperature corresponding to this enthalpy is between 2600 and 2650°K. The actual temperature must lie below this value. A summary of several trial calculations in the vicinity of the final answer is shown below.

|  | 2350°K | 2400°K |
|---|---|---|
| $4h_T(CO_2)$ | 488,400 | 500,600 |
| $5h_T(H_2O)$ | 504,200 | 517,500 |
| $24.4h_T(N_2)$ | 1,890,900 | 1,935,400 |
| $\sum h_T(\text{prod})$ | 2,883,500 | 2,935,500 |

Since the summation of enthalpies for the products should equal 2,934,500 kJ, the preceding table indicates that the adiabatic-flame temperature is very nearly 2400°K.

It is interesting to compare the above result with that obtained in Example 14-13M. Even though the air in that case is preheated to 600°K, the adiabatic-combustion temperature is only around 1220°K when 400 percent theoretical air is used. This is some 1200 degrees less than the value for stoichiometric combustion. In the original example the excess nitrogen and oxygen molecules are a large sink for the energy released by the reaction. The presence of this excess gas greatly reduces the combustion temperature. It should also be pointed out that temperatures in the range of 2300 to 2500°K, as found in this example, are fairly typical for the adiabatic combustion of many hydrocarbons with the theoretical or stoichiometric quantity of air.

**Example 14-14** Determine the adiabatic-flame temperature for the reaction of liquid butane $(C_4H_{10})$ with the theoretical air requirements at 1 atm if all the reactants enter the combustion chamber at 77°F in steady flow.

SOLUTION The chemical equation for the reaction is

$$C_4H_{10}(l) + 6.5O_2(g) + 24.4N_2(g) \longrightarrow 4CO_2(g) + 5H_2O(g) + 24.4N_2(g)$$

The energy balance for the steady-flow process again reduces to

$$\sum_{\text{reac}} N_i(\Delta h_f^\circ + h_T - h_{537})_i = \sum_{\text{prod}} N_i(\Delta h_f^\circ + h_T - h_{537})_i$$

The only reactant terms in this case are $\Delta h_f^\circ$ and the enthalpy of vaporization for butane. Substitution of tabular data yields

$$1(-54,270 - 9060) = 4(-169,300 + h_{T, CO_2} - 4028) + 5(-104,040 + h_{T, H_2O} - 4258)$$

$$+ 24.4(h_{T, N_2} - 3730)$$

Therefore,

$$4h_{T, CO_2} + 5h_{T, H_2O} + 24.4h_{T, N_2} = 1,262,500 \text{ Btu}$$

To shorten the iteration process, we again assume that all the products are nitrogen. This leads to the approximation,

$$h_{T, N_2} = 1,262,500/33.4 = 37,800 \text{ Btu/lb·mol}$$

From the nitrogen table, A-6, the temperature corresponding to this enthalpy is around 4740°R. The actual temperature must lie below this value. A summary of several trial calculations in the vicinity of the final answer is shown below.

|  | 4300°R | 4340°R |
|---|---|---|
| $4h_T(CO_2)$ | 214,500 | 216,900 |
| $5h_T(H_2O)$ | 221,400 | 223,900 |
| $24.4h_T(N_2)$ | 828,100 | 836,700 |
| $\Sigma h_T(\text{prod})$ | 1,264,000 | 1,277,500 |

Since the summation of enthalpies for the products should equal 1,262,500 Btu, the table above indicates that the adiabatic-flame temperature is very nearly 4300°R.

It is interesting to compare the above result with that obtained in Example 14-13. Even though the air in that case is preheated to 1080°R, the adiabatic-combustion temperature is only around 2180°R when 400 percent theoretical air is used. This is some 2100 degrees less than the value for stoichiometric combustion. In the original example the excess nitrogen and oxygen molecules are a large sink for the energy released by the reaction. The presence of this excess gas greatly reduces the combustion temperature. It should also be pointed out that temperatures in the range of 4200 to 4500°R, as found in this example, are fairly typical for the adiabatic combustion of many hydrocarbons with the theoretical or stoichiometric quantity of air.

# 14-6 THERMOCHEMICAL ANALYSIS OF CONSTANT-VOLUME SYSTEMS

If a chemical reaction occurs in a closed system maintained at constant volume, then the basic energy balance is

$$Q = \Delta U = U_{\text{prod}} - U_{\text{reac}} \tag{14-13}$$

The internal energy change associated with any chemical reaction for which the products are returned to the original temperature and volume of the reactants is called the *internal energy of reaction*, $\Delta u_R$, or the heat of reaction at constant volume. Like the enthalpy of reaction, the standard-state value, $\Delta u_R^\circ$, is measured at 25°C or 77°F. The effect of pressure on the internal energy of many substances may be neglected. In such a case the enthalpy of reaction and the internal energy of reaction are related by the expression

$$\Delta h_R^\circ = \Delta u_R^\circ + P \, \Delta V.$$

Furthermore, the volume of solid and liquid components in a chemical reaction is usually negligible in comparison to the volume occupied by the gaseous components. Hence, $\Delta V$ in the above equation refers primarily to the gas phase. If the gas phase can be considered an ideal-gas mixture, then, for a constant-pressure reaction at temperature $T$, $P \, \Delta V = \Delta NRT$. In this form of the ideal-gas equation $\Delta V$ and $\Delta N$ represent the difference in $V$ and $N$ between products and reactants in the gas phase. As a result the above equation for $\Delta h_R^\circ$ becomes

$$\Delta h_R^\circ = \Delta u_R^\circ + \Delta NRT \tag{14-14}$$

It is important to keep in mind, when applying Eq. (14-14) to a chemical reaction, that $N$ represents only moles of gaseous products and reactants. In general, $\Delta N$ may be positive, negative, or zero in value.

Now consider a constant-volume reaction for which the temperatures of the reactants and products are different from the standard-state value of 298°K or 537°R. In a manner analogous to the development of Eq. (14-11), Eq. (14-13) can be written in the form

$$Q = \Delta u_R^\circ + \sum_{\text{prod}} N_i(u_T - u_{298})_i - \sum_{\text{reac}} N_i(u_T - u_{298})_i \qquad (14\text{-}15)$$

where $\Delta u_R^\circ$ is evaluated by Eq. (14-14). When one is working in USCS units, $u_{298}$ is replaced by $u_{537}$.

**Example 14-15M** Carbon monoxide with 200 percent theoretical air at 25°C reacts completely in a constant-volume reaction chamber. After a given time interval a temperature measurement indicates a gas temperature within the chamber of 1200°K. Determine the heat loss from the chamber, in kJ/kg·mol of carbon monoxide.

SOLUTION The overall chemical reaction is given by

$$CO(g) + O_2(g) + 3.76N_2(g) \longrightarrow CO_2(g) + \tfrac{1}{2}O_2(g) + 3.76N_2(g)$$

The heat loss is determined by means of Eq. (14-15),

$$Q = \Delta u_R^\circ + \sum [N_i(u_T - u_{298})_i]_{\text{prod}} - \sum [N_i(u_T - u_{298})_i]_{\text{reac}}$$

The internal energy of reaction is evaluated by means of Eq. (14-14). The enthalpy of reaction (combustion) of CO is found in Table A-23M to be $-282{,}990$ kJ/kg·mol. This value applies to the basic chemical reaction,

$$CO(g) + \tfrac{1}{2}O_2(g) \longrightarrow CO_2(g)$$

Since all the chemical species in the reaction are gases, $\Delta N = 1 - 1\tfrac{1}{2} = -\tfrac{1}{2}$. The use of Eq. (14-14) then yields

$$\Delta u_R^\circ = \Delta h_R^\circ - \Delta NRT$$

$$= -282{,}990 - (-\tfrac{1}{2})(8.315)(298)$$

$$= -281{,}750 \text{ kJ/kg·mol}$$

Note that there is little difference between $\Delta u_R^\circ$ and $\Delta h_R^\circ$. This is frequently true, unless the energy released by the reaction is extremely small, since the value of $\Delta NRT$ for most reactions is less than 5000 kJ.

The remaining $u$ data for the energy balance come from Table A-6M through A-9M. Substitution of proper data leads to

$$Q = -281{,}750 + 1(43{,}871 - 6885) + \tfrac{1}{2}(28{,}469 - 6203)$$

$$+ 3.76(27{,}799 - 6190) - 1(0) - 1(0) - 3.76(0)$$

$$= -156{,}140 \text{ kJ/kg·mol CO}$$

Since the initial reactants are at 25°C, the last three terms in the above energy balance are zero in value.

**Example 14-15** Carbon monoxide with 200 percent theoretical air at 77°F reacts completely in a constant-volume reaction chamber. After a given time interval a temperature measurement indicates a gas temperature within the chamber of 2100°R. Determine the heat loss from the chamber, in Btu/lb·mol of carbon monoxide.

SOLUTION The overall chemical reaction is given by

$$CO(g) + O_2(g) + 3.76N_2(g) \longrightarrow CO_2(g) + \tfrac{1}{2}O_2(g) + 3.76N_2(g)$$

The heat loss is determined by means of Eq. (14-15),

$$Q = \Delta u_R^\circ + \sum [N_i(u_T - u_{537})_i]_{\text{prod}} - \sum [N_i(u_T - u_{537})_i]_{\text{reac}}$$

The internal energy of reaction is evaluated by means of Eq. (14-14). The enthalpy of reaction (combustion) of CO is found in Table A-23 to be $-121,750$ Btu/lb·mol. This value applies to the basic chemical reaction,

$$CO(g) + \tfrac{1}{2}O_2(g) \longrightarrow CO_2(g)$$

Since all the chemical species in the reaction are gases, $\Delta N = 1 - 1\tfrac{1}{2} = -\tfrac{1}{2}$. The use of Eq. (14-14) then yields

$$\Delta h_R^\circ = \Delta h_R^\circ - \Delta NRT$$

$$= -121,750 - (-\tfrac{1}{2})(1.986)(537)$$

$$= -121,220 \text{ Btu/lb·mol}$$

Note that there is little difference between $\Delta u_R^\circ$ and $\Delta h_R^\circ$. This is frequently true, unless the energy released by the reaction is extremely small, since the value of $\Delta NRT$ for most reactions is less than 3000 Btu.

The remaining $u$ data for the energy balance come from Tables A-6 through A-9. Substitution of proper data leads to

$$Q = -121,220 + 1(18,182 - 2964) + \tfrac{1}{2}(11,841 - 2659)$$

$$+ 3.76(11,164 - 2664) - 1(0) - 1(0) - 3.76(0)$$

$$= -69,450 \text{ Btu/lb·mol CO}$$

Since the initial reactants are at 77°F, the last three terms in the above energy balance are zero in value.

When chemical reactions occur under constant-volume conditions, not only elevated temperatures but also higher pressures usually result. It is of practical interest to the engineer to estimate the maximum possible pressures that might develop. This would occur theoretically under adiabatic conditions, with the simultaneous achievement of the adiabatic-flame temperature. From a more practical viewpoint the heat release would have to be nearly instantaneous and conditions of internal equilibrium also would have to be met. For calculation purposes these criteria will be assumed to be valid. This will lead to an upper limit, which is of real interest and importance. Dissociation effects at these high temperatures will be neglected at this time.

In general, two methods of calculation present themselves. One requires a knowledge of the initial volume; the other requires information on the initial bomb pressure. In order to simplify both methods, we shall assume that the gases involved will behave ideally. If one applies the ideal-gas relation to the resultant gases after a chemical reaction at constant volume, the expression for the final pressure is

$$P_f = \frac{N_f R_u T_f}{V}$$

where $T_f$ = adiabatic-flame temperature
$\qquad N_f$ = total number of gaseous moles of products
$\qquad V$ = bomb volume

In this case $V$ will be known through measurement, and $T_f$ may be calculated by methods previously outlined. The value of $N_f$ may be determined from the basic chemical equation representing the reaction.

**Example 14-16M** Two and one half grams of liquid benzene ($C_6H_6$) are placed in a 0.030-$m^3$ constant-volume bomb. The initial temperature of the fuel and oxidizer is 25°C. If the fuel is burned with 20 percent excess air and the reaction goes to completion, determine the maximum pressure in the bomb, in bars.

SOLUTION The chemical equation for the reaction is

$$C_6H_6(l) + 9O_2(g) + 33.8N_2(g) \longrightarrow 6CO_2(g) + 3H_2O(g) + 1.5O_2(g) + 33.8N_2(g)$$

The enthalpy of reaction in Table A-23M for gaseous benzene and liquid water in the products is $-3,301,500$ kJ/kg·mol. The correction for the enthalpy of vaporization of benzene is 33,820 kJ/kg·mol, and the correction to the lower heating value is 3(43,980), or 131,940 kJ. Both of these corrections lower the amount of energy released; consequently, the enthalpy of combustion for the above reaction is $-3,135,740$ kJ/kg·mol of benzene. The process is at constant volume; hence we need the internal energy of combustion, which is

$$\Delta u_c^\circ = \Delta h_c^\circ - \Delta NRT = -3,135,740 - 1.5(8.315)(298)$$

$$= -3,139,460 \text{ kJ/kg·mol } C_6H_6$$

The adiabatic-combustion temperature is evaluated next, on the basis of sensible internal-energy data. For this problem Eq. (14-15) reduces to the internal energy of combustion plus the sensible internal-energy change of the products equals zero. The tabulated internal energy data from Tables A-6M to A-10M are used. The final form of the energy balance is

$$6u_{T,CO_2} + 3u_{T,H_2O} + 1.5u_{T,O_2} + 33.8u_{T,N_2} = 3,421,570 \text{ kJ}$$

A temperature must now be found which satisfies the above equation. Results of the iteration process are summarized below.

|  | 2650°K | 2700°K |
| --- | --- | --- |
| $3u(H_2O)$ | 284,870 | 291,800 |
| $6u(CO_2)$ | 711,000 | 727,030 |
| $1.5u(O_2)$ | 106,320 | 108,650 |
| $33.8u(N_2)$ | 2,246,180 | 2,294,340 |
| $\sum u(\text{prod})$ | 3,348,370 | 3,421,830 |

Results of the table indicate that the maximum combustion temperature is essentially 2700°K. Finally, there remains the determination of $N_f$, the number of kilogram-moles of gaseous products at equilibrium. From the chemical equation it is noted that 44.3 mol of gaseous products are formed for each mole of benzene. The initial quantity of 2.5 g of benzene is equivalent to $3.21 \times 10^{-5}$ kg·mol. The number of kilogram-moles of gaseous products then is $44.3(3.21 \times 10^{-5}) = 0.00142$ kg·mol. Therefore the maximum combustion pressure is

$$P_f = \frac{N_f R_u T_f}{V} = \frac{0.00142(0.08315)(2700)}{0.030} = 10.6 \text{ bars}$$

**Example 14-16** Two and one half grams of liquid benzene ($C_6H_6$) is placed in a 1.0-ft³ constant-volume bomb. The initial temperature of the fuel and oxidant is 77°F. If the fuel is burned with 20 percent excess air, determine the explosion pressure in psia.

SOLUTION The balanced chemical equation for the reaction is

$$C_6H_6(l) + 9O_2(g) + 33.8N_2(g) \longrightarrow 6CO_2(g) + 3H_2O(g) + 33.8N_2(g) + 1.5O_2(g)$$

The enthalpy of reaction for gaseous benzene and liquid water in the products is listed as −1,420,300 Btu/lb·mol in Table A-23. The correction for the enthalpy of vaporization of benzene is 14,550 Btu/lb·mol, and the correction to the lower heating value is 3(18,920), or 56,760 Btu. Both of these corrections lower the amount of energy released; consequently, the enthalpy of combustion for the above reaction is −1,348,990 Btu/lb·mol of benzene. The process is at constant volume; hence we need the internal energy of combustion, which is

$$\Delta u_c = \Delta h_c - \Delta N R_u T = -1,348,990 - 1.5(1.986)(537)$$

$$= -1,349,590 \text{ Btu/lb·mol } C_6H_6$$

The adiabatic-combustion temperature is now evaluated on the basis of sensible internal-energy data since, for an adiabatic, constant-volume vessel, $\Delta U = 0$. The internal energy of reaction equals the sensible internal-energy change of the products. The tabulated data from the gas tables A-6 to A-10 are used, and the results of the iteration process are summarized below.

|  | 537°R | 4800°R | 4900°R |
|---|---|---|---|
| $3u(H_2O)$ | 9600 | 123,500 | 127,000 |
| $6u(CO_2)$ | 17,800 | 309,000 | 317,000 |
| $33.8u(N_2)$ | 90,000 | 974,000 | 996,000 |
| $1.5u(O_2)$ | 4000 | 46,000 | 47,100 |
| $\sum u(\text{prod})$ | 121,400 | 1,452,500 | 1,487,100 |

Hence, for $T_f = 4800°R$, $\sum N \, \Delta u = 1,452,500 - 121,400 = 1,331,100$ Btu, and at 4900°R, $\sum N \, \Delta u = 1,487,100 - 121,400 = 1,365,700$ Btu. Since the actual energy released is 1,349,590 Btu, the maximum combustion temperature by interpolation is 4850°R.
Finally, there remains the determination of $N_f$, the number of pound-moles of gaseous products at equilibrium. From the chemical equation it is noted that 44.3 moles of gaseous products are formed for each mole of benzene in the reaction. The initial quantity of 2.5 g of benzene is equivalent to $7.06 \times 10^{-5}$ lb·mol. Therefore the explosion pressure becomes

$$P_f = \frac{N_f R_u T_f}{V} = \frac{(7.06 \times 10^{-5})(44.3)(1545)(4850)}{1} = 23,400 \text{ psfa} = 162.5 \text{ psia}$$

A second method of determining explosion pressures is based on a knowledge of the initial pressure in the reaction vessel. By applying the ideal-gas relation to conditions both before and after the reaction takes place, the following relation results:

$$P_f = \frac{P_i N_f T_f}{N_i T_i}$$

where the subscripts $f$ and $i$ indicate the final and initial conditions, respectively.

**Example 14-17M** Consider again the combustion of 2.5 g of liquid benzene with 20 percent excess air, both initially at 25°C. In this case, however, the volume of the bomb will be such that the initial pressure $P_i$ is 1.0 bar. Determine the maximum bomb pressure under this new condition.

SOLUTION The following data are known from Example 14-16M: $N_i/N_f = 42.8/44.3$, $T_i = 298°$K, $T_f = 2700°$K. In addition, $P_i = 1.0$ bar. Therefore

$$P_f = \frac{P_i N_f T_f}{N_i T_i} = \frac{1.0(44.3)(2700)}{42.8(298)} = 9.38 \text{ bars}$$

**Example 14-17** Consider again the combustion of 2.5 g of liquid benzene with 20 percent excess air, both initially at 77°F. In this case the volume of the bomb will be sufficient so that the initial pressure $P_i$ is 14.7 psia. Now determine the explosion pressure under this new condition.

SOLUTION The following data are known from Example 14-16: $P_i = 14.7$ psia, $N_i = 42.8(7.06 \times 10^{-5})$ lb·mol, $N_f = 44.3(7.06 \times 10^{-5})$ lb·mol, $T_i = 537°$R, $T_f = 4850°$R. Therefore

$$P_f = \frac{14.7(44.3)(4850)}{42.8(537)} = 137.5 \text{ psia}$$

## 14-7 ENTROPY CHANGES FOR REACTING MIXTURES

In the discussion up to this point we have relied solely upon the conservation of energy principle, in conjunction with the conservation of atomic species, in analyzing reactive systems. As in the case of systems of fixed compositions, the second law and the evaluation of the change of the entropy for processes is of particular value in the study of chemical reactions. In the application of the first law, we found it necessary to assign arbitrary reference values to the enthalpy of stable elements at 77°F and 1 atm pressure, in order to establish the absolute enthalpy values for pertinent chemical species. Only by considering both sensible- and chemical-enthalpy changes could the change in the enthalpy of a reacting system be obtained in a consistent and straightforward manner. The same problem of a reference state occurs also in evaluating the entropy change of a process for a chemical reaction, since to date we have established equations only for the change in entropy of pure substances or ideal-gas mixtures of fixed composition. Similar methods of assigning arbitrary reference values for the entropy of particular elements or compounds might be employed. In this case, however, a more fundamental approach is available for establishing the entropy values of substances, based on what is known as the third law of thermodynamics.

The *third law* of thermodynamics, on the basis of experimental data, states that the entropy of a pure crystalline substance may be taken to be zero at the absolute zero thermodynamic temperature, that is, 0°K or 0°R. This law was fairly well accepted by around 1920, following the work of Boltzmann, Planck, Nernst, Einstein, Lewis, and others in the two or three decades preceding this time. It may also be shown that the entropy of crystalline substances at absolute zero temperature is not a function of pressure; that is $(\partial s/\partial P)_{T=0} = 0$. At temperatures above absolute zero, however, the entropy of a substance is a function of the pressure.

Because of this the tabulated values of the absolute entropy are usually given at 1 atm pressure. From a knowledge of specific-heat and latent-heat (enthalpy) data for a substance at 1 atm, the absolute entropy can be evaluated for any desired temperature at that pressure. Corrections for pressures other than 1 atm can also be made. The absolute entropy of a pure substance can also be obtained from statistical, or microscopic, considerations, based on methods introduced in Chap. 10. Considerable insight into the behavior of matter has been gained in the past by determining the absolute entropy of a substance by these two independent methods. The absolute entropy values of a large number of substances at 1 atm and a wide range of temperatures are available in the literature, for example, in the *JANAF* tables.

The evaluation of entropy changes for reactive mixtures requires a means of determining the entropy of a substance, either pure or in a mixture, at pressures other than that generally used in reference tables. In order to simplify the presentation, the following discussion will be restricted to ideal gases. The entropy of a pure ideal gas at temperatures and pressures other than those found in tables based on the third law are easily obtained by employing the equation derived in Chap. 7, namely,

$$ds = c_p \frac{dT}{T} - R \frac{dP}{P} \tag{7-15}$$

where the integration would be carried out from the reference state to the desired state. In the case of ideal-gas mixtures, it may be recalled from Chap. 11 that, based on the Gibbs–Dalton law, each constituent of a mixture of ideal gases behaves as if it existed alone at the temperature and volume of the mixture. Thus, in a gas mixture, an ideal gas exerts a pressure equal to its partial pressure. Therefore the entropy of a gas in a mixture can be found by integrating Eq. (7-15) for $ds$ from a reference state where the gas is pure and at 1 atm pressure to the mixture state where its pressure is its partial pressure $p_i$. Hence

$$(s_i)_{T, P} = s^0_{i, T} - R \ln \frac{p_i}{P_0}$$

where the reference state of the pure ideal gas at 1 atm pressure has been denoted by the superscript 0 on $s_i$ and by a subscript 0 on the initial pressure and temperature. Recall that the partial pressure $p_i = x_i P_{tot}$.

The ideal-gas tables, A-5M through A-11M and A-5 through A-11, list values of $s^0_{i, T}$ for a wide range of temperatures at 1 atm. The absolute entropy of an ideal gas at any other pressure and the given temperature is found from the equation

$$s_{i, T, P} = s^0_{i, T} - R \ln p_i \tag{14-16}$$

where $p_i$ must be measured in atmospheres. For a mixture of ideal gases the total entropy is

$$S_{m, T, P} = \sum_i N_i s_i(T, p_i) \tag{14-17}$$

and for a chemical reaction the change in entropy is given by

$$\Delta S = \sum_i (N_i s_i)_{\text{prod}} - \sum_i (N_i s_i)_{\text{reac}} \qquad (14\text{-}18)$$

where $N_i$ is the number of moles of the $i$th component in the mixture, and $s_i$ is the absolute entropy of the pure substances at the required temperature and partial pressure.

**Example 14-18M** Carbon monoxide and oxygen in a stoichiometric ratio at 25°C and 1 atm react to form carbon dioxide at the same pressure and temperature. If the reaction goes to completion, evaluate the entropy change in kJ/°K per kilogram-mole of CO.

SOLUTION The basic reaction is $CO + \frac{1}{2}O_2 \rightarrow CO_2$ in the gas phase. The overall entropy change is given by Eq. (14-18), namely,

$$\Delta S = \sum (N_i s_i)_{\text{prod}} - \sum (N_i s_i)_{\text{reac}}$$

The absolute entropies of the three reacting components are calculated below on the basis of Eq. (14-16). The values of $s_i^0$ at 25°C are from Tables A-7M, A-8M, and A-9M. In units of kJ/(kg·mol)(°K),

$$s(CO_2) = 213.685 - 8.315 \ln (1.0) = 213.685$$

$$s(CO) = 197.543 - 8.315 \ln (2/3) = 200.914$$

$$s(O_2) = 205.033 - 8.315 \ln (1/3) = 214.168$$

Therefore the total change in entropy for the complete reaction is

$$\Delta S = 213.685 - 200.914 - \tfrac{1}{2}(214.168) = -94.463 \text{ kJ/°K}$$

The negative change in entropy for the gas-phase reaction is not a violation of the second law. To keep the mixture at 25°C the enthalpy of combustion is released to the surroundings. If the surroundings are at 25°C, they undergo an entropy change of

$$\Delta S_{\text{surr}} = Q/T = -\Delta h_c/T = -(-282,990)/298 = 949.6 \text{ kJ/°K}$$

Hence the increase in entropy of the surroundings overshadows the decrease in entropy of the chemical reaction.

**Example 14-18** Carbon monoxide and oxygen in a stoichiometric ratio at 77°F and 1 atm react to form carbon dioxide at the same pressure and temperature. If the reaction goes to completion, evaluate the entropy change in Btu/°R per pound-mole of CO.

SOLUTION The basic reaction is $CO + \frac{1}{2}O_2 \rightarrow CO_2$ in the gas phase. The overall entropy change is given by Eq. (14-18), namely,

$$\Delta S = \sum (N_i s_i)_{\text{prod}} - \sum (N_i s_i)_{\text{reac}}$$

The absolute entropies of the three reacting components are calculated below on the basis of Eq. (14-16). The values of $s_i^0$ at 77°F are from Tables A-7, A-8, and A-9. In units of Btu/(lb·mol)(°R),

$$s(CO_2) = 51.032 - 1.986 \ln (1.0) = 51.032$$

$$s(CO) = 47.272 - 1.986 \ln (2/3) = 48.077$$

$$s(O_2) = 48.982 - 1.986 \ln (1/3) = 51.164$$

Therefore the total change in entropy for the complete reaction is

$$\Delta S = 51.032 - 48.077 - \tfrac{1}{2}(51.164) = -22.627 \text{ Btu/}^\circ\text{R}$$

The negative change in entropy for the gas-phase reaction is not a violation of the second law. To keep the mixture at 77°F the enthalpy of combustion is released to the surroundings. If the surroundings are at 77°F, they undergo an entropy change of

$$\Delta S_{surr} = Q/T = -\Delta h_c/T = -(-121{,}750)/537 = 226.7 \text{ Btu/}^\circ\text{R}$$

Hence the increase in entropy of the surroundings overshadows the decrease in entropy of the chemical reaction.

**Example 14-19M** A 1 : 1 molar ratio of gaseous carbon monoxide and water vapor enters a steady-flow, adiabatic reactor at 400°K and 1 atm. If it is assumed that the pressure is constant and that the reaction goes to completion according to the following equation,

$$CO(g) + H_2O(g) \longrightarrow CO_2(g) + H_2(g)$$

determine the overall entropy change in kJ/°K per kilogram-mole of CO.

SOLUTION To evaluate the entropy change, we must first determine the final temperature. By neglecting kinetic- and potential-energy changes, the steady-flow energy balance reduces to

$$0 = \sum_{prod} N_i(\Delta h_f^\circ + h_T - h_{298})_i - \sum_{reac} N_i(\Delta h_f^\circ + h_T - h_{298})_i$$

Using $\Delta h_f^\circ$ data from Table A-22M and sensible-enthalpy data for ideal gases from Tables A-8M through A-11M, we find that

$$0 = 1(-393{,}520 - h_{T,CO_2} - 9364) + 1(0 + h_{T,H_2} - 8468) - 1(-110{,}530 + 11{,}644 - 8669)$$
$$- 1(-241{,}820 + 13{,}356 - 9904)$$

Upon rearrangement the equation becomes

$$h_{T,CO_2} + h_{T,H_2} = 65{,}430 \text{ kJ}$$

From Tables A-9M and A-11M at 920°K, it is found that

$$h_{T,CO_2} + h_{T,H_2} = 38{,}467 + 26{,}747 = 65{,}210 \text{ kJ}$$

Similarly, at 960°K, it is found that

$$h_{T,CO_2} + h_{T,H_2} = 40{,}607 + 27{,}947 = 68{,}555 \text{ kJ}$$

Therefore the final temperature of the products is very close to 920°K. The entropy change is found by evaluating $s_{i,T}^\circ - R \ln p_i$ for each chemical species in the reaction. Since the mole fraction of each reactant and each product is $\tfrac{1}{2}$ and the total pressure is constant, the partial pressure in each constituent is the same. Hence the value of $R \ln p_i$ is the same for each, and the terms will cancel out when the change in entropy is calculated. As a consequence, Eq. (14-18) for this problem reduces to

$$\Delta S = (s_{CO_2} + s_{H_2})_{920\,^\circ K} - (s_{CO} + s_{H_2O})_{400\,^\circ K}$$
$$= 264.728 + 163.607 - 206.125 - 198.673$$
$$= 23.537 \text{ kJ/}^\circ\text{K}$$

**Example 14-19** A 1 : 1 molar ratio of gaseous carbon monoxide and water vapor enters a steady-flow, adiabatic reactor at 260°F and 1 atm. If it is assumed that the pressure is constant and that the reaction goes to completion according to the following equation,

$$CO(g) + H_2O(g) \longrightarrow CO_2(g) + H_2(g)$$

determine the overall entropy change in Btu/°R per pound-mole of CO.

SOLUTION In order to evaluate the entropy change, we must first determine the final temperature. By neglecting kinetic- and potential-energy changes, the steady-flow energy balance reduces to

$$0 = \sum_{\text{prod}} N_i (\Delta h_f^\circ + h_T - h_{537})_i - \sum_{\text{reac}} N_i (\Delta h_f^\circ + h_T - h_{537})_i$$

Using $\Delta h_f^\circ$ data from Table A-22 and sensible-enthalpy data for ideal gases from Tables A-8 through A-11, we find that

$$0 = 1(-169,300 + h_{T,CO_2} - 4028) + 1(0 + h_{T,H_2} - 3640) - 1(-47,540 + 5006 - 3725)$$

$$- 1(-104,040 + 5,739 - 4258)$$

Upon rearrangement the equation becomes

$$h_{T,CO_2} + h_{T,H_2} = 28,150 \text{ Btu}$$

From Tables A-9 and A-11 at $1600°R$, it is found that

$$h_{T,CO_2} + h_{T,H_2} = 15,829 + 11,093 = 26,920 \text{ Btu}$$

Similarly, at $1700°R$, it is found that

$$h_{T,CO_2} + h_{T,H_2} = 17,101 + 11,807 = 28,910 \text{ Btu}$$

Therefore, the final temperature of the products is very close to $1660°R$. The entropy change is found by evaluating $s_{i,T}^\circ - R \ln p_i$ for each chemical species in the reaction. Since the mole fraction of each reactant and each product is $\frac{1}{2}$ and the total pressure is constant, the partial pressure in each constituent is the same. Hence the value of $R \ln p_i$ is the same for each, and the terms will cancel out when the change in entropy is calculated. As a consequence, Eq. (14-18) for this problem reduces to

$$\Delta S = (s_{CO_2} + s_{H_2})_{1660°R} - (s_{CO} + s_{H_2O})_{720°R}$$

$$= 63.250 + 39.090 - 49.317 - 47.450$$

$$= 5.573 \text{ Btu}/°R$$

The positive answer is in accord with the second law.

In the two preceding examples it is noted that the entropy change is positive for this adiabatic reaction, in accordance with the second-law requirement. However, this does not necessarily mean that the reaction as proposed is possible. If the entropy of the system attains a maximum value at some other concentration of the four chemical species present, the reaction will stop at that point. That is, whatever composition of reactants and products leads to the highest entropy value will be the composition at the equilibrium state. In order to determine the equilibrium composition of a reacting mixture of CO and $H_2O$ initially at $260°F$ ($400°K$) in an adiabatic-reaction vessel at 1 atm pressure, we need to evaluate the total absolute entropy of the system for different degrees or stages of reaction. Recall that the stoichiometric equation is

$$CO(g) + H_2O(g) \longrightarrow CO_2(g) + H_2(g)$$

We shall evaluate the total absolute entropy of the reactive gases under those conditions for which 0, 0.2, 0.4, 0.6, 0.7, 0.8, 0.9, and 1.0 mole of $CO_2$ are formed under adiabatic conditions. Only one actual calculation will be shown, since the same method is used throughout. For example, if 0.8 mole of $CO_2$ is formed from the initial $1:1$ molar mixture of CO and $H_2O$, the chemical reaction proceeds accordingly as

$$CO(g) + H_2O(g) \longrightarrow 0.8CO_2(g) + 0.8H_2(g) + 0.2CO(g) + 0.2H_2O(g)$$

As in Examples 14-19M and 14-19, one must employ the first law to determine the final temperature that would be achieved for this reaction, which only goes to 80 percent completion. By equating the enthalpy of the products with that of the reactants, it is found by an iteration process that the final temperature of the four product gases is 834°K, or roughly 1040°F. This is approximately 30°C or 60°F cooler than for complete combustion, but this is to be expected, since the total enthalpy of reaction is not released. The absolute entropy of each gas may now be found from

$$S_i = N_i(s_{i,T}^0 - R \ln p_i)$$

Recall that $s_{i,T}^0$ is the value of the absolute entropy at 1 atm and the desired temperature. The values of $s_{i,T}^0$ used in the calculation below are based on data from the ideal-gas tables in the Appendix. The following calculations are in terms of metric data, with the USCS results shown in parentheses. Hence

$$S_{CO_2} = 0.8(259.560 - 8.315 \ln 0.4) = 213.743 \text{ kJ/°K} \quad (51.035 \text{ Btu/°R})$$

$$S_{H_2} = 0.8(160.673 - 8.315 \ln 0.4) = 134.634 \text{ kJ/°K} \quad (32.154 \text{ Btu/°R})$$

$$S_{CO} = 0.2(228.493 - 8.315 \ln 0.1) = \phantom{0}49.528 \text{ kJ/°K} \quad (11.846 \text{ Btu/°R})$$

$$S_{H_2O} = 0.2(225.311 - 8.315 \ln 0.1) = \phantom{0}48.891 \text{ kJ/°K} \quad (11.676 \text{ Btu/°R})$$

$$\overline{\phantom{xxxxxxxxxxxxxxxxxxxxxxxxxxx}}$$

$$S_{tot} = 446.796 \text{ kJ/°K} \quad (106.711 \text{ Btu/°R})$$

Thus the total entropy, when the reaction goes to 80 percent completion, is 446.796 kJ/°K or 106.711 Btu/°R per mole of carbon monoxide. The total entropy for the system for no reaction and for complete reaction can be found from data used in Examples 14-19M and 14-19. For metric data

$$S_{init} = 206.125 + 198.673 - 2(8.315)(\ln 0.5) = 416.325 \text{ kJ/°K}$$

$$S_{complete} = 264.728 + 163.607 - 2(8.315)(\ln 0.5) = 439.862 \text{ kJ/°K}$$

For the same calculation in USCS units,

$$S_{init} = 49.317 + 47.450 - 2(1.986)(\ln 0.5) = 99.520 \text{ Btu/°R}$$

$$S_{complete} = 63.250 + 39.090 - 2(1.986)(\ln 0.5) = 105.093 \text{ Btu/°R}$$

From these data, in either metric or USCS, it is seen that the entropy for 80 percent completion is greater than that for 100 percent completion, and both values are greater than the initial value. This indicates that the entropy function is a maximum before the reaction reaches completion. To ascertain the actual state of equilibrium, one needs to carry out similar calculations at other degrees of completion. A summary of these computations appears in the table below. The

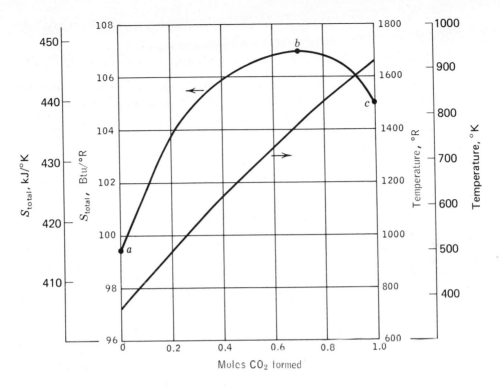

**Figure 14-5** The total entropy and the combustion temperature for the adiabatic reaction of an equimolar mixture of CO and $H_2O$ initially at $400°K$ ($720°R$) as a function of the extent of the reaction, for the chemical reaction $CO(g) + H_2O(g) \rightleftharpoons CO_2(g) + H_2(g)$

system temperature at each degree of completion is also indicated. The data from this table are plotted in Fig. 14-5. It is found that the state of maximum entropy occurs when approximately 70 percent of the CO and $H_2O$ have reacted. This, then, is the state at equilibrium.

| Moles $CO_2$ formed | Metric | | USCS | |
|---|---|---|---|---|
| | $S_{tot}$, kJ/°K | $T_{final}$, °K | $S_{tot}$, Btu/°R | $T_{final}$, °R |
| 0 | 416.325 | 400 | 99.520 | 720 |
| 0.2 | 433.870 | 520 | 103.759 | 940 |
| 0.4 | 442.525 | 630 | 105.828 | 1140 |
| 0.6 | 447.218 | 740 | 106.815 | 1330 |
| 0.7 | 447.752 | 790 | 106.935 | 1420 |
| 0.8 | 446.796 | 834 | 106.711 | 1500 |
| 1.0 | 439.862 | 920 | 105.093 | 1660 |

The foregoing discussion illustrates the fact that a second-law analysis is frequently as important as a first-law analysis when dealing with reactive systems.

In the absence of the second law it is possible to arrive at conclusions which are incorrect. In the next chapter we shall extend our analysis of reactive systems in the light of the second law. We shall find a more direct method of ascertaining the equilibrium state of a chemical reaction.

## 14-8 BATTERIES AND FUEL CELLS

There are a number of conventional devices which convert thermal energy into some form of work. Sources of thermal energy include solar radiation, solid, liquid, and gaseous fuels, and nuclear reactions. The work produced commonly is in the form of a rotating shaft or electrical work. The maximum work obtainable from a cyclic device which receives heat from a high-temperature source and rejects to the environment is limited to the Carnot efficiency. If the theoretical thermal efficiency is to be relatively large, the temperature of the source must be fairly high, e.g., upwards to 1500–1800°K or 3000–3500°R. Because of the presence of irreversibilities in actual practice, modern heat engines seldom achieve thermal efficiencies greater than about 40 percent. A considerable fraction of the energy supplied to the cyclic device is discharged as heat to the local surroundings. It would be highly desirable if other methods of energy conversion were available which did not rely on the conversion of heat into work, since such methods would not be limited to the Carnot efficiency.

One well-known device which bypasses the heat-work energy-conversion step is the conventional battery. By means of a controlled chemical reaction, a battery converts energy stored in the chemical bonds of the reactants into electrical work. Usually a small quantity of heat is also discharged by the battery, but the operation is nearly isothermal. A fuel cell is a form of a battery with several important changes. The electrode materials are not consumed during its operation, but remain invariant. As a result, the fuel and oxidizer must be supplied continuously from an outside source. In addition, a means of eliminating the products of the reaction must be provided. The conventional battery is a closed system, whereas the fuel cell operates as an open system.

The electrical-work output of a battery or fuel cell can be ascertained from a thermodynamic analysis of either a closed or open system. Consider the battery shown in Fig. 14-6a. We shall assume that heat transfer to or from the battery results in an isothermal process. In addition to electrical work, boundary work in the amount $P \, dV$ may also be present, since the system is maintained at a constant pressure. An energy balance on the closed system is of the form

$$\delta q + \delta w_{elec} + (-P \, dv) = du$$

or

$$\delta w_{elec} = du + P \, dv - \delta q = dh - \delta q$$

Finally, for a closed system in general, $\delta q \leq T \, ds$. When this inequality is substituted into the above expression, we find that

$$\delta w_{elec} \leq dh - T \, ds$$

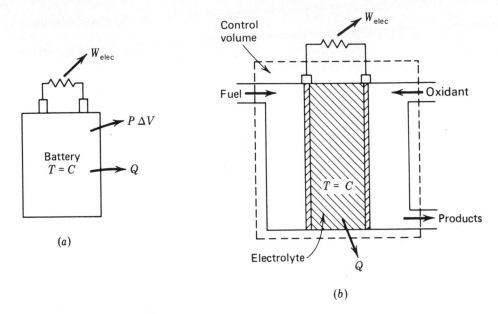

**Figure 14-6** Schematic of (*a*) a battery, and (*b*) a fuel cell.

However, since $g = h - Ts$, then under isothermal conditions we see that $dg_T = dh - T\,ds$. Therefore, for the closed system described in Fig. 14-6a,

$$\delta w_{\text{elec}} \leqslant dg_T \qquad (14\text{-}19)$$

or, for a finite change of state,

$$w_{\text{elec}} \leqslant \Delta g_T \qquad \text{(closed, isothermal)} \qquad (14\text{-}20)$$

The equality sign applies to an internally reversible process, while the inequality sign applies when irreversibilities within the battery are present.

A schematic of a steady-state fuel cell is shown in Fig. 14-6b. Again we shall assume isothermal operation. An energy balance on the control volume around the fuel cell shows that

$$\delta q + \delta w_{\text{elec}} = dh + d\text{KE} + d\text{PE}$$

In the operation of most fuel cells the changes in kinetic and potential energy may be neglected. In addition, for a unit mass passing through the control volume, $\delta q \leqslant T\,ds$. As a result, we obtain the identical result that was developed above for a closed system such as a battery, namely,

$$w_{\text{elec}} \leqslant \Delta g_T \qquad \text{(steady flow, isothermal)} \qquad (14\text{-}20)$$

Hence the maximum electrical work output of a battery or fuel cell operating under isothermal conditions is measured by the decrease in the Gibbs function for the overall process.

There are several different efficiencies frequently defined for the operation of batteries and fuel cells. One possible standard of performance is the ratio of the maximum-useful-work output to the energy input. We have seen above that the maximum-useful-work output is given by $\Delta g_R$. The energy input is the enthalpy of reaction $\Delta h_R$ released by the overall chemical reaction. The ideal efficiency or effectiveness of a battery or fuel cell, therefore, is defined by the relation

$$\eta_i = \frac{\Delta g_R}{\Delta h_R} \qquad (14\text{-}21)$$

where both $\Delta g_R$ and $\Delta h_R$ normally have negative values. For isothermal operation it is also true that $\Delta g = \Delta h - T \, \Delta s$. Therefore, the ideal efficiency frequently is seen in the form

$$\eta_i = 1 - \frac{T \, \Delta s_R}{\Delta h_R} \qquad (14\text{-}22)$$

For isothermal, reversible processes the quantity $T \, \Delta s$ represents the heat transfer to or from the system. When the reversible heat transfer is out of the system ($\Delta s$ is negative), then the ideal efficiency is less than unity.

Recall from Sec. 14-4 that the enthalpy of reaction $\Delta h_R$ can be found from the relation

$$\Delta h_R = \sum_i (v_i \, \Delta h_{f, i})_{\text{prod}} - \sum_i (v_i \, \Delta h_{f, i})_{\text{reac}} \qquad (14\text{-}10)$$

A similar equation holds for the Gibbs function. That is,

$$\Delta g_R = \sum_i (v_i \, \Delta g_{f, i})_{\text{prod}} - \sum_i (v_i \, \Delta g_{f, i})_{\text{reac}} \qquad (14\text{-}23)$$

Typical values of the Gibbs function of formation and the enthalpy of formation of compounds containing carbon, hydrogen, and oxygen are found in Tables A-22M and A-22. Similar data for substances commonly found in batteries are found in Table 14-1 for conditions of 1 atm and 25°C (77°F). Thus the change in the Gibbs function and the enthalpy for any reaction can be calculated if suitable data similar to those found in these tables are available.

In addition to the ideal efficiency of a battery or fuel cell, another important parameter is the ideal open circuit voltage developed by the cell. We have already seen that the Gibbs-function change for the reversible case is a measure of the electrical work produced by the cell. The electrical work is the product of the amount of charge $Q_e$ that passes from the cell per mole of reacting fuel and the ideal electrostatic potential $V_i$ developed. That is, $w_{\text{elec}} = -Q_e V_i$. The quantity of charge $Q_e$ is equal to the number of moles of electrons $j$ produced by the cell reaction per mole of reacting fuel multiplied by the number of coulombs (C) per mole of electrons, $\mathscr{F}$. Hence $Q_e = j\mathscr{F}$. Therefore

$$w_{\text{elec}} = -j\mathscr{F} V_i = \Delta g_R$$

or
$$V_i = \frac{-\Delta g_R}{j\mathscr{F}} \qquad (14\text{-}24)$$

**Table 14-1 Some values of the enthalpy of formation, Gibbs function of formation, and the absolute entropy at 25°C and 1 atm**

| Substance and state* | | $\Delta h_f^\circ$ kJ/kg·mol | $\Delta g_f^\circ$ kJ/kg·mol | $s^\circ$ kJ/(kg·mol)°K |
|---|---|---|---|---|
| Hg | (l) | 0 | 0 | 76.02 |
| HgO | (c) | −90,210 | −58.400 | 73.2 |
| Mn | (c) | 0 | 0 | 31.8 |
| $MnO_2$ | (c) | −520,030 | −465,180 | 53.05 |
| $Mn_2O_3$ | (c) | −958,970 | −881,150 | 110.5 |
| Pb | (c) | 0 | 0 | 64.81 |
| $PbO_2$ | (c) | −277,400 | −217,360 | 68.6 |
| $PbSO_4$ | (c) | −919,940 | −813,200 | 148.57 |
| Zn | (c) | 0 | 0 | 41.63 |
| ZnO | (c) | −348,280 | −318,320 | 43.64 |
| $H_2SO_4$ | (l) | −813,990 | −690,100 | 156.90 |
| $H_2SO_4$ | (aq, m = 1) | −909,270 | −744,630 | 20.1 |

\* (c), crystalline solid; (l), liquid; (aq, m = 1), aqueous solution, unit molality

The quantity $\Delta g_R$ is negative for battery and fuel-cell reactions, so that the ideal voltage $V_i$ is a positive value.

The quantity of charge $\mathscr{F}$ is called a faraday. Its value is

$$\mathscr{F} = \frac{6.023 \times 10^{23} \text{ electrons}}{\text{g·mol electrons}} \times \frac{1.602 \times 10^{-19} \text{ C}}{\text{electron}}$$

$$= 96,487 \text{ C/g·mol electrons}$$

If this value is multiplied by the identity, $1\text{ J} = 1\text{V·C}$, then

$$\mathscr{F} = 96,487 \text{ kJ/(V)(kg·mol electrons)} \tag{14-25}$$

Therefore Eq. (14-24) becomes

$$V_i = \frac{-\Delta g_R}{96,487j} \tag{14-26}$$

where $\Delta g_R$ is expressed in kilojoules per kilogram mole and $V_i$ is in volts.

The method of evaluation of the quantity $j$ in Eq. (14-26) is made more clear by examining two specific examples. Figure 14-7a is a schematic of the well-known lead-acid storage battery. This cell consists of two electrodes: one of metallic lead and the other of lead oxide, $PbO_2$. The electrode plates are immersed in a water solution of sulfuric acid. In the solution, called the *electrolyte*, the sulfuric acid dissociates into hydrogen and sulfate ions. The overall reaction during discharge of the cell is

$$Pb + PbO_2 + 2H_2SO_4 \longrightarrow 2PbSO_4 + 2H_2O \tag{14-27}$$

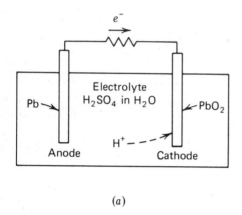

(a)

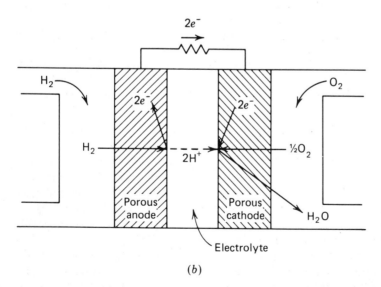

(b)

**Figure 14-7** Schematic of (a) a lead-acid battery, and (b) a hydrogen-oxygen fuel cell.

Thus, as the cell is discharged, both of the plates are converted to lead sulfate and the sulfuric acid is consumed while excess water is produced. It is now convenient to express the overall chemical reaction in terms of the individual reactions which occur at each electrode. The reactions that occur at the anode and cathode are called the *half-cell* reactions. For the lead-acid battery the anode half-cell reaction is

$$Pb + SO_4^{--} \longrightarrow PbSO_4 + 2e^-$$

The electrons released at the anode now pass through an external circuit in order to complete the reaction. By placing an external load in the circuit, useful work can be produced by the process, even though the reaction is isothermal. Since chemical energy is converted directly to electrical energy, there is no Carnot limitation of the conversion process. At the cathode the reaction is

$$PbO_2 + SO_4^{--} + 4H^+ + 2e^- \longrightarrow PbSO_4 + 2H_2O$$

This reaction requires the electrons that were released at the anode, and which pass through the external circuit. In this particular case it is seen that $j$ has a value of 2. That is, 2 electrons are released when a molecule of reactant at the anode is consumed.

A fuel cell operating on hydrogen and oxygen has already been used on space missions to the moon. The overall reaction at room temperature is

$$H_2(g) + \tfrac{1}{2}O_2(g) \longrightarrow H_2O(l) \tag{14-28}$$

A schematic of the cell operating with an acidic electrolyte is shown in Fig. 14-7$b$. The half-cell reactions are:

Anode: $\qquad\qquad\qquad H_2 \longrightarrow 2e^- + 2H^+$

Cathode: $\quad 2H^+ + 2e^- + \tfrac{1}{2}O_2 \longrightarrow H_2O(l)$

Thus, during the process hydrogen ions migrate through the electrolyte from the anode to the cathode, while electrons pass through the external circuit. In this case $j$ is again 2, since 2 electrons are liberated at the anode for every molecule of hydrogen consumed.

One important consideration in the analysis of any fuel cell is the effect of temperature on performance. First of all, the ideal efficiency may be affected by temperature. This depends upon the variation of $\Delta g_R$ and $\Delta h_R$ with temperature. These properties are fairly sensitive to temperature for the hydrogen-oxygen cell, while it is found that they are not affected significantly for the carbon-oxygen cell, for example. The influence of temperature on the rate of reaction is also important. Cells employing carbon or hydrocarbons as the fuel frequently must be operated at elevated temperatures in order to achieve sufficient rates.

Finally, we need to consider the factors which influence the performance of actual fuel cells. In the actual operation there are a number of irreversible effects within the cell which greatly reduce the terminal voltage of the cell. The losses within a fuel cell are generally spoken of as overvoltages or polarization effects.

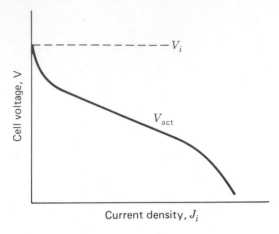

**Figure 14-8** A typical voltage-current density diagram for a fuel cell.

These frequently are grouped into three classes: (1) resistance or ohmic polarization, (2) activation or chemical polarization, and (3) concentration polarization. The magnitude of each of these effects is a function of the current density $J_i$. The first of these is caused by the internal resistance of the cell, and it includes losses in the electrolyte as well as in the electrodes. The activation polarization arises from chemical changes occurring at the surface of the electrodes, as well as adsorption and desorption effects on the surface. The concentration polarization is caused by the concentration gradients set up in the electrolyte and in the gas streams in the vicinity of the electrodes.

The net result of the three polarization effects or overvoltages is a reduction in the terminal voltage at a given current density. Figure 14-8 presents a typical diagram for the cell voltage versus the current density. Current densities typically range from 0 to 1 A/cm². The rapid drop at low current is due to the activation or chemical polarization. In the middle range neither the activation or concentration effects are changing appreciably, and the curve varies linearly due to the ohmic drop. Finally, at high current densities the concentration polarization drops the cell output sharply.

## PROBLEMS (METRIC)

**Fuel-air combustion analysis**

**14-1M** Ethane ($C_2H_6$) is burned with air in a mass ratio of 1 : 18. Compute (a) the percent excess air used, (b) the equivalence ratio used, and (c) the percent $CO_2$ by volume in the product gases, if combustion is complete.

**14-2M** Same as Prob. 14-1M, except that the fuel is ethylene ($C_2H_4$).

**14-3M** Same as Prob. 14-1M, except that the fuel is acetylene ($C_2H_2$).

**14-4M** Same as Prob. 14-1M, except that the fuel is propylene ($C_3H_6$).

**14-5M** One mole of ethane ($C_2H_6$) is burned with 20 percent excess air. Determine (a) the air-fuel ratio, (b) the equivalence ratio used, and (c) the mole percent of $N_2$ in the product gas, if combustion is complete.

**14-6M** Same as Prob. 14-5M, except that the fuel is ethylene $(C_2H_4)$.

**14-7M** Same as Prob. 14-5M, except that the fuel is acetylene $(C_2H_2)$.

**14-8M** Same as Prob. 14-5M, except that the fuel is propylene $(C_3H_6)$.

**14-9M** Determine the composition (mole percent on a dry basis) of the products formed by the complete combustion (with 20 percent excess of air) of a casinghead gas with the following volumetric composition: $CH_4$, 60 percent; $C_2H_6$, 30 percent; $N_2$, 10 percent.

**14-10M** A gaseous fuel having a volumetric analysis of 65 percent $CH_4$, 25 percent $C_2H_6$, 5 percent $CO_2$, and 5 percent $N_2$ is burned with 30 percent excess air. Determine the air-fuel ratio used.

**14-11M** Liquid benzene is burned with (a) 200 percent, and (b) 150 percent, theoretical-air requirements. Compute the air-fuel ratio and the equivalence ratio used, assuming complete combustion.

**14-12M** The flammable range for propane $(C_3H_8)$ at normal temperature and pressure is from 2.2 to 9.5 percent fuel vapor by volume. Determine the percent excess air present when combustion occurs at (a) the lowest possible vapor content, and (b) the highest possible vapor content.

**14-13M** Determine the volume flow rate in $m^3/min$ of air at 1 bar and 27°C required to burn 1 kg/min of fuel for the fuel-air mixtures specified in (a) Prob. 14-1M, (b) Prob. 14-2M, (c) Prob. 14-3M, (d) Prob. 14-4M, (e) Prob. 14-5M, (f) Prob. 14-6M, (g) Prob. 14-7M, and (h) Prob. 14-8M.

### Dew-point temperature evaluation

**14-14M** Determine the dew-point temperature in °C for the products of the reaction specified in (a) Prob. 14-1M, (b) Prob. 14-2M, (c) Prob. 14-3M, (d) Prob. 14-4M, (e) Prob. 14-5M, (f) Prob. 14-6M, (g) Prob. 14-7M, and (h) Prob. 14-8M. The pressure of the exit gas stream is 1 bar.

**14-15M** Pentane gas $(C_5H_{12})$ is burned to completion with the stoichiometric amount of air. If the total pressure is 1.03 bars, find the dew-point temperature of the products, in °C.

**14-16M** Methane is burned with 50 percent excess air. The products of combustion are at 0.95 bar and 400°K. Determine the dew-point temperature of the products, in °C.

**14-17M** Ethyl alcohol $(C_2H_5OH)$ is burned with (a) 25 percent excess air, and (b) 50 percent excess air. Determine the dew-point temperature, in °C, for the product gas if the total pressure is 1 bar.

**14-18M** If ethane $(C_2H_6)$ is burned with (a) stoichiometric air and (b) 50 percent excess air, and the products are cooled to 25°C and 1 bar, how many moles of water vapor, per mole of fuel burned, would condense?

**14-19M** If butane $(C_4H_{10})$ is burned with (a) stoichiometric air and (b) 50 percent excess air, and the products are cooled to 25°C and 1 bar, how many moles of water vapor, per mole of fuel burned, would condense?

### Analysis of product gases

**14-20M** Octane fuel $(C_8H_{18})$ is burned with air in a combustion test, and a volumetric analysis reveals the following composition of the products: 11.0 percent $CO_2$, 3.6 percent $O_2$, and 1.5 percent CO. Compute the actual air-fuel ratio used during the test and the percent theoretical air used.

**14-21M** The combustion of methane gas with air leads to the following volumetric analysis during a test: 9.7 percent $CO_2$, 0.5 percent CO, and 3.0 percent $O_2$. Determine the moles of air used per mole of fuel in the test, and compare with the same ratio for stoichiometric combustion.

**14-22M** Solid carbon is burned with air in a combustion test. A volumetric analysis reveals that the products of combustion include 3.5 percent CO, 13.8 percent $CO_2$, and 5.2 percent $O_2$. Determine the air-fuel ratio used during the test and the percent theoretical air used.

**14-23M** A fuel gas with the volumetric analysis 60 percent CO, 20 percent $H_2$, 20 percent $N_2$ is burned in a steady-flow process with dry air to give combustion gases at 500°F. A volumetric analysis of these combustion gases gives 20.0 percent $CO_2$, 5.0 percent CO, 2.8 percent $O_2$. Calculate the pounds of air supplied for each pound of fuel gas burned.

**14-24M** A gaseous fuel is composed of 20 percent $CH_4$, 40 percent $C_2H_6$, 40 percent $C_3H_8$, where all percentages are by volume. The volumetric analysis of dry products of combustion of this fuel gives

10.6 percent $CO_2$, 3 percent $O_2$, 1 percent CO. Determine (a) the gravimetric analysis of fuel, (b) the theoretical air-fuel ratio, and (c) the percent excess air.

**14-25M** A gas has the following volumetric analysis by percent: $CH_4$, 80.62; $C_2H_6$, 5.41; $C_3H_8$, 1.87; $C_4H_{10}$, 1.60; $N_2$, 10.50. A volumetric analysis of the products of combustion shows 7.8 percent $CO_2$, 7.0 percent $O_2$, 0.2 percent CO. Calculate the actual fuel-air ratio.

**14-26M** Sulfur dioxide ($SO_2$) is burned with air catalytically to form $SO_3$. The products of combustion reveal 12.8 percent $SO_3$, 2.0 percent $SO_2$, and 12.9 percent $O_2$, by volume. The remaining gas in the products is nitrogen. Determine the percent excess air used.

**14-27M** Ethylene ($C_2H_4$) is burned with 33 percent excess air. An analysis of the products of combustion on a dry basis reveals 6.06 percent $O_2$ by volume. The remaining data for the product analysis are missing. What percent of the carbon in the fuel was converted to CO instead of $CO_2$?

**14-28M** Solid carbon is reacted with air, and a volumetric analysis of the products reveals 17.85 percent $CO_2$, 1.14 percent CO, and 2.47 percent $O_2$. Determine the air-fuel ratio used in the test.

**14-29M** The volumetric analysis of the dry products of combustion of a hydrocarbon fuel described by the general formula $C_xH_y$ is: $CO_2$, 13.6 percent; $O_2$, 0.4 percent; CO, 0.8 percent; $CH_4$, 0.4 percent; and $N_2$, 84.8 percent. Determine the values of $x$ and $y$ for the fuel on the basis of 13.6 mol of $CO_2$ in the products of combustion.

**14-30M** Methane ($CH_4$) is burned with 80 percent of the theoretical air requirements.
   (a) What percent of the carbon in the fuel is converted to $CO_2$?
   (b) What equivalence ratio was used?

**14-31M** Propane ($C_3H_8$) reacts with a deficiency of air so that no $CO_2$ is formed, only CO.
   (a) What percent theoretical air would be used?
   (b) What would be the equivalence ratio used?

**14-32M** Pentane ($C_5H_{12}$) is reacted with a deficiency of air so that only (a) 68 percent, and (b) 12 percent, of the carbon in the fuel is converted to $CO_2$. The remaining carbon appears as CO. Determine (1) the percent theoretical air used and (2) the equivalence ratio used.

### Energy analysis with heat effects

**14-33M** Liquid octane ($n$-$C_8H_{18}$) and air in 50 percent excess enter a combustion chamber in steady flow at 1 atm and 27°C. Determine the amount of heat per mole of fuel which must be removed to cool the combustion products down to 1000°K.

**14-34M** Ethane gas ($C_2H_6$) at 25°C and air at 227°C enter a combustion chamber in steady flow at 1 bar pressure. The products of combustion leave at 1100°K. If the percent excess air used is 25 percent, and combustion is complete, determine the heat loss, in kJ/kg·mol of fuel.

**14-35M** Liquid octane ($n$-$C_8H_{18}$) at 25°C is sprayed into a combustor of a jet engine where it is burned continuously with (a) 400 percent theoretical air, and (b) 500 percent theoretical air from the compressor at 500°K. The products of combustion leave at 1000°K. Determine the heat transfer from the combustor, in kJ/kg·mol of fuel.

**14-36M** If propane is burned in a steady-flow system with 100 percent excess air (both originally at 25°C) and the products are cooled to 600°K, how much energy is given up as heat transfer, in kJ/kg·mol of fuel?

**14-37M** The combustor of a gas turbine is supplied with air at 500°K and liquid $n$-heptane at 25°C. The products of combustion leave at 1200°K. If the air-fuel ratio used is 18 : 1, determine the heat transfer, in kJ/kg·mol of fuel, for complete combustion.

**14-38M** Carbon monoxide is burned with air at an initial state of 25°C. The final temperature of the products is (a) 1100°K, and (b) 1000°K, and during the process 38,000 kJ of heat per kilogram-mole of CO are removed. Determine the percent excess air used if combustion is complete.

**14-39M** In the reaction of solid carbon and $O_2$ gas at 25°C and a molar ratio of 1 : 2, the analysis of the products of combustion indicated that 90 percent of the carbon burned to $CO_2$, and the remaining 10 percent appeared as CO. If the final temperature is 2500°K, how much heat, in kJ/kg·mol of carbon, was added or rejected during the process?

**14-40M** Hydrogen is burned with (*a*) stoichiometric air, and (*b*) 100 percent excess air at 25°C and 1 bar. During the process, a heat loss of 35,000 kJ/kg·mol of fuel occurs. Find the final gas temperature in °K.

**14-41M** Acetylene gas ($C_2H_2$) and air enter a combustion chamber at 25°C in such a ratio that 1 mole of CO per mole of $CO_2$ is formed in the product gas. If the products are at 800°K, find the quantity of heat transferred, in kJ/kg·mol of fuel.

**14-42M** Propane gas ($C_3H_8$) with 20 percent excess air enters a combustion chamber in steady flow at 1 bar and 25°C. In the final products, which leave the chamber at 900°K, the percent of carbon converted to $CO_2$ is (*a*) 94 percent, and (*b*) 96 percent. The remaining carbon appears as CO. Determine the heat transferred, in kJ/kg·mol of fuel.

### Enthalpy of formation and reaction evaluation

**14-43M** The enthalpy of reaction for the combustion of hydrazine ($N_2H_4$) with oxygen is 624,800 kJ/kg·mol for the specific reaction $N_2H_4(l) + O_2(g) \rightarrow N_2(g) + 2H_2O(l)$. Determine the enthalpy of formation of liquid hydrazine at 25°C in kJ/kg·mol.

**14-44M** Consider the use of hydrogen peroxide ($H_2O_2$) and methane as the oxidant and fuel in a propulsion system. The overall reaction is $4H_2O_2(l) + CH_4(g) \rightarrow 6H_2O(g) + CO_2(g)$. Determine the enthalpy of reaction for the above reaction at 25°C if the enthalpy of formation of $H_2O_2(l)$ is $-197,400$ kJ/kg·mol.

**14-45M** The enthalpy of combustion of isobutane is $-2,872,600$ kJ/kg·mol at 25°C. Determine the enthalpy of formation of this isomer, in kJ/kg·mol.

**14-46M** The lower heating value of gaseous butene ($C_4H_8$) at 25°C is 45,220 kJ/kg. Determine the enthalpy of formation of butene in kJ/kg·mol. Also determine the enthalpy of hydrogenation in kJ/kg·mol associated with the reaction $C_4H_8(g) + H_2(g) \rightarrow C_4H_{10}(g)$.

### Adiabatic-combustion processes

**14-47M** With reference to Prob. 14-44M, if the stoichiometric ratio of fuel to oxidant were introduced at 25°C into a steady-flow combustion device, what would the adiabatic-combustion temperature be in °K?

**14-48M** If ethylene ($C_2H_4$) at 25°C is burned with 300 percent excess air supplied at 400°K and the reaction occurs at constant pressure in the gas phase, estimate the maximum combustion temperature, in °K.

**14-49M** Hydrogen gas at 25°C is reacted with (*a*) 500 percent, (*b*) 600 percent, and (*c*) 700 percent theoretical oxygen requirements which enter at 500°K. Estimate the maximum combustion temperature in °K.

**14-50M** Propane gas ($C_3H_8$) is burned at constant pressure with (*a*) 20 percent, and (*b*) 40 percent excess air starting at 25°C. Determine the maximum combustion temperature in °K.

**14-51M** Determine the maximum theoretical combustion temperature for the reaction of ethane with (*a*) 30 percent, and (*b*) 50 percent excess air in a steady-flow process. The reactants enter at 25°C and the reaction goes to completion.

**14-52M** Determine the maximum temperature when methane is burned with the stoichiometric amount of air, both entering at 25°C. Assume that 20 percent of the carbon is burned only to CO.

**14-53M** Carbon monoxide undergoes adiabatic combustion with the stoichiometric amount of air. Neglecting dissociation of products, determine the maximum combustion temperature, in °K, if the initial reactants are (*a*) at 25°C, and (*b*) at 1000°K.

**14-54M** The gas turbine combustor is supplied with air at 500°K and liquid octane at 25°C. The products of combustion leave at 1400°K. Calculate the air-fuel ratio used if the flow is steady, the combustion is complete, and the heat loss is negligible.

**14-55M** Determine the adiabatic-combustion temperature for the reaction of liquid methanol ($CH_3OH$) with the theoretical air requirements at 1 bar if all the reactants enter the combustion chamber at 25°C in steady flow. Assume complete combustion.

**14-56M** The decomposition of tetranitromethane may proceed according to the reaction $C(NO_2)_4 \rightarrow CO_2 + 2N_2 + 3O_2$. The enthalpy of reaction is $-250,600$ kJ/kg·mol. Estimate the adiabatic-combustion temperature, in °K, if the reactant enters at 25°C.

**14-57M** One possible reaction for the catalytic decomposition of hydrazine is $2N_2H_4 \rightarrow N_2 + H_2 + 2NH_3$. Its enthalpy of reaction at 25°C is $-151,300$ kJ/kg·mol. Estimate the maximum possible temperature of the product gases for adiabatic decomposition.

**14-58M** Consider the adiabatic decomposition of gaseous hydrogen peroxide according to the reaction $H_2O_2 \rightarrow H_2O + \frac{1}{2}O_2$. Determine the maximum theoretical temperature of the product gases, in °K, if the reactant enters at 25°C.

### Constant-volume combustion

**14-59M** Determine the adiabatic-combustion temperature, in °K, and the explosion pressure in bars for the constant-volume combustion of CO with 50 percent excess air. Initial conditions are 1 bar and 27°C.

**14-60M** An equimolar mixture of hydrogen and carbon monoxide, together with the theoretical amount of air for complete combustion, is ignited in a constant-volume bomb. The initial conditions are 3 bars and 25°C. Estimate the maximum temperature and pressure that would be attained, assuming complete combustion.

**14-61M** Two cubic centimeters of liquid benzene ($C_6H_6$) are placed in a 28.3-l constant-volume bomb at 25°C. If the fuel is burned with the stoichiometric amount of air initially at 1 bar, determine the explosion pressure in bars.

**14-62M** Same as Prob. 14-61M, except the fuel is liquid methyl alcohol ($CH_3OH$).

**14-63M** A very strong tank holds a large supply of oxygen and a small amount of a fuel at 25°C and 1 bar. The fuel is ignited and the combustion products are brought back to 25°C. Determine the heat transfer in kilojoules if the fuel is (a) 0.1 g·mol of CO, and (b) 0.1 g·mol of solid carbon.

### Entropy changes for reacting mixtures

**14-64M** One kilogram-mole of methane gas is burned with the stoichiometric amount of air initially at 25°C and 1 atm. If the complete products of combustion are cooled to 25°C and 1 atm, determine the entropy change of the process, in kJ/°K per kilogram-mole of fuel.

**14-65M** Same as Prob. 14-64M, except that the fuel is gaseous ethane.

**14-66M** Same as Prob. 14-64M, except that the fuel is gaseous ethylene.

**14-67M** Same as Prob. 14-64M, except that the fuel is gaseous ethane.

**14-68M** Same as Prob. 14-64M, except that the fuel is gaseous propane.

**14-69M** Same as Prob. 14-64M, except that the fuel is liquid methyl alcohol.

**14-70M** Gaseous carbon monoxide and water vapor react at 25°C and 1 atm according to the reaction $CO + H_2O \rightarrow CO_2 + H_2$. If it is assumed that the reaction goes to completion, determine the entropy change in kJ/°K per kilogram-mole of CO.

**14-71M** Consider the process described in Prob. 14-50M, part (a). Evaluate the entropy change in kJ/°K per kilogram-mole of fuel.

**14-72M** Consider the process described in Prob. 14-52M. Evaluate the entropy change in kJ/°K per kilogram-mole of fuel.

**14-73M** Consider the process described in Prob. 14-53M, part (a). Evaluate the entropy change in kJ/°K per kilogram-mole of fuel.

**14-74M** Consider the process described in Prob. 14-55M. Evaluate the entropy change in kJ/°K per kilogram-mole of fuel.

### Batteries and fuel cells

**14-75M** The composition of a gaseous mixture of CO, $H_2$, and air is 1 : 1 : 5.71 on a mole basis. The reactants enter and the products leave at 25°C and 1 atm. Water is a liquid in the products. Determine the maximum work output for complete reaction in a steady-flow process, in kJ/kg·mol of CO and of $H_2$.

**14-76M** Methyl alcohol and stoichiometric air in the gas phase at 25°C react in steady flow and 1 atm to form complete products also at 25°C. Determine the maximum work possible in kJ/kg·mol of fuel if the water in the products is a liquid.

**14-77M** Determine the maximum possible work output in kJ/kg·mol of fuel for the reaction of gaseous methane with stoichiometric air at 25°C and 1 atm. The water in the products is a liquid.

**14-78M** Same as Prob. 14-77M, except that the fuel is ethane.

**14-79M** Same as Prob. 14-77M, except that the fuel is ethylene.

**14-80M** On the basis of Prob. 14-76M, determine the ideal efficiency and the ideal voltage if the reaction occurred in a fuel cell.

**14-81M** On the basis of Prob. 14-77M, determine the ideal efficiency and the ideal voltage if the reaction occurred in a fuel cell.

**14-82M** On the basis of Prob. 14-78M, determine the ideal efficiency and the ideal voltage if the reaction occurred in a fuel cell.

**14-83M** On the basis of Prob. 14-79M, determine the ideal efficiency and the ideal voltage if the reaction occurred in a fuel cell.

# PROBLEMS (USCS)

### Fuel-air combustion analysis

**14-1–14-12** Use Probs. 14-1M through 14-12M above.

**14-13** Determine the volume flow rate in ft³/min of air at 1 atm and 80°F required to burn 1 lb/min of fuel for the fuel-air mixtures specified in (a) Prob. 14-1, (b) Prob. 14-2, (c) Prob. 14-3, (d) Prob. 14-4, (e) Prob. 14-5, (f) Prob. 14-6, (g) Prob. 14-7, and (h) Prob. 14-8.

### Dew-point temperature evaluation

**14-14** Determine the dew-point temperature, in °F, for the products of the reaction specified in (a) Prob. 14-1, (b) Prob. 14-2, (c) Prob. 14-3, (d) Prob. 14-4, (e) Prob. 14-5, (f) Prob. 14-6, (g) Prob. 14-7, and (h) Prob. 14-8. The pressure of the exit stream is 14.7 psia.

**14-15** Pentane gas is burned to completion with the stoichiometric amount of air. If the total pressure is 15.2 psia, find the dew-point temperature of the products in °F.

**14-16** Methane is burned with 50 percent excess air. The products of combustion then exist at 14.5 psia and 500°F. Determine the dew-point temperature of the water vapor in the products of combustion.

**14-17** Ethyl alcohol ($C_2H_5OH$) is burned with 50 percent excess air. Determine (a) the dew-point temperature for the water vapor in the products of combustion if the total pressure is 1 atm, and (b) the air-fuel ratio.

**14-18** If ethane ($C_2H_6$) is burned with (a) stoichiometric air, and (b) 50 percent excess air, and the products are cooled to 80°F and 1 atm, how many moles of water vapor, per mole of fuel burned, would condense?

**14-19** If butane ($C_4H_{10}$) is burned with (a) stoichiometric air, and (b) 50 percent excess air, and the products are cooled to 80°F and 1 atm, how many moles of water vapor, per mole of fuel burned, would condense?

### Analysis of product gases

**14-20–14-32** Use Probs. 14-20M through 14-32M above.

### Energy analysis with heat effects

**14-33** Liquid octane ($n$-$C_8H_{18}$) and air in 50 percent excess enter a combustion chamber in steady flow at 1 atm and 80°F. Determine the amount of heat per mole which must be removed to cool the combustion products down to 1800°R.

**14-34** Ethane gas ($C_2H_6$) at 77°F and air at 540°F enter a combustion chamber in steady flow at 1 atm pressure. The products of combustion leave at 2000°R and 1 atm. If the percent excess air is 25 percent, determine the heat loss during the process, in Btu/mol of fuel, assuming complete combustion.

**14-35** Liquid octane ($n$-$C_8H_{18}$) at 77°F is sprayed into the combustor of a jet engine where it is burned continuously with 400 percent theoretical air from the compressor at 900°R. The products of combustion leave at 1800°R. Determine the heat loss from the combustor per mole of octane burned. Neglect kinetic- and potential-energy changes. $h_{fg}$ is 17,800 Btu/mol for liquid octane at 77°F.

**14-36** If propane is burned in an open-steady-flow system with 100 percent excess air (both originally at 77°F) and the products are cooled to 620°F, how much energy, in Btu/mole of fuel, is given up to the surroundings in the form of heat?

**14-37** The combustor of a gas turbine is supplied with dry air at 440°F and liquid $n$-heptane at 77°F. The products of combustion leave at 1700°F. Calculate the required air-fuel ratio. Assume that the flow is steady, the heat loss to the surroundings is negligible, and the combustion is complete.

**14-38** Carbon monoxide is burned with air at an initial state of 25°C. The final temperature of the products is found to be 2000°R, and during the process 17,000 Btu of heat are removed per mole of CO. Determine the percent excess air used if combustion is complete.

**14-39** In the reaction of $C(s)$ and $2O_2(g)$, both initially at 77°F, the analysis of the products of combustion at 4500°R indicated that 90 percent of the carbon burned to $CO_2$, and the remaining appeared as CO. How much heat, in Btu/lb·mol of $C(s)$, was added or rejected during the reaction?

**14-40** Hydrogen is burned with (a) stoichiometric air, and (b) 100 percent excess air at 77°F and 1 atm. During the process a heat loss of 17000 Btu/lb·mol of fuel occurs. Determine the final gas temperature, in °R.

**14-41** Acetylene gas ($C_2H_2$) and air enter a combustion chamber at 77°F in such a ratio that 1 mole of CO is formed per mole of $CO_2$ in the product gas. If the products are at 1440°R, find the quantity of heat transferred, in Btu/lb·mol of fuel.

**14-42** Propane gas ($C_3H_8$) with 20 percent excess air enters a combustion chamber in steady flow at 1 atm and 77°F. In the final products, which leave the chamber at 1200°F, the percent of carbon converted to $CO_2$ is 97 percent. The remaining carbon appears as CO. Determine the heat transferred in Btu/lb·mol of fuel.

### Enthalpy of formation and reaction evaluation

**14-43** The enthalpy of reaction for the combustion of hydrazine ($N_2H_4$) with oxygen is $-268,600$ Btu/lb·mol for the specific reaction $N_2H_4(l) + O_2(g) \rightarrow N_2(g) + 2H_2O(l)$. Determine the enthalpy of formation of liquid hydrazine at 77°F.

**14-44** Consider the use of hydrogen peroxide ($H_2O_2$) and methane ($CH_4$) as the oxidant and fuel in a propulsion system. The overall reaction is $4H_2O_2(l) + CH_4(g) \rightarrow 6H_2O(g) + CO_2(g)$. Determine the enthalpy of reaction for the above reaction at 77°F if the enthalpy of formation of $H_2O_2(l)$ is $-84,850$ Btu/lb·mol.

**14-45** The enthalpies of combustion of $n$-butane and isobutane are $-1,237,800$ and $-1,234,850$ Btu/lb·mol, respectively. Calculate the enthalpy of formation of each of these isomers from its stable elements, and also compute the enthalpy of isomerization for the reaction $n$-butane $\rightarrow$ isobutane in the gas phase.

**14-46** The lower heating value of gaseous butene ($C_4H_8$) is 19,483 Btu/lb at 77°F and 14.7 psia.

   (a) Determine the enthalpy of formation of butene in Btu/mol.

   (b) Determine the enthalpy of hydrogenation that is associated with the reaction $C_4H_8(g) + H_2(g) \rightarrow C_4H_{10}(g)$.

**Adiabatic-combustion processes**

**14-47** With reference to Prob. 14-44, if the stoichiometric ratio of fuel to oxidant were introduced at 77°F into a steady-flow combustion device, what would the adiabatic-combustion temperature be in °R?

**14-48** If ethylene ($C_2H_4$) at 77°F is burned with 300 percent excess air supplied at 260°F and the gaseous reaction occurs at constant pressure, estimate the maximum combustion temperature in °R.

**14-49** Hydrogen gas at 77°F is reacted with 700 percent theoretical oxygen requirements which enter at 1200°F. Assuming complete combustion, determine the adiabatic-flame temperature in °F.

**14-50** Propane gas ($C_3H_8$) is burned at a constant pressure of 1 atm with 20 percent excess air starting at 77°F. Assuming no dissociation, determine the maximum adiabatic temperature.

**14-51** Determine the maximum theoretical combustion temperature for the reaction of ethane with 50 percent excess air in a steady-flow process. The ethane and the air both enter the combustor at 1 atm and 77°F, and it is assumed that the reaction goes to completion.

**14-52** Determine the maximum flame temperature when methane gas is burned with the stoichiometric amount of air at a constant pressure of 1 atm, both reactants entering at 25°C. Assume that the combustion with respect to carbon is only 80 percent complete; that is, 20 percent of the carbon in methane is burned only to CO. Dissociation of the products at high temperatures may be neglected.

**14-53** One mole of carbon monoxide undergoes adiabatic combustion with the stoichiometric amount of air. Assuming no dissociation of the products, determine the maximum combustion temperature in °R if (*a*) the initial reactants are at 77°F, and (*b*) the initial reactants are at 1800°R.

**14-54** The combustor of a gas-turbine power plant is supplied with air at 440°F and liquid octane at 77°F. The products of combustion leave at 2060°F. Calculate the air-fuel ratio used if the flow is steady, the combustion is complete, and the heat loss is negligible.

**14-55** Determine the adiabatic-combustion temperature for the reaction of liquid methanol ($CH_3OH$) with the theoretical air requirements at 1 atm if all the reactants enter the combustion chamber at 77°F in steady flow. Assume complete combustion.

**14-56** The decomposition of tetranitromethane may proceed according to the reaction $C(NO_2)_4 \rightarrow CO_2 + 2N_2 + 3O_2$. The enthalpy of reaction is $-107,800$ Btu/lb·mol. Estimate the adiabatic-combustion temperature, in °R, if the reactants enter at 77°F.

**14-57** One possible reaction for the catalytic decomposition of hydrazine is $2N_2H_4 \rightarrow N_2 + H_2 + 2NH_3$. The enthalpy of reaction at 77°F is $-65,100$ Btu/lb·mol $N_2H_4$. Estimate the maximum possible temperature of the product gases for adiabatic decomposition in °R.

**14-58** Consider the adiabatic decomposition of gaseous hydrogen peroxide according to the reaction $H_2O_2 \rightarrow H_2O + \frac{1}{2}O_2$. Determine the maximum theoretical temperature of the product gases, in °R, if the reactant enters at 77°F.

**Constant volume combustion**

**14-59** Determine the adiabatic-flame temperature and the explosion pressure for the constant-volume combustion of CO with 50 percent excess air. The initial conditions are 1 atm and 27°C.

**14-60** An equimolar mixture of hydrogen and carbon monoxide, together with the theoretical amount of air for complete combustion, is ignited in a constant-volume bomb. The initial pressure and temperature are 3 atm and 77°F, respectively. Estimate the maximum temperature and pressure that would be attained, assuming complete combustion and ideal-gas behavior.

**14-61** Two cubic centimeters of liquid benzene ($C_6H_6$) are placed in a 1.0-ft³ constant-volume bomb. The initial temperature of the reactants is 25°C. If the fuel is burned with the stoichiometric amount of air, determine the explosion pressure in psia.

**14-62** Two cubic centimeters of liquid methyl alcohol ($CH_3OH$) are placed in a 1.0-ft³ constant-volume bomb. The initial reactant temperature is 77°F. If the fuel is burned with the stoichiometric amount of air, determine the explosion pressure in psia.

**14-63** A very strong tank holds a large supply of oxygen and a small amount of a fuel at 77°F and 1 atm. The fuel is ignited and the combustion products are brought back to 77°F. Determine the heat transfer, in Btu, if the fuel is (a) 0.1 lb·mol of CO, and (b) 0.1 lb·mol of carbon.

### Entropy changes for reacting mixtures

**14-64** One pound-mole of methane gas is burned with the stoichiometric amount of air initially at 77°F and 1 atm. If the complete products of combustion are cooled to 77°F and 1 atm, determine the entropy change of the process, in Btu/°R per pound-mole of fuel.

**14-65** Same as Prob. 14-64, except that the fuel is gaseous ethane.

**14-66** Same as Prob. 14-64, except that the fuel is gaseous ethylene.

**14-67** Same as Prob. 14-64, excpet that the fuel is gaseous acetylene.

**14-68** Same as Prob. 14-64, except that the fuel is gaseous propane.

**14-69** Same as Prob. 14-64, except that the fuel is liquid methyl alcohol.

**14-70** Gaseous carbon monoxide and water vapor react at 77°F and 1 atm according to the reaction $CO + H_2O \rightarrow CO_2 + H_2$. If it is assumed that the reaction goes to completion, determine the entropy change, in Btu/°R per pound-mole of CO.

**14-71** Consider the process described in Prob. 14-50. Evaluate the entropy change, in Btu/°R per pound-mole of fuel.

**14-72** Consider the process described in Prob. 14-52. Evaluate the entropy change, in Btu/°R per pound-mole of fuel.

**14-73** Consider the process described in Prob. 14-53, part (a). Evaluate the entropy change, in Btu/°R per pound-mole of fuel.

**14-74** Consider the process described in Prob. 14-55. Evaluate the entropy change, in Btu/°R per pound-mole of fuel.

# FIFTEEN

## CHEMICAL EQUILIBRIUM

The study of energy transformations which result from chemical reactions was presented in the preceding chapter from the viewpoint of the first law of thermodynamics. On the basis of either the theoretical chemical equation for complete combustion or the experimental analysis of the products of the actual reaction, two types of calculations are possible. If the initial and final states of the reaction are known, the net heat and/or work interactions during the process can be computed. In the case of adiabatic process, a theoretical maximum temperature can be determined. Either type of calculation may be made on open or closed systems.

A first-law computation based on a theoretical chemical equation is often quite misleading, since the actual products frequently do not match those predicted by the theoretical reaction. In the absence of experimental measurements on the product gas, a better theoretical prediction of actual behavior can be achieved through the use of the second law of thermodynamics. Although the second-law approach described in this chapter leads to a definite improvement in the thermodynamic analysis of reactive systems, the results still may not match actual behavior. One major reason is that many chemical reactions are rate controlled. Since thermodynamics is directed toward equilibrium situations, it does not account for rate-controlled phenomena. Consequently, in numerous studies of reactive systems the chemical kinetics of the reaction must also be studied. Nevertheless, a second-law analysis of reactive systems is of major importance in the study of engineering processes which involve chemical reactions.

## 15-1 INTRODUCTION

In conjunction with the third law of thermodynamics, which establishes a base, or reference state, for the evaluation of absolute entropy values, the entropy change for any theoretical chemical reaction may be computed. This type of calculation for reactive ideal-gas mixtures was illustrated at the end of the preceding chapter. In general, this entropy change could be either positive or negative, depending upon the initial and final states of the system. However, if a reaction occurs in an isolated or an adiabatic, closed system, the second law of thermodynamics requires that the value of the entropy of the system must increase for all spontaneous, or irreversible, processes. In view of this requirement, the entropy of the system must reach a maximum value at the final equilibrium state, consistent with the external constraints on the system. In Sec. 14-7 the variation of the entropy for the reactive system, $CO + H_2O \rightarrow CO_2 + H_2$, was illustrated. For an equimolar mixture of reactants, it was found that the entropy of the system increased for the complete formation of $CO_2$ and $H_2$ in the products. Although this agrees with the basic statement of the second law for adiabatic, closed systems, the final state for the theoretical reaction is not the state of maximum entropy. As shown in Fig. 14-5 for initial reactants at $400°K$ $(260°F)$, the entropy is a maximum for the reactive mixture when approximately 0.7 mol of $CO_2$ (and $H_2$) is formed from 1 mole each of $CO$ and $H_2O$. A process from state $b$ on the diagram to state $c$ is impossible, since it violates the second law of thermodynamics. For any change of state of an adiabatic, closed system, the entropy can never decrease in value. Consequently, the process on a macroscopic basis will stop at state $b$, where the number of moles of $CO_2$ formed is considerably less than unity. Hence the degree to which any reaction reaches completion, as given by the theoretical equation, depends upon the composition of the system which leads to a maximum value of the entropy under these conditions.

Although the entropy function is quite useful for the analysis of reactive systems under adiabatic conditions, it is impractical for the majority of systems of engineering interest. For example, consider the flow of a reactive gas mixture through a reaction vessel. The pressure may be fairly uniform during the flow process, but the temperature may be changing because of both the energy released by the reaction and heat losses to the surroundings. More generally, we wish to determine the extent of a chemical reaction for a given temperature and pressure. What is a suitable criterion for chemical equilibrium in this case? Any such criterion must be established in terms of the properties of the system alone, since information on the surroundings generally will not be known. In this chapter we shall first investigate the use of the second law in establishing a suitable criterion for determining the state of equilibrium of a system at a specified temperature and pressure. Such a criterion will then be employed to ascertain the equilibrium composition of a reactive mixture. Although we shall restrict ourselves to ideal-gas mixtures, the major points presented are of general usefulness.

## 15-2 EQUILIBRIUM CRITERIA

For an isolated system the second law of thermodynamics requires that

$$(dS)_{isol} \geq 0$$

Any simple, compressible system which is isolated is one of constant internal energy, since both heat and work effects are prohibited. It is also one of constant volume, since work is zero, which negates expansion or compression processes. Hence it is quite common to find the above equation written in the following notation:

$$(dS)_{U,V,N} > 0 \qquad \text{spontaneous process} \qquad (15\text{-}1a)$$

$$(dS)_{U,V,N} = 0 \qquad \text{reversible process} \qquad (15\text{-}1b)$$

$$(dS)_{U,V,N} < 0 \qquad \text{unnatural process} \qquad (15\text{-}1c)$$

The subscript $N$ is employed to indicate that the system is one of fixed mass. The set of equations above simply states that, for a spontaneous, irreversible process of a system of fixed mass and constant internal energy and volume, the change in entropy must be positive. In the limit of a reversible change, the variation of $S$ is zero. Under these same conditions a process involving a negative change in $S$ is deemed unnatural, or impossible. We may conclude that when the entropy reaches a maximum value such that, for a further infinitesimal change, the value of $dS$ is zero, then the system is in thermodynamic equilibrium. No further change will occur in the system unless some further interaction between the system and the surroundings is permitted.

Important criteria for other experimental conditions may be obtained by combining the second law with the definitions of the Helmholtz and Gibbs functions. From the definition of the Helmholtz function, $A = U - TS$, it is seen that a differential change in $A$ is given by

$$dA = dU - T\,dS - S\,dT \qquad (15\text{-}2)$$

At constant temperature this reduces to

$$dA_T = dU - T\,dS \qquad (15\text{-}2a)$$

If we combine this equation with the first law for a closed system, then

$$dA_T = \delta Q + \delta W - T\,dS \qquad (15\text{-}2b)$$

There exists a unique relationship between the first and third terms on the right-hand side of Eq. (15-2b), namely, $\delta Q \leq T\,dS$ for a closed system. Rearrangement of Eq. (15-2b) and use of this inequality yields

$$-dA_T + \delta W = T\,dS - \delta Q \geq 0 \qquad (15\text{-}3)$$

In the study of a chemical reaction in a constant-volume vessel there normally would be no work interactions to consider. The constant-volume condition will be denoted by a subscript $V$. In this special case Eq. (15-3) becomes

$$dA_{T,V,N} \le 0 \tag{15-4}$$

This result may be expressed in an equivalent form by stating that

$$(dA)_{T,V,N} < 0 \qquad \text{spontaneous process} \tag{15-5a}$$

$$(dA)_{T,V,N} = 0 \qquad \text{reversible process} \tag{15-5b}$$

$$(dA)_{T,V,N} > 0 \qquad \text{unnatural process} \tag{15-5c}$$

Thus the Helmholtz function must decrease when a spontaneous process occurs in a closed system at constant volume and temperature, in the absence of all forms of work. Ultimately, it reaches a minimum value and the system has reached a state of thermodynamic equilibrium.

A set of equations analogous to Eqs. (15-5) may be derived in terms of the Gibbs function. Since the Gibbs function is defined as $G = H - TS$, it is easily shown that the Gibbs and Helmholtz functions are related by

$$dG = dA + d(PV)$$

At constant pressure the differential Gibbs-function change would be

$$dG_P = dA + P\,dV$$

If an additional restriction of constant temperature is added, then $dA$ may be expressed by Eq. (15-3). Consequently, we find that

$$dG_{T,P} - P\,dV = dA_T \le \delta W$$

Since $P\,dV$ represents boundary work and $\delta W$ is the total-work term, the expression for the Gibbs-function change becomes

$$dG_{T,P} \le \delta W_{\text{tot}} - \delta W_{P\,dV} = \delta W_{\text{net}} \tag{15-6}$$

In Eq. (15-6) $\delta W_{\text{net}}$ represents the sum of all work interactions except those associated with a volume change for the system.

Now consider a closed system for which all forms of work interactions are absent except boundary work. In such a case the net-work term in Eq. (15-6) reduces to zero, and the equation takes on a new form and meaning, namely,

$$dG_{T,P} \le 0 \tag{15-7}$$

This equation simply states that the Gibbs function always decreases for a spontaneous, isothermal, isobaric change of a closed system in the absence of all work effects except boundary work. As a process approaches equilibrium, the Gibbs function attains a minimum value, and in this limiting case of equilibrium, $dG$ is zero. For comparison with the two previous sets of equations, (15-1) and (15-5), one may write

$$dG_{T,P,N} < 0 \qquad \text{spontaneous process} \qquad\qquad (15\text{-}8a)$$

$$dG_{T,P,N} = 0 \qquad \text{reversible process} \qquad\qquad (15\text{-}8b)$$

$$dG_{T,P,N} > 0 \qquad \text{unnatural process} \qquad\qquad (15\text{-}8c)$$

This set of equations would be directly applicable, for example, to a chemically reactive system at a given pressure and temperature.

A graphical interpretation of Eq. (15-7) may be obtained in a manner similar to that used in Sec. 14-7 for the entropy function. Consider the gas-phase reaction $CO + H_2O \rightarrow CO_2 + H_2$. The temperature will be chosen to be $1000°K$, and the pressure is low enough so that the gases are essentially ideal. On the basis of 1 mole each of the initial reactants, the reaction might proceed until 1 mole each of $CO_2$ and $H_2$ is formed. As CO and $H_2O$ are consumed, the composition of the system continually changes. Consequently, the Gibbs function for the total system also changes. Employing the concepts of absolute-enthalpy and absolute-entropy values for each component introduced in Chap. 14, one may calculate the value of $G_{tot}$ at the given temperature and pressure for various compositions of the system. These compositions depend, of course, upon the extent of the reaction. The values of $G_{tot}$ (in kJ/kg·mol) for various values of the number of moles of $CO_2$ formed are summarized for a temperature of $1000°K$:

| Moles $CO_2$ | 0 | 0.2 | 0.4 | 0.5 | 0.6 | 0.7 | 0.8 | 1.0 |
|---|---|---|---|---|---|---|---|---|
| $G_{tot}$ | $-783.2$ | $-792.2$ | $-795.6$ | $-796.3$ | $-796.3$ | $-795.5$ | $-793.8$ | $-786.3$ |

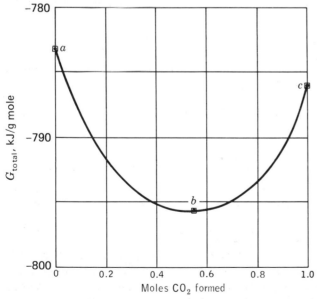

**Figure 15-1** The total Gibbs function for the reaction of an equimolar mixture of CO and $H_2O$ at $1000°K$ as a function of the extent of the reaction, for the reaction

$$CO(g) + H_2O(g) \ \rightleftharpoons \ CO_2(g) + H_2O(g)$$

These data are plotted in Fig. 15-1. When an equimolar mixture of CO and $H_2O$ is allowed to react, the Gibbs function of the system decreases until approximately 0.55 mol of $CO_2$ is formed, for a temperature of 1000°K. A further change in composition represented by the process from state $b$ to state $c$ is impossible, since this requires that the Gibbs function increase for a process at constant temperature and pressure. Although state $c$ has a lower value of $G$ than state $a$, it is not the minimum value based on the composition of the original system. Consequently, only state $b$ will eventually prevail. It should also be noted that if an equimolar mixture of $CO_2$ and $H_2$ is heated to 1000°K, it will react, and the quantity of $CO_2$ will decrease to 0.55 mol for each initial mole in the mixture. Hence the reaction from state $c$ to state $b$ is also possible, since this is not a violation of the criterion established by Eq. (15-7). It must be kept in mind that the application of the Gibbs function as a criterion for equilibrium is equivalent to the usual second-law statement in terms of the entropy function. The use of the Gibbs function, however, has the advantage that easily controlled properties such as the temperature and the pressure are involved in its application to reactive systems.

## 15-3 EQUILIBRIUM AND THE CHEMICAL POTENTIAL

The equilibrium criterion established in the preceding section in terms of the Gibbs function is valid for any type of homogeneous or heterogeneous system. For the present time, however, it is more meaningful to restrict the initial use of this criterion to a study of chemical reactions which occur in the gas phase. The condition of equilibrium for a mixture of gases undergoing a chemical reaction at a given $T$ and $P$ is $dG_{T,P} = 0$. To find the equilibrium composition we need to develop a general expression for $dG$ in terms of the moles of reactants and products present at any given time. This expression is then set equal to zero, in accordance with the above criterion. Equating the derivative $dG$ to zero is equivalent to setting the total Gibbs function of the mixture at its minimum value. The solution to the resulting equation leads to an evaluation of the equilibrium composition.

In Chap. 13 a basic equation is presented which relates the change in the Gibbs function in terms of changes in the pressure and temperature of the system. This equation is

$$dG = V\,dP - S\,dT \tag{13-10}$$

This equation is applicable to changes in simple, compressible systems between equilibrium states in the absence of chemical reactions. For a system of variable composition, however, the Gibbs function is also a function of the number of moles of each of the substances present at a given time. This is expressed mathematically by

$$G = G(T, P, N_1, N_2, \ldots, N_i) \tag{15-9}$$

where the quantities $N_i$ represent the numbers of moles of each chemical species within the system at some instant. A change of state may be brought about by altering the temperature or the pressure or the composition. The overall change in $G$ is given by the total differential of $G$, that is,

$$dG = \left(\frac{\partial G}{\partial P}\right)_{T,N_i} dP + \left(\frac{\partial G}{\partial T}\right)_{P,N_i} dT + \sum_i \left(\frac{\partial G}{\partial N_i}\right)_{P,T,N_j} dN_i \qquad (15\text{-}10)$$

In Eq. (15-10) the subscript $N_i$ on the first two partial derivatives implies that the moles of every species are held fixed during the change in either $P$ or $T$. In the summation the subscript $N_j$ on the partial indicates that the moles of every component are held constant except one, along with fixed values of $P$ and $T$. When the composition is held fixed, the last term of Eq. (15-10) is zero. Under this condition Eq. (15-10) should be identical with Eq. (13-10); consequently, we find that

$$\left(\frac{\partial G}{\partial P}\right)_{T,N_i} = V \qquad \left(\frac{\partial G}{\partial T}\right)_{P,N_i} = -S$$

In addition, the partial derivative which appears in the summation in Eq. (15-10) will be defined as the *chemical potential* $\mu_i$. Hence

$$\mu_i = \left(\frac{\partial G}{\partial N_i}\right)_{P,T,N_j} \qquad (15\text{-}11)$$

On this basis Eq. (15-10) can now be written as

$$dG = V\,dP - S\,dT + \sum_i \mu_i\,dN_i \qquad (15\text{-}12)$$

This equation is a fundamental relationship which relates changes in the Gibbs function to changes in the pressure, temperature, and composition of any phase of a simple, compressible system.

At this point we are specifically interested in obtaining an expression for the change in the Gibbs function of a reactive system when infinitesimal amounts of reactants are converted into products. Consider a generalized chemical reaction represented by the equation

$$v_A A + v_B B \leftrightarrows v_E E + v_F F \qquad (15\text{-}13)$$

The $v$ symbols represent the stoichiometric coefficients for the balanced chemical equation and the uppercase letters stand for the chemical species involved in the reaction. Although the reaction we have chosen has two reactants and two products, it must be realized that the format of the resulting equations will be valid for any number of reactants and products. The values of $v_A$, $v_B$, $v_E$, etc., are not related in any way to the actual numbers of moles $N_i$ of each component actually placed in the reaction vessel, or the actual numbers of moles of products produced by the reaction. The values of the stoichiometric coefficients $v_i$ are always known, once the reaction is specified.

Consider now that a reactive mixture having a reaction represented by

Eq. (15-13) undergoes an infinitesimally small change of composition. If the temperature and pressure remain the same, then the Gibbs function change for the overall reaction, in terms of Eq. (15-12), is

$$dG_{T,P} = \sum_i \mu_i \, dN_i$$

$$= \mu_A \, dN_A + \mu_B \, dN_B + \mu_E \, dN_E + \mu_F \, dN_G \qquad (15\text{-}14)$$

The $dN_i$ terms may be positive or negative, depending upon whether the reaction is proceeding from left to right or right to left. The $dN_i$ terms in Eq. (15-14) are not independent of each other, since the change in the moles of reactants and products is always in proportion to the stoichiometric coefficients. Thus we may write for the forward direction of the reaction

$$dN_A = -kv_A \qquad dN_B = -kv_B \qquad dN_E = kv_E \qquad dN_F = kv_F$$

where $k$ is a proportionality constant and is an extremely small number. The negative signs are necessary for the two reactant terms since the stoichiometric coefficients such as $v_A$ and $v_B$ are always considered to be positive in value. Substitution of the above expressions for $dN_i$ into Eq. (15-14) yields an equation for the Gibbs-function change in terms of the stoichiometric coefficients. The result is

$$dG_{T,P} = (-\mu_A v_A - \mu_B v_B + \mu_E v_E + \mu_F v_F)k$$

However, if such an infinitesimal reaction occurred essentially at the point of chemical equilibrium, the net change in the Gibbs function would be zero. This is the particular case in which we are interested. Therefore, for $dG_{T,P} = 0$ at equilibrium, the criterion for equilibrium is

$$v_E \mu_E + v_F \mu_F - v_A \mu_A - v_B \mu_B = 0 \qquad (15\text{-}15)$$

The proportionality constant $k$ has dropped out since it appears in each term of the equation as a multiplying factor. When Eq. (15-15) is satisfied, the Gibbs function of a system at a given temperature and pressure will be a minimum, in line with the general criterion set up earlier.

Equation (15-15) is known as the *equation of reaction equilibrium*, and it relates intensive properties of the reactants and products. The equation is valid for any chemical reaction, regardless of the phases of the reacting species. To establish the equilibrium composition of a reacting mixture we must first determine expressions for the chemical potential of a given component as a function of the temperature, pressure, and composition. In the next section this is done for a mixture of ideal gases.

## 15-4 THE CHEMICAL POTENTIAL OF AN IDEAL GAS

According to the Gibbs-Dalton rule introduced in Chap. 11, an ideal gas in a mixture of gases behaves as if it alone occupies the volume of the system at the given temperature. Under this circumstance the gas exerts a pressure equal to its component or partial pressure $p_i$. Therefore the chemical potential of an ideal gas in a mixture of gases is the same as that for the pure component maintained at a pressure equal to its partial pressure in the mixture. Symbolically,

$$\mu_{i,\text{ mixture, }P} = \mu_{i,\text{ pure, }p_i}$$

In addition, the chemical potential of a pure substance is equal to its specific Gibbs function $g$ at that state. That is,

$$\mu_{i,\text{ pure}} = \left(\frac{\partial G}{\partial N_i}\right)_{T,P,N_j} = \left(\frac{\partial G_i}{\partial N_i}\right)_{T,p_i} = g_{i,\text{ pure}} \tag{15-16}$$

Hence the chemical potential of an ideal gas in a mixture may be evaluated in terms of the specific Gibbs function of the pure component at a pressure $p_i$ and temperature $T$. The specific Gibbs function of a substance at temperature $T$ is, by definition,

$$g_{i,T} = h_{i,T} - Ts_{i,T} \tag{a}$$

The enthalpy and the entropy of an ideal gas may be evaluated in terms of the standard reference state discussed in Chap. 14. Recall that the standard state of an ideal gas is taken as 1 atm, and that it is symbolized by the superscript $\circ$. Since the enthalpy of an ideal gas is not a function of pressure, we may write

$$h_{i,T} = h^{\circ}_{i,T} \tag{b}$$

where $h^{\circ}_{i,T}$ accounts for the enthalpy of formation as well as the enthalpy difference between 298°K (537°R) and the specified temperature $T$. Also, the absolute entropy at temperature $T$ and pressure $p_i$ is given by Eq. (14-16), namely,

$$s_{i,T} = s^{\circ}_{i,T} - R \ln p_i \tag{c}$$

The chemical potential of a component in an ideal-gas mixture is then found by combining Eqs. (a), (b), and (c) with Eq. (15-16). The result is

$$\mu_{i,T} = g_{i,T} = h^{\circ}_{i,T} - Ts^{\circ}_{i,T} + RT \ln p_i$$

With reference to Eq. (a), it is noted that the first two terms on the right-hand side of the equation above is the standard-state Gibbs function, $g^{\circ}_{i,T}$. Therefore the chemical potential of an ideal gas at temperature $T$ and pressure $p_i$ is given by

$$\mu_{i,\text{ ideal}} = g^{\circ}_i + RT \ln p_i \tag{15-17}$$

where the quantity $p_i$, *must* be measured in atmospheres. This relation, in conjunction with the equation of reaction equilibrium, enables us to determine the equilibrium composition of a reacting ideal-gas mixture.

## 15-5 THE EQUILIBRIUM CONSTANT $K_p$

The equilibrium composition of a reacting ideal-gas mixture is determined by means of (1) the equation of reaction equilibrium and (2) the equation for the chemical potential of an ideal gas. In review, these equations are

$$v_E \mu_E + v_F \mu_F - v_A \mu_A - v_B \mu_B = 0 \tag{15-15}$$

and
$$\mu_i = g_i^\circ + RT \ln p_i \tag{15-18}$$

where $A$ and $B$ represent reactants and $E$ and $F$ are products. By using Eq. (15-17) to replace the $\mu_i$ terms in Eq. (15-15), we find that

$$v_E(g_E^\circ + RT \ln p_E) + v_F(g_F^\circ + RT \ln p_F) - v_A(g_A^\circ + RT \ln p_A)$$
$$- v_B(g_B^\circ + RT \ln p_B) = 0 \tag{15-19}$$

At this point it is convenient to rearrange Eq. (15-19) and collect common terms. This leads to

$$(v_E g_E^\circ + v_F g_F^\circ - v_A g_A^\circ - v_B g_B^\circ) + v_E RT \ln p_E + v_F RT \ln p_F - v_A RT \ln p_A$$
$$- v_B RT \ln p_B = 0 \tag{15-20}$$

The quantity in the parentheses is called the standard-state Gibbs-function change for a reaction and is given the symbol $\Delta G_T^\circ$. From the definition of the Gibbs function, it is easily seen that

$$\Delta G_T^\circ = \Delta H_T^\circ - T \Delta S_T^\circ \tag{15-21}$$

Recall that the standard state for ideal gases is defined as 1 atm pressure. Equation (15-21) then indicates that the standard-state Gibbs-function change can be evaluated from a knowledge of the enthalpy of reaction and the entropy change for the stoichiometric reaction occurring at 1 atm pressure and temperature $T$. Methods for computing these quantities were introduced in the preceding chapter. Consequently, the value of $\Delta G_T^\circ$ is known once the stoichiometric chemical equation and the temperature are specified.

We shall now substitute this definition of $\Delta G_T^\circ$ into Eq. (15-20). At the same time the terms containing logarithms can be combined into a single term. Thus we find that

$$-\Delta G_T^\circ = v_E RT \ln p_E + v_F RT \ln p_F - v_A RT \ln p_A - v_B RT \ln p_B$$
$$= RT[\ln (p_E)^{v_E} + \ln (p_F)^{v_F} - \ln (p_A)^{v_A} - \ln (p_B)^{v_B}]$$
$$= RT \ln \frac{(p_E)^{v_E}(p_F)^{v_F}}{(p_A)^{v_A}(p_B)^{v_B}} \tag{15-22}$$

where the $p_i$ values are the actual component pressures of the reacting gases at equilibrium. The exponents of the component pressures are the stoichiometric coefficients based on the balanced theoretical chemical equation. One final change in the form of the relationship is now necessary. The left-hand side of expression (15-22) is solely a function of temperature for a given reaction. Hence its value could be tabulated in standard tables of thermodynamic data. It is more convenient, however, to define first an *equilibrium constant* $K_p$ for ideal-gas reactions by the relation

$$K_p \equiv \frac{(p_E)^{\nu_E}(p_F)^{\nu_F}}{(p_A)^{\nu_A}(p_B)^{\nu_B}} \tag{15-23}$$

In general, of course, the number of terms in the numerator and denominator depends upon the number of product and reactant species in the theoretical equation. Based on this definition for an ideal-gas reaction, Eq. (15-22) becomes

$$\Delta G_T^\circ = -RT \ln K_p$$

or

$$K_p = e^{-\Delta G_T^\circ/RT} \tag{15-24}$$

As a result of the introduction of the term $K_p$, it is more common to see tabulations of $K_p$ or $\log_{10} K_p$ in thermodynamic tables than $\Delta G_T^\circ$ itself. Since $\Delta G_T^\circ$ is solely a function of temperature, the values of $K_p$ or $\log_{10} K_p$ are formally tabulated against the temperature. The use of the base 10 on the logarithm, rather than the base $e$, is one of convenience. A tabulation of $\log_{10} K_p$ values over a range of temperatures for some common ideal-gas reactions is presented in Table A-29M. For the purposes of this text we shall assume that the values of $K_p$ for any reaction of interest are always available.

Although the definition and the subsequent evaluation of $K_p$ values appear straightforward, the following items need to be emphasized.

1. As noted above, the value of $K_p$ is independent of the pressure during the actual reaction. The reason for this is that the standard-state Gibbs-function change $\Delta G_T^\circ$ is defined as the value in the standard state of 1 atm pressure, regardless of the actual conditions.
2. The equilibrium constant based on component pressures $K_p$ is defined with products in the numerator of the definition and reactants in the denominator. This is fairly standard procedure, but there are authors who invert the definition. Some care must be taken in this respect when obtaining $K_p$ values from tables which are unfamiliar.
3. The value of $K_p$ is a function of the method of writing a chemical equation for particular reactants and products. For example, consider the following three reactions at a given temperature:

$$CO + \tfrac{1}{2}O_2 \longrightarrow CO_2 \tag{a}$$

$$2CO + O_2 \longrightarrow 2CO_2 \tag{b}$$

$$CO_2 \longrightarrow CO + \tfrac{1}{2}O_2 \tag{c}$$

The standard state Gibbs-function change for reaction $(b)$ is twice that for reaction $(a)$. Consequently, doubling the stoichiometric equation in effect will square the value of $K_p$, that is, $(K_p)_b = (K_p)_a^2$. On the other hand, reversal of the direction of a chemical equation reverses the sign on $\Delta G_T^\circ$. Thus the $K_p$ value for reaction $(c)$ is the reciprocal of the value for reaction $(a)$. These results stem from the exponential relationship between $K_p$ and $\Delta G_T^\circ$. One can conclude from these results that tabulated values of $K_p$ as a function of temperature are meaningless unless the chemical equation to which they refer is also cited.

4. The difference in $dG_{T,P}$ and $\Delta G_T^\circ$ should be kept in mind. The former term is the criterion for equilibrium at any given temperature and pressure. It must be zero at equilibrium under any conditions. The second quantity is the standard-state Gibbs-function change at the required temperature, but at 1 atm pressure. Its value is usually finite (positive or negative) and is zero at only one particular temperature for each possible reaction. When $\Delta G_T^\circ$ is zero, it merely means that $K_p$ is unity. The value of $\Delta G_T^\circ$ (and hence $K_p$) enables one to calculate the equilibrium composition, as we shall see shortly.

## 15-6 CALCULATION OF EQUILIBRIUM COMPOSITIONS

In theory, the standard-state Gibbs-function change $\Delta G_T^\circ$ can be computed for any desired reaction. Therefore the value of $K_p$ for any reaction involving ideal gases is also known, once the temperature is known. Moreover, the equilibrium constant $K_p$ is related to the component pressures of the gases at chemical equilibrium by Eq. (15-23), namely,

$$K_p = \frac{(p_E)^{v_E}(p_F)^{v_F} \cdots}{(p_A)^{v_A}(p_B)^{v_B} \cdots} \tag{15-23}$$

where the dots indicate that, in general, any number of products and reactants may be involved. The equilibrium composition of the reacting species is uniquely related by expression (15-23). This relationship is more meaningful and useful to us if it is written in terms of the number of moles of each constituent present at equilibrium, rather than the partial pressures. Since the component pressure of any ideal gas is defined by

$$p_i = y_i P = \frac{N_i}{N_m} P$$

the expression for $K_p$ may be modified to the following form for a reaction with two reactants and two products:

$$K_p = \frac{(N_E)^{v_E}(N_F)^{v_F}}{(N_A)^{v_A}(N_B)^{v_B}} \left(\frac{P}{N_m}\right)^{\Delta v} \tag{15-25}$$

In Eq. (15-25) $N_m$ is the sum of the total number of moles of mixture present in the reaction vessel at equilibrium, $P$ is the total system pressure, and $\Delta v$ by definition is the sum of the stoichiometric coefficients of the products minus the sum of the stoichiometric coefficients of the reactants all from the balanced chemical equation. In Eq. (15-25) $\Delta v$ equals $v_E + v_F - v_A - v_B$.

At a given temperature and pressure for a reaction, the only unknowns in Eq. (15-25) are the values of the $N_i$, that is, the number of moles of each reacting chemical species present at chemical equilibrium. The value of $N_m$ is related, of course, to the $N_i$ values by

$$N_m = N_A + N_B + \cdots + N_E + N_F + \cdots + N_{\text{inerts}} \qquad (15\text{-}26)$$

It is important not to omit the last term in this expression. The presence of inert gases certainly affects the component pressure of each reacting species. If we again restrict ourselves for the moment to a reaction with two reactants and two products, Eqs. (15-25) and (15-26) constitute two equations with five unknowns. The remaining equations necessary for the solution are based on the principle of the conservation of atomic species previously used, for example, in determining equations for chemical reactions based on product gas analyses, in Chap. 14. In practice, it will be found that there are sufficient mass balances to make a solution possible. The method employed in obtaining the equilibrium composition of a reacting mixture of ideal gases is illustrated below by several pertinent examples.

**Examples 15-1M and 15-1** Carbon monoxide (CO) and oxygen ($O_2$) in equimolar propor tions are allowed to attain equilibrium at 1 atm and $3000°K$ ($5400°R$). Determine the composition of the equilibrium mixture.

SOLUTION The main reaction to consider is

$$CO(g) + \tfrac{1}{2}O_2(g) \rightleftharpoons CO_2(g)$$

Another possible reaction might be the further dissociation of $O_2$ into atomic oxygen (O). It is a matter of experience that appreciable dissociation of $O_2$ requires temperatures much higher than $3000°K$. Consequently, we shall assume that the only chemical species present at equilibrium are $CO$, $O_2$, and $CO_2$. The equilibrium constant $K_p$ at this temperature is listed as $3.06$ in Table A-29M. Therefore

$$K_p = \frac{(N_{CO_2})^1}{(N_{O_2})^{1/2}(N_{CO})^1} \left(\frac{P}{N_m}\right)^{1-1-1/2}$$

or

$$3.06 = \frac{N_{CO_2}}{(N_{O_2})^{1/2} N_{CO}} \left(\frac{1}{N_m}\right)^{-1/2}$$

The actual chemical reaction itself, which does not go to completion, may be written as

$$1CO + 1O_2 \longrightarrow xCO + yO_2 + zCO_2$$

where $x$, $y$, and $z$ represent the numbers of moles of CO, $O_2$, and $CO_2$ present in the mixture at equilibrium. Since $N_m = x + y + z$, the expression for $K_p$ becomes, upon rearrangement,

$$3.06 = \frac{z(x + y + z)^{1/2}}{x(y)^{1/2}}$$

In addition to this relationship, which contains three unknowns, two mass balances may be made on the carbon and oxygen atoms. Hence

C balance:          $1 = x + z$

O balance:          $3 = x + 2y + 2z$

Solving for $y$ and $z$, we find that

$$z = 1 - x$$

and

$$y = \tfrac{1}{2}(3 - x - 2z) = \tfrac{1}{2}(1 + x)$$

Also

$$N_m = x + y + z = x + (1 - x) + \tfrac{1}{2}(1 + x) = \tfrac{1}{2}(3 + x)$$

We have chosen to evaluate the numbers of moles of $O_2$ and $CO_2$ in terms of CO. This is an arbitrary choice, and either $y$ or $z$ could have been selected as the remaining unknown. Substitution of the equations for $y$, $z$, and $N_m$ into the expression for $K_p$ yields

$$3.06 = \frac{(1 - x)[(3 + x)/2]^{1/2}}{x[(1 + x)/2]^{1/2}} = \frac{(1 - x)(3 + x)^{1/2}}{x(1 + x)^{1/2}}$$

This equation for $x$ may be solved by iteration, synthetic division, Newton's method, or other suitable technique. A value of 0.34 satisfies the equation within the desired accuracy. Therefore the correct chemical equation for the reaction becomes

$$1CO + 1O_2 \longrightarrow 0.34CO + 0.67O_2 + 0.66CO_2$$

This compares with the theoretical reaction for complete combustion as

$$1CO + 1O_2 \longrightarrow CO_2 + \tfrac{1}{2}O_2$$

For the particular reactant mixture and final pressure and temperature, the $CO_2$ formed is, roughly, two-thirds that expected for complete combustion.

**Examples 15-2M and 15-2** An equimolar mixture of carbon monoxide and oxygen is allowed to attain equilibrium at 3000°K and 5 atm pressure. Determine the composition of the equilibrium mixture.

SOLUTION On the basis of the analysis made in the preceding example, we may write the following set of equations:

$$1CO + 1O_2 \longrightarrow xCO + yO_2 + zCO_2$$

$$z = 1 - x$$

$$y = \tfrac{1}{2}(1 + x)$$

$$N_m = \tfrac{1}{2}(3 + x)$$

$$K_p = \frac{z(5)^{-1/2}}{x(y)^{1/2}(x + y + z)^{-1/2}} = 3.06$$

or, upon rearrangement,

$$3.06(5)^{1/2} = \frac{(1 - x)(3 + x)^{1/2}}{x(1 + x)^{1/2}}$$

The method of solution is exactly the same as in the preceding example. However, we see that the change in pressure does affect the numerical answer. The right-hand side of the equation is the same as before, but the numerical value of the left-hand side has increased by the square root of the pressure, expressed in atmospheres. The value of $x$ which satisfies the above relation is approximately 0.193 mol. Thus 0.807 mol of $CO_2$ is formed at equilibrium. When the pressure is 1 atm, the $CO_2$ formed is 0.66 mol per mole of initial CO. Apparently, the effect of pressure on the equilibrium composition is quite pronounced for this reaction.

As a generalization, any time the exponent of the pressure term in the $K_p$

expression is negative (i.e., the sum $v_E + v_F - v_A - v_B$ is negative), an increase in the pressure will always increase the number of moles of products formed at a given temperature, with a concomitant decrease in the number of moles of reactants present at equilibrium. The opposite result is true when the exponent on the pressure is positive. When the exponent is zero, the pressure has no effect on the equilibrium composition.

**Examples 15-3M and 15-3** In order to determine the effect of the presence of inert gases on the equilibrium composition, compute the mixture composition at $3000°K$ ($5400°R$) and 1 atm pressure for a mixture initially composed of 1 mol of carbon monoxide and 4.76 mol of air.

SOLUTION The approach is basically the same as in the two preceding examples, except that the equation for the total number of moles $N_m$ must be revised. The chemical equation for the ideal-gas reaction now becomes

$$1CO + 1O_2 + 3.76N_2 \longrightarrow xCO + yO_2 + zCO_2 + 3.76N_2$$

Again, from the carbon and oxygen balances, $z = 1 - x$ and $y = \frac{1}{2}(1 + x)$. However, the total number of moles of mixture at equilibrium is now

$$N_m = x + y + z + 3.76 = \frac{1}{2}(10.52 + x)$$

The expression for $K_p$ is then

$$3.06 = \frac{(1 - x)(1)^{-1/2}}{x[(1 + x)/2]^{1/2}[(10.52 + x)/2]^{-1/2}} = \frac{(1 - x)(10.52 + x)^{1/2}}{x(1 + x)^{1/2}}$$

A suitable solution for $x$ from this equation is 0.47 mol of CO at the equilibrium state of $3000°K$ and 1 atm pressure. Compared with the original example, we find that the presence of inert nitrogen has decreased the $CO_2$ formed from 0.66 to 0.53 mol for the same pressure and temperature. It is now apparent that the pressure of the system and the presence of inert gases must be taken into account when evaluating the equilibrium composition of a reacting ideal-gas mixture at a given temperature.

The preceding examples have illustrated the general method for evaluating the equilibrium composition of an ideal-gas mixture when the final pressure and temperature are known. In order to make any computations for a reacting mixture, the identity of the chemical species expected in the equilibrium mixture must be assumed. This is necessary, because the presence of various species determines which equilibrium reactions must be considered. To date we have examined mixtures for which only one equilibrium reaction was significant. Other reactions undoubtedly would occur, such as the dissociation of $O_2$ into atomic oxygen (O) or $N_2$ into atomic nitrogen (N), but the extent of these reactions was assumed to be negligible. Experience enables one to predict which reactions can be safely neglected and which reactions should be taken into account. The magnitude of the equilibrium constant $K_p$ is often a good clue to the importance of a reaction. As a rule of thumb, when $K_p$ is less than 0.001 (or $\log_{10} K_p$ is less than $-3.0$), the extent of the reaction is usually not significant. When $K_p$ is greater than 1000 (or $\log_{10} K_p$ is greater than $+3.0$), the reaction probably proceeds very close to completion. For example, consider the reaction $CO + \frac{1}{2}O_2 \rightleftharpoons CO_2$. The equilibrium constant $K_p$ at $3600°R$ ($2000°K$) is, roughly, 730. For an equimolar ratio of

CO and $O_2$ at 1 atm pressure, the CO present at equilibrium at this temperature is approximately 0.0024 mol per initial mole of CO. The reaction essentially goes to completion. Consequently, for temperatures below, roughly, 4000°R, the dissociation of $CO_2$ into CO and $O_2$ may usually be neglected, except at pressures very much below atmospheric. At higher temperatures the effects of dissociation must be considered, as shown by the examples previously given, since the value of $K_p$ falls rapidly with increasing temperatures above 4000°R. (See $K_p$ data in Table A-29M.) We shall prove that the value of $K_p$ always decreases with increasing temperatures for all exothermic reactions.

The concept of the dissociation of a compound into two or more smaller particles can be extended to ionization effects. Elements, for example, will ionize into a positively charged ion and an electron. At elevated temperatures diatomic nitrogen dissociates into its monatomic form according to the equation

$$N_2 \rightleftharpoons 2N$$

As the gas is heated to higher temperatures the following reaction also occurs:

$$N \rightleftharpoons N^+ + e^-$$

It is reasonable to assume in many situations that the positive ions and the electrons behave as ideal-gas particles. Consequently the $K_p$ expression [Eq. (15-25)] is equally valid for a mixture of neutral particles, ions, and electrons at a given temperature and pressure. (In the presence of electric fields the temperature of the electrons may not necessarily be the same as the temperature of the ions and neutral particles. In the presence of moderate fields, one can assume that the temperature of all particles is the same.) In general, the degree of ionization increases as the temperature is raised and the pressure is lowered.

**Examples 15-4M and 15-4** A hypothetical monatomic gas species A ionizes according to the relation $A \rightleftharpoons A^+ + e^-$. If at some temperature $T$ the value of $K_p$ is 0.1 and the pressure is 0.1 atm, determine the percent ionization.

SOLUTION The $K_p$ expression for the reaction is

$$K_p = \frac{N_{A^+} N_{e^-}}{N_A} \frac{P}{N_m}$$

where $N_m = N_{A^+} + N_{e^-} + N_A$. In addition, since charge must be conserved, $N_{A^+} = N_{e^-}$. Also, every ion must come originally from a neutral particle. Therefore, on the basis of 1 initial mole of A, $1 - N_A = N_{A^+}$. If these latter three equalities are substituted into the $K_p$ expression, we find that

$$K_p = 0.1 = \frac{(1 - N_A)(1 - N_A)}{N_A} \frac{0.1}{2 - N_A}$$

or

$$N_A(2 - N_A) = (1 - N_A)^2$$

The solution to this equation is

$$N_A = 0.293$$

Therefore the degree of ionization is close to 70 percent.

## 15-7 SIMULTANEOUS REACTIONS

In the preceding discussion we have focused our attention on determining the equilibrium state for a single chemical reaction at a given temperature and pressure. To attain equilibrium, the total Gibbs function of all the reacting species must be minimized. It is quite common, however, for two or more reactions to occur simultaneously in a reacting mixture. In addition, some of the reacting species will appear in several of the competing chemical reactions. As might be anticipated, the correct analysis of this more complex process requires the use of the equation of reaction equilibrium for every independent reaction which occurs in the mixture.

As a generalization, consider a system for which there are $R$ *independent* reactions. The solution to this problem of simultaneous reactions requires that $R$ equations of the form

$$\sum_{\text{prod}} v_i \mu_i - \sum_{\text{reac}} v_i \mu_i = 0$$

be written. By substituting the chemical potential of an ideal gas for $\mu_i$ in these $R$ equations, there will arise $R$ different expressions for $K_p$. Each $K_p$ is evaluated from the general relationship

$$K_p = \exp(\ \Delta G_T^\circ / RT)$$

where a different $\Delta G_T^\circ$ value exists for each reaction at the given temperature. In addition, atomic balances are written in terms of the initial state of the reacting mixture. The $K_p$ expressions plus the atomic balances will provide sufficient information to determine the equilibrium composition of the ideal-gas mixture.

It has been stressed that the $R$ chemical reactions must be independent, in order to achieve a valid solution. As an example, consider the situation where the reacting species include $CO$, $CO_2$, $H_2$, $H_2O$, and $O_2$. In this case we might write

$$CO + \tfrac{1}{2}O_2 \ \rightleftharpoons \ CO_2 \qquad\qquad (a)$$

$$CO + H_2O \ \rightleftharpoons \ CO_2 + H_2 \qquad\qquad (b)$$

A third reaction we also might have written is

$$H_2 + \tfrac{1}{2}O_2 \ \rightleftharpoons \ H_2O \qquad\qquad (m)$$

These three reactions are not independent, however, since Eq. ($m$) can be formed by subtracting Eq. ($b$), algebraically, from Eq. ($a$). Hence, only two $K_p$ expressions should be written in this case, although which two is a matter of choice.

**Examples 15-5M and 15-5** One mole of CO and one mole of $H_2O$ are heated to $2500°K(4500°R)$ and 1 atm. Determine the equilibrium composition, if it is assumed that only $CO$, $CO_2$, $H_2$, $H_2O$, and $O_2$ are present and that all these gases are ideal gases.

SOLUTION The theoretical equations we shall choose are

$$CO + \tfrac{1}{2}O_2 \;\rightleftharpoons\; CO_2 \qquad (a)$$

and

$$CO + H_2O \;\rightleftharpoons\; CO_2 + H_2 \qquad (b)$$

From Table A-29M we find the $K_p$ values for these reactions as written are 27.5 and 0.164, respectively. Thus

$$K_{p,a} = 27.4 = \frac{N_{CO_2}}{N_{CO}(N_{O_2})^{1/2}} \left(\frac{P}{N_T}\right)^{-1/2}$$

and

$$K_{p,b} = 0.164 = \frac{(N_{CO_2})(N_{H_2})}{(N_{CO})(N_{H_2O})} \left(\frac{P}{N_T}\right)^0$$

where

$$N_T = N_{CO} + N_{CO_2} + N_{O_2} + N_{H_2} + N_{H_2O}.$$

The overall chemical equation relating the reactants and the products is

$$1CO + 1H_2O \;\longrightarrow\; N_{CO} + N_{CO_2} + N_{O_2} + N_{H_2} + N_{H_2O}$$

This relationship can be used to write mass balances on the atomic species present in the overall reaction. These are

C balance: $\qquad\qquad\qquad\qquad 1 = N_{CO} + N_{CO_2}$

O balance: $\qquad\qquad\qquad\qquad 2 = N_{CO} + 2N_{CO_2} + N_{H_2O} + 2N_{O_2}$

H$_2$ balance: $\qquad\qquad\qquad\qquad 1 = N_{H_2} + N_{H_2O}$

The two equilibrium-constant expressions and the above three species balances constitute five equations with five unknowns. Hence a unique numerical solution exists. First, we solve the three atomic balances for three of the $N_i$ quantities in terms of the remaining two independent $N_i$ values. As an arbitrary choice, let $N_{CO_2}$ and $H_{H_2O}$ be the two independent variables. Then the expressions for the three dependent variables, and $N_T$, are

$$N_{CO} = 1 - N_{CO_2} \qquad\qquad N_{H_2} = 1 - N_{H_2O}$$

$$N_{O_2} = \tfrac{1}{2}(1 - N_{CO_2} - N_{H_2O}) \qquad N_T = \tfrac{1}{2}(5 - N_{CO_2} - N_{H_2O})$$

Substitution of these four relations into the $K_p$ relations yields

$$27.4 = \frac{N_{CO_2}(5 - N_{CO_2} - N_{H_2O})^{1/2}}{(1 - N_{CO_2})(1 - N_{CO_2} - N_{H_2O})^{1/2}}$$

and

$$0.164 = \frac{N_{CO_2}(1 - N_{H_2O})}{(1 - N_{CO_2})(N_{H_2O})}$$

By suitable numerical techniques the simultaneous solution of the two preceding equations may be carried out. The table below summarizes the results.

| Species | Moles initially | Moles finally |
|---------|-----------------|---------------|
| CO      | 1               | 0.71          |
| H$_2$O  | 1               | 0.71          |
| CO$_2$  | 0               | 0.29          |
| H$_2$   | 0               | 0.29          |
| O$_2$   | 0               | 0.005         |

From these data it is seen that nearly 30 percent of both CO and H$_2$O have reacted at the state of equilibrium.

# 15-8 FIRST-LAW ANALYSIS OF REACTING IDEAL-GAS MIXTURES

In Chap. 14 the conservation of energy principle was applied to chemical reactions which were assumed to proceed to completion. Two types of processes were analyzed. In the first of these, the initial and final states of the reactants and products, respectively, were known. On the basis of this information, the heat interactions could be evaluated. In the second case, the process was assumed to be adiabatic. From a knowledge of the initial state of the reactants, the maximum theoretical reaction temperature (the adiabatic-flame temperature) was computed. These two types of calculations may now be repeated in the light of the second-law restrictions on chemical reactions. We realize now that in actual combustion processes two things may occur. Either the principal reaction itself may not go to completion, or products of the principal reaction may dissociate into other chemical species not initially present. Hence computations for chemical reactions should take into account these two factors if it is thought that they might be important. If sufficient heat is removed, for example, during a combustion process, it is quite possible that neither effect need be considered. The two examples that follow illustrate the use of both the first and second laws in the analysis of chemical reactions of ideal-gas mixtures.

**Examples 15-6M and 15-6** An equimolar mixture of carbon monoxide and water vapor enters a steady-flow device at 400°K or 260°F and 1 atm pressure. The water-gas reaction occurs, and the final products leave at 1500°K or 2240°F. Compute the quantity and direction of the heat transferred per mole of initial CO in the reactants.

SOLUTION The theoretical water-gas reaction is given by

$$CO(g) + H_2O(g) \rightleftharpoons CO_2(g) + H_2(g)$$

In lieu of the expected equilibrium reaction among the constituents, the actual reaction must be written as

$$CO + H_2O \longrightarrow wCO + xCO_2 + yH_2O + zH_2$$

The use of mass balances leads to relationships among $w$, $x$, $y$, and $z$:

O$_2$ balance: $\qquad\qquad\qquad\qquad 2 = w + 2x + y$

C balance: $\qquad\qquad\qquad\qquad 1 = w + x$

H$_2$ balance: $\qquad\qquad\qquad\qquad 2 = 2y + 2z$

A further relationship among the four variables is provided by the equilibrium-constant expression for the water-gas reaction:

$$K_p = \frac{(N_{CO_2})(N_{H_2})}{(N_{CO})(N_{H_2O})}\left(\frac{P}{N_m}\right)^0$$

At 1500°K, the values of $K_p$ for this reaction is listed in Table A-29M as 0.390. Therefore we may write

$$0.390 = \frac{xz}{wy}$$

These four equations can now be solved simultaneously. The solution in terms of the actual chemical equation is

$$1CO + 1H_2O \longrightarrow 0.615CO + 0.385CO_2 + 0.615H_2O + 0.385H_2$$

Thus the reaction goes to only 38.5 percent of completion. With this information the value of the heat interaction may now be calculated either in metric or USCS units.

(a) METRIC SOLUTION The conservation of energy principle for this steady-flow process is

$$Q = \sum_{prod} N_i h_i - \sum_{reac} N_i h_i$$

$$= \sum_{prod} N_i(\Delta h_f^\circ + h_T - h_{298})_i - \sum_{reac} N_i(\Delta h_f^\circ + h_T - h_{298})_i$$

where the enthalpy of formation of each constituent is evaluated at 298°K. Substituting $\Delta h_f$ data from Table A-22M and $h$ data from Tables A-8M to A-11M, we find that

$$Q = 0.615(-110{,}530 + 47{,}517 - 8669) + 0.385(-393{,}520 + 71{,}078 - 9364)$$

$$+ 0.615(-241{,}810 + 57{,}999 - 9904) + 0.385(0 + 44{,}738 - 8468)$$

$$- 1(-110{,}530 + 11{,}644 - 8669) - 1(-241{,}810 + 13{,}356 - 9904)$$

$$= 69{,}910 \text{ kJ/kg·mol CO}$$

The positive value of $Q$ indicates that 69,910 kJ of heat would have to be added per initial kilogram-mole of CO to permit the products to reach 1500°K. At this temperature, dissociation of $CO_2$ and $H_2O$ according to the chemical equations, $CO_2 \rightleftharpoons CO + \frac{1}{2}O_2$ and $H_2O \rightleftharpoons H_2 + \frac{1}{2}O_2$, may be neglected.

(b) USCS SOLUTION The conservation of energy principle for this steady-flow process is

$$Q = \sum_{prod} N_i h_i - \sum_{reac} N_i h_i$$

$$= \sum_{prod} N_i(\Delta h_f^\circ + h_T - h_{537})_i - \sum_{reac} N_i(\Delta h_f^\circ + h_T - h_{537})_i$$

where the enthalpy of formation of each constituent is evaluated at 537°R. Substituting $\Delta h_f$ data from Table A-22 and $h$ data from Tables A-8 to A-11, we find that

$$Q = 0.615(-47{,}540 + 20{,}434 - 3725) + 0.385(-169{,}300 + 30{,}581 - 4028)$$

$$+ 0.615(-104{,}040 + 24{,}957 - 4258) + 0.385(0 + 19{,}238 - 3640)$$

$$- 1(-47{,}540 + 5006 - 3725) - 1(-104{,}040 + 5739 - 4258)$$

$$= 29{,}700 \text{ Btu/lb·mol CO}$$

The positive value of $Q$ indicates that 29,700 Btu of heat would have to be added per initial pound·mole of CO in order for the products to reach 2700°R. At this temperature, dissociation of $CO_2$ and $H_2O$ according to the chemical equations, $CO_2 \rightleftharpoons CO + \frac{1}{2}O_2$ and $H_2O \rightleftharpoons H_2 + \frac{1}{2}O_2$, may be neglected.

**Examples 15-7M and 15-7** One mole of carbon monoxide and 220 percent theoretical oxygen requirements, both initially at 25°C or 77°F, undergo a reaction in a steady-flow process at 1 atm pressure. Neglecting dissociation of $O_2$, determine the final equilibrium composition and the final temperature if the process is adiabatic.

SOLUTION The theoretical equation for complete combustion would be

$$CO + 2.2(0.5)O_2 \quad \longrightarrow \quad CO_2 + 0.6O_2$$

In this particular case, however, when the effects of chemical equilibrium are to be considered, we must write

$$CO + 1.1O_2 \quad \longrightarrow \quad xCO_2 + yCO + zO_2$$

The mass balances on the equation are

O balance: $\qquad\qquad\qquad\qquad\qquad 1 + 2.2 = 2z + y + 2z$

C balance: $\qquad\qquad\qquad\qquad\qquad 1 = x + y$

The expression for $K_p$ for the reaction $CO + \frac{1}{2}O_2 \rightleftharpoons CO_2$ is

$$K_p = \frac{N_{CO_2}}{(N_{CO})(N_{O_2})^{1/2}}\left(\frac{P}{N_m}\right)^{-1/2} = \frac{x(x + y + z)^{1/2}}{y(z)^{1/2}}$$

At this point we have three equations and four unknowns. The unknowns are $x$, $y$, $z$, and $K_p$. The value of $K_p$, of course, is solely a function of the temperature. The additional required equation is the conservation of energy principle, which in this case reduces to

$$H_{reac} = H_{prod}$$

The energy balance may be evaluated in either metric (SI) or USCS units.

(a) METRIC SOLUTION The above energy balance can be written more explicitly as

$$\sum_{reac} N_i(\Delta h_f^\circ + h_T - h_{298})_i = \sum_{prod} N_i(\Delta h_f^\circ + h_T - h_{298})_i$$

In terms of the above reaction, this can be written as

$$1(-110{,}530 + 0) + 1.1(0) = x(-393{,}520 + h_{T,\,CO_2} - 9364) + y(-110{,}530 + h_{T,\,CO} - 8669)$$
$$+ z(0 + h_{T,\,O_2} - 8682)$$

where the enthalpy terms are to be evaluated at the final but as yet unknown temperature. This equation introduces no more unknowns; hence we now have a sufficient number of equations for a solution. The method of solution, however, requires an iteration, or trial-and-error, technique. Before starting the iteration process, the mass balances and the $K_p$ expression may be combined into the form

$$K_p = \frac{x(2.1 - 0.5x)^{1/2}}{(1 - x)(1.1 - 0.5x)^{1/2}}$$

At the same time a combination of the mass balances and the energy equation leads to

$$-110{,}530 = x(-402{,}884 + h_{T,\,CO_2}) + (1 - x)(-119{,}199 + h_{T,\,CO}) + (1.1 - 0.5x)(h_{T,\,O_2} - 8682)$$

Thus we have two equations with two unknowns, primarily $x$ and $T$.

As a first approximation we shall let $T$ equal 2900°K. The sensible-enthalpy values in the energy balance may now be obtained from gas tables A-7M to A-9M, and the linear energy equation can be solved for $x$. This yields $x = 0.707$. This value of $x$ is now substituted into the expression for $K_p$. Thus

$$K_p = \frac{0.707(2.1 - 0.354)^{1/2}}{0.293(1.1 - 0.354)^{1/2}} = 3.69$$

At 2900°K, however, the equilibrium constant is found to be 4.7 by interpolation in Table A-29M. As a second guess, we shall let $T$ equal 3000°K The energy balance then gives $x = 0.739$. Substitution of this quantity into the relation for $K_p$ yields a value of 4.36. At 3000°K the value of

$K_p$ given in Table A-29M is 3.06. Since this latter value is smaller than the value calculated from the value of $x$, we have now guessed too high a value for the temperature. The correct answer must lie between 2900 and $3000°$ K. We need not refine the method further, since the general approach has been sufficiently demonstrated.

(b) USCS SOLUTION The energy balance can be written more explicitly as

$$\sum_{\text{reac}} N_i(\Delta h_f^\circ + h_T - h_{537})_i = \sum_{\text{prod}} N_i(\Delta h_f^\circ + h_T - h_{537})_i$$

In terms of the above reaction, this can be written as

$$1(-47,540 + 0) + 1.1(0) = x(-169,300 + h_{T,CO_2} - 4028) + y(-47,540 + h_{T,CO} - 3725)$$
$$+ z(0 + h_{T,O_2} - 3725)$$

where the enthalpy terms are to be evaluated at the final but as yet unknown temperature. This equation introduces no more unknowns; hence we now have a sufficient number of equations for a solution. The method of solution, however, requires an iteration, or trial-and-error, technique. Before starting the iteration process, the mass balances and the $K_p$ expression may be combined into the form

$$K_p = \frac{x(2.1 - 0.5x)^{1/2}}{(1 - x)(1.1 - 0.5x)^{1/2}}$$

At the same time a combination of the mass balances and the energy equations leads to

$$-47,540 = x(-173,328 + h_{T,CO_2}) + (1 - x)(-51,265 + h_{T,CO}) + (1.1 - 0.5x)(h_{T,O_2} - 3725)$$

Thus we have two equations with two unknowns, primarily $x$ and $T$.

As a first-approximation we shall let $T$ equal $5220°$R ($2900°$K). The sensible-enthalpy values in the energy balance may now be obtained by linear interpolation from gas tables A-7 to A-9 and the energy equation can be solved for $x$. This yields $x = 0.704$. This value of $x$ is now substituted into the $K_p$ expression. Thus

$$K_p = \frac{0.704(2.1 - 0.352)^{1/2}}{0.296(1.1 - 0.352)^{1/2}} = 3.64$$

At $2900°$K, however, the equilibrium constant is found to be 4.7 by interpolation in Table A-28. As a second guess we might try $T$ equal $5400°$R. The energy equation then gives $x = 0.739$. Substitution of this value into the relation for $K_p$ yields a $K_p$ value of 4.36. At $5400°$R (or $3000°$K) the value of $K_p$ is given in the tables as 3.06. Since this tabulated value is smaller than the value calculated from the value of $x$, we have now guessed too high a value of the temperature. The correct answer must lie between 5220 and $5400°$R, and closer to $5300°$R. We need not refine the method further, since the general approach has been sufficiently demonstrated.

The foregoing example is fairly simple because only one equilibrium reaction is involved. In actual practice there may be a number of equilibrium reactions involved for a given set of initial reactions. Nevertheless, the method of solution is exactly the same, whether the process is adiabatic or nonadiabatic. Mass balances based on the conservation of atomic species, the conservation of energy principle, and the second law in the form of $K_p$ expressions must be employed. Care must be taken to assure that all the $K_p$ expressions are independent of each other. In addition, it is important to keep in mind that our study of chemical reactions involving the use of $K_p$ requires ideal-gas behavior. The study of chemical equilibrium in systems containing real gases, liquids, and solids is beyond the aims of this text.

## 15-9 A RELATIONSHIP BETWEEN $K_p$ AND THE ENTHALPY OF REACTION

In a preceding section the equilibrium constant $K_p$ was related to the standard-state Gibbs-function change $\Delta G_T^\circ$. Since this latter quantity is a function of the enthalpy of reaction and the entropy of reaction at 1 atm pressure and a given temperature, its value theoretically can be determined for any desired reaction from basic thermodynamic data. Consequently, the values of $K_p$ are also known as a function of temperature. It is informative to derive a general expression for the variation of $\ln K_p$ with temperature.

From Eq. (15-24) it is seen that $-R \ln K_p = \Delta G_T^\circ/T$. If the relation $\Delta G_T = \Delta H_T - T \Delta S_T$ is substituted and the resulting expression differentiated with respect to temperature, we find that

$$-R \frac{d \ln K_p}{dT} = -\frac{\Delta H_T}{T^2} + \frac{d(\Delta H_T)}{T \, dT} - \frac{d(\Delta S_T)}{dT}$$

However, if one applies the basic equation $T \, dS = dH - V \, dP$ to a chemical reaction at constant pressure, then $T \, d(\Delta S)/dT = d(\Delta H)/dT$. Hence the second and third terms on the right cancel, and the above equation reduces to

$$\frac{d \ln K_p}{dT} = \frac{\Delta H_T^\circ}{RT^2} = \frac{\Delta h_R^\circ}{RT^2} \tag{15-27}$$

The expression for the change in $\ln K_p$ with temperature may be written also as

$$\frac{d \ln K_p}{d(1/T)} = \frac{-\Delta h_R^\circ}{R} \tag{15-28}$$

Either of these equations is referred to as the van't Hoff isobar equation.

In order to integrate these equations, the functional relationship between $\Delta h_R^\circ$ and $T$ must be known. The enthalpy of reaction for many reactions is nearly independent of temperature. Consequently, it is often possible to assume that $\Delta h_R^\circ$ is constant over the range of temperatures of interest. If some average or initial value for the enthalpy of reaction is chosen, integration of Eq. (15-28) leads directly to

$$\ln \frac{K_{p2}}{K_{p1}} = -\frac{\Delta h_R^\circ}{R} \left( \frac{1}{T_2} - \frac{1}{T_1} \right) \tag{15-29}$$

This approximation is generally quite good for small temperature intervals.

Equation (15-29) leads to an interesting qualitative result. If a reaction is exothermic (heat released), $\Delta h_R^\circ$ is negative by convention. In addition, when $T_2 > T_1$ for such a reaction, the right-hand side of Eq. (15-29) is negative. Hence

$K_{p2}$ must be less than $K_{p1}$. Thus, for exothermic reactions, the equilibrium constant $K_p$ decreases with increasing temperature. As $K_p$ decreases, the tendency for the reaction to proceed to completion is reduced. Under adiabatic conditions, then, the energy released by an exothermic reaction is diminished since the reaction does not proceed to completion.

**Examples 15-8M and 15-8** The equilibrium constant $K_p$ for the gas-phase reaction $CO + \frac{1}{2}O_2 \rightleftharpoons CO_2$ is found to be 3.055 at 3000°K (5400°R). Estimate the value at 2000°K, based on the van't Hoff isobar equation.

SOLUTION The enthalpies of reaction at 2000 and 3000°K are $-119,500$ and $-117,200$ Btu/mole, respectively. Since the change in $\Delta h°$ is relatively small, Eq. (15-25) should lead to a fairly good evaluation of $K_p$ at the lower temperature. The average value of $\Delta h°$ is $-118,350$ Btu/mol. Therefore

$$\ln \frac{K_{p2}}{K_{p1}} = \frac{118,350}{1.986}\left(\frac{1}{3600} - \frac{1}{5400}\right) = 59,590(0.000093) = 5.54$$

Hence, $K_{p2}/K_{p1} = 254$, or $K_{p2} = 776$. The tabulated value found in Table A-33 is 766. The error is 1.3 percent, but this is for a temperature difference of 1000°K. The calculation does indicate that reasonable accuracy is obtained, in the absence of direct experimental information.

**Examples 15-9M and 15-9** At high temperature the potassium atom is ionized according to the equation $K \rightleftharpoons K^+ + e^-$. The values of equilibrium constant $K_p$ for this gas-phase reaction at 3000°K and 3500°K are $8.33 \times 10^{-6}$ and $1.33 \times 10^{-4}$, respectively. Estimate the average enthalpy of reaction in the given temperature range, in joules per gram·mole and electronvolts per molecule.

SOLUTION The average enthalpy of reaction can be determined for the van't Hoff isobar equation. Use of Eq. (15-29) yields

$$\ln \frac{1.33 \times 10^{-4}}{8.33 \times 10^{-6}} = -\frac{\Delta h_R°}{8.315}\left(\frac{1}{3500} - \frac{1}{3000}\right)$$

or $\qquad \Delta h_R° = -2.77(8.315)/(-4.76 \times 10^{-5}) = +483,700$ J·g·mol

The conversion of this quantity into electronvolts per molecule is carried out as follows:

$$\Delta h_R° = \frac{483,700 \text{ J}}{\text{g·mol}} \times \frac{\text{g·mol}}{6.023 \times 10^{23} \text{ molecule}} \times \frac{\text{eV}}{1.06 \times 10^{-19} \text{ J}}$$

$$= +7.58 \text{ eV/molecule}$$

This value of the enthalpy of reaction is also known as the *ionization potential* of the potassium atom. Note that the reaction is highly endothermic.

## PROBLEMS*

### Calculation of $K_p$ data

**15-1M** The standard-state Gibbs function of an ideal gas is given by $g° = \Delta h_f° - Ts°$ when $T$ is 298°K. Data for $\Delta h_f°$ and $s°$ at 298°K are given in gas tables A-6M through A-11M in the Appendix. Now consider the reaction $H_2 + \frac{1}{2}O_2 \rightarrow H_2O$.

* All problems for this chapter are metric.

(a) Calculate the value of $g°$ for each gas in the reaction at 298°K, in kJ/kg·mol.

(b) Evaluate $\Delta g_T°$ for the stoichiometric reaction, also in kilojoules.

(c) On the basis of Eq. (15-24a), calculate ln $K_p$.

(d) Compare this value to that given in Table A-29M, after first converting the answer in part (c) to the base 10 logarithm.

**15-2M** Same as Prob. 15-1M, except that we shall consider the reaction to be $CO + \frac{1}{2}O_2 \rightarrow CO_2$.

**15-3M** Same as Prob. 15-1M, except that we shall consider the reaction to be $CO_2 + H_2 \rightarrow CO + H_2O$.

**15-4M** Same as Prob. 15-1M, except that we shall consider the reaction to be $\frac{1}{2}H_2 + OH \rightarrow H_2O$. The enthalpy of formation and absolute entropy of hydroxyl (OH) are found in Table A-22M.

**15-5M** A gas mixture of 10 atm consists of 0.1 mol of CO, 0.6 mol of $CO_2$, 0.3 mol of $O_2$, and 2 mol of $N_2$. The mixture is in equilibrium for the reaction $CO + \frac{1}{2}O_2 \leftrightharpoons CO_2$. Determine (a) the value of $K_p$, and (b) the mixture temperature in °K.

**15-6M** At what temperature in °K will CO be 10 percent of the total products of combustion if CO is burned with the stoichiometric amount of $O_2$ at a total pressure of 2 atm?

**15-7M** Consider the chemical reaction

$$CO(g) + 3H_2(g) \rightleftharpoons CH_4(g) + H_2O(g)$$

Initially, a constant-pressure reaction vessel is filled with 2 mol of CO, 5 mol of $H_2$, and 2 moles of nitrogen (an inert gas) at temperature $T$. For the reaction given above, it is found that, at chemical equilibrium, the products include 0.5 mol CO at 9 atm total pressure and temperature $T$. Determine the equilibrium constant $K_p$ for the reaction.

**15-8M** At what temperature is 10 percent of $CO_2$ dissociated at 2 atm pressure?

**15-9M** A system consisting initially of 1 mol of water vapor is heated at 1 atm until it is 10 percent dissociated. Determine the final temperature of the mixture if the sole products of dissociation are $H_2$ and $O_2$.

**15-10M** What temperature is necessary to dissociate oxygen ($O_2$) to a state where the monatomic species (O) comprises 20 percent of the total number of moles of the equilibrium mixture at a pressure of 0.25 atm?

**Equilibrium-composition calculations**

**15-11M** One mole of diatomic oxygen at 25°C and 1 atm is heated to 3500°K. If $O_2$ and O are the only constituents present at equilibrium, determine the number of moles of O present at 3500°K if the final pressure is (a) 1 atm and (b) 10 atm.

**15-12M** The equilibrium constant for the dissociation of $O_2$ into atomic oxygen is found to be 0.391 at 3600°K. If a mole of molecular oxygen is placed in a constant-pressure reaction vessel at 1.09 atm and heated to 3600°K, what percent of the original oxygen will remain at equilibrium?

**15-13M** At elevated temperatures iodine vapor ($I_2$) dissociated according to the equation $I_2 \rightleftharpoons 2I$. At 1500°K the equilibrium constant $K_p$ for this reaction is known to be equal to 1.22. If 1 mol of diatomic iodine is heated to 1500°K and 0.5 atm pressure, determine the number of moles of $I_2$ in the reaction vessel at equilibrium.

**15-14M** The standard-state Gibbs-function change for the dissociation of diatomic fluorine into the monatomic species at 1200°K is 12,550 kJ/kg·mol. Compute the percent dissociation of $F_2$ at this temperature and (a) 1 atm and (b) 0.1 atm.

**15-15M** At 300°K the equilibrium constant for the reaction $N_2O_4 \rightleftharpoons 2NO_2$ is $K_p = 0.18$. What system pressure will be required for (a) 20 percent and (b) 15 percent of the initially pure $N_2O_4$ to dissociate at this temperature?

**15-16M** Gaseous nitrogen ($N_2$) is heated until there are equal numbers of diatomic and monatomic species present at equilibrium at a given temperature. Determine the ratio of atoms (N) to molecules ($N_2$) at the given temperature if now the pressure is (a) doubled, and (b) tripled.

**15-17M** Pure diatomic hydrogen ($H_2$) is placed in a reaction vessel at 298°K. Determine the mole percent of $H_2$ in the vessel after it is heated to 4000°K at a pressure of (a) 0.5 atm, and (b) 2 atm.

**15-18M** The dissociation reaction $NO \rightleftharpoons \frac{1}{2}O_2 + \frac{1}{2}N_2$ has an equilibrium-constant value of $K_p = 4$ at temperature $T_1$. Find the equilibrium-mixture composition at $T_1$ and 2 atm pressure if initially you had 2.0 moles of NO.

**15-19M** One mole of water is heated to 3000°K at a pressure of 0.90 atm. If it is assumed that the reaction $H_2O \rightleftharpoons \frac{1}{2}H_2 + OH$ represents the only dissociation reaction at this temperature, determine the moles of water present at equilibrium.

**15-20M** One mole of water at 3000°K dissociates according to the reaction $H_2O \rightleftharpoons H_2 + \frac{1}{2}O_2$.

    (a) Determine the percent dissociation at this temperature and 1 atm pressure.

    (b) Repeat the calculation for the same temperature and pressure if the initial reactants include 1 mole of water and 1 mole of nitrogen.

    (c) Repeat the calculation for the same temperature and reactants as part (a), but with the pressure 0.5 atm.

**15-21M** The equilibrium constant $K_p$ for the reaction $\frac{3}{2}H_2 + \frac{1}{2}N_2 \rightleftharpoons NH_3$ is 0.0068 at 450°C. Determine the equilibrium composition at 450°C if 3 mol of $H_2$ and 1 mol of $N_2$ are mixed at a pressure of (a) 20 atm and (b) 50 atm.

**15-22M** A gas mixture enters a reaction vessel with the following composition by volume: $SO_2$, 7.8 percent; $O_2$, 10.8 percent; $N_2$, 81.4 percent. The pressure and temperature are 1 atm and 500°C. If the $K_p$ value is 56, then determine the number of moles of $SO_3$ per mole of reactant gas at equilibrium for the overall reaction $SO_2 + \frac{1}{2}O_2 \rightleftharpoons SO_3$.

**15-23M** A reacting mixture consisting of 1 mole of $O_2$, 2 mol of $N_2$, and 1 mol of Ar reach equilibrium at 1 atm and 3000°K. For the reaction $N_2 + O_2 \rightleftharpoons 2NO$, determine the number of moles of NO present at equilibrium.

**15-24M** A mixture consisting of 1 mol of $H_2$, 0.6 mol of $O_2$, and 1 mol of $N_2$ react at 3000°K and 0.5 atm. For the reaction $H_2 + \frac{1}{2}O_2 \rightleftharpoons H_2O$, determine the number of moles of $H_2O$ present at equilibrium.

**15-25M** A mixture consisting initially of 2 moles $CO_2$, 2 moles $H_2$, and 0.5 mole $N_2$ is heated to 1500°K. The pressure is constant at 5 atm. Determine the equilibrium mixture, assuming that no monatomic species are present.

**15-26M** One mole of $CO_2$ is mixed with one mole of $H_2$. Determine the equilibrium composition for the reaction $CO_2 + H_2 \rightleftharpoons CO + H_2O$ if the temperature is (a) 1000°K and (b) 2000°K.

**15-27M** Repeat Prob. 15-26M if the initial composition is 1 mol of $CO_2$ and 2 mol of $H_2$.

**15-28M** Initially, a reaction vessel contains 2 mol of $CO_2$, 1 mol of CO, and 2 mol of $H_2O$ at 1500°K and 2 atm pressure. If the basic reaction is $CO_2 + H_2 \rightleftharpoons CO + H_2O$, what will be the percent hydrogen present in the equilibrium mixture?

**Ionization reactions**

**15-29M** The equilibrium constant for the ionization reaction for argon ($Ar \rightleftharpoons Ar^+ + e^-$) at 10,000°K is 0.00042. Determine the mole fraction of ionized argon atoms at (a) 0.01 atm and (b) 0.05 atm.

**15-30M** Cesium vapor ionizes at elevated temperatures according to the reaction $Cs \rightleftharpoons Cs^+ + e^-$. Determine the percent ionization at (a) 1400°K, (b) 1800°K, and (c) 2000°K, if the corresponding $\log_{10} K_p$ values are $-1.01$, 0.609, and 1.19, respectively, and the pressure is 1 atm.

**15-31M** For the ionization reaction $Na \rightleftharpoons Na^+ + e^-$ the value of $\log_{10} K_p$ at 2000°K is $-0.175$. Determine the percent ionization at this temperature at (a) 1 atm and (b) 0.5 atm.

**15-32M** The ionization equilibrium constant for the reaction $N \rightleftharpoons N^+ + e^-$ is $6.26 \times 10^{-4}$ at 10,000°K. What system pressure is necessary for 5 percent of N to ionize, assuming that this is the sole species present initially.

**Simultaneous reactions**

**15-33M** A mixture of 1 mole of $CO_2$, 1 mole of $H_2$, and $\frac{1}{2}$ mole of $O_2$ is raised to 3000°K at 1 atm.

Assume that the mixture consists of $CO_2$, $H_2$, $O_2$, CO, and $H_2O$. Determine the equilibrium composition, assuming H, O, and OH are absent.

**15-34M** The reaction of 1 mol of $CO_2$, 1 mol of $H_2$, and 0.5 mol of $O_2$ yields an equilibrium mixture of 0.5 mol of $CO_2$, 0.11 mol of $H_2$, 0.305 mol of $O_2$, 0.5 mol of CO, and 0.89 mol of $H_2O$. Determine the temperature, in °K, at which this equilibrium condition exists, if the pressure is 30.3 atm.

**15-35M** One mole of $H_2O$ vapor is heated to (a) 2800°K, and (b) 3000°K. Determine the equilibrium composition assuming that $H_2O$, $H_2$, $O_2$, and OH are present.

### Energy analysis of equilibrium mixtures

**15-36M** Diatomic oxygen ($O_2$) is heated at a constant pressure of 0.1 atm from 298 to 3000°K. How much heat is transferred to or from the reaction vessel in kJ/kg·mol?

**15-37M** Carbon monoxide is mixed with the theoretically correct amount of air and burned adiabatically at the constant pressure of 1 atm. Assume that the reaction products consist of CO, $O_2$, $CO_2$ and $N_2$. Calculate the temperature attained if the pressure is 1 atm and the initial temperature is 330°K.

**15-38M** One mole of carbon monoxide and 80 percent stoichiometric air initially at 25°C are allowed to react adiabatically at a pressure of 1 atm. Compute the maximum combustion temperature with dissociation.

**15-39M** Carbon monoxide is reacted with a 10 percent excess of the stoichiometric amount of air, both initially at 500°K.
　(a) Determine the adiabatic-flame temperature without dissociation.
　(b) Determine the final equilibrium-flame temperature and the percent dissociation for the same initial conditions if the pressure is (1) 2 atm, (2) 5 atm, and (3) 10 atm.

**15-40M** Carbon monoxide is burned at constant volume, using the theoretical amount of air for combustion to $CO_2$. The initial reactant pressure and temperature at 1 atm and 500°K, respectively. Determine the maximum combustion temperature if dissociation is considered.

**15-41M** Hydrogen gas at 25°C is being burned with an enriched air-oxygen mixture, which enters at 500°K. The enriched oxidant mixture is 70 percent air with 30 percent $O_2$ by volume. Determine the adiabatic-flame temperature with dissociation, in °K, if 1.4 mol of oxidizer mixture are used per mole of hydrogen. The pressure is 1 atm.

**15-42M** One mole of carbon monoxide is burned with 20 percent excess air at a pressure of 0.5 atm. If the original temperature of the reactants is 25°C, determine the adiabatic-combustion temperature with dissociation.

**15-43M** Carbon monoxide is burned with 120 percent of the theoretical air requirement, both being supplied at 25°C and burned at a pressure of 0.1 atm. It is desired to supply the products of the reaction at 2500°K, assuming that only $CO_2$, CO, $O_2$, and $N_2$ are present in the products.
　(a) Determine the equilibrium composition at 2500°K.
　(b) Determine the heat which must be added or removed from the product gases (per mole of initial CO) in order to achieve the required final temperature.

**15-44M** Same as Prob. 15-43M, except that the initial temperature is 500°K, the pressure is 0.2 atm, and the final products of reaction are at 2800°K. Assume that only $CO_2$, CO, $O_2$, and $N_2$ are present in the products.

**15-45M** One kilogram·mole of $CO_2$ and two kilogram·moles of $O_2$, both initially at 25°C, are heated to an unknown temperature at a pressure of 1 atm.
　(a) Determine the temperature in °K at which CO represents 10 percent of the total products of combustion, assuming that only $CO_2$, CO, and $O_2$ are present.
　(b) Determine the quantity of heat transferred, in kilojoules.

**15-46M** Carbon monoxide and water vapor in a 1 : 1 molar ratio enter a reaction vessel maintained at 1 atm at 500°K. The mixture is heated at 1500°K. Determine the magnitude and direction of any heat transfer in kJ/mol of initial CO, if the only gases present are CO, $H_2O$, $CO_2$, and $H_2$.

**15-47M** On the basis of the van't Hoff isobar equation, estimate at $2000°K$ the enthalpy of formation of $O_2$ from $O$ per mole of $O_2$. Is the reaction endothermic or exothermic?

**15-48M** From the equilibrium-constant data for the reaction $H_2 \rightleftharpoons 2H$ estimate the energy required to break down $H_2$ into its monatomic species at $1500°K$ and 1 atm in kJ/kg·mol $H_2$.

**15-49M** Calculate $\Delta h_R$ at $2000°K$ for the reaction $CO + \frac{1}{2}O_2 \rightleftharpoons CO_2$ by using the relationship between $\Delta h_R$ and $\ln K_p$. Compare with tabulated data.

**15-50M** For the reaction $SO_3 \rightleftharpoons SO_2 + \frac{1}{2}O_2$, the following experimental data are available:

| $T$, °K | $K_p$ |
|---------|-------|
| 900     | 0.146 |
| 1000    | 0.518 |
| 1100    | 1.45  |

Estimate $\Delta h_R$ at $1000°K$ in kJ/kg·mol.

**15-51M** The equilibrium constant for the gas-phase reaction $H_2 + \frac{1}{2}O_2 \rightleftharpoons H_2O$ is given approximately by $\ln K_p = 29,600[(1/T) - (1/4200)]$, where $T$ is in °K. Estimate the enthalpy of reaction in kJ/kg·mol.

**15-52M** For the reaction $N_2O_4 \rightleftharpoons 2NO_2$, the heat of reaction at $300°K$ is $57,930$ kJ/kg·mol. Will the amount of $N_2O_4$ in the equilibrium mixture increase or decrease with (a) an increase in temperature at constant pressure and (b) an increase in pressure at constant temperature?

**15-53M** For the dissociation of $SO_3$ according to the equation $SO_3 \rightleftharpoons SO_2 + \frac{1}{2}O_2$, the equilibrium constant is given by the approximate relation, $\ln K_p = -11,400/T + 10.75$, where $T$ is in °K.

(a) Estimate the enthalpy of reaction in kJ/kg·mol.

(b) If 0.01 mol of $SO_3$ is sealed in a rigid, but uninsulated, vessel at 1 atm and $300°K$, determine the temperature to which the system must be heated to permit 5 percent of the $SO_3$ to dissociate.

(c) For the final conditions of part (b), determine the final vessel pressure in atm.

**15-54M** The equilibrium-constant data for the reaction $Cs \rightleftharpoons Cs^+ + e^-$ are

| $T$, °K         | 1400    | 1600    | 1800  | 2000  |
|-----------------|---------|---------|-------|-------|
| $\log_{10} K_p$ | -1.010  | -0.108  | 0.609 | 1.194 |

Estimate the enthalpy of ionization for the above reaction at (a) $1600°K$, and (b) $1800°K$, in kJ/kg·mol.

**15-55M** The equilibrium-constant data for the reaction $Na \rightleftharpoons Na^+ + e^-$ are

| $T$, °K         | 1600    | 1800    | 2000    | 2200  |
|-----------------|---------|---------|---------|-------|
| $\log_{10} K_p$ | -1.819  | -0.913  | -0.175  | 0.438 |

Estimate the enthalpy of ionization for the above reaction at (a) $1800°K$, and (b) $2000°K$, in kJ/kg·mol.

# SIXTEEN

## GAS CYCLES

An important use of thermodynamics is made in the study of cyclic devices for power production and refrigeration. In this chapter we shall restrict ourselves to devices which employ a gas as the working fluid. Modern automotive, truck, and gas-turbine engines are examples of the extremely fruitful application of thermodynamic analysis. Air-refrigeration processes are used aboard aircraft for cooling of the passenger cabin. In the next chapter the study will be directed toward power and refrigeration processes which involve the presence of two phases during the course of the process.

## 16-1 THE AIR-STANDARD CYCLE

Because of the complexities of the actual processes, it is profitable in the initial study of gas-power and refrigeration cycles to examine the general characteristics of each cycle without going into a detailed analysis. The advantage of a simple model is that the main parameters which govern the cycle are made more apparent. By stripping the actual process of all its complications and retaining only a bare minimum of detail, the engineer is able to examine the influence of major operating variables on the performance of the device. However, it must be kept in mind that any numerical values calculated from such models may not be strictly representative of the actual process. Thus modeling is an important tool in engineering analysis, but at times it is highly qualitative.

Gas cycles are those in which the working fluid remains a gas throughout the cycle. In actual gas-power cycles the fluid consists mainly of air, plus the products of combustion such as carbon dioxide and water vapor. Since the gas is predominantly air, especially in gas-turbine cycles, it is convenient to examine gas-power cycles in terms of an air-standard cycle. An *air-standard* cycle is an idealized cycle based on the following approximations: (1) the working fluid is taken to be solely air throughout the cycle, and the air behaves like an ideal gas; (2) any combustion process which might appear in actual practice is replaced by a heat-addition process from an external source; (3) a heat-rejection process to the surroundings is used to restore the fluid to its initial state and complete the cycle; and (4) all processes are internally reversible.

In applying the air-standard cycle restrictions to various processes, it is sometimes customary to place additional constraints on the property values of air. For example, in the cold air-standard cycle the specific-heat quantities $c_v$ and $c_p$ and the specific-heat ratio $k$ are assumed to have constant values, and these are measured at room temperature. This approach is frequently used, but the numerical results may be considerably different from those obtained by accounting for variable specific heats. This is due to the wide temperature variation in most gas-power cycles, which severely alters the values of $c_v$ and $c_p$ throughout the cycle. In actual practice, of course, it would be desirable to employ information on the actual gases which result from the combustion of hydrocarbon fuels mixed with air.

## 16-2 THE AIR-STANDARD CARNOT CYCLE

In Chap. 6 it is pointed out that the maximum thermal efficiency of any heat engine operating between two fixed temperature levels is the Carnot efficiency. The relation for the Carnot efficiency is

$$\eta_{\text{Carnot}} = 1 - \frac{T_L}{T_H} \tag{6-4}$$

where $T_L$ and $T_H$ represent the sink temperature and the source temperature, respectively. A Carnot heat engine, as described in Sec. 7-2, undergoes a cyclic process composed of two isothermal, reversible processes and two adiabatic, reversible processes. The $PV$ and $TS$ diagrams for air undergoing a Carnot cycle are shown in Fig. 16-1. This cycle may take place in a closed system, such as a reciprocating piston-cylinder device, or in a steady-flow device. The equipment required for a steady-flow Carnot cycle using air as the working fluid is shown in Fig. 16-2. Steady-flow turbines are required for the isothermal, reversible and adiabatic, reversible expansion processes (steps 1-2 and 2-3). Similarly, steady-flow compressors are needed for the isothermal, reversible and adiabatic, reversible compression processes (steps 3-4 and 4-1.). Heat addition $Q_H$ occurs to the fluid in the isothermal turbine, while heat removal $Q_L$ from the air occurs in the isothermal compressor.

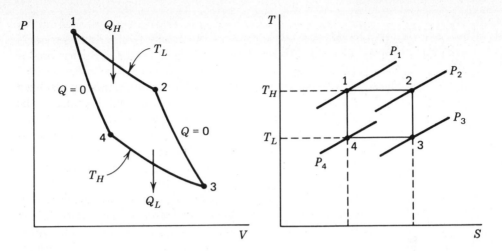

**Figure 16-1** The $PV$ and $TS$ diagrams for a Carnot cycle.

To approach the Carnot efficiency given by Eq. (6-4), an actual engine must be relatively free of dissipative effects, such as friction. In addition, the fluid temperature should be constant during the heat-addition and heat-removal processes. In practice these restrictions, among others, are impossible to meet. The achievement of an engine which approximates the Carnot cycle is not practical. Hence the efficiency of an actual engine is always considerably less than that for a Carnot engine operating between the same maximum and minimum temperatures. Nevertheless the performance of a Carnot heat engine operating on an air-standard cycle is an important standard against which actual engines may be compared.

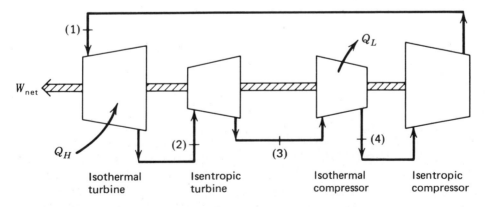

**Figure 16-2** Steady-flow Carnot engine employing air as the working fluid.

## 16-3 SOME INTRODUCTORY NOMENCLATURE FOR RECIPROCATING DEVICES

A number of applications make use of a piston-cylinder arrangement in which the piston is observed to undergo cycles, or revolutions. The *bore* of the piston is its diameter, and the distance the piston moves in one direction is known as the *stroke*. When the piston has moved to a position such that a minimum volume of fluid is left in the cylinder, the piston is said to be at top dead center (TDC). This minimum volume is called the *clearance volume*. When the piston has moved the distance of the stroke so that the fluid now occupies the maximum volume, the piston is in the bottom dead center (BDC) position. The volume displaced by the piston as it moves the distance of the stroke between TDC and BDC is the displacement volume. The clearance volume is frequently cited in terms of the *percent clearance*, which is the percent of the piston displacement equal to the clearance volume. The *compression ratio r* of a reciprocating device is defined as the volume of the fluid at BDC divided by the volume of the fluid at TDC; that is,

$$r = \frac{V_{BDC}}{V_{TDC}} = \frac{\text{clearance volume} + \text{displacement volume}}{\text{clearance volume}} \tag{16-1}$$

The compression ratio is always expressed in terms of a volume ratio.

The *mean effective pressure* (MEP) is a useful parameter in the study of reciprocating devices that are used for power production. It is defined as an average pressure which, if it acted on the piston during the entire power or outward stroke, would produce the same work output as the net work output for the actual cyclic process. It is seen that the work per cycle is given by

$$W_{cycle} = (MEP)(\text{piston area})(\text{stroke})$$

$$= (MEP)(\text{displacement volume}) \tag{16-2}$$

For reciprocating engines of comparable size, a larger mean effective pressure is an indication of better performance in terms of power produced at the same rated speed. For the purpose of illustration, consider the hypothetical clockwise cycle 1-2-3-4-5-1 in Fig. 16-3. The net work produced is represented by the enclosed area on the diagram. The mean effective pressure of the cycle is shown by the horizontal line, and the area under this line equals the enclosed area of the actual cycle.

## 16-4 THE AIR-STANDARD OTTO CYCLE

The four-stroke spark-ignition engine is an important component in the operation of a technology to meet the modern needs of society. Although it is undergoing some modifications in order to meet pollution standards for mobile equipment,

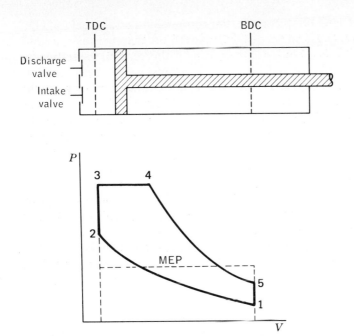

**Figure 16-3** Interpretation of mean effective pressure on a *PV* diagram.

this engine will undoubtedly continue to play a significant role as a device for producing relatively small quantities of power. A typical *PV* diagram for such an engine at wide-open throttle is shown in Fig. 16-4. The series of events includes the intake stroke *ab*, the compression stroke *bc*, the expansion or power stroke *cd*, and finally the exhaust stroke *da*. The intake and exhaust strokes occur essentially at atmospheric pressure. The process lines *ab* and *da* do not lie on top of each other, but since Fig. 16-4 is drawn to scale it is difficult to show the separation between the intake- and exhaust-process lines, except near the BDC position.

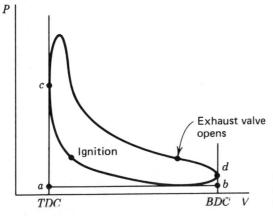

**Figure 16-4** Typical *PV* diagram at wide-open throttle for a 4-stroke spark-ignition engine.

Normally the point of ignition occurs on the compression stroke before TDC position, since flame propagation across the combustion chamber takes a finite time. For a given engine, the point of ignition can be altered until the setting for maximum power is determined. Note also that the exhaust valve is opened before the piston reaches BDC. This allows the pressure of the exhaust gases to nearly reach atmospheric pressure before the exhaust stoke begins.

As discussed in Sec. 16-1, the initial step in analyzing the performance of a four-stroke reciprocating spark-ignition engine is to prepare a simple model of the overall process. While such a model is of value only in a qualitative sense, it does provide some information on the influence of major operating variables on performance.

A theoretical cycle of interest in analyzing the behavior of reciprocating spark-ignition engines is the Otto cycle. A four-stroke Otto cycle is composed of four internally reversible processes, plus an intake and an exhaust portion of the cycle. Both $PV$ and $TS$ diagrams for the theoretical cycle are shown in Fig. 16-5. Consider a piston-cylinder assembly containing air and the piston situated at the bottom dead center position. This is shown as point 1 on the diagrams. As the piston moves to the top dead center position, compression of the air occurs adiabatically. Since the processes are reversible, the compression process is isentropic, ending at state 2. Heat is then added to the air instantaneously, so that both the pressure and the temperature rise to high values during a constant-volume process (2-3). As the piston now moves toward the BDC position once more, the expansion is carried out adiabatically and internally reversibly, i.e., isentropically, to state 4. Now with the piston in its BDC position, heat is rejected at constant volume until the initial state is achieved.

At this point the fluid, theoretically, could begin to go through another cycle. In order to make the cycle somewhat more realistic, the following sequence might be considered before resuming the cyclic pattern: In actual practice, the gases would contain the products of hydrocarbon combustion, so that an exhaust stroke would be necessary. Consequently, the exhaust valve now opens, and the piston moves from BDC to TDC, expelling the gases to the ambient surroundings. Then

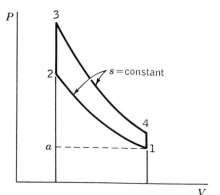
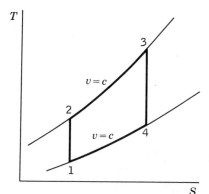

**Figure 16-5** Air-standard Otto cycle.

the exhaust valve closes and the intake valve opens, while the piston returns to the BDC position. During this suction stroke the cylinder is filled with fresh air for the next cycle. In the air-standard cycle this recharging of the cylinder is not necessary, since the same fluid continually undergoes the cyclic variations. Note that the work required to push the charge from the cylinder is the same in magnitude but opposite in sign to that required to suck in the new charge. Hence these two portions of the theoretical cycle do not affect the net work done by the cycle, and it is only the cycle 1-2-3-4-1 which is important in the thermodynamic analysis. In review, the theoretical Otto cycle is composed of the following internally reversible processes:

1. Adiabatic compression, 1-2
2. Constant-volume heat addition, 2-3
3. Adiabatic expansion, 3-4
4. Constant-volume heat rejection, 4-1

In addition, one may consider an exhaust stroke, 1-$a$, and an intake stroke, $a$-1, for completeness, although this is not necessary.

Since the air acts as a closed system, the conservation of energy principle, when applied to the various processes, leads to the following equations: For the adiabatic compression and expansion processes, since $q = 0$,

$$w = \Delta u$$

For the constant-volume heat input and rejection processes, since $w = 0$,

$$q = \Delta u$$

At this point it is informative to analyze the Otto cycle on the basis of the cold air-standard cycle, since it provides some insight as to the important parameters that determine the thermal efficiency of the cycle. Employing the cold air-standard cycle, we find that

$$q_{in} = q_{23} = u_3 - u_2 = c_v(T_3 - T_2)$$

and

$$q_{out} = -q_{41} = u_4 - u_1 = c_v(T_4 - T_1)$$

Since the net work is the sum of $q_{23}$ and $q_{41}$, the thermal efficiency is given by

$$\eta_{th} = \frac{w_{net}}{q_{in}} = \frac{c_v(T_3 - T_2) - c_v(T_4 - T_1)}{c_v(T_3 - T_2)}$$

$$= 1 - \frac{T_4 - T_1}{T_3 - T_2} = 1 - \left(\frac{T_1}{T_2}\right)\frac{T_4/T_1 - 1}{T_3/T_2 - 1}$$

Note that $V_2 = V_3$ and $V_1 = V_4$. Since the isentropic relations show that

$$\frac{T_2}{T_1} = \left(\frac{V_1}{V_2}\right)^{k-1} \qquad \text{and} \qquad \frac{T_3}{T_4} = \left(\frac{V_4}{V_3}\right)^{k-1} = \left(\frac{V_1}{V_2}\right)^{k-1}$$

then $T_2/T_1 = T_3/T_4$, or $T_4/T_1 = T_3/T_2$. When this result is substituted into the equation for the thermal efficiency of an air-standard Otto cycle, it is seen that

$$\eta_{\text{th, Otto}} = 1 - \frac{T_1}{T_2} = 1 - \left(\frac{V_2}{V_1}\right)^{k-1} = 1 - \frac{1}{r^{k-1}} \qquad (16\text{-}3)$$

where $r$ is the *compression ratio* for the theoretical cycle. Equation (16-3) indicates that the major parameters governing the thermal efficiency of an Otto cycle are the compression ratio and the specific-heat ratio. The influence of these factors on the thermal efficiency is demonstrated in Fig. 16-6. For a given specific-heat ratio, the value of the thermal efficiency increases with increasing compression ratio. It should be noted, however, that the curves flatten out at compression ratios above 10 or so. Hence the advantage of operating at high compression ratios lessens rapidly. From a practical viewpoint, the compression ratio is limited by the occurrence of preignition or engine knock when the compression ratio rises much above 10, for common hydrocarbon fuels. Figure 16-6 also shows that the thermal efficiency increases with increasing specific-heat ratio. This means that higher values of the thermal efficiency are obtained when simple monatomic or diatomic molecules are present in the working fluid. Unfortunately, the presence of carbon dioxide and water vapor and other heavier molecules in actual practice makes the attainment of high specific-heat-ratio values a practical impossibility.

To account for variable specific-heat values, the thermal efficiency must be determined by the relation

$$\eta_{\text{th}} = 1 - \frac{u_4 - u_1}{u_3 - u_2} \qquad \cdot \qquad (16\text{-}4)$$

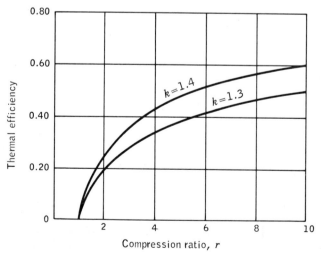

**Figure 16-6** Thermal efficiency of the air-standard Otto cycle as a function of compression ratio and specific-heat ratio.

In addition, the $u$ values must be read from the air table in the Appendix at the corresponding temperatures around the cycle. The temperatures at states 2 and 4 are found from the isentropic relations

$$v_{r2} = v_{r1}\left(\frac{V_2}{V_1}\right) = \frac{v_{r2}}{r} \quad \text{and} \quad v_{r4} = v_{r3}\left(\frac{V_4}{V_3}\right) = rv_{r3}$$

Recall that the $v_r$ data are solely a function of temperature.

**Example 16-1M** The initial conditions for an air-standard Otto cycle operating with a compression ratio of 8 : 1 are 0.95 bar and 17°C. At the beginning of the compression stroke, the cylinder volume is 3.80 l, and 7.5 kJ of heat are added to the gas during the constant-volume heating process. Calculate the pressure and temperature at the end of each process of the cycle, and determine the thermal efficiency and mean effective pressure of the cycle.

SOLUTION By denoting the states in the same manner as in Fig. 16-4, we let $P_1 = 0.95$ bar and $T_1 = 290°$K. From the ideal-gas equation of state

$$v_1 = \frac{RT}{P} = \frac{0.08315(290)}{29(0.95)} = 0.875 \text{ m}^3/\text{kg}$$

As a first approximation we shall assume $c_v$ and $k$ of air are constant, and shall use the room-temperature values of 0.717 kJ/(kg)(°K) and 1.40, respectively. On this basis the temperature and pressure after isentropic compression is found by

$$T_{2s} = T_1\left(\frac{V_1}{V_2}\right)^{k-1} = 290(8)^{0.40} = 666°\text{K}$$

$$P_{2s} = P_1\left(\frac{V_1}{V_2}\right)^{k} = 0.95(8)^{1.40} = 17.5 \text{ bars}$$

It might be noted that $P_2$ can also be calculated from the relation $P_2 = P_1(V_1/V_2)(T_2/T_1)$. To determine state 3, the amount of heat added per kilogram of air must be found.

$$q = \frac{Q}{m} = \frac{Qv_1}{V_1} = \frac{(7.5 \text{ kJ})(0.875 \text{ m}^3/\text{kg})}{3.8 \times 10^{-3}\text{m}^3} = 1727 \text{ kJ/kg}$$

This quantity of heat equals the change in the internal energy. That is, $q_{in} = u_3 - u_2 = c_v(T_3 - T_2)$. Therefore

$$T_3 = T_2 + \frac{q}{c_v} = 666 + \frac{1727}{0.717} = 3075°\text{K}$$

The pressure at state 3 is

$$P_3 = P_2\left(\frac{T_3}{T_2}\right) = 17.5(3075/666) = 80.8 \text{ bars}$$

State 4 is again found from isentropic relations.

$$T_4 = T_3\left(\frac{V_3}{V_4}\right)^{k-1} = 3075(1/8)^{0.4} = 1338°\text{K}$$

and

$$P_4 = P_3\left(\frac{V_3}{V_4}\right)^{k} = 80.8(1/8)^{1.4} = 4.40 \text{ bars}$$

The thermal efficiency is found from the net work output and the heat supplied. The heat rejected is

$$u_4 - u_1 = c_v(T_4 - T_1) = 0.717(1338 - 290) = 751 \text{ kJ/kg}$$

The net work, then, is

$$q_{in} - q_{out} = 1727 - 751 = 976 \text{ kJ/kg}$$

Consequently,

$$\eta_{th} = w_{net}/q_{in} = 976/1727 = 0.565 \qquad \text{(or 56.5 percent)}$$

Since we have assumed that $k$ is a constant, the thermal efficiency may also be determined by

$$\eta_{th} = 1 - \frac{1}{r^{k-1}} = 1 - \frac{1}{8^{0.4}} = 1 - 0.435 = 0.565 \qquad \text{(or 56.5 percent)}$$

Finally, the mean effective pressure of the cycle is

$$\text{MEP} = \frac{w_{net}}{v_1 - v_2} = \frac{976 \text{ J/g}}{(0.875 - 0.109)\text{m}^3/\text{g}} \times \frac{\text{bar·m}^2}{10^5 \text{N}} \times \frac{10^3 \text{g}}{\text{kg}} = 12.7 \text{ bars}$$

where $v_2$ is one-eighth of $v_1$. The values obtained above would change somewhat if the variation of the specific heats with temperature were taken into consideration.

**Example 16-1** The intake conditions for a theoretical Otto cycle operating with a compression ratio of 8 : 1 are 14.4 psia and 60°F. At the beginning of the compression stroke, the cylinder volume is 270 in³, and 9 Btu of heat are added to the gas during the constant-volume heating process. Calculate the pressure and temperature at the end of each process of the cycle, and determine the thermal efficiency and the mean effective pressure of the cycle.

SOLUTION By denoting the states in the same manner as in Fig. 16-5 we let $P_1 = 14.4$ psia and $T_1 = 520°$R. From the ideal-gas equation of state

$$v_1 = \frac{RT}{P} = \frac{1545(520)}{29(14.4)(144)} = 13.33 \text{ ft}^3/\text{lb}$$

The state after isentropic compression is found by using the $v_r$ and $p_r$ data in gas table A-5 for air. At 520°R, $v_r = 158.58$ and $p_r = 1.2147$. Hence

$$v_{r2} = v_{r1} \frac{v_2}{v_1} = 158.58\left(\frac{1}{8}\right) = 19.82$$

Therefore $T_2 = 1170°$R $= 710°$F, and $P_2 = P_1(p_{r2}/p_{r1}) = 14.4(21.86/1.2147) = 259$ psia. In order to determine state 3, the amount of heat added per pound of air must be found. This is determined from knowledge of the initial cylinder volume.

$$q = Q/m = Qv/V = 9(13.3)(1728)/270 = 777 \text{ Btu}$$

This quantity of heat equals the change in internal energy, $u_3 - u_2$. At 1170°R, $u_2 = 203.5$ Btu/lb; hence

$$u_3 = u_2 + q = 203.5 + 777 = 980.5 \text{ Btu/lb}$$

From Table A-5, it is found that $T_3$ is 4708°R and $v_{r3} = 0.2584$. The pressure at state 3 is

$$P_3 = P_2 \frac{T_3}{T_2} = 259\left(\frac{4708}{1170}\right) = 1041 \text{ psia}$$

State 4 is again found from isentropic relations.

$$v_{r4} = v_{r3} \frac{v_4}{v_3} = 0.2584(8) = 2.067$$

Therefore $T_4 = 2522°R$, and $P_4 = P_3(p_{r4}/p_{r3}) = 1041(\frac{452}{6749}) = 69.8$ psia. The thermal efficiency is found from the net work output and the heat supplied. The heat rejected is

$$u_4 - u_1 = 479.2 - 88.6 = 390.6 \text{ Btu/lb}$$

The net work, then, is

$$q_{in} - q_{out} = 777 - 390.6 = 386.4 \text{ Btu/lb}$$

Consequently,

$$\eta_{th} = \frac{w_{net}}{q_{in}} = \frac{386.4}{777} = 0.497 \qquad \text{(or 49.7 percent)}$$

Finally, the mean effective pressure of the cycle is

$$\text{MEP} = \frac{w_{net}}{v_1 - v_2} = \frac{386.4(778)}{13.33 - 1.67} = 25,900 \text{ psf} = 179.5 \text{ psi}$$

## 16-5 THE AIR-STANDARD DIESEL CYCLE AND THE DUAL CYCLE

In a spark-ignition engine the fuel is ignited by energy supplied from an external source. An alternative method for initiating the combustion process in a recipro-cating engine is to raise the fuel-air mixture above its autoignition temperature. An engine built on this principle is called a compression-ignition (CI) engine. By using compression ratios in the range of 14 : 1 to 24 : 1 and using diesel fuel instead of gasoline, the temperature of the air within the cylinder will exceed the ignition temperature at the end of the compression stroke. If the fuel were premixed with the air, as in a spark-ignition engine, combustion would begin throughout the mixture when the ignition temperature is reached. As a result we would have no control on the timing of the combustion process. To overcome this difficulty, the fuel is injected into the cylinder in a separate operation. Injection begins when the piston is near TDC position. Thus the CI engine differs from the spark-ignition (SI) engine primarily in the method of achieving combustion and in the adjustment of the timing of the combustion process. The rest of the four-stroke CI cycle is similar to that of the SI cycle discussed in Sec. 16-4.

A typical $PV$ diagram for a compression-ignition engine at rated load is shown in Fig. 16-7. Note that the modern CI engine, frequently called a diesel engine, has a $PV$ diagram very similar to that of a spark-ignition engine at full load. In the earlier history of this engine the combustion part of the cycle was somewhat flatter, so that the initial section of the expansion process was closer to a constant-pressure process. As a result the compression-ignition engine was modeled early in its history by a theoretical cycle known as a Diesel cycle. The modern compression-ignition engine is better modeled by the Otto cycle discussed in Sec. 16-4 or the dual cycle discussed later in this section. Since the theoretical Diesel cycle is of limited use, only its basic characteristics will be outlined.

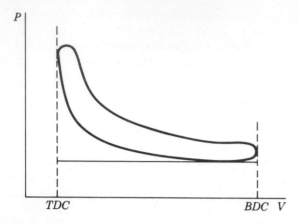

**Figure 16-7** Typical $PV$ diagram for a CI engine at rated load.

The theoretical Diesel cycle for a reciprocating engine is shown in Fig. 16-8 on both $PV$ and $Ts$ diagrams. This cycle, like the Otto cycle, is composed of four internally reversible processes. The only difference between the two cycles is that a Diesel cycle models the combustion as occurring at constant pressure, while the Otto cycle assumes constant-volume heat addition. A useful analysis of the Diesel cycle is made possible by an air-standard cycle based on constant specific heats. Under this circumstance the heat input and output for the cycle are given by

$$q_{in} = c_p(T_3 - T_2) \quad \text{and} \quad q_{out} = c_v(T_4 - T_1)$$

Consequently,

$$\eta_{th,\,Diesel} = \frac{q_{in} - q_{out}}{q_{in}} = \frac{c_p(T_3 - T_2) - c_v(T_4 - T_1)}{c_p(T_3 - T_2)}$$

$$= 1 - \frac{T_4 - T_1}{k(T_3 - T_2)}$$

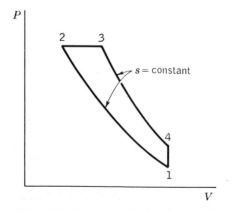

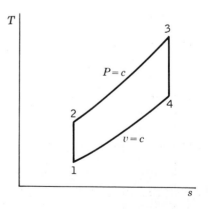

**Figure 16-8** The air-standard Diesel cycle.

The above equation is more informative if one introduces the concept of the *cutoff ratio* $r_c$, which is defined as $V_3/V_2$. By recalling that the compression ratio $r$ is defined as $V_1/V_2$, it can be shown that the preceding equation containing temperatures can be expressed in terms of volumes in the following manner:

$$\eta_{\text{th, Diesel}} = 1 - \frac{1}{r^{k-1}}\left[\frac{r_c^k - 1}{k(r_c - 1)}\right] \tag{16-5}$$

This equation indicates that the theoretical Diesel cycle is primarily a function of the compression ratio $r$, the cutoff ratio $r_c$, and the specific-heat ratio $k$.

It may be demonstrated that the term in the brackets in Eq. (16-5) is always equal to or greater than unity. Hence the thermal efficiency of a Diesel cycle is always less than that of an Otto cycle for the same compression ratio when $r_c$ is greater than unity. In fact, an increase in the cutoff ratio has a drastic effect on the thermal efficiency of an air-standard Diesel cycle. This effect is shown in Fig. 16-9 for one particular compression ratio. Hence manufacturers of modern compression-ignition engines strive to design their engines so that the performance more nearly equals that of an Otto cycle.

In the case where variable specific-heat values are to be considered, the equation for the thermal efficiency of the diesel cycle becomes

$$\eta_{\text{th, Diesel}} = 1 - \frac{u_4 - u_1}{h_3 - h_2} \tag{16-6}$$

To use Eq. (16-6) it is necessary to evaluate the $u$ and $h$ data from the air table in the Appendix. In this case the temperatures at states 2 and 4 are found from the isentropic relations

$$v_{r2} = v_{r1}\frac{V_2}{V_1} = \frac{v_{r2}}{r} \quad \text{and} \quad v_{r4} = v_{r3}\frac{V_4}{V_3} = rv_{r3}$$

Recall that the $v_r$ data in the air table are solely a function of temperature.

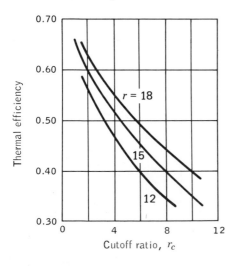

Figure **16-9** Air-standard Diesel cycle—thermal efficiency versus cutoff ratio and compression ratio for $k = 1.4$.

A theoretical cycle which comes closer than the Diesel cycle to matching the actual performance of modern compression-ignition engines is the *dual* cycle. As shown in Fig. 16-10, a short heat-addition process $(2 - x)$ at constant volume is followed by a second heat addition process $(x - 3)$ at constant pressure. The other three parts of the cycle are similar to those found in the Otto and Diesel cycles. The dual cycle is also called a mixed, or limited-pressure, cycle. Note that the use of a two-step heat-addition process allows the theoretical dual cycle to model fairly closely the upper left-hand portion of the actual performance curve shown in Fig. 16-7 for a compression-ignition engine.

The thermal efficiency of the air-standard dual cycle is a function of the heat-input and heat-output quantities. The heat added during the constant-volume process $(2 - x)$ is

$$q_{in} = c_v(T_x - T_2)$$

if we assume that the specific heats are constant. For the constant-pressure path $(x - 3)$ the heat addition is

$$q_{in} = c_p(T_3 - T_x)$$

Therefore the total heat input for the dual cycle is

$$q_{in, total} = c_v(T_x - T_2) + c_p(T_3 - T_x)$$

The heat rejected along the constant-volume path is

$$q_{out} = c_v(T_4 - T_1)$$

Consequently, the thermal efficiency of an air-standard dual cycle for constant-specific-heat data is

$$\eta_{th, dual} = \frac{q_{in} - q_{out}}{q_{in}} = 1 - \frac{c_v(T_4 - T_1)}{c_v(T_x - T_2) + c_p(T_3 - T_x)}$$

$$= 1 - \frac{T_4 - T_1}{(T_x - T_2) + k(T_3 - T_x)} \tag{16-7}$$

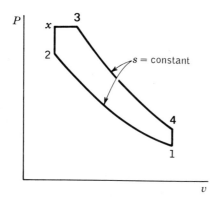
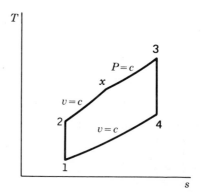

**Figure 16-10** The air-standard dual cycle.

where $k$ is the specific-heat ratio. Recall that the compression ratio $r = V_1/V_2$ and the cutoff ratio $r_c = V_3/V_x = V_3/V_2$. If, in addition, we define the pressure ratio $r_{p,v}$ during the constant-volume combustion process as

$$r_{p,v} = \frac{P_x}{P_2} = \frac{P_3}{P_2}$$

then the thermal efficiency as evaluated by Eq. (16-7) may be expressed solely as a function of $r$, $r_c$, $r_{p,v}$, and $k$. The result is

$$\eta_{th} = 1 - \frac{1}{r^{k-1}}\left[\frac{r_{p,v}r_c^k - 1}{kr_{p,v}(r_c - 1) + r_{p,v} - 1}\right] \qquad (16\text{-}8)$$

When $r_{p,v}$ is unity, Eq. (16-8) reduces to the Diesel efficiency equation given by Eq. (16-5). Also, Eq. (16-8) reduces to Eq. (16-3) for the Otto cycle when $r_c$ is unity.

On the basis of the same heat input and compression ratio, the thermal efficiency of the three theoretical cycles decreases in the following order: Otto cycle, dual cycle, Diesel cycle. This is a major reason that modern CI engines are designed to operate close to the Otto- or dual-cycle models, rather than the Diesel cycle.

**Example 16-2M** The intake conditions for an air-standard dual cycle operating with a compression ratio of 15 : 1 are 0.95 bar and 17°C. The pressure ratio during constant-volume heating is 1.5 : 1 and the volume ratio during the constant-pressure part of the heating process is 2 : 1. Use a cold-air cycle for which $c_v$ and $c_p$ are 0.717 and 1.00 kJ/(kg)(°K), respectively, and $k$ is 1.40. Calculate the temperatures and pressures around the cycle, the heat input and heat rejection, and the thermal efficiency.

SOLUTION We shall use the notation of Fig. 16-10. The pressure and temperature after isentropic compression are

$$P_{2s} = P_1\left(\frac{V_1}{V_2}\right)^k = 0.95(15)^{1.40} = 42.1 \text{ bars}$$

$$T_{2s} = T_1\left(\frac{V_1}{V_2}\right)^{k-1} = 290(15)^{0.40} = 857°K$$

The pressure after constant-volume heating is simply

$$P_x = 1.5P_2 = 1.5(42.1) = 63.2 \text{ bars}$$

On the basis of the ideal-gas equation,

$$T_x = T_2\frac{P_x}{P_2} = 857(1.5) = 1286°K$$

Therefore the heat input during the constant-volume process is

$$q_{in} = c_v(T_x - T_2) = 0.717(1286 - 857) = 308 \text{ kJ/kg}$$

The pressure at state 3 is the same as that at state $x$, namely, 63.2 bars. The temperature at state 3 is

$$T_3 = T_x\frac{V_3}{V_x} = 1286(2) = 2572°K$$

The heat input during the constant-pressure process $x - 3$ is

$$q_{in} = c_p(T_3 - T_x) = 1.00(2572 - 1286) = 1286 \text{ kJ/kg}$$

The total heat input is

$$q_{in, \text{total}} = 308 + 1286 = 1594 \text{ kJ/kg}$$

State 4 is determined from isentropic relations. First, however, we need to determine $V_3/V_4$. This ratio is: $V_3/V_4 = V_3/V_1 = (V_3/V_2)(V_2/V_1) = r_c/r$. Therefore,

$$T_{4s} = T_3\left(\frac{r_c}{r}\right)^{k-1} = 2572\left(\frac{2}{15}\right)^{0.40} = 1149°\text{K}$$

$$P_{4s} = P_3\left(\frac{r_c}{r}\right)^{k} = 63.2\left(\frac{2}{15}\right)^{1.40} = 3.76 \text{ bars}$$

The heat rejected from the cycle then is

$$q_{out} = c_v(T_4 - T_1) = 0.717(1149 - 290) = 616 \text{ kJ/kg}$$

Finally, the thermal efficiency for this cold-air-standard dual cycle is

$$\eta_{th} = 1 - \frac{q_{out}}{q_{in}} = 1 - \frac{616}{1594} = 0.614 \qquad (\text{or } 61.4 \text{ percent})$$

Since this is a cold-air cycle, the efficiency value calculated above may be checked against Eq. (16-7), where $r = 15$, $r_{p,v} = 1.5$, and $r_c = 2$.

**Example 16-2** The intake conditions for an air-standard dual cycle operating with a compression ratio of $15 : 1$ are 14.4 psia and 60°F. The pressure ratio during constant-volume heating is $1.5 : 1$ and the volume ratio during the constant-pressure part of the heating process is $2 : 1$. Calculate the temperatures and pressures around the cycle, the heat input and heat rejection, and the thermal efficiency. Use a cold-air cycle for which $c_v$ and $c_p$ are 0.171 and 0.240 Btu/(lb)(°F), respectively, and $k$ is 1.40.

SOLUTION The notation of Fig. 16-10 will be used. The pressure and temperature after isentropic compression are

$$P_{2s} = P_1\left(\frac{V_1}{V_2}\right)^{k} = 14.4(15)^{1.40} = 638 \text{ psia}$$

$$T_{2s} = T_1\left(\frac{V_1}{V_2}\right)^{k-1} = 520(15)^{0.40} = 1536°\text{R}$$

The pressure after constant-volume heating is simply

$$P_x = 1.5P_2 = 1.5(638) = 957 \text{ psia}$$

On the basis of the ideal-gas equation,

$$T_x = T_2\frac{P_x}{P_2} = 1536(1.5) = 2304°\text{R}$$

Therefore the heat input during the constant-volume process is

$$q_{in} = c_v(T_x - T_2) = 0.171(2304 - 1536) = 131 \text{ Btu/lb}$$

The pressure at state 3 is the same as that at state $x$, namely, 957 psia. The temperature at state 3 is

$$T_3 = T_x \frac{V_3}{V_x} = 2304(2) = 4608°R$$

The heat input during the constant-pressure process $x - 3$ is

$$q_{in} = c_p(T_3 - T_x) = 0.240(4608 - 2304) = 553 \text{ Btu/lb}$$

The total heat input is

$$q_{in, total} = 131 + 553 = 684 \text{ Btu/lb}$$

State 4 is determined from isentropic relations. First, however, we need to determine $V_3/V_4$. This ratio is: $V_3/V_4 = V_3/V_1 = (V_3/V_2)(V_2/V_1) = r_c/r$. Therefore,

$$T_{4s} = T_3 \left(\frac{r_c}{r}\right)^{k-1} = 4608\left(\frac{2}{15}\right)^{0.40} = 2058°R$$

$$P_{4s} = P_3 \left(\frac{r_c}{r}\right)^{k} = 957\left(\frac{2}{15}\right)^{1.40} = 57.0 \text{ psia}$$

The heat rejected from the cycle then is

$$q_{out} = c_v(T_4 - T_1) = 0.171(2058 - 520) = 263 \text{ Btu/lb}$$

Finally, the thermal efficiency for this cold-air-standard dual cycle is

$$\eta_{th} = 1 - \frac{q_{out}}{q_{in}} = 1 - \frac{263}{684} = 0.615 \qquad \text{(or 61.5 percent)}$$

Since this is a cold-air cycle, the efficiency calculated above may be checked against Eq. (16-7), where $r = 15$, $r_{p,v} = 1.5$, and $r_c = 2$.

## 16-6 THE AIR-STANDARD BRAYTON CYCLE

In a simple gas-turbine-power cycle, separate equipment is used for the various processes of the cycle. Initially, air is compressed adiabatically in a rotating axial or centrifugal compressor. At the end of this process the air enters a combustion chamber where fuel is injected and burned at essentially constant pressure. The products of combustion are then expanded through a turbine until they reach the ambient pressure of the surroundings. A cycle composed of these three steps is called an open cycle, because the cycle is not actually completed. Actual gas-turbine cycles are open cycles, since new air must continually be introduced into the compressor. If one wishes to examine a closed cycle, the products of combustion which have expanded through the turbine must be sent through a heat exchanger, where heat is rejected from the gas until the initial temperature is attained. The open and closed gas-turbine cycles are shown in Fig. 16-11. In the analysis of gas-turbine cycles it is useful in the beginning to employ an air-standard cycle. An air-standard gas-turbine cycle with isentropic compression and expansion is called a Brayton cycle. In the Brayton cycle it is necessary to replace the actual combustion process by a heat-addition process. The use of air as the sole working medium throughout the cycle is a fairly good model since a relatively high air-fuel ratio of, roughly, at least 50 : 1 on a mass basis is quite common in actual operation with conventional hydrocarbon fuels.

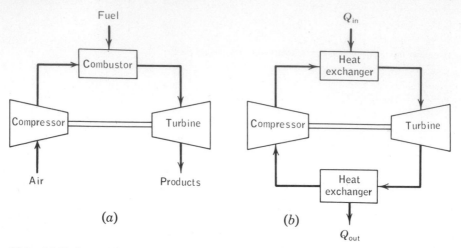

**Figure 16-11** Gas turbine operating on the (a) open, and (b) closed Brayton cycle.

In the Brayton cycle the compression and expansion processes are assumed to be isentropic, and the heat-addition and heat-removal processes occur at constant pressure. Both $Pv$ and $Ts$ diagrams for this ideal cycle are shown in Fig. 16-12. For processes 1-2 and 3-4, which are isentropic, the conservation of energy principle reduces to

$$w = \Delta h + \Delta KE$$

A steady-flow energy balance for the heat exchangers used in processes 2-3 and 4-1 is of the form

$$q = \Delta h + \Delta KE$$

In the absence of appreciable kinetic-energy changes, the thermal efficiency of the ideal Brayton cycle is given by

$$\eta_{th, \text{Brayton}} = 1 - \frac{q_{out}}{q_{in}} = 1 - \frac{h_{4s} - h_1}{h_3 - h_{2s}} \tag{16-9}$$

The enthalpy values may be obtained from the tabular properties of air, or from average $c_p$ data. Assuming that the specific heats of air are constant and recalling that the enthalpy of an ideal gas is given by $c_p T$, we have Eq. (16-9) in the form

$$\eta_{th, \text{Brayton}} = 1 - \frac{T_{4s} - T_1}{T_3 - T_{2s}} = 1 - \frac{1}{r^{k-1}} = 1 - \frac{1}{r_p^{(k-1)/k}} \tag{16-10}$$

where $r$ is defined as $v_1/v_2$ and $r_p$ as $P_2/P_1$. Thus the thermal efficiency of an air-standard Brayton cycle is primarily a function of the overall pressure ratio.

The use of constant-specific-heat data, which led to Eq. (16-10), is quite useful in the initial modeling of a gas turbine power cycle. However, to obtain reasonable values for the heat and work terms in a cycle analysis, it is necessary to account for the variation of $c_p$ with temperature. This requires the use of the air tables in the Appendix. For the isentropic compression and expansion processes in the Brayton cycle, it is convenient to make use of relative-pressure data. Recall that for ideal

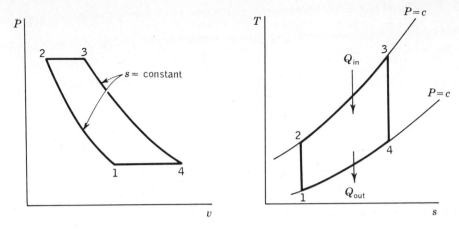

**Figure 16-12** Typical $PV$ and $Ts$ diagrams of the air-standard Brayton cycle.

gases undergoing isentropic changes of state, $P_2/P_1 = p_{r2}/p_{r1}$, where the relative pressure $p_r$ is solely a function of the temperature. The use of $p_r$ data automatically accounts for the variation of $c_p$ with temperature. For a review of the concept of the relative pressure, see Sec. 7-9.

**Example 16-3M** An ideal air-standard Brayton cycle operates with air entering the compressor at 0.95 bar and 22°C. The pressure ratio $r_p$ is 6 : 1, and the air leaves the combustion chamber at 1100°K. Compute the compressor-work input, the turbine-work output, and the thermal efficiency of the cycle, based on the data of Table A-5M.

SOLUTION The compressor outlet condition can be determined from isentropic relationships. For the given inlet state, it is found from Table A-5M that $h_1 = 295.17$ kJ/kg and $p_{r1} = 1.3068$. The relative pressure at the compressor outlet is

$$p_{r2} = p_{r1} \frac{P_2}{P_1} = 1.3068(6) = 7.841$$

The enthalpy and temperature at state 2 are then approximated by interpolation of the $p_{r2}$ data in Table A-5M. We find that $T_{2s} = 490°$K and $h_{2s} = 492.7$ kJ/kg. In addition, since $T_3$ is given as 1100°K, the value of $h_3 = 1161.1$ kJ/kg and the relative pressure $p_{r3} = 167.1$. Hence, for state 4 we find that

$$p_{r4} = p_{r3} \frac{P_4}{P_3} = 167.1(1/6) = 27.85$$

Again interpolating in Table A-5M, we estimate that $T_{4s} = 694°$K and $h_{4s} = 706.5$ kJ/kg. On the basis of these data, the following quantities are determined:

$$w_{C,\,in} = h_{2s} - h_1 = 492.7 - 295.2 = 197.5 \text{ kJ/kg}$$

$$w_{T,\,out} = h_3 - h_{4s} = 1161.1 - 706.5 = 454.6 \text{ kJ/kg}$$

$$q_{in} = h_3 - h_{2s} = 1161.1 - 492.7 = 668.4 \text{ kJ/kg}$$

$$\eta_{th} = \frac{w_T - w_C}{q_{in}} = \frac{454.6 - 197.5}{668.4} = \frac{257.1}{668.4} = 0.385 \qquad \text{(or 38.5 percent)}$$

**Example 16-3** An ideal Brayton cycle operates with air entering the compressor at 14.5 psia and 80°F. The pressure ratio is 6 : 1, and the air leaves the combustion chamber at 1540°F. Compute the compressor-work input, the turbine-work output, and the thermal efficiency of the cycle, based on an air-standard cycle and employing Table A-15.

SOLUTION The compressor outlet conditions can be determined from isentropic relationships. For the given inlet state, it is found that $h_1 = 129.06$ Btu/lb and $p_{r1} = 1.386$. The relative pressure at the compressor outlet, then, is

$$p_{r2} = p_{r1} \frac{P_2}{P_1} = 1.386(6) = 8.316$$

The enthalpy and temperature at state 2 can be approximated by interpolation in the abridged tables in the Appendix. If the unabridged Gas Tables are used, one finds that $h_{2s} = 215.52$ Btu/lb and $T_{2s} = 897°$R. In addition, since $T_3$ is given as 1540°F (2000°R), the value of $h_3 = 504.71$ Btu/lb, and the relative pressure $p_{r3} = 174.00$. Hence

$$p_{r4} = p_{r3} \frac{P_4}{P_3} = 174.00\left(\frac{1}{6}\right) = 29.00$$

Again employing the unabridged tables for accuracy, one finds that $h_{4s} = 307.25$ Btu/lb and $T_{4s} = 1262°$R (802°F). On the basis of these data, the following quantities are determined:

$$w_{C,\,in} = h_{2s} - h_1 = 215.52 - 129.06 = 86.46 \text{ Btu/lb}$$

$$w_{T,\,out} = h_3 - h_{4s} = 504.71 - 307.25 = 197.46 \text{ Btu/lb}$$

$$q_{in} = h_3 - h_{2s} = 504.71 - 215.52 = 289.19 \text{ Btu/lb}$$

$$\eta_{th} = \frac{w_T - w_c}{q_{in}} = \frac{197.46 - 86.46}{289.19} = \frac{111.0}{289.19} = 0.384(38.4 \text{ percent})$$

The data of the preceding example illustrate an important feature of the gas-turbine power cycle. In the ideal Brayton cycle analyzed above, the ratio of the compressor work to the turbine work is roughly 0.44, or 44 percent. This quantity, $w_C/w_T$, is known as the *back work ratio* of a power cycle. In practice this ratio typically ranges from 40 to 80 percent. As the next several sections will show, the effect of irreversibilities in the compressor and turbine is to greatly increase the back work ratio. Obviously, the ratio cannot exceed unity, for in that limiting case the net work output becomes zero. In the case of vapor-power cycles, discussed in Chap. 17, it is found that the back work ratio of pump work to turbine work is several percent or less.

Figure 16-13 illustrates the effect of the pressure ratio $r_p$ and the combustor-outlet temperature $T_3$ on the Brayton cycle. Equation (16-10) shows that the thermal efficiency increases with an increasing pressure ratio, but as $r_p$ increases, it is found that $T_3$ increases for the same heat rejection. For example, increasing the compressor-outlet pressure from state 2 to state 5 on the figure requires that the combustor-outlet temperature increase from state 3 to state 6. The heat rejections for the two cycles 1-2-3-4-1 and 1-5-6-4-1 are exactly the same, as given by the area under the curve 4-1. Increasing $T_3$ to $T_6$ may be disadvantageous, since the temperature may now exceed the maximum allowable at the inlet to the turbine. To overcome this difficulty, it might be proposed that $r_p$ be increased but that the combustor-outlet temperature be restricted to a value not exceeding $T_3$. Such a

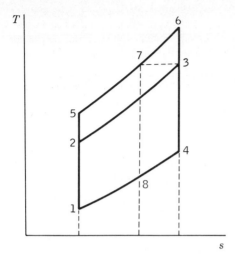

**Figure 16-13** Effect of the pressure ratio and the combustor-outlet temperature on the thermal efficiency of the Brayton cycle.

cycle would be represented by 1-5-7-8-1. The efficiency of this cycle would be greater than that for the basic cycle 1-2-3-4-1, but the figure easily demonstrates that the net work output would be much less. To achieve the same power output as the basic cycle, a larger mass flow rate would be required, which in turn means a larger physical system would be required. This may be undesirable. In the limit, as $r_p$ increases and the combustor-outlet temperature is restricted to a value equivalent to $T_3$, the net work approaches zero. It may also be shown that as $r_p$ decreases for a fixed value of $T_3$, the net work decreases. Consequently, there must be a value of $T_2$ (and hence $P_2$) which leads to a maximum net work output. This is found by differentiating the equation for the net work with respect to $T_2$, for fixed values of $T_1$ and $T_3$. It is found that the maximum net work for the Brayton cycle under these restrictions occurs when $T_2 = (T_1 T_3)^{1/2}$. The lower curve in Fig. 16-14 shows the variation of thermal efficiency with pressure ratio for an ideal air-standard cycle. The data in this figure are based on constant-specific-heat values and a turbine-inlet to compressor-inlet temperature ratio of 4 : 1. If the net work were plotted on this figure, the function would maximize at a pressure ratio around 12 : 1.

Compared with the Otto cycle, the Brayton cycle operates over a wider range of volume but a smaller range of pressures and temperatures. This feature makes the Brayton cycle unadaptable for use with reciprocating machinery. Instead, one makes use of turbomachinery, which comprises steady-flow devices. The Brayton cycle is a standard for the practical design of modern gas-turbine power plants. It should be kept in mind that irreversibilities occur in the compressor and turbine in actual operation. Hence the compressor and turbine work quantities are not equal to the isentropic enthalpy change across each device. The effect of these irreversibilities is to decrease the heat input in the combustor and increase the heat output in the heat-rejection part of the cycle. The overall result of this is a decrease in the thermal efficiency of the cycle. Other differences appear between

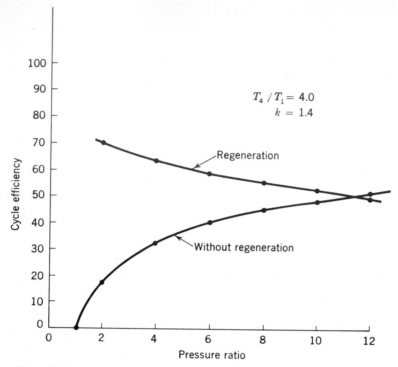

**Figure 16-14** Brayton-cycle efficiency as a function of pressure ratio for the ideal cycle, without and with regeneration.

the actual and the theoretical performance of the cycle. In actuality, $P_3$ does not equal $P_2$, and $P_4$ does not equal $P_1$, because of pressure drops in the combustor and the heat exchanger. In the cycle analysis presented here, these pressure drops are ignored. Heat losses from the compressor and turbine may also be present.

The effect of irreversibilities in the compressor and turbine is frequently accounted for by the use of a term called the adiabatic efficiency of these devices. Without actual test data on a particular model, the adiabatic efficiency can be predicted fairly well from experience with similar devices. The adiabatic efficiency permits one to evaluate the actual work quantities of turbines and compressors, as will be seen in the next section. It will be found that the thermal efficiency of a gas-turbine power cycle is quite sensitive to the efficiencies of the turbine and the compressor.

## 16-7 THE ADIABATIC EFFICIENCY OF WORK DEVICES

Since the maximum work output, or the minimum work input, occurs when a process is internally reversible, it is apparent that these conditions are achieved for an adiabatic process when the process is isentropic. For those processes which are close to being adiabatic in practice, the isentropic work associated with a change in state is a goal, or standard, with which all actual processes may be compared. For example, when turbines and compressors operate adiabatically, the isentropic work output or input of these devices is a standard with which their actual performance should be compared. In order to describe the actual performance of work-producing devices which are essentially adiabatic, we shall define an *adiabatic turbine efficiency* $\eta_T$ by

$$\eta_T \equiv \frac{\text{actual work output}}{\text{isentropic work output}} = \frac{W_a}{W_s} \qquad (16\text{-}11)$$

This quantity is frequently termed, simply, the turbine efficiency. For devices which require work input, the *adiabatic compressor efficiency* is defined as

$$\eta_C \equiv \frac{\text{isentropic work input}}{\text{actual work input}} = \frac{W_s}{W_a} \qquad (16\text{-}12)$$

It is convenient, often, to speak simply of the compressor efficiency. For a given device, the initial state used in evaluating both $W_a$ and $W_s$ is the same, of course. The final state in each case has the same pressure. Other properties in the final state, such as temperature, enthalpy, etc., will be different for the isentropic versus the actual process. Figure 16-15 illustrates reversible and irreversible adiabatic processes for both turbines and compressors. It is noted in the figure that the specific entropy always increases for the actual adiabatic processes, as dictated by consequences of the second law. If the changes in kinetic and potential energies are neglected, the work terms may be evaluated as a function of the enthalpy changes for a steady-flow process. Use of the nomenclature from Fig. 16-15a allows the adiabatic efficiency of a work-producing device such as a turbine to be written as

$$\eta_T = \frac{h_1 - h_{2a}}{h_1 - h_{2s}} \qquad (16\text{-}13)$$

where the subscript $a$ represents the actual outlet condition and the subscript $s$

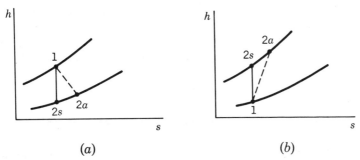

(a)                                      (b)

**Figure 16-15** Process diagrams for adiabatic, reversible and adiabatic, irreversible work devices. (a) Turbine; (b) compressor.

represents the isentropic outlet state. Similarly for the adiabatic compressor, the efficiency becomes

$$\eta_C = \frac{h_{2s} - h_1}{h_{2a} - h_1} \tag{16-14}$$

Slightly different definitions for these efficiencies are employed sometimes; so the student should not accept them as having universal applicability. For instance, the kinetic-energy change may be included in either the numerator or denominator, or both, of the efficiency expression. This is a matter of choice. In the absence of kinetic-energy effects it is seen that the efficiencies defined above are ratios of vertical distances taken from the Mollier diagram. Efficiencies of 80 to 85 percent are not uncommon for modern turbines and compressors. The effect of the irreversibilities is to require a larger compressor-work input, which is available from a smaller turbine output. Hence the compressor in actual practice may consume 40 to 70 percent of the turbine output.

**Example 16-4M** An air-standard gas-turbine cycle operates under the conditions used in Example 16-3M, except that the compressor and turbine are only 82 and 85 percent efficient, respectively. Compute the compressor- and turbine-work quantities and the thermal efficiency for this cycle, again using the data of Table A-5M for air.

SOLUTION The data for the compressor- and turbine-inlet states remain the same as before. The isentropic values previously calculated for the outlet states of these two devices are employed to find the actual states, in conjunction with the definitions of compressor and turbine efficiencies. For the compressor, in the absence of kinetic-energy effects,

$$\Delta h_a = \frac{\Delta h_s}{\eta_C} = \frac{197.5}{0.82} = 240.9 \text{ kJ/kg}$$

Since $\Delta h_a = h_{2a} - h_1$, the actual compressor-outlet enthalpy is

$$h_{2a} = 240.9 + 295.2 = 536.1 \text{ kJ/kg}$$

From Table A-5M it is seen that this corresponds to an outlet temperature of 532°K. Consequently, the irreversibilities within the compressor have resulted in a temperature rise of 42°K over that for the isentropic compression. For the turbine,

$$\Delta h_a = \eta_T \, \Delta h_s = 0.85(-454.6) = -386.4 \text{ kJ/kg}$$

Since $\Delta h_a = h_{4a} - h_3$, the actual turbine-outlet enthalpy is

$$h_{4a} = 1161.1 - 386.4 = 774.7 \text{ kJ/kg}$$

This value of $h$ corresponds to a temperature of 757°K, which is 63°K above that for the isentropic expansion. The thermal efficiency becomes

$$\eta_{th} = \frac{w_{T,a} - w_{C,a}}{h_3 - h_{2a}} = \frac{386.4 - 240.9}{1161.1 - 536.1} = 0.233 \qquad \text{(or 23-3 percent)}$$

Therefore the presence of irreversibilities within the compressor and turbine has led to a decrease in the thermal efficiency from 38.5 to 23.3 percent, which is quite significant. In addition, irreversibilities have increased the ratio of compressor work to turbine work (back work ratio) to 0.62 from the value of 0.44 for isentropic flow. Relatively small changes in adiabatic efficiencies greatly influence the net work output and the back work ratio of the device.

**Example 16-4** A gas-turbine cycle operates under the conditions used in Example 16-3, except that the compressor and turbine are only 82 and 85 percent efficient, respectively. Compute the compressor- and turbine-work quantities and the thermal efficiency for this cycle, again using Table A-5, the table of properties for air.

SOLUTION The data for the compressor and turbine-inlet states remain the same as before. The isentropic values for the outlet states of these two devices are employed to find the actual states, in conjunction with the definitions of compressor and turbine efficiencies. For the compressor, in the absence of kinetic-energy effects,

$$\Delta h_a = \frac{\Delta h_s}{\eta_C} = \frac{86.46}{0.82} = 105.4 \text{ Btu/lb}$$

Since $\Delta h_a = h_{2a} - h_1$, the actual compressor-outlet enthalpy is

$$h_{2a} = 105.4 + 129.06 = 234.5 \text{ Btu/lb}$$

From Table A-5 it is seen that this corresponds to an outlet temperature of $974°$R. Consequently, the irreversibilities within the compressor have resulted in a temperature rise of $77°$F over that for the isentropic compression. For the turbine,

$$\Delta h_a = \eta_T \Delta h_s = 0.85(-197.46) = -167.8 \text{ Btu/lb}$$

Since $\Delta h_a = h_{4a} - h_3$, the actual turbine-outlet enthalpy is

$$h_{4a} = 504.71 - 167.8 = 336.9 \text{ Btu/lb}$$

This value corresponds to a temperature which is $115°$F above that for the isentropic expansion. The thermal efficiency becomes

$$\eta_{th} = \frac{167.8 - 105.4}{504.7 - 234.5} = \frac{62.4}{270.2} = 0.232 \qquad \text{(or 23.2 percent)}$$

Therefore the presence of fluid friction within the compressor and turbine has led to a decrease in the thermal efficiency from 38.4 to 23.2 percent, which is quite significant. In addition, irreversibilities have increased the ratio of compressor work to turbine work to 0.63 from the value of 0.44 for isentropic flow. Small changes in adiabatic efficiencies greatly influence the net work output of the device.

## 16-8 THE REGENERATIVE GAS-TURBINE CYCLE

The basic gas turbine cycle can be modified in several important ways to increase its overall efficiency. One of these ways is based on the concept of regeneration. In the preceding examples for the Brayton cycle, it is noted that the gases leaving the turbine are at a relatively high temperature. In many cases the outlet turbine temperature is higher than the outlet compressor temperature. It is possible, then, to reduce the amount of fuel injected into the combustor by reheating the air leaving the compressor with energy taken from the turbine exhaust gases. The exchange of heat between the two flow streams takes place in a heat exchanger usually called a regenerator. A flow diagram for the regenerative gas-turbine cycle is shown in Fig. 16-16a. In the ideal situation it is assumed that the flow through the regenerator occurs at constant pressure. If an internally reversible heat exchanger is assumed, the heat transfer between the two flow streams may be represented as an area on a $Ts$ diagram. Since the heat flow from the turbine-exhaust

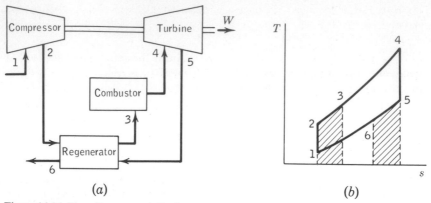

Figure 16-16 Flow diagram and $Ts$ diagram for an ideal regenerative gas-turbine cycle.

flow stream must equal the heat input into the compressor-outlet stream, the two crosshatched areas in Fig. 16-16$b$ must be equal in magnitude; that is, $Q_{23} = -Q_{56}$. Note that $T_5$ is considerably higher than $T_2$.

If the regenerator were ideal in its operation, it would be possible to preheat the compressor-outlet stream up to the temperature of the turbine-outlet stream; i.e., ideally, $T_3 = T_5$. This is impractical, though, since a very large surface area is required for heat transfer as the temperature difference between two systems approaches zero. As a measure of the approach to this limiting ideal condition, it is convenient to introduce the effectiveness of a regenerator. The regenerator effectiveness, $\eta_{eff}$, is defined as

$$\eta_{eff} \equiv \frac{h_3 - h_2}{h_5 - h_2} \qquad (16\text{-}15)$$

where the temperature corresponding to $h_3$ is somewhat lower than the temperature corresponding to $h_5$. In the presence of a regenerator, we find that $q_{in} = h_4 - h_3$, and $q_{out} = h_6 - h_1$. Consequently, the thermal efficiency becomes

$$\eta_{th} = 1 - \frac{h_6 - h_1}{h_4 - h_3} \qquad . \qquad (16\text{-}16)$$

If the gases behave ideally and the specific heats can be considered constant, it may be shown that the thermal efficiency reduces to

$$\eta_{th} = 1 - \frac{T_1}{T_4}\left(\frac{P_2}{P_1}\right)^{(k-1)/k} \qquad (16\text{-}17)$$

and at the same time the effectiveness becomes

$$\eta_{eff} = \frac{T_3 - T_2}{T_5 - T_2} \qquad (16\text{-}18)$$

Thus the value of the thermal efficiency for a regenerative-type gas-turbine cycle is also a function of the ratio of the maximum to minimum temperatures of the cycle, in addition to the pressure ratio. Note from Fig. 16-14 that the efficiency decreases with increasing $r_p$, contrary to the basic Brayton-cycle efficiency. To increase the thermal efficiency, the value of $h_3$, and hence $T_3$, should be as high as possible, since the heat input is reduced but the net work remains the same. The usual value of the effectiveness is somewhat less than 0.7 for gas-turbine stationary power-plant applications. To increase it beyond this value usually leads to equipment costs which negate any advantage of the higher thermal efficiency. Also, a higher effectiveness requires a greater heat-transfer area. However, this leads to an increased pressure drop through the regenerator, which is a loss in terms of cycle efficiency.

**Example 16-5M** Employing the data used in Example 16-3M for an ideal air-standard Brayton cycle, determine the effect on the thermal efficiency of the cycle if an ideal regenerator is inserted into the cycle.

SOLUTION The data for the enthalpies at the inlet and outlet of the compressor and turbine remain the same. Employing the symbols shown in Fig. 16-16, we note that $h_1 = 295.2$, $h_{2s} = 492.7$, $h_4 = 1161.1$, and $h_{5s} = 706.5$, all the values listed in kJ/kg. For an ideal regenerator, the exit temperature of the fluid heated in the regenerator is the same as the outlet temperature of the turbine. Consequently, the enthalpy of the fluid leaving the regenerator $h_3$ must be equal to $h_{5s}$. The heat input saved because of the regenerator is

$$q_{saved} = h_3 - h_{2s} = h_5 - h_2 = 706.5 - 492.7 = 213.8 \text{ kJ/kg}$$

The required heat input now becomes

$$q_{in, regen} = h_4 - h_3 = h_4 - h_5 = 1161.1 - 706.5 = 454.6 \text{ kJ/kg}$$

The percent saved over the case without regeneration is

$$\% q_{saved} = \frac{213.8}{668.4} 100 = 32.0 \text{ percent}$$

Since the turbine and compressor work are exactly the same as in Example 16-3M,

$$\eta_{th} = \frac{257.1}{454.6} = 0.565 \quad \text{(or 56.5 percent)}$$

This compares with an efficiency of 38.5 percent without regeneration. Thus the effect of regeneration on the thermal efficiency is considerable, although in actual cases the effect would not be quite as large, since ideal regeneration is not possible. In addition, internal irreversibilities and heat losses will reduce the efficiency further.

**Example 16-5** Employing the data used in Example 16-3 for an ideal Brayton cycle, determine the effect on the thermal efficiency of the cycle if an ideal regenerator is inserted into the cycle.

SOLUTION The data for the enthalpies at the inlet and outlet of the compressor and turbine remain the same. Employing the symbols shown in Fig. 16-16, we note that $h_1 = 129.06$, $h_{2s} = 215.52$, $h_4 = 504.71$, and $h_{5s} = 307.25$, all the values listed in Btu/lb. For an ideal regenerator, the exit temperature of the fluid heated in the regenerator is the same as the outlet temperature of the turbine. Consequently, the enthalpy of the fluid leaving the regenerator $h_3$ must be equal to $h_{5s}$. The heat input no longer required because of the regenerator is

$$q_{saved} = h_3 - h_2 = h_5 - h_2 = 307.25 - 215.52 = 91.73 \text{ Btu/lb}$$

The required heat input now becomes

$$q_{in, regen} = h_4 - h_3 = h_4 - h_5 = 504.71 - 307.25 = 197.46 \text{ Btu/lb}$$

The percent heat saved over the case without regeneration is

$$\% \, q_{saved} = \frac{91.73}{289.19} \, 100 = 31.7 \text{ percent}$$

Since the turbine and compressor work are exactly the same as before,

$$\eta_{th} = \frac{111.0}{197.46} = 0.562 \qquad \text{(or 56.2 percent)}$$

This compares with an efficiency of 38.4 percent without regeneration. It is seen that the effect of regeneration on the thermal efficiency is considerable, although in actual cases the effect would not be quite as large since ideal regeneration is not possible. In addition, internal irreversibilities and heat losses will reduce the efficiency further.

**Example 16-6M** Employing the data of Example 16-4, determine the effect on the thermal efficiency if a regenerator of 70 percent effectiveness is inserted into the cycle.

SOLUTION In Example 16-4 the compressor and turbine efficiencies are 82 and 85 percent, respectively. The data for the enthalpies at the inlet and outlet of the compressor and turbine remain the same. Using the symbols shown in Fig. 16-17, we note that $h_1 = 295.2$, $h_{2a} = 536.1$, $h_4 = 1161.1$, and $h_{5a} = 774.7$, all the values listed in kJ/kg. Employment of the regenerator-effectiveness equation enables us to determine the heat added between states 2 and 3. Use of Eq. (16-12) yields

$$0.70 = \frac{h_3 - h_2}{h_5 - h_2} = \frac{h_3 - h_2}{774.7 - 536.1}$$

or

$$h_3 - h_2 = 0.7(238.6) = 167.0 \text{ kJ/kg}$$

This quantity represents the heat input saved by regeneration. Without regeneration the heat input would need to be

$$q_{without \, regen} = h_4 - h_{2a} = 1161.1 - 536.1 = 625.0 \text{ kJ/kg}$$

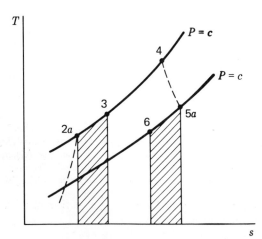

**Figure 16-17** The $Ts$ diagram for the regenerative gas-turbine cycle in Examples 16-6M and 16-6.

The percent heat input saved over the case without regeneration

is
$$\% \, q_{saved} = \frac{167.0}{625.0} \, 100 = 26.7 \text{ percent}$$

Since the turbine and compressor work are exactly the same as in Example 16-4,

$$\eta_{th} = \frac{386.4 - 240.9}{625.0 - 167.0} = \frac{145.5}{458.0} = 0.318 \qquad (\text{or } 31.8 \text{ percent})$$

This compares to an efficiency of 23.3 percent without regeneration. It is also a more realistic value than the 56.5 percent value found in Example 16-5M, because reasonable turbine and compressor efficiencies and regenerator effectiveness have been assumed.

**Example 16-6** Employing the data of Example 16-4, determine the effect on the thermal efficiency if a regenerator of 70 percent effectiveness is inserted into the cycle.

SOLUTION The data for the enthalpies at the inlet and outlet of the compressor and turbine remain the same. Employing the symbols shown in Fig. 16-17, we note that $h_1 = 129.1$, $h_{2a} = 234.5$, $h_4 = 504.7$, and $h_{5a} = 336.9$, all the values listed in Btu/lb. Employment of the regenerator-effectiveness equation enables us to determine the heat added between states 2 and 3. Use of Eq. (16-12) yields

$$0.70 = \frac{h_3 - h_2}{336.9 - 234.5}$$

or
$$h_3 - h_2 = 0.7(102.4) - 71.7 \text{ Btu/lb}$$

This quantity represents the heat input saved by regeneration. Without regeneration the heat input would need to be

$$q_{without \ regen} = h_4 - h_{2a} = 504.7 - 234.5 = 270.2 \text{ Btu/lb}$$

The percent heat input saved over the case without regeneration is

$$\% \, q_{saved} = \frac{71.7}{270.2} \, 100 = 26.5 \text{ percent}$$

Since the turbine and compressor work are exactly the same as in Example 16-4,

$$\eta_{th} = \frac{62.4}{270.2 - 71.7} = \frac{62.4}{198.5} = 0.315 \qquad (\text{or } 31.5 \text{ percent})$$

This compares to an efficiency of 23.2 percent without regeneration. It is also a more realistic value than the 56.2 percent value found in Example 16-5, since realistic turbine and compressor efficiencies and regenerator effectiveness have been assumed.

## 16-9 THE POLYTROPIC PROCESS

When an ideal gas is expanded or compressed in an internally reversible, adiabatic manner, the relationship between $P$ and $v$ is given by the isentropic equation

$$Pv^k = \text{constant} \tag{7-29b}$$

This process equation is a reasonably good model when the heat transfer during an actual process is quite small, compared to the work interaction. Even if the heat transfer is not negligible during a process, the format of Eq. (7-29b) can be retained by writing the equation in the form

$$Pv^n = \text{constant} \tag{16-19}$$

A process which follows this relation between equilibrium states is called a *polytropic* process. The equation is restricted to a quasistatic process for a simple, compressible gaseous system. If the relation $Pv = RT$ is valid, then it is easily shown that two other equations are equally valid for a polytropic process. Analogous to equations developed for adiabatic, reversible processes, we find that for polytropic processes,

$$T(v)^{n-1} = \text{constant} \tag{16-20}$$

and

$$TP^{-(n-1)/n} = \text{constant} \tag{16-21}$$

The simplicity of the polytropic equations is a distinct advantage when deriving mathematical relationships for various thermodynamic functions such as $\delta w$, $\delta q$, and $du$, if the validity of the model is justified.

One of the notable features of the polytropic process is that, when $n$ takes on certain specific values, the process may be identified with a more familiar change of state. For example, when the process is adiabatic, then $n = k$. The reader should identify that when $n = 1$ and 0, the processes are isothermal and isobaric, respectively. The table below summarizes the situation for four familiar cases.

| Process | F | n |
|---|---|---|
| Isentropic | 0 | k |
| Isothermal ($T = c$) | 1 | 1 |
| Isobaric ($P = c$) | $k/(k-1)$ | 0 |
| Isovolumic ($v = c$) | $\infty$ | $-\infty$ |

When $n$ is negative for a process, both $P$ and $v$ will increase (or decrease) during the process. Physically, this implies that heat must be added at a sufficient rate that the pressure will rise during an expansion process in spite of the increase

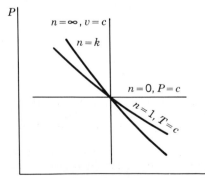

$v$    **Figure 16-18** Polytropic processes.

in volume. The limiting case is when heat is added and no work is performed, in which case $n$ is negative infinity. (A constant-volume process is a special case of the polytropic process for values of $n$ of either positive or negative infinity.) Figure 16-18 illustrates the polytropic process for the specific values of $n$ discussed above. In some instances the polytropic expression $Pv^n = c$ is a good approximation for the process equation for real gases, even though $Pv \neq RT$. Under such circumstances it must be recognized that Eqs. (16-20) and (16-21) will no longer be valid, since they require that $Pv = RT$.

**Example 16-7M** Air is compressed polytropically along a path for which $n = 1.30$ in a closed system. The initial temperature and pressure are 17°C and 1 bar, respectively, and the final pressure is 5 bars. Find (a) the final temperature in °C, (b) the work done on the gas in kJ/kg, and (c) the heat transferred in kJ/kg.

SOLUTION  Air is essentially an ideal gas in the range of pressures and temperatures involved. Consequently, we may use Eq. (16-21) to determine the final temperature.

(a)
$$T_2 = T_1 \left(\frac{P_2}{P_1}\right)^{(n-1)/n} = 290(5/1)^{(1.30-1)/1.30}$$

$$- 290(1.45) = 420°K = 147°C$$

(b) The first-law statement for a closed system is $q + w = \Delta u$. Both $q$ and $w$ are unknown. Since we have no equations for calculating $q$ independently, we must first evaluate $w$ by means of the integral of $-P\,dv$. Substitution of the process equation $Pv^n = c$ and subsequent integration leads to the following equation for the work associated with polytropic processes within a closed system:

$$w - -\int P\,dv = -\int cv^{-n}\,dv = \frac{c(v_2)^{1-n}c(v_1)^{1-n}}{n-1}$$

$$= \frac{P_2 v_2 - P_1 v_1}{n-1} = \frac{R(T_2 - T_1)}{n-1}$$

Employing the proper values, we find that

$$w = \frac{8.315(420 - 290)}{29(1.3 - 1)} = 124.2 \text{ kJ/kg}$$

(c) The heat transferred is determined by the conservation of energy statement for the closed system, $q = \Delta u - w$. The specific internal-energy change for an ideal gas is simply $c_v \Delta T$, and for the temperature range involved the average value of $c_v$ is close to 0.723 J/(g)(°K). Hence the heat transferred is

$$q = 0.723(420 - 290) - 124.2 = -30.2 \text{ kJ/kg}$$

The negative sign indicates that heat is removed from the gas during the compression process. The ratio of $q$ to $w$ for this specific polytropic process is $30.2/124.1 = 0.24$.

**Example 16-7** Air is compressed polytropically along a path for which $n = 1.3$ in a closed system. The initial temperature is 40°F, the initial pressure is 15 psia, and the final pressure is 75 psia. Find (a) the final temperature, (b) the work done on the gas, and (c) the heat transferred.

SOLUTION  Air will follow closely the ideal-gas equation of state in the range of pressures and temperatures involved. Consequently, we may employ Eq. (16-21) to determine the final temperature.

(a)
$$T_2 = T_1\left(\frac{P_2}{P_1}\right)^{(n-1)/n} = 500\left(\frac{75}{15}\right)^{(1.3-1)/1.3}$$

$$= 500(1.45) = 725°R = 265°F$$

(b) The work done on the gas normally might be computed from the first-law statement for a closed system, namely, $q + w = \Delta u$. However, in the absence of a means of calculating the value of $q$ independently, we must turn to the analytical expression for boundary work and determine the value of $w$ independent of other quantities. The boundary work is given by the integral of $-P\,dv$, and the relationship between $P$ and $v$ is simply $Pv^n = c$. Subsequent substitution and definite integration leads to the following equation for the work associated with polytropic processes:

$$w = -\int P\,dv = -\int cv^{-n}\,dv = \frac{c(v_2)^{1-n} - c(v_1)^{1-n}}{n-1}$$

$$= \frac{P_2 v_2 - P_1 v_1}{n-1} = \frac{R(T_2 - T_1)}{n-1}$$

Employing the proper numbers, we find that

$$w = \frac{1545(265 - 40)}{29(1.3 - 1)} = +40,000 \text{ ft·lb}_f/\text{lb}_m$$

$$= +51.5 \text{ Btu/lb}_m$$

(c) The heat transferred is determined by the conservation of energy statement for the closed system, $q = \Delta u - w$. The specific internal-energy change for an ideal gas is simply $c_v\,\Delta T$, and in the temperature range involved the average value of $c_v$ is, roughly, 0.172 Btu/$(\text{lb}_m)(°F)$. Hence the heat transferred per pound of gas is

$$q = 0.172(265 - 40) - (51.5) = -12.8 \text{ Btu/lb}$$

The negative sign indicates that heat was removed from the gas during the compression.

## 16-10 STEADY-FLOW ANALYSIS OF COMPRESSORS

The steady-flow equation for the conservation of energy with regard to a compressor reduces to

$$w = h_2 - h_1 + KE_2 - KE_1 - q$$

Often the kinetic-energy difference is small; hence

$$w = (h_2 - h_1) - q$$

For ideal gases the enthalpy values can be found in the ideal-gas tables or calculated from the integral of $c_p\,dT$. For real gases the enthalpy can be obtained from tables or calculated from generalized equations introduced in Chap. 13. The heat-transfer quantity $q$ must be known from other experimental measurements, although in the case of polytropic processes it can be calculated. In analyzing any compression process it is convenient to consider three particular cases. We shall restrict ourselves in the following discussion to ideal gases.

## 1 Adiabatic Compression with Constant Specific Heats

For an adiabatic process the conservation of energy principle reduces to

$$w = (h_2 - h_1) = c_p(T_2 - T_1)$$

If, in addition, the process is reversible, then $T_2/T_1 = (P_2/P_1)^{(k-1)/k}$. It may be shown that, under these circumstances, the steady-flow mechanical work becomes

$$w_{sf} = \frac{kRT_1[1 - (P_2/P_1)^{(k-1)/k}]}{1 - k}$$

## 2 Isothermal Compression

In the case of an ideal gas, the enthalpy change for an isothermal process is zero. Hence the energy balance for a simple substance becomes $w = -q$. An analytical expression for the steady-flow mechanical work of a frictionless process can be obtained from the integral of $v\ dP$; thus

$$w - \int v\ dP = \int \frac{RT}{P}\ dP = RT \ln \frac{P_2}{P_1} = RT \ln \frac{v_1}{v_2}$$

## 3 Polytropic Compression

In Sec. 16-9 it was shown that, under certain circumstances, a quasistatic process with heat transfer may follow a path represented by the equation $Pv^n = $ constant. In this situation it is found that the steady-flow work is given by

$$w_{sf} = \int v\ dP - \Delta KE = \frac{n(P_2 v_2 - P_1 v_1)}{n - 1} - \Delta KE \qquad (16\text{-}22)$$

In addition, a specific equation for the heat transfer may be obtained from the conservation of energy principle for steady-flow processes and relation (16-22) for steady-flow work. Since $q + w = \Delta h + \Delta KE$, then

$$q = -w_{sf} + \Delta h + \Delta KE = \Delta h + \frac{n(P_2 v_2 - P_1 v_1)}{1 - n}$$

$$= c_p(T_2 - T_1) + \frac{nR(T_2 - T_1)}{1 - n} = \left(c_v + \frac{R}{1 - n}\right)(T_2 - T_1)$$

$$= \frac{c_v(k - n)(T_2 - T_1)}{1 - n} = c_n(T_2 - T_1) \qquad (16\text{-}23a)$$

where the polytropic specific heat has been defined as

$$c_n = \frac{c_v(k - n)}{1 - n} \qquad (16\text{-}23b)$$

It is helpful to compare the frictionless steady-flow mechanical work for the three types of processes just described. In the absence of significant kinetic-energy changes, the work of compression for a steady-flow process is evaluated by the integral of $v \, dP$. A graphical representation of such a process is shown in Fig. 16-19 for a fixed pressure ratio $P_2/P_1$. Since the area to the left of the process curve is a measure of the steady-flow work, it is seen that the maximum frictionless work is required for an adiabatic compression, where $n = k$. As the value of $n$ is decreased by removal of heat during the process, the work requirements also decrease. When sufficient heat is removed during a process that $n$ is unity, the process is isothermal. Figure 16-19 illustrates the simple fact that the removal of heat during a compression process is advantageous with regard to the work-input requirements. This result is equally true for both reciprocating and rotating (axial and centrifugal) compressors. Consequently, it might be practical to use cooling jackets around the casing of a compressor in order to lower the value of the polytropic coefficient during the compression process. For air the specific-heat ratio at normal temperatures is 1.4, but by external cooling the value of $n$ for the process may be reduced to the neighborhood of 1.35. The practice of cooling a fluid during compression depends upon the use to which the fluid will be placed. If a gas at high pressure is the primary purpose of a compression process, cooling is beneficial. If a fluid of high enthalpy is desired at a high pressure, adiabatic operation is more desirable.

Although the practice of cooling the gas passing through a compressor has its advantages, in many cases it is either not possible or practical to have much heat transfer through the compressor casing. To achieve the benefits of cooling, another physical arrangement is frequently used, called multistage compression with intercooling. This is especially effective when a comparatively large pressure change is desired. In multistage compression, the fluid is first compressed to some intermediate pressure which lies between $P_1$ and $P_2$. The fluid then is passed through an intercooler (a heat exchanger), where it is cooled down to a lower temperature at essentially constant pressure. In some instances this lower temperature could be the initial temperature $T_1$. Then it passes through another state of

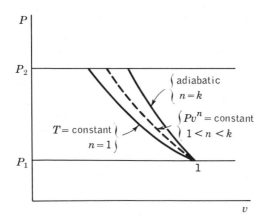

**Figure 16-19** The effect of cooling on the required work of compression for a steady-flow process between fixed pressure limits.

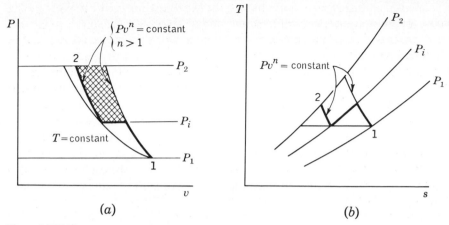

**Figure 16-20** Two-stage compression with intercooling to the initial temperature.

the compressor, where its pressure is further raised. This would be followed by another intercooler process, and then another staging of the compressor, until the final high pressure is achieved. The overall result is, of course, a lowering of the net work input required for a given pressure ratio.

The effect of intercooling on a two-stage compressor is demonstrated in Fig. 16-20. In this special case it has been assumed that the intercooler cools the fluid down to the initial temperature before it enters the second stage. The compression processes themselves between pressures $P_1$ to $P_i$ and $P_i$ to $P_2$ are of the type $Pv^n = $ constant, where the value of $n$ depends upon the amount of heat transfer that occurs during compression. The crosshatched area on the $Pv$ diagram represents the work input that is saved by the process of intercooling. Diagrams similar to the $Pv$ and $Ts$ diagrams shown in Fig. 16-20 could be drawn for compressors with a larger number of stages. The compression processes on the $Ts$ diagram have been sketched for a polytropic process, where $n$ lies between 1 and $k$.

It is of theoretical and practical interest to determine the conditions under which the work input would be minimized, that is, the crosshatched area in Fig. 16-20a would be maximized. For a two-stage compressor, both the intermediate pressure and temperature can be varied, which leads to an indefinitely large number of ways of carrying out the compression process. In order to simplify the calculation, we shall assume that it is possible to design an intercooler that will reduce the temperature of the fluid to the initial temperature of the system before it enters the second state. By using Eq. (16-22), the total work of compression for the two stages, neglecting kinetic-energy changes, is

$$w = \frac{nRT_1\left[\left(\dfrac{P_x}{P_1}\right)^{(n-1)/n} - 1\right]}{n - 1} + \frac{nRT_1\left[\left(\dfrac{P_2}{P_x}\right)^{(n-1)/n} - 1\right]}{n - 1} \qquad (16\text{-}24)$$

The work is minimized by differentiating the work quantity with respect to the pressure $P_x$ and setting the resulting equation equal to zero. When this is done, it is found that the minimum work input is obtained when the pressures are related by the expression

$$P_x = (P_1 P_2)^{1/2} \qquad \text{or} \qquad \frac{P_x}{P_1} = \frac{P_2}{P_x} \tag{16-25}$$

If this result is substituted into the equation for the work, it is noted that the work done by each stage is equal; that is, $w_{1x} = w_{x2}$. It is easily demonstrated by use of Fig. 16-20$a$ that, as $P_x$ approaches either $P_1$ or $P_2$, the work saved by intercooling is extremely small, if $T_i$ is assumed to equal $T_1$. Consequently, the most effective value of $P_x$ must lie at some intermediate value between $P_1$ and $P_2$.

## 16-11 GAS-TURBINE CYCLES WITH INTERCOOLING AND REHEATING

Another method of increasing the overall efficiency of a gas-turbine cycle is to decrease the work input to the compression process and/or increase the work output of the turbine. The effect of either of these procedures is to increase the net work output. In Sec. 16-10, it was pointed out that the work of a compressor operating between fixed pressure levels can be reduced by intercooling between stages of the compressor. Figure 16-21 is a $Ts$ diagram for a cycle with intercooling between a two-stage compressor. Note that the final temperature $T_2$ is less than the temperature which would occur without intercooling. As a consequence of this low temperature, a gas-turbine cycle with intercooling is especially adaptable to regeneration. In fact, intercooling is promising only if a regenerator is employed at the same time; otherwise a considerable amount of heat must be supplied to the cycle at a relatively low temperature. It must also be kept in mind that a considerably larger regenerator will be required than in the case of no intercooling.

In conjunction with intercooling, it is often found effective to use turbine staging as well. Instead of expanding directly through a single turbine, the gases

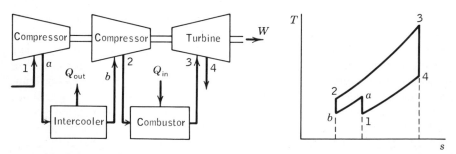

**Figure 16-21** The air-standard gas-turbine cycle with intercooling.

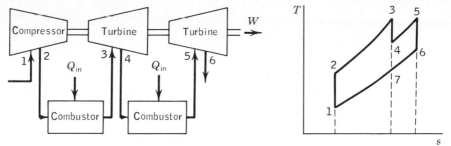

**Figure 16-22** The air-standard gas-turbine cycle with reheat.

are allowed to expand only partially before they are returned to another combustion chamber, where heat ideally is added at constant pressure until the limiting temperature is reached. Then, in the case of a two-stage turbine, further expansion occurs until the ambient pressure is attained. Figure 16-22 illustrates the process path for a cycle with reheating on a $Ts$ diagram. The use of reheat increases the turbine-work output without changing either the compressor work or the maximum limiting temperature. It is noted that the use of reheat requires a substantial increase in the heat input to the cycle. However, the final turbine-exhaust temperature $T_6$ shown in the figure is somewhat above the outlet-turbine temperature without reheat, $T_7$. Consequently, reheating is quite effective when used in conjunction with regeneration, since the quantity of heat exchanged in the regenerator can be greatly increased.

Under the condition of ideal reheating ($T_3$ and $T_5$ are equal in Fig. 16-22), the total work output from the two stages depends upon the pressure ratio across each stage. Analogous to two-stage compression with ideal intercooling, the maximum work output is found by differentiation of a general equation for the work of the two stages. Equation (16-24) again applies, except $T_1$ in this equation becomes $T_3$ and the pressure ratios in the equation now represent those across the two turbine stages. Since the derivation is the same, the answer for maximum work output is the same as minimum work input for a two-stage compressor. That is, the intermediate pressure is the geometric mean between the inlet pressure and the outlet pressure, or the pressure ratio across each stage is the square root of the overall pressure ratio. As a result, the work output of each stage will be equal for the case of ideal reheating. A $Ts$ diagram for an ideal-gas-turbine cycle employing intercooling, reheat, and regeneration is shown in Fig. 16-23.

From the foregoing discussion it should be apparent that, from practical considerations, a gas-turbine-power cycle is improved most when a combination of intercooling and reheating is employed with regeneration. Under any circumstances, the effects of irreversibilities in the turbine and compressor, or pressure losses in the combustor, etc., must be considered in predicting the actual performance of a gas-turbine cycle. One should not conclude, however, that intercooling and reheating without regeneration improve the thermal efficiency. In fact, the effect without regeneration is always to decrease the thermal efficiency. The reason

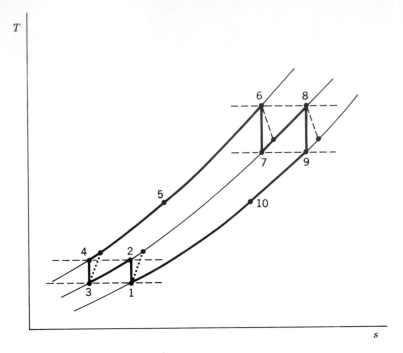

**Figure 16-23** An ideal gas-turbine cycle with intercooling, reheat, and regeneration.

is that intercooling and reheating used alone decrease the average temperature of the heat supply and increase the average temperature of heat rejection. This argument can be seen qualitatively from a $Ts$ diagram such as Fig. 16-23. The major purpose in employing intercooling and reheating is to increase the effective use of a regenerator.

**Example 16-8M** Ideal intercooling and reheating are added to the cycle proposed in Example 16-6M. In addition, the pressure ratios across the two-stage compressor and two-stage turbine are adjusted to provide the minimum work input, and the maximum work output, respectively. Determine the effect of this additional equipment on the thermal efficiency.

SOLUTION In order to minimize the compressor work and maximize the turbine work, the pressure ratio across each stage must be the same. This requires the pressure ratio across each stage to be the square root of the overall pressure ratio. Since this latter value is 6, the ratio for each stage (compressor or turbine) is 2.45. At the inlet to the compressor's first stage $h_1 = 295.2$ kJ/kg and $p_{r1} = 1.3068$. The relative pressure at the outlet of this stage for isentropic flow would be

$$p_{r2} = p_{r1} \frac{P_2}{P_1} = 1.3068(2.45) = 3.202$$

Interpolation in the abridged air table A-5M indicates that $h_{2s} = 381.8$ kJ/kg and $T_{2s} = 381°$K. The actual compressor-enthalpy rise is

$$\Delta h_{C, a} = \frac{\Delta h_{C, s}}{\eta_C} = \frac{381.8 - 295.2}{0.82} = 105.6 \text{ kJ/kg}$$

The actual compressor work is twice this value, that is,

$$w_C = 2(105.6) = 211.2 \text{ kJ/kg}$$

The actual outlet enthalpy at state 4 (see dotted lines in Fig. 16-23) for the compressor is

$$h_{4a} = h_3 + \Delta h_{C,a} = h_1 + \Delta h_{C,a}$$

$$= 295.2 + 105.6 = 400.8 \text{ kJ/kg}$$

For the first stage of the turbine, $h_6 = 1161.1 \text{ kJ/kg}$, $p_{r6} = 167.1$, and

$$p_{r7} = p_{r6} \frac{P_7}{P_6} = 167.1 \frac{1}{2.45} = 68.2$$

From the abridged air table $h_{7s} = 907.6 \text{ kJ/kg}$ and $T_{7s} = 877°\text{K}$. The actual turbine-enthalpy drop is

$$\Delta h_{T,a} = \Delta h_{T,s} \eta_T = (1161.1 - 907.6)(0.85)$$

$$= 253.5(0.85) = 215.5 \text{ kJ/kg}$$

The total turbine output is

$$w_T = 2(215.5) = 431.0 \text{ kJ/kg}$$

The actual enthalpy at state 9 for the turbine outlet is

$$h_{9a} = h_8 - \Delta h_{T,a} = h_6 - \Delta h_{T,a}$$

$$= 1161.1 - 215.5 = 945.6 \text{ kJ/kg}$$

For the regenerator, $\eta_{\text{eff}} = (h_5 - h_4)/(h_9 - h_4)$. Hence

$$0.70 = \frac{h_5 - h_4}{945.6 - 400.8}$$

or

$$h_5 - h_4 = q_{\text{saved}} = 0.7(544.8) = 381.4 \text{ kJ/kg}$$

The heat input required in the combustor becomes

$$q_{\text{comb}} = (h_6 - h_4) - (h_5 - h_4)$$

$$= (1161.1 - 400.8) - 381.4 = 378.9 \text{ kJ/kg}$$

The total heat input, combustor plus reheat, is

$$q_{\text{in}} = q_{\text{comb}} + (h_8 - h_{7a}) = q_{\text{comb}} + (h_6 - h_{7a})$$

$$= q_{\text{comb}} + w_{T,a} = 378.9 + 215.5 = 594.4 \text{ kJ/kg}$$

The thermal efficiency then is

$$\eta_{\text{th}} = \frac{w_T - w_C}{q_{\text{in}}} = \frac{431.0 - 211.2}{594.4} = 0.370 \quad \text{(or 37.0 percent)}$$

In comparison to Example 16-6M, the addition of intercooling and reheating to the cycle with regeneration increases the cycle thermal efficiency from 31.8 percent to 37.0 percent.

**Example 16-8** Ideal intercooling and reheating are added to the cycle proposed in Example 16-6. In addition, the pressure-ratios across the two-stage compressor and two-stage turbine are adjusted to provide the minimum work input, and maximum work output, respectively. Determine the effect on the thermal efficiency.

SOLUTION In order to minimize the compressor work and maximize the turbine work, the pressure ratio across each stage must be the same. This requires the pressure ratio across each stage to be the square root of the overall pressure ratio. Since this latter value is 6, the ratio for each stage (compressor or turbine) is 2.45. At the inlet to the compressor's first stage $h_1 = 129.1$ Btu/lb and $p_{r1} = 1.386$. The relative pressure at the outlet of this stage for isentropic flow would be

$$p_{r2} = p_{r1} \frac{P_2}{P_1} = 1.386(2.45) = 3.40$$

Interpolation in the abridged tables indicates that $h_{2s} = 167$ Btu/lb and $T_{2s} = 697°$R. The actual compressor-enthalpy rise is

$$\Delta h_{C,a} = \frac{\Delta h_s}{\eta_C} = \frac{167 - 129}{0.82} = 46.4 \text{ Btu/lb}$$

The total compressor work is twice this value, that is,

$$w_C = 2(46.4) = 92.8 \text{ Btu/lb}$$

The actual outlet enthalpy at state 4 (see Fig. 16-23) for the compressor is

$$h_{4a} = h_3 + \Delta h_{C,a} = h_1 + \Delta h_{C,a}$$

$$= 129.1 + 46.4 = 175.5 \text{ Btu/lb}$$

For the first stage of the turbine, $h_6 = 54.7$, $p_{r6} = 174.00$, and

$$p_{r7} = p_{r6} \frac{P_7}{P_6} = 174.0 \frac{1}{2.45} = 71.0$$

From the abridged air tables $h_{7s} = 394.7$ and $T_{7s} = 1596°$R. The actual turbine-enthalpy drop is

$$\Delta h_{T,a} = \Delta h_{T,s} \eta_T = (504.7 - 394.7)(.85)$$

$$= 110(0.85) = 93.5 \text{ Btu/lb}$$

The total turbine output is

$$w_T = 2(93.5) = 187.0 \text{ Btu/lb}$$

The actual enthalpy at state 9 for the turbine outlet is

$$h_{9a} = h_8 - \Delta h_{T,a} = h_6 - \Delta h_{T,a}$$

$$= 504.7 - 93.5 = 411.2 \text{ Btu/lb}$$

For the regenerator, $\eta_{eff} = (h_5 - h_4)/(h_9 - h_4)$. Hence,

$$0.70 = \frac{h_5 - h_4}{411.2 - 175.5}$$

or $$h_5 - h_4 = q_{saved} = 0.7(235.7) = 165.0 \text{ Btu/lb}$$

The heat input required in the combustor becomes

$$q_{comb} = (h_6 - h_4) - (h_5 - h_4)$$

$$= (504.7 - 175.5) - (165.0)$$

$$= 164.2 \text{ Btu/lb}$$

The total heat input, combustor plus reheat, is

$$q = q_{comb} + (h_8 + h_{7a}) = q_{comb} + (h_6 - h_{7a})$$

$$= q_{comb} + w_{T,a} = 164.2 + 93.5 = 257.7 \text{ Btu/lb}$$

The thermal efficiency then is

$$\eta_{th} = \frac{w_T - w_C}{q_{in}} = \frac{187.0 - 92.8}{257.7} = 0.365 \qquad \text{(or 36.5 percent)}$$

In comparison to Example 16-6 the addition of intercooling and reheating to a cycle with regeneration increases the cycle efficiency from 31.5 percent to 36.5 percent.

## 16-12 THE ERICSSON CYCLE

In the preceding section it was demonstrated that the composite effect of intercooling, reheat, and regeneration is an increase in the thermal efficiency of a gas-turbine-power cycle. It is interesting to examine the situation where the number of stages of both intercooling and reheat become indefinitely large. Figure 16-24 shows the $Ts$ diagram for the case with five stages of isentropic compression and expansion, and ideal intercooling and reheating. Line $a$-$d$ represents the average temperature during compression and intercooling, and line $b$-$c$ represents the average temperature during expansion and reheating. It is apparent from

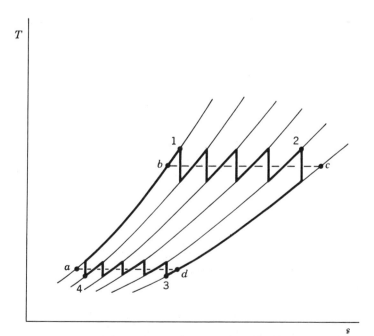

**Figure 16-24** Gas-turbine cycle with many stages of compression and expansion, with ideal intercooling and reheating.

the diagram that as the number of stages increases, line $a$-$d$ approaches line 3-4, and line $b$-$c$ approaches line 1-2. In the limit, the compression and expansion processes become isothermal, rather than isentropic ones. That is, for an indefinitely large number of stages the cycle can be represented by two constant-temperature processes and two constant-pressure processes, with regeneration. Such a cycle is called an Ericsson cycle.

Figure 16-25 shows the $PV$ and $TS$ diagrams for the cycle and also a schematic diagram of an Ericsson engine operating as a steady-flow device. The fluid expands isothermally through the turbine from state 1 to 2. Work is produced and heat is absorbed reversibly from a reservoir at $T_H$. The fluid is then cooled at constant pressure in a regenerator (heat exchanger). From state 3 to state 4 the fluid is compressed isothermally. This requires work input and reversible heat rejection to a reservoir at a temperature $T_L$. Finally, the fluid is heated at constant pressure to the initial state by passing it in a countercurrent direction through the regenerator and picking up the heat rejected to the regenerator during the constant-pressure cooling process. In this manner the temperature difference across the heat exchanger is always infinitesimal throughout the length of the exchanger, as required for reversible heat transfer. The regenerator again acts as an energy-storage unit within the system. For any complete cycle the net energy storage is zero. Since the only external changes are heat transfer with heat reservoirs and the processes described are all reversible, the thermal efficiency of

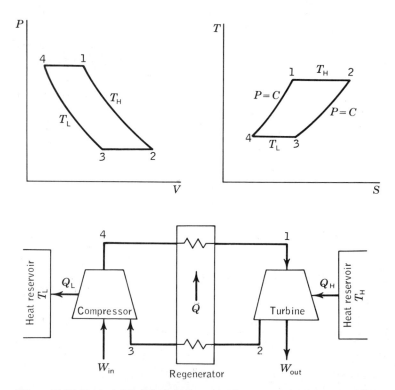

**Figure 16-25** Typical $PV$, $TS$, and schematic diagrams of the Ericsson cycle.

the Ericsson cycle equals that of the Carnot cycle. The original engine constructed by Ericsson (1803–1889) was a nonflow device, but the thermodynamic cycle is the same as for the steady-flow device described above. By increasing the number of compressor and turbine stages indefinitely, with ideal regeneration, the efficiency of such a gas-turbine-power plant would approach the Carnot efficiency, given by $1 - T_L/T_H$. For an inlet temperature of around 300°K or 540°R and a turbine limiting temperature of around 1200°K or 2200°R, the theoretical efficiency would approach 75 percent. Such a design is impractical, however, in terms of intercooling and reheating, since the cost and size requirements are prohibitive.

## 16-13 TURBOJET PROPULSION

The gas-turbine-power cycle has been effectively adapted for the propulsion of aircraft. Utilization of the power cycle for this purpose requires only that a diffuser and a nozzle be placed in the cycle at the appropriate place. Figure 16-26 is a schematic of the equipment for a turbojet engine. The center section of the engine contains the three components already discussed—compressor, combustor, and turbine. A diffuser is placed in front of the compressor to slow down the entering fluid to a reasonable value before it enters the compressor. A pressure rise of a few decibars (or psi) accompanies the decrease in kinetic energy, as shown ideally in the $Ts$ diagram of Fig. 16-27. The ideal compressor and combustor processes are similar to those discussed in preceding sections. The major modification in turbine operation is that no net work output is required beyond that needed to drive the compressor and auxiliary equipment. Hence the gas does not expand back to ambient pressure in the turbine. The pressure ratio $P_4/P_5$ may be two or more. The final expansion to ambient condition occurs in the nozzle, where the fluid is accelerated to a relatively high velocity. As a result of the momentum change of the gases as they pass through the engine, a forward thrust is exerted on the engine. If we neglect the effect of the fuel added to the air passing through the

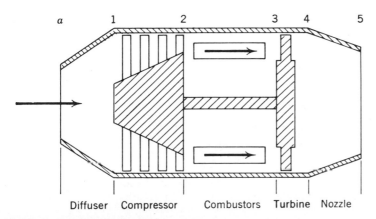

Diffuser  Compressor  Combustors  Turbine  Nozzle

**Figure 16-26** Turbojet engine schematic.

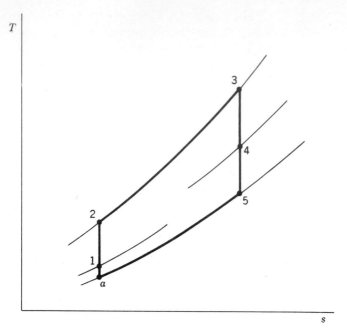

**Figure 16-27** A $Ts$ diagram for an ideal turbojet engine.

engine (about 2 percent of the total flow), then the reaction force or thrust $F$ of the engine is given by

$$F = \frac{\dot{m}_{\text{air}}}{g_c} (V_5 - V_a)$$

where $V_5$ and $V_a$ are measured relative to the engine.

In actual practice irreversibilities occur in the engine, such as a pressure drop in the combustor. In addition, the compressor, turbine, diffuser, and nozzle are not isentropic. These irreversibilities modify the basic cycle shown in Fig. 16-27.

Compared to stationary gas-turbine-power plants, turbojet engines operate at higher pressure ratios and higher inlet-turbine temperatures. Pressure ratios from $8:1$ to $25:1$ are common. Inlet-turbine temperatures are in the range of $900$–$1300°$K or $1500$–$2300°$F. The following example demonstrates the basic thermodynamic calculations for a turbojet engine. The calculated values are somewhat unrealistic, since idealized behavior is assumed in order to shorten the calculations.

**Example 16-9M** A turbojet aircraft flies at 260 m/s at an altitude of 5,000 m, where the atmospheric pressure is 0.60 bar and the temperature is $250°$K. The compressor pressure ratio is $8:1$, and the maximum temperature at the turbine inlet is $1300°$K. Assuming ideal performance of the various components of the engine, determine the compressor-work input, the pressures and temperatures throughout the cycle. and the exit-jet velocity.

SOLUTION The values calculated will correspond to the states shown in Fig. 16-27. For the diffuser process, $1 - a$, we shall assume that the outlet velocity is negligible compared to the inlet value. At 250°K the enthalpy is 250.1 kJ/kg and $p_{r1} = 0.7329$ (Table A-5M). For isentropic flow through the diffuser, we find that

$$h_a + \frac{V_a^2}{2} = h_1 + \frac{V_1^2}{2}$$

$$250.1 + \frac{(260)^2}{2(1000)} = h_1 + 0$$

$$h_1 = 250.1 + 33.8 = 283.9 \text{ kJ/kg}$$

From the air table, $T_1 = 284°K$ and $p_{r1} = 1.141$. Hence

$$P_1 = P_a \frac{p_{r1}}{p_{a1}} = 0.60 \frac{1.141}{0.7329} = 0.934 \text{ bar}$$

Thus there is a 0.33-bar pressure rise in the diffuser. For the compressor, with $r_p = 8$, $P_2 = 8(0.934) = 7.47$ bars. Therefore

$$p_{r2} = p_{r1} \frac{P_2}{P_1} = 1.141(8) = 9.13$$

From the air table A-5M, $T_{2s} = 511°K$, $h_{2s} = 514.9$ kJ/kg. The isentropic compressor work is

$$w_{C,s} = h_{2s} - h_1 = 514.9 - 283.9 = 231.0 \text{ kJ/kg}$$

At the turbine inlet, where $T_3 = 1300°K$, we find that $h_3 = 1396.0$ kJ/kg and $p_{r3} = 330.9$. Since the turbine work must equal the compressor work,

$$h_{4s} = h_3 - w_C = 1396.0 - 231.0 = 1165.0 \text{ kJ/kg}$$

For this value of the enthalpy, Table A-5M shows that $T_{4s} = 1102°K$ and $p_{r4} = 169.3$. Hence

$$P_4 = P_3 \frac{p_{r4}}{p_{r3}} = 7.47 \frac{169.3}{330.9} = 3.82 \text{ bars}$$

Note that the turbine-outlet pressure is considerably above the atmospheric value, and hence a reasonably large pressure drop still exists across the nozzle. The nozzle-outlet state is found from

$$p_{r5} = p_{r4} \frac{P_5}{P_4} = 169.3 \frac{0.60}{3.82} = 26.6$$

At this condition, $T_{5s} = 685°K$ and $h_{5s} = 697.4$ kJ/kg. An energy balance on the nozzle yields, neglecting the inlet velocity,

$$h_4 + \frac{V_4^2}{2} = h_5 + \frac{V_5^2}{2}$$

$$1165.0 + 0 = 697.4 + \frac{V_5^2}{2(1000)}$$

$$V_5^2 = 2(1000)(1165.0 - 697.4) = 935{,}200$$

$$V_5 = 967 \text{ m/s}$$

This velocity is supersonic, with a Mach number around 1.8. A more meaningful calculation would include compressor, turbine, and nozzle efficiencies, and a diffuser pressure coefficient.

**Example 16-9** A turbojet aircraft flies at 575 mph at an altitude of 16,500 ft, where $P_a = 8.7$ psia and $T_a = 460°R$. The compressor pressure ratio is 8, and the maximum temperature in the combustion chamber is 1840°F. Assuming ideal performance of the various components of the engine, determine the compressor-work input, the pressures and temperatures throughout the cycle, and the exit-jet velocity.

SOLUTION The values calculated will correspond to the states shown on Fig. 16-27. For the diffuser process, $1 - a$, we shall assume that the outlet velocity is negligible compared to the inlet value. At $460°R$ the enthalpy is 109.9 Btu and $p_{ra} = 0.7913$ (Table A-5). Assuming isentropic flow through the diffuser, we find that

$$h_a + \frac{V_a^2}{2} = h_1 + \frac{V_1^2}{2}$$

$$109.9 + \frac{(843)^2}{2(32.2)(778)} = h_1 + 0$$

$$h_1 = 109.9 + 14.2 = 124.1 \text{ Btu/lb}$$

From the air table, $T_1 = 519°R$ and $p_{r1} = 1.21$. Hence

$$P_1 = P_a \frac{p_{r1}}{p_{ra}} = 8.7 \frac{1.21}{0.7913} = 13.3 \text{ psia}$$

There is a 4.6-psi pressure rise in the diffuser. For the compressor, with $r_p = 8$, $P_2 = 8(13.3) = 106.4$ psia.

$$p_{r2} = p_{r1} \frac{P_2}{P_1} = 1.21(8) = 9.68$$

From the air table, $T_2 = 936°R$, $h_2 = 225.1$ Btu/lb. The isentropic compressor work is

$$w_{C.s} = 225.1 - 124.1 = 101.0 \text{ Btu/lb}$$

At the turbine inlet, where $T_3 = 2300°R$, we find that $h_3 = 588.8$ Btu/lb and $p_{r3} = 308.1$. Since the turbine work must equal the compressor work,

$$h_{4s} = h_3 - w_C = 588.8 - 101.0 = 487.8 \text{ Btu/lb}$$

Table A-5 shows that $T_{4s} = 1939°R$ and $p_{r4} = 153.6$. Hence

$$P_4 = P_3 \frac{p_{r4}}{p_{r3}} = 106.4 \frac{153.6}{308.1} = 53.0 \text{ psia}$$

Note that a considerable drop still exists across the nozzle. The nozzle-outlet state is found from

$$p_{r5} = p_{r4} \frac{P_5}{P_4} = 153.6 \frac{8.7}{53.0} = 25.2$$

At this condition, $T_5 = 1216°R$ and $h_5 = 295.3$ Btu/lb. An energy balance on the nozzle yields, neglecting the inlet velocity,

$$h_4 + \frac{V_4^2}{2} = h_5 + \frac{V_5^2}{2}$$

$$487.9 + 0 = 295.3 + \frac{V_5^2}{2} \frac{1}{32.2(778)}$$

$$V_5^2 = 2(32.2)(778)(487.8 - 295.3) = 9.62 \times 10^6$$

$$V_5 = 3100 \text{ ft/s}$$

This velocity is supersonic, with a Mach number around 1.8. A more meaningful calculation would include compressor, turbine, and nozzle efficiencies, and a diffuser pressure coefficient.

# 16-14 GAS-TURBINE SPACE-POWER SYSTEM

The modern gas turbine engine has two major applications: (1) as a power source for the production of electrical power when coupled with an electrical generator and (2) as a power source for the propulsion of aircraft. The device is also applicable as a power source for ground-level propulsion equipment, such as passenger cars, trucks, and trains. The thermodynamic fundamentals of the basic air-breathing gas turbine cycle have been discussed in Secs. 16-6 through 16-8. The Brayton cycle may also be used as a power system for orbiting vehicles in the space environment outside our atmosphere. In this case the space-power system is being used to generate electrical power aboard the spacecraft, and not to propel the vehicle. The major features of the cycle described previously remain the same for space applications. Typical turbine expansion ratios are around 2 : 1 or less, and cycle temperature ratios (minimum to maximum temperature) are around 0.25 : 0.30. However three major modifications must be made to permit operation in a space environment. These modifications are:

1. The device must operate as a closed cycle, since the working fluid must be conserved for reuse. Air is not available as in terrestrial applications of open gas-turbine cycles.
2. The heat source must be of a different nature than the conventional hydrocarbons which undergo combustion directly in the cycle. This is necessary since the cycle loop is closed, and hence accumulation of products of combustion cannot be permitted.
3. The temperature of the gas prior to reentering the compressor in the cycle must be reduced to its initial value. In an open cycle relatively hot gases are simply thrown out into the atmosphere, even if a regenerator (recuperator) has been employed to preheat the gases between the compressor and the heat source.

The first modification to a closed cycle increases the physical complexity of the system. However, the closed-loop operation is advantageous, since it permits the use of working fluids which have properties more desirable than those of air. The use of inert gases, for example, reduces the problem of corrosion. Fundamental studies of turbomachinery indicate that the number of turbine blade rows decreases rapidly as the molecular weight increases. An application requiring 10 stages using helium (molar mass of 4) requires two stages with neon (molar mass of 20) and one stage with argon (molar mass of 40). For gases of even higher molecular weight, a single stage could be used at lower speeds than for argon, and hence at lower stresses. The turbomachinery diameter also increases as the molecular weight increases. For the relatively low-power turbomachinery used in space applications, this also is an advantage. For power levels in the range of 10–50 kW, gases under study have molecular weights in the range of 40–80. However, high-molecular-weight gases do have disadvantages. A major one is discussed when the third modification of the basic system is presented.

The energy sources of principal interest are solar radiation and nuclear reactors, although radioisotopes are also a possibility. Nuclear reactors are compact and require no special orientation with respect to the space environment. However, they are hazardous. A solar heat source is not hazardous, but it does require special means of orientation with respect to the sun. In addition, a relatively large solar concentrator (mirror) with good collection efficiency, is required. For example, for a 10-kW Brayton power cycle with a maximum temperature of 2000°R, a 25-ft concentrator may be required. Whichever energy source is used, the heat generated must be transferred to the working fluid by means of a heat exchanger. This heat exchanger is shown as the "energy-source heat exchanger" in Fig. 16-28. The regenerator is, of course, a second heat exchanger in the loop. Regenerators working with argon have been built with an effectiveness of 90 percent and a 2 percent pressure drop.

The third modification for a Brayton space power system involves the "heat-sink heat exchanger" shown in Fig. 16-28. After passing through the regenerator (recuperator), the working fluid is cooled further by passing through this third heat exchanger. The energy discharged from the working fluid to a secondary liquid in the radiator loop is then radiated into space. The radiator size is significantly reduced by increasing the turbine-inlet temperature. This third heat-exchange process reduces the temperature of the primary working fluid to the required value at the compression inlet. A disadvantage of the liquid loop is the added complexity. Passage of the inert gas (primary fluid) directly through the radiator, without a liquid loop, could be considered. However, for a given radiator size the extra gas-pressure loss in the radiator is found to be greater than the equivalent power required to pump the liquid.

It has been mentioned that high-molecular-weight gases (molecular weight greater than air) have properties favorable to turbomachinery design. Unfortunately, the thermal conductivity of a gas falls rapidly with increasing molecular

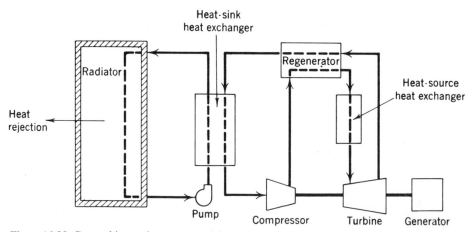

**Figure 16-28** Gas-turbine-cycle space-power system.

weight. The conductivity of krypton (molecular weight of 83) is roughly $\frac{1}{15}$ of that of helium (molecular weight of 4). This implies a severe penalty in heat-exchanger size. Recall from Fig. 16-28 that the working fluid passes through three different heat exchangers. A heat exchanger designed for krypton would be about 10 times the size of one designed for helium. The increased weight and volume suggest that the choice of a working fluid is a compromise between turbomachinery needs and heat-exchanger requirements.

An interesting engineering-design feature is that the benefits of high molecular weight to turbomachinery design can be met in a subtle way without sacrificing heat-exchanger size. This is done by using a mixture of inert gases, rather than a pure gas. For example, helium and krypton could be mixed to give a gas which has the same molecular weight as argon. However, the average thermal conductivity of the gas mixture is greatly increased by the presence of helium. The result is a reduction in heat-exchanger size of about one-half of that for pure argon.

In summary, the Brayton cycle for space-power applications appears quite promising for several reasons:

1. It has a relatively high cycle efficiency (which reduces the required size of a solar collector or space radiator.
2. It can cover a range of power levels for a given system.
3. It can be used with a number of different energy sources.
4. It can operate over extended periods of time.

## 16-15 GAS REFRIGERATION CYCLES

The adiabatic expansion of gases can be used to produce a refrigeration effect. In its simplest form this is accomplished by reversing the Brayton cycle used as a standard for gas-power cycles. Figure 16-29 is a $Ts$ diagram for the ideal reversed Brayton cycle. The gas is first compressed isentropically from state 1 to state 2, which is at a relatively high temperature compared with the ambient temperature

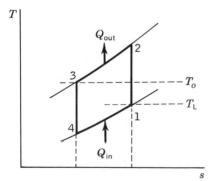

**Figure 16-29** Gas-refrigeration cycle using a reversed Brayton cycle.

$T_0$. It then passes through a heat exchanger, where in the limit it can be cooled to the ambient temperature. (In practice, state 3 may be 10° or so above the ambient condition.) The gas then enters an adiabatic-expansion device, such as a turbine, where work is extracted and the gas is cooled to a temperature $T_4$, which is considerably below the temperature $T_L$ of the cold region. Consequently, heat may be transferred from the cold region to the gas through a heat exchanger, and ideally the gas will be heated to the temperature $T_L$, which is also $T_1$ in the cycle. The fluid in the ideal-gas refrigeration cycle is assumed to pass through the cycle in an internally reversible manner. For steady-flow operation, and in the absence of kinetic- and potential-energy changes, the heat and work interactions for each piece of equipment may be evaluated in terms of the enthalpy change during the process. On the $Ts$ diagram in Fig. 16-29 the area under the curve 4-1 represents the heat removed from the cold region, the area enclosed by the cycle 1-2-3-4-1 represents the net work input, and the ratio of these areas is a measure of the coefficient of performance (COP) of the device. This latter quantity is defined in Chap. 7 as

$$\text{COP} = \frac{q_{\text{refrig}}}{w_{\text{net, in}}} \qquad (7\text{-}3)$$

The reversed Brayton cycle has the disadvantage of large temperature variations of the fluid during the heat-addition and heat-removal processes. Hence the coefficient of performance of this cycle is considerably below that for a reversed Carnot engine operating between the same minimum temperatures, $T_0$ and $T_L$.

Modifications of the reversed Brayton cycle lead to some useful applications of gas refrigeration cycles. For example, in Fig. 16-29, it is noted that the temperature of the fluid after picking up heat from the cold region is below that of state 3, where the fluid enters the expansion engine. If the gas at state 1 could be used to cool the gas further, below the temperature of state 3, subsequent expansion would lead to a temperature below that of state 4. In this manner extremely low temperatures might be achieved. Physically, this is accomplished by inserting a heat exchanger internally into the cycle, as shown in Fig. 16-30. Heat transfer external to the cycle results in the usual temperature drop from state 2 to state 3 in

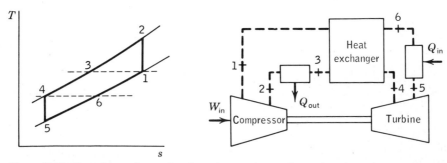

**Figure 16-30** Gas-refrigeration cycle using a heat exchanger internal to the cycle.

the figure. The additional heat exchanger, however, permits the gas to be cooled further to state 4 in the ideal case. The gas, after expansion, receives heat from the cold region from state 5 to state 6, and receives a further quantity of heat in the heat exchanger from state 6 to state 1. Such use of heat exchangers internal to the cycle is important in processes for the liquefaction of gases. The gas refrigeration cycle in the form of an open cycle is employed for the purpose of aircraft cooling, since it has a definite weight advantage over vapor-compression refrigeration. There are a number of modifications of the technique, depending, for example, on the type of aircraft. In general, however, air is compressed, cooled by rejecting heat to the ambient atmosphere, and then expanded through a turbine. The cool air leaving the turbine goes directly into the cabin of the aircraft.

**Example 16-10M** A reversed Brayton cycle is proposed to operate between $T_0 = 27°C$ and $T_L = -8°C$ (see Fig. 16-29 for notation). The compression and expansion ratios are 3 : 1. Determine the coefficient of performance.

SOLUTION From the air table (A-5M) for $T_1 = -8°C$, $h_1 = 265.1$ kJ/kg and $p_{r1} = 0.900$. State 2 is found by noting that

$$p_{r2} = p_{r1} \frac{P_2}{P_1} = 0.900(3) = 2.70$$

For this $p_r$ value, the air table indicates that $T_{2s} = 363°K$ (90°C) and $h_{2s} = 363.6$ kJ/kg. A similar calculation can be carried out for process 3-4. At 27°C, $h_3 = 300.2$ kJ/kg and $p_{r3} = 1.386$. Consequently, $p_{r4} = 1.386/3 = 0.462$, $T_{4s} = 219°K$ $(-54°C)$, and $h_{4s} = 219.0$ kJ/kg. The compressor work, the turbine work, and the heat removed from the cold region are found by

$$w_{C, in} = h_{2s} - h_1 = 363.6 - 265.1 = 98.5 \text{ kJ/kg}$$

$$w_{T, out} = h_3 - h_{4s} = 300.2 - 219.0 = 81.2 \text{ kJ/kg}$$

$$q_{in} = h_1 - h_{4s} = 265.1 - 219.0 = 46.1 \text{ kJ/kg}$$

Therefore the coefficient of performance is

$$\text{COP} = \frac{q_{in}}{w_C - w_T} = \frac{46.1}{98.5 - 81.2} = 2.66$$

This coefficient of performance is somewhat high, since we have not accounted for the irreversibilities in the turbine and compressor, among other things.

**Example 16-10** A reversed Brayton cycle is proposed to operate between $T_0 = 80°F$ and $T_L = 20°F$ (see Fig. 16-29 for notation). The compression and expansion ratios are 3 : 1. Determine the coefficient of performance.

SOLUTION From the air table (A-5) for $T_1 = 20°F$, $h_1 = 114.7$ Btu/lb and $p_r = 0.9182$. State 2 is found by noting that

$$p_{r2} = p_{r1} \frac{P_2}{P_1} = 0.9182(3) = 2.755$$

For this $p_r$ value, $T_{2s} = 655°R$ (195°F) and $h_{2s} = 157.0$ Btu/lb. A similar calculation can be carried out for process 3-4. At 80°F, $h_3 = 129.1$ Btu/lb and $p_{r3} = 1.386$. Consequently

$p_{r4} = 1.386/3 = 0.462$, $T_{4s} = 395°R$ ($-65°F$), and $h_{4s} = 94.3$ Btu/lb. The compressor work, the turbine work, and the heat removed from the cold region are found by

$$w_{C, in} = h_2 - h_1 = 157.0 - 114.7 = 42.3 \text{ Btu/lb}$$

$$w_{T, out} = h_3 - h_4 = 129.1 - 94.3 = 34.8 \text{ Btu/lb}$$

$$q_{in} = h_1 - h_4 = 144.7 - 94.3 = 20.4 \text{ Btu/lb}$$

Therefore the coefficient of performance is

$$\text{COP} = \frac{q_{in}}{w_C - w_T} = \frac{20.4}{42.3 - 34.8} = 2.72$$

## 16-16 THE STIRLING CYCLE

In addition to the Carnot cycle, other totally reversible cycles may be conceived which equal the Carnot cycle in its theoretical thermal efficiency. The sole requirement is that all heat transfer across the system boundaries occur reversibly and isothermally. One of the earliest devised cycles of this nature was proposed by Robert Stirling (1790–1878). The Stirling cycle is composed of two isothermal, reversible processes and two constant-volume, reversible processes. The $PV$ and $TS$ diagrams for the cycle are shown in Fig. 16-31.

From an initial state 1 the gas is expanded isothermally to state 2 and heat is added reversibly from a reservoir at $T_H$. From state 2 to state 3 heat is removed at constant volume until the temperature of the fluid reaches $T_L$. The volume is then reduced to its original value isothermally, and heat is removed reversibly to a second reservoir at $T_L$. Finally, heat is added at constant volume from state 4 to state 1. The cycle would operate between two fixed-temperature reservoirs if the heat quantities for the processes 2-3 and 4-1 could be kept within the system. Application of an energy balance on the closed system for these two processes shows that the two heat quantities are equal in magnitude. This is shown by the cross-hatched areas under process curves 4-1 and 2-3 on the $TS$ plot. What is needed is simply a means of storing the heat given up by process 2-3 and then

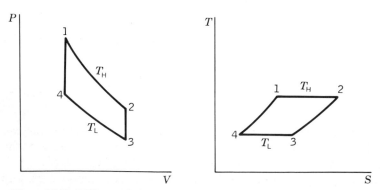

**Figure 16-31** Stirling cycle for a gaseous substance.

supplying this same energy to the working medium during the process 4-1. Moreover, this exchange of heat must be done across a differential temperature difference.

This requirement of energy storage within the system necessitates the use of a regenerator. A regenerator can be visualized as a series of heat reservoirs the temperatures of which are varied differentially. In this particular case the temperature range of the reservoirs is from $T_H$ to $T_L$. During the constant-volume cooling process (2-3), increments of heat removed are stored as energy in the reservoirs, the temperatures of which correspond to that of the system at that particular moment. Thus the heat transfer to the set of reservoirs is always reversible. During the constant-volume heating process (4-1), the energy stored is returned reversibly to the fluid as heat beginning at the low-temperature end of the regenerator. The actual physical arrangement for a practical regenerator is not important at the moment. The use of a regenerator offers a means of operating a heat engine on the Stirling cycle so that the only effect external to the system during each cycle is heat exchange between two fixed-temperature heat reservoirs.

The theoretical thermal efficiency of the Stirling cycle is found by using exactly the same equations for the first and second laws as were used for the Carnot cycle. Consequently, the efficiency must be the same. This can also be seen from the general definition of thermal efficiency. Referring again to Fig. 16-31, one notes that, for the Stirling engine,

$$\eta_{th} = \frac{Q_H - Q_L}{Q_H} = \frac{T_H(S_2 - S_1) - T_L(S_3 - S_4)}{T_H(S_2 - S_1)} \qquad (16\text{-}26)$$

Since the heat transfer to the regenerator is reversible, the entropy change of the working medium and the regenerator must be equal during either process, 2-3 or 4-1. But the change in entropy of the regenerator for any cycle is zero, since it is returned to its initial state. Consequently, $(S_3 - S_2) + (S_1 - S_4)$ must be zero for the working medium. It is also true that

$$(S_1 - S_2) + (S_2 - S_3) + (S_3 - S_4) + (S_4 - S_1) = 0$$

Since the second and fourth terms add up to zero,

$$S_3 - S_4 = S_2 - S_1$$

The substitution of this result into Eq. (16-26) shows that the efficiency of a Stirling engine with regeneration is given by

$$\eta_{th}(\text{Stirling}) = \frac{T_H - T_L}{T_H} \qquad (16\text{-}27)$$

For many years the Stirling cycle has been of only theoretical interest. However, since the early 1950s considerable work has been carried out toward the development of a practical engine working on the Stirling cycle. Although it matches the Carnot cycle in thermal efficiency, it is difficult to build a Stirling engine without introducing some inherent disadvantages. For example, it operates at very high pressures, and the most suitable working fluids are helium and hydrogen. Its weight-to-horsepower ratio is not too favorable, except possibly for large vehicles such as trucks and buses. The high temperature in the cycle also leads to problems, since the pistons are not lubricated to avoid fouling of the regenerator. However, one major advantage of the Stirling engine is its excellent emission quality. This engine is an "external-combustion engine," as opposed to the common internal-combustion type for automotive use. Hence the combustion process is much more complete than in an internal-combustion type in terms of carbon monoxide, hydrocarbon, and nitrogen oxide content in the exhaust.

An interesting proposed application of the Stirling engine is as a power source for pumping blood through the circulatory system. The energy source for the engine would be a radioisotope, and source, engine, and pump would be totally implanted in the body. A major modification to the cycle is that the Stirling engine's output would be in the form of a pressurized gas instead of mechanical shaft work. The pressurized gas is used in an expander, such as a turbine, to produce power. Hence the piston of the engine is a "free piston" type.

Although there are certain limitations to the Stirling-cycle engine for power production for vehicles, the Stirling cycle has some inherent advantages when operated in the reverse direction so that it produces a refrigeration effect. It is particularly effective for achieving temperatures in the range of $-100$ to $-200°C$. A discussion of the Stirling cycle as a refrigeration technique appears in Sec. 17-10.

## 16-17 THE COMBINED CYCLE

The thermal efficiency of both the gas-turbine-power cycle and the vapor-power cycle (discussed in Chap. 17) is typically less than 40 percent in practice. Although techniques such as reheating and regeneration do improve the cycle performance, the rejected or waste energy is still a large fraction of the energy input in either case. One possible way of achieving further improvement is by an arrangement called a combined cycle. One possible combined cycle involves the combination of a gas-turbine (Brayton) and a vapor-turbine (Rankine) cycle. The basic thermodynamic principles for gas-turbine and Rankine-vapor cycles are still applicable.

In a gas-turbine cycle the exhaust stream leaving the turbine is relatively hot. As discussed in Sec. 16-8, one method of utilizing this high-temperature energy is through the concept of regeneration. A portion of the energy is returned to the air as it passes from the compressor outlet to the combustor inlet. An alternative possibility is shown in Fig. 16-32. In place of regeneration, the gas-turbine exhaust stream is used as the energy source in the boiler of a conventional steam-power cycle. Although this is not illustrated, the steam cycle would probably employ feedwater heaters. However, reheat in the turbine section of the steam cycle would not normally be possible. The energy in the hot gas-turbine exhaust is needed primarily for the boiling and superheating of the steam. The thermodynamic analysis of this combined cycle would follow the procedures established in sections on the Brayton and Rankine cycles.

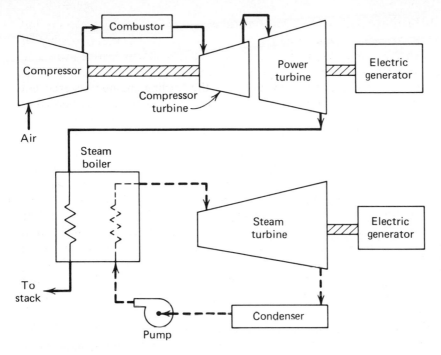

**Figure 16-32** Schematic of a gas-steam–turbine combined cycle.

The practical development of the combined gas-steam–turbine cycle was delayed until modern technology provided the means of building gas-turbine-power plants which operated at relatively high pressure ratios and with turbine-inlet temperatures which exceeded 1400°K (2500°R). Pressure ratios of 10 : 1 to 13 : 1 are typical for this temperature. When turbine-inlet temperatures around 1600°K (2900°R) are possible on commercially available units, pressure ratios in the vicinity of 20 : 1 may be used. At the time of this writing commercial gas-steam–turbine combined power plants were available on the market and several had been installed. As gas-turbine technology improves, the advantages of this combined cycle will improve in relation to conventional steam-power plants. This cycle appears to be extremely useful in conjunction with coal gasification. Not only is cycle efficiency improved by the combined cycle itself, but coal gasification as a source of fuel for the gas-turbine combustor offers the additional advantage of removing potential air pollutants before they enter the combustion process.

·

## REFERENCES

Buck, K. E., An Implantable Artificial Heart, *Mech. Eng.*, **91**(9): 20–25 (1969).

Hesse, W. J., and N. V. S. Mumford, Jr.: "Jet Propulsion for Aerospace Applications," Pitman, New York, 1964.

Taylor, C. F., and E. S. Taylor: "The Internal Combustion Engine," International Textbook, Scranton, Pa., 1962.

# PROBLEMS (METRIC)

### Air standard cycles

**16-1M** An air-standard cycle is proposed which begins at an initial state of 27°C, 1 bar, and 0.860 m³/kg. Process 1-2 is constant-volume heating to 2 bar; process 2-3 is constant-pressure heating to 1200°K and 1.72 m³/kg; process 3-4 is isentropic expansion to 1 bar; process 4-1 is constant-pressure cooling to the initial state.

(a) Sketch $Pv$ and $Ts$ diagrams for the cycle.

(b) Determine the total amount of heat added, in kJ/kg.

(c) Determine the net work output, in kJ/kg.

(d) Compute the thermal efficiency.

(e) What is the Carnot efficiency for a heat engine with the same maximum and minimum temperatures as the above cycle? Use $u$ and $h$ data from the air table.

**16-2M** An air-standard cycle is proposed which begins at an initial state of 37°C, 1.2 bars, and 0.741 m³/kg. Process 1-2 is constant-volume heating to 3.0 bars; process 2-3 is constant-pressure heating to 1550°K and 1.48 m³/kg; process 3-4 is isentropic expansion to 1.2 bars; process 4-1 is constant-pressure cooling to the initial state.

(a) Sketch $Pv$ and $Ts$ diagrams for the cycle.

(b) Determine the total amount of heat added, in kJ/kg.

(c) Find the net work output, in kJ/kg.

(d) Compute the thermal efficiency.

(e) What is the Carnot efficiency for a heat engine with the same maximum and minimum temperatures as the above cycle? Use $u$ and $h$ data from the air table.

**16-3M** A proposed air-standard cycle begins at 1 bar, 27°C, and 0.860 m³/kg. Process 1-2 is isentropic compression to 2.75 bar; process 2-3 is constant-pressure heating to 1200°K and 1.25 m³/kg; process 3-4 is constant-volume cooling to 1 bar; process 4-1 is constant-pressure cooling to the initial state.

(a) Sketch $Pv$ and $Ts$ diagrams of the cycle.

(b) Determine the amount of heat added and the net work output, in kJ/kg.

(c) Compute the thermal efficiency.

(d) What is the Carnot efficiency for a heat engine with the same maximum and minimum temperatures as the above cycle? Use air-table data.

**16-4M** A proposed air-standard cycle begins at 1.1 bars, 37°C, and 0.808 m³/kg. Process 1-2 is isentropic compression to 3.2 bars; process 2-3 is constant-pressure heating to 1050°K and 0.940 m³/kg; process 3-4 is constant-volume cooling to 1.1 bars; process 4-1 is constant-pressure cooling to the initial state.

(a) Sketch $Pv$ and $Ts$ diagrams of the cycle.

(b) Determine the amount of heat added and the net work output, in kJ/kg.

(c) Compute the thermal efficiency.

(d) What is the Carnot efficiency for a heat engine with the same maximum and minimum temperatures as the above cycle? Use air-table data.

### Air-standard Carnot cycles

**16-5M** An air-standard Carnot cycle rejects 100 kJ/kg to a sink at 300°K. The minimum and maximum pressures in the cycle are 1 bar and 174 bars. On the basis of the air table, determine (a) the pressure after isothermal compression, (b) the temperature of heat supply, and (c) the thermal efficiency.

**16-6M** An air-standard Carnot cycle is supplied with 200 kJ/kg from a source at 1000°K. The minimum and maximum pressures in the cycle are 1 bar and 69.3 bar. On the basis of the air table, determine (a) the pressure after isothermal heat addition, (b) the temperature of heat rejection, and (c) the thermal efficiency.

**16-7M** An air-standard Carnot cycle operates between temperatures of 300 and 900°K. The minimum pressure in the cycle is 1 bar. Determine the maximum pressure in the cycle, in atmospheres, if the heat rejection is (a) 60 kJ/kg, and (b) 40 kJ/kg.

**16-8M** An air-standard Carnot cycle operates between temperatures of 300 and 1100°K. The minimum pressure in the cycle is 1 bar. Determine the maximum pressure in the cycle, in atmospheres, if the amount of heat supplied is (a) 120 kJ/kg, and (b) 150 kJ/kg.

### Compression ratio and mean effective pressure

**16-9M** For the data in Prob. 16-5M, determine the compression ratio and the mean effective pressure for a reciprocating device.

**16-10M** For the data in Prob. 16-6M, determine the compression ratio and the mean effective pressure for a reciprocating device.

**16-11M** For the data in Prob. 16-7M, determine the compression ratio and the mean effective pressure for a reciprocating device if the heat rejection is (a) 60 kJ/kg, and (b), 40 kJ/kg.

**16-12M** For the data in Prob. 16-8M, determine the compression ratio and the mean effective pressure for a reciprocating device if the amount of heat supplied is (a) 120 kJ/kg, and (b) 150 kJ/kg.

### Otto cycle

**16-13M** The compression ratio of an Otto cycle is 8 : 1. Before the compression stroke of the cycle begins, the pressure is 0.98 bar and the temperature is 27°C. The amount of heat added to the air per cycle is 1430 kJ/kg. On the basis of the air-standard cycle and employing air table A-5M, determine (a) the pressure and temperature at the end of each process of the cycle, (b) the theoretical thermal efficiency, and (c) the mean effective pressure for the cycle, in bars.

**16-14M** The air at the beginning of the compression stroke of an air-standard Otto cycle is at 0.95 bar and 22°C, and the cylinder volume is 5600 cm³. The compression ratio is 9, and 8.6 kJ are added during the heat-addition process. Calculate (a) the temperature and pressure after the compression and heat-addition processes, and (b) the thermal efficiency.

**16-15M** An air-standard Otto cycle operates with a compression ratio of 8.55. At the beginning of compression the air is at 0.98 bar and 32°C, and during the heat-addition process the pressure is tripled.

(a) Calculate the thermal efficiency of the cycle on the basis of variable specific heats.

(b) Determine the thermal efficiency of a Carnot engine operating between the same overall temperature limits.

**16-16M** Consider an air-standard Otto cycle that has a compression ratio of 8.3 and a heat addition of 1456 kJ/kg of air. If the pressure and temperature at the beginning of the compression process are 0.95 bar and 7°C, determine (a) the maximum pressure and temperature for the cycle, (b) the thermal efficiency, and (c) the mean effective pressure, on the basis of Table A-5M.

**16-17M** At the beginning of compression of an air-standard Otto cycle the pressure is 0.96 bar and the temperature is 27°C. At the end of compression the pressure has risen by a factor of 15.3, and during the combustion process 1160 kJ/kg of heat are added. Using the air table, determine (a) the compression ratio, (b) the temperature at the end of compression, in °K, (c) the temperature at the end of expansion, in °K, (d) the work output during the expansion, in kJ/kg, (e) the work input during the compression, in kJ/kg, and (f) the thermal efficiency of the cycle.

**16-18M** Same as Prob. 16-13M, except that the amount of heat added is 1188 kJ/kg.

**16-19M** Same as Prob. 16-14M, except that the amount of heat added is 7.07 kJ.

**16-20M** Same as Prob. 16-16M, except that the amount of heat added is 1310 kJ/kg.

## Diesel cycle

**16-21M** An air-standard Diesel cycle operates with a compression ratio of 14.8 and a cutoff ratio of 2. At the beginning of compression the air temperature and pressure are 37°C and 1 bar, respectively. Determine (a) the maximum temperature in the cycle, in °K, and (b) the heat input per cycle, in kJ/kg.

**16-22M** An engine operates on the theoretical Diesel cycle with a compression ratio of 15 : 1, and fuel is injected for 10 percent of the stroke. The pressure and temperature of the air entering the cylinder are 0.98 bar and 17°C, respectively.

    (a) Determine the cutoff ratio.

    (b) Determine the temperature, in °K, at the end of the compression process and at the end of the heat-addition process.

    (c) Compute the heat input, in kJ/kg. Use Table A-5M.

**16-23M** An air-standard Diesel cycle operates with a compression ratio of 16.3 and a cutoff ratio of 2. At the beginning of compression the air temperature and pressure are 32°C and 1 bar, respectively. Determine the maximum temperature in the cycle, in °K.

**16-24M** A Diesel air-standard cycle has a compression ratio of 15 : 1. The pressure and temperature at the beginning of compression are 1 bar and 17°C, respectively. If the maximum temperature of the cycle is 2250°K, determine (a) the cutoff ratio, (b) the thermal efficiency, and (c) the mean effective pressure, in bars.

**16-25M** An air-standard Diesel cycle is supplied with 1535 kJ/kg of air per cycle. The pressure and temperature at the beginning of compression are 0.95 bar and 27°C, respectively, and the pressure after compression is 36 bars. Determine (a) the compression ratio, and (b) the cutoff ratio.

**16-26M** The intake conditions for an air-standard Diesel cycle operating with a compression ratio of 15 : 1 are 0.95 bar and 17°C. At the beginning of the compression stroke the cylinder volume is 3.80 l, and 7.5 kJ of heat are added to the gas during the constant-pressure heating process. Calculate the pressure and temperature at the end of each process of the cycle, and determine the thermal efficiency and the mean effective pressure of the cycle.

## Dual cycle

**16-27M** An air-standard dual cycle operates with a compression ratio of 15 : 1. At the beginning of compression the conditions are 17°C, 0.95 bar, and 3.80 l. The amount of heat added is 6.6 kJ, of which one-third is added at constant volume and the remainder at constant pressure. Determine (a) the pressure after the constant-volume heat-addition process, (b) the temperature before and after the constant-pressure heat-addition process, (c) the temperature after isentropic expansion, and (d) the thermal efficiency.

**16-28M** An air-standard dual cycle operates with a compression ratio of 14 : 1. At the beginning of the isentropic compression the conditions are 27°C and 0.96 bar. The total heat addition is 1680 kJ/kg, of which one-fourth is added at constant volume and the remainder at constant pressure. Determine (a) the temperatures at the end of each process around the cycle, in °K, (b) the thermal efficiency, and (c) the mean effective pressure.

**16-29M** Same as Prob. 16-27M, except that the heat addition is 8.0 kJ, of which 30 percent is added at constant volume.

**16-30M** Same as Prob. 16-28M, except that the heat addition is 1800 kJ/kg, of which one-third is added at constant volume.

## Ideal open gas-turbine cycle

**16-31M** A gas-turbine-power plant operates under the following conditions: (a) Air enters the compressor at 1 bar and 22°C and leaves at 3.46 bars, and (b) air leaves the combustion chamber at 747°C. Calculate the thermal efficiency of the Brayton cycle using the data of Table A-5M.

**16-23M** A stationary gas-turbine-power plant delivers 10,000 kW to an electric generator. The maximum and minimum cycle temperatures are 827 and 27°C, respectively. The pressure ratio is 5.

(*a*) What is the gross turbine-power output, in kilowatts?

(*b*) What fraction of the turbine output is used to drive the compressor?

(*c*) What mass flow rate of air is required, in kg/min?

(*d*) What is the thermal efficiency of the cycle?

**16-33M** The pressure ratio of an air-standard Brayton cycle is 4.5, and the inlet conditions to the compressor are 1 bar and 27°C. The turbine is limited to a temperature of 827°C, and the mass flow rate is 5 kg/s. Determine (*a*) the compressor work, in kJ/kg, (*b*) the turbine work, in kJ/kg, (*c*) the thermal efficiency, and (*d*) the net power output, in kilowatts.

**16-34M** Prove that the maximum net work for a simple Brayton cycle, for fixed compressor- and turbine-inlet temperatures, occurs when $T_2 = (T_1 T_3)^{1/2}$ if the specific heats are constant.

**16-35M** Show that the pressure ratio which yields the largest net work for a simple Brayton cycle with fixed inlet-compressor and turbine temperatures is given by $P_2/P_1 = (T_3/T_1)^n$, where $n = k/[2(k-1)]$ and $k$ is a constant.

**16-36M** The inlet conditions for an air-standard Brayton cycle are 1 bar and 22°C. The inlet-gas temperature to the turbine is limited to 877°C. What pressure ratio should be employed in the ideal compressor so that the cycle produces the maximum work output, assuming (*a*) variable $k$ data, and (*b*) constant $k$ data.

**16-37M** Consider an ideal gas-turbine cycle. The maximum allowable temperature to the turbine is 747°C. Find the maximum pressure ratio for which this cycle will just deliver power if the inlet compressor temperature is 27°C.

(*a*) Use the air table.

(*b*) Let $k = 1.40 = $ constant.

**16-38M** An air-standard gas-turbine cycle operates on an ideal Brayton cycle, with inlet conditions of 1 bar and 17°C. The turbine-inlet temperature is limited to 1000°K. Calculate the percent change in net work output and thermal efficiency if the pressure ratio is increased from 4 : 1 to 6 : 1.

**16-39M** A gas-turbine-power plant operates on an air-standard cycle between pressure limits of 1 bar and 6.4 bars. The inlet-air temperature is 22°C, and the temperature limitation on the turbine is 807°C. Calculate the net work output and the thermal efficiency if the cycle is ideal.

**16-40M** A gas-turbine-power plant operates on an air-standard cycle between pressure limits of 1 bar and 6 bars. The inlet-air temperature to the compressor is 17°C, and the inlet temperature to the turbine is 1080°K. Calculate the net work output and the thermal efficiency if the cycle is ideal. Use the air table.

**Nonideal simple gas-turbine-power cycle**

**16-41M** Using the data of Prob. 16-31M, calculate the thermal efficiency if the compressor and turbine adiabatic efficiencies are 83 and 86 percent, respectively.

**16-42M** Using the data of Prob. 16-32M, calculate the required quantities if the adiabatic efficiencies of the compressor and turbine are 81 and 86 percent, respectively.

**16-43M** Using the data of Prob. 16-33M, calculate the required quantities if the compressor and turbine adiabatic efficiencies are 78 and 84 percent, respectively.

**16-44M** A turbine operates with inlet conditions of 567°C and 4 bars. If the outlet conditions are 447°C and 2 bars, what is the turbine adiabatic efficiency?

**16-45M** A gas-turbine cycle, with compressor and turbine inlet temperatures of 22°C and 827°C, respectively, uses a 6 : 1 pressure ratio. If the compressor efficiency is 78 percent, what minimum turbine adiabatic efficiency will lead to zero work output?

**16-46M** A gas-turbine-power cycle operates with a pressure ratio of 12 : 1. The compressor and turbine adiabatic efficiencies are 85 and 90 percent, respectively. The compressor-inlet temperature is 22°C, and the turbine-inlet temperature is 1027°C. For a mass flow rate of 1 kg/s, find the power generated by the cycle. Now double the pressure ratio, and find the power output in this latter case, in kilowatts.

**16-47M** Air enters a gas-turbine-power plant at 17°C and 1 bar. The compressor and turbine both operate with a pressure ratio of 8, and the adiabatic efficiencies of each are 80 and 85 percent, respectively. Gases enter the combustor at 887°C, and the mass flow rate is 10 kg/s. Determine what percent of the total output of the turbine may be used to drive devices other than the compressor, on the basis of data in Table A-5M.

**16-48M** A gas turbine cycle operates with a pressure ratio of 10. The inlet temperatures to the compressor and turbine are 7 and 1027°C, respectively. The adiabatic efficiencies of the turbine and compressor are 88 and 82 percent, respectively. Calculate (a) the net work, in kJ/kg and (b) the thermal efficiency.

**16-49M** Using the data of Prob. 16-39M, calculate the required quantities if the adiabatic efficiencies of the compressor and turbine are 82 and 85 percent, respectively.

**16-50M** Using the data of Prob. 16-40M, rework the problem if the adiabatic efficiencies of the compressor and turbine are 82 and 85 percent, respectively.

**16-51M** A gas-turbine-power plant producing 6700 kW of power operates under the following test data: (1) entering compressor, 0.95 bar and 22°C; (2) leaving compressor, 4.20 bars and 207°C; (3) entering turbine, 4.10 bars and 727°C; (4) leaving turbine, 0.97 bar and 457°C.

    (a) Compute the adiabatic compressor efficiency.
    (b) Find the adiabatic turbine efficiency.
    (c) Determine the gross output of the turbine, in kilowatts.

**Regenerative gas-turbine cycle**

**16-52M** Using the data of Prob. 16-41M, compute the thermal efficiency if a regenerator is installed with an effectiveness of (a) 80 percent, and (b) 65 percent.

**16-53M** Using the data of Prob. 16-42M, determine the thermal efficiency if a regenerator is added with an effectiveness of (a) 70 percent, and (b) 50 percent.

**16-54M** Using the data of Prob. 16-43M, determine the thermal efficiency of the cycle if a regenerator is used with an effectiveness of (a) 100 percent, (b) 80 percent, and (c) 60 percent.

**16-55M** Using the data from Prob. 16-49M, determine the percentage of the fuel costs which could be saved if a regenerator with an effectiveness of (a) 50 percent, and (b) 70 percent, were installed.

**16-56M** Using the data from Prob. 16-50M, determine the thermal efficiency if a regenerator with 70 percent effectiveness were installed.

**16-57M** In an air-standard gas-turbine cycle with regeneration, the compressor is driven directly by the turbine. The following enthalpy data were taken on a test of the gas turbine with a pressure ratio of 5.41 : 1. Determine (a) the thermal efficiency of the actual cycle, (b) the effectiveness of the regenerator, (c) the adiabatic efficiency of the compressor, and (d) the adiabatic efficiency of the turbine. Data are in kJ/kg.

| System | Entering | Leaving |
|---|---|---|
| Compressor | 290.2 | 505.0 |
| Regenerator | 505.0 | 629.4 |
| Combustor | 629.4 | 1046.0 |
| Turbine | 1046.0 | 713.7 |
| Regenerator | 713.7 | 590.1 |

**16-58M** An automobile manufacture is contemplating the use of a regenerative-type gas turbine for a power source of a new model. Fresh air enters the compressor at 22°C and 1 bar, with a flow rate of 1 kg/s, and is compressed at a ratio of 4 : 1. Air from the compressor enters the regenerator, where it is circulated until its temperature is approximately 537°C. It then passes to the combustion chamber, where it is heated to 927°C. After expanding through the two-stage turbine, the hot gases pass through the other half of the regenerator and then are exhausted to the atmosphere. Assuming ideal compression and expansion and negligible pressure losses through the burner and regenerator, determine (a) the thermal efficiency of the unit, (b) the net power output, in kilowatts, (c) the regenerator effectiveness, and (d) the exhaust stream temperature, in °C.

**16-59M** Air enters the compressor of a gas-turbine cycle at 1 bar and 12°C. The compressor and turbine are ideal, and the pressure ratio is 6.2 : 1. A regenerator is used which has an effectiveness of 70 percent. The gas stream leaving the compressor is heated 125°C as it passes through the regenerator. Determine the turbine-inlet temperature, in °C.

**16-60M** With reference to Prob. 16-51M, compute the percentage of fuel saved if a regenerator with 50 percent effectiveness is installed.

## Polytropic processes

**16-61M** Air is compressed in a piston-cylinder device from 1 bar, 7°C, to a pressure of 5 bars. Calculate the work required, in kJ/kg, if the process is (a) polytropic with $n = 1.30$, (b) isentropic, and (c) isothermal.

**16-62M** Work Prob. 16-1 with the same data except that the final pressure is 6 bars.

**16-63M** Air is compressed polytropically in steady flow, with $n = 1.35$, from 1 bar and 37°C to 5 bars, at a rate of 2 kg/s. Calculate (a) the required power input, in kilowatts, and (b) the heat removed, in kJ/kg.

**16-64M** Air is compressed in a steady-flow polytropic process from 1 bar and 7°C, to 6 bars. If the polytropic constant is 1.30, calculate the work required and the heat transferred, in kJ/kg.

**16-65M** An air compressor operates between an inlet condition of 1 bar and 17°C, and an outlet pressure of 5 bars. Determine the work input, in kJ/kg, required for the following steady-flow reversible processes: (a) isothermal, (b) $n = 1.30$, and (c) adiabatic.

**16-66M** The work output of a polytropic process involving air is twice the value of the heat input. The initial temperature and pressure are 300°K and 2 bars, respectively, and it is assumed that the specific-heat values are constant. Determine the value of the polytropic constant $n$ for the process, and the shaft work, in kJ/kg, if the process is steady-flow and the exhaust pressure is 1 bar.

**16-67M** A reciprocating air compressor is designed to have a displacement volume of 4000 cm$^3$ and a clearance volume of 500 cm$^3$. If the cycle is adiabatic, determine the overall pressure ratio which will reduce the mass flow through the compressor to zero.

**16-68M** Find the power requirements, in kilowatts, if 7 m$^3$/min of air at 0.95 bar and 12°C are compressed polytropically ($n = 1.3$) to 7.6 bars. Cooling water runs through the compressor jacket at a rate of 14 kg/min and has a temperature rise of 6°C. The air velocity at the inlet is small, but it is 160 m/s at the outlet.

## Two-stage compression with intercooling

**16-69M** The intake conditions of a two-stage air compressor are 1 bar and 27°C. The outlet pressure is 4 bars, and the pressure ratio across each stage is the same. If an intercooler cools the air to the initial inlet temperature and the stages are isentropic, determine the work required, in kJ/kg. Compare with the work required for a single-stage compressor.

**16-70M** Five kilograms of air per second are compressed isentropically from 1 bar and 27°C to 5.6 bars.

(a) Using Table A-5M, determine the power requirement, in kilowatts, for a single-stage compressor.

(b) Using the same data, compute the minimum kilowatt requirement for two-stage compression when an intercooler cools the air to the initial temperature between stages.

**16-71M** The inlet conditions of a two-stage steady-flow compressor are 0.95 bar and 27°C. The outlet pressure is 9.5 bars, and the pressure ratio across each stage is the same. If an intercooler cools the air to the initial temperature and the stages are isentropic, (a) determine the work input, in kJ/kg, and (b) compare the result of (a) with the work required for a single-stage compressor.

**16-72M** Same as Prob. 16-71M, except that the inlet conditions are 1.05 bars and 37°C, and the final outlet pressure is 6.3 bars.

### Gas-turbine cycle with intercooling and reheating

**16-73M** Reconsider Prob. 16-52M. Determine the net work output and the thermal efficiency if two-stage compression and expansion are used with a regenerator of (a) 80 percent, and (b) 65 percent effectiveness. The compressor and turbine efficiencies are given in Prob. 16-41M. The intercooling and reheating are ideal.

**16-74M** Reconsider Prob. 16-53M. Determine the net work output and the thermal efficiency if two-stage compression and expansion are used with a regenerator effectiveness of (a) 70 percent, and (b) 50 percent. The compressor and turbine efficiencies are given in Prob. 16-42M. The intercooling and reheating are ideal.

**16-75M** Reconsider Prob. 16-54M. Determine the net work output and the thermal efficiency if two-stage compression and expansion are used with a regenerator effectiveness of (a) 80 percent, and (b) 60 percent. The compressor and turbine efficiencies are given in Prob. 16-43M. The intercooling and reheating are ideal.

**16-76M** Reconsider Prob. 16-55M. Determine the net work output and the thermal efficiency if two-stage compression and expansion are used with a regenerator effectiveness of (a) 50 percent, and (b) 70 percent. The compressor and turbine efficiencies are given in Prob. 16-49M. The intercooling and reheating are ideal.

**16-77M** A gas-turbine cycle operates with two stages of compression and two stages of expansion. The pressure ratio across each stage is 2. The inlet temperature and pressure to the compressor are 22°C and 1 bar, and the inlet turbine temperature to each stage is 827°C. The compressor operates with an ideal intercooler, and an ideal regenerator also is used. Determine (a) the compressor work and turbine work, in kJ/kg, and (b) the thermal efficiency.

**16-78M** Using the data for Prob. 16-77M, calculate the same required quantities if the compressor and turbine efficiencies are 78 and 84 percent, respectively, and the regenerator has an effectiveness of 70 percent.

**16-79M** A gas-turbine-power plant employs two-stage compression and expansion, with intercooling, reheating, and regeneration. The temperature at the outlet of the compressor second stage is 390°K, and the combustor inlet temperature is 750°K. The turbine-inlet temperature is limited to 1180°K.

    (a) Determine the gross maximum work output from the two-stage turbine if the overall pressure ratio is 6 : 1.

    (b) Determine the regenerator effectiveness.

**16-80M** Consider the data for Prob. 16-33M, but employ two-stage compression and two-stage expansion with intercooling and reheating. If the pressure ratio across each stage is 2.12, determine (a) the compressor work, in kJ/kg, (b) the turbine work, and (c) the thermal efficiency.

### Ericsson cycle

**16-81M** An Ericsson cycle operates on air, and the minimum volume of the cycle is 0.01 m³. The maximum pressure is 6 bars, and at the end of constant-pressure expansion the volume is 0.02 m³. Draw a $PV$ diagram, and compute, for each 40 kJ of heat added, the amount of heat that must be rejected.

**16-82M** Same as Prob. 16-81M, except that the minimum volume is 5 l, the maximum pressure is 5 bars, and the volume at the end of constant-pressure expansion is 12 l. The heat added is 20 kJ.

### Turbojet cycles

**16-83M** Rework Example 16-9 in the text under the following conditions: (1) The actual pressure rise in the diffuser is 92 percent of theoretical. (2) The compressor efficiency is 82 percent. (3) The turbine efficiency is 86 percent. (4) The nozzle efficiency is 95 percent. Determine (a) the pressure and temperature throughout the cycle, (b) the compressor work, and (c) the exit-jet velocity, in m/s.

**16-84M** The air speed of a turbojet aircraft is 300 m/s in still air at 0.25 bar and 220°K. The compressor pressure ratio is 9, and the maximum temperature in the cycle is 1220°K. Assume ideal performance of the various components. Determine (a) the temperatures and pressures throughout the cycle, (b) the compressor work required, and (c) the exit-jet velocity, in m/s.

**16-85M** Rework Prob. 16-84M under the following conditions: (1) The actual pressure rise in the diffuser is 90 percent of the theoretical. (2) The compressor and turbine efficiencies are 83 and 87 percent, respectively. (3) The nozzle efficiency is 96 percent. Determine the quantities specified in Prob. 16-84M.

**16-86M** The air speed of a jet aircraft is 250 m/s in still air at 0.50 bar and 250°K. The pressure ratio across the compressor is 7, and the maximum permissible temperature is 1160°K. Assume ideal performance of the various components. Determine (a) the compressor work required, and (b) the exit-jet velocity, in m/s.

**16-87M** Rework Prob. 16-86M under the following conditions: (1) The actual pressure rise is 92 percent of the theoretical. (2) The compressor and turbine efficiencies are 84 and 87 percent, respectively and (3) the nozzle efficiency is 94 percent.

**16-88M** The nozzle of a turbojet engine receives gas from the turbine at 2.4 bars and 940°K and with a velocity of 40 m/s. The nozzle expands the air adiabatically to 0.60 bar.

    (a) Find the discharge velocity if the nozzle efficiency is 92 percent.

    (b) The aircraft is flying at 240 m/s in still air and the air enters the intake at this speed and at a rate of 55 kg/s. What is the thrust of the engine, in newtons?

**16-89M** The nozzle of a turbojet engine receives gas from the turbine exhaust at 2.0 bars and 980°K and with a velocity of 40 m/s. The nozzle expands the air adiabatically to 0.75 bar.

    (a) Determine the exit velocity if the nozzle efficiency is 90 percent.

    (b) The aircraft flies at 300 m/s, and air enters at this speed and at a rate of 75 kg/s. What is the thrust of the engine, in newtons?

**16-90M** A simple gas-turbine cycle operates so that there is no net work output. The inlet pressure and temperature to the isentropic compressor are 1 bar and 22°C, and the compressor pressure ratio is 6. If the limiting temperature of the isentropic turbine is 1120°K, determine the turbine-outlet pressure and temperature.

**16-91M** Consider a simple gas-turbine cycle with no net work output, as in a jet aircraft. The inlet pressure and temperature to the compressor are 1 bar and 27°C, and the pressure ratio is 7. The limiting temperature in the turbine is 1160°K. If the compressor and turbine efficiencies are 81 and 85 percent, respectively, determine the turbine-outlet pressure and temperature.

**16-92M** Same as Prob. 16-91M, except that the compressor inlet conditions are 0.7 bar and 250°K.

**16-93M** Same as Prob. 16-90M, except that the compressor and turbine efficiencies are 80 and 85 percent, respectively.

## Closed gas-turbine cycles

**16-94M** A closed-cycle gas-turbine-power plant operates with helium as the working fluid. The compressor and turbine efficiencies are each 85 percent. The compressor-inlet state is 5 bars and 47°C, while the turbine-inlet state is 12 bars and 980°K. If the mass flow rate is 5 kg/s, determine (a) the net power output, in kilowatts, and (b) the rate of heat removed from the cycle, in kJ/min.

**16-95M** A closed-cycle gas-turbine-power plant operates with argon as the working fluid. The compressor and turbine efficiencies are 0.83 and 0.86, respectively. Compressor-inlet conditions are 8 bars and 340°K, while the turbine-inlet state is 15 bars and 1000°K. If the mass flow rate is 7 kg/s, find (a) the net power output in kilowatts, and (b) the rate of heat removed from the cycle, in kJ/min.

**16-96M** A closed-cycle gas-turbine-power plant operates with argon as the working fluid. The compressor and turbine efficiencies are 84 and 86 percent, respectively. The compressor-inlet conditions are 6 bars and 37°C, while the turbine-inlet conditions are 21 bars, and 652°C. If the net power output is 3500 kW, determine (a) the required mass flow rate in kg/s, (b) the gross kilowatt output of the turbine, and (c) the rate of heat removed from the cycle and radiated to outer space, in kJ/min.

## Gas refrigeration cycles

**16-97M** A reversed Brayton cycle with a pressure ratio of 3 is used to produce a refrigeration effect. The compressor-inlet temperature is 27°C, and the turbine-inlet temperature is 7°C.

(a) What are the maximum and minimum temperatures in the cycle, in °C?

(b) What is the coefficient of performance?

**16-98M** An air-standard Brayton refrigerator compresses air at 1 bar and 280°K to 5 bars. The turbine-inlet temperature is 360°K. Calculate the coefficient of performance for the cycle.

**16-99M** An ideal, reversed Brayton cycle operates with air entering the compressor at 1 bar and 7°C and leaving at 4 bars. Air enters the turbine at 27°C. Calculate (a) the coefficient of performance, and (b) the mass flow rate of air required, in kg/min, to remove 211 kJ/min of heat from the cold region.

**16-100M** A reversed Brayton refrigeration cycle operates with air between the pressure limits of 5 and 10 bars. The fluid enters the turbine at 27°C, and it leaves the cold heat exchanger at 7°C. Determine (a) the coefficient of performance, and (b) the mass flow rate in kg/min required to cool the cold region at a rate of 211 kJ/min.

**16-101M** Using the data of Prob. 16-100M, consider a heat exchanger placed in the cycle similar to that shown in Fig. 16-30. Assume that the fluid entering the turbine is precooled down to the compressor-inlet temperature. Determine (a) the turbine-outlet temperature, in °C, and (b) the coefficient of performance.

**16-102M** An ideal reversed Brayton cycle operates with a pressure ratio of 3. Air enters the compressor at 7°C, and enters the turbine normally at 27°C. The cycle is now modified similarly to that shown in Fig. 16-30. The heat exchanger allows the air entering the turbine to be precooled down to the compressor-inlet temperature. Determine (a) the old and new turbine-outlet temperatures, in °K, and (b) the old and new COP values.

**16-103M** An air-refrigerating machine operating between 3 and 12 bars is to be used to heat a building. The air temperature entering the compressor is 7°C, and the air temperature entering the expander is 27°C. If the expansion and compression are reversible, determine (a) the kilojoules input in work for each kilojoule received as heat to the building, and (b) the coefficient of performance of the device.

**16-104M** Rework Prob. 16-97M if the compressor- and turbine-adiabatic efficiencies are 85 and 88 percent, respectively.

**16-105M** Rework Prob. 16-98M if the compressor- and turbine-adiabatic efficiencies are 84 and 87 percent, respectively.

**16-106M** Rework Prob. 16-99M if the compressor- and turbine-adiabatic efficiencies are 84 and 88 percent, respectively.

## Stirling cycle

**16-107M** A Stirling cycle operates with air, and at the beginning of isothermal expansion the state is 447°C and 4 bars. The minimum pressure in the cycle is 1 bar, and at the end of isothermal compression the volume is 60 percent of the maximum volume. Determine (a) the thermal efficiency of the cycle, and (b) the mean effective pressure.

**16-108M** A Stirling cycle operates with air, and at the beginning of isothermal compression the state is 77°C and 2 bars. The maximum pressure is 6 bars, and during isothermal expansion the volume increases by 40 percent. Determine (a) the thermal efficiency, and (b) the mean effective pressure of the cycle.

**16-109M** A Stirling cycle operates with air, and at the beginning of the isothermal expansion the pressure is 5 bars and the temperature is 257°C. The thermal efficiency is 44 percent, and at the end of the isothermal compression the volume is two-thirds the maximum volume. Determine the mean effective pressure of the cycle.

### Combined gas steam turbine cycle

**16-110M** A gas-turbine-power plant operates on a pressure ratio of 12 : 1 with compressor- and turbine-inlet temperatures of 300 and 1400°K, respectively. The adiabatic efficiencies of the compressor and turbine are 85 and 87 percent, respectively. The turbine exhaust, used as the energy source for a steam cycle, leaves the boiler at 480°K. The inlet conditions to the 85 percent efficient turbine in the steam cycle are 140 bars and 520°C. The condenser pressure is 0.10 bar and the pump is 75 percent efficient. Determine (a) the required heat input, in kJ/kg of air, (b) the mass-flow-rate ratio of air to steam, (c) the net work output of the gas turbine cycle, in kJ/kg of air, (d) the net work output of the steam cycle, in kJ/kg of steam, and (e) the overall thermal efficiency of the combined cycle.

**16-111M** Consider the data of Prob. 16-110M. In addition, an open feedwater heater is operated in the steam cycle with steam bled from the turbine at 7 bars. For this modified cycle, determine the quantities that were required in Prob. 16-110M.

**16-112M** A gas-turbine-power plant operates on a pressure ratio of 13 : 1 with compressor- and turbine-inlet temperatures of 290 and 1440°K, respectively. The adiabatic efficiencies of the compressor and turbine are 84 and 88 percent, respectively. The turbine exhaust, used as the energy source for a steam cycle, leaves the boiler at 500°K. The inlet conditions to the 86 percent efficient turbine in the steam cycle are 160 bars and 560°C. The condenser pressure is is 0.08 bar and the pump is 70 percent efficient. Determine (a) the required heat input, in kJ/kg of air, (b) the mass-flow-rate ratio of air to steam, (c) the net work output of the gas-turbine cycle, in kJ/kg of air, (d) the net work output of the steam cycle, in kJ/kg of steam, and (e) the overall thermal efficiency of the combined cycle.

**16-113M** Consider the data of Prob. 16-112M. In addition, an open feedwater heater is operated in the steam cycle with steam bled to the heater from the turbine at 10 bars. For this modified cycle, determine the quantities that were required in Prob. 16-112M.

**16-114M** A gas-turbine-power plant operates on a 10 : 1 pressure ratio with compressor- and turbine-inlet temperatures of 300 and 1340°K, respectively. The adiabatic efficiencies of the compressor and turbine are 84 and 87 percent, respectively. The turbine exhaust, used as the energy source for a steam cycle, leaves the boiler at 460°K. The inlet conditions to the 86 percent efficient turbine in the steam cycle are 100 bars and 480°C. The condenser pressure is 0.08 bar and the pump is 70 percent efficient. Determine (a) the required heat input, in kJ/kg of air, (b) the mass-flow-rate ratio of air to steam, (c) the net work output of the gas-turbine cycle, in kJ/kg of air, (d) the net work output of the steam cycle, in kJ/kg of steam, and (e) the overall thermal efficiency of the combined cycle.

**16-115M** Determine the quantities that were required in Prob. 16-114M if in addition, an open feedwater heater is added to the steam cycle with steam bled from the turbine to the heater at 7 bars.


# PROBLEMS (USCS)


### Air standard cycles

**16-1** An air-standard cycle is proposed which begins at an initial state of 70°F, 14.7 psia, and 13.35 ft³/lb. Process 1-2 is constant-volume heating to 29.4 psia; process 2-3 is constant-pressure heating to 1660°F and 26.7 ft³/lb; process 3-4 is isentropic expansion to 14.7 psia; process 4-1 is constant-pressure cooling to the initial state.

    (a) Sketch $Pv$ and $Ts$ diagrams for the cycle.

    (b) Determine the total heat added, in Btu/lb.

    (c) Determine the net work output, in Btu/lb.

    (d) Compute the thermal efficiency.

    (e) What is the Carnot efficiency for a heat engine with the same maximum and minimum temperature as the above cycle? Use $u$ and $h$ data from the air table.

**16-2** An air-standard cycle is proposed which begins at an initial state of 100°F, 16 psia, and 12.95 ft³/lb. Process 1-2 is constant-volume heating to 42 psia; process 2-3 is constant-pressure heating to 2340°F and 25.90 ft³/lb; process 3-4 is isentropic expansion to 16 psia; process 4-1 is constant-pressure cooling to the initial state.

(a) Sketch Pv and Ts diagrams for the cycle.

(b) Determine the total heat added, in Btu/lb.

(c) Find the net work output, in Btu/lb.

(d) Compute the thermal efficiency.

(e) What is the Carnot efficiency for a heat engine with the same maximum and minimum temperatures as the above cycle? Use u and h data from the air table.

**16-3** A proposed air-standard cycle begins at 15 psia, 80°F, and 13.32 ft³/lb. Process 1-2 is isentropic compression to 41.2 psia; process 2-3 is constant-pressure heating to 1700°F and 19.41 ft³/lb; process 3-4 is constant-volume cooling to 15 psia; process 4-1 is constant-pressure cooling to the initial state.

(a) Sketch Pv and Ts diagrams of the cycle.

(b) Determine the heat added and the net work output, in Btu/lb.

(c) Compute the thermal efficiency.

(d) What is the Carnot efficiency for a heat engine with the same maximum and minimum temperatures as the above cycle? Use air table data.

**16-4** A proposed air-standard cycle begins at 16 psia, 100°F, and 12.95 ft³/lb. Process 1-2 is isentropic compression to 35.0 psia; process 2-3 is constant-pressure heating to 1290°F and 18.5 ft³/lb; process 3-4 is constant-volume cooling to 16 psia; process 4-1 is constant-pressure cooling to the initial state.

(a) Sketch Pv and Ts diagrams of the cycle.

(b) Determine the heat added and the net work output, in Btu/lb.

(c) Compute the thermal efficiency.

(d) What is the Carnot efficiency for a heat engine with the same maximum and minimum temperatures as the above cycle? Use air table data.

### Air-standard Carnot cycles

**16-5** An air-standard Carnot cycle rejects 50 Btu/lb to a sink at 500°R. The minimum and maximum pressures in the cycle are 1 and 292 atm. On the basis of the air table, determine (a) the pressure after isothermal compression, (b) the temperature of the heat supply, and (c) the thermal efficiency.

**16-6** An air-standard Carnot cycle is supplied 100 Btu/lb from a source at 1200°R. The minimum and maximum pressures in the cycle are 1 and 88 atm. On the basis of the air table, determine (a) the pressure after isothermal heat addition, (b) the temperature of heat rejection, and (c) the thermal efficiency.

**16-7** An air-standard Carnot cycle operates between temperatures of 500 and 1500°R. The minimum pressure in the cycle is 1 atm. Determine the maximum pressure in the cycle, in atm, if the heat rejection is (a) 30 Btu/lb, and (b) 20 Btu/lb.

**16-8** An air-standard Carnot cycle operates between temperatures of 540 and 2000°R. The minimum pressure in the cycle is 1 atm. Determine the maximum pressure in the cycle, in atm, if the heat supplied is (a) 80 Btu/lb, and (b) 70 Btu/lb.

### Compression ratio and mean effective pressure

**16-9** For the data in Prob. 16-5, determine the compression ratio and the mean effective pressure for a reciprocating device.

**16-10** For the data in Prob. 16-6, determine the compression ratio and the mean effective pressure for a reciprocating device.

**16-11** For the data in Prob. 16-7, determine the compression ratio and the mean effective pressure for a reciprocating device if the heat rejection is (a) 30 Btu/lb, and (b) 20 Btu/lb.

**16-12** For the data in Prob. 16-8, determine the compression ratio and the mean effective pressure for a reciprocating device if the heat supplied is (a) 80 Btu/lb, and (b) 70 Btu/lb.

## Otto cycle

**16-13** The compression ratio of an Otto cycle is 8 : 1. Before the compression stroke of the cycle begins, the pressure is 14.5 psia and the temperature is 80°F. The heat added to the air per cycle is 888 Btu/lb of air. On the basis of the air standard cycle, and employing air table A-5, determine (a) the pressure and temperature at the end of each process of the cycle, (b) the theoretical thermal efficiency, and (c) the mean effective pressure for the cycle, in lb/in$^2$.

**16-14** The air at the beginning of the compression stroke of an air-standard Otto cycle is at 14 psia and 80°F, and the cylinder volume is 0.2 ft$^3$. The compression ratio is 9, and 9.2 Btu are added during the heat-addition process. Calculate (a) the temperatures and pressures after the compression and heat-addition processes, and (b) the thermal efficiency.

**16-15** An air-standard Otto cycle operates with a compression ratio of 8.5. At the beginning of compression the air is at 14.5 psia and 90°F, and during the heat-addition process the pressure is tripled.

(a) Calculate the thermal efficiency of the cycle.

(b) Determine the efficiency of a Carnot engine operating between the same overall temperature limits.

**16-16** Consider an air-standard Otto cycle that has a compression ratio of 9 and a heat addition of 870 Btu/lb of air. If the pressure and temperature at the beginning of the compression process are 14 psia and 40°F, determine (a) the maximum pressure and temperature for the cycle, (b) the thermal efficiency, and (c) the mean effective pressure.

**16-17** At the beginning of compression of an air-standard Otto cycle the pressure is 14 psia and the temperature is 80°F. At the end of compression the pressure has risen by a factor of 15.3, and during the combustion process 497 Btu/lb of heat are added. Using the air tables, determine (a) the compression ratio, (b) the temperature at the end of compression, in °R, (c) the temperature at the end of expansion, in °R, (d) the work output during expansion, in Btu/lb, (e) the work input during the compression, in Btu/lb, and (f) the thermal efficiency of the cycle.

**16-18** Same as Prob. 16-13, except that the amount of heat added is 792 Btu/lb.

**16-19** Same as Prob. 16-14, except that the amount of heat added is 8.3 Btu.

**16-20** Same as Prob. 16-16, except that the amount of heat added is 750 Btu/lb.

## Diesel cycle

**16-21** An air-standard Diesel cycle operates with a compression ratio of 14.8 and a cutoff ratio of 2. At the beginning of compression the air temperature and pressure are 100°F and 14.5 psia, respectively. Determine (a) the maximum temperature in the cycle, in °R, and (b) the heat input per cycle, in Btu/lb.

**16-22** An engine operates on the theoretical Diesel cycle with a compression ratio of 12 : 1, and fuel is injected for 10 percent of the stroke. The pressure of the air entering the cylinder is 14.0 psia. and its temperature is 60°F.

(a) Determine the cutoff ratio.

(b) Determine the temperature, in °R, at the end of the compression process and at the end of the heat-addition process.

(c) Compute the heat input, in Btu/lb.

**16-23** An air-standard Diesel cycle operates with a compression ratio of 10 and a cutoff ratio of 2. At the beginning of compression the air temperature and pressure are 90°F and 14.5 psia. Determine the maximum temperature in the cycle, in °F.

**16-24** A Diesel cycle operating on an air-standard cycle has a compression ratio of 15. The pressure and temperature at the beginning of compression are 15 psia and 60°F. If the maximum temperature of the cycle is 4200°R, determine (a) the thermal efficiency, and (b) the mean effective pressure.

**16-25** An air-standard Diesel cycle is supplied 700 Btu/lb of air per cycle. The pressure and temperature at the beginning of compression are 14 psia and 80°F, and the pressure after compression is 540 psia. Determine (a) the compression ratio, and (b) the cutoff ratio.

**16-26** The intake conditions for a theoretical Diesel cycle operating with a compression ratio of 15 are 14.4 psia and 60°F. At the beginning of the compression stroke the cylinder volume is 270 in³, and 9 Btu of heat are added to the gas during the constant-pressure heating process.

    (*a*) Calculate the pressure and temperature at the end of each process of the cycle.

    (*b*) Determine the thermal efficiency and the mean effective pressure.

### Dual cycle

**16-27** An air-standard dual cycle operates with a compression ratio of 15 : 1. At the beginning of compression the conditions are 60°F, 14.6 psia, and 230 in³. The amount of heat added is 7.5 Btu, of which one-third is added at constant volume and the remainder at constant pressure. Determine (*a*) the pressure after the constant-volume heat-addition process, (*b*) the temperature before and after the constant-pressure heat-addition process, (*c*) the temperature after isentropic expansion, and (*d*) the thermal efficiency.

**16-28** An air-standard dual cycle operates with a compression ratio of 14 : 1. At the beginning of isentropic compression the conditions are 80°F and 14.5 psia. The total heat addition is 800 Btu/lb, of which one-fourth is added at constant volume and the remainder at constant pressure. Determine (*a*) the temperatures at the end of each process around the cycle, in °R, (*b*) the thermal efficiency, and (*c*) the mean effective pressure.

**16-29** Same as Prob. 16-27, except that the heat addition is 8.1 Btu, of which 30 percent is added at constant volume.

**16-30** Same as Prob. 16-28 that the heat addition is 900 Btu/lb, of which one-third is added at constant volume.

### Ideal gas-turbine open cycle

**16-31** A gas-turbine-power plant operates under the following conditions: (*a*) Air enters the compressor at 14.7 psia and 60°F and leaves the compressor at 50.7 psia, and (*b*) air leaves combustion chamber at 1120°F. Calculate the thermal efficiency of the Brayton cycle.

**16-32** An air-standard gas turbine operates on the Brayton cycle between pressure limits of 14.7 and 67 psia. The inlet-air temperature to the compressor is 60°F, and the air entering the turbine is at a temperature of 1240°F. The net work and efficiency are 79.3 Btu/lb and 34.8 percent, respectively. The upper pressure limit is now raised to 94 psia. Calculate the net work output and the efficiency of the cycle at the new pressure ratio, for the temperature values of the original cycle. Use air table A-5.

**16-33** The pressure ratio of an air-standard Brayton cycle is 4, and the inlet conditions to the compressor are 15 psia and 60°F. The turbine is limited to a temperature of 1540°F, and the flow rate is 10 lb/s. Determine (*a*) the compressor work, in Btu/lb, (*b*) the turbine work, and (*c*) the thermal efficiency.

**16-34** Prove that the maximum net work for a simple Brayton cycle, for fixed compressor- and turbine-inlet temperatures, occurs when $T_2 = (T_1 T_3)^{1/2}$.

**16-35** Show that the pressure ratio which yields the largest net work for a simple Brayton cycle with fixed inlet-compressor and turbine temperatures is given by $P_2/P_1 = (T_3/T_1)^{k/[2(k-1)]}$.

**16-36** The inlet conditions for an air-standard Brayton cycle are 15 psia and 60°F. The inlet-gas temperature to the turbine is limited to 1540°F. What pressure ratio should be employed in the ideal compressor so that the cycle produces the maximum work output, assuming (*a*) variable *k* data and (*b*) constant *k* data.

**16-37** Consider an ideal gas-turbine cycle. The maximum allowable inlet temperature to the turbine is 1400°F. Find the maximum pressure ratio for which this cycle will just deliver power if the inlet-compressor temperature is 80°F.

    (*a*) Use the air table.

    (*b*) Let *k* = 1.40 = constant.

**16-38** An ideal gas-turbine-power plant delivers 10,000 hp to an electric generator. The maximum and minimum cycle temperatures are 1540 and 80°F, respectively, and the pressure ratio is 5. Determine (*a*) the fraction of the gross power output used to drive the generator, (b) the mass flow rate of air to the compressor required, in lb/min, and (c) the thermal efficiency. Use the air table.

**16-39** An air-standard Brayton cycle compresses atmospheric air at 15 psia and 40°F by a factor of six. The turbine-inlet temperature is 1440°F, and the heated air is expanded back to 15 psia.
    (a) Determine the compressor work, the turbine work, and the amount of heat added, in Btu/lb.
    (b) Compute the thermal efficiency.

**16-40** A gas-turbine-power plant operates on an air-standard cycle between pressure limits of 14.7 and 94 psia. The inlet-air temperature to the compressor is 60°F, and the temperature limitation on the turbine is 1480°F. Calculate the new work output and the thermal efficiency if the cycle is ideal. Use the air table.

### Non-ideal simple gas-turbine power cycle

**16-41** Using the data of Prob. 16-31, calculate the thermal efficiency if the compressor- and turbine-adiabatic efficiencies are 83 and 86 percent, respectively.

**16-42** Using the data of Prob. 16-33, calculate the required quantities if the compressor and turbine efficiencies are 78 and 84 percent, respectively.

**16-43** A turbine operates with inlet conditions of 1000°R and 50 psia. If the outlet conditions are 900°R and 30 psia, what is the adiabatic-turbine efficiency?

**16-44** A gas-turbine cycle, with compressor- and turbine-inlet temperatures of 60°F and 1540°F, uses 4 : 1 pressure ratio. If the compressor efficiency is 78 percent, what minimum turbine efficiency will lead to zero work output?

**16-45** A gas-turbine power cycle operates with a pressure ratio of 12 : 1. The compressor and turbine efficiencies are 85 and 90 percent, respectively. The compressor-inlet temperature is 60°F, and the turbine-inlet temperature is 2200°F. For a mass flow rate of 1 lb/s, find the power generated by the cycle. Now double the pressure ratio. What is the power output in this case?

**16-46** Air enters a gas-turbine-power plant at 60°F and 14.5 psia. The compressor and turbine operate with a pressure ratio of 8, and because of fluid friction, the efficiency of each is 80 and 85 percent, respectively. Gases leave the combustor at 1540°F, and the mass flow rate of inlet air is 20 lb/s. Part of the turbine output is used to drive the compressor. Determine what percent of the total output of the turbine may be used to drive other than the compressor, on the basis of air-table calculations.

**16-47** A simple gas-turbine cycle operates with a pressure ratio of 10. The inlet temperatures to the compressor and turbine are 40 and 2000°F, respectively. The adiabatic efficiencies of the turbine and compressor are 0.88 and 0.82, respectively. Calculate (a) the net work, in Btu/lb and (b) the thermal efficiency.

**16-48** Using the data of Prob. 16-38, rework the problem if the adiabatic efficiencies of the compressor and turbine are 81 and 86 percent, respectively.

**16-49** Using the data of Prob. 16-39, rework the problem if the adiabatic efficiencies of the compressor and turbine are 80 and 85 percent, respectively.

**16-50** Using the data of Prob. 16-40, rework the problem if the adiabatic efficiencies of the compressor and turbine are 82 and 85 percent, respectively.

**16-51** A stationary gas-turbine-power plant producing 5000 hp operates under the following test data: (1) entering compressor, 14.25 psia and 60°F; (2) leaving compressor, 64.80 psia and 380°F; (3) entering turbine, 61.0 psia and 1340°F; (4) leaving turbine, 14.5 psia and 860°F.
    (a) Compute the adiabatic-compressor efficiency.
    (b) Compute the turbine efficiency.
    (c) Find the gross output of the turbine.

### Regenerative gas turbine cycle

**16-52** Using the data from Prob. 16-42, determine the thermal efficiency of the cycle if a regenerator is used with an effectiveness of (a) 100 percent, (b) 80 percent, and (c) 60 percent.

**16-53** Using the data of Prob. 16-41, compute the thermal efficiency if a regenerator is installed with an effectiveness of (a) 80 percent and (b) 65 percent.

**16-54** Using the data of Prob. 16-48, determine the thermal efficiency if a regenerator is added with an effectiveness of 70 percent.

**16-55** Using the data of Prob. 16-50, determine the thermal efficiency and the percentage of the fuel costs saved if a regenerator with 50 percent effectiveness were to be installed.

**16-56** Using the data from Prob. 16-49, determine the thermal efficiency if a regenerator with 70 percent effectiveness were to be installed.

**16-57** In an air-standard gas-turbine cycle with regeneration, the compressor is driven directly by the turbine. The following enthalpy data were taken on a test of a gas turbine with a pressure ratio of 5.41. Determine the thermal efficiency of the actual cycle and the regenerator effectiveness.

| System | Entering | Leaving |
|---|---|---|
| Compressor | 124.3 | 216.3 |
| Regenerator | 216.3 | 266.4 |
| Combustor | 266.4 | 449.7 |
| Turbine | 449.7 | 306.7 |
| Regenerator | 306.7 | 256.6 |

**16-58** An automotive manufacturer is contemplating the use of a regenerative-type gas turbine for the power source of a new model. Fresh air enters the turbine compressor at 60°F and 1 atm, with a flow rate of 2.2 lb/s, and is compressed at a ratio of 4 : 1. Air from the compressor enters the regenerator, where it is circulated until its temperature is approximately 1000°F. It then passes to the combustion chamber, where it is heated to 1700°F. After expanding through the two-stage turbine, the hot gases pass through the other half of the regenerator and are then exhausted to the atmosphere. Assuming ideal compression and expansion and negligible pressure losses through the burner and regenerator, determine (a) the thermal efficiency of the unit, (b) the power output, in horsepower, (c) the regenerator effectiveness, and (d) the exhaust stream temperature, in °F.

**16-59** Air enters the compressor of a gas turbine cycle at 15 psia and 40°F. The compressor and turbine are ideal, and the pressure ratio is 6.2 : 1. A regenerator is used which has an effectiveness of 70 percent. The gas stream leaving the compressor is heated 225°F as it passes through the regenerator. Determine the turbine-inlet temperature, in °F.

**16-60** With reference to Prob. 16-51, compute the percentage of fuel saved if a regenerator of 50 percent effectiveness is installed.

**Polytropic processes**

**16-61** Air is compressed in a piston-cylinder device from 14 psia, 40°F, to a pressure of 70 psia. Calculate the work required, in Btu/lb, if the process is (a) polytropic with $n = 1.30$, (b) isentropic, and (c) isothermal.

**16-62** Work Prob. 16-61 with the same data except the final pressure is 84 psia, and list the answers in $ft/lb_f/lb_m$.

**16-63** Air is compressed polytropically in steady flow, with $n = 1.35$, from 15 psia, 100°F, to 75 psia at a rate of 2 lb/s. Calculate (a) the required power input, and (b) the amount of heat removed per pound.

**16-64** Air is compressed in a steady-flow polytropic process from 15 psia, 40°F, to 90 psia. If the polytropic constant is 1.30, calculate the work required and the amount of heat transferred.

**16-65** An air compressor operates between an inlet condition of 15 psia and 60°F and an outlet pressure of 75 psia. Determine the work input required for the following reversible processes: (a) isothermal, (b) $n = 1.30$, and (c) adiabatic.

**16-66** The work output of a polytropic process involving air is twice the value of the heat input. The initial temperature and pressure are 540°R and 30 psia, respectively, and it is assumed that the specific-heat values are constant. Determine the value of the polytropic constant $n$ for the process, and the shaft work developed in ft·lb$_f$/lb$_m$ if the process is steady-flow and the exhaust pressure is 15 psia.

**16-67** A reciprocating air compressor is designed to have a displacement volume of 0.9 ft$^3$ and a clearance volume of 0.1 ft$^3$. If the cycle is adiabatic, determine the overall pressure ratio which will reduce the mass flow through the compressor to zero.

**16-68** Find the compressor horsepower requirements if 220 ft$^3$/min of air at 14 psia and 50°F are compressed polytropically ($n = 1.3$) to 112 psia. Cooling water runs through the compressor jacket at a rate of 28.5 lb/min and has a temperature rise of 10°F. The air velocity at inlet is small, but it is 500 ft/s at the outlet.

### Two-stage compression with intercooling

**16-69** The intake conditions of a two-stage air compressor are 14.5 psia and 80°F. The outlet pressure is 58 psia, and the pressure ratio across each stage is the same. If an intercooler cools the air to the initial inlet temperature and the stages are isentropic, determine the work required per pound. Compare with the work required for a single-stage compressor.

**16-70** Ten pounds of air per second are compressed isentropically from 15 psia and 80°F to 84 psia.

(a) Using air table A-5, determine the horsepower requirements for a single-stage compressor.

(b) Using the same data, compute the minimum horsepower requirements for two-stage compression when an intercooler cools the air to the initial temperature after the first stage.

**16-71** The intake conditions of a two-stage steady-flow compressor are 14 psia and 80°F. The outlet pressure is 140 psia, and the pressure ratio across each stage is the same. If an intercooler cools the air to the initial inlet temperature and the stages are isentropic, (a) determine the work required per pound, and (b) compare to the work required for a single-stage compressor.

**16-72** Same as Prob. 16-71, except that the inlet conditions are 16 psia and 100°F, and the final outlet pressure is 96 psia.

### Gas-turbine cycle with intercooling and reheating

**16-73** Reconsider the data of Prob. 16-55. Determine the net work output and the thermal efficiency if two-stage compression and expansion are used with a regenerator of 50 percent effectiveness. The compressor and turbine efficiencies are given in Prob. 16-50. The intercooling and reheating are ideal.

**16-74** Reconsider Prob. 16-52. Determine the net work output and the thermal efficiency if two-stage compression and expansion are used with a regenerator effectiveness of (a) 60 percent, and (b) 80 percent. The compressor and turbine efficiencies are given in Prob. 16-42. The intercooling and reheating are ideal.

**16-75** Reconsider Prob. 16-53. Determine the net work output and the thermal efficiency if two-stage compression and expansion are used with a regenerator effectiveness of (a) 80 percent, and (b) 65 percent. The compressor and turbine efficiencies are given in Prob. 16-41. The intercooling and reheating are ideal.

**16-76** Reconsider Prob. 16-54. Determine the net work output and the thermal efficiency if two-stage compression and expansion are used with a regenerator effectiveness of 70 percent. The compressor and turbine efficiencies are given in Prob. 16-48. The intercooling and reheating are ideal.

**16-77** A gas-turbine cycle operates with two stages of compression and two stages of expansion. The pressure ratio across each stage is 2. The inlet temperature and pressure to the compressor are 60°F and 15 psia, and the inlet turbine temperature to each stage is 1540°F. The compressor operates with an ideal intercooler, and an ideal regenerator also is used. Determine the compressor work, the turbine work, and the thermal efficiency.

**16-78** Using the data for Prob. 16-77, calculate the same required quantities if the compressor and turbine efficiencies are 78 and 84 percent, respectively, and the regenerator has an effectiveness of 70 percent.

**16-79** A gas-turbine-power plant employs two-stage compression and expansion, with intercooling, reheating, and regeneration. The temperature at the outlet of the compressor second stage is 700°R, and the combustor-inlet temperature is 1360°R. The turbine-inlet temperature is limited to 2100°R.

(a) Determine the total maximum work output from the two-stage turbine if the overall pressure ratio is 6:1.

(b) Determine the regenerator effectiveness.

**16-80** Consider the data for Prob. 16-33, but employ two-stage compression and two-stage expansion with intercooling and reheating. If the pressure ratio across each stage is 2, determine (a) the compressor work, (b) the turbine work, and (c) the thermal efficiency.

### Ericsson cycle

**16-81** An Ericsson cycle operates on air, and the minimum volume of the cycle is 1 ft³. The maximum pressure is 100 psia, and at the end of the constant pressure expansion process the volume is 2 ft³. Draw a $PV$ diagram, and compute, for each 100 Btu of heat added, the amount of heat that must be rejected.

**16-82** Same as Prob. 16-81, except that the minimum volume is 0.2 ft³, the maximum pressure is 120 psia, and the volume at the end of constant-pressure expansion is 0.44 ft³. The amount of heat added is 24 Btu.

### Turbojet cycles

**16-83** Rework Example 16-9 in the text under the following conditions: (1) the actual pressure rise in the diffuser is 92 percent of theoretical, (2) the compressor efficiency is 82 percent, (3) the turbine efficiency is 86 percent, and (4) the nozzle efficiency is 95 percent. Determine (a) the pressures and temperatures throughout the cycle, (b) the compressor work input, and (c) the exit-jet velocity, in ft/s.

**16-84** The air speed of a turbojet aircraft is 900 ft/s in still air at 4 psia and −60°F. The compressor pressure ratio is 9, and the maximum temperature at the turbine inlet is 1740°F. Assume ideal performance of the various components. Determine (a) the temperatures and pressures throughout the cycle, (b) the compressor work input required, and (c) the exit-jet velocity, in ft/s.

**16-85** Rework Prob. 16-84 under the following conditions: (1) the actual pressure rise in the diffuser is 90 percent of theoretical, (2) the compressor efficiency is 83 percent, (3) the turbine efficiency is 87 percent, and (4) the nozzle efficiency is 96 percent. Determine the quantities specified in Prob. 16-84.

**16-86** The air speed of a jet aircraft is 800 ft/s in still air at 8 psia and 0°F. The pressure ratio across the compressor is 7, and the turbine-inlet temperature is 1640°F. Assume ideal performance of the various components. Determine (a) the compressor work input required, in Btu/lb, and (b) the exit-jet velocity, in ft/s.

**16-87** Rework Prob. 16-86 under the following conditions: (1) the actual pressure rise in the diffuser is 92 percent of theoretical, (2) the compressor efficiency is 84 percent, (3) the turbine efficiency is 87 percent, and (4) the nozzle efficiency is 94 percent.

**16-88** The nozzle of a turbojet engine receives gas from the turbine at 35 psia and 1240°F and with a velocity of 120 ft/s. The nozzle expands the gas adiabatically to 9 psia.

(a) Find the discharge velocity if the nozzle efficiency is 92 percent.

(b) The aircraft is flying at 800 ft/s in still air and air enters the intake at this speed and at a rate of 120 lb/s. What is the thrust of the engine?

**16-89** The nozzle of a turbojet engine receives gas from the turbine exhaust at 27 psia and 1300°F and with a velocity of 100 ft/s. The nozzle expands the gas adiabatically to 10.5 psia.

(a) Determine the discharge velocity if the nozzle efficiency is 90 percent.

(b) The engine is mounted on an aircraft flying at 600 mph (880 ft/s) and air enters the intake at this velocity and at a rate of 150 lb/s. What is the thrust of the engine?

**16-90** A simple gas-turbine cycle operates so that there is no net work output. The inlet pressure and temperature to the isentropic compressor are 15 psia and 60°F, and the compressor pressure ratio is 6. If the limiting temperature of the isentropic turbine is 1540°F, determine the turbine-outlet pressure and temperature.

**16-91** Consider a simple gas-turbine cycle with no net work output, as in a jet aircraft. The inlet pressure and temperature to the compressor are 15 psia and 80°F, and the pressure ratio is 7. The limiting temperature in the turbine is 1640°F. If the compressor and turbine efficiencies are 81 and 85 percent, respectively, find the turbine-outlet pressure and temperature.

**16-92** Same as Prob. 16-91, except that the compressor-inlet conditions are 10 psia and 0°F.

**16-93** Same as Prob. 16-90, except that the compressor and turbine adiabatic efficiencies are 80 and 85 percent, respectively.

## Closed gas-turbine cycles

**16-94** A closed-cycle gas-turbine-power plant operates with helium as the working fluid. The compressor and turbine efficiencies are each 85 percent. The compressor-inlet state is 80 psia and 120°F, while the turbine-inlet state is 180 psia and 1300°F. If the mass flow rate is 10 lb/s, determine (a) the net power output, in horsepower, and (b) the rate of heat removed from the cycle, in Btu/min.

**16-95** A closed-cycle gas-turbine-power plant operates with argon as the working fluid. The compressor and turbine efficiencies are 83 and 86 percent, respectively. Compressor-inlet conditions are 120 psia and 150°F, while the turbine-inlet state is 250 psia and 1400°F. If the mass flow rate is 15 lb/s, determine (a) the net power output, in horsepower, and (b) the rate of heat removed from the cycle, in Btu/min.

**16-96** A closed-cycle gas-turbine-power plant operates with argon as the working fluid. The compressor and turbine efficiencies are 0.86 and 0.84, respectively. The compressor-inlet conditions are 100 psia and 100°F, while the turbine-inlet conditions are 350 psia and 1220°F. If the net power output is 2500 hp, determine (a) the required mass flow rate in lb/s, (b) the horsepower output of the turbine, and (c) the rate of heat removed from the cycle and radiated to outer space, in Btu/min.

## Gas refrigeration cycles

**16-97** A reversed Brayton cycle with a pressure ratio of 3 is used to produce a refrigeration effect. The compressor-inlet temperature is 80°F, and the turbine-inlet temperature is 40°F.

(a) What are the maximum and minimum temperatures in the cycle in °F?

(b) What is the coefficient of performance?

**16-98** An air-standard closed-cycle Brayton refrigeraton compresses air at 15 psia and 500°R by a factor of 5. The turbine-inlet temperature is 650°R. Calculate the coefficient of performance for the cycle.

**16-99** An ideal, reversed Brayton refrigeration cycle operates with air entering the compressor at 14.5 psia and 20°F and leaving at 58 psia. Air enters the turbine at 80°F. Calculate (a) the coefficient of performance and (b) the mass flow rate of air required to produce 1 ton of refrigeration, in lb/min.

**16-100** A reversed Brayton refrigeration cycle operates with air between the pressure limits of 75 and 150 psia. The fluid enters the turbine at 80°F; it leaves the cold heat exchanger at 40°F. Determine (a) the coefficient of performance and (b) the mass flow rate required for a 1-ton unit.

**16-101** Using the data of Prob. 16-100, consider a heat exchanger placed in the cycle similar to that shown in Fig. 16-30. Assume that the fluid entering the turbine is precooled down to the compressor-inlet temperature. Determine (a) the turbine-outlet temperature, in °F and (b) the coefficient of performance.

**16-102** An air refrigerating machine operating between 50 and 200 psia is to be used to heat a building. The air temperature entering the compressor is 33°F, and the air temperature entering the expander is 96°F. If the expansion and compression are reversible and adiabatic, determine (a) the Btus input in work for each 1000 Btu received as heat, and (b) the coefficient of performance of the device.

**16-103** An ideal, reversed Brayton cycle operates with a pressure ratio of 3. Air enters the compressor at 45°F, and normally would enter the turbine at 80°F. However, the cycle is modified similarly to that shown in Fig. 16-30. The heat exchanger allows the air entering the turbine to be precooled down to the compressor-inlet temperature. Determine (a) the old and new turbine-outlet temperatures, in °F, and (b) the old and new values of the coefficient of performance.

**16-104** Rework Prob. 16-97 if the compressor- and turbine-adiabatic efficiencies are 85 and 88 percent, respectively.

**16-105** Rework Prob. 16-98 if the compressor- and turbine-adiabatic efficiencies are 84 and 87 percent, respectively.

**16-106** Rework Prob. 16-99 if the compressor- and turbine-adiabatic efficiencies are 84 and 88 percent, respectively.

**Stirling cycle**

**16-107** A Stirling cycle operates with air, and at the beginning of isothermal expansion the state is 740°F and 80 psia. The minimum pressure in the cycle is 20 psia, and at the end of isothermal compression the volume is 60 percent of the maximum volume. Determine (a) the thermal efficiency, and (b) the mean effective pressure of the cycle.

**16-108** A Stirling cycle operates with air, and at the beginning of isothermal compression the state is 240°F and 25 psia. The maximum pressure is 75 psia, and during isothermal expansion the volume increases by 40 percent. Determine (a) the thermal efficiency, and (b) the mean effective pressure of the cycle.

**16-109** A Stirling cycle operates with air, and at the beginning of the isothermal expansion the pressure is 60 psia and the temperature is 500°F. The thermal efficiency is 43.7 percent, and at the end of the isothermal compression the volume is two-thirds the maximum volume. Assuming constant specific heats of air, determine the mean effective pressure of the cycle.

**Combined gas-steam turbine cycle**

**16-110** A gas-turbine-power plant operates on a pressure ratio of 12 : 1 with compressor- and turbine-inlet temperatures of 80 and 2040°F, respectively. The adiabatic efficiencies of the compressor and turbine are 85 and 87 percent, respectively. The turbine exhaust, used as the energy source for a steam cycle, leaves the boiler at 400°F. The inlet conditions to the 85 percent efficient turbine in the steam cycle are 2000 psia and 1000°F. The condenser pressure is 1 psia and the pump is 75 percent efficient. Determine (a) the required heat input, in Btu/lb air, (b) the mass-flow-rate ratio of air to steam, (c) the net work output of the gas-turbine cycle, Btu/lb of air, (d) the net work output of the steam cycle, Btu/lb of steam, and (e) the overall thermal efficiency of the combined cycle.

**16-111** Consider the data of Prob. 16-110. In addition, an open feedwater heater is operated in the steam cycle with steam bled from the turbine at 100 psia. For this modified cycle determine the quantities that were required in Prob. 16-110.

**16-112** A gas-turbine-power plant operates on a pressure ratio of 13 : 1 with compressor- and turbine-inlet temperatures of 60 and 2140°F, respectively. The adiabatic efficiencies of the compressor and turbine are 84 and 88 percent, respectively. The turbine exhaust, used as the energy source for a steam cycle, leaves the boiler at 440°F. The inlet conditions to the 86 percent efficient turbine in the steam cycle are 2500 psia and 1000°F. The condenser pressure is 0.8 psia and the pump is 70 percent efficient. Determine (a) the required heat input, in Btu/lb of air, (b) the mass-flow-rate ratio of air to steam, (c) the net work output of the gas-turbine cycle, in Btu/lb of air, (d) the net work output of the steam cycle, in Btu/lb of steam, and (e) the overall thermal efficiency of the combined cycle.

**16-113** Consider the data of Prob. 16-112. In addition, an open feedwater heater is operated in the steam cycle with steam bled from the turbine at 150 psia. For this modified cycle determine the quantities that were required in Prob. 16-112.

**16-114** A gas-turbine-power plant operates on a 10 : 1 pressure ratio with compressor- and turbine-inlet temperatures of 80 and 1940°F, respectively. The adiabatic efficiencies of the compressor and turbine are 84 and 87 percent, respectively. The turbine exhaust, used as the energy source for a steam cycle, leaves the boiler at 380°F. The inlet conditions to the 86 percent efficient turbine in the steam cycle are 1600 psia and 900°F. The condenser pressure is 1 psia and the pump is 70 percent efficient. Determine (a) the required heat input, in Btu/lb of air, (b) the mass-flow-rate ratio of air to steam, (c) the net work output of the gas-turbine cycle, in Btu/lb of air, (d) the net work output of the steam cycle, in Btu/lb of steam, and (e) the overall thermal efficiency of the combined cycle.

# SEVENTEEN

## VAPOR CYCLES

Two large and basic industries of interest to the engineer are electric-power generation and refrigeration. A major portion of the electricity generated commercially is through the use of steam-power plants. These power plants operate essentially on the same basic cycle, whether the input energy is from the combustion of fuels or from a fission process in a nuclear reactor. Steam-power cycles and refrigeration cycles have one feature in common. The substance which undergoes a cyclic process appears in both the vapor and liquid phases during the process. Hence these cycles are excellent examples of the application of thermodynamic principles presented earlier. To a large degree the following discussion deals with idealized cycles. More comprehensive analyses of vapor cycles may be obtained from standard textbooks on these subjects.

## 17-1 THE RANKINE CYCLE

The model cycle for a steam-power plant is called the Rankine cycle. It is a modification of a Carnot cycle, which uses a vapor as its working fluid. Consider a Carnot engine which operates with steam through a cycle illustrated by Fig. 17-1. Wet steam at state 1 is compressed isentropically to saturated liquid at state 2. At this elevated pressure, energy is added at constant pressure until the water is completely evaporated to saturated vapor at state 3. It is then allowed to expand isentropically through a turbine to state 4. The wet steam leaving the steam turbine is then partially condensed at constant pressure back to state 1. The

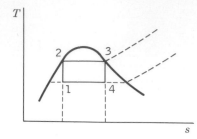

**Figure 17-1** Carnot cycle using a vapor as the working substance.

thermal efficiency of the cycle is, of course, the highest for any engine operating between temperatures $T_1$ and $T_2$, and is given by $(T_2 - T_1)/T_2$. The Carnot cycle is impractical to use, however, with fluids which undergo phase changes. For example, it is difficult, first of all, to compress a two-phase mixture isentropically, as required by process 1-2. Second, the condensing process 4-1 would have to be controlled very accurately to end up with the desired quality of state 1. Third, the efficiency of a Carnot cycle is greatly affected by the temperature $T_2$ at which energy is added. For steam, the critical temperature is only 705°F. Therefore, if the cycle is to be operated within the wet region, the maximum possible temperature is severely limited.

In part, these objections can be eliminated by a slight modification of the basic Carnot vapor cycle discussed above. Instead of condensing from state 4 to a low-quality vapor, the condensation process is carried out so that the wet steam leaving the turbine is condensed to saturated liquid at the turbine-outlet pressure. The compression process is now handled by a liquid pump, which isentropically compresses the liquid leaving the condenser to the pressure desired in the heat-addition process. The heat added may come from the combustion of conventional fuels or be supplied from a nuclear reactor. The ideal cycle of a simple *Rankine* steam power cycle then consists of:

1. Isentropic compression in a pump
2. Constant-pressure heat addition in a boiler
3. Isentropic expansion in a turbine
4. Constant-pressure heat removal in a condenser

The basic cycle is presented schematically and on a $Ts$ diagram in Fig. 17-2. If the changes in potential and kinetic energies can be neglected, the heat transfer to the fluid in the boiler is represented on the $Ts$ diagram by the area enclosed by states 2-2'-3-b-a-2. The area enclosed by states 1-4-b-a-1 then represents the heat removed from the fluid in the condenser. The first law for a closed cyclic process indicates that the net heat effect equals the net work effect. Hence the net work is represented by the difference in the areas for the heat input and heat rejection, i.e., area 1-2-2'-3-4-1. The thermal efficiency for the cycle is again defined as $W_{net}/Q_{in}$.

Expressions for the work and heat interactions in the ideal cycle are found by applying the steady-flow energy equation to each separate piece of equipment. If

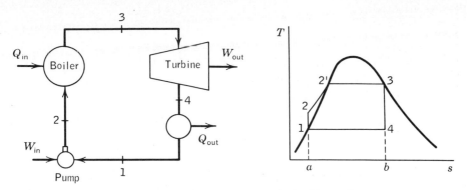

**Figure 17-2** The simple steam-power plant operating on the Rankine cycle.

we may neglect kinetic- and potential-energy changes, the basic equation for each process reduces to $q + w = h_2 - h_1$. The isentropic pump work is given by

$$w_{in, pump} = h_2 - h_1$$

However, the isentropic pump work may also be computed from the steady-flow mechanical-work equation

$$w = \int v \, dP \qquad (7\text{-}25)$$

Since the change in the specific volume of liquid water from the saturation state to pressures in the compressed-liquid state normally encountered in steam-power plants is less than 1 percent, the fluid in the pump may be considered incompressible. Consequently, the pump work is frequently determined within the desired accuracy from the relation

$$w_{in, pump} = v_f(P_2 - P_1) \qquad s_1 = s_2$$

where $v_f$ is the saturated-liquid specific volume at state 1. The heat input, the isentropic work output from the turbine, and the heat rejection in the condenser, all on a unit-mass basis, are

$$q_{in, boiler} = h_3 - h_2 \qquad P_3 = P_2$$

$$w_{out, turbine} = h_3 - h_4 \qquad s_3 = s_4$$

$$q_{out, condenser} = h_4 - h_1 \qquad P_4 = P_1$$

The thermal efficiency of an ideal Rankine cycle then may be written as

$$\eta_{th} = \frac{w_T - w_P}{q_{in}} = \frac{(h_3 - h_4) - v_{f,1}(P_2 - P_1)}{h_3 - h_2}$$

Typical examples are given below.

**Example 17-1M** Compute the thermal efficiency of an ideal Rankine cycle for which steam leaves the boiler as saturated vapor at 30 bars and is condensed at 1.0 bar.

SOLUTION The evaluation of the thermal efficiency requires calculating the turbine and pump work and the heat input. Using the notation employed in Fig. 17-2, we find the following data from the steam tables:

$$h_3 = h_g \text{ at } 30 \text{ bars} = 2804.2 \text{ kJ/kg}$$

$$h_1 = h_f \text{ at } 1.0 \text{ bar} = 417.5 \text{ kJ/kg}$$

The pump work is

$$w_{\text{in, pump}} = w_P = \frac{1.04 \text{ cm}^3}{g} \times 29.0 \text{ bars} \times \frac{10^5 \text{ N}}{\text{bar·m}^2} \times \frac{\text{m}^3}{10^6 \text{ cm}^3}$$

$$= 3.0 \text{ kJ/kg}$$

Therefore $\qquad h_2 = h_1 + w_P = 417.5 + 3.0 = 420.5 \text{ kJ/kg}$

The enthalpy at state 4 is the value for a pressure of 1.0 bar and for $s_4 = s_3 = s_g$ at 25 bars. Since $s_3 = 6.2575 \text{ kJ/(kg)(°K)}$, $6.2575 = (s_f + xs_{fg})_{\text{at 1.0 bar}} = 1.3026 + x(7.3594 - 1.3026)$

or $\qquad\qquad x = \dfrac{4.9549}{6.0568} = 0.818 \qquad \text{(or 81.8 percent)}$

Then the turbine-outlet enthalpy can be found.

$$h_4 = h_f + xh_{fg} = 417.5 + 0.818(2258.0) = 2264.7 \text{ kJ/kg}$$

Consequently,

$$w_{\text{T, out}} = h_3 - h_4 = 2804.2 - 2264.7 = 539.5 \text{ kJ/kg}$$

$$q_{\text{in}} = h_3 - h_2 = 2804.2 - 420.0 = 2384.2 \text{ kJ/kg}$$

The thermal efficiency is

$$\eta_{\text{th}} = \frac{539.5 - 3.0}{2384.2} = 0.225 \qquad \text{(or 22.5 percent)}$$

**Example 17-1** Compute the thermal efficiency of an ideal Rankine cycle for which steam leaves the boiler as saturated vapor at 400 psia and is condensed at 14.7 psia.

SOLUTION The evaluation of the thermal efficiency requires calculating the turbine and pump work and the heat input. Using the notation employed in Fig. 17-2, we find the following data from the steam tables:

$$h_3 = h_g \text{ at } 400 \text{ psia} = 1205.5 \text{ Btu/lb}$$

$$h_1 = h_f \text{ at } 14.7 \text{ psia} = 180.2 \text{ Btu/lb}$$

The pump work is

$$w_{\text{in, pump}} = w_P = \frac{0.01614(400 - 14.7)(144)}{778} = 1.2 \text{ Btu/lb}$$

Therefore $\qquad h_2 = h_1 + w_P = 180.2 + 1.2 = 181.4 \text{ Btu/lb}$

The enthalpy at state 4 is the value for a pressure of 1 psia and for $s_4 = s_3 = s_g$ at 400 psia. Since $s_3 = 1.4856$ Btu/(lb)(°R),

$$1.4856 = (s_f + xs_{fg})_{\text{at } 14.7 \text{ psia}} = 0.3121 + x(1.4446)$$

or

$$x = \frac{1.1735}{1.4446} = 0.812 \quad \text{(or 81.2 percent)}$$

Then

$$h_4 = 180.2 + 0.812(970.4) = 968.2 \text{ Btu/lb}$$

Consequently,

$$w_{T, \text{ out}} = h_3 - h_4 = 1205.5 - 968.2 = 237.3 \text{ Btu/lb}$$

$$q_{\text{in}} = h_3 - h_2 = 1205.5 - 181.4 = 1024.1 \text{ Btu/lb}$$

The thermal efficiency is

$$\eta_{\text{th}} = \frac{237.3 - 1.2}{1024.1} = 0.231 \quad \text{(or 23.1 percent)}$$

It is interesting to note from the preceding examples that the pump work is a very small fraction of the turbine work. In the analysis of the gas-turbine cycle introduced in Chap. 16, it is found that the compressor consumes a large fraction of the turbine output. The ratio of work input to work output is known as the back work ratio, and for the ideal Rankine cycle analyzed above this ratio is less than 0.01. This major difference between a Rankine vapor-power cycle and a Brayton gas-turbine-power cycle is due to the difference between the work required to compress a liquid and the work needed to compress a gas. Since the work required to compress a fluid reversibly over a given increment of pressure is directly proportional to the specific volume of the fluid [Eq. (7-25)], it takes considerably less work to compress a liquid than to compress a gas under normal conditions.

The efficiency of the simple Rankine cycle discussed above may be increased, on the basis of the theoretical Carnot cycle, either by decreasing the temperature at which heat is rejected or by increasing the average temperature at which heat is added. The first of these effects is accomplished by lowering the exhaust pressure, which in turn lowers the value of $T_4$ (or $T_1$) shown in Fig. 17-2. This point is illustrated by Fig. 17-3. The crosshatched area enclosed by the states 1-2-2″-1″-4″-4-1 is a measure of the decrease in heat rejected when the exhaust

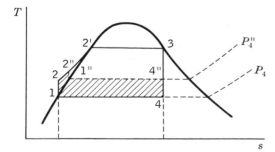

**Figure 17-3** Effect of turbine-exhaust pressure on the efficiency of an ideal Rankine cycle.

pressure is lowered from $P_4''$ to $P_4$. This area is also a measure of the increase in net work output. The increase in heat input is measured by the total area under the curve 2-2″. As a result of the lower exhaust pressure, both the quantity of heat input required and the net work output increase. Since these two quantities are about equal in magnitude, the thermal efficiency increases. This point is illustrated in the following examples.

**Example 17-2M** Compute the thermal efficiency of an ideal Rankine cycle for which the steam leaves the boiler as saturated vapor at 30 bars and is condensed to 0.10 bar.

SOLUTION The evaluation of the thermal efficiency is carried out in a manner similar to that shown in Example 17-1M. The notation is the same as that used in Fig. 17-3, where the important states are 1, 2, 3, and 4. A summary of the calculations follows:

$$h_3 = h_g \text{ at 30 bars} = 2804.2 \text{ kJ/kg}$$

$$h_1 = h_f \text{ at 0.10 bar} = 191.8 \text{ kJ/kg}$$

$$w_{in, pump} = v_f(P_2 - P_1) = 1.01(29.9)/10 = 3.0 \text{ kJ/kg}$$

$$h_2 = h_1 + w_P = 191.8 + 3.0 = 194.8 \text{ kJ/kg}$$

$$s_3 = 6.2575 = (s_f + xs_{fg})_{at\ 0.1\ bar} = 0.6493 + x(8.1502 - 0.6493)$$

$$x = \frac{5.6082}{7.5009} = 0.748 \quad \text{(or 74.8 percent)}$$

$$h_4 = h_f + xh_{fg} = 191.8 + 0.748(2392.8) = 1981.8 \text{ kJ/kg}$$

$$w_{T, out} = h_3 - h_4 = 2804.2 - 1981.8 = 822.4 \text{ kJ/kg}$$

$$q_{in} = h_3 - h_2 = 2804.2 - 194.8 = 2609.4 \text{ kJ/kg}$$

$$\eta_{th} = \frac{822.4 - 3.0}{2609.4} = 0.314 \quad \text{(or 31.4 percent)}$$

In comparison to Example 17-1M, the thermal efficiency has increased from 22.5 to 31.4 percent, and the turbine-work output has increased from 539.5 kJ/kg to 822.4 kJ/kg. This is accomplished by lowering the exhaust pressure from 1 to 0.1 bar, which is equivalent to lowering the exhaust temperature from 100 to 46°C.

**Example 17-2** Compute the thermal efficiency of an ideal Rankine cycle for which steam leaves the boiler as saturated vapor at 400 psia and is condensed at 1 psia.

SOLUTION The evaluation of the thermal efficiency requires calculating the turbine and pump work and the heat input. Using the notation seen in Fig. 17-3, we find the following data:

$$h_3 = h_g \text{ at 400 psia} = 1205.5 \text{ Btu/lb}$$

$$h_1 = h_f \text{ at 1 psia} = 69.7$$

The pump work is

$$w_{in, pump} = w_P = \frac{0.01614(400 - 1)(144)}{778} = 1.2 \text{ Btu/lb}$$

Therefore

$$h_2 = h_1 + w_P = 69.7 + 1.2 = 70.9 \text{ Btu/lb}$$

The enthalpy at state 4 is the value for a pressure of 1 psia and for $s_4 = s_3 = s_g$ at 400 psia. Since $s_3 = 1.4856$ Btu/(lb)(°R),

$$1.4856 = (s_f + x s_{fg})_{\text{at 1 psia}} = 0.1327 + x(1.8453)$$

or

$$x = \frac{1.3529}{1.8453} = 0.733 \qquad \text{(or 73.3 percent)}$$

Then

$$h_4 = 69.70 + 0.733(1036.0) = 829.3 \text{ Btu/lb}$$

Consequently,

$$w_{\text{T, out}} = h_3 - h_4 = 1205.5 - 829.3 = 376.2 \text{ Btu/lb}$$

$$q_{\text{in}} = h_3 - h_2 = 1205.5 - 70.85 = 1134.6 \text{ Btu/lb}$$

The thermal efficiency is

$$\eta_{\text{th}} = \frac{376.3 - 1.2}{1134.6} = 0.33 \qquad \text{(or 33 percent)}$$

In comparison to Example 17-1, the thermal efficiency has increased from 23.1 to 33 percent, and the turbine-work output has increased from 237.3 to 376.2 Btu/lb. This is accomplished by lowering the exhaust pressure from 14.7 to 1 psia, which is equivalent to lowering the exhaust temperature from 212 to 102°F.

The preceding examples illustrate the large effect of the exhaust pressure on the work output and thermal efficiency of a simple Rankine cycle. There is a limit, however, to the minimum pressure in the condenser. Heat is transferred from the condensing steam to cooling water or atmospheric air. The temperature of the cooling water or air normally would vary only over a very narrow range. Typically this might be 15 to 30°C, or 60 to 90°F. A temperature differential of 10 to 15°C, or 15 to 25°F, must exist across the heat-transfer surface in order to maintain adequate heat-transfer rates. Hence the minimum condensing temperature for the steam would range from 25 to 45°C or 75 to 115°F. From the saturated-steam tables, the saturation pressures corresponding to these ranges of temperature are roughly from 0.03 to 0.10 bar or 0.5 to 1.5 psia. Thus, fairly low pressures, well below atmospheric values, are possible in the condensers of modern steam-power plants. However, this pressure varies over a fairly small range.

Although the effect of lowering the exhaust pressure is advantageous from the standpoint of increasing the thermal efficiency, it has the great disadvantage of increasing the moisture content (decreasing the quality) of the fluid leaving the turbine. This increased moisture content throughout the turbine decreases the efficiency of an actual turbine. In addition, the impingement of liquid droplets on the turbine blades may lead to a serious erosion problem. In practice, it is desirable to keep the moisture content less than about 10 percent at the low-pressure end of the turbine. It was also mentioned previously that increasing the average temperature at which heat is supplied would increase the efficiency of the Rankine cycle. Both the increase of the efficiency of the cycle by raising the temperature of the fluid entering the turbine and the removal of the moisture problem in the turbine can be met by the addition of a superheater to the simple Rankine cycle

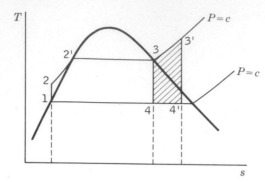

**Figure 17-4** Effect of superheating on the ideal Rankine cycle.

already presented. The process of superheating leads to a higher temperature at the turbine inlet without increasing the maximum pressure in the cycle. After the saturated vapor leaves the boiler, the fluid passes through another heat-input section, where the temperature is increased theoretically at constant pressure. The steam leaves the superheater at a temperature which is usually restricted only by metallurgical effects. Temperatures in the range of 540 to 600°C, or 1000 to 1100°F, are generally permissible. A $Ts$ diagram for a Rankine cycle with super-heating is shown in Fig. 17-4. The shaded area represents the additional net work output, and the area beneath the curve 3-3' represents the additional heat added in the superheater section. Note that the average temperature at which heat is added during process 3-3' is greater than that during the heat-addition process in the boiler section, whereas the temperature of rejection is still the same. On the basis of a Carnot-engine analysis, the efficiency of the cycle would be expected to have increased. The other important point to note is that the quality at state 4' is considerably higher than that at state 4. Hence the moisture problem in the turbine has been alleviated.

**Example 17-3M** Determine the thermal efficiency and the change in the steam quality leaving the turbine of an ideal Rankine cycle with superheating, for which the turbine-inlet conditions are 30 bars and 500°C and the condenser pressure is 0.1 bar.

SOLUTION This example contains the same pressure conditions as Example 17-2M. However, the steam entering the turbine has been superheated from the saturation state of 234 to 500°C. The pump work will remain the same, i.e., 3.0 kJ/kg, but all other heat and work quantities are altered. The turbine-inlet enthalpy is now 3456.5 kJ/kg, and the entropy at this state is 7.2338 kJ/(kg)(°K). Hence, using $h_2$ from the preceding example, we find that

$$q_{in} = h_3' - h_2 = 3456.5 - 194.8 = 3261.7 \text{ kJ/kg}$$

where the notation of Fig. 17-4 is being used. Calculation of the turbine work requires that the final isentropic state 4' be found first. Since $s_3' = s_4' = 7.2338$ kJ/(kg)(°K),

$$7.2338 = (s_f + xs_{fg})_{\text{at } 0.1 \text{ bar}} = 0.6493 + x(7.5009)$$

or

$$x = \frac{6.5845}{7.5009} = 0.878 \quad \text{(or 87.8 percent)}$$

$$h_4' = (h_f + xh_{fg})_{\text{at } 0.1 \text{ bar}} = 191.8 + 0.878(2392.8) = 2292.7 \text{ kJ/kg}$$

Therefore $\qquad w_{T, in} = h_3' - h_4' = 3456.5 - 2292.7 = 1163.8 \text{ kJ/kg}$

The thermal efficiency becomes

$$\eta_{th} = \frac{1163.8 - 3.0}{3261.7} = 0.356 \qquad \text{(or 35.6 percent)}$$

The result of superheating is an increase in the thermal efficiency from 31.4 to 35.6 percent, and an increase in the quality of the steam leaving the turbine from 74.8 to 87.8 percent. Although this latter value of the quality is still objectionable, it must be remembered that the above calculations are based on isentropic flow through the turbine. The presence of irreversibilities in the flow, although decreasing the work output of the turbine, will be found to increase the quality at the outlet.

**Example 17.3** Determine the thermal efficiency of an ideal Rankine cycle with superheating, for which the turbine-inlet conditions are 400 psia and 900°F and the condenser pressure is 1 psia.

SOLUTION This example contains the same pressure conditions as Example 17-2. However, the steam entering the turbine has been superheated from the saturation state of 444.6 to 900°F. The pump work will remain the same, i.e., 1.2 Btu/lb, but all other heat and work quantities are altered. The turbine-inlet enthalpy is now 1470.1 Btu/lb, and the entropy in this state is 1.7252 Btu/(lb)(°R). Hence, using $h_2$ from the preceding example, we find that

$$q_{in, tot} = h_3' - h_2 = 1470.1 - 70.9 = 1399.2 \text{ Btu/lb}$$

Calculation of the turbine work requires that the final isentropic state at 4′ be found first. Since $s_3' = s_4' = 1.7252 \text{ Btu/(lb)(°R)}$,

$$1.7252 = (s_f + x s_{fg})_{\text{at 1 psia}} = 0.1327 + x(1.8453)$$

or $\qquad x = 0.863 \qquad \text{(or 86.3 percent)}$

$$h_4' = (h_f + x h_{fg})_{\text{at 1 psia}} = 69.7 + 0.863(1036.0) = 964.0 \text{ Btu/lb}$$

Therefore $\qquad w_{T, in} = h_3' - h_4' = 1470.1 - 964.0 = 506.1 \text{ Btu/lb}$

The thermal efficiency becomes

$$\eta_{th} = \frac{506.1 - 1.2}{1399.2} = 0.361 \qquad \text{(or 36.1 percent)}$$

The result of superheating is an increase in the thermal efficiency from 33 to 36 percent and an increase in the quality of steam leaving the turbine from 73.3 to 86.3 percent. Although this latter quality is still objectionable, it must be remembered that the calculations above are based on isentropic flow through the turbine. The presence of irreversibilities in the flow, although decreasing the work output of the turbine, will be found to increase the outlet enthalpy even further, thus eliminating the moisture problem to a large extent.

## 17-2 THE REHEAT CYCLE

In the ideal Rankine cycle the efficiency may be increased by the use of a super-heater section. The process of superheating in general raises the average temperature at which heat is supplied to the cycle, thus raising the theoretical efficiency. An equivalent gain in the average temperature during the heat-input process may be accomplished by raising the maximum pressure of the cycle, i.e., the boiler

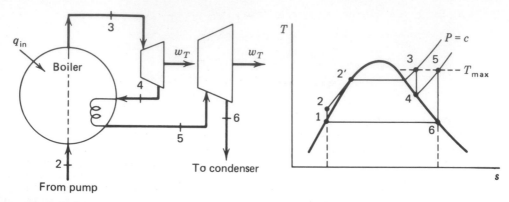

**Figure 17-5** The ideal reheat cycle.

pressure. This may result in a higher initial cost of the steam generator (boiler plus superheater), because of the higher pressure that must be contained, but over a period of years the higher efficiency of the overall unit may more than compensate for this factor. However, for a given maximum temperature in the steam generator, an increase in the generator pressure results in a decrease in the quality of the steam leaving the turbine. To avoid the erosion problem and still take advantage of the higher temperatures made available to increasing the boiler pressure, the reheat cycle has been developed.

In the reheat cycle the steam is not allowed to expand completely to the condenser pressure in a single stage. After partial expansion the steam is withdrawn from the turbine and reheated at constant pressure. Then it is returned to the turbine for further expansion to the exhaust pressure. The turbine may be considered to consist of two stages, a high-pressure one and a low-pressure one. Figure 17-5 illustrates the reheat cycle on a $Ts$ diagram. Since the temperature of the reheat process 4-5 is not much different from that of the heat-addition process 2-3, the thermal efficiency is not greatly affected. Nevertheless, the reheat cycle does remove the objectionable feature of high moisture content at the turbine exhaust. In computing the thermal efficiency of a reheat cycle, one must remember to account for the work output of both the high-pressure and low-pressure turbines, as well as the heat inputs to both the boiler-superheater section and the reheat section.

**Example 17-4M** In a steam-power plant utilizing the reheat cycle, the turbine-inlet condition is 30 bars and 500°C. After expansion to 5.0 bar, the steam is reheated to 500°C and then expanded to the condenser pressure of 0.1 bar. Compute the cycle efficiency and the state of the steam at the outlet of the turbine.

SOLUTION The problem is based on the data of Example 17-3M, except for the addition of the reheat section. From the preceding example, the following data are still valid. The subscripts on the property values are consistent with the notation in Fig. 17-5.

$$h_1 = 191.8 \text{ kJ/kg}$$

$$h_2 = 194.8 \text{ kJ/kg}$$

$$h_3 = 3456.5 \text{ kJ/kg}$$

$$w_P = h_2 - h_1 = 3.0 \text{ kJ/kg}$$

$$q_{\text{boiler-superheater}} = h_3 - h_2 = 3261.7 \text{ kJ/kg}$$

$$s_3 = 7.2338 \text{ kJ/(kg)(}^\circ\text{K)}$$

From the last value we can estimate the enthalpy at state 4, where the pressure is 5 bars. The entropy of saturated vapor at 5 bars is 6.8213 kJ/(kg)(°K). Therefore state 4 is still in the super-heat region. From superheat table A-14M it is found by linear interpolation that $T_4$ is close to 241°C and that $h_4$ is 2941.6 kJ/kg. State 5 is at 5 bars and 500°C, for which $h_5 = 3483.9$ kJ/kg and $s_5 = 8.0873$ kJ/(kg)(°K). Finally, state 6 is the result of isentropic expansion from state 5 to 0.1 bar. From the saturation table at 0.1 bar,

$$8.0873 = (s_f + x s_{fg})_{\text{at 0.1 bar}} = 0.6493 + x(8.5009)$$

or

$$x = \frac{8.0873 - 0.6493}{7.5009} = 0.992 \quad \text{(or 99.2 percent)}$$

Hence

$$h_6 = h_f + x h_{fg} = 191.8 + 0.992(2392.8) = 2565.5 \text{ kJ/kg}$$

Consequently,

$$w_{T,\text{ high pressure}} = h_3 - h_4 = 3456.5 - 2941.6 = 514.9 \text{ kJ/kg}$$

$$w_{T,\text{ low pressure}} = h_5 - h_6 = 3483.9 - 2565.5 = 918.4 \text{ kJ/kg}$$

$$q_{\text{reheat}} = h_5 - h_4 = 3483.9 - 2941.6 = 542.3 \text{ kJ/kg}$$

The thermal efficiency, then, is

$$\eta_{\text{th}} = \frac{514.9 + 918.4 - 3.0}{3261.7 + 542.3} = 0.376 \quad \text{(or 37.6 percent)}$$

Compared with Example 17-3M, without reheat, the thermal efficiency has increased only from 35.6 to 37.6 percent. However, the outlet quality of the steam at the turbine exhaust has increased from 87.8 to 99.2 percent.

**Example 17-4** In a steam-power plant utilizing the reheat cycle, the turbine-inlet condition is 400 psia and 900°F. After expansion to 60 psia, the steam is reheated to 900°F, and then expanded to the condenser pressure of 1 psia. Compute the cycle efficiency and the net work output per pound of steam.

SOLUTION The problem is based on the data of Example 17-3, except for the addition of the reheat section. From the preceding example, the following data are still valid. The subscripts on the property values are consistent with the notation in Fig. 17-5.

$$h_1 = 69.7 \text{ Btu/lb}$$

$$h_2 = 70.9 \text{ Btu/lb}$$

$$h_3 = 1470.1 \text{ Btu/lb}$$

$$w_P = h_2 - h_1 = 1.2 \text{ Btu/lb}$$

$$q_{\text{boiler-superheater}} = h_3 - h_2 = 1399.2 \text{ Btu/lb}$$

$$s_3 = 1.7252 \text{ Btu/(lb)(}^\circ\text{R)}$$

From the last value we can estimate the enthalpy at state 4, where the pressure is 60 psia. The entropy of saturated vapor at 60 psia is 1.6443 Btu/(lb)(°R). Therefore state 4 is still in the superheat region. From Table A-14 it is found that $T_4$ is close to 422°F and that $h_4$ is 1244.2 Btu/lb. State 5 is at 60 psia and 900°F, for which $h_5 = 1481.8$ Btu/lb and $s_5 = 1.9408$ Btu/(lb)(°R). Finally, state 6 is the result of isentropic expansion from state 5 to 1 psia. From the saturation tables at 1 psia,

$$1.9408 = s_f + xs_{fg} = 0.1327 + x(1.8453)$$

or
$$x = 0.980 \quad \text{(or 98 percent)}$$

Hence
$$h_6 = h_f + xh_{fg} = 69.7 + 0.98(1036.0) = 1085.0 \text{ Btu/lb}$$

Consequently,

$$w_{T,\text{ high pressure}} = h_3 - h_4 = 1470.1 - 1244.2 = 225.9 \text{ Btu/lb}$$

$$w_{T,\text{ low pressure}} = h_5 - h_6 = 1481.8 - 1085.0 = 396.8 \text{ Btu/lb}$$

$$q_{\text{reheat}} = h_5 - h_4 = 1481.8 - 1244.2 = 237.6 \text{ Btu/lb}$$

The thermal efficiency, then, is

$$\eta_{\text{th}} = \frac{225.9 + 396.8 - 1.2}{1399.2 + 237.6} = 0.379 \quad \text{(or 37.9 percent)}$$

Compared to Example 17-3, without reheat, the efficiency has increased only from 36.1 to 37.9 percent. However, the outlet quality of the steam at the turbine exhaust has increased from 86.3 to 98 percent.

## 17-3 THE REGENERATIVE CYCLE

Referring to Fig. 17-4 for the simple Rankine cycle with superheating reveals a serious disadvantage of the basic cycle. For the portion of the heat-addition process 2-2', the average temperature is much below the temperature of the vaporization and superheating process 2'-3-3'. From the viewpoint of the second law, the cycle efficiency is greatly reduced as a result of this relatively low-temperature heat-addition process. If the average temperature for this portion of the heat-addition process could be raised, the efficiency of the cycle would more nearly approach that of the Carnot cycle. One practical method of accomplishing this is by the use of a regeneration process internal to the overall cycle.

The ideal regeneration vapor-power cycle, shown in Fig. 17-6, is accomplished in the following way: Part of the superheated steam which enters the turbine at state 3 is bled or extracted from the turbine at state 4, which is an intermediate state in the turbine expansion process. The extracted steam is directed into a feedwater heater. The portion of the steam which is not extracted expands completely to the condenser pressure (state 5), and it is then condensed to saturated liquid at state 6. A pump then increases the pressure of the liquid leaving the condenser isentropically to the same pressure as that of the extracted steam. The compressed liquid at state 7 then enters the feedwater heater, where it mixes directly with the flow stream extracted from the turbine. Because of this direct-mixing process, the feedwater heater in Fig. 17-6 is called an *open* or *direct-*

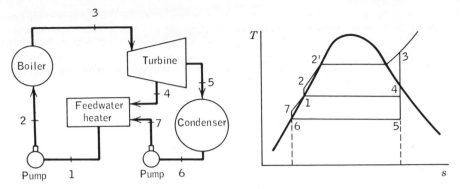

**Figure 17-6** Ideal regenerative cycle with one open feedwater heater.

*contact* type of heater. In the ideal situation the mass flow rates for the two streams entering the heater are adjusted so that the state of the mixture leaving the heater is a saturated liquid at the heater pressure (state 1). A second pump then isentropically raises the pressure of the liquid to state 2, which corresponds to the steam-generator pressure.

For the ideal, internally reversible cycle the areas beneath the curves on the $Ts$ diagram in Fig. 17-6 represent heat-addition or heat-rejection quantities. One must keep in mind, however, that the mass flow rates through certain parts of the cycle are not equal to the total mass flow rate through the steam generator. For example, the area under line 5-6 is a measure of the heat rejected in the condenser, but only for each pound which passes through the condenser. Before presenting a numerical example for an ideal regenerative cycle, it is convenient to employ the conservation of mass and energy principles to a control volume which lies just inside the open feedwater heater. On the basis of Fig. 17-6 we may write

$$\dot{m}_1 = \dot{m}_4 + \dot{m}_7$$

and
$$\dot{m}_1 h_1 = \dot{m}_4 h_4 + \dot{m}_7 h_7$$

Combining the two equations so that $\dot{m}_7$ is eliminated, one observes that

$$\dot{m}_1 h_1 = \dot{m}_4 h_4 + (\dot{m}_1 - \dot{m}_4)h_7$$

or
$$h_1 = \frac{\dot{m}_4}{\dot{m}_1} h_4 + \left(1 - \frac{\dot{m}_4}{\dot{m}_1}\right) h_7$$

If the fraction of the steam extracted from the turbine $\dot{m}_4/\dot{m}_1$ is represented by $y$, then

$$h_1 = yh_4 + (1 - y)h_7$$

or
$$y = \frac{h_1 - h_7}{h_4 - h_7} \tag{17-1}$$

Since the enthalpies of the flow streams entering and leaving the open feedwater heater are known for the ideal cycle, the above relation permits the evaluation of the quantity of steam which must be extracted from the turbine at state 4 to ensure that the flow stream leaving the heater will be a saturated liquid.

**Example 17-5M** An ideal regenerative steam-power cycle operates so that steam enters the turbine at 30 bars and 500°C and exhausts at 0.1 bar. A single, open feedwater heater is employed which operates at 5 bars. Compute the thermal efficiency of the cycle.

SOLUTION The $Ts$ and flow diagrams for the cycle are given in Fig. 17-6. From the preceding examples the following data are still valid:

$$P_3 = 30 \text{ bars}, \ T_3 = 500°C, \ h_3 = 3456.5 \text{ kJ/kg}$$

$$P_4 = 5 \text{ bars}, \ s_4 = s_3, \ h_4 = 2941.6 \text{ kJ/kg}$$

$$P_5 = 0.1 \text{ bar}, \ s_5 = s_3, \ h_5 = 2292.7 \text{ kJ/kg}$$

$$P_6 = 0.1 \text{ bar, saturated liquid}, \ h_6 = 191.8 \text{ kJ/kg}$$

In addition, state 1 is a saturated-liquid state at 5 bars; thus $h_1 = h_f$ at 5 bars = 640.2 kJ/kg. The enthalpies at states 2 and 7 can be found by adding the pump work $(w_P)$ to the enthalpies at states 1 and 6, respectively. The pump work is approximated by the value of $v \, \Delta P$. Hence

$$w_{P, \ 1\text{-}2} = 1.09(25) \frac{10^5}{10^6} = 2.7 \text{ kJ/kg}$$

$$h_2 = h_1 + w_{P, \ 1\text{-}2} = 640.2 + 2.7 = 642.9 \text{ kJ/kg}$$

$$w_{P, \ 6\text{-}7} = 1.01(4.9) \frac{10^5}{10^6} = 0.5 \text{ kJ/kg}$$

$$h_7 = h_6 + w_{P, \ 6\text{-}7} = 191.8 + 0.5 = 192.3 \text{ kJ/kg}$$

In order to determine the turbine work output, it is necessary to find, first, the fraction of the steam which is extracted at 5 bars. Substitution of the enthalpy values into Eq. (17-1) yields

$$y = \frac{h_1 - h_7}{h_4 - h_7} = \frac{640.2 - 192.3}{2941.6 - 192.3} = 0.174$$

Therefore 0.174 kg of steam is extracted for every kilogram which enters the turbine. The work output of the turbine becomes

$$w_{T, \ out} = h_3 - h_4 + (1 - y)(h_4 - h_5)$$

$$= (3456.5 - 2941.6) + 0.826(2941.6 - 2292.7)$$

$$= 514.9 + 536.0 = 1050.9 \text{ kJ/kg}$$

The total pump work is

$$w_{P, \ tot} = 2.7 + 0.826(0.5) = 3.1 \text{ kJ/kg}$$

and the heat input is

$$q_{in} = h_3 - h_2 = 3456.5 - 642.9 = 2813.6 \text{ kJ/kg}$$

Consequently, the thermal efficiency for the cycle is

$$\eta_{th} = \frac{w_{net}}{q_{in}} = \frac{1050.9 - 3.1}{2813.6} = 0.372 \qquad \text{(or 37.2 percent)}$$

In Example 17-3 the conditions given were the same as for the above cycle, except that no regeneration was used. In that case the thermal efficiency was 35.6 percent. Thus the inclusion of one open feedwater heater in the ideal cycle has increased the cycle efficiency by some 4.5 percent. Note, however, that the work output of the turbine has been lowered by nearly 10 percent as compared to the case without regeneration.

**Example 17-5** An ideal regenerative steam-power cycle operates so that steam enters the turbine at 400 psia and 900°F and exhausts at 1 psia. A single, open feedwater heater is employed which operates at 60 psia. Compute the thermal efficiency of the cycle.

SOLUTION The $Ts$ and flow diagrams for the cycle are given in Fig. 17-6. From the preceding examples the following data are still valid:

$$P_3 = 400 \text{ psia}, \ T_3 = 900°F, \ h_3 = 1470.1 \text{ Btu/lb}$$

$$P_4 = 60 \text{ psia}, \ s_4 = s_3, \ h_4 = 1244.2 \text{ Btu/lb}$$

$$P_5 = 1 \text{ psia}, \ s_5 = s_3, \ h_5 = 964.0 \text{ Btu/lb}$$

$$P_6 = 1 \text{ psia, saturated liquid}, \ h_6 = 69.7 \text{ Btu/lb}$$

In addition, state 1 is a saturated-liquid state at 60 psia; thus $h_1 = 262.2$ Btu/lb. The enthalpies at states 2 and 7 can be found by adding the pump work to the enthalpies at states 1 and 6, respectively. The pump work ($w_P$) can be approximated again by the value of $v \ \Delta P$. Hence

$$w_{P, 1-2} = \frac{0.01727(400 - 60)(144)}{778} = 1.1 \text{ Btu/lb}$$

$$h_2 = h_1 + w_{P, 1-2} = 262.2 + 1.1 = 263.3 \text{ Btu/lb}$$

$$w_{P, 6-7} = \frac{0.01614(60 - 1)(144)}{778} = 0.2 \text{ Btu/lb}$$

$$h_7 = h_6 + w_{P, 6-7} = 69.7 + 0.2 = 69.9 \text{ Btu/lb}$$

In order to determine the turbine work output, it is necessary to find, first, the fraction of the steam which is extracted at 60 psia. Substitution of the enthalpy values into the equation derived above yields

$$h_1 = yh_4 + (1 - y)h_7$$

$$262.2 = y(1244.2) + (1 - y)(69.9)$$

$$y = 0.164$$

Therefore 0.164 lb of steam is extracted for every pound which enters the turbine. The work output of the turbine becomes

$$w_{T, \text{ out}} = h_3 - h_4 + (1 - y)(h_4 - h_5)$$

$$= (1470.1 - 1244.2) + 0.836(1244.2 - 964.0)$$

$$= 225.9 + 234.2 = 460.1 \text{ Btu/lb}$$

The total pump work is

$$w_{P, \text{ tot}} = 1.1 + 0.2(0.836) = 1.3 \text{ Btu/lb}$$

and the heat input is

$$q_{\text{in}} = h_3 - h_2 = 1470.1 - 263.3 = 1206.8 \text{ Btu/lb}$$

Consequently, the thermal efficiency for the cycle is

$$\eta_{th} = \frac{w_{net}}{q_{in}} = \frac{460.1 - 1.3}{1206.8} = 0.381 \qquad \text{(or 38.1 percent)}$$

In Example 17-3 the conditions given were the same as for the above cycle, except that no regeneration was used. In that case the thermal efficiency was 36.1 percent. Thus the inclusion of one open feedwater heater in the ideal cycle has increased the cycle efficiency by some 5.5 percent. In modern high-pressure steam-power plants several open feedwater heaters operating at different pressures may be employed.

A modification of the regenerative cycle described above occurs when closed feedwater heaters are used to preheat the water being returned from the condenser to the boiler. As the term might imply, in a closed feedwater heater the two entering flow streams do not mix directly. The feedwater leaving the condenser flows inside tubes that pass through the heater. The steam extracted from the turbine enters the heater and condenses on the outside of the tubes which carry the feedwater. A schematic diagram of a closed feedwater heater is shown in Fig. 17-7. Two alternatives are shown for removal of the condensate. A pump may be used to return the condensate directly back into the feedwater line (line *a*), or the condensate may be collected in a trap which permits only liquid to flow to another region of lower pressure (line *b*). This lower-pressure region may be the condenser itself or another feedwater heater.

In the ideal case the condensate leaving the feedwater heater is a saturated liquid at the turbine extraction pressure, and the feedwater leaves as a compressed liquid at the same temperature as the condensate. In practice, of course, the feedwater would necessarily have to be 10° or so lower in temperature than the condensate, because of heat-transfer requirements. As in the case of open feedwater heaters, modern high-pressure steam-power plants usually employ several closed feedwater heaters in a given cycle. Since open feedwater heaters usually operate at fairly low pressures, they are less expensive. Another advantage is that they bring the feedwater up to its saturation temperature at the heater pressure. However, they require a separate pump for each heater.

The various descriptions of steam-power cycles presented above are for plants in which the total output is for power purposes. It must be recognized that there

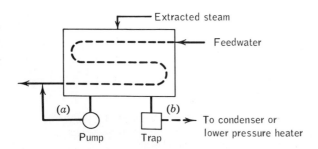

Figure 17-7 Schematic of a closed feedwater heater.

are many industrial situations where energy is also required for heating purposes, possibly in addition to power requirements. A power plant on a college campus is another example of this situation. Consequently, further modifications of the various cycles already introduced may be made. For example, the steam might first be passed through the turbine to some low pressure, say, 30 psia. Then it might be used for heating purposes with subsequent condensation, before being returned to the steam generator. In another case it might be more practical to extract steam from the low-pressure end of the turbine for heating purposes and then combine the two flow streams before they reenter the steam generator. Other designs are possible, depending upon the particular requirements.

## 17-4 EFFECT OF IRREVERSIBILITIES ON VAPOR-POWER-CYCLE PERFORMANCE

It must be kept in mind that the various cycles which have been presented earlier in this chapter are ideal cycles, and irreversibilities have not been considered. In all cases there are frictional losses in the piping which lead to pressure drops. Heat-transfer losses occur throughout the equipment. Irreversibilities caused by the flow of the fluid are especially important in the turbine and the pump. If heat losses are negligible in a turbine or pump, the adiabatic efficiency of these pieces of equipment must be considered to bring the ideal analysis more in line with actual performance. The concept of an adiabatic efficiency for various flow devices was introduced in Sec.7-11. In review, for turbines and pumps

$$\eta_T = \frac{w_{act}}{w_{isen}} \quad \text{and} \quad \eta_P = \frac{w_{isen}}{w_{act}}$$

Fig. 17-8 shows the change in the position of the exit states for the turbine and pump when irreversibilities are present.

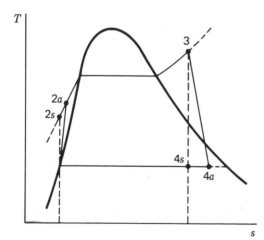

**Figure 17-8** $Ts$ diagram of a simple steam-power cycle with irreversible turbine and pump performance.

**Example 17-6M** The turbine-inlet conditions in a simple steam-power-cycle are 30 bars and 500°C, and the turbine-exhaust condition is 0.1 bar. The turbine operates adiabatically with an efficiency of 82 percent, and the efficiency of the pump is 78 percent. Compute the thermal efficiency of the cycle.

SOLUTION Losses other than those in the turbine and pump will be neglected. From Example 17-3M, it is known that the inlet enthalpy $h_3$ and entropy $s_3$ to the turbine are 3456.5 kJ/kg and 7.2338 kJ/(kg)(°K), respectively. Also, isentropic expansion to 0.1 bar gives a turbine-outlet enthalpy $h_{4s}$ of 2292.7 kJ/kg. The actual work output is the isentropic enthalpy change times the turbine-adiabatic efficiency. Hence

$$w_{T, \text{actual}} = 0.82(3456.5 - 2292.7) = 954.3 \text{ kJ/kg}$$

$$= h_3 - h_{4a}$$

Since $h_3$ is 3456.5 kJ/kg, $h_{4a} = 3456.5 - 954.3 = 2502.2$ kJ/kg. Since $h_g$ at 0.1 bar is 2584.7 kJ/kg, the steam leaving the turbine is in the wet region. The exit quality is, roughly, 96 percent. The pump work for an isentropic process was previously determined to be 3.0 kJ/kg. For the irreversible process this becomes

$$w_{P, \text{actual}} = \frac{3.0}{0.78} = 3.8 \text{ kJ/kg}$$

Therefore the enthalpy of the water leaving the pump is

$$h_2 = h_1 + w_P = 191.8 + 3.8 = 195.6 \text{ kJ/kg}$$

The thermal efficiency then becomes

$$\eta_{th} = \frac{w_T - w_P}{h_3 - h_2} = \frac{954.3 - 3.8}{3456.5 - 195.6} = 0.291 \quad \text{(or 29.1 percent)}$$

Thus irreversibilities lead to a considerable decrease from the ideal Rankine-cycle efficiency of 35.6 percent obtained in Example 17-3M.

**Example 17-6** The turbine inlet conditions in a simple steam power cycle are 400 psia and 900°F, and the exhaust condition is 1 psia. The turbine operates adiabatically with an efficiency of 82 percent, and the efficiency of the pump is 78 percent. Compute the thermal efficiency of the cycle.

SOLUTION Losses other than those in the turbine and pump will be neglected. From Example 17-3 it is known that the inlet enthalpy $h_3$ and entropy $s_3$ to the turbine are 1470.1 Btu/lb and 1.7252 Btu/(lb)(°R), respectively. Also, isentropic expansion to 1 psia gives the turbine-outlet condition as 964.0 Btu/lb for the enthalpy $h_{4s}$. The actual work output is the isentropic enthalpy change times the adiabatic efficiency. Hence

$$w_{T, \text{actual}} = 0.82(1470.1 - 964.0) = 415.0 \text{ Btu/lb}$$

$$= h_3 - h_{4a}$$

Since $h_3$ is 1470.1 Btu/lb, $h_{4a} = 1470.1 - 415.0 = 1055.1$ Btu/lb. This corresponds to an exit steam quality of, roughly, 95 percent. The pump work for an isentropic process was previously determined to be 1.2 Btu/lb. For the irreversible process this becomes

$$w_{P, \text{actual}} = \frac{1.2}{0.78} = 1.5 \text{ Btu/lb}$$

Therefore the enthalpy of the steam leaving the pump is

$$h_2 = h_1 + w_P = 69.7 + 1.5 = 71.2 \text{ Btu/lb}$$

The thermal efficiency then becomes

$$\eta_{th} = \frac{w_T - w_P}{h_3 - h_2} = \frac{415.0 - 1.5}{1470.1 - 71.2} = 0.296 \qquad \text{(or 29.6 percent)}$$

This is a considerable decrease from the ideal Rankine-cycle efficiency of 36.1 percent obtained in Example 17-3.

## 17-5 SUPERCRITICAL RANKINE CYCLE

In the preceding discussion of the Rankine cycle, the final pressure at the outlet of the pump preceding the heat-addition section of the cycle was chosen to be below the critical pressure of the fluid. That is, the maximum pressure was less than 221 bars, or 3200 psia, when the working fluid was water. A number of modern steam-power plants operate on a cycle for which the turbine-inlet pressure is supercritical. A typical value might be 250 to 325 bars, or 3500 to 5000 psia.

A Rankine cycle with a supercritical turbine-inlet pressure is shown in Fig. 17-9. Note that during the heat-addition process (4-1) a phase change does not occur. The pressurized fluid enters a bank of tubes (rather than a boiler), and gradually expands in volume with no bubbling occurring as it passes through the tubes. The fluid behaves like an extremely dense gas. At supercritical pressures the change in specific volume from subcooled water to superheated steam is smaller than in the case of subcritical pressure cycles. Inlet turbine temperatures may approach 620°C or 1150°F. For large power units working under these conditions, two reheat stages are often employed, and as many as five to eight feedwater heaters are commonly required. The thermal efficiency of supercritical units is around 40 percent.

The theoretical Rankine cycle does not require expansion to a low pressure followed by condensation of the working fluid. Two other possibilities are shown in Fig. 17-10. In Fig. 17-10a, the upper and lower pressures are each supercritical

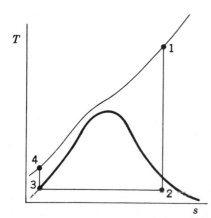

**Figure 17-9** Rankine cycle with supercritical heat addition.

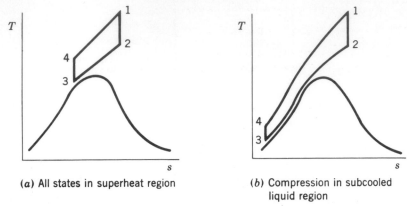

(a) All states in superheat region

(b) Compression in subcooled liquid region

**Figure 17-10** Two possible totally supercritical Rankine cycles.

and the lowest temperature in the cycle represents a state that could be considered still in the superheat region. In Fig. 17-10b, the lowest temperature now represents a state which is in the subcooled liquid region, although all pressures are still supercritical. In these totally supercritical power cycles, it is also not necessary for the working fluid to be steam. Carbon dioxide, for example, has been proposed. Conservative calculations for this type of power cycle indicate that competitive performance with present equipment will be attained only if a turbine-inlet temperature above 650°C or 1200°F is used.

## 17-6 RANKINE-CYCLE SPACE POWER SYSTEM

The Rankine cycle is the basic thermodynamic cycle for steam-power generation. In addition, the cycle may be used for the generation of power aboard a spacecraft. Although the basic thermodynamic principles of the cycle remain the same as presented in Sec. 17-1, the following modifications for space power systems must be considered:

1. The working fluid is a liquid metal, such as potassium, sodium, a potassium-sodium mixture, or even mercury. Major characteristics of liquid metals are their high heat-transfer coefficients and their high heat capacity (mass times specific heat). As a result any heat-exchanger size is significantly reduced.
2. The energy source typically is a nuclear reactor. As a result of possible radioactive contamination, a separate reactor loop is included, as shown in Fig. 17-11.
3. As in a conventional power plant, a condenser or heat-rejection loop must be provided. In this case, the fluid in this loop (also a liquid metal) passes through a radiator for the final heat rejection to space.
4. In the boiling and condensing steps of the cycle, two-phase flow occurs. The zero-gravity environment of space introduces problems in the separation of phases, which directly affects the design of the heat-transfer equipment.

The maximum temperature for a Rankine cycle operating with potassium, for

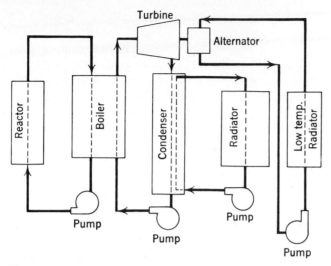

**Figure 17-11** Schematic of a potassium Rankine cycle.

example, may be as high as 1200°C or 2200°F. The degree of superheating at the turbine inlet typically might be 30 to 80°C, or 50 to 150°F. The vapor pressure of potassium at 1150°C is roughly 13.5 bars, which is equivalent to 200 psia at 2100°F. The condensing temperature of potassium may be as low as 600°C, with a vapor pressure of 0.17 bar. This is equivalent to 1100°F, with a vapor pressure of 2.5 psia. These temperature limits give a turbine expansion ratio of 80, which is relatively high when only two stages of expansion are used. Hence the expansion ratio places a practical lower limit on the cycle-temperature ratio (minimum to maximum temperature). Typical space-power Rankine cycles are constrained to operate with cycle-temperature ratios on the order of 0.65 to 0.75. Consequently the cycle efficiency is relatively low, probably 25 percent or less.

The low cycle-temperature ratio does have one advantage in terms of space applications. Since the condensing temperature is high, the temperature of the fluid in the heat-rejection loop is also high. A high radiator rejection temperature leads to a small radiator size, since the radiant-heat-transfer per unit area varies as the fourth power of the temperature. Tabular data for potassium in the saturation and superheat regions are presented in Tables A-24M and A-24.

## REFRIGERATION CYCLES

## 17-7 THE REVERSED CARNOT CYCLE

In the study of cyclic devices which operate for the purpose of removing heat continuously from a low-temperature source, it is useful to recall the features of the reversed Carnot cycle introduced in Chap. 7. A diagram of a reversed Carnot-cycle engine operating as a heat pump or refrigerator is shown in Fig. 17-12a. A quantity of heat $Q_L$ is transferred reversibly from a low-temperature source of temperature $T_L$ to the reversed heat engine. The reversed heat engine operates through a cycle during which net work $W$ is added to the engine and a quantity of heat $Q_H$ is transferred reversibly to a higher-temperature sink of temperature $T_H$. On the basis of the first law for a closed cyclic process, $Q_L + W = Q_H$. From the second law for a totally reversible process it is found that $T_H/T_L = Q_H/Q_L$. The reversed Carnot heat engine is useful as a standard of comparison because it requires the minimum work input for a given refrigeration effect between two given bodies of fixed temperature.

In place of the thermal efficiency, which is used as a criterion in the analysis of heat engines, the standard for refrigeration processes is the coefficient of performance. The definition of the coefficient of performance depends upon whether the reversed cycle is operating as a heat pump or as a refrigerator. The differentiation between a heat pump and a refrigerator is based upon the primary use of the device. The objective of a heat pump is to provide heat to a building or other relatively high-temperature body by removing heat from the colder surroundings. Hence the coefficient of performance for a heat pump is defined as

$$\text{COP}_{\text{heat pump}} = \frac{Q_H}{W_{\text{in}}} \tag{17-2}$$

For a Carnot heat pump, it is easily shown that

$$\text{COP}_{\text{heat pump, Carnot}} = \frac{T_H}{T_H - T_L} \tag{17-3}$$

On the other hand, the objective of a refrigerator is to remove heat from a

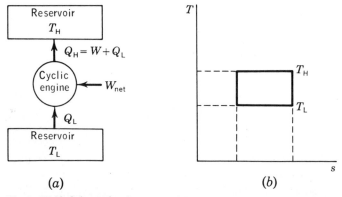

(a)　　　　　　　　　　　　(b)

**Figure 17-12** Schematic of a reversed Carnot engine and the $Ts$ diagram for the cycle.

low-temperature source in order to maintain the temperature at a desired value. Consequently, for a refrigerator,

$$COP_{refrig} = \frac{Q_L}{W_{in}} \tag{17-4}$$

and

$$COP_{refrig,\,Carnot} = \frac{T_L}{T_H - T_L} \tag{17-5}$$

It is important to note that the value of the COP can exceed unity, and, in fact, does so for a well-designed unit. Also note that the coefficient of performance increases as the temperature difference $T_H - T_L$ decreases.

## 17-8 THE VAPOR-COMPRESSION REFRIGERATION CYCLE

Although the reversed Carnot cycle is a standard with which all actual cycles may be compared, it is not a practical device for refrigeration purposes. It would be highly desirable, however, to approximate the constant-temperature heat-addition and rejection processes in order to achieve the highest coefficient of performance possible. This is accomplished in a large degree by operating a refrigeration device on a vapor-compression cycle. The schematic of the equipment for the cycle, along with $Ts$ and $Ph$ diagrams of the ideal cycle, is shown in Fig. 17-13. Saturated vapor at state 1 is compressed isentropically to a superheated vapor at state 2. The refrigerant vapor then enters a condenser, where heat is removed at constant pressure until the fluid becomes a saturated liquid at state 3. In order to return the fluid to a lower pressure, it is expanded adiabatically through a valve or a capillary tube to state 4. Process 3-4 is a throttling process, and $h_3 = h_4$. At state 4 the refrigerant is a low-quality wet mixture. Finally, it passes through the evaporator at constant pressure. Heat enters the evaporator from the cold-temperature source and vaporizes the fluid to the saturated-vapor state. Thus the cycle is completed. Note that all of process 4-1 and a large portion of process 2-3 occur at constant temperature. Unlike many other ideal cycles, the vapor-compression cycle modeled in Fig. 17-13 contains an irreversible process within it, the throttling process. All other portions of the cycle are assumed to be reversible.

The rating of refrigeration systems is frequently given on the basis of the tons of refrigeration provided by the unit when operating at design conditions. A *ton of refrigeration* is defined as a heat-removal rate from the cold region (or heat-absorption rate by the fluid passing through the evaporator) of 211 kJ/min or 200 Btu/min. On the basis of this definition, one possible measure of refrigeration-cycle performance is the power input required per ton of refrigeration capacity. In the United States this parameter is frequently expressed as horsepower per ton, and its value is in the neighborhood of unity. Another parameter frequently cited for a refrigeration device is the effective displacement of the compressor. This quantity will be defined as the volume flow rate at the compressor inlet.

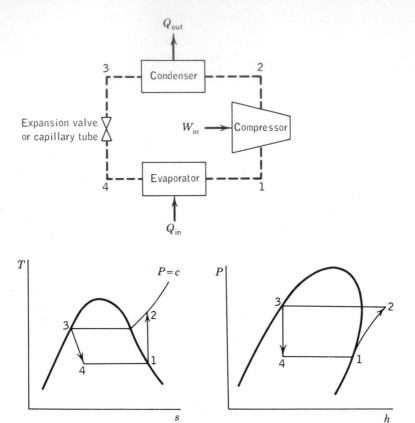

**Figure 17-13** Vapor-compression refrigeration cycle.

**Example 17-7M** An ideal vapor-compression refrigeration cycle with refrigerant 12 as the working fluid operates with an evaporator temperature of $-20°C$ and a condenser pressure of 9.0 bars. The mass flow rate of the refrigerant is 3 kg/min through the cycle. Compute the coefficient of performance, the tons of refrigeration, and the coefficient of performance of a Carnot reversed heat engine operating under the same maximum and minimum temperatures as the actual cycle.

SOLUTION Tables A-16M to A-18M provide the following data for the states illustrated in Fig. 17-13:

$$h_1 = h_{g, \text{ at } -20°C} = 178.74 \text{ kJ/kg} \qquad s_1 = s_g = 0.7087 \text{ kJ/(kg)(°K)}$$

In addition, the fluid pressure in the evaporator is roughly 1.5 bars. At 9 bars the entropy of saturated vapor is 0.6832 kJ/(kg)(°K). Hence isentropic compression to state 2 leads to a superheated vapor. Interpolation in the superheat table at 9 bars indicates that $T_2$ is, roughly, 43°C and $h_{2s} = 208.2$ kJ/kg. At the end of the condensation process

$$h_3 = h_{f. \text{ at } 9 \text{ bars}} = 71.93 \text{ kJ/kg} \qquad T_3 = T_{\text{sat}} = 37.37°C$$

The enthalpy at state 3 is also the enthalpy at state 4, after the fluid passes through the throttling process. Consequently

$$\text{COP} = \frac{h_1 - h_4}{h_2 - h_1} = \frac{178.74 - 71.93}{208.2 - 178.74} = 3.63$$

$$\text{Tons of refrigeration} = \frac{3(178.74 - 71.93)}{211} = 1.52$$

$$\text{COP}_{\text{Carnot}} = \frac{T_1}{T_2 - T_1} = 253/63 = 4.02$$

**Example 17-7** An ideal vapor-compression refrigeration cycle with refrigerant 12 as the working fluid operates with an evaporator temperature of 0°F and a condenser pressure of 100 psia. The mass flow rate of the refrigerant is 6 lb/min. Compute the coefficient of performance, the tons of refrigeration, the horsepower input per ton of refrigeration, and the coefficient of performance of a reversed Carnot heat engine operating under the same maximum and minimum temperatures.

SOLUTION Tables A-16 to A-18 provide the following data for the states illustrated in Fig. 17-13:

$$h_1 = h_{g, \text{ at } 0\,°F} = 77.27 \text{ Btu/lb} \qquad s_1 = s_g = 0.1689 \text{ Btu/(lb)(°R)}$$

In addition, the fluid pressure in the evaporator is roughly 24 psia. At 100 psia the entropy of saturated vapor is 0.1639 Btu/(lb)(°R). Hence isentropic compression to state 2 leads to a slightly superheated vapor. Interpolation in the superheat table at 100 psia indicates that $T_{2s}$ is, roughly, 96.6°F and $h_{2s} = 88.09$ Btu/lb. At the end of the condensation process

$$h_3 = h_{f, \text{ at } 100 \text{ psia}} = 26.54 \text{ Btu/lb} \qquad T_3 = T_{\text{sat}} = 80.76\,°F$$

The enthalpy at state 3 is also the enthalpy at state 4, after the fluid passes through the throttling process. Consequently

$$\text{COP} = \frac{h_1 - h_4}{h_2 - h_1} = \frac{77.27 - 26.54}{88.09 - 77.27} = 4.69$$

$$\text{Tons of refrigeration} = \frac{6(77.27 - 26.54)}{200} = 1.52$$

$$\text{hp/ton} = \frac{h_2 - h_1}{42.4} \frac{200}{h_1 - h_4} = \frac{4.715}{\text{COP}} = 1.01$$

$$\text{COP}_{\text{Carnot}} = \frac{T_1}{T_2 - T_1} = 460/96.6 = 4.76$$

In actual operation, a refrigeration cycle differs from the ideal cycle in a number of ways. The presence of fluid friction results in pressure drops throughout the cycle and in nonisentropic flow through the compressor. In addition, the presence of heat transfer must also be taken into account. Since it is not possible to control exactly the state of the fluid leaving the evaporator, it usually leaves as a superheated vapor, instead of as the saturated vapor found in the ideal cycle. Irreversibilities in the flow through the compressor lead to an increase in the entropy of the fluid during the process and a concomitant increase in the final temperature over that found in the ideal case. If heat losses from the compressor are great enough, however, the actual entropy of the fluid at the compressor exit can be less than that at the inlet. Even if pressure losses in the condenser are small,

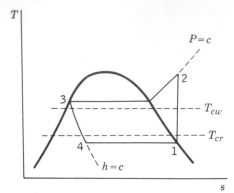

**Figure 17-14** Effect of irreversible heat transfer on the vapor-compression refrigeration cycle.

the fluid will probably leave the condenser as a subcooled liquid rather than as the saturated-liquid state assumed in the ideal cycle. This is a beneficial effect, since the lower enthalpy which results from the subcooling effect permits a larger quantity of heat to be absorbed by the fluid during the evaporation process.

The evaluation of certain parameters of interest in refrigeration cycles has been based on the saturation temperatures of the refrigerant in the evaporator and condenser. Nevertheless, the operating temperatures in an actual cycle are really established by the temperature desired to be maintained in the cold region and the temperature of the cooling water or air that is available for use in the condenser. Figure 17-14 illustrates this point. In order to obtain sufficient heat-transfer rates, the temperature difference between two fluids must be at least on the order of 10°C or 20°F. In the evaporator, heat is transferred from a cold region to the refrigerant, which undergoes a phase change at constant temperature. If the temperature of the cold region ($T_{cr}$ on the figure) is to be $-18$°C (0°F), for example, the refrigerant would have to be maintained at a saturation temperature corresponding to, say, $-25$°C ($-15$°F) for effective heat transfer. At the same time the refrigerant is being condensed in the condenser by the transfer of heat to a coolant flow stream external to the cycle. Cooling water and atmospheric air are two typical coolants which might be passed over the condenser tubes. Since these two substances usually would be available at temperatures from 15 to 30°C or 60 to 90°F, roughly ($T_{cw}$ in the figure), the saturation temperature of the refrigerant in the condenser must be above this range.

**Example 17-8M** Reconsider the ideal vapor-compression refrigeration cycle introduced in Example 17-7M. The cold region is to be maintained at $-20$°C. To allow for a finite temperature difference between the cold region and the evaporator fluid, the evaporator will be set to operate at $-30$°C. The air which passes over the condenser coils in the preceding example is at the saturation temperature corresponding to 9.0 bars, or 37.37°C. Again, to allow for a finite temperature difference, the condenser pressure will be raised to 12.0 bars, for which the saturation temperature is 49.31°C. Determine the coefficient of performance of the cycle under the new conditions, and compare with the result of Example 17-7M.

SOLUTION Tables A-16M to A-18M provide data for the problem. At the evaporator outlet

$$h_1 = h_{g,\,\text{at}\,-30\,°C} = 174.20 \text{ kJ/kg} \qquad s_1 = s_g = 0.7170 \text{ kJ/(kg)(°K)}$$

State 2 at the compressor outlet is found from knowledge of $P_2$ and the fact that $s_2 = s_1$. By interpolation at 12 bars in the superheat table, we find that $T_{2s}$ is, roughly, 64.6°C and $h_{2s} = 218.56$ kJ/kg. At the end of the condensation process

$$h_3 = h_{f,\,\text{at}\,12\,\text{bars}} = 84.21 \text{ kJ/kg}$$

The enthalpy at state 3 is also the enthalpy at state 4, after the fluid passes through the throttling process. Consequently

$$\text{COP} = \frac{h_1 - h_4}{h_2 - h_1} = \frac{174.20 - 84.21}{218.56 - 174.20} = 2.03$$

By allowing for a 10°C temperature differential at the evaporator end of the cycle and an 11.9°C difference at the condenser, the COP has decreased from 3.63 to 2.03. Although these calculations were based on an idealized cycle, the trend in the actual case would be similar. The coefficient of performance is severely affected by the requirement of finite temperature differences across the condenser and evaporator heat-transfer surfaces.

**Example 17-8** Reconsider the ideal vapor-compression refrigeration cycle introduced in Example 17-7. The cold region is to be maintained at 0°F. To allow for a finite temperature difference between the cold region and the evaporator fluid, the evaporator will be set to operate at −20°F. The air which passes over the condenser coil in the preceding example is at the saturation temperature corresponding to 100 psia, or 80.76°F. Again, to allow for a finite temperature difference, the condenser pressure will be raised to 120 psia, for which the saturation temperature is 93.29°F. Determine the coefficient of performance for the cycle under the new conditions, and compare with the result of Example 17-7.

SOLUTION Tables A-16 to A-18 provide data for the problem. At the evaporator outlet

$$h_1 = h_{g,\,\text{at}\,-20\,°F} = 75.11 \text{ Btu/lb} \qquad s_1 = s_g = 0.1710 \text{ Btu/(lb)(°R)}$$

State 2 at the compressor outlet is found from knowledge of $P_2$ and the fact that $s_2 = s_1$. By interpolation at 120 psia in the superheat table, we find that $T_{2s}$ is, roughly, 117.4°F and $h_{2s} = 90.78$ Btu/lb. At the end of the condensation process

$$h_3 = h_{f,\,\text{at}\,120\,\text{psia}} = 29.49 \text{ Btu/lb}$$

The enthalpy at state 3 is also the enthalpy at state 4, after the fluid passes through the throttling process. Consequently

$$\text{COP} = \frac{h_1 - h_4}{h_2 - h_1} = \frac{75.11 - 29.49}{90.78 - 75.11} = 2.91$$

By allowing for a 20°F temperature differential at the evaporator end of the cycle and a 12.5°F difference at the condenser, the COP has decreased from 4.69 to 2.91. Although these calculations were based on an idealized cycle, the trend in the actual case would be similar. The coefficient of performance is severely affected by the requirement of finite temperature differences across the condenser and evaporator heat-transfer surfaces.

In the vapor refrigeration cycle the two desired saturation temperatures for the evaporation and condensation processes determine the operating pressures of the cycle for a given refrigerant. Consequently, the choice of a refrigerant partially depends upon the saturation pressure-temperature relationship in the range of interest. Normally, the minimum pressure in the cycle should be above 1 atm to avoid leakage into the equipment, but maximum pressures over 150 to 200 psia (10 to 15 bars) would be undesirable. In addition, the fluid needs to be nontoxic, stable, and of low cost and to have a relatively high enthalpy of vaporization. These restrictions and others limit the number of compounds which are suitable for refrigerants. In fact, because of the range of applicability of refrigeration cycles, no one fluid is suitable for all cases. Even with a proper choice of the refrigerant, a number of modifications may be made on the basic cycle to improve the coefficient of performance. Such changes are discussed in standard textbooks and handbooks on refrigeration.

**Example 17-9M** Reconsider the vapor-compression cycle analyzed in Example 17-8M. The evaporator operates at $-30°C$, and the condenser pressure is 12.0 bars, for which the saturation temperature is 49.31°C. To further modify the cycle, we shall consider that the fluid leaves the evaporator superheated by 10°C, and that it leaves the condenser subcooled by 1.31°C. In addition, the compressor is essentially adiabatic, with an efficiency of 75 percent. Calculate the coefficient of performance.

SOLUTION The saturation pressure at $-30°C$ is very close to 1.0 bar. Thus the fluid leaving the evaporator is at 1 bar and $-20°C$. From the superheat table the enthalpy $h_1$ and the entropy $s_1$ are found to be 179.99 kJ/kg and 0.7406 kJ/(kg)(°K), respectively. The isentropic enthalpy $h_{2s}$ at the compressor outlet (12 bars) is found from the superheat table to be 227.86 kJ/kg at roughly 76.5°C. Consequently, the actual outlet enthalpy is

$$h_{2a} = h_1 + \frac{w_s}{\eta_C} = 179.99 + \frac{228.86 - 179.99}{0.75} = 243.82 \text{ kJ/kg}$$

Since the fluid at the condenser outlet is subcooled, its enthalpy at state 3 is approximated by the saturated-liquid enthalpy at $T_3 = 49.31 - 1.31 = 48.0°C$. Hence $h_3 = 82.83$ kJ/kg, and this value is also $h_4$. Therefore

$$\text{COP} = \frac{h_1 - h_4}{h_2 - h_1} = \frac{179.99 - 82.83}{243.82 - 179.99} = 1.52$$

The effects of superheating at the evaporator outlet, subcooling at the condenser outlet, and irreversibilities in the compressor lower the COP found in Example 17-8M from 2.03 to 1.52.

**Example 17-9** Reconsider the vapor-compression cycle analyzed in Example 17-8. The evaporator operates at $-20°F$, and the condenser pressure is 120 psia, for which the saturation temperature is 93.29°F. To further modify the cycle, we shall consider that the fluid leaves the evaporator superheated by 10°F, and that it leaves the condenser subcooled by 3.29°F. In addition, the compressor is essentially adiabatic, with an efficiency of 75 percent. Calculate the coefficient of performance.

SOLUTION The saturation pressure at $-20°F$ is 15.267 psia. Thus the fluid leaving the evaporator is at 15.267 psia and $-10°F$. To approximate the data at state 1, we shall use the pressure of 15.0 psia in the superheat table. By interpolation between the saturation temperature of $-20.75°F$ and 0°F, we find that the enthalpy $h_1$ and the entropy $s_1$ are 76.52 Btu/lb and 0.1744 Btu/(lb)(°R), respectively. The isentropic enthalpy $h_{2s}$ at the compressor outlet (120 psia) is found from the superheat table to be 92.75 Btu/lb at roughly 124°F. Consequently, the actual outlet enthalpy is

$$h_{2a} = h_1 + \frac{w_s}{\eta_C} = 76.52 + \frac{92.75 - 76.52}{0.75} = 98.16 \text{ Btu/lb}$$

Since the fluid at the condenser outlet is subcooled, its enthalpy at state 3 is approximated by the saturated-liquid enthalpy at $T_3 = 93.29 - 3.29 = 90°F$. Hence $h_3 = 28.71$ Btu/lb, and this value is also $h_4$. Therefore

$$\text{COP} = \frac{h_1 - h_4}{h_2 - h_1} = \frac{76.52 - 28.71}{98.16 - 76.52} = 2.21$$

The effects of superheating at the evaporator outlet, subcooling at the condenser outlet, and irreversibilities in the compressor lower the COP found in Example 17-8 from 2.91 to 2.21.

## 17-9 LIQUEFACTION AND SOLIDIFICATION OF GASES

In modern technology the preparation of liquids at temperatures below $-100°F$ is quite important. In the study of the properties and the behavior of substances at low temperatures, the materials are placed in baths composed of liquefied gases. Mixtures of gases may be separated by liquefaction techniques. For example, liquid oxygen and nitrogen are separated from air in this manner, and in the same manner rarefied gases such as helium may be obtained. Liquefied gases are also successful rocket propellants. Liquid helium and hydrogen are especially useful in research studies in the temperature range from 2 to 30°K, such as those of super-conductivity and superfluidity. The following discussion provides a brief introduction to the thermodynamic cycles employed in the liquefaction and solidification of gases.

In general, two criteria must be met if a process is to alter a gas into the liquid or solid phase successfully. First, the gas must be cooled below its critical temperature, since only below the critical temperature can one make a distinction between the liquid and gas phases. The critical temperatures of a number of substances are given in Table A-3. For common monatomic and diatomic gases, the critical temperature is considerably below atmospheric values. Hence these substances exist in the liquid phase only at extremely low temperatures. The second criterion involves the maximum inversion temperature of a substance, since liquefaction processes rely on throttling to achieve a portion of the cooling effect. In Sec. 13-6 it was pointed out that cooling could be obtained from a Joule–Thompson expansion only if the initial state of the fluid was within the inversion curve. Under this circumstance $(\partial T/\partial P)_h$ would be positive. The maximum inversion temperatures of most gases are well above room temperature, except for hydrogen and helium. Hence precooling of these two gases is necessary before one can take advantage of the Joule–Thomson effect.

The simplest cycle for liquefying gases from a theoretical viewpoint is illustrated in Fig. 17-15. Makeup gas at a conventional pressure and temperature enters the steady-flow system, and is compressed to an elevated pressure and temperature (state 1). Note that the compression is shown not to be isentropic on the $Ts$ diagram. In fact, in practice, multistaging with intercooling might be used, so that the compression process is closer to isothermal than isentropic. Before undergoing a throttling process, the gas is cooled to state 2, without a significant pressure drop, by passing it through a heat exchanger. The gas is now throttled to

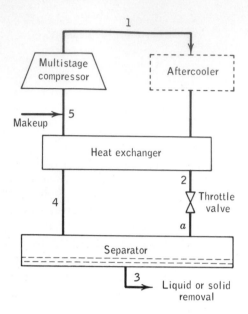

**Figure 17-15** Basic cycle for liquefying gases.

state $a$, which must lie within the wet region (or solid-gas region) if a phase change is to occur. The liquid is collected in the bottom of the separation chamber as a saturated liquid at state 3, and the remaining gas is passed back through the heat exchanger. Since the gas leaving the separator as saturated vapor at state 4 may be considerably colder than the gas at state 1, it may be used effectively to cool the gas stream passing from the compressor to the throttling valve. The gas now at state 5 is mixed with makeup gas, and the cycle is repeated. The process is traced on the $Ts$ diagram in Fig. 17-15. (The use of the aftercooler shown in the figure is discussed in a subsequent paragraph of this section.)

Consider a control volume drawn around the heat exchanger, throttling valve, and separator shown in Fig. 17-15. It is assumed that these three pieces of equipment are adiabatic and the changes in kinetic and potential energies are negligible. The steady-state energy equation for this control volume consequently reduces to $\Delta H = 0$. Let a unit mass enter the control volume and let $y$ equal the fraction of liquid or solid formed in the separator per unit mass entering. Therefore, on the basis of symbols shown in Fig. 17-15,

$$yh_3 + (1 - y)h_5 = 1(h_1)$$

Solving for $y$, we find that

$$y = \frac{h_5 - h_1}{h_5 - h_3} \tag{17-6}$$

The value of $h_3$ is set by the pressure in the separator, $P_3$. At the same time $h_5$ is set by the design of the heat exchanger. The enthalpy $h_1$ is fixed by $T_1$ and $P_1$; however, $T_1$ is set by the design of the separator and heat exchanger. Con-

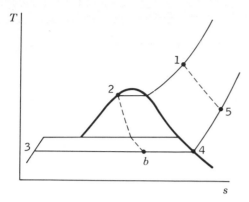

**Figure 17-16** Cycle for the production of solid carbon dioxide.

sequently, the compressor-outlet pressure $P_1$ is the main variable controlling the fraction of the gas which is liquefied or solidified. It is noted that $y$ is made larger by decreasing the value of $h_1$. It may be shown that the maximum degree of liquefaction occurs when the initial value of $P_1$ for a given temperature of the gas entering the heat exchanger is the inversion pressure at that temperature.

An interesting application of the apparatus described in Fig. 17-16 is in the production of solid carbon dioxide, or Dry Ice. A $Ts$ diagram for carbon dioxide appears as Fig. A-26. Figure 17-16 is a representation of the approximate path of the process on a $Ts$ diagram. Carbon dioxide at atmospheric conditions (state 5) is compressed by multistaging to a high pressure (roughly, 1000 psia), represented by state 1. It then passes through the heat exchanger, where it is cooled and condensed to a saturated liquid at state 2. The corresponding temperature in this liquid state is room temperature. The liquid is then expanded through a throttling valve to atmospheric pressure in the separator. The resulting mixture is solid and gas at state $b$. The corresponding saturation states are states 3 and 4 on the figure. Solid is formed because the triple state of carbon dioxide is, roughly, 5 atm and $-70°F$. Consequently, liquid carbon dioxide cannot exist at a pressure of 1 atm. The sublimation temperature of carbon dioxide at atmospheric pressure is approximately $-110°F$. Solid carbon dioxide will be formed by cooling to this temperature by throttling. The solid phase is pressed into bars, and the gas phase at state 4 and the required makeup gas return to the compressor at state 5.

A number of modifications of the basic cycle lead to improved operation and efficiency. One of these is the use of an aftercooler, which appears between the compressor and the heat exchanger. This is indicated by the dashed box in Fig. 17-15. In practice, a separate refrigerating system is used to precool the gas as it passes through the aftercooler. Any auxiliary refrigeration device such as this is inherently more efficient in cooling the gas, compared with a throttling process, because the refrigeration device can be made more reversible than a throttling process. The inherent irreversibilities of a throttling process can be partially overcome by another method of modifying the basic liquefaction cycle discussed above. In addition to cooling by the throttling process, a gas can be cooled by

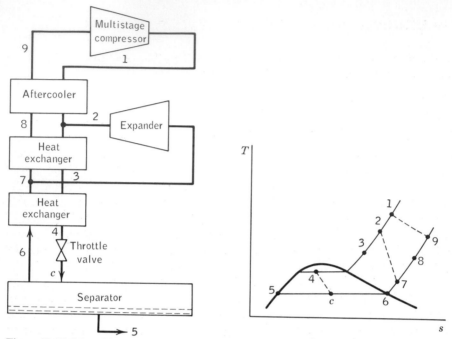

**Figure 17-17** Schematic and $Ts$ diagram for gas-liquefaction process involving a turbine expander.

expansion through a device such as a turbine. This was discussed in Sec. 16-15 with respect to gas-refrigeration cycles. If the throttling device is simply replaced by an expander, operating difficulties will ensue, because of the complications introduced by two-phase flow through a turbine, which is highly undesirable. This is overcome in the Claude system for liquefying gases by employing both a throttling valve and an expansion engine. Figure 17-17 illustrates the process schematically and on a $Ts$ diagram. Multistage compression with intercooling changes the gas from state 9 to state 1. An auxiliary refrigerating system cools the gas further to state 2, at which point the flow stream splits. A small portion continues through another heat exchanger, and then through the throttling valve into the separator. The other portion passes through an expansion engine, where it is cooled, and it then rejoins the flow stream coming from the separator. After passing through additional heat exchangers, the main flow returns to the compressor and begins the cycle again. By placing the expander early in the cycle, a cooling effect is achieved, but the fluid, even after expansion, still lies outside the wet region. Thus two-phase flow in the expansion device is avoided. The work output of the expander can be used as work input to the compressor section of the cycle. The overall result is a smaller expenditure of net work input for a given amount of liquefaction.

Other modifications for systems which liquefy gases at low temperatures are possible. These are discussed fully in texts and handbooks on the subject.

## 17-10 STIRLING REFRIGERATION CYCLE

The attainment of extremely low temperatures (less than 200°K or −100°F) is usually achieved by three well-established methods.

1. Vaporization of a liquid.
2. Joule–Thomson effect by isenthalpic expansion
3. Adiabatic expansion in an engine with production of work

Since the early 1950s considerable work has been done on developing a practical refrigeration device based on the Stirling cycle. (See Sec. 16-16 for a discussion of the Stirling heat engine.) Devices arising from this work have proved useful in the temperature range of 100 to 200°K. Other methods for maintaining temperatures in this range are not too plentiful. As noted in Sec. 16-16, the Stirling cycle is composed to two constant-temperature processes and two constant-volume processes. Figure 17-18 is a repeat of Fig. 16-31 and it shows the $PV$ and $TS$ diagrams for the cycle. If the cycle is reversible, then the heat quantities for the cycle are represented by areas on the $TS$ diagram. In the presence of an ideal regenerator in the cycle the heat quantities $Q_{14}$ and $Q_{32}$, which are equal in magnitude but opposite in sign, are exchanged between fluid streams within the device. Hence the only external heat transfer occurs in processes 4-3 and 2-1 at constant temperatures $T_L$ and $T_H$. Consequently the coefficient of performance of the Stirling refrigerator theoretically equals that of the Carnot refrigerator, namely, $T_L/(T_H − T_L)$. The COP will be quite small if $T_L$ is small, since $T_H − T_L$ will also be quite large. For example, if $T_L$ is 100°K and $T_H$ is 300°K, a value for the COP of 0.5 results. Thus considerable work must be performed per unit of heat removal, compared to a household refrigerator or air-conditioning unit.

An ideal reciprocating Stirling refrigeration unit is shown in Fig. 17-19. A cold temperature is produced by a reversible expansion of a gas in region $E$, while the gas is heated by compression in region $D$. Particles of the gas oscillate between the two spaces which are connected by a regenerator $F$. The regenerator must be composed of a material with a high heat capacity. Cylinder $A$ encloses the regular

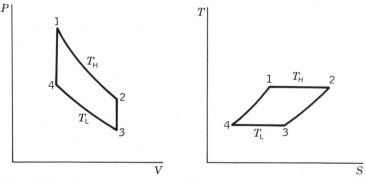

**Figure 17-18** Stirling cycle for a gaseous substance.

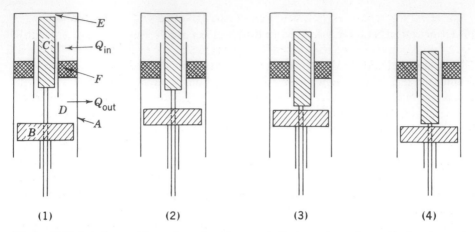

**Figure 17-19** Relative positions of regular piston and displacer piston in an ideal reciprocating Stirling refrigeration cycle.

piston $B$ and a displacer piston $C$. The shaft of the displacer piston passes through piston $B$. Consider the piston initially in position 1 shown in Fig. 17-19. Four distinct ideal processes take place in the cycle.

1. The piston moves upward, compressing the gas in region $D$. To keep the process isothermal at temperature $T_H$, heat is removed through the cylinder walls during the compression process. This is equivalent to process 2-1 in Fig. 17-19.
2. The displacer piston now moves downward, forcing part of the gas through the regenerator to space $E$. The gas is cooled as it passes through the regenerator, the energy being stored in the regenerator material. This creates a temperature gradient in the regenerator, the temperature increasing from region $E$ to region $D$. Since piston $B$ does not move, this transfer of part of the gas from $E$ to $D$, with energy storage in $E$, occurs at constant volume. This corresponds to process 1-4.
3. Both pistons now move downward, with the resulting expansion of the gas in region $E$. The expansion process would tend to cool the gas, but heat added from an outside source keeps the temperature at $T_L$. This heat removal at a very low $T_L$ values constitutes the refrigeration effect. For example, a gas circulated around the outside of region $E$ could be liquefied. This process is equivalent to process 4-3 in Fig. 17-19.
4. Finally, the displacer piston moves upward to its initial position, forcing gas from region $D$ to region $E$. As the gas passes through the regenerator from the cold to the hot side, it is reheated. This internal heat-addition process corresponds to process 3-2. Note that the temperature gradient in the regenerator makes the internal heat transfer between the gas and the refrigerator material reversible.

Several discrepancies between theory and practice should be pointed out. First, the entire gas within the system does not follow the process path shown in Fig. 17-18. A considerable fraction of the gas remains either in region $D$ or region $E$. Hence the $TS$ and $PV$ diagrams do not have their usual clear-cut representation of the process. The system at any instant is nonhomogeneous, and therefore cannot be represented by a single state. Different gas particles describe different cycles. Secondly, the simple piston displacements described above would be difficult to match, since they occur discontinuously. In an actual device the reciprocating motions of the pistons would be harmonic. To match more nearly the theoretical cycle, the displacer and regular pistons are placed out of phase by some angle $\phi$ (on the same drive shaft), so that the compression region $D$ lags in phase with respect to the expansion space $E$. This compromise tends to blur even further the four steps of the theoretical cycle.

As noted earlier, the working fluid for a practical Stirling refrigeration unit is either helium or hydrogen. Maximum and minimum pressures may range around 35 and 15 bars (or 500 and 200 psia), respectively, with an engine speed of 1500 rpm. A number of applications include:

1. Means for cooling electronic equipment and superconducting research magnets
2. Freeze-drying of materials
3. A precooler for production of liquid hydrogen and helium
4. An air liquefier
5. Gas separation, e.g., as a liquid nitrogen generator from air

## 17-11 ABSORPTION REFRIGERATION

In any refrigeration process the energy removed from the cold region eventually must be rejected to another region which is at a considerably higher temperature. This second region is usually the surrounding environment. In order to carry out the heat-rejection process, the temperature of the fluid within the refrigeration cycle must be raised to a value above that of the environment. In a vapor-compression refrigeration cycle, as discussed in Sec. 17-8, the temperature of the vapor leaving the evaporator is raised by a compression process. The work input required in an ideal steady-flow compression process is given by the integral of $v\, dP$. The pressure limits on the integral are set by the saturation temperatures required in the evaporator and condenser of the cycle. Once the pressure range is determined for a given refrigerant, the main variable which controls the amount of work input is the specific volume of the fluid, In a vapor-compression refrigeration cycle the value of $v$ is relatively large, since the fluid is in the superheat region throughout the compression process. Therefore, the work input is also relatively large. One method of overcoming this disadvantage is to design a refrigeration cycle in which the fluid is a liquid during the compression process. Then the work input will be significantly smaller.

The technique of absorption refrigeration is based on this approach. To accomplish this, however, the overall cycle becomes physically more complex. In addition, a two-component mixture, such as ammonia and water or lithium bromide and water, must be used as the circulating fluid in part of the cycle, rather than the single component used in a vapor-compression cycle. Two-component fluids have an important characteristic which must be recognized. When two phases are present at equilibrium, the composition of a given component is not the same in the two phases. The vapor phase will contain more of that component which is more volatile at the given temperature. For example, consider an ammonia-water mixture. At 43°C (110°F) the saturation pressure of ammonia is 17 bars (247 psia), while that of water is 0.09 bar (1.3 psia). Therefore ammonia has a much greater tendency to vaporize at a given temperature than does water. Hence for an ammonia-water solution, the vapor phase contains much more ammonia (is richer in ammonia) than the liquid phase in equilibrium with it. This fact is extremely important when making mass and energy balances on equipment used in absorption refrigeration.

A schematic diagram of a simple absorption-refrigeration cycle is shown in Fig. 17-20. A condenser, throttle valve, and evaporator are shown on the left-hand side of the diagram. These three pieces of equipment also are used in a conventional vapor-compression cycle, as shown earlier in Fig. 17-13. The compressor in that cycle, however, is now replaced by four pieces of equipment. These are an absorber, a pump, a generator, and a valve. For the purpose of discussion, we shall consider ammonia and water as the two components in the cycle. Essentially pure ammonia vapor leaves the evaporator at state 1 and enters the absorber. The absorbing medium is a weak solution (low ammonia concentration) of ammonia and water which continually enters at state 5. The process of absorption releases energy, hence cooling water must be circulated through the absorber in order to

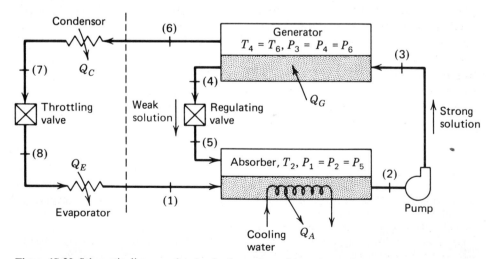

**Figure 17-20** Schematic diagram of a simple absorption-refrigeration cycle.

keep the solution at a constant temperature. The temperature of the absorbing fluid must be kept at as low a temperature as possible, since the amount of pure refrigerant (ammonia) which can be absorbed decreases as the temperature increases. However, the absorber must operate at 10 to 20° above the cooling water temperature to allow for adequate heat-transfer rates. The liquid which leaves the absorber at state 2 is a rich or strong solution (high ammonia concentration). This binary liquid mixture is now compressed by a pump to state 3, which is at the desired condenser pressure.

The temperature rise of the binary mixture due to the pump work usually is quite small. Thus the strong solution is subcooled liquid as it enters the generator shown in Fig. 17-20. Heat $Q_G$ must now be added to the solution in the generator to warm the incoming liquid to the saturation temperature and to drive out of solution some of the ammonia. This nearly-pure ammonia passes to the condenser at state 6, and eventually returns to the absorber at state 1. The weak solution left in the generator (state 4) now flows through a regulating valve, which drops the pressure of the solution to that in the absorber. It then mixes with the solution remaining in the absorber, and cold vapor coming from the evaporator is added to the overall liquid solution. The strong solution is cooled, as before, and the cycle is repeated. Hence the ammonia-water solution which cycles through the absorber, pump, generator, and valve merely serves as a transport medium for the ammonia refrigerant. Note that any absorption-refrigeration device requires an external heat source for the generation of refrigerant vapor. Thus absorption refrigeration is especially attractive if a low-temperature (100–200°C, 200–400°F) source of thermal energy is readily available.

In actual practice, absorption-refrigeration units have at least two modifications not shown in Fig. 17-20. First, the cold, strong solution at state 3 needs warming before it enters the generator, and the warm, weak solution at state 4 needs cooling before it enters the absorber. Consequently, a heat exchanger is placed between the absorber and generator, which permits heat transfer from the weak solution to the strong solution. Second, a major requirement is that the ammonia from the generator should be essentially free of water as it passes through the condenser–throttling-valve–evaporator loop. Any traces of water will freeze in the expansion valve and evaporator. Hence the vapor leaving the generator passes through a device called a rectifier before it enters the condenser. The rectifier separates any remaining water vapor from the vapor stream leaving the generator, and returns the water to the generator.

## REFERENCES

American Society of Heating, Refrigerating, and Air Conditioning Engineers Guide and Data Book, 1964.

"Combustion Engineering," Combustion Engineering, Inc., New York, 1957.

Potter, J. H.: The Totally Supercritical Steam Cycle," *J. Eng. Power*, 91A(2); 113–120 (April 1969).

"Steam, Its Generation and Uses," The Babcock and Wilcox Company, New York, 1972.

Stocker, W. F.: "Refrigeration and Air Conditioning," McGraw-Hill, New York, 1958.

# PROBLEMS (METRIC)

## The Rankine cycle

**17-1M** A power plant operating on the Rankine cycle has steam entering the turbine at 40 bars and 440°C. If the turbine output is 10,000 kW, determine the thermal efficiency of the cycle and the mass flow rate of steam for condenser pressures of (a) 0.08 bar and (b) 0.04 bar.

**17-2M** A Rankine cycle has an exhaust pressure of 0.08 bar. Determine the moisture content of the fluid leaving the turbine and the thermal efficiency of the cycle which has a turbine-inlet pressure of 60 bars and a turbine-inlet temperature of (a) 540°C, (b) 440°C, and (c) 320°C.

**17-3M** A Rankine cycle has an exhaust pressure of 0.08 bar. Determine the moisture content of the fluid leaving the turbine and the thermal efficiency of the cycle which has a turbine-inlet temperature of 600°C and a turbine-inlet pressure of (a) 120 bars, (b) 60 bars, and (c) 30 bars.

**17-4M** An ideal steam turbine operates initially with inlet conditions of 30 bars and 360°C, and an exhaust pressure of 0.1 bar. In order to reduce the work output, the steam may first be throttled to a lower pressure and then expanded to the same exhaust pressure. What pressure must the inlet steam be throttled to, in order to reduce the work output to (a) 94 percent and (b) 84 percent of its normal output?

**17-5M** An ideal Rankine-cycle steam-power plant generates steam at 140 bars and 560°C, and condenses steam at 0.06 bar. Neglecting pump work, determine the thermal efficiency. If the plant generates 10 MW, determine the mass flow rate of steam in kg/h.

**17-6M** An ideal Rankine steam-power cycle which produces 125 MW of turbine power has the following operating conditions: turbine inlet, 80 bars and 600°C; condenser pressure, 0.06 bar. Compute the required mass flow rate and the cycle efficiency.

**17-7M** A Rankine steam-power cycle operates with turbine-inlet conditions of 60 bars and 600°C. The condenser pressure is 0.08 bar. Determine the thermal efficiency.

**17-8M** A steam-generating unit has a throughput of 23,800 kg/h. The unit, operating on the Rankine cycle, is furnished with $50 \times 10^6$ kJ/h of heat in the boiler and superheater. Water enters the boiler at 30 bars and 40°C, and it leaves the condenser at 0.1 bar and 40°C. Cooling water in the condenser is circulated at the rate of $1.31 \times 10^6$ kg/h, and it experiences a temperature rise of 8.3°C.

    (a) Determine the enthalpy of the steam leaving the boiler-superheat unit, in kJ/kg.
    (b) Determine the enthalpy of steam entering the condenser, in kJ/kg.
    (c) Calculate the power output of the ideal turbine, in kilowatts.

**17-9M** A Rankine power cycle uses solar energy for the heat input and refrigerant 12 as the working fluid. Fluid enters the pump as a saturated liquid at 9 bars, and is pumped to 16 bars. It then passes through the steam-generator section at this pressure, and enters the turbine at 120°C. The mass flow rate is 1000 kg/h. Determine (a) the net work output, in kJ/kg, (b) the thermal efficiency, and (c) the area, in square meters, of the solar collector needed if the collector picks up 650 $J/(m^2)(s)$.

## Reheat cycles

**17-10M** Modify the data of Prob. 17-1M in the following manner. The steam in the turbine is expanded to 7 bars, reheated to 440°C, and then expanded to (a) 0.08 bar, and (b) 0.04 bar. Neglect the pump work and compute the thermal efficiency of the cycle.

**17-11M** Modify the data of Prob. 17-3M in the following manner. The steam in the turbine is expanded to 10 bars, reheated to 540°C, and then expanded to 0.08 bar. Neglect pump work and calculate the thermal efficiency for turbine-inlet pressures of (a) 120 bars, (b) 60 bars, and (c) 30 bars.

**17-12M** The data of Prob. 17-5M are modified in the following manner. The steam in the turbine is expanded to 15 bars, reheated to 540°C, and then expanded to 0.06 bar. Neglect the pump work and compute the thermal efficiency.

**17-13M** An ideal reheat cycle which produces 125,000 kW of turbine power has the following given conditions: turbine inlet, 80 bars and 600°C; reheat at 15 bars to 540°C; condenser pressure, 0.06 bar. Compute the required mass flow rate of steam in kg/h and the cycle efficiency.

**17-14M** The Rankine cycle in Prob. 17-7M is modified in the following manner. The steam is first expanded to 10 bars, reheated to 540°C, and then expanded to 0.08 bar. Determine the thermal efficiency.

### Regenerative cycles

**17-15M** Modify the data of Prob. 17-1M as follows. The steam is expanded to 7 bars, where a portion is bled to a single, open feedwater heater operating at this pressure. Neglecting pump work, determine the thermal efficiency of the cycle for a condenser pressure of (a) 0.08 bar, and (b) 0.04 bar.

**17-16M** The data of Prob. 17-3M are modified in the following manner. The steam is expanded in the turbine to 10 bars, where a portion is bled to a single, open feedwater heater operating at this pressure. The remaining portion of the turbine flow expands to 0.08 bar. Determine the thermal efficiency for a turbine-inlet pressure of (a) 120 bars, (b) 60 bars, and (c) 30 bars.

**17-17M** Consider the following modification of Prob. 17-5M. Steam is expanded to 15 bars, where a portion is bled to (a) a single, open feedwater heater operating at this pressure, and (b) a single, closed feedwater heater followed by a pump. The remaining portion of the turbine flow expands to 0.06 bar. Determine the thermal efficiency of the cycle.

**17-18M** Modify Prob. 17-6M in the following manner. The steam is expanded in the turbine until a portion is bled off to (a) a single, open feedwater heater operating at 5 bars, and (b) a single, closed feedwater heater with a bleedoff pressure of 15 bars, and followed by a pump. The remaining portion of the turbine flow expands to 0.06 bar. Determine the thermal efficiency.

**17-19M** Modify the data of Prob. 17-7M as follows. The steam is expanded to 10 bars, where a portion is bled to a single, closed feedwater heater which is followed by a pump. The remaining portion of the turbine flow is expanded to 0.08 bar. Determine the thermal efficiency.

**17-20M** Steam enters a turbine at 140 bars and 560°C and expands isentropically. A portion is bled at 15 bars for use in a closed feedwater heater, where it condenses to saturated liquid at this pressure. The other flow stream to the heater enters at 15 bars and 100°C and leaves at the same pressure. Compute the ratio of the mass flow rate from the bleed-point in the turbine to the mass flow rate of the other flow stream.

**17-21M** A direct-contact type of feedwater heater is to be used to supply 50,000 kg/h of heated water at the maximum possible temperature when steam is supplied at 1.5 bars and 90 percent quality. The condensate entering the heater is at a temperature of 40°C. Determine the amount of steam required, in kg/h.

**17-22M** An ideal regenerative cycle operates with steam entering the turbine at 40 bars and 440°C. The exhaust pressure is 0.06 bar. If a single, open feedwater heater is operated at 3 bars, compute the fraction of the steam bled from the turbine and the thermal efficiency of the cycle.

**17-23M** An ideal regenerative cycle operates with a turbine-inlet condition of 30 bars, 400°C, a single, open feedwater heater at 3 bars, and a condenser pressure of 0.06 bar. Calculate (a) the fraction of steam bled from the turbine for feedwater heating, (b) the work output of the turbine, in kJ/kg, and (c) the thermal efficiency of the cycle.

**17-24M** Steam enters the turbine of an ideal regenerative cycle at 40 bars and 500°C. Steam is then extracted at 7 and 3 bars and introduced into two open feedwater heaters that are in series. Appropriate pumps are employed after the condenser, which operates at 0.06 bar, and after each heater. Determine the fraction of the overall flow stream which is bled off into the 7-bar heater.

**17-25M** Steam is generated at 140 bars and 520°C. The steam is expanded through the first stage of a turbine to 40 bars. Part of the exhaust stream from this turbine is supplied to a closed feedwater heater, and the remainder is reheated to 520°C. The reheated steam is then expanded in a second stage of the turbine to an exhaust pressure of 0.08 bar. Some steam is bled from the second turbine at 7 bars for use in an open feedwater heater. The steam bled from the first turbine condenses as it passes through the closed heater and is throttled back into the open heater at 7 bars. The saturated liquid leaving the open heater first passes through a pump to 140 bars and then through the other side of the closed heater to the boiler. If we assume an ideal cycle and neglect pump work, determine (a) the percent of the steam entering the first turbine which goes to the closed heater, (b) the percent of the steam entering the second turbine which is bled to the open heater, and (c) the thermal efficiency.

### Effect of irreversibilities on vapor-cycle performance

**17-26M** The adiabatic efficiency of the turbine in a simple steam-power cycle is 85 percent. On this basis determine the thermal efficiency for the data given in (a) Prob. 17-1Ma, (b) Prob. 17-3Ma, (c) Prob. 17-3Mb, (d) Prob. 17-5M, (e) Prob. 17-6M, and (f) Prob. 17-7M.

**17-27M** The adiabatic efficiency of the turbine in a reheat cycle is 85 percent. On this basis determine the thermal efficiency for the data given in (a) Prob. 17-10Ma, (b) Prob. 17-11Ma, (c) Prob. 17-11Mb, (d) Prob. 17-12M, (e) Prob. 17-13M, and (f) Prob. 17-14M.

**17-28M** The adiabatic efficiency of the turbine in a regenerative cycle is 85 percent. On this basis determine the thermal efficiency for the data given in (a) Prob. 17-15Ma, (b) Prob. 17-16Ma, (c) Prob. 17-16Mb, (d) Prob. 17-17M, (e) Prob. 17-18M, and (f) Prob. 17-19M.

**17-29M** A steam-power cycle operates with a turbine-inlet temperature of 600°C and an exhaust pressure of 0.06 bar. If the turbine efficiency is 85 percent, what is the maximum turbine-inlet pressure which can be used if the exhaust moisture content is not to exceed 7.5 percent?

**17-30M** Steam at 60 bars enters a 20,000-kW condensing turbine and exhausts at 0.06 bar. The steam loses 2227 kJ/kg to the condenser cooling water, which increases 8°C in temperature while passing through the condenser. The cycle thermal efficiency is 34 percent, and the turbine adiabatic efficiency is 84 percent. Assuming that the steam condensate leaves the condenser as a liquid subcooled 1.16°C, determine the temperature of the steam entering the turbine, in °C.

**17-31M** Steam enters a boiler at 80 bars and 45°C and leaves the superheater at 80 bars and 520°C. It then expands through an adiabatic turbine with an adiabatic efficiency of 80 percent to 0.06 bar. Determine the thermal efficiency if the flow stream is reheated at 3 bars to 500°C during its passage through the turbine.

**17-32M** A steam cycle operates between a turbine-inlet condition of 40 bars and 500°C and an exhaust pressure of 0.06 bar. The efficiency of the turbine is 80 percent. The radiation loss for the turbine is 30 kJ/kg of steam entering the turbine.

(a) Determine the quality of steam leaving the turbine.

(b) If steam is extracted at 3 bars to heat the feedwater in an open heater, determine the number of kilograms of steam extracted per kilogram of steam entering the turbine. Assume constant engine efficiency in the turbine and 20 kJ/kg of radiation loss up to that point in the turbine.

(c) Determine the saving in heat supplied (from an outside source) in kJ/kg of steam entering the turbine. Disregard pump work.

**17-33M** A simple steam-power cycle uses solar energy for the heat input. Water in the cycle enters the pump as a saturated liquid at 40°C, and is pumped to 2 bars. It then evaporates in the boiler at this pressure, and enters the turbine as saturated vapor. At the turbine exhaust the conditions are 40°C and 8 percent moisture. The mass flow rate is 1500 kg/h, and the pump is rated at 0.3 kW. Determine (a) the net work output, in kJ/kg, (b) the energy-conversion efficiency, and (c) the area, in square meters of the solar collector needed if the collector picks up 700 J/(m²)(s).

### Reheat-regenerative cycles

**17-34M** An ideal reheat-regenerative steam cycle is operated with steam entering the high-pressure turbine at 200 bars and 600°C, and it is expanded to 10 bars. Part of the steam is extracted to heat the feedwater in an open feedwater heater (at 10 bars), and the remaining steam is reheated to 600°C and returned to a low-pressure turbine, where it is expanded to 0.04 bar. Determine the number of kilograms of steam which must be supplied per hour by the steam generator for an output of 100 MW by the turbine.

**17-35M** Consider a combined reheat-regenerative steam-turbine cycle where inlet conditions to the high-pressure turbine are 140 bars, and 600°C, and reheat is at 15 bars with a temperature at the low-pressure turbine inlet of 540°C. Closed feedwater heaters operate at pressures of 15 and 3 bars, with the trap outlet of the 3-bar heater connected to the condenser drain. There is just one feedwater pump located at the inlet to the 3-bar heater.

(a) Determine the enthalpy change of the feedwater as it passes through the 15-bar heater.

(b) Determine the ratio of the mass of steam bled at 15 bars to the mass of steam entering the high-pressure turbine.

**17-36M** Consider the following modification of Probs. 17-10M and 17-15M. The steam in the turbine is expanded to 7 bars, where a portion is bled to an open feedwater heater operating at this pressure. The remaining portion of the turbine flow is reheated to 500°C. Neglecting pump work, determine the thermal efficiency if the turbine exhaust is expanded to (a) 0.08 bar, and (b) 0.04 bar.

**17-37M** Consider the following modification of Probs. 17-12M and 17-17M. The steam is expanded to 15 bars, where a portion is bled to a single, open feedwater heater operating at this pressure. The remaining portion of the turbine flow is reheated to 540°C and then expanded to 0.06 bar. Neglecting pump work, determine the thermal efficiency.

**17-38M** An ideal, reheat-regenerative steam-power plant generates steam at 160 bars and 560°C which expands isentropically through the high-pressure side of the turbine to 15 bars. Steam is then extracted to heat the feedwater from the condenser in an open heater at 15 bars. The remaining part of the flow is reheated to 500°C and then expanded through the low-pressure turbine to 0.06 bar, isentropically. A low-pressure pump raises the pressure to 15 bars, and a high-pressure pump raises the pressure to 160 bars. Determine (a) the fraction of the total flow which is extracted to the heater, (b) the net work output in kJ/kg, and (c) the amount of heat added during reheating between turbine stages, in kJ/kg of total flow.

**17-39M** An ideal, reheat-regenerative cycle operates with conditions at the turbine inlet of 140 bars, and 600°C, and reheat is at 7 bars to 500°C. A closed feedwater heater operates at 15 bars, and the drain from the closed heater is trapped back into an open feedwater heater which operates at 3 bars. The condenser pressure is 0.06 bar. Determine (a) the thermal efficiency of the cycle, and (b) the mass flow rate through the steam generator required for a turbine output of 100,000 kW.

**17-40M** A steam-power generating station operates on a reheat-regeneration cycle, with part of the steam extracted for heating the feedwater and the rest of the steam reheated before being expanded in the low-pressure turbine. Steam enters the high-pressure turbine at 100 bars and 560°C and leaves at 10 bars. Part is then extracted for heating the feedwater in an open heater at 10 bars, and the rest is superheated to 540°C before entering the low-pressure turbine, where it is expanded to 0.06 bar. Pump work in the low-pressure pump is 5 kJ/kg, while pump work for the high-pressure pump is 50 kJ/kg. Assume reversible-adiabatic expansion in each turbine. Determine (a) the number of kilograms of steam extracted for heating the feedwater per kilogram of steam which entered the high-pressure turbine, and (b) the thermal efficiency of the cycle.

**Supercritical steam-power cycle**

**17-41M** A supercritical steam-power cycle operates with reheat and regeneration. The steam enters the turbine at 240 bars and 600°C and expands to 0.06 bar. Steam leaves the first stage at 30 bars, where part of it enters a closed heater, while the rest is reheated to 540°C. Both sections of the turbine have an adiabatic efficiency of 88 percent. There is a condensate pump between the condenser and the heater. Another pump for the extracted steam that is condensed lies between the heater and the condensate line from the heater. The total feedwater enters the steam generator at 280 bars. Determine (a) the enthalpies at the ten states around the cycle, (b) the fraction of the total mass flow rate which is extracted to the heater, and (c) the thermal efficiency of the cycle.

**17-42M** In a supercritical steam-power cycle steam expands ideally from 320 bars and 600°C to 240 bars. It is then cooled at constant pressure to 400°C, followed by isentropic compression to 320 bars. Heat is then added until the temperature reaches 600°C. Determine (a) the amount of heat added, in kJ/kg, (b) the amount of heat rejected, and (c) the thermal efficiency.

**17-43M** A supercritical steam-power plant operates with reheat and regeneration. The steam enters the turbine at 240 bars and 560°C and expands to 0.08 bar. Steam leaves the first stage at 40 bars. where part of it enters a closed heater, while the rest is reheated to 540°C. Both sections of the turbine have an adiabatic efficiency of 87 percent. There is a condensate pump between the main condenser and the closed heater. Another pump for the extracted steam that is condensed lies between the heater outlet and the condenser-outlet line from the heater. Determine (a) the enthalpies at the 10 states around the cycle, (b) the fraction of the total mass flow which is extracted to the heater, and (c) the thermal efficiency of the cycle.

### Steam cycles for heating and power

**17-44M** A steam plant is used for the supply of both heat and power. Steam enters the turbine at 30 bars and 320°C and expands to 0.06 bar. After expansion to 2 bars, steam is removed for heating purposes, but 2000 kg/h must always go through the low-pressure end of the turbine to keep the blades from overheating. The adiabatic efficiency of the turbine is 80 percent and its output is 3000 kW. Determine the maximum number of kg/h of steam which can be supplied for heating.

**17-45M** A steam-power plant is being used for both heating and power purposes. Steam enters the adiabatic turbine at 40 bars and 440°C, and extraction for heating purposes occurs at 3 bars. The condensate from the extraction flow stream enters a pump at 50°C, which pumps it back directly to the steam generator. The heating system provides 10 million kJ/h, and the turbine output is 5000 kW. If the turbine is 75 percent efficient and the exhaust pressure is 0.06 bar, determine the flow rate through the steam generator, in kg/h.

**17-46M** A steam-power station is being used for both power production and heating purposes. The power load is 20 million kJ/h, and the heating load is 45 million kJ/h. Steam is generated at 40 bars and 500°C and is expanded isentropically through a turbine to a condenser at 0.06 bar. The condensate leaves the condenser as saturated liquid. The heating load is supplied by bleeding steam off the turbine at 2 bars. The steam for heating is cooled by the external load down to saturated liquid at 2 bars, then pumped back to the boiler. Compute (a) the total steam leaving the boiler, in kg/h, (b) the heat input to the boiler, in kJ/h, and (c) the heat rejected to the condenser, in kJ/h.

### Rankine space power cycle

**17-47M** A Rankine cycle operating on potassium as the working fluid has a turbine-inlet state of 1325°K and 6 atm, and the temperature of the condensing fluid is 950°K.

(a) Determine the thermal efficiency for an ideal cycle.

(b) If the turbine and pump efficiencies are 85 and 60 percent, respectively, determine the thermal efficiency.

**17-48M** A Rankine cycle using potassium as the working fluid has a turbine inlet state of 1450°K and 10 atm, and the condenser operates at 900°K.

(a) Determine the quality at the exit of the turbine and the thermal efficiency for an ideal cycle.

(b) Repeat part (a) if the turbine and pump efficiencies are 82 and 50 percent, respectively.

**17-49M** A potassium Rankine cycle operates with a turbine-inlet state of 1200°K and 3 atm, and the condenser temperature is 900°K.

(a) Determine the thermal efficiency for an ideal cycle.

(b) Repeat part (a) if the turbine and pump efficiencies are 86 and 65 percent, respectively.

### Ideal vapor-compression refrigeration cycle

**17-50M** Refrigerant 12 is used as the working fluid in a heat pump which follows a reversed Carnot cycle and operates between 2 and 8 bars. In the condenser the refrigerant changes from dry saturated vapor to saturated liquid. Heat is absorbed in the evaporator at constant pressure. Determine (a) the coefficient of performance, and (b) the quality of the fluid leaving the evaporator.

**17-51M** Dry, saturated refrigerant 12 enters the compressor of a vapor-compression refrigerator at 1.6 bars; saturated-liquid refrigerant 12 enters the expansion valve at 7 bars. Assuming no pressure change in the condenser or evaporator, compute (a) the temperature of the fluid leaving the compressor (assume adiabatic reversible compression), (b) the change in specific entropy during the free-expansion process, and (c) the coefficient of performance.

**17-52M** An ideal vapor-compression refrigeration cycle uses refrigerant 12 and operates between 1.4 and 8 bars. Entering the compressor, the refrigerant 12 is superheated 1.9°C, and the fluid leaves the condenser as a saturated liquid. Determine (a) the coefficient of performance, (b) the maximum temperature in the cycle, and (c) the tons of refrigeration, if the input of the compressor is 3.7 kW.

**17-53M** An ideal vapor-compression refrigeration cycle uses refrigerant 12 and operates between 1.6 and 8 bars. Entering the compressor, the fluid is a saturated vapor. Determine (a) the coefficient of performance, and (b) the effective displacement of the compressor in l/min for a nominal 7-ton refrigeration plant.

**17-54M** An ideal vapor-compression refrigeration uses refrigerant 12 and operates between 2.8 and 8 bars. Entering the compressor, the fluid is superheated 2.9°C. Determine (a) the coefficient of performance, and (b) the effective displacement of the compressor in l/min for a 5-ton-capacity unit.

**17-55M** An ideal vapor-compression refrigeration cycle operates between evaporating and condensing temperatures of 0 and 36°C. Fluid leaves the evaporator as saturated vapor and enters the expansion valve as saturated liquid. For a refrigeration capacity of 3 tons, compute the coefficient of performance and the required power input (in kilowatts) if the fluid is (a) refrigerant 12, and (b) water.

**17-56M** The pressures in the evaporator and condenser of a 5-ton refrigeration plant operating on refrigerant 12 are 2 and 7 bars, respectively. For the ideal cycle the fluid enters the compressor as dry, saturated vapor, and no subcooling occurs in the condenser. Determine the power input to the compressor, in kilowatts.

### Nonideal vapor-compression refrigeration cycles

**17-57M** A refrigeration cycle operates with refrigerant 12 with an evaporator temperature of −18.49°C and a condenser temperature of 32.74°C. The vapor is at a temperature of 60°C at the end of compression, and there is no subcooling of the liquid leaving the condenser. The compressor is cooled externally, and the compressor cooling water removes 10 kJ/min. For 2 kg of fluid per minute, compute (a) the coefficient of performance, and (b) the condenser cooling water required, in kg/min, if the water-temperature rise is 6°C.

**17-58M** Refrigerant 12 leaves the evaporator of a vapor-compression refrigeration plant at 1.4 bars and −20°C, and is compressed to 7 bars and 50°C. The temperature of the fluid leaving the condenser is 24°C. Determine (a) the coefficient of performance, (b) the effective displacement of the compressor in l/min per ton of refrigeration, and (c) the compressor efficiency.

**17-59M** In a vapor-compression refrigeration cycle, refrigerant 12 leaves the evaporator as saturated vapor. The evaporator and condenser pressures are 1.4 and 10 bars, respectively. The fluid entering the condenser is at 80°C, and the refrigeration capacity is 10 tons.

(a) Determine the mass flow rate required in kg/min.

(b) Determine the power input in kilowatts.

(c) Determine the compressor efficiency.

**17-60M** A refrigeration cycle operates with refrigerant 12 at evaporator and condenser pressure of 1.6 and 9 bars, respectively. Saturated vapor enters the compressor. There is a negligible drop in pressure in the piping, but the adiabatic compressor is only 75 percent efficient. The rates capacity of the unit is 10 tons. The condenser is water-cooled, with the water entering at 18°C and leaving at 25°C.

(a) Sketch the cycle on a Ts diagram.

Then determine (b) the coefficient of performance, (c) the refrigerant mass flow rate, in kg/min, (d) the rate of heat rejection, in kJ/min, (e) the flow rate of cooling water, in kg/min, (f) the kilowatt requirement of the compressor, and (g) the fraction of the liquid entering the throttle valve that flashes to vapor.

**17-61M** In a vapor-compression refrigeration cycle which circulates refrigerant 12 at a rate of 6 kg/min, the refrigerant enters the compressor superheated 1 9°C at 1.4 bars and leaves at 7 bars. The fluid leaves the condenser as a saturated liquid. The compressor-adiabatic efficiency is 66.7 percent.

(a) Determine the coefficient of performance of the cycle.

(b) Compute the tons of refrigeration produced by the cycle.

(c) If an ideal turbine were used instead of an expansion valve or a capillary tube in the cycle, would the refrigeration capacity be increased or decreased for the same mass flow rate? Use a $Ts$ diagram to explain the effect qualitatively.

**17-62M** In a refrigerant 12 refrigeration cycle the evaporator and condenser pressures are 1.6 and 9 bars, respectively. Saturated vapor enters the compressor, and vapor leaves at 80°C. Saturated liquid leaves the condenser. When the refrigerating effect is 20 tons, cooling water passing through the compressor-cylinder water jacket picks up heat at the rate of 200 kJ/min. Calculate (a) the refrigerant flow rate, in kg/min, (b) the power input to the compressor, in kilowatts, and (c) the coefficient of performance.

**17-63M** A refrigeration system uses refrigerant 12 and operates with evaporator and condenser temperatures of −25 and 32°C, respectively. Saturated liquid leaves the condenser. Saturated vapor enters the compressor, where its pressure and temperature are increased to 8 bars and 60°C. Calculate (a) the coefficient of performance, (b) the refrigerant flow rate for a 10-ton unit, and (c) the adiabatic efficiency of the compressor.

**17-64M** In a vapor-compression refrigeration cycle, refrigerant 12 leaves the evaporator at 1.826 bars as a saturated vapor, and is compressed to 9 bars and 60°C. Heat loss from the compressor amounts to 6.0 kJ/kg. The refrigerant enters the expansion valve at 36°C, and the cycle delivers 3.5 tons of refrigeration. Compute (a) the coefficient of performance, and (b) the required compressor power input, in kilowatts.

**17-65M** A building requires 200,000 kJ/h of heat to maintain the interior air supply at 35°C when the outside temperature is − 10°C. The heat will be supplied by a heat-pump refrigeration system using refrigerant 12. The evaporator operates at a temperature 10° below the outside air temperature and the condenser operates at 10 bars. The compressor has an adiabatic efficiency of 75 percent. The fluid leaving the evaporator is a saturated vapor, and it is a saturated liquid leaving the condenser.

(a) What is the pressure in the evaporator?

(b) What is the minimum temperature difference in the condenser between the refrigerant and the heated air supply?

(c) What is the temperature of the fluid at the compressor outlet, in °C?

(d) Determine the quality of the fluid as it leaves the throttle valve.

### Liquefaction and solidification

**17-66M** Consider the basic process for liquefying gases shown in Fig. 17-15, without the compressor and aftercooler. Refrigerant 12 enters the heat exchanger at 16 bars and 80°C and is cooled by saturated vapor which is withdrawn from the heavily insulated separator. The high-pressure cooled gas leaves the heat exchanger and passes through a throttling valve into the separator. The separator contains liquid and vapor in equilibrium at 1.4 bars. The vapor leaving at low pressure from the heat exchanger has a temperature of 60°C. As a first approximation neglect all pressure losses except that across the throttling valve. Determine (a) the fraction of the inlet high-pressure gas which is liquefied, and (b) the temperature (if superheated) or the quality (if saturated) of the flow stream entering the valve.

**17-67M** Work Prob. 17-66M under the following set of conditions. Refrigerant 12 enters the heat exchanger at 14 bars and 60°C. The separator operates at 1.8 bars, and the vapor leaving the heat exchanger at low pressure has a temperature of 30°C.

### Stirling refrigeration cycle

**17-68M** A Stirling cycle is used for refrigeration purposes. Air at 230°K and 10 bars is expanded isothermally to 1 bar. It is then heated to 315°K at constant volume. Compression at a constant temperature of 315°K follows, and the cycle is completed by constant-volume heat removal. Compute the coefficient of performance for the cycle and the quantity $q_L$, in kJ/kg.

**17-69M** Air enters a Stirling cycle used for refrigeration purposes at 240°K and 8 bars and expands isothermally to 1.2 bars. It is then heated to 320°K at constant volume. Compression at a constant temperature of 320°K follows, and the cycle is completed by constant-volume heat removal. Compute the coefficient of performance for the cycle and the quantity $q_L$, in kJ/kg.

**17-70M** Work Prob. 17-68M if the working fluid is carbon dioxide and it is assumed to behave as an ideal gas.

**17-71M** Work Prob. 17-69M if the working fluid is argon.

# PROBLEMS (USCS)

### The Rankine cycle

**17-1** A power plant operating on the Rankine cycle has steam entering the turbine at 550 psia and 800°F. If the turbine output is equivalent to 10,000 kW, determine the efficiency of the cycle and the mass flow rate of steam for condenser pressures of (a) 1 psia and (b) 0.6 psia.

**17-2** A Rankine cycle has an exhaust pressure of 1 psia. Determine the moisture content leaving the turbine and the thermal efficiency of a cycle which has a turbine-inlet pressure of 800 psia and a turbine-inlet temperature of (a) 1000°F, (b) 800°F, and (c) 600°F.

**17-3** A Rankine cycle has an exhaust pressure of 1 psia. Determine the moisture content leaving the turbine and the thermal efficiency of a cycle which has a turbine-inlet temperature of 1000°F and a turbine-inlet pressure of (a) 1600 psia, (b) 800 psia, and (c) 400 psia.

**17-4** An ideal steam turbine operates initially with inlet conditions of 400 psia and 600°F and with an exhaust pressure of 2.0 psia. In order to reduce the work output, the steam may first be throttled to a lower pressure and then expanded to the same exhaust pressure. What pressure must the inlet steam be throttled to, in order to reduce the work output to (a) 80 percent and (b) 70 percent of its normal output?

**17-5** An ideal Rankine-cycle steam-power plant generates steam at 2000 psia and 1000°F, and condenses steam at 1.0 psia.

    (a) Determine the thermal efficiency.

    (b) If the plant produces 100 MW of power, determine the mass flow rate of steam, in lb/h.

**17-6** An ideal Rankine steam-power cycle which produces 125 MW of turbine power has the following operating conditions: turbine inlet, 1000 psia and 1000°F; condenser pressure, 1.0 psia. Compute the required mass flow rate and the cycle efficiency.

**17-7** A Rankine steam-power cycle operates with turbine-inlet conditions of 800 psia and 1050°F. The condenser pressure is 1 psia. Determine the thermal efficiency.

**17-8** A steam-generating unit has a throughput of 50,000 lb of steam per hour. This unit, running on the Rankine cycle, is composed only of a boiler and superheater; the boiler receives $507.3 \times 10^5$ Btu/h. Water enters the boiler at 450 psia and 200°F. After passing through the superheater, the steam goes to a turbine, which produces net work. The steam is then condensed in the condenser, where $2.89 \times 10^6$ lb/h of cooling water is circulated with a temperature rise of 15°F. The steam leaves the condenser at 1.4 psia and 110°F.

    (a) Determine the enthalpy of the steam leaving the boiler, in Btu/lb.

    (b) Determine the enthalpy of steam entering the condenser, in Btu/lb.

    (c) From a Mollier diagram, determine approximately the quality of steam entering the condenser.

    (d) Using a Mollier diagram, determine the enthalpy of the steam entering the turbine, assuming isentropic expansion through the turbine.

    (e) Calculate the horsepower output of the ideal turbine.

**17-9** A Rankine power cycle uses solar energy for the heat input and refrigerant 12 as the working fluid. Fluid enters the pump as a saturated liquid at 120 psia, and is pumped to 400 psia. It then passes through the steam-generator section at this pressure, and enters the turbine at 220°F. The mass flow rate is 1000 lb/h. Determine (a) the net work output in Btu/lb, (b) the thermal efficiency, and (c) the area, in square feet, of solar collector needed if the collector absorbs 220 Btu/(ft$^2$)(h).

### Reheat cycle

**17-10** Modify the data of Prob. 17-1 in the following manner. The steam to the turbine is expanded to 100 psia, reheated to 800°F, and then expanded to (a) 1 psia, and (b) 0.6 psia. Neglect the pump work and compute the thermal efficiency of the cycle.

**17-11** Modify the data of Prob. 17-3 in the following manner. The steam in the turbine is expanded to 140 psia, reheated to 1000°F, and then expanded to 1 psia. Neglect pump work and calculate the thermal efficiency for turbine-inlet pressures of (a) 1600 psia, (b) 800 psia, and (c) 400 psia.

**17-12** The data of Prob. 17-5 are modified in the following manner. The steam in the turbine is expanded to 800 psia, reheated to 1000°F, and then expanded to 1 psia. Neglect the pump work and compute the thermal efficiency.

**17-13** An ideal reheat cycle which produces 125,000 kW of turbine power has the following given conditions: turbine inlet, 1000 psia and 1000°F; reheat, 200 psia, to 1000°F; condenser pressure, 1 psia. Compute the required mass flow rate of steam, in lb/h, and the cycle efficiency.

**17-14** The Rankine cycle in Prob. 17-7 is modified in the following manner. The steam is first expanded to 200 psia, reheated to 1000°F, and then expanded to 1 psia. Determine the thermal efficiency.

### Regenerative cycles

**17-15** Modify the data of Prob. 17-1 as follows. The steam is expanded to 100 psia, where a portion is bled to a single, open feedwater heater operating at this pressure. Determine the thermal efficiency for a condenser pressure of (a) 1 psia, and (b) 0.6 psia.

**17-16** The data of Prob. 17-3 are modified as follows. The steam is expanded in the turbine to 140 psia, where a portion is bled to a single, open feedwater heater. The remaining turbine flow expands to 1 psia. Determine the thermal efficiency for turbine-inlet pressures of (a) 1600 psia, (b) 800 psia, and (c) 400 psia.

**17-17** Consider the following modification of Prob. 17-5. Steam is expanded to 800 psia, where a portion is bled off to a single, closed feedwater heater. After condensing, this flow stream is pumped back into the main flow leaving the condenser pump. The remaining portion of the turbine flow is expanded to 1 psia. Determine the thermal efficiency.

**17-18** Modify Prob. 17-6 in the following manner. The steam is expanded in the turbine until a portion is bled off at 200 psia to a single, closed feedwater heater. The condensed steam is pumped back to the main flow coming from the condenser pump. The remaining portion of the turbine flow expands to 1 psia. Determine the thermal efficiency.

**17-19** Modify the data of Prob. 17-7 as follows. The steam expands in the turbine to 200 psia, where a portion is bled to a single, closed feedwater heater which is followed by a pump. The remaining portion of the turbine flow is expanded to 1 psia. Determine the thermal efficiency.

**17-20** Steam enters a turbine at 2000 psia and 1000°F, and expands isentropically. A portion is bled at 200 psia for use in a closed feedwater heater, where it condenses to saturated liquid at this pressure. The other flow stream to the heater enters at 200 psi and 235°F, and leaves at the same pressure. Compute the ratio of the mass flow rates of the two streams.

**17-21** A direct-contact type of feedwater heater is to be used to supply 100,000 lb/h of heated water at the maximum temperature possible when supplied with steam at 20 psia and 90 percent quality. The condensate entering the heater is at a temperature of 100°F. Determine the number of pounds of steam required per hour.

**17-22** An ideal regenerative cycle operates with steam entering the turbine at 500 psia and 800°F. The exhaust condition is 1 psia. If a single, open feedwater heater is operated at 40 psia, compute the efficiency of the cycle and the fraction of the steam bled from the turbine.

**17-23** An ideal regenerative cycle operates with throttle conditions at 400 psia and 700°F; a single, open feedwater heater at 80 psia; and a condenser pressure of 1 psia. Calculate (a) the fraction of steam bled from the turbine for feedwater heating, (b) the work output of the turbine, in Btu/lb, and (c) the thermal efficiency of the cycle.

**17-24** Steam enters the turbine of an ideal regenerative cycle at 500 psia and 900°F. Steam is then extracted at 90 and 30 psia and introduced into two open feedwater heaters that are in series. Appropriate pumps are employed after the condenser, which operates at 1 psia, and, after each heater. Determine the fraction of the flow stream which is bled off into the 90-psia heater.

**17-25** Steam is generated at 2000 psia and 1000°F. The steam is expanded through the first stage of a turbine to 600 psia. Part of the exhaust stream from this turbine is supplied to a closed-feedwater heater, and the remainder is reheated to 1000°F. The reheated steam is then expanded in a second turbine to an exhaust pressure of 1 psia. Some steam is bled from the second turbine at 100 psia for use in an open feedwater heater. The steam bled from the first turbine condenses as it passes through the closed heater and is throttled back into the open heater at 100 psia. The saturated liquid leaving the open heater first passes through a pump to 2000 psia and then through the other side of the closed heater to the boiler-superheater. If we assume an ideal cycle and neglect pump work, determine (a) the percent of the steam entering the first turbine which goes to the closed heater, (b) the percent of the steam entering the second turbine which is bled to the open heater, and (c) the thermal efficiency.

### Effect of irreversibilities on vapor cycle performance

**17-26** The adiabatic efficiency of the turbine in a simple steam-power cycle is 85 percent. On this basis determine the thermal efficiency for the data given in (a) Prob. 17-1a, (b) Prob. 17-3a, (c) Prob. 17-3b, (d) Prob. 17-5, (e) Prob. 17-6, and (f) Prob. 17-7.

**17-27** The adiabatic efficiency of the turbine in a reheat cycle is 85 percent. On this basis determine the thermal efficiency for the data given in (a) Prob. 17-10a, (b) Prob. 17-11a, (c) Prob. 17-11b, (d) Prob. 17-12, (e) Prob. 17-13, and (f) Prob. 17-14.

**17-28** The adiabatic efficiency of the turbine in a regenerative cycle is 85 percent. On this basis determine the thermal efficiency for the data given in (a) Prob. 17-15a, (b) Prob. 17-16a, (c) Prob. 17-16b, (d) Prob. 17-17, (e) Prob. 17-18, and (f) Prob. 17-19.

**17-29** A steam-power cycle operates with a turbine-inlet temperature of 1050°F and an exhaust pressure of 1 psia. If the turbine efficiency is 85 percent, what is the maximum inlet turbine pressure which can be used if the exhaust moisture content is not to exceed 8 percent?

**17-30** Steam at 800 psia enters a 20,000-kW condensing turbine and exhausts at 1.0 psia. The steam loses 956 Btu/lb to the condenser cooling water, which increases 15°F in temperature while passing through the condenser. The Rankine-cycle efficiency is 32.5 percent; the turbine-adiabatic efficiency is 82 percent. Assuming that the steam condensate leaves the condenser as a liquid subcooled 3.6°F, determine the temperature of the steam entering the turbine.

**17-31** Steam enters a boiler at 800 psia and 110°F and leaves the superheater at 800 psia and 900°F. It then expands through an adiabatic turbine with efficiency of 80 percent to 1 psia. Determine the thermal efficiency for the same steam-generator and turbine-exhaust conditions if the flow stream is reheated at 50 psia to 900°F.

**17-32** A steam cycle operates between 600 psia, 900°F, and 1 psia. The efficiency of the turbine is 75 percent. Radiation loss is 15 Btu/lb$_m$ of steam entering.

(a) Determine the quality of steam entering the turbine.

(b) If steam is extracted at 40 psia to heat the feedwater in an open heater, determine the number of pounds of steam extracted per pound of steam entering the turbine. Assume constant engine efficiency in the turbine and 10 Btu/lb$_m$ heat loss due to radiation up to that point.

(c) Determine the saving in heat supplied (from an outside source), in Btu/lb$_m$ of steam entering the turbine. Disregard pump work.

**17-33** A simple steam-power cycle uses solar energy for the heat input. Water in the cycle enters the pump as a saturated liquid at 120°F, and is pumped to 30 psia. It then evaporates in the boiler at this pressure, and enters the turbine as saturated vapor. At the turbine exhaust the conditions are 120°F and 6 percent moisture. The flow rate is 300 lb/h, and the pump is rated at $\frac{1}{2}$ hp. Determine (a) the net work output, (b) the energy-conversion efficiency, and (c) the area, in square feet, of solar collector needed if the collectors pick up 225 Btu/(sq ft)(h).

## Reheat-regenerative cycles

**17-34** An ideal reheat-regenerative steam cycle is operated with steam entering the high-pressure turbine at 3000 psia and 1100°F, and it is expanded to 100 psia. Part of the steam is extracted to heat the feedwater in an open feedwater heater (at 100 psia), and the remaining steam is reheated to 1100°F and returned to a low-pressure turbine, where it is expanded to 0.5 psia back pressure. Determine the number of pounds of steam which must be supplied per hour by the steam generator for an output of 50,000 hp by the turbine.

**17-35** Consider a combined reheat-regenerative steam-turbine cycle where inlet conditions to the high-pressure turbine are 2000 psia and 1100°F and reheat is at 200 psia with a temperature at the low-pressure turbine inlet of 1000°F. Closed feedwater heaters operate at pressures of 200 and 50 psia, with the trap outlet of the 50-psia heater connected to the condenser drain. There is just one feedwater pump located at the inlet to the 50-psia heater.

    (a) Determine the enthalpy change of the feedwater as it passes through the 200-psia heater.

    (b) Determine the ratio of the mass of steam bled at 200 psia to the mass of steam entering the high-pressure turbine.

**17-36** Consider the following modification of Probs. 17-10 and 17-15. The steam is expanded to 100 psia, where a portion is bled to a single, open feedwater heater operating at this pressure. The remaining portion of the turbine flow is reheated to 800°F. Neglecting pump work, determine the thermal efficiency if the turbine exhaust is expanded to (a) 1 psia, and (b) 0.6 psia.

**17-37** Consider the following modification to Probs. 17-12 and 17-17. The steam is expanded to 800 psia, where a portion is bled to a single, closed feedwater heater. The remaining portion of the turbine flow is reheated to 1000°F and then expanded to 1 psia. The condensed steam leaving the heater is pumped back to the main flow leaving the condenser pump. Determine the thermal efficiency.

**17-38** An ideal reheat-regenerative steam power plant generates steam at 2000 psia and 1000°F, which expands isentropically through the high-pressure side of the turbine to 200 psia. Steam is then extracted to heat the feedwater from the condenser in an open heater at 200 psia. The remaining part of the flow is reheated to 900°F and then expanded through the low-pressure turbine to 1 psia, isentropically. A low-pressure pump raises the pressure to 200 psia, and a high-pressure pump raises the pressure to 2000 psia. Determine the fraction of the total flow which is extracted to the heater.

**17-39** An ideal reheat-regenerative cycle operates with throttle conditions at the turbine inlet of 2000 psia and 1100°F, and reheat is at 100 psia to 1000°F. A closed feedwater heater operates at 200 psia, and the drain from the closed heater is trapped back into an open feedwater heater which operates at 30 psia. The condenser pressure is 1 psia. Determine (a) the efficiency of the cycle, and (b) the required mass flow rate through the steam generator for a turbine output equivalent to 100,000 kW.

**17-40** A steam-power generating station operates on the reheat-regeneration cycle, with part of the steam extracted for heating the feedwater and the rest of the steam reheated before being expanded to the back pressure in the low-pressure turbine. Steam enters the high-pressure turbine at 1300 psia, 900°F, and leaves the high-pressure turbine at 100 psia. Part is extracted for heating the feedwater in an open heater at 100 psia, and the rest is superheated to 900°F before entering the low-pressure turbine, where it is expanded to 1 psia. Pump work in the low-pressure pump is 0.3 Btu/lb$_m$. Pump work in the high-pressure pump is 4.0 Btu/lb$_m$. Assume reversible, adiabatic expansion in each turbine. Determine (a) the number of pounds mass of steam extracted for heating per pound mass of steam which entered the high-pressure turbine, and (b) the thermal efficiency of the cycle.

## Supercritical steam power cycle

**17-41** A supercritical steam-power cycle operates with regeneration and reheat. The steam enters the turbine at 3500 psia, 1200°F, and expands to 1 psia. Steam leaves the first stage at 200 psia, where part of it enters a closed heater, while the rest is reheated to 1000°F. Both sections of the turbine have an adiabatic efficiency of 89 percent. There is a condensate pump between the condenser and the heater.

Another pump for the extracted steam that is condensed lies between the heater and the condensate-outlet line from the heater. The total feedwater enters the steam generator at 4000 psia. Determine (a) the enthalpies at the 10 states around the cycle, (b) the fraction of the total mass flow rate which is extracted to the heater, and (c) the thermal efficiency of the cycle.

**17-42** In a supercritical steam-power cycle steam expands ideally from 4800 psia and 1100°F to 3500 psia. It is then cooled at constant pressure to 650°F, followed by compression to 4800 psia. Heat is then added until the temperature reaches 1100°F. Determine (a) the amount of heat added, in Btu/lb, (b) the amount of heat rejected, and (c) the thermal efficiency.

**17-43** A supercritical steam-power plant operates with reheat and regeneration. The steam enters the turbine at 4400 psia, 1100°F, and expands to 1 psia. Steam leaves the first stage at 400 psia, where part of it enters a closed heater, while the rest is reheated to 1000°F. Both sections of the turbine have an adiabatic efficiency of 88 percent. There is a condensate pump between the main condenser and the heater. Another pump for the extracted steam that is condensed lies between the heater and the condensate-outlet line from the heater. Determine (a) the enthalpies at the 10 states around the cycle, (b) the fraction of the total mass flow which is extracted to the heater, and (c) the thermal efficiency of the cycle.

## Steam cycles for heating and power

**17-44** A steam plant is used for supplying both heat and power. Steam enters the turbine at 225 psia and 440°F, and expands to 1 psia. After expansion to 30 psia, steam is removed for heating purposes, but 3000 lb/h must always go through the low-pressure end of the turbine to keep the blades from overheating. The efficiency of the turbine is 80 percent and its output is 3000 kW. Determine the maximum number of pounds per hour of steam which can be supplied for heating.

**17-45** A steam-power plant is being used for both heating and power purposes. Steam enters the adiabatic turbine at 400 psia and 700°F, and extraction for heating purposes occurs at 20 psia. The condensate from the extraction flow stream enters a pump at 125°F, which pumps it back directly to the steam generator. The heating system provides 10 million Btu/h, and the turbine output is 8000 kW. If the turbine is 70 percent efficient and the exhaust pressure is 1 psia, determine the flow rate through the steam generator, in lb/h.

**17-46** A steam-generating power station is being used for both power production and heating purposes. The power load is 20 million Btu/h, and the heating load is 45 million Btu/h. Steam is generated at 500 psia and 900°F, and is expanded isentropically through a turbine to a condenser at 1 psia. The condensate leaves the condenser as saturated liquid. The heating load is supplied by bleeding steam off the turbine at 30 psia. The steam for heating is cooled by the external load down to saturated liquid at 30 psia, then pumped back to the boiler. Compute (a) the total number of pounds of steam leaving the boiler, in lb/h, (b) the heat input to the boiler in Btu/h, and (c) the amount of heat rejected to the condenser, in Btu/h.

## Rankine space power cycle

**17-47** A Rankine cycle with potassium as the working fluid has a turbine-inlet state of 2400°R and 90 psia, and the temperature of the condensing fluid is 1600°R.

(a) Determine the thermal efficiency for an ideal cycle.

(b) Determine the thermal efficiency if the turbine- and pump-adiabatic efficiencies are 85 and 60 percent, respectively.

**17-48** A Rankine cycle using potassium as the working fluid has a turbine-inlet state of 2600°R and 150 psia, and the condenser operates at 1700°R.

(a) Determine the quality at the turbine exit and the thermal efficiency for an ideal cycle.

(b) Repeat part (a) if the turbine and pump efficiencies are 84 and 50 percent, respectively.

**17-49** A Potassium-Rankine cycle operates with a turbine-inlet state of 2200°R and 50 psia, and the condenser temperature is 1600°R.

(a) Determine the thermal efficiency for an ideal cycle.

(b) Repeat part (a) if the turbine and pump efficiencies are 86 and 65 percent, respectively.

### Ideal vapor-compression refrigeration cycles

**17-50** Refrigerant 12 is used as a working fluid in a heat pump which follows the Carnot cycle (reversed) and operates between 40 and 160 psia. In the condenser the refrigerant 12 changes from dry, saturated vapor to saturated liquid. Heat is absorbed in the evaporator at constant pressure. Determine (a) the coefficient of performance, and (b) the quality of the fluid leaving the evaporator.

**17-51** Dry saturated refrigerant 12 enters the compressor of a vapor-compression refrigerator at 20 psia; saturated liquid enters the expansion valve at 160 psia. Assuming no pressure change in the condenser or evaporator, compute (a) the temperature of the fluid leaving the compressor (assume reversible, adiabatic compression), (b) the change in specific entropy during the free-expansion process, and (c) the coefficient of performance.

**17-52** An ideal vapor-compression refrigeration cycle uses refrigerant 12 and operates between 20 and 180 psia. Entering the compressor, the vapor is superheated 8°F, and the fluid leaves the condenser as a saturated liquid. Determine (a) the coefficient of performance, (b) the maximum temperature in the cycle, and (c) the number of tons of refrigeration, if the input to the compressor is 5 hp.

**17-53** An ideal vapor-compression refrigeration cycle uses refrigerant 12 and operates between 40 and 160 psia. Entering the compressor, the vapor is superheated 4.1°F. Determine (a) the coefficient of performance, and (b) the effective displacement of the compressor in $ft^3$/min for a nominal 7-ton refrigeration plant.

**17-54** An ideal vapor refrigeration cycle uses refrigerant 12 and operates between 30 and 200 psia. Entering the compressor, the vapor is superheated 8.9°. Determine (a) the coefficient of performance, and (b) the effective displacement of the compressor in $ft^3$/min for a 5-ton-capacity unit.

**17-55** An ideal vapor-compression refrigerating cycle operates between evaporating and condensing temperatures of 40 and 100°F. Fluid leaves the evaporator as saturated vapor and enters the expansion valve as saturated liquid. For a refrigeration capacity of 3 tons, compute the coefficient of performance and the required power input to the compressor if the fluid is (a) refrigerant 12, and (b) water.

**17-56** The pressures in the evaporator and condenser of a 5-ton refrigeration plant operating on refrigerant 12 are 29.335 and 100 psia, respectively. For the ideal cycle the fluid enters the compressor as dry, saturated vapor, and no subcooling occurs in the condenser. Determine the power input to the compressor.

### Nonideal vapor compression refrigeration cycles

**17-57** A refrigeration cycle operates with refrigerant 12 with an evaporator temperature of $-8.13$°F and a condenser pressure of 120 psia. The vapor is at a temperature of 150°F at the end of compression, and there is no subcooling of the liquid leaving the condenser. The compressor is cooled externally, and the compressor cooling water removes 10 Btu/min. For a flow rate of 4 lb/min, compute (a) the coefficient of performance, and (b) the flow rate of condenser cooling water required, in lb/min, if the water-temperature rise is 10°F.

**17-58** Refrigerant 12 leaves the evaporator of a vapor-compression refrigeration plant at 20 psia, and 0°F and is compressed to 100 psia and 120°F. The temperature leaving the condenser is 80°F. Determine (a) the coefficient of performance, and (b) the effective displacement per minute of the compressor per nominal ton of refrigeration.

**17-59** In a vapor-compression refrigeration cycle, refrigerant 12 leaves the evaporator as saturated, dry vapor. The evaporator and condenser pressures are 20 and 180 psia, respectively. The fluid entering the condenser is at 200°F. The refrigeration capacity is 10 tons.

(a) Determine the mass flow rate required, in lb/min.
(b) Determine the power input, in horsepower.
(c) Determine the compressor efficiency.

**17-60** A refrigeration cycle is operating at an evaporator pressure of 20 psia and a condenser pressure of 180 psia. Saturated vapor enters the compressor. There is a negligible drop in pressure in the piping, but the adiabatic compressor is only 75 percent efficient. The rated capacity of the unit is 10 tons. The condenser is water-cooled, with the water entering at 65°F and leaving at 80°F.

(a) Sketch the cycle on a $Ts$ diagram.

Then determine (b) the coefficient of performance, (c) the refrigerant 12 mass flow rate in lb/min, (d) the rate of heat rejection in Btu/min, (e) the flow rate of cooling water in lb/min, (f) the horsepower requirement of the compressor, and (g) the fraction of the liquid entering the throttle valve that flashes to vapor.

**17-61** In a vapor-compression refrigeration cycle, which circulates refrigerant 12 at a rate of 12 lb/min, the refrigerant enters the compressor superheated 8.1°F and 20 psia and leaves at 100 psia and 150°F. The fluid leaves the condenser as a saturated liquid

(a) Determine the coefficient of performance of the cycle.

(b) Compute the number of tons of refrigeration produced by the cycle.

(c) Determine the compressor efficiency.

(d) If an ideal turbine were used instead of an expansion valve or a capillary tube in the cycle, would the refrigeration capacity be increased or decreased for the same mass flow rate? Use a $Ts$ diagram of the cycle to explain the effect qualitatively.

**17-62** In a compression refrigeration cycle, the evaporator and condenser pressures are 20 and 180 psia, respectively. Dry, saturated vapor enters the compressor, and vapor at 160°F leaves. Saturated liquid leaves the condenser. When the refrigerating effect is 20 tons, cooling water passing through the compressor-cylinder water jacket picks up heat at a rate of 250 Btu/min. Calculate (a) the refrigerant 12 flow rate, in lb/min, (b) the power input to the compressor, in Btu/min, and (c) the coefficient of performance.

**17-63** A cooling system using refrigerant 12 operates with an evaporator temperature of −30°F and a condenser temperature of 100°F. Saturated liquid leaves the condenser. Dry, saturated vapor enters the compressor, where its pressure and temperature are increased to 140 psia and 140°F. Calculate (a) the coefficient of performance, and (b) the flow rate for a 10-ton capacity unit, assuming ideal throttling.

**17-64** In a vapor-compression refrigerator cycle, refrigerant 12 leaves the evaporator at 23.85 psia as a saturated vapor, and is compressed to 100 psia and 140°F. Heat loss from the compressor amounts to 3.5 Btu/lb of fluid. The refrigerant enters the expansion valve at 70°F, and the cycle delivers an equivalent of 3.5 tons of refrigeration. Compute (a) the coefficient of performance, and (b) the power input to the compressor.

**17-65** A building requires 200,000 Btu/h to maintain the interior at 95°F when the outside temperature is 10°F. The heat will be supplied by a heat-pump refrigeration system using refrigerant 12. The evaporator operates at a temperature 20° below the outside-air temperature and the condenser operates at 160 psia. The compressor has an adiabatic efficiency of 75 percent. The fluid leaving the evaporator is a saturated vapor, and it is a saturated liquid leaving the condenser.

(a) What is the pressure in the evaporator?

(b) What is the minimum temperature difference in the condenser between the refrigerant and the heated air?

(c) What is the temperature of the fluid at the compressor outlet, in °F?

(d) Determine the quality of the fluid as it leaves the throttle valve.

## Liquefaction and solidification

**17-66** Consider the equipment diagram shown in Fig. 17-15 and the $Ts$ diagram for the solidification of $CO_2$ shown in Fig. 17-16. Omitting the aftercooler, state 1 has a pressure of 1000 psia and a temperature of 90°F. After passing through the heat exchanger, the fluid is throttled to 50 psia. Employing the data from Fig. A-26, determine (a) the percent solid in the separation process, and (b) the temperature at state 5, in °F. The fluid at state 2 is 50 percent liquid at 1000 psia.

**17-67** Solid carbon dioxide is to be produced by throttling the fluid under the following steady-flow conditions:

(a) The gas is compressed to 600 psia and then precooled to 100°F. A pressure drop to 10 psi is incurred in the precooler. The fluid is passed through a heat exchanger and then throttled to 20 psia into a separator. The gas phase in the separator is passed back through the heat exchanger and discharged at 15 psia and 80°F. The solid $CO_2$ is removed at 20 psia. Determine (1) the mass fraction of solid $CO_2$ produced, and (2) the temperature and condition of the $CO_2$ entering the throttle device.

(b) Now the same process is carried out, except that the fluid is precooled another 60° down to 40°F before it enters the heat exchanger. The outlet temperature at 15 psia is also 60° lower than before. Repeat parts (1) and (2).

**17-68** Consider the basic process for liquefying gases as shown in Fig. 17-15, without the compressor and aftercooler. Refrigerant 12 enters the heat exchanger at 300 psia and 180°F and is cooled by saturated vapor which is withdrawn from the heavily insulated separator. The high-pressure cooled gas leaves the heat exchanger and passes through a throttling valve into the separator. The separator contains liquid and vapor refrigerant in equilibrium at 20 psia. The vapor leaving at low pressure from the heat exchanger has a temperature of 160°F. As a first approximation neglect all pressure losses except that across the throttling valve. Determine (a) the fraction of the inlet high-pressure gas which is liquefied, and (b) the temperature (if superheated) or the quality (if saturated) of the flow stream entering the valve.

**17-69** Work Prob. 17-68 under the following set of conditions. Refrigerant 12 enters the heat exchanger at 200 psia and 160°F. The separator operates at 20 psia and the vapor leaving the heat exchanger from the separator has a temperature of 150°F.

### Stirling refrigeration cycle

**17-70** Consider the use of a Stirling cycle for refrigeration purposes. Air at $-40$°F and 10 atm is expanded isothermally to 1 atm. It is then heated to 100°F at constant volume. Compression to 10 atm at a constant temperature of 100°F follows, and the cycle is completed by constant-volume heat removal. Compute the coefficient of performance for the cycle, using the air table.

**17-71** A Stirling cycle is used for refrigeration purposes. Air at $-20$°F and 8 atm is expanded isothermally to 1.2 atm. It is then heated to 120°F at constant volume. Compression to 8 atm at a constant temperature of 120°F follows, and the cycle is completed by constant-volume heat removal. Compute the coefficient of performance for the cycle, using the air table.

# THE METHODS OF LAGRANGIAN UNDETERMINED MULTIPLIERS

For a function represented by $F(x, y, z)$, the calculus requires that, at an extremum point such as a maximum or minimum, the function satisfy the relation

$$dF = \left(\frac{\partial F}{\partial x}\right)_{y, z} dx + \left(\frac{\partial F}{\partial y}\right)_{x, z} dy + \left(\frac{\partial F}{\partial z}\right)_{x, y} dz = 0 \qquad \text{(A-1)}$$

When $x$, $y$, and $z$ are independent variables, the coefficients of $dx$, $dy$, and $dz$ are required to be zero at the extremum of the function in order for Eq. (A-1) to be zero. Hence, at a maximum or minimum point,

$$\left(\frac{\partial F}{\partial x}\right)_{y, z} = 0 \qquad \left(\frac{\partial F}{\partial y}\right)_{x, z} = 0 \qquad \left(\frac{\partial F}{\partial z}\right)_{x, y} = 0 \qquad \text{(A-2)}$$

The partial derivatives in Eqs. (A-2) are a function of $x$, $y$, and $z$, in general. This set of equations gives us the required number of equations to solve for the values of the independent variables at the extremum condition.

In some cases the approach to finding a maximum or minimum point is complicated somewhat since there may be one or more constraints on the possible values of $x$, $y$, and $z$. That is, the values of the independent variables at the extremum must satisfy one or more relations of the form $G_n(x, y, z) = c$, where $c$ is a constant and $n$ is the number of constraining equations. The number of constraining equations must be less than the number of independent variables; otherwise an extremum point cannot be found for the function.

In this new situation a conditional maximum or minimum exists at some state $(x_0, y_0, z_0)$. Equation (A-1) is still valid, but we realize that $dx$, $dy$, and $dz$ are no longer independent of one another, since relations of the type $G_n(x, y, z) = c$ must also be fulfilled. Since these functional relationships equal different constants, the total derivatives of the constraining equations must be zero. Hence we may write for one such equation

$$dG = \left(\frac{\partial G}{\partial x}\right)_{y, z} dx + \left(\frac{\partial G}{\partial y}\right)_{x, z} dy + \left(\frac{\partial G}{\partial z}\right)_{x, y} dz = 0 \qquad \text{(A-3)}$$

This shows that $dx$, $dy$, and $dz$ are interrelated by Eq. (A-3) as well as Eq. (A-1). One straightforward method of solving for such a conditional extremum is as follows: For the case of one constraining equation the relationship $G(x, y, z) = c$ may be used to eliminate one of the variables from the function $F(x, y, z)$. This function can be maximized with respect to the remaining two variables, which are now independent of each other. By setting the coefficients of the two remaining variables equal to zero, the values of $x_0$, $y_0$, and $z_0$ at the extremum condition are determined.

A more general method of solution for the foregoing type of problem was developed by Lagrange. It is termed the method of undetermined multipliers. In general, we wish to formulate an equation like (A-1) in which the variables $dx$, $dy$, and $dz$ are independent of one another. If any additional constraints are present, this normally will not be true. Once the variables are made independent, however, their coefficients may then be set equal to zero. This is accomplished in the following manner.

At the conditional extremum both $dF$ and $dG$ equal zero. Then it is also true that $A\, dG = 0$, where $A$ is some arbitrary undetermined constant multiplier. Thus we may write

$$dF - A\, dG = 0 \qquad \text{(A-4)}$$

The quantities $dF$ and $dG$ are now replaced by Eqs. (A-1) and (A-3), and common terms are collected. The result is

$$\left[\left(\frac{\partial F}{\partial x}\right) - A\left(\frac{\partial G}{\partial x}\right)\right] dx + \left[\left(\frac{\partial F}{\partial y}\right) - A\left(\frac{\partial G}{\partial y}\right)\right] dy + \left[\left(\frac{\partial F}{\partial z}\right) - A\left(\frac{\partial G}{\partial z}\right)\right] dz = 0$$
$$\text{(A-5)}$$

The subscripts on the partial derivatives have been omitted to simplify the equation. In expression (A-5) still only two of the variations $dx$, $dy$, and $dz$ are independent. However, one term in Eq. (A-5) can be eliminated by the proper choice of the constant $A$, since $A$ is any arbitrary value. One proper choice of $A$ is that

$$A = \frac{(\partial F/\partial z)_{x_0, y_0, z_0}}{(\partial G/\partial z)_{x_0, y_0, z_0}} \qquad \text{(A-6)}$$

The subscripts $x_0$, $y_0$, and $z_0$ again simply denote that the partials must be evaluated at the maximum point. Since the location of the extremum is yet unknown, the value of $A$ likewise is not yet known. With the assignment to $A$ of the value given by Eq. (A-6), the final term in Eq. (A-5) drops out. At this point the remaining variables $dx$ and $dy$ can be varied independently. Thus we are required to set the coefficients of $dx$ and $dy$ equal to zero in order to satisfy Eq. (A-5), since $dx$ and $dy$ can take on nonzero values. Hence, as a solution,

$$\left(\frac{\partial F}{\partial x}\right)_{y,z} - A\left(\frac{\partial G}{\partial x}\right)_{y,z} = 0 \quad \text{and} \quad \left(\frac{\partial F}{\partial y}\right)_{x,z} - A\left(\frac{\partial G}{\partial y}\right)_{x,z} = 0 \quad \text{(A-7)}$$

The two expressions in Eq. (A-7) constitute in this case two equations with three unknowns, namely, $x_0$, $y_0$, and $A$. Equation (A-6) gives a third equation, but at the same time introduces a fourth unknown, $z_0$. The additional relationship we need is given by the constraining equation itself, $G(x_0, y_0, z_0) = c$. Such an expression introduces no other unknown quantities than those already sought.

In actual practice, one follows a less formal procedure once the mathematical justification is recognized. An equation like Eq. (A-5) is set up which contains $m$ differential terms with coefficients. All the $m$ coefficients are set equal to zero. In addition, there are $n$ constraining equations. These $m + n$ equations are solved for the $m + n$ unknowns, where $m$ is the number of independent variables and $n$ is the number of lagrangian multipliers. The following example illustrates the general method.

It is desired to find the minimum distance from the line $ax + by + c = 0$ to the origin. It should be clear from geometric considerations that such a minimum exists. If a circle of radius $r$ is drawn with the origin at its center, the distance we seek is the radius of a circle which is just tangent to the line. In such a case the radius will be perpendicular to the line and hence will represent the minimum distance. Thus we must determine the minimum radius of a circle such that a single point on it also lies on the line $ax + by + c = 0$. Let the function to be minimized by given by

$$F(x, y) = x^2 + y^2 = r^2$$

subject to the constraint

$$G(x, y) = ax + by + c = 0$$

By differentiating each of these equations we obtain

$$dF = 2x\,dx + 2y\,dy \quad \text{and} \quad dG = a\,dx + b\,dy$$

The expression for $dG$ is now multiplied by the multiplier $A$, and the result is subtracted from $dF$. In order to be a minimum radius, the final expression must equal zero; hence

$$2x\,dx + 2y\,dy - Aa\,dx - Ab\,dy = 0$$

Separation of variables yields

$$(2x - Aa)\,dx + (2y - Ab)\,dy = 0$$

By proper selection of the value of $A$, the coefficient of $dy$ becomes zero. Consequently, $x$ then becomes an independent variable, and the coefficient of $dx$ may be set equal to zero. Hence $A = 2x/a$. By substituting this result into the coefficient of $dy$, which was arbitrarily made equal to zero, we obtain $b = 2y/A = 2ya/2x = ya/x$. This relation, in conjunction with the constraining equation $ax + by + c = 0$, enables one to determine $x$ and $y$ in terms of the constants $a$, $b$, and $c$, and hence to evaluate the radius $r$. It may easily be verified that the simultaneous solution of these two equations leads to

$$x = \frac{-ac}{a^2 + b^2} \qquad y = \frac{-bc}{a^2 + b^2} \qquad r = \frac{c}{(a^2 + b^2)^{1/2}}$$

This may be confirmed in an actual example by considering the line passing through the points $(2, 0)$ and $(0, 2)$ on cartesian rectangular coordinates and showing mathematically and graphically that $r = (2)^{1/2}$.

# DERIVATION OF EQ. (9-11)

In Sec. 9-7 it is desired to develop an expression for $(\partial \ln W_{mp})/\partial n_i$. Mathematically, this requires the maximization of the function, $\ln W$, in conjunction with certain constraining equations. This type of problem is usually solved in the following manner. First, the constraining equations are substituted into the general equation, thus eliminating several of the independent variables. For example, if there are two constraining equations, one can use these equations to eliminate two of the independent variables in the general equation. Following this substitution, the general equation is differentiated with respect to the remaining independent variables, and the total differential is set equal to zero. Then by equating to zero the coefficients in front of the derivative of each independent variable, one determines the values of the independent variables which maximize the function. Our approach will be basically the same as this, with one slight modification. Instead of substituting the constraining equations and then differentiating the general equation, we shall first differentiate both the general equation and the constraining equations and then substitute the differentiated forms of the general equation. Using this combination of equations, we then set the coefficients in front of the remaining independent variables equal to zero. The end result is exactly the same as the method above, since the order of the mathematical steps is immaterial.

First we differentiate $\ln W$ with respect to $n_i$ for each energy level, and then set the resulting total differential equal to zero. Therefore,

$$d \ln W = \sum_i \frac{\partial \ln W}{\partial n_i} dn_i = 0 \tag{A-8}$$

where the summation is over all the [...]
maximizing the function, $\ln W$, the part [...]
solely to the most probable macrostate. [...]
subscript "mp" has not been placed on the [...]
striction. However, this restriction must be [...]
constraining equations. In past discussions w [...]
$U$ and the total number of particles $n$ be conse [...]
These constraining equations are given by

2, 3, etc. Sinc[...]
he above equ[...]s re-
mplify the [.]duce the
[.]atives to [...]
nd. No[...]e total energy
equire[.]ering macrostates.
[...]hen [.]

$$n = \sum_i n_i \tag{A-9}$$

and

$$U = \sum_i n_i \epsilon_i \tag{A-10}$$

If these equations are differentiated, holding the $\epsilon_i$ values constant during the differentiation, we find that

$$dn_1 + dn_2 + \cdots + dn_a + dn_b + \cdots + dn_i + \cdots = 0$$

and

$$\epsilon_1\, dn_1 + \epsilon_2\, dn_2 + \cdots + \epsilon_a\, dn_a + \epsilon_b\, dn_b + \cdots + \epsilon_i\, dn_i + \cdots = 0$$

Two of the energy levels have been given special identifying subscripts $a$ and $b$. These two summations can be written more compactly as

$$dn_a + dn_b + \sum_j dn_j = 0 \tag{A-11}$$

and

$$\epsilon_a\, dn_a + \epsilon_b\, dn_b + \sum_j \epsilon_j\, dn_j = 0 \tag{A-12}$$

where the sum of $j$ terms signifies a summation over all energy levels except the two levels denoted by $a$ and $b$, which are any two arbitrarily chosen levels. In a similar fashion we may rewrite Eq. (A-8) in the form

$$\frac{\partial \ln W}{\partial n_a}\, dn_a + \frac{\partial \ln W}{\partial n_b}\, dn_b + \sum_j \frac{\partial \ln W}{\partial n_j}\, dn_j = 0 \tag{A-13}$$

Next we solve Eqs. (A-11) and (A-12) for $dn_a$ and $dn_b$. Hence,

$$dn_a = \frac{\sum_j \epsilon_j\, dn_j - \epsilon_b \sum_j dn_j}{\epsilon_b - \epsilon_a}$$

and

$$dn_b = \frac{\epsilon_a \sum_j dn_j - \sum_j \epsilon_j\, dn_j}{\epsilon_b - \epsilon_a}$$

Insertion of these two expressions into Eq. (A-13) yields, after rearrangement,

$$\sum_j \left[ \frac{\partial \ln W}{\partial n_j} + \left( \frac{\epsilon_a \dfrac{\partial \ln W}{\partial n_b} - \epsilon_b \dfrac{\partial \ln W}{\partial n_a}}{\epsilon_b - \epsilon_a} \right) + \epsilon_j \frac{\dfrac{\partial \ln W}{\partial n_a} - \dfrac{\partial \ln W}{\partial n_b}}{\epsilon_b - \epsilon_a} \right] dn_j = 0 \tag{A-14}$$

In effect, we have used the two constraining equations to eliminate two of the variables in Eq. (A-13). Since two $dn_i$ terms have been arbitrarily eliminated in the above equation, the remaining $(n - 2)$ $dn_i$ terms are independent of one another. In our new notation, these remaining terms are indicated by a sum over $j$ terms.

Now note that all the quantities in the two bracketed terms in Eq. (A-14) refer only to energy levels $a$ and $b$. Since the choice of these two levels is completely arbitrary, the two bracketed terms must be constants. If they were not constants, we would get different solutions to Eq. (A-14) depending upon which energy levels were represented by $a$ and $b$. But there can be only one solution, since we are dealing with the most probable macrostate. We shall represent these constants by $-A$ and $-B$. Consequently Eq. (A-14) can be written as

$$\sum_j \left( \frac{\partial \ln W_{mp}}{\partial n_j} - A - B\epsilon_j \right) dn_j = 0$$

where we have reintroduced the subscript "mp" for the most probable macrostate. Now, if the $dn_j$ quantities can be varied independently, the only way for this summation to equal zero is for the coefficients of all the $dn_j$ terms to be zero. These coefficients all have the same form, so that

$$\frac{\partial \ln W_{mp}}{\partial n_j} - A - B\epsilon_j = 0$$

or

$$\frac{\partial \ln W_{mp}}{\partial n_j} = A + B\epsilon_j \tag{A-15}$$

Although Eq. (A-15) was derived for $j$ terms, the expression actually holds for energy levels $a$ and $b$ as well. This is easily proved by setting the two bracketed terms in Eq. (A-14) equal to $-A$ and $-B$, respectively, and solving. This proof is left as an exercise. Therefore Eq. (A-15) really applies to $(j + 2)$ levels, or all $i$ energy levels. That is, Eq. (A-15) is a general solution. This equation appears as Eq. (9-11) in the text, and is the desired result.

# SUPPLEMENTARY TABLES AND FIGURES (METRIC UNITS)

$g_c = 8.315 \; \dfrac{KJ}{Kg \, mol \, K}$

$g_c = .08315 \; \dfrac{bar \cdot m}{Kg \, mol \, K}$

ADIABATIC WORK

P.E. $W = \dfrac{mg}{g_c}(z_2 - z_1)$

K.E.

$W = \dfrac{m}{2g_c}(V_2^2 - V_1^2)$

### Table A-1M  Physical constants and conversion factors

#### Physical constants

| | |
|---|---|
| Avogadro's number | $N_A = 6.024 \times 10^{26}$ atoms/kg·mol |
| Universal gas constant | $R_u = 0.08205$ l·atm/(g·mol)(°K) |
| | $= 8.315$ kJ/(kg·mol)(°K) |
| | $= 0.08315$ bar·m³/(kg·mol)(°K) |
| Planck's constant | $h = 6.625 \times 10^{-34}$ J·s/molecule |
| Boltzmann's constant | $k = 1.380 \times 10^{-23}$ J/(°K)(molecule) |
| Speed of light | $c = 2.998 \times 10^{10}$ cm/s |
| Standard gravity | $g = 9.80665$ m/s² |

CONST. VOL. GAS THERMOMETER

$\downarrow P_2$

$\Delta P = -\rho g \Delta z$

ISOTHERMAL

boundary $W = -N R_U T \ln\left(\dfrac{V_2}{V_1}\right)$

SOLID & LIQUID

$W = \dfrac{N K_T}{2}(P_2^2 - P_1^2)$

#### Conversion factors

| | |
|---|---|
| Length | 1 cm = 0.3937 in = $10^4$ $\mu m$ = $10^8$ Å |
| | 1 km = 0.6215 m = 3281 ft |
| Mass | 1 kg = 2.2005 $lb_m$ |
| Force | 1 N = 1 kg·m/s² = 0.2248 $lb_f$ |
| Pressure | 1 bar = $10^5$ N/m² = 0.9869 atm |
| | = 100 kilopascals (kPa) |
| | 1 torr = 1 mm Hg at 0°C = 1.333 mbar |
| | = $1.933 \times 10^{-2}$ psi |
| | 1 mbar = 0.402 in $H_2O$ |
| Volume | 1 l = 0.0353 ft³ = 0.2642 gal = 61.025 in³ = $10^{-3}$ m³ |
| Density | 1 g/cm³ = 1 kg/l = 62.4 $lb_m$/ft³ = $10^3$ kg/m³ |
| Energy | 1 J = 1 N·m = 1 V·C |
| | = 0.7375 ft·$lb_f$ = 10 bar·cm³ = $0.624 \times 10^{19}$ eV |
| | 1 kJ = 0.948 Btu = 737.6 ft·$lb_f$ |
| | 1 kJ/kg = 0.431 Btu/lb |
| Power | 1 W = 1 J/s = 3.413 Btu/h |
| | 1 kW = 1.3405 hp = 737.3 ft·$lb_f$/s |
| Velocity | 1 m/s = 2.237 mi/h = 3.60 km/h = 3.281 ft/s |

## Table A-2M Values of the molar mass (molecular weight) of some common elements and compounds

$\frac{Kg}{Kg\text{-}mol}$

| Substance | Formula | Molar Mass | Substance | Formula | Molar Mass |
|---|---|---|---|---|---|
| Argon | Ar | 39.94 | Water | $H_2O$ | 18.02 |
| Aluminum | Al | 26.97 | Hydrogen | | |
| Carbon | C | 12.01 | peroxide | $H_2O_2$ | 34.02 |
| Copper | Cu | 63.54 | Ammonia | $NH_3$ | 17.04 |
| Helium | He | 4.003 | Hydroxyl | $-OH$ | 17.01 |
| Hydrogen | $H_2$ | 2.018 | Methane | $CH_4$ | 16.04 |
| Iron | Fe | 55.85 | Acetylene | $C_2H_2$ | 26.04 |
| Lead | Pb | 207.2 | Ethylene | $C_2H_4$ | 28.05 |
| Mercury | Hg | 200.6 | Ethane | $C_2H_6$ | 30.07 |
| Nitrogen | $N_2$ | 28.008 | Propylene | $C_3H_6$ | 42.08 |
| Oxygen | $O_2$ | 32.00 | Propane | $C_3H_8$ | 44.09 |
| Potassium | K | 39.096 | n-Butane | $C_4H_{10}$ | 58.12 |
| Silver | Ag | 107.88 | n-Pentane | $C_5H_{12}$ | 72.15 |
| Sodium | Na | 22.997 | n-Octane | $C_8H_{18}$ | 114.22 |
| Air | | 28.97 | Benzene | $C_6H_6$ | 78.11 |
| Carbon | | | Methyl alcohol | $CH_3OH$ | 32.05 |
| monoxide | CO | 28.01 | Ethyl alcohol | $C_2H_5OH$ | 46.07 |
| Carbon | | | Refrigerant 12 | $CCl_2F_2$ | 120.92 |
| dioxide | $CO_2$ | 44.01 | | | |

## Table A-3M Critical properties and van der Waals constants

| Substance | $T_c$, °K | $P_c$, bar | $v_c$ $\frac{m^3}{kg \cdot mol}$ | $Z_c = \frac{P_c v_c}{RT_c}$ | van der Waals $a$ $bar\left(\frac{m^3}{kg \cdot mol}\right)^2$ | van der Waals $b$ $\frac{m^3}{kg \cdot mol}$ |
|---|---|---|---|---|---|---|
| Acetylene ($C_2H_2$) | 309 | 62.8 | 0.112 | 0.274 | 4.410 | 0.0510 |
| Air (equivalent) | 133 | 37.7 | 0.0829 | 0.284 | 1.358 | 0.0364 |
| Ammonia ($NH_3$) | 406 | 112.8 | 0.0723 | 0.242 | 4.233 | 0.0373 |
| Benzene ($C_6H_6$) | 562 | 49.3 | 0.256 | 0.274 | 18.63 | 0.1181 |
| n-Butane ($C_4H_{10}$) | 425.2 | 38.0 | 0.257 | 0.274 | 13.80 | 0.1196 |
| Carbon dioxide ($CO_2$) | 304.2 | 73.9 | 0.0941 | 0.276 | 3.643 | 0.0427 |
| Carbon monoxide (CO) | 133 | 35.0 | 0.0928 | 0.294 | 1.463 | 0.0394 |
| Refrigerant 12 ($CCl_2F_2$) | 385 | 40.1 | 0.214 | 0.270 | 10.78 | 0.0998 |
| Ethane ($C_2H_6$) | 305.4 | 48.8 | 0.221 | 0.273 | 5.575 | 0.0650 |
| Ethylene ($C_2H_4$) | 283 | 51.2 | 0.143 | 0.284 | 4.563 | 0.0574 |
| Helium (He) | 5.2 | 2.3 | 0.0579 | 0.300 | 0.0341 | 0.0234 |
| Hydrogen ($H_2$) | 33.2 | 13.0 | 0.0648 | 0.304 | 0.247 | 0.0265 |
| Methane ($CH_4$) | 190.7 | 46.4 | 0.0991 | 0.290 | 2.285 | 0.0427 |
| Nitrogen ($N_2$) | 126.2 | 33.9 | 0.0897 | 0.291 | 1.361 | 0.0385 |
| Oxygen ($O_2$) | 154.4 | 50.5 | 0.0741 | 0.290 | 1.369 | 0.0315 |
| Propane ($C_3H_8$) | 370 | 42.7 | 0.195 | 0.276 | 9.315 | 0.0900 |
| Sulfur dioxide ($SO_2$) | 431 | 78.7 | 0.124 | 0.268 | 6.837 | 0.0568 |
| Water ($H_2O$) | 647.3 | 220.9 | 0.0558 | 0.230 | 5.507 | 0.0304 |

Source: Adapted from the data in Table A-3.

## Table A-4M Zero-pressure specific heats for various gases, kJ/(kg)(°C)

| Temp. °K | $c_p$ | $c_v$ | $k$ | $c_p$ | $c_v$ | $k$ | $c_p$ | $c_v$ | $k$ | Temp. °K |
|---|---|---|---|---|---|---|---|---|---|---|
| | | Air | | | Carbon dioxide, $CO_2$ | | | Carbon monoxide, CO | | |
| 250 | 1.003 | 0.716 | 1.401 | 0.791 | 0.602 | 1.314 | 1.039 | 0.743 | 1.400 | 250 |
| 300 | 1.005 | 0.718 | 1.400 | 0.846 | 0.657 | 1.288 | 1.040 | 0.744 | 1.399 | 300 |
| 350 | 1.008 | 0.721 | 1.398 | 0.895 | 0.706 | 1.268 | 1.043 | 0.746 | 1.398 | 350 |
| 400 | 1.013 | 0.726 | 1.395 | 0.939 | 0.750 | 1.252 | 1.047 | 0.751 | 1.395 | 400 |
| 450 | 1.020 | 0.733 | 1.391 | 0.978 | 0.790 | 1.239 | 1.054 | 0.757 | 1.392 | 450 |
| 500 | 1.029 | 0.742 | 1.387 | 1.014 | 0.825 | 1.229 | 1.063 | 0.767 | 1.387 | 500 |
| 600 | 1.051 | 0.764 | 1.376 | 1.075 | 0.886 | 1.213 | 1.087 | 0.790 | 1.376 | 600 |
| 700 | 1.075 | 0.788 | 1.364 | 1.126 | 0.937 | 1.202 | 1.113 | 0.816 | 1.364 | 700 |
| 800 | 1.099 | 0.812 | 1.354 | 1.169 | 0.980 | 1.193 | 1.139 | 0.842 | 1.353 | 800 |
| | | Hydrogen, $H_2$ | | | Nitrogen, $N_2$ | | | Oxygen, $O_2$ | | |
| 250 | 14.051 | 9.927 | 1.416 | 1.039 | 0.742 | 1.400 | 0.913 | 0.653 | 1.398 | 250 |
| 300 | 14.307 | 10.183 | 1.405 | 1.039 | 0.743 | 1.400 | 0.918 | 0.658 | 1.395 | 300 |
| 350 | 14.427 | 10.302 | 1.400 | 1.041 | 0.744 | 1.399 | 0.928 | 0.668 | 1.389 | 350 |
| 400 | 14.476 | 10.352 | 1.398 | 1.044 | 0.747 | 1.397 | 0.941 | 0.681 | 1.382 | 400 |
| 450 | 14.501 | 10.377 | 1.398 | 1.049 | 0.752 | 1.395 | 0.956 | 0.696 | 1.373 | 450 |
| 500 | 14.513 | 10.389 | 1.397 | 1.056 | 0.759 | 1.391 | 0.972 | 0.712 | 1.365 | 500 |
| 600 | 14.546 | 10.422 | 1.396 | 1.075 | 0.778 | 1.382 | 1.003 | 0.743 | 1.350 | 600 |
| 700 | 14.604 | 10.480 | 1.394 | 1.098 | 0.801 | 1.371 | 1.031 | 0.771 | 1.337 | 700 |
| 800 | 14.695 | 10.570 | 1.390 | 1.121 | 0.825 | 1.360 | 1.054 | 0.794 | 1.327 | 800 |

*Monatomic gases.* Over a wide range of temperatures at low pressures the specific heats $c_v$ and $c_p$ of all monatomic gases are essentially independent of temperature and pressure. On a molar basis all monatomic gases have the same value for either $c_v$ or $c_p$, and these values may be taken as

$$c_v = 12.5 \text{ kJ/(kg)(°C)} \quad \text{and} \quad c_p = 20.8 \text{ kJ/(kg)(°C)}$$

Data adapted from "Tables of Thermal Properties of Gases," NBS Circular 564, 1955.

## Table A-5M Ideal-gas properties of air $T$, °K; $h$, kJ/kg; $u$, kJ/kg; $s^0$, kJ/(kg)(°K)

| $T$ | $h$ | $p_r$ | $u$ | $v_r$ | $s^0$ | $T$ | $h$ | $p_r$ | $u$ | $v_r$ | $s^0$ |
|---|---|---|---|---|---|---|---|---|---|---|---|
| 200 | 199.97 | 0.3363 | 142.56 | 1707. | 1.29559 | 305 | 305.22 | 1.4686 | 217.67 | 596.0 | 1.71865 |
| 210 | 209.97 | 0.3987 | 149.69 | 1512. | 1.34444 | 310 | 310.24 | 1.5546 | 221.25 | 572.3 | 1.73498 |
| 220 | 219.97 | 0.4690 | 156.82 | 1346. | 1.39105 | 315 | 315.27 | 1.6442 | 224.85 | 549.8 | 1.75106 |
| 230 | 230.02 | 0.5477 | 164.00 | 1205. | 1.43557 | 320 | 320.29 | 1.7375 | 228.43 | 528.6 | 1.76690 |
| 240 | 240.02 | 0.6355 | 171.13 | 1084. | 1.47824 | 325 | 325.31 | 1.8345 | 232.02 | 508.4 | 1.78249 |
| 250 | 250.05 | 0.7329 | 178.28 | 979. | 1.51917 | 330 | 330.34 | 1.9352 | 235.61 | 489.4 | 1.79783 |
| 260 | 260.09 | 0.8405 | 185.45 | 887.8 | 1.55848 | 340 | 340.42 | 2.149 | 242.82 | 454.1 | 1.82790 |
| 270 | 270.11 | 0.9590 | 192.60 | 808.0 | 1.59634 | 350 | 350.49 | 2.379 | 250.02 | 422.2 | 1.85708 |
| 280 | 280.13 | 1.0889 | 199.75 | 738.0 | 1.63279 | 360 | 360.58 | 2.626 | 257.24 | 393.4 | 1.88543 |
| 285 | 285.14 | 1.1584 | 203.33 | 706.1 | 1.65055 | 370 | 370.67 | 2.892 | 264.46 | 367.2 | 1.91313 |
| 290 | 290.16 | 1.2311 | 206.91 | 676.1 | 1.66802 | 380 | 380.77 | 3.176 | 271.69 | 343.4 | 1.94001 |
| 295 | 295.17 | 1.3068 | 210.49 | 647.9 | 1.68515 | 390 | 390.88 | 3.481 | 278.93 | 321.5 | 1.96633 |
| 300 | 300.19 | 1.3860 | 214.07 | 621.2 | 1.70203 | 400 | 400.98 | 3.806 | 286.16 | 301.6 | 1.99194 |

## Table A-5M  (Continued)

| T | h | $p_r$ | u | $v_r$ | $s^0$ | T | h | $p_r$ | u | $v_r$ | $s^0$ |
|---|---|---|---|---|---|---|---|---|---|---|---|
| 410 | 411.12 | 4.153 | 293.43 | 283.3 | 2.01699 | 1000 | 1046.04 | 114.0 | 758.94 | 25.17 | 2.96770 |
| 420 | 421.26 | 4.522 | 300.69 | 266.6 | 2.04142 | 1020 | 1068.89 | 123.4 | 776.10 | 23.72 | 2.99034 |
| 430 | 431.43 | 4.915 | 307.99 | 251.1 | 2.06533 | 1040 | 1091.85 | 133.3 | 793.36 | 22.39 | 3.01260 |
| 440 | 441.61 | 5.332 | 315.30 | 236.8 | 2.08870 | 1060 | 1114.86 | 143.9 | 810.62 | 21.14 | 3.03449 |
| 450 | 451.80 | 5.775 | 322.62 | 223.6 | 2.11161 | 1080 | 1137.89 | 155.2 | 827.88 | 19.98 | 3.05608 |
| 460 | 462.02 | 6.245 | 329.97 | 211.4 | 2.13407 | 1100 | 1161.07 | 167.1 | 845.33 | 18.896 | 3.07732 |
| 470 | 472.24 | 6.742 | 337.32 | 200.1 | 2.15604 | 1120 | 1184.28 | 179.7 | 862.79 | 17.886 | 3.09825 |
| 480 | 482.49 | 7.268 | 344.70 | 189.5 | 2.17760 | 1140 | 1207.57 | 193.1 | 880.35 | 16.946 | 3.11883 |
| 490 | 492.74 | 7.824 | 352.08 | 179.7 | 2.19876 | 1160 | 1230.92 | 207.2 | 897.91 | 16.064 | 3.13916 |
| 500 | 503.02 | 8.411 | 359.49 | 170.6 | 2.21952 | 1180 | 1254.34 | 222.2 | 915.57 | 15.241 | 3.15916 |
| 510 | 513.32 | 9.031 | 366.92 | 162.1 | 2.23993 | 1200 | 1277.79 | 238.0 | 933.33 | 14.470 | 3.17888 |
| 520 | 523.63 | 9.684 | 374.36 | 154.1 | 2.25997 | 1220 | 1301.31 | 254.7 | 951.09 | 13.747 | 3.19834 |
| 530 | 533.98 | 10.37 | 381.84 | 146.7 | 2.27967 | 1240 | 1324.93 | 272.3 | 968.95 | 13.069 | 3.21751 |
| 540 | 544.35 | 11.10 | 389.34 | 139.7 | 2.29906 | 1260 | 1348.55 | 290.8 | 986.90 | 12.435 | 3.23638 |
| 550 | 554.74 | 11.86 | 396.86 | 133.1 | 2.31809 | 1280 | 1372.24 | 310.4 | 1004.76 | 11.835 | 3.25510 |
| 560 | 565.17 | 12.66 | 404.42 | 127.0 | 2.33685 | 1300 | 1395.97 | 330.9 | 1022.82 | 11.275 | 3.27345 |
| 570 | 575.59 | 13.50 | 411.97 | 121.2 | 2.35531 | 1320 | 1419.76 | 352.5 | 1040.88 | 10.747 | 3.29160 |
| 580 | 586.04 | 14.38 | 419.55 | 115.7 | 2.37348 | 1340 | 1443.60 | 375.3 | 1058.94 | 10.247 | 3.30959 |
| 590 | 596.52 | 15.31 | 427.15 | 110.6 | 2.39140 | 1360 | 1467.49 | 399.1 | 1077.10 | 9.780 | 3.32724 |
| 600 | 607.02 | 16.28 | 434.78 | 105.8 | 2.40902 | 1380 | 1491.44 | 424.2 | 1095.26 | 9.337 | 3.34474 |
| 610 | 617.53 | 17.30 | 442.42 | 101.2 | 2.42644 | 1400 | 1515.42 | 450.5 | 1113.52 | 8.919 | 3.36200 |
| 620 | 628.07 | 18.36 | 450.09 | 96.92 | 2.44356 | 1420 | 1539.44 | 478.0 | 1131.77 | 8.526 | 3.37901 |
| 630 | 638.63 | 19.48 | 457.78 | 92.84 | 2.46048 | 1440 | 1563.51 | 506.9 | 1150.13 | 8.153 | 3.39586 |
| 640 | 649.22 | 20.64 | 465.50 | 88.99 | 2.47716 | 1460 | 1587.63 | 537.1 | 1168.49 | 7.801 | 3.41247 |
| 650 | 659.84 | 21.86 | 473.25 | 85.34 | 2.49364 | 1480 | 1611.79 | 568.8 | 1186.95 | 7.468 | 3.42892 |
| 660 | 670.47 | 23.13 | 481.01 | 81.89 | 2.50985 | 1500 | 1635.97 | 601.9 | 1205.41 | 7.152 | 3.44516 |
| 670 | 681.14 | 24.46 | 488.81 | 78.61 | 2.52589 | 1520 | 1660.23 | 636.5 | 1223.87 | 6.854 | 3.46120 |
| 680 | 691.82 | 25.85 | 496.62 | 75.50 | 2.54175 | 1540 | 1684.51 | 672.8 | 1242.43 | 6.569 | 3.47712 |
| 690 | 702.52 | 27.29 | 504.45 | 72.56 | 2.55731 | 1560 | 1708.82 | 710.5 | 1260.99 | 6.301 | 3.49276 |
| 700 | 713.27 | 28.80 | 512.33 | 69.76 | 2.57277 | 1580 | 1733.17 | 750.0 | 1279.65 | 6.046 | 3.50829 |
| 710 | 724.04 | 30.38 | 520.23 | 67.07 | 2.58810 | 1600 | 1757.57 | 791.2 | 1298.30 | 5.804 | 3.52364 |
| 720 | 734.82 | 32.02 | 528.14 | 64.53 | 2.60319 | 1620 | 1782.00 | 834.1 | 1316.96 | 5.574 | 3.53879 |
| 730 | 745.62 | 33.72 | 536.07 | 62.13 | 2.61803 | 1640 | 1806.46 | 878.9 | 1335.72 | 5.355 | 3.55381 |
| 740 | 756.44 | 35.50 | 544.02 | 59.82 | 2.63280 | 1660 | 1830.96 | 925.6 | 1354.48 | 5.147 | 3.56867 |
| 750 | 767.29 | 37.35 | 551.99 | 57.63 | 2.64737 | 1680 | 1855.50 | 974.2 | 1373.24 | 4.949 | 3.58335 |
| 760 | 778.18 | 39.27 | 560.01 | 55.54 | 2.66176 | 1700 | 1880.1 | 1025 | 1392.7 | 4.761 | 3.5979 |
| 780 | 800.03 | 43.35 | 576.12 | 51.64 | 2.69013 | 1750 | 1941.6 | 1161 | 1439.8 | 4.328 | 3.6336 |
| 800 | 821.95 | 47.75 | 592.30 | 48.08 | 2.71787 | 1800 | 2003.3 | 1310 | 1487.2 | 3.944 | 3.6684 |
| 820 | 843.98 | 52.49 | 608.59 | 44.84 | 2.74504 | 1850 | 2065.3 | 1475 | 1534.9 | 3.601 | 3.7023 |
| 840 | 866.08 | 57.60 | 624.95 | 41.85 | 2.77170 | 1900 | 2127.4 | 1655 | 1582.6 | 3.295 | 3.7354 |
| 860 | 888.27 | 63.09 | 641.40 | 39.12 | 2.79783 | 1950 | 2189.7 | 1852 | 1630.6 | 3.022 | 3.7677 |
| 880 | 910.56 | 68.98 | 657.95 | 36.61 | 2.82344 | 2000 | 2252.1 | 2068 | 1678.7 | 2.776 | 3.7994 |
| 900 | 932.93 | 75.29 | 674.58 | 34.31 | 2.84856 | 2050 | 2314.6 | 2303 | 1726.8 | 2.555 | 3.8303 |
| 920 | 955.38 | 82.05 | 691.28 | 32.18 | 2.87324 | 2100 | 2377.4 | 2559 | 1775.3 | 2.356 | 3.8605 |
| 940 | 977.92 | 89.28 | 708.08 | 30.22 | 2.89748 | 2150 | 2440.3 | 2837 | 1823.8 | 2.175 | 3.8901 |
| 960 | 1000.55 | 97.00 | 725.02 | 28.40 | 2.92128 | 2200 | 2503.2 | 3138 | 1872.4 | 2.012 | 3.9191 |
| 980 | 1023.25 | 105.2 | 741.98 | 26.73 | 2.94468 | 2250 | 2566.4 | 3464 | 1921.3 | 1.864 | 3.9474 |

*Source:*  Adapted from Keenan, J. H., and J. Kaye, "Gas Tables," Wiley, New York, 1945.

## Table A-6M Ideal-gas enthalpy, internal energy, and absolute entropy of diatomic nitrogen, N

$$\Delta h_f = 0 \text{ kJ/kg·mol}$$

$T$, °K; $h$ and $u$, kJ/kg·mol; $s$, kJ/(kg·mol)(°K)

| $T$ | $h$ | $u$ | $s^0$ | $T$ | $h$ | $u$ | $s^0$ |
|---|---|---|---|---|---|---|---|
| 0 | 0 | 0 | 0 | 600 | 17,563 | 12,574 | 212.066 |
| 220 | 6,391 | 4,562 | 182.639 | 610 | 17,864 | 12,792 | 212.564 |
| 230 | 6,683 | 4,770 | 183.938 | 620 | 18,166 | 13,011 | 213.055 |
| 240 | 6,975 | 4,979 | 185.180 | 630 | 18,468 | 13,230 | 213.541 |
| 250 | 7,266 | 5,188 | 186.370 | 640 | 18,772 | 13,450 | 214.018 |
| 260 | 7,558 | 5,396 | 187.514 | 650 | 19,075 | 13,671 | 214.489 |
| 270 | 7,849 | 5,604 | 188.614 | 660 | 19,380 | 13,892 | 214.954 |
| 280 | 8,141 | 5,813 | 189.673 | 670 | 19,685 | 14,114 | 215.413 |
| 290 | 8,432 | 6,021 | 190.695 | 680 | 19,991 | 14,337 | 215.866 |
| 298 | 8,669 | 6,190 | 191.502 | 690 | 20,297 | 14,560 | 216.314 |
| 300 | 8,723 | 6,229 | 191.682 | 700 | 20,604 | 14,784 | 216.756 |
| 310 | 9,014 | 6,437 | 192.638 | 710 | 20,912 | 15,008 | 217.192 |
| 320 | 9,306 | 6,645 | 193.562 | 720 | 21,220 | 15,234 | 217.624 |
| 330 | 9,597 | 6,853 | 194.459 | 730 | 21,529 | 15,460 | 218.059 |
| 340 | 9,888 | 7,061 | 195.328 | 740 | 21,839 | 15,686 | 218.472 |
| 350 | 10,180 | 7,270 | 196.173 | 750 | 22,149 | 15,913 | 218.889 |
| 360 | 10,471 | 7,478 | 196.995 | 760 | 22,460 | 16,141 | 219.301 |
| 370 | 10,763 | 7,687 | 197.794 | 770 | 22,772 | 16,370 | 219.709 |
| 380 | 11,055 | 7,895 | 198.572 | 780 | 23,085 | 16,599 | 220.113 |
| 390 | 11,347 | 8,104 | 199.331 | 790 | 23,398 | 16,830 | 220.512 |
| 400 | 11,640 | 8,314 | 200.071 | 800 | 23,714 | 17,061 | 220.907 |
| 410 | 11,932 | 8,523 | 200.794 | 810 | 24,027 | 17,292 | 221.298 |
| 420 | 12,225 | 8,733 | 201.499 | 820 | 24,342 | 17,524 | 221.684 |
| 430 | 12,518 | 8,943 | 202.189 | 830 | 24,658 | 17,757 | 222.067 |
| 440 | 12,811 | 9,153 | 202.863 | 840 | 24,974 | 17,990 | 222.447 |
| 450 | 13,105 | 9,363 | 203.523 | 850 | 25,292 | 18,224 | 222.822 |
| 460 | 13,399 | 9,574 | 204.170 | 860 | 25,610 | 18,459 | 223.194 |
| 470 | 13,693 | 9,786 | 204.803 | 870 | 25,928 | 18,695 | 223.562 |
| 480 | 13,988 | 9,997 | 205.424 | 880 | 26,248 | 18,931 | 223.927 |
| 490 | 14,285 | 10,210 | 206.033 | 890 | 26,568 | 19,168 | 224.288 |
| 500 | 14,581 | 10,423 | 206.630 | 900 | 26,890 | 19,407 | 224.647 |
| 510 | 14,876 | 10,635 | 207.216 | 910 | 27,210 | 19,644 | 225.002 |
| 520 | 15,172 | 10,848 | 207.792 | 920 | 27,532 | 19,883 | 225.353 |
| 530 | 15,469 | 11,062 | 208.358 | 930 | 27,854 | 20,122 | 225.701 |
| 540 | 15,766 | 11,277 | 208.914 | 940 | 28,178 | 20,362 | 226.047 |
| 550 | 16,064 | 11,492 | 209.461 | 950 | 28,501 | 20,603 | 226.389 |
| 560 | 16,363 | 11,707 | 209.999 | 960 | 28,826 | 20,844 | 226.728 |
| 570 | 16,662 | 11,923 | 210.528 | 970 | 29,151 | 21,086 | 227.064 |
| 580 | 16,962 | 12,139 | 211.049 | 980 | 29,476 | 21,328 | 227.398 |
| 590 | 17,262 | 12,356 | 211.562 | 990 | 29,803 | 21,571 | 227.728 |

**Table A-6M**  (*Continued*)

| T | h | u | s⁰ | T | h | u | s⁰ |
|---|---|---|---|---|---|---|---|
| 1000 | 30,129 | 21,815 | 228.057 | 1760 | 56,227 | 41,594 | 247.396 |
| 1020 | 30,784 | 22,304 | 228.706 | 1780 | 56,938 | 42,139 | 247.798 |
| 1040 | 31,442 | 22,795 | 229.344 | 1800 | 57,651 | 42,685 | 248.195 |
| 1060 | 32,101 | 23,288 | 229.973 | 1820 | 58,363 | 43,231 | 248.589 |
| 1080 | 32,762 | 23,782 | 230.591 | 1840 | 59,075 | 43,777 | 248.979 |
| 1100 | 33,426 | 24,280 | 231.199 | 1860 | 59,790 | 44,324 | `249.365 |
| 1120 | 34,092 | 24,780 | 231.799 | 1880 | 60,504 | 44,873 | 249.748 |
| 1140 | 34,760 | 25,282 | 232.391 | 1900 | 61,220 | 45,423 | 250.128 |
| 1160 | 35,430 | 25,786 | 232.973 | 1920 | 61,936 | 45,973 | 250.502 |
| 1180 | 36,104 | 26,291 | 233.549 | 1940 | 62,654 | 46,524 | 250.874 |
| 1200 | 36,777 | 26,799 | 234.115 | 1960 | 63,381 | 47,075 | 251.242 |
| 1220 | 37,452 | 27,308 | 234.673 | 1980 | 64,090 | 47,627 | 251.607 |
| 1240 | 38,129 | 27,819 | 235.223 | 2000 | 64,810 | 48,181 | 251.969 |
| 1260 | 38,807 | 28,331 | 235.766 | 2050 | 66,612 | 49,567 | 252.858 |
| 1280 | 39,488 | 28,845 | 236.302 | 2100 | 68,417 | 50,957 | 253.726 |
| 1300 | 40,170 | 29,361 | 236.831 | 2150 | 70,226 | 52,351 | 254.578 |
| 1320 | 40,853 | 29,878 | 237.353 | 2200 | 72,040 | 53,749 | 255.412 |
| 1340 | 41,539 | 30,398 | 237.867 | 2250 | 73,856 | 55,149 | 256.227 |
| 1360 | 42,227 | 30,919 | 238.376 | 2300 | 75,676 | 56,553 | 257.027 |
| 1380 | 42,915 | 31,441 | 238.878 | 2350 | 77,496 | 57,958 | 257.810 |
| 1400 | 43,605 | 31,964 | 239.375 | 2400 | 79,320 | 59,366 | 258.580 |
| 1420 | 44,295 | 32,489 | 239.865 | 2450 | 81,149 | 60,779 | 259.332 |
| 1440 | 44,988 | 33,014 | 240.350 | 2500 | 82,981 | 62,195 | 260.073 |
| 1460 | 45,682 | 33,543 | 240.827 | 2550 | 84,814 | 63,613 | 260.799 |
| 1480 | 46,377 | 34,071 | 241.301 | 2600 | 86,650 | 65,033 | 261.512 |
| 1500 | 47,073 | 34,601 | 241.768 | 2650 | 88,488 | 66,455 | 262.213 |
| 1520 | 47,771 | 35,133 | 242.228 | 2700 | 90,328 | 67,880 | 262.902 |
| 1540 | 48,470 | 35,665 | 242.685 | 2750 | 92,171 | 69,306 | 263.577 |
| 1560 | 49,168 | 36,197 | 243.137 | 2800 | 94,014 | 70,734 | 264.241 |
| 1580 | 49,869 | 36,732 | 243.585 | 2850 | 95,859 | 72,163 | 264.895 |
| 1600 | 50,571 | 37,268 | 244.028 | 2900 | 97,705 | 73,593 | 265.538 |
| 1620 | 51,275 | 37,806 | 244.464 | 2950 | 99,556 | 75,028 | 266.170 |
| 1640 | 51,980 | 38,344 | 244.896 | 3000 | 101,407 | 76,464 | 266.793 |
| 1660 | 52,686 | 38,884 | 245.324 | 3050 | 103,260 | 77,902 | 267.404 |
| 1680 | 53,393 | 39,424 | 245.747 | 3100 | 105,115 | 79,341 | 268.007 |
| 1700 | 54,099 | 39,965 | 246.166 | 3150 | 106,972 | 80,782 | 268.601 |
| 1720 | 54,807 | 40,507 | 246.580 | 3200 | 108,830 | 82,224 | 269.186 |
| 1740 | 55,516 | 41,049 | 246.990 | 3250 | 110,690 | 83,668 | 269.763 |

Based on data from the JANAF Thermochemical Tables, NSRDS-NBS-37, 1971.

## Table A-7M  Ideal-gas enthalpy, internal energy, and absolute entropy of diatomic oxygen, $O_2$

$$\Delta h_f = 0 \text{ kJ/kg·mol}$$

$T$, °K; $h$ and $u$, kJ/kg·mol; $s$, kJ/(kg·mol)(°K)

| $T$ | $h$ | $u$ | $s^0$ | $T$ | $h$ | $u$ | $s^0$ |
|---|---|---|---|---|---|---|---|
| 0 | 0 | 0 | 0 | 600 | 17,929 | 12,940 | 226.346 |
| 220 | 6,404 | 4,575 | 196.171 | 610 | 18,250 | 13,178 | 226.877 |
| 230 | 6,694 | 4,782 | 197.461 | 620 | 18,572 | 13,417 | 227.400 |
| 240 | 6,984 | 4,989 | 198.696 | 630 | 18,895 | 13,657 | 227.918 |
| 250 | 7,275 | 5,197 | 199.885 | 640 | 19,219 | 13,898 | 228.429 |
| 260 | 7,566 | 5,405 | 201.027 | 650 | 19,544 | 14,140 | 228.932 |
| 270 | 7,858 | 5,613 | 202.128 | 660 | 19,870 | 14,383 | 229.430 |
| 280 | 8,150 | 5,822 | 203.191 | 670 | 20,197 | 14,626 | 229.920 |
| 290 | 8,443 | 6,032 | 204.218 | 680 | 20,524 | 14,871 | 230.405 |
| 298 | 8,682 | 6,203 | 205.033 | 690 | 20,854 | 15,116 | 230.885 |
| 300 | 8,736 | 6,242 | 205.213 | 700 | 21,184 | 15,364 | 231.358 |
| 310 | 9,030 | 6,453 | 206.177 | 710 | 21,514 | 15,611 | 231.827 |
| 320 | 9,325 | 6,664 | 207.112 | 720 | 21,845 | 15,859 | 232.291 |
| 330 | 9,620 | 6,877 | 208.020 | 730 | 22,177 | 16,107 | 232.748 |
| 340 | 9,916 | 7,090 | 208.904 | 740 | 22,510 | 16,357 | 233.201 |
| 350 | 10,213 | 7,303 | 209.765 | 750 | 22,844 | 16,607 | 233.649 |
| 360 | 10,511 | 7,518 | 210.604 | 760 | 23,178 | 16,859 | 234.091 |
| 370 | 10,809 | 7,733 | 211.423 | 770 | 23,513 | 17,111 | 234.528 |
| 380 | 11,109 | 7,949 | 212.222 | 780 | 23,850 | 17,364 | 234.960 |
| 390 | 11,409 | 8,166 | 213.002 | 790 | 24,186 | 17,618 | 235.387 |
| 400 | 11,711 | 8,384 | 213.765 | 800 | 24,523 | 17,872 | 235.810 |
| 410 | 12,012 | 8,603 | 214.510 | 810 | 24,861 | 18,126 | 236.230 |
| 420 | 12,314 | 8,822 | 215.241 | 820 | 25,199 | 18,382 | 236.644 |
| 430 | 12,618 | 9,043 | 215.955 | 830 | 25,537 | 18,637 | 237.055 |
| 440 | 12,923 | 9,264 | 216.656 | 840 | 25,877 | 18,893 | 237.462 |
| 450 | 13,228 | 9,487 | 217.342 | 850 | 26,218 | 19,150 | 237.864 |
| 460 | 13,535 | 9,710 | 218.016 | 860 | 26,559 | 19,408 | 238.264 |
| 470 | 13,842 | 9,935 | 218.676 | 870 | 26,899 | 19,666 | 238.660 |
| 480 | 14,151 | 10,160 | 219.326 | 880 | 27,242 | 19,925 | 239.051 |
| 490 | 14,460 | 10,386 | 219.963 | 890 | 27,584 | 20,185 | 239.439 |
| 500 | 14,770 | 10,614 | 220.589 | 900 | 27,928 | 20,445 | 239.823 |
| 510 | 15,082 | 10,842 | 221.206 | 910 | 28,272 | 20,706 | 240.203 |
| 520 | 15,395 | 11,071 | 221.812 | 920 | 28,616 | 20,967 | 240.580 |
| 530 | 15,708 | 11,301 | 222.409 | 930 | 28,960 | 21,228 | 240.953 |
| 540 | 16,022 | 11,533 | 222.997 | 940 | 29,306 | 21,491 | 241.323 |
| 550 | 16,338 | 11,765 | 223.576 | 950 | 29,652 | 21,754 | 241.689 |
| 560 | 16,654 | 11,998 | 224.146 | 960 | 29,999 | 22,017 | 242.052 |
| 570 | 16,971 | 12,232 | 224.708 | 970 | 30,345 | 22,280 | 242.411 |
| 580 | 17,290 | 12,467 | 225.262 | 980 | 30,692 | 22,544 | 242.768 |
| 590 | 17,609 | 12,703 | 225.808 | 990 | 31,041 | 22,809 | 243.120 |

**Table A-7M** *(Continued)*

| T | h | u | $s^0$ | T | h | u | $s^0$ |
|---|---|---|---|---|---|---|---|
| 1000 | 31,389 | 23,075 | 243.471 | 1760 | 58,880 | 44,247 | 263.861 |
| 1020 | 32,088 | 23,607 | 244.164 | 1780 | 59,624 | 44,825 | 264.283 |
| 1040 | 32,789 | 24,142 | 244.844 | 1800 | 60,371 | 45,405 | 264.701 |
| 1060 | 33,490 | 24,677 | 245.513 | 1820 | 61,118 | 45,986 | 265.113 |
| 1080 | 34,194 | 25,214 | 246.171 | 1840 | 61,866 | 46,568 | 265.521 |
| 1100 | 34,899 | 25,753 | 246.818 | 1860 | 62,616 | 47,151 | 265.925 |
| 1120 | 35,606 | 26,294 | 247.454 | 1880 | 63,365 | 47,734 | 266.326 |
| 1140 | 36,314 | 26,836 | 248.081 | 1900 | 64,116 | 48,319 | 266.722 |
| 1160 | 37,023 | 27,379 | 248.698 | 1920 | 64,868 | 48,904 | 267.115 |
| 1180 | 37,734 | 27,923 | 249.307 | 1940 | 65,620 | 49,490 | 267.505 |
| 1200 | 38,447 | 28,469 | 249.906 | 1960 | 66,374 | 50,078 | 267.891 |
| 1220 | 39,162 | 29,018 | 250.497 | 1980 | 67,127 | 50,665 | 268.275 |
| 1240 | 39,877 | 29,568 | 251.079 | 2000 | 67,881 | 51,253 | 268.655 |
| 1260 | 40,594 | 30,118 | 251.653 | 2050 | 69,772 | 52,727 | 269.588 |
| 1280 | 41,312 | 30,670 | 252.219 | 2100 | 71,668 | 54,208 | 270.504 |
| 1300 | 42,033 | 31,224 | 252.776 | 2150 | 73,573 | 55,697 | 271.399 |
| 1320 | 42,753 | 31,778 | 253.325 | 2200 | 75,484 | 57,192 | 272.278 |
| 1340 | 43,475 | 32,334 | 253.868 | 2250 | 77,397 | 58,690 | 273.136 |
| 1360 | 44,198 | 32,891 | 254.404 | 2300 | 79,316 | 60,193 | 273.981 |
| 1380 | 44,923 | 33,449 | 254.932 | 2350 | 81,243 | 61,704 | 274.809 |
| 1400 | 45,648 | 34,008 | 255.454 | 2400 | 83,174 | 63,219 | 275.625 |
| 1420 | 46,374 | 34,567 | 255.968 | 2450 | 85,112 | 64,742 | 276.424 |
| 1440 | 47,102 | 35,129 | 256.475 | 2500 | 87,057 | 66,271 | 277.207 |
| 1460 | 47,831 | 35,692 | 256.978 | 2550 | 89,004 | 67,802 | 277.979 |
| 1480 | 48,561 | 36,256 | 257.474 | 2600 | 90,956 | 69,339 | 278.738 |
| 1500 | 49,292 | 36,821 | 257.965 | 2650 | 92,916 | 70,883 | 279.485 |
| 1520 | 50,024 | 37,387 | 258.450 | 2700 | 94,881 | 72,433 | 280.219 |
| 1540 | 50,756 | 37,952 | 258.928 | 2750 | 96,852 | 73,987 | 280.942 |
| 1560 | 51,490 | 38,520 | 259.402 | 2800 | 98,826 | 75,546 | 281.654 |
| 1580 | 52,224 | 39,088 | 259.870 | 2850 | 100,808 | 77,112 | 282.357 |
| 1600 | 52,961 | 39,658 | 260.333 | 2900 | 102,793 | 78,682 | 283.048 |
| 1620 | 53,696 | 40,227 | 260.791 | 2950 | 104,785 | 80,258 | 283.728 |
| 1640 | 54,434 | 40,799 | 261.242 | 3000 | 106,780 | 81,837 | 284.399 |
| 1660 | 55,172 | 41,370 | 261.690 | 3050 | 108,778 | 83,419 | 285.060 |
| 1680 | 55,912 | 41,944 | 262.132 | 3100 | 110,784 | 85,009 | 285.713 |
| 1700 | 56,652 | 42,517 | 262.571 | 3150 | 112,795 | 86,601 | 286.355 |
| 1720 | 57,394 | 43,093 | 263.005 | 3200 | 114,809 | 88,203 | 286.989 |
| 1740 | 58,136 | 43,669 | 263.435 | 3250 | 116,827 | 89,804 | 287.614 |

Based on data from the JANAF Thermochemical Tables, NSRDS-NBS-37, 1971.

## Table A-8M Ideal-gas enthalpy, internal energy, and absolute entropy of carbon monoxide, CO

$$\Delta h_f = -110{,}530 \text{ kJ/kg·mol}$$

$T$, °K; $h$ and $u$, kJ/kg·mol; $s$, kJ/(kg·mol)(°K)

| $T$ | $h$ | $u$ | $s^0$ | $T$ | $h$ | $u$ | $s^0$ |
|---|---|---|---|---|---|---|---|
| 0 | 0 | 0 | 0 | 600 | 17,611 | 12,622 | 218.204 |
| 220 | 6,391 | 4,562 | 188.683 | 610 | 17,915 | 12,843 | 218.708 |
| 230 | 6,683 | 4,771 | 189.980 | 620 | 18,221 | 13,066 | 219.205 |
| 240 | 6,975 | 4,979 | 191.221 | 630 | 18,527 | 13,289 | 219.695 |
| 250 | 7,266 | 5,188 | 192.411 | 640 | 18,833 | 13,512 | 220.179 |
| 260 | 7,558 | 5,396 | 193.554 | 650 | 19,141 | 13,736 | 220.656 |
| 270 | 7,849 | 5,604 | 194.654 | 660 | 19,449 | 13,962 | 221.127 |
| 280 | 8,140 | 5,812 | 195.713 | 670 | 19,758 | 14,187 | 221.592 |
| 290 | 8,432 | 6,020 | 196.735 | 680 | 20,068 | 14,414 | 222.052 |
| 298 | 8,669 | 6,190 | 197.543 | 690 | 20,378 | 14,641 | 222.505 |
| 300 | 8,723 | 6,229 | 197.723 | 700 | 20,690 | 14,870 | 222.953 |
| 310 | 9,014 | 6,437 | 198.678 | 710 | 21,002 | 15,099 | 223.396 |
| 320 | 9,306 | 6,645 | 199.603 | 720 | 21,315 | 15,328 | 223.833 |
| 330 | 9,597 | 6,854 | 200.500 | 730 | 21,628 | 15,558 | 224.265 |
| 340 | 9,889 | 7,062 | 201.371 | 740 | 21,943 | 15,789 | 224.692 |
| 350 | 10,181 | 7,271 | 202.217 | 750 | 22,258 | 16,022 | 225.115 |
| 360 | 10,473 | 7,480 | 203.040 | 760 | 22,573 | 16,255 | 225.533 |
| 370 | 10,765 | 7,689 | 203.842 | 770 | 22,890 | 16,488 | 225.947 |
| 380 | 11,058 | 7,899 | 204.622 | 780 | 23,208 | 16,723 | 226.357 |
| 390 | 11,351 | 8,108 | 205.383 | 790 | 23,526 | 16,957 | 226.762 |
| 400 | 11,644 | 8,319 | 206.125 | 800 | 23,844 | 17,193 | 227.162 |
| 410 | 11,938 | 8,529 | 206.850 | 810 | 24,164 | 17,429 | 227.559 |
| 420 | 12,232 | 8,740 | 207.549 | 820 | 24,483 | 17,665 | 227.952 |
| 430 | 12,526 | 8,951 | 208.252 | 830 | 24,803 | 17,902 | 228.339 |
| 440 | 12,821 | 9,163 | 208.929 | 840 | 25,124 | 18,140 | 228.724 |
| 450 | 13,116 | 9,375 | 209.593 | 850 | 25,446 | 18,379 | 229.106 |
| 460 | 13,412 | 9,587 | 210.243 | 860 | 25,768 | 18,617 | 229.482 |
| 470 | 13,708 | 9,800 | 210.880 | 870 | 26,091 | 18,858 | 229.856 |
| 480 | 14,005 | 10,014 | 211.504 | 880 | 26,415 | 19,099 | 230.227 |
| 490 | 14,302 | 10,228 | 212.117 | 890 | 26,740 | 19,341 | 230.593 |
| 500 | 14,600 | 10,443 | 212.719 | 900 | 27,066 | 19,583 | 230.957 |
| 510 | 14,898 | 10,658 | 213.310 | 910 | 27,392 | 19,826 | 231.317 |
| 520 | 15,197 | 10,874 | 213.890 | 920 | 27,719 | 20,070 | 231.674 |
| 530 | 15,497 | 11,090 | 214.460 | 930 | 28,046 | 20,314 | 232.028 |
| 540 | 15,797 | 11,307 | 215.020 | 940 | 28,375 | 20,559 | 232.379 |
| 550 | 16,097 | 11,524 | 215.572 | 950 | 28,703 | 20,805 | 232.727 |
| 560 | 16,399 | 11,743 | 216.115 | 960 | 29,033 | 21,051 | 233.072 |
| 570 | 16,701 | 11,961 | 216.649 | 970 | 29,362 | 21,298 | 233.413 |
| 580 | 17,003 | 12,181 | 217.175 | 980 | 29,693 | 21,545 | 233.752 |
| 590 | 17,307 | 12,401 | 217.693 | 990 | 30,024 | 21,793 | 234.088 |

## Table A-8M (*Continued*)

| T | h | u | $s^0$ | T | h | u | $s^0$ |
|---|---|---|---|---|---|---|---|
| 1000 | 30,355 | 22,041 | 234.421 | 1760 | 56,756 | 42,123 | 253.991 |
| 1020 | 31,020 | 22,540 | 235.079 | 1780 | 57,473 | 42,673 | 254.398 |
| 1040 | 31,688 | 23,041 | 235.728 | 1800 | 58,191 | 43,225 | 254.797 |
| 1060 | 32,357 | 23,544 | 236.364 | 1820 | 58,910 | 43,778 | 255.194 |
| 1080 | 33,029 | 24,049 | 236.992 | 1840 | 59,629 | 44,331 | 255.587 |
| 1100 | 33,702 | 24,557 | 237.609 | 1860 | 60,351 | 44,886 | 255.976 |
| 1120 | 34,377 | 25,065 | 238.217 | 1880 | 61,072 | 45,441 | 256.361 |
| 1140 | 35,054 | 25,575 | 238.817 | 1900 | 61,794 | 45,997 | 256.743 |
| 1160 | 35,733 | 26,088 | 239.407 | 1920 | 62,516 | 46,552 | 257.122 |
| 1180 | 36,406 | 26,602 | 239.989 | 1940 | 63,238 | 47,108 | 257.497 |
| 1200 | 37,095 | 27,118 | 240.663 | 1960 | 63,961 | 47,665 | 257.868 |
| 1220 | 37,780 | 27,637 | 241.128 | 1980 | 64,684 | 48,221 | 258.236 |
| 1240 | 38,466 | 28,426 | 241.686 | 2000 | 65,408 | 48,780 | 258.600 |
| 1260 | 39,154 | 28,678 | 242.236 | 2050 | 67,224 | 50,179 | 259.494 |
| 1280 | 39,844 | 29,201 | 242.780 | 2100 | 69,044 | 51,584 | 260.370 |
| 1300 | 40,534 | 29,725 | 243.316 | 2150 | 70,864 | 52,988 | 261.226 |
| 1320 | 41,226 | 30,251 | 243.844 | 2200 | 72,688 | 54,396 | 262.065 |
| 1340 | 41,919 | 30,778 | 244.366 | 2250 | 74,516 | 55,809 | 262.887 |
| 1360 | 42,613 | 31,306 | 244.880 | 2300 | 76,345 | 57,222 | 263.692 |
| 1380 | 43,309 | 31,836 | 245.388 | 2350 | 78,178 | 58,640 | 264.480 |
| 1400 | 44,007 | 32,367 | 245.889 | 2400 | 80,015 | 60,060 | 265.253 |
| 1420 | 44,707 | 32,900 | 246.385 | 2450 | 81,852 | 61,482 | 266.012 |
| 1440 | 45,408 | 33,434 | 246.876 | 2500 | 83,692 | 62,906 | 266.755 |
| 1460 | 46,110 | 33,971 | 247.360 | 2550 | 85,537 | 64,335 | 267.485 |
| 1480 | 46,813 | 34,508 | 247.839 | 2600 | 87,383 | 65,766 | 268.202 |
| 1500 | 47,517 | 35,046 | 248.312 | 2650 | 89,230 | 67,197 | 268.905 |
| 1520 | 48,222 | 35,584 | 248.778 | 2700 | 91,077 | 68,628 | 269.596 |
| 1540 | 48,928 | 36,124 | 249.240 | 2750 | 92,930 | 70,066 | 270.285 |
| 1560 | 49,635 | 36,665 | 249.695 | 2800 | 94,784 | 71,504 | 270.943 |
| 1580 | 50,344 | 37,207 | 250.147 | 2850 | 96,639 | 72,945 | 271.602 |
| 1600 | 51,053 | 37,750 | 250.592 | 2900 | 98,495 | 74,383 | 272.249 |
| 1620 | 51,763 | 38,293 | 251.033 | 2950 | 100,352 | 75,825 | 272.884 |
| 1640 | 52,472 | 38,837 | 251.470 | 3000 | 102,210 | 77,267 | 273.508 |
| 1660 | 53,184 | 39,382 | 251.901 | 3050 | 104,073 | 78,715 | 274.123 |
| 1680 | 53,895 | 39,927 | 252.329 | 3100 | 105,939 | 80,164 | 274.730 |
| 1700 | 54,609 | 40,474 | 252.751 | 3150 | 107,802 | 81,612 | 275.326 |
| 1720 | 55,323 | 41,023 | 253.169 | 3200 | 109,667 | 83,061 | 275.914 |
| 1740 | 56,039 | 41,572 | 253.582 | 3250 | 111,534 | 84,513 | 276.494 |

Based on data from the JANAF Thermochemical Tables, NSRDS-NBS-37, 1971.

## Table A-9M Ideal-gas enthalpy, internal energy, and absolute entropy of carbon dioxide, $CO_2$

$$\Delta h_f = -393{,}520 \text{ kJ/kg·mol}$$

$T$, °K; $h$ and $u$, kJ/kg·mol; $s$, kJ/(kg·mol)(°K)

| $T$ | $h$ | $u$ | $s^0$ | $T$ | $h$ | $u$ | $s^0$ |
|---|---|---|---|---|---|---|---|
| 0 | 0 | 0 | 0 | 600 | 22,280 | 17,291 | 243.199 |
| 220 | 6,601 | 4,772 | 202.966 | 610 | 22,754 | 17,683 | 243.983 |
| 230 | 6,938 | 5,026 | 204.464 | 620 | 23,231 | 18,076 | 244.758 |
| 240 | 7,280 | 5,285 | 205.920 | 630 | 23,709 | 18,471 | 245.524 |
| 250 | 7,627 | 5,548 | 207.337 | 640 | 24,190 | 18,869 | 246.282 |
| 260 | 7,979 | 5,817 | 208.717 | 650 | 24,674 | 19,270 | 247.032 |
| 270 | 8,335 | 6,091 | 210.062 | 660 | 25,160 | 19,672 | 247.773 |
| 280 | 8,697 | 6,369 | 211.376 | 670 | 25,648 | 20,078 | 248.507 |
| 290 | 9,063 | 6,651 | 212.660 | 680 | 26,138 | 20,484 | 249.233 |
| 298 | 9,364 | 6,885 | 213.685 | 690 | 26,631 | 20,894 | 249.952 |
| 300 | 9,431 | 6,939 | 213.915 | 700 | 27,125 | 21,305 | 250.663 |
| 310 | 9,807 | 7,230 | 215.146 | 710 | 27,622 | 21,719 | 251.368 |
| 320 | 10,186 | 7,526 | 216.351 | 720 | 28,121 | 22,134 | 252.065 |
| 330 | 10,570 | 7,826 | 217.534 | 730 | 28,622 | 22,552 | 252.755 |
| 340 | 10,959 | 8,131 | 218.694 | 740 | 29,124 | 22,972 | 253.439 |
| 350 | 11,351 | 8,439 | 219.831 | 750 | 29,629 | 23,393 | 254.117 |
| 360 | 11,748 | 8,752 | 220.948 | 760 | 30,135 | 23,817 | 254.787 |
| 370 | 12,148 | 9,068 | 222.044 | 770 | 30,644 | 24,242 | 255.452 |
| 380 | 12,552 | 9,392 | 223.122 | 780 | 31,154 | 24,669 | 256.110 |
| 390 | 12,960 | 9,718 | 224.182 | 790 | 31,665 | 25,097 | 256.762 |
| 400 | 13,372 | 10,046 | 225.225 | 800 | 32,179 | 25,527 | 257.408 |
| 410 | 13,787 | 10,378 | 226.250 | 810 | 32,694 | 25,959 | 258.048 |
| 420 | 14,206 | 10,714 | 227.258 | 820 | 33,212 | 26,394 | 258.682 |
| 430 | 14,628 | 11,053 | 228.252 | 830 | 33,730 | 26,829 | 259.311 |
| 440 | 15,054 | 11,393 | 229.230 | 840 | 34,251 | 27,267 | 259.934 |
| 450 | 15,483 | 11,742 | 230.194 | 850 | 34,773 | 27,706 | 260.551 |
| 460 | 15,916 | 12,091 | 231.144 | 860 | 35,296 | 28,125 | 261.164 |
| 470 | 16,351 | 12,444 | 232.080 | 870 | 35,821 | 28,588 | 261.770 |
| 480 | 16,791 | 12,800 | 233.004 | 880 | 36,347 | 29,031 | 262.371 |
| 490 | 17,232 | 13,158 | 233.916 | 890 | 36,876 | 29,476 | 262.968 |
| 500 | 17,678 | 13,521 | 234.814 | 900 | 37,405 | 29,922 | 263.559 |
| 510 | 18,126 | 13,885 | 235.700 | 910 | 37,935 | 30,369 | 264.146 |
| 520 | 18,576 | 14,253 | 236.575 | 920 | 38,467 | 30,818 | 264.728 |
| 530 | 19,029 | 14,622 | 237.439 | 930 | 39,000 | 31,268 | 265.304 |
| 540 | 19,485 | 14,996 | 238.292 | 940 | 39,535 | 31,719 | 265.877 |
| 550 | 19,945 | 15,372 | 239.135 | 950 | 40,070 | 32,171 | 266.444 |
| 560 | 20,407 | 15,751 | 239.962 | 960 | 40,607 | 32,625 | 267.007 |
| 570 | 20,870 | 16,131 | 240.789 | 970 | 41,145 | 33,081 | 267.566 |
| 580 | 21,337 | 16,515 | 241.602 | 980 | 41,685 | 33,537 | 268.119 |
| 590 | 21,807 | 16,902 | 242.405 | 990 | 42,226 | 33,995 | 268.670 |

## Table A-9M  (*Continued*)

| T | h | u | $s^0$ | T | h | u | $s^0$ |
|---|---|---|---|---|---|---|---|
| 1000 | 42,769 | 34,455 | 269.215 | 1760 | 86,420 | 71,787 | 301.543 |
| 1020 | 43,859 | 35,378 | 270.293 | 1780 | 87,612 | 72,812 | 302.217 |
| 1040 | 44,953 | 36,306 | 271.354 | 1800 | 88,806 | 73,840 | 302.884 |
| 1060 | 46,051 | 37,238 | 272.400 | 1820 | 90,000 | 74,868 | 303.544 |
| 1080 | 47,153 | 38,174 | 273.430 | 1840 | 91,196 | 75,897 | 304.198 |
| 1100 | 48,258 | 39,112 | 274.445 | 1860 | 92,394 | 76,929 | 304.845 |
| 1120 | 49,369 | 40,057 | 275.444 | 1880 | 93,593 | 77,962 | 305.487 |
| 1140 | 50,484 | 41,006 | 276.430 | 1900 | 94,793 | 78,996 | 306.122 |
| 1160 | 51,602 | 41,957 | 277.403 | 1920 | 95,995 | 80,031 | 306.751 |
| 1180 | 52,724 | 42,913 | 278.361 | 1940 | 97,197 | 81,067 | 307.374 |
| 1200 | 53,848 | 43,871 | 279.307 | 1960 | 98,401 | 82,105 | 307.992 |
| 1220 | 54,977 | 44,834 | 280.238 | 1980 | 99,606 | 83,144 | 308.604 |
| 1240 | 56,108 | 45,799 | 281.158 | 2000 | 100,804 | 84,185 | 309.210 |
| 1260 | 57,244 | 46,768 | 282.066 | 2050 | 103,835 | 86,791 | 310.701 |
| 1280 | 58,381 | 47,739 | 282.962 | 2100 | 106,864 | 89,404 | 312.160 |
| 1300 | 59,522 | 48,713 | 283.847 | 2150 | 109,898 | 92,023 | 313.589 |
| 1320 | 60,666 | 49,691 | 284.722 | 2200 | 112,939 | 94,648 | 314.988 |
| 1340 | 61,813 | 50,672 | 285.586 | 2250 | 115,984 | 97,277 | 316.356 |
| 1360 | 62,963 | 51,656 | 286.439 | 2300 | 119,035 | 99,912 | 317.695 |
| 1380 | 64,116 | 52,643 | 287.283 | 2350 | 122,091 | 102,552 | 319.011 |
| 1400 | 65,271 | 53,631 | 288.106 | 2400 | 125,152 | 105,197 | 320.302 |
| 1420 | 66,427 | 54,621 | 288.934 | 2450 | 128,219 | 107,849 | 321.566 |
| 1440 | 67,586 | 55,614 | 289.743 | 2500 | 131,290 | 110,504 | 322.808 |
| 1460 | 68,748 | 56,609 | 290.542 | 2550 | 134,368 | 113,166 | 324.026 |
| 1480 | 69,911 | 57,606 | 291.333 | 2600 | 137,449 | 115,832 | 325.222 |
| 1500 | 71,078 | 58,606 | 292.114 | 2650 | 140,533 | 118,500 | 326.396 |
| 1520 | 72,246 | 59,609 | 292.888 | 2700 | 143,620 | 121,172 | 327.549 |
| 1540 | 73,417 | 60,613 | 292.654 | 2750 | 146,713 | 123,849 | 328.684 |
| 1560 | 74,590 | 61,620 | 294.411 | 2800 | 149,808 | 126,528 | 329.800 |
| 1580 | 76,767 | 62,630 | 295.161 | 2850 | 152,908 | 129,212 | 330.896 |
| 1600 | 76,944 | 63,741 | 295.901 | 2900 | 156,009 | 131,898 | 331.975 |
| 1620 | 78,123 | 64,653 | 296.632 | 2950 | 159,117 | 134,589 | 333.037 |
| 1640 | 79,303 | 65,668 | 297.356 | 3000 | 162,226 | 137,283 | 334.084 |
| 1660 | 80,486 | 66,592 | 298.072 | 3050 | 165,341 | 139,982 | 335.114 |
| 1680 | 81,670 | 67,702 | 298.781 | 3100 | 168,456 | 142,681 | 336.126 |
| 1700 | 82,856 | 68,721 | 299.482 | 3150 | 171,576 | 145,385 | 337.124 |
| 1720 | 84,043 | 69,742 | 300.177 | 3200 | 174,695 | 148,089 | 338.109 |
| 1740 | 85,231 | 70,764 | 300.863 | 3250 | 177,822 | 150,801 | 339.069 |

Based on data from the JANAF Thermochemical Tables, NSRDS-NBS-37, 1971.

## Table A-10M Ideal-gas enthalpy, internal energy, and absolute entropy of water, $H_2O$

$$\Delta h_f = -241,810 \text{ kJ/kg·mol}$$

$T$, °K; $h$ and $u$, kJ/kg·mol; $s$, kJ/(kg·mol)(°K)

| $T$ | $h$ | $u$ | $s^o$ | $T$ | $h$ | $u$ | $s^o$ |
|---|---|---|---|---|---|---|---|
| 0 | 0 | 0 | 0 | 600 | 20,402 | 15,413 | 212.920 |
| 220 | 7,295 | 5,466 | 178.576 | 610 | 20,765 | 15,693 | 213.529 |
| 230 | 7,628 | 5,715 | 180.054 | 620 | 21,130 | 15,975 | 214.122 |
| 240 | 7,961 | 5,965 | 181.471 | 630 | 21,495 | 16,257 | 214.707 |
| 250 | 8,294 | 6,215 | 182.831 | 640 | 21,862 | 16,541 | 215.285 |
| 260 | 8,627 | 6,466 | 184.139 | 650 | 22,230 | 16,826 | 215.856 |
| 270 | 8,961 | 6,716 | 185.399 | 660 | 22,600 | 17,112 | 216.419 |
| 280 | 9,296 | 6,968 | 186.616 | 670 | 22,970 | 17,399 | 216.976 |
| 290 | 9,631 | 7,219 | 187.791 | 680 | 23,342 | 17,688 | 217.527 |
| 298 | 9,904 | 7,425 | 188.720 | 690 | 23,714 | 17,978 | 218.071 |
| 300 | 9,966 | 7,472 | 188.928 | 700 | 24,088 | 18,268 | 218.610 |
| 310 | 10,302 | 7,725 | 190.030 | 710 | 24,464 | 18,561 | 219.142 |
| 320 | 10,639 | 7,978 | 191.098 | 720 | 24,840 | 18,854 | 219.668 |
| 330 | 10,976 | 8,232 | 192.136 | 730 | 25,218 | 19,148 | 220.189 |
| 340 | 11,314 | 8,487 | 193.144 | 740 | 25,597 | 19,444 | 220.707 |
| 350 | 11,652 | 8,742 | 194.125 | 750 | 25,977 | 19,741 | 221.215 |
| 360 | 11,992 | 8,998 | 195.081 | 760 | 26,358 | 20,039 | 221.720 |
| 370 | 12,331 | 9,255 | 196.012 | 770 | 26,741 | 20,339 | 222.221 |
| 380 | 12,672 | 9,513 | 196.920 | 780 | 27,125 | 20,639 | 222.717 |
| 390 | 13,014 | 9,771 | 197.807 | 790 | 27,510 | 20,941 | 223.207 |
| 400 | 13,356 | 10,030 | 198.673 | 800 | 27,896 | 21,245 | 223.693 |
| 410 | 13,699 | 10,290 | 199.521 | 810 | 28,284 | 21,549 | 224.174 |
| 420 | 14,043 | 10,551 | 200.350 | 820 | 28,672 | 21,855 | 224.651 |
| 430 | 14,388 | 10,813 | 201.160 | 830 | 29,062 | 22,162 | 225.123 |
| 440 | 14,734 | 11,075 | 201.955 | 840 | 29,454 | 22,470 | 225.592 |
| 450 | 15,080 | 11,339 | 202.734 | 850 | 29,846 | 22,779 | 226.057 |
| 460 | 15,428 | 11,603 | 203.497 | 860 | 30,240 | 23,090 | 226.517 |
| 470 | 15,777 | 11,869 | 204.247 | 870 | 30,635 | 23,402 | 226.973 |
| 480 | 16,126 | 12,135 | 204.982 | 880 | 31,032 | 23,715 | 227.426 |
| 490 | 16,477 | 12,403 | 205.705 | 890 | 31,429 | 24,029 | 227.875 |
| 500 | 16,828 | 12,671 | 206.413 | 900 | 31,828 | 24,345 | 228.321 |
| 510 | 17,181 | 12,940 | 207.112 | 910 | 32,228 | 24,662 | 228.763 |
| 520 | 17,534 | 13,211 | 207.799 | 920 | 32,629 | 24,980 | 229.202 |
| 530 | 17,889 | 13,482 | 208.475 | 930 | 33,032 | 25,300 | 229.637 |
| 540 | 18,245 | 13,755 | 209.139 | 940 | 33,436 | 25,621 | 230.070 |
| 550 | 18,601 | 14,028 | 209.795 | 950 | 33,841 | 25,943 | 230.499 |
| 560 | 18,959 | 14,303 | 210.440 | 960 | 34,247 | 26,265 | 230.924 |
| 570 | 19,318 | 14,579 | 211.075 | 970 | 34,653 | 26,588 | 231.347 |
| 580 | 19,678 | 14,856 | 211.702 | 980 | 35,061 | 26,913 | 231.767 |
| 590 | 20,039 | 15,134 | 212.320 | 990 | 35,472 | 27,240 | 232.184 |

## Table A-10M  (*Continued*)

| $T$ | $h$ | $u$ | $s^0$ | $T$ | $h$ | $u$ | $s^0$ |
|------|--------|--------|---------|------|---------|---------|---------|
| 1000 | 35,882 | 27,568 | 232.597 | 1760 | 70,535 | 55,902 | 258.151 |
| 1020 | 36,709 | 28,228 | 233.415 | 1780 | 71,523 | 56,723 | 258.708 |
| 1040 | 37,542 | 28,895 | 234.223 | 1800 | 72,513 | 57,547 | 259.262 |
| 1060 | 38,380 | 29,567 | 235.020 | 1820 | 73,507 | 58,375 | 259.811 |
| 1080 | 39,223 | 30,243 | 235.806 | 1840 | 74,506 | 59,207 | 260.357 |
| 1100 | 40,071 | 30,925 | 236.584 | 1860 | 75,506 | 60,042 | 260.898 |
| 1120 | 40,923 | 31,611 | 237.352 | 1880 | 76,511 | 60,880 | 261.436 |
| 1140 | 41,780 | 32,301 | 238.110 | 1900 | 77,517 | 61,720 | 261.969 |
| 1160 | 42,642 | 32,997 | 238.859 | 1920 | 78,527 | 62,564 | 262.497 |
| 1180 | 43,509 | 33,698 | 239.600 | 1940 | 79,540 | 63,411 | 263.022 |
| 1200 | 44,380 | 34,403 | 240.333 | 1960 | 80,555 | 64,259 | 263.542 |
| 1220 | 45,256 | 35,112 | 241.057 | 1980 | 81,573 | 65,111 | 264.059 |
| 1240 | 46,137 | 35,827 | 241.773 | 2000 | 82,593 | 65,965 | 264.571 |
| 1260 | 47,022 | 36,546 | 242.482 | 1050 | 85,156 | 68,111 | 265.838 |
| 1280 | 47,912 | 37,270 | 243.183 | 2100 | 87,735 | 70,275 | 267.081 |
| 1300 | 48,807 | 38,000 | 243.877 | 2150 | 90,330 | 72,454 | 268.301 |
| 1320 | 49,707 | 38,732 | 244.564 | 2200 | 92,940 | 74,649 | 269.500 |
| 1340 | 50,612 | 39,470 | 245.243 | 2250 | 95,562 | 76,855 | 270.679 |
| 1360 | 51,521 | 40,213 | 245.915 | 2300 | 98,199 | 79,076 | 271.839 |
| 1380 | 52,434 | 40,960 | 246.582 | 2350 | 100,846 | 81,308 | 272.978 |
| 1400 | 53,351 | 41,711 | 247.241 | 2400 | 103,508 | 83,553 | 274.098 |
| 1420 | 54,273 | 42,466 | 247.895 | 2450 | 106,183 | 85,811 | 275.201 |
| 1440 | 55,198 | 43,226 | 248.543 | 2500 | 108,868 | 88,082 | 276.286 |
| 1460 | 56,128 | 43,989 | 249.185 | 2550 | 111,565 | 90,364 | 277.354 |
| 1480 | 57,062 | 44,756 | 249.820 | 2600 | 114,273 | 92,656 | 278.407 |
| 1500 | 57,999 | 45,528 | 250.450 | 2650 | 116,991 | 94,958 | 279.441 |
| 1520 | 58,942 | 46,304 | 251.074 | 2700 | 119,717 | 97,269 | 280.462 |
| 1540 | 59,888 | 47,084 | 251.693 | 2750 | 122,453 | 99,588 | 281.464 |
| 1560 | 60,838 | 47,868 | 252.305 | 2800 | 125,198 | 101,917 | 282.453 |
| 1580 | 61,792 | 48,655 | 252.912 | 2850 | 127,952 | 104,256 | 283.429 |
| 1600 | 62,748 | 49,445 | 253.513 | 2900 | 130,717 | 106,605 | 284.390 |
| 1620 | 63,709 | 50,240 | 254.111 | 2950 | 133,486 | 108,959 | 285.338 |
| 1640 | 64,675 | 51,039 | 254.703 | 3000 | 136,264 | 111,321 | 286.273 |
| 1660 | 65,643 | 51,841 | 255.290 | 3050 | 139,051 | 113,692 | 287.194 |
| 1680 | 66,614 | 52,646 | 255.873 | 3100 | 141,846 | 116,072 | 288.102 |
| 1700 | 67,589 | 53,455 | 256.450 | 3150 | 144,648 | 118,458 | 288.999 |
| 1720 | 68,567 | 54,267 | 257.022 | 3200 | 147,457 | 120,851 | 289.884 |
| 1740 | 69,550 | 55,083 | 257.589 | 3250 | 150,272 | 123,250 | 290.756 |

Based on data from the JANAF Thermochemical Tables, NSRDS-NBS-37, 1971.

## Table A-11M  Ideal-gas enthalpy, internal energy, and absolute entropy of diatomic hydrogen, $H_2$

$$\Delta h_f = 0 \text{ kJ/kg·mol}$$

$T$, °K; $h$ and $u$, kJ/kg·mol; $s$, kJ/(kg·mol)(°K)

| $T$ | $h$ | $u$ | $s^0$ | $T$ | $h$ | $u$ | $s^0$ |
|---|---|---|---|---|---|---|---|
| 0 | 0 | 0 | 0 | 1440 | 42,808 | 30,835 | 177.410 |
| 260 | 7,370 | 5,209 | 126.636 | 1480 | 44,091 | 31,786 | 178.291 |
| 270 | 7,657 | 5,412 | 127.719 | 1520 | 45,384 | 32,746 | 179.153 |
| 280 | 7,945 | 5,617 | 128.765 | 1560 | 46,683 | 33,713 | 179.995 |
| 290 | 8,233 | 5,822 | 129.775 | 1600 | 47,990 | 34,687 | 180.820 |
| 298 | 8,468 | 5,989 | 130.574 | 1640 | 49,303 | 35,668 | 181.632 |
| 300 | 8,522 | 6,027 | 130.754 | 1680 | 50,622 | 36,654 | 182.428 |
| 320 | 9,100 | 6,440 | 132.621 | 1720 | 51,947 | 37,646 | 183.208 |
| 340 | 9,680 | 6,853 | 134.378 | 1760 | 53,279 | 38,645 | 183.973 |
| 360 | 10,262 | 7,268 | 136.039 | 1800 | 54,618 | 39,652 | 184.724 |
| 380 | 10,843 | 7,684 | 137.612 | 1840 | 55,962 | 40,663 | 185.463 |
| 400 | 11,426 | 8,100 | 139.106 | 1880 | 57,311 | 41,680 | 186.190 |
| 420 | 12,010 | 8,518 | 140.529 | 1920 | 58,668 | 42,705 | 186.904 |
| 440 | 12,594 | 8,936 | 141.888 | 1960 | 60,031 | 43,735 | 187.607 |
| 460 | 13,179 | 9,355 | 143.187 | 2000 | 61,400 | 44,771 | 188.297 |
| 480 | 13,764 | 9,773 | 144.432 | 2050 | 63,119 | 46,074 | 189.148 |
| 500 | 14,350 | 10,193 | 145.628 | 2100 | 64,847 | 47,386 | 189.979 |
| 520 | 14,935 | 10,611 | 146.775 | 2150 | 66,584 | 48,708 | 190.796 |
| 560 | 16,107 | 11,451 | 148.945 | 2200 | 68,328 | 50,037 | 191.598 |
| 600 | 17,280 | 12,291 | 150.968 | 2250 | 70,080 | 51,373 | 192.385 |
| 640 | 18,453 | 13,133 | 152.863 | 2300 | 71,839 | 52,716 | 193.159 |
| 680 | 19,630 | 13,976 | 154.645 | 2350 | 73,608 | 54,069 | 193.921 |
| 720 | 20,807 | 14,821 | 156.328 | 2400 | 75,383 | 55,429 | 194.669 |
| 760 | 21,988 | 15,669 | 157.923 | 2450 | 77,168 | 56,798 | 195.403 |
| 800 | 23,171 | 16,520 | 159.440 | 2500 | 78,960 | 58,175 | 196.125 |
| 840 | 24,359 | 17,375 | 160.891 | 2550 | 80,755 | 59,554 | 196.837 |
| 880 | 25,551 | 18,235 | 162.277 | 2600 | 82,558 | 60,941 | 197.539 |
| 920 | 26,747 | 19,098 | 163.607 | 2650 | 84,368 | 62,335 | 198.229 |
| 960 | 27,948 | 19,966 | 164.884 | 2700 | 86,186 | 63,737 | 198.907 |
| 1000 | 29,154 | 20,839 | 166.114 | 2750 | 88,008 | 65,144 | 199.575 |
| 1040 | 30,364 | 21,717 | 167.300 | 2800 | 89,838 | 66,558 | 200.234 |
| 1080 | 31,580 | 22,601 | 168.449 | 2850 | 91,671 | 67,976 | 200.885 |
| 1120 | 32,802 | 23,490 | 169.560 | 2900 | 93,512 | 69,401 | 201.527 |
| 1160 | 34,028 | 24,384 | 170.636 | 2950 | 95,358 | 70,831 | 202.157 |
| 1200 | 35,262 | 25,284 | 171.682 | 3000 | 97,211 | 72,268 | 202.778 |
| 1240 | 36,502 | 26,192 | 172.698 | 3050 | 99,065 | 73,707 | 203.391 |
| 1280 | 37,749 | 27,106 | 173.687 | 3100 | 100,926 | 75,152 | 203.995 |
| 1320 | 39,002 | 28,027 | 174.652 | 3150 | 102,793 | 76,604 | 204.592 |
| 1360 | 40,263 | 28,955 | 175.593 | 3200 | 104,667 | 78,061 | 205.181 |
| 1400 | 41,530 | 29,889 | 176.510 | 3250 | 106,545 | 79,523 | 205.765 |

## Table A-11M (*Continued*) Ideal-gas enthalpy, internal energy, and absolute entropy for monatomic oxygen, O

$$\Delta h_f = 249,190 \text{ kJ/kg·mol}$$

| T | h | u | $s^0$ | T | h | u | $s^0$ |
|---|---|---|---|---|---|---|---|
| 0 | 0 | 0 | 0 | 2400 | 50,894 | 30,940 | 204.932 |
| 298 | 6,852 | 4,373 | 160.944 | 2450 | 51,936 | 31,566 | 205.362 |
| 300 | 6,892 | 4,398 | 161.079 | 2500 | 52,979 | 32,193 | 205.783 |
| 500 | 11,197 | 7,040 | 172.088 | 2550 | 54,021 | 32,820 | 206.196 |
| 1000 | 21,713 | 13,398 | 186.678 | 2600 | 55,064 | 33,447 | 206.601 |
| 1500 | 32,150 | 19,679 | 195.143 | 2650 | 56,108 | 34,075 | 206.999 |
| 1600 | 34,234 | 20,931 | 196.488 | 2700 | 57,152 | 34,703 | 207.389 |
| 1700 | 36,317 | 22,183 | 197.751 | 2750 | 58,196 | 35,332 | 207.772 |
| 1800 | 38,400 | 23,434 | 198.941 | 2800 | 59,241 | 35,961 | 208.148 |
| 1900 | 40,482 | 24,685 | 200.067 | 2850 | 60,286 | 36,590 | 208.518 |
| 2000 | 42,564 | 25,935 | 201.135 | 2900 | 61,332 | 37,220 | 208.882 |
| 2050 | 43,605 | 26,560 | 201.649 | 2950 | 62,378 | 37,851 | 209.240 |
| 2100 | 44,646 | 27,186 | 202.151 | 3000 | 63,425 | 38,482 | 209.592 |
| 2150 | 45,687 | 27,811 | 202.641 | 3100 | 65,520 | 39,746 | 210.279 |
| 2200 | 46,728 | 28,436 | 203.119 | 3200 | 67,619 | 41,013 | 210.945 |
| 2250 | 47,769 | 29,062 | 203.588 | 3300 | 69,720 | 42,283 | 211.592 |
| 2300 | 48,811 | 29,688 | 204.045 | 3400 | 71,824 | 43,556 | 212.220 |
| 2350 | 49,852 | 30,314 | 204.493 | 3500 | 73,932 | 44,832 | 212.831 |

## Ideal-gas enthalpy, internal energy, and absolute entropy for hydroxyl, OH

$$\Delta h_f = 39,040 \text{ kJ/kg·mol}$$

| T | h | u | $s^0$ | T | h | u | $s^0$ |
|---|---|---|---|---|---|---|---|
| 0 | 0 | 0 | 0 | 2400 | 77,015 | 57,061 | 248.628 |
| 298 | 9,188 | 6,709 | 183.594 | 2450 | 78,801 | 58,431 | 249.364 |
| 300 | 9,244 | 6,749 | 183.779 | 2500 | 80,592 | 59,806 | 250.088 |
| 500 | 15,181 | 11,024 | 198.955 | 2550 | 82,388 | 61,186 | 250.799 |
| 1000 | 30,123 | 21,809 | 219.624 | 2600 | 84,189 | 62,572 | 251.499 |
| 1500 | 46,046 | 33,575 | 232.506 | 2650 | 85,995 | 63,962 | 252.187 |
| 1600 | 49,358 | 36,055 | 234.642 | 2700 | 87,806 | 65,358 | 252.864 |
| 1700 | 52,706 | 38,571 | 236.672 | 2750 | 89,622 | 66,757 | 253.530 |
| 1800 | 56,089 | 41,123 | 238.606 | 2800 | 91,442 | 68,162 | 254.186 |
| 1900 | 59,505 | 43,708 | 240.453 | 2850 | 93,266 | 69,570 | 254.832 |
| 2000 | 62,952 | 46,323 | 242.221 | 2900 | 95,095 | 70,983 | 255.468 |
| 2050 | 64,687 | 47,642 | 243.077 | 2950 | 96,927 | 72,400 | 256.094 |
| 2100 | 66,428 | 48,968 | 243.917 | 3000 | 98,763 | 73,820 | 256.712 |
| 2150 | 68,177 | 50,301 | 244.740 | 3100 | 102,447 | 76,673 | 257.919 |
| 2200 | 69,932 | 51,641 | 245.547 | 3200 | 106,145 | 79,539 | 259.093 |
| 2250 | 71,694 | 52,987 | 246.338 | 3300 | 109,855 | 82,418 | 260.235 |
| 2300 | 73,462 | 54,339 | 247.116 | 3400 | 113,578 | 85,309 | 261.347 |
| 2350 | 75,236 | 55,697 | 247.879 | 3500 | 117,312 | 88,212 | 262.429 |

Based on data from the JANAF Thermochemical Tables, NSRDS-NBS-37, 1971.

## Table A-12M Properties of saturated water: temperature table

$v$, cm$^3$/g; $u$, kJ/kg; $h$, kJ/kg; $s$, kJ/(kg)($^\circ$K)

| Temp. °C $T$ | Press. bars $P$ | Specific volume | | Internal energy | | Enthalpy | | | Entropy | |
|---|---|---|---|---|---|---|---|---|---|---|
| | | Sat. liquid $v_f$ | Sat. vapor $v_g$ | Sat. liquid $u_f$ | Sat. vapor $u_g$ | Sat. liquid $h_f$ | Evap. $h_{fg}$ | Sat. vapor $h_g$ | Sat. liquid $s_f$ | Sat vapor $s_g$ |
| 0 | .00611 | 1.0002 | 206278 | −.03 | 2375.4 | −.02 | 2501.4 | 2501.3 | −.0001 | 9.1565 |
| 5 | .00872 | 1.0001 | 147120 | 20.97 | 2382.3 | 20.98 | 2489.6 | 2510.6 | .0761 | 9.0257 |
| 10 | .01228 | 1.0004 | 106379 | 42.00 | 2389.2 | 42.01 | 2477.7 | 2519.8 | .1510 | 8.9008 |
| 15 | .01705 | 1.0009 | 77926 | 62.99 | 2396.1 | 62.99 | 2465.9 | 2528.9 | .2245 | 8.7814 |
| 20 | .02339 | 1.0018 | 57791 | 83.95 | 2402.9 | 83.96 | 2454.1 | 2538.1 | .2966 | 8.6672 |
| 25 | .03169 | 1.0029 | 43360 | 104.88 | 2409.8 | 104.89 | 2442.3 | 2547.2 | .3674 | 8.5580 |
| 30 | .04246 | 1.0043 | 32894 | 125.78 | 2416.6 | 125.79 | 2430.5 | 2556.3 | .4369 | 8.4533 |
| 35 | .05628 | 1.0060 | 25216 | 146.67 | 2423.4 | 146.68 | 2418.6 | 2565.3 | .5053 | 8.3531 |
| 40 | .07384 | 1.0078 | 19523 | 167.56 | 2430.1 | 167.57 | 2406.7 | 2574.3 | .5725 | 8.2570 |
| 45 | .09593 | 1.0099 | 15258 | 188.44 | 2436.8 | 188.45 | 2394.8 | 2583.2 | .6387 | 8.1648 |
| 50 | .1235 | 1.0121 | 12032 | 209.32 | 2443.5 | 209.33 | 2382.7 | 2592.1 | .7038 | 8.0763 |
| 55 | .1576 | 1.0146 | 9568 | 230.21 | 2450.1 | 230.23 | 2370.7 | 2600.9 | .7679 | 7.9913 |
| 60 | .1994 | 1.0172 | 7671 | 251.11 | 2456.6 | 251.13 | 2358.5 | 2609.6 | .8312 | 7.9096 |
| 65 | .2503 | 1.0199 | 6197 | 272.02 | 2463.1 | 272.06 | 2346.2 | 2618.3 | .8935 | 7.8310 |
| 70 | .3119 | 1.0228 | 5042 | 292.95 | 2469.6 | 292.98 | 2333.8 | 2626.8 | .9549 | 7.7553 |
| 75 | .3858 | 1.0259 | 4131 | 313.90 | 2475.9 | 313.93 | 2321.4 | 2635.3 | 1.0155 | 7.6824 |
| 80 | .4739 | 1.0291 | 3407 | 334.86 | 2482.2 | 334.91 | 2308.8 | 2643.7 | 1.0753 | 7.6122 |
| 85 | .5783 | 1.0325 | 2828 | 355.84 | 2488.4 | 355.90 | 2296.0 | 2651.9 | 1.1343 | 7.5445 |
| 90 | .7014 | 1.0360 | 2361 | 376.85 | 2494.5 | 376.92 | 2283.2 | 2660.1 | 1.1925 | 7.4791 |
| 95 | .8455 | 1.0397 | 1982 | 397.88 | 2500.6 | 397.96 | 2270.2 | 2668.1 | 1.2500 | 7.4159 |
| 100 | 1.014 | 1.0435 | 1673. | 418.94 | 2506.5 | 419.04 | 2257.0 | 2676.1 | 1.3069 | 7.3549 |
| 110 | 1.433 | 1.0516 | 1210. | 461.14 | 2518.1 | 461.30 | 2230.2 | 2691.5 | 1.4185 | 7.2387 |
| 120 | 1.985 | 1.0603 | 891.9 | 503.50 | 2529.3 | 503.71 | 2202.6 | 2706.3 | 1.5276 | 7.1296 |
| 130 | 2.701 | 1.0697 | 668.5 | 546.02 | 2539.9 | 546.31 | 2174.2 | 2720.5 | 1.6344 | 7.0269 |
| 140 | 3.613 | 1.0797 | 508.9 | 588.74 | 2550.0 | 589.13 | 2144.7 | 2733.9 | 1.7391 | 6.9299 |

## Table A-12M  (*Continued*)

| Temp. °C $T$ | Press. bars $P$ | Specific volume | | Internal energy | | Enthalpy | | | Entropy | |
|---|---|---|---|---|---|---|---|---|---|---|
| | | Sat. liquid $v_f$ | Sat. vapor $v_g$ | Sat. liquid $u_f$ | Sat. vapor $u_g$ | Sat. liquid $h_f$ | Evap. $h_{fg}$ | Sat. vapor $h_g$ | Sat. liquid $s_f$ | Sat. vapor $s_g$ |
| 150 | 4.758 | 1.0905 | 392.8 | 631.68 | 2559.5 | 632.20 | 2114.3 | 2746.5 | 1.8418 | 6.8379 |
| 160 | 6.178 | 1.1020 | 307.1 | 674.86 | 2568.4 | 675.55 | 2082.6 | 2758.1 | 1.9427 | 6.7502 |
| 170 | 7.917 | 1.1143 | 242.8 | 718.33 | 2576.5 | 719.21 | 2049.5 | 2768.7 | 2.0419 | 6.6663 |
| 180 | 10.02 | 1.1274 | 194.1 | 762.09 | 2583.7 | 763.22 | 2015.0 | 2778.2 | 2.1396 | 6.5857 |
| 190 | 12.54 | 1.1414 | 156.5 | 806.19 | 2590.0 | 807.62 | 1978.8 | 2786.4 | 2.2359 | 6.5079 |
| 200 | 15.54 | 1.1565 | 127.4 | 850.65 | 2595.3 | 852.45 | 1940.7 | 2793.2 | 2.3309 | 6.4323 |
| 210 | 19.06 | 1.1726 | 104.4 | 895.53 | 2599.5 | 897.76 | 1900.7 | 2798.5 | 2.4248 | 6.3585 |
| 220 | 23.18 | 1.1900 | 86.19 | 940.87 | 2602.4 | 943.62 | 1858.5 | 2802.1 | 2.5178 | 6.2861 |
| 230 | 27.95 | 1.2088 | 71.58 | 986.74 | 2603.9 | 990.12 | 1813.8 | 2804.0 | 2.6099 | 6.2146 |
| 240 | 33.44 | 1.2291 | 59.76 | 1033.2 | 2604.0 | 1037.3 | 1766.5 | 2803.8 | 2.7015 | 6.1437 |
| 250 | 39.73 | 1.2512 | 50.13 | 1080.4 | 2602.4 | 1085.4 | 1716.2 | 2801.5 | 2.7927 | 6.0730 |
| 260 | 46.88 | 1.2755 | 42.21 | 1128.4 | 2599.0 | 1134.4 | 1662.5 | 2796.9 | 2.8838 | 6.0019 |
| 270 | 54.99 | 1.3023 | 35.64 | 1177.4 | 2593.7 | 1184.5 | 1605.2 | 2789.7 | 2.9751 | 5.9301 |
| 280 | 64.12 | 1.3321 | 30.17 | 1227.5 | 2586.1 | 1236.0 | 1543.6 | 2779.6 | 3.0668 | 5.8571 |
| 290 | 74.36 | 1.3656 | 25.57 | 1278.9 | 2576.0 | 1289.1 | 1477.1 | 2766.2 | 3.1594 | 5.7821 |
| 300 | 85.81 | 1.4036 | 21.67 | 1332.0 | 2563.0 | 1344.0 | 1404.9 | 2749.0 | 3.2534 | 5.7045 |
| 320 | 112.7 | 1.4988 | 15.49 | 1444.6 | 2525.5 | 1461.5 | 1238.6 | 2700.1 | 3.4480 | 5.5362 |
| 340 | 145.9 | 1.6379 | 10.80 | 1570.3 | 2464.6 | 1594.2 | 1027.9 | 2622.0 | 3.6594 | 5.3357 |
| 360 | 186.5 | 1.8925 | 6.945 | 1725.2 | 2351.5 | 1760.5 | 720.5 | 2481.0 | 3.9147 | 5.0526 |
| 374.14 | 220.9 | 3.155 | 3.155 | 2029.6 | 2029.6 | 2099.3 | 0 | 2099.3 | 4.4298 | 4.4298 |

*Source:*  Keenan, J. H., F. G. Keyes, P. G. Hill, and J. G. Moore, "Steam Tables," Wiley, New York, 1969.

## Table A-13M  Properties of saturated water: pressure table

$v$, cm$^3$/g; $u$, kJ/kg; $h$, kJ/kg; $s$, kJ/(kg)($^\circ$K)

| Press. bars $P$ | Temp. °C $T$ | Specific volume Sat. liquid $v_f$ | Specific volume Sat. vapor $v_g$ | Internal energy Sat. liquid $u_f$ | Internal energy Sat. vapor $u_g$ | Enthalpy Sat. liquid $h_f$ | Enthalpy Evap. $h_{fg}$ | Enthalpy Sat. vapor $h_g$ | Entropy Sat. liquid $s_f$ | Entropy Sat. vapor $s_g$ |
|---|---|---|---|---|---|---|---|---|---|---|
| .040 | 28.96 | 1.0040 | 34800 | 121.45 | 2415.2 | 121.46 | 2432.9 | 2554.4 | .4226 | 8.4746 |
| .060 | 36.16 | 1.0064 | 23739 | 151.53 | 2425.0 | 151.53 | 2415.9 | 2567.4 | .5210 | 8.3304 |
| .080 | 41.51 | 1.0084 | 18103 | 173.87 | 2432.2 | 173.88 | 2403.1 | 2577.0 | .5926 | 8.2287 |
| 0.10 | 45.81 | 1.0102 | 14674 | 191.82 | 2437.9 | 191.83 | 2392.8 | 2584.7 | .6493 | 8.1502 |
| 0.20 | 60.06 | 1.0172 | 7649. | 251.38 | 2456.7 | 251.40 | 2358.3 | 2609.7 | .8320 | 7.9085 |
| 0.30 | 69.10 | 1.0223 | 5229. | 289.20 | 2468.4 | 289.23 | 2336.1 | 2625.3 | .9439 | 7.7686 |
| 0.40 | 75.87 | 1.0265 | 3993. | 317.53 | 2477.0 | 317.58 | 2319.2 | 2636.8 | 1.0259 | 7.6700 |
| 0.50 | 81.33 | 1.0300 | 3240. | 340.44 | 2483.9 | 340.49 | 2305.4 | 2645.9 | 1.0910 | 7.5939 |
| 0.60 | 85.94 | 1.0331 | 2732. | 359.79 | 2489.6 | 359.86 | 2293.6 | 2653.5 | 1.1453 | 7.5320 |
| 0.70 | 89.95 | 1.0360 | 2365. | 376.63 | 2494.5 | 376.70 | 2283.3 | 2660.0 | 1.1919 | 7.4797 |
| 0.80 | 93.50 | 1.0380 | 2087. | 391.58 | 2498.8 | 391.66 | 2274.1 | 2665.8 | 1.2329 | 7.4346 |
| 0.90 | 96.71 | 1.0410 | 1869. | 405.06 | 2502.6 | 405.15 | 2265.7 | 2670.9 | 1.2695 | 7.3949 |
| 1.00 | 99.63 | 1.0432 | 1694. | 417.36 | 2506.1 | 417.46 | 2258.0 | 2675.5 | 1.3026 | 7.3594 |
| 1.50 | 111.4 | 1.0528 | 1159. | 466.94 | 2519.7 | 467.11 | 2226.5 | 2693.6 | 1.4336 | 7.2233 |
| 2.00 | 120.2 | 1.0605 | 885.7 | 504.49 | 2529.5 | 504.70 | 2201.9 | 2706.7 | 1.5301 | 7.1271 |
| 2.50 | 127.4 | 1.0672 | 718.7 | 535.10 | 2537.2 | 535.37 | 2181.5 | 2716.9 | 1.6072 | 7.0527 |
| 3.00 | 133.6 | 1.0732 | 605.8 | 561.15 | 2543.6 | 561.47 | 2163.8 | 2725.3 | 1.6718 | 6.9919 |
| 3.50 | 138.9 | 1.0786 | 524.3 | 583.95 | 2548.9 | 584.33 | 2148.1 | 2732.4 | 1.7275 | 6.9405 |
| 4.00 | 143.6 | 1.0836 | 462.5 | 604.31 | 2553.6 | 604.74 | 2133.8 | 2738.6 | 1.7766 | 6.8959 |
| 4.50 | 147.9 | 1.0882 | 414.0 | 622.77 | 2557.6 | 623.25 | 2120.7 | 2743.9 | 1.8207 | 6.8565 |
| 5.00 | 151.9 | 1.0926 | 374.9 | 639.68 | 2561.2 | 640.23 | 2108.5 | 2748.7 | 1.8607 | 6.8213 |
| 6.00 | 158.9 | 1.1006 | 315.7 | 669.90 | 2567.4 | 670.56 | 2086.3 | 2756.8 | 1.9312 | 6.7600 |
| 7.00 | 165.0 | 1.1080 | 272.9 | 696.44 | 2572.5 | 697.22 | 2066.3 | 2763.5 | 1.9922 | 6.7080 |
| 8.00 | 170.4 | 1.1148 | 240.4 | 720.22 | 2576.8 | 721.11 | 2048.0 | 2769.1 | 2.0462 | 6.6628 |
| 9.00 | 175.4 | 1.1212 | 215.0 | 741.83 | 2580.5 | 742.83 | 2031.1 | 2773.9 | 2.0946 | 6.6226 |
| 10.0 | 179.9 | 1.1273 | 194.4 | 761.68 | 2583.6 | 762.81 | 2015.3 | 2778.1 | 2.1387 | 6.5863 |
| 15.0 | 198.3 | 1.1539 | 131.8 | 843.16 | 2594.5 | 844.89 | 1947.3 | 2792.2 | 2.3150 | 6.4448 |
| 20.0 | 212.4 | 1.1767 | 99.63 | 906.44 | 2600.3 | 908.79 | 1890.7 | 2799.5 | 2.4474 | 6.3409 |
| 25.0 | 224.0 | 1.1973 | 79.98 | 959.11 | 2603.1 | 962.11 | 1841.0 | 2803.1 | 2.5547 | 6.2575 |
| 30.0 | 233.9 | 1.2165 | 66.68 | 1004.8 | 2604.1 | 1008.4 | 1795.7 | 2804.2 | 2.6457 | 6.1869 |
| 35.0 | 242.6 | 1.2347 | 57.07 | 1045.4 | 2603.7 | 1049.8 | 1753.7 | 2803.4 | 2.7253 | 6.1253 |
| 40.0 | 250.4 | 1.2522 | 49.78 | 1082.3 | 2602.3 | 1087.3 | 1714.1 | 2801.4 | 2.7964 | 6.0701 |
| 45.0 | 257.5 | 1.2692 | 44.06 | 1116.2 | 2600.1 | 1121.9 | 1676.4 | 2798.3 | 2.8610 | 6.0199 |
| 50.0 | 264.0 | 1.2859 | 39.44 | 1147.8 | 2597.1 | 1154.2 | 1640.1 | 2794.3 | 2.9202 | 5.9734 |
| 60.0 | 275.6 | 1.3187 | 32.44 | 1205.4 | 2589.7 | 1213.4 | 1571.0 | 2784.3 | 3.0267 | 5.8892 |
| 70.0 | 285.9 | 1.3513 | 27.37 | 1257.6 | 2580.5 | 1267.0 | 1505.1 | 2772.1 | 3.1211 | 5.8133 |
| 80.0 | 295.1 | 1.3842 | 23.52 | 1305.6 | 2569.8 | 1316.6 | 1441.3 | 2758.0 | 3.2068 | 5.7432 |
| 90.0 | 303.4 | 1.4178 | 20.48 | 1350.5 | 2557.8 | 1363.3 | 1378.9 | 2742.1 | 3.2858 | 5.6772 |
| 100. | 311.1 | 1.4524 | 18.03 | 1393.0 | 2544.4 | 1407.6 | 1317.1 | 2724.7 | 3.3596 | 5.6141 |
| 110. | 318.2 | 1.4886 | 15.99 | 1433.7 | 2529.8 | 1450.1 | 1255.5 | 2705.6 | 3.4295 | 5.5527 |
| 120. | 324.8 | 1.5267 | 14.26 | 1473.0 | 2513.7 | 1491.3 | 1193.6 | 2684.9 | 3.4962 | 5.4924 |
| 130. | 330.9 | 1.5671 | 12.78 | 1511.1 | 2496.1 | 1531.5 | 1130.7 | 2662.2 | 3.5606 | 5.4323 |
| 140. | 336.8 | 1.6107 | 11.49 | 1548.6 | 2476.8 | 1571.1 | 1066.5 | 2637.6 | 3.6232 | 5.3717 |
| 150. | 342.2 | 1.6581 | 10.34 | 1585.6 | 2455.5 | 1610.5 | 1000.0 | 2610.5 | 3.6848 | 5.3098 |
| 160. | 347.4 | 1.7107 | 9.306 | 1622.7 | 2431.7 | 1650.1 | 930.6 | 2580.6 | 3.7461 | 5.2455 |
| 170. | 352.4 | 1.7702 | 8.364 | 1660.2 | 2405.0 | 1690.3 | 856.9 | 2547.2 | 3.8079 | 5.1777 |
| 180. | 357.1 | 1.8397 | 7.489 | 1698.9 | 2374.3 | 1732.0 | 777.1 | 2509.1 | 3.8715 | 5.1044 |
| 190. | 361.5 | 1.9243 | 6.657 | 1739.9 | 2338.1 | 1776.5 | 688.0 | 2464.5 | 3.9388 | 5.0228 |
| 200. | 365.8 | 2.036 | 5.834 | 1785.6 | 2293.0 | 1826.3 | 583.4 | 2409.7 | 4.0139 | 4.9269 |
| 220.9 | 374.1 | 3.155 | 3.155 | 2029.6 | 2029.6 | 2099.3 | 0 | 2099.3 | 4.4298 | 4.4298 |

CRITICAL PRESSURE

*Source:*  Keenan et al., "Steam Tables," Wiley, New York, 1969.

### Table A-14M Properties of water: superheated-vapor table
$v$, cm$^3$/g; $u$, kJ/kg; $h$, kJ/kg; $s$, kJ/(kg)($^\circ$K)

| Temp. °C | $v$ | $u$ | $h$ | $s$ | $v$ | $u$ | $h$ | $s$ |
|---|---|---|---|---|---|---|---|---|
| | 0.06 bar (36.16°C) | | | | 0.35 bar (72.69°C) | | | |
| Sat. | 23739 | 2425.0 | 2546.4 | 8.3304 | 4526. | 2473.0 | 2631.4 | 7.7158 |
| 80 | 27132 | 2487.3 | 2650.1 | 8.5804 | 4625. | 2483.7 | 2645.6 | 7.7564 |
| 120 | 30219 | 2544.7 | 2726.0 | 8.7840 | 5163. | 2542.4 | 2723.1 | 7.9644 |
| 160 | 33302 | 2602.7 | 2802.5 | 8.9693 | 5696. | 2601.2 | 2800.6 | 8.1519 |
| 200 | 36383 | 2661.4 | 2879.7 | 9.1398 | 6228. | 2660.4 | 2878.4 | 8.3237 |
| 240 | 39462 | 2721.0 | 2957.8 | 9.2982 | 6758. | 2720.3 | 2956.8 | 8.4828 |
| 280 | 42540 | 2781.5 | 3036.8 | 9.4464 | 7287. | 2780.9 | 3036.0 | 8.6314 |
| 320 | 45618 | 2843.0 | 3116.7 | 9.5859 | 7815. | 2842.5 | 3116.1 | 8.7712 |
| 360 | 48696 | 2905.5 | 3197.7 | 9.7180 | 8344. | 2905.1 | 3197.1 | 8.9034 |
| 400 | 51774 | 2969.0 | 3279.6 | 9.8435 | 8872. | 2968.6 | 3279.2 | 9.0291 |
| 440 | 54851 | 3033.5 | 3362.6 | 9.9633 | 9400. | 3033.2 | 3362.2 | 9.1490 |
| 500 | 59467 | 3132.3 | 3489.1 | 10.134 | 10192. | 3132.1 | 3488.8 | 9.3194 |
| | 0.70 bar (89.95°C) | | | | 1.0 bar (99.63°C) | | | |
| Sat. | 2365. | 2494.5 | 2660.0 | 7.4797 | 1694. | 2506.1 | 2675.5 | 7.3594 |
| 100 | 2434. | 2509.7 | 2680.0 | 7.5341 | 1696. | 2506.7 | 2676.2 | 7.3614 |
| 120 | 2571. | 2539.7 | 2719.6 | 7.6375 | 1793. | 2537.3 | 2716.6 | 7.4668 |
| 160 | 2841. | 2599.4 | 2798.2 | 7.8279 | 1984. | 2597.8 | 2796.2 | 7.6597 |
| 200 | 3108. | 2659.1 | 2876.7 | 8.0012 | 2172. | 2658.1 | 2875.3 | 7.8343 |
| 240 | 3374. | 2719.3 | 2955.5 | 8.1611 | 2359. | 2718.5 | 2954.5 | 7.9949 |
| 280 | 3640. | 2780.2 | 3035.0 | 8.3162 | 2546. | 2779.6 | 3034.2 | 8.1445 |
| 320 | 3905. | 2842.0 | 3115.3 | 8.4504 | 2732. | 2841.5 | 3114.6 | 8.2849 |
| 360 | 4170. | 2904.6 | 3196.5 | 8.5828 | 2917. | 2904.2 | 3195.9 | 8.4175 |
| 400 | 4434. | 2968.2 | 3278.6 | 8.7086 | 3103. | 2967.9 | 3278.2 | 8.5435 |
| 440 | 4698. | 3032.9 | 3361.8 | 8.8286 | 3288. | 3032.6 | 3361.4 | 8.6636 |
| 500 | 5095. | 3131.8 | 3488.5 | 8.9991 | 3565. | 3131.6 | 3488.1 | 8.8342 |
| | 1.5 bars (111.37°C) | | | | 3.0 bars (133.55°C) | | | |
| Sat. | 1159. | 2519.7 | 2693.6 | 7.2233 | 606. | 2543.6 | 2725.3 | 6.9919 |
| 120 | 1188. | 2533.3 | 2711.4 | 7.2693 | | | | |
| 160 | 1317. | 2595.2 | 2792.8 | 7.4665 | 651. | 2587.1 | 2782.3 | 7.1276 |
| 200 | 1444. | 2656.2 | 2872.9 | 7.6433 | 716. | 2650.7 | 2865.5 | 7.3115 |
| 240 | 1570. | 2717.2 | 2952.7 | 7.8052 | 781. | 2713.1 | 2947.3 | 7.4774 |
| 280 | 1695. | 2778.6 | 3032.8 | 7.9555 | 844. | 2775.4 | 3028.6 | 7.6299 |
| 320 | 1819. | 2840.6 | 3113.5 | 8.0964 | 907. | 2838.1 | 3110.1 | 7.7722 |
| 360 | 1943. | 2903.5 | 3195.0 | 8.2293 | 969. | 2901.4 | 3192.2 | 7.9061 |
| 400 | 2067. | 2967.3 | 3277.4 | 8.3555 | 1032. | 2965.6 | 3275.0 | 8.0330 |
| 440 | 2191. | 3032.1 | 3360.7 | 8.4757 | 1094. | 3030.6 | 3358.7 | 8.1538 |
| 500 | 2376. | 3131.2 | 3487.6 | 8.6466 | 1187. | 3130.0 | 3486.0 | 8.3251 |
| 600 | 2685. | 3301.7 | 3704.3 | 8.9101 | 1341. | 3300.8 | 3703.2 | 8.5892 |

## Table A-14M  (*Continued*)

| Temp. °C | $v$ | $u$ | $h$ | $s$ | $v$ | $u$ | $h$ | $s$ |
|---|---|---|---|---|---|---|---|---|
| | 5.0 bars (151.86°C) | | | | 7.0 bars (164.97°C) | | | |
| Sat. | 374.9 | 2561.2 | 2748.7 | 6.8213 | 272.9 | 2572.5 | 2763.5 | 6.7080 |
| 180 | 404.5 | 2609.7 | 2812.0 | 6.9656 | 284.7 | 2599.8 | 2799.1 | 6.7880 |
| 200 | 424.9 | 2642.9 | 2855.4 | 7.0592 | 299.9 | 2634.8 | 2844.8 | 6.8865 |
| 240 | 464.6 | 2707.6 | 2939.9 | 7.2307 | 329.2 | 2701.8 | 2932.2 | 7.0641 |
| 280 | 503.4 | 2771.2 | 3022.9 | 7.3865 | 357.4 | 2766.9 | 3017.1 | 7.2233 |
| 320 | 541.6 | 2834.7 | 3105.6 | 7.5308 | 385.2 | 2831.3 | 3100.9 | 7.3697 |
| 360 | 579.6 | 2898.7 | 3188.4 | 7.6660 | 412.6 | 2895.8 | 3184.7 | 7.5063 |
| 400 | 617.3 | 2963.2 | 3271.9 | 7.7938 | 439.7 | 2960.9 | 3268.7 | 7.6350 |
| 440 | 654.8 | 3028.6 | 3356.0 | 7.9152 | 466.7 | 3026.6 | 3353.3 | 7.7571 |
| 500 | 710.9 | 3128.4 | 3483.9 | 8.0873 | 507.0 | 3126.8 | 3481.7 | 7.9299 |
| 600 | 804.1 | 3299.6 | 3701.7 | 8.3522 | 573.8 | 3298.5 | 3700.2 | 8.1956 |
| 700 | 896.9 | 3477.5 | 3925.9 | 8.5952 | 640.3 | 3476.6 | 3924.8 | 8.4391 |
| | 10.0 bars (179.91°C) | | | | 15.0 bars (198.32°C) | | | |
| Sat. | 194.4 | 2583.6 | 2778.1 | 6.5865 | 131.8 | 2594.5 | 2792.2 | 6.4448 |
| 200 | 206.0 | 2621.9 | 2827.9 | 6.6940 | 132.5 | 2598.1 | 2796.8 | 6.4546 |
| 240 | 227.5 | 2692.9 | 2920.4 | 6.8817 | 148.3 | 2676.9 | 2899.3 | 6.6628 |
| 280 | 248.0 | 2760.2 | 3008.2 | 7.0465 | 162.7 | 2748.6 | 2992.7 | 6.8381 |
| 320 | 267.8 | 2826.1 | 3093.9 | 7.1962 | 176.5 | 2817.1 | 3081.9 | 6.9938 |
| 360 | 287.3 | 2891.6 | 3178.9 | 7.3349 | 189.9 | 2884.4 | 3169.2 | 7.1363 |
| 400 | 306.6 | 2957.3 | 3263.9 | 7.4651 | 203.0 | 2951.3 | 3255.8 | 7.2690 |
| 440 | 325.7 | 3023.6 | 3349.3 | 7.5883 | 216.0 | 3018.5 | 3342.5 | 7.3940 |
| 500 | 354.1 | 3124.4 | 3478.5 | 7.7622 | 235.2 | 3120.3 | 3473.1 | 7.5698 |
| 540 | 372.9 | 3192.6 | 3565.6 | 7.8720 | 247.8 | 3189.1 | 3560.9 | 7.6805 |
| 600 | 401.1 | 3296.8 | 3697.9 | 8.0290 | 266.8 | 3293.9 | 3694.0 | 7.8385 |
| 640 | 419.8 | 3367.4 | 3787.2 | 8.1290 | 279.3 | 3364.8 | 3783.8 | 7.9391 |
| | 20.0 bars (212.42°C) | | | | 30.0 bars (233.90°C) | | | |
| Sat. | 99.6 | 2600.3 | 2799.5 | 6.3409 | 66.7 | 2604.1 | 2804.2 | 6.1869 |
| 240 | 108.5 | 2659.6 | 2876.5 | 6.4952 | 68.2 | 2619.7 | 2824.3 | 6.2265 |
| 280 | 120.0 | 2736.4 | 2976.4 | 6.6828 | 77.1 | 2709.9 | 2941.3 | 6.4462 |
| 320 | 130.8 | 2807.9 | 3069.5 | 6.8452 | 85.0 | 2788.4 | 3043.4 | 6.6245 |
| 360 | 141.1 | 2877.0 | 3159.3 | 6.9917 | 92.3 | 2861.7 | 3138.7 | 6.7801 |
| 400 | 151.2 | 2945.2 | 3247.6 | 7.1271 | 99.4 | 2932.8 | 3230.9 | 6.9212 |
| 440 | 161.1 | 3013.4 | 3335.5 | 7.2540 | 106.2 | 3002.9 | 3321.5 | 7.0520 |
| 500 | 175.7 | 3116.2 | 3467.6 | 7.4317 | 116.2 | 3108.0 | 3456.5 | 7.2338 |
| 540 | 185.3 | 3185.6 | 3556.1 | 7.5434 | 122.7 | 3178.4 | 3546.6 | 7.3474 |
| 600 | 199.6 | 3290.9 | 3690.1 | 7.7024 | 132.4 | 3285.0 | 3682.3 | 7.5085 |
| 640 | 209.1 | 3362.2 | 3780.4 | 7.8035 | 138.8 | 3357.0 | 3773.5 | 7.6106 |
| 700 | 223.2 | 3470.9 | 3917.4 | 7.9487 | 148.4 | 3466.5 | 3911.7 | 7.7571 |

## Table A-14M  (Continued)

| Temp. °C | $v$ | $u$ | $h$ | $s$ | $v$ | $u$ | $h$ | $s$ |
|---|---|---|---|---|---|---|---|---|
| | 40 bars (250.40°C) | | | | 60 bars (275.64°C) | | | |
| Sat. | 49.78 | 2602.3 | 2801.4 | 6.0701 | 32.44 | 2589.7 | 2784.3 | 5.8892 |
| 280 | 55.46 | 2680.0 | 2901.8 | 6.2568 | 33.17 | 2605.2 | 2804.2 | 5.9252 |
| 320 | 61.99 | 2767.4 | 3015.4 | 6.4553 | 38.76 | 2720.0 | 2952.6 | 6.1846 |
| 360 | 67.88 | 2845.7 | 3117.2 | 6.6215 | 43.31 | 2811.2 | 3071.1 | 6.3782 |
| 400 | 73.41 | 2919.9 | 3213.6 | 6.7690 | 47.39 | 2892.9 | 3177.2 | 6.5408 |
| 440 | 78.72 | 2992.2 | 3307.1 | 6.9041 | 51.22 | 2970.0 | 3277.3 | 6.6853 |
| 500 | 86.43 | 3099.5 | 3445.3 | 7.0901 | 56.65 | 3082.2 | 3422.2 | 6.8803 |
| 540 | 91.45 | 3171.1 | 3536.9 | 7.2056 | 60.15 | 3156.1 | 3517.0 | 6.9999 |
| 600 | 98.85 | 3279.1 | 3674.4 | 7.3688 | 65.25 | 3266.9 | 3658.4 | 7.1677 |
| 640 | 103.7 | 3351.8 | 3766.6 | 7.4720 | 68.59 | 3341.0 | 3752.6 | 7.2731 |
| 700 | 111.0 | 3462.1 | 3905.9 | 7.6198 | 73.52 | 3453.1 | 3894.1 | 7.4234 |
| 740 | 115.7 | 3536.6 | 3999.6 | 7.7141 | 76.77 | 3528.3 | 3989.2 | 7.5190 |
| | 80 bars (295.06°C) | | | | 100 bars (311.06°C) | | | |
| Sat. | 23.52 | 2569.8 | 2758.0 | 5.7432 | 18.03 | 2544.4 | 2724.7 | 5.6141 |
| 320 | 26.82 | 2662.7 | 2877.2 | 5.9489 | 19.25 | 2588.8 | 2781.3 | 5.7103 |
| 360 | 30.89 | 2772.7 | 3019.8 | 6.1819 | 23.31 | 2729.1 | 2962.1 | 6.0060 |
| 400 | 34.32 | 2863.8 | 3138.3 | 6.3634 | 26.41 | 2832.4 | 3096.5 | 6.2120 |
| 440 | 37.42 | 2946.7 | 3246.1 | 6.5190 | 29.11 | 2922.1 | 3213.2 | 6.3805 |
| 480 | 40.34 | 3025.7 | 3348.4 | 6.6586 | 31.60 | 3005.4 | 3321.4 | 6.5282 |
| 520 | 43.13 | 3102.7 | 3447.7 | 6.7871 | 33.94 | 3085.6 | 3425.1 | 6.6622 |
| 560 | 45.82 | 3178.7 | 3545.3 | 6.9072 | 36.19 | 3164.1 | 3526.0 | 6.7864 |
| 600 | 48.45 | 3254.4 | 3642.0 | 7.0206 | 38.37 | 3241.7 | 3625.3 | 6.9029 |
| 640 | 51.02 | 3330.1 | 3738.3 | 7.1283 | 40.48 | 3318.9 | 3723.7 | 7.0131 |
| 700 | 54.81 | 3443.9 | 3882.4 | 7.2812 | 43.58 | 3434.7 | 3870.5 | 7.1687 |
| 740 | 57.29 | 3520.4 | 3978.7 | 7.3782 | 45.60 | 3512.1 | 3968.1 | 7.2670 |
| | 120 bars (324.75°C) | | | | 140 bars (336.75°C) | | | |
| Sat. | 14.26 | 2513.7 | 2684.9 | 5.4924 | 11.49 | 2476.8 | 2637.6 | 5.3717 |
| 360 | 18.11 | 2678.4 | 2895.7 | 5.8361 | 14.22 | 2617.4 | 2816.5 | 5.6602 |
| 400 | 21.08 | 2798.3 | 3051.3 | 6.0747 | 17.22 | 2760.9 | 3001.9 | 5.9448 |
| 440 | 23.55 | 2896.1 | 3178.7 | 6.2586 | 19.54 | 2868.6 | 3142.2 | 6.1474 |
| 480 | 25.76 | 2984.4 | 3293.5 | 6.4154 | 21.57 | 2962.5 | 3264.5 | 6.3143 |
| 520 | 27.81 | 3068.0 | 3401.8 | 6.5555 | 23.43 | 3049.8 | 3377.8 | 6.4610 |
| 560 | 29.77 | 3149.0 | 3506.2 | 6.6840 | 25.17 | 3133.6 | 3486.0 | 6.5941 |
| 600 | 31.64 | 3228.7 | 3608.3 | 6.8037 | 26.83 | 3215.4 | 3591.1 | 6.7172 |
| 640 | 33.45 | 3307.5 | 3709.0 | 6.9164 | 28.43 | 3296.0 | 3694.1 | 6.8326 |
| 700 | 36.10 | 3425.2 | 3858.4 | 7.0749 | 30.75 | 3415.7 | 3846.2 | 6.9939 |
| 740 | 37.81 | 3503.7 | 3957.4 | 7.1746 | 32.25 | 3495.2 | 3946.7 | 7.0952 |

## Table A-14M  (Continued)

| Temp. °C | $v$ | $u$ | $h$ | $s$ | $v$ | $u$ | $h$ | $s$ |
|---|---|---|---|---|---|---|---|---|
| | | 160 bars (347.44°C) | | | | 180 bars (357.06°C) | | |
| Sat. | 9.31 | 2431.7 | 2580.6 | 5.2455 | 7.49 | 2374.3 | 2509.1 | 5.1044 |
| 360 | 11.05 | 2539.0 | 2715.8 | 5.4614 | 8.09 | 2418.9 | 2564.5 | 5.1922 |
| 400 | 14.26 | 2719.4 | 2947.6 | 5.8175 | 11.90 | 2672.8 | 2887.0 | 5.6887 |
| 440 | 16.52 | 2839.4 | 3103.7 | 6.0429 | 14.14 | 2808.2 | 3062.8 | 5.9428 |
| 480 | 18.42 | 2939.7 | 3234.4 | 6.2215 | 15.96 | 2915.9 | 3203.2 | 6.1345 |
| 520 | 20.13 | 3031.1 | 3353.3 | 6.3752 | 17.57 | 3011.8 | 3378.0 | 6.2960 |
| 560 | 21.72 | 3117.8 | 3465.4 | 6.5132 | 19.04 | 3101.7 | 3444.4 | 6.4392 |
| 600 | 23.23 | 3201.8 | 3573.5 | 6.6399 | 20.42 | 3188.0 | 3555.6 | 6.5696 |
| 640 | 24.67 | 3284.2 | 3678.9 | 6.7580 | 21.74 | 3272.3 | 3663.6 | 6.6905 |
| 700 | 26.74 | 3406.0 | 3833.9 | 6.9224 | 23.62 | 3396.3 | 3821.5 | 6.8580 |
| 740 | 28.08 | 3486.7 | 3935.9 | 7.0251 | 24.83 | 3478.0 | 3925.0 | 6.9623 |
| | | 200 bars (365.81°C) | | | | 240 bars | | |
| Sat. | 5.83 | 2293.0 | 2409.7 | 4.9269 | | | | |
| 400 | 9.94 | 2619.3 | 2818.1 | 5.5540 | 6.73 | 2477.8 | 2639.4 | 5.2393 |
| 440 | 12.22 | 2774.9 | 3019.4 | 5.8450 | 9.29 | 2700.6 | 2923.4 | 5.6506 |
| 480 | 13.99 | 2891.2 | 3170.8 | 6.0518 | 11.00 | 2838.3 | 3102.3 | 5.8950 |
| 520 | 15.51 | 2992.0 | 3302.2 | 6.2218 | 12.41 | 2950.5 | 3248.5 | 6.0842 |
| 560 | 16.89 | 3085.2 | 3423.0 | 6.3705 | 13.66 | 3051.1 | 3379.0 | 6.2448 |
| 600 | 18.18 | 3174.0 | 3537.6 | 6.5048 | 14.81 | 3145.2 | 3500.7 | 6.3875 |
| 640 | 19.40 | 3260.2 | 3648.1 | 6.6286 | 15.88 | 3235.5 | 3616.7 | 6.5174 |
| 700 | 21.13 | 3386.4 | 3809.0 | 6.7993 | 17.39 | 3366.4 | 3783.8 | 6.6947 |
| 740 | 22.24 | 3469.3 | 3914.1 | 6.9052 | 18.35 | 3451.7 | 3892.1 | 6.8038 |
| 800 | 23.85 | 3592.7 | 4069.7 | 7.0544 | 19.74 | 3578.0 | 4051.6 | 6.9567 |
| | | 280 bars | | | | 320 bars | | |
| 400 | 3.83 | 2223.5 | 2330.7 | 4.7494 | 2.36 | 1980.4 | 2055.9 | 4.3239 |
| 440 | 7.12 | 2613.2 | 2812.6 | 5.4494 | 5.44 | 2509.0 | 2683.0 | 5.2327 |
| 480 | 8.85 | 2780.8 | 3028.5 | 5.7446 | 7.22 | 2718.1 | 2949.2 | 5.5968 |
| 520 | 10.20 | 2906.8 | 3192.3 | 5.9566 | 8.53 | 2860.7 | 3133.7 | 5.8357 |
| 560 | 11.36 | 3015.7 | 3333.7 | 6.1307 | 9.63 | 2979.0 | 3287.2 | 6.0246 |
| 600 | 12.41 | 3115.6 | 3463.0 | 6.2823 | 10.61 | 3085.3 | 3424.6 | 6.1858 |
| 640 | 13.38 | 3210.3 | 3584.8 | 6.4187 | 11.50 | 3184.5 | 3552.5 | 6.3290 |
| 700 | 14.73 | 3346.1 | 3758.4 | 6.6029 | 12.73 | 3325.4 | 3732.8 | 6.5203 |
| 740 | 15.58 | 3433.9 | 3870.0 | 6.7153 | 13.50 | 3415.9 | 3847.8 | 6.6361 |
| 800 | 16.80 | 3563.1 | 4033.4 | 6.8720 | 14.60 | 3548.0 | 4015.1 | 6.7966 |
| 900 | 18.73 | 3774.3 | 4298.8 | 7.1084 | 16.33 | 3762.7 | 4285.1 | 7.0372 |

*Source:*  Keenan, J. H., F. G. Keyes, P. G. Hill, and J. G. Moore, "Steam Tables," Wiley, New York, 1969.

## Table A-15M Properties of water: compressed liquid table
$v$, cm$^3$/g; $u$, kJ/kg; $h$, kJ/kg; $s$, kJ/(kg)($^\circ$K)

| Temp. $^\circ$C | $v$ | $u$ | $h$ | $s$ | $v$ | $u$ | $h$ | $s$ |
|---|---|---|---|---|---|---|---|---|
| | 25 bars (223.99$^\circ$C) | | | | 50 bars (263.99$^\circ$C) | | | |
| 20 | 1.0006 | 83.80 | 86.30 | .2961 | .9995 | 83.65 | 88.65 | .2956 |
| 40 | 1.0067 | 167.25 | 169.77 | .5715 | 1.0056 | 166.95 | 171.97 | .5705 |
| 80 | 1.0280 | 334.29 | 336.86 | 1.0737 | 1.0268 | 333.72 | 338.85 | 1.0720 |
| 120 | 1.0590 | 502.68 | 505.33 | 1.5255 | 1.0576 | 501.80 | 507.09 | 1.5233 |
| 160 | 1.1006 | 673.90 | 676.65 | 1.9404 | 1.0988 | 672.62 | 678.12 | 1.9375 |
| 200 | 1.1555 | 849.9 | 852.8 | 2.3294 | 1.1530 | 848.1 | 848.1 | 2.3255 |
| 220 | 1.1898 | 940.7 | 943.7 | 2.5174 | 1.1866 | 938.4 | 944.4 | 2.5128 |
| Sat. | 1.1973 | 959.1 | 962.1 | 2.5546 | 1.2859 | 1147.8 | 1154.2 | 2.9202 |
| | 75 bars (290.59$^\circ$C) | | | | 100 bars (311.06$^\circ$C) | | | |
| 20 | .9984 | 83.50 | 90.99 | .2950 | .9972 | 83.36 | 93.33 | .2945 |
| 40 | 1.0045 | 166.64 | 174.18 | .5696 | 1.0034 | 166.35 | 176.38 | .5686 |
| 80 | 1.0256 | 333.15 | 340.84 | 1.0704 | 1.0245 | 332.59 | 342.83 | 1.0688 |
| 100 | 1.0397 | 416.81 | 424.62 | 1.3011 | 1.0385 | 416.12 | 426.50 | 1.2992 |
| 140 | 1.0752 | 585.72 | 593.78 | 1.7317 | 1.0737 | 584.68 | 595.42 | 1.7292 |
| 180 | 1.1219 | 758.13 | 766.55 | 2.1308 | 1.1199 | 756.65 | 767.84 | 2.1275 |
| 220 | 1.1835 | 936.2 | 945.1 | 2.5083 | 1.1805 | 934.1 | 945.9 | 2.5039 |
| 260 | 1.2696 | 1124.4 | 1134.0 | 2.8763 | 1.2645 | 1121.1 | 1133.7 | 2.8699 |
| Sat. | 1.3677 | 1282.0 | 1292.2 | 3.1649 | 1.4524 | 1393.0 | 1407.6 | 3.3596 |
| | 150 bars (342.24$^\circ$C) | | | | 200 bars (365.81$^\circ$C) | | | |
| 20 | .9950 | 83.06 | 97.99 | .2934 | .9928 | 82.77 | 102.62 | .2923 |
| 40 | 1.0013 | 165.76 | 180.78 | .5666 | .9992 | 165.17 | 185.16 | .5646 |
| 100 | 1.0361 | 414.75 | 430.28 | 1.2955 | 1.0337 | 413.39 | 434.06 | 1.2917 |
| 180 | 1.1159 | 753.76 | 770.50 | 2.1210 | 1.1120 | 750.95 | 773.20 | 2.1147 |
| 220 | 1.1748 | 929.9 | 947.5 | 2.4953 | 1.1693 | 925.9 | 949.3 | 2.4870 |
| 260 | 1.2550 | 1114.6 | 1133.4 | 2.8576 | 1.2462 | 1108.6 | 1133.5 | 2.8459 |
| 300 | 1.3770 | 1316.6 | 1337.3 | 3.2260 | 1.3596 | 1306.1 | 1333.3 | 3.2071 |
| Sat. | 1.6581 | 1585.6 | 1610.5 | 3.6848 | 2.036 | 1785.6 | 1826.3 | 4.0139 |
| | 250 bars | | | | 300 bars | | | |
| 20 | .9907 | 82.47 | 107.24 | .2911 | .9886 | 82.17 | 111.84 | .2899 |
| 40 | .9971 | 164.60 | 189.52 | .5626 | .9951 | 164.04 | 193.89 | .5607 |
| 100 | 1.0313 | 412.08 | 437.85 | 1.2881 | 1.0290 | 410.78 | 441.66 | 1.2844 |
| 200 | 1.1344 | 834.5 | 862.8 | 2.2961 | 1.1302 | 831.4 | 865.3 | 2.2893 |
| 300 | 1.3442 | 1296.6 | 1330.2 | 3.1900 | 1.3304 | 1287.9 | 1327.8 | 3.1741 |

*Source:* Keenan, J. H., F. G. Keyes, P. G. Hill, and J. G. Moore, "Steam Tables," Wiley, New York, 1969.

### Table A-16M Properties of saturated refrigerant 12, $CCl_2F_2$: temperature table
$v$, cm$^3$/g; $u$, kJ/kg; $h$, kJ/kg; $s$, kJ/(kg)($^\circ$K)

| Temp. °C $T$ | Press. bar(s) $P$ | Specific volume | | Internal energy | | Enthalpy | | | Entropy | |
|---|---|---|---|---|---|---|---|---|---|---|
| | | Sat. liquid $v_f$ | Sat. vapor $v_g$ | Sat. liquid $u_f$ | Sat. vapor $u_g$ | Sat. liquid $h_f$ | Evap. $h_{fg}$ | Sat. vapor $h_g$ | Sat. liquid $s_f$ | Sat. vapor $s_g$ |
| −40 | .6417 | .6595 | 241.91 | −0.04 | 154.07 | 0 | 169.59 | 169.59 | 0 | .7274 |
| −35 | .8071 | .6656 | 195.40 | 4.37 | 156.13 | 4.42 | 167.48 | 171.90 | .0187 | .7219 |
| −30 | 1.0041 | .6720 | 159.38 | 8.79 | 158.20 | 8.86 | 165.33 | 174.20 | .0371 | .7170 |
| −25 | 1.2368 | .6786 | 131.17 | 13.25 | 160.26 | 13.33 | 163.15 | 176.48 | .0552 | .7126 |
| −20 | 1.5093 | .6855 | 108.85 | 17.72 | 162.31 | 17.82 | 160.92 | 178.74 | .0731 | .7087 |
| −15 | 1.8260 | .6926 | 91.02 | 22.20 | 164.35 | 22.33 | 158.64 | 180.97 | .0906 | .7051 |
| −10 | 2.1912 | .7000 | 76.65 | 26.72 | 166.39 | 26.87 | 156.31 | 183.19 | .1080 | .7019 |
| −5 | 2.6096 | .7078 | 64.96 | 31.27 | 168.42 | 31.45 | 153.93 | 185.37 | .1251 | .6991 |
| 0 | 3.0861 | .7159 | 55.39 | 35.83 | 170.44 | 36.05 | 151.48 | 187.53 | .1420 | .6965 |
| 4 | 3.5124 | .7227 | 48.95 | 39.51 | 172.04 | 39.76 | 149.47 | 189.23 | .1553 | .6946 |
| 8 | 3.9815 | .7297 | 43.40 | 43.21 | 173.63 | 43.50 | 147.41 | 190.91 | .1686 | .6929 |
| 12 | 4.4962 | .7370 | 38.60 | 46.93 | 175.20 | 47.26 | 145.30 | 192.56 | .1817 | .6913 |
| 16 | 5.0591 | .7446 | 34.42 | 50.67 | 176.78 | 51.05 | 143.14 | 194.19 | .1948 | .6898 |
| 20 | 5.6729 | .7525 | 30.78 | 54.44 | 178.32 | 54.87 | 140.91 | 195.78 | .2078 | .6884 |
| 24 | 6.3405 | .7607 | 27.59 | 58.25 | 179.85 | 58.73 | 138.61 | 197.34 | .2207 | .6871 |
| 28 | 7.0648 | .7694 | 24.78 | 62.09 | 181.36 | 62.63 | 136.24 | 198.87 | .2335 | .6859 |
| 32 | 7.8485 | .7785 | 22.31 | 65.96 | 182.85 | 66.57 | 133.79 | 200.36 | .2463 | .6847 |
| 36 | 8.6948 | .7880 | 20.12 | 69.86 | 184.31 | 70.55 | 131.25 | 201.80 | .2591 | .6836 |
| 40 | 9.6066 | .7980 | 18.17 | 73.82 | 185.74 | 74.59 | 128.61 | 203.20 | .2718 | .6825 |
| 44 | 10.587 | .8086 | 16.44 | 77.82 | 187.13 | 78.68 | 125.87 | 204.54 | .2845 | .6814 |
| 48 | 11.639 | .8199 | 14.88 | 81.88 | 188.51 | 82.83 | 123.00 | 205.83 | .2973 | .6802 |
| 52 | 12.766 | .8318 | 13.49 | 86.00 | 189.83 | 87.06 | 119.99 | 207.05 | .3101 | .6791 |
| 56 | 13.972 | .8445 | 12.24 | 90.18 | 191.10 | 91.36 | 116.84 | 208.20 | .3229 | .6779 |
| 60 | 15.259 | .8581 | 11.11 | 94.43 | 192.31 | 95.74 | 113.52 | 209.26 | .3358 | .6765 |
| 112 | 41.155 | 1.792 | 1.79 | 175.98 | 175.98 | 183.35 | 0 | 183.35 | .5687 | .5687 |

*Source:* Based on data supplied by Freon Products Division, E. I. du Pont de Nemours & Company, 1969.

**Table A-17M Properties of saturated refrigerant 12, $CCl_2F_2$: pressure table**
$v$, cm$^3$/g; $u$, kJ/kg; $h$, kJ/kg; $s$, kJ/(kg)(°K)

| Press. bar(s) $P$ | Temp. °C $T$ | Specific volume | | Internal energy | | Enthalpy | | | Entropy | |
|---|---|---|---|---|---|---|---|---|---|---|
| | | Sat. liquid $v_f$ | Sat. vapor $v_g$ | Sat. liquid $u_f$ | Sat. vapor $u_g$ | Sat. liquid $h_f$ | Evap. $h_{fg}$ | Sat. vapor $h_g$ | Sat. liquid $s_f$ | Sat. vapor $s_g$ |
| 0.6 | −41.42 | 0.6578 | 257.5 | −1.29 | 153.49 | −1.25 | 170.19 | 168.94 | −0.0054 | 0.7290 |
| 1.0 | −30.10 | .6719 | 160.0 | 8.71 | 158.15 | 8.78 | 165.37 | 174.15 | 0.0368 | .7171 |
| 1.2 | −25.74 | .6776 | 134.9 | 12.58 | 159.95 | 12.66 | 163.48 | 176.14 | .0526 | .7133 |
| 1.4 | −21.91 | .6828 | 116.8 | 15.99 | 161.52 | 16.09 | 161.78 | 177.87 | .0663 | .7102 |
| 1.6 | −18.49 | .6876 | 103.1 | 19.07 | 162.91 | 19.18 | 160.23 | 179.41 | .0784 | .7076 |
| 1.8 | −15.38 | .6921 | 92.25 | 21.86 | 164.19 | 21.98 | 158.82 | 180.80 | .0893 | .7054 |
| 2.0 | −12.53 | .6962 | 83.54 | 24.43 | 165.36 | 24.57 | 157.50 | 182.07 | .0992 | .7035 |
| 2.4 | −7.42 | .7040 | 70.33 | 29.06 | 167.44 | 29.23 | 155.09 | 184.32 | .1168 | .7004 |
| 2.8 | −2.93 | .7111 | 60.76 | 33.15 | 169.26 | 33.35 | 152.92 | 186.27 | .1321 | .6980 |
| 3.2 | 1.11 | .7177 | 53.51 | 36.85 | 170.88 | 37.08 | 150.92 | 188.00 | .1457 | .6960 |
| 4.0 | 8.15 | .7299 | 43.21 | 43.35 | 173.69 | 43.64 | 147.33 | 190.97 | .1691 | .6928 |
| 5.0 | 15.60 | .7438 | 34.82 | 50.30 | 176.61 | 50.67 | 143.35 | 194.02 | .1935 | .6899 |
| 6.0 | 22.00 | .7566 | 29.13 | 56.35 | 179.09 | 56.80 | 139.77 | 196.57 | .2142 | .6878 |
| 7.0 | 27.65 | .7686 | 25.01 | 61.75 | 181.23 | 62.29 | 136.45 | 198.74 | .2324 | .6860 |
| 8.0 | 32.74 | .7802 | 21.88 | 66.68 | 183.13 | 67.30 | 133.33 | 200.63 | .2487 | .6845 |
| 9.0 | 37.37 | .7914 | 19.42 | 71.22 | 184.81 | 71.93 | 130.36 | 202.29 | .2634 | .6832 |
| 10.0 | 41.64 | .8023 | 17.44 | 75.46 | 186.32 | 76.26 | 127.50 | 203.76 | .2770 | .6820 |
| 12.0 | 49.31 | .8237 | 14.41 | 83.22 | 188.95 | 84.21 | 122.03 | 206.24 | .3015 | .6799 |
| 14.0 | 56.09 | .8448 | 12.22 | 90.28 | 191.11 | 91.46 | 116.76 | 208.22 | .3232 | .6778 |
| 16.0 | 62.19 | .8660 | 10.54 | 96.80 | 192.95 | 98.19 | 111.62 | 209.81 | .3329 | .6758 |

*Source:* Based on data supplied by Freon Products Division, E. I. du Pont de Nemours & Company, 1969.

$$PV = \frac{NRT}{P} =$$

## Table A-18M Properties of superheated refrigerant 12 ($CCl_2F_2$)

$v$, cm$^3$/g; $u$, kJ/kg; $h$, kJ/kg; $s$, kJ/(kg)($^\circ$K)

| Temp. °C | $v$ | $u$ | $h$ | $s$ | $v$ | $u$ | $h$ | $s$ |
|---|---|---|---|---|---|---|---|---|
| | | 0.6 bar ($-41.42^\circ$C) | | | | 1.0 bar ($-30.10^\circ$C) | | |
| Sat. | 257.5 | 153.49 | 168.94 | 0.7290 | 160.0 | 158.15 | 174.15 | 0.7171 |
| $-40$ | 259.3 | 154.16 | 169.72 | .7324 | | | | |
| $-20$ | 283.8 | 163.91 | 180.94 | .7785 | 167.7 | 163.22 | 179.99 | .7406 |
| 0 | 307.9 | 174.05 | 192.52 | .8225 | 182.7 | 173.50 | 191.77 | .7854 |
| 10 | 319.8 | 179.26 | 198.45 | .8439 | 190.0 | 178.77 | 197.77 | .8070 |
| 20 | 331.7 | 184.57 | 204.47 | .8647 | 197.3 | 184.12 | 203.85 | .8281 |
| 30 | 343.5 | 189.96 | 210.57 | .8852 | 204.5 | 189.57 | 210.02 | .8488 |
| 40 | 355.2 | 195.46 | 216.77 | .9053 | 211.7 | 195.09 | 216.26 | .8691 |
| 50 | 367.0 | 201.02 | 223.04 | .9251 | 218.8 | 200.70 | 222.58 | .8889 |
| 60 | 378.7 | 206.69 | 229.41 | .9444 | 226.0 | 206.38 | 228.98 | .9084 |
| 80 | 402.0 | 218.25 | 242.37 | .9822 | 240.1 | 218.00 | 242.01 | .9464 |
| | | 1.4 bars ($-21.91^\circ$C) | | | | 1.8 bars ($-15.38^\circ$C) | | |
| Sat. | 116.8 | 161.52 | 177.87 | 0.7102 | 92.2 | 164.20 | 180.80 | 0.7054 |
| $-20$ | 117.9 | 162.50 | 179.01 | .7147 | | | | |
| 0 | 128.9 | 172.94 | 190.99 | .7602 | 99.1 | 172.37 | 190.21 | .7408 |
| 10 | 134.3 | 178.28 | 197.08 | .7821 | 103.4 | 177.77 | 196.38 | .7630 |
| 20 | 139.7 | 183.67 | 203.23 | .8035 | 107.6 | 183.23 | 202.60 | .7846 |
| 30 | 144.9 | 189.17 | 209.46 | .8243 | 111.8 | 188.77 | 208.89 | .8057 |
| 40 | 150.2 | 194.72 | 215.75 | .8447 | 116.0 | 194.35 | 215.23 | .8263 |
| 50 | 155.3 | 200.38 | 222.12 | .8648 | 120.1 | 200.02 | 221.64 | .8464 |
| 60 | 160.5 | 206.08 | 228.55 | .8844 | 124.1 | 205.78 | 228.12 | .8662 |
| 80 | 170.7 | 217.74 | 241.64 | .9225 | 132.2 | 217.47 | 241.27 | .9045 |
| 100 | 180.9 | 229.67 | 255.00 | .9593 | 140.2 | 229.45 | 254.69 | .9414 |
| | | 2.0 bars ($-12.53^\circ$C) | | | | 2.4 bars ($-7.42^\circ$C) | | |
| Sat. | 83.5 | 165.37 | 182.07 | 0.7035 | 70.3 | 167.45 | 184.32 | 0.7004 |
| 0 | 88.6 | 172.08 | 189.80 | .7325 | 72.9 | 171.49 | 188.99 | .7177 |
| 10 | 92.6 | 177.50 | 196.02 | .7548 | 76.3 | 176.98 | 195.29 | .7404 |
| 20 | 96.4 | 183.00 | 202.28 | .7766 | 79.6 | 182.53 | 201.63 | .7624 |
| 30 | 100.2 | 188.56 | 208.60 | .7978 | 82.8 | 188.14 | 208.01 | .7838 |
| 40 | 104.0 | 194.17 | 214.97 | .8184 | 86.0 | 193.80 | 214.44 | .8047 |
| 50 | 107.7 | 199.86 | 221.40 | .8387 | 89.2 | 199.51 | 220.92 | .8251 |
| 60 | 111.4 | 205.62 | 227.90 | .8585 | 92.3 | 205.31 | 227.46 | .8450 |
| 80 | 118.7 | 217.35 | 241.09 | .8969 | 98.5 | 217.07 | 240.71 | .8836 |
| 100 | 125.9 | 229.35 | 254.53 | .9339 | 104.5 | 229.12 | 254.20 | .9208 |
| 120 | 133.1 | 241.59 | 268.21 | .9696 | 110.5 | 241.41 | 267.93 | .9566 |

**Table A-18M**  (*Continued*)

| Temp. °C | $v$ | $u$ | $h$ | $s$ | $v$ | $u$ | $h$ | $s$ |
|---|---|---|---|---|---|---|---|---|
| | | 2.8 bars | $(-2.93°C)$ | | | 3.2 bars | $(1.11°C)$ | |
| Sat. | 60.76 | 169.26 | 186.27 | 0.6980 | 53.51 | 170.88 | 188.00 | 0.6960 |
| 0 | 61.66 | 170.89 | 188.15 | .7049 | | | | |
| 10 | 64.64 | 176.45 | 194.55 | .7279 | 55.90 | 175.90 | 193.79 | 0.7167 |
| 20 | 67.55 | 182.06 | 200.97 | .7502 | 58.52 | 181.57 | 200.30 | .7393 |
| 30 | 70.40 | 187.71 | 207.42 | .7718 | 61.06 | 187.28 | 206.82 | .7612 |
| 40 | 73.19 | 193.42 | 213.91 | .7928 | 63.55 | 193.02 | 213.36 | .7824 |
| 50 | 75.94 | 199.18 | 220.44 | .8134 | 66.00 | 198.82 | 219.94 | .8031 |
| 60 | 78.65 | 205.00 | 227.02 | .8334 | 68.41 | 204.68 | 226.57 | .8233 |
| 80 | 83.99 | 216.82 | 240.34 | .8722 | 73.14 | 216.55 | 239.96 | .8623 |
| 100 | 89.24 | 228.89 | 253.88 | .9095 | 77.78 | 228.66 | 253.55 | .8997 |
| 120 | 94.43 | 241.21 | 267.65 | .9455 | 82.36 | 241.00 | 267.36 | .9358 |
| | | 4.0 bars | $(8.15°C)$ | | | 5.0 bars | $(15.60°C)$ | |
| Sat. | 43.21 | 173.69 | 190.97 | 0.6928 | 34.82 | 176.61 | 194.02 | 0.6899 |
| 10 | 43.63 | 174.76 | 192.21 | .6972 | | | | |
| 20 | 45.84 | 180.57 | 198.91 | .7204 | 35.65 | 179.26 | 197.08 | 0.7004 |
| 30 | 47.97 | 186.39 | 205.58 | .7428 | 37.46 | 185.23 | 203.96 | .7235 |
| 40 | 50.05 | 192.23 | 212.25 | .7645 | 39.22 | 191.20 | 210.81 | .7457 |
| 50 | 52.07 | 198.11 | 218.94 | .7855 | 40.91 | 197.19 | 217.64 | .7672 |
| 60 | 54.06 | 204.03 | 225.65 | .8060 | 42.57 | 203.20 | 224.48 | .7881 |
| 80 | 57.91 | 216.03 | 239.19 | .8454 | 45.78 | 215.32 | 238.21 | .8281 |
| 100 | 61.73 | 228.20 | 252.89 | .8831 | 48.89 | 227.61 | 252.05 | .8662 |
| 120 | 65.46 | 240.61 | 266.79 | .9194 | 51.93 | 240.10 | 266.06 | .9028 |
| 140 | 69.13 | 253.23 | 280.88 | .9544 | 54.92 | 252.77 | 280.23 | .9379 |
| | | 6.0 bars | $(22.00°C)$ | | | 7.0 bars | $(27.65°C)$ | |
| Sat. | 29.13 | 179.09 | 196.57 | 0.6878 | 25.01 | 181.23 | 198.74 | 0.6860 |
| 30 | 30.42 | 184.01 | 202.26 | .7068 | 25.35 | 182.72 | 200.46 | .6917 |
| 40 | 31.97 | 190.13 | 209.31 | .7297 | 26.76 | 189.00 | 207.73 | .7153 |
| 50 | 33.45 | 196.23 | 216.30 | .7516 | 28.10 | 195.23 | 214.90 | .7378 |
| 60 | 34.89 | 202.34 | 223.27 | .7729 | 29.39 | 201.45 | 222.02 | .7595 |
| 80 | 37.65 | 214.61 | 237.20 | .8135 | 31.84 | 213.88 | 236.17 | .8008 |
| 100 | 40.32 | 227.01 | 251.20 | .8520 | 34.19 | 226.40 | 250.33 | .8398 |
| 120 | 42.91 | 239.57 | 265.32 | .8889 | 36.46 | 239.05 | 264.57 | .8769 |
| 140 | 45.45 | 252.31 | 279.58 | .9243 | 38.67 | 251.85 | 278.92 | .9125 |
| 160 | 47.94 | 265.25 | 294.01 | .9584 | 40.85 | 264.83 | 293.42 | .9468 |

## Table A-18M  (Continued)

| Temp. °C | v | u | h | s | v | u | h | s |
|---|---|---|---|---|---|---|---|---|
| | | 8.0 bars | (32.74°C) | | | 9.0 bars | (37.37°C) | |
| Sat. | 21.88 | 183.13 | 200.63 | 0.6845 | 19.42 | 184.81 | 202.29 | 0.6832 |
| 40 | 22.83 | 187.81 | 206.07 | .7021 | 19.74 | 186.55 | 204.32 | .6897 |
| 50 | 24.07 | 194.19 | 213.45 | .7253 | 20.91 | 193.10 | 211.92 | .7136 |
| 60 | 25.25 | 200.52 | 220.72 | .7474 | 22.01 | 199.56 | 219.37 | .7363 |
| 80 | 27.48 | 213.13 | 235.11 | .7894 | 24.07 | 212.37 | 234.03 | .7790 |
| 100 | 29.59 | 225.77 | 249.44 | .8289 | 26.01 | 225.13 | 248.54 | .8190 |
| 120 | 31.62 | 238.51 | 263.81 | .8664 | 27.85 | 237.97 | 263.03 | .8569 |
| 140 | 33.59 | 251.39 | 278.26 | .9022 | 29.64 | 250.90 | 277.58 | .8930 |
| 160 | 35.52 | 264.41 | 292.83 | .9367 | 31.38 | 263.99 | 292.23 | .9276 |
| 180 | 37.42 | 277.60 | 307.54 | .9699 | 33.09 | 277.23 | 307.01 | .9609 |
| | | 10.0 bars | (41.64°C) | | | 12.0 bars | (49.31°C) | |
| Sat. | 17.44 | 186.32 | 203.76 | 0.6820 | 14.41 | 188.95 | 206.24 | 0.6799 |
| 50 | 18.37 | 191.95 | 210.32 | .7026 | 14.48 | 189.43 | 206.81 | .6816 |
| 60 | 19.41 | 198.56 | 217.97 | .7259 | 15.46 | 196.41 | 214.96 | .7065 |
| 80 | 21.34 | 211.57 | 232.91 | .7695 | 17.22 | 209.91 | 230.57 | .7520 |
| 100 | 23.13 | 224.48 | 247.61 | .8100 | 18.81 | 223.13 | 245.70 | .7937 |
| 120 | 24.84 | 237.41 | 262.25 | .8482 | 20.30 | 236.27 | 260.63 | .8326 |
| 140 | 26.47 | 250.43 | 276.90 | .8845 | 21.72 | 249.45 | 275.51 | .8696 |
| 160 | 28.07 | 263.56 | 291.63 | .9193 | 23.09 | 263.70 | 290.41 | .9048 |
| 180 | 29.63 | 276.84 | 306.47 | .9528 | 24.43 | 276.05 | 305.37 | .9385 |
| 200 | 31.16 | 290.26 | 321.42 | .9851 | 25.74 | 289.55 | 320.44 | .9711 |
| | | 14.0 bars | (56.09°C) | | | 16.0 bars | (62.19°C) | |
| Sat. | 12.22 | 191.11 | 208.22 | 0.6778 | 10.54 | 192.95 | 209.81 | 0.6758 |
| 60 | 12.58 | 194.00 | 211.61 | .6881 | | | | |
| 80 | 14.25 | 208.11 | 228.06 | .7360 | 11.98 | 206.17 | 225.34 | 0.7209 |
| 100 | 15.71 | 221.70 | 243.69 | .7791 | 13.37 | 220.19 | 241.58 | .7656 |
| 120 | 17.05 | 235.09 | 258.96 | .8189 | 14.61 | 233.84 | 257.22 | .8065 |
| 140 | 18.32 | 248.43 | 274.08 | .8564 | 15.77 | 247.38 | 272.61 | .8447 |
| 160 | 19.54 | 261.80 | 289.16 | .8921 | 16.86 | 260.90 | 287.88 | .8808 |
| 180 | 20.71 | 275.27 | 304.26 | .9262 | 17.92 | 274.47 | 303.14 | .9152 |
| 200 | 21.86 | 288.84 | 319.44 | .9589 | 18.95 | 288.11 | 318.43 | .9482 |
| 220 | 22.99 | 302.51 | 334.70 | .9905 | 19.96 | 301.84 | 333.78 | .9800 |

*Source:* Based on data supplied by Freon Products Division, E. I. du Pont de Nemours & Company, 1969.

## Table A-19M  Specific heats of some common liquids and solids
$c_p$, kJ/(kg)(°C)

### A.  Liquids

| Substance | State | $c_p$ | Substance | State | $c_p$ |
|-----------|-------|-------|-----------|-------|-------|
| Water | 1 atm, 0°C | 4.213 | Glycerin | 1 atm, 10°C | 2.32 |
|  | 1 atm, 25°C | 4.177 |  | 1 atm, 50°C | 2.58 |
|  | 1 atm, 50°C | 4.178 | Bismuth | 1 atm, 425°C | 0.144 |
|  | 1 atm, 100°C | 4.213 |  | 1 atm, 760°C | 0.164 |
| Ammonia | sat., −20°C | 4.52 | Mercury | 1 atm, 10°C | 0.138 |
|  | sat., 50°C | 5.10 |  | 1 atm, 315°C | 0.134 |
| Refrigerant 12 | sat., −40°C | 0.883 | Sodium | 1 atm, 95°C | 1.38 |
|  | sat., −20°C | 0.908 |  | 1 atm, 540°C | 1.26 |
|  | sat., 50°C | 1.02 | Propane | 1 atm, 0°C | 2.41 |
| Benzene | 1 atm, 15°C | 1.80 | Ethyl |  |  |
|  | 1 atm, 65°C | 1.92 | alcohol | 1 atm, 25°C | 2.43 |

### B.  Solids

| Substance | $T$, °C | $c_p$ | Substance | $T$, °C | $c_p$ |
|-----------|---------|-------|-----------|---------|-------|
| Ice | −200 | 0.678 | Lead | −270 | 0.0033 |
|  | −140 | 1.096 |  | −259 | 0.0305 |
|  | −60 | 1.640 |  | −100 | 0.118 |
|  | −11 | 2.033 |  | 0 | 0.124 |
|  | −2.2 | 1.682 |  | 100 | 0.134 |
| Aluminum | −250 | 0.0163 |  | 300 | 0.149 |
|  | −200 | 0.318 | Copper | −223 | 0.0967 |
|  | −100 | 0.699 |  | −173 | 0.252 |
|  | 0 | 0.870 |  | −100 | 0.328 |
|  | 100 | 0.941 |  | −50 | 0.361 |
|  | 300 | 1.04 |  | 0 | 0.381 |
| Iron | 20 | 0.448 |  | 27 | 0.385 |
| Silver | 20 | 0.233 |  | 100 | 0.393 |
|  | 500 | 0.243 |  | 200 | 0.403 |

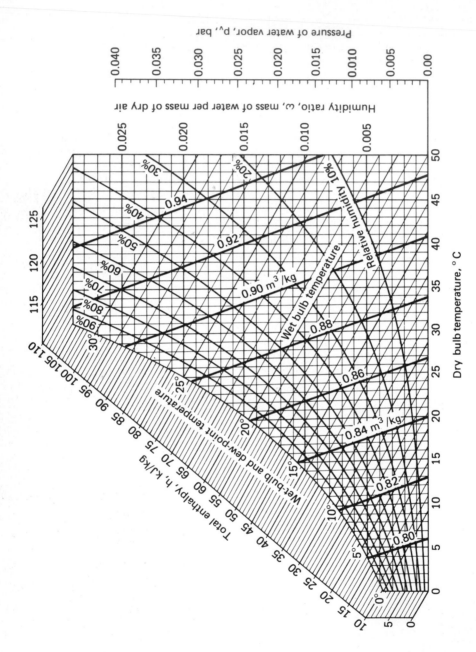

**Figure A-20M** Psychrometric chart, metric units, barometric pressure 1.01 bar.

832

## Table A-21M  Constants for the Benedict-Webb-Rubin and Redlich-Kwong equations of state

1. Benedict-Webb-Rubin; Units are bar(s), $m^3/kg\cdot mol$, and $°K$

| Constants | n-Butane, $C_4H_{10}$ | Carbon dioxide, $CO_2$ | Carbon monoxide, CO | Methane, $CH_4$ | Nitrogen, $N_2$ |
|---|---|---|---|---|---|
| $a$ | 1.9068 | 0.1386 | 0.0371 | 0.0500 | 0.0254 |
| $A_0$ | 10.216 | 2.7730 | 1.3587 | 1.8791 | 1.0673 |
| $b$ | 0.039998 | 0.007210 | 0.002632 | 0.003380 | 0.002328 |
| $B_0$ | 0.12436 | 0.04991 | 0.05454 | 0.04260 | 0.04074 |
| $c$ | $3.205 \times 10^5$ | $1.511 \times 10^4$ | $1.054 \times 10^3$ | $2.578 \times 10^3$ | $7.379 \times 10^2$ |
| $C_0$ | $1.006 \times 10^6$ | $1.404 \times 10^5$ | $8.673 \times 10^3$ | $2.286 \times 10^4$ | $8.164 \times 10^3$ |
| $\alpha$ | $1.101 \times 10^{-3}$ | $8.470 \times 10^{-5}$ | $1.350 \times 10^{-4}$ | $1.244 \times 10^{-4}$ | $1.272 \times 10^{-4}$ |
| $\gamma$ | 0.0340 | 0.00539 | 0.0060 | 0.0060 | 0.0053 |

*Source:*  Cooper, H. W., and J. C. Goldfrank, *Hydrocarbon Processing*, **46** (12): 141 (1967).

2. Redlich-Kwong: Units are bar(s), $m^3/kg\cdot mol$, and $°K$

| Substance | $a$ | $b$ |
|---|---|---|
| Carbon dioxide, $CO_2$ | 64.64 | 0.02969 |
| Carbon monoxide, CO | 17.26 | 0.02743 |
| Methane, $CH_4$ | 32.19 | 0.02969 |
| Nitrogen, $N_2$ | 15.59 | 0.02681 |
| Oxygen, $O_2$ | 17.38 | 0.02199 |
| Propane, $C_3H_8$ | 183.07 | 0.06269 |
| Refrigerant 12, $CCl_2F_2$ | 214.03 | 0.06913 |
| Sulfur dioxide, $SO_2$ | 144.49 | 0.03939 |
| Water, $H_2O$ | 142.64 | 0.02110 |

*Source:*  Computed from critical data.

## Table A-22M  Values of the enthalpy of formation, Gibbs function of formation, and absolute entropy at 25°C and 1 atm

$\Delta h_f^0$ and $\Delta g_f^0$ in $kJ/kg\cdot mol$ and $s^0$ in $kJ/(kg\cdot mol)(°K)$

| Substance | Formula | $\Delta h_f^0$ | $\Delta g_f^0$ | $s^0$ |
|---|---|---|---|---|
| Carbon | C(s) | 0 | 0 | 5.74 |
| Hydrogen | $H_2(g)$ | 0 | 0 | 130.57 |
| Nitrogen | $N_2(g)$ | 0 | 0 | 191.50 |
| Oxygen | $O_2(g)$ | 0 | 0 | 205.03 |
| Carbon monoxide | CO(g) | −110,530 | −137,150 | 197.56 |
| Carbon dioxide | $CO_2(g)$ | −393,520 | −394,360 | 213.64 |
| Water | $H_2O(g)$ | −241,820 | −228,590 | 188.72 |

## Table A-22M  (Continued)

| Substance | Formula | $\Delta h_f^0$ | $\Delta g_f^0$ | $s^0$ |
|---|---|---|---|---|
| Water | $H_2O(l)$ | $-285{,}830$ | $-237{,}180$ | 69.92 |
| Hydrogen peroxide | $H_2O_2(g)$ | $-136{,}310$ | $-105{,}600$ | 232.63 |
| Ammonia | $NH_3(g)$ | $-46{,}190$ | $-16{,}590$ | 192.33 |
| Methane | $CH_4(g)$ | $-74{,}850$ | $-50{,}790$ | 186.16 |
| Acetylene | $C_2H_2(g)$ | $+226{,}730$ | $+209{,}170$ | 200.85 |
| Ethylene | $C_2H_4(g)$ | $+52{,}280$ | $+68{,}120$ | 219.83 |
| Ethane | $C_2H_6(g)$ | $-84{,}680$ | $-32{,}890$ | 229.49 |
| Propylene | $C_3H_6(g)$ | $+20{,}410$ | $+62{,}720$ | 266.94 |
| Propane | $C_3H_8(g)$ | $-103{,}850$ | $-23{,}490$ | 269.91 |
| n-Butane | $C_4H_{10}(g)$ | $-126{,}150$ | $-15{,}710$ | 310.12 |
| n-Octane | $C_8H_{18}(g)$ | $-208{,}450$ | $+16{,}530$ | 466.73 |
| n-Octane | $C_8H_{18}(l)$ | $-249{,}950$ | $+6{,}610$ | 360.79 |
| Benzene | $C_6H_6(g)$ | $+82{,}930$ | $+129{,}660$ | 269.20 |
| Methyl alcohol | $CH_3OH(g)$ | $-200{,}670$ | $-162{,}000$ | 239.70 |
| Methyl alcohol | $CH_3OH(l)$ | $-238{,}660$ | $-166{,}360$ | 126.80 |
| Ethyl alcohol | $C_2H_5OH(g)$ | $-235{,}310$ | $-168{,}570$ | 282.59 |
| Ethyl alcohol | $C_2H_5OH(l)$ | $-277{,}690$ | $-174{,}890$ | 160.70 |
| Oxygen | $O(g)$ | $+249{,}190$ | $+231{,}770$ | 160.95 |
| Hydrogen | $H(g)$ | $+218{,}000$ | $+203{,}290$ | 114.61 |
| Nitrogen | $N(g)$ | $+472{,}650$ | $+455{,}510$ | 153.19 |
| Hydroxyl | $OH(g)$ | $+39{,}460$ | $+34{,}280$ | 183.75 |

*Sources:*  From the JANAF Thermochemical Tables, Dow Chemical Co., 1971; *Selected Values of Chemical Thermodynamic Properties*, NBS Tech. Note 270-3, 1968; and *API Research Project 44*, Carnegie Press, 1953.

## Table A-23M  Values of the enthalpy of combustion $\Delta h_c^0$ and the enthalpy of vaporization at 25°C and 1 atm

Water appears as a liquid in the products of combustion.

| Substance | Formula | $\Delta h_c^0$ kJ/kg·mol | $h_{fg}$ kJ/kg·mole |
|---|---|---|---|
| Hydrogen | $H_2(g)$ | $-285{,}840$ | |
| Carbon | $C(s)$ | $-393{,}520$ | |
| Carbon monoxide | $CO(g)$ | $-282{,}990$ | |
| Methane | $CH_4(g)$ | $-890{,}360$ | |
| Acetylene | $C_2H_2(g)$ | $-1{,}299{,}600$ | |
| Ethylene | $C_2H_4(g)$ | $-1{,}410{,}970$ | |
| Ethane | $C_2H_6(g)$ | $-1{,}559{,}900$ | |
| Propylene | $C_3H_6(g)$ | $-2{,}058{,}500$ | |
| Propane | $C_3H_8(g)$ | $-2{,}220{,}000$ | 15,060 |
| n-Butane | $C_4H_{10}(g)$ | $-2{,}877{,}100$ | 21,060 |
| n-Pentane | $C_5H_{12}(g)$ | $-3{,}536{,}100$ | 26,410 |
| n-Hexane | $C_6H_{14}(g)$ | $-4{,}194{,}800$ | 31,530 |

## Table A-23M   (Continued)

| Substance | Formula | $\Delta h_c^o$ kJ/kg·mol | $h_{fg}$ kJ/kg·mole |
|---|---|---|---|
| n-Heptane | $C_7H_{16}(g)$ | −4,853,500 | 36,520 |
| n-Octane | $C_8H_{18}(g)$ | −5,512,200 | 41,460 |
| Benzene | $C_6H_6(g)$ | −3,301,500 | 33,830 |
| Toluene | $C_7H_8(g)$ | −3,947,900 | 39,920 |
| Methyl alcohol | $CH_3OH(g)$ | −764,540 | 37,900 |
| Ethyl alcohol | $C_2H_5OH(g)$ | −1,409,300 | 42,340 |

*Sources:* Primarily from "Selected Values of Chemical Thermodynamic Properties," NBS Circular 500, 1952, and "Selected Values of Physical and Thermodynamic Properties of Hydrocarbons and Related Compounds," API Research Project 44, Carnegie Institute of Technology, Pittsburgh, Pa., 1953.

## Table A-24M  Thermodynamic properties of potassium
$T$, °K; $P$, atm; $v$, l/kg; $h$, kJ/kg; $s$, kJ/(kg)(°K)

A. Saturation temperature table

| $T$ | $P$ | $v_f$ | $v_g$ | $h_f$ | $h_g$ | $s_f$ | $s_g$ |
|---|---|---|---|---|---|---|---|
| 900 | 0.251 | 1.438 | 7180 | 731.0 | 2739.7 | 2.6874 | 4.9175 |
| 950 | 0.447 | 1.462 | 4204 | 771.4 | 2750.4 | 2.7313 | 4.8129 |
| 1000 | 0.753 | 1.493 | 2592 | 812.6 | 2760.4 | 2.7732 | 4.7196 |

B. Superheat table

| $P$ | $v$ | $h$ | $s$ | $P$ | $v$ | $h$ | $s$ |
|---|---|---|---|---|---|---|---|
| 1075°K (1.494 atm) | | | | 1200°K (3.860 atm) | | | |
| Sat. | 1.375 | 2772.9 | 4.5977 | Sat. | 0.573 | 2795.1 | 4.4367 |
| 1.0 | 2.120 | 2813.6 | 4.7144 | 3.0 | 0.759 | 2831.1 | 4.5144 |
| 0.8 | 2.684 | 2830.0 | 4.7745 | 2.0 | 1.178 | 2874.0 | 4.6295 |
| 1325°K (8.293 atm) | | | | 1450°K (15.55 atm) | | | |
| Sat. | 0.285 | 2820.4 | 4.3163 | Sat. | 0.161 | 2849.4 | 4.2255 |
| 6.0 | 0.411 | 2873.7 | 4.4162 | 10.0 | 0.267 | 2928.5 | 4.3599 |
| 4.0 | 0.641 | 2923.1 | 4.5316 | 8.0 | 0.343 | 2959.2 | 4.4233 |
| 3.0 | 0.873 | 2948.7 | 4.6078 | 6.0 | 0.469 | 2991.3 | 4.5011 |

*Sources:* Data adapted from Naval Research Laboratory Report 6233, 1965, and Air Force Aero Propulsion Laboratory Technical Report 66-104, 1966.

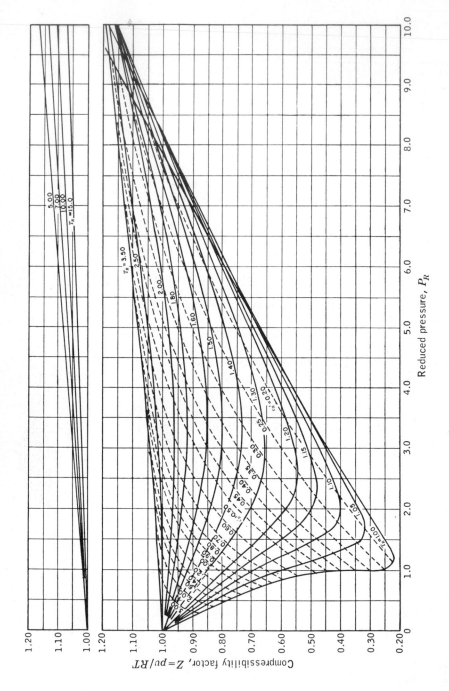

**Figure A-25M** Compressibility chart, low-pressure range. *Source: V. M. Faires, "Problems on Thermodynamics," Macmillan, New York, 1962. Data from L. C. Nelson and E. F. Obert, Generalized Compressibility Charts, Chem. Eng.,* **61** : 203 (1954).

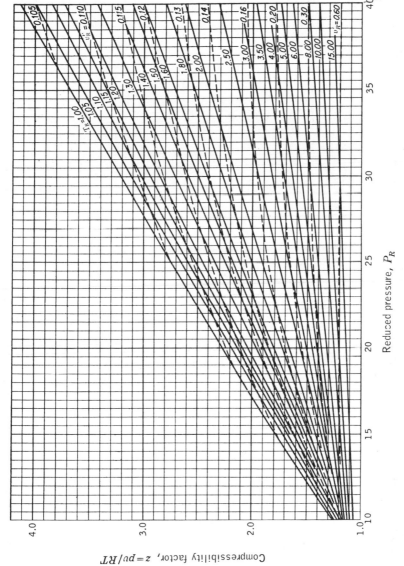

**Figure A-26M** Compressibility chart, high-pressure range. Adapted from E. F. Obert, "Concepts of Thermodynamics," McGraw-Hill, New York, 1960.

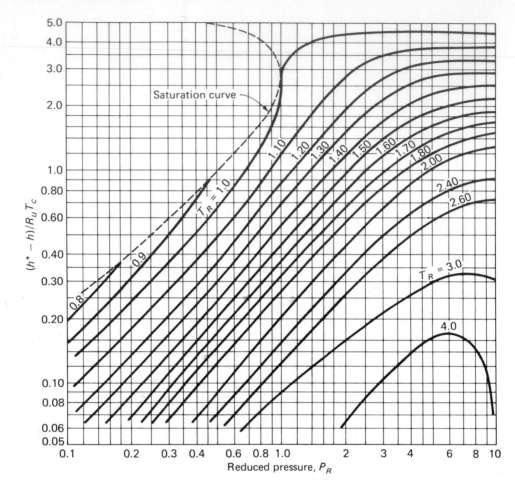

**Figure A-27M** Generalized enthalpy chart. *Source:* Based on data from A. L. Lydersen, R. A. Greenkorn, and O. A. Hougen, "Engineering Experiment Station Report No. 4," University of Wisconsin, 1955. $T_R = T/T_c$ = reduced temperature, $P_R = P/P_c$ = reduced pressure, $T_c$ = critical temperature, $P_c$ = critical pressure, $h^*$ = enthalpy of an ideal gas, $h$ = enthalpy of an actual gas.

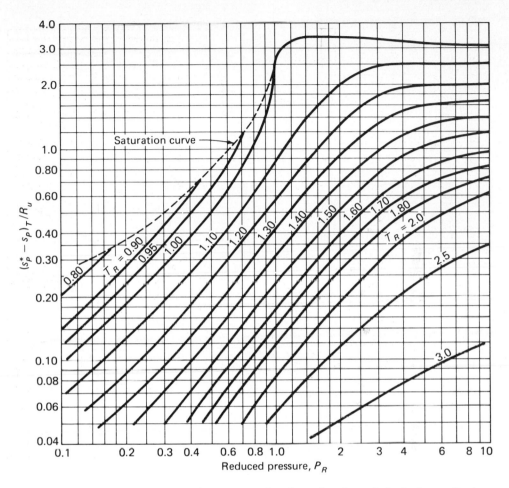

**Figure A-28M** Generalized entropy chart. *Source:* Based on data from A. L. Lydersen, R. A. Greenkorn, and O. A. Hougen, "Engineering Experiment Station Report No. 4," University of Wisconsin, 1955. $T_R = T/T_c$ = reduced temperature, $P_R = P/P_c$ = reduced pressure, $T_c$ = critical temperature, $P_c$ = critical pressure, $s_P^*$ = entropy of an ideal gas, $s_P$ = entropy of an actual gas.

**Table A-29M Logarithms to the base 10 of the equilibrium constant $K_p$**

$$K_p = \frac{(p_E)^{v_E}(p_F)^{v_F}}{(p_A)^{v_A}(p_B)^{v_B}} \text{ for the reaction } v_A A + v_B B \rightleftharpoons v_E E + v_F F$$

Numbered reactions

(1) $H_2 \rightleftharpoons 2H$

(2) $O_2 \rightleftharpoons 2O$

(3) $N_2 \rightleftharpoons 2N$

(4) $\frac{1}{2}O_2 + \frac{1}{2}N_2 \rightleftharpoons NO$

(5) $H_2O \rightleftharpoons H_2 + \frac{1}{2}O_2$

(6) $H_2O \rightleftharpoons OH + \frac{1}{2}H_2$

(7) $CO_2 \rightleftharpoons CO + \frac{1}{2}O_2$

(8) $CO_2 + H_2 \rightleftharpoons CO + H_2O$

| Temp. °K | $\log_{10} K_p$ values for reactions numbered above | | | | | | | |
|---|---|---|---|---|---|---|---|---|
| | (1) | (2) | (3) | (4) | (5) | (6) | (7) | (8) |
| 298 | −71.224 | −81.208 | −159.600 | −15.171 | −40.048 | −46.054 | −45.066 | −5.018 |
| 500 | −40.316 | −45.880 | −92.672 | −8.783 | −22.886 | −26.130 | −25.025 | −2.139 |
| 1000 | −17.292 | −19.614 | −43.056 | −4.062 | −10.062 | −11.280 | −10.221 | −0.159 |
| 1200 | −13.414 | −15.208 | −34.754 | −3.275 | −7.899 | −8.811 | −7.764 | +0.135 |
| 1400 | −10.630 | −12.054 | −28.812 | −2.712 | −6.347 | −7.021 | −6.014 | +0.333 |
| 1600 | −8.532 | −9.684 | −24.350 | −2.290 | −5.180 | −5.677 | −4.706 | +0.474 |
| 1700 | −7.666 | −8.706 | −22.512 | −2.116 | −4.699 | −5.124 | −4.169 | +0.530 |
| 1800 | −6.896 | −7.836 | −20.874 | −1.962 | −4.270 | −4.613 | −3.693 | +0.577 |
| 1900 | −6.204 | −7.058 | −19.410 | −1.823 | −3.886 | −4.190 | −3.267 | +0.619 |
| 2000 | −5.580 | −6.356 | −18.092 | −1.699 | −3.540 | −3.776 | −2.884 | +0.656 |
| 2100 | −5.016 | −5.720 | −16.898 | −1.586 | −3.227 | −3.434 | −2.539 | +0.688 |
| 2200 | −4.502 | −5.142 | −15.810 | −1.484 | −2.942 | −3.091 | −2.226 | +0.716 |
| 2300 | −4.032 | −4.614 | −14.818 | −1.391 | −2.682 | −2.809 | −1.940 | +0.742 |
| 2400 | −3.600 | −4.130 | −13.908 | −1.305 | −2.443 | −2.520 | −1.679 | +0.764 |
| 2500 | −3.202 | −3.684 | −13.070 | −1.227 | −2.224 | −2.270 | −1.440 | +0.784 |
| 2600 | −2.836 | −3.272 | −12.298 | −1.154 | −2.021 | −2.038 | −1.219 | +0.802 |
| 2800 | −2.178 | −2.536 | −10.914 | −1.025 | −1.658 | −1.624 | −0.825 | +0.833 |
| 3000 | −1.606 | −1.898 | −9.716 | −0.913 | −1.343 | −1.265 | −0.485 | +0.858 |
| 3200 | −1.106 | −1.340 | −8.664 | −0.815 | −1.067 | −0.951 | −0.189 | +0.878 |
| 3400 | −0.664 | −0.846 | −7.736 | −0.729 | −0.824 | −0.687 | −0.071 | +0.895 |

*Source:* Based on data from the JANAF Tables, NSRDS-NBS-37, 1971.

# SUPPLEMENTARY TABLES AND FIGURES (USCS UNITS)

## Table A-1 Physical constants and conversion factors

### Physical constants

| | |
|---|---|
| Avogadro's number | $N_A = 6.024 \times 10^{23}$ atoms/g·mol |
| Universal gas constant | $R_u = 0.08205$ l·atm/(g·mol)(°K) |
| | $= 8.315$ J/(g·mol)(°K) |
| | $= 0.08315$ bar·m³/(kg·mol)(°K) |
| | $= 1545$ ft·lb$_f$/(lb·mol)(°R) |
| | $= 1.986$ Btu/(lb·mol)(°R) |
| | $= 0.730$ atm·ft³/(lb·mol)(°R) |
| | $= 10.73$ psia·ft³/(lb·mol)(°R) |
| Planck's constant | $h = 6.625 \times 10^{-34}$ J·s/molecule |
| Boltzmann's constant | $k = 1.380 \times 10^{-23}$ J/(°K)(molecule) |
| Speed of light | $c = 2.998 \times 10^{10}$ cm/s |

### Conversion factors

| | |
|---|---|
| Length | 1 cm = 0.3937 in = $10^4$ μm = $10^8$ Å |
| | 1 in = 2.540 cm |
| | 1 ft = 30.48 cm |
| Mass | 1 lb$_m$ = 453.59 g$_m$ = 7000 grains |
| | 1 kg$_m$ = 2.2005 lb$_m$ |
| Force | 1 lb$_f$ = 32.174 lb$_m$·ft/s² |
| | 1 dyne = 1 g$_m$·cm/s² |
| | 1 N = 1 kg$_m$·m/s² |
| | 1 lb$_f$ = 444,800 dyn = 4.448 N |
| Pressure | 1 lb$_f$/in² = 2.036 in Hg at 32°F |
| | 1 in Hg = 33,864 dyn/cm² = 0.0334 atm = 0.491 lb$_f$/in² |
| | 1 atm = 14.696 lb$_f$/in² = 760 mm Hg at 32°F |
| | $= 29.92$ in Hg at 32°F |
| | 1 bar = 0.9869 atm = $10^5$ N/m² = 100 kPa |
| | 1 torr = 1 mm Hg at 0°C = $10^3$ μm |
| | $= 1.933 \times 10^{-2}$ psi |
| Volume | 1 l = 0.0353 ft³ = 0.2642 gal = 61.025 in³ |
| | 1 ft³ = 28.316 l = 7.4805 gal = 0.02832 m³ |
| | 1 in³ = 16.387 cm³ |
| Density | 1 lb$_m$/ft³ = 0.01602 g$_m$/cm³ |
| Energy | 1 Btu = 778.16 ft·lb$_f$ = 252.16 cal = 1055 J |
| | 1 Btu/lb = 2.32 kJ/kg |
| | 1 J = 1 N·m = 1 V·C |
| | 1 eV = $1.602 \times 10^{-19}$ J |
| Power | 1 W = 1 J/s = 860.42 cal/h = 3.413 Btu/h |
| | 1 hp = 746 W = 550 ft·lb$_f$/s = 2545 Btu/h |
| Velocity | 1 m/h = 0.447 m/s |

## Table A-2 Values of the molar mass (molecular weight) of some common elements and compounds

| Substance | Formula | Molar mass |
|---|---|---|
| Argon | Ar | 39.94 |
| Aluminum | Al | 26.97 |
| Carbon | C | 12.01 |
| Copper | Cu | 63.54 |
| Helium | He | 4.003 |
| Hydrogen | $H_2$ | 2.018 |
| Iron | Fe | 55.85 |
| Lead | Pb | 207.2 |
| Mercury | Hg | 200.6 |
| Nitrogen | $N_2$ | 28.008 |
| Oxygen | $O_2$ | 32.00 |
| Potassium | K | 39.096 |
| Silver | Ag | 107.88 |
| Sodium | Na | 22.997 |
| Air | | 28.97 |
| Carbon monoxide | CO | 28.01 |
| Carbon dioxide | $CO_2$ | 44.01 |
| Water | $H_2O$ | 18.02 |
| Hydrogen peroxide | $H_2O_2$ | 34.02 |
| Ammonia | $NH_3$ | 17.04 |
| Hydroxyl | $-OH$ | 17.01 |
| Methane | $CH_4$ | 16.04 |
| Acetylene | $C_2H_2$ | 26.04 |
| Ethylene | $C_2H_4$ | 28.05 |
| Ethane | $C_2H_6$ | 30.07 |
| Propylene | $C_3H_6$ | 42.08 |
| Propane | $C_3H_8$ | 44.09 |
| n-Butane | $C_4H_{10}$ | 58.12 |
| n-Pentane | $C_5H_{12}$ | 72.15 |
| n-Octane | $C_8H_{18}$ | 114.22 |
| Benzene | $C_6H_6$ | 78.11 |
| Methyl alcohol | $CH_3OH$ | 32.05 |
| Ethyl alcohol | $C_2H_5OH$ | 46.07 |
| Refrigerant 12 | $CCl_2F_2$ | 120.92 |

## Table A-3 Critical properties and van der Waals constants

| Substance | $T_c$ , °R | $p_c$ , atm | $\bar{v}_c$, ft³/lb·mol | $Z_c = \dfrac{p_c v_c}{RT_c}$ | van der Waals $a$ atm(ft³/lb·mol)² | $b$ ft³/lb·mol |
|---|---|---|---|---|---|---|
| Acetylene ($C_2H_2$) | 556 | 62 | 1.80 | 0.274 | 1121 | 0.818 |
| Air (equivalent) | 239 | 37.2 | 1.33 | 0.284 | 345.2 | 0.585 |
| Ammonia ($NH_3$) | 730 | 111.3 | 1.16 | 0.242 | 1076 | 0.598 |
| Benzene ($C_6H_6$) | 1013 | 48.7 | 4.11 | 0.274 | 4736 | 1.896 |
| n-Butane ($C_4H_{10}$) | 765 | 37.5 | 4.13 | 0.274 | 3508 | 1.919 |
| Carbon dioxide ($CO_2$) | 548 | 72.9 | 1.51 | 0.276 | 926 | 0.686 |
| Carbon monoxide (CO) | 239 | 34.5 | 1.49 | 0.294 | 372 | 0.632 |
| Refrigerant 12 ($CCl_2F_2$) | 693 | 39.6 | 3.43 | 0.270 | 2718 | 1.595 |
| Ethane ($C_2H_6$) | 549 | 48.2 | 3.55 | 0.273 | 1410 | 1.041 |
| Ethylene ($C_2H_4$) | 510 | 50.5 | 2.29 | 0.284 | 1158 | 0.922 |
| Helium (He) | 9.33 | 2.26 | 0.93 | 0.300 | 8.66 | 0.376 |
| n-Heptane ($C_7H_{16}$) | 972 | 27 | 6.86 | 0.26 | 7866 | 3.298 |
| Hydrogen ($H_2$) | 59.8 | 12.8 | 1.04 | 0.304 | 62.8 | 0.426 |
| Methane ($CH_4$) | 344 | 45.8 | 1.59 | 0.290 | 581 | 0.685 |
| Methyl chloride ($CH_3Cl$) | 749 | 65.8 | 2.29 | 0.276 | 1917 | 1.040 |
| Nitrogen ($N_2$) | 227 | 33.5 | 1.44 | 0.291 | 346 | 0.618 |
| Nonane ($C_9H_{20}$) | 1071 | 22.86 | 8.86 | 0.250 | | |
| n-Octane ($C_8H_{18}$) | 1025 | 24.6 | 7.82 | 0.258 | 9601 | 3.76 |
| Oxygen ($O_2$) | 278 | 49.8 | 1.19 | 0.290 | 348 | 0.506 |
| Propane ($C_3H_8$) | 666 | 42.1 | 3.13 | 0.276 | 2368 | 1.445 |
| Sulfur dioxide ($SO_2$) | 775 | 77.7 | 1.99 | 0.268 | 1738 | 0.911 |
| Water ($H_2O$) | 1165 | 218.2 | 0.896 | 0.230 | 1400 | 0.488 |

Most of the values of $T_c$ and $p_c$ are adapted from Nelson and E. F. Obert, Generalized Compressibility Charts, *Chem. Eng.*, **61** : 203 (1954); others from International Critical Tables. Most values of $\bar{v}_c$ are from Physical Tables, Smithsonian Institution, 1954; others from International Critical Tables. The values of $Z_c$, $a$, and $b$ are computed from these critical data; $a = 27R^2T_c^2/64p_c$ and $b = RT_c/8p_c$.

*Source:* Faires, V. M., "Problems on Thermodynamics," 4th ed., Macmillan, New York, 1962.

## Table A-4 Zero-pressure specific heats for various gases, Btu/(lb)(°F)

| | $c_p$ | $c_v$ | $k$ | $c_p$ | $c_v$ | $k$ | $c_p$ | $c_v$ | $k$ | |
|---|---|---|---|---|---|---|---|---|---|---|
| Temp. °F | | Air | | | Carbon dioxide, $CO_2$ | | | Carbon monoxide, CO | | Temp. °F |
| 40 | 0.240 | 0.171 | 1.401 | 0.195 | 0.150 | 1.300 | 0.248 | 0.177 | 1.400 | 40 |
| 100 | 0.240 | 0.172 | 1.400 | 0.205 | 0.160 | 1.283 | 0.249 | 0.178 | 1.399 | 100 |
| 200 | 0.241 | 0.173 | 1.397 | 0.217 | 0.172 | 1.262 | 0.249 | 0.179 | 1.397 | 200 |
| 300 | 0.243 | 0.174 | 1.394 | 0.229 | 0.184 | 1.246 | 0.251 | 0.180 | 1.394 | 300 |
| 400 | 0.245 | 0.176 | 1.389 | 0.239 | 0.193 | 1.233 | 0.253 | 0.182 | 1.389 | 400 |
| 500 | 0.248 | 0.179 | 1.383 | 0.247 | 0.202 | 1.223 | 0.256 | 0.185 | 1.384 | 500 |
| 600 | 0.250 | 0.182 | 1.377 | 0.255 | 0.210 | 1.215 | 0.259 | 0.188 | 1.377 | 600 |
| 700 | 0.254 | 0.185 | 1.371 | 0.262 | 0.217 | 1.208 | 0.262 | 0.191 | 1.371 | 700 |
| 800 | 0.257 | 0.188 | 1.365 | 0.269 | 0.224 | 1.202 | 0.266 | 0.195 | 1.364 | 800 |
| 900 | 0.259 | 0.191 | 1.358 | 0.275 | 0.230 | 1.197 | 0.269 | 0.198 | 1.357 | 900 |
| 1000 | 0.263 | 0.195 | 1.353 | 0.280 | 0.235 | 1.192 | 0.273 | 0.202 | 1.351 | 1000 |
| 1500 | 0.276 | 0.208 | 1.330 | 0.298 | 0.253 | 1.178 | 0.287 | 0.216 | 1.328 | 1500 |
| 2000 | 0.286 | 0.217 | 1.312 | 0.312 | 0.267 | 1.169 | 0.297 | 0.226 | 1.314 | 2000 |
| | | Hydrogen, $H_2$ | | | Nitrogen, $N_2$ | | | Oxygen, $O_2$ | | |
| 40 | 3.397 | 2.412 | 1.409 | 0.248 | 0.177 | 1.400 | 0.219 | 0.156 | 1.397 | 40 |
| 100 | 3.426 | 2.441 | 1.404 | 0.248 | 0.178 | 1.399 | 0.220 | 0.158 | 1.394 | 100 |
| 200 | 3.451 | 2.466 | 1.399 | 0.249 | 0.178 | 1.398 | 0.223 | 0.161 | 1.387 | 200 |
| 300 | 3.461 | 2.476 | 1.398 | 0.250 | 0.179 | 1.396 | 0.226 | 0.164 | 1.378 | 300 |
| 400 | 3.466 | 2.480 | 1.397 | 0.251 | 0.180 | 1.393 | 0.230 | 0.168 | 1.368 | 400 |
| 500 | 3.469 | 2.484 | 1.397 | 0.254 | 0.183 | 1.388 | 0.235 | 0.173 | 1.360 | 500 |
| 600 | 3.473 | 2.488 | 1.396 | 0.256 | 0.185 | 1.383 | 0.239 | 0.177 | 1.352 | 600 |
| 700 | 3.477 | 2.492 | 1.395 | 0.260 | 0.189 | 1.377 | 0.242 | 0.181 | 1.344 | 700 |
| 800 | 3.494 | 2.509 | 1.393 | 0.262 | 0.191 | 1.371 | 0.246 | 0.184 | 1.337 | 800 |
| 900 | 3.502 | 2.519 | 1.392 | 0.265 | 0.194 | 1.364 | 0.249 | 0.187 | 1.331 | 900 |
| 1000 | 3.513 | 2.528 | 1.390 | 0.269 | 0.198 | 1.359 | 0.252 | 0.190 | 1.326 | 1000 |
| 1500 | 3.618 | 2.633 | 1.374 | 0.283 | 0.212 | 1.334 | 0.263 | 0.201 | 1.309 | 1500 |
| 2000 | 3.758 | 2.773 | 1.355 | 0.293 | 0.222 | 1.319 | 0.270 | 0.208 | 1.298 | 2000 |

Monatomic gases. Over a wide range of temperatures at low pressures the specific heats $c_v$ and $c_p$ of all monatomic gases are essentially independent of temperature. On a molar basis all monatomic gases have the same value for either $c_v$ or $c_p$, and these values may be taken as

$$c_v = 3.0 \text{ Btu/(lb)(°F)} \quad \text{and} \quad c_p = 5.0 \text{ Btu/(lb)(°F)}$$

*Source:* Data adapted from "Tables of Thermal Properties of Gases," NBS Circ. 564, 1955.

## Table A-5 Ideal-gas properties of air

| $T$ °R | $h$ Btu/lb | $p_r$ | $u$ Btu/lb | $v_r$ | $s^0$ Btu/(lb)(°R) | $T$ °R | $h$ Btu/lb | $p_r$ | $u$ Btu/lb | $v_r$ | $s^0$ Btu/(lb)(°R) |
|---|---|---|---|---|---|---|---|---|---|---|---|
| 360 | 85.97 | 0.3363 | 61.29 | 396.6 | 0.50369 | 860 | 206.45 | 7.149 | 147.50 | 44.57 | 0.71323 |
| 380 | 90.75 | 0.4061 | 64.70 | 346.6 | 0.51663 | 880 | 211.35 | 7.761 | 151.02 | 42.01 | 0.71886 |
| 400 | 95.53 | 0.4858 | 68.11 | 305.0 | 0.52890 | 900 | 216.25 | 8.411 | 154.57 | 39.64 | 0.72438 |
| 420 | 100.32 | 0.5760 | 71.52 | 270.1 | 0.54058 | 920 | 221.13 | 9.102 | 158.12 | 37.44 | 0.72979 |
| 440 | 105.11 | 0.6776 | 74.93 | 240.6 | 0.55172 | 940 | 226.11 | 9.834 | 161.68 | 35.41 | 0.73509 |
| 460 | 109.90 | 0.7913 | 78.36 | 215.33 | 0.56235 | 960 | 231.05 | 10.61 | 165.26 | 33.52 | 0.74030 |
| 480 | 114.69 | 0.9182 | 81.77 | 193.65 | 0.57255 | 980 | 236.02 | 11.43 | 168.83 | 31.76 | 0.74540 |
| 500 | 119.48 | 1.0590 | 85.20 | 174.90 | 0.58233 | 1000 | 240.93 | 12.30 | 172.43 | 30.12 | 0.75042 |
| 520 | 124.27 | 1.2147 | 88.62 | 158.58 | 0.59173 | 1040 | 250.95 | 14.18 | 179.66 | 27.17 | 0.76019 |
| 537 | 128.10 | 1.3593 | 91.53 | 146.34 | 0.59945 | 1080 | 260.97 | 16.28 | 186.93 | 24.58 | 0.76964 |
| 540 | 129.06 | 1.3860 | 92.04 | 144.32 | 0.60078 | 1120 | 271.03 | 18.60 | 194.25 | 22.30 | 0.77880 |
| 560 | 133.86 | 1.5742 | 95.47 | 131.78 | 0.60950 | 1160 | 281.14 | 21.18 | 201.63 | 20.29 | 0.78767 |
| 580 | 138.66 | 1.7800 | 98.90 | 120.70 | 0.61793 | 1200 | 291.30 | 24.01 | 209.05 | 18.51 | 0.79628 |
| 600 | 143.47 | 2.005 | 102.34 | 110.88 | 0.62607 | 1240 | 301.52 | 27.13 | 216.53 | 16.93 | 0.80466 |
| 620 | 148.28 | 2.249 | 105.78 | 102.12 | 0.63395 | 1280 | 311.79 | 30.55 | 224.05 | 15.52 | 0.81280 |
| 640 | 153.09 | 2.514 | 109.21 | 94.30 | 0.64159 | 1320 | 322.11 | 34.31 | 231.63 | 14.25 | 0.82075 |
| 660 | 157.92 | 2.801 | 112.67 | 87.27 | 0.64902 | 1360 | 332.48 | 38.41 | 239.25 | 13.12 | 0.82848 |
| 680 | 162.73 | 3.111 | 116.12 | 80.96 | 0.65621 | 1400 | 342.90 | 42.88 | 246.93 | 12.10 | 0.83604 |
| 700 | 167.56 | 3.446 | 119.58 | 75.25 | 0.66321 | 1440 | 353.37 | 47.75 | 254.66 | 11.17 | 0.84341 |
| 720 | 172.39 | 3.806 | 123.04 | 70.07 | 0.67002 | 1480 | 363.89 | 53.04 | 262.44 | 10.34 | 0.85062 |
| 740 | 177.23 | 4.193 | 126.51 | 65.38 | 0.67665 | 1520 | 374.47 | 58.78 | 270.26 | 9.578 | 0.85767 |
| 760 | 182.08 | 4.607 | 129.99 | 61.10 | 0.68312 | 1560 | 385.08 | 65.00 | 278.13 | 8.890 | 0.86456 |
| 780 | 186.94 | 5.051 | 133.47 | 57.20 | 0.68942 | 1600 | 395.74 | 71.73 | 286.06 | 8.263 | 0.87130 |
| 800 | 191.81 | 5.526 | 136.97 | 53.63 | 0.69558 | 1650 | 409.13 | 80.89 | 296.03 | 7.556 | 0.87954 |
| 820 | 196.69 | 6.033 | 140.47 | 50.35 | 0.70160 | 1700 | 422.59 | 90.95 | 306.06 | 6.924 | 0.88758 |
| 840 | 201.56 | 6.573 | 143.98 | 47.34 | 0.70747 | 1750 | 436.12 | 101.98 | 316.16 | 6.357 | 0.89542 |

**Table A-5** (*Continued*)

| T °R | h Btu/lb | $p_r$ | u Btu/lb | $v_r$ | $s^0$ Btu/(lb)(°R) |
|---|---|---|---|---|---|
| 1800 | 449.71 | 114.0 | 326.32 | 5.847 | 0.90308 |
| 1850 | 463.37 | 127.2 | 336.55 | 5.388 | 0.91056 |
| 1900 | 477.09 | 141.5 | 346.85 | 4.974 | 0.91788 |
| 1950 | 490.88 | 157.1 | 357.20 | 4.598 | 0.92504 |
| 2000 | 504.71 | 174.0 | 367.61 | 4.258 | 0.93205 |
| 2050 | 518.61 | 192.3 | 378.08 | 3.949 | 0.93891 |
| 2100 | 532.55 | 212.1 | 388.60 | 3.667 | 0.94564 |
| 2150 | 546.54 | 233.5 | 399.17 | 3.410 | 0.95222 |
| 2200 | 560.59 | 256.6 | 409.78 | 3.176 | 0.95919 |
| 2250 | 574.69 | 281.4 | 420.46 | 2.961 | 0.96501 |
| 2300 | 588.82 | 308.1 | 431.16 | 2.765 | 0.97123 |
| 2350 | 603.00 | 336.8 | 441.91 | 2.585 | 0.97732 |
| 2400 | 617.22 | 367.6 | 452.70 | 2.419 | 0.98331 |
| 2450 | 631.48 | 400.5 | 463.54 | 2.266 | 0.98919 |
| 2500 | 645.78 | 435.7 | 474.40 | 2.125 | 0.99497 |
| 2550 | 660.12 | 473.3 | 485.31 | 1.996 | 1.00064 |
| 2600 | 674.49 | 513.5 | 496.26 | 1.876 | 1.00623 |
| 2650 | 688.90 | 556.3 | 507.25 | 1.765 | 1.01172 |
| 2700 | 703.35 | 601.9 | 518.26 | 1.662 | 1.01712 |
| 2750 | 717.83 | 650.4 | 529.31 | 1.566 | 1.02244 |
| 2800 | 732.33 | 702.0 | 540.40 | 1.478 | 1.02767 |
| 2850 | 746.88 | 756.7 | 551.52 | 1.395 | 1.03282 |
| 2900 | 761.45 | 814.8 | 562.66 | 1.318 | 1.03788 |
| 2950 | 776.05 | 876.4 | 573.84 | 1.247 | 1.04288 |
| 3000 | 790.68 | 941.4 | 585.04 | 1.180 | 1.04779 |
| 3050 | 805.34 | 1011 | 596.28 | 1.118 | 1.05264 |
| 3100 | 820.03 | 1083 | 607.53 | 1.060 | 1.05741 |
| 3150 | 834.75 | 1161 | 618.82 | 1.006 | 1.06212 |
| 3200 | 849.48 | 1242 | 630.12 | 0.955 | 1.06676 |
| 3250 | 864.24 | 1328 | 641.46 | 0.907 | 1.07134 |

| T °R | h Btu/lb | $p_r$ | u Btu/lb | $v_r$ | $s^0$ Btu/(lb)(°R) |
|---|---|---|---|---|---|
| 3300 | 879.02 | 1418 | 652.81 | .8621 | 1.07585 |
| 3350 | 893.83 | 1513 | 664.20 | .8202 | 1.08031 |
| 3400 | 908.66 | 1613 | 675.60 | .7807 | 1.08470 |
| 3450 | 923.52 | 1719 | 687.04 | .7436 | 1.08904 |
| 3500 | 938.40 | 1829 | 698.48 | .7087 | 1.09332 |
| 3550 | 953.30 | 1946 | 709.95 | .6759 | 1.09755 |
| 3600 | 968.21 | 2068 | 721.44 | .6449 | 1.10172 |
| 3650 | 983.15 | 2196 | 732.95 | .6157 | 1.10584 |
| 3700 | 998.11 | 2330 | 744.48 | .5882 | 1.10991 |
| 3750 | 1013.1 | 2471 | 756.04 | .5621 | 1.11393 |
| 3800 | 1028.1 | 2618 | 767.60 | .5376 | 1.11791 |
| 3850 | 1043.1 | 2773 | 779.19 | .5143 | 1.12183 |
| 3900 | 1058.1 | 2934 | 790.80 | .4923 | 1.12571 |
| 3950 | 1073.2 | 3103 | 802.43 | .4715 | 1.12955 |
| 4000 | 1088.3 | 3280 | 814.06 | .4518 | 1.13334 |
| 4050 | 1103.4 | 3464 | 825.72 | .4331 | 1.13709 |
| 4100 | 1118.5 | 3656 | 837.40 | .4154 | 1.14079 |
| 4150 | 1133.6 | 3858 | 849.09 | .3985 | 1.14446 |
| 4200 | 1148.7 | 4067 | 860.81 | .3826 | 1.14809 |
| 4300 | 1179.0 | 4513 | 884.28 | .3529 | 1.15522 |
| 4400 | 1209.4 | 4997 | 907.81 | .3262 | 1.16221 |
| 4500 | 1239.9 | 5521 | 931.39 | .3019 | 1.16905 |
| 4600 | 1270.4 | 6089 | 955.04 | .2799 | 1.17575 |
| 4700 | 1300.9 | 6701 | 978.73 | .2598 | 1.18232 |
| 4800 | 1331.5 | 7362 | 1002.5 | .2415 | 1.18876 |
| 4900 | 1362.2 | 8073 | 1026.3 | .2248 | 1.19508 |
| 5000 | 1392.9 | 8837 | 1050.1 | .2096 | 1.20129 |
| 5100 | 1423.6 | 9658 | 1074.0 | .1956 | 1.20738 |
| 5200 | 1454.4 | 10539 | 1098.0 | .1828 | 1.21336 |
| 5300 | 1485.3 | 11481 | 1122.0 | .1710 | 1.21923 |

*Source:* Data abridged from J. H. Keenan and J. Kaye, "Gas Tables," John Wiley, New York, 1945.

**Table A-6 Ideal-gas enthalpy, internal energy, and absolute entropy of diatomic nitrogen, $N_2$**

$$\Delta h_f = 0 \ \text{Btu/lb·mole}$$

$\bar{h}$ and $\bar{u}$, Btu/lb·mol; $\bar{s}$, Btu/(lb·mole)(°R)

| $T$, °R | $\bar{h}^{\bullet}$ | $\bar{u}^{\bullet}$ | $s^0$ | $T$, °R | $\bar{h}^{\bullet}$ | $\bar{u}^{\bullet}$ | $s^0$ |
|---|---|---|---|---|---|---|---|
| 300 | 2082.0 | 1486.2 | 41.695 | 1080 | 7551.0 | 5406.2 | 50.651 |
| 320 | 2221.0 | 1585.5 | 42.143 | 1100 | 7695.0 | 5510.5 | 50.783 |
| 340 | 2360.0 | 1684.8 | 42.564 | 1120 | 7839.3 | 5615.2 | 50.912 |
| 360 | 2498.9 | 1784.0 | 42.962 | 1140 | 7984.0 | 5720.1 | 51.040 |
| 380 | 2638.0 | 1883.4 | 43.337 | 1160 | 8129.0 | 5825.4 | 51.167 |
| 400 | 2777.0 | 1982.6 | 43.694 | 1180 | 8274.4 | 5931.0 | 51.291 |
| 420 | 2916.1 | 2082.0 | 44.034 | 1200 | 8420.0 | 6037.0 | 51.413 |
| 440 | 3055.1 | 2181.3 | 44.357 | 1220 | 8566.1 | 6143.4 | 51.534 |
| 460 | 3194.1 | 2280.6 | 44.665 | 1240 | 8712.6 | 6250.1 | 51.653 |
| 480 | 3333.1 | 2379.9 | 44.962 | 1260 | 8859.3 | 6357.2 | 51.771 |
| 500 | 3472.2 | 2479.3 | 45.246 | 1280 | 9006.4 | 6464.5 | 51.887 |
| 520 | 3611.3 | 2578.6 | 45.519 | 1300 | 9153.9 | 6572.3 | 52.001 |
| 537 | 3729.5 | 2663.1 | 45.743 | 1320 | 9301.8 | 6680.4 | 52.114 |
| 540 | 3750.3 | 2678.0 | 45.781 | 1340 | 9450.0 | 6788.9 | 52.225 |
| 560 | 3889.5 | 2777.4 | 46.034 | 1360 | 9598.6 | 6897.8 | 52.335 |
| 580 | 4028.7 | 2876.9 | 46.278 | 1380 | 9747.5 | 7007.0 | 52.444 |
| 600 | 4167.9 | 2976.4 | 46.514 | 1400 | 9896.9 | 7116.7 | 52.551 |
| 620 | 4307.1 | 3075.9 | 46.742 | 1420 | 10046.6 | 7226.7 | 52.658 |
| 640 | 4446.4 | 3175.5 | 46.964 | 1440 | 10196.6 | 7337.0 | 52.763 |
| 660 | 4585.8 | 3275.2 | 47.178 | 1460 | 10347.0 | 7447.6 | 52.867 |
| 680 | 4725.3 | 3374.9 | 47.386 | 1480 | 10497.8 | 7558.7 | 52.969 |
| 700 | 4864.9 | 3474.8 | 47.588 | 1500 | 10648.0 | 7670.1 | 53.071 |
| 720 | 5004.5 | 3574.7 | 47.785 | 1520 | 10800.4 | 7781.9 | 53.171 |
| 740 | 5144.3 | 3674.7 | 47.977 | 1540 | 10952.2 | 7893.9 | 53.271 |
| 760 | 5284.1 | 3774.9 | 48.164 | 1560 | 11104.3 | 8006.4 | 53.369 |
| 780 | 5424.2 | 3875.2 | 48.345 | 1580 | 11256.9 | 8119.2 | 53.465 |
| 800 | 5564.4 | 3975.7 | 48.522 | 1600 | 11409.7 | 8232.3 | 53.561 |
| 820 | 5704.7 | 4076.3 | 48.696 | 1620 | 11562.8 | 8345.7 | 53.656 |
| 840 | 5845.3 | 4177.1 | 48.865 | 1640 | 11716.4 | 8459.6 | 53.751 |
| 860 | 5985.9 | 4278.1 | 49.031 | 1660 | 11870.2 | 8573.6 | 53.844 |
| 880 | 6126.9 | 4379.4 | 49.193 | 1680 | 12024.3 | 8688.1 | 53.936 |
| 900 | 6268.1 | 4480.8 | 49.352 | 1700 | 12178.9 | 8802.9 | 54.028 |
| 920 | 6409.6 | 4582.6 | 49.507 | 1720 | 12333.7 | 8918.0 | 54.118 |
| 940 | 6551.2 | 4684.5 | 49.659 | 1740 | 12488.8 | 9033.4 | 54.208 |
| 960 | 6693.1 | 4786.7 | 49.808 | 1760 | 12644.3 | 9149.2 | 54.297 |
| 980 | 6835.4 | 4889.3 | 49.955 | 1780 | 12800.2 | 9265.3 | 54.385 |
| 1000 | 6977.9 | 4992.0 | 50.099 | 1800 | 12956.3 | 9381.7 | 54.472 |
| 1020 | 7120.7 | 5095.1 | 50.241 | 1820 | 13112.7 | 9498.4 | 54.559 |
| 1040 | 7263.8 | 5198.5 | 50.380 | 1840 | 13269.5 | 9615.5 | 54.645 |
| 1060 | 7407.2 | 5302.2 | 50.516 | 1860 | 13426.5 | 9732.8 | 54.729 |

## Table A-6 Ideal-gas enthalpy, internal energy, and absolute entropy of diatomic nitrogen, $N_2$ (*Continued*)

| $T$, °R | $\bar{h}$ | $\bar{u}$ | $s^0$ | $T$, °R | $\bar{h}$ | $\bar{u}$ | $\bar{s}^0$ |
|---|---|---|---|---|---|---|---|
| 1900 | 13,742 | 9,968 | 54.896 | 3500 | 27,016 | 20,065 | 59.944 |
| 1940 | 14,058 | 10,205 | 55.061 | 3540 | 27,359 | 20,329 | 60.041 |
| 1980 | 14,375 | 10,443 | 55.223 | 3580 | 27,703 | 20,593 | 60.138 |
| 2020 | 14,694 | 10,682 | 55.383 | 3620 | 28,046 | 20,858 | 60.234 |
| 2060 | 15,013 | 10,923 | 55.540 | 3660 | 28,391 | 21,122 | 60.328 |
| 2100 | 15,334 | 11,164 | 55.694 | 3700 | 28,735 | 21,387 | 60.422 |
| 2140 | 15,656 | 11,406 | 55.846 | 3740 | 29,080 | 21,653 | 60,515 |
| 2180 | 15,978 | 11,649 | 55.995 | 3780 | 29,425 | 21,919 | 60.607 |
| 2220 | 16,302 | 11,893 | 56.141 | 3820 | 29,771 | 22,185 | 60.698 |
| 2260 | 16,626 | 12,138 | 56.286 | 3860 | 30,117 | 22,451 | 60.788 |
| 2300 | 16,951 | 12,384 | 56.429 | 3900 | 30,463 | 22,718 | 60.877 |
| 2340 | 17,277 | 12,630 | 56.570 | 3940 | 30,809 | 22,985 | 60.966 |
| 2380 | 17,604 | 12,878 | 56.708 | 3980 | 31,156 | 23,252 | 61.053 |
| 2420 | 17,932 | 13,126 | 56.845 | 4020 | 31,503 | 23,520 | 61.139 |
| 2460 | 18,260 | 13,375 | 56.980 | 4060 | 31,850 | 23,788 | 61.225 |
| 2500 | 18,590 | 13,625 | 57.112 | 4100 | 32,198 | 24,056 | 61.310 |
| 2540 | 18,919 | 13,875 | 57.243 | 4140 | 32,546 | 24,324 | 61.395 |
| 2580 | 19,250 | 14,127 | 57.372 | 4180 | 32,894 | 24,593 | 61.479 |
| 2620 | 19,582 | 14,379 | 57.499 | 4220 | 33,242 | 24,862 | 61.562 |
| 2660 | 19,914 | 14,631 | 57.625 | 4260 | 33,591 | 25,131 | 61.644 |
| 2700 | 20,246 | 14,885 | 57.750 | 4300 | 33,940 | 25,401 | 61.726 |
| 2740 | 20,580 | 15,139 | 57.872 | 4340 | 34,289 | 25,670 | 61.806 |
| 2780 | 20,914 | 15,393 | 57.993 | 4380 | 34,638 | 25,940 | 61.887 |
| 2820 | 21,248 | 15,648 | 58.113 | 4420 | 34,988 | 26,210 | 61.966 |
| 2860 | 21,584 | 15,904 | 58.231 | 4460 | 35,338 | 26,481 | 62.045 |
| 2900 | 21,920 | 16,161 | 58.348 | 4500 | 35,688 | 26,751 | 62.123 |
| 2940 | 22,256 | 16,417 | 58.463 | 4540 | 36,038 | 27,022 | 62.201 |
| 2980 | 22,593 | 16,675 | 58.576 | 4580 | 36,389 | 27,293 | 62.278 |
| 3020 | 22,930 | 16,933 | 58.688 | 4620 | 36,739 | 27,565 | 62.354 |
| 3060 | 23,268 | 17,192 | 58.800 | 4660 | 37,090 | 27,836 | 62.429 |
| 3100 | 23,607 | 17,451 | 58.910 | 4700 | 37,441 | 28,108 | 62.504 |
| 3140 | 23,946 | 17,710 | 59.019 | 4740 | 37,792 | 28,379 | 62.578 |
| 3180 | 24,285 | 17,970 | 59.126 | 4780 | 38,144 | 28,651 | 62.652 |
| 3220 | 24,625 | 18,231 | 59.232 | 4820 | 38,495 | 28,924 | 62.725 |
| 3260 | 24,965 | 18,491 | 59.338 | 4860 | 38,847 | 29,196 | 62.798 |
| 3300 | 25,306 | 18,753 | 59.442 | 4900 | 39,199 | 29,468 | 62.870 |
| 3340 | 25,647 | 19,014 | 59.544 | 5000 | 40,080 | 30,151 | 63.049 |
| 3380 | 25,989 | 19,277 | 59.646 | 5100 | 40,962 | 30,834 | 63.223 |
| 3420 | 26,331 | 19,539 | 59.747 | 5200 | 41,844 | 31,518 | 63.395 |
| 3460 | 26,673 | 19,802 | 59.846 | 5300 | 42,728 | 32,203 | 63.563 |

*Source:* Data abridged from J. H. Keenan and J. Kaye, "Gas Tables," Wiley, New York, 1945.

## Table A-7 Ideal-gas enthalpy, internal energy, and absolute entropy of diatomic oxygen, $O_2$

$$\Delta h_f = 0 \text{ Btu/lb·mole}$$

$\bar{h}$ and $\bar{u}$, Btu/lb·mol; $\bar{s}$, Btu/(lb·mol)(°R)

| $T$, °R | $\bar{h}$ | $\bar{u}$ | $\bar{s}^0$ | $T$, °R | $\bar{h}$ | $\bar{u}$ | $s^0$ |
|---|---|---|---|---|---|---|---|
| 300 | 2073.5 | 1477.8 | 44.927 | 1080 | 7696.8 | 5552.1 | 54.064 |
| 320 | 2212.6 | 1577.1 | 45.375 | 1100 | 7850.4 | 5665.9 | 54.204 |
| 340 | 2351.7 | 1676.5 | 45.797 | 1120 | 8004.5 | 5780.3 | 54.343 |
| 360 | 2490.8 | 1775.9 | 46.195 | 1140 | 8159.1 | 5895.2 | 54.480 |
| 380 | 2630.0 | 1875.3 | 46.571 | 1160 | 8314.2 | 6010.6 | 54.614 |
| 400 | 2769.1 | 1974.8 | 46.927 | 1180 | 8469.8 | 6126.5 | 54.748 |
| 420 | 2908.3 | 2074.3 | 47.267 | 1200 | 8625.8 | 6242.8 | 54.879 |
| 440 | 3047.5 | 2173.8 | 47.591 | 1220 | 8782.4 | 6359.6 | 55.008 |
| 460 | 3186.9 | 2273.4 | 47.900 | 1240 | 8939.4 | 6476.9 | 55.136 |
| 480 | 3326.5 | 2373.3 | 48.198 | 1260 | 9096.7 | 6594.5 | 55.262 |
| 500 | 3466.2 | 2473.2 | 48.483 | 1280 | 9254.6 | 6712.7 | 55.386 |
| 520 | 3606.1 | 2573.4 | 48.757 | 1300 | 9412.9 | 6831.3 | 55.508 |
| 537 | 3725.1 | 2658.7 | 48.982 | 1320 | 9571.6 | 6950.2 | 55.630 |
| 540 | 3746.2 | 2673.8 | 49.021 | 1340 | 9730.7 | 7069.6 | 55.750 |
| 560 | 3886.6 | 2774.5 | 49.276 | 1360 | 9890.2 | 7189.4 | 55.867 |
| 580 | 4027.3 | 2875.5 | 49.522 | 1380 | 10050.1 | 7309.6 | 55.984 |
| 600 | 4168.3 | 2976.8 | 49.762 | 1400 | 10210.4 | 7430.1 | 56.099 |
| 620 | 4309.7 | 3078.4 | 49.993 | 1420 | 10371.0 | 7551.1 | 56.213 |
| 640 | 4451.4 | 3180.4 | 50.218 | 1440 | 10532.0 | 7672.4 | 56.326 |
| 660 | 4593.5 | 3282.9 | 50.437 | 1460 | 10693.3 | 7793.9 | 56.437 |
| 680 | 4736.2 | 3385.8 | 50.650 | 1480 | 10855.1 | 7916.0 | 56.547 |
| 700 | 4879.3 | 3489.2 | 50.858 | 1500 | 11017.1 | 8038.3 | 56.656 |
| 720 | 5022.9 | 3593.1 | 51.059 | 1520 | 11179.6 | 8161.1 | 56.763 |
| 740 | 5167.0 | 3697.4 | 51.257 | 1540 | 11342.4 | 8284.2 | 56.869 |
| 760 | 5311.4 | 3802.2 | 51.450 | 1560 | 11505.4 | 8407.4 | 56.975 |
| 780 | 5456.4 | 3907.5 | 51.638 | 1580 | 11668.8 | 8531.1 | 57.079 |
| 800 | 5602.0 | 4013.3 | 51.821 | 1600 | 11832.5 | 8655.1 | 57.182 |
| 820 | 5748.1 | 4119.7 | 52.002 | 1620 | 11996.6 | 8779.5 | 57.284 |
| 840 | 5894.8 | 4226.6 | 52.179 | 1640 | 12160.9 | 8904.1 | 57.385 |
| 860 | 6041.9 | 4334.1 | 52.352 | 1660 | 12325.5 | 9029.0 | 57.484 |
| 880 | 6189.6 | 4442.0 | 52.522 | 1680 | 12490.4 | 9154.1 | 57.582 |
| 900 | 6337.9 | 4550.6 | 52.688 | 1700 | 12655.6 | 9279.6 | 57.680 |
| 920 | 6486.7 | 4659.7 | 52.852 | 1720 | 12821.1 | 9405.4 | 57.777 |
| 940 | 6636.1 | 4769.4 | 53.012 | 1740 | 12986.9 | 9531.5 | 57.873 |
| 960 | 6786.0 | 4879.5 | 53.170 | 1760 | 13153.0 | 9657.9 | 57.968 |
| 980 | 6936.4 | 4990.3 | 53.326 | 1780 | 13319.2 | 9784.4 | 58.062 |
| 1000 | 7087.5 | 5101.6 | 53.477 | 1800 | 13485.8 | 9911.2 | 58.155 |
| 1020 | 7238.9 | 5213.3 | 53.628 | 1820 | 13652.5 | 10038.2 | 58.247 |
| 1040 | 7391.0 | 5325.7 | 53.775 | 1840 | 13819.6 | 10165.6 | 58.339 |
| 1060 | 7543.6 | 5438.6 | 53.921 | 1860 | 13986.8 | 10293.1 | 58.428 |

**Table A-7 Ideal-gas enthalpy, internal energy, and absolute entropy of diatomic oxygen, $O_2$** (*Continued*)

| $T, °R$ | $\bar{h}$ | $\bar{u}$ | $\bar{s}^0$ | $T, °R$ | $\bar{h}$ | $\bar{u}$ | $\bar{s}^0$ |
|---|---|---|---|---|---|---|---|
| 1900 | 14,322 | 10,549 | 58.607 | 3500 | 28,273 | 21,323 | 63.914 |
| 1940 | 14,658 | 10,806 | 58.782 | 3540 | 28,633 | 21,603 | 64.016 |
| 1980 | 14,995 | 11,063 | 58.954 | 3580 | 28,994 | 21,884 | 64.114 |
| 2020 | 15,333 | 11,321 | 59.123 | 3620 | 29,354 | 22,165 | 64.217 |
| 2060 | 15,672 | 11,581 | 59.289 | 3660 | 29,716 | 22,447 | 64.316 |
| 2100 | 16,011 | 11,841 | 59.451 | 3700 | 30,078 | 22,730 | 64.415 |
| 2140 | 16,351 | 12,101 | 59.612 | 3740 | 30,440 | 23,013 | 64.512 |
| 2180 | 16,692 | 12,363 | 59.770 | 3780 | 30,803 | 23,296 | 64.609 |
| 2220 | 17,036 | 12,625 | 59.926 | 3820 | 31,166 | 23,580 | 64.704 |
| 2260 | 17,376 | 12,888 | 60.077 | 3860 | 31,529 | 23,864 | 64.800 |
| 2300 | 17,719 | 13,151 | 60.228 | 3900 | 31,894 | 24,149 | 64.893 |
| 2340 | 18,062 | 13,416 | 60.376 | 3940 | 32,258 | 24,434 | 64.986 |
| 2380 | 18,407 | 13,680 | 60.522 | 3980 | 32,623 | 24,720 | 65.078 |
| 2420 | 18,572 | 13,946 | 60.666 | 4020 | 32,989 | 25,006 | 65.169 |
| 2460 | 19,097 | 14,212 | 60.808 | 4060 | 33,355 | 25,292 | 65.260 |
| 2500 | 19,443 | 14,479 | 60.946 | 4100 | 33,722 | 25,580 | 65.350 |
| 2540 | 19,790 | 14,746 | 61.084 | 4140 | 34,089 | 25,867 | 65.439 |
| 2580 | 20,138 | 15,014 | 61.220 | 4180 | 34,456 | 26,155 | 65.527 |
| 2620 | 20,485 | 15,282 | 61.354 | 4220 | 34,824 | 26,444 | 65.615 |
| 2660 | 20,834 | 15,551 | 61.486 | 4260 | 35,192 | 26,733 | 65.702 |
| 2700 | 21,183 | 15,821 | 61.616 | 4300 | 35,561 | 27,022 | 65.788 |
| 2740 | 21,533 | 16,091 | 61.744 | 4340 | 35,930 | 27,312 | 65.873 |
| 2780 | 21,883 | 16,362 | 61.871 | 4380 | 36,300 | 27,602 | 65.958 |
| 2820 | 22,232 | 16,633 | 61.996 | 4420 | 36,670 | 27,823 | 66.042 |
| 2860 | 22,584 | 16,905 | 62.120 | 4460 | 37,041 | 28,184 | 66.125 |
| 2900 | 22,936 | 17,177 | 62.242 | 4500 | 37,412 | 28,475 | 66.208 |
| 2940 | 23,288 | 17,450 | 62.363 | 4540 | 37,783 | 28,768 | 66.290 |
| 2980 | 23,641 | 17,723 | 62.483 | 4580 | 38,155 | 29,060 | 66.372 |
| 3020 | 23,994 | 17,997 | 62.599 | 4620 | 38,528 | 29,353 | 66.453 |
| 3060 | 24,348 | 18,271 | 62.716 | 4660 | 38,900 | 29,646 | 66.533 |
| 3100 | 24,703 | 18,546 | 62.831 | 4700 | 39,274 | 29,940 | 66.613 |
| 3140 | 25,057 | 18,822 | 62.945 | 4740 | 39,647 | 30,234 | 66.691 |
| 3180 | 25,413 | 19,098 | 63.057 | 4780 | 40,021 | 30,529 | 66.770 |
| 3220 | 25,769 | 19,374 | 63.169 | 4820 | 40,396 | 30,824 | 66.848 |
| 3260 | 26,175 | 19,651 | 63.279 | 4860 | 40,771 | 31,120 | 66.925 |
| 3300 | 26,412 | 19,928 | 63.386 | 4900 | 41,146 | 31,415 | 67.003 |
| 3340 | 26,839 | 20,206 | 63.494 | 5000 | 42,086 | 32,157 | 67.193 |
| 3380 | 27,197 | 20,485 | 63.601 | 5100 | 43,021 | 32,901 | 67.380 |
| 3420 | 27,555 | 20,763 | 63.706 | 5200 | 43,974 | 33,648 | 67.562 |
| 3460 | 27,914 | 21,043 | 63.811 | 5300 | 44,922 | 34,397 | 67.743 |

*Source:* Data abridged from J. H. Keenan and J. Kaye, "Gas Tables," Wiley, New York, 1945.

## Table A-8 Ideal-gas enthalpy, internal energy, and absolute entropy of carbon monoxide, CO

$\Delta h_f = -47,540$ Btu/lb·mole

$\bar{h}$ and $\bar{u}$, Btu/lb·mole; $\bar{s}$, Btu/(lb·mole)(°R)

| $T$, °R | $\bar{h}$ | $\bar{u}$ | $\bar{s}^0$ | $T$, °R | $\bar{h}$ | $\bar{u}$ | $\bar{s}^0$ |
|---|---|---|---|---|---|---|---|
| 300 | 2081.9 | 1486.1 | 43.223 | 1080 | 7571.1 | 5426.4 | 52.203 |
| 320 | 2220.9 | 1585.4 | 43.672 | 1100 | 7716.8 | 5532.3 | 52.337 |
| 340 | 2359.9 | 1684.7 | 44.093 | 1120 | 7862.9 | 5638.7 | 52.468 |
| 360 | 2498.8 | 1783.9 | 44.490 | 1140 | 8009.2 | 5745.4 | 52.598 |
| 380 | 2637.9 | 1883.3 | 44.866 | 1160 | 8156.1 | 5852.5 | 52.726 |
| 400 | 2776.9 | 1982.6 | 45.223 | 1180 | 8303.3 | 5960.0 | 52.852 |
| 420 | 2916.0 | 2081.9 | 45.563 | 1200 | 8450.8 | 6067.8 | 52.976 |
| 440 | 3055.0 | 2181.2 | 45.886 | 1220 | 8598.8 | 6176.0 | 53.098 |
| 460 | 3194.0 | 2280.5 | 46.194 | 1240 | 8747.2 | 6284.7 | 53.218 |
| 480 | 3333.0 | 2379.8 | 46.491 | 1260 | 8896.0 | 6393.8 | 53.337 |
| 500 | 3472.1 | 2479.2 | 46.775 | 1280 | 9045.0 | 6503.1 | 53.455 |
| 520 | 3611.2 | 2578.6 | 47.048 | 1300 | 9194.6 | 6613.0 | 53.571 |
| 537 | 3725.1 | 2663.1 | 47.272 | 1320 | 9344.6 | 6723.2 | 53.685 |
| 540 | 3750.3 | 2677.9 | 47.310 | 1340 | 9494.8 | 6833.7 | 53.799 |
| 560 | 3889.5 | 2777.4 | 47.563 | 1360 | 9645.5 | 6944.7 | 53.910 |
| 580 | 4028.7 | 2876.9 | 47.807 | 1380 | 9796.6 | 7056.1 | 54.021 |
| 600 | 4168.0 | 2976.5 | 48.044 | 1400 | 9948.1 | 7167.9 | 54.129 |
| 620 | 4307.4 | 3076.2 | 48.272 | 1420 | 10100.0 | 7280.1 | 54.237 |
| 640 | 4446.9 | 3175.9 | 48.494 | 1440 | 10252.2 | 7392.6 | 54.344 |
| 660 | 4586.5 | 3275.8 | 48.709 | 1460 | 10404.8 | 7505.4 | 54.448 |
| 680 | 4726.2 | 3375.8 | 48.917 | 1480 | 10557.8 | 7618.7 | 54.552 |
| 700 | 4866.0 | 3475.9 | 49.120 | 1500 | 10711.1 | 7732.3 | 54.655 |
| 720 | 5006.1 | 3576.3 | 49.317 | 1520 | 10864.9 | 7846.4 | 54.757 |
| 740 | 5146.4 | 3676.9 | 49.509 | 1540 | 11019.0 | 7960.8 | 54.858 |
| 760 | 5286.8 | 3777.5 | 49.697 | 1560 | 11173.4 | 8075.4 | 54.958 |
| 780 | 5427.4 | 3878.4 | 49.880 | 1580 | 11328.2 | 8190.5 | 55.056 |
| 800 | 5568.2 | 3979.5 | 50.058 | 1600 | 11483.4 | 8306.0 | 55.154 |
| 820 | 5709.4 | 4081.0 | 50.232 | 1620 | 11638.9 | 8421.8 | 55.251 |
| 840 | 5850.7 | 4182.6 | 50.402 | 1640 | 11794.7 | 8537.9 | 55.347 |
| 860 | 5992.3 | 4284.5 | 50.569 | 1660 | 11950.9 | 8654.4 | 55.441 |
| 880 | 6134.2 | 4386.6 | 50.732 | 1680 | 12107.5 | 8771.2 | 55.535 |
| 900 | 6276.4 | 4489.1 | 50.892 | 1700 | 12264.3 | 8888.3 | 55.628 |
| 920 | 6419.0 | 4592.0 | 51.048 | 1720 | 12421.4 | 9005.7 | 55.720 |
| 940 | 6561.7 | 4695.0 | 51.202 | 1740 | 12579.0 | 9123.6 | 55.811 |
| 960 | 6704.9 | 4798.5 | 51.353 | 1760 | 12736.7 | 9241.6 | 55.900 |
| 980 | 6848.4 | 4902.3 | 51.501 | 1780 | 12894.9 | 9360.0 | 55.990 |
| 1000 | 6992.2 | 5006.3 | 51.646 | 1800 | 13053.2 | 9478.6 | 56.078 |
| 1020 | 7136.4 | 5110.8 | 51.788 | 1820 | 13212.0 | 9597.7 | 56.166 |
| 1040 | 7281.0 | 5215.7 | 51.929 | 1840 | 13371.0 | 9717.0 | 56.253 |
| 1060 | 7425.9 | 5320.9 | 52.067 | 1860 | 13530.2 | 9836.5 | 56.339 |

**Table A-8 Ideal-gas enthalpy, internal energy, and absolute entropy of carbon monoxide, CO** (*Continued*)

| $T$, °R | $\bar{h}$ | $\bar{u}$ | $\bar{s}^0$ | $T$, °R | $\bar{h}$ | $\bar{u}$ | $\bar{s}^0$ |
|---|---|---|---|---|---|---|---|
| 1900 | 13,850 | 10,077 | 56.509 | 3500 | 27,262 | 20,311 | 61.612 |
| 1940 | 14,170 | 10,318 | 56.677 | 3540 | 27,608 | 20,576 | 61.710 |
| 1980 | 14,492 | 10,560 | 56.841 | 3580 | 27,954 | 20,844 | 61.807 |
| 2020 | 14,815 | 10,803 | 57.007 | 3620 | 28,300 | 21,111 | 61.903 |
| 2060 | 15,139 | 11,048 | 57.161 | 3660 | 28,647 | 21,378 | 61.998 |
| 2100 | 15,463 | 11,293 | 57.317 | 3700 | 28,994 | 21,646 | 62.093 |
| 2140 | 15,789 | 11,539 | 57.470 | 3740 | 29,341 | 21,914 | 62.186 |
| 2180 | 16,116 | 11,787 | 57.621 | 3780 | 29,688 | 22,182 | 62.279 |
| 2220 | 16,443 | 12,035 | 57.770 | 3820 | 30,036 | 22,450 | 62.370 |
| 2260 | 16,772 | 12,284 | 57.917 | 3860 | 30,384 | 22,719 | 62.461 |
| 2300 | 17,101 | 12,534 | 58.062 | 3900 | 30,733 | 22,988 | 62.511 |
| 2340 | 17,431 | 12,784 | 58.204 | 3940 | 31,082 | 23,257 | 62.640 |
| 2380 | 17,762 | 13,035 | 58.344 | 3980 | 31,431 | 23,527 | 62.728 |
| 2420 | 18,093 | 13,287 | 58.482 | 4020 | 31,780 | 23,797 | 62.816 |
| 2460 | 18,426 | 13,541 | 58.619 | 4060 | 32,129 | 24,067 | 62.902 |
| 2500 | 18,759 | 13,794 | 58.754 | 4100 | 32,479 | 24,337 | 62.988 |
| 2540 | 19,093 | 14,048 | 58.885 | 4140 | 32,829 | 24,608 | 63.072 |
| 2580 | 19,427 | 14,303 | 59.016 | 4180 | 33,179 | 24,878 | 63.156 |
| 2620 | 19,762 | 14,559 | 59.145 | 4220 | 33,530 | 25,149 | 63.240 |
| 2660 | 20,098 | 14,815 | 59.272 | 4260 | 33,880 | 25,421 | 63.323 |
| 2700 | 20,434 | 15,072 | 59.398 | 4300 | 34,231 | 25,692 | 63.405 |
| 2740 | 20,771 | 15,330 | 59.521 | 4340 | 34,582 | 25,934 | 63.486 |
| 2780 | 21,108 | 15,588 | 59.644 | 4380 | 34,934 | 26,235 | 63.567 |
| 2820 | 21,446 | 15,846 | 59.765 | 4420 | 35,285 | 26,508 | 63.647 |
| 2860 | 21,785 | 16,105 | 59.884 | 4460 | 35,637 | 26,780 | 63.726 |
| 2900 | 22,124 | 16,365 | 60.002 | 4500 | 35,989 | 27,052 | 63.805 |
| 2940 | 22,463 | 16,225 | 60.118 | 4540 | 36,341 | 27,325 | 63.883 |
| 2980 | 22,803 | 16,885 | 60.232 | 4580 | 36,693 | 27,598 | 63.960 |
| 3020 | 23,144 | 17,146 | 60.346 | 4620 | 37,046 | 27,871 | 64.036 |
| 3060 | 23,485 | 17,408 | 60.458 | 4660 | 37,398 | 28,144 | 64.113 |
| 3100 | 23,826 | 17,670 | 60.569 | 4700 | 37,751 | 28,417 | 64.188 |
| 3140 | 24,168 | 17,932 | 60.679 | 4740 | 38,104 | 28,691 | 64.263 |
| 3180 | 24,510 | 18,195 | 60.787 | 4780 | 38,457 | 28,965 | 64.337 |
| 3220 | 24,853 | 18,458 | 60.894 | 4820 | 38,811 | 29,239 | 64.411 |
| 3260 | 25,196 | 18,722 | 61.000 | 4860 | 39,164 | 29,513 | 64.484 |
| 3300 | 25,539 | 18,986 | 61.105 | 4900 | 39,518 | 29,787 | 64.556 |
| 3340 | 25,883 | 19,250 | 61.209 | 5000 | 40,403 | 30,473 | 64.735 |
| 3380 | 26,227 | 19,515 | 61.311 | 5100 | 41,289 | 31,161 | 64.910 |
| 3420 | 26,572 | 19,780 | 61.412 | 5200 | 42,176 | 31,849 | 65.082 |
| 3460 | 26,917 | 20,045 | 61.513 | 5300 | 43,063 | 32,538 | 65.252 |

*Source:* Data abridged from J. H. Keenan and J. Kaye, "Gas Tables," Wiley, New York, 1945.

## Table A-9 Ideal-gas enthalpy, internal energy, and absolute entropy of carbon dioxide, $CO_2$

$\Delta h_f = -169{,}290$ Btu/lb·mol

$\bar{h}$ and $\bar{u}$, Btu/lb·mole; $\bar{s}$, Btu/(lb·mole)(°R)

| $T$, °R | $\bar{h}$ | $\bar{u}$ | $\bar{s}^0$ | $T$, °R | $\bar{h}$ | $\bar{u}$ | $\bar{s}^0$ |
|---|---|---|---|---|---|---|---|
| 300 | 2108.2 | 1512.4 | 46.353 | 1080 | 9575.8 | 7431.1 | 58.072 |
| 320 | 2256.6 | 1621.1 | 46.832 | 1100 | 9802.6 | 7618.1 | 58.281 |
| 340 | 2407.3 | 1732.1 | 47.289 | 1120 | 10030.6 | 7806.4 | 58.485 |
| 360 | 2560.5 | 1845.6 | 47.728 | 1140 | 10260.1 | 7996.2 | 58.689 |
| 380 | 2716.4 | 1961.8 | 48.148 | 1160 | 10490.6 | 8187.0 | 58.889 |
| 400 | 2874.7 | 2080.4 | 48.555 | 1180 | 10722.3 | 8379.0 | 59.088 |
| 420 | 3035.7 | 2201.7 | 48.947 | 1200 | 10955.3 | 8572.3 | 59.283 |
| 440 | 3199.4 | 2325.6 | 49.329 | 1220 | 11189.4 | 8766.6 | 59.477 |
| 460 | 3365.7 | 2452.2 | 49.698 | 1240 | 11424.6 | 8962.1 | 59.668 |
| 480 | 3534.7 | 2581.5 | 50.058 | 1260 | 11661.0 | 9158.8 | 59.858 |
| 500 | 3706.2 | 2713.3 | 50.408 | 1280 | 11898.4 | 9356.5 | 60.044 |
| 520 | 3880.3 | 2847.7 | 50.750 | 1300 | 12136.9 | 9555.3 | 60.229 |
| 537 | 4027.5 | 2963.8 | 51.032 | 1320 | 12376.4 | 9755.0 | 60.412 |
| 540 | 4056.8 | 2984.4 | 51.082 | 1340 | 12617.0 | 9955.9 | 60.593 |
| 560 | 4235.8 | 3123.7 | 51.408 | 1360 | 12858.5 | 10157.7 | 60.772 |
| 580 | 4417.2 | 3265.4 | 51.726 | 1380 | 13101.0 | 10360.5 | 60.949 |
| 600 | 4600.9 | 3409.4 | 52.038 | 1400 | 13344.7 | 10564.5 | 61.124 |
| 620 | 4786.8 | 3555.6 | 52.343 | 1420 | 13589.1 | 10769.2 | 61.298 |
| 640 | 4974.9 | 3704.0 | 52.641 | 1440 | 13834.5 | 10974.8 | 61.469 |
| 660 | 5165.2 | 3854.6 | 52.934 | 1460 | 14080.8 | 11181.4 | 61.639 |
| 680 | 5357.6 | 4007.2 | 53.225 | 1480 | 14328.0 | 11388.9 | 61.800 |
| 700 | 5552.0 | 4161.9 | 53.503 | 1500 | 14576.0 | 11597.2 | 61.974 |
| 720 | 5748.4 | 4318.6 | 53.780 | 1520 | 14824.9 | 11806.4 | 62.138 |
| 740 | 5946.8 | 4477.3 | 54.051 | 1540 | 15074.7 | 12016.5 | 62.302 |
| 760 | 6147.0 | 4637.9 | 54.319 | 1560 | 15325.3 | 12227.3 | 62.464 |
| 780 | 6349.1 | 4800.1 | 54.582 | 1580 | 15576.7 | 12439.0 | 62.624 |
| 800 | 6552.9 | 4964.2 | 54.839 | 1600 | 15829.0 | 12651.6 | 62.783 |
| 820 | 6758.3 | 5129.9 | 55.093 | 1620 | 16081.9 | 12864.8 | 62.939 |
| 840 | 6965.7 | 5297.6 | 55.343 | 1640 | 16335.7 | 13078.9 | 63.095 |
| 860 | 7174.7 | 5466.9 | 55.589 | 1660 | 16590.2 | 13293.7 | 63.250 |
| 880 | 7385.3 | 5637.7 | 55.831 | 1680 | 16845.5 | 13509.2 | 63.403 |
| 900 | 7597.6 | 5810.3 | 56.070 | 1700 | 17101.4 | 13725.4 | 63.555 |
| 920 | 7811.4 | 5984.4 | 56.305 | 1720 | 17358.1 | 13942.4 | 63.704 |
| 940 | 8026.8 | 6160.1 | 56.536 | 1740 | 17615.5 | 14160.1 | 63.853 |
| 960 | 8243.8 | 6337.4 | 56.765 | 1760 | 17873.5 | 14378.4 | 64.001 |
| 980 | 8462.2 | 6516.1 | 56.990 | 1780 | 18132.2 | 14597.4 | 64.147 |
| 1000 | 8682.1 | 6696.2 | 57.212 | 1800 | 18391.5 | 14816.9 | 64.292 |
| 1020 | 8903.4 | 6877.8 | 57.432 | 1820 | 18651.5 | 15037.2 | 64.435 |
| 1040 | 9126.2 | 7060.9 | 57.647 | 1840 | 18912.2 | 15258.2 | 64.578 |
| 1060 | 9350.3 | 7245.3 | 57.861 | 1860 | 19173.4 | 15479.7 | 64.719 |

**Table A-9 Ideal-gas enthalpy, internal energy, and absolute entropy of carbon dioxide, $CO_2$** (*Continued*)

| $T, °R$ | $\bar{h}$ | $\bar{u}$ | $\bar{s}^0$ | $T, °R$ | $\bar{h}$ | $\bar{u}$ | $\bar{s}^0$ |
|---|---|---|---|---|---|---|---|
| 1900 | 19,698 | 15,925 | 64.999 | 3500 | 41,965 | 35,015 | 73.462 |
| 1940 | 20,224 | 16,372 | 65.272 | 3540 | 42,543 | 35,513 | 73.627 |
| 1980 | 20,753 | 16,821 | 65.543 | 3580 | 43,121 | 36,012 | 73.789 |
| 2020 | 21,284 | 17,273 | 65.809 | 3620 | 43,701 | 36,512 | 73.951 |
| 2060 | 21,818 | 17,727 | 66.069 | 3660 | 44,280 | 37,012 | 74.110 |
| 2100 | 22,353 | 18,182 | 66.327 | 3700 | 44,861 | 37,513 | 74.267 |
| 2140 | 22,890 | 18,640 | 66.581 | 3740 | 45,442 | 38,014 | 74.423 |
| 2180 | 23,429 | 19,101 | 66.830 | 3780 | 46,023 | 38,517 | 74.578 |
| 2220 | 23,970 | 19,561 | 67.076 | 3820 | 46,605 | 39,019 | 74.732 |
| 2260 | 24,512 | 20,024 | 67.319 | 3860 | 47,188 | 39,522 | 74.884 |
| 2300 | 25,056 | 20,489 | 67.557 | 3900 | 47,771 | 40,026 | 75.033 |
| 2340 | 25,602 | 20,955 | 67.792 | 3940 | 48,355 | 40,531 | 75.182 |
| 2380 | 26,150 | 21,423 | 68.025 | 3980 | 48,939 | 41,035 | 75.330 |
| 2420 | 26,699 | 21,893 | 68.253 | 4020 | 49,524 | 41,541 | 75.477 |
| 2460 | 27,249 | 22,364 | 68.479 | 4060 | 50,109 | 42,047 | 75.622 |
| 2500 | 27,801 | 22,837 | 68.702 | 4100 | 50,695 | 42,553 | 75.765 |
| 2540 | 28,355 | 23,310 | 68.921 | 4140 | 51,282 | 43,060 | 75.907 |
| 2580 | 28,910 | 23,786 | 69.138 | 4180 | 51,868 | 43,568 | 76.048 |
| 2620 | 29,465 | 24,262 | 69.352 | 4220 | 52,456 | 44,075 | 76.188 |
| 2660 | 30,023 | 24,740 | 69.563 | 4260 | 53,044 | 44,584 | 76.327 |
| 2700 | 30,581 | 25,220 | 69.771 | 4300 | 53,632 | 45,093 | 76.464 |
| 2740 | 31,141 | 25,701 | 69.977 | 4340 | 54,221 | 45,602 | 76.601 |
| 2780 | 31,702 | 26,181 | 70.181 | 4380 | 54,810 | 46,112 | 76.736 |
| 2820 | 32,264 | 26,664 | 70.382 | 4420 | 55,400 | 46,622 | 76.870 |
| 2860 | 32,827 | 27,148 | 70.580 | 4460 | 55,990 | 47,133 | 77.003 |
| 2900 | 33,392 | 27,633 | 70.776 | 4500 | 56,581 | 47,645 | 77.135 |
| 2940 | 33,957 | 28,118 | 70.970 | 4540 | 57,172 | 48,156 | 77.266 |
| 2980 | 34,523 | 28,605 | 71.160 | 4580 | 57,764 | 48,668 | 77.395 |
| 3020 | 35,090 | 29,093 | 71.350 | 4620 | 58,356 | 49,181 | 77.581 |
| 3060 | 35,659 | 29,582 | 71.537 | 4660 | 58,948 | 49,694 | 77.652 |
| 3100 | 36,228 | 30,072 | 71.722 | 4700 | 59,541 | 50,208 | 77.779 |
| 3140 | 36,798 | 30,562 | 71.904 | 4740 | 60,134 | 50,721 | 77.905 |
| 3180 | 37,369 | 31,054 | 72.085 | 4780 | 60,728 | 51,236 | 78.029 |
| 3220 | 37,941 | 31,546 | 72.264 | 4820 | 61,322 | 51,750 | 78.153 |
| 3260 | 38,513 | 32,039 | 72.441 | 4860 | 61,916 | 52,265 | 78.276 |
| 3300 | 39,087 | 32,533 | 72.616 | 4900 | 62,511 | 52,781 | 78.398 |
| 3340 | 39,661 | 33,028 | 72.788 | 5000 | 64,000 | 54,071 | 78.698 |
| 3380 | 40,236 | 33,524 | 72.960 | 5100 | 65,491 | 55,363 | 78.994 |
| 3420 | 40,812 | 34,020 | 73.129 | 5200 | 66,984 | 56,658 | 79.284 |
| 3460 | 41,388 | 34,517 | 73.297 | 5300 | 68,471 | 57,954 | 79.569 |

*Source:* Data abridged from J. H. Keenan and J. Kaye, "Gas Tables," Wiley, New York, 1945.

**Table A-10 Ideal-gas enthalpy, internal energy, and absolute entropy of water, $H_2O$**

$\Delta h_f = -104{,}040$ Btu/lb·mol

$\bar{h}$ and $\bar{u}$, Btu/lb·mol; $\bar{s}$, Btu/(lb·mol)($°R$)

| $T$, $°R$ | $\bar{h}$ | $\bar{u}$ | $\bar{s}^0$ | $T$, $°R$ | $\bar{h}$ | $\bar{u}$ | $\bar{s}^0$ |
|---|---|---|---|---|---|---|---|
| 300 | 2367.6 | 1771.8 | 40.439 | 1080 | 8768.2 | 6623.5 | 50.854 |
| 320 | 2526.8 | 1891.3 | 40.952 | 1100 | 8942.0 | 6757.5 | 51.013 |
| 340 | 2686.0 | 2010.8 | 41.435 | 1120 | 9116.4 | 6892.2 | 51.171 |
| 360 | 2845.1 | 2130.2 | 41.889 | 1140 | 9291.4 | 7027.5 | 51.325 |
| 380 | 3004.4 | 2249.8 | 42.320 | 1160 | 9467.1 | 7163.5 | 51.478 |
| 400 | 3163.8 | 2369.4 | 42.728 | 1180 | 9643.4 | 7300.1 | 51.630 |
| 420 | 3323.2 | 2489.1 | 43.117 | 1200 | 9820.4 | 7437.4 | 51.777 |
| 440 | 3482.7 | 2608.9 | 43.487 | 1220 | 9998.0 | 7575.2 | 51.925 |
| 460 | 3642.3 | 2728.8 | 43.841 | 1240 | 10176.1 | 7713.6 | 52.070 |
| 480 | 3802.0 | 2848.8 | 44.182 | 1260 | 10354.9 | 7852.7 | 52.212 |
| 500 | 3962.0 | 2969.1 | 44.508 | 1280 | 10534.4 | 7992.5 | 52.354 |
| 520 | 4122.0 | 3089.4 | 44.821 | 1300 | 10714.5 | 8132.9 | 52.494 |
| 537 | 4258.0 | 3191.9 | 45.079 | 1320 | 10895.3 | 8274.0 | 52.631 |
| 540 | 4282.4 | 3210.0 | 45.124 | 1340 | 11076.6 | 8415.5 | 52.768 |
| 560 | 4442.8 | 3330.7 | 45.415 | 1360 | 11258.7 | 8557.9 | 52.903 |
| 580 | 4603.7 | 3451.9 | 45.696 | 1380 | 11441.4 | 8700.9 | 53.037 |
| 600 | 4764.7 | 3573.2 | 45.970 | 1400 | 11624.8 | 8844.6 | 53.168 |
| 620 | 4926.1 | 3694.9 | 46.235 | 1420 | 11808.8 | 8988.9 | 53.299 |
| 640 | 5087.8 | 3816.8 | 46.492 | 1440 | 11993.4 | 9133.8 | 53.428 |
| 660 | 5250.0 | 3939.3 | 46.741 | 1460 | 12178.8 | 9279.4 | 53.556 |
| 680 | 5412.5 | 4062.1 | 46.984 | 1480 | 12364.8 | 9425.7 | 53.682 |
| 700 | 5575.4 | 4185.3 | 47.219 | 1500 | 12551.4 | 9572.7 | 53.808 |
| 720 | 5738.8 | 4309.0 | 47.450 | 1520 | 12738.8 | 9720.3 | 53.932 |
| 740 | 5902.6 | 4433.1 | 47.673 | 1540 | 12926.8 | 9868.6 | 54.055 |
| 760 | 6066.9 | 4557.6 | 47.893 | 1560 | 13115.6 | 10017.6 | 54.177 |
| 780 | 6231.7 | 4682.7 | 48.106 | 1580 | 13305.0 | 10167.3 | 54.298 |
| 800 | 6396.9 | 4808.2 | 48.316 | 1600 | 13494.9 | 10317.6 | 54.418 |
| 820 | 6562.6 | 4934.2 | 48.520 | 1620 | 13685.7 | 10468.6 | 54.535 |
| 840 | 6728.9 | 5060.8 | 48.721 | 1640 | 13877.0 | 10620.2 | 54.653 |
| 860 | 6895.6 | 5187.8 | 48.916 | 1660 | 14069.2 | 10772.7 | 54.770 |
| 880 | 7062.9 | 5315.3 | 49.109 | 1680 | 14261.9 | 10925.6 | 54.886 |
| 900 | 7230.9 | 5443.6 | 49.298 | 1700 | 14455.4 | 11079.4 | 54.999 |
| 920 | 7399.4 | 5572.4 | 49.483 | 1720 | 14649.5 | 11233.8 | 55.113 |
| 940 | 7568.4 | 5701.7 | 49.665 | 1740 | 14844.3 | 11388.9 | 55.226 |
| 960 | 7738.0 | 5831.6 | 49.843 | 1760 | 15039.8 | 11544.7 | 55.339 |
| 980 | 7908.2 | 5962.0 | 50.019 | 1780 | 15236.1 | 11701.2 | 55.449 |
| 1000 | 8078.9 | 6093.0 | 50.191 | 1800 | 15433.0 | 11858.4 | 55.559 |
| 1020 | 8250.4 | 6224.8 | 50.360 | 1820 | 15630.6 | 12016.3 | 55.668 |
| 1040 | 8422.4 | 6357.1 | 50.528 | 1840 | 15828.7 | 12174.7 | 55.777 |
| 1060 | 8595.0 | 6490.0 | 50.693 | 1860 | 16027.6 | 12333.9 | 55.884 |

**Table A-10  Ideal-gas enthalpy, internal energy, and absolute entropy of water, H$_2$O** (*Continued*)

| $T$, °R | $\bar{h}$ | $\bar{u}$ | $\bar{s}^0$ | $T$, °R | $\bar{h}$ | $\bar{u}$ | $\bar{s}^0$ |
|---|---|---|---|---|---|---|---|
| 1900 | 16,428 | 12,654 | 56.097 | 3500 | 34,324 | 27,373 | 62.876 |
| 1940 | 16,830 | 12,977 | 56.307 | 3540 | 34,809 | 27,779 | 63.015 |
| 1980 | 17,235 | 13,303 | 56.514 | 3580 | 35,296 | 28,187 | 63.153 |
| 2020 | 17,643 | 13,632 | 56.719 | 3620 | 35,785 | 28,596 | 63.288 |
| 2060 | 18,054 | 13,963 | 56.920 | 3660 | 36,274 | 29,006 | 63.423 |
| 2100 | 18,467 | 14,297 | 57.119 | 3700 | 36,765 | 29,418 | 63.557 |
| 2140 | 18,883 | 14,633 | 57.315 | 3740 | 37,258 | 29,831 | 63.690 |
| 2180 | 19,301 | 14,972 | 57.509 | 3780 | 37,752 | 30,245 | 63.821 |
| 2220 | 19,722 | 15,313 | 57.701 | 3820 | 38,247 | 30,661 | 63.952 |
| 2260 | 20,145 | 15,657 | 57.889 | 3860 | 38,743 | 31,077 | 64.082 |
| 2300 | 20,571 | 16,003 | 58.077 | 3900 | 39,240 | 31,495 | 64.210 |
| 2340 | 20,999 | 16,352 | 58.261 | 3940 | 39,739 | 31,915 | 64.338 |
| 2380 | 21,429 | 16,703 | 58.445 | 3980 | 40,239 | 32,335 | 64.465 |
| 2420 | 21,862 | 17,057 | 58.625 | 4020 | 40,740 | 32,757 | 64.591 |
| 2460 | 22,298 | 17,413 | 58.803 | 4060 | 41,242 | 33,179 | 64.715 |
| 2500 | 22,735 | 17,771 | 58.980 | 4100 | 41,745 | 33,603 | 64.839 |
| 2540 | 23,175 | 18,131 | 59.155 | 4140 | 42,250 | 34,028 | 64.962 |
| 2580 | 23,618 | 18,494 | 59.328 | 4180 | 42,755 | 34,454 | 65.084 |
| 2620 | 24,062 | 18,859 | 59.500 | 4220 | 43,267 | 34,881 | 65.204 |
| 2660 | 24,508 | 19,226 | 59.669 | 4260 | 43,769 | 35,310 | 65.325 |
| 2700 | 24,957 | 19,595 | 59.837 | 4300 | 44,278 | 35,739 | 65.444 |
| 2740 | 25,408 | 19,967 | 60.003 | 4340 | 44,788 | 36,169 | 65.563 |
| 2780 | 25,861 | 20,340 | 60.167 | 4380 | 45,298 | 36,600 | 65.680 |
| 2820 | 26,316 | 20,715 | 60.330 | 4420 | 45,810 | 37,032 | 65.797 |
| 2860 | 26,773 | 21,093 | 60.490 | 4460 | 46,322 | 37,465 | 65.913 |
| 2900 | 27,231 | 21,472 | 60.650 | 4500 | 46,836 | 37,900 | 66.028 |
| 2940 | 27,692 | 21,853 | 60.809 | 4540 | 47,350 | 38,334 | 66.142 |
| 2980 | 28,154 | 22,237 | 60.965 | 4580 | 47,866 | 38,770 | 66.255 |
| 3020 | 28,619 | 22,621 | 61.120 | 4620 | 48,382 | 39,207 | 66.368 |
| 3060 | 29,085 | 23,085 | 61.274 | 4660 | 48,899 | 39,645 | 66.480 |
| 3100 | 29,553 | 23,397 | 61.426 | 4700 | 49,417 | 40,083 | 66.591 |
| 3140 | 30,023 | 23,787 | 61.577 | 4740 | 49,936 | 40,523 | 66.701 |
| 3180 | 30,494 | 24,179 | 61.727 | 4780 | 50,455 | 40,963 | 66.811 |
| 3220 | 30,967 | 24,572 | 61.874 | 4820 | 50,976 | 41,404 | 66.920 |
| 3260 | 31,442 | 24,968 | 62.022 | 4860 | 51,497 | 41,856 | 67.028 |
| 3300 | 31,918 | 25,365 | 62.167 | 4900 | 52,019 | 42,288 | 67.135 |
| 3340 | 32,396 | 25,763 | 62.312 | 5000 | 53,327 | 43,398 | 67.401 |
| 3380 | 32,876 | 26,164 | 62.454 | 5100 | 54,640 | 44,512 | 67.662 |
| 3420 | 33,357 | 26,565 | 62.597 | 5200 | 55,957 | 45,631 | 67.918 |
| 3460 | 33,839 | 26,968 | 62.738 | 5300 | 57,279 | 46,754 | 68.172 |

*Source:* Data abridged from J. H. Keenan and J. Kaye, "Gas Tables," Wiley, New York, 1945.

### Table A-11 Ideal-gas enthalpy, internal energy, and absolute entropy of diatomic hydrogen, $H_2$

$\Delta h_f = 0$ Btu/lb·mol

$\bar{h}$ and $\bar{u}$, Btu/lb·mol; $\bar{s}$, Btu/(lb·mol)(°R)

| $T$, °R | $\bar{h}$ | $\bar{u}$ | $\bar{s}^0$ | $T$, °R | $\bar{h}$ | $\bar{u}$ | $\bar{s}^0$ |
|---|---|---|---|---|---|---|---|
| 300 | 2063.5 | 1467.7 | 27.337 | 1400 | 9673.8 | 6893.6 | 37.883 |
| 320 | 2189.4 | 1553.9 | 27.742 | 1500 | 10381.5 | 7402.7 | 38.372 |
| 340 | 2317.2 | 1642.0 | 28.130 | 1600 | 11092.5 | 7915.1 | 38.830 |
| 360 | 2446.8 | 1731.9 | 28.501 | 1700 | 11807.4 | 8431.4 | 39.264 |
| 380 | 2577.8 | 1823.2 | 28.856 | 1800 | 12526.8 | 8952.2 | 39.675 |
| 400 | 2710.2 | 1915.8 | 29.195 | 1900 | 13250.9 | 9477.8 | 40.067 |
| 420 | 2843.7 | 2009.6 | 29.520 | 2000 | 13980.1 | 10008.4 | 40.441 |
| 440 | 2978.1 | 2104.3 | 29.833 | 2100 | 14714.5 | 10544.2 | 40.799 |
| 460 | 3113.5 | 2200.0 | 30.133 | 2200 | 15454.4 | 11085.5 | 41.143 |
| 480 | 3249.4 | 2296.2 | 30.424 | 2300 | 16199.8 | 11632.3 | 41.475 |
| 500 | 3386.1 | 2393.2 | 30.703 | 2400 | 16950.6 | 12184.5 | 41.794 |
| 520 | 3523.3 | 2490.6 | 30.972 | 2500 | 17707.3 | 12742.6 | 42.104 |
| 537 | 3640.3 | 2573.9 | 31.194 | 2600 | 18469.7 | 13306.4 | 42.403 |
| 540 | 3660.9 | 2588.5 | 31.232 | 2700 | 19237.8 | 13876.0 | 42.692 |
| 560 | 3798.8 | 2686.7 | 31.482 | 2800 | 20011.8 | 14451.4 | 42.973 |
| 580 | 3937.1 | 2785.3 | 31.724 | 2900 | 20791.5 | 15032.5 | 43.247 |
| 600 | 4075.6 | 2884.1 | 31.959 | 3000 | 21576.9 | 15619.3 | 43.514 |
| 620 | 4214.3 | 2983.1 | 32.187 | 3100 | 22367.7 | 16211.5 | 43.773 |
| 640 | 4353.1 | 3082.1 | 32.407 | 3200 | 23164.1 | 16809.3 | 44.026 |
| 660 | 4492.1 | 3181.4 | 32.621 | 3300 | 23965.5 | 17412.1 | 44.273 |
| 680 | 4631.1 | 3280.7 | 32.829 | 3400 | 24771.9 | 18019.9 | 44.513 |
| 700 | 4770.2 | 3380.1 | 33.031 | 3500 | 25582.9 | 18632.4 | 44.748 |
| 720 | 4909.5 | 3479.6 | 33.226 | 3600 | 26398.5 | 19249.4 | 44.978 |
| 740 | 5048.8 | 3579.2 | 33.417 | 3700 | 27218.5 | 19870.8 | 45.203 |
| 760 | 5188.1 | 3678.8 | 33.603 | 3800 | 28042.8 | 20496.5 | 45.423 |
| 780 | 5327.6 | 3778.6 | 33.784 | 3900 | 28871.1 | 21126.2 | 45.638 |
| 800 | 5467.1 | 3878.4 | 33.961 | 4000 | 29703.5 | 21760.0 | 45.849 |
| 820 | 5606.7 | 3978.3 | 34.134 | 4100 | 30539.8 | 22397.7 | 46.056 |
| 840 | 5746.3 | 4078.2 | 34.302 | 4200 | 31379.8 | 23039.2 | 46.257 |
| 860 | 5885.9 | 4178.0 | 34.466 | 4300 | 32223.5 | 23684.3 | 46.456 |
| 880 | 6025.6 | 4278.0 | 34.627 | 4400 | 33070.9 | 24333.1 | 46.651 |
| 900 | 6165.3 | 4378.0 | 34.784 | 4500 | 33921.6 | 24985.2 | 46.842 |
| 920 | 6305.1 | 4478.1 | 34.938 | 4600 | 34775.7 | 25640.7 | 47.030 |
| 940 | 6444.9 | 4578.1 | 35.087 | 4700 | 35633.0 | 26299.4 | 47.215 |
| 960 | 6584.7 | 4678.3 | 35.235 | 4800 | 36493.4 | 26961.2 | 47.396 |
| 980 | 6724.6 | 4778.4 | 35.379 | 4900 | 37356.9 | 27626.1 | 47.574 |
| 1000 | 6864.5 | 4878.6 | 35.520 | 5000 | 38223.3 | 28294.0 | 47.749 |
| 1100 | 7564.6 | 5380.1 | 36.188 | 5100 | 39092.8 | 28964.9 | 47.921 |
| 1200 | 8265.8 | 5882.8 | 36.798 | 5200 | 39965.1 | 29638.6 | 48.090 |
| 1300 | 8968.7 | 6387.1 | 37.360 | 5300 | 40840.2 | 30315.1 | 48.257 |

*Source:* Data abridged from J. H. Keenan and J. Kaye, "Gas Tables," Wiley, New York, 1945.

## Table A-12 Properties of saturated water: temperature table
$v$, ft$^3$/lb; $h$, Btu/lb; $s$, Btu/(lb)(°R)

| Temp., °F $T$ | Press., psia $P$ | Specific volume Sat. liquid $v_f$ | Sat. vapor $v_g$ | Internal energy Sat. liquid $u_f$ | Sat. vapor $u_g$ | Enthalpy Sat. liquid $h_f$ | Evap. $h_{fg}$ | Sat. vapor $h_g$ | Entropy Sat. liquid $s_f$ | Evap. $s_{fg}$ | Sat. vapor $s_g$ | Temp., °F $T$ |
|---|---|---|---|---|---|---|---|---|---|---|---|---|
| 32 | 0.0886 | 0.01602 | 3305 | -.01 | 1021.2 | -.01 | 1075.4 | 1075.4 | -.00003 | 2.1870 | 2.1870 | 32 |
| 35 | 0.0999 | 0.01602 | 2948 | 2.99 | 1022.2 | 3.00 | 1073.7 | 1076.7 | 0.00607 | 2.1704 | 2.1764 | 35 |
| 40 | 0.1217 | 0.01602 | 2445 | 8.02 | 1023.9 | 8.02 | 1070.9 | 1078.9 | 0.01617 | 2.1430 | 2.1592 | 40 |
| 45 | 0.1475 | 0.01602 | 2037 | 13.04 | 1025.5 | 13.04 | 1068.1 | 1081.1 | 0.02618 | 2.1162 | 2.1423 | 45 |
| 50 | 0.1780 | 0.01602 | 1704 | 18.06 | 1027.2 | 18.06 | 1065.2 | 1083.3 | 0.03607 | 2.0899 | 2.1259 | 50 |
| 55 | 0.2140 | 0.01603 | 1431 | 23.07 | 1028.8 | 23.07 | 1062.4 | 1085.5 | 0.04586 | 2.0641 | 2.1099 | 55 |
| 60 | 0.2563 | 0.01604 | 1207 | 28.08 | 1030.4 | 28.08 | 1059.6 | 1087.7 | 0.05555 | 2.0388 | 2.0943 | 60 |
| 65 | 0.3057 | 0.01604 | 1022 | 33.09 | 1032.1 | 33.09 | 1056.8 | 1089.9 | 0.06514 | 2.0140 | 2.1791 | 65 |
| 70 | 0.3632 | 0.01605 | 867.7 | 38.09 | 1033.7 | 38.09 | 1054.0 | 1092.0 | 0.07463 | 1.9896 | 2.0642 | 70 |
| 75 | 0.4300 | 0.01606 | 739.7 | 43.09 | 1035.4 | 43.09 | 1051.1 | 1094.2 | 0.08402 | 1.9657 | 2.0497 | 75 |
| 80 | 0.5073 | 0.01607 | 632.8 | 48.08 | 1037.0 | 48.09 | 1048.3 | 1096.4 | 0.09332 | 1.9423 | 2.0356 | 80 |
| 85 | 0.5964 | 0.01609 | 543.1 | 53.08 | 1038.6 | 53.08 | 1045.5 | 1098.6 | 0.1025 | 1.9193 | 2.0218 | 85 |
| 90 | 0.6988 | 0.01610 | 467.7 | 58.07 | 1040.2 | 58.07 | 1042.7 | 1100.7 | 0.1117 | 1.8966 | 2.0083 | 90 |
| 95 | 0.8162 | 0.01611 | 404.0 | 63.06 | 1041.9 | 63.06 | 1039.8 | 1102.9 | 0.1207 | 1.8744 | 1.9951 | 95 |
| 100 | 0.9503 | 0.01613 | 350.0 | 68.04 | 1043.5 | 68.05 | 1037.0 | 1105.0 | 0.1296 | 1.8526 | 1.9822 | 100 |
| 110 | 1.276 | 0.01617 | 265.1 | 78.02 | 1046.7 | 78.02 | 1031.3 | 1109.3 | 0.1473 | 1.8101 | 1.9574 | 110 |
| 120 | 1.695 | 0.01621 | 203.0 | 87.99 | 1049.9 | 88.00 | 1025.5 | 1113.5 | 0.1647 | 1.7690 | 1.9336 | 120 |
| 130 | 2.225 | 0.01625 | 157.2 | 97.97 | 1053.0 | 97.98 | 1019.8 | 1117.8 | 0.1817 | 1.7292 | 1.9109 | 130 |
| 140 | 2.892 | 0.01629 | 122.9 | 107.95 | 1056.2 | 107.96 | 1014.0 | 1121.9 | 0.1985 | 1.6907 | 1.8892 | 140 |
| 150 | 3.722 | 0.01634 | 97.0 | 117.95 | 1059.3 | 117.96 | 1008.1 | 1126.1 | 0.2150 | 1.6533 | 1.8684 | 150 |
| 160 | 4.745 | 0.01640 | 77.2 | 127.94 | 1062.3 | 127.96 | 1002.2 | 1130.1 | 0.2313 | 1.6171 | 1.8484 | 160 |
| 170 | 5.996 | 0.01645 | 62.0 | 137.95 | 1065.4 | 137.97 | 996.2 | 1134.2 | 0.2473 | 1.5819 | 1.8293 | 170 |
| 180 | 7.515 | 0.01651 | 50.2 | 147.97 | 1068.3 | 147.99 | 990.2 | 1138.2 | 0.2631 | 1.5478 | 1.8109 | 180 |
| 190 | 9.343 | 0.01657 | 41.0 | 158.00 | 1071.3 | 158.03 | 984.1 | 1142.1 | 0.2787 | 1.5146 | 1.7932 | 190 |
| 200 | 11.529 | 0.01663 | 33.6 | 168.04 | 1074.2 | 168.07 | 977.9 | 1145.9 | 0.2940 | 1.4822 | 1.7762 | 200 |
| 210 | 14.13 | 0.01670 | 27.82 | 178.1 | 1077.0 | 178.1 | 971.6 | 1149.7 | 0.3091 | 1.4508 | 1.7599 | 210 |
| 212 | 14.70 | 0.01672 | 26.80 | 180.1 | 1077.6 | 180.2 | 970.3 | 1150.5 | 0.3121 | 1.4446 | 1.7567 | 212 |
| 220 | 17.19 | 0.01677 | 23.15 | 188.2 | 1079.8 | 188.2 | 965.3 | 1153.5 | 0.3241 | 1.4201 | 1.7441 | 220 |
| 230 | 20.78 | 0.01685 | 19.39 | 198.3 | 1082.6 | 198.3 | 958.8 | 1157.1 | 0.3388 | 1.3901 | 1.7289 | 230 |
| 240 | 24.97 | 0.01692 | 16.33 | 208.4 | 1085.3 | 208.4 | 952.3 | 1160.7 | 0.3534 | 1.3609 | 1.7143 | 240 |

| Temp | | | | | | | | | | | Press | Temp |
|---|---|---|---|---|---|---|---|---|---|---|---|---|
| 250 | 1.7001 | 1.3324 | 0.3677 | 1164.2 | 945.6 | 218.6 | 1087.9 | 218.5 | 13.83 | 0.01700 | 29.82 | 250 |
| 260 | 1.6864 | 1.3044 | 0.3819 | 1167.6 | 938.8 | 228.8 | 1090.5 | 228.6 | 11.77 | 0.01708 | 35.42 | 260 |
| 270 | 1.6731 | 1.2771 | 0.3960 | 1170.9 | 932.0 | 239.0 | 1093.0 | 238.8 | 10.07 | 0.01717 | 41.85 | 270 |
| 280 | 1.6602 | 1.2504 | 0.4099 | 1174.1 | 924.9 | 249.2 | 1095.4 | 249.0 | 8.65 | 0.01726 | 49.18 | 280 |
| 290 | 1.6477 | 1.2241 | 0.4236 | 1177.2 | 917.8 | 259.4 | 1097.7 | 259.3 | 7.47 | 0.01735 | 57.53 | 290 |
| 300 | 1.6356 | 1.1984 | 0.4372 | 1180.2 | 910.4 | 269.7 | 1100.0 | 269.5 | 6.472 | 0.01745 | 66.98 | 300 |
| 310 | 1.6238 | 1.1731 | 0.4507 | 1183.0 | 903.0 | 280.1 | 1102.1 | 279.8 | 5.632 | 0.01755 | 77.64 | 310 |
| 320 | 1.6123 | 1.1483 | 0.4640 | 1185.8 | 895.3 | 290.4 | 1104.2 | 290.1 | 4.919 | 0.01765 | 89.60 | 320 |
| 330 | 1.6010 | 1.1238 | 0.4772 | 1188.4 | 887.5 | 300.8 | 1106.2 | 300.5 | 4.312 | 0.01776 | 103.00 | 330 |
| 340 | 1.5901 | 1.0997 | 0.4903 | 1190.8 | 879.5 | 311.3 | 1108.0 | 310.9 | 3.792 | 0.01787 | 117.93 | 340 |
| 350 | 1.5793 | 1.0760 | 0.5033 | 1193.1 | 871.3 | 321.8 | 1109.8 | 321.4 | 3.346 | 0.01799 | 134.53 | 350 |
| 360 | 1.5688 | 1.0526 | 0.5162 | 1195.2 | 862.9 | 332.4 | 1111.4 | 331.8 | 2.961 | 0.01811 | 152.92 | 360 |
| 370 | 1.5585 | 1.0295 | 0.5289 | 1197.2 | 854.2 | 343.0 | 1112.9 | 342.4 | 2.628 | 0.01823 | 173.23 | 370 |
| 380 | 1.5483 | 1.0067 | 0.5416 | 1199.0 | 845.4 | 353.6 | 1114.3 | 353.0 | 2.339 | 0.01836 | 195.60 | 380 |
| 390 | 1.5383 | 0.9841 | 0.5542 | 1200.6 | 836.2 | 364.3 | 1115.6 | 363.6 | 2.087 | 0.01850 | 220.2 | 390 |
| 400 | 1.5284 | 0.9617 | 0.5667 | 1202.0 | 826.8 | 375.1 | 1116.6 | 374.3 | 1.866 | 0.01864 | 247.1 | 400 |
| 410 | 1.5187 | 0.9395 | 0.5792 | 1203.1 | 817.2 | 386.0 | 1117.6 | 385.0 | 1.673 | 0.01878 | 276.5 | 410 |
| 420 | 1.5091 | 0.9175 | 0.5915 | 1204.1 | 807.2 | 396.9 | 1118.3 | 395.8 | 1.502 | 0.01894 | 308.5 | 420 |
| 430 | 1.4995 | 0.8957 | 0.6038 | 1204.8 | 796.9 | 407.9 | 1118.9 | 406.7 | 1.352 | 0.01909 | 343.3 | 430 |
| 440 | 1.4900 | 0.8740 | 0.6161 | 1205.3 | 786.3 | 419.0 | 1119.3 | 417.6 | 1.219 | 0.01926 | 381.2 | 440 |
| 450 | 1.4806 | 0.8523 | 0.6282 | 1205.6 | 775.4 | 430.2 | 1119.5 | 428.6 | 1.1011 | 0.01943 | 422.1 | 450 |
| 460 | 1.4712 | 0.8308 | 0.6404 | 1205.5 | 764.1 | 441.4 | 1119.6 | 439.7 | 0.9961 | 0.01961 | 466.3 | 460 |
| 470 | 1.4618 | 0.8093 | 0.6525 | 1205.2 | 752.4 | 452.8 | 1119.4 | 450.9 | 0.9025 | 0.01980 | 514.1 | 470 |
| 480 | 1.4524 | 0.7878 | 0.6646 | 1204.6 | 740.3 | 464.3 | 1118.9 | 462.2 | 0.8187 | 0.02000 | 565.5 | 480 |
| 490 | 1.4430 | 0.7663 | 0.6767 | 1203.7 | 727.8 | 475.9 | 1118.3 | 473.6 | 0.7436 | 0.02021 | 620.7 | 490 |
| 500 | 1.4335 | 0.7448 | 0.6888 | 1202.5 | 714.8 | 487.7 | 1117.4 | 485.1 | 0.6761 | 0.02043 | 680.0 | 500 |
| 520 | 1.4145 | 0.7015 | 0.7130 | 1198.9 | 687.3 | 511.7 | 1114.8 | 508.5 | 0.5605 | 0.02091 | 811.4 | 520 |
| 540 | 1.3950 | 0.6576 | 0.7374 | 1193.8 | 657.5 | 536.4 | 1111.0 | 532.6 | 0.4658 | 0.02145 | 961.5 | 540 |
| 560 | 1.3749 | 0.6129 | 0.7620 | 1187.0 | 625.0 | 562.0 | 1105.8 | 548.4 | 0.3877 | 0.02207 | 1131.8 | 560 |
| 580 | 1.3540 | 0.5668 | 0.7872 | 1178.0 | 589.3 | 588.6 | 1098.9 | 583.1 | 0.3225 | 0.02278 | 1324.3 | 580 |
| 600 | 1.3317 | 0.5187 | 0.8130 | 1166.4 | 549.7 | 616.7 | 1090.0 | 609.9 | 0.2677 | 0.02363 | 1541.0 | 600 |
| 620 | 1.3075 | 0.4677 | 0.8398 | 1151.4 | 505.0 | 646.4 | 1078.5 | 638.3 | 0.2209 | 0.02465 | 1784.4 | 620 |
| 640 | 1.2803 | 0.4122 | 0.8681 | 1131.9 | 453.4 | 673.6 | 1063.2 | 668.7 | 0.1805 | 0.02593 | 2057.1 | 640 |
| 660 | 1.2483 | 0.3493 | 0.8990 | 1105.5 | 391.1 | 714.4 | 1042.3 | 702.3 | 0.1446 | 0.02767 | 2362 | 660 |
| 680 | 1.2068 | 0.2718 | 0.9350 | 1066.7 | 309.8 | 755.9 | 1011.0 | 741.7 | 0.1113 | 0.03032 | 2705 | 680 |
| 700 | 1.1346 | 0.1444 | 0.9902 | 990.2 | 167.5 | 822.7 | 947.7 | 801.7 | 0.0744 | 0.03666 | 3090 | 700 |
| 705.4 | 1.0580 | 0 | 1.0580 | 902.5 | 0 | 902.5 | 872.6 | 872.6 | 0.05053 | 0.05053 | 3204 | 705.4 |

*Source:* Keenan, J. H., F. G. Keyes, P. G. Hill, and J. G. Moore, "Steam Tables," Wiley, New York, 1969.

## Table A-13 Properties of saturated water: pressure table
$v$, ft$^3$/lb; $h$, Btu/lb; $s$, Btu/(lb)($°$R)

| Abs. press., psi $p$ | Temp., $°$F $t$ | Specific volume | | Internal energy | | Enthalpy | | | Entropy | | | Abs. press., psi $p$ |
|---|---|---|---|---|---|---|---|---|---|---|---|---|
| | | Sat. liquid $v_f$ | Sat. vapor $v_g$ | Sat. liquid $u_f$ | Sat. vapor $u_g$ | Sat. liquid $h_f$ | Evap. $h_{fg}$ | Sat. vapor $h_g$ | Sat. liquid $s_f$ | Evap. $s_{fg}$ | Sat. vapor $s_g$ | |
| 0.4 | 72.84 | 0.01606 | 792.0 | 40.94 | 1034.7 | 40.94 | 1052.3 | 1093.3 | 0.0800 | 1.9760 | 2.0559 | 0.4 |
| 0.6 | 85.19 | 0.01609 | 540.0 | 53.26 | 1038.7 | 53.27 | 1045.4 | 1098.6 | 0.1029 | 1.9184 | 2.0213 | 0.6 |
| 0.8 | 94.35 | 0.01611 | 411.7 | 62.41 | 1041.7 | 62.41 | 1040.2 | 1102.6 | 0.1195 | 1.8773 | 1.9968 | 0.8 |
| 1.0 | 101.70 | 0.01614 | 333.6 | 69.74 | 1044.0 | 69.74 | 1036.0 | 1105.8 | 0.1327 | 1.8453 | 1.9779 | 1.0 |
| 1.2 | 107.88 | 0.01616 | 280.9 | 75.90 | 1046.0 | 75.90 | 1032.5 | 1108.4 | 0.1436 | 1.8190 | 1.9626 | 1.2 |
| 1.5 | 115.65 | 0.01619 | 227.7 | 83.65 | 1048.5 | 83.65 | 1028.0 | 1111.7 | 0.1571 | 1.7867 | 1.9438 | 1.5 |
| 2.0 | 126.04 | 0.01623 | 173.75 | 94.02 | 1051.8 | 94.02 | 1022.1 | 1116.1 | 0.1750 | 1.7448 | 1.9198 | 2.0 |
| 3.0 | 141.43 | 0.01630 | 118.72 | 109.38 | 1056.6 | 109.39 | 1013.1 | 1122.5 | 0.2009 | 1.6852 | 1.8861 | 3.0 |
| 4.0 | 152.93 | 0.01636 | 90.64 | 120.88 | 1060.2 | 120.89 | 1006.4 | 1127.3 | 0.2198 | 1.6426 | 1.8624 | 4.0 |
| 5.0 | 162.21 | 0.01641 | 73.53 | 130.15 | 1063.0 | 130.17 | 1000.9 | 1131.0 | 0.2349 | 1.6093 | 1.8441 | 5.0 |
| 6.0 | 170.03 | 0.01645 | 61.98 | 137.98 | 1065.4 | 138.00 | 996.2 | 1134.2 | 0.2474 | 1.5819 | 1.8292 | 6.0 |
| 7.0 | 176.82 | 0.01649 | 53.65 | 144.78 | 1067.4 | 144.80 | 992.1 | 1136.9 | 0.2581 | 1.5585 | 1.8167 | 7.0 |
| 8.0 | 182.84 | 0.01653 | 47.35 | 150.81 | 1069.2 | 150.84 | 988.4 | 1139.3 | 0.2675 | 1.5383 | 1.8058 | 8.0 |
| 9.0 | 188.26 | 0.01656 | 42.41 | 156.25 | 1070.8 | 156.27 | 985.1 | 1141.4 | 0.2760 | 1.5203 | 1.7963 | 9.0 |
| 10 | 193.19 | 0.01659 | 38.42 | 161.20 | 1072.2 | 161.23 | 982.1 | 1143.3 | 0.2836 | 1.5041 | 1.7877 | 10 |
| 14.696 | 211.99 | 0.01672 | 26.80 | 180.10 | 1077.6 | 180.15 | 970.4 | 1150.5 | 0.3121 | 1.4446 | 1.7567 | 14.696 |
| 15 | 213.03 | 0.01672 | 26.29 | 181.14 | 1077.9 | 181.19 | 969.7 | 1150.9 | 0.3137 | 1.4414 | 1.7551 | 15 |
| 20 | 227.96 | 0.01683 | 20.09 | 196.19 | 1082.0 | 196.26 | 960.1 | 1156.4 | 0.3358 | 1.3962 | 1.7320 | 20 |
| 25 | 240.08 | 0.01692 | 16.31 | 208.44 | 1085.3 | 208.52 | 952.2 | 1160.7 | 0.3535 | 1.3607 | 1.7142 | 25 |
| 30 | 250.34 | 0.01700 | 13.75 | 218.84 | 1088.0 | 218.93 | 945.4 | 1164.3 | 0.3682 | 1.3314 | 1.6996 | 30 |
| 35 | 259.30 | 0.01708 | 11.90 | 227.93 | 1090.3 | 228.04 | 939.3 | 1167.4 | 0.3809 | 1.3064 | 1.6873 | 35 |
| 40 | 267.26 | 0.01715 | 10.50 | 236.03 | 1092.3 | 236.16 | 933.8 | 1170.0 | 0.3921 | 1.2845 | 1.6767 | 40 |
| 45 | 274.46 | 0.01721 | 9.40 | 243.37 | 1094.0 | 243.51 | 928.8 | 1172.3 | 0.4022 | 1.2651 | 1.6673 | 45 |
| 50 | 281.03 | 0.01727 | 8.52 | 250.08 | 1095.6 | 250.24 | 924.2 | 1174.4 | 0.4113 | 1.2476 | 1.6589 | 50 |
| 55 | 287.10 | 0.01733 | 7.79 | 256.28 | 1097.0 | 256.46 | 919.9 | 1176.3 | 0.4196 | 1.2317 | 1.6513 | 55 |

| | | | | | | | | | | | | |
|---|---|---|---|---|---|---|---|---|---|---|---|---|
| 60 | 1.6443 | 1.2170 | 0.4273 | 1173.0 | 915.8 | 262.2 | 1098.3 | 262.1 | 7.177 | 0.01738 | 292.73 | 60 |
| 65 | 1.6380 | 1.2035 | 0.4345 | 1179.6 | 911.9 | 267.7 | 1099.5 | 267.5 | 6.647 | 0.01743 | 298.00 | 65 |
| 70 | 1.6321 | 1.1909 | 0.4412 | 1181.0 | 908.3 | 272.8 | 1100.6 | 272.6 | 6.209 | 0.01748 | 302.96 | 70 |
| 75 | 1.6265 | 1.1790 | 0.4475 | 1182.4 | 904.8 | 277.6 | 1101.6 | 277.4 | 5.818 | 0.01752 | 307.63 | 75 |
| 80 | 1.6213 | 1.1679 | 0.4534 | 1183.6 | 901.4 | 282.2 | 1102.6 | 282.0 | 5.474 | 0.01757 | 312.07 | 80 |
| 85 | 1.6165 | 1.1574 | 0.4591 | 1184.8 | 898.2 | 286.6 | 1103.5 | 286.3 | 5.170 | 0.01761 | 316.29 | 85 |
| 90 | 1.6119 | 1.1475 | 0.4644 | 1185.9 | 895.1 | 290.8 | 1104.3 | 290.5 | 4.898 | 0.01766 | 320.31 | 90 |
| 95 | 1.6075 | 1.1380 | 0.4695 | 1186.9 | 892.1 | 294.8 | 1105.0 | 294.5 | 4.654 | 0.01770 | 324.16 | 95 |
| 100 | 1.6034 | 1.1290 | 0.4744 | 1187.8 | 889.2 | 298.6 | 1105.8 | 298.3 | 4.434 | 0.01774 | 327.86 | 100 |
| 110 | 1.5958 | 1.1122 | 0.4836 | 1189.6 | 883.7 | 305.9 | 1107.1 | 305.5 | 4.051 | 0.01781 | 334.82 | 110 |
| 120 | 1.5886 | 1.0966 | 0.4920 | 1191.1 | 878.5 | 312.7 | 1108.3 | 312.3 | 3.730 | 0.01789 | 341.30 | 120 |
| 130 | 1.5821 | 1.0822 | 0.4999 | 1192.5 | 873.5 | 319.0 | 1109.4 | 318.6 | 3.457 | 0.01796 | 347.37 | 130 |
| 140 | 1.5761 | 1.0688 | 0.5073 | 1193.8 | 868.7 | 325.1 | 1110.3 | 324.6 | 3.221 | 0.01802 | 353.08 | 140 |
| 150 | 1.5704 | 1.0562 | 0.5142 | 1194.9 | 864.2 | 330.8 | 1111.2 | 330.2 | 3.016 | 0.01809 | 358.48 | 150 |
| 160 | 1.5651 | 1.0443 | 0.5208 | 1196.0 | 859.8 | 336.2 | 1112.0 | 335.6 | 2.836 | 0.01815 | 363.60 | 160 |
| 170 | 1.5600 | 1.0330 | 0.5270 | 1196.9 | 855.6 | 341.3 | 1112.7 | 340.8 | 2.676 | 0.01821 | 368.47 | 170 |
| 180 | 1.5552 | 1.0223 | 0.5329 | 1197.8 | 851.5 | 346.3 | 1113.4 | 345.7 | 2.533 | 0.01827 | 373.13 | 180 |
| 190 | 1.5508 | 1.0122 | 0.5386 | 1198.6 | 847.5 | 351.0 | 1114.0 | 350.4 | 2.405 | 0.01833 | 377.59 | 190 |
| 200 | 1.5465 | 1.0025 | 0.5440 | 1199.3 | 843.7 | 355.6 | 1114.6 | 354.9 | 2.289 | 0.01839 | 381.86 | 200 |
| 250 | 1.5274 | 0.9594 | 0.5680 | 1202.1 | 825.8 | 376.2 | 1116.7 | 375.4 | 1.845 | 0.01865 | 401.04 | 250 |
| 300 | 1.5115 | 0.9232 | 0.5883 | 1203.9 | 809.8 | 394.1 | 1118.2 | 393.0 | 1.544 | 0.01890 | 417.43 | 300 |
| 350 | 1.4977 | 0.8917 | 0.6060 | 1204.9 | 795.0 | 409.9 | 1119.0 | 408.7 | 1.327 | 0.01912 | 431.82 | 350 |
| 400 | 1.4856 | 0.8638 | 0.6218 | 1205.5 | 781.2 | 424.2 | 1119.5 | 422.8 | 1.162 | 0.01934 | 444.70 | 400 |
| 450 | 1.4745 | 0.8385 | 0.6360 | 1205.6 | 768.2 | 437.4 | 1119.6 | 435.7 | 1.033 | 0.01955 | 456.39 | 450 |
| 500 | 1.4644 | 0.8154 | 0.6490 | 1205.3 | 755.8 | 449.5 | 1119.4 | 447.7 | 0.928 | 0.01975 | 467.13 | 500 |
| 550 | 1.4551 | 0.7941 | 0.6611 | 1204.8 | 743.9 | 460.9 | 1119.1 | 458.9 | 0.842 | 0.01994 | 477.07 | 550 |
| 600 | 1.4464 | 0.7742 | 0.6723 | 1204.1 | 732.4 | 471.7 | 1118.6 | 469.4 | 0.770 | 0.02013 | 486.33 | 600 |
| 700 | 1.4305 | 0.7378 | 0.6927 | 1202.0 | 710.5 | 491.5 | 1117.0 | 488.9 | 0.656 | 0.02051 | 503.23 | 700 |
| 800 | 1.4160 | 0.7050 | 0.7110 | 1199.3 | 689.6 | 509.7 | 1115.0 | 506.6 | 0.569 | 0.02087 | 518.36 | 800 |
| 900 | 1.4027 | 0.6750 | 0.7277 | 1196.0 | 669.5 | 526.6 | 1112.6 | 523.0 | 0.501 | 0.02123 | 532.12 | 900 |

**Table A-13** (*Continued*)

| Abs. press psi $p$ | Temp., °F $t$ | Specific volume | | Internal energy | | Enthalpy | | | Entropy | | | Abs. press., psi $p$ |
|---|---|---|---|---|---|---|---|---|---|---|---|---|
| | | Sat. liquid $v_f$ | Sat. vapor $v_g$ | Sat. liquid $u_f$ | Sat. vapor $u_g$ | Sat. liquid $h_f$ | Evap. $h_{fg}$ | Sat. vapor $h_g$ | Sat. liquid $s_f$ | Evap. $s_{fg}$ | Sat. vapor $s_g$ | |
| 1000 | 544.75 | 0.02159 | 0.446 | 538.4 | 1109.9 | 542.4 | 650.0 | 1192.4 | 0.7432 | 0.6471 | 1.3903 | 1000 |
| 1100 | 556.45 | 0.02195 | 0.401 | 552.9 | 1106.8 | 557.4 | 631.0 | 1188.3 | 0.7576 | 0.6209 | 1.3786 | 1100 |
| 1200 | 567.37 | 0.02232 | 0.362 | 566.7 | 1103.5 | 571.7 | 612.3 | 1183.9 | 0.7712 | 0.5961 | 1.3673 | 1200 |
| 1300 | 577.60 | 0.02269 | 0.330 | 579.9 | 1099.8 | 585.4 | 593.8 | 1179.2 | 0.7841 | 0.5724 | 1.3565 | 1300 |
| 1400 | 587.25 | 0.02307 | 0.302 | 592.7 | 1096.0 | 598.6 | 575.5 | 1174.1 | 0.7964 | 0.5497 | 1.3461 | 1400 |
| 1500 | 596.39 | 0.02346 | 0.277 | 605.0 | 1091.8 | 611.5 | 557.2 | 1168.7 | 0.8082 | 0.5276 | 1.3359 | 1500 |
| 1600 | 605.06 | 0.02386 | 0.255 | 616.9 | 1087.4 | 624.0 | 538.9 | 1162.9 | 0.8196 | 0.5062 | 1.3258 | 1600 |
| 1700 | 613.32 | 0.02428 | 0.236 | 628.6 | 1082.7 | 636.2 | 520.6 | 1156.9 | 0.8307 | 0.4852 | 1.3159 | 1700 |
| 1800 | 621.21 | 0.02472 | 0.218 | 640.0 | 1077.7 | 648.3 | 502.1 | 1150.4 | 0.8414 | 0.4645 | 1.3060 | 1800 |
| 1900 | 628.76 | 0.02517 | 0.203 | 651.3 | 1072.3 | 660.1 | 483.4 | 1143.5 | 0.8519 | 0.4441 | 1.2961 | 1900 |
| 2000 | 636.00 | 0.02565 | 0.188 | 662.4 | 1066.6 | 671.9 | 464.4 | 1136.3 | 0.8623 | 0.4238 | 1.2861 | 2000 |
| 2250 | 652.90 | 0.02698 | 0.157 | 689.9 | 1050.6 | 701.1 | 414.8 | 1115.9 | 0.8876 | 0.3728 | 1.2604 | 2250 |
| 2500 | 668.31 | 0.02860 | 0.131 | 717.7 | 1031.0 | 730.9 | 360.5 | 1091.4 | 0.9131 | 0.3196 | 1.2327 | 2500 |
| 2750 | 682.46 | 0.03077 | 0.107 | 747.3 | 1005.9 | 763.0 | 297.4 | 1060.4 | 0.9401 | 0.2604 | 1.2005 | 2750 |
| 3000 | 695.52 | 0.03431 | 0.084 | 783.4 | 968.8 | 802.5 | 213.0 | 1015.5 | 0.9732 | 0.1843 | 1.1575 | 3000 |
| 3203.6 | 705.44 | 0.05053 | 0.0505 | 872.6 | 872.6 | 902.5 | 0 | 902.5 | 1.0580 | 0 | 1.0580 | 3203.6 |

*Source:* Keenan, J. H., F. G. Keyes, P. G. Hill, and J. G. Moore, "Steam Tables," Wiley, New York, 1969.

## Table A-14 Properties of water: superheated-vapor table
$v$, ft$^3$/lb; $h$, Btu/lb; $s$, Btu/(lb)($^\circ$R)

| Temp., °F | $v$ | $u$ | $h$ | $s$ | $v$ | $u$ | $h$ | $s$ |
|---|---|---|---|---|---|---|---|---|
| | 1 psia (101.7°F) | | | | 5 psia (162.2°F) | | | |
| Sat. | 333.6 | 1044.0 | 1105.8 | 1.9779 | 73.53 | 1063.0 | 1131.0 | 1.8441 |
| 150 | 362.6 | 1060.4 | 1127.5 | 2.0151 | | | | |
| 200 | 392.5 | 1077.5 | 1150.1 | 2.0508 | 78.15 | 1076.0 | 1148.6 | 1.8715 |
| 250 | 422.4 | 1094.7 | 1172.8 | 2.0839 | 84.21 | 1093.8 | 1171.7 | 1.9052 |
| 300 | 452.3 | 1112.0 | 1195.7 | 2.1150 | 90.24 | 1111.3 | 1194.8 | 1.9367 |
| 400 | 511.9 | 1147.0 | 1241.8 | 2.1720 | 102.24 | 1146.6 | 1241.2 | 1.9941 |
| 500 | 571.5 | 1182.8 | 1288.5 | 2.2235 | 114.20 | 1182.5 | 1288.2 | 2.0458 |
| 600 | 631.1 | 1219.3 | 1336.1 | 2.2706 | 126.15 | 1219.1 | 1335.8 | 2.0930 |
| 700 | 690.7 | 1256.7 | 1384.5 | 2.3142 | 138.08 | 1256.5 | 1384.3 | 2.1367 |
| 800 | 750.3 | 1294.4 | 1433.7 | 2.3550 | 150.01 | 1294.7 | 1433.5 | 2.1775 |
| 900 | 809.9 | 1333.9 | 1483.8 | 2.3932 | 161.94 | 1333.8 | 1483.7 | 2.2158 |
| 1000 | 869.5 | 1373.9 | 1534.8 | 2.4294 | 173.86 | 1373.9 | 1534.7 | 2.2520 |
| | 10 psia (193.2°F) | | | | 14.7 psia (212.0°F) | | | |
| Sat. | 38.42 | 1072.2 | 1143.3 | 1.7877 | 26.80 | 1077.6 | 1150.5 | 1.7567 |
| 200 | 38.85 | 1074.7 | 1146.6 | 1.7927 | | | | |
| 250 | 41.95 | 1092.6 | 1170.2 | 1.8272 | 28.42 | 1091.5 | 1168.8 | 1.7832 |
| 300 | 44.99 | 1110.4 | 1193.7 | 1.8592 | 30.52 | 1109.6 | 1192.6 | 1.8157 |
| 400 | 51.03 | 1146.1 | 1240.5 | 1.9171 | 34.67 | 1145.6 | 1239.9 | 1.8741 |
| 500 | 57.04 | 1182.2 | 1287.7 | 1.9690 | 38.77 | 1181.8 | 1287.3 | 1.9263 |
| 600 | 63.03 | 1218.9 | 1335.5 | 2.0164 | 42.86 | 1218.6 | 1335.2 | 1.9737 |
| 700 | 69.01 | 1256.3 | 1384.0 | 2.0601 | 46.93 | 1256.1 | 1383.8 | 2.0175 |
| 800 | 74.98 | 1294.6 | 1433.3 | 2.1009 | 51.00 | 1294.4 | 1433.1 | 2.0584 |
| 900 | 80.95 | 1333.7 | 1483.5 | 2.1393 | 55.07 | 1333.6 | 1483.4 | 2.0967 |
| 1000 | 86.91 | 1373.8 | 1534.6 | 2.1755 | 59.13 | 1373.7 | 1534.5 | 2.1330 |
| 1100 | 92.88 | 1414.7 | 1586.6 | 2.2099 | 63.19 | 1414.6 | 1586.4 | 2.1674 |
| | 20 psia (228.0°F) | | | | 40 psia (267.3°F) | | | |
| Sat. | 20.09 | 1082.0 | 1156.4 | 1.7320 | 10.50 | 1093.3 | 1170.0 | 1.6767 |
| 250 | 20.79 | 1090.3 | 1167.2 | 1.7475 | | | | |
| 300 | 22.36 | 1108.7 | 1191.5 | 1.7805 | 11.04 | 1105.1 | 1186.8 | 1.6993 |
| 350 | 23.90 | 1126.9 | 1215.4 | 1.8110 | 11.84 | 1124.2 | 1211.8 | 1.7312 |
| 400 | 25.43 | 1145.1 | 1239.2 | 1.8395 | 12.62 | 1143.0 | 1236.4 | 1.7606 |
| 500 | 28.46 | 1181.5 | 1286.8 | 1.8919 | 14.16 | 1180.1 | 1284.9 | 1.8140 |
| 600 | 31.47 | 1218.4 | 1334.8 | 1.9395 | 15.69 | 1217.3 | 1333.4 | 1.8621 |
| 700 | 34.47 | 1255.9 | 1383.5 | 1.9834 | 17.20 | 1255.1 | 1382.4 | 1.9063 |
| 800 | 37.46 | 1294.3 | 1432.9 | 2.0243 | 18.70 | 1293.7 | 1432.1 | 1.9474 |
| 900 | 40.45 | 1333.5 | 1483.2 | 2.0627 | 20.20 | 1333.0 | 1482.5 | 1.9859 |
| 1000 | 43.44 | 1373.5 | 1534.3 | 2.0989 | 21.70 | 1373.1 | 1533.8 | 2.0223 |
| 1100 | 46.42 | 1414.5 | 1586.3 | 2.1334 | 23.20 | 1414.2 | 1585.9 | 2.0568 |

## Table A-14 Properties of water: superheated-vapor table *(Continued)*

| Temp., °F | $v$ | $u$ | $h$ | $s$ | $v$ | $u$ | $h$ | $s$ |
|---|---|---|---|---|---|---|---|---|
| | 60 psia (292.7°F) | | | | 80 psia (312.1°F) | | | |
| Sat. | 7.17 | 1098.3 | 1178.0 | 1.6444 | 5.47 | 1102.6 | 1183.6 | 1.6214 |
| 300 | 7.26 | 1101.3 | 1181.9 | 1.6496 | | | | |
| 350 | 7.82 | 1121.4 | 1208.2 | 1.6830 | 5.80 | 1118.5 | 1204.3 | 1.6476 |
| 400 | 8.35 | 1140.8 | 1233.5 | 1.7134 | 6.22 | 1138.5 | 1230.6 | 1.6790 |
| 500 | 9.40 | 1178.6 | 1283.0 | 1.7678 | 7.02 | 1177.2 | 1281.1 | 1.7346 |
| 600 | 10.43 | 1216.3 | 1332.1 | 1.8165 | 7.79 | 1215.3 | 1330.7 | 1.7838 |
| 700 | 11.44 | 1254.4 | 1381.4 | 1.8609 | 8.56 | 1253.6 | 1380.3 | 1.8285 |
| 800 | 12.45 | 1293.0 | 1431.2 | 1.9022 | 9.32 | 1292.4 | 1430.4 | 1.8700 |
| 900 | 13.45 | 1332.5 | 1481.8 | 1.9408 | 10.08 | 1332.0 | 1481.2 | 1.9087 |
| 1000 | 14.45 | 1372.7 | 1533.2 | 1.9773 | 10.83 | 1372.3 | 1532.6 | 1.9453 |
| 1100 | 15.45 | 1413.8 | 1585.4 | 2.0119 | 11.58 | 1413.5 | 1584.9 | 1.9799 |
| 1200 | 16.45 | 1455.8 | 1638.5 | 2.0448 | 12.33 | 1455.5 | 1638.1 | 2.0130 |
| | 100 psia (327.8°F) | | | | 120 psia (341.3°F) | | | |
| Sat. | 4.434 | 1105.8 | 1187.8 | 1.6034 | 3.730 | 1108.3 | 1191.1 | 1.5886 |
| 350 | 4.592 | 1115.4 | 1200.4 | 1.6191 | 3.783 | 1112.2 | 1196.2 | 1.5950 |
| 400 | 4.934 | 1136.2 | 1227.5 | 1.6517 | 4.079 | 1133.8 | 1224.4 | 1.6288 |
| 450 | 5.265 | 1156.2 | 1253.6 | 1.6812 | 4.360 | 1154.3 | 1251.2 | 1.6590 |
| 500 | 5.587 | 1175.7 | 1279.1 | 1.7085 | 4.633 | 1174.2 | 1277.1 | 1.6868 |
| 600 | 6.216 | 1214.2 | 1329.3 | 1.7582 | 5.164 | 1213.2 | 1327.8 | 1.7371 |
| 700 | 6.834 | 1252.8 | 1379.2 | 1.8033 | 5.682 | 1252.0 | 1378.2 | 1.7825 |
| 800 | 7.445 | 1291.8 | 1429.6 | 1.8449 | 6.195 | 1291.2 | 1428.7 | 1.8243 |
| 900 | 8.053 | 1331.5 | 1480.5 | 1.8838 | 6.703 | 1330.9 | 1479.8 | 1.8633 |
| 1000 | 8.657 | 1371.9 | 1532.1 | 1.9204 | 7.208 | 1371.5 | 1531.5 | 1.9000 |
| 1100 | 9.260 | 1413.1 | 1584.5 | 1.9551 | 7.711 | 1412.8 | 1584.0 | 1.9348 |
| 1200 | 9.861 | 1455.2 | 1637.7 | 1.9882 | 8.213 | 1454.9 | 1637.3 | 1.9679 |
| | 140 psia (353.1°F) | | | | 160 psia (363.6°F) | | | |
| Sat. | 3.221 | 1110.3 | 1193.8 | 1.5761 | 2.836 | 1112.0 | 1196.0 | 1.5651 |
| 400 | 3.466 | 1131.4 | 1221.2 | 1.6088 | 3.007 | 1128.8 | 1217.8 | 1.5911 |
| 450 | 3.713 | 1152.4 | 1248.6 | 1.6399 | 3.228 | 1150.5 | 1246.1 | 1.6230 |
| 500 | 3.952 | 1172.7 | 1275.1 | 1.6682 | 3.440 | 1171.2 | 1273.0 | 1.6518 |
| 550 | 4.184 | 1192.5 | 1300.9 | 1.6945 | 3.646 | 1191.3 | 1299.2 | 1.6785 |
| 600 | 4.412 | 1212.1 | 1326.4 | 1.7191 | 3.848 | 1211.1 | 1325.0 | 1.7034 |
| 700 | 4.860 | 1251.2 | 1377.1 | 1.7648 | 4.243 | 1250.4 | 1376.0 | 1.7494 |
| 800 | 5.301 | 1290.5 | 1427.9 | 1.8068 | 4.631 | 1289.9 | 1427.0 | 1.7916 |
| 900 | 5.739 | 1330.4 | 1479.1 | 1.8459 | 5.015 | 1329.9 | 1478.4 | 1.8308 |
| 1000 | 6.173 | 1371.0 | 1531.0 | 1.8827 | 5.397 | 1370.6 | 1530.4 | 1.8677 |
| 1100 | 6.605 | 1412.4 | 1583.6 | 1.9176 | 5.776 | 1412.1 | 1583.1 | 1.9026 |
| 1200 | 7.036 | 1454.6 | 1636.9 | 1.9507 | 6.154 | 1454.3 | 1636.5 | 1.9358 |

## Table A-14 Properties of water: superheated-vapor table  *(Continued)*

| Temp., °F | $v$ | $u$ | $h$ | $s$ | $v$ | $u$ | $h$ | $s$ |
|---|---|---|---|---|---|---|---|---|
| | 180 psia (373.1°F) | | | | 200 psia (381.8°F) | | | |
| Sat. | 2.533 | 1113.4 | 1197.8 | 1.5553 | 2.289 | 1114.6 | 1199.3 | 1.5464 |
| 400 | 2.648 | 1126.2 | 1214.4 | 1.5749 | 2.361 | 1123.5 | 1210.8 | 1.5600 |
| 450 | 2.850 | 1148.5 | 1243.4 | 1.6078 | 2.548 | 1146.4 | 1240.7 | 1.5938 |
| 500 | 3.042 | 1169.6 | 1270.9 | 1.6372 | 2.724 | 1168.0 | 1268.8 | 1.6239 |
| 550 | 3.228 | 1190.0 | 1297.5 | 1.6642 | 2.893 | 1188.7 | 1295.7 | 1.6512 |
| 600 | 3.409 | 1210.0 | 1323.5 | 1.6893 | 3.058 | 1208.9 | 1322.1 | 1.6767 |
| 700 | 3.763 | 1249.6 | 1374.9 | 1.7357 | 3.379 | 1248.8 | 1373.8 | 1.7234 |
| 800 | 4.110 | 1289.3 | 1426.2 | 1.7781 | 3.693 | 1288.6 | 1425.3 | 1.7660 |
| 900 | 4.453 | 1329.4 | 1477.7 | 1.8174 | 4.003 | 1328.9 | 1477.1 | 1.8055 |
| 1000 | 4.793 | 1370.2 | 1529.8 | 1.8545 | 4.310 | 1369.8 | 1529.3 | 1.8425 |
| 1100 | 5.131 | 1411.7 | 1582.6 | 1.8894 | 4.615 | 1411.4 | 1582.2 | 1.8776 |
| 1200 | 5.467 | 1454.0 | 1636.1 | 1.9227 | 4.918 | 1453.7 | 1635.7 | 1.9109 |
| | 250 psia (401.0°F) | | | | 300 psia (417.4°F) | | | |
| Sat. | 1.845 | 1116.7 | 1202.1 | 1.5274 | 1.544 | 1118.2 | 1203.9 | 1.5115 |
| 450 | 2.002 | 1141.1 | 1233.7 | 1.5632 | 1.636 | 1135.4 | 1226.2 | 1.5365 |
| 500 | 2.150 | 1163.8 | 1263.3 | 1.5948 | 1.766 | 1159.5 | 1257.5 | 1.5701 |
| 550 | 2.290 | 1185.3 | 1291.3 | 1.6233 | 1.888 | 1181.9 | 1286.7 | 1.5997 |
| 600 | 2.426 | 1206.1 | 1318.3 | 1.6494 | 2.004 | 1203.2 | 1314.5 | 1.6266 |
| 700 | 2.688 | 1246.7 | 1371.1 | 1.6970 | 2.227 | 1244.0 | 1368.3 | 1.6751 |
| 800 | 2.943 | 1287.0 | 1423.2 | 1.7301 | 2.442 | 1285.4 | 1421.0 | 1.7187 |
| 900 | 3.193 | 1327.6 | 1475.3 | 1.7799 | 2.653 | 1326.3 | 1473.6 | 1.7589 |
| 1000 | 3.440 | 1368.7 | 1527.9 | 1.8172 | 2.860 | 1367.7 | 1526.5 | 1.7964 |
| 1100 | 3.685 | 1410.5 | 1581.0 | 1.8524 | 3.066 | 1409.6 | 1579.8 | 1.8317 |
| 1200 | 3.929 | 1453.0 | 1634.8 | 1.8858 | 3.270 | 1452.2 | 1633.8 | 1.8653 |
| 1300 | 4.172 | 1496.3 | 1689.3 | 1.9177 | 3.473 | 1495.6 | 1688.4 | 1.8973 |
| | 350 psia (431.8°F) | | | | 400 psia (444.7°F) | | | |
| Sat. | 1.327 | 1119.0 | 1204.9 | 1.4978 | 1.162 | 1119.5 | 1205.5 | 1.4856 |
| 450 | 1.373 | 1129.2 | 1218.2 | 1.5125 | 1.175 | 1122.6 | 1209.6 | 1.4901 |
| 500 | 1.491 | 1154.9 | 1251.5 | 1.5482 | 1.284 | 1150.1 | 1245.2 | 1.5282 |
| 550 | 1.600 | 1178.3 | 1281.9 | 1.5790 | 1.383 | 1174.6 | 1277.0 | 1.5605 |
| 600 | 1.703 | 1200.3 | 1310.6 | 1.6068 | 1.476 | 1197.3 | 1306.6 | 1.5892 |
| 700 | 1.898 | 1242.5 | 1365.4 | 1.6562 | 1.650 | 1240.4 | 1362.5 | 1.6397 |
| 800 | 2.085 | 1283.8 | 1418.8 | 1.7004 | 1.816 | 1282.1 | 1416.6 | 1.6844 |
| 900 | 2.267 | 1325.0 | 1471.8 | 1.7409 | 1.978 | 1323.7 | 1470.1 | 1.7252 |
| 1000 | 2.446 | 1366.6 | 1525.0 | 1.7787 | 2.136 | 1365.5 | 1523.6 | 1.7632 |
| 1100 | 2.624 | 1408.7 | 1578.6 | 1.8142 | 2.292 | 1407.8 | 1577.4 | 1.7989 |
| 1200 | 2.799 | 1451.5 | 1632.8 | 1.8478 | 2.446 | 1450.7 | 1631.8 | 1.8327 |
| 1300 | 2.974 | 1495.0 | 1687.6 | 1.8799 | 2.599 | 1494.3 | 1686.8 | 1.8648 |

## Table A-14 Properties of water: superheated-vapor table  (*Continued*)

| Temp., °F | $v$ | $u$ | $h$ | $s$ | $v$ | $u$ | $h$ | $s$ |
|---|---|---|---|---|---|---|---|---|
| | 450 psia (456.4°F) | | | | 500 psia (467.1°F) | | | |
| Sat. | 1.033 | 1119.6 | 1205.6 | 1.4746 | 0.928 | 1119.4 | 1205.3 | 1.4645 |
| 500 | 1.123 | 1145.1 | 1238.5 | 1.5097 | 0.992 | 1139.7 | 1231.5 | 1.4923 |
| 550 | 1.215 | 1170.7 | 1271.9 | 1.5436 | 1.079 | 1166.7 | 1266.6 | 1.5279 |
| 600 | 1.300 | 1194.3 | 1302.5 | 1.5732 | 1.158 | 1191.1 | 1298.3 | 1.5585 |
| 700 | 1.458 | 1238.2 | 1359.6 | 1.6248 | 1.304 | 1236.0 | 1356.7 | 1.6112 |
| 800 | 1.608 | 1280.5 | 1414.4 | 1.6701 | 1.441 | 1278.8 | 1412.1 | 1.6571 |
| 900 | 1.752 | 1322.4 | 1468.3 | 1.7113 | 1.572 | 1321.0 | 1466.5 | 1.6987 |
| 1000 | 1.894 | 1364.4 | 1522.2 | 1.7495 | 1.701 | 1363.3 | 1520.7 | 1.7471 |
| 1100 | 2.034 | 1406.9 | 1576.3 | 1.7853 | 1.827 | 1406.0 | 1575.1 | 1.7731 |
| 1200 | 2.172 | 1450.0 | 1630.8 | 1.8192 | 1.952 | 1449.2 | 1629.8 | 1.8072 |
| 1300 | 2.308 | 1493.7 | 1685.9 | 1.8515 | 2.075 | 1493.1 | 1685.1 | 1.8395 |
| 1400 | 2.444 | 1538.1 | 1741.7 | 1.8823 | 2.198 | 1537.6 | 1741.0 | 1.8704 |
| | 600 psia (486.3°F) | | | | 700 psia (503.2°F) | | | |
| Sat. | 0.770 | 1118.6 | 1204.1 | 1.4464 | 0.656 | 1117.0 | 1202.0 | 1.4305 |
| 500 | 0.795 | 1128.0 | 1216.2 | 1.4592 | | | | |
| 550 | 0.875 | 1158.2 | 1255.4 | 1.4990 | 0.728 | 1149.0 | 1243.2 | 1.4723 |
| 600 | 0.946 | 1184.5 | 1289.5 | 1.5320 | 0.793 | 1177.5 | 1280.2 | 1.5081 |
| 700 | 1.073 | 1231.5 | 1350.6 | 1.5872 | 0.907 | 1226.9 | 1344.4 | 1.5661 |
| 800 | 1.190 | 1275.4 | 1407.6 | 1.6343 | 1.011 | 1272.0 | 1402.9 | 1.6145 |
| 900 | 1.302 | 1318.4 | 1462.9 | 1.6766 | 1.109 | 1315.6 | 1459.3 | 1.6576 |
| 1000 | 1.411 | 1361.2 | 1517.8 | 1.7155 | 1.204 | 1358.9 | 1514.9 | 1.6970 |
| 1100 | 1.517 | 1404.2 | 1572.7 | 1.7519 | 1.296 | 1402.4 | 1570.2 | 1.7337 |
| 1200 | 1.622 | 1447.7 | 1627.8 | 1.7861 | 1.387 | 1446.2 | 1625.8 | 1.7682 |
| 1300 | 1.726 | 1491.7 | 1683.4 | 1.8186 | 1.476 | 1490.4 | 1681.7 | 1.8009 |
| 1400 | 1.829 | 1536.5 | 1739.5 | 1.8497 | 1.565 | 1535.3 | 1738.1 | 1.8321 |
| | 800 psia (518.3°F) | | | | 900 psia (532.1°F) | | | |
| Sat. | 0.569 | 1115.0 | 1199.3 | 1.4160 | 0.501 | 1112.6 | 1196.0 | 1.4027 |
| 550 | 0.615 | 1138.8 | 1229.9 | 1.4469 | 0.527 | 1127.5 | 1215.2 | 1.4219 |
| 600 | 0.677 | 1170.1 | 1270.4 | 1.4861 | 0.587 | 1162.2 | 1260.0 | 1.4652 |
| 650 | 0.732 | 1197.2 | 1305.6 | 1.5186 | 0.639 | 1191.1 | 1297.5 | 1.4999 |
| 700 | 0.783 | 1222.1 | 1338.0 | 1.5471 | 0.686 | 1217.1 | 1331.4 | 1.5297 |
| 800 | 0.876 | 1268.5 | 1398.2 | 1.5969 | 0.772 | 1264.9 | 1393.4 | 1.5810 |
| 900 | 0.964 | 1312.9 | 1455.6 | 1.6408 | 0.851 | 1310.1 | 1451.9 | 1.6257 |
| 1000 | 1.048 | 1356.7 | 1511.9 | 1.6807 | 0.927 | 1354.5 | 1508.9 | 1.6662 |
| 1100 | 1.130 | 1400.5 | 1567.8 | 1.7178 | 1.001 | 1398.7 | 1565.4 | 1.7036 |
| 1200 | 1.210 | 1444.6 | 1623.8 | 1.7526 | 1.073 | 1443.0 | 1621.7 | 1.7386 |
| 1300 | 1.289 | 1489.1 | 1680.0 | 1.7854 | 1.144 | 1487.8 | 1687.3 | 1.7717 |
| 1400 | 1.367 | 1534.2 | 1736.6 | 1.8167 | 1.214 | 1533.0 | 1735.1 | 1.8031 |

## Table A-14 Properties of water: superheated-vapor table (*Continued*)

| Temp., °F | v | u | h | s | v | u | h | s |
|---|---|---|---|---|---|---|---|---|
| | | 1000 psia | (544.7°F) | | | 1200 psia | (567.4°F) | |
| Sat. | 0.446 | 1109.0 | 1192.4 | 1.3903 | 0.362 | 1103.5 | 1183.9 | 1.3673 |
| 600 | 0.514 | 1153.7 | 1248.8 | 1.4450 | 0.402 | 1134.4 | 1223.6 | 1.4054 |
| 650 | 0.564 | 1184.7 | 1289.1 | 1.4822 | 0.450 | 1170.9 | 1270.8 | 1.4490 |
| 700 | 0.608 | 1212.0 | 1324.6 | 1.5135 | 0.491 | 1201.3 | 1310.2 | 1.4837 |
| 800 | 0.688 | 1261.2 | 1388.5 | 1.5665 | 0.562 | 1253.7 | 1378.4 | 1.5402 |
| 900 | 0.761 | 1307.3 | 1448.1 | 1.6120 | 0.626 | 1301.5 | 1440.4 | 1.5876 |
| 1000 | 0.831 | 1352.2 | 1505.9 | 1.6530 | 0.685 | 1347.5 | 1499.7 | 1.6297 |
| 1100 | 0.898 | 1396.8 | 1562.9 | 1.6908 | 0.743 | 1393.0 | 1557.9 | 1.6682 |
| 1200 | 0.963 | 1441.5 | 1619.7 | 1.7261 | 0.798 | 1438.3 | 1615.5 | 1.7040 |
| 1300 | 1.027 | 1486.5 | 1676.5 | 1.7593 | 0.853 | 1483.8 | 1673.1 | 1.7377 |
| 1400 | 1.091 | 1531.9 | 1733.7 | 1.7909 | 0.906 | 1529.6 | 1730.7 | 1.7696 |
| 1600 | 1.215 | 1624.4 | 1849.3 | 1.8499 | 1.011 | 1622.6 | 1847.1 | 1.8290 |
| | | 1400 psia | (587.2°F) | | | 1600 psia | (605.1°F) | |
| Sat. | 0.302 | 1096.0 | 1174.1 | 1.3461 | 0.255 | 1087.4 | 1162.9 | 1.3258 |
| 600 | 0.318 | 1110.9 | 1193.1 | 1.3641 | | | | |
| 650 | 0.367 | 1155.5 | 1250.5 | 1.4171 | 0.303 | 1137.8 | 1227.4 | 1.3852 |
| 700 | 0.406 | 1189.6 | 1294.8 | 1.4562 | 0.342 | 1177.0 | 1278.1 | 1.4299 |
| 800 | 0.471 | 1245.8 | 1367.9 | 1.5168 | 0.403 | 1237.7 | 1357.0 | 1.4953 |
| 900 | 0.529 | 1295.6 | 1432.5 | 1.5661 | 0.456 | 1289.5 | 1424.4 | 1.5468 |
| 1000 | 0.582 | 1342.8 | 1493.5 | 1.6094 | 0.504 | 1338.0 | 1487.1 | 1.5913 |
| 1100 | 0.632 | 1389.1 | 1552.8 | 1.6487 | 0.549 | 1385.2 | 1547.7 | 1.6315 |
| 1200 | 0.681 | 1435.1 | 1611.4 | 1.6851 | 0.592 | 1431.8 | 1607.1 | 1.6684 |
| 1300 | 0.728 | 1481.1 | 1669.6 | 1.7192 | 0.634 | 1478.3 | 1666.1 | 1.7029 |
| 1400 | 0.774 | 1527.2 | 1727.8 | 1.7513 | 0.675 | 1524.9 | 1724.8 | 1.7354 |
| 1600 | 0.865 | 1620.8 | 1844.8 | 1.8111 | 0.755 | 1619.0 | 1842.6 | 1.7955 |
| | | 1800 psia | (621.2°F) | | | 2000 psia | (636.0°F) | |
| Sat. | 0.218 | 1077.7 | 1150.4 | 1.3060 | 0.188 | 1066.6 | 1136.3 | 1.2861 |
| 650 | 0.251 | 1117.0 | 1200.4 | 1.3517 | 0.206 | 1091.1 | 1167.2 | 1.3141 |
| 700 | 0.291 | 1163.1 | 1259.9 | 1.4042 | 0.249 | 1147.7 | 1239.8 | 1.3782 |
| 750 | 0.322 | 1198.6 | 1305.9 | 1.4430 | 0.280 | 1187.3 | 1291.1 | 1.4216 |
| 800 | 0.350 | 1229.1 | 1345.7 | 1.4753 | 0.307 | 1220.1 | 1333.8 | 1.4562 |
| 900 | 0.399 | 1283.2 | 1416.1 | 1.5291 | 0.353 | 1276.8 | 1407.6 | 1.5126 |
| 1000 | 0.443 | 1333.1 | 1480.7 | 1.5749 | 0.395 | 1328.1 | 1474.1 | 1.5598 |
| 1100 | 0.484 | 1381.2 | 1542.5 | 1.6159 | 0.433 | 1377.2 | 1537.2 | 1.6017 |
| 1200 | 0.524 | 1428.5 | 1602.9 | 1.6534 | 0.469 | 1425.2 | 1598.6 | 1.6398 |
| 1300 | 0.561 | 1475.5 | 1662.5 | 1.6883 | 0.503 | 1472.7 | 1659.0 | 1.6751 |
| 1400 | 0.598 | 1522.5 | 1721.8 | 1.7211 | 0.537 | 1520.2 | 1718.8 | 1.7082 |
| 1600 | 0.670 | 1617.2 | 1840.4 | 1.7817 | 0.602 | 1615.4 | 1838.2 | 1.7692 |

## Table A-14 Properties of water: superheated-vapor table (*Continued*)

| Temp., °F | $v$ | $u$ | $h$ | $s$ | $v$ | $u$ | $h$ | $s$ |
|---|---|---|---|---|---|---|---|---|
| | 2500 psia (668.3°F) | | | | 3000 psia (695.5°F) | | | |
| Sat. | 0.1306 | 1031.0 | 1091.4 | 1.2327 | 0.0840 | 968.8 | 1015.5 | 1.1575 |
| 700 | 0.1684 | 1098.7 | 1176.6 | 1.3073 | 0.0977 | 1003.9 | 1058.1 | 1.1944 |
| 750 | 0.2030 | 1155.2 | 1249.1 | 1.3686 | 0.1483 | 1114.7 | 1197.1 | 1.3122 |
| 800 | 0.2291 | 1195.7 | 1301.7 | 1.4112 | 0.1757 | 1167.6 | 1265.2 | 1.3675 |
| 900 | 0.2712 | 1259.9 | 1385.4 | 1.4752 | 0.2160 | 1241.8 | 1361.7 | 1.4414 |
| 1000 | 0.3069 | 1315.2 | 1457.2 | 1.5262 | 0.2485 | 1301.7 | 1439.6 | 1.4967 |
| 1100 | 0.3393 | 1366.8 | 1523.8 | 1.5704 | 0.2772 | 1356.2 | 1510.1 | 1.5434 |
| 1200 | 0.3696 | 1416.7 | 1587.7 | 1.6101 | 0.3086 | 1408.0 | 1576.6 | 1.5848 |
| 1300 | 0.3984 | 1465.7 | 1650.0 | 1.6465 | 0.3285 | 1458.5 | 1640.9 | 1.6224 |
| 1400 | 0.4261 | 1514.2 | 1711.3 | 1.6804 | 0.3524 | 1508.1 | 1703.7 | 1.6571 |
| 1500 | 0.4531 | 1562.5 | 1772.1 | 1.7123 | 0.3754 | 1557.3 | 1765.7 | 1.6896 |
| 1600 | 0.4795 | 1610.8 | 1832.6 | 1.7424 | 0.3978 | 1606.3 | 1827.1 | 1.7201 |
| | 3500 psia | | | | 4000 psia | | | |
| 650 | 0.0249 | 663.5 | 679.7 | 0.8630 | 0.0245 | 657.7 | 675.8 | 0.8574 |
| 700 | 0.0306 | 759.5 | 779.3 | 0.9506 | 0.0287 | 742.1 | 763.4 | 0.9345 |
| 750 | 0.1046 | 1058.4 | 1126.1 | 1.2440 | 0.0633 | 960.7 | 1007.5 | 1.1395 |
| 800 | 0.1363 | 1134.7 | 1223.0 | 1.3226 | 0.1052 | 1095.0 | 1172.9 | 1.2740 |
| 900 | 0.1763 | 1222.4 | 1336.5 | 1.4096 | 0.1462 | 1201.5 | 1309.7 | 1.3789 |
| 1000 | 0.2066 | 1287.6 | 1421.4 | 1.4699 | 0.1752 | 1272.9 | 1402.6 | 1.4449 |
| 1100 | 0.2328 | 1345.2 | 1496.0 | 1.5193 | 0.1995 | 1333.9 | 1481.6 | 1.4973 |
| 1200 | 0.2566 | 1399.2 | 1565.3 | 1.5624 | 0.2213 | 1390.1 | 1553.9 | 1.5423 |
| 1300 | 0.2787 | 1451.1 | 1631.7 | 1.6012 | 0.2414 | 1443.7 | 1622.4 | 1.5823 |
| 1400 | 0.2997 | 1501.9 | 1696.1 | 1.6368 | 0.2603 | 1495.7 | 1688.4 | 1.6188 |
| 1500 | 0.3199 | 1552.0 | 1759.2 | 1.6699 | 0.2784 | 1546.7 | 1752.8 | 1.6526 |
| 1600 | 0.3395 | 1601.7 | 1821.6 | 1.7010 | 0.2959 | 1597.1 | 1816.1 | 1.6841 |
| | 4400 psia | | | | 4800 psia | | | |
| 650 | 0.0242 | 653.6 | 673.3 | 0.8535 | 0.0237 | 649.8 | 671.0 | 0.8499 |
| 700 | 0.0278 | 732.7 | 755.3 | 0.9257 | 0.0271 | 725.1 | 749.1 | 0.9187 |
| 750 | 0.0415 | 870.8 | 904.6 | 1.0513 | 0.0352 | 832.6 | 863.9 | 1.0154 |
| 800 | 0.0844 | 1056.5 | 1125.3 | 1.2306 | 0.0668 | 1011.2 | 1070.5 | 1.1827 |
| 900 | 0.1270 | 1183.7 | 1287.1 | 1.3548 | 0.1109 | 1164.8 | 1263.4 | 1.3310 |
| 1000 | 0.1552 | 1260.8 | 1387.2 | 1.4260 | 0.1385 | 1248.3 | 1317.4 | 1.4078 |
| 1100 | 0.1784 | 1324.7 | 1469.9 | 1.4809 | 0.1608 | 1315.3 | 1458.1 | 1.4653 |
| 1200 | 0.1989 | 1382.8 | 1544.7 | 1.5274 | 0.1802 | 1375.4 | 1535.4 | 1.5133 |
| 1300 | 0.2176 | 1437.7 | 1614.9 | 1.5685 | 0.1979 | 1431.7 | 1607.4 | 1.5555 |
| 1400 | 0.2352 | 1490.7 | 1682.3 | 1.6057 | 0.2143 | 1485.7 | 1676.1 | 1.5934 |
| 1500 | 0.2520 | 1542.7 | 1747.6 | 1.6399 | 0.2300 | 1538.2 | 1742.5 | 1.6282 |
| 1600 | 0.2681 | 1593.4 | 1811.7 | 1.6718 | 0.2450 | 1589.8 | 1807.4 | 1.6605 |

*Source:* Keenan, J. H., F. G. Keyes, P. G. Hill, and J. G. Moore, "Steam Tables," Wiley, New York, 1969.

## Table A-15 Properties of water: compressed liquid table
$v$, ft$^3$/lb; $h$, Btu/lb; $s$, Btu/(lb)($^\circ$R)

| Temp., $^\circ$F | 500 psia ($T_{sat} = 467.1^\circ$F) | | | | 1000 psia ($T_{sat} = 544.7^\circ$F) | | | |
|---|---|---|---|---|---|---|---|---|
| | $v$ | $u$ | $h$ | $s$ | $v$ | $u$ | $h$ | $s$ |
| 32 | 0.015994 | 0.00 | 1.49 | 0.00000 | 0.015967 | 0.03 | 2.99 | 0.00005 |
| 50 | 0.015998 | 18.02 | 19.50 | 0.03599 | 0.015972 | 17.99 | 20.94 | 0.03592 |
| 100 | 0.016106 | 67.87 | 69.36 | 0.12932 | 0.016082 | 67.70 | 70.68 | 0.12901 |
| 150 | 0.016318 | 117.66 | 119.17 | 0.21457 | 0.016293 | 117.38 | 120.40 | 0.21410 |
| 200 | 0.016608 | 167.65 | 169.19 | 0.29341 | 0.016580 | 167.26 | 170.32 | 0.29281 |
| 300 | 0.017416 | 268.92 | 270.53 | 0.43641 | 0.017379 | 268.24 | 271.46 | 0.43552 |
| 400 | 0.018608 | 373.68 | 375.40 | 0.56604 | 0.018550 | 372.55 | 375.98 | 0.56472 |
| Sat. | 0.019748 | 447.70 | 449.53 | 0.64904 | 0.021591 | 538.39 | 542.38 | 0.74320 |

| Temp., $^\circ$F | 1500 psia ($T_{sat} = 596.4^\circ$F) | | | | 2000 psia ($T_{sat} = 636.0^\circ$F) | | | |
|---|---|---|---|---|---|---|---|---|
| 32 | 0.015939 | 0.05 | 4.47 | 0.00007 | 0.015912 | 0.06 | 5.95 | 0.00008 |
| 50 | 0.015946 | 17.95 | 22.38 | 0.03584 | 0.015920 | 17.91 | 23.81 | 0.03575 |
| 100 | 0.016058 | 67.53 | 71.99 | 0.12870 | 0.016034 | 67.37 | 73.30 | 0.12839 |
| 150 | 0.016268 | 117.10 | 121.62 | 0.21364 | 0.016244 | 116.83 | 122.84 | 0.21318 |
| 200 | 0.016554 | 166.87 | 171.46 | 0.29221 | 0.016527 | 166.49 | 172.60 | 0.29162 |
| 300 | 0.017343 | 267.58 | 272.39 | 0.43463 | 0.017308 | 266.93 | 273.33 | 0.43376 |
| 400 | 0.018493 | 371.45 | 376.59 | 0.56343 | 0.018439 | 370.38 | 377.21 | 0.56216 |
| 500 | 0.02024 | 481.8 | 487.4 | 0.6853 | 0.02014 | 479.8 | 487.3 | 0.6832 |
| Sat. | 0.02346 | 605.0 | 611.5 | 0.8082 | 0.02565 | 662.4 | 671.9 | 0.8623 |

| Temp., $^\circ$F | 3000 psia ($T_{sat} = 695.5^\circ$F) | | | | 4000 psia | | | |
|---|---|---|---|---|---|---|---|---|
| 32 | 0.015859 | 0.09 | 8.90 | 0.00009 | 0.015807 | 0.10 | 11.80 | 0.00005 |
| 50 | 0.015870 | 17.84 | 26.65 | 0.03555 | 0.015821 | 17.76 | 29.47 | 0.03534 |
| 100 | 0.015987 | 67.04 | 75.91 | 0.12777 | 0.015942 | 66.72 | 78.52 | 0.12714 |
| 150 | 0.016196 | 116.30 | 125.29 | 0.21226 | 0.016150 | 115.77 | 127.73 | 0.21136 |
| 200 | 0.016476 | 165.74 | 174.89 | 0.29046 | 0.016425 | 165.02 | 177.18 | 0.28931 |
| 300 | 0.017240 | 265.66 | 275.23 | 0.43205 | 0.017174 | 264.43 | 277.15 | 0.43038 |
| 400 | 0.018334 | 368.32 | 378.50 | 0.55970 | 0.018235 | 366.35 | 379.85 | 0.55734 |
| 500 | 0.019944 | 476.2 | 487.3 | 0.6794 | 0.019766 | 472.9 | 487.5 | 0.6758 |
| Sat. | 0.034310 | 783.5 | 802.5 | 0.9732 | | | | |

*Source:* Keenan, J. H., F. G. Keyes, P. G. Hill, and J. G. Moore, "Steam Tables," Wiley, New York, 1969.

### Table A-16  Properties of saturated refrigerant 12 ($CCl_2F_2$): temperature table
$v$, ft$^3$/lb; $u$, Btu/lb; $h$, Btu/lb; $s$, Btu/(lb)($°$R)

| Temp. °F $T$ | Press. psi $P$ | Specific volume | | Internal energy | | Enthalpy | | | Entropy | |
|---|---|---|---|---|---|---|---|---|---|---|
| | | Sat. liquid $v_f$ | Sat. vapor $v_g$ | Sat. liquid $u_f$ | Sat. vapor $u_g$ | Sat. liquid $h_f$ | Evap. $h_{fg}$ | Sat. vapor $h_g$ | Sat. liquid $s_f$ | Sat. vapor $s_g$ |
| −40 | 9.308 | 0.01056 | 3.8750 | −0.02 | 66.24 | 0 | 72.91 | 72.91 | 0 | 0.1737 |
| −30 | 11.999 | 0.01067 | 3.0585 | 1.93 | 67.22 | 2.11 | 71.90 | 74.01 | 0.0050 | 0.1723 |
| −20 | 15.267 | 0.01079 | 2.4429 | 4.21 | 68.21 | 4.24 | 70.87 | 75.11 | 0.0098 | 0.1710 |
| −10 | 19.189 | 0.01091 | 1.9727 | 6.33 | 69.19 | 6.37 | 69.82 | 76.19 | 0.0146 | 0.1699 |
| 0 | 23.849 | 0.01103 | 1.6089 | 8.47 | 70.17 | 8.52 | 68.75 | 77.27 | 0.0193 | 0.1689 |
| 10 | 29.335 | 0.01116 | 1.3241 | 10.62 | 71.15 | 10.68 | 67.65 | 78.33 | 0.0240 | 0.1680 |
| 20 | 35.736 | 0.01130 | 1.0988 | 12.79 | 72.12 | 12.86 | 66.52 | 79.38 | 0.0285 | 0.1672 |
| 30 | 43.148 | 0.01144 | 0.9188 | 14.97 | 73.08 | 15.06 | 65.36 | 80.42 | 0.0330 | 0.1665 |
| 40 | 51.667 | 0.01159 | 0.7736 | 17.16 | 74.04 | 17.27 | 64.16 | 81.43 | 0.0375 | 0.1659 |
| 50 | 61.394 | 0.01175 | 0.6554 | 19.51 | 74.99 | 19.51 | 62.93 | 82.44 | 0.0418 | 0.1653 |
| 60 | 72.433 | 0.01191 | 0.5584 | 21.61 | 75.92 | 21.77 | 61.64 | 83.41 | 0.0462 | 0.1648 |
| 70 | 84.888 | 0.01209 | 0.4782 | 23.86 | 76.85 | 24.05 | 60.31 | 84.36 | 0.0505 | 0.1643 |
| 80 | 98.870 | 0.01228 | 0.4114 | 26.14 | 77.76 | 26.37 | 58.92 | 85.29 | 0.0548 | 0.1639 |
| 90 | 114.49 | 0.01248 | 0.3553 | 28.45 | 78.65 | 28.71 | 57.46 | 86.17 | 0.0590 | 0.1635 |
| 100 | 131.86 | 0.01269 | 0.3079 | 30.79 | 79.51 | 31.10 | 55.93 | 87.03 | 0.0632 | 0.1632 |
| 110 | 151.11 | 0.01292 | 0.2677 | 33.16 | 80.36 | 33.53 | 54.31 | 87.84 | 0.0675 | 0.1628 |
| 120 | 172.35 | 0.01317 | 0.2333 | 35.59 | 81.17 | 36.01 | 52.60 | 88.61 | 0.0717 | 0.1624 |
| 140 | 221.32 | 0.01375 | 0.1780 | 40.60 | 82.68 | 41.16 | 48.81 | 89.97 | 0.0802 | 0.1616 |
| 160 | 279.82 | 0.01445 | 0.1360 | 45.88 | 83.96 | 46.63 | 44.37 | 91.00 | 0.0889 | 0.1605 |
| 180 | 349.00 | 0.01536 | 0.1033 | 51.57 | 84.89 | 52.56 | 39.00 | 91.56 | 0.0980 | 0.1590 |
| 200 | 430.09 | 0.01666 | 0.0767 | 57.87 | 85.17 | 59.20 | 32.08 | 91.28 | 0.1079 | 0.1565 |
| 233.6 | 596.9 | 0.0287 | 0.0287 | 75.69 | 75.69 | 78.86 | 0 | 78.86 | 0.1359 | 0.1359 |

Data from Freon Products Division, E. I. du Pont de Nemours & Company, 1956.

**Table A-17 Properties of saturated refrigerant 12 ($CCl_2F_2$): pressure table**
$v$, ft$^3$/lb; $u$, Btu/lb; $h$, Btu/lb; $s$, Btu/(lb)($^\circ$R)

| | | Specific volume | | Internal energy | | Enthalpy | | | Entropy | |
|---|---|---|---|---|---|---|---|---|---|---|
| Press. psi $P$ | Temp. $^\circ$F $T$ | Sat. liquid $v_f$ | Sat. vapor $v_g$ | Sat. liquid $u_f$ | Sat. vapor $u_g$ | Sat. liquid $h_f$ | Evap. $h_{fg}$ | Sat. vapor $h_g$ | Sat. liquid $s_f$ | Sat. vapor $s_g$ |
| 5 | −62.35 | 0.0103 | 6.9069 | −4.69 | 64.04 | −4.68 | 75.11 | 70.43 | −0.0114 | 0.1776 |
| 10 | −37.23 | 0.0106 | 3.6246 | 0.56 | 66.51 | 0.58 | 72.64 | 73.22 | 0.0014 | 0.1733 |
| 15 | −20.75 | 0.0108 | 2.4835 | 4.05 | 68.13 | 4.08 | 70.95 | 75.03 | 0.0095 | 0.1711 |
| 20 | −8.13 | 0.0109 | 1.8977 | 6.73 | 69.37 | 6.77 | 69.63 | 76.40 | 0.0155 | 0.1697 |
| 30 | 11.11 | 0.0112 | 1.2964 | 10.86 | 71.25 | 10.93 | 67.53 | 78.45 | 0.0245 | 0.1679 |
| 40 | 25.93 | 0.0114 | 0.9874 | 14.08 | 72.69 | 14.16 | 65.84 | 80.00 | 0.0312 | 0.1668 |
| 50 | 38.15 | 0.0116 | 0.7982 | 16.75 | 73.86 | 16.86 | 64.39 | 81.25 | 0.0366 | 0.1660 |
| 60 | 48.64 | 0.0117 | 0.6701 | 19.07 | 74.86 | 19.20 | 63.10 | 82.30 | 0.0413 | 0.1654 |
| 70 | 57.90 | 0.0119 | 0.5772 | 21.13 | 75.73 | 21.29 | 61.92 | 83.21 | 0.0453 | 0.1649 |
| 80 | 66.21 | 0.0120 | 0.5068 | 23.00 | 76.50 | 23.18 | 60.82 | 84.00 | 0.0489 | 0.1645 |
| 90 | 73.79 | 0.0122 | 0.4514 | 24.72 | 77.20 | 24.92 | 59.79 | 84.71 | 0.0521 | 0.1642 |
| 100 | 80.76 | 0.0123 | 0.4067 | 26.31 | 77.82 | 26.54 | 58.81 | 85.35 | 0.0551 | 0.1639 |
| 120 | 93.29 | 0.0126 | 0.3389 | 29.21 | 78.93 | 29.49 | 56.97 | 86.46 | 0.0604 | 0.1634 |
| 140 | 104.35 | 0.0128 | 0.2896 | 31.82 | 79.89 | 32.15 | 55.24 | 87.39 | 0.0651 | 0.1630 |
| 160 | 114.30 | 0.0130 | 0.2522 | 34.21 | 80.71 | 34.59 | 53.59 | 88.18 | 0.0693 | 0.1626 |
| 180 | 123.38 | 0.0133 | 0.2228 | 36.42 | 81.44 | 36.86 | 52.00 | 88.86 | 0.0731 | 0.1623 |
| 200 | 131.74 | 0.0135 | 0.1989 | 38.50 | 82.08 | 39.00 | 50.44 | 89.44 | 0.0767 | 0.1620 |
| 220 | 139.51 | 0.0137 | 0.1792 | 40.48 | 82.08 | 41.03 | 48.90 | 89.94 | 0.0816 | 0.1616 |
| 240 | 146.77 | 0.0140 | 0.1625 | 42.35 | 83.14 | 42.97 | 47.39 | 90.36 | 0.0831 | 0.1613 |
| 260 | 153.60 | 0.0142 | 0.1483 | 44.16 | 83.58 | 44.84 | 45.88 | 90.72 | 0.0861 | 0.1609 |
| 280 | 160.06 | 0.0145 | 0.1359 | 45.90 | 83.97 | 46.65 | 44.36 | 91.01 | 0.0890 | 0.1605 |
| 300 | 166.18 | 0.0147 | 0.1251 | 47.59 | 84.30 | 48.41 | 42.83 | 91.24 | 0.0917 | 0.1601 |

Data from Freon Products Division, E. I. du Pont de Nemours & Company, 1956.

## Table A-18  Properties of superheated refrigerant 12 ($CCl_2F_2$)
$v$, ft$^3$/lb; $u$, Btu/lb; $h$, Btu/lb; $s$, Btu/(lb)($^\circ$R)

| Temp. °F | $v$ | $u$ | $h$ | $s$ | $v$ | $u$ | $h$ | $s$ |
|---|---|---|---|---|---|---|---|---|
| | 10 psia ($-37.23^\circ$F) | | | | 15 psia ($-20.75^\circ$F) | | | |
| Sat. | 3.6246 | 66.512 | 73.219 | 0.1733 | 2.4835 | 68.134 | 75.028 | 0.1711 |
| 0 | 3.9809 | 70.879 | 78.246 | 0.1847 | 2.6201 | 70.629 | 77.902 | 0.1775 |
| 20 | 4.1691 | 73.299 | 81.014 | 0.1906 | 2.7494 | 73.080 | 80.712 | 0.1835 |
| 40 | 4.3556 | 75.768 | 83.828 | 0.1964 | 2.8770 | 75.575 | 83.561 | 0.1893 |
| 60 | 4.5408 | 78.286 | 86.689 | 0.2020 | 3.0031 | 78.115 | 86.451 | 0.1950 |
| 80 | 4.7248 | 80.853 | 89.596 | 0.2075 | 3.1281 | 80.700 | 89.383 | 0.2005 |
| 100 | 4.9079 | 83.466 | 92.548 | 0.2128 | 3.2521 | 83.330 | 92.357 | 0.2059 |
| 120 | 5.0903 | 86.126 | 95.546 | 0.2181 | 3.3754 | 86.004 | 95.373 | 0.2112 |
| 140 | 5.2720 | 88.830 | 98.586 | 0.2233 | 3.4981 | 88.719 | 98.429 | 0.2164 |
| 160 | 5.4533 | 91.578 | 101.669 | 0.2283 | 3.6202 | 91.476 | 101.525 | 0.2215 |
| 180 | 5.6341 | 94.367 | 104.793 | 0.2333 | 3.7419 | 94.274 | 104.661 | 0.2265 |
| 200 | 5.8145 | 97.197 | 107.957 | 0.2381 | 3.8632 | 97.112 | 107.835 | 0.2314 |
| | 20 psia ($-8.13^\circ$F) | | | | 30 psia ($11.11^\circ$F) | | | |
| Sat. | 1.8977 | 69.374 | 76.397 | 0.1697 | 1.2964 | 71.255 | 78.452 | 0.1679 |
| 20 | 2.0391 | 72.856 | 80.403 | 0.1783 | 1.3278 | 72.394 | 79.765 | 0.1707 |
| 40 | 2.1373 | 75.379 | 83.289 | 0.1842 | 1.3969 | 74.975 | 82.730 | 0.1767 |
| 60 | 2.2340 | 77.942 | 86.210 | 0.1899 | 1.4644 | 77.586 | 85.716 | 0.1826 |
| 80 | 2.3295 | 80.546 | 89.168 | 0.1955 | 1.5306 | 80.232 | 88.729 | 0.1883 |
| 100 | 2.4241 | 83.192 | 92.164 | 0.2010 | 1.5957 | 82.911 | 91.770 | 0.1938 |
| 120 | 2.5179 | 85.879 | 95.198 | 0.2063 | 1.6600 | 85.627 | 94.843 | 0.1992 |
| 140 | 2.6110 | 88.607 | 98.270 | 0.2115 | 1.7237 | 88.379 | 97.948 | 0.2045 |
| 160 | 2.7036 | 91.374 | 101.380 | 0.2166 | 1.7868 | 91.166 | 101.086 | 0.2096 |
| 180 | 2.7957 | 94.181 | 104.528 | 0.2216 | 1.8494 | 93.991 | 104.258 | 0.2146 |
| 200 | 2.8874 | 97.026 | 107.712 | 0.2265 | 1.9116 | 96.852 | 107.464 | 0.2196 |
| 220 | 2.9789 | 99.907 | 110.932 | 0.2313 | 1.9735 | 99.746 | 110.702 | 0.2244 |
| | 40 psia ($25.93^\circ$F) | | | | 50 psia ($38.15^\circ$F) | | | |
| Sat. | 0.9874 | 72.691 | 80.000 | 0.1668 | 0.7982 | 73.863 | 81.249 | 0.1660 |
| 40 | 1.0258 | 74.555 | 82.148 | 0.1711 | 0.8025 | 74.115 | 81.540 | 0.1666 |
| 60 | 1.0789 | 77.220 | 85.206 | 0.1771 | 0.8471 | 76.838 | 84.676 | 0.1727 |
| 80 | 1.1306 | 79.908 | 88.277 | 0.1829 | 0.8903 | 79.574 | 87.811 | 0.1786 |
| 100 | 1.1812 | 82.624 | 91.367 | 0.1885 | 0.9322 | 82.328 | 90.953 | 0.1843 |
| 120 | 1.2309 | 85.369 | 94.480 | 0.1940 | 0.9731 | 85.106 | 94.110 | 0.1899 |
| 140 | 1.2798 | 88.147 | 97.620 | 0.1993 | 1.0133 | 87.910 | 97.286 | 0.1953 |
| 160 | 1.3282 | 90.957 | 100.788 | 0.2045 | 1.0529 | 90.743 | 100.485 | 0.2005 |
| 180 | 1.3761 | 93.800 | 103.985 | 0.2096 | 1.0920 | 93.604 | 103.708 | 0.2056 |
| 200 | 1.4236 | 96.674 | 107.212 | 0.2146 | 1.1307 | 96.496 | 106.958 | 0.2106 |
| 220 | 1.4707 | 99.583 | 110.469 | 0.2194 | 1.1690 | 99.419 | 110.235 | 0.2155 |
| 240 | 1.5176 | 102.524 | 113.757 | 0.2242 | 1.2070 | 102.371 | 113.539 | 0.2203 |

## Table A-18 Properties of superheated refrigerant 12 ($CCl_2F_2$)  (*Continued*)

| Temp., °F | $v$ | $u$ | $h$ | $s$ | $v$ | $u$ | $h$ | $s$ |
|---|---|---|---|---|---|---|---|---|
| | 60 psia (48.64°F) | | | | 70 psia (57.90°F) | | | |
| Sat. | 0.6701 | 74.859 | 82.299 | 0.1654 | 0.5772 | 75.729 | 83.206 | 0.1649 |
| 60 | 0.6921 | 76.442 | 84.126 | 0.1689 | 0.5809 | 76.027 | 83.552 | 0.1656 |
| 80 | 0.7296 | 79.229 | 87.330 | 0.1750 | 0.6146 | 78.871 | 86.832 | 0.1718 |
| 100 | 0.7659 | 82.024 | 90.528 | 0.1808 | 0.6469 | 81.712 | 90.091 | 0.1777 |
| 120 | 0.8011 | 84.836 | 93.731 | 0.1864 | 0.6780 | 84.560 | 93.343 | 0.1834 |
| 140 | 0.8355 | 87.668 | 96.945 | 0.1919 | 0.7084 | 87.421 | 96.597 | 0.1889 |
| 160 | 0.8693 | 90.524 | 100.776 | 0.1972 | 0.7380 | 90.302 | 99.862 | 0.1943 |
| 180 | 0.9025 | 93.406 | 103.427 | 0.2023 | 0.7671 | 93.205 | 103.141 | 0.1995 |
| 200 | 0.9353 | 96.315 | 106.700 | 0.2074 | 0.7957 | 96.132 | 106.439 | 0.2046 |
| 220 | 0.9678 | 99.252 | 109.997 | 0.2123 | 0.8240 | 99.083 | 109.756 | 0.2095 |
| 240 | 0.9998 | 102.217 | 113.319 | 0.2171 | 0.8519 | 102.061 | 113.096 | 0.2144 |
| 260 | 1.0318 | 105.210 | 116.666 | 0.2218 | 0.8796 | 105.065 | 116.459 | 0.2191 |
| | 80 psia (66.21°F) | | | | 90 psia (73.79°F) | | | |
| Sat. | 0.5068 | 76.500 | 84.003 | 0.1645 | 0.4514 | 77.194 | 84.713 | 0.1642 |
| 80 | 0.5280 | 78.500 | 86.316 | 0.1689 | 0.4602 | 78.115 | 85.779 | 0.1662 |
| 100 | 0.5573 | 81.389 | 89.640 | 0.1749 | 0.4875 | 81.056 | 89.175 | 0.1723 |
| 120 | 0.5856 | 84.276 | 92.945 | 0.1807 | 0.5135 | 83.984 | 92.536 | 0.1782 |
| 140 | 0.6129 | 87.169 | 96.242 | 0.1863 | 0.5385 | 86.911 | 95.879 | 0.1839 |
| 160 | 0.6394 | 90.076 | 99.542 | 0.1917 | 0.5627 | 89.845 | 99.216 | 0.1894 |
| 180 | 0.6654 | 93.000 | 102.851 | 0.1970 | 0.5863 | 92.793 | 102.557 | 0.1947 |
| 200 | 0.6910 | 95.945 | 106.174 | 0.2021 | 0.6094 | 95.755 | 105.905 | 0.1998 |
| 220 | 0.7161 | 98.912 | 109.513 | 0.2071 | 0.6321 | 98.739 | 109.267 | 0.2049 |
| 240 | 0.7409 | 101.904 | 112.872 | 0.2119 | 0.6545 | 101.743 | 112.644 | 0.2098 |
| 260 | 0.7654 | 104.919 | 116.251 | 0.2167 | 0.6766 | 104.771 | 116.040 | 0.2146 |
| 280 | 0.7898 | 107.960 | 119.652 | 0.2214 | 0.6985 | 107.823 | 119.456 | 0.2192 |
| | 100 psia (80.76°F) | | | | 120 psia (93.29°F) | | | |
| Sat. | 0.4067 | 77.824 | 85.351 | 0.1639 | 0.3389 | 78.933 | 86.459 | 0.1634 |
| 100 | 0.4314 | 80.711 | 88.694 | 0.1700 | 0.3466 | 79.978 | 87.675 | 0.1656 |
| 120 | 0.4556 | 83.685 | 92.116 | 0.1760 | 0.3684 | 83.056 | 91.237 | 0.1718 |
| 140 | 0.4788 | 86.647 | 95.507 | 0.1817 | 0.3890 | 86.098 | 94.736 | 0.1778 |
| 160 | 0.5012 | 89.610 | 98.884 | 0.1873 | 0.4087 | 89.123 | 98.199 | 0.1835 |
| 180 | 0.5229 | 92.580 | 102.257 | 0.1926 | 0.4277 | 92.144 | 101.642 | 0.1889 |
| 200 | 0.5441 | 95.564 | 105.633 | 0.1978 | 0.4461 | 95.170 | 105.076 | 0.1942 |
| 220 | 0.5649 | 98.564 | 109.018 | 0.2029 | 0.4640 | 98.205 | 108.509 | 0.1993 |
| 240 | 0.5854 | 101.582 | 112.415 | 0.2078 | 0.4816 | 101.253 | 111.948 | 0.2043 |
| 260 | 0.6055 | 104.622 | 115.828 | 0.2126 | 0.4989 | 104.317 | 115.396 | 0.2092 |
| 280 | 0.6255 | 107.684 | 119.258 | 0.2173 | 0.5159 | 107.401 | 118.857 | 0.2139 |
| 300 | 0.6452 | 110.768 | 122.707 | 0.2219 | 0.5327 | 110.504 | 122.333 | 0.2186 |

## Table A-18  Properties of superheated refrigerant 12 ($CCl_2F_2$)  (*Continued*)

| Temp. °F | $v$ | $u$ | $h$ | $s$ | $v$ | $u$ | $h$ | $s$ |
|---|---|---|---|---|---|---|---|---|
| | 140 psia (104.35°F) | | | | 160 psia (114.30°F) | | | |
| Sat. | 0.2896 | 79.886 | 87.389 | 0.1630 | 0.2522 | 80.713 | 88.180 | 0.1626 |
| 120 | 0.3055 | 82.382 | 90.297 | 0.1681 | 0.2576 | 81.656 | 89.283 | 0.1645 |
| 140 | 0.3245 | 85.516 | 93.923 | 0.1742 | 0.2756 | 84.899 | 93.059 | 0.1709 |
| 160 | 0.3423 | 88.615 | 97.483 | 0.1801 | 0.2922 | 88.080 | 96.732 | 0.1770 |
| 180 | 0.3594 | 91.692 | 101.003 | 0.1857 | 0.3080 | 91.221 | 100.340 | 0.1827 |
| 200 | 0.3758 | 94.765 | 104.501 | 0.1910 | 0.3230 | 94.344 | 103.907 | 0.1882 |
| 220 | 0.3918 | 97.837 | 107.987 | 0.1963 | 0.3375 | 97.457 | 107.450 | 0.1935 |
| 240 | 0.4073 | 100.918 | 111.470 | 0.2013 | 0.3516 | 100.570 | 110.980 | 0.1986 |
| 260 | 0.4226 | 104.008 | 114.956 | 0.2062 | 0.3653 | 103.690 | 114.506 | 0.2036 |
| 280 | 0.4375 | 107.115 | 118.449 | 0.2110 | 0.3787 | 106.820 | 118.033 | 0.2084 |
| 300 | 0.4523 | 110.235 | 121.953 | 0.2157 | 0.3919 | 109.964 | 121.567 | 0.2131 |
| 320 | 0.4668 | 113.376 | 125.470 | 0.2202 | 0.4049 | 113.121 | 125.109 | 0.2177 |
| | 180 psia (123.38°F) | | | | 200 psia (131.74°F) | | | |
| Sat. | 0.2228 | 81.436 | 88.857 | 0.1623 | 0.1989 | 82.077 | 89.439 | 0.1620 |
| 140 | 0.2371 | 84.238 | 92.136 | 0.1678 | 0.2058 | 83.521 | 91.137 | 0.1648 |
| 160 | 0.2530 | 87.513 | 95.940 | 0.1741 | 0.2212 | 86.913 | 95.100 | 0.1713 |
| 180 | 0.2678 | 90.727 | 99.647 | 0.1800 | 0.2354 | 90.211 | 98.921 | 0.1774 |
| 200 | 0.2818 | 93.904 | 103.291 | 0.1856 | 0.2486 | 93.451 | 102.652 | 0.1831 |
| 220 | 0.2952 | 97.063 | 106.896 | 0.1910 | 0.2612 | 96.659 | 106.325 | 0.1886 |
| 240 | 0.3081 | 100.215 | 110.478 | 0.1961 | 0.2732 | 99.850 | 109.962 | 0.1939 |
| 260 | 0.3207 | 103.364 | 114.046 | 0.2012 | 0.2849 | 103.032 | 113.576 | 0.1990 |
| 280 | 0.3329 | 106.521 | 117.610 | 0.2061 | 0.2962 | 106.214 | 117.178 | 0.2039 |
| 300 | 0.3449 | 109.686 | 121.174 | 0.2108 | 0.3073 | 109.402 | 120.775 | 0.2087 |
| 320 | 0.3567 | 112.863 | 124.744 | 0.2155 | 0.3182 | 112.598 | 124.373 | 0.2134 |
| 340 | 0.3683 | 116.053 | 128.321 | 0.2200 | 0.3288 | 115.805 | 127.974 | 0.2179 |
| | 300 psia (166.18°F) | | | | 400 psia (192.93°F) | | | |
| Sat. | 0.1251 | 84.295 | 91.240 | 0.1601 | 0.0856 | 85.178 | 91.513 | 0.1576 |
| 180 | 0.1348 | 87.071 | 94.556 | 0.1654 | | | | |
| 200 | 0.1470 | 90.816 | 98.975 | 0.1722 | 0.0910 | 86.982 | 93.718 | 0.1609 |
| 220 | 0.1577 | 94.379 | 103.136 | 0.1784 | 0.1032 | 91.410 | 99.046 | 0.1689 |
| 240 | 0.1676 | 97.835 | 107.140 | 0.1842 | 0.1130 | 95.371 | 103.735 | 0.1757 |
| 260 | 0.1769 | 101.225 | 111.043 | 0.1897 | 0.1216 | 99.102 | 108.105 | 0.1818 |
| 280 | 0.1856 | 104.574 | 114.879 | 0.1950 | 0.1295 | 102.701 | 112.286 | 0.1876 |
| 300 | 0.1940 | 107.899 | 118.670 | 0.2000 | 0.1368 | 106.217 | 116.343 | 0.1930 |
| 320 | 0.2021 | 111.208 | 122.430 | 0.2049 | 0.1437 | 109.680 | 120.318 | 0.1981 |
| 340 | 0.2100 | 114.512 | 126.171 | 0.2096 | 0.1503 | 113.108 | 124.235 | 0.2031 |
| 360 | 0.2177 | 117.814 | 129.900 | 0.2142 | 0.1567 | 116.514 | 128.112 | 0.2079 |

Data from Freon Products Division, E. I. du Pont de Nemours & Company, 1956.

## Table A-19 Specific heats of some common liquids and solids

### A. Liquids

| Substance | State | $c_p$, Btu/(lb)(°F) | Substance | State | $c_p$, Btu/(lb)(°F) |
|---|---|---|---|---|---|
| Water | 1 atm, 32°F | 1.007 | Glycerin | 1 atm, 50°F | 0.554 |
| | 1 atm, 77°F | 0.998 | | 1 atm, 120°F | 0.617 |
| | 1 atm, 212°F | 1.007 | Bismuth | 1 atm, 800°F | 0.0345 |
| Ammonia | sat., 0°F | 1.08 | | 1 atm, 1400°F | 0.0393 |
| | sat., 120°F | 1.22 | Mercury | 1 atm, 50°F | 0.033 |
| Refrigerant 12 | sat., −40°F | 0.211 | | 1 atm, 600°F | 0.032 |
| | sat., 0°F | 0.217 | Sodium | 1 atm, 200°F | 0.33 |
| | sat., 120°F | 0.244 | | 1 atm, 1000°F | 0.30 |
| Benzene | 1 atm, 60°F | 0.43 | Propane | 1 atm, 32°F | 0.576 |
| | 1 atm, 150°F | 0.46 | | | |

### B. Solids

| Substance | $T$, °F | $c_p$ | Substance | $T$, °F | $c_p$ |
|---|---|---|---|---|---|
| Ice | −330 | 0.162 | Lead | −455 | 0.0008 |
| | −220 | 0.262 | | −435 | 0.0073 |
| | −75 | 0.392 | | −150 | 0.0283 |
| | 12 | 0.486 | | 32 | 0.0297 |
| | 28 | 0.402 | | 210 | 0.0320 |
| Aluminum | −420 | 0.0039 | | 570 | 0.0356 |
| | −330 | 0.076 | Copper | −370 | 0.0231 |
| | −150 | 0.167 | | −240 | 0.0674 |
| | 32 | 0.208 | | −150 | 0.0783 |
| | 210 | 0.225 | | −60 | 0.0862 |
| | 570 | 0.248 | | 32 | 0.0910 |
| Iron | 68 | 0.107 | | 68 | 0.0921 |
| Silver | 68 | 0.0558 | | 210 | 0.0939 |
| | 930 | 0.0581 | | 390 | 0.0963 |

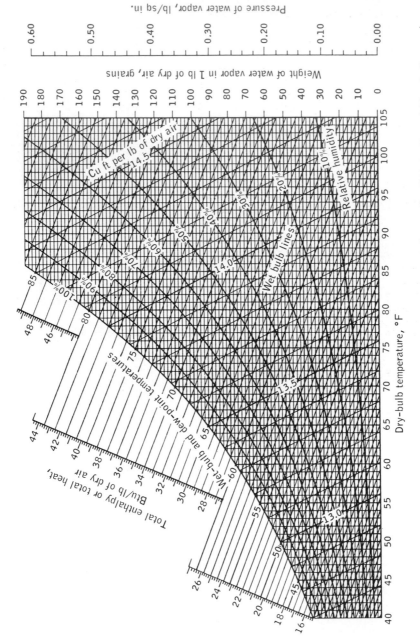

**Figure A-20** General Electric psychrometric chart, barometric pressure 14.696 psia. (*Copyright, 1942, by General Electric Company*)

## Table A-21 Constants for the Benedict-Webb-Rubin and Redlich-Kwong equations of state

1. Benedict-Webb-Rubin: Units are atm, $ft^3$/lb·mol, and °R

| Constants | $n$-Butane $C_4H_{10}$ | Carbon dioxide, $CO_2$ | Carbon monoxide, CO | Methane $CH_4$ | Nitrogen $N_2$ |
|---|---|---|---|---|---|
| $a$ | 7747 | 563.1 | 150.7 | 203.1 | 103.2 |
| $A_0$ | 2590 | 703.0 | 344.5 | 476.4 | 270.6 |
| $b$ | 10.27 | 1.852 | 0.676 | 0.868 | 0.598 |
| $B_0$ | 1.993 | 0.7998 | 0.8740 | 0.6827 | 0.6529 |
| $c$ | $4.219 \times 10^9$ | $1.989 \times 10^8$ | $1.387 \times 10^7$ | $3.393 \times 10^7$ | $9.713 \times 10^6$ |
| $C_0$ | $8.263 \times 10^8$ | $1.153 \times 10^8$ | $7.124 \times 10^6$ | $1.878 \times 10^7$ | $6.706 \times 10^6$ |
| $\alpha$ | 4.531 | 0.3486 | 0.5556 | 0.5120 | 0.5235 |
| $\gamma$ | 8.732 | 1.384 | 1.541 | 1.541 | 1.361 |

*Source:* Cooper, H. W., and J. C. Goldfrank, *Hydrocarbon Processing,* **46** (12): 141 (1967).

2. Redlich-Kwong: Units are atm, $ft^3$/lb·mol, and °R

| Substance | $a$ | $b$ |
|---|---|---|
| Carbon dioxide, $CO_2$ | 21,970 | 0.4757 |
| Carbon monoxide, CO | 5,870 | 0.4395 |
| Methane, $CH_4$ | 10,930 | 0.4757 |
| Nitrogen, $N_2$ | 5,300 | 0.4294 |
| Oxygen, $O_2$ | 5,900 | 0.3522 |
| Propane, $C_3H_8$ | 62,190 | 1.0040 |
| Refrigerant 12, $CCl_2F_2$ | 72,710 | 1.1080 |
| Water, $H_2O$ | 48,460 | 0.3381 |

*Source:* Computed from critical data.

**Table A-22  Values of the enthalpy of formation, Gibbs function of formation, and absolute entropy at 25°C (77°F) and 1 atm**
$\Delta h_f^o$ and $\Delta g_f^o$ in Btu/lb·mol, $s^o$ in Btu/(lb·mol)(°R)

| Substance | Formula | $\Delta h_f^o$ | $\Delta g_f^o$ | $s^o$ |
|---|---|---|---|---|
| Carbon | $C(s)$ | 0 | 0 | 1.36 |
| Hydrogen | $H_2(g)$ | 0 | 0 | 31.21 |
| Nitrogen | $N_2(g)$ | 0 | 0 | 45.77 |
| Oxygen | $O_2(g)$ | 0 | 0 | 49.00 |
| Carbon monoxide | $CO(g)$ | −47,540 | −59,010 | 47.21 |
| Carbon dioxide | $CO_2(g)$ | −169,300 | −169,680 | 51.07 |
| Water | $H_2O(g)$ | −104,040 | −98,350 | 45.11 |
| Water | $H_2O(l)$ | −122,970 | −102,040 | 16.71 |
| Hydrogen peroxide | $H_2O_2(g)$ | −58,640 | −45,430 | 55.60 |
| Ammonia | $NH_3(g)$ | −19,750 | −7,140 | 45.97 |
| Methane | $CH_4(g)$ | −32,210 | −21,860 | 44.49 |
| Acetylene | $C_2H_2(g)$ | +97,540 | +87,990 | 48.00 |
| Ethylene | $C_2H_4(g)$ | +22,490 | +29,306 | 52.54 |
| Ethane | $C_2H_6(g)$ | −36,420 | −14,150 | 54.85 |
| Propylene | $C_3H_6(g)$ | +8,790 | +26,980 | 63.80 |
| Propane | $C_3H_8(g)$ | −44,680 | −10,105 | 64.51 |
| n-Butane | $C_4H_{10}(g)$ | −54,270 | −6,760 | 74.11 |
| n-Octane | $C_8H_{18}(g)$ | −89,680 | +7,110 | 111.55 |
| n-Octane | $C_8H_{18}(l)$ | −107,530 | +2,840 | 86.23 |
| Benzene | $C_6H_6(g)$ | +35,680 | +55,780 | 64.34 |
| Methyl alcohol | $CH_3OH(g)$ | −86,540 | −69,700 | 57.29 |
| Methyl alcohol | $CH_3OH(l)$ | −102,670 | −71,570 | 30.30 |
| Ethyl alcohol | $C_2H_5OH(g)$ | −101,230 | −72,520 | 67.54 |
| Ethyl alcohol | $C_2H_5OH(g)$ | −119,470 | −75,240 | 38.40 |
| Oxygen | $O(g)$ | +107,210 | +99,710 | 38.47 |
| Hydrogen | $H(g)$ | +93,780 | +87,460 | 27.39 |
| Nitrogen | $N(g)$ | +203,340 | +195,970 | 36.61 |
| Hydroxyl | $OH(g)$ | +16,790 | +14,750 | 43.92 |

*Sources:*  From the JANAF Thermochemical Tables, NSRDS-NBS-37, 1971; Selected Values of Chemical Thermodynamic Properties, *NBS Tech. Note* 270-3, 1968; and API Res. Project 44, Carnegie Press, Carnegie Institute of Technology, Pittsburgh, Pa., 1953.

## Table A-23 Values of the enthalpy of combustion $\Delta h_c^o$ and the enthalpy of vaporization $h_{fg}$ at 77°F and 1 atm

Water appears as a liquid in the products of combustion.

| Substance | Formula | $\Delta h_c^o$ Btu/lb mole | $h_{fg}$ Btu/lb mole |
|-----------|---------|------------|----------|
| Hydrogen | $H_2(g)$ | −122,970 | |
| Carbon | $C(s)$ | −169,290 | |
| Carbon monoxide | $CO(g)$ | −121,750 | |
| Methane | $CH_4(g)$ | −383,040 | |
| Acetylene | $C_2H_2(g)$ | −559,120 | |
| Ethylene | $C_2H_4(g)$ | −607,010 | |
| Ethane | $C_2H_6(g)$ | −671,080 | |
| Propylene | $C_3H_6(g)$ | −885,580 | |
| Propane | $C_3H_8(g)$ | −955,070 | 6,480 |
| n-Butane | $C_4H_{10}(g)$ | −1,237,800 | 9,060 |
| n-Pentane | $C_5H_{12}(g)$ | −1,521,300 | 11,360 |
| n-Hexane | $C_6H_{14}(g)$ | −1,804,600 | 13,563 |
| n-Heptane | $C_7H_{16}(g)$ | −2,088,000 | 15,713 |
| n-Octane | $C_8H_{18}(g)$ | −2,371,400 | 17,835 |
| Benzene | $C_6H_6(g)$ | 1,420,300 | 14,552 |
| Toluene | $C_7H_8(g)$ | −1,698,400 | 17,176 |
| Methyl alcohol | $CH_3OH(g)$ | −328,700 | 16,092 |
| Ethyl alcohol | $C_2H_5OH(g)$ | −606,280 | 18,216 |

*Sources:* Primarily from Selected Values of Chemical Thermodynamic Properties, *NBS Circ.* 500, 1952, and Selected Values of Physical and Thermodynamic Properties of Hydrocarbons and Related Compounds, *API Res. Project* 44, Carnegie Press, Carnegie Institute of Technology, Pittsburgh, Pa., 1953.

## Table A-24 Thermodynamic properties of potassium
$T$, °R; $P$, psia; $v$, ft$^3$/lb; $h$, Btu/lb; $s$, Btu/(lb)(°R)

A. Saturation temperature table

| $T$ | $P$ | $v_f$ | $v_g$ | $h_f$ | $h_g$ | $s_f$ | $s_g$ |
|------|--------|--------|--------|-------|--------|--------|--------|
| 1500 | 1.523  | 0.0225 | 261.13 | 291.7 | 1170.8 | 0.6279 | 1.2140 |
| 1600 | 3.210  | 0.0229 | 130.62 | 310.4 | 1176.6 | 0.6400 | 1.1813 |
| 1700 | 6.185  | 0.0234 | 71.09  | 329.6 | 1181.8 | 0.6516 | 1.1529 |
| 1800 | 11.060 | 0.0239 | 41.49  | 349.3 | 1186.6 | 0.6628 | 1.1280 |

B. Superheat table

| $P$ | $v$ | $h$ | $s$ | $P$ | $v$ | $h$ | $s$ |
|------|-------|--------|--------|------|-------|--------|--------|
| \multicolumn{4}{} 2000°R (29.58 psia) | | | | 2200°R (65.83 psia) | | | |
| Sat. | 16.71 | 1195.0 | 1.0867 | Sat. | 8.01  | 1203.5 | 1.0547 |
| 20.0 | 25.58 | 1213.7 | 1.1143 | 50.0 | 10.89 | 1220.8 | 1.0749 |
| 14.0 | 37.34 | 1225.5 | 1.1372 | 40.0 | 13.89 | 1232.0 | 1.0904 |
| 8.0  | 66.73 | 1237.3 | 1.1705 | 30.0 | 18.91 | 1243.5 | 1.1092 |
| 2400°R (127.7 psia) | | | | 2600°R (222.9 psia) | | | |
| Sat. | 4.37  | 1214.0 | 1.0303 | Sat. | 2.63 | 1224.4 | 1.0108 |
| 110  | 5.19  | 1225.2 | 1.0415 | 150  | 4.16 | 1255.2 | 1.0398 |
| 90   | 6.50  | 1238.5 | 1.0559 | 130  | 4.88 | 1264.3 | 1.0497 |
| 70   | 8.57  | 1252.3 | 1.0732 | 110  | 5.87 | 1273.5 | 1.0608 |
| 50   | 12.32 | 1266.5 | 1.0949 | 90   | 7.30 | 1283.1 | 1.0738 |

*Source:* Air Force Aero Propulsion Laboratory Technical Report 66-104, 1966.

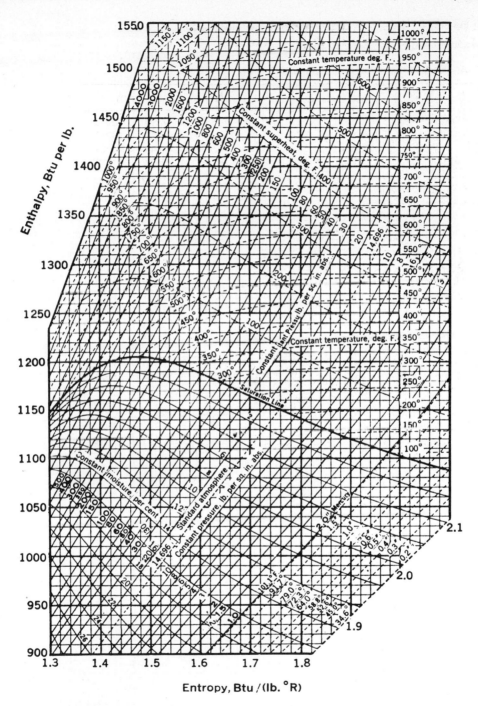

**Figure A-25** Mollier diagram for steam. *Source*: J. H. Keenan and J. Keyes, "Thermodynamic Properties of Steam," Wiley, New York, 1936.

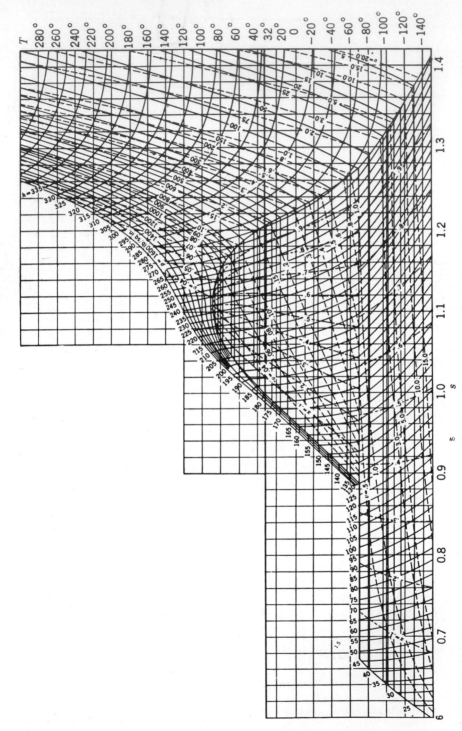

**Figure A-26** Temperature-entropy diagram for carbon dioxide ($CO_2$). $T$ is in °F, $h$ in Btu/lb, $v$ in ft³/lb, $s$ in Btu/(lb)(°R).

# SYMBOLS

| | |
|---|---|
| $A$ | Area |
| | Helmholtz function, $A = U - TS$ |
| | Lagrangian undetermined multiplier |
| $a$ | Specific Helmholtz function, $a = u - Ts$ |
| | Virial coefficient |
| | Acceleration |
| $B$ | Lagrangian undetermined multiplier, $B = 1/kT$ |
| $b$ | Virial coefficient |
| $C$ | Number of components (in Gibbs's phase rule) |
| | A constant |
| $CV$ | Control volume |
| $c$ | Virial coefficient |
| | Velocity of a particle |
| | Speed of light |
| $c_v$ | Specific heat at constant volume, $(\partial u/\partial T)_v$ |
| $c_p$ | Specific heat at constant pressure, $(\partial h/\partial T)_P$ |
| $c_H$ | Specific heat in a constant applied magnetic field |
| $c_M$ | Specific heat at constant magnetization |
| $d$ | An infinitesimal increase in a point function |
| $E$ | Stored energy |
| $\mathbf{E}$ | Electric field strength |
| $e$ | Specific stored energy |
| $F$ | Force |
| | Variance or degrees of freedom (in phase rule) |
| $F_k$ | Measurable generalized force |
| $F_{k,\,eq}$ | Equilibrium generalized force |

| | |
|---|---|
| $f$ | Degrees of freedom of a particle |
| $G$ | Gibbs function, $G = H - TS$ |
| $\Delta G$ | Gibbs-function change for a unit reaction |
| $\Delta G_T^0$ | Standard-state Gibbs-function change for a unit reaction |
| $g$ | Local acceleration of gravity |
| $g_c$ | Gravitational constant in Newton's law, $F = (1/g_c)ma$ |
| $g_i$ | Degeneracy, or statistical weight |
| $H$ | Enthalpy, $H = U + PV$ |
| $\mathbf{H}$ | Magnetic field strength |
| $\Delta h_R$ | Enthalpy of reaction |
| $\Delta h_f$ | Enthalpy of formation |
| $\Delta h_c$ | Enthalpy of combustion |
| $h$ | Specific enthalpy, $h = u + Pv$ |
| | Planck's constant |
| $I$ | Irreversibility |
| | Moment of inertia |
| $j$ | Rotational quantum number |
| KE | Kinetic energy |
| $K_T$ | Isothermal coefficient of compressibility |
| $K_p$ | Equilibrium constant for ideal-gas reactions |
| $k$ | Specific-heat ratio, $c_p/c_v$ |
| | Translational quantum number |
| | Boltzmann's constant |
| $k_s$ | Spring constant |
| $L$ | Length |
| $M$ | Mass, as a dimension |
| | Molar mass, or molecular weight |
| $\mathbf{M}$ | Magnetization per unit volume |
| MEP | Mean effective pressure |
| $m$ | Mass of a substance |
| $\dot{m}$ | Mass flow rate |
| $m'$ | Mass of a particle |
| $N$ | Number of moles |
| $N_A$ | Avogadro's number |
| $n$ | Number of particles |
| | Polytropic constant |
| $P$ | Pressure |
| | Number of phases (in the phase rule) |
| $\mathbf{P}$ | Polarization |
| PE | Potential energy |
| $P_m$ | Permutations |
| $P_R$ | Reduced pressure |
| $p_r$ | Relative pressure |
| $p_i$ | Probability, $n_i/n$ |
| | Component pressure |
| $p_i'$ | Partial pressure |
| $Q$ | Heat interaction |
| $q$ | Heat interaction per unit mass |
| $Q_A$ | Available energy |

| $Q_U$ | Unavailable energy |
|---|---|
| $Q_e$ | Electric charge |
| $R$ | Gas constant, $R_u/M$ |
| $R_u$ | Universal gas constant |
| $r$ | Distance between masses |
| | Compression ratio |
| $r_c$ | Cutoff ratio |
| $r_p$ | Pressure ratio |
| $S$ | Entropy |
| | Surface tension |
| $s$ | Specific entropy |
| | Distance |
| $T$ | Temperature |
| $t$ | Time |
| $U$ | Internal energy |
| $u$ | Specific internal energy |
| | Velocity of particle in $x$ direction |
| $\Delta u_R$ | Internal energy of reaction |
| $V$ | Volume |
| | Velocity |
| | Electrostatic potential |
| $v$ | Specific volume |
| | Vibrational quantum number |
| $v_r$ | Relative volume |
| $W$ | Work interaction |
| | Weight |
| | Thermodynamic probability |
| $w$ | Specific weight |
| | Vibrational wave number |
| $X_k$ | Generalized displacement |
| $x$ | Quality |
| | Mole fraction |
| | Cartesian coordinate |
| | Statistical parameter, $hcw/kT$ |
| $y$ | Mole fraction in the vapor phase |
| | Cartesian coordinate |
| $Z$ | Compressibility factor, $Z = Pv/RT$ |
| $z$ | Height |
| | Cartesian coordinate |
| | Molecular partition function |

### Greek symbols and special notation

| $\beta$ | Isobaric compressibility |
|---|---|
| | Coefficient of performance |
| $\Delta$ | A finite increase in a point function or property |
| $\delta$ | Symbol for an infinitesimal increase in a path function |
| $\epsilon$ | Energy of a particle |
| | Strain |

| | |
|---|---|
| $\mathscr{E}$ | emf, electromotive force |
| $\eta$ | Efficiency |
| $\mathscr{F}$ | Faraday constant |
| $\theta$ | Temperature function |
| $\theta_D$ | Debye temperature, $hv_m/k$ |
| $\theta_R$ | Rotational temperature, $h^2/8h^2Ik$ |
| $\theta_v$ | Vibrational temperature, $hcw/k$ |
| $\mu_{JT}$ | Joule-Thomson coefficient |
| $\mu_i$ | Chemical potential of the $i$th component |
| $v$ | Fundamental frequency of an oscillator |
| | Stoichiometric coefficient |
| $\Delta v$ | Change in the stoichiometric coefficients for a unit reaction |
| $\rho$ | Density |
| $\sigma$ | Symmetry number |
| | Stress |
| $\tau$ | Time |
| | Force |
| $\phi$ | Closed system availability |
| | Equivalence ratio |
| | Relative humidity |
| $\Psi$ | Time-dependent wave function |
| | Total stream availability |
| $\psi$ | Time-independent wave function |
| | Stream availability per unit mass |
| $\omega$ | Angular velocity |
| | Specific humidity, humidity ratio |
| $\mathscr{V}$ | Volume (Sec. 5-2 only) |
| $\sum_i x_i$ | The sum $x_1 + x_2 + \cdots + x_n$ |
| $\prod_i x_i$ | The product $x_1 x_2 x_3 \cdots x_n$ |
| $\equiv$ | Identity symbol, used to specify a definition |

**Subscripts**

| | |
|---|---|
| $a$ | End state of actual process |
| $c$ | Critical state |
| $f$ | Saturated-liquid value |
| $fg$ | Change in value between saturated-liquid and saturated-vapor phases |
| $g$ | Saturated-vapor value |
| H | High temperature (as in $T_H$ and $Q_H$) |
| L | Low temperature (as in $T_L$ and $Q_L$) |
| $m$ | Mixture value |
| $r$ | Relative value |
| $s$ | End state of isentropic process |
| $R$ | Reduced state |
| $v$ | Vapor state |
| $x$ | Property value in wet region |
| $\sigma$ | System designation |
| mp | Most probable macrostate |
| rot | Rotational energy mode |
| tr | Translational energy mode |

| vib | Vibrational energy mode |
|-----|------------------------|
| BE  | Bose-Einstein          |
| FD  | Fermi–Dirac            |
| MB  | Maxwell–Boltzmann      |

# SELECTED PROBLEM ANSWERS (METRIC)

**1-3M** (*a*) 32.5 percent, (*b*) 70.3 percent

**1-6M** (*a*) 35.9 percent, (*b*) 2250 kW

**1-9M** (*a*) 25°C, 1 bar, 0.855 $m^3$/kg; 3 $m^3$, 3.51 kg; (*b*) 11.35 $N/m^3$

**1-12M** (*a*) 200 $m/s^2$, (*b*) 330 $m/s^2$

**1-15M** (*a*) 59.1 km, (*b*) 118 km

**1-18M** (*a*) 22.1 $m/s^2$, (*b*) 2.9 $m/s^2$

**1-21M** (*a*) 10.2m, (*b*) 12.9 m

**1-24M** (*a*) 1.57 bar, (*b*) 0.23 bar, (*c*) 30 mbar, (*d*) 130 kPa

**1-27M** $-0.65$ bar

**1-30M** (*a*) 552 m, (*b*) 969 m, (*c*) 1385 m

**1-33M** 30,200 m

**1-36M** (*a*) $T_N = T_K - 173.15$, (*b*) $-173.15$

**1-39M** 3.48 $m^3$

**1-42M** (*a*) 4.3 kg, (*b*) 6.4 kg

**1-45M** 131°C

**1-48M** (*a*) 276°K, (*b*) 304°K

**1-51M** 398°K

**2-3M** 878,000 kJ

**2-6M** 21 kJ, (in)

**2-9M** (*a*) $-7$, $-11$; (*b*) 2, $-5$

**2-12M** (*a*) $-17,600$ J, (*c*) $-15,600$ J

**2-15M** $-150$ J

**2-18M** $-2,220$ N·m

**2-21M** (*a*) 11,880 N·m, (*b*) 14,400 N·m

**2-24M** 4.07 kJ

**2-27M** (*a*) 261°K, (*b*) 312°K

**2-30M** $3.56 \times 10^{-3}$ kJ/kg

**2-33M** (*a*) 0.053 $KTL_0$, (*b*) 0.194 $KTL_0$, (*c*) 0.292 $KTL_0$

**2-36M** $6.75 \times 10^{-3}$ kJ

**2-39M** 25.2 N/cm

**2-42M** (*a*) $10^{-4}$ C, (*b*) $10^{-3}$ J, (*c*) $5 \times 10^{-4}$ J

**2-45M** (*a*) 239 N·m, (*b*) 286 N·m

**2-48M** (*a*) 124.2 kJ, (*b*) 310.5 kJ

**2-51M** $-17.6$ kJ (out)

**2-54M** (*a*) 82.5 J/g, (*b*) 38.8 J/g, (*c*) 61.3 N·m/g

**3-3M** 5850 J

**3-6M** $-14.3$ kJ/kg

**3-9M**  (*a*) 6036, 6034, 0.22 percent; (*b*) 12,448, 12,448, 0.21 percent

**3-12M**  (*a*) 1563 kJ/kg, (*b*) 2600 kJ/kg

**3-15M**  229°C

**3-18M**  −10.1 kJ

**3-21M**  12.9 kJ, 0

**3-24M**  4.9 kJ

**3-27M**  (*a*) 15,200 N·m, (*b*) 9100 J

**3-30M**  (*a*) 94,500 N·m, (*b*) −94.5 kJ, (*c*) 0

**3-33M**  (*a*) 20.1 kJ, (*b*) 20.8 kJ

**3-36M**  (*a*) 1765 J, 445 J, 1320 J; (*b*) 1498 J, 178 J, 1320 J

**3-39M**  0.657 kJ/(kg)(°C)

**3-42M**  (*a*) 19.7, (*b*) 26.1, (*c*) 28.3 kJ/kg·mol

**3-45M**  60°C

**4-3M**  (*a*) 4.758, 2746.5, 2559.5, 1.0; (*b*) 99.4, 3230.9, 2932.8;
(*c*) 1.014, 1319, 2066, 0.789; (*d*) 275.6, 2409, 2260, 0.762;
(*e*) 1.0256, 340.84, 333.15; (*f*) 151.2, 3247.6, 2945.2;
(*g*) 151.9, 337, 2537, 2369; (*h*) 12.54, 2786.4, 2590, 1.0;
(*i*) 1.985, 658, 2132, 0.739; (*j*) 4.758, 1.0905, 631.68, 0.0;
(*k*) 158.9, 1.1006, 670.56, 0.0

**4-6M**  (*a*) 5.6729, 0.7525; (*b*) 32.74, 200.63; (*c*) 28.10, 195.23;
(*d*) 8.15, 13.48; (*e*) 0.723, 39.76; (*f*) 2.1912, 58.13;
(*g*) 80, 212.37; (*h*) 0.80, 13.31; (*i*) 181.36, 24.78;
(*j*) 4, 212.25; (*k*) 54.87, 0.7525

**4-9M**  (*a*) 27,200, (*b*) 219, (*c*) 376, (*d*) 285, (*e*) 43.0

**4-12M**  0.464

**4-15M**  $1.134 \times 10^6$ J

**4-18M**  (*a*) 0.0757, (*b*) 0.147

**4-21M**  0.418 percent

**4-24M**  (*a*) −49.0 l, (*b*) 165.0°C

**4-27M**  (*a*) 15.54 bars, 2621; (*b*) 500°C, 3456.5

**4-30M**  434 kJ

**4-33M**  (*a*) 3.94, (*b*) 7.92, (*c*) 15.89 kJ/kg

**4-36M**  (*a*) 419.2, 421.18, 0.47 percent; (*b*) 417.12, 421.18, 0.97 percent;
(*c*) 413.13, 421.18, 1.95 percent

**4-45M**  (*a*) 1.366 kg, (*b*) 31.6 kJ, (*c*) −27.35 kJ

**4-48M**  1.40 min

**4-51M**  (*a*) 110.3, (*b*) 76.7 kJ/kg

**4-54M**  (*a*) 33.44 bars, (*b*) −1054 kJ/kg

**4-57M**  (*a*) 200°C, (*b*) 12.3 kJ

**4-60M**  1.25 g

**4-63M**  48.8 kJ/kg

**4-66M**  (*a*) 250.12, 251.47, 0.54 percent; (*b*) 591.46, 595.65, 0.71 percent

**4-69M**  (*a*) 229, 227, 226 cm³/g; (*b*) 16.4, 12.7, 12.58 cm³/g

**4-72M**  (*a*) 45°C, (*b*) 98–102°C

**4-75M**  (*a*) 73–78°C, (*b*) 80°C

**4-78M**  71.9°C

**4-81M** (*a*) 420, 0 percent; (*b*) 420, 0.2 percent; (*c*) 420, 0.75 percent

**5-3M** (*a*) $2.336 \times 10^6$ kg/h, (*b*) 218.5 m/s

**5-6M** (*a*) 47.1 kg/s, (*b*) 1.62 m²

**5-9M** (*a*) 0.904 kg/s, (*b*) 0.053 m²

**5-12M** (*a*) 15.6 kg/s, (*b*) 0.738 bar

**5-15M** 2950 cm²

**5-18M** (*a*) 65°C, (*b*) 875 cm²

**5-21M** (*a*) 26 m/s, (*b*) 5.69 kg/s

**5-24M** 484 m/s

**5-27M** 27.5 cm², 312°C

**5-30M** −54.3 kJ/kg

**5-33M** 5000 kW

**5-36M** (*a*) 238 kW, (*b*) 0.346/1

**5-39M** −720 kJ/h

**5-42M** 9530 kJ/s

**5-45M** (*a*) 91.8 kJ/kg, (*b*) 0.465 kg/s, (*c*) 42.7 kW

**5-48M** 0.954, 109/1

**5-51M** (*a*) 3.5124 bars, 0.233; (*b*) 1.826 bars, 0.329

**5-54M** Positive

**5-57M** $0.129 \times 10^6$ kg/h

**5-60M** 515 kg/h

**5-63M** 2.84 kg/min

**5-66M** (*a*) 4.62 kg/min, (*b*) 0.373 m/s

**5-69M** 148 m/s, 426 kg/min

**5-72M** 530 m/s

**5-75M** 297 kJ/kg, (in)

**5-78M** (*a*) 0.684 kJ/kg, (*b*) 0.946 kW

**5-81M** 71.6 kJ

**5-84M** (*a*) 274°K, 15 bars; 396°K, 3.5 bars; (*b*) 3.87 kg

**5-87M** 68°C

**5-90M** (*a*) 341°K, (*b*) 355°K

**5-93M** $T_R = k/[1 + P_R(k - 1)]$

**5-96M** (*a*) 348°K, (*b*) 407°K

**6-3M** (*a*) 679°C, (*b*) 847°C

**6-6M** 17°C, 250 kJ

**6-9M** Yes

**6-12M** (*a*) 3.5714, (*b*) 3.5714 kJ/(min)(°K)

**7-3M** 500, 350 kJ

**7-6M** (*a*) 327, (*b*) 247°C

**7-9M** (*a*) 20, 0, −10, 0 kJ; −20, −98, 10, 98 kJ; (*b*) 537, 269°K; (*c*) 0.143 m³

**7-12M** (*a*) 20.9 percent, (*b*) 326 kJ/kg

**7-15M** (*a*) 42.4 percent, (*b*) 200 kW

**7-18M** −42°C

**7-21M** 8.39

**7-24M** (a) 2.03 kW, (b) $1.70

**7-27M** $1.55

**7-30M** 1.386, 2.360 kJ

**7-36M** 20.0 percent

**7-39M** (a) $-0.182$, 0.357 kJ/°K; (b) $-0.222$, 0.357 kJ/°K

**7-42M** 3000 kJ of heat

**7-45M** 530 kJ of work

**7-48M** $-0.075$ kJ/(kg)(°K)

**7-15M** (a) $-5.10$; (b) $-5.10$ kJ/°K

**7-54M** $-0.135$ kJ/(kg)(°K)

**7-57M** $-0.0209$ kJ/°K

**7-60M** 23.3 kJ/°K

**7-63M** (a) 5.21 bars, (b) $-0.237$ kJ/°K

**7-66M** (a) 60°C, (b) 0.106 kJ/°K

**7-69M** (a) 3.1 bars, (b) 2.1 bars

**7-72M** (a) 0.4369, (b) 6.7080, (c) 7.1962, (d) 6.325
(e) 1.0688, (f) 5.202, (g) 6.8381, (h) 4.433
(i) 7.3549, (j) 7.2233, (k) 0.5646 kJ/(kg)(°K)

**7-75M** 339 J/g

**7-78M** 35.3 kJ/kg

**7-81M** (a) $-48.4$, (b) 37.4 kJ/kg

**7-84M** $-0.0313$ kJ/°K

**7-87M** 1.41 kJ/°K

**7-90M** 0.215 kJ/°K

**7-93M** (a) $-3.422$, (b) 5.826 kJ/°K, (c) Yes

**7-96M** (a) 0, (b) 7.76 J/°K

**7-99M** (a) 323.8 kJ/kg, (b) Irrev, (c) 265°C

**7-102M** 0.0105 kJ/(kg)(°K)

**7-105M** (a) $-0.0244$, (b) 0.0405 kJ/(kg)(°K), (c) Yes

**7-108M** 88 kJ

**7-111M** 496 m/s, 1.10 kg/s

**7-114M** (a) 295°K, 0.818 bar; (b) 1.603

**7-117M** 2640 kW

**7-120M** 292 m/s

**7-123M** (a) 1.4 bars, (b) 69 m/s

**7-126M** 42.7 kJ/kg

**7-129M** (a) 280°K, (b) 323.5 kJ/kg, (c) 900 kW

**7-132M** 0.476 kW

**7-135M** 282 kJ/kg

**7-138M** 655 m/s

**7-141M** 83.0 percent

**7-144M** (a) 77.1 percent, (b) 81 percent

**8-3M** (a) 1095, (b) 315 kJ

**8-6M** 29.3 kJ/kg, 0.1009 kJ/(kg)(°K)

**8-9M** (a) 21; (b) 882, (c) 1054, (d) 1312; 1312 kJ/kg

**8-12M** (*a*) 187.4, (*b*) 0.475 kJ/°K, (*c*) 54.4 kJ

**8-15M** (*a*) 0.0975, (*b*) 21,200 kJ/h

**8-18M** (*a*) 104.2, (*b*) 19.4 kJ/kg

**8-21M** 230 kJ

**8-24M** 975 kJ

**8-27M** (*a*) 0.88, (*b*) 0.33 m$^3$

**8-30M** (*a*) 13.8, (*b*) 189, (*c*) 12.8 kJ

**8-33M** (*a*) $-26.65, -19.7$ kJ; (*b*) 3.57 kJ

**8-36M** 97.3, 17.1 kJ/kg

**8-39M** (*a*) 2.60, (*b*) 3.20 kJ/(kg)(°K)

**9-3M** (*a*) 210, (*b*) 105

**9-6M** (*a*) 0.247, (*b*) 0.0823

**9-9M** C

**9-12M** 3, 10

**9-15M** (*a*) 4, (*b*) 3490, (*c*) 2520, (*d*) 5, 3, 2

**9-18M** (*a*) 8, 4852, (*b*) No

**9-21M** 2, 3, 4

**9-24M** $-4$, 2, 8

**9-27M** (*a*) 554, 292, 154; (*b*) 333, 333, 333; (*c*) 154, 292, 554

**9-30M** 231°C

**9-33M** 7.17 kW, 156°C

**9-36M** 0.444 kW, 224°C

**9-39M** $-0.862$, 0.862 kJ/°K

**10-3M** (*a*) 250 m/s, (*b*) 750 m/s

**10-9M** FD = 6, BE = 10, MB = 16

**10-12M** $1.5 \times 10^{-17}$ J/g·mol

**10-15M** (*a*) 18,910, (*b*) 52,720, (*c*) 6,710 J/g·mol

**10-18M** (*a*) $7.9 \times 10^{30}$, (*b*) $4.32 \times 10^{31}$

**10-21M** (*a*) 1.025

**10-24M** 162°K

**10-27M** (*a*) 0.932, 0.063, 0.004; (*c*) 0.554, 0.248, 0.110, 0.049, 0.022, 0.010, 0.002

**10-30M** (*a*) $10^{-6}$, (*b*) 0.141, (*c*) 0.388

**10-33M** 0.009, 0.0434, 0.0688, 0.0813, 0.0805, 0.0702, 0.0550, 0.0387, 0.0248, 0.0146, 0.0075

**10-36M** 34.7, 37.3 J/(g·mol)(°K)

**10-39M** 170.92 J/(g·mol)(°K)

**10-42M** (*a*) 0, 0.590, 5.795; (*b*) 4.653, 7.858, 8.255

**10-45M** 58.62, 308.24 J/(g·mol)(°K)

**10-48M** 95.2 percent

**10-51M** 22.64 J/(g·mol)(°K)

**10-54M** 32,800°K

**11-3M** (*a*) 0.620, 0.279, 0.101; (*b*) 0.2 bar; (*c*) 17.12 kg

**11-6M** (*a*) 6.236 bars, (*b*) 0.05 m$^3$, (*c*) 12.47 bars, (*d*) 0.231

**11-9M** (*a*) 60 percent $O_2$, (*b*) 1.08, (*c*) 33.80, (*d*) 56.8 percent

**11-12M** (*a*) 0.96 bar, (*b*) 50 percent, (*c*) 0.0119, (*d*) 3.12

**11-15M** 103 mbar, 58.7 kJ/kg

**11-18M** (*a*) 0.711, (*b*) 0.938

**11-21M** 14.45

**11-24M** (*a*) 273.7, (*b*) 274.8, (*c*) 179.8

**11-27M** 457 kJ

**11-30M** 533 kJ

**11-33M** (*a*) 343°K, (*b*) 2.66 bars, (*c*) 0.0029, 0.1916, 0.1945

**11-36M** $-3.425$ kJ, $-2.593$ kJ, $-9.836$ kJ/(kg·mol)(°K)

**11-39M** (*a*) 0.252 m³, (*b*) 370.4 mbar

**11-42M** (*a*) 78.9 percent, (*b*) 21°C, (*c*) 16.3 g/kg

**11-45M** (*a*) 9.51 mbar, (*b*) 6.0 g/kg, (*c*) 6.1°C

**11-48M** 35, 8, 19

**11-51M** (*a*) 13.34 g/kg, (*b*) 58.9 percent, (*c*) 61.17 kJ/kg

**11-54M** (*a*) 0.0243 bar, (*b*) 0.015, (*c*) 0.89, (*d*) 57 percent, (*e*) 68.9

**11-57M** (*a*) 39 percent, (*b*) 0.0148 bar, (*c*) 51.7 kJ/kg, (*d*) 0.873 m³/kg, (*e*) 12.5°C

**11-60M** $-3265$ kJ/h

**11-63M** (*a*) $-23,600$, (*b*) 8120, (*c*) 5.25

**11-66M** (*a*) 0.0143, (*b*) $-58.8$ kJ/kg, (*c*) 14.6 kJ/kg

**11-69M** (*a*) 16.1°C; (*b*) 0.0034, 27.4°C; 0.0054, 22.5°C

**11-72M** (*a*) 0.0052 kg/kg, 9.5 kJ/kg; (*b*) 0.0068 kg/kg, 14.0 kJ/kg

**11-75M** (*a*) 22.8°C, (*b*) 16.6°C, (*c*) 0.0094, (*d*) 55 percent

**11-78M** 40.5°C

**12-6M** (*a*) 0.400, (*b*) 0.3176, (*c*) 0.310, (*d*) 0.311, (*e*) 0.310 m³/kg·mol

**12-9M** (*a*) 20.2, (*b*) 18.9, (*c*) 16.1, (*d*) 16.5, (*e*) 16.0 bars

**12-12M** (*a*) 134, (*b*) 65.3, (*c*) 76.2, (*d*) 67.7, (*e*) 66 bars

**12-15M** (*a*) 69.6, (*b*) 68.6, (*c*) 68.7 bars

**12-18M** (*a*) 0.759, (*b*) 0.76

**12-21M** 645°K

**12-24M** 116 percent

**12-27M** (*a*) 117, (*b*) 106 cm³/g·mol

**12-30M** 119.5, 113.0 cm³/g·mol

**12-33M** 2600 kJ/kg·mol

**12-36M** 0.7, 1.1 m³

**12-39M** (*a*) 54.7, (*b*) 42, (*c*) 51.6; 62.9 bars

**13-9M** (*a*) 0.692, 0.698 kJ/(kg)(°C); (*b*) 0.765, 0.769 kJ/(kg)(°C)

**13-15M** (*a*) 0.331, 0.343 cm³/(g)(°C); (*b*) 0.0816, 0.0822 cm³/(g)(°C)

**13-18M** $[3a/2vT^{1/2}(v+b)]\,dv$

**13-21M** $R \ln [(v_2 - b)/(v_1 - b)] - (a/T^2)[(1/v_2) - (1/v_1)]$

**13-24M** $(2aR/T)\,\Delta P$

**13-27M** 0.3987 percent

**13-30M** (*a*) 1.248, (*b*) 5.02 percent

**13-33M** 208 kJ/kg

**13-36M** (*a*) $-1.86$, (*b*) $-3.70$, (*c*) $-5.53$°C

**13-39M** (*a*) 71.7, (*b*) 0.479, (*c*) $2.1 \times 10^{-5}$

**13-42M** (*a*) 1.125, (*b*) 1.29, (*c*) 0.708

**13-45M** (*a*) 1.55, (*b*) 1.63, (*c*) 1.69

**13-48M** about 650°K

**13-51M** (*a*) 93, (*b*) 101

**13-54M** 56°C, 60°C

**13-57M** (*a*) 1677, (*b*) −5980

**13-60M** (*a*) 3520 kJ/kg·mol, 227 bars; (*b*) 3120, 175

**13-63M** (*a*) −80.0, (*b*) −9,080, (*c*) −27,200, (*d*) 18,120

**13-66M** 58°C, 155 m/s

**14-3M** (*a*) 35.6 percent, (*b*) 0.737, (*c*) 12.0 percent

**14-6M** (*a*) 17.74 kg/kg, (*b*) 0.833, (*c*) 57.5 percent

**14-9M** 10.1 percent $CO_2$, 3.8 percent $O_2$, 86.1 percent $N_2$

**14-12M** (*a*) 86.8 percent, (*b*) −60 percent

**14-15M** 55°C

**14-18M** (*a*) 2.50, (*b*) 2.23

**14-21M** 10.8, 9.52

**14-24M** (*a*) 9.75 percent $CH_4$, (*b*) 16.0, (*c*) 15.2 percent

**14-27M** 9.5 percent

**14-30M** (*a*) 20 percent, (*b*) 1.25

**14-33M** −2,914,000 kJ/kg·mol

**14-36M** −1,582,600 kJ/kg·mol

**14-39M** −165,700 kJ/kg·mol

**14-42M** −1,391,000 kJ/kg·mol

**14-45M** −130,600 kJ/kg·mol

**14-48M** 1100°K

**14-51M** (*a*) 1650°K, (*b*) 1825°K

**14-54M** 40.6 kg/kg

**14-57M** 2000°K

**14-60M** 3000°K, 25.8 bars

**14-63M** (*a*) −28.18 kJ, (*b*) −39.35 kJ

**14-66M** −312.76 kJ/°K

**14-69M** −88.00 kJ/°K

**14-72M** 760.9 kJ/°K

**14-75M** 481,400 kJ

**14-78M** 1,383,000 kJ

**14-81M** 0.90, 1.04 V

**15-3M** (*a*) −169,400, −298,050; −457,200; −38,910; (*b*) −28,660; (*c*) 11.567

**15-6M** 2375°K

**15-9M** 2825°K

**15-12M** 70 percent

**15-15M** (*a*) 1.08 atm, (*b*) 1.96 atm

**15-18M** 0.222 mol NO

**15-21M** 0.156 mol $NH_3$

**15-24M** 0.80

**15-27M** (*a*) 0.625 mol CO, (*b*) 0.865 mol CO

**15-30M** (*a*) 29.8 percent, (*b*) 89.6 percent, (*c*) 96.9 percent

**15-33M** 0.11 mol $H_2$, 0.50 mol CO

**15-36M** 188,000 kJ/kg·mol

**15-39M** (*a*) 2675°K; (*b*) (1) 2450°K, 10.7 percent; (2) 2495°K, 8.9 percent; (3) 2520°K, 7.5 percent

**15-42M** 2275°K

**15-45M** (*a*) 3040°K, (*b*) 440,000 kJ

**15-48M** 445,000 kJ/kg·mol

**15-51M** −246,400 kJ/kg·mol

**15-54M** (*a*) 195,300, (*b*) 199,500

**16-3M** (*b*) 877, −128.6; (*c*) 14.7 percent; (*d*) 75 percent

**16-6M** (*a*) 34.5 bars, (*b*) 315°K, (*c*) 68.5 percent

**16-9M** 58/1, 2.37 bars

**16-12M** (*a*) 48/1, 1.04 bars; (*b*) 52.9/1, 1.29 bars

**16-15M** (*a*) 52 percent, (*b*) 85 percent

**16-18M** (*a*) 673°K, 17.6 bars; 2000°K, 52.3 bars; 1043°K, 3.41 bars;
(*b*) 0.510; (*c*) 7.88 bars

**16-21M** (*a*) 1728°K, (*b*) 1022 kJ/kg

**16-24M** (*a*) 2.87, (*b*) 0.50, (*c*) 11.8 bars

**16-27M** (*a*) 69.1 bars; (*b*) 1400°K, 2225°K; (*c*) 1112°K; (*d*) 57.5 percent

**16-30M** (*a*) 823, 1408, 2233, 1122°K; (*b*) 57 percent; (*c*) 10.5 bars

**16-33M** (*a*) 161, (*b*) 395, (*c*) 33 percent, (*d*) 1165 kW

**16-36M** (*b*) 10.85

**16-39M** 252 kJ/kg, 39.6 percent

**16-42M** (*a*) 25,300 kW, (*b*) 0.60, (*c*) 4225 kg/min, (*d*) 22 percent

**16-45M** 56 percent

**16-48M** (*a*) 258, (*b*) 32.3 percent

**16-51M** (*a*) 84 percent, (*b*) 87 percent, (*c*) 17,790 kW

**16-54M** (*a*) 37.3 percent, (*b*) 31.3 percent, (*c*) 27.0 percent

**16-57M** (*a*) 27.4 percent, (*b*) 59 percent, (*c*) 83.7 percent, (*d*) 84.3 percent

**16-60M** 23.4 percent

**16-63M** (*a*) 354 kW, (*b*) −17

**16-66M** 1.2, 57

**16-69M** 132, 146.5

**16-72M** (*a*) 194, (*b*) 208

**16-75M** (*a*) 176 kJ/kg, 36.7 percent; (*b*) 176 kJ/kg, 29.5 percent

**16-78M** 167, 338 kJ/kg; 32.8 percent

**16-81M** 20 kJ

**16-84M** (*a*) 265°K, 0.48 bar; 495°K, 4.32 bars; 1020°K, 2.09 bars; 584°K, 0.25 bar; (*b*) 233 kJ/kg; (*c*) 977

**16-87M** (*a*) 250 kJ/kg, (*b*) 765 m/s

**16-90M** 3.06 bars, 947°K

**16-93M** 2.15 bars, 903°K

**16-96M** (*a*) 92, (*b*) 15,000, (*c*) 720,000

**16-99M** (*a*) 2.07, (*b*) 2.7

**16-102M** (*a*) 219, 204°K; (*b*) 2.72, 2.00

**16-105M** 0.46

**16-108M** (*a*) 53.3 percent, (*b*) 2.68 bars

**16-111M** (*a*) 850, (*b*) 6.95, (*c*) 285, (*d*) 1031, (*e*) 50.9 percent

**16-114M** (*a*) 811, (*b*) 7.95, (*c*) 255, (*d*) 1099, (*e*) 48.5 percent

**17-3M** (*a*) 0.81, 43 percent; (*b*) 0.86, 40.5 percent; (*c*) 0.906, 37.9 percent

**17-6M** 304,000 kg/h, 42.3 percent

**17-9M** (*a*) 12.6 kJ/kg, (*b*) 6.8 percent, (*c*) 79 m$^2$

**17-12M** 45 percent

**17-15M** (*a*) 38.8 percent, (*b*) 40.8 percent

**17-18M** (*a*) 45.0 percent, (*b*) 46.2 percent

**17-21M** 6490 kg/h

**17-24M** 0.057

**17-27M** (*a*) 32.9 percent, (*b*) 38.7 percent, (*c*) 39.2 percent

**17-30M** 540°C

**17-33M** (*a*) 325, (*b*) 12.8 percent, (*c*) 1510 m$^2$

**17-36M** (*a*) 40.4 percent, (*b*) 42.3 percent

**17-39M** (*a*) 47.5 percent, (*b*) 224,000 kg/h

**17-42M** (*a*) 742, (*b*) 692, (*c*) 6.7 percent

**17-45M** 22,100 kg/h

**17-48M** (*a*) 0.75, 28.1 percent; (*b*) 0.803, 23.0 percent

**17-51M** (*a*) 36.7°C, (*b*) 0.0150 kJ/(kg)(°K), (*c*) 4.5

**17-54M** (*a*) 6.4, (*b*) 538

**17-57M** (*a*) 2.42, (*b*) 12.2

**17-60M** (*b*) 2.63, (*c*) 19.63, (*d*) 2910, (*e*) 99.5, (*f*) 13.4, (*g*) 0.33

**17-63M** (*a*) 2.48, (*b*) 19.2, (*c*) 74.4 percent

**17-66M** (*a*) 1.5 percent, (*b*) 69.2 percent

**17-69M** 3.0, 130.6 kJ/kg

# SELECTED PROBLEM ANSWERS (USCS)

**1-3** (*a*) 32.4 percent, (*b*) 68.8 percent

**1-6** (*a*) 35.6 percent, 302 hp

**1-9** (*a*) 70°F, 14.6 psia, 0.0745 lb/ft$^3$; 2 ft$^3$, 0.149 lb;
(*b*) 0.0722 lb$_f$/ft$^3$

**1-12** 1.61 ft/s$^2$

**1-15** 36.5 m

**1-18** (*a*) 76.1 ft/s$^2$ (down), (*b*) 13.9 ft/s$^2$ (up)

**1-21** (*a*) 33.9 ft, (*b*) 43.0 ft

**1-24** (*a*) 20.0 psig; (*b*) 10.55 in Hg, 5.19 psia; (*c*) 10.20 in Hg;
(*d*) 24.85 psia

**1-27** −9.8 psig

**1-30** (*a*) 1147 ft, (*b*) 2255 ft, (*c*) 4128 ft

**1-33** 100,200 ft

**1-36** (*a*) 38°C, (*b*) 318°K, (*c*) 285°F, (*d*) 522°R, (*e*) 134°F

**1-39** 122 ft$^3$

**1-42**  8.5 lb

**1-45**  141°C

**1-48**  504°R

**1-51**  772°R

**2-3**  $6.15 \times 10^8$ ft·lb$_f$

**2-6**  26 Btu

**2-9**  50; 40; −90, 60

**2-12**  12,280 ft·lb$_f$

**2-15**  3,685,000 ft·lb$_f$/lb·mol

**2-18**  (a) 17,100 ft·lb$_f$, (b) 13,610 ft·lb$_f$

**2-21**  (a) 1230 ft·lb$_f$

**2-24**  119,500 ft·lb$_f$

**2-27**  (a) 1580, (b) 1527 ft·lb$_f$

**2-30**  (a) 0.593, (b) 2.35, (c) 5.25 ft·lb$_f$

**2-33**  (a) 13.4, (b) 333 ft·lb$_f$

**2-36**  8.2 in

**2-39**  0.121

**2-45**  (a) 42.8, (b) 71.4

**2-48**  502

**2-51**  −22.1 Btu

**3-3**  3940

**3-6**  −6.80

**3-9**  (a) 6825, 6806, 0.34 percent; (b) 15,220, 15,210, 2.1 percent;
(c) 7025, 7015, 1.1 percent

**3-12**  675, 1125

**3-15**  (a) 44.3, (b) 1.60

**3-18**  −6.6 Btu (out)

**3-21**  28.5 Btu/lb, 0

**3-24**  4.65

**3-27**  (a) 5050 ft·lb$_f$, (b) 3030 ft·lb$_f$

**3-30**  (a) 78.4 Btu, (b) −78.4 Btu, 0

**3-33**  (a) 4.70 Btu, (b) 4.98 Btu

**3-36**  (a) 691, 148, 543; (b) 617, 73.7, 543

**3-39**  0.156

**3-42**  (a) 80°F, (b) 30 psia

**3-45**(a) 167°F, (b) 0

**4-3**  (a) 400, 1205.5, 1.0; (b) 1.650, 1362.5; (c) 24.97, 13.58, 0.832;
(d) 467.13, 1031, 0.769; (e) 0.01658, 170.32; (f)700, 0.783;
(g) 327.86, 3.992, 1099; (h) 134.53, 1193.1, 1.0;
(i) 11.529, 25.16, 0.748; (j) 274.1, 0.01864, 0.0

**4-6**  (a) 84.888, 0.01209; (b) 93.29, 86.459; (c) 0.4788, 86.647;
(d) 48.64, 0.2092; (e) 0.01116, 10.68; (f) 23.849, 25.71;
(g) 140, 86.098; (h) 0.760, 0.237; (i) 80.36, 0.2677;
(j) 160, 93.059; (k) 17.27, 0.01159

**4-9**  (a) 394, 392.5; (b) 2.565, 2.361; (c) 4.35, 4.31; (d) 0.346, 0.349; (e) 0.614, 0.602

**4-12**  (a) 213°F, (b) 0.309

**4-15** 1115

**4-18** (a) 21.4 percent, (b) 36.5 percent

**4-21** 0.40 percent

**4-24** (a) 1.02 ft$^3$, (b) 327.86°F

**4-27** (a) 400°F, 948; (b) 600°F, 1302.5

**4-30** 242,000

**4-33** (a) 1.21, 0; (b) 3.66, 0; (c) 7.33, 0

**4-36** (a) 100.88, 101.46, 0.57 percent; (b) 99.54, 101.46, 1.93 percent;
(c) 97.62, 101.46, 3.93 percent

**4-45** (a) 0.753, (b) 13.98 Btu, (c) −7.09 Btu

**4-48** 72 s

**4-51** 58.8

**4-54** −640 (out)

**4-57** (a) 400, (b) 11.1

**4-60** 0.0025 lb

**4-63** −25 Btu/lb

**4-66** (a) 99.24, 99.9, 0.67 percent; (b) 250.01, 251.64, 0.65 percent

**4-69** (a) 3.67, 3.62, 3.6202; (b) 0.275, 0.227, 0.2212

**4-72** (a) 550, (b) 662

**4-75** (a) 195°F, (b) 200°F

**4-78** 140°F

**4-81** (a) 102.24, 0.21 percent; (b) 105.39, 1.03 percent; (c) 110.10, 2.74 percent

**5-3** (a) $6.0 \times 10^6$ lb/h, (b) 4900 ft/s

**5-6** (a) 168, (b) 357

**5-9** (a) 1.405, (b) 0.535 ft$^2$

**5-12** (a) 0.283, (b) 9.31

**5-15** 402

**5-18** (a) 160, (b) 0.879

**5-21** (a) 76 ft/s, (b) 12.2

**5-24** 1590 ft/s

**5-27** 0.0265, 112°F

**5-30** −27.9 Btu/lb (out)

**5-33** 27,700

**5-36** 9500

**5-39** 102,000

**5-42** 3880

**5-45** 220°F

**5-48** 154/1

**5-51** (a) 23.849, 0.329; (b) 29.335, 0.302

**5-54** Positive

**5-57** 203,000

**5-60** 814

**5-63** 5.54

**5-66** (a) 8.43, (b) 1.0

**5-69** 413, 922

**5-72** 1975 ft/s

**5-75** $-14.4$ (out)

**5-78** (a) 242, (b) 1.35

**5-81** 20.7

**5-84** (a) 570°R, 733°R, 30 psia; (b) 0.63 lb; (c) 0.132 Btu/°R

**5-87** 160°F

**5-90** 825°R

**5-93** $k/[1 + P_R(k - 1)]$

**5-96** 188°F

**6-3** 1100°F

**6-6** 50°F, 250 Btu

**6-9** Yes

**6-12** $-2.0$, 2.0

**7-3** (a) 50 percent, (b) 500, (c) 250

**7-6** (a) 590°F, (b) 457°F

**7-9** (a) 20, $-10$, 92.4; (b) 80, 620°F; (c) 5.24 ft$^3$

**7-12** (a) 21.9 percent, (b) 145

**7-15** (a) 41.7 percent, (b) 132 hp

**7-18** $-10$°F

**7-21** 9

**7-24** $1.97

**7-27** 9.55¢/h

**7-30** 5.31, 3.80

**7-36** 20.5 percent

**7-39** (a) 0.2, $-0.1$; Yes; (b) 0.20, $-0.125$; Yes

**7-42** 3000 Btu at 200°F

**7-45** 40,000 ft·lb$_f$

**7-48** 0.975 Btu/(lb·mol)(°F)

**7-51** (a) $-0.385$, (b) $-0.385$ Btu/°R

**7-54** $-0.0197$

**7-57** $-0.0120$

**7-60** 10.1

**7-63** 75 psia, 0.11 Btu/°R

**7-66** 140°F, 0.0252

**7-69** (a) 60, (b) 70

**7-72** (a) 0.1296, (b) 1.6034, (c) 1.7085, (d) 1.4902, (e) 0.1296, (f) 1.2399, (g) 1.6642, (h) 1.0612, (i) 1.7567, (j) 1.7085, (k) 0.29162

**7-75** 240 Btu

**7-78** 4.47

**7-81** $-20.7$, 16.1

**7-84** $-0.0276$

**7-87** 0.84

**7-90** 215

**7-93** (a) $-1.0992$, (b) 1.51, (c) Yes

**7-96** (*a*) 0, (*b*) 0.00268

**7-99** (*a*) 144 Btu/lb, (*b*) Irrev.

**7-102** 0.0020

**7-105** 0.0027

**7-108** 833 Btu

**7-111** 2090 ft/s, 1.45 lb/s

**7-114** 60°F, 10.6 psia, 1.64

**7-117** 2370

**7-120** 950

**7-123** 65, 390

**7-126** − 11.45

**7-129** (*a*) 695, (*b*) 184.5, (*c*) 1450

**7-132** 0.545

**7-135** 100,000 ft-lb$_f$/lb$_m$

**7-138** 2140 ft/s

**7-141** 79.6 percent

**7-144** 81.8 percent

**8-3** (*a*) 513, (*b*) 167, (*c*) 346 Btu

**8-6** 13.7 Btu/lb, 0.026 Btu/(lb)(°R)

**8-9** (*a*) 0, (*b*) 382, (*c*) 505, (*d*) 588 Btu

**8-12** (*a*) 81.3, (*b*) 0.1132, (*c*) 56.6, (*d*) 23.9

**8-15** (*a*) 0.1277, (*b*) 21,600 Btu/h

**8-18** (*a*) 49.8, (*b*) 0, (*c*) 9.8 Btu/lb

**8-21** 244.5 Btu

**8-24** 434 Btu/lb

**8-27** 14.45 ft$^3$

**8-30** (*a*) 29.46, 24.62 Btu; (*b*) 22.5 Btu

**8-33** 42.1, 10.0 Btu/lb

**8-36** (*a*) 1.36, (*b*) 1.76 Btu/lb

**11-3** (*a*) 62 percent $N_2$, (*b*) 3 psia, (*c*) 33.1

**11-6** (*a*) 49.1 psia, (*b*) 1.5 ft$^3$, (*c*) 98.2 psia, (*d*) 42.9

**11-9** (*a*) 60 percent $O_2$, (*b*) 15.3 psia, (*c*) 30.4, (*d*) 63.2 percent $O_2$

**11-12** (*a*) 14 psia, (*b*) 50 percent, (*c*) 220.6, (*d*) 0.464

**11-15** 1.55 psia, 1880 Btu

**11-18** (*a*) 0.163, (*b*) 0.217

**11-21** 5.46

**11-24** 192,000

**11-27** 1.7 × 10$^6$

**11-30** 6.24

**11-33** (*a*) 81°F, (*b*) 36.7 psia, (*c*) 0.0279 Btu/°R

**11-36** 477 Btu, 358 Btu, − 0.868 Btu/°R

**11-39** 11.64, 4.071

**11-42** (*a*) 69 percent, (*b*) 69, (*c*) 0.0153

**11-45** 46.5 gr/lb, 0.152 psia, 46°F

**11-48** 97, 46, 66

**11-51** (a) 0.01456, (b) 62.1 percent, (c) 35.66

**11-54** (a) 0.36 psia, (b) 109 gr/lb, (c) 14.07 ft$^3$/lb,
(d) 60 percent, (e) 37.55 Btu/lb

**11-57** (a) 32 percent, (b) 0.22 psia, (c) 32.3 Btu/lb, (d) 14.06 ft$^3$/lb
(e) 56°F

**11-60** 3110

**11-63** (a) 2100, (b) 690, (c) 0.96

**11-66** (a) 0.0149 lb, (b) 26.1 Btu/lb, (c) 6.3 Btu/lb

**11-69** (a) 58°F, (b) (1) 0.004, 72.5°F; (2) 0.006, 63.5°F

**11-72** (a) 39.5 gr/lb, 3.7 Btu; (b) 52.5 gr/lb, 5.7 Btu

**11-75** (a) 74.6°F, (b) 63.5°F, (c) 69.6 gr/lb, (d) 55 percent

**11-78** 98°F

**12-3** (a) 0.362, (b) 0.447, (c) 0.375, (d) 0.368 ft$^3$/lb

**12-6** (a) 254, (b) 226, (c) 234, (d) 229, (e) 220 psia

**12-9** 0.54 ft$^3$/lb

**12-12** (a) 21,970, 0.476; (b) 2.91 ft$^3$/lb·mol; (c) 9.05 ft$^3$/lb·mol

**12-15** (a) 1218, (b) 1740

**12-18** (a) 0.0775, (b) 3000

**12-21** 116.1, 100.2 cm$^3$/g·mol

**12-24** (a) 529, (b) 871, (c) 1675, (d) 1000 atm

**12-27** (a) 5 ft$^3$, (b) 7 ft$^3$

**13-3** (b) $T(\partial P/\partial T)_v - P$

**13-9** 0.394 Btu/(lb)(°F)

**13-15** 0.00345 ft$^3$/(lb)(°F)

**13-18** $\Delta s = R \ln \left[(v_2 - b)/(v_1 - b)\right] - \left[(a/T^2)(1/v_2 - 1/v_1)\right]$

**13-21** $\Delta T + (B/2)(T_2^2 - T_1^2) + AR\Delta P$

**13-24** (a) 0.82, (b) 14 percent

**13-27** (a) 60.45, (b) 71.1; 60.31 Btu/lb

**13-30** 830 Btu/lb

**13-33** Approx. $T_2 = -10°C$

**13-36** No

**13-39** 755

**13-42** (a) 0.395, (b) 0.083, (c) -0.0254 °F/atm

**13-45** 4070

**13-48** (a) 750 Btu/lb·mol, (b) -1850 Btu/lb·mol

**13-51** (a) ≈5.5 Btu/lb, (b) ~5.4 Btu/lb

**13-54** (a) $\left[-aT^2/(T-b)^2\right]\Delta P$, (b) decrease, (c) yes

**13-57** (a) 748, (b) -2390

**13-60** (a) 62.4, (b) 53.3 Btu/lb

**13-63** -0.566, -126, -330, -206

**13-66** 125°F, 430 ft/s

**14-3** (a) 35.6 percent, (b) 0.737, (c) 12 percent

**14-6** (a) 17.75, (b) 0.833, (c) 57.5 percent

**14-9** 10.1 percent $CO_2$, 3.8 percent $O_2$, 86.1 percent $N_2$

**14-12** (a) 86.8 percent, (b) -60 percent

**14-15** 130°F

**14-18** (a) 2.46, (b) 2.16

**14-21** 10.8, 9.52

**14-24** (a) 9.75 percent $CH_4$, (b) 16.0, (c) 15.2 percent

**14-27** 9.5 percent

**14-30** (a) 20 percent, (b) 1.25

**14-33** $1.26 \times 10^6$ Btu

**14-36** 681,000

**14-39** $-71,230$

**14-42** $-714,700$

**14-45** $-54,210$, $-57,160$, $-2950$ Btu/lb·mol

**14-48** 2000

**14-51** 3240°R

**14-54** 40.5 lb/lb

**14-57** 3400°R

**14-60** 5600°R, 26.6 atm

**14-63** (a) $-12.120$, (b) $-16,930$

**14-66** $-74.71$

**14-69** $-21.01$

**14-72** 182.3

**16-3** (b) 376.9, 50.8; (c) 13.5 percent; (d) 75 percent

**16-6** (a) 26.1 atm, (b) 480°R, (c) 60 percent

**16-9** 91.2/1, 47.7 psi

**16-12** (a) 60.8/1, 23.6 psi; (b) 56.5/1, 20.7 psi

**16-15** (a) 51.8 percent, (b) 85.4 percent

**16-18** (a) 1208°R, 259 psia; 4800°R, 1030 psia; 2555°R, 68.5 psia; (b) 50.4 percent; (c) 179 psi

**16-21** (a) 3120, (b) 441

**16-24** (a) 49.9 percent, (b) 177 psia

**16-27** (a) 1120 psia; (b) 2654, 4320°R; (c) 2195°R; (d) 56.9 percent

**16-30** (a) 1130, 2600, 4610, 2480°R; (b) 57.9 percent; (c) 220 psi

**16-33** (a) 60.6, (b) 160.5, (c) 31.3 percent

**16-36** (a) 10.9, (b) 10.6

**16-39** (a) 85.4, 187.3, 272.2; (b) 37.4 percent

**16-42** (a) 77.7, (b) 134.9, (c) 18.9 percent

**16-45** 218 hp, 205 hp

**16-48** (a) 0.40, (b) 6820, (c) 22.1 percent

**16-51** (a) 87.3 percent, (b) 86.1 percent, (c) 12,700

**16-54** 32.1 percent

**16-57** 27.8 percent, 55.5 percent

**16-60** 24.3 percent

**16-63** (a) 217 hp, (b) 6.3 Btu

**16-66** 1.17; 19,100

**16-69** 56.6, 62.5 Btu/lb

**16-72** (a) 78.2 Btu, (b) 89.6 Btu

**16-75** (*a*) 34.0 percent, (*b*) 27.1 percent

**16-78** 70.0, 146.6, 33.6 percent

**16-81** 50 Btu

**16-84** (*a*) 468°R, 6.92 psia; 873°R, 62.3 psia; 1847°R, 30.7 psia; 1085°R, 4 psia; (*b*) 98.0; (*c*) 3170

**16-87** (*a*) 108.9, (*b*) 2485

**16-90** 46.5 psia, 1695°R

**16-93** 32.7 psia, 1623°R

**16-96** (*a*) 119 lb/s; (*b*) 11,400; (*c*) 504,000

**16-99** (*a*) 2.06, (*b*) 7.14

**16-102** (*a*) 325 Btu, (*b*) 3.05

**16-105** 0.425

**16-108** (*a*) 53.3 percent, (*b*) 33.7 psia

**16-111** (*a*) 360, (*b*) 7.19, (*c*) 120, (*d*) 455, (*e*) 50.9 percent

**16-114** (*a*) 345, (*b*) 8.21, (*c*) 108, (*d*) 477, (*e*) 48.2 percent

**17-3** (*a*) 79 percent, 42 percent; (*b*) 84 percent,39.6 percent; (*c*) 88 percent, 37.0 percent

**17-6** (*a*) 732,000 lb/h, (*b*) 40.5 percent

**17-9** (*a*) 8.83, (*b*) 12.8 percent, (*c*) 313

**17-12** 44.4 percent

**17-15** (*a*) 38.7 percent, (*b*) 40.8 percent

**17-18** 42.6 percent

**17-21** 12,900

**17-24** 0.069

**17-27** (*b*) 38.2 percent, (*c*) 36.2 percent, (*d*) 38.2 percent, (*e*) 37.4 percent

**17-30** 940°F

**17-33** (*a*) 108 Btu/lb, (*b*) 10.1 percent, (*c*) 1285 ft$^2$

**17-36** (*a*) 39.6 percent, (*b*) 41.1 percent

**17-39** (*a*) 47.9 percent, (*b*) 486,000 lb/h

**17-42** (*a*) 772 Btu/lb, (*b*) 735 Btu/lb, (*c*) 4.8 percent

**17-45** 91,000 lb/h

**17-48** (*a*) 28.6 percent, (*b*) 24.0 percent

**17-51** (*a*) 136°F, (*b*) 0.0078 Btu/(lb)(°R), (*c*) 2.62

**17-54** (*a*) 4.64, (*b*) 26.4 ft$^3$/min

**17-57** (*a*) 2.08, (*b*) 1.0

**17-60** (*b*) 1.76, (*c*) 50.6, (*d*) 3140, (*e*) 209, (*f*) 26.9, (*g*) 0.432

**17-63** (*a*) 2.16, (*b*) 46.6 lb/min

**17-66** (*a*) 24 percent, (*b*) 210°F

**17-69** (*a*) 0.051, (*b*) 57.1 percent

# INDEX